Table of Integrals

ELEMENTARY FORMS

1. $\int [f(u) + g(u)]\, du = \int f(u)\, du + \int g(u)\, du$

2. $\int af(u)\, du = a \int f(u)\, du$

3. $\int u^n\, du = \dfrac{1}{n+1} u^{n+1} + C, \quad n \neq -1$

4. $\int \dfrac{du}{u} = \ln |u| + C$

5. $\int e^u\, du = e^u + C$

6. $\int a^u\, du = \dfrac{1}{\ln a} a^u + C$

7. $\int \sin u\, du = -\cos u + C$

8. $\int \cos u\, du = \sin u + C$

9. $\int \sec^2 u\, du = \tan u + C$

10. $\int \csc^2 u\, du = -\cot u + C$

11. $\int \sec u \tan u\, du = \sec u + C$

12. $\int \csc u \cot u\, du = -\csc u + C$

13. $\int \tan u\, du = \ln |\sec u| + C$

14. $\int \cot u\, du = \ln |\sin u| + C$

15. $\int \sec u\, du = \ln |\sec u + \tan u| + C$

16. $\int \csc u\, du = \ln |\csc u - \cot u| + C$

17. $\int \dfrac{du}{\sqrt{a^2 - u^2}} = \operatorname{Sin}^{-1} \dfrac{u}{a} + C$

18. $\int \dfrac{du}{a^2 + u^2} = \dfrac{1}{a} \operatorname{Tan}^{-1} \dfrac{u}{a} + C$

19. $\int \dfrac{du}{u\sqrt{u^2 - a^2}} = \dfrac{1}{a} \operatorname{Sec}^{-1} \dfrac{u}{a} + C$

20. $\int u\, dv = uv - \int v\, du$

RATIONAL FORMS

21. $\int \dfrac{du}{a + bu} = \dfrac{1}{b} \ln |a + bu| + C$

22. $\int \dfrac{u\, du}{a + bu} = \dfrac{1}{b^2}(a + bu - a \ln |a + bu|) + C$

23. $\int \dfrac{u\, du}{(a + bu)^2} = \dfrac{a}{b^2(a + bu)} + \dfrac{1}{b^2} \ln |a + bu| + C$

24. $\int \dfrac{u^2\, du}{(a + bu)^2} = \dfrac{1}{b^3}\left(a + bu - \dfrac{a^2}{a + bu} - 2a \ln |a + bu|\right) + C$

25. $\int \dfrac{du}{u(a + bu)^2} = \dfrac{1}{a(a + bu)} - \dfrac{1}{a^2} \ln \left|\dfrac{a + bu}{u}\right| + C$

26. $\int \dfrac{du}{u^2(a + bu)} = -\dfrac{1}{au} + \dfrac{b}{a^2} \ln \left|\dfrac{a + bu}{u}\right| + C$

27. $\int u(a + bu)^n\, du = \dfrac{(a + bu)^{n+1}}{b^2}\left(\dfrac{a + bu}{n + 2} - \dfrac{a}{n + 1}\right) + C, \quad n \neq -1, -2$

28. $\int \dfrac{u\, du}{(a + bu)(c + du)} = \dfrac{1}{bc - ad}\left(-\dfrac{a}{b} \ln |a + bu| + \dfrac{c}{d} \ln |c + du|\right) + C, \quad bc - ad \neq 0$

29. $\int \dfrac{u\, du}{(a + bu)^2(c + du)} = \dfrac{1}{bc - ad}\left[\dfrac{a}{b(a + bu)} + \dfrac{c}{bc - ad} \ln \left|\dfrac{a + bu}{c + du}\right|\right] + C, \quad bc - ad \neq 0$

30. $\int \dfrac{du}{a^2 - u^2} = \dfrac{1}{2a} \ln \left|\dfrac{u + a}{u - a}\right| + C$

31. $\int \dfrac{du}{u^2 - a^2} = \dfrac{1}{2a} \ln \left|\dfrac{u - a}{u + a}\right| + C$

32. $\int \dfrac{du}{(a^2 \pm u^2)^n} = \dfrac{1}{2(n - 1)a^2}\left[\dfrac{u}{(a^2 \pm u^2)^{n-1}} + (2n - 3) \int \dfrac{du}{(a^2 \pm u^2)^{n-1}}\right], \quad n \neq 1$

33. $\int \dfrac{du}{(u^2 - a^2)^n} = \dfrac{1}{2(n - 1)a^2}\left[-\dfrac{u}{(u^2 - a^2)^{n-1}} - (2n - 3) \int \dfrac{du}{(u^2 - a^2)^{n-1}}\right], \quad n \neq 1$

FORMS CONTAINING $\sqrt{a + bu}$

34. $\int u\sqrt{a + bu}\, du = \dfrac{2}{15b^2}(3bu - 2a)(a + bu)^{3/2} + C$

35. $\int u^n\sqrt{a + bu}\, du = \dfrac{2}{b(2n + 3)}\left[u^n(a + bu)^{3/2} - na\int u^{n-1}\sqrt{a + bu}\, du\right]$

36. $\int \dfrac{u\, du}{\sqrt{a + bu}} = \dfrac{2}{3b^2}(bu - 2a)\sqrt{a + bu} + C$

37. $\int \dfrac{u^2\, du}{\sqrt{a + bu}} = \dfrac{2}{15b^3}(8a^2 + 3b^2u^2 - 4abu)\sqrt{a + bu} + C$

38. $\int \dfrac{u^n\, du}{\sqrt{a + bu}} = \dfrac{2u^n\sqrt{a + bu}}{b(2n + 1)} - \dfrac{2na}{b(2n + 1)}\int \dfrac{u^{n-1}\, du}{\sqrt{a + bu}}$

39. $\int \dfrac{du}{u\sqrt{a + bu}} = \begin{cases} \dfrac{1}{\sqrt{a}}\ln\left|\dfrac{\sqrt{a + bu} - \sqrt{a}}{\sqrt{a + bu} + \sqrt{a}}\right| + C, & a > 0 \\[3mm] \dfrac{2}{\sqrt{-a}}\tan^{-1}\sqrt{\dfrac{a + bu}{-a}} + C, & a < 0 \end{cases}$

40. $\int \dfrac{du}{u^n\sqrt{a + bu}} = -\dfrac{\sqrt{a + bu}}{a(n - 1)u^{n-1}} - \dfrac{b(2n - 3)}{2a(n - 1)}\int \dfrac{du}{u^{n-1}\sqrt{a + bu}}$

41. $\int \dfrac{\sqrt{a + bu}}{u}\, du = 2\sqrt{a + bu} + a\int \dfrac{du}{u\sqrt{a + bu}}$

42. $\int \dfrac{\sqrt{a + bu}}{u^2}\, du = -\dfrac{\sqrt{a + bu}}{u} + \dfrac{b}{2}\int \dfrac{du}{u\sqrt{a + bu}}$

FORMS CONTAINING $\sqrt{u^2 \pm a^2}$

43. $\int \sqrt{u^2 \pm a^2}\, du = \dfrac{u}{2}\sqrt{u^2 \pm a^2} \pm \dfrac{a^2}{2}\ln\left|u + \sqrt{u^2 \pm a^2}\right| + C$

44. $\int u\sqrt{u^2 \pm a^2}\, du = \dfrac{1}{3}(u^2 \pm a^2)^{3/2} + C$

45. $\int u^2\sqrt{u^2 \pm a^2}\, du = \dfrac{u}{8}(2u^2 \pm a^2)\sqrt{u^2 \pm a^2} - \dfrac{a^4}{8}\ln\left|u + \sqrt{u^2 \pm a^2}\right| + C$

46. $\int \dfrac{\sqrt{u^2 + a^2}}{u}\, du = \sqrt{u^2 + a^2} - a\ln\left|\dfrac{a + \sqrt{u^2 + a^2}}{u}\right| + C$

47. $\int \dfrac{\sqrt{u^2 - a^2}}{u}\, du = \sqrt{u^2 - a^2} - a\,\mathrm{Sec}^{-1}\left|\dfrac{u}{a}\right| + C$

48. $\int \dfrac{\sqrt{u^2 \pm a^2}}{u^2}\, du = -\dfrac{\sqrt{u^2 \pm a^2}}{u} + \ln\left|u + \sqrt{u^2 \pm a^2}\right| + C$

49. $\int \dfrac{du}{\sqrt{u^2 \pm a^2}} = \ln\left|u + \sqrt{u^2 \pm a^2}\right| + C$

50. $\int \dfrac{u^2\, du}{\sqrt{u^2 \pm a^2}} = \dfrac{u}{2}\sqrt{u^2 \pm a^2} \mp \dfrac{a^2}{2}\ln\left|u + \sqrt{u^2 \pm a^2}\right| + C$

51. $\int \dfrac{du}{u\sqrt{u^2 + a^2}} = -\dfrac{1}{a}\ln\left|\dfrac{a + \sqrt{u^2 + a^2}}{u}\right| + C$

52. $\int \dfrac{du}{u\sqrt{u^2 - a^2}} = \dfrac{1}{a}\,\mathrm{Sec}^{-1}\left|\dfrac{u}{a}\right| + C$

53. $\int \dfrac{du}{u^2\sqrt{u^2 \pm a^2}} = \mp\dfrac{\sqrt{u^2 \pm a^2}}{a^2 u} + C$

54. $\int (u^2 \pm a^2)^{3/2}\, du = \dfrac{u}{8}(2u^2 \pm 5a^2)\sqrt{u^2 \pm a^2} + \dfrac{3a^4}{8}\ln\left|u + \sqrt{u^2 \pm a^2}\right| + C$

55. $\int \dfrac{du}{(u^2 \pm a^2)^{3/2}} = \pm\dfrac{u}{a^2\sqrt{u^2 \pm a^2}} + C$

(continued inside back cover)

Calculus and Analytic Geometry
Third Edition

Calculus and Analytic Geometry
Third Edition

Douglas F. Riddle
St. Joseph's University

Wadsworth Publishing Company, Inc., Belmont, California

ISBN 0-534-00626-4

Library of Congress Cataloging in Publication Data

Riddle, Douglas F.
  Calculus and analytic geometry.

  Includes index.
  1. Calculus. 2. Geometry, Analytic. I. Title.
QA303.R53   1979   515'.15   78-23206
ISBN 0-534-00626-4

Printed in the United States of America

The Table of Trigonometric Functions in the Appendix is from *The Calculus with Analytic Geometry* by Louis Leithold. Copyright © 1968 by Louis Leithold. Reprinted by permission of the publishers, Harper & Row, Publishers, Inc.

The tables of Exponential Functions, Common Logarithms, Natural Logarithms of Numbers, and Squares, Square Roots, and Prime Factors are reprinted from *College Algebra,* 2nd ed., by Edwin F. Beckenbach, Irving Drooyan, and William Wooton, Copyright © 1964, 1968 by Wadsworth Publishing Company, Inc., Belmont, California.

5 6 7 8 9 10—85 84 83

Contents

Preface

As in the previous editions, we have striven to present calculus in a way that is as easy as possible for the student to understand as well as easy for the instructor to present. Many changes, both large and small, have been made in this edition to further these aims. One of the most important of these changes is in the problem sets. Over half of the problems have been altered, and new ones have been added. The new problems are, for the most part, problems that present the student with something of a challenge. Many of the alterations in the problems are minor changes in the numbers; this has been done to prevent the students from making use of "files," assembled from previous editions. In addition, the problems have been graded for difficulty by separating them (in most sections) into three parts, labeled A, B, and C. The part labeled A consists of routine problems that every student can be expected to do. Those problems labeled B are less routine but still not a great challenge. The average student can be expected to work most of these. The C problems present something of a challenge; only the better students can be expected to work them. Of course, any such system of classification must be very subjective; it must be considered as a rough guide only.

The approach to vectors has been reviewed and completely revised. The previously used definition of a vector as an equivalence class of directed line segments was felt to be too abstract for the level set by the rest of the book. Thus, in this edition, vectors are approached from an algebraic point of view that simplifies their introduction.

In addition, a more conventional proof of the fundamental theorem of calculus has been given.

However, the greatest change has occurred in the last few chapters on multivariate calculus. Chapters 20, 21, and 22 have been extensively revised and a chapter on line and surface integrals has been added.

Retained from the previous editions are the wide variety of applications and the important area of rapid curve sketching *without* the use of calculus. The recent trend toward the consideration of curve sketching only in conjunction with calculus is, in our view, unfortunate. A sketch is necessary for setting up most integrals; and generally only a very rough sketch (without locating relative maxima, minima, or points of inflection) is needed. It is felt that a student who cannot make such a sketch without resorting to differentiation is at a distinct disadvantage. Thus, several sections are devoted to the sketching of curves without any consideration of the derivative.

As in earlier editions, we have frequently been faced with direct opposition between what is mathematically proper and what is pedagogically proper. As mathematicians, we feel that we should use proper mathematics, but as teachers we feel that to say that one is teaching when no one is learning is like saying that one is selling when no one is buying. The view that we must use "proper" mathematics at all costs is responsible for the current wave of ultrarigorous texts that begin with an epsilon-delta definition of limits, introduce the mean-value theorem at an early stage, and give proofs of all theorems. The hoped-for results—students who really understand the underlying concepts of calculus—

simply have not been attained with this level of sophistication. Thus, we have tried to steer a middle course between the extremes of rigor and nonrigor.

Limits and continuity are presented at three levels; an intuitive introduction in Chapter 3, rigorous definitions in a geometric setting in Chapter 10, and finally the epsilon-delta definitions in Chapter 17. An early, rigorous introduction to limits seems to be an exercise in futility. The beginning student is in no position to recognize the need for such definitions, nor does he or she possess the mathematical maturity needed for their understanding.

The intuitive introduction gives the student a working knowledge of limits and allows evaluation of simple limits. The geometric definitions, due to R. L. Moore, give the student rigorous definitions in forms that provide greater insight into the meanings of limits and continuity than do the traditional epsilon-delta definitions. Professor Moore had used them successfully with his students for many years to provide the pictures behind the epsilon-delta definitions. Finally, the epsilon-delta definitions are introduced with the feeling that more than one or two students will understand them.

Despite our convictions about the teaching of calculus, it would be foolish not to recognize the great diversity of opinion that exists on the subject. Therefore we have taken pains to make the three-level approach to limits and continuity versatile enough to allow the instructor to tailor the treatment to his or her own taste and time limitations as well as to the needs and abilities of the students. Certainly there are those who will not find the geometric treatment of limits to their taste. Others simply do not have time to consider both the geometric and analytic treatments. The instructor who feels that an intuitive introduction to limits is all that should be presented in a first course can easily skip the other two levels. And the instructor who feels that the only proper introduction to calculus begins with the epsilon-delta treatment of limits, can simply replace the intuitive introduction (Section 3.3) with the epsilon-delta definition (Sections 17.1 and 17.2). Other combinations between these two extremes are possible.

Finally, the mean-value theorem is treated rather late in the text, since it depends upon a careful consideration of limits and continuity. This is proper, we think, because it is not an easy theorem to understand, prove, or use. Proofs of earlier theorems requiring the mean-value theorem are omitted in the text proper; however, they are supplied in Appendix B, and an instructor who wants to state the theorem early and prove the theorems based upon it will find it easy to do so.

We gratefully thank the following individuals who contributed their suggestions and thoughts about how the text is and should be used. We have not always agreed with them, but they have made us reexamine many of our thoughts and statements. This itself is especially helpful.

Dennis Alber, Palm Beach Junior College; Frank D. Anger, University of Puerto Rico; Leon E. Arnold, Delaware County Community College; Richard W. Ball, Auburn University; Jerry Beehler, Tri-State University; R. E. Briney, Salem State College; Joe K. Bryant, Monterey Peninsula College; Paul Bugl, University of Hartford; Edward W. Chillak, Drew University; D. M. Crystal, Rochester Institute of Technology; Dale Ewen, Parkland Community College; Dodd M. Fisher, West Liberty State College; Jay E. Folkert, Hope College; Gary Ford, Radford College; Brian Garman, University of Kentucky; Kialynn Glubrecht, Spokane Falls Community College; Robert M. Gravina, University of Lowell; Mark P. Hale, Jr., University of Florida; Allen Hansen, Riverside Community College; Jack L. Hennington, Johnson County Community College; Stephen Hinthorne, Palomar College; Marvin Y. Johnson, College of Lake County;

Richard Johnsonbaugh, Chicago State University; W. D. Kaigh, University of Texas—El Paso; Kenneth Kalmanson, Montclaire State University; Eleanor Killam, University of Massachusetts; Joseph F. Krebs, Boston College; Howard B. Lambert, East Texas State University; Eric Liban, York College; Richard Luebbe, Triton College; Joseph Malhevitch, York College–CUNY; Jay P. Morgan, Wright State University; Don Nelson, Western Michigan University; L. N. Nigam, Quinnipiac College; N. B. Patterson, Pennsylvania State University; Larry H. Potter, Memphis State University; James J. Reynolds, Pennsylvania State University—Beaver Campus; F. Schaffer, University of Lethbridge; Kenneth Shabell, Riverside Community College; Ron Smit, University of Portland; Karl Smith, Santa Rosa Junior College; Robert E. Spencer, Virginia Polytechnic Institute and State University; L. L. Stearley, Mott Community College; Robert C. Steinbach, Grossmont College; Marion J. Stokes, University of Richmond; Allen C. Utterback, Cabrillo College; A. R. Van Cleave, Columbus College; Abe Weinstein, Nassau Community College; Harry Whitcomb, Philadelphia College of Pharmacy and Science; Dennis Williams, Salem State College; Jimmie D. Woods, U.S. Coast Guard Academy.

In particular, we are indebted to several professors who have taught courses using the second edition and have given us a considerable amount of detailed criticism based upon their teaching experiences: Les Birdsall, Russ DiPrizio, Art Dull, Michael Freeman, Dan Henry, Raymond McGivney, Dave Sanders, and Lin Wyant.

1

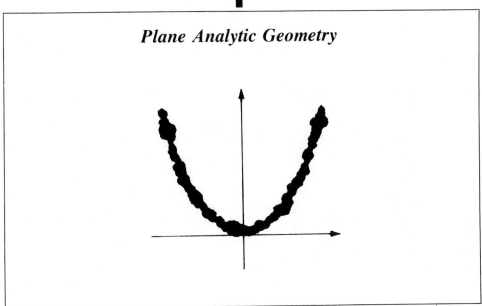

Plane Analytic Geometry

1.1

Introduction: Distance and Midpoint Formulas

Analytic geometry provides a bridge between algebra and geometry that makes it possible for geometric problems to be solved algebraically (or analytically). It also allows us to solve algebraic problems geometrically, but the former capability is far more important, especially when numbers are assigned to essentially geometric concepts. Consider, for instance, the length of a line segment or the angle between two lines. Even if the lines and points in question are accurately known, the number representing the length of a segment or the angle between two lines can be determined only approximately by measurement. Algebraic methods provide an exact determination of the number.

The association between the algebra and geometry is made by assigning numbers to points. Suppose we look at this assignment of numbers to the points on a line. First of all, we select a pair of points, O and P, on the line, as shown in Figure 1.1. The point O, which we call the origin, is assigned the number zero, and the point P is assigned the number one. Using $\overline{OP}$ as our unit of length,† we assign numbers to all other points on the line in the following way; Q on the P side of the origin is assigned the positive number x if and only if its

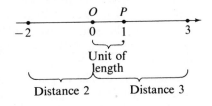

Figure 1.1

---

† We shall use the notation OP for the line segment joining the points O and P, and $\overline{OP}$ for the length of the segment.

distance from the origin is x. A point Q on the opposite side of the origin is assigned the negative number $-x$ if and only if its distance from the origin is x. In this way every point on the line is assigned a real number and, for each real number, there corresponds a point on the line.

Thus a *scale* is established on the line, which we now call a *coordinate line*. The number representing a given point is called the *coordinate* of that point, and the point is called the *graph* of the number.

Just as points on a line (a one-dimensional space) are represented by single numbers, so points in a plane (a two-dimensional space) can be represented by pairs of numbers. Later we shall see that points in a three-dimensional space can be represented by triples of numbers.

In order to represent points in a plane by pairs of numbers, we select two intersecting lines and establish a scale on each line, as shown in Figure 1.2. The point of intersection is the origin. These two lines, called the axes, are distinguished by identifying symbols (usually by the letters x and y). For a given point P in the plane, there corresponds a point P_x on the x axis. It is the point of intersection of the x axis and the line parallel to the y axis that contains P. (If P is on the y axis, this line coincides with the y axis.) Similarly, there exists a point P_y on the y axis which is the point of intersection of the y axis and the line through P that parallels (or is) the x axis. The coordinates of these two points on the axes are the *coordinates* of P. If a is the coordinate of P_x and b is the coordinate of P_y, then the point

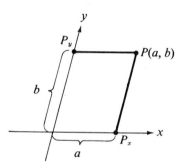

Figure 1.2

P is represented by (a, b). In this example, a is called the x *coordinate*, or *abscissa*, of P and b is the y *coordinate*, or *ordinate*, of P.

In a coordinate plane, the following conventions normally apply:

(1) the axes are taken to be perpendicular to each other;
(2) the x axis is a horizontal line with the positive coordinates to the right of the origin, and the y axis is a vertical line with the positive coordinates above the origin;
(3) the same scale is used on both axes.

These conventions, of course, need not be followed when others are more convenient. We shall violate the third rather frequently when considering figures that would be very difficult to sketch if we insisted upon using the same scale on both axes. In such cases, we shall feel free to use different scales, remembering that we have distorted the figures in the process. Unless a departure from convention is specifically stated or is obvious from the context, we shall always follow the first two conventions.

We can now identify the coordinates of the points in Figure 1.3. Note that all points on the x axis have the y coordinate zero, while those on the y axis have the x coordinate zero. The origin has both coordinates zero, since it is on both axes.

The axes separate the plane into four regions, called *quadrants*. It is convenient to identify them by the numbers shown in Figure 1.4. The points on the axes are not in any quadrant.

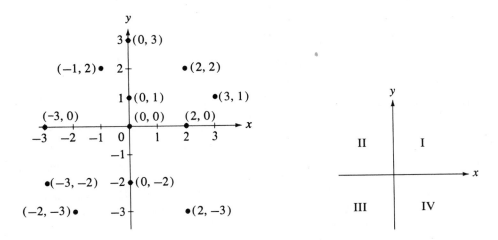

Figure 1.3   Figure 1.4

A point's coordinates, determined in this way, are sometimes called Cartesian co-ordinates, after the French mathematician and philosopher René Descartes. In the appendix of a book published in 1637, Descartes gave the first description of analytic geometry. From it came the developments that eventually led to the invention of the calculus.

Now that we have assigned coordinates to points, let us see how we may use the co-ordinates to solve geometric problems. We begin with the distance between two points. Suppose we consider the distance between two points on a coordinate line. Let P_1 and P_2 be two points on a line, and let them have coordinates x_1 and x_2, respectively. If P_1 and P_2 are both to the right of the origin, with P_2 farther right than P_1 (as in Figure 1.5 (a)), then

$$\overline{P_1P_2} = \overline{OP_2} - \overline{OP_1} = x_2 - x_1.$$

Expressing the distance between two points is only slightly more complicated if the origin is on the right of one or both of the points. In Figure 1.5(b),

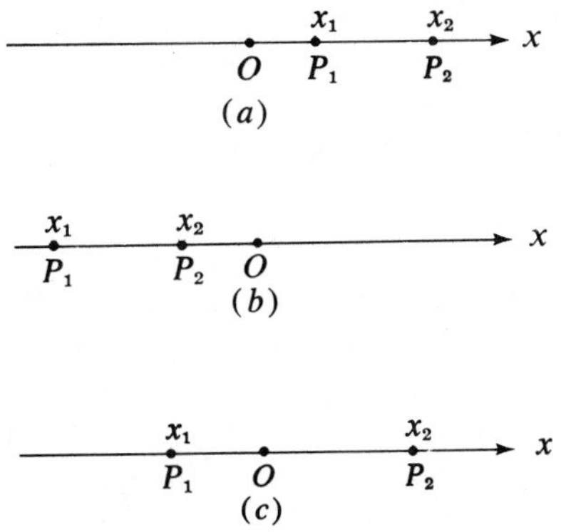

$$\overline{P_1P_2} = \overline{P_1O} - \overline{P_2O} = -x_1 - (-x_2) = x_2 - x_1,$$

and in Figure 1.5(c),

$$\overline{P_1P_2} = \overline{P_1O} + \overline{OP_2} = -x_1 + x_2 = x_2 - x_1.$$

Figure 1.5

Thus, we see that $\overline{P_1P_2} = x_2 - x_1$ in all three of these cases in which P_2 is to the right of P_1. If P_2 were to the left of P_1, we would have

$$\overline{P_1P_2} = x_1 - x_2,$$

as you can easily verify. Thus $\overline{P_1P_2}$ can always be represented as the larger coordinate minus the smaller. Since $x_2 - x_1$ and $x_1 - x_2$ differ only in that one is the negative of the other and since distance is always nonnegative, we see that $\overline{P_1P_2}$ is the difference that is nonnegative. Thus

$$\overline{P_1P_2} = |x_2 - x_1|.$$

This form is especially convenient when the relative positions of P_1 and P_2 are unknown. However, since absolute values are sometimes rather bothersome, they will be avoided whenever the relative positions of P_1 and P_2 are known.

Let us now turn our attention to the more difficult problem of finding the distance between two points in the plane. Suppose we are interested in the distance between P_1 (x_1, y_1) and P_2 (x_2, y_2) (see Figure 1.6). A vertical line through P_1 and a horizontal

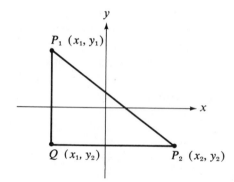

Figure 1.6

line through P_2 intersect at a point Q (x_1, y_2). Assuming P_1 and P_2 are not on the same horizontal or vertical line, P_1P_2Q forms a right triangle with the right angle at Q. Now we can use the theorem of Pythagoras to determine the length of P_1P_2. By the previous discussion,

$$\overline{QP_2} = |x_2 - x_1| \quad \text{and} \quad \overline{P_1Q} = |y_2 - y_1|$$

(the absolute values are retained here, since we want the resulting formula to hold for *any* choice of P_1 and P_2, not merely for the one shown in Figure 1.6). Now by the Pythagorean theorem,

$$\overline{P_1P_2} = \sqrt{|x_2 - x_1|^2 + |y_2 - y_1|^2}.$$

But, since $|x_2 - x_1|^2 = (x_2 - x_1)^2 = (x_1 - x_2)^2$, the absolute values may be dropped at this stage and we have

$$\overline{P_1P_2} = \sqrt{(x_2 - x_1)^2 + (y_2 - y_1)^2}.$$

Thus we have proved the following theorem.

Theorem 1.1

The distance between two points P_1 (x_1, y_1) and P_2 (x_2, y_2) is

$$\overline{P_1 P_2} = \sqrt{(x_2 - x_1)^2 + (y_2 - y_1)^2}.$$

In deriving this formula, we assumed that P_1 and P_2 are not on the same horizontal or vertical line; however, the formula would hold even in these cases. For example, if P_1 and P_2 are on the same horizontal line, then $y_1 = y_2$ and $y_2 - y_1 = 0$. Thus,

$$\overline{P_1 P_2} = \sqrt{(x_2 - x_1)^2} = |x_2 - x_1|.$$

Note that $\sqrt{(x_2 - x_1)^2}$ is *not always* $x_2 - x_1$. Since the symbol $\sqrt{}$ indicates the non-negative square root, we see that if $x_2 - x_1$ is negative, then $\sqrt{(x_2 - x_1)^2}$ is not equal to $x_2 - x_1$ but, rather, equals $|x_2 - x_1|$. Suppose, for example, that $x_2 - x_1 = -5$. Then

$$\sqrt{(x_2 - x_1)^2} = \sqrt{(-5)^2} = \sqrt{25} = 5 = |x_2 - x_1|.$$

Example 1

Find the distance between P_1 $(1, 4)$ and P_2 $(-3, 2)$.

$$\overline{P_1 P_2} = \sqrt{(-3 - 1)^2 + (2 - 4)^2} = 2\sqrt{5}.$$

Example 2

Determine whether or not A $(1, 7)$, B $(0, 3)$, and C $(-2, -5)$ are collinear.

$$\overline{AB} = \sqrt{(0 - 1)^2 + (3 - 7)^2} = \sqrt{17},$$
$$\overline{BC} = \sqrt{(-2 - 0)^2 + (-5 - 3)^2} = \sqrt{68} = 2\sqrt{17},$$
$$\overline{AC} = \sqrt{(-2 - 1)^2 + (-5 - 7)^2} = \sqrt{153} = 3\sqrt{17}.$$

Since $\overline{AC} = \overline{AB} + \overline{BC}$, the three points must be collinear (if they were not, they would form a triangle and any one side would be less than the sum of the other two).

Example 3

Show that $(1, 2)$, $(4, 7)$, $(-6, 13)$, and $(-9, 8)$ are the vertices of a rectangle.

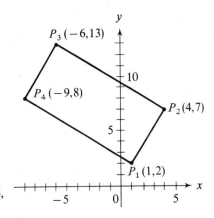

The points are plotted in Figure 1.7. Let us check lengths.

$$\overline{P_1 P_2} = \sqrt{(4 - 1)^2 + (7 - 2)^2} = \sqrt{34},$$

$$\overline{P_3 P_4} = \sqrt{(-9 + 6)^2 + (8 - 13)^2} = \sqrt{34},$$

$$\overline{P_2 P_3} = \sqrt{(-6 - 4)^2 + (13 - 7)^2} = \sqrt{136},$$

$$\overline{P_4 P_1} = \sqrt{(1 + 9)^2 + (2 - 8)^2} = \sqrt{136}.$$

Figure 1.7

Although $\overline{P_1 P_2} = \overline{P_3 P_4}$ and $\overline{P_2 P_3} = \overline{P_4 P_1}$, we are not justified in saying that we have a rectangle; we can merely conclude that we have a parallelogram. But if the diagonals of a parallelogram are equal, then the parallelogram is a rectangle (see Problem 34). Let us then consider the lengths of the diagonals,

$$\overline{P_1 P_3} = \sqrt{(-6-1)^2 + (13-2)^2} = \sqrt{170},$$

$$\overline{P_2 P_4} = \sqrt{(-9-4)^2 + (8-7)^2} = \sqrt{170}.$$

Now that we see we have a parallelogram with equal diagonals, we may conclude that it is a rectangle. SMART !

Another useful formula that is easily found with Cartesian coordinates is the midpoint of a line segment. Suppose P is the midpoint of AB in Figure 1.8. Since BB_1 and PP_1 are vertical lines, they are parallel, and $\triangle APP_1 \sim \triangle ABB_1$. Therefore,

$$\frac{\overline{AP_1}}{\overline{AB_1}} = \frac{\overline{AP}}{\overline{AB}} = \frac{1}{2}$$

and

$$\overline{AP_1} = \frac{1}{2}\,\overline{AB_1}.$$

Then

$$x = x_1 + \overline{AP_1}$$

$$= x_1 + \frac{1}{2}\,\overline{AB_1}$$

$$= x_1 + \frac{1}{2}(x_2 - x_1)$$

$$= \frac{x_1 + x_2}{2}.$$

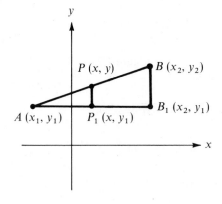

Figure 1.8

A similar argument can be used to show that

$$y = \frac{y_1 + y_2}{2}.$$

This proves the following theorem.

Theorem 1.2

If P is the midpoint of AB, then the coordinates of P are:

$$x = \frac{x_1 + x_2}{2}, \qquad y = \frac{y_1 + y_2}{2}.$$

Thus to find the midpoint of a segment AB, we merely average both the x and y coordinates of the given points. A moment of thought will reveal how reasonable this is; the average of two grades is halfway between them, the average of two temperatures is halfway between them, and so forth.

Example 4

Find the midpoint of the segment AB where $A = (1, 5)$ and $B = (-3, -1)$.

$$x = \frac{x_1 + x_2}{2} \qquad y = \frac{y_1 + y_2}{2}$$

$$= \frac{1 - 3}{2} \qquad = \frac{5 - 1}{2}$$

$$= -1; \qquad = 2.$$

Thus $P = (-1, 2)$.

When we use the methods of analytic geometry to prove geometric theorems, such proofs are called analytic proofs. When carrying out analytic proofs, we should recall that a plane does not come fully equipped with coordinate axes—they are imposed upon the plane to make the transition from geometry to algebra. Thus we are free to place the axes in any position we choose in relation to the given figure. We place them in a way that makes the algebra as simple as possible.

Example 5

Prove analytically that the diagonals of a rectangle are equal.

First we place the axes in a convenient position. Let us put the x axis on one side of the rectangle and the y axis on another, as illustrated in Figure 1.9. Since we have a rectangle, the coordinates of B and D determine those of C.

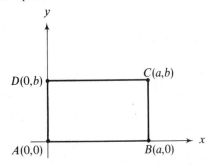

$$\overline{AC} = \sqrt{(a - 0)^2 + (b - 0)^2} = \sqrt{a^2 + b^2}$$

$$\overline{BD} = \sqrt{(0 - a)^2 + (b - 0)^2} = \sqrt{a^2 + b^2}$$

Since $\overline{AC} = \overline{BD}$, the theorem is proved.

Figure 1.9

Example 6

Prove analytically that the segment joining the midpoints of two sides of a triangle is parallel to the third side and one-half its length.

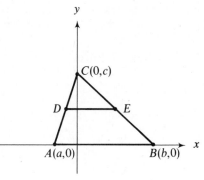

Figure 1.10

Let us place the axes as indicated in Figure 1.10 and let D and E be the midpoints of AC and BC, respectively. By the midpoint formula, $D = (a/2, c/2)$ and $E = (b/2, c/2)$. Since D and E have identical y coordinates, DE is horizontal and therefore parallel to AB. Finally $\overline{DE} = b/2 - a/2 = (b - a)/2$ and $\overline{AB} = b - a$; thus $\overline{DE} = \overline{AB}/2$.

Problems

A

In Problems 1–4, find the distance between the given points.

1. $(1, -3), (2, 5)$.
2. $(5, -3), (5, 4)$.
3. $(1/2, 2), (-3/2, 1/2)$.
4. $(\sqrt{3}, 3\sqrt{2}), (3\sqrt{3}, -\sqrt{2})$.

In Problems 5 and 6, find the midpoint of the segment AB.

5. $A = (4, -1), B = (3, 3)$.
6. $A = (-3, 3), B = (1, 5)$.

In Problems 7–10, find the unknown quantity.

7. $P_1 = (5, -2), \quad P_2 = (x, -5), \quad \overline{P_1 P_2} = 5$.
8. $P_1 = (-4, y), \quad P_2 = (8, 5), \quad \overline{P_1 P_2} = 13$.
9. $P_1 = (x, x), \quad P_2 = (1, 4), \quad \overline{P_1 P_2} = \sqrt{5}$.
10. $P_1 = (x, 2x), \quad P_2 = (2x, 1), \quad \overline{P_1 P_2} = \sqrt{34}$.

In Problems 11–16, determine whether or not the three given points are collinear.

11. $(-2, 3), (7, -2), (2, 5)$.
12. $(-3, 4), (0, 2), (6, -2)$.
13. $(1, -1), (3, 4), (-1, -6)$.
14. $(1, 2\sqrt{2}), (-1, 5\sqrt{2}), (3, -2\sqrt{2})$.
15. $(-3, 3), (2, -1), (7, -5)$.
16. $(2, 3), (1, -2), (-1, 11)$.

In Problems 17 and 18, determine whether or not the three given points are the vertices of a right triangle.

17. $(0, 2), (-2, 4), (1, 3)$.
18. $(\sqrt{3} - 3, 2\sqrt{3} + 1), \quad (\sqrt{3} - 1, \sqrt{3} + 1), \quad (2\sqrt{3} - 1, \sqrt{3} + 2)$.
19. If $P = (4, -1)$ is the midpoint of the segment AB, where $A = (2, 5)$, find B.
20. If $P = (4, 1)$ is the midpoint of the segment AB, where $A = (5, -2)$, find B.

B

21. Show that $(5, 2)$ is on the perpendicular bisector of the segment AB where $A = (1, 3)$ and $B = (4, -2)$.
22. Show that $(-2, 4), (2, 0), (2, 8)$, and $(6, 4)$ are the vertices of a square.
23. Find the center and radius of the circle circumscribed about the right triangle with vertices $(-2, -1), (3, -1)$, and $(3, 11)$.
24. Determine whether each of the following points is inside, on, or outside the circle with center $(-2, 3)$ and radius 5: $(1, 7), (-3, 8), (2, 0), (-5, 7), (0, -1), (-5, -1), (-6, 6), (4, 2)$.
25. Show that $(1, 1), (4, 1), (3, -2)$, and $(0, -2)$ are the vertices of a parallelogram.
26. Find all possible values for y so that $(5, 8), (-4, 11)$, and $(2, y)$ are the vertices of a right triangle.
27. Prove analytically that the diagonals of a parallelogram bisect each other.

28. Find the point of intersection of the diagonals of the parallelogram with vertices $(-3, 2)$, $(2, 2)$, $(1, 5)$, and $(6, 5)$.
29. The point $(1, 4)$ is at a distance 5 from the midpoint of the segment joining $(3, -2)$ and $(x, 4)$. Find x.
30. The midpoints of the sides of a triangle are $(-1, 6)$, $(4, -2)$, and $(10, 1)$. Find the vertices.

C

31. Three vertices of a parallelogram are $(1, -1)$, $(3, 3)$, and $(6, -3)$. Find the fourth vertex. (*Hint:* There is more than one solution. Sketch all possible parallelograms using the three given vertices.)
32. Find the center and radius of the circle circumscribed about the triangle with vertices $(4, 1)$, $(2, 5)$, and $(0, -1)$.
33. Show that a triangle with vertices (x_1, y_1), (x_2, y_2), and (x_3, y_3) has area

$$\frac{1}{2}|x_1 y_2 + x_2 y_3 + x_3 y_1 - x_1 y_3 - x_2 y_1 - x_3 y_2|.$$

(*Hint:* Consider the smallest rectangle with sides parallel to the coordinate axes and containing the vertices of the triangle.)

34. Prove analytically that if the diagonals of a parallelogram are equal, then the parallelogram is a rectangle. (*Hint:* Place the axes as shown in Figure 1.11 and show that $\overline{AC} = \overline{BD}$ implies that A is the origin.)

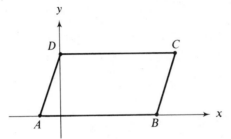

Figure 1.11

35. Prove analytically that the sum of the lengths of two sides of a triangle is greater than the length of the third side.

1.2

Inclination and Slope

An important concept in the description of a line and one that is used quite extensively throughout calculus has to do with the inclination of a line. First let us recall a convention from trigonometry: angles measured in the counterclockwise direction are positive; those measured in the clockwise direction are negative.

Definition

The **inclination** of a line that intersects the x axis is the measure of the smallest nonnegative angle which the line makes with the positive end of the x axis. The inclination of a line parallel to the x axis is 0.

We shall use the symbol θ to represent an inclination. The inclination of a line is always less than 180°, or π radians, and every line has an inclination. Thus, for any line,

$$0° \leq \theta < 180° \quad \text{or} \quad 0 \leq \theta < \pi.$$

Figure 1.12 shows several lines with their inclinations. Note that the angular measure is given in both degrees and radians. Although there is no reason to show preference for one over the other at this time, we shall see when working with calculus that radian measure is the more natural way of representing an angle.

While the inclination of a line may seem like a simple representation, we cannot, in general, find a simple relationship between the inclination of a line and the co-ordinates of points on it without resorting to tables of trigonometric functions. Thus we consider another expression related to the inclination—namely, the slope of a line.

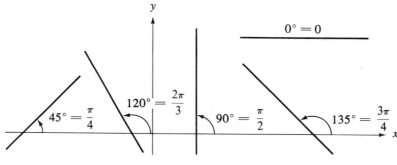

Figure 1.12

Definition

*The **slope** m of a line is the tangent of the inclination; thus*

$$m = \tan \theta.$$

While it is possible for two different angles to have the same tangent, it is not possible for lines having two different inclinations to have the same slope. The reason for this is the restriction on the inclination, $0° \leq \theta < 180°$. Nevertheless, one minor problem does arise from the use of slope, since the tangent of 90° is nonexistent. Thus vertical lines have inclination 90° but no slope. *Do not confuse "no slope" with "zero slope."* A horizontal line definitely has a slope and that slope is the number 0, but there is no number at all (not even 0) which is the slope of a vertical line. Some might object to this nonexistence of tan 90° by saying that it is "infinity," or "∞." However, infinity is not a number. Also, while the symbol ∞ is quite useful in calculus when dealing with limits, as you will see later, its use in algebra and an algebraic development of trigonometry leads to trouble.

While the nonexistence of the slope of certain lines is somewhat bothersome, it is more than counterbalanced by the simple relationship between the slope and the coordinates of a pair of points on the line. Recall that if θ is in either of the positions shown in Figure 1.13, then

$$\tan \theta = \frac{y}{x}.$$

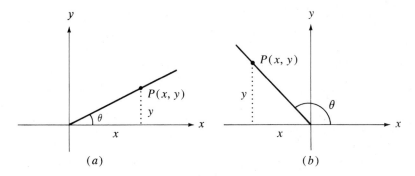

Figure 1.13

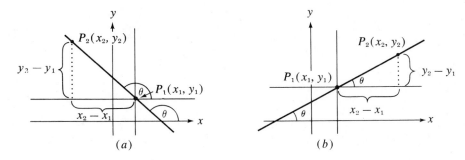

Figure 1.14

Unfortunately, the lines we must deal with are not always so conveniently placed. Suppose we have a line with a pair of points, P_1 (x_1, y_1) and P_2 (x_2, y_2), on it (see Figure 1.14). If we place a pair of axes parallel to the old axes, with P_1 as the new origin, then the coordinates of P_2 with respect to this new coordinate system are $x = x_2 - x_1$ and $y = y_2 - y_1$. Now θ is situated in a position that allows us to use the definition of $\tan \theta$ and state the following theorem.

Theorem 1.3

A line through P_1 (x_1, y_1) and P_2 (x_2, y_2), where $x_1 \neq x_2$, has slope

$$m = \frac{y_2 - y_1}{x_2 - x_1} = \frac{y_1 - y_2}{x_1 - x_2}.$$

Example 1

Find the slope of the line containing P_1 $(1, 5)$ and P_2 $(7, -7)$.

$$m = \frac{y_2 - y_1}{x_2 - x_1} = \frac{-7 - 5}{7 - 1} = \frac{-12}{6} = -2.$$

Since a vertical line has no slope, Theorem 1.3 does not hold for it; however, $x_1 = x_2$ for any pair of points on a vertical line, and the right-hand side of the slope formula is also nonexistent. Thus there is no slope when the right-hand side of the slope formula does not exist.

If two nonvertical lines are parallel, they must have the same inclination and, thus, the same slope (see Figure 1.15). If two parallel lines are vertical, then neither one has slope. Similarly, if $m_1 = m_2$ or if neither line has slope, then the two lines are parallel. Thus, two lines are parallel if and only if $m_1 = m_2$ or neither line has slope.

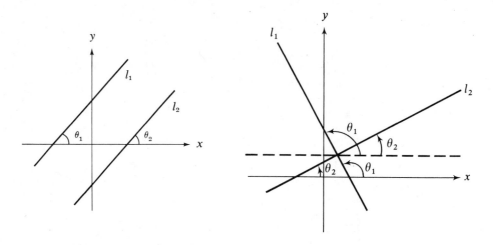

Figure 1.15   Figure 1.16

If two nonvertical lines l_1 and l_2 with the respective inclinations θ_1 and θ_2 are perpendicular (see Figure 1.16), then (assuming l_1 is the line with the larger inclination)

$$\theta_1 - \theta_2 = 90°,$$

or

$$\theta_1 = \theta_2 + 90°.$$

Thus

$$\tan \theta_1 = \tan (\theta_2 + 90°) = -\cot \theta_2 = -\frac{1}{\tan \theta_2},$$

or

$$m_1 = -\frac{1}{m_2}.$$

On the other hand, if $m_1 = -1/m_2$, the argument can be traced backward to show that the difference of the inclinations is $90°$ and the lines are perpendicular. Therefore we have the following theorem.

Theorem 1.4

Two lines l_1 and l_2 with slopes m_1 and m_2, respectively, are

(a) parallel if and only if $m_1 = m_2$,
(b) perpendicular if and only if $m_1 m_2 = -1$.

Example 2

Find the slopes of l_1 containing (1, 5) and (3, 8) and l_2 containing $(-4, 1)$ and (0, 7); determine whether l_1 and l_2 are parallel, coincident, perpendicular, or none of these.

$$m_1 = \frac{8-5}{3-1} = \frac{3}{2}, \qquad m_2 = \frac{7-1}{0+4} = \frac{6}{4} = \frac{3}{2}.$$

We now know that l_1 and l_2 are either parallel or coincident. To determine which, let us find the slope of the line joining (1, 5) on l_1 and $(-4, 1)$ on l_2.

$$m_3 = \frac{5-1}{1+4} = \frac{4}{5}.$$

Since $m_3 \neq m_1$, the point $(-4, 1)$ cannot be on l_1. Thus l_1 and l_2 are not coincident; they must be parallel. Figure 1.17 shows the actual situation.

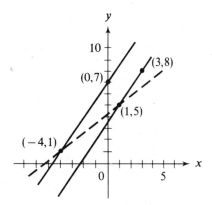

Figure 1.17

Problems

A

In Problems 1–8, find the slope (if any) and the inclination of the line through the given points.

1. (2, 3), (5, 8).
2. $(3, -2), (5, 1)$.
3. $(-3, -2), (3, 2)$.
4. (4, 0), (2, 5).
5. $(-4, 2), (-4, 5)$.
6. $(5, 0), (0, -3)$.
7. $(a, a), (b, b)$, where $a \neq b$.
8. $(a, a), (-a, 2a)$, where $a \neq 0$.

In Problems 9–16, find the slopes of the lines through the two pairs of points; then determine whether the lines are parallel, coincident, perpendicular, or none of these.

9. $(1, -2), (-2, -11)$;   (2, 8), (0, 2).
10. $(3, 4), (1, -2)$; $(-5, -4), (4, -1)$.
11. (3, 5), (2, 1); $(6, 1), (-2, 3)$.
12. $(1, 6), (-4, 3)$; $(8, 1), (3, -2)$.
13. $(1, 1), (4, -1)$; $(-2, 3), (7, -3)$.
14. $(5, 5), (4, -1)$; $(6, 3), (2, -2)$.
15. $(2, 2), (-2, 7)$; $(0, 4), (6, -5)$.
16. $(3, 7), (-3, -1)$; $(-1, -2), (-5, 1)$.

B

17. If the line through $(x, 5)$ and (4, 3) is parallel to a line with slope 3, find x.
18. If the line through $(x, 4)$ and (3, 2) is perpendicular to a line with slope 3, find x.
19. If the line through $(x, 1)$ and $(0, y)$ is coincident with the line through $(5, -1)$ and $(-1, 3)$, find x and y.
20. If the line through (2, 7) and $(0, y)$ is perpendicular to the line through (1, 3) and $(x, 2)$, find a relationship between x and y.
21. If the line through $(x, 4)$ and (3, 7) is parallel to the line through $(x, -1)$ and (5, 1), find x.

22. Show by means of slopes that $(1, 1)$, $(4, 1)$, $(3, -2)$, and $(0, -2)$ are the vertices of a parallelogram.
23. Show by means of slopes that $(-2, 4)$, $(2, 0)$, $(2, 8)$, and $(6, 4)$ are the vertices of a square.
24. Prove analytically that the diagonals of a square intersect at right angles.
25. Prove analytically that the diagonals of a rhombus intersect at right angles.
26. Prove analytically that one median of an isosceles triangle is an altitude.
27. Prove analytically that the medians of an equilateral triangle are altitudes.

C

28. Given $A = (-1, 2)$ and $B = (3, 2)$, find a point P such that $\angle APB$ is a right angle. Describe the set of all such points.
29. Show that if θ is the acute angle of intersection of two lines with slopes m_1 and m_2, then

$$\tan \theta = \frac{|m_1 - m_2|}{1 + m_1 m_2}.$$

1.3

Graphs and Points of Intersection

The graph of an equation in two variables x and y is simply the set of all points (x, y) in the plane whose coordinates satisfy the given equation. The determination of the graph of an equation is one of the principal problems of analytic geometry. Although we shall consider other methods in Chapter 5, we consider only point-by-point plotting here. To do this, we assign a value to either x or y, substitute the assigned value into the given equation, and solve for the other.

Example 1

Graph $x^2 + y^2 = 25$.

x	y
0	± 5
± 1	$\pm 2\sqrt{6}$
± 2	$\pm \sqrt{21}$
± 3	± 4
± 4	± 3
± 5	0

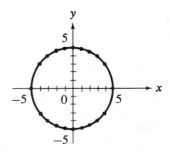

Figure 1.18

Example 2

Graph $y = |x| = \begin{cases} x & \text{if } x \geq 0, \\ -x & \text{if } x < 0. \end{cases}$

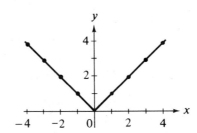

x	y
0	0
± 1	1
± 2	2
± 3	3
± 4	4

Figure 1.19

One obvious question that arises is, How many points must one plot before drawing the graph? There is no specific answer—just plot as many as are needed for a reasonable idea of what the graph looks like.

Since each point of a graph satisfies the given equation, a point of intersection of two graphs is simply a point that satisfies both equations. Thus, any such point can be found by solving the two equations simultaneously.

Example 3

Find all points of intersection of $x^2 + y^2 = 25$ and $x + y = 2$.

Solving the second equation for y and substituting into the first, we have

$$x^2 + (2 - x)^2 = 25,$$

$$2x^2 - 4x - 21 = 0,$$

$$x = \frac{4 \pm \sqrt{16 + 168}}{4} = \frac{2 \pm \sqrt{46}}{2} = 1 \pm \frac{1}{2}\sqrt{46},$$

$$y = 2 - x = 1 \mp \frac{1}{2}\sqrt{46}.$$

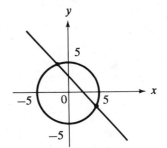

Figure 1.20

Thus, the two points of intersection are

$$\left(1 + \frac{1}{2}\sqrt{46},\, 1 - \frac{1}{2}\sqrt{46}\right) \quad \text{and} \quad \left(1 - \frac{1}{2}\sqrt{46},\, 1 + \frac{1}{2}\sqrt{46}\right).$$

Another interesting graph is represented by the equation $y = [\![x]\!]$, where $[\![x]\!]$ denotes the largest integer less than or equal to x. This is called the greatest integer function. For instance, $[\![1]\!] = 1$, $[\![3/2]\!] = 1$, $[\![7/4]\!] = 1$, $[\![-1/2]\!] = -1$, and so on. The graph of $y = [\![x]\!]$ is given in Figure 1.21. Note that each horizontal segment includes the left-hand end point but not the right-hand one. The relation between the weight of a letter and the amount of postage required is similar to this one (see Problem 31).

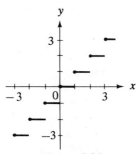

Figure 1.21

Problems

A

Plot the graphs of the equations in Problems 1–12.

1. $3x - 5y = 2.$
2. $y = 3x + 2.$
3. $x^2 + y^2 = 4.$
4. $x^2 - y^2 = 4.$
5. $x^2 - y^2 = -4.$
6. $x^2 + y^2 = 0.$
7. $9x^2 + y^2 = 9.$
8. $x^2 = y^3.$
9. $y = |x| + 2.$
10. $y = |x + 2|.$
11. $y = |x| - 1.$
12. $y = \sqrt{x - 1}.$

In Problems 13–16, find the points of intersection and sketch the graphs of the equations.

13. $3x - 5y = 2$
    $4x + 2y = 1.$
14. $x + y = 2$
    $x^2 + y^2 = 1.$
15. $x + y = 1$
    $x^2 + y^2 = 5.$
16. $y = x^2 + 1$
    $x + y = 2.$

B

Plot the graphs of the equations in Problems 17–26.

17. $|x| + |y| = 1.$
18. $|x + y| = 1.$
19. $y = [\![-x]\!].$
20. $y = [\![x - 1]\!].$
21. $y = [\![|x|]\!].$
22. $y = |[\![x]\!]|.$
23. $y = x + [\![x]\!].$
24. $\sqrt{x} + \sqrt{y} = 4.$
25. $y = \dfrac{x}{x - 2}.$

26. $y = \dfrac{x - 2}{x}.$

In Problems 27 and 28, find the points of intersection and sketch the graph.

27. $x^2 + y^2 = 1$
    $(x - 2)^2 + y^2 = 5.$
28. $y = \sqrt{4 - (x - 2)^2}$
    $x = \sqrt{4 - y^2}.$

29. Graph on one set of axes: $y = x,\ \ y = x^2,\ \ y = x^3,\ \ y = x^4,\ \ y = x^5,$ and $y = x^6.$
30. Graph $y = x,\ \ y = -x,\ \ y = |x|,$ and $y = \sqrt{x^2}.$ Compare.
31. Sketch the "postage stamp graph," giving the amount of postage in terms of the weight. What is its equation? (See the last paragraph of this section.)

C

32. Sketch $y = \{x\},$ where $\{x\} = \min(x - [\![x]\!], 1 - x + [\![x]\!]).$ Interpret the values of y geometrically. *Note:* $\min(x - [\![x]\!], 1 - x + [\![x]\!])$ means the smaller of the two numbers $x - [\![x]\!]$ and $1 - x + [\![x]\!].$

Review Problems

A

1. Use distances to determine whether or not the three points $(1, 5), (-2, -1),$ and $(4, 10)$ are collinear. Check your work by using slopes.
2. Determine x so that $(x, 1)$ is on the line joining $(0, 4)$ and $(4, -2).$
3. Find the lengths of the medians of the triangle with vertices $(-3, 4), (5, 5),$ and $(3, -2).$
4. Use distances to determine whether or not the points $(1, 6), (5, 3),$ and $(3, 1)$ are the vertices of a right triangle. Check your work by using slopes.

5. Find the point of intersection of $2x + y = 5$ and $x - 3y = 7$.   Sketch.

6. Line l_1 contains the points (4, 7) and (2, 3), while l_2 contains (5, 6) and $(-3, 4)$. Are l_1 and l_2 parallel, perpendicular, coincident, or none of these?

7. Find the slopes of the altitudes of the triangle with vertices $(-2, 4)$, (3, 3), and $(-5, -2)$.

B

8. If $A = (1, 7)$ and $B = (-2, 3)$,   find the end points of the segment half the length of AB and centered inside AB.

9. Prove analytically that the lines joining the midpoints of adjacent sides of a quadrilateral form a parallelogram.

10. Find the points of intersection of $x - 7y + 2 = 0$ and $x^2 + y^2 - 4x + 6y - 12 = 0$. Sketch.

11. Sketch the graph of $y = |x + 1| - 2$.

12. If the line l has slope 3 and contains the point $(-1, 1)$, at what points does it cross the coordinate axes?

13. Sketch the graph of $y = 2[\![x - 3]\!]$.

14. The point $(5, -2)$ is at a distance $\sqrt{13}$ from the midpoint of the segment joining (5, y) and $(-1, 1)$. Find y.

C

15. Find the center of the circle circumscribed about the triangle with vertices $(-1, 1)$, (6, 2), and $(7, -5)$.

16. Two vertices of an equilateral triangle are $(a, -a)$ and $(-a, a)$. Find the third.

17. A square has all its vertices in the first quadrant and one of its sides joins (3, 1) and (6, 3). Find the other two vertices.

18. A parallelogram has three vertices (3, 4), (6, 3), and (1, 0) and the fourth vertex in the first quadrant. Find the fourth vertex.

2

Equations of Lines and Circles

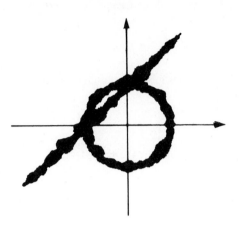

2.1

Lines: Point-Slope and Two-Point Forms

The last section of Chapter 1 dealt with one of the two principal problems of analytic geometry, namely, sketching the graph of an equation. Now we shall begin considering the reverse problem—that is, finding the equation of a curve from its description. We shall restrict ourselves to lines at first.

The simplest ways of determining a line use either a pair of points or one point and the slope. Thus, if a line is described in either of these ways, we should be able to give an equation for it. We begin with a line described by its slope and a point on it.

Theorem 2.1

(*Point-slope form of a line.*) *A line that has slope m and contains the point* (x_1, y_1) *has equation*

$$y - y_1 = m(x - x_1).$$

Proof

Let (x, y) be any point different from (x_1, y_1) on the given line (see Figure 2.1). Since the line has slope, it is not vertical. Thus $x \neq x_1$, which gives

$$m = \frac{y - y_1}{x - x_1}$$

and

$$y - y_1 = m(x - x_1).$$

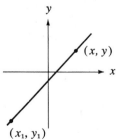

Figure 2.1

Although the formula was derived only for points on the line different from the given point (x_1, y_1), it is easily seen that (x_1, y_1) also satisfies the equation. Thus, every point on the line satisfies the equation. Suppose now that the point (x_2, y_2) satisfies the equation—that is,

$$y_2 - y_1 = m(x_2 - x_1).$$

If $x_2 = x_1$, then $y_2 - y_1 = 0$, or $y_2 = y_1$. In this case, $(x_2, y_2) = (x_1, y_1)$, which is on the line. If $x_2 \neq x_1$, then

$$\frac{y_2 - y_1}{x_2 - x_1} = m.$$

Thus, the slope of the line joining (x_1, y_1) and (x_2, y_2) is m, and this line has the point (x_1, y_1) in common with the given line. Since there can be only one line with slope m containing (x_1, y_1), we see that (x_2, y_2) is on the given line.

Example 1

Find an equation of the line through $(3, -2)$ with slope 4.

$$y - y_1 = m(x - x_1),$$
$$y - (-2) = 4(x - 3),$$
$$4x - y - 14 = 0.$$

Of course vertical lines cannot be represented by the point-slope form, since they have no slope. Again, remember that "no slope" does not mean "zero slope." A horizontal line has $m = 0$, and it can be represented by the point-slope form, which gives $y - y_1 = 0$. There is no x in the resulting equation! But the points on a horizontal line satisfy the condition that all have the same y coordinate, no matter what the x coordinate is. Similarly, the points on a vertical line satisfy the condition that all have the same x coordinate. Thus, if (x_1, y_1) is one point on a vertical line, then $x = x_1$, or $x - x_1 = 0$ for every point (x, y) on the line.

Example 2

Find an equation of the vertical line through $(5, -2)$.

Since the x coordinate of the given point is 5, all points on the line have x coordinates 5. Thus

$$x = 5 \quad \text{or} \quad x - 5 = 0.$$

Theorem 2.2

(Two-point form of a line.) A line through (x_1, y_1) and (x_2, y_2), $x_1 \neq x_2$, has equation

$$y - y_1 = \frac{y_2 - y_1}{x_2 - x_1}(x - x_1).$$

It might be noted that this result is often stated in the form

$$\frac{y - y_1}{x - x_1} = \frac{y_2 - y_1}{x_2 - x_1}.$$

While the symmetry of this form is appealing, the form has one serious defect—the point (x_1, y_1) is on the desired line, but it does not satisfy this equation. It does satisfy the equation of Theorem 2.2.

The proof of Theorem 2.2 follows directly from Theorem 2.1 and the fact that $m = (y_2 - y_1)/(x_2 - x_1)$, provided $x_1 \neq x_2$. Actually this follows so easily from Theorem 2.1 that you may prefer to use the earlier theorem after finding the slope from the two given points. Of course, the designation of the two points as "point 1" and "point 2" is quite arbitrary.

Example 3

Find an equation of the line through $(4, 1)$ and $(-2, 3)$.

$$y - y_1 = \frac{y_2 - y_1}{x_2 - x_1} (x - x_1),$$

$$y - 1 = \frac{3 - 1}{-2 - 4} (x - 4),$$

$$x + 3y - 7 = 0.$$

Example 4

Find the perpendicular bisector of the segment joining $(5, -3)$ and $(1, 7)$.

First let us find the midpoint.

$$x = \frac{5 + 1}{2} = 3, \qquad y = \frac{-3 + 7}{2} = 2.$$

The midpoint is $(3, 2)$.

The slope of the line joining $(5, -3)$ and $(1, 7)$ is

$$m = \frac{-3 - 7}{5 - 1} = -\frac{5}{2};$$

the slope of the perpendicular line is $m = 2/5$.

Now we merely need to use the point-slope formula, using $(3, 2)$ and $m = 2/5$.

$$y - 2 = \frac{2}{5} (x - 3),$$

$$2x - 5y + 4 = 0.$$

Problems

A

In Problems 1–16, find an equation of the line indicated and sketch the graph.

1. Through $(2, -4)$; $m = -2$.
2. Through $(-2, 1)$; $m = 3$.
3. Through $(5, 1)$; $m = -4$.
4. Through $(2, -3)$; $m = -2$.
5. Through $(0, 0)$; $m = 1$.
6. Through $(0, -2)$; $m = -4$.
7. Through $(-3, 1)$; $m = 0$.
8. Through $(-4, -5)$; no slope.

9. Through $(1, 4)$ and $(3, 5)$.

10. Through $(4, 3)$ and $(6, -1)$.

11. Through $(-2, 4)$ and $(3, 5)$.

12. Through $(5, 5)$ and $(-1, -1)$.

13. Through $(0, 0)$ and $(1, 5)$.

14. Through $(0, 4)$ and $(8, 0)$.

15. Through $(4, 5)$ and $(4, 8)$.

16. Through $(2, 4)$ and $(-1, 4)$.

17. Find equations of the three sides of the triangle with vertices $(1, 4)$, $(3, 0)$, and $(-1, -2)$.

18. Find equations of the medians of the triangle of Problem 17.

19. Find equations of the altitudes of the triangle of Problem 17.

20. Find the vertices of the triangle with sides $x - 5y + 8 = 0$, $4x - y - 6 = 0$, and $3x + 4y + 5 = 0$.

21. Find equations of the medians of the triangle of Problem 20.

22. Find equations of the altitudes of the triangle of Problem 20.

23. Find an equation of the chord of the circle $x^2 + y^2 = 25$ that joins $(-4, -3)$ and $(0, 5)$. Sketch the circle and its chord.

24. Find an equation of the chord of the parabola $y = x^2$ that joins $(1, 1)$ and $(-3, 9)$. Sketch the curve and its chord.

B

25. Find an equation of the perpendicular bisector of the segment joining $(5, 4)$ and $(-3, 0)$.

26. Find an equation of the line through the points of intersection of the circles

$$x^2 + y^2 + 2x - 19 = 0 \quad \text{and} \quad x^2 + y^2 - 10x - 12y + 41 = 0.$$

Look over your work. Is there any easier way?

27. Repeat Problem 26 for the circles

$$x^2 + y^2 - 4x + 2y + 1 = 0 \quad \text{and} \quad x^2 + y^2 - 16x - 4y + 52 = 0.$$

What is wrong?

28. Find an equation of the line through the centers of the two circles of Problem 26.

29. What condition must the coordinates of a point satisfy in order that it be equidistant from $(2, 5)$ and $(4, -1)$?

30. Find the center and radius of the circle through the points $(-1, 3)$, $(2, 2)$, and $(3, -5)$.

31. Consider the triangle with vertices $A = (3, 1)$, $B = (0, 5)$, and $C = (7, 4)$. Find equations of the altitude and the median from A. What do your results tell you about the triangle?

32. The pressure within a partially evacuated container is being measured by means of an open-end manometer. This gives the difference between the pressure in the container and atmospheric pressure. It is known that a difference of 0 mm of mercury corresponds to a pressure of 1 atmosphere and that if the pressure in the container were reduced to 0 atmospheres, a difference of 760 mm of mercury would be observed. Assuming that the difference D in mm of mercury and the pressure P in atmospheres are related by a linear relation, determine what such a relation is.

33. Knowing that water freezes at $0°C$, or $32°F$, that it boils at $100°C$, or $212°F$, and that the relation between the temperature in degrees centigrade C and in degrees Fahrenheit F is linear, find that relation.

34. The amount of a given commodity that consumers are willing to buy at a given price is called the demand for that commodity corresponding to the given price; the relationship between the price and the demand is called a demand equation. Similarly the amount that manufacturers are willing to offer for sale at a given price is called the supply corresponding to the given price, and the relationship between the price and the supply is called a supply equation. Market equilibrium exists when the supply and demand are equal. The demand and supply equations for a given commodity are

$$2p + x - 100 = 0 \quad \text{and} \quad p - x + 10 = 0,$$

respectively, where p is the price of the commodity and x is its supply or demand. At what price will there be market equilibrium? Graph both equations with p on the vertical axis. What happens to the demand as the price increases? What happens to the supply as the price increases?

C

35. Show that a line through points (x_1, y_1) and (x_2, y_2) can be represented by

$$\begin{vmatrix} x & y & 1 \\ x_1 & y_1 & 1 \\ x_2 & y_2 & 1 \end{vmatrix} = 0.$$

The expression on the left-hand side of this equation is a determinant. Some authors use the notation

$$\det \begin{bmatrix} x & y & 1 \\ x_1 & y_1 & 1 \\ x_2 & y_2 & 1 \end{bmatrix}$$

for this determinant.

36. Show that the points (x_1, y_1), (x_2, y_2), (x_3, y_3) are collinear if and only if

$$\begin{vmatrix} x_1 & y_1 & 1 \\ x_2 & y_2 & 1 \\ x_3 & y_3 & 1 \end{vmatrix} = 0.$$

37. Show that if no pair of the equations

$$A_1 x + B_1 y + C_1 = 0$$
$$A_2 x + B_2 y + C_2 = 0$$
$$A_3 x + B_3 y + C_3 = 0$$

represents parallel lines, then the lines are concurrent if and only if

$$\begin{vmatrix} A_1 & B_1 & C_1 \\ A_2 & B_2 & C_2 \\ A_3 & B_3 & C_3 \end{vmatrix} = 0.$$

2.2

Lines: Slope-Intercept and Intercept Forms

The x and y intercepts of a line are the points at which the line crosses the x and y axes, respectively. These points are of the form $(a, 0)$ and $(0, b)$ (see Figure 2.2), but they are usually represented simply by a and b, since the 0's are understood by their position on the axes. We shall continue using the convention that the x and y intercepts of a line are represented by the symbols a and b, respectively. It might be noted that lines parallel to the x axis have no x intercept and those parallel to the y axis have no y intercept. While a line on the x axis has infinitely many points in common with the x axis, we shall adopt the convention that it has no x intercept. Similarly, a line on the y axis has no y intercept. Thus no horizontal line has an x intercept and no vertical line has a y intercept. One other special case is that of a line through the origin which is neither horizontal nor vertical; it has a single point (the origin) which is both its x intercept and y intercept. In this case $a = b = 0$. With these special points defined, we now introduce two more forms of a line.

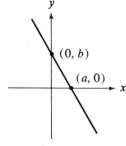

Figure 2.2

Theorem 2.3

(*Slope-intercept form of a line.*) *A line with slope m and y intercept b has equation*

$$y = mx + b.$$

Proof

Since the y intercept is really the point $(0, b)$, the use of the point-slope form gives

$$y - b = m(x - 0) \quad \text{or} \quad y = mx + b.$$

Theorem 2.4

(*Intercept form of a line.*) *A line with nonzero intercepts a and b has equation*

$$\frac{x}{a} + \frac{y}{b} = 1.$$

Proof

Since the intercepts are the points $(a, 0)$ and $(0, b)$, the line has slope

$$m = -\frac{b}{a}.$$

Using the slope-intercept form, we have

$$y = -\frac{b}{a}x + b.$$

Dividing through by b gives

$$\frac{y}{b} = \frac{-x}{a} + 1, \quad \text{or} \quad \frac{x}{a} + \frac{y}{b} = 1.$$

It might be noted that these two forms are merely special cases of the point-slope and two-point forms; thus, the earlier forms may be used in place of these at any time. However, these variations, especially the slope-intercept form, are so convenient to use that it is well to remember them. We shall see an example of their use shortly.

Example 1

Find an equation of the line with slope 2 and y intercept 5.

$$y = mx + b,$$
$$y = 2x + 5,$$
$$2x - y + 5 = 0.$$

There is no commonly used special form for a line with a given slope and x intercept. Although one can easily be derived (see Problem 34), it has not proved as convenient as the slope-intercept form. If you know the slope and the x intercept, simply use the point-slope form, the point being $(a, 0)$.

Example 2

Find an equation of the line with x and y intercepts 5 and -2, respectively.

$$\frac{x}{a} + \frac{y}{b} = 1,$$

$$\frac{x}{5} + \frac{y}{-2} = 1,$$

$$-2x + 5y = -10,$$

$$2x - 5y - 10 = 0.$$

Just as it was true that vertical lines could not be represented by the point-slope form, we see that vertical lines cannot be represented by the slope-intercept form, since vertical lines have neither slope nor y intercept. The intercept form of a line is even more restrictive, accommodating neither horizontal nor vertical lines, because a horizontal line has no x intercept and a vertical line has no y intercept. Furthermore, no line through the origin can be put into the intercept form, since $a = b = 0$ gives 0's in the denominators.

In all of the examples considered so far, we used the special forms only as a starting point; the final form was always $Ax + By + C = 0$. The question arises, Can every equation representing a line be put into such a form and does every equation in such a form represent a line?

Theorem 2.5

(*General form of a line.*) *Every line can be represented by an equation of the form*

$$Ax + By + C = 0,$$

where A and B are not both zero, and any such equation represents a line.

Proof

Any line we consider is either vertical or can be put into slope-intercept form. Thus any line can be represented by either

$$x = k \quad \text{or} \quad y = mx + b.$$

Thus any line is in the form

$$x - k = 0 \quad \text{or} \quad mx - y + b = 0.$$

Both are special cases of $Ax + By + C = 0$.

Suppose we have an equation of the form $Ax + By + C = 0$, where A and B are not both 0. Let us consider two cases.

Case I: $B = 0$. Then

$$Ax + C = 0 \quad \text{and} \quad x = -\frac{C}{A}.$$

(since $B = 0$ and A and B are not both 0, we know that $A \neq 0$ and we may divide by A). This represents an equation of a vertical line.

Case II: $B \neq 0$. Solving $Ax + By + C = 0$ for y, we have

$$y = -\frac{A}{B}x - \frac{C}{B}$$

(since $B \neq 0$, we may divide by B). This represents an equation of a line with slope $-A/B$ and y intercept $-C/B$.

Theorem 2.5 has the following implication for graphing: any equation of the form $Ax + By + C = 0$ represents a line, and its graph can be determined by two of its points. Since the intercepts are so easily located, finding the line through these two points (if there are two) is the quickest way of sketching a line. Of course, vertical or horizontal lines do not have two intercepts, but these are easily sketched. The only problem comes from lines through the origin. The origin is both the x and y intercept; so just find a second point in any convenient way.

Example 3

Sketch the line $2x - 3y - 6 = 0$.

When $y = 0$, $x = 3$, and when $x = 0$, $y = -2$. We did not put the equation into intercept form in order to determine the intercepts, although we might have done so; however, we can find the intercepts by inspection if we set y and x equal to zero in turn and solve for the other. Actually this represents a convenient way of putting the line into intercept form. Since $a = 3$ and $b = -2$, the intercept form of $2x - 3y - 6 = 0$ is

$$\frac{x}{3} + \frac{y}{-2} = 1.$$

The graph of this equation is given in Figure 2.3.

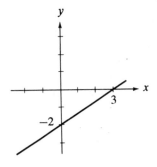

Figure 2.3

Problems

A

In Problems 1–18, find an equation of the line described and express it in general form with integer coefficients. Sketch the line.

1. $m = 4, b = 2$.
2. $m = -2, b = 4$.
3. $m = 3, b = 2/3$.
4. $m = 1/4. b = 3/4$.
5. $m = 3/4, b = 2/3$.
6. $m = 1/4, b = -2/5$.
7. $m = 3, a = 4$.
8. $m = 3, a = -2$.
9. $a = 4, b = 2$.
10. $a = -2, b = 5$.
11. $a = 1/3, b = 4$.
12. $a = 1/3, b = 1/3$.
13. $a = 2/3, b = -2/5$.
14. $a = -2/5, b = 3/4$.
15. $a = b = 0$, through $(5, 1)$.
16. $a = b = 0$, through $(-3, 2)$.
17. $a = 4$, no b.
18. No a, $b = -5$.

19. Find an equation of the line parallel to $3x - 2y + 1 = 0$ and containing the point (5, 1).
20. Find an equation of the line parallel to $2x + y - 3 = 0$ with y intercept -5.
21. Find an equation of the line perpendicular to $x + 2y - 5 = 0$ and containing the point (4, 1).
22. Find an equation of the line perpendicular to $3x + y - 5 = 0$ with x intercept 4.

B

23. Find the center of the circle circumscribed about the triangle with vertices (1, 3), (4, -2), and (-2, 1).
24. Find the center of the circle circumscribed about the triangle with sides $x + y = 2$, $x - y = 0$,   and $2x - y = 4$.
25. Find the orthocenter (point of concurrency of the altitudes) of the triangle with vertices (1. 4), (7, 3), and (2, -3).
26. Prove analytically that the altitudes of a triangle are concurrent.
27. For what value(s) of m does the line $y = mx - 2$ have x intercept 5?
28. For what value(s) of m does the line $y = mx + 2$ contain the point (5, 3)?
29. For what value(s) of a does the line $(x/a) - (y/2) = 1$ have slope 5?
30. For what value(s) of b does the line $(x/3) + (y/b) = 1$ have slope -3?
31. Plot the graph of $x^2 - y^2 = 0$.
32. Plot the graph of $xy = 0$.
33. Plot the graph of $x^2 - 5x + 6 = 0$.
34. Find an equation of the line with slope m and x intercept a.

C

35. The relationship between the vapor pressure P of a liquid and its absolute temperature T is given by the Clausius-Clapeyron equation

$$2.303 \log_{10} P = \frac{-\Delta H}{R} \cdot \frac{1}{T} + C,$$

where ΔH is the molar heat of vaporization of the liquid and R is the ideal gas constant, 1.987 calories degree^{-1} mole^{-1}. Measurements of the vapor pressure of a liquid were made at several temperatures and $\log_{10} P$ as ordinate was plotted against $1/T$ as abscissa. The resulting set of points determined a line with slope -0.0155. What is the molar heat of vaporization of the liquid?

36. The Freundlich equation for adsorption is

$$y = kC^{1/n},$$

where y represents the weight in grams of substance adsorbed, C the concentration in moles/liter of the solute. In logarithmic form, the equation is

$$\log_{10} y = \log_{10} k + \frac{1}{n} \log_{10} C.$$

Freundlich experimented with the adsorption of acetic acid from water solutions by charcoal and plotted $\log_{10} C$ as abscissa against $\log_{10} y$ as ordinate. He found that the points determined a line with slope 0.431 and "$\log_{10} y$" intercept -0.796. What are k and n?

37. Work Problem 35 of the previous section without expanding the determinant. (*Hint:* Use Theorem 2.5.)

38. One vertex of a rectangle is (4, 1); the diagonals intersect at (1, 5); and one side has slope 7. Find the other three vertices.

39. One vertex of a parallelogram is (2, 1); the diagonals intersect at (1/2, 5/2); and the sides have slopes 5/2 and 2/5. Find the other three vertices.

2.3

Distance from a Point to a Line

Before considering the distance from a point to a line, let us recall some simple facts from the preceding sections. $Ax + By + C_1 = 0$ and $Ax + By + C_2 = 0$ must be parallel, since they give

$$y = -\frac{A}{B}x - \frac{C_1}{B} \quad \text{and} \quad y = -\frac{A}{B}x - \frac{C_2}{B}$$

when $B \neq 0$, and they represent two vertical lines when $B = 0$. Moreover, if we are given the line $Ax + By + C = 0$ and the point (x_1, y_1), then the line through (x_1, y_1) and parallel to the given line is

$$Ax + By - (Ax_1 + By_1) = 0.$$

Also, $Ax + By + C_1 = 0$ and $Bx - Ay + C_2 = 0$ are perpendicular, since they give

$$y = -\frac{A}{B}x - \frac{C_1}{B} \quad \text{and} \quad y = \frac{B}{A}x + \frac{C_2}{A}$$

when neither A nor B is 0; and they give horizontal and vertical lines when either $A = 0$ or $B = 0$. Furthermore,

$$Bx - Ay - (Bx_1 - Ay_1) = 0$$

contains (x_1, y_1) and is perpendicular to $Ax + By + C = 0$.

Theorem 2.6

The distance from the point (x_1, y_1) to the line $Ax + By + C = 0$ is

$$d = \frac{|Ax_1 + By_1 + C|}{\sqrt{A^2 + B^2}}.$$

Proof

Given the line

$$Ax + By + C = 0$$

and the point (x_1, y_1), then

$$Bx - Ay - (Bx_1 - Ay_1) = 0$$

is perpendicular to the given line and contains (x_1, y_1) (see Figure 2.4). The distance we seek is the distance between (x_1, y_1) and the point of intersection of this line with the given line. This point of intersection is

$$\left(-\frac{AC - B^2x_1 + ABy_1}{A^2 + B^2}, \quad -\frac{BC + ABx_1 - A^2y_1}{A^2 + B^2} \right).$$

Using the distance formula, we have

$$d = \sqrt{\left(x_1 + \frac{AC - B^2x_1 + ABy_1}{A^2 + B^2}\right)^2 + \left(y_1 + \frac{BC + ABx_1 - A^2y_1}{A^2 + B^2}\right)^2}$$

$$= \sqrt{\left(\frac{A^2x_1 + AC + ABy_1}{A^2 + B^2}\right)^2 + \left(\frac{B^2y_1 + BC + ABx_1}{A^2 + B^2}\right)^2}$$

$$= \sqrt{\left(\frac{A(Ax_1 + By_1 + C)}{A^2 + B^2}\right)^2 + \left(\frac{B(Ax_1 + By_1 + C)}{A^2 + B^2}\right)^2}$$

$$= \sqrt{(A^2 + B^2)\left(\frac{Ax_1 + By_1 + C}{A^2 + B^2}\right)^2}$$

$$= \sqrt{\frac{(Ax_1 + By_1 + C)^2}{A^2 + B^2}}$$

$$= \frac{|Ax_1 + By_1 + C|}{\sqrt{A^2 + B^2}}.$$

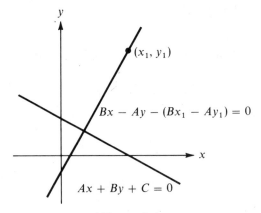

Figure 2.4

Example 1

Find the distance from the point $(1, 4)$ to the line $3x - 5y + 2 = 0$.

$$d = \frac{|Ax_1 + By_1 + C|}{\sqrt{A^2 + B^2}}$$

$$= \frac{|3 \cdot 1 - 5 \cdot 4 + 2|}{\sqrt{3^2 + (-5)^2}}$$

$$= \frac{15}{\sqrt{34}}.$$

The absolute value in the distance formula is sometimes very inconvenient in practice. We could get rid of it if we knew whether $Ax_1 + By_1 + C$ were positive or negative. The following theorem gives us a method of determining this.

Theorem 2.7

If $P(x_1, y_1)$ is a point not on the line $Ax + By + C = 0$ $(B \neq 0)$, then

(a) B and $Ax_1 + By_1 + C$ agree in sign if P is above the line;
(b) B and $Ax_1 + By_1 + C$ have opposite signs if P is below the line.

Proof

Case I: $B > 0$. Let Q be the point on the given line with abscissa x_1 (see Figure 2.5). If P is above the line, then $y_1 > y$. $By_1 > By$, since $B > 0$. Therefore,

$$Ax_1 + By_1 + C > Ax_1 + By + C.$$

Since (x_1, y) is on the line,

$$Ax_1 + By + C = 0 \quad \text{and} \quad Ax_1 + By_1 + C > 0.$$

If P is below the line, all of the above inequalities are reversed and

$$Ax_1 + By_1 + C < 0.$$

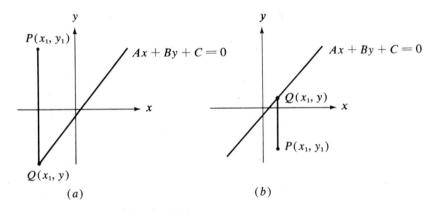

Figure 2.5

Case II: $B < 0$. If P is above the line, then $y_1 > y$. $By_1 < By$, since $B < 0$. Thus,

$$Ax_1 + By_1 + C < Ax_1 + By + C.$$

Again

$$Ax_1 + By + C = 0 \quad \text{and} \quad Ax_1 + By_1 + C < 0.$$

As with Case I, all of these inequalities are reversed if P is below the line, and

$$Ax_1 + By_1 + C > 0.$$

If $B = 0$, the line is vertical and there is no "above" nor "below." Theorem 2.7 does not apply to this case, but the distance from a point to a vertical line is easily found without using Theorem 2.6. Other methods of determining the sign of $Ax_1 + By_1 + C$ are given in Problems 34 and 35.

Example 2

Find an equation of the line bisecting the obtuse angle between $3x - 4y - 3 = 0$ and $5x + 12y + 1 = 0$.

If (x, y) is any point on the desired line (see Figure 2.6), then it is equidistant from the two given lines. By Theorem 2.6,

$$\frac{|5x + 12y + 1|}{\sqrt{5^2 + 12^2}} = \frac{|3x - 4y - 3|}{\sqrt{3^2 + (-4)^2}},$$

$$5|5x + 12y + 1| = 13|3x - 4y - 3|.$$

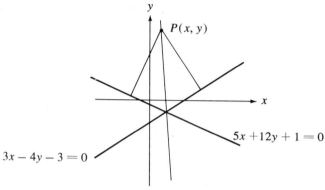

Figure 2.6

Now let us apply Theorem 2.7. Since P is above $5x + 12y + 1 = 0$ and the coefficient of y is positive, $5x + 12y + 1$ is also positive. Similarly, since P is above $3x - 4y - 3 = 0$ and B is negative,

$$3x - 4y - 3 < 0.$$

Thus

$$5(5x + 12y + 1) = -13(3x - 4y - 3) \quad \text{or} \quad 32x + 4y - 17 = 0.$$

Perhaps you object to the designation of P above both lines. Not every point on the bisector is above them. While this is true, the points on the bisector that are not above both are below both. Thus, we still have one expression positive and the other negative, and the result is the same.

It might be noted that we can avoid the use of Theorem 2.7 by considering both cases; that is, $5x + 12y + 1$ and $3x - 4y - 3$ either agree in sign or have opposite signs. We then get two answers, and Figure 2.6 indicates which is correct.

Problems

A

In Problems 1–10, find the distance from the given point to the given line.

1. $x + y - 5 = 0$, (2, 5).
2. $2x - 4y + 3 = 0$, (1, 3).
3. $4x + 3y - 5 = 0$, (−2, 4).
4. $x - 3y + 5 = 0$, (1, −2).
5. $3x + 4y - 5 = 0$, (1, 1).
6. $5x + 12y + 13 = 0$, (−2, 3).
7. $2x - 5y = 3$, (−2, 5).
8. $2x + y = 5$, (4, 2).
9. $3x + 4 = 0$, (2, 4).
10. $y = 3$, (4, −3).

11. Find the altitudes of the triangle with vertices $(1, 2)$, $(5, 5)$, and $(-1, 7)$.

12. Find the altitudes of the triangle with sides $x + y - 3 = 0$, $x - 2y + 4 = 0$, and $2x + 3y = 5$.

B

In Problems 13–18, find an equation of the line bisecting the acute angle between the given lines.

13. $3x - 4y - 2 = 0$, $4x - 3y + 4 = 0$.   14. $8x + 15y - 5 = 0$, $12x + 5y + 3 = 0$.

15. $24x - 7y - 1 = 0$, $3x + 4y - 5 = 0$.   16. $12x + 35y + 1 = 0$, $8x + 15y + 9 = 0$.

17. $x + y - 2 = 0$, $2x - 3 = 0$.   18. $2x + y + 3 = 0$, $y + 5 = 0$.

In Problems 19–24, find the distance between the given parallel lines.

19. $2x - 5y + 5 = 0$, $2x - 5y + 8 = 0$.   20. $x + 2y - 1 = 0$, $x + 2y + 4 = 0$.

21. $2x + y + 2 = 0$, $4x + 2y - 3 = 0$.   22. $4x - y + 6 = 0$, $12x - 3y + 1 = 0$.

23. $2x - y + 2 = 0$, $2x - y - 8 = 0$.   24. $3x + 2y = 0$, $6x + 4y - 13 = 0$.

25. Find the area of the triangle of Problem 11.

26. Find the area of the triangle of Problem 12.

27. The center of the circle inscribed in a triangle is the incenter of the triangle. The center of a circle that is tangent to one side and the extensions of the other two sides is an excenter of the triangle. Find the incenter and the three excenters of the triangle with vertices $(3, 1)$, $(5, 6)$, and $(-9/4, 31/10)$.

28. For what value(s) of m is the line $y = mx + 13$ at a distance 5 from the origin?

29. For what value(s) of m is the line $y = mx + 1$ at a distance 3 from $(4, 1)$?

30. For what value(s) of a is the line $(x/a) + (y/2) = 1$ at a distance 2 from the point $(4, 0)$?

31. For what value(s) of b is the line $(x/5) + (y/b) = 1$ at a distance 4 from the origin.

32. Find the center of the circle inscribed in the triangle with vertices $(0, 0)$, $(4, 0)$, and $(0, 3)$.

33. A board leaning against a fence makes an angle of $30°$ with the horizontal. If the board is 4 ft long (see Figure 2.7), what is the diameter of the largest pipe that will fit between the board, the fence, and the ground?

C

34. Prove that if $P(x_1, y_1)$ is a point not on the line $Ax + By + C = 0$ $(A \neq 0)$, then
    (a) A and $Ax_1 + By_1 + C$ agree in sign if P is to the right of the line;
    (b) A and $Ax_1 + By_1 + C$ have opposite signs if P is to the left of the line.

35. Prove that if $P(x_1, y_1)$ is a point not on the line $Ax + By + C = 0$ $(C \neq 0)$, then
    (a) C and $Ax_1 + By_1 + C$ agree in sign if P and the origin are on the same side of the line;
    (b) C and $Ax_1 + By_1 + C$ have opposite signs if P and the origin are on opposite sides of the line.

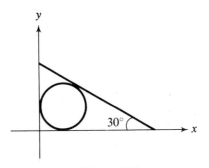

Figure 2.7

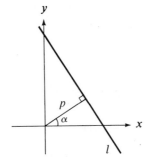

Figure 2.8

36. Suppose that α is the inclination of a line perpendicular (or normal) to the line l and p is the directed distance of l from the origin, p being positive if l is above the origin and negative if l is below (see Figure 2.8). Show that the equation of l can be put into the form

$$x \cos \alpha + y \sin \alpha - p = 0.$$

This is called the normal form of the line.

2.4

The Circle

The standard form for an equation of a circle is a direct consequence of the definition and the length formula.

Definition

A ***circle*** *is the set of all points in a plane at a fixed positive distance (radius) from a fixed point (center).*

Theorem 2.8

A circle with center (h, k) and radius r has equation in standard form

$$(x - h)^2 + (y - k)^2 = r^2.$$

Proof

If (x, y) is any point on the circle, then the distance from the center (h, k) to (x, y) is r (see Figure 2.9):

$$r = \sqrt{(x - h)^2 + (y - k)^2}.$$

Squaring, we have

$$(x - h)^2 + (y - k)^2 = r^2.$$

Since the steps above are reversible (why? see Problem 42), we see that every point satisfying the equation of Theorem 2.8 is on the circle described.

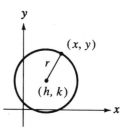

Figure 2.9

Example 1

Give an equation for the circle with center $(3, -5)$ and radius 2.

From Theorem 2.8, an equation is

$$(x - 3)^2 + [y - (-5)]^2 = 2^2,$$

or

$$(x - 3)^2 + (y + 5)^2 = 4.$$

Although the above form is convenient, in that it shows at a glance the center and radius of the circle, another form is usually used. Called the general form, it is comparable to the general form of a line. Let us first illustrate it with the result of Example 1. Squaring the two binomials and combining similar terms, we have

$$(x - 3)^2 + (y + 5)^2 = 4,$$
$$x^2 - 6x + 9 + y^2 + 10y + 25 = 4,$$
$$x^2 + y^2 - 6x + 10y + 30 = 0.$$

Normally an equation of a circle will be given in this form. Let us now repeat the process, starting with the standard form of Theorem 2.8.

$$(x - h)^2 + (y - k)^2 = r^2,$$
$$x^2 - 2hx + h^2 + y^2 - 2ky + k^2 = r^2,$$
$$x^2 + y^2 - 2hx - 2ky + (h^2 + k^2 - r^2) = 0,$$

which is in the form

$$x^2 + y^2 + D'x + E'y + F' = 0.$$

Multiplying by a nonzero constant, A, gives

$$Ax^2 + Ay^2 + Dx + Ey + F = 0 \quad (A \neq 0),$$

as the following theorem states.

Theorem 2.9

Every circle can be represented in the general form

$$Ax^2 + Ay^2 + Dx + Ey + F = 0 \quad (A \neq 0).$$

It is a simple matter to take an equation of a circle in the standard form and reduce it to the general form. We have already seen an example of this. However, it is somewhat more difficult to go from the general form to the standard form. The latter is accomplished by the process of "completing the square." To see how this is done, suppose we consider

$$(x + a)^2 = x^2 + 2ax + a^2.$$

The constant term a^2 and the coefficient of x have a definite relationship; namely, the constant term is the square of one-half the coefficient of x. Thus,

$$a^2 = \left[\frac{1}{2}(2a)\right]^2.$$

Note, however, that this relationship holds only when the coefficient of x^2 is 1. If the coefficients of x^2 and y^2 are not 1, make them 1 by division. Group the x terms and the y terms on one side of the equation and take the constant to the other side. Then complete the square on both the x and the y terms.

Example 2

Express $2x^2 + 2y^2 - 2x + 6y - 3 = 0$ in the standard form.

$$2x^2 + 2y^2 - 2x + 6y - 3 = 0,$$

$$x^2 + y^2 - x + 3y - \frac{3}{2} = 0,$$

$$(x^2 - x \quad) + (y^2 + 3y \quad) = \frac{3}{2},$$

$$\left(x^2 - x + \frac{1}{4}\right) + \left(y^2 + 3y + \frac{9}{4}\right) = \frac{3}{2} + \frac{1}{4} + \frac{9}{4},$$

$$\left(x - \frac{1}{2}\right)^2 + \left(y + \frac{3}{2}\right)^2 = 4.$$

Thus, the original equation represents a circle with center $(1/2, -3/2)$ and radius 2.

The next two examples show that the converse of Theorem 2.9 is not true: that is, an equation of the form

$$Ax^2 + Ay^2 + Dx + Ey + F = 0$$

does not necessarily represent a circle.

Example 3

Express $x^2 + y^2 + 4x - 6y + 13 = 0$ in standard form.

$$x^2 + y^2 + 4x - 6y + 13 = 0,$$
$$(x^2 + 4x \quad) + (y^2 - 6y \quad) = -13,$$
$$(x^2 + 4x + 4) + (y^2 - 6y + 9) = -13 + 4 + 9,$$
$$(x + 2)^2 + (y - 3)^2 = 0.$$

Since neither of the two expressions on the left-hand side of the last equation can be negative, their sum can be zero only if both expressions are zero. This is possible only when $x = -2$ and $y = 3$. Thus, the point $(-2, 3)$ is the only point in the plane that satisfies the original equation.

Example 4

Express $x^2 + y^2 + 2x + 8y + 19 = 0$ in standard form.

$$x^2 + y^2 + 2x + 8y + 19 = 0,$$
$$(x^2 + 2x \quad) + (y^2 + 8y \quad) = -19,$$
$$(x^2 + 2x + 1) + (y^2 + 8y + 16) = -19 + 1 + 16,$$
$$(x + 1)^2 + (y + 4)^2 = -2.$$

Again, since neither expression on the left-hand side of the last equation can be negative, their sum cannot possibly be negative. There is no point in the plane satisfying this equation. It has no graph.

The results illustrated by the last three examples are stated in the next theorem.

Theorem 2.10

Every equation of the form

$$Ax^2 + Ay^2 + Dx + Ey + F = 0 \quad (A \neq 0)$$

represents either a circle, a point, or no graph. (The last two cases are called the degenerate cases of a circle.)

Problems

A

In Problems 1–12, write an equation of the circle in both the standard form and the general form. Sketch.

1. Center $(1, 3)$; radius 5.
2. Center $(-2, 1)$, radius 2.
3. Center $(3, -2)$; radius 3.
4. Center $(0, 0)$; radius $1/2$.
5. Center $(1/2, -3/2)$; radius 2.
6. Center $(-1/4, 3/4)$; radius $1/4$.
7. Center $(5, 1)$; $(2, -3)$ on the circle.
8. Center $(2, -4)$; $(5, 1)$ on the circle.
9. $(2, -3)$ and $(-2, 0)$ are the end points of a diameter.
10. $(4, -1)$ and $(8, 3)$ are the end points of a diameter.
11. Radius 5; in the second quadrant and tangent to both axes.
12. Radius 2; in the third quadrant and tangent to both axes.

In Problems 13–16, express the equation in standard form. Sketch if there is a graph.

13. $x^2 + y^2 - 2x - 4y + 1 = 0$.
14. $x^2 + y^2 - 4x + 10y + 20 = 0$.
15. $x^2 + y^2 + 2x - 6y + 10 = 0$.
16. $x^2 + y^2 - 4y = 0$.

B

In Problems 17–20, write an equation of the circle described in both the standard form and the general form. Sketch.

17. Radius 2; tangent to $x = 2$ and $y = -1$ and above and to the right of these lines.
18. Radius 5; tangent to $x = 3$ and $y = -2$ and below and to the left of these lines.
19. Tangent to both axes at $(-3, 0)$ and $(0, 3)$.
20. Tangent to $x = -3$ and $y = 2$ at $(-3, 0)$ and $(-1, 2)$.

In Problems 21–28, express the equation in standard form. Sketch if there is a graph.

21. $4x^2 + 4y^2 - 4x - 12y + 1 = 0$.
22. $9x^2 + 9y^2 - 6x - 12y - 4 = 0$.
23. $2x^2 + 2y^2 - 2x + 6y + 5 = 0$.
24. $9x^2 + 9y^2 + 12x + 24y - 16 = 0$.
25. $9x^2 + 9y^2 - 6x + 18y + 11 = 0$.
26. $36x^2 + 36y^2 - 36x - 24y - 59 = 0$.
27. $16x^2 + 16y^2 - 16x + 8y + 21 = 0$.
28. $8x^2 + 8y^2 + 12x - 4y - 27 = 0$.

29. Find the point(s) of intersection of
$$x^2 + y^2 - x - 3y - 6 = 0 \quad \text{and} \quad 4x - y - 9 = 0.$$

30. Find the point(s) of intersection of
$$x^2 + y^2 + 6x - 12y + 5 = 0 \quad \text{and} \quad 2x + 3y + 6 = 0.$$

31. Find the point(s) of intersection of
$$x^2 + y^2 - 6x + 2y - 15 = 0 \quad \text{and} \quad x^2 + y^2 + 6x - 22y + 45 = 0.$$

32. Find the point(s) of intersection of
$$x^2 + y^2 + x + 12y + 8 = 0 \quad \text{and} \quad 2x^2 + 2y^2 - 4x + 9y + 4 = 0.$$

33. What happens when we try to solve simultaneously
$$x^2 + y^2 - 2x + 4y + 1 = 0 \quad \text{and} \quad x - 2y + 2 = 0?$$
Interpret geometrically.

34. What happens when we try to solve simultaneously
$$x^2 + y^2 - 4x - 2y + 1 = 0 \quad \text{and} \quad x^2 + y^2 + 6x - 6y + 14 = 0?$$
Interpret geometrically.

35. Find the line through the points of intersection of
$$x^2 + y^2 - 4x - 2y + 1 = 0 \quad \text{and} \quad x^2 + y^2 - 8x + 2y + 16 = 0.$$

C

36. For what value(s) of k is the line $x + 2y + k = 0$ tangent to the circle
$$x^2 + y^2 - 2x + 4y + 1 = 0?$$

37. Prove analytically that if P_1 and P_2 are the ends of a diameter of a circle and Q is any point on the circle ($Q \neq P_1, P_2$), then $\angle P_1 Q P_2$ is a right angle.

38. Suppose that $A = (\sqrt{3}, 0)$ and $B = (-\sqrt{3}, 0)$. Consider the set of all points P such that AP and BP intersect at right angles. What equation must all such points satisfy? Is every point that satisfies this equation in the above set?

39. Suppose that $A = (\sqrt{3}, 0)$ and $B = (-\sqrt{3}, 0)$. Consider the set of all points P such that AP and BP intersect at an angle of 60°. What equation(s) must all such points satisfy? Is every point that satisfies such equation(s) in the above set? (*Hint:* See Problem 29 in Section 1.2.)

40. A set of points in the plane has the property that every point in it is twice as far from $(-1, 2)$ as it is from $(2, 4)$. What equation must be satisfied by every point (x, y) in the set?

41. Find the relation between A, D, E, and F of Theorem 2.10 in order that the equation represent (a) a circle, (b) a point, (c) no graph. If the equation represents a circle, find h, k, and r in terms of A, D, E, and F.

42. In general, squaring both sides of an equation is not reversible (if $x = 2$, then $x^2 = 4$; but if $x^2 = 4$, then $x = \pm 2$). Yet, in the proof of Theorem 2.8, the argument was declared to be reversible even though both sides of an equation were squared. Why?

Review Problems

A

1. Write an equation (in general form with integer coefficients) for each of the following lines.
    (a) The line through $(1, 5)$ and $(-2, 3)$.
    (b) The line with slope 2 and x intercept 3.
    (c) The line with inclination 135° and y intercept 1/3.
    (d) The line through $(2, 3)$ and $(2, 8)$.

2. Write an equation (in general form with integer coefficients) for each of the following lines.

    (a) The line through (4, 2) and parallel to $3x - y + 4 = 0$.
    (b) The line with x intercept $1/2$ and y intercept $-5/4$.
    (c) The horizontal line through $(3, -2)$.
    (d) The line with y intercept $2/3$ and perpendicular to $2x - y + 3 = 0$.

3. Find an equation of the perpendicular bisector of the segment joining (4, 1) and $(0, -3)$.
4. A triangle has vertices (1, 5), $(-2, 3)$, and $(4, -1)$. Find equations for the three altitudes.
5. Find equations for the three medians of the triangle of Problem 4.
6. Find the slope and the intercepts of each of the following lines.

    (a) $x - 4y + 1 = 0$.   (b) $2x + 3y + 5 = 0$.
    (c) $5x + 2y = 0$.   (d) $3x + 1 = 0$.

7. Put the following equations into standard form, and identify each as a circle, a point, or having no graph. If it is a circle, give its center and radius. If it is a point, give its coordinates.

    (a) $x^2 + y^2 - 10x + 4y + 13 = 0$.   (b) $36x^2 + 36y^2 - 24x + 108y + 85 = 0$.

B

8. The distance between $3x + By - 2 = 0$ and $(5, -3)$ is 5. Find B.
9. Find the lengths of the altitudes of the triangle of Problem 4.
10. Find the area of the triangle with vertices (3, 1), (5, 3), and (1, 4).
11. Find the line through the point of intersection of the circles $x^2 + y^2 - 2x - 4y + 1 = 0$ and $x^2 + y^2 + 4x - 6y + 9 = 0$.
12. Find an equation of the circle with center (4, 1) and tangent to $3x + 4y - 2 = 0$.
13. Find an equation of the bisector of the acute angle between $x + y - 4 = 0$ and $x - 7y + 2 = 0$.
14. Sketch $x^2 - xy + 3x - 3y = 0$.
15. Find an equation of the circle inscribed in the triangle with vertices (5, 4), $(-15, -1)$, and $(23/3, -20/3)$.
16. Find an equation of the circle of radius 4 that is tangent to the x axis and has its center on $x - 2y = 2$.
17. Prove analytically that the perpendicular bisector of a chord of a circle contains the center.

C

18. Find an equation of the line tangent to $x^2 + y^2 = 25$ at $(4, -3)$.
19. Suppose that l_1 and l_2 are the intersecting lines

$$A_1x + B_1y + C_1 = 0 \quad \text{and} \quad A_2x + B_2y + C_2 = 0,$$

respectively. Show that for any constant k

$$A_1x + B_1y + C_1 + k(A_2x + B_2y + C_2) = 0$$

is a line containing the point of intersection of l_1 and l_2. If l_3 is a line containing the point of intersection of l_1 and l_2, can a value of k always be found so that the above equation represents l_3? Find the line through (1, 2) and the point of intersection of $3x + 2y - 5 = 0$ and $x - 3y + 4 = 0$.

3

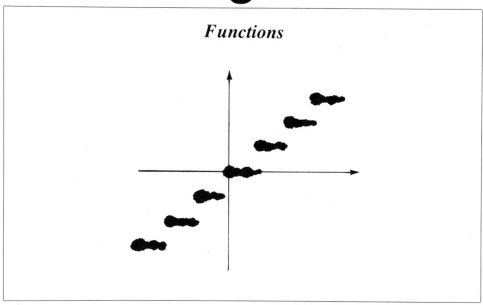

Functions

3.1

Definition of a Function

Recall that the graph of an equation is the set of all points whose coordinates satisfy the equation. For example, the equation $y = x^2$ has the graph shown in Figure 3.1. Points on the graph are identified by their coordinates, which are ordered pairs of real numbers. If we disregard the graph, we see that the equation is represented by a set of ordered pairs of real numbers.

The example $y = x^2$ has the special property of a *function:* for each value of x there is only one value of y. That is to say, we cannot find two ordered pairs in the set that have the same first numbers but different second ones. This property may be expressed geometrically by saying that no vertical line contains two points of the graph. Note, however, that the property does not exclude the possibility of having two ordered pairs with the same second numbers but different first ones. For example, in Figure 3.1 we see the ordered pairs (2, 4) and $(-2, 4)$ with the same second numbers; but we do not have another ordered pair with 2 as its first number.

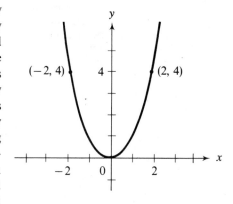

Figure 3.1

A functional relationship between x and y values can exist even when it cannot be expressed in terms of a single equation (or any combination of equations). An example is the relationship between time and temperature at a given place. For any value of time there corresponds one and only one temperature. Although no equation is given to represent this relationship, there is no doubt that the relationship does exist. Thus the set of ordered pairs, rather than the equation, is the important concept in the idea of a function.

From the foregoing discussion we now abstract a definition.

Definition

*A **function** f is a set of ordered pairs of objects such that no two ordered pairs of the set have the same first object but different second ones. The set of all first terms of the ordered pairs is the **domain** of f. The set of all second terms is the **range** of f.*

Note that a function is defined as a set of ordered pairs of "objects," rather than ordered pairs of real numbers. We do not wish to restrict the term "function" to refer only to a certain relationship between numbers. Although we shall be mainly interested in those functions that relate real numbers, we shall deal with others as well.

Example 1

The set of all ordered pairs (x, y) such that $y = x^2$ and x is a real number is a function. Its domain is the set of all real numbers. Its range is the set of all nonnegative real numbers. The graph of this function is given in Figure 3.1.

Example 2

The set of all ordered pairs (x, y) such that $y = x^2$ and $x \geq 0$ is a function with both domain and range the set of all nonnegative real numbers. Although the relationship between the first and second terms of the ordered pairs is the same as that of Example 1, this is a different function, since there are many ordered pairs of Example 1 that are not in this set—for example, $(-2, 4)$. This difference is easily seen if we compare the graph of this function, given in Figure 3.2, with that of Figure 3.1.

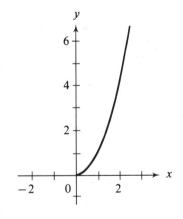

Figure 3.2

Example 3

Consider the correspondence between circles in a plane and their areas. For each circle there is one and only one area. Thus the function involved is the set of all ordered pairs in which the first terms are circles in the given plane and each second term is the area of the circle given in the first term. The domain is the set of all circles in the given plane and the range is the set of all positive real numbers.

Example 4

The set $\{(1, 3), (2, 5), .(4, 8), (5, 4), (6, 3)\}$ is a function with domain $\{1, 2, 4, 5, 6\}$ and range $\{3, 4, 5, 8\}$. Its graph is given in Figure 3.3. Note that no equation is given to relate the first and second terms of the ordered pairs, and none is necessary.

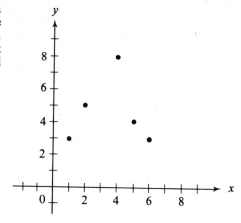

Figure 3.3

Example 5

The set of all ordered pairs (x, y) such that $x^2 + y^2 = 4$ is *not* a function, because the solution contains ordered pairs with the same first number but different second numbers. For example, it contains the pairs $(0, 2)$, $(0, -2)$ and $(1, \sqrt{3})$, $(1, -\sqrt{3})$. Geometrically speaking, there exist vertical lines containing more than one point of the graph of $x^2 + y^2 = 4$ (see Figure 3.4(a)). If we solve for y, we have

$$y = \pm \sqrt{4 - x^2},$$

which easily shows that y is not a function of x. It also gives a convenient way of expressing the original equation as a combination of two equations that do represent functions: namely, $y = \sqrt{4 - x^2}$ and $y = -\sqrt{4 - x^2}$. The graphs of these equations are given in Figures 3.4(b) and (c).

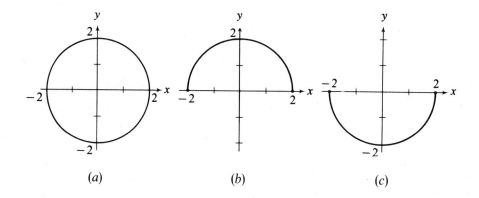

(a)   (b)   (c)

Figure 3.4

The descriptions in the foregoing examples are rather cumbersome because we have no convenient notation for representing a function. In most of the cases we consider, the relationship between the first and second terms of the ordered pairs can be expressed in the form of an equation (or a combination of equations). For example, the equation $y = x^2$ expresses the relationship between the first and second terms of the ordered pairs given in Examples 1 and 2. If the ordered pair (x, y) is a member of some function named f, then y is called the value of the function at x, and is often represented by $f(x)$. The function of Example 1 is represented by

$$f:f(x) = x^2, \quad x \text{ real},$$

which is read "f is the function whose value at x is x^2 and whose domain is the set of all real numbers," or, more simply,

$$f(x) = x^2, \quad x \text{ real},$$

which is read "f of x is x^2, x real." Similarly, the function of Example 2 is represented by

$$f(x) = x^2, \quad x \text{ real}, x \geq 0.$$

Note that f is the function and $f(x)$ is the value of the function at x. Thus, a function is identified by its value at x and its domain. If we wish to talk about a function without specifying its value at x, we use the symbol f rather than $f(x)$.

Since we shall be dealing almost exclusively with real functions of real variables— that is, the ranges and domains of our functions will be subsets of the set of real numbers—we shall omit the statement "x real." *Therefore, when the domain is not stated, it is understood to be the set of all real numbers for which $f(x)$ is real.* Thus,

$$f(x) = \frac{1}{x}$$

describes a function whose domain is the set of all real numbers except 0, and

$$f(x) = \sqrt{x}$$

describes a function whose domain is the set of all nonnegative real numbers.

The advantage of the notation $f(x) = x^2$ over $y = x^2$ can be seen in the following table.

$f(x) = x^2$	$y = x^2$
$f(1) = 1^2 = 1$	If $x = 1$, then $y = 1^2 = 1$.
$f(-1) = (-1)^2 = 1$	If $x = -1$, then $y = (-1)^2 = 1$.
$f(2) = 2^2 = 4$	If $x = 2$, then $y = 2^2 = 4$.
$f(2x + 3) = (2x + 3)^2$	If x is replaced by $2x + 3$, then $y = (2x + 3)^2$.

Note that the function notation, on the left, is shorter throughout the table. While this difference is slight for the first three lines, the last line is very cumbersome in the old notation. Note that, while we can say $x = 1$ on the first line, we cannot say $x = 2x + 3$ on the last; this is not a proper symbolic representation for "x is replaced by $2x + 3$." Furthermore, the type of expression given in the last line occurs rather frequently.

Example 6

Given

$$f(x) = x^2 + x + 1,$$

then

$$f(0) = 0^2 + 0 + 1 = 1,$$
$$f(1) = 1^2 + 1 + 1 = 3,$$
$$f(-1) = (-1)^2 + (-1) + 1 = 1,$$
$$f(z) = z^2 + z + 1,$$
$$f(x + h) = (x + h)^2 + (x + h) + 1.$$

Example 7

Given

$$f(x) = \begin{cases} 0 & \text{if } x \le 0, \\ x & \text{if } x > 0, \end{cases}$$

then

$$f(0) = 0, \qquad f(1) = 1,$$
$$f(-1) = 0, \qquad f(3) = 3,$$
$$f(-10) = 0, \qquad f(0.1) = 0.1;$$

$$f(1 + h) = \begin{cases} 1 + h & \text{if } h > -1, \\ 0 & \text{if } h \le -1. \end{cases}$$

Functions that are expressed in terms of more than one equation, like the one above, are called *compound functions*. Rather than describing an inherent property of the function, this name only describes the representation. For example,

$$f(x) = |x| \quad \text{and} \quad f(x) = \begin{cases} x & \text{if } x \ge 0, \\ -x & \text{if } x < 0 \end{cases}$$

represent the same function, the first in simple form and the second in compound form.

In an equation relating x and y, the symbols x and y are called *variables*, because they take on many different values. We use the word "variables" whether the equation represents a function or not. If y is a function of x, we say that x is the *independent variable* and y the *dependent variable*.

Problems

A

In Problems 1–10, indicate whether or not the given equation determines y as a function of x.

1. $x + y = 1.$

2. $x^2 + 2y = 1.$

3. $2x + y^2 = 1.$

4. $x = 2y - 5.$

5. $y^2 = x^2.$

6. $x = y^3.$

7. $y = \sqrt{x}.$

8. $x^2 - y^2 = 1.$

9. $y = \pm\sqrt{x}.$

10. $x^2 + y^2 + 2x - 6y + 2 = 0.$

In Problems 11–18, indicate the domain of the given function.

11. $f(x) = \dfrac{x}{x-2}$.

12. $f(x) = \dfrac{x+1}{x-3}$.

13. $f(x) = \dfrac{1}{x^2-1}$.

14. $f(x) = \dfrac{1}{x^2+3x+2}$.

15. $f(x) = \sqrt{x^2+x-6}$.

16. $f(x) = \sqrt{6+x-x^2}$.

17. $f(x) = \sqrt{\dfrac{x-1}{x}}$.

18. $f(x) = \sqrt{\dfrac{x-3}{x+4}}$.

In Problems 19–30, indicate the domain and range of the given function.

19. $f(x) = x^2 + 1, \quad x \geq 0$.

20. $f(x) = x^2 + 1, \quad x < 0$.

21. $f(x) = \sqrt{x-2}$.

22. $f(x) = \sqrt{x+4}$.

23. $f(x) = x, \quad x = 1, 2, 3, \ldots$.

24. $f(x) = x^2, \quad x = 0, \pm 1, \pm 2, \pm 3, \ldots$.

25. $f(x) = \dfrac{1}{x}, \quad x > 0$.

26. $f(x) = x + 1, \quad 0 \leq x < 3$.

27. $f(x) = \begin{cases} x & \text{if } x < 0, \\ 1 & \text{if } x \geq 0. \end{cases}$

28. $f(x) = \begin{cases} \dfrac{1}{x} & \text{if } x \neq 0, \\ 0 & \text{if } x = 0. \end{cases}$

29. $f(x) = \begin{cases} x & \text{if } x < 0, \\ 1 & \text{if } 0 < x \leq 2, \\ x-1 & \text{if } 2 < x. \end{cases}$

30. $f(x) = \begin{cases} x^2 & \text{if } x < 0, \\ 1 & \text{if } x = 0, \\ x & \text{if } 0 < x < 2. \end{cases}$

In Problems 31–38, express the given function by giving the relation between x and $f(x)$. Give the domain and range.

31. $\{(x, f(x)) \mid f(x) = x^3, 0 \leq x \leq 1\}$.

32. $\{(x, f(x)) \mid f(x) = 1 - x, x \geq 0\}$.

33. $\{(x, f(x)) \mid f(x) = x - 3, 0 < x \leq 6\}$.

34. $\{(x, f(x)) \mid f(x) = 2x + 1, x \text{ real}\}$.

35. $\{(0, 0), (1, 1), (2, 2), (3, 3), (4, 4), (5, 5)\}$.

36. $\{(0, 0), (1, 1), (2, 4), (3, 9)\}$.

37. $\{(1, 2), (2, 4), (3, 6), \ldots, (n, 2n), \ldots\}$.

38. $\{(1, -1), (2, -2), (3, -3)\}$.

39. If $f(x) = 3x + 2$ find $f(0), f(1), f(5),$ and $f(-3)$.

40. If $f(x) = 4x - 3,$ find $f(-1), f(0), f(3),$ and $f(5)$.

41. If $f(x) = 1/x,$ find $f(-1), f(0), f(2),$ and $f(x+1)$.

42. If $f(x) = 1/(x+1),$ find $f(-1), f(0), f(1),$ and $f(x+1)$.

43. If $f(x) = \sqrt{x-1},$ find $f(0), f(4), f(x^2),$ and $f(x+h)$.

44. If $f(x) = \sqrt{x+1}/x,$ find $f(0), f(3), f(x^2),$ and $f(x+h)$.

45. If $f(x) = x^2 + 1,$ find $f(y)$ and $f(x+h)$.

46. If $f(x) = (x+2)/(x+1),$ find $f(x+1)$ and $f(2+h)$.

B

47. If
$$f(x) = \begin{cases} 0 & \text{if } x < 0, \\ 2x & \text{if } 0 \leq x \leq 1, \\ 2 & \text{if } 1 < x, \end{cases}$$

find $f(-2), f(1/2),$ and $f(3)$.

48. If
$$f(x) = \begin{cases} 3x & \text{if } x \neq 0, \\ 1 & \text{if } x = 0, \end{cases}$$

find $f(1), f(0),$ and $f(1+h)$.

49. Express the area of a circle as a function of its radius; the circumference as a function of its radius.

50. Express the area of a circle as a function of its diameter; the circumference as a function of its diameter.

51. A rectangle is inscribed in a circle of radius R. Express the area of the rectangle as a function of one of its sides.

52. A trapezoid is inscribed in a circle of radius 2 cm with one base a diameter. Express the area of the trapezoid as a function of its altitude.

53. The height of a cone is twice the radius of the base. Express the volume of the cone as a function of the radius of the base.

C

54. The function f is odd if $f(-x) = -f(x)$ for all x in the domain of f; and it is even if $f(-x) = f(x)$. Give an example of each of the following.
    (a) An odd function.
    (b) An even function.
    (c) A function that is neither odd nor even.
    (d) A function that is both odd and even.

55. What are the common characteristics of the graphs of all odd functions? Of all even functions? (See Problem 54 for the definitions of odd and even functions.)

56. Express $x^2 + y^2 = 4$ as a combination of two functions different from those given in Example 5.

57. Given $f(x) = ax + b$, where a and b are nonzero real numbers, find all real numbers t such that $f(2t + 5) = f(t^2 - 3)$.

58. Given $f(x) = x^2 - x$, find all real numbers t such that $f(3t - 2) = f(t - 1)$.

59. For what values of x are the functions below both defined?

$$f(x) = \sqrt{x^2 + 3x - 4} \quad \text{and} \quad g(x) = \sqrt{\frac{x - 3}{x + 2}}.$$

60. For what values of x are the functions below both defined?

$$f(x) = \sqrt{1 - x^2} \quad \text{and} \quad g(x) = \sqrt{\frac{x - 4}{x + 1}}.$$

3.2

Functional Expressions and Combinations of Functions

One big advantage of function notation is the simplicity it provides for representing expressions involving several different values of the same function. For example, we shall frequently encounter the expression

$$\frac{f(x + h) - f(x)}{h},$$

in which we want the values of the function f at both $x + h$ and x.

Example 1

Given $f(x) = x^2$, evaluate $\dfrac{f(x+h) - f(x)}{h}$.

$f(x+h) = (x+h)^2$ and $f(x) = x^2$.

Thus

$$\frac{f(x+h) - f(x)}{h} = \frac{(x+h)^2 - x^2}{h} = \frac{x^2 + 2hx + h^2 - x^2}{h} = \frac{2hx + h^2}{h} = 2x + h$$

$$(h \neq 0).$$

While the expression of Example 1 is very important, it is only one of many that we might consider. The expressions might even involve two or more functions.

Example 2

Given $f(x) = x^2 + 1$, evaluate

$$\frac{f(x) - f(a)}{x - a} \quad \text{and} \quad f(x) - f(y).$$

$$\frac{f(x) - f(a)}{x - a} = \frac{(x^2 + 1) - (a^2 + 1)}{x - a} = \frac{x^2 - a^2}{x - a} = \frac{(x+a)(x-a)}{x - a} = x + a \quad (x \neq a).$$

$$f(x) - f(y) = (x^2 + 1) - (y^2 + 1) = x^2 - y^2.$$

Example 3

If $f(x) = x^2 + 1$ and $g(x) = x^2 - 1$, evaluate

$$\frac{f(x+h) - g(x+h)}{f(x) - g(x)} \quad \text{and} \quad \frac{f(x+h) + g(x+h)}{f(x) + g(x)}.$$

$$\frac{f(x+h) - g(x+h)}{f(x) - g(x)} = \frac{[(x+h)^2 + 1] - [(x+h)^2 - 1]}{(x^2 + 1) - (x^2 - 1)} = \frac{2}{2} = 1,$$

$$\frac{f(x+h) + g(x+h)}{f(x) + g(x)} = \frac{[(x+h)^2 + 1] + [(x+h)^2 - 1]}{(x^2 + 1) + (x^2 - 1)} = \frac{2(x+h)^2}{2x^2} = \frac{(x+h)^2}{x^2}.$$

Frequently several functions are combined to give new functions. The procedure is basically no different from what we just did in Example 3. The obvious combinations are the sum, difference, product, and quotient of two functions. Note, however, that these combinations exist only at those values of x which are in the domains of *both* of the given functions. If at least one of the two given functions is not defined at a given value of x, then the sum, difference, product, or quotient is not defined there. (One exception to this rule occurs when we have some combination of complex numbers giving a real result. For examples of this see Example 1 in Section 5.3.) Furthermore, the quotient is not defined at any value of x for which the denominator is zero.

These four combinations are sometimes represented by $f + g$, $f - g$, fg, and f/g, with their values at x given by $(f+g)(x)$, $(f-g)(x)$, $fg(x)$, and $(f/g)(x)$.

Example 4

Given $f(x) = x^2 + x - 1$ for $x \geq -2$ and $g(x) = x^2 - x$ for $x < 5$, find $f + g$, $f - g$, fg, and f/g.

Note that to find $f + g$ we need to find $(f + g)(x)$ and determine its domain.

$$(f + g)(x) = f(x) + g(x) = (x^2 + x - 1) + (x^2 - x) = 2x^2 - 1 \quad (-2 \leq x < 5)$$
$$(f - g)(x) = f(x) - g(x) = (x^2 + x - 1) - (x^2 - x) = 2x - 1 \quad (-2 \leq x < 5)$$
$$fg(x) = f(x)g(x) = (x^2 + x - 1)(x^2 - x) = x^4 - 2x^2 + x \quad (-2 \leq x < 5)$$
$$\frac{f}{g}(x) = \frac{f(x)}{g(x)} = \frac{x^2 + x - 1}{x^2 - x} \quad (-2 \leq x < 5 \text{ and } x \neq 0, 1)$$

Another important combination of two functions is a function of a function. Suppose f is a function and x is in its domain; then $f(x)$ is in the range of f. Suppose now that g is a function and that the number $y = f(x)$ is in the domain of g. Then we can find $g(y) = g(f(x))$. This, of course, can be done only for those values of x in the domain of f for which $f(x)$ is in the domain of g.

The notation used for a function of a function is $g \circ f$ (do not confuse this with the product gf). Thus $(g \circ f)(x) = g(f(x))$. Note that the function f operates on x first and that g operates on the result. The domain of $g \circ f$ is the set of all numbers x in the domain of f for which $f(x)$ is in the domain of g. Since the notation $g(f(x))$ is more suggestive than $(g \circ f)(x)$, we shall normally use it. However when we are interested in considering the function itself, apart from its value at x, we shall use $g \circ f$. The function $g \circ f$ is a *composite function* formed by the *composition* of f by g.

Example 5

If $f(x) = x^2 + x - 1$ for $x \geq -2$ and $g(x) = x^2 - x$ for $x < 5$, find $g \circ f$ and $f \circ g$.

Again, to find $g \circ f$ we need to find $(g \circ f)(x)$ and the domain of $g \circ f$.

$$(g \circ f)(x) = g(f(x)) = g(x^2 + x - 1)$$
$$= (x^2 + x - 1)^2 - (x^2 + x - 1)$$
$$= x^4 + 2x^3 - 2x^2 - 3x + 2.$$

The domain of $g \circ f$ is the set of all numbers $x \geq -2$ such that $f(x) < 5$. Let us first find all x such that $f(x) = 5$.

$$x^2 + x - 1 = 5,$$
$$x^2 + x - 6 = 0,$$
$$(x + 3)(x - 2) = 0;$$
$$x = -3, \quad x = 2.$$

Now $f(x) < 5$ either for those values of x between -3 and 2 or for those which are either greater than 2 or less than -3. A simple check shows that $f(x) < 5$ if $-3 < x < 2$. Then the domain of $g \circ f$ is the set of all x in the domain of f for which $-3 < x < 2$. This is $\{x \mid -2 \leq x < 2\}$.

We now consider $f \circ g$.

$$
\begin{aligned}
(f \circ g)(x) = f(g(x)) &= f(x^2 - x) \\
&= (x^2 - x)^2 + (x^2 - x) - 1 \\
&= x^4 - 2x^3 + 2x^2 - x - 1.
\end{aligned}
$$

The domain of $f \circ g$ is the set of all numbers x such that $x < 5$ and $g(x) \geq -2$. Let us proceed as above and find all x such that $g(x) = -2$.

$$
x^2 - x = -2,
$$
$$
x^2 - x + 2 = 0,
$$
$$
x = \frac{1 \pm \sqrt{-7}}{2}.
$$

Since $g(x) = -2$ for no real value of x, either $g(x) < -2$ for all x or $g(x) > -2$ for all x. Again a simple check shows that $g(x) > -2$ for all x. Thus the domain of $f \circ g$ is the domain of g, $\{x \mid x < 5\}$.

You might wonder at this point what should have happened in the last example if $g(x)$ were always less than -2. In that case the domain of $f \circ g$ is empty, which implies that there simply is no function $f \circ g$.

Problems

A

1. Given $f(x) = x^2 + 1$, find $\dfrac{f(x + h) - f(x)}{h}$ $(h \neq 0)$.

2. Given $f(x) = x^3 - 1$, find $\dfrac{f(x + h) - f(x)}{h}$ $(h \neq 0)$.

3. Given $f(x) = x^2 + 2x - 3$, find $\dfrac{f(x + h) - f(x)}{h}$ $(h \neq 0)$.

4. Given $f(x) = \dfrac{1}{x}$, find $\dfrac{f(x + h) - f(x)}{h}$ $(h \neq 0)$.

5. Given $f(x) = (2x + 3)^2$, find $\dfrac{f(2 + h) - f(2)}{h}$ $(h \neq 0)$.

6. Given $f(x) = \dfrac{x - 1}{x + 1}$, find $\dfrac{f(1 + k) - f(1)}{k}$ $(k \neq 0)$.

7. Given $f(x) = x^2 + 1$, find $\dfrac{f(x) - f(a)}{x - a}$ $(x \neq a)$.

8. Given $f(x) = (x - 3)^3$, find $\dfrac{f(x) - f(a)}{x - a}$ $(x \neq a)$.

9. Given $f(x) = \dfrac{1}{x}$, find $\dfrac{f(x) - f(1)}{x - 1}$ $(x \neq 1)$.

10. Given $f(x) = \sqrt{x}$, find $\dfrac{f(x^2) - f(4)}{x - 2}$ $(x \neq 2)$.

B

11.   Given $f(x) = \begin{cases} x^2 & \text{if } x \neq 0, \\ 1 & \text{if } x = 0, \end{cases}$ find $\dfrac{f(1 + h) - f(1)}{h}$ $(h \neq 0)$.

12.   Given $f(x) = \begin{cases} \sin(1/x) & \text{if } x \neq 0, \\ 0 & \text{if } x = 0, \end{cases}$ find $\dfrac{f(h) - f(0)}{h}$ $(h \neq 0)$.

13.   Given $f(x) = \begin{cases} x \sin(1/x) & \text{if } x \neq 0, \\ 0 & \text{if } x = 0, \end{cases}$ find $\dfrac{f(h) - f(0)}{h}$ $(h \neq 0)$.

14.   Given $f(x) = \begin{cases} x^2 & \text{if } x \geq 0, \\ 1 & \text{if } x < 0, \end{cases}$ find $\dfrac{f(h) - f(0)}{h}$ $(h \neq 0)$.

15.   If $f(x) = 3x^2 - 2$,   find $f(xy) - f(y)$.

16.   If $f(x) = \dfrac{x + 1}{x - 1}$,   find $f(x + h) - f(x + k)$.

17.   If $f(x) = \dfrac{1}{x}$,   find $\dfrac{1}{f(x) + f(y)}$.

18.   If $f(x) = \sqrt{x + 1}$,   find $[f(x) + f(y)]^2$.

In Problems 19–26, find $f + g, f - g, fg, f/g$, and $g \circ f$ and their domains for the given pairs of functions.

19. $f(x) = x^2$ and $g(x) = 1/x^2$.
20. $f(x) = x^2$ and $g(x) = 2x - 1$.
21. $f(x) = \sqrt{x}$ and $g(x) = \sqrt{1 - x}$.
22. $f(x) = \sqrt{x - 1}$ and $g(x) = \sqrt{1 - x}$.
23. $f(x) = 3x + 1$ for $x > -3$ and $g(x) = x - 3$ for $x < 4$.
24. $f(x) = x^2 + 3$ for $x > 1$ and $g(x) = 2x - 3$ for $x < 5$.
25. $f(x) = x^3$ for $x < 2$ and $g(x) = x + 4$ for $x \geq 0$.
26. $f(x) = 3x - 1$ for $x \geq 1$ and $g(x) = x^2 + 1$ for $x \leq 5$.

27.   Given $f(x) = 1 + 2x$ and $(f + g)(x) = 4x - 2$,   find $g(x)$.
28.   Given $f(x) = 2x^2 - 1$ and $(f - g)(x) = x - 4$,   find $g(x)$.
29.   Given $f(x) = 1/x$ and $(fg)(x) = x, x \neq 0$,   find $g(x)$.
30.   Given $f(x) = 3x + 2$ and $(f \circ g)(x) = x - 5$,   find $g(x)$.

C

31.   Given $f(x) = ax + b$,   find the values of a and b such that $ff = f \circ f$.
32.   Given $f(x) = ax^2 + bx + c$,   find the values of a, b, and c such that

$$f(x + y) = f(x) + f(y).$$

33.   Suppose f and g are odd functions and F and G are even functions (see Problem 54 of Section 3.1 for definitions of odd and even functions), all defined for all real numbers. Indicate whether each of the following is odd, even, or neither.

(a) fg (b) fG (c) FG

(d) $f + g$ (e) $f + G$ (f) $F + G$

(g) $f \circ g$ (h) $f \circ G$ (i) $F \circ G$

3.3

Introduction to Limits

The concept of a limit of a function is one of the most important ideas of calculus. Unfortunately it is also a rather sophisticated concept. Thus we shall consider it here only from an intuitive point of view, leaving actual definitions to Chapters 10 and 17.

Suppose that y is a function of x; for example, suppose $y = x^2$. Now what can be said about the value of y when x is near 2? In order to answer that question let us consider the following tables.

x	y	x	y
1	1	3	9
1.5	2.25	2.5	6.25
1.9	3.61	2.1	4.41
1.99	3.9601	2.01	4.0401
1.999	3.996001	2.001	4.004001

We can see that as the value of x gets closer and closer to 2, the value of y approaches 4. In fact, no matter how close to 4 we might insist that y be, we can find a value of x close enough to 2 so that the corresponding value of y is within the desired range. We express this by saying that the limit of y (or x^2) as x approaches 2 is 4. In symbols,

$$\lim_{x \to 2} y = 4.$$

Perhaps you noticed above that when $x = 2$, $y = 4$. Thus you might feel it reasonable to expect y to be near 4 when x is near 2. This is not necessarily the case, as we can see from the following example.

Example 1

Does $\lim_{x \to 1} [\![x]\!]$ exist? If so, what is the limit? (The greatest integer function $[\![x]\!]$ was defined on page 15 as the greatest integer that is less than or equal to x. Its graph is shown in Figure 1.21.)

Let us make tables as before.

x	$[\![x]\!]$	x	$[\![x]\!]$
0	0	2	2
0.5	0	1.5	1
0.9	0	1.1	1
0.99	0	1.01	1
0.999	0	1.001	1

Now we see that as x approaches 1 from the left ($x < 1$), $[\![x]\!]$ becomes and remains 0; but as x approaches 1 from the right ($x > 1$), $[\![x]\!]$ becomes and remains 1. Thus, since $[\![x]\!]$ does not approach any one number as $x \to 1$, the given limit does not exist.

We see in the foregoing example that, in spite of the fact that $[\![x]\!] = 1$ when $x = 1$, $\lim_{x \to 1} [\![x]\!]$ does not exist. We are interested in the behavior of the function *near* $x = 1$, not *at* $x = 1$.

Example 2

Does $\lim\limits_{x \to 1} \dfrac{x^2 - 1}{x - 1}$ exist? If so, what is the limit?

If we substitute $x = 1$ into

$$f(x) = \frac{x^2 - 1}{x - 1},$$

we get 0/0, which is meaningless. The number 1 is simply not in the domain of this function. However

$$\frac{x^2 - 1}{x - 1} = \frac{(x + 1)(x - 1)}{x - 1} = x + 1$$

if $x \neq 1$. Since we have equality for all values of x except $x = 1$, $(x^2 - 1)/(x - 1)$ and $x + 1$ are the same for all x near $x = 1$ (see Problem 32). Thus

$$\lim_{x \to 1} \frac{x^2 - 1}{x - 1} = \lim_{x \to 1} (x + 1) = 2.$$

Example 3

Does $\lim\limits_{h \to 0} \dfrac{(x + h)^2 - x^2}{h}$ exist? If so, what is the limit?

Since we are interested in the behavior of the function when h is near 0, the x is regarded as a constant. Again a simple substitution, $h = 0$, gives the meaningless expression 0/0. But

$$\frac{(x + h)^2 - x^2}{h} = \frac{x^2 + 2hx + h^2 - x^2}{h}$$

$$= \frac{2hx + h^2}{h}$$

$$= 2x + h \quad (h \neq 0).$$

Thus

$$\lim_{h \to 0} \frac{(x + h)^2 - x^2}{h} = \lim_{h \to 0} (2x + h) = 2x.$$

Example 4

Does $\lim\limits_{h \to 0} \dfrac{\dfrac{1}{x + h} - \dfrac{1}{x}}{h}$ exist? If so, what is the limit?

Again, a simple substitution, $h = 0$, gives the meaningless expression $0/0$. But

$$\frac{\dfrac{1}{x+h} - \dfrac{1}{x}}{h} = \frac{\dfrac{x-(x+h)}{x(x+h)}}{h}$$

$$= \frac{-h}{hx(x+h)}$$

$$= \frac{-1}{x(x+h)} \quad (h \neq 0).$$

Thus

$$\lim_{h \to 0} \frac{\dfrac{1}{x+h} - \dfrac{1}{x}}{h} = \lim_{h \to 0} \frac{-1}{x(x+h)} = \frac{-1}{x^2}.$$

The following theorem on limits is used in the next few chapters. We make no attempt to prove it here; the proof can be found in Chapter 10 and in Chapter 17.

Theorem 3.1

If $\lim_{x \to a} f(x) = L$ and $\lim_{x \to a} g(x) = M$, then

(a) $\lim_{x \to a} [f(x) + g(x)] = L + M$,

(b) $\lim_{x \to a} [f(x) - g(x)] = L - M$,

(c) $\lim_{x \to a} f(x)\, g(x) = LM$,

(d) $\lim_{x \to a} \dfrac{f(x)}{g(x)} = \dfrac{L}{M}$, provided $M \neq 0$.

Problems

A

In Problems 1–18, does the limit exist? If so, what is the limit?

1. $\lim_{x \to 1} x^2$.

2. $\lim_{x \to 2} (3x - 2)$.

3. $\lim_{x \to 0} \dfrac{2x^2 - x}{x}$.

4. $\lim_{x \to 3} \dfrac{x^2 - x - 6}{x - 3}$.

5. $\lim_{x \to 2} \dfrac{x^2 - 1}{x - 1}$.

6. $\lim_{x \to -1} \dfrac{x^2 - 1}{x - 1}$.

7. $\lim_{x \to 0} [\![x]\!]$.

8. $\lim_{x \to -1/2} [\![x]\!]$.

9. $\lim_{x \to 2} |x|$.

10. $\lim_{x \to 0} |x|$.

11. $\lim_{x \to 0} f(x)$, where $f(x) = \begin{cases} 1 & \text{if } x \leq 0, \\ x & \text{if } x > 0. \end{cases}$

12. $\lim\limits_{x \to 1} f(x)$, where $f(x) = \begin{cases} 1 & \text{if } x \leq 0, \\ x & \text{if } x > 0. \end{cases}$

13. $\lim\limits_{x \to 1} f(x)$, where $f(x) = \begin{cases} x + 1 & \text{if } x \neq 1, \\ 3 & \text{if } x = 1. \end{cases}$

14. $\lim\limits_{x \to 2} f(x)$, where $f(x) = \begin{cases} 3x - 1 & \text{if } x \neq 2, \\ 4 & \text{if } x = 2. \end{cases}$

15. $\lim\limits_{h \to 0} \dfrac{3(x + h) - 2 - (3x - 2)}{h}$.

16. $\lim\limits_{h \to 0} \dfrac{3(x + h)^2 - 3x^2}{h}$.

17. $\lim\limits_{h \to 0} \dfrac{(x + h + 1)^2 - (x + 1)^2}{h}$.

18. $\lim\limits_{h \to 0} \dfrac{(x + h + 1)(x + h + 2) - (x + 1)(x + 2)}{h}$.

B

In Problems 19–30, does the limit exist? If so, what is the limit?

19. $\lim\limits_{h \to 0} \dfrac{\dfrac{1}{x + h - 1} - \dfrac{1}{x - 1}}{h}$.

20. $\lim\limits_{h \to 0} \dfrac{\dfrac{x + h + 1}{x + h} - \dfrac{x + 1}{x}}{h}$.

21. $\lim\limits_{h \to 0} \dfrac{\sqrt{x + h} - \sqrt{x}}{h}$. (*Hint:* Rationalize the numerator.)

22. $\lim\limits_{h \to 0} \dfrac{\sqrt{2x + 2h + 1} - \sqrt{2x + 1}}{h}$. (*Hint:* Rationalize the numerator.)

23. $\lim\limits_{h \to 0} \dfrac{|h|}{h}$. (*Hint:* Consider positive and negative values of h separately.)

24. $\lim\limits_{h \to 0} \dfrac{|2 + h| - 2}{h}$.

25. $\lim\limits_{h \to 0} \dfrac{|-1 + h| - 1}{h}$.

26. $\lim\limits_{x \to a} \dfrac{x^3 - a^3}{x - a}$.

27. $\lim\limits_{x \to a} \dfrac{x^4 - a^4}{x - a}$.

28. $\lim\limits_{x \to a} \dfrac{\dfrac{1}{x^2} - \dfrac{1}{a^2}}{x - a}$.

29. $\lim\limits_{x \to a} \dfrac{\sqrt{x} - \sqrt{a}}{x - a}$. (*Hint:* Rationalize the numerator.)

30. $\lim\limits_{x \to a} \dfrac{|x| - |a|}{x - a}$. (*Hint:* Consider the following three cases: $a > 0$, $a < 0$, $a = 0$.)

C

31. For $f(x) = x^2$, evaluate $\lim\limits_{h \to 0} \dfrac{f(1+h) - f(1)}{h}$ and $\lim\limits_{x \to 1} \dfrac{f(x) - f(1)}{x-1}$.

   Compare. Can you give a geometric interpretation for

   $$\frac{f(1+h) - f(1)}{h} \quad \text{and} \quad \frac{f(x) - f(1)}{x-1}?$$

32. Sketch the graph of $f(x) = (x^2 - 1)/(x-1)$, given in Example 2. How does it differ from the graph of $g(x) = x + 1$?

3.4

Continuity

The idea of continuity is closely related to that of a limit. We saw in the last section that there is a basic difference in the limits

$$\lim_{x \to 2} x^2 \quad \text{and} \quad \lim_{x \to 1} [\![x]\!].$$

Furthermore, as we can see by an inspection of Figure 3.5, there is a basic difference in the graphs of $f(x) = x^2$ and $f(x) = [\![x]\!]$. While the graph of $f(x) = x^2$ is all of one piece, the graph of $f(x) = [\![x]\!]$ has many vertical jumps. In particular, the vertical jump at $x = 1$ is responsible for the nonexistence of $\lim_{x \to 1} [\![x]\!]$. We

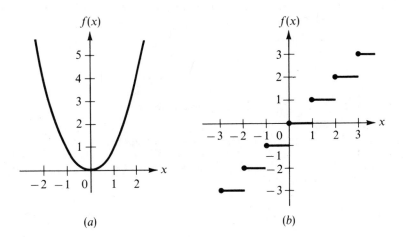

(a)   (b)

Figure 3.5

saw that $f(x) = 0$ for values of x less than but near 1, while $f(x) = 1$ for values of x greater than but near 1. On the other hand, we do not have this separation of the values of $f(x) = x^2$ when x is near 2. In this case, the values of $f(x)$ approach the same number as x approaches 2 from either side.

We distinguish between these two cases by saying that $f(x) = x^2$ is *continuous* at $x = 2$ while $f(x) = [\![x]\!]$ is *discontinuous* at $x = 1$. Intuitively, a function is discontinuous at $x = a$ if its graph has a vertical jump at $x = a$, requiring one to pick up one's pencil at $x = a$ when drawing the graph. While we might use this to give a rough idea of continuity and discontinuity, let us be somewhat more precise. In Chapters 10 and 17 precise definitions of continuity are given. While these definitions are independent of the limit, the relationship between limits and continuity is given in Theorem 10.3. We use a simplified statement of this theorem to show that relationship.

Let f be a function whose domain includes all numbers in some interval $c \leq x \leq d$ ($c \neq d$), which contains a. Then f is continuous at $x = a$ if and only if

$$\lim_{x \to a} f(x) = f(a).$$

Of course, this implies that both $\lim_{x \to a} f(x)$ and $f(a)$ exist (actually the existence of $f(a)$ is assured by the assumption that a is in an interval belonging to the domain of f). Let us consider some examples.

Example 1

Show that $f(x) = x^2$ is continuous at $x = 2$.

We saw in the last section that

$$\lim_{x \to 2} x^2 = 4.$$

Since $f(2) = 2^2 = 4$, it follows that $f(x) = x^2$ is continuous at $x = 2$.

It might be noted that $f(x) = x^2$ is continuous not only at $x = 2$, but at every value of x. This is in keeping with our intuitive idea of continuity since there are no vertical jumps in the graph of $f(x) = x^2$ (see Figure 3.5(a)).

Example 2

Show that $f(x) = \begin{cases} 0 & \text{if } x \leq 0, \\ 1 & \text{if } x > 0 \end{cases}$ is discontinuous at $(0, 0)$.

Since $f(x) = 0$ if $x < 0$, $f(x)$ remains 0 as x approaches 0 from the left (see Figure 3.6). On the other hand, since $f(x) = 1$ if $x > 0$, $f(x)$ remains 1 as x approaches 0 from the right. Thus $\lim_{x \to 0} f(x)$ does not exist. This is sufficient to assure us that f is discontinuous at $(0, 0)$.

Example 3

Show that $f(x) = \begin{cases} x^2 & \text{if } x \neq 0, \\ 1 & \text{if } x = 0 \end{cases}$ is discontinuous at $(0, 1)$.

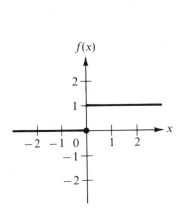

Figure 3.6

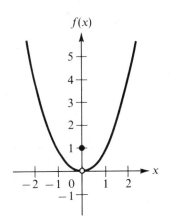

Figure 3.7

When considering $\lim_{x \to 0} f(x)$, we are concerned only with the values of $f(x)$ for x near 0 and not for $x = 0$. Since $f(x) = x^2$ for all x near 0, we see that $f(x)$ approaches 0 as x approaches 0 from either side. Thus

$$\lim_{x \to 0} f(x) = 0.$$

But $f(0) = 1$. Thus

$$\lim_{x \to 0} f(x) \neq f(0),$$

and f is discontinuous at $(0, 1)$. Again this agrees with our intuitive idea of a discontinuity giving a vertical jump in the graph, as it does at $x = 0$ (see Figure 3.7).

We see that a function f that is continuous at $x = a$ has the point $(a, f(a))$, roughly speaking, where we expect it to be. This contrasts with the function of Example 3 in which the point $(0, 1)$ is displaced from its expected position $(0, 0)$ or with the function of Example 2 in which the behavior of the function to the left and right of $x = 0$ gives us no clear expectation of what $f(0)$ "should" be. It is this property of continuity that makes it so important. For example, knowing that $f(x) = x^2$ is continuous at $x = 2$, we then have the following by a simple substitution.

$$\lim_{x \to 2} x^2 = 2^2 = 4.$$

It appears that we are arguing in circles here. We see that $\lim_{x \to a} f(x)$ is simply $f(a)$ if we know that f is continuous at $x = a$; we know that f is continuous at $x = a$ because $\lim_{x \to a} f(x) = f(a)$. Two elementary theorems on continuity allow us to break into this circle.

Theorem 3.2

If f and g are continuous at $x = a$, then so are $f + g$, $f - g$, fg, and, if $g(a) \neq 0$, f/g.

Theorem 3.3

If $f(x) = g(h(x))$, h is continuous at $x = a$, and g is continuous at $x = h(a)$, then f is continuous at $x = a$.

These theorems, together with the continuity of certain simple functions, allow us to assert the continuity of a broad range of functions. For example, it is not difficult to establish that $f(x) = k$ (where k is a constant) and $f(x) = x$ are continuous for every real number a. With that and Theorem 3.2, we may show that any polynomial is continuous for every real number and that any rational function (quotient of two polynomials) is continuous throughout its domain (see Problems 29 and 30). In fact, most of the common functions with which we deal are continuous throughout their domains. The only functions we have encountered that have at least one discontinuity are those involving $[\![x]\!]$ and some (not all) functions defined by more than one equation. For example, we have seen that

$$f(x) = \begin{cases} 0 & \text{if } x \le 0, \\ 1 & \text{if } x > 0 \end{cases} \quad \text{and} \quad f(x) = \begin{cases} x^2 & \text{if } x \ne 0, \\ 1 & \text{if } x = 0 \end{cases}$$

are both discontinuous at $x = 0$.

Since most of the functions we deal with are continuous throughout their domains, evaluating limits for these functions is simply a matter of substitution. But you must remember that a function is not continuous where it is not defined. For example,

$$f(x) = \frac{x^2 - 1}{x - 1}$$

is continuous throughout its domain since it is a quotient of two polynomials; but its domain consists of all real numbers except $x = 1$. Thus

$$\lim_{x \to 2} \frac{x^2 - 1}{x - 1}$$

can be evaluated simply by substitution since $x = 2$ is in the domain of the given function. But

$$\lim_{x \to 1} \frac{x^2 - 1}{x - 1}$$

cannot be so easily evaluated, because $x = 1$ is not in the domain. As was noted in the preceding section,

$$\lim_{x \to 1} \frac{x^2 - 1}{x - 1} = \lim_{x \to 1} (x + 1)$$

since

$$f(x) = \frac{x^2 - 1}{x - 1} \quad \text{and} \quad g(x) = x + 1$$

are identical for all values of x except $x = 1$. Knowing that g is continuous for all values of x, including $x = 1$, $\lim_{x \to 1} (x + 1)$ can be evaluated by a simple substitution. Thus

$$\lim_{x \to 1} \frac{x^2 - 1}{x - 1} = \lim_{x \to 1} (x + 1) = 2.$$

Example 4

Does $\lim\limits_{h \to 0} \dfrac{(2 + h)^2 - 4}{h}$ exist? If so, what is the limit?

The function

$$f(h) = \frac{(2 + h)^2 - 4}{h}$$

is continuous throughout its domain. However $h = 0$ is not in that domain. Let us simplify.

$$\begin{aligned} f(h) &= \frac{(2 + h)^2 - 4}{h} \\ &= \frac{4 + 4h + h^2 - 4}{h} \\ &= \frac{4h + h^2}{h} \end{aligned}$$

Finally, we divide the numerator and denominator by h. The resulting function is not f. It is a new function,

$$g(h) = 4 + h,$$

that is equal to f for all values of h except $h = 0$. Thus

$$\lim_{h \to 0} f(h) = \lim_{h \to 0} g(h).$$

Furthermore, since g is continuous for all real numbers h, we evaluate $\lim_{h \to 0} g(h)$ by substituting $h = 0$. We now see that the given limit exists, and

$$\lim_{h \to 0} \frac{(2 + h)^2 - 4}{h} = \lim_{h \to 0} \frac{4h + h^2}{h} = \lim_{h \to 0} (4 + h) = 4.$$

Problems

A

In Problems 1–12, indicate whether the given function is continuous or discontinuous at the given value of x by comparing $f(a)$ and $\lim_{x \to a} f(x)$.

1. $f(x) = x^3$; $a = 0$.

2. $f(x) = 4x^2 - x + 2$; $a = 1$.

3. $f(x) = [\![x]\!]$; $a = 1/2$.

4. $f(x) = \begin{cases} 0 & \text{if } x \le 0, \\ 1 & \text{if } x > 0; \end{cases}$ $a = 1$.

5. $f(x) = \begin{cases} 0 & \text{if } x \le 0, \\ x + 1 & \text{if } x > 0; \end{cases}$ $a = 0$.

6. $f(x) = \begin{cases} 0 & \text{if } x \le 0, \\ x + 1 & \text{if } x > 0; \end{cases}$ $a = 1$.

7. $f(x) = x - [\![x]\!]$; $a = 0$.

8. $f(x) = x - |x|$; $a = 0$.

9. $f(x) = \begin{cases} 0 & \text{if } x \le 0, \\ x & \text{if } x > 0; \end{cases}$ $a = 0$.

10. $f(x) = \begin{cases} x^2 & \text{if } x \ne 1, \\ 0 & \text{if } x = 1; \end{cases}$ $a = 1$.

11. $f(x) = \begin{cases} x + 1 & \text{if } x > 0, \\ 0 & \text{if } x = 0, \\ x - 1 & \text{if } x < 0; \end{cases}$ $a = 0$.

12. $f(x) = \begin{cases} x + 1 & \text{if } x \ge 0, \\ x - 1 & \text{if } x < 0; \end{cases}$ $a = 0$.

In Problems 13–20, use the results of this section to evaluate the given limits.

13. $\lim_{x \to 0} x^3$.

14. $\lim_{x \to 1} x^3$.

15. $\lim_{x \to 1/2} [\![x]\!]$.

16. $\lim_{x \to 1} \sqrt{x}$.

17. $\lim_{x \to 2} \dfrac{x^2 + x - 6}{x - 2}$.

18. $\lim_{x \to 1} \dfrac{x^2 - x}{x - 1}$.

19. $\lim_{h \to 0} \dfrac{(1 + h)^2 - 1}{h}$.

20. $\lim_{h \to 0} \dfrac{(1 + h)^2 - 3(1 + h) + 2}{h}$.

B

In Problems 21–24, indicate whether the given function is continuous or discontinuous at the given value of x by comparing $f(a)$ and $\lim_{x \to a} f(x)$.

21. $f(x) = \begin{cases} 0 & \text{if } x \text{ is rational,} \\ 1 & \text{if } x \text{ is irrational;} \end{cases}$ $a = 0.$

22. $f(x) = \begin{cases} 0 & \text{if } x \text{ is rational,} \\ 1 & \text{if } x \text{ is irrational;} \end{cases}$ $a = \sqrt{2}.$

23. $f(x) = \begin{cases} 0 & \text{if } x \text{ is rational,} \\ x & \text{if } x \text{ is irrational;} \end{cases}$ $a = 0.$

24. $f(x) = \begin{cases} 0 & \text{if } x \text{ is rational,} \\ x & \text{if } x \text{ is irrational;} \end{cases}$ $a = 1.$

In Problems 25–28, use the results of this section to evaluate the given limits.

25. $\lim_{h \to 0} \dfrac{\dfrac{1}{2 + h} - \dfrac{1}{2}}{h}$.

26. $\lim_{h \to 0} \dfrac{\sqrt{1 + h} - 1}{h}$.

27. $\lim_{h \to 0} \dfrac{\dfrac{1}{(x + h)^2} - \dfrac{1}{x^2}}{h}$.

28. $\lim_{h \to 0} \dfrac{\sqrt{x + h} - \sqrt{x}}{h}$.

C

29. Use Theorem 3.2 and the fact that $f(x) = k$ and $g(x) = x$ are continuous to show that any polynomial is continuous for all real numbers.

30. Use Theorem 3.2 and the fact that $f(x) = k$ and $g(x) = x$ are continuous to show that any rational function is continuous throughout its domain.

Review Problems

A

1. Give the domain and range of each of the following functions.

   (a) $f(x) = x^2 + 2$.

   (b) $f(x) = \sqrt{x - 4}$.

   (c) $f(x) = \dfrac{1}{x + 5}$.

   (d) $f(x) = \dfrac{x^2}{x^2 + 1}$.

   (e) $f(x) = \sqrt[3]{x + 1}$.

   (f) $f(x) = \sqrt{x(x - 2)}$.

2. Give the domain and range of each of the following functions.

   (a) $\{(1, 5), (3, 1), (4, -2), (6, 1), (10, 2)\}$.

   (b) $\{(2, 3), (-1, 4), (0, 1), (-3, 4)\}$.

3. For $f(x) = x^2 - 3x$, find

    (a) $f(0)$,   (b) $f(1)$,   (c) $f(1/2)$,   (d) $f(x + y)$,

    (e) $\dfrac{f(x + h) - f(x)}{h}$,   (f) $\dfrac{f(x) - f(a)}{x - a}$.

4. For $f(x) = \dfrac{x + 3}{x - 1}$, find

    (a) $f(0)$,   (b) $f(1)$,   (c) $f(-1)$,   (d) $f(x - 1)$,

    (e) $\dfrac{f(2 + h) - f(2)}{h}$,   (f) $f(x) + f(x + 1)$.

5. For $f(x) = \sqrt{x}$, find

    (a) $f(0)$,   (b) $f(4)$,   (c) $f(-1)$,   (d) $f(x^2)$,

    (e) $\dfrac{f(1 + h) - f(1)}{h}$.

6. Given $f(x) = 2x - 1$ for $x \le 5$ and $g(x) = x^2 + 1$ for $x > -1$, find $f + g$, $f - g$, fg, and f/g.

7. Given $f(x) = x^2$ for $-4 < x < 4$ and $g(x) = x + 2$ for $x \ge 0$, find $f + g$, $f - g$, fg, and f/g.

B

8. Given $f(x) = 2x^2$ for $x \ge 0$ and $g(x) = 2 - x^2$ for $x \ge 0$, find $f \circ g$.

9. Given $f(x) = 2x + 3$ for $x \ge 0$ and $g(x) = -x$ for $x \le 0$, find $f \circ g$ and $g \circ f$.

10. In each of the following expressions, indicate whether or not the limit exists. If there is a limit, give its value.

    (a) $\lim\limits_{x \to 1} \dfrac{x + 1}{x}$.

    (b) $\lim\limits_{x \to 0} f(x)$, where $f(x) = \begin{cases} 1 & \text{if } x < 0, \\ x & \text{if } x \ge 0. \end{cases}$

    (c) $\lim\limits_{h \to 0} \dfrac{(1 + h)^2 + 2(1 + h) - 3}{h}$.

    (d) $\lim\limits_{h \to 0} \dfrac{\dfrac{1}{3 + h} - \dfrac{1}{3}}{h}$.

11. For each of the following functions show whether it is continuous or discontinuous at the indicated value of x by comparing $f(a)$ and $\lim_{x \to a} f(x)$.

    (a) $f(x) = x^3 + 1$; $a = 2$.

    (b) $f(x) = \begin{cases} |x|/x & \text{if } x \ne 0, \\ 0 & \text{if } x = 0; \end{cases}$ $a = 0$.

    (c) $f(x) = \begin{cases} 2x + 1 & \text{if } x \le 1, \\ 3x & \text{if } x > 1; \end{cases}$ $a = 1$.

12. In each of the following cases, is there a value of $f(1)$ that will make f continuous at $x = 1$? If so, what is it? If not, why not?

    (a) $f(x) = \begin{cases} x^2 - 2x & \text{if } x < 1, \\ x^3 & \text{if } x > 1. \end{cases}$

    (b) $f(x) = \begin{cases} [\![x]\!] & \text{if } x < 1, \\ x - 1 & \text{if } x > 1. \end{cases}$

13. If $f(x) = (ax + b)/(cx + d)$, where $c \ne 0$, show that $f(x) = f_3(f_2(f_1(x)))$, where

$$f_1(x) = cx + d, \qquad f_2(x) = \frac{1}{x}, \qquad f_3(x) = \frac{a}{c} + \frac{bc - ad}{c}x.$$

C

14. A fixed point of a function f is a number x such that $f(x) = x$. Show that $f(x) = (ax + b)/(cx + d)$, where $ad - bc \neq 0$, has at most two fixed points in general.† Find the fixed point(s), if any, in each of the following cases.

    (a) $a = d, c \neq 0$;   (b) $a \neq d, c = 0$;   (c) $a = d, c = 0$.

15. Show that an arbitrary function f can be expressed as the sum of an odd function and an even function. (*Hint:* See Problem 54 of Section 3.1 for definitions of odd and even functions. Consider $f(x) + f(-x)$.)

---

†Note that there are possibilities of zero, one, two, or an infinite number of fixed points. Generally, there are two, but exceptions occur in case (b), where there is one fixed point, and in case (c), where there may be zero or an infinite number of fixed points.

4

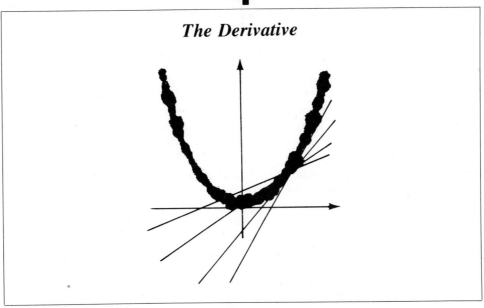

The Derivative

4.1

The Slope of a Curve

Suppose we have the graph of $y = f(x)$ and we are presented with the problem of finding the slope of the curve at the point A (Figure 4.1). Before asking, "What is the solution?" we might first ask, "What is the question?"

If our original curve were a line, it would be quite a simple problem. The slope of a line is simply the tangent of the inclination; or, given a pair of points on the line, we could find the slope by taking the difference between the y coordinates and dividing by the difference between the x coordinates taken in the same order. Since finding the slope of a line is so easy, we might try to relate the slope of a curve at a given point to the slope of some line. Let us consider Figure 4.2. Since the curve and the line have the same "direction" at the point A, they must have the same slope. What is the relationship between the curve and the line? It appears that the line

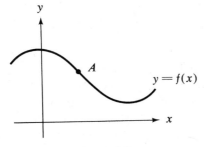

Figure 4.1

Figure 4.2

is tangent to the curve at point A. If it is, our problem of finding the slope of the curve at a given point reduces to one of finding the slope of the line tangent to the curve at the point.

But what is meant by the tangent to a curve at a given point? When we are dealing with circles, any line having exactly one point in common with the circle is defined as a tangent line, while any line having two points in common with it is called a secant line. But suppose we consider the two lines in Figure 4.3 and ask which of these is tangent to the curve at point A. If we use the same definition we used for a circle, we must conclude that m is tangent to the curve, while l is not. But this is exactly the reverse of what we intuitively think of as the tangent line. Perhaps we should cast off our intuitive idea and use this definition. However, it is not difficult to see that there are several lines that have only the point A in common with the curve of Figure 4.3. Thus it is easily seen that this definition of a tangent line—though quite suitable for circles—does not give us what we really want in the case of more general curves. What, then, do we mean by the tangent line to a curve at a given point? The fact that the line l intersects the curve at B as well as A seems, intuitively, to be rather unimportant when considering l as a tangent to the curve at A. Thus we seem to be interested only in the portion of the curve "near" A. Perhaps we can get the idea of a tangent line by considering secant lines joining A to the points of the curve "near" A. In Figure 4.4,

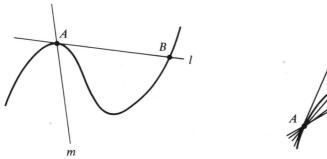

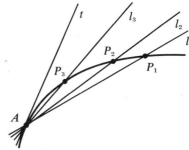

Figure 4.3   Figure 4.4

t is what we intuitively consider to be the tangent to the curve at A. Note that when we take the points P_1, P_2, P_3 closer and closer to A, the secant lines joining A to P_1, P_2, P_3, are getting closer and closer to the line t. In fact, we can get as "close" as we please to the line t, provided we take a point P close enough to A. Thus we might define the tangent to a curve at a given point to be the limiting position of a secant line joining A to some point P of the curve, different from A, as P moves closer and closer to A. We might note that P can approach A from two sides. We are assuming that the secant line approaches the same limiting position as the point on the curve approaches A from either side (see Problem 25).

Now that we know what the question is, we have gone a long way toward finding an answer. If we want the slope of the curve at a given point, we are interested in the slope of the tangent line. If we want the tangent line, we must go through a limiting process with secant lines. Thus suppose we return to our original problem. Given the graph of $y = f(x)$ with the point A, suppose we identify the abscissa (x coordinate) of A by a (see Figure 4.5). The ordinate (y coordinate) of A is then $f(a)$. In order to get a secant line, we must select a point P of the curve different from A. Suppose we do it

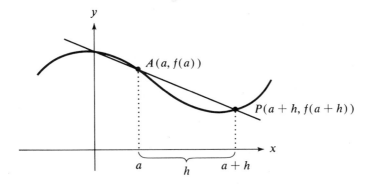

Figure 4.5

by choosing some number h and take P as the point of the curve with abscissa $a + h$. Note that while Figure 4.5 shows h to be positive, it need not be so—it may be negative; however, it may not be zero, since P is to be different from A. The slope of the line AP is easily found to be

$$m_{AP} = \frac{f(a+h) - f(a)}{(a+h) - a} = \frac{f(a+h) - f(a)}{h}.$$

If h is taken closer and closer to zero (but not equal to zero), the line AP then approaches the tangent to the curve at A. It would seem reasonable that if the line AP approaches the tangent line desired, then the slope of AP also approaches the slope of the tangent and thus the slope of the curve. Hence the slope of the curve at A is the limit of the slope of AP as P approaches A or, equivalently, as h approaches zero; in symbols,

$$m = \lim_{P \to A} m_{AP} = \lim_{h \to 0} \frac{f(a+h) - f(a)}{h}.$$

Example 1

Find the slope of $f(x) = x^2$ at the point (2, 4).

If we let P be the point with abscissa $2 + h$, its ordinate, as indicated by the equation of the curve, is $(2 + h)^2$. Thus

$$m_{AP} = \frac{(2+h)^2 - 2^2}{h}.$$

If we attempt to take the limit immediately, we see that both the numerator and denominator of this fraction approach zero as h approaches zero. However, *the fact that both the numerator and denominator are approaching zero tells us nothing about what the fraction is approaching.* On simplifying the numerator, we have

$$m_{AP} = \frac{h^2 + 4h}{h}$$

$$= h + 4.$$

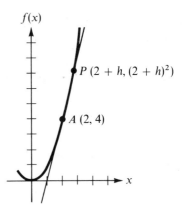

Figure 4.6

Note that the last two expressions are the same for all values of h except $h = 0$, but zero is the one value that h cannot have. Now

$$m = \lim_{P \to A} m_{AP} = \lim_{h \to 0} (h + 4) = 4,$$

which is the desired slope.

Example 2

Find the slope of $f(x) = x^2$ at the point $(0, 0)$.

Before using the method given above, note that the x axis is what we intuitively think of as the tangent to $y = x^2$ at $(0, 0)$, and it has slope 0. Thus, we know the answer before we start. This gives us a way of checking the validity of our new method—at least for this case. Letting $P = (h, h^2)$, we have

$$m_{AP} = \frac{h^2 - 0}{h} = h$$

and

$$m = \lim_{P \to A} m_{AP} = \lim_{h \to 0} h = 0.$$

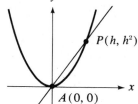

Figure 4.7

Once we have the slope of the tangent line at a given point we can use the point-slope form to find an equation of the tangent line. Also of interest is the *normal line* at a point A. It is simply the line through A which is perpendicular, or normal, to the curve.

Example 3

Find equations for the tangent and normal lines to $y = \dfrac{1}{x - 2}$ at $(4, 1/2)$.

First we find the slope of the curve at $(4, 1/2)$. Letting $P = \left(4 + h, \dfrac{1}{2 + h}\right)$, we have

$$m_{AP} = \frac{\dfrac{1}{2 + h} - \dfrac{1}{2}}{h} = \frac{\dfrac{-h}{2(2 + h)}}{h} = \frac{-1}{2(2 + h)}$$

and

$$m = \lim_{P \to A} m_{AP} = \lim_{h \to 0} \frac{-1}{2(2 + h)} = -\frac{1}{4}.$$

Now the tangent line has slope $m = -1/4$ and contains the point $(4, 1/2)$; its equation is

$$y - \frac{1}{2} = -\frac{1}{4}(x - 4),$$

$$x + 4y - 6 = 0.$$

The normal line has slope $m = 4$ (by Theorem 1.4(b)) and also contains $(4, 1/2)$; its equation is

$$y - \frac{1}{2} = 4(x - 4),$$

$$8x - 2y - 31 = 0.$$

Problems

A

In Problems 1–8, find the slope of the curve at the point indicated.

1. $f(x) = x^2$,   at $(-1, 1)$.
2. $f(x) = x^2$,   at $(1, 1)$.
3. $f(x) = x^3$,   at $(2, 8)$.
4. $f(x) = x^2 + 3x$,   at $(1, 4)$.
5. $f(x) = 3x^2 - 5x + 2$,   at $(2, 4)$.
6. $f(x) = (x - 2)(x + 1)$,   at $(2, 0)$.
7. $f(x) = (3x + 1)(x - 2)$,   at $(0, -2)$.
8. $f(x) = (x^2 + 1)(x - 3)$,   at $(2, -5)$.

In Problems 9–12, find equations of the tangent and normal lines to the given curve at the point indicated.

9. $f(x) = x^2$,   at $(2, 4)$.
10. $f(x) = 2x^2 - 3$,   at $(1, -1)$.
11. $f(x) = 2x^3 - 3x^2$,   at $(2, 4)$.
12. $f(x) = (x + 3)(2x - 3)$,   at $(2, 5)$.

B

In Problems 13–16, find the slope of the curve at the point indicated.

13. $f(x) = \dfrac{x + 1}{x - 1}$,   at $(0, -1)$.
14. $f(x) = \dfrac{x - 3}{2x - 1}$,   at $(4, 1/7)$.

15. $f(x) = \sqrt{x}$,   at $(0, 0)$; what restrictions must be put on h here?
16. $f(x) = \sqrt{x}$,   at $(4, 2)$; what restrictions must be put on h here?

In Problems 17 and 18, find equations of the tangent and normal lines to the given curve at the point indicated.

17. $f(x) = \dfrac{x - 2}{x + 1}$,   at $(2, 0)$.
18. $f(x) = \sqrt{x}$,   at $(0, 0)$.

19. What is the slope of the curve $f(x) = |x|$ at $x = 1, 3, 8, -2, -5, -7, 0$? Sketch this curve.
20. At what point does $f(x) = x^2 + 6x$ have slope 0?
21. At what point does $f(x) = x^2 - 4x$ have a horizontal tangent?
22. At what point does $f(x) = \sqrt{x}$ have no slope? (Do not confuse "no slope" with "slope 0.") Sketch the curve and the tangent line.
23. Use the method of this section to find the slope of $f(x) = 4x + 3$ at $(2, 11)$. Repeat for the point $(0, 3)$. Repeat for the point with abscissa x. How does this compare with what you already know about the curve?

C

24. Find the slope of $f(x) = x^2$ at the point with abscissa x. What is the value of this slope when $x = 2$? When $x = 0$? Compare these results with those of Examples 1 and 2.
25. We noted earlier that P can approach A from two sides and that we were assuming the line AP to approach the same limiting position in either case. Give an example in which there are two different limiting positions, depending upon the side from which A is approached by P.

4.2

The Derivative of a Function

In the preceding section we found the slope of a curve by finding the slope of a secant line and using a certain limiting process. We saw that if the curve is represented by the function f and we are interested in the point with abscissa a, then

$$m_{AP} = \frac{f(a + h) - f(a)}{h}$$

and

$$m = \lim_{P \to A} m_{AP} = \lim_{h \to 0} \frac{f(a + h) - f(a)}{h}.$$

It is often convenient to consider the slope of the curve not merely at some particular point $(a, f(a))$, but at any one of its points $(x, f(x))$. The result is a function, which we shall represent by f', defined by

$$f'(x) = \lim_{h \to 0} \frac{f(x + h) - f(x)}{h}.$$

Of course, the value of f' at $x = a$ is still the slope of the graph at $(a, f(a))$, but this is only one of several interpretations we shall encounter. For this reason, we give f' a name that does not suggest any one particular interpretation.

Definition

*The **derivative** (or **derived function**) f' of a function f is given by*

$$f'(x) = \lim_{h \to 0} \frac{f(x + h) - f(x)}{h},$$

*provided this limit exists. If $f'(a)$ exists, then f is said to be **differentiable** at $x = a$. The process of finding the derivative is called **differentiation**.*

If $y = f(x)$, then the derivative of y (or f) at x is represented by any one of the following notations:

$$y', \quad f'(x), \quad \frac{dy}{dx}, \quad \frac{d}{dx}f(x), \quad Dy, \quad \text{or} \quad Df(x).$$

These are all used interchangeably, and all mean exactly the same thing. It might be noted here that the notation dy/dx is particularly useful in certain cases. It is sometimes read "the derivative of y with respect to x." If, for example, $y = u^2$, where $u = x^2 + 1$, then y is a function of u, but at the same time it is, indirectly, a function of x; that is $y = (x^2 + 1)^2$. Thus it is possible to take two derivatives of y, one with respect to x, which is dy/dx, and another with respect to u, which is dy/du. Whenever this ambiguity exists, we shall use the notation dy/dx, which is called the differential notation.

Example 1

Differentiate $f(x) = 1/x^2$, and find an equation of the tangent line at $x = 2$.

$$f'(x) = \lim_{h \to 0} \frac{f(x+h) - f(x)}{h}$$

$$= \lim_{h \to 0} \frac{\dfrac{1}{(x+h)^2} - \dfrac{1}{x^2}}{h}$$

$$= \lim_{h \to 0} \frac{x^2 - (x+h)^2}{hx^2(x+h)^2}$$

$$= \lim_{h \to 0} \frac{-2hx - h^2}{hx^2(x+h)^2}$$

$$= \lim_{h \to 0} \frac{-2x - h}{x^2(x+h)^2}$$

$$= -\frac{2x}{x^4} = -\frac{2}{x^3}.$$

Note that the derivative is the *limit* of $(f(x+h) - f(x))/h$; it is not merely $(f(x+h) - f(x))/h$. The "$\lim_{h \to 0}$" must be retained until the limit is actually taken at the last step. For instance,

$$f'(x) \neq \frac{-2x - h}{x^2(x+h)^2};$$

rather,

$$f'(x) = \lim_{h \to 0} \frac{-2x - h}{x^2(x+h)^2} = -\frac{2}{x^3}.$$

The "$\lim_{h \to 0}$" is dropped in the last expression, because the limit now has been taken and you no longer need a symbol that tells you to take the limit.

Now that we have the derivative, the slope of the curve at $x = 2$ is

$$m = f'(2) = -\frac{2}{2^3} = -\frac{1}{4}.$$

Note here that the slope of the tangent line at $x = 2$ is the constant $f'(2)$, *not* the function $f'(x)$. The point of the curve for which $x = 2$ is $(2, f(2))$ or $(2, 1/4)$. Thus the tangent line is

$$y - \frac{1}{4} = -\frac{1}{4}(x - 2),$$

$$x + 4y - 3 = 0.$$

We have already seen one interpretation of the derivative: namely, the slope of a curve. Let us now consider another. Suppose the motion of an object in a straight-line path is carefully observed and it is determined that the object satisfies the equation

$$s = 8t^2 + 4t, \quad 0 \le t \le 5,$$

where s represents the position of the object, expressed in the number of feet to the right of the starting position, and t is the time, in seconds, after the starting time. This is called an *equation of motion*. Now, if we want to determine the average velocity throughout the 5 seconds, we simply divide the total distance, 220 feet (since $s = 0$ when $t = 0$ and $s = 220$ when $t = 5$), by the total time, 5 seconds, to get 44 feet per second. If we want the average velocity during the last two seconds, we divide the distance covered in those two seconds, which is $220 - 84 = 136$ feet (since $s = 84$ when $t = 3$ and $s = 220$ when $t = 5$), by 2 seconds, to obtain 68 feet per second. For the average velocity between $t = 3$ and $t = 4$, we have (letting $s = f(t)$)

$$\frac{f(4) - f(3)}{4 - 3} = \frac{144 - 84}{1} = 60 \text{ ft/sec.}$$

In fact, for the average velocity between any two times (within the domain of f), t and $t + h$, we have

$$v_{av} = \frac{f(t + h) - f(t)}{(t + h) - t} = \frac{f(t + h) - f(t)}{h}.$$

Of course, we cannot use this method to find the velocity at any particular instant, since the denominator would be zero. But we may think of these average velocities as approximations of an instantaneous velocity at some time within the interval under consideration; and, in general, we expect better approximations with shorter intervals. With this in mind it is reasonable to define the velocity at time t to be

$$v = \lim_{h \to 0} \frac{f(t + h) - f(t)}{h},$$

which is the derivative of s.

Example 2

For an object with the equation of motion $s = 8t^2 + 4t$, $0 \le t \le 5$, find the position and velocity at $t = 4$.

The position is found simply by substitution into the equation of motion.

$$s = 8 \cdot 4^2 + 4 \cdot 4 = 144.$$

Since the velocity is the derivative, we have

$$v = \lim_{h \to 0} \frac{f(t + h) - f(t)}{h}$$

$$= \lim_{h \to 0} \frac{[8(t + h)^2 + 4(t + h)] - [8t^2 + 4t]}{h}$$

$$= \lim_{h \to 0} \frac{8t^2 + 16ht + 8h^2 + 4t + 4h - 8t^2 - 4t}{h}$$

$$= \lim_{h \to 0} \frac{16ht + 8h^2 + 4h}{h}$$

$$= \lim_{h \to 0} (16t + 8h + 4)$$

$$= 16t + 4.$$

When $t = 4$, $v = 16 \cdot 4 + 4 = 68$ ft/sec.

In the foregoing example the position and velocity are always nonnegative. This need not be the case. The position is given by reference to a coordinate axis for which the zero point is arbitrarily set. In some cases it may be set in such a way that the position of the object is sometimes negative. In the same way, the velocity may be either positive or negative, depending upon the direction of motion; positive or negative velocities merely indicate motion in the positive or negative direction, respectively. Note that a negative velocity does not necessarily imply a negative position; it implies motion in the direction of decreasing numbers. In this connection we introduce another term, the *speed* of the object. The speed is the absolute value of the velocity. Thus speed gives only the rate at which the object is moving, while the velocity gives both rate and direction. Thus the velocity is a vector and the speed a scalar; we shall say more about these terms in Chapter 8.

Example 3

An arrow is shot upward from ground level ($s = 0$) resulting in the equation of motion $s = 128t - 16t^2$, $0 \leq t \leq ?$. How high does it go? When, and at what speed, does it hit the ground?

At its highest point the arrow stops going up ($v > 0$) and starts coming down ($v < 0$); the velocity must be zero at this point.

$$
\begin{aligned}
v &= \lim_{h \to 0} \frac{f(t+h) - f(t)}{h} \\
&= \lim_{h \to 0} \frac{[128(t+h) - 16(t+h)^2] - [128t - 16t^2]}{h} \\
&= \lim_{h \to 0} \frac{128t + 128h - 16t^2 - 32ht - 16h^2 - 128t + 16t^2}{h} \\
&= \lim_{h \to 0} \frac{128h - 32ht - 16h^2}{h} \\
&= \lim_{h \to 0} (128 - 32t - 16h) \\
&= 128 - 32t.
\end{aligned}
$$

If $v = 0$, then

$$128 - 32t = 0 \quad \text{and} \quad t = 4 \text{ sec.}$$

When $t = 4$, $s = 256$ ft.

The arrow hits the ground when $s = 0$; thus

$$
\begin{aligned}
128t - 16t^2 &= 0, \\
16t(8 - t) &= 0; \\
t = 0, \qquad t &= 8.
\end{aligned}
$$

Thus the arrow left the ground when $t = 0$ and hit the ground 8 seconds later. At that time the velocity was

$$v = 128 - 32 \cdot 8 = -128,$$

indicating that the arrow hit at 128 ft/sec. (Note that this was the initial speed of the arrow.)

We now have a second interpretation of the derivative—as a velocity. This interpretation can be generalized somewhat. We have a velocity when we take a derivative of an equation of motion, which gives the position as a function of time. Thus we say that velocity is a derivative of position with respect to time. But velocity is a rate of change of position with respect to time; so that if $y = f(x)$, we may interpret dy/dx to be the rate of change of y with respect to x. This interpretation of a derivative as velocity, or, more broadly, rate of change, has very wide application in such areas as rate of flow of fluids or electricity, rate of change of temperature, rate of growth, and so on.

Economists often refer to the derivative of a function as the *marginal value* of that function. For example, the cost function for some item gives the cost C of producing a number x of items as a function of the number of items produced; $C = f(x)$. The marginal cost, dC/dx, is interpreted as the rate of increase of the cost corresponding to a small increase in the number of items produced.

Example 4

The ABC Company has determined that the cost of producing x widgets per week is

$$C = 2000 + 100x - 0.1x^2.$$

Find the cost of producing 100 widgets per week. What is the marginal cost when production is at that level?

The cost of producing 100 widgets is

$$C = 2000 + 100(100) - 0.1(100)^2 = \$11,000.$$

The marginal cost is

$$\frac{dC}{dx} = \lim_{h \to 0} \frac{[2000 + 100(x + h) - 0.1(x + h)^2] - [2000 + 100x - 0.1x^2]}{h}$$

$$= \lim_{h \to 0} \frac{100h - 0.2hx - 0.1h^2}{h}$$

$$= \lim_{h \to 0} (100 - 0.2x - 0.1h)$$

$$= 100 - 0.2x.$$

When $x = 100$, the marginal cost is

$$\frac{dC}{dx} = 100 - 0.2(100) = \$80.$$

This indicates that the cost of producing one more widget is approximately \$80.

We can consider marginal revenue and marginal profit in exactly the same way. The revenue function gives the total revenue (income) received as a function of the number of items sold; $R = f(x)$. The profit is simply the total revenue minus the cost of production; $P = R - C$.

Our first two approaches to the derivative are the same ones taken by the two inventors of the calculus. Although there was initially a great deal of controversy over who had priority, it is now generally agreed that the calculus was invented independently by Isaac Newton and Gottfried Wilhelm Leibniz around 1675. Leibniz, who was well known as a mathematician, philosopher, and linguist, interpreted a derivative as the slope of a curve. Newton looked upon the derivative (which he called a fluxion) as a rate of change. He applied calculus to mechanics and astronomy, using it to show the consistency of his law of universal gravitation with Kepler's laws of planetary motion.

Much of our present-day notation is due to Leibniz, who introduced the notation dy/dx for the derivative. Newton used the symbol $\dot{s}$ for the derivative of s; this notation is still used in mechanics when taking a derivative with respect to time.

Problems

A
In Problems 1–6, differentiate the given function.

1. $f(x) = 2x^2 - 4x + 1.$

2. $y = 4x^3 - x.$

3. $f(x) = \dfrac{1}{x^3}.$

4. $f(z) = \dfrac{z - 3}{z + 3}.$

5. $f(s) = (s^2 + 1)^2.$

6. $y = \sqrt{x}.$

In Problems 7–10, find the derivative at the indicated points.

7. $f(x) = x^3 - 2x,$ at $x = 1.$

8. $y = 4x^3 + 3x^2,$ at $x = 2;$ at $x = 1.$

9. $f(v) = \dfrac{v - 4}{v + 1},$ at $v = 0;$ at $v = 2.$

10. $p = (q - 2)^3,$ at $q = 0.$

11. Find an equation of the line tangent to $f(x) = x^3$ at $(-1, -1).$

12. Find an equation of the line tangent to $y = 3x^2 - 2x + 1$ at $(-1, 6).$

In Problems 13–22, find the position and velocity for the given value of t.

13. $s = t^2 + t - 1;$ $t = 4.$

14. $s = 3t^2 + 2t - 4;$ $t = 1.$

15. $s = t^3 - 3t^2 + 5;$ $t = 0.$

16. $s = t^3 + t^2 - t - 1;$ $t = 1.$

17. $s = t^4 - 1;$ $t = 2.$

18. $s = t^5 - t^3,$ $t = 2.$

19. $s = (t + 2)(t^2 + 1);$ $t = 2.$

20. $s = (t + 1)(t^2 - 1);$ $t = -1.$

21. $s = t^2(t + 1)^2;$ $t = 3.$

22. $s = \dfrac{t}{t^2 + 1};$ $t = 0.$

B
In Problems 23–26, find the points at which the curve has horizontal tangents.

23. $y = x^2 - 6x + 2.$

24. $y = 2x^3 + 6x.$

25. $f(x) = \dfrac{x^2 + 1}{x}.$

26. $y = x^2(x^2 - 2).$

27. Find the points of $y = x^3$ at which the slope is 1.

28. Find the points of $y = x^3$ at which the slope is 12.

29. Find an equation(s) of the line(s) with slope 4 tangent to $y = 3x^2 - 2x + 4.$

30. Find an equation(s) of the line(s) with slope 4 tangent to $y = x^3 + 3x^2 - 5x + 6.$

31. For a ball thrown upward, we have the equation of motion $s = -16t^2 + 16t + 32,$ $0 \le t \le ?.$ How high does the ball go? When, and at what speed does it hit the ground $(s = 0)$?

32. If the arrow of Example 3 were shot upward from the moon, its equation of motion would be $s = 128t - 2.65t^2$, $0 \leq t \leq$?. How high would it go? When, and with what speed, would it hit the moon?

C

33. Suppose an object thrown upward has the equation of motion $s = -At^2 + Bt + C$, $0 \leq t \leq$?, where $A > 0$, $B > 0$, $C \geq 0$. How high does the object go? When, and at what speed, does it hit the ground?

34. An object propelled upward from ground level ($s = 0$) with velocity v_0 has equation of motion $s = -16t^2 + v_0 t$, $0 \leq t \leq$?. In Example 3 it was noted that the arrow hit the ground with a speed equal to the initial speed. Show that this is always the case.

In Problems 35–38, use the interpretation of the derivative as the slope of the graph to sketch the function f' if the following are graphs of f.

35. $f(x)$ 36. $f(x)$ 37. $f(x)$ 38. $f(x)$

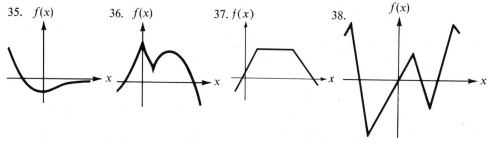

39. Given the cost function $C = 4000 + 120x - 0.5x^2 + 0.001x^3$,   find the cost and marginal cost at $x = 300$;   at $x = 400$.

40. For the revenue function $R = 150x - 5000$,   find the revenue and the marginal revenue at $x = 300$;   at $x = 400$.

41. Using the cost and revenue functions of Problems 39 and 40, find the profit and marginal profit at $x = 300$;   at $x = 400$. What advice would you give the manufacturer in each of these two cases?

4.3

Derivative Formulas: Sum and Product Formulas

Before looking at other applications, it would be well for us to achieve a greater facility in finding derivatives of functions. Until now we have relied on the definition of a derivative to find derivatives of functions, and the process has been rather tedious even for the relatively simple functions with which we have been dealing. Imagine what a problem it would be to find the derivative of

$$y = \frac{x\sqrt{x+1}}{x^2 - 4} \quad \text{or} \quad y = \left(\frac{x^2 + 1}{x^2 - 1}\right)^{2/3}$$

by the definition! Obviously we need some easier method of taking derivatives. We now consider some short cuts.

Theorem 4.1

If f is a function such that $f(x) = c$, then $f'(x) = 0$. (In other words, the derivative of a constant is zero.)

> **Proof**
>
> $$f'(x) = \lim_{h \to 0} \frac{f(x+h) - f(x)}{h}$$
>
> $$= \lim_{h \to 0} \frac{c - c}{h}$$
>
> $$= \lim_{h \to 0} 0$$
>
> $$= 0.$$
>
> This result is quite obvious graphically. The graph of $f(x) = c$ is a horizontal line, which has slope 0 at every point.

Theorem 4.2

If f is a function such that $f(x) = x^n$, where n is a positive integer, then $f'(x) = nx^{n-1}$. (In other words, the derivative of x to a power is the exponent times x to the next lower power.)

> **Proof**
>
> $$f'(x) = \lim_{h \to 0} \frac{f(x+h) - f(x)}{h}$$
>
> $$= \lim_{h \to 0} \frac{(x+h)^n - x^n}{h}$$
>
> $$= \lim_{h \to 0} \frac{x^n + nx^{n-1}h + [n(n-1)/2!]x^{n-2}h^2 + \cdots + h^n - x^n}{h} \, ,$$
>
> by the binomial theorem (see Appendix C.1 (2)),
>
> $$= \lim_{h \to 0} \left(nx^{n-1} + \frac{n(n-1)}{2!} x^{n-2}h + \cdots + h^{n-1} \right)$$
>
> $$= nx^{n-1}.$$

With this formula we can see that if $y = x^2$, then $y' = 2x$, a fact we proved earlier with much more difficulty. Actually, this formula holds whether n is restricted to positive integers or not. We have already proved it for the case $n = 0$. If $y = x^0 = 1$ ($x \neq 0$), then $y' = 0 \cdot x^{0-1} = 0$. This same formula will be derived later with fewer restrictions on n.

Theorem 4.3

If u and v are differentiable at x and f is a function such that $f(x) = u(x) + v(x)$, then $f'(x) = u'(x) + v'(x)$. (That is to say, the derivative of a sum is the sum of the derivatives.)

Proof

$$f'(x) = \lim_{h \to 0} \frac{f(x+h) - f(x)}{h}$$

$$= \lim_{h \to 0} \frac{[u(x+h) + v(x+h)] - [u(x) + v(x)]}{h}$$

$$= \lim_{h \to 0} \left(\frac{u(x+h) - u(x)}{h} + \frac{v(x+h) - v(x)}{h} \right)$$

$$= \lim_{h \to 0} \frac{u(x+h) - u(x)}{h} + \lim_{h \to 0} \frac{v(x+h) - v(x)}{h}, \quad \text{by Theorem 3.1(a),}$$

$$= u'(x) + v'(x).$$

A similar result can be proved for the sum of three or more functions as well as for the difference of two functions (see Problems 32, 33, and 36). Thus, if

$$f(x) = x^3 - x^2 + x - 3,$$

then

$$f'(x) = 3x^2 - 2x + 1.$$

Although it might be assumed that the same type of formula would work for products and quotients, it is easily seen that this is not the case; for if

$$u(x) = x,\ v(x) = x \quad \text{and} \quad f(x) = u(x) \cdot v(x) = x^2,$$

then

$$f'(x) = 2x; \quad \text{but} \quad u'(x) \cdot v'(x) = 1 \cdot 1 = 1 \neq f'(x).$$

Thus the derivative of a product is *not* merely the product of the derivatives.

Theorem 4.4

(*Product rule*) *If u and v are differentiable at x and f is a function such that*

$$f(x) = u(x) \cdot v(x),$$

then

$$f'(x) = u(x) \cdot v'(x) + v(x) \cdot u'(x).$$

(*The derivative of a product is the first factor times the derivative of the second plus the second factor times the derivative of the first.*)

Proof

$$f'(x) = \lim_{h \to 0} \frac{f(x+h) - f(x)}{h}$$

$$= \lim_{h \to 0} \frac{u(x+h) \cdot v(x+h) - u(x) \cdot v(x)}{h}$$

$$= \lim_{h \to 0} \frac{u(x+h) \cdot v(x+h) - u(x+h) \cdot v(x) + u(x+h) \cdot v(x) - u(x) \cdot v(x)}{h}$$

(See Note 1.)

$$= \lim_{h \to 0} \left[u(x + h) \, \frac{v(x + h) - v(x)}{h} + v(x) \, \frac{u(x + h) - u(x)}{h} \right]$$

$$= \left[\lim_{h \to 0} u(x + h) \right] \left[\lim_{h \to 0} \frac{v(x + h) - v(x)}{h} \right] + \left[\lim_{h \to 0} v(x) \right] \left[\lim_{h \to 0} \frac{u(x + h) - u(x)}{h} \right],$$

by Theorems 3.1(a) and (c),

$$= u(x) \cdot v'(x) + v(x) \cdot u'(x). \quad \text{(See Note 2.)}$$

Note 1: All that was done here was to add and subtract $u(x + h) \cdot v(x)$ in the numerator. You might well ask, "How did you know to do that?" The answer "It works" is not very satisfactory. Let us look for a more reasonable answer. At the second step we had

$$\lim_{h \to 0} \frac{u(x + h) \cdot v(x + h) - u(x) \cdot v(x)}{h}.$$

In its place, we would like something that can be handled easily and looks almost like the expression we have. This expression would be much easier to handle if the $u(x)$ were, instead, $u(x + h)$, to match the $u(x + h)$ in the other term. That way the $u(x + h)$ could be factored out, and the limit of the expression which remains is a derivative. (See Problem 34 for another approach.)

Of course, if we want the new expression to be equal to the former one, we must compensate for it in some way. We do this by putting it in a second time, with the opposite sign. This same idea will be repeated several times later on. To summarize, when we do not have what we want, we simply put in what we want and compensate for it to have equality.

Note 2: In taking the limit, we had $\lim_{h \to 0} u(x + h)$, which we assumed to be $u(x)$. Although $\lim_{h \to 0} u(x + h)$ is not always $u(x)$ (see Problem 35), it is $u(x)$ if u is continuous at x. This was noted in Section 3.4. As we shall see in Chapter 10, the differentiability of u at x implies its continuity at x. We shall continue to make this assumption about continuity throughout this chapter.

Theorem 4.5

If u is differentiable at x and f is a function such that $f(x) = c \cdot u(x)$, then $f'(x) = c \cdot u'(x)$. (In other words, the derivative of a constant times a function of x is the constant times the derivative of the function.)

Proof

$$f'(x) = c \cdot u'(x) + u(x) \cdot \frac{d}{dx} c, \qquad \text{by Theorem 4.4,}$$

$$= c \cdot u'(x) + u(x) \cdot 0, \qquad \text{by Theorem 4.1,}$$

$$= c \cdot u'(x).$$

Example 1

Differentiate $y = 6x^3$.

$$y' = 6 \cdot 3x^2 = 18x^2.$$

Example 2

Differentiate $y = 4x^3 + 3x^2 - 2x + 2$.

$$y' = 4 \cdot 3x^2 + 3 \cdot 2x - 2 \cdot 1 + 0$$
$$= 12x^2 + 6x - 2.$$

With very little practice you should be able to omit the intermediate step and go directly to the final answer.

Example 3

Differentiate $y = (x^2 + x + 1)(x + 1)$.

There are two methods. We can either differentiate by the product rule and simplify the result, or we can multiply first and differentiate the result. Both methods are given here.

$$y' = (x^2 + x + 1)1 + (x + 1)(2x + 1)$$
$$= x^2 + x + 1 + 2x^2 + 3x + 1$$
$$= 3x^2 + 4x + 2,$$

or

$$y = (x^2 + x + 1)(x + 1)$$
$$= x^3 + 2x^2 + 2x + 1$$
$$y' = 3x^2 + 4x + 2.$$

Problems

A

In Problems 1–8, differentiate the given function.

1. $y = 3x^2 + 5x - 2$.
2. $y = 3x^3 - 4x^2 + 2x - 5$.
3. $y = x^7 + 7x^6$.
4. $y = x^6 - x^3 + 1$.
5. $y = 7x^5 + 5x^3 + x - 3$.
6. $u = 4v^6 - 6v^4$.
7. $s = 5t^3 + 2t^2 - 6t$.
8. $y = (x^2 + 4)(x - 2)$.

In Problems 9–12, find the derivative at the indicated point.

9. $y = 3x^2 - 5x + 2$, at $(1, 0)$.
10. $y = 3x^4 - 3x^2$, at $(-1, 0)$.
11. $y = x^3 - 2x$, at $(2, 4)$.
12. $y = x^4 + 4x^3$, at $x = 2$.

B

In Problems 13–16, differentiate the given function.

13. $y = (x^2 + 2x - 1)(x^2 - 2x + 1)$.
14. $y = (x^3 + 1)(x^3 - 1)$.
15. $p = (q^2 + 4)(q^2 + 1)$.
16. $y = x(2x + 1)(2x - 1)$.

In Problems 17 and 18, find the derivative at the indicated point.

17. $y = (x^4 + 1)(x^2 - 1)$, at $(1, 0)$.
18. $y = (2x^2 + 1)(x^2 + 1)$, at $x = -1$.

19. At what point(s) does the tangent to $y = x^3 + 2x^2 + x$ have slope 1?
20. At what point(s) does the tangent to $y = (x^2 + 5)(x - 5)$ have slope 2?

21. At what point(s) does $y = x^3 - 12x^2 + 45x - 55$ have a horizontal tangent?
22. At what point(s) does $y = x^4 - 4x^2$ have a horizontal tangent?
23. Find an equation of the line tangent to $y = x^3 - 3x + 4$ at the point $(1, 2)$.
24. Find an equation of the line tangent to $y = x^4$ at the point $(2, 16)$.
25. A moving object has equation of motion $s = 5 - 3t + 4t^2 - t^3$.   Find the position, velocity, and speed of the object when $t = 1$;   when $t = 4$.
26. A ball is thrown upward from an elevated platform, giving an equation of motion $s = -16t^2 + 8t + 8$.   How high does it go? When, and with what speed, does it hit the ground $(s = 0)$?
27. Prove that if $f(x) = u(x)v(x)w(x)$,   then

$$f'(x) = u(x)v(x)w'(x) + u(x)v'(x)w(x) + u'(x)v(x)w(x).$$

   If $f(x)$ is the product of four functions, what is $f'(x)$?

In Problems 28–31, use the results of Problem 27 to find the derivative.

28. $y = x(x + 2)(x - 1)$.
29. $y = (x + 1)(x + 2)(x + 3)$.
30. $y = (x^2 + 1)(x^2 + 2)(x^2 + 3)$.
31. $y = (x + 1)(x + 2)(x + 3)(x + 4)$.

C

32. Prove that if $f(x) = u(x) + v(x) + w(x)$, then $f'(x) = u'(x) + v'(x) + w'(x)$.
33. Prove that if $f(x) = u(x) - v(x)$, then $f'(x) = u'(x) - v'(x)$.
34. Prove Theorem 4.4 by adding and subtracting $u(x)v(x + h)$ instead of $u(x + h) \cdot v(x)$.
35. Give an example of a function $f(x)$ such that $\lim_{h \to 0} f(x + h) \neq f(x)$ for some value of x.
36. Prove that if $f(x) = u_1(x) + u_2(x) + u_3(x) + \cdots + u_n(x)$, where n is a positive integer, then

$$f'(x) = u_1'(x) + u_2'(x) + u_3'(x) + \cdots + u_n'(x).$$

4.4

Derivative Formulas: Quotient Rule and Chain Rule

Theorem 4.6

(*Quotient rule*)   *If u and v are differentiable at x, $v(x) \neq 0$, and f is a function such that $f(x) = u(x)/v(x)$,   then*

$$f'(x) = \frac{v(x) \cdot u'(x) - u(x) \cdot v'(x)}{[v(x)]^2}.$$

(*The derivative of a quotient is the denominator times the derivative of the numerator minus the numerator times the derivative of the denominator, all divided by the denominator squared.*)

Proof

$$f'(x) = \lim_{h \to 0} \frac{f(x+h) - f(x)}{h}$$

$$= \lim_{h \to 0} \frac{\dfrac{u(x+h)}{v(x+h)} - \dfrac{u(x)}{v(x)}}{h}$$

$$= \lim_{h \to 0} \frac{v(x) \cdot u(x+h) - v(x+h) \cdot u(x)}{v(x)v(x+h)h}$$

$$= \lim_{h \to 0} \frac{\dfrac{v(x) \cdot u(x+h) - v(x+h) \cdot u(x)}{h}}{v(x)v(x+h)}$$

$$= \lim_{h \to 0} \frac{\dfrac{v(x) \cdot u(x+h) - v(x) \cdot u(x) + v(x) \cdot u(x) - v(x+h)u(x)}{h}}{v(x)v(x+h)}$$

(See Note 1.)

$$= \lim_{h \to 0} \frac{v(x)\dfrac{u(x+h) - u(x)}{h} - u(x)\dfrac{v(x+h) - v(x)}{h}}{v(x)v(x+h)}$$

$$= \frac{\left[\lim_{h \to 0} v(x)\right]\left[\lim_{h \to 0} \dfrac{u(x+h) - u(x)}{h}\right] - \left[\lim_{h \to 0} u(x)\right]\left[\lim_{h \to 0} \dfrac{v(x+h) - v(x)}{h}\right]}{\left[\lim_{h \to 0} v(x)\right]\left[\lim_{h \to 0} v(x+h)\right]},$$

by Theorems 3.1(b), (c), and (d),

$$= \frac{v(x) \cdot u'(x) - u(x) \cdot v'(x)}{[v(x)]^2}.$$

(See Note 2.)

Note 1: (Note 1 of Theorem 4.4 also applies here.) We add and subtract $v(x) \cdot u(x)$. This particular expression is selected because we want a $v(x)$ to match the one in the first term so it can be factored out and a $u(x)$ so that

$$\frac{u(x+h) - u(x)}{h}$$

remains after the factoring.

Note 2: We again assume that v is continuous at x and $\lim_{h \to 0} v(x+h) = v(x)$.

Example 1

Differentiate $y = \dfrac{x+1}{x-2}$.

$$y' = \frac{(x-2) \cdot 1 - (x+1) \cdot 1}{(x-2)^2} = \frac{-3}{(x-2)^2}.$$

With Theorem 4.6, we can now extend Theorem 4.2.

Theorem 4.7

If f is a function such that $f(x) = x^n$, where n is an integer, then $f'(x) = nx^{n-1}$.

Proof

It has already been proved for the cases in which n is positive or zero. The only remaining case is that in which n is negative. In that case, there is a positive integer k such that $n = -k$. Thus,

$$f(x) = x^{-k} = 1/x^k.$$

By Theorems 4.6 and 4.2,

$$f'(x) = \frac{x^k \cdot 0 - 1 \cdot kx^{k-1}}{x^{2k}} = -kx^{-k-1} = nx^{n-1}.$$

Before considering other differentiation formulas, it is convenient to have another form for the derivative of a function—that is, a form different from the definition. To do so let us look back at the geometric motivation for the derivative. Compare (a) and (b) of Figure 4.8; basically, they represent the same thing, but they use different notation. Similarly, the slope of the line AP can be represented by

$$m_{AP} = \frac{f(x+h) - f(x)}{x+h-x} = \frac{f(x+h) - f(x)}{h}$$

or

$$m_{AP} = \frac{f(t) - f(x)}{t - x} = \frac{f(x) - f(t)}{x - t},$$

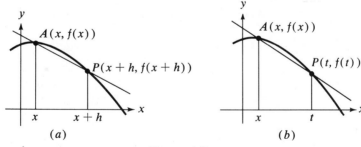

Figure 4.8

and as P approaches A, h approaches 0 and t approaches x. Thus, when we defined the derivative as

$$f'(x) = \lim_{h \to 0} \frac{f(x+h) - f(x)}{h},$$

we could just as easily have used

$$f'(x) = \lim_{t \to x} \frac{f(t) - f(x)}{t - x} = \lim_{t \to x} \frac{f(x) - f(t)}{x - t},$$

which is the statement of the next theorem.

Theorem 4.8

$$f'(x) = \lim_{t \to x} \frac{f(t) - f(x)}{t - x} = \lim_{t \to x} \frac{f(x) - f(t)}{x - t}.$$

This theorem also can be proved directly from the definition of derivative without recourse to the above geometric argument, by making the substitution $t = x + h$. For some purposes, this new form for the derivative is more convenient than the definition. Some authors use this as the definition and derive the other form from it.

Our next theorem uses the result of Theorem 4.8, but first let us take up a question touched upon earlier in connection with notation. Suppose y is a function of u while u is in turn a function of x, in which case y is a function of x as well as a function of u. Thus, if $y = f(u)$ and $u = g(x)$, then

$$y = f(g(x)) = F(x).$$

We may now consider two derivatives of y—the derivative of y with respect to u, which is dy/du, and the derivative of y with respect to x, which is dy/dx. Now when these derivatives are presented in the limit forms, how do we distinguish one from another? Basically it is done merely by noting that there are two different functions, $f(u)$ and $F(x)$. The important thing to note here is not the difference between u and x but the difference between the *functions* f and F. Thus

$$\frac{dy}{du} = \lim_{h \to 0} \frac{f(u + h) - f(u)}{h} = \lim_{t \to u} \frac{f(t) - f(u)}{t - u},$$

while

$$\frac{dy}{dx} = \lim_{h \to 0} \frac{F(x + h) - F(x)}{h} = \lim_{t \to x} \frac{F(t) - F(x)}{t - x}.$$

Since $u = g(x)$ and $F(x) = f(g(x))$, both can be written in other forms. Thus we get

$$\frac{dy}{du} = \lim_{t \to u} \frac{f(t) - f(u)}{t - u} = \lim_{g(s) \to g(x)} \frac{f(g(s)) - f(g(x))}{g(s) - g(x)}$$

simply by substituting $u = g(x)$ and $t = g(s)$. Note that the letters used here are immaterial; if t had not been used for something else, it might have been used instead of s, thus:

$$\frac{dy}{du} = \lim_{g(t) \to g(x)} \frac{f(g(t)) - f(g(x))}{g(t) - g(x)}.$$

Similarly,

$$\frac{dy}{dx} = \lim_{t \to x} \frac{F(t) - F(x)}{t - x} = \lim_{t \to x} \frac{f(g(t)) - f(g(x))}{t - x}.$$

Theorem 4.9

(*Chain rule*)   *If f and g are functions such that g is differentiable at x and f is differentiable at $u = g(x)$ and F is a function such that $F(x) = f(g(x))$,   then*

$$\frac{d}{dx} F(x) = \frac{d}{du} f(u) \cdot \frac{d}{dx} g(x).$$

A simpler, but less accurate statement of this theorem is: If $y = f(u)$ and $u = g(x)$, then

$$\frac{dy}{dx} = \frac{dy}{du} \cdot \frac{du}{dx}.$$

Note here that the differential notation is an aid to the memory. While you are reminded not to think of dy/du and du/dx as fractions, they seem to behave as if they were.

Proof

Since $y = f(u)$ and $u = g(x)$, $y = f(g(x)) = F(x)$.

$$\frac{dy}{dx} = \lim_{t \to x} \frac{F(t) - F(x)}{t - x}$$

$$= \lim_{t \to x} \frac{f(g(t)) - f(g(x))}{t - x}.$$

Now let us stop and take stock. Given

$$\lim_{t \to x} \frac{f(g(t)) - f(g(x))}{t - x},$$

we prefer to have

$$\lim_{t \to x} \frac{g(t) - g(x)}{t - x}.$$

Thus we multiply both numerator and denominator by $g(t) - g(x)$ to get

$$\frac{dy}{dx} = \lim_{t \to x} \frac{f(g(t)) - f(g(x))}{g(t) - g(x)} \cdot \frac{g(t) - g(x)}{t - x}$$

$$= \left[\lim_{t \to x} \frac{f(g(t)) - f(g(x))}{g(t) - g(x)} \right] \left[\lim_{t \to x} \frac{g(t) - g(x)}{t - x} \right]$$

$$= \frac{dy}{du} \frac{du}{dx}. \qquad \text{(See Note.)}$$

Actually there is a problem here. We have no assurance that $g(t) - g(x) \neq 0$. We can remedy the situation by splitting the argument into two cases.

Case I: $g(t) \neq g(x)$ for any t within a distance H of x. By choosing t close enough to x, we are assured that $g(t) - g(x) \neq 0$. Thus, for $|t - x|$ small enough ($|t - x| < H$),

$$\frac{dy}{dx} = \lim_{t \to x} \frac{f(g(t)) - f(g(x))}{g(t) - g(x)} \cdot \frac{g(t) - g(x)}{t - x}$$

$$= \frac{df(u)}{du} \cdot \frac{dg(x)}{dx} \qquad \text{(See Note.)}$$

$$= \frac{dy}{du} \cdot \frac{du}{dx}.$$

Case II: No matter how small H is, there is a number t within a distance H of x such that $g(t) = g(x)$. Note that this does not imply that $g(t) = g(x)$ for *all* t

within a distance H of x; it does imply that $g(t) = g(x)$ for infinitely many values of t within a distance H of x. In any case

$$\lim_{t \to x} \frac{g(t) - g(x)}{t - x} = 0,$$

since we know that this limit (which is $g'(x)$) exists and

$$\frac{g(t) - g(x)}{t - x} = 0$$

for values of t arbitrarily close to x. Now let us consider

$$\lim_{t \to x} \frac{F(t) - F(x)}{t - x}.$$

First let M be the set of all numbers t of the domain of F such that t is within a distance H of x and $g(t) \neq g(x)$; let N be the set of all numbers t of the domain of F such that t is within a distance H of x and $g(t) = g(x)$. If M contains numbers arbitrarily close to x, then (by Case I)

$$\lim_{\substack{t \to x \\ t \text{ in } M}} \frac{F(t) - F(x)}{t - x} = \frac{dy}{du} \cdot \frac{du}{dx} = \frac{dy}{du} \cdot 0 = 0.$$

If M does not contain numbers arbitrarily close to x, we need not consider this limit. In addition,

$$\lim_{\substack{t \to x \\ t \text{ in } N}} \frac{F(t) - F(x)}{t - x} = \lim_{\substack{t \to x \\ t \text{ in } N}} \frac{f(g(t)) - f(g(x))}{t - x} = \lim_{\substack{t \to x \\ t \text{ in } N}} \frac{f(g(x)) - f(g(x))}{t - x}$$

$$= 0.$$

Thus the combination of these two limits gives

$$\frac{dy}{dx} = \lim_{t \to x} \frac{F(t) - F(x)}{t - x} = 0 = \frac{dy}{du} \cdot \frac{du}{dx}.$$

Note: While we had indicated that

$$\frac{dy}{du} = \lim_{g(t) \to g(x)} \frac{f(g(t)) - f(g(x))}{g(t) - g(x)},$$

the above limit is

$$\lim_{t \to x} \frac{f(g(t)) - f(g(x))}{g(t) - g(x)}.$$

We are assuming again that as t approaches x, $g(t)$ approaches $g(x)$. Here we are using the fact, to be proved later, that the existence of $g'(x)$ implies the continuity of $g(x)$. Thus these two limits are equivalent.

Example 2

If $y = u^2$ and $u = x^2 - 4x + 3$, find dy/dx.

By the chain rule, $\dfrac{dy}{dx} = \dfrac{dy}{du} \cdot \dfrac{du}{dx}$

$$= 2u(2x - 4)$$
$$= 4(x^2 - 4x + 3)(x - 2).$$

Example 3

Find y' for $y = \dfrac{1}{(x^2 + 3x - 5)^3}$.

Let us make the substitution $u = x^2 + 3x - 5$. Then $y = u^{-3}$; and, by the chain rule,

$$\frac{dy}{dx} = \frac{dy}{du} \cdot \frac{du}{dx}$$

$$= -3u^{-4}(2x + 3)$$

$$= \frac{-3(2x + 3)}{(x^2 + 3x - 5)^4}.$$

The chain rule is very important and we shall use it to prove the next two theorems, as well as several others.

Theorem 4.10

If f is a function such that $f(x) = x^n$, where n is any rational number, then $y' = nx^{n-1}$.

Proof

First, $n = p/q$, where p and q are integers and q is positive. Now we shall split the argument into two parts. We shall show first that $y' = nx^{n-1}$ if $n = 1/q$ (where q is a positive integer); then we shall use this together with the chain rule to prove the general case.

Suppose that $y = x^{1/q}$, where q is a positive integer. Then

$$y' = \lim_{t \to x} \frac{t^{1/q} - x^{1/q}}{t - x}.$$

Using the substitutions $z = x^{1/q}$ and $s = t^{1/q}$, we get

$$y' = \lim_{s^q \to z^q} \frac{s - z}{s^q - z^q}$$

$$= \lim_{s \to z} \frac{s - z}{(s - z)(s^{q-1} + s^{q-2}z + s^{q-3}z^2 + \cdots + sz^{q-2} + z^{q-1})}$$

$$= \lim_{s \to z} \frac{1}{(s^{q-1} + s^{q-2}z + s^{q-3}z^2 + \cdots + sz^{q-2} + z^{q-1})}$$

$$= \frac{1}{qz^{q-1}}$$

$$= \frac{1}{q} z^{1-q}$$

$$= \frac{1}{q} (x^{1/q})^{1-q}$$

$$= \frac{1}{q} x^{(1/q)-1}.$$

Thus, if $n = 1/q$ and $y = x^n$, then $y' = nx^{n-1}$.

Now suppose $y = x^{p/q}$, where p and q are integers and q is positive. Then $y = (x^{1/q})^p$. By substituting $u = x^{1/q}$ we have $y = u^p$.

$$\frac{dy}{dx} = \frac{dy}{du} \cdot \frac{du}{dx}$$

$$= pu^{p-1} \cdot \frac{1}{q} x^{(1/q)-1}$$

$$= \frac{p}{q} (x^{1/q})^{p-1} x^{(1/q)-1}$$

$$= \frac{p}{q} x^{(p/q)-(1/q)} x^{(1/q)-1}$$

$$= \frac{p}{q} x^{(p/q)-1}.$$

Thus, the formula which was first stated for positive integers in Theorem 4.2 was extended to all integers in Theorem 4.7 and now applies to all rational numbers. We shall not be in a position to prove it for all real numbers until Chapter 12 (see Theorem 12.6). Nevertheless, the extensions we have made allow us to find derivatives of a wide range of functions.

Example 4

Differentiate $y = \sqrt{x}$.

$$y = \sqrt{x} = x^{1/2}, \qquad y' = \frac{1}{2} x^{-1/2} = \frac{1}{2\sqrt{x}}.$$

Example 5

Differentiate $y = \dfrac{x+2}{\sqrt{x}}$.

There are several possible methods. Two are illustrated.

$$y = \frac{x+2}{x^{1/2}}, \qquad\qquad\qquad y = x^{1/2} + 2x^{-1/2},$$

$$y' = \frac{x^{1/2} \cdot 1 - (x+2)\frac{1}{2}x^{-1/2}}{x} \qquad\qquad y' = \frac{1}{2} x^{-1/2} - x^{-3/2}$$

$$= \frac{\sqrt{x} - \dfrac{x+2}{2\sqrt{x}}}{x} \qquad\qquad\qquad = \frac{1}{2x^{1/2}} - \frac{1}{x^{3/2}}$$

$$= \frac{2x - (x+2)}{2x^{3/2}} \qquad\qquad\qquad = \frac{x-2}{2x^{3/2}}.$$

$$= \frac{x-2}{2x^{3/2}}.$$

In the second method of Example 5 we have avoided the relatively complicated quotient formula by carrying out the division and using negative exponents. While this method is not universally recommended, it sometimes simplifies a problem considerably.

Example 6

Differentiate $y = \dfrac{1}{x}$.

$$y' = \frac{x \cdot 0 - 1 \cdot 1}{x^2} = \frac{-1}{x^2}.$$

However, by writing the original problem as $y = x^{-1}$, we get

$$y' = -x^{-2} = -\frac{1}{x^2}.$$

The use of negative exponents makes the problem simple enough to do in your head. This method can be used to advantage when the denominator is very simple or when the numerator is a constant.

Problems

A

In Problems 1–12, differentiate.

1. $y = \dfrac{1}{x^2}$.

2. $y = \dfrac{1}{x^5}$.

3. $y = \dfrac{x^2 + 1}{x^2}$.

4. $y = \dfrac{x^2 - 4}{x}$.

5. $y = x^{2/3} - a^{2/3}$ (*a* is a constant).

6. $y = \dfrac{x}{x^2 + 4}$.

7. $y = \dfrac{x^3}{x - 1}$.

8. $y = \dfrac{x(x + 2)}{x + 1}$.

9. $s = t^{2/3} - t^{-1/3}$.

10. $s = \dfrac{3}{4} t^{4/3} + 3t^{1/3} + \dfrac{3}{2} t^{-2/3}$.

11. $u = \dfrac{v^2 + 2v - 2}{v^2 - 2v + 2}$.

12. $u = \dfrac{v^2 + 4a^2}{v^2 - 4a^2}$ (*a* is a constant).

In Problems 13–16, use the chain rule to find dy/dx.

13. $y = \sqrt{u}$, $u = 2x^2 - 3$.

14. $y = u^2 + 1$, $u = 2x + 5$.

15. $y = u^3 + u$, $u = 3x - 2$.

16. $y = u\sqrt{u}$, $u = 2x + 3$.

In Problems 17–22, find the derivative at the point indicated.

17. $y = 2x^{1/2} - 3x^{1/3}$,   at $(1, -1)$.

18. $y = \dfrac{x - 1}{x + 2}$,   at $(1, 0)$.

19. $y = \dfrac{x^2 + 1}{4x^2 - 9}$,   at $x = 2$.

20. $y = \sqrt[3]{x} - 1$,   at $(8, 1)$.

21. $y = \dfrac{x^3 + 1}{x^2 - 2}$,   at $(-1, 0)$.

22. $y = \dfrac{\sqrt{x} - 1}{\sqrt[3]{x} + 1}$,   at $x = 64$.

B

In Problems 23–26, differentiate.

23. $p = \dfrac{q^{1/3} - a^{1/3}}{q^{1/3} + a^{1/3}}$ (a is a constant).

24. $p = \dfrac{q^{2/3} - q^{1/3}}{q^{-2/3} - q^{-1/3}}$.

25. $y = \dfrac{(x^2 + 1)(2x - 3)}{x}$.

26. $y = \dfrac{(2x + 5)(3x - 2)}{x + 1}$.

In Problems 27 and 28, use the chain rule to find dy/dx.

27. $y = \sqrt{x^2 + 4x - 3}$.

28. $y = \dfrac{4x^2 + 12x + 10}{\sqrt{2x + 3}}$.   (*Hint:* Express $4x^2 + 12x + 10$ in the form $(2x + 3)^2 + k$.)

29. At what point(s) does the tangent to $y = 2x^3 - 3x^2 + 1$ have slope 12?
30. At what point(s) does the tangent to $y = x^3 + 3x^2 - 9x$ have slope 0?

31. At what point(s) does $y = \dfrac{x^2}{x^2 + 1}$ have a horizontal tangent?

32. At what point(s) does $y = \sqrt[3]{x - 1}$ have a vertical tangent?
33. Find an equation of the line tangent to $y = \sqrt{x}$ at the point (4, 2).

34. Find an equation of the line tangent to $y = \dfrac{2x + 1}{2x - 1}$ at the point (1, 3).

35. If $\dfrac{d}{dx} \sin x = \cos x$ and $\dfrac{d}{dx} \cos x = -\sin x$,   find

 (a) $\dfrac{d}{dx} \tan x$,

 (b) $\dfrac{d}{dx} \cot x$,

 (c) $\dfrac{d}{dx} \sec x$,

 (d) $\dfrac{d}{dx} \csc x$.

36. If $\dfrac{d}{dx} \sin x = \cos x$ and $\dfrac{d}{dx} \cos x = -\sin x$,   find

 (a) $\dfrac{d}{dx} \sin x^2$,

 (b) $\dfrac{d}{dx} \cos (2x + 1)$,

 (c) $\dfrac{d}{dx} \sin (3x^2 + 4x)$,

 (d) $\dfrac{d}{dx} \cos \sqrt{x}$.

C

37. An ideal gas satisfies the equation $PV = k$,   where P is the pressure, V is the volume, and k is a constant. A gas occupies 10 liters at 1 atmosphere of pressure. If the pressure satisfies the equation $P = 1 + t$,   where P is in atmospheres and t in minutes, find the rate of change of volume when $t = 9$.

38. Suppose f is a function such that $f(2) = 3$ and

$$\lim_{x \to 2} \frac{f(x) - 3}{x - 2} = 7.$$

If $g(x) = (x^2 - 7)\, f(x)$,   find an equation of the line tangent to the curve $y = g(x)$ at the point with abscissa 2.

39. Prove Theorem 4.6 by adding and subtracting $v(x + h) \cdot u(x + h)$ instead of $v(x) \cdot u(x)$.
40. Prove Theorem 4.2 by using Theorem 4.8 instead of the definition of derivative.

4.5

Derivative Formulas: Power Rule

We now extend Theorem 4.10 and, in proving the new theorem, illustrate a second use of the chain rule. It will be used again to extend other theorems.

Theorem 4.11

(*Power rule*)   *If u is a function that is differentiable at x and f is a function such that $f(x) = [u(x)]^n$, where n is a rational number, then*

$$f'(x) = n[u(x)]^{n-1} \cdot u'(x).$$

(*More simply, if $y = u^n$, then $dy/dx = nu^{n-1} \cdot du/dx$.*)

Proof

By the chain rule, we have

$$\frac{dy}{dx} = \frac{dy}{du} \cdot \frac{du}{dx} = nu^{n-1} \cdot \frac{du}{dx}.$$

Example 1

Differentiate $y = (x^2 + 1)^2$.

$y' = 2(x^2 + 1) \cdot 2x = 4x(x^2 + 1)$.
Compare this with the power-rule formula. The "u" here is $x^2 + 1$. Thus nu^{n-1} is $2(x^2 + 1)^1$, while $u' = 2x$. The derivative can be found here without using Theorem 4.11.

$$y = (x^2 + 1)^2 = x^4 + 2x^2 + 1,$$
$$y' = 4x^3 + 4x = 4x(x^2 + 1).$$

Example 2

Differentiate $y = (x^2 + 1)^{2/3}$.

$$y' = \frac{2}{3}(x^2 + 1)^{-1/3}2x = \frac{4x}{3(x^2 + 1)^{1/3}}.$$

In this case, we cannot resort to the method of expansion by the binomial theorem—we must use Theorem 4.11. If you protest that the binomial theorem can be used even in this case, when the exponent is not a positive integer, a few things should be noted about that situation. First of all, the expansion cannot be given in finite terms—the result is an infinite series. This subject is better postponed until we are in a position to answer some of the questions that it brings up. What is an infinite series or infinite sum? Does the sum always exist? Does it ever exist? You might also look back to the place you first saw the binomial theorem for exponents other than positive integers. Very likely it appeared in an algebra book and was stated after the binomial theorem for positive integers; and while a proof was given for the binomial theorem for positive integers, none was given for the other cases. Even assuming that the expansion can be carried out, there is the problem of extending Theorem 4.3 to an infinite sum.

Of course, even if we determined that the expansion could be carried out and Theorem 4.3 could be extended to the infinite case, the question still arises, " Do we really want to do it this way when Theorem 4.11 gives us such a simple way of taking the derivative?" This question might arise in connection with the problem of finding the derivative of $y = (2x^3 - 5)^{12}$. While it is true that this expression can be expanded and differentiated term by term, Theorem 4.11 offers a *far simpler* method of differentiation.

Example 3

Differentiate $y = \dfrac{(x+1)^2}{(x-2)^3}$.

By the quotient rule,

$$y' = \frac{(x-2)^3 \dfrac{d}{dx}(x+1)^2 - (x+1)^2 \dfrac{d}{dx}(x-2)^3}{(x-2)^6}$$

$$= \frac{(x-2)^3 2(x+1) - (x+1)^2 3(x-2)^2}{(x-2)^6}$$

$$= \frac{(x+1)(x-2)^2[2(x-2) - 3(x+1)]}{(x-2)^6}$$

$$= \frac{(x+1)(-x-7)}{(x-2)^4}$$

$$= -\frac{(x+1)(x+7)}{(x-2)^4}.$$

Example 4

Differentiate $y = \sqrt{\dfrac{2x-1}{2x+1}}$.

$$y = \left(\frac{2x-1}{2x+1}\right)^{1/2},$$

$$y' = \frac{1}{2}\left(\frac{2x-1}{2x+1}\right)^{-1/2} \frac{(2x+1)2 - (2x-1)2}{(2x+1)^2}$$

$$= \frac{(2x+1)^{1/2} \cdot 4}{2(2x-1)^{1/2}(2x+1)^2}$$

$$= \frac{2}{(2x-1)^{1/2}(2x+1)^{3/2}}.$$

We have proved the power rule only for rational exponents. However, with that exception, Theorems 4.1–4.11 allow us to differentiate any algebraic function, that is, any function formed by addition, subtraction, multiplication, division, or raising to a constant power. For convenience of reference, the results are summarized here. It is understood that u and v are functions of x, and c and n are constants (n rational).

$$\frac{d}{dx}c = 0$$

$$\frac{d}{dx}x^n = nx^{n-1}$$

$$\frac{d}{dx}(u+v) = \frac{du}{dx} + \frac{dv}{dx}$$

$$\frac{d}{dx}(uv) = u\frac{dv}{dx} + v\frac{du}{dx}$$

$$\frac{d}{dx}\left(\frac{u}{v}\right) = \frac{v\dfrac{du}{dx} - u\dfrac{dv}{dx}}{v^2}$$

$$\frac{d}{dx}u^n = nu^{n-1}\frac{du}{dx}$$

$$\frac{dy}{dx} = \frac{dy}{du} \cdot \frac{du}{dx}$$

Problems

A

In Problems 1–16, differentiate.

1. $y = (x+1)^4$.

2. $y = (x^2 + 2)^4$.

3. $y = (4x+3)^5$.

4. $y = (3x-2)^4$.

5. $y = \sqrt{4x+2}$.

6. $y = \sqrt[3]{3x-5}$.

7. $p = (q^3 - 8)^{2/3}$.

8. $p = (3q-1)^{-1/3}$.

9. $y = \sqrt{\dfrac{x+1}{x-1}}$.

10. $y = \dfrac{\sqrt[3]{x^3+1}}{(3x-2)^2}$.

11. $u = \dfrac{(v+1)^{1/3}}{(v-1)^{2/3}}$.

12. $u = \dfrac{v^2-1}{(v+1)^2}$.

13. $y = \left(\dfrac{2x-1}{2x+1}\right)^{2/3}$.

14. $y = (x^{2/3} - a^{2/3})^{4/3}$ (a is a constant).

15. $y = x + \sqrt{2x+1}$.

16. $y = x - \sqrt{2x+1}$.

In Problems 17–20, find the derivative at the point indicated.

17. $y = (x^2+1)^4$,   at $x = 1$.

18. $y = (x^{1/3}+1)^3$,   at $(8, 27)$.

19. $y = \dfrac{(x-1)^3}{(x+2)^2}$,   at $(-1, -8)$.

20. $y = x - \sqrt{x^2-3}$,   at $(2, 1)$.

B

In Problems 21–26, differentiate.

21. $y = (x^{-2} + x)^{-3}$.

22. $y = (x^{-2} + x^{-1})^{-2}$.

23. $y = \dfrac{x^{-2} + x^{-1}}{x^{-2} - x^{-1}}$.

24. $y = \sqrt{x + \sqrt{x + \sqrt{x}}}$.

25. $y = \dfrac{x\sqrt{x+1}}{x^2 - 4}$.

26. $y = \dfrac{\sqrt{x^2+1}}{(2x-3)^2}$.

In Problems 27 and 28, find the derivative at the point indicated.

27. $y = (x^{-2} + x^{-1})^{-1}$, at $x = 1$.

28. $y = \dfrac{\sqrt{x + 2}}{\sqrt[3]{3x + 2}}$, at $x = 2$.

29. For what value(s) of x does the tangent to $y = \dfrac{x + 2}{x - 3}$ have slope -5?

30. For what value(s) of x does the tangent to $y = (x - 2)^4$ have slope 4?

31. For what value(s) of x does $y = (x - 3)^2(x - 2)^3$ have a horizontal tangent?

32. For what value(s) of x does $y = \dfrac{1}{(x - 2)^2}$ have a vertical tangent?

33. Find an equation of the line tangent to $y = (x - 4)^2$ at the point $(1, 9)$.

34. Find an equation of the line tangent to $y = (x + 3)^{2/3}$ at the point $(5, 4)$.

C

35. Find the derivative of $y = \sqrt{x^2}$. What is the derivative for $x = 1, 3, 8, -2, -5, -7, 0$? Can you express y in a different form?

36. Repeat Problem 35 for $y = \sqrt{(x - 1)^2}$.

37. Show that if $f(x) = (x - a)^n \cdot P(x)$, then $f'(x) = (x - a)^{n-1} \cdot Q(x)$. What does the graph of $f(x) = (x - 1)^2(x + 3)$ look like at $x = 1$?

4.6

Derivatives of Implicit Functions

Until now, all of the functions we have considered in this chapter have been stated explicitly in the form, $y = f(x)$. However, many functions are only implied by an equation. For example, the equation of a parabola $(x - 1)^2 = 4(y + 1)$ implies that y is some function of x, namely,

$$y = \frac{1}{4}(x - 1)^2 - 1,$$

although it is not explicitly stated. Unfortunately, the problem is not always so simple; even some relatively simple equations imply a combination of several functions, rather than a single function. For instance, the equation $x^2 + y^2 = 4$ implies that $y = \pm\sqrt{4 - x^2}$. It is easily seen that y is not a function of x, since some values of x determine two values of y. However, y can be represented as a combination of the two functions

$$\sqrt{4 - x^2} \quad \text{and} \quad -\sqrt{4 - x^2},$$

the first representing the top half of the original circle and the second representing the bottom half (see Figure 4.9). Thus an equation may imply that one of the variables can be represented as a function—or combination of functions—of the other. These are referred to as *implicit* functions.

Now suppose that we have a function (or combination of functions) defined implicitly by an equation and we wish to find the derivative of it. How do we proceed? One obvious method would be to determine the function or functions and proceed to find its derivative as in the past.

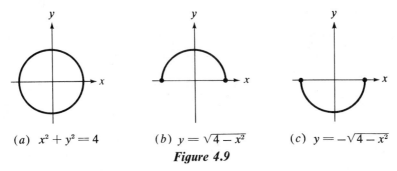

(a) $x^2 + y^2 = 4$ (b) $y = \sqrt{4 - x^2}$ (c) $y = -\sqrt{4 - x^2}$

Figure 4.9

Example 1

Solve $x^2 + y^2 = 4$ for y and differentiate both functions.

We have already solved for y and found

$$y = \sqrt{4 - x^2} \quad \text{or} \quad y = -\sqrt{4 - x^2}.$$

The differentiation is now quite routine.

$$y' = \frac{1}{2}(4 - x^2)^{-1/2}(-2x) \quad \text{or} \quad y' = -\frac{1}{2}(4 - x^2)^{-1/2}(-2x)$$

$$= \frac{-x}{\sqrt{4 - x^2}} \qquad\qquad\qquad = \frac{x}{\sqrt{4 - x^2}}.$$

Now if we want the derivative at a given point on the circle, we merely determine which of the two halves of the circle contains the point and use the derivative formula corresponding to that one. Thus if we want the derivative at the point $(1, -\sqrt{3})$, we note that this point corresponds to

$$y = -\sqrt{4 - x^2} \quad \text{and} \quad y' = \frac{x}{\sqrt{4 - x^2}}$$

at $x = 1$, which gives $y' = 1/\sqrt{3}$.

This is all quite simple. But suppose we try to repeat this process with the equation

$$x^3 + 3x^2y + y^3 = 1.$$

We have considerably more trouble solving for y as a function of x. While we might be able to do so for this problem, we can easily imagine problems for which the solution is much more difficult or even impossible to find in the form we want. Nevertheless we feel that y can be represented as *some* function (or functions) of x. The fact that we cannot find it does not deny its existence. If we want the slope of the curve represented by some equation, must we give up if we cannot solve for y as a function of x, or is it still possible to find the slope without finding the function explicitly?

There is a way out. Since the general form is

$$F(x, y) = G(x, y),$$

where y is some function of x, we may look upon both sides of the equation as an expression in x. Since the two sides of the equation are equal, their derivatives are equal. Thus we simply differentiate both sides with respect to x. Of course, since we have not solved for y as a function of x, we merely know that $y = f(x)$, but we do not know what $f(x)$ is. Thus, whenever we come to finding the derivative of y, we must simply leave it as dy/dx, or y'. Thus y' appears in the expression we get after taking derivatives of both $F(x, y)$ and $G(x, y)$, and sometimes we can solve for y' in terms of x and y.

Example 2

Assuming y to be a function (or a combination of functions) of x, differentiate both sides of $x^2 + y^2 = 4$ and solve for y'.

Differentiating both sides (keep in mind that y is a function of x), we get

$$2x + 2yy' = 0, \quad \text{or} \quad y' = -\frac{x}{y}.$$

This process is called *implicit differentiation*. Now this result does not look at all like the result we had when solving for y and taking derivatives of the functions so obtained. Suppose we compare them.

$$\left. \begin{array}{ll} y' = -\dfrac{x}{\sqrt{4 - x^2}} & \text{when } y = \sqrt{4 - x^2} \\[4mm] y' = \dfrac{x}{\sqrt{4 - x^2}} & \text{when } y = -\sqrt{4 - x^2} \end{array} \right\} \text{from } x^2 + y^2 = 4$$

and

$$y' = -\frac{x}{y} \quad \text{when } x^2 + y^2 = 4.$$

The first thing to notice is that there are two forms for the derivative in the first case but only one in the other. Can the second result possibly tell as much as the first? Yes, because the second expression for the derivative has a "y" in it which is seen to be either $\sqrt{4 - x^2}$ or $-\sqrt{4 - x^2}$. When these are substituted for y, the second result looks like the first.

Note that there is a certain advantage in using $y' = -x/y$,   in that it gives us the derivatives of both functions at the same time. When we wanted the derivative at the point $(1, -\sqrt{3})$, we had to select the one of the two functions satisfied by the given point and substitute $x = 1$ into the derivative formula corresponding to that function. Now, since we have only one expression for the derivative and it has both x and y in it, we merely substitute $x = 1$ and $y = -\sqrt{3}$ into $y' = -x/y$ to get $y' = 1/\sqrt{3}$,   the same result as before. Thus not only are we spared the chore of solving for y, but we have a simpler result to use. Suppose we repeat this process for more complicated equations.

Example 3

Find dy/dx for $x^3 + 3x^2y + y^3 = 1$.

$$3x^2 + 3(x^2y' + y \cdot 2x) + 3y^2y' = 0,$$
$$x^2 + 2xy + (x^2 + y^2)y' = 0,$$
$$y' = -\frac{x^2 + 2xy}{x^2 + y^2}.$$

As you can see, implicit differentiation is much simpler than first trying to solve for y and then differentiating.

Problems

A

In Problems 1–4, find dy/dx at the point indicated by (a) solving for y as a function(s) of x and finding the derivative of an explicit function, and (b) using implicit differentiation.

1. $x^2 + 2y = 1$, $(1, 0)$.
2. $3x^2 + 2y - 5 = 0$, $(-1, 1)$.
3. $y^2 = x + 3$, $(1, 2)$.
4. $y^2 = 2x + 3$, $(3, -3)$.

In Problems 5–14, find dy/dx.

5. $x^3 + y^3 = 5$.
6. $3x^2 - 4y^3 = 1$.
7. $x^{4/3} + y^{4/3} = 1$.
8. $\sqrt{x} + \sqrt{y} = 1$.
9. $x^{1/3} - y^{1/3} = 1$.
10. $x^{-2/3} + y^{-2/3} = 1$.
11. $xy^2 = 4$.
12. $3x^2 - 2xy + y^2 = 5$.
13. $2xy - y^2 = 1$.
14. $(x + y)^2 = (2x - y)^3$.

B

In Problems 15–18, find dy/dx at the point indicated by (a) solving for y as a function(s) of x and finding the derivative of an explicit function, and (b) using implicit differentiation.

15. $x^2 + y^2 + 2y = 0$, $(0, -2)$.
16. $3x^2 - 4xy + y^2 = 15$, $(1, -2)$.
17. $3x^2 + 4xy + y^2 + x - 2y + 7 = 0$, $(-1, 3)$.
18. $x^3 + 2x^2y + y^2 - 2x + 3 = 0$, $(2, -1)$.

In Problems 19–26, find dy/dx.

19. $(x + y)^3 = (x - y + 1)^2$.
20. $(x + y)^{2/3} = \sqrt{x^2 + y^2}$.
21. $x^3 + 3x^2y + y^3 = 8$.
22. $x^4 + 2x^2y^2 - y^4 = 8$.
23. $(x^2 - y^2)^2 = 2x^2 + y^2$.
24. $xy(2x - 3y) = x^2 + y^2$.
25. $\dfrac{x + y}{x - y} = x^2 + y^2$.
26. $3x + 2\sqrt{xy} - y = 4$.

27. At what point(s) does $3x^2 + y^2 + 4x - 6y + 1 = 0$ have a horizontal tangent? A vertical tangent?
28. At what point(s) does $\sqrt{x} + \sqrt{y} = \sqrt{a}$ (a is a positive constant) have a horizontal tangent? A vertical tangent?
29. Find an equation of the line tangent to $x^2 + y^2 - 6x - 8y = 0$ at the point $(6, 0)$.
30. Show that the line tangent to $y^2 = 4ax$ at the point (x_0, y_0) is $yy_0 = 2a(x + x_0)$.
31. Show that the line tangent to $x^2/a^2 + y^2/b^2 = 1$ at the point (x_0, y_0) is

$$\frac{xx_0}{a^2} + \frac{yy_0}{b^2} = 1.$$

32. Show that the tangent to $x^2/a^2 - y^2/b^2 = 1$ at the point (x_0, y_0) is

$$\frac{xx_0}{a^2} - \frac{yy_0}{b^2} = 1.$$

C

33. Find equations of the lines tangent to $x^2 + y^2 - 6x - 4y - 12 = 0$ and containing the point $(-4, 3)$.
34. Find equations of the lines tangent to $y^2 - 3x + 4y + 4 = 0$ and containing the origin.
35. If $4x^2 + 9y^2 + 36 = 0$, find dy/dx. What is the graph of the given equation? What conclusion can be drawn?

36. In Theorem 4.10, we extended the result of Theorem 4.2 to the case in which n is a rational number. If $y = x^{p/q}$, where p and q are integers, then $y^q = x^p$. Use this fact to prove Theorem 4.10 by implicit differentiation.

37. Find dy/dx at the points $(2, 4)$, $(2, 1)$, $(3, 6)$, $(3, 0)$, and $(1, 2)$ of

$$2x^2 + xy - y^2 - 6x + 3y = 0.$$

Sketch the graph of the equation.

38. Show that the line tangent to $Ax^2 + Bxy + Cy^2 + Dx + Ey + F = 0$ at (x_1, y_1) is

$$(2Ax_1 + By_1 + D)x + (Bx_1 + 2Cy_1 + E)y + (Dx_1 + Ey_1 + 2F) = 0.$$

4.7

Derivatives of Higher Order

The derivative of a function is also a function. For example, if

$$f(x) = x^4,$$

then

$$f'(x) = 4x^3.$$

Since f' is itself a function, we can find its derivative (if it exists). The result, called the second derivative of f, is written f''. Thus, for the function $f(x) = x^4$,

$$f''(x) = 12x^2.$$

We may continue indefinitely, taking the third, fourth, fifth, etc., derivatives, which are written f''', $f^{(4)}$, $f^{(5)}$, etc. Again, for the original function,

$$f'''(x) = 24x,$$
$$f^{(4)}(x) = 24,$$
$$f^{(5)}(x) = 0,$$
$$f^{(6)}(x) = 0,$$

and so forth. The various notations used are as follows:

$$y', \quad y'', \quad y''', \quad y^{(4)} \quad \text{or} \quad y^{(\mathrm{IV})}, \quad \text{etc.}$$

$$f'(x), \quad f''(x), \quad f'''(x), \quad f^{(4)} \quad \text{or} \quad f^{(\mathrm{IV})}(x), \quad \text{etc.}$$

$$\frac{dy}{dx}, \quad \frac{d^2y}{dx^2}, \quad \frac{d^3y}{dx^3}, \quad \frac{d^4y}{dx^4}, \quad \text{etc.}$$

$$\frac{d}{dx}f(x), \quad \frac{d^2}{dx^2}f(x), \quad \frac{d^3}{dx^3}f(x), \quad \frac{d^4}{dx^4}f(x), \quad \text{etc.}$$

$$Dy, \quad D^2y, \quad D^3y, \quad D^4y, \quad \text{etc.}$$

$$Df(x), \quad D^2f(x), \quad D^3f(x), \quad D^4f(x), \quad \text{etc.}$$

Example 1

Find the first three derivatives of $y = x^2 + x^{1/2}$.

$$y' = 2x + \frac{1}{2} x^{-1/2} = \frac{4x^{3/2} + 1}{2\sqrt{x}},$$

$$y'' = 2 - \frac{1}{4} x^{-3/2} = \frac{8x^{3/2} - 1}{4x^{3/2}},$$

$$y''' = \frac{3}{8} x^{-5/2} = \frac{3}{8x^{5/2}}.$$

Example 2

Find d^2y/dx^2 for $x^2 + xy - y^2 = 2$.

$$2x + xy' + y - 2yy' = 0,$$

$$(2y - x)y' = 2x + y,$$

$$y' = \frac{2x + y}{2y - x};$$

$$y'' = \frac{(2y - x)(2 + y') - (2x + y)(2y' - 1)}{(2y - x)^2}$$

$$= \frac{(2y - x)\left(2 + \frac{2x + y}{2y - x}\right) - (2x + y)\left(2\frac{2x + y}{2y - x} - 1\right)}{(2y - x)^2}$$

$$= \frac{(2y - x)[2(2y - x) + (2x + y)] - (2x + y)[2(2x + y) - (2y - x)]}{(2y - x)^3}$$

$$= \frac{(2y - x)(5y) - (2x + y)(5x)}{(2y - x)^3}$$

$$= \frac{5(2y^2 - 2xy - 2x^2)}{(2y - x)^3}$$

$$= \frac{-10(x^2 + xy - y^2)}{(2y - x)^3}$$

$$= \frac{-20}{(2y - x)^3}, \qquad \text{since } x^2 + xy - y^2 = 2.$$

The questions of geometric interpretations and uses of the higher derivatives still remain. Of course, the first derivative represents the slope of the original graph and the second derivative gives the slope of the graph of the first derivative, and so on; but we have not answered the question of the relationship between the second derivative and the original function. This will be discussed in the next chapter.

There is a simple interpretation of the second derivative of the position of a moving object as a function of time. We have already seen that the first derivative, interpreted as the rate of change of the position with respect to time, is the velocity. The second derivative then is the rate of change of velocity with respect to time which, by definition, is the acceleration. The acceleration gives the rate at which the object is speeding up or slowing down. If the acceleration agrees in sign with the velocity, the moving object is speeding up; if it has the opposite sign, the object is slowing down. Of course, zero acceleration means no change in the velocity.

Example 3

Analyze the motion of an object with the equation of motion $s = t^3/3 - 3t^2 + 8t$, where t is in seconds, s is in feet, and the positive direction is to the right.

$$s = t^3/3 - 3t^2 + 8t,$$

$$\frac{ds}{dt} = v = t^2 - 6t + 8,$$

$$\frac{d^2s}{dt^2} = a = 2t - 6.$$

Let us find the times at which s, v, and a are zero. By use of the quadratic formula (after factoring out t), we see that $s = 0$ only if $t = 0$; $v = 0$ if $t = 2$ or $t = 4$; and $a = 0$ if $t = 3$. Beyond $t = 4$, s, v, and a are all positive. Hence we have the following table.

t	s	v	a
0	0	8	-6
2	20/3	0	-2
3	6	-1	0
4	16/3	0	2
>4	$+$	$+$	$+$

Now at $t = 0$, the object is at the zero position, moving to the right at 8 ft/sec, but slowing down at the rate of 6 ft/sec/sec. After 2 seconds, the object is $6\frac{2}{3}$ ft to the right of the zero position. It has stopped moving to the right, but is still accelerating to the left at the rate of 2 ft/sec/sec. The object is moving to the left at 1 ft/sec when $t = 3$ sec. It has moved back to $s = 6$, and a is now 0. After $t = 3$, the acceleration becomes positive. Thus the speed of the object decreases (v and a have opposite signs) until the object again stops when $t = 4$. At this time it is $5\frac{1}{3}$ ft to the right of the zero point and accelerating to the right at the rate of 2 ft/sec/sec. Beyond $t = 4$, the object moves to the right with increasing speed (the signs of v and a agree). This is indicated schematically by Figure 4.10.

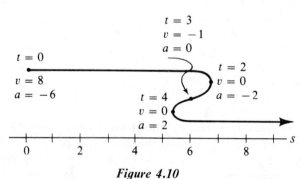

Figure 4.10

Problems

A

In Problems 1–6, find the indicated derivative.

1. $y = 7x^2 + 2x - 5$, $\quad y'''$.

2. $p = 3q^4 - 4q^2 + 3$, $\quad d^3p/dq^3$.

3. $f(v) = (v^2 + 1)^4$, $\quad f'''(v)$.

4. $f(x) = 2x^{3/2} - 6x^{1/2}$, $\quad f''(x)$.

5. $f(x) = \dfrac{1}{x^2 - 1}$, $\quad f''(x)$.

6. $g(p) = \sqrt{p + 2}$, $\quad g'''(p)$.

B

In Problems 7–12, find the indicated derivative and express the result in terms of x and y.

7. $2xy - y^2 = 5$, d^2y/dx^2.
8. $4x^2 - 9y^2 = 36$, d^3y/dx^3.
9. $x^{2/3} + y^{2/3} = 1$, d^2y/dx^2.
10. $3x^2 - 2xy - y^2 = 1$, d^2y/dx^2.
11. $(x + y)^2 = 2xy + 5$, d^2y/dx^2.
12. $\sqrt{x} + \sqrt{y} = 1$, d^2y/dx^2.

In Problems 13–16, find $f(x), f'(x)$, and $f''(x)$ for the indicated value of x.

13. $f(x) = 4x^2 - 2x + 1$, $x = 1$.
14. $f(x) = (3x - 5)^3$, $x = 2$.

15. $f(x) = \sqrt{3x - 2}$, $x = 2$.
16. $f(x) = \dfrac{x - 1}{x + 1}$, $x = 1$.

In Problems 17–20, find dy/dx and d^2y/dx^2 at the indicated point.

17. $x^2 + y^2 = 25$, $(3, -4)$.
18. $3x^2 + 4xy - y^2 - 6 = 0$, $(1, 3)$.
19. $\sqrt{x} + \sqrt{y} = 1$, $(1/4, 1/4)$.
20. $(x + y)^2 = x - y$, $(3, -1)$.

In Problems 21–28, equations of motion are given with s in feet, $t \geq 0$ in seconds, and the positive direction to the right. **Analyze the motion as in Example 3, and give a schematic diagram.**

21. $s = t^3 - 2t^2 - 4t - 8$.
22. $s = t^3 - 3t$.
23. $s = (t - 3)^2$.
24. $s = (t - 3)^3$.
25. $s = 2t^3 - 15t^2 + 24t$.
26. $s = t^4 - 8t^3 + 22t^2 - 24t$.

27. $s = t + \dfrac{16}{t + 2}$.
28. $s = \dfrac{1}{t^2 - 2t + 2}$.

In Problems 29–32, sketch the graphs of f' and f'' if the following are graphs of f.

29.

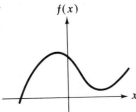

$f(x)$

30.

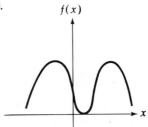

$f(x)$

31.

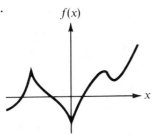

$f(x)$

32.

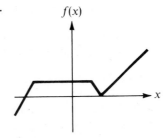

$f(x)$

C

33. Show that if $f(x) = u(x) \cdot v(x)$, then
$$f''(x) = u(x) \cdot v''(x) + 2u'(x) \cdot v'(x) + u''(x) \cdot v(x)$$
and
$$f'''(x) = u(x) \cdot v'''(x) + 3u'(x) \cdot v''(x) + 3u''(x) \cdot v'(x) + u'''(x) \cdot v(x).$$
Compare with $(a + b)^2$ and $(a + b)^3$. Give an expression for $f^{(4)}(x)$.

34. If $f(x) = u(x) \cdot v(x)$, express $f^{(n)}(x)$ in terms of $u(x), v(x)$, and their derivatives of any order. (*Hint:* See Problem 33, and consider a proof of the binomial theorem.)

Review Problems

A

In Problems 1–12, find dy/dx and simplify.

1. $y = \dfrac{4}{x^3}$.

2. $y = \dfrac{3x + 2}{2x - 1}$.

3. $y = (x^3 - 2)^4$.

4. $y = \sqrt{4x^2 + 1}$.

5. $y = \dfrac{\sqrt{x^2 + 1}}{x}$.

6. $y = (x - 2)^3(x + 4)^2$.

7. $y = \left(\dfrac{2x - 3}{2x + 3}\right)^{3/2}$

8. $y = 2x - \sqrt{x^2 + 4}$.

9. $y = \dfrac{(2x + 1)(x - 3)}{x + 4}$.

10. $y = x^2\sqrt{x^2 + 4}$.

11. $x^2 + 4xy - 3y^2 = 5$.

12. $(x + y)^2 = x^3 + y^3$.

B

In Problems 13–16, find dy/dx and simplify.

13. $y = \dfrac{(x + 1)^2}{2x - 1}$.

14. $y = (x - 2)^{1/5}(3x + 4)^{3/5}$.

15. $y = \dfrac{\sqrt{x^2 + 4}}{(2x + 3)^3}$.

16. $y^2 = \dfrac{x^3 + 1}{x^3 - 1}$.

17. Given $y = u^2 + 1$ and $u = x + \sqrt{x + 1}$, find dy/du and dy/dx.
18. Given $y = \sqrt{4x + 5}$, find y''.
19. Given $x^3 + 12x^2y = 4$, find d^2y/dx^2.
20. Use the definition of a derivative to differentiate $y = 3x^3 - x^2 + 2x - 5$.

21. Use the definition of a derivative to differentiate $y = \dfrac{x + 1}{2x - 3}$.

22. Find equations of the tangent and normal lines to $y = x^4 + 3x^2 - 6$ at $(1, -2)$.
23. An object has equation of motion $s = (t^2 - 4)^3$, $-3 \le t \le 3$. Analyze its motion.

24. For $(x - h)^2 + (y - k)^2 = r^2$, show that $\dfrac{d^2y}{dx^2} = -\dfrac{r^2}{(y - k)^3}$.

25. At what point of $y = x^3 - x$ is the slope 2?
26. Find the points of intersection of $x^2 + y^2 = 4$ and $(x - 4)^2 + y^2 = 12$. Show that these circles intersect at right angles—that is, that their tangents at either point of intersection are perpendicular.

C

27. Find an equation of the circle that is tangent to $y = x^3$ at $(1, 1)$ and has the same second derivative there.
28. Suppose that $f(x) = (x - a)^n g(x)$, where $g(a) \neq 0$. Show that $f^{(n)}(a)$ is the first non-zero derivative of f at $x = a$. What is $f^{(n)}(a)$?

5

Curve Sketching

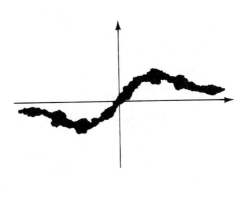

5.1

Intercepts and Asymptotes

In the first chapter we sketched the graph of an equation by the tedious process of point-by-point plotting—a method that sometimes causes one to overlook some "interesting" portions of the graph or to sketch certain portions incorrectly. Suppose for example, you are asked to sketch the graph of

$$y = \frac{10x(x + 8)}{(x + 10)^2}.$$

x	y
−5	−6.00
−4	−4.44
−3	−3.06
−2	−1.88
−1	−0.86
0	0.00
1	0.74
2	1.39
3	1.95
4	2.45
5	2.89

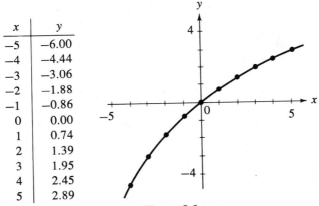

Figure 5.1

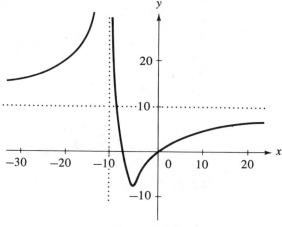

Figure 5.2

The methods of Chapter 1 might lead you to the graph of Figure 5.1. A better sketch of the graph is given in Figure 5.2. While the earlier method produced correct results for the portion we were sketching, it provided no means for determining which portions of the curve are most "interesting."

Let us consider one more example. Suppose we want to graph

$$y = \frac{2x(2x - 1)}{4x - 1}.$$

The methods of Chapter 1 lead us to the set of points shown in Figure 5.3. Now, what does the graph look like? How would you join the points? You might join them as indicated in (a) of Figure 5.4. The correct graph is shown in (b). These examples demonstrate the need for better methods of sketching curves.

We begin with noncalculus methods in the first three sections of this chapter. They give us only a general idea of the graph, but in many cases this is sufficient for the purposes we have in mind. In cases in which these methods are not sufficient, we can use calculus to get a much more detailed picture of the graph. This is done in the later sections of this chapter.

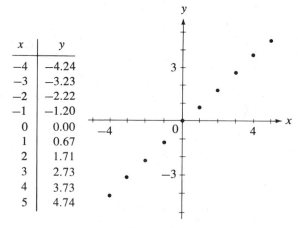

x	y
-4	-4.24
-3	-3.23
-2	-2.22
-1	-1.20
0	0.00
1	0.67
2	1.71
3	2.73
4	3.73
5	4.74

Figure 5.3

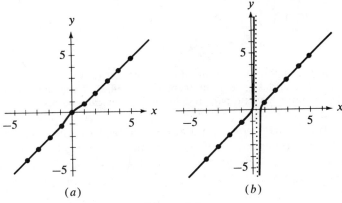

Figure 5.4

We begin the noncalculus methods of sketching with a consideration of intercepts and asymptotes. The intercepts of a curve are simply the points of the curve that lie on the coordinate axes; those on the x axis are the x intercepts, and those on the y axis are the y intercepts (the origin is both an x intercept and a y intercept). We determine the intercepts (if any) by setting x and y equal to zero in turn and solving for y and x, respectively.

Example 1

Find the intercepts of $\dfrac{x^2}{4} + \dfrac{y^2}{9} = 1$.

When $y = 0$, we have $x^2 = 4$ and $x = \pm 2$. If $x = 0$, then $y^2 = 9$ and $y = \pm 3$. Thus, the intercepts are $(2, 0)$, $(-2, 0)$, $(0, 3)$, and $(0, -3)$.

An equation frequently encountered is one of the type $y = P(x)$ or $y = P(x)/Q(x)$, where $P(x)$ and $Q(x)$ are polynomials having no common factor. If $P(x)$ can be factored in the form

$$P(x) = c(x - a_1)^{n_1}(x - a_2)^{n_2} \cdots (x - a_k)^{n_k},$$

where $c, a_1, \ldots, a_k$ are real numbers, then the x intercepts are $(a_1, 0)$, $(a_2, 0)$, $\ldots$, $(a_k, 0)$. The y intercept (an equation in this form has at most one) is still found by setting x equal to zero.

Example 2

Find the intercepts of $y = (x + 1)^2(x - 3)$.

The x intercepts can be taken from the two factors: $(-1, 0)$ from $(x + 1)^2$ and $(3, 0)$ from $(x - 3)$. When $x = 0$,

$$y = 1^2(-3) = -3.$$

Thus the y intercept is $(0, -3)$.

Example 3

Find the intercepts of $y = \dfrac{(x-2)^2(x+1)}{(x-3)(x-1)^2}$.

From the factors $(x-2)^2$ and $(x+1)$, we get $(2, 0)$ and $(-1, 0)$. When $x = 0$, $y = -4/3$, and so the y intercept is $(0, -4/3)$. Note that the factors of the denominator have no part in determining the x intercepts.

Let us now turn to *asymptotes* (you are encouraged to study the spelling of that word). Rather than attempting a definition of an asymptote, which is rather difficult to define properly, let us consider some examples. In Figure 5.2, the lines $x = -10$ and $y = 10$ are asymptotes. We see that portions of the curve approach $x = -10$ and $y = 10$. This is the main feature to be considered in determining asymptotes. Note that the curve contains the point $(-25/3, 10)$ of the line $y = 10$. This does not prevent $y = 10$ from being an asymptote. A curve *can* have one or more (even infinitely many) points in common with its asymptote; however, a line is not an asymptote of itself, nor does $y = |x|$ have an asymptote. In Figure 5.4(b), the line $x = 1/4$ is an asymptote. Again we see that portions of the curve approach this line.

Let us now begin our study of asymptotes by considering vertical asymptotes. In Figure 5.2 you can see that as x approaches -10 from either side, y approaches no definite number but gets larger and larger. This is written

$$\lim_{x \to -10} y = +\infty.$$

As x approaches $1/4$ from the right in Figure 5.4(b), y gets large and negative; and as x approaches $1/4$ from the left, y gets large and positive. These statements are symbolized as

$$\lim_{x \to 1/4^+} y = -\infty \quad \text{and} \quad \lim_{x \to 1/4^-} y = +\infty.$$

To find vertical asymptotes, we are not concerned with whether y gets large and positive or large and negative. We are interested only in determining values of x for which y gets large in absolute value. If the equation is in the form

$$y = \frac{P(x)}{Q(x)},$$

then, as x approaches a, y gets large in absolute value if $Q(x)$ approaches zero and $P(x)$ does not. Thus we need determine only the values of x which make $Q(x) = 0$ and $P(x) \neq 0$.

Example 4

Determine the vertical asymptotes of $y = \dfrac{(x+1)(x-3)}{(2x-1)(x+2)^2}$.

The denominator is zero when either one of the two factors is zero.

$$2x - 1 = 0 \quad \text{gives} \quad x = 1/2.$$
$$x + 2 = 0 \quad \text{gives} \quad x = -2.$$

Since neither value of x gives zero for the numerator, $x = 1/2$ and $x = -2$ are the vertical asymptotes. Figure 5.5 shows the graph with the vertical asymptotes. The method of sketching the graph is deferred until the next section.

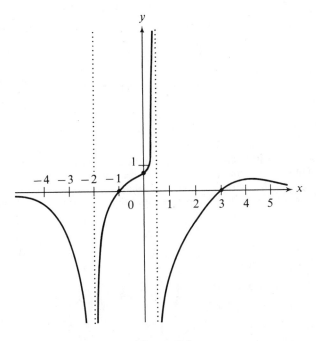

Figure 5.5

Let us now consider horizontal asymptotes. Of course, if the given equation is in the form

$$x = \frac{P(y)}{Q(y)},$$

or can easily be put into that form, we can determine the horizontal asymptotes by the methods given for vertical asymptotes. We merely reverse the role of the x and y here. Unfortunately, it is often difficult or impossible to solve for x as a function of y (consider the equation of Example 4); so another method must be found.

If $y = k$ is a horizontal asymptote for $y = f(x)$, then the distance between a point of the graph of $y = f(x)$ and the line $y = k$ must approach zero as x gets large in absolute value. That is,

$$\lim_{x \to +\infty} [f(x) - k] = 0 \quad \text{or} \quad \lim_{x \to -\infty} [f(x) - k] = 0,$$

from which we have

$$\lim_{x \to +\infty} f(x) = k \quad \text{or} \quad \lim_{x \to -\infty} f(x) = k.$$

Thus, we simply evaluate the limits; if one of them equals some number k, then $y = k$ is a horizontal asymptote. Although it is often true that both of the limits equal the same number k, it sometimes happens that one of the limits is a number k, while the other limit does not exist (see Problems 31–34), or the two limits equal two different numbers (we shall see examples of this in Section 5.3).

Example 5

Determine the horizontal asymptote of $\quad y = \dfrac{x^2 - 4}{x^2 + 3x}$.

One of the limits we are interested in here is

$$\lim_{x \to +\infty} y = \lim_{x \to +\infty} \frac{x^2 - 4}{x^2 + 3x}.$$

As x becomes large and positive, both the numerator and denominator also become large and positive. This fact alone tells us nothing about what value the quotient is approaching (just as in the case in which both numerator and denominator approach zero). We must evaluate this limit by means of some trick. Suppose we divide both numerator and denominator by x^2. Then

$$y = \frac{x^2 - 4}{x^2 + 3x} = \frac{1 - 4/x^2}{1 + 3/x}.$$

Now as x gets large,

$$\frac{4}{x^2} \to 0 \quad \text{and} \quad \frac{3}{x} \to 0.$$

Thus

$$\lim_{x \to +\infty} y = \lim_{x \to +\infty} \frac{1 - 4/x^2}{1 + 3/x} = 1,$$

and $y = 1$ is the horizontal asymptote. By the same argument,

$$\lim_{x \to -\infty} y = 1.$$

Thus the asymptote $y = 1$ is approached by the curve in both directions. Note in Figure 5.6 that the graph crosses the horizontal asymptote at $(-4/3, 1)$.

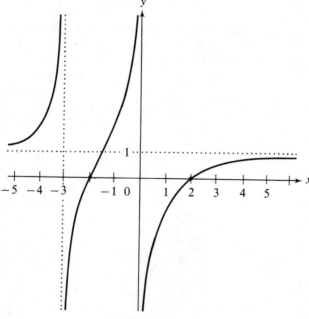

Figure 5.6

In evaluating the limit of a quotient we first divided both numerator and denominator by the highest power of x (x^2 in Example 5) in the given expression. This trick often helps in evaluating limits of the forms

$$\lim_{x \to +\infty} f(x) \quad \text{and} \quad \lim_{x \to -\infty} f(x).$$

Example 6

Determine the horizontal asymptote of $\quad y = \dfrac{(x+1)(x-3)}{(2x-1)(x+2)^2}.$

If we multiplied out the numerator, the highest power of x would be x^2; in the denominator it would be x^3. Thus we shall divide the numerator and denominator by x^3 (do *not* divide the numerator by x^2 and the denominator by x^3; the result would *not* equal y).

$$y = \frac{(x+1)(x-3)}{(2x-1)(x+2)^2} = \frac{\left(\dfrac{x+1}{x}\right)\left(\dfrac{x-3}{x}\right)\left(\dfrac{1}{x}\right)}{\left(\dfrac{2x-1}{x}\right)\left(\dfrac{x+2}{x}\right)^2} = \frac{\left(1+\dfrac{1}{x}\right)\left(1-\dfrac{3}{x}\right)\left(\dfrac{1}{x}\right)}{\left(2-\dfrac{1}{x}\right)\left(1+\dfrac{2}{x}\right)^2}.$$

As x gets large, all of the expressions with x in the denominator approach zero and

$$\lim_{x \to +\infty} y = \lim_{x \to -\infty} y = \frac{(1+0)(1-0)(0)}{(2-0)(1+0)^2} = 0.$$

Thus $y = 0$ is the only horizontal asymptote.

The graph was given in Figure 5.5. Again note that the graph crosses the horizontal asymptote at $(-1, 0)$ and $(3, 0)$.

Example 7

Find the horizontal asymptote of $\quad y = \dfrac{x(x+1)(x-2)}{(x-4)(x+2)}.$

The highest power of x in this expression is x^3. Dividing numerator and denominator by x^3, we have

$$y = \frac{x(x+1)(x-2)}{(x-4)(x+2)} = \frac{\left(\dfrac{x}{x}\right)\left(\dfrac{x+1}{x}\right)\left(\dfrac{x-2}{x}\right)}{\left(\dfrac{x-4}{x}\right)\left(\dfrac{x+2}{x}\right)\left(\dfrac{1}{x}\right)} = \frac{(1)\left(1+\dfrac{1}{x}\right)\left(1-\dfrac{2}{x}\right)}{\left(1-\dfrac{4}{x}\right)\left(1+\dfrac{2}{x}\right)\left(\dfrac{1}{x}\right)}.$$

As $x \to \pm\infty$, the numerator is approaching 1 and the denominator, 0. Thus the fraction becomes arbitrarily large as $x \to \pm\infty$, and so neither $\lim_{x \to +\infty} y$ nor $\lim_{x \to -\infty} y$ exists. There is no horizontal asymptote. The graph is given in Figure 5.7.

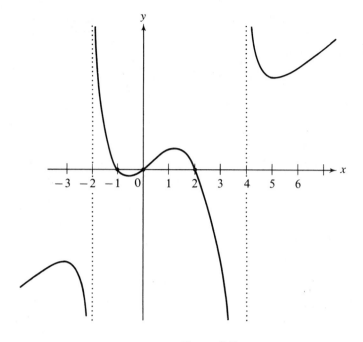

Figure 5.7

Let us now consider slant asymptotes—that is, asymptotes that are neither horizontal nor vertical. Suppose we have an equation of the form

$$y = \frac{P(x)}{Q(x)},$$

where $P(x)$ and $Q(x)$ are polynomials and the degree of $Q(x)$ is n and the degree of $P(x)$ is $n + 1$. We may then carry out the division to get

$$y = \frac{P(x)}{Q(x)} = ax + b + \frac{R(x)}{Q(x)},$$

where the degree of $R(x)$ is less than that of $Q(x)$. By dividing the numerator and denominator by x^n, we find that

$$\lim_{x \to \pm\infty} \frac{R(x)}{Q(x)} = 0.$$

Thus for large values of $|x|$, the original curve is near the line $y = ax + b$; in other words, $y = ax + b$ is a slant asymptote.

Note that we get a slant asymptote only when the degree of the numerator exceeds the degree of the denominator by one. If the degree of the numerator is less than or equal to that of the denominator, we get a horizontal asymptote (see Problem 43); if the degree of the numerator exceeds that of the denominator by more than one, the quotient is no longer linear. The situation is far more complicated if $P(x)$ and $Q(x)$ are not polynomials. Thus we shall only consider quotients of polynomials.

Example 8

Find the slant asymptote of $y = \dfrac{x^2 - 2x - 3}{x + 2}$.

Carrying out the division, we have

$$y = \frac{x^2 - 2x - 3}{x + 2} = x - 4 + \frac{5}{x + 2}.$$

Since

$$\lim_{x \to \pm \infty} \frac{5}{x + 2} = 0,$$

the slant asymptote is $y = x - 4$. The graph of this equation is given in Figure 5.8.

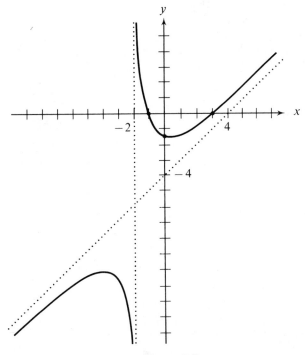

Figure 5.8

Problems

A

In Problems 1–16, find the intercepts.

1. $y = x^2 + 3x$.
2. $y = x^2 - x - 6$.
3. $y = (x + 3)(x^2 - 4)$.
4. $y = (2x - 3)^2(3x + 4)^3$.
5. $y = (4x + 1)(x - 2)(2x + 3)^2$.
6. $y = (3x - 1)^3(3x + 2)^2(x - 2)$.
7. $y = (x + 2)(x^2 + 4)$.
8. $y = (2x - 5)(x^2 + x + 2)$.
9. $y = (3x - 1)^2(x^2 + 2)^3(2x + 1)^4$.
10. $y = (2x + 3)(x^2 - x + 1)^3(x^2 + 2)$.

11. $y = \dfrac{x}{x+1}$.

12. $y = \dfrac{(x-1)^2(x+2)^2}{(3x+2)^2}$.

13. $y = \dfrac{(x-3)^2}{2x+1}$.

14. $y = \dfrac{2x-1}{3x+2}$.

15. $y = \dfrac{2}{x^3}$.

16. $y = \dfrac{(x+2)^2(x^2-1)}{x^2}$.

In Problems 17–20, find all horizontal and vertical asymptotes.

17. $y = (x+1)(x-2)$.

18. $y = (4x+3)(x-2)$.

19. $y = \dfrac{1}{x-3}$.

20. $y = \dfrac{4x-3}{x-1}$.

B

In Problems 21–34, find all horizontal and vertical asymptotes.

21. $y = \dfrac{x}{x+3}$.

22. $y = \dfrac{(x-3)^2}{2x(x+4)}$.

23. $y = \dfrac{3x(x+2)}{(x-1)^2}$.

24. $y = \dfrac{(x-2)(x+4)^2}{(x-3)^2}$.

25. $y = \dfrac{(2x-3)(x-2)}{(x+1)(x-3)^2}$.

26. $y = \dfrac{(3x+2)(x+5)^2}{(x-1)(x-2)(x-3)}$.

27. $y = \dfrac{(2x+1)^2}{2x^2+1}$.

28. $y = \dfrac{(2x-1)^2(x+3)^2}{x(4x+3)}$.

29. $y = \dfrac{(3x+2)^3(x-4)}{(2x+3)^2(x+1)^3}$.

30. $y = \dfrac{(2x+1)^2(x-3)^3}{x(2x-3)^2}$.

31. $y = x - \sqrt{x^2+4}$.
    (*Hint:* Rationalize the numerator.)

32. $y = 2x + \sqrt{4x^2+1}$.

33. $y = 2x - \sqrt{4x^2+3}$.

34. $y = 4x + \sqrt{16x^2-1}$.

In Problems 35–40, find all asymptotes.

35. $y = \dfrac{4x^2-1}{x}$.

36. $y = \dfrac{(x-3)^2}{x}$.

37. $y = \dfrac{(x-1)(x+2)}{x}$.

38. $y = \dfrac{(2x+1)(x-2)}{x-1}$.

39. $y = \dfrac{(x-2)^3}{x^2}$.

40. $y = \dfrac{x^2(x-2)}{(x+1)^2}$.

41. Under certain conditions, the size of a colony of bacteria is related to time by

$$x = 40{,}000 \,\frac{2t+1}{t+1} \quad (t \geq 0),$$

where x is the number of bacteria and t is the time in hours. To what number does the size of the colony tend over a long period of time? In how many hours will the colony have achieved 95% of its ultimate size?

42. A certain chemical decomposition proceeds according to the formula

$$x = \frac{15}{15+t} \quad (t \geq 0),$$

where t is the time in minutes and x is the proportion of undecomposed chemical; that is, x is the ratio of the weight of undecomposed chemical to the weight of the chemical when $t = 0$. Show that the chemical never decomposes entirely. How long does it take for 99% of the chemical to decompose?

C

43.   If

$$y = \frac{a_n x^n + a_{n-1} x^{n-1} + \cdots + a_1 x + a_0}{b_m x^m + b_{m-1} x^{m-1} + \cdots + b_1 x + b_0},$$

where $a_n \neq 0$ and $b_m \neq 0$, what can be said about horizontal asymptotes in case (a) $n < m$? (b) $n = m$?   (c) $n > m$?

5.2

Symmetry, Sketching

Another characteristic that helps in sketching a curve is symmetry. There are two types: symmetry about a line and symmetry about a point. If a curve is symmetric about a line, then one-half of it is the mirror image of the other half, with the mirror as the line of symmetry. More precisely, for every point P of the curve, on one side of the line there is another point P' of the curve such that PP' is perpendicular to the line of symmetry and is bisected by it. An example of this type of symmetry occurs with the graph of $y = 1/x^2$, in which the y axis is the line of symmetry (see Figure 5.9).

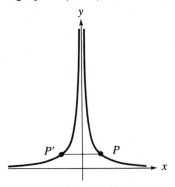

Symmetry about the y axis.
Figure 5.9

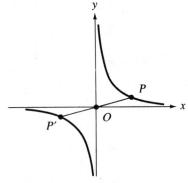

Symmetry about the origin.
Figure 5.10

A curve is symmetric about a point O if for every point $P \neq O$ of the curve, there corresponds a point P' such that PP' is bisected by the point O. The origin is the point of symmetry of the graph of $y = 1/x$ (see Figure 5.10).

While there is no restriction on the lines or points that may be lines or points of symmetry (see Problem 38), we shall consider here only symmetry about the axes and about the origin. We begin with symmetry about the y axis. If a curve is symmetric about the y axis, then, corresponding to every point $P(x, y)$ on the curve, there is a point $P'(-x, y)$ (see Figure 5.9) with the same y coordinate and an x coordinate that is the negative of the x coordinate of P. In this situation, we get the same value for y whether we substitute a positive number x into the equation or its negative, $-x$.

Theorem 5.1

If every x in an equation is replaced by $-x$ and the resulting equation is equivalent to the original (has the same graph), then its graph is symmetric about the y axis.

Example 1

Use Theorem 5.1 to show that $y = 1/x^2$ is symmetric about the y axis.

Replacing x by $-x$, we have

$$y = \frac{1}{(-x)^2}.$$

Since $(-x)^2 = x^2$, we see that the substitution has produced an equation equivalent to the original equation, proving symmetry about the y axis.

An argument similar to the one preceding Theorem 5.1 can be used to obtain the following theorem.

Theorem 5.2

If every y in an equation is replaced by $-y$ and the resulting equation is equivalent to the original, then its graph is symmetric about the x axis.

If a curve is symmetric about the origin, then, for every point $P(x, y)$ on the curve, there is a point $P'(-x, -y)$ on the curve (see Figure 5.10). This is the statement of the next theorem.

Theorem 5.3

If every x in an equation is replaced by $-x$ and every y by $-y$ and the resulting equation is equivalent to the original, then its graph is symmetric about the origin.

Example 2

Use Theorem 5.3 to show that $y = 1/x$ is symmetric about the origin.

Replacing x by $-x$ and y by $-y$ gives

$$-y = \frac{1}{-x}.$$

This is equivalent to the original equation, since we get the original equation if we multiply both sides by -1.

Before using what we have observed about intercepts, asymptotes, and symmetry to sketch a curve, we shall look at a characteristic of curves represented by equations of the form

$$y = \frac{P(x)}{Q(x)},$$

where $P(x)$ and $Q(x)$ are polynomials in reduced form—that is, $P(x)$ and $Q(x)$ have no common factor. The factors of $P(x)$ determine the x intercepts, and the factors of

$Q(x)$ determine the vertical asymptotes of the curve. These are the only two places at which y can change from positive to negative or from negative to positive. This is not to say that the value of y *must* change there—only that it cannot do so elsewhere. We can easily determine whether or not the change occurs at a given intercept or asymptote by considering the exponent on the factor that produces it.

Theorem 5.4

Given an equation of the form

$$y = \frac{P(x)}{Q(x)}$$

in reduced form, if $(x - a)^n$ (where n is a positive integer) is a factor of either $P(x)$ or $Q(x)$ and if $(x - a)^{n+1}$ is a factor of neither, then

(a) *the graph crosses the x axis at $x = a$ if and only if n is odd, and*
(b) *the graph stays on the same side of the x axis at $x = a$ if and only if n is even.*

The expression "the graph crosses the x axis" does not necessarily imply that the graph has a point in common with the x axis. It means that the graph is above (or below) the x axis for $c < x < a$ and below (or above) for $a < x < d$ for some c and d. Of course the graph does contain a point of the x axis at an x intercept, but it "crosses the x axis" at an asymptote by "hopping" over and not touching the axis. This is illustrated in Figure 5.11(a) in which the graph crosses the x axis at both intercepts and both asymptotes.

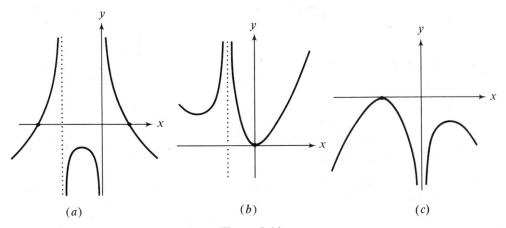

(a)   (b)   (c)

Figure 5.11

Similarly the graph can contain a point of the x axis and still stay on the same side of it. This is illustrated in Figures 5.11(b) and (c) in which the graph stays on the same side of the x axis at the intercept and asymptote in each case.

Although the following discussion does not constitute a proof of this theorem, it serves to show why the theorem works. Let us consider the case in which $(x - a)^n$ is a factor of $P(x)$ (a similar argument can be used for the other case). Then

$$y = \frac{P(x)}{Q(x)} = \frac{R(x)}{Q(x)}(x - a)^n.$$

For all values of x at and "near" $x = a$, $R(x)/Q(x)$ is either positive throughout or negative throughout, not making any sign change. But

$$x - a < 0 \quad \text{for} \quad x < a,$$
$$x - a = 0 \quad \text{for} \quad x = a,$$
$$x - a > 0 \quad \text{for} \quad x > a.$$

In other words, $x - a$ changes sign at $x = a$. If n is odd, then $(x - a)^n$ also changes sign; thus y changes sign at $x = a$. If n is even, then $(x - a)^n$ is positive whether $x < a$ or $x > a$; that is, $(x - a)^n$ does not change sign and y does not change sign.

Table 5.1 summarizes our methods up to this point. We consider only two forms of equations: those of the form $y = P(x)$, where $P(x)$ is a factored polynomial, and those of the form $y = P(x)/Q(x)$, where $P(x)$ and $Q(x)$ are factored polynomials with no common factors. Let us now use all this information to sketch the graph of an equation.

Table 5.1

	$y = P(x)$ Polynomial	$y = P(x)/Q(x)$ Rational Function
x intercepts	Set each factor equal to 0 and solve for x. Note whether exponent on the factor is odd or even.	Set each factor of the numerator equal to 0 and solve for x. Note whether exponent on the factor is odd or even.
y intercepts	Set $x = 0$ and solve for y.	
Vertical asymptotes	None.	Set each factor of the denominator equal to 0 and solve for x. Note whether exponent on the factor is odd or even.
Horizontal asymptotes	None.	Find $\lim_{x \to \pm \infty} y$. Begin by dividing numerator and denominator by the highest power of x in the expression.
Slant asymptotes	None.	Only present when the degree of the numerator exceeds that of the denominator by 1. Express in the form $y = ax + b + R(x)/Q(x)$ by division; $y = ax + b$ is the asymptote.
Symmetry about x axis	Replace y by $-y$ and see whether the result is equivalent to the original.	
Symmetry about y axis	Replace x by $-x$ and see whether the result is equivalent to the original.	
Symmetry about origin	Replace x by $-x$ and y by $-y$ and see whether the result is equivalent to the original.	
Odd intercept or asymptote	Graph crosses the x axis.	
Even intercept or asymptote	Graph stays on the same side of the x axis.	
General remarks	Every vertical line contains one and only one point of the graph.	A vertical line contains one and only one point of the graph unless it is a vertical asymptote, which contains no point of the graph.

Example 3

Sketch $y = (x - 3)(x + 1)^2$.

From the "numerator," we get x intercepts $(3, 0)$ with an odd exponent and $(-1, 0)$ with an even exponent. If $x = 0$, then $y = -3$, which gives $(0, -3)$. Since the denominator is 1, there are no vertical asymptotes; and

$$\lim_{x \to +\infty} y = +\infty \quad \text{and} \quad \lim_{x \to -\infty} y = -\infty,$$

which means that there are no horizontal asymptotes. It is easy to see that no symmetry exists about either axis or the origin. Summing up, we have:

Intercepts: $(3, 0)$ odd, $(-1, 0)$ even, $(0, -3)$;

No asymptotes: $\lim_{x \to +\infty} y = +\infty$, $\lim_{x \to -\infty} y = -\infty$,

No symmetry.

All of this is indicated in Figure 5.12. Let us sketch the graph, starting at the far left and working to the right (this choice is quite arbitrary; we might just as well go from right to left or start in the middle and work outward). We keep in

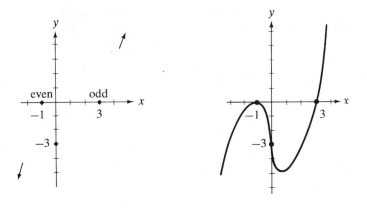

Figure 5.12   **Figure 5.13**

mind that the curve must go through all intercepts and that y is a *function* of x; that is, it is single-valued. Since $\lim_{x \to -\infty} y = -\infty$, we start in the lower left-hand corner. Going to the right, we first reach the intercept $(-1, 0)$. Since it is an even intercept, the graph merely touches the x axis but does not cross it. Next, the graph goes through $(0, -3)$ and then turns back up in order to go through $(3, 0)$. Since $(3, 0)$ is an odd intercept, the graph crosses the x axis there and proceeds upward. The result is given in Figure 5.13.

Note that the lowest point of the "dip" is at approximately $x = 1$. How did we know to put it there? We didn't. We made no attempt to locate it—we simply guessed. Without further work, the best we can say is that it is between $x = -1$ and $x = 3$. Furthermore, how do we know that the graph does not have some extra "turns" and "wiggles" and perhaps look like Figure 5.14? Again, we don't. As a general rule, unless there is some special reason to put in some extra "turn" or "wiggle," we shall leave it out. This rule will not necessarily give us the correct graph every

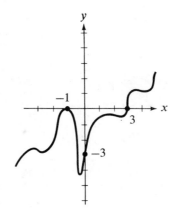

Figure 5.14

time but there is no point in needlessly complicating the situation. These methods give only a general idea of the graph. In Sections 5.4–5.7 we shall see how derivatives may be used to answer the questions raised here.

Note that, with the exception of the three intercepts, we have not plotted a single point. Yet we have some idea (within the restrictions noted above) of the main features of the curve. With a little practice, you should be able to sketch such curves quite quickly and thus achieve the principal aim here.

Example 4

Sketch $\quad y = \dfrac{(2x - 1)(x + 2)^2}{(x + 1)^2(x - 3)}$.

Intercepts:

$$(1/2, 0), \text{ odd}; \quad (-2, 0), \text{ even}; \quad (0, 4/3).$$

Asymptotes:
From the denominator:

$$x = -1, \text{ even}; \quad x = 3, \text{ odd};$$

$$\lim_{x \to \pm \infty} \frac{(2x - 1)(x + 2)^2}{(x + 1)^2(x - 3)} = \lim_{x \to \pm \infty} \frac{\left(\dfrac{2x - 1}{x}\right)\left(\dfrac{x + 2}{x}\right)^2}{\left(\dfrac{x + 1}{x}\right)^2\left(\dfrac{x - 3}{x}\right)} = \lim_{x \to \pm \infty} \frac{\left(2 - \dfrac{1}{x}\right)\left(1 + \dfrac{2}{x}\right)^2}{\left(1 + \dfrac{1}{x}\right)^2\left(1 - \dfrac{3}{x}\right)}$$

$$= \frac{2 \cdot 1}{1 \cdot 1} = 2.$$

Thus $y = 2$ is the horizontal asymptote.
No symmetry.

This information is summarized in Figure 5.15. If we begin sketching at one end or the other, we have the problem of not knowing whether the curve is approaching the asymptote from above or below. Similar problems exist at the vertical asymptotes and x intercepts. Suppose, then, we start at (0, 4/3). Going to the right, we first come to (1/2, 0). Since it is an odd intercept, the graph crosses the x axis there and then goes down to the vertical asymptote $x = 3$ (it cannot go up, since it cannot cross the x axis anywhere between $x = 1/2$ and $x = 3$). Since this asymptote is also odd, the graph now jumps above the x axis. Finally it comes down to the horizontal asymptote $y = 2$.

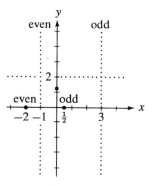

Figure 5.15

Going back to (0, 4/3) and proceeding to the left, we see that the graph must go up to the vertical asymptote $x = -1$ (remember there is nothing to prevent the graph from crossing a horizontal asymptote). Since $x = -1$ is an even asymptote, the curve stays above the x axis. It must then proceed down to the intercept $(-2, 0)$. This is also even, so the graph again remains above the x axis, finally going up to the horizontal asymptote. Thus, we have the graph indicated in Figure 5.16.

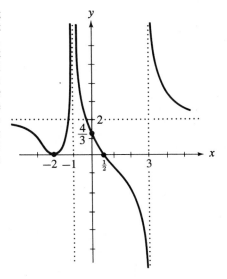

Figure 5.16

Problems

A

In Problems 1–10, check for symmetry about both axes and the origin.

1. $y = x^4 - x^2$.

2. $y = x^4 - x$.

3. $y = x^2 - 1$.

4. $\dfrac{x^2}{4} + \dfrac{y^2}{9} = 1$.

5. $y^2 = \dfrac{x + 1}{x}$.

6. $y^3 = \dfrac{(x + 1)^2}{x}$.

7. $xy^2 = 1$.

8. $x^2y^3 = 1$.

9. $y = \dfrac{x}{x^2 + 1}$.

10. $y = \dfrac{(x + 1)(x - 1)}{x^2}$.

In Problems 11–18, use the methods of this and the preceding section to sketch the graph. Do not plot the graph point-by-point.

11. $y = (x - 2)(x + 3)$.

12. $y = (x - 3)(x + 1)^2$.

13. $y = x^2 - 5x - 6$.

14. $y = x^3 - 2x^2 - 3x$.

15. $y = x^5 - 4x^3$.

16. $y = x^3 + x^2$.

17. $y = \dfrac{x + 1}{x}$.

18. $y = \dfrac{x + 3}{x - 1}$.

B

In Problems 19–30, use the methods of this and the preceding section to sketch the graph. Do not plot the graph point-by-point.

19. $y = \dfrac{(3x - 2)(x - 1)^2}{(x + 1)^3}$.

20. $y = \dfrac{x + 2}{(x - 1)(x + 3)}$.

21. $y = \dfrac{(x + 2)(x - 4)}{x - 1}$.

22. $y = \dfrac{(x - 2)^2}{(x - 1)(x + 2)}$.

23. $y = \dfrac{(x - 2)^2(x + 1)}{x + 4}$.

24. $y = \dfrac{x^2}{x^2 + 1}$.

25. $y = \dfrac{x^2 + 1}{x}$.

26. $xy + y = 2x - 3$.

27. $xy = 3x + 2$.

28. $x^2y - 4y = x^2$.

29. $x^2y - y = x^3$.

30. $x^2y + y = x$.

31. Show that if two perpendicular lines are lines of symmetry of a given curve, then their point of intersection is a point of symmetry.

32. An even function f is one for which $f(-x) = f(x)$; an odd function is one for which $f(-x) = -f(x)$. What can we say concerning the symmetry of the graph of an even function? Of an odd function?

33. A graph (with at least one point not on the x axis) is symmetric about the x axis. Can it be the graph of a function? Explain.

C

34. Show that if a graph has any two of the three types of symmetry—about the x axis, about the y axis, about the origin—then it must have the third.

35. Give an example of a curve with exactly two lines of symmetry.

36. Give an example of a curve with infinitely many lines of symmetry.

37. Can a graph have two or more points of symmetry?

38. Show that if every x in an equation is replaced by $2k - x$ and the resulting equation is equivalent to the original, then its graph is symmetric about the line $x = k$.

5.3

Radicals and the Domain of the Equation

Recall that two things can keep us from getting a value for y when we substitute a value of x into an equation: a zero in the denominator and an even root of a negative number. A zero in the denominator gives a vertical asymptote (provided the numerator is not also zero for the same value of x). Even roots of negative numbers simply cause gaps in the domain of the equation.

Example 1

Sketch $y = \dfrac{2x}{\sqrt{x^2 - 4}}$.

Using the previous methods for determining intercepts, asymptotes, and symmetry, we have:

Intercepts:

$$(0, 0) \text{ odd.}$$

Asymptotes:

$$x = 2, \quad x = -2.$$

The radical is equivalent to the one-half power, which is neither odd nor even. We have a special problem in finding the horizontal asymptotes. The highest power of x in the numerator is clearly x. The highest power in the denominator appears to be x^2. But since it is under the radical, the highest power is really $(x^2)^{1/2} = x$. Thus we shall want to divide the numerator and denominator by x. But we shall want to put the x under the radical in the denominator, which leads to further complications. The symbol $\sqrt{}$ means the *nonnegative* square root. Thus $x = \sqrt{x^2}$ is true only when $x \geq 0$; when $x < 0$, $\sqrt{x^2} = -x$ (note that, since x itself is negative, $-x$ is positive), and we have two cases to consider:

$$\frac{2x}{\sqrt{x^2 - 4}} = \frac{\dfrac{2x}{x}}{\sqrt{\dfrac{x^2 - 4}{x^2}}} = \frac{2}{\sqrt{1 - \dfrac{4}{x^2}}} \quad \text{when } x > 0,$$

$$\frac{2x}{\sqrt{x^2 - 4}} = \frac{\dfrac{2x}{-x}}{\sqrt{\dfrac{x^2 - 4}{x^2}}} = \frac{-2}{\sqrt{1 - \dfrac{4}{x^2}}} \quad \text{when } x < 0.$$

Thus,

$$\lim_{x \to +\infty} \frac{2x}{\sqrt{x^2 - 4}} = \lim_{x \to +\infty} \frac{2}{\sqrt{1 - \dfrac{4}{x^2}}} = 2,$$

$$\lim_{x \to -\infty} \frac{2x}{\sqrt{x^2 - 4}} = \lim_{x \to -\infty} \frac{-2}{\sqrt{1 - \dfrac{4}{x^2}}} = -2,$$

giving two horizontal asymptotes: $y = 2$, which is approached on the right, and $y = -2$, which is approached on the left. Replacing x by $-x$ and y by $-y$ gives

$$-y = \frac{2(-x)}{\sqrt{(-x)^2 - 4}} = \frac{-2x}{\sqrt{x^2 - 4}},$$

which is equivalent to the original equation. Thus we have symmetry about the origin.

Finally, $\sqrt{x^2 - 4}$ represents a real number only when $x^2 - 4 \geq 0$, which gives

$$x^2 \geq 4 \quad \text{or} \quad \begin{cases} x \geq 2 \\ x \leq -2. \end{cases}$$

But y is real for one additional value of x, namely, $x = 0$. If $x = 0$, y equals zero divided by a complex number, which is still zero. Thus the domain is

$$\{x \mid x > 2 \text{ or } x < -2 \text{ or } x = 0\}.$$

We see here that $(0, 0)$ is an isolated point of the graph.   (See Note.)

All of this information is represented graphically in Figure 5.17, which shows the intercept as an isolated point; the fact that it is odd is of no use. Note one thing more: Since $\sqrt{x^2 - 4}$ is never negative, y is positive whenever x is positive, and negative whenever x is negative. This additional information makes it easy for us to sketch the curve (see Figure 5.18).

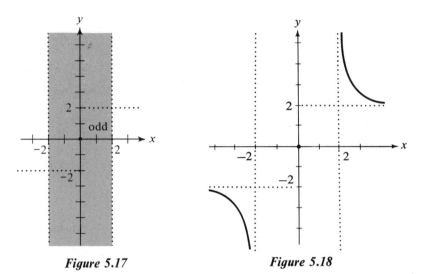

Figure 5.17   Figure 5.18

Note: The existence of an isolated point of this graph at the origin is open to controversy. On one hand are those who say that there is no point of the graph at $(0, 0)$. Their reasoning is as follows: Whenever we restrict ourselves to real functions of real variables, we, in effect, say that imaginary numbers do not exist. Thus, instead of having zero over an imaginary number, we have zero divided by no number at all, which yields no number. By this line of reasoning, every "part" of the equation must be real in order to yield a valid result.

On the other hand are those who maintain that the result is independent of the means of obtaining it. The mere fact that we go from one real number to another by way of imaginary numbers, they say, does not invalidate the result. Exactly the same controversy arose when Cardan published his solution of a cubic equation in 1545. His rule for the solutions of $x^3 = 15x + 4$ leads to

$$x = \sqrt[3]{2 + \sqrt{-121}} + \sqrt[3]{2 - \sqrt{-121}}.$$

This simplifies to $x = 4$, which is the only positive root of the given equation.† It was decided then that the excursion into complex numbers did not invalidate the result.

We shall take the latter point of view throughout this text, mostly because it is easier to discard an unwanted point than it is to produce one that is not present. Nevertheless, it is well to bear in mind the controversial nature of the problem.

†In this connection, see Carl B. Boyer, *A History of Mathematics* (New York: John Wiley, 1968), pp. 310–316.

Example 2

Sketch $y^2 = x^4 - x^2$.

To graph this equation, we use the following device: Since $y = \pm\sqrt{x^4 - x^2}$, we first graph $z = x^4 - x^2$ and then, from the values of z, get $y = \pm\sqrt{z}$. Using our previous methods for determining the intercepts, asymptotes, and symmetry of

$$z = x^4 - x^2 = x^2(x^2 - 1) = x^2(x + 1)(x - 1),$$

we have

Intercepts: $(0, 0)$, even, $(1, 0)$, odd, $(-1, 0)$, odd.

No asymptotes.
Symmetry about the z axis. In Figure 5.19, the dotted curve is the graph of $z = x^4 - x^2$.

We see on the graph that, for each value of x, we have a value of $z = x^4 - x^2$. Now let us find the corresponding values for $y = \pm\sqrt{z}$. But first, we note the following points.

(1) $\sqrt{z} = z$ if $z = 0$ or $z = 1$,

(2) $\sqrt{z} > z$ if $0 < z < 1$,

(3) $\sqrt{z} < z$ if $z > 1$,

(4) $\sqrt{z}$ is not real if $z < 0$.

These four points have the following implications for the graph of the original equation: The upper half of the graph of $y^2 = x^4 - x^2$ crosses the dotted curve when $z = 0$ or $z = 1$; it is above the dotted curve when $0 < z < 1$; it is below the dotted curve when $z > 1$; and there is no graph for those values of x for which the dotted curve is below the x axis.

The final result is given by the solid graph of Figure 5.19. Note that while the auxiliary curve is symmetric only about the z axis, the original curve is symmetric about both axes and the origin. The origin is again an isolated point of this graph.

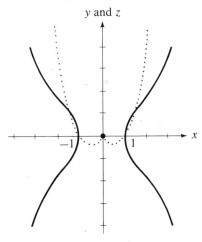

Figure 5.19

The same method could be used to sketch $y = \sqrt{x^4 - x^2}$.   The only difference would be that we would have only the top half of the result in Figure 5.19. We might also have used this method in Example 1, starting with

$$y^2 = \frac{4x^2}{x^2 - 4}.$$

In that case, we would have to be careful which branch we chose; we would have to choose the top portion when x is positive and the bottom portion when x is negative.

One final point. Let us recall that when we had an equation of the form

$$y = \frac{P(x)}{Q(x)},$$

we noted that x intercepts come from factors in the numerator and vertical asymptotes come from factors in the denominator, *provided there is no value of x for which both numerator and denominator are zero.* In the examples we have been considering, this is equivalent to the provision that there is no factor common to both numerator and denominator. What happens if there *are* common factors? The answer is simple. We simply cancel the common factors and sketch the resulting equation. But remember that if we cancel the factor $x - a$, the original equation is not defined at $x = a$ (it gives 0/0) and there is no point on the graph with x coordinate a.

Example 3

Sketch $y = \dfrac{x^2 - 1}{x - 1}$.

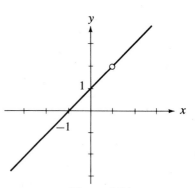

Since the numerator and denominator have the common factor $x - 1$, we cancel them (if $x \neq 1$) to get

$$y = x + 1,$$

which gives a straight line. But recall that the original equation gives no value of y when $x = 1$. Thus the point $(1, 2)$ should be deleted from the graph, as in Figure 5.20.

Figure 5.20

Problems

A

In Problems 1–10, sketch the graphs of the equations.

1. $y = x\sqrt{x^2 - 1}$.

2. $y = \dfrac{x - 2}{\sqrt{x + 1}}$.

3. $y^2 = \dfrac{x + 1}{x}$.

4. $y^2 = \dfrac{x^2}{(x - 1)(x + 2)}$.

5. $y^2 = (x - 1)(x - 3)^2$.

6. $y^2 = -(x - 1)(x - 3)^2$.

7. $y = \dfrac{x^2 - 4}{x + 2}$.

8. $y = \dfrac{x^2 + x}{x + 1}$.

9. $y = \dfrac{x^2 + x}{x^2}$.

10. $y = \dfrac{x^3 + x^2}{x}$.

B

In Problems 11–28, sketch the graphs of the equations.

11. $y = \dfrac{-x}{\sqrt{x^2 - 4}}$.

12. $y = \dfrac{x - 3}{\sqrt{x(x + 2)}}$.

13. $y = x + \sqrt{x^2 - 1}$.

14. $y = x - \sqrt{x^2 - 1}$.

15. $y^2 = \dfrac{(x^2 - 4)^2}{x - 1}.$

16. $y^2 = (1 - x)^2(3 - x).$

17. $y^2 = \dfrac{2x}{(x - 1)^2}.$

18. $y^2 = \dfrac{(x + 2)^2}{x + 1}.$

19. $y^2 = \dfrac{x(x + 2)}{(x + 1)^2}.$

20. $y^2 = \dfrac{x(x + 2)^2}{x + 1}.$

21. $y^2 = \dfrac{x(x + 1)}{(x + 1)^2}.$

22. $y^2 = \dfrac{x(1 - x)}{(x + 2)^2}.$

23. $y = \dfrac{x(x + 1)^3}{(x - 1)(x + 1)^2}.$

24. $y = \dfrac{2x(x + 1)^2}{x(x - 1)}.$

25. $y = \dfrac{1 - (1 + h)^2}{h}.$

26. $y = \dfrac{-1 - [(1 + h)^2 - 2(1 + h)]}{h}.$

   (*Hint:* Simplify the numerator.)

27. $y = \dfrac{2 - \dfrac{2 + h}{1 + h}}{h}.$

28. $y = \dfrac{1 - \sqrt{1 + h}}{h}.$

C

In Problems 29–32, sketch the graphs of the equations.

29. $y^3 = \dfrac{x - 4}{x}.$

30. $y^4 = \dfrac{(x - 3)^2}{x + 1}.$

31. $x^2 = \dfrac{y^2 - 1}{y}.$

32. $x = y\sqrt{y^2 - 1}.$

5.4

Relative Maxima and Minima

Until now we have avoided using calculus in sketching curves. We have also noted that our methods have left some things undetermined (see page 114). Whether or not we want to determine these things depends upon the use we plan for our sketch. Often we are not interested in locating highest or lowest points and we may not care how may "turns" or "wiggles" the graph has. But if we are interested in such things, we must use calculus to determine them. In fact, the problem of locating the highest and lowest points of a graph prompted some of the early work in calculus and led Pierre de Fermat to the derivative in 1629. Although the Greeks considered tangents in a very limited way, Fermat is looked upon as the first to find tangents for a very broad range of functions. Furthermore, there is some evidence to indicate that his ideas of analytic geometry predate those of Descartes. While Fermat is regarded as one of the leading mathematicians of the seventeenth century, mathematics was only a leisure time activity for him—he made his living as a lawyer!

Before going on, we need some new notation.

Definition

*If a and b are real numbers, a < b, then the **open interval** (a, b) is the set {x | a < x < b}
and the **closed interval** [a, b] is the set {x | a ≤ x ≤ b}. Similarly, [a, b) = {x | a ≤ x < b}
and (a, b] = {x | a < x ≤ b}; the latter two are **half-open intervals**.*

The notations for a point in the plane and for an open interval are the same, but it
is easy to distinguish between them by context.

Definition

*The function f is said to be **increasing** (or **decreasing**) on the interval (a, b) if, when
a < c < d < b,*

$$f(c) < f(d) \quad (or\ f(c) > f(d)).$$

*The function f is said to be **nondecreasing** (or **nonincreasing**) on the interval (a, b) if,
when a < c < d < b,*

$$f(c) \le f(d) \quad (or\ f(c) \ge f(d)).$$

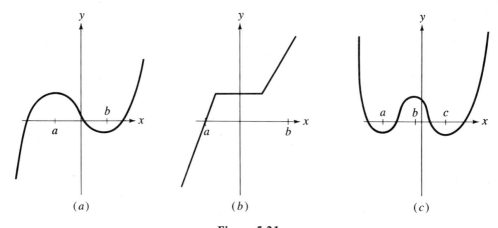

(a)   (b)   (c)

Figure 5.21

In (a) of Figure 5.21, the function is decreasing on the interval (a, b). In part (b),
it is nondecreasing on the interval (a, b). In part (c), it is increasing on (a, b) and
decreasing on (b, c). Now let us consider the relationship between increasing and
decreasing functions and the derivative.

Theorem 5.5

*If f is a function such that f'(x) > 0 for every x on (a, b), then f is increasing on (a, b);
if f'(x) < 0 for every x on (a, b), then f is decreasing on (a, b).*

We shall not attempt to prove this theorem; a proof (see Appendix B) requires the
use of the mean-value theorem, which we shall consider later (page 573). But we can
see that a positive derivative implies a tangent line with an inclination between 0° and
90°, which, in turn, implies an increasing function. On the other hand, a negative
derivative implies a tangent line with an inclination between 90° and 180° and, thus,

a decreasing function. It might be noted that the converse of this theorem is not true; that is, it does not follow that if f is increasing on (a, b), then $f'(x) > 0$ for every x on (a, b). Consider the function

$$f(x) = x^3.$$

It is increasing on the interval $(-1, 1)$ but $f'(0) = 0$. However, it is true that if f is increasing on (a, b), then for every x on (a, b), either $f'(x) \geq 0$ or $f'(x)$ does not exist.

Definition

*The point $A(a, f(a))$ is termed a **relative maximum** (or **minimum**) of the graph of the function f when there is an interval (b, c) containing a such that $f(a) \geq f(x)$ (or $f(a) \leq f(x)$) for any value of x that is both in the domain of f and in the interval (b, c).*

Definition

*The point $A(a, f(a))$ is termed an **absolute maximum** (or **minimum**) of the graph of the function f when $f(a) \geq f(x)$ (or $f(a) \leq f(x)$) for any value of x in the domain of f.*

Let us note, first of all, that we said *an* absolute maximum (or minimum) rather than *the* absolute maximum (or minimum). There may be several "highest" (or "lowest") points of the graph, all having the same y coordinate (see Figure 5.22). Roughly speaking, an absolute maximum (or minimum) is a highest (or lowest) point of the graph (in the sense that there is none higher (or lower)), while a relative maximum (or minimum) is at least as high (or low) as those points "near" it. Of course, an absolute maximum (or minimum) is also a relative maximum (or minimum); in fact, if a graph has an absolute maximum (or minimum), then it is the highest (or lowest) of the relative maxima (or minima). Note the qualifier, "if a

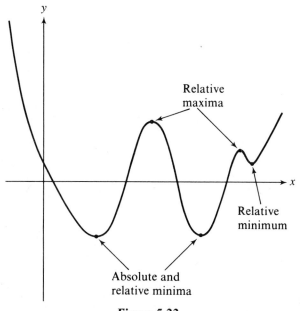

Relative maxima

Relative minimum

Absolute and relative minima

Figure 5.22

graph has an absolute maximum (or minimum)." Do not just find a highest (or lowest) relative maximum (or minimum) and conclude that it is an absolute maximum (or minimum)—there may not be any absolute maximum (or minimum) (see parts (a) and (c) of Figure 5.21).

Now, let us suppose that the function f is defined on (b, c) and there is a number a in (b, c) such that $f'(x) > 0$ for x in (b, a) and $f'(x) < 0$ for x in (a, c). By Theorem 5.5, f is increasing on (b, a) and decreasing on (a, c); $(a, f(a))$ is a relative maximum. Since $f'(x)$ changes from positive to negative, either $f'(a) = 0$ or $f'(a)$ does not exist. A similar argument shows that, if $f'(x) < 0$ for x in (b, a) and $f'(x) > 0$ for x in (a, c), then $(a, f(a))$ is a relative minimum and either $f'(a) = 0$ or $f'(a)$ does not exist. Thus a first step in finding relative maxima and minima is to find those values of x that give either a zero derivative or no derivative.

Definition

The point $(x, f(x))$ is a **critical point** of f if $f(x)$ is defined and $f'(x)$ is either zero or undefined. The abscissa x of a critical point is a **critical number**, or **critical value**.

A critical point is not necessarily a relative maximum or minimum—it may be neither. We can easily determine whether a given critical point is a relative maximum or minimum or neither by considering whether the graph is increasing or decreasing on each side of the point. The results are summarized below:

Type of Critical Point	Left of Point	Right of Point	Graphically
Relative maximum	Increasing	Decreasing	/ \
Relative minimum	Decreasing	Increasing	\ /
Neither	Increasing	Increasing	/ /
	Decreasing	Decreasing	\ \

Example 1

Find all relative maxima and minima of $y = x^3 - 3x^2 - 9x + 9$. Indicate which, if any, are absolute maxima or minima.

$$y' = 3x^2 - 6x - 9 = 3(x + 1)(x - 3).$$

This gives critical values $x = -1$ and $x = 3$. Substituting back into the original equation to get the corresponding values of y, we have the points $(-1, 14)$ and $(3, -18)$.

Now we shall see where the graph is increasing and where it is decreasing in order to determine, in each case, whether the critical point is a relative maximum or minimum or neither. The only place that y' can change from positive to negative (or negative to positive) is at values of x for which $y' = 0$ or y' is nonexistent (compare this with a similar statement about y on page 111). Thus we need only determine the sign of y' to the left of -1, between -1 and 3, and to the right of 3; we obtain $+$, $-$, $+$, respectively. This is represented graphically in Figure 5.23, which shows that $(-1, 14)$ is a relative maximum and $(3, -18)$ is a relative minimum. There is neither an absolute maximum nor an absolute minimum, since $\lim_{x \to +\infty} y = +\infty$ and $\lim_{x \to -\infty} y = -\infty$.

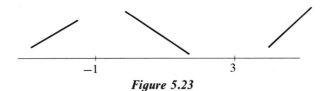

Figure 5.23

Example 2

Find all relative maxima and minima of $y = (x^2 + 1)/x$. Indicate which if, any, are absolute maxima or minima.

$$y' = \frac{x \cdot 2x - (x^2 + 1)}{x^2} = \frac{x^2 - 1}{x^2}.$$

The derivative is 0 when $x = \pm 1$ and there is no derivative when $x = 0$. These are *not* all critical values. A critical value is the x coordinate of a critical *point*. When we put the values of x back into the original equation in order to find the y coordinates, we get $(-1, -2)$ and $(1, 2)$; but putting $x = 0$ into the original equation gives no value at all for y. Thus there is no value of y' at $x = 0$ because there is no value for y there, and $x = 0$ is a vertical asymptote.

Now in checking to see where the graph is increasing and where it is decreasing, we must take into consideration the fact that a change may occur at *any* value at which $y' = 0$ or y' does not exist. This includes the vertical asymptote $x = 0$. Thus we must check to the left of -1, between -1 and 0, between 0 and 1, and to the right of 1. The results are summarized graphically in Figure 5.24. Thus $(-1, -2)$ is a relative maximum and $(1, 2)$ is a relative minimum. Again there is neither an absolute maximum nor an absolute minimum, since $\lim_{x \to +\infty} y = +\infty$ and $\lim_{x \to -\infty} y = -\infty$.

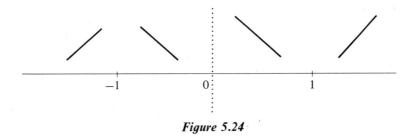

Figure 5.24

One thing about this result that may seem startling is that the lower of the two critical points is the relative maximum, while the higher one is the relative minimum. Thus the y coordinates give no hint as to which point is the relative maximum and which is the relative minimum! The graph of this equation is given in Figure 5.25. We shall say more about sketching the graph in Section 5.7.

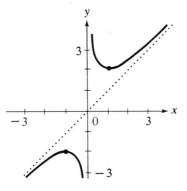

Figure 5.25

Example 3

Find all relative maxima and minima of $y = x^{2/3}$. Indicate which, if any, are absolute maxima or minima.

$$y' = \frac{2}{3} x^{-1/3} = \frac{2}{3x^{1/3}}.$$

We see that this derivative is never 0, but there is no derivative for $x = 0$. This time we see that it is a critical value, being the x coordinate of $(0, 0)$. Checking values of y' to the left and right of $x = 0$ gives us the result shown in Figure 5.26, so $(0, 0)$ is a relative minimum and, indeed, an absolute minimum (by reference to Figure 5.26). There is neither a relative nor an absolute maximum.

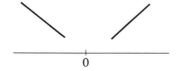

Figure 5.26

Example 4

Find all relative maxima and minima of $y = x^3$. Indicate which, if any, are absolute maxima or minima.

$$y' = 3x^2.$$

The only critical point is $(0, 0)$. Checking the values of y' to the left and right of $x = 0$ gives the result that $(0, 0)$ is neither a relative maximum nor minimum (see Figure 5.27). The graph $y = x^3$ is given in Figure 5.28. Note that, while $(0, 0)$ is

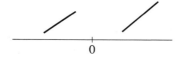

Figure 5.27

neither a relative maximum nor minimum, it does have a horizontal tangent at $(0, 0)$. Since there is no relative maximum nor minimum, there cannot be an absolute maximum or minimum.

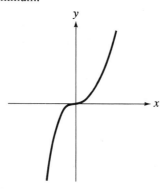

Figure 5.28

Problems

A

In Problems 1–12, find all relative maxima and minima. Indicate which, if any, are absolute maxima or minima.

1. $y = x^2 + 2x - 3.$
2. $y = 3x^2 + 6x - 5.$
3. $y = -x^2 + 8x + 2.$
4. $y = -2x^2 + 12x - 4.$
5. $y = x^3 - 3x^2 + 1.$
6. $y = x^3 - 3x^2 + 3x.$
7. $y = x^3 - 12x + 2.$
8. $y = 2x^3 + 9x^2 + 12x + 1.$
9. $y = 3x^4 + 4x^3.$
10. $y = x^4 - 18x^2 + 1.$
11. $y = 3x^4 + 16x^3 + 24x^2 + 32.$
12. $y = x^5 - 5x^4.$

B

In Problems 13–34, find all relative maxima and minima. Indicate which, if any, are absolute maxima or minima.

13. $y = \dfrac{x}{x + 1}.$

14. $y = \dfrac{x^2}{x^2 + 1}.$

15. $y = \dfrac{x^2 + 1}{x^2}.$

16. $y = \dfrac{x^2 - 1}{x^3}.$

17. $y = \dfrac{1}{x^2 - 1}.$

18. $y = \dfrac{x^3}{x^2 - 1}.$

19. $y = \dfrac{x^2}{(x - 1)^3}.$

20. $y = \dfrac{x^3 - 1}{x^3 + 1}.$

21. $y = (x + 1)^2(x - 2).$
22. $y = x^3(x + 1)^4.$
23. $y = (x + 4)^4(x - 2)^3.$
24. $y = x^3(x - 2)^2.$
25. $y = x^{1/3}.$
26. $y = x^{4/3}.$
27. $y = x^{2/5} - 1.$
28. $y = (x - 1)^{2/5}.$

29. $y = (x - 1)^{1/3}(x + 2)^{2/3}.$

30. $y = \dfrac{(x - 1)^{1/3}}{(x + 2)^{2/3}}.$

31. $y = x^{3/2}(x - 3)^{1/2}.$

32. $y = \dfrac{x^{3/2}}{(x - 3)^{1/2}}.$

33. $y = x^{4/3} - x^{2/3}.$

34. $y = \dfrac{x^{4/3} - x^{2/3}}{x^{4/3} + x^{2/3}}.$

C

35. Show that if $(x - a)^b$, where $b > 0$ and $b \neq 1$, is a factor of $P(x)$ but not of $Q(x)$ in $f(x) = P(x)/Q(x)$, then $(a, 0)$ is a critical point.

36. Show that if $(x - a)^n$, where n is a positive even integer, is a factor of $P(x)$ but not of $Q(x)$ in $f(x) = P(x)/Q(x)$, then $(a, 0)$ is either a relative maximum or a relative minimum.

37. Find the critical value of $y = Ax^2 + Bx + C$. For what conditions on the coefficients is it a relative maximum? A relative minimum?

38. Find the critical values of $y = Ax^3 + Bx^2 + Cx + D$. For what conditions on the coefficients are there two critical values? One? None? If there is exactly one critical value, show that it is neither a relative maximum nor a relative minimum.

5.5

Second-Derivative Test and Points of Inflection

When we introduced higher-order derivatives, we did not attempt a geometrical interpretation of them. Let us do so now for the second derivative. Figure 5.29 shows three graphs; the top one gives y as a function of x, the middle one gives the derivative y' as a function of x, and the bottom one gives the second derivative, y'', as a function of x. How was the middle one derived from the top one? At the far left we see that y is increasing very rapidly; thus the corresponding value of y' is large and positive. As x increases, y increases more and more slowly until finally, at $x = a$, y stops increasing and the graph has a horizontal tangent; thus the corresponding values of y' become smaller and smaller and finally reach zero at $x = a$. As x goes from $x = a$ to $x = c$, the graph is decreasing, slowly at first, then faster and faster until it reaches $x = b$, at which time the rate of decrease begins slowing until it again reaches zero at $x = c$. The corresponding values of y', starting from zero at $x = a$, go negative, decreasing until, at $x = b$, y' attains the minimum value. Then the value of y' increases until it reaches zero at $x = c$. Beyond $x = c$, y increases faster and faster and y' gets larger and larger in the positive direction. The bottom graph is derived from the middle one in exactly the same way.

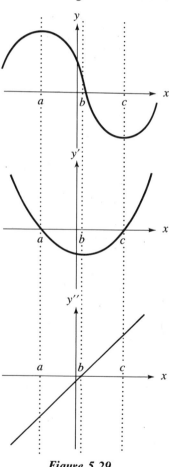

Now let us compare the second derivative with the original. We see that the second derivative is negative to the left of $x = b$, where the original graph is concave downward, and it is positive to the right of $x = b$, where the original graph is concave upward. Note that $y'' = 0$ at $x = b$, and the original graph changes from concave downward to concave upward. Now let us give definitions of concave upward and concave downward and express the foregoing result in the form of a theorem.

Figure 5.29

Definition

*A function f is **concave upward (downward)** at $x = a$ if $f(a)$ is defined and the graph of f lies on or above (below) the tangent line to f at $x = a$ for all x in some open interval (b, c) containing a.*

Theorem 5.6

If f is a function such that $f''(x)$ is positive (negative) for all x on (a, b), then the graph of f is concave upward (downward) on (a, b).

Of course the foregoing discussion does not constitute a proof of this theorem and we shall not attempt one. A proof can be given (see Appendix B) using the mean-value theorem (see page 573). Again, as in Theorem 5.5, the converse is not true; the most that can be said is that if the graph is concave upward on (a, b), then $f''(x) \geq 0$ for all x on (a, b) at which $f''(x)$ exists. This gives us another method of testing a critical point to determine whether it is a relative maximum or minimum.

Theorem 5.7

(Second-derivative test for maxima and minima.)   If $A(a, f(a))$ is a point of the graph of f such that $f'(a) = 0$ and $f''(a) < 0$, then A is a relative maximum; if $f'(a) = 0$ and $f''(a) > 0$, then A is a relative minimum.

Note that the theorem mentions only the cases in which the second derivative is either positive or negative. Therefore, it does *not* justify a conclusion that neither a relative maximum nor minimum exists if the second derivative either is zero or does not exist—*one must resort to the first-derivative test* or some other analysis.

Example 1

Find all relative maxima and minima of $y = x^3 - 3x^2 - 9x + 9$.

$$y' = 3x^2 - 6x - 9 = 3(x + 1)(x - 3),$$
$$y'' = 6x - 6 = 6(x - 1).$$

From the first derivative we have critical values $x = -1$ and $x = 3$. From the second derivative we see that $f''(-1) = -12$ and $f''(3) = 12$. Thus $(-1, 14)$ is a relative maximum, and $(3, -18)$ is a relative minimum. This is the same result that we obtained in Example 1 of the previous section.

This second-derivative test is a nice, simple test when it works; unfortunately it does not work in every case. You can be sure that the second-derivative test will fail for critical points at which the first derivative does not exist. If the first derivative does not exist for a given value of x, then the second derivative (which is the derivative of the first derivative) cannot possibly exist. Of course, there is no obligation to use this test even when it does work. If, after looking at the first derivative, you feel that you would rather do almost anything than take another derivative, then don't take another derivative—use the first-derivative test instead.

The second derivative also is useful for locating those points at which the graph changes from concave upward to downward or vice versa.

Definition

*If $A(a, f(a))$ is a point of the graph of the function f such that the concavity of f changes at A, then A is a **point of inflection**.*

Points of inflection are determined in much the same way as are critical points, by using the second derivative instead of the first. The second derivative can change from positive to negative or negative to positive only at those points at which it either is zero or does not exist. Again, as with critical points, the points of the graph at which the second derivative is either zero or nonexistent are only *possible* points of inflection. You must check to see whether or not the sign of the second derivative really does change by considering values to the left and right of the point in question. This is summed up in the following theorem.

Theorem 5.8

(*Second-derivative test for points of inflection.*) *A function f is concave upward when f" > 0 and concave downward when f" < 0. A point of inflection is a point at which f" changes sign; this may occur when f" is either zero or undefined.*

Can one use the third derivative to make the check? Yes, you can, but it might be noted that the third-derivative test is not so valuable as it might seem at first glance. In most problems you will be interested in finding not only points of inflection, but also relative maxima and minima. When using the second-derivative test to determine relative maxima and minima, you would check several values of the second derivative. Thus it sometimes happens that no further checking is needed when you consider points of inflection; when further checking is needed, it is usually minimal.

If you do want to use the third-derivative test, the procedure is much like the second-derivative test. If $f''(a) = 0$ and $f'''(a) \neq 0$, then $(a, f(a))$ is a point of inflection, with the graph changing from concave upward to downward if $f'''(a)$ is negative and from concave downward to upward if $f'''(a)$ is positive. If $f'''(a) = 0$, the test fails.

Example 2

Find all points of inflection and determine the concavity of the graph of $y = x^4 - 6x^2$.

$$y' = 4x^3 - 12x,$$
$$y'' = 12x^2 - 12 = 12(x - 1)(x + 1).$$

We see that $y'' = 0$ if $x = \pm 1$, and so we have two *possible* points of inflection, $(1, -5)$ and $(-1, -5)$.

To determine concavity we can use one of two methods. Checking the values of the second derivative to the left of -1, between -1 and 1, and to the right of 1 gives us Figure 5.30, which shows that the two points are really points of inflection, with the graph concave downward between them and concave upward elsewhere.

Figure 5.30

The third-derivative test gives us $y''' = 24x$, which is clearly negative at $x = -1$ and positive at $x = 1$. Therefore, the graph changes from concave upward to downward at $x = -1$ and back to concave upward at $x = 1$.

Had we been asked to determine relative maxima and minima, there would be no point in using the third-derivative test. The critical values are $x = \pm\sqrt{3}$ and $x = 0$. When using the second-derivative test to determine relative maxima and minima, you would check values of the second derivative to the left of -1 (at $-\sqrt{3}$), between -1 and 1 (at 0), and to the right of 1 (at $\sqrt{3}$). Thus no further checking would be needed when you consider points of inflection.

Problems

A

In Problems 1–8, use the second-derivative test where possible to determine relative maxima and minima. Indicate which, if any, are absolute maxima or minima.

1. $y = x^2 - x - 6$.
2. $y = x^2 - 4x - 5$.
3. $y = 2x^3 - 3x^2 - 12x + 18$.
4. $y = x^3 - 3x^2 - 9x + 27$.
5. $y = 3x^4 - 4x^3 - 12x^2 + 24$.
6. $y = 3x^4 - 4x^3 - 6x^2 + 12x$.
7. $y = 4x^5 - 5x^4$.
8. $y = 4x^5 - x^4$.

In Problems 9–14, find all points of inflection and determine the concavity.

9. $y = x^2 - 2x - 1$.
10. $y = 2 - 5x - 3x^2$.
11. $y = x^3 - 3x^2 - 4x + 5$.
12. $y = x^3 + 6x^2 - 4x + 1$.
13. $y = x^4 - 6x^3$.
14. $y = x^4 + 2x^3 - 12x^2 - 15x + 21$.

B

In Problems 15–18, use the second-derivative test where possible to determine relative maxima and minima. Indicate which, if any, are absolute maxima or minima.

15. $y = \dfrac{x^2 + 1}{x}$.
16. $y = \dfrac{(x + 1)^2}{x^2 + 1}$.

17. $y = \dfrac{x^2}{x + 1}$.
18. $y = \dfrac{x}{(x + 4)^3}$.

In Problems 19–22, find all points of inflection and determine the concavity.

19. $y = \dfrac{3}{x + 2}$.
20. $y = \dfrac{x - 2}{x + 3}$.

21. $y = \dfrac{x}{x^2 + 1}$.
22. $y = \dfrac{1}{x^2 + 3}$.

In Problems 23–32, find all relative maxima and minima and points of inflection. Indicate which, if any, are absolute maxima or minima.

23. $y = x^3 - 3x^2$.
24. $y = x^3 + 6x^2 - 24$.
25. $y = 4x^3 - 15x^2 - 18x + 10$.
26. $y = x^3 - 9x^2 + 15x - 5$.

27. $y = 2x^4 - x.$

28. $y = x^3 - 3x.$

29. $y = x^5 - 5x^4.$

30. $y = x^4 - 4x^3.$

31. $y = \dfrac{x - 1}{x^2}.$

32. $y = \dfrac{4}{x^2 + 3}.$

C

33. Show that the graph of $y = (ax + b)/(cx + d)$ has no critical point and no point of inflection.

34. For what values of a, b, and c does the graph of $y = x^3 + ax^2 - 4bx + c$ have a relative minimum at $(-2, -19)$ and contain the point $(0, 1)$?

35. Use derivatives to show that $x^2 - 6x + 10 > 0$ for all values of x. Can you show it without the use of derivatives?

36. Use derivatives to show that $3x^4 - 4x^3 - 12x^2 + 36 > 0$ for all values of x.

37. Suppose that f has a critical point at $x = a$, but it is neither a relative maximum nor relative minimum. Show that $(a, f(a))$ must be a point of inflection.

5.6

End-Point Extrema

Sometimes a relative maximum or minimum occurs at some point other than a critical point. Let us consider the function

$$f(x) = x^2, \quad -2 \le x \le 2.$$

We can easily see that it has a critical point $(0, 0)$ which is an absolute minimum. Furthermore, this is its only critical point. But it has a pair of absolute maxima at $(-2, 4)$ and $(2, 4)$, as we can see in Figure 5.31.

In general, for cases such as this, we need to consider not only critical points, but also end points of the domain. Therefore, relative maxima or minima that occur at points other than critical points are called end-point extrema.

Note that the end point must actually belong to the domain in order to have an end-point extremum. The function

$$f(x) = x^2, \quad -2 < x < 2,$$

has no relative or absolute maximum. Since $(2, 4)$ and $(-2, 4)$ are not points of the graph, a highest point would be one that is nearest $(2, 4)$ or $(-2, 4)$. Of course, there is no such nearest point—there is no maximum.

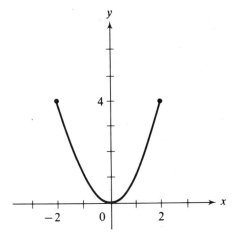

Figure 5.31

Example 1

Find all relative maxima and minima of $y = x^2 - x$
$0 \leq x \leq 3$.   Indicate which, if any, are absolute maxima
or minima.

Let us first consider critical points;

$$y' = 2x - 1, \qquad y'' = 2.$$

We have a relative minimum at $(1/2, -1/4)$.

Now the two end points are $(0, 0)$ and $(3, 6)$.
Since $y' < 0$ for $x < 1/2$ and $y' > 0$ for
$x > 1/2$, the graph decreases from $(0, 0)$ to
$(1/2, -1/4)$ and increases thereafter to $(3, 6)$.
Thus $(0, 0)$ and $(3, 6)$ are both relative maxima.
Furthermore, $(3, 6)$ is the absolute maximum,
and $(1/2, -1/4)$ is the absolute minimum.
The graph is given in Figure 5.32.

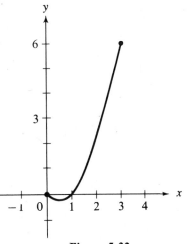

Figure 5.32

Example 2

Find all relative maxima and minima of $y = 1/x$, $x \geq 1$.   Indicate which, if any, are absolute
maxima or minima.

We find that $y' = -1/x^2$, $y'' = 2/x^3$.

There is no critical point; furthermore, the graph is decreasing and concave upward for $x \geq 1$. Thus the end point $(1, 1)$ is an absolute maximum, while there is
no minimum (see Figure 5.33).

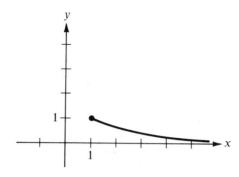

Figure 5.33

When dealing with functions defined by several equations, the ordinary methods of finding derivatives cannot be used at the "joints," where the value of the function is determined by one equation on the left and another on the right. To be specific, let us suppose that

$$f(x) = \begin{cases} g_1(x) & \text{if } x < a, \\ g_2(x) & \text{if } x \geq a. \end{cases}$$

Then $f'(x) = g_1'(x)$ for $x < a$ and $f'(x) = g_2'(x)$ for $x > a$, but special care must be taken for $f'(a)$. We cannot simply say that since $f(x) = g_2(x)$ at $x = a$, it follows that $f'(a) = g_2'(a)$. The reason for this is that

$$f'(a) = \lim_{x \to a} \frac{f(x) - f(a)}{x - a}.$$

Since the limit depends upon the value of f near $x = a$ rather than at $x = a$, we are concerned with f on both sides of $x = a$. Thus $f'(a)$ depends upon both g_1 and g_2.

There are two things necessary for the existence of $f'(a)$:

(1) $\quad \lim_{x \to a^-} g_1(x) = \lim_{x \to a^+} g_2(x)$

and

(2) $\quad \lim_{x \to a^-} g_1'(x) = \lim_{x \to a^+} g_2'(x).$

The first of these conditions assures us that f is continuous at $x = a$, and the second that the two halves of the graph approach the same slope at $x = a$. Assuming that g_1, g_2, g_1', and g_2' are all continuous at $x = a$, the two conditions become

(1) $\quad g_1(a) = g_2(a)$

and

(2) $\quad g_1'(a) = g_2'(a).$

Example 3

Find all relative maxima and minima of

$$y = \begin{cases} -x & \text{if } x < 0, \\ 2x - x^2 & \text{if } x \geq 0. \end{cases}$$

Indicate which, if any, are absolute maxima or minima.

The portion $y = -x$ for $x < 0$ is a line; it has no critical point. For the other portion, $y = 2x - x^2$ if $x \geq 0$.

$$y' = 2 - 2x, \qquad y'' = -2.$$

Thus $(1, 1)$ is a relative maximum.

Now let us consider the point $(0, 0)$. Since

$$g_1(x) = -x = 0$$

and

$$g_2(x) = 2x - x^2 = 0$$

at $x = 0$, the graph is continuous at $x = 0$. However,

$$g_1'(x) = -1$$

and

$$g_2'(x) = 2 - 2x = 2$$

at $x = 0$. Since $g_1'(0) \neq g_2'(0)$, the two halves of the curve do not have the same slope at $x = 0$. This implies the nonexistence of y' at $x = 0$, and $(0, 0)$ is a critical point. Since the slope changes from negative to positive there, $(0, 0)$ is a relative minimum. There is no absolute maximum or minimum, since

$$\lim_{x \to -\infty} y = +\infty$$

and

$$\lim_{x \to +\infty} y = -\infty.$$

The graph of this function is given in Figure 5.34.

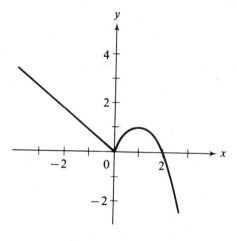

Figure 5.34

Example 4

Find all relative maxima and minima of

$$y = \begin{cases} x & \text{if } x < 0, \\ \dfrac{x}{x^2 + 1} & \text{if } x \geq 0. \end{cases}$$

Indicate which, if any, are absolute maxima or minima.

There is no critical point for $x < 0$. This portion is a straight line. For $x > 0$,

$$y = \frac{x}{x^2 + 1},$$

$$y' = \frac{(x^2 + 1) \cdot 1 - x \cdot 2x}{(x^2 + 1)^2} = \frac{1 - x^2}{(x^2 + 1)^2},$$

$$y'' = \frac{(x^2 + 1)^2(-2x) - (1 - x^2)2(x^2 + 1)2x}{(x^2 + 1)^4} = \frac{2x(x^2 - 3)}{(x^2 + 1)^3}.$$

There is a critical point at $(1, 1/2)$ which is a relative maximum. (The point $(-1, -1/2)$ is not a critical point since we are only considering values of x greater than zero.) Now consider the point $(0, 0)$. Since $g_1(x) = x = 0$ and $g_2(x) = x/(x^2 + 1) = 0$ at $x = 0$, the graph is continuous at $(0, 0)$. Since $g'_1(x) = 1$ and $g'_2(x) = (1 - x^2)/(x^2 + 1)^2 = 1$ at $x = 0$, $y' = 1$ at $x = 0$. Thus $(0, 0)$ is not a critical point.

Thus $(1, 1/2)$ is the only relative maximum and there is no relative minimum. Furthermore, since $y = 0$ is a horizontal asymptote and there is no critical point to the right of $(1, 1/2)$, the relative maximum is the absolute maximum. Of course, there is no absolute minimum, since there is no relative minimum. The graph of this function is given in Figure 5.35.

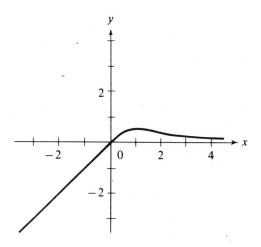

Figure 5.35

Problems

A

Find all relative maxima and minima. Indicate which, if any, are absolute maxima or minima.

1. $y = x^2 - 2$, $-1 \le x \le 2$.
2. $y = x^2 - 2$, $2 \le x \le 5$.
3. $y = 4x^3 - 9x^2 - 12x$, $-2 \le x \le 1$.
4. $y = x^3 + 3x^2 - 9x$, $-1 \le x \le 2$.
5. $y = \dfrac{1}{x^2}$, $-1 \le x \le 1$.
6. $y = \dfrac{x}{x + 1}$, $-2 \le x \le 1$.
7. $y = (x^2 - 1)^2$, $-2 \le x \le 3$.
8. $y = x^2(x - 1)$, $-1 \le x \le 1$.
9. $y = \begin{cases} x^2 & \text{if } x < 0, \\ x^3 & \text{if } x \ge 0. \end{cases}$
10. $y = \begin{cases} 1 - x^2 & \text{if } x < 0, \\ x^3 - 1 & \text{if } x \ge 0. \end{cases}$

11. $y = \begin{cases} -x & \text{if } x < 0, \\ 2x^3 + 3x^2 & \text{if } x \geq 0. \end{cases}$

12. $y = \begin{cases} 1 + x^2 & \text{if } x < 0, \\ 1 - x^2 & \text{if } x \geq 0. \end{cases}$

B

Find all relative maxima and minima. Indicate which, if any, are absolute maxima or minima.

13. $y = x^3 - 3x, \quad -\sqrt{3} \leq x \leq 2.$

14. $y = \dfrac{x^2}{x^2 + 1}, \quad -1 \leq x \leq 2.$

15. $y = \sqrt{1 - x^2}.$

16. $y = \sqrt{9 - x^2}.$

17. $y = \begin{cases} -x - 1 & \text{if } x < -1, \\ 1 - x^2 & \text{if } -1 \leq x \leq 1, \\ x - 1 & \text{if } x > 1. \end{cases}$

18. $y = \begin{cases} -x - 1 & \text{if } x < -1, \\ x^2 - 1 & \text{if } -1 \leq x \leq 1, \\ 2x - 2 & \text{if } x > 1. \end{cases}$

19. $y = \begin{cases} -4(3x + 4) & \text{if } x < -2, \\ x^4 - 2x^2 & \text{if } -2 \leq x \leq 2, \\ 4(3x - 4) & \text{if } x > 2. \end{cases}$

20. $y = \begin{cases} -8x + 4 & \text{if } x < -2, \\ x^3 - 2x & \text{if } -2 \leq x \leq 2, \\ 4x - 4 & \text{if } x > 2. \end{cases}$

21. $y = \begin{cases} 2x + 6 & \text{if } x < -2, \\ -x & \text{if } -2 \leq x \leq 0, \\ 2x - x^2 & \text{if } x > 0. \end{cases}$

22. $y = \begin{cases} x + 1 & \text{if } x \leq -1, \\ -x - x^2 & \text{if } -1 < x \leq 0, \\ 2x - x^2 & \text{if } 0 < x \leq 2. \\ 2x - 4 & \text{if } x > 2. \end{cases}$

C

23. A widget can be manufactured by either of two methods, A and B. The manufacturing expense E per widget depends upon the total daily output x in the following way:

$$\text{Method A: } E = 0.01x^2 - 15x + 17{,}000,$$

$$\text{Method B: } E = 0.005x^2 - 12x + 19{,}000.$$

Show that method A is more efficient (gives a lower expense) when the daily output x is less than 1000, and method B is more efficient for $x > 1000$. Thus, by the proper choice of methods, the manufacturing expense is

$$E = \begin{cases} 0.01x^2 - 15x + 17{,}000, & \text{if } 0 \leq x \leq 1000, \\ 0.005x^2 - 12x + 19{,}000, & \text{if } x > 1000. \end{cases}$$

What daily output minimizes E? What is the minimum expense per widget?

5.7

The Derivative As an Aid in Sketching Curves

With the use of calculus we are now in a much better position to sketch curves. Table 5.2 shows the role of the first and second derivatives in sketching the graph of $y = f(x)$.

Table 5.2

$f(x)$ Defined		
$f'(x)$	$f''(x)$	At $(x, f(x))$
+		Graph is increasing.
−		Graph is decreasing.
	+	Graph is concave upward.
	−	Graph is concave downward.
0	+	Graph has a relative minimum with a horizontal tangent.
0	−	Graph has a relative maximum with a horizontal tangent.
0	0 or undefined	Graph has a horizontal tangent and a relative maximum, relative minimum, or inflection point.
Undefined	Undefined	Graph has a possible vertical tangent and a possible relative maximum, relative minimum, or inflection point.
	0 or undefined	Graph has a possible inflection point.

Example 1

Sketch $y = (x - 3)(x + 1)^2$.

This is the same as Example 3 of Section 5.2. The information we found there is summed up graphically in Figure 5.12. Let us now determine maxima, minima, and points of inflection.

$$y' = (x + 1)(3x - 5), \qquad y'' = 2(3x - 1).$$

From these derivatives we have:

$(-1, 0)$ is a relative maximum,
$(5/3, -256/27)$ is a relative minimum,
$(1/3, -128/27)$ is a point of inflection of the curve, which has slope $-16/3$ there.

We now see that the curve must look like Figure 5.36.

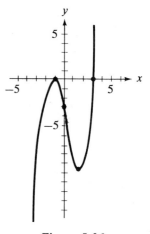

Figure 5.36

You might compare the graph of this example with our best guess in Figure 5.13. You might also recall that we were not sure, when we first sketched the graph, that it did not look like Figure 5.14. We now know that it cannot have any more "turns," since we know that there cannot be any other relative maximum or minimum. Furthermore, there cannot even be a little "wiggle," since we found only one point of inflection. We now have a very accurate picture of the graph.

Example 2

Sketch $y = x^3 - 6x^2 + 9x + 1$.

Without calculus we know almost nothing about the graph. In order to find the x intercept(s), we would have to solve the equation $x^3 - 6x^2 + 9x + 1 = 0$! This seems to be too much work. Let's see if we can get along without it. The y intercept is (0, 1). There are no asymptotes of any kind, but

$$\lim_{x \to +\infty} y = +\infty \quad \text{and} \quad \lim_{x \to -\infty} y = -\infty.$$

Taking derivatives we have

$$y' = 3x^2 - 12x + 9 = 3(x-1)(x-3),$$
$$y'' = 6x - 12 = 6(x-2).$$

These give the following results.

(1, 5) is a relative maximum,
(3, 1) is a relative minimum,
(2, 3) is a point of inflection, slope -3.

The resulting graph is given in Figure 5.37.

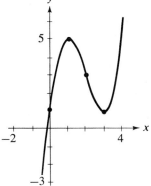

Figure 5.37

Example 3

Sketch $y = (x-1)^{1/3}(x+2)^{2/3}$.

The intercepts are (1, 0) odd, (−2, 0) even, and $(0, -\sqrt[3]{4})$. Even though the exponents are not integers, we may assign odd and even designations according to the numerator of the exponents. There are no asymptotes and no symmetry, but

$$\lim_{x \to +\infty} y = +\infty \quad \text{and} \quad \lim_{x \to -\infty} y = -\infty.$$

This information alone gives us a rough idea of the graph which has the general shape given in Figure 5.38. We already know that there is a relative maximum

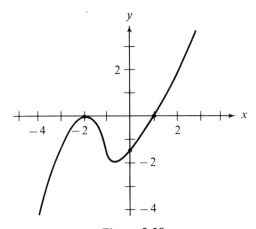

Figure 5.38

at $(-2, 0)$, but we have only guessed at the location of the relative minimum. Furthermore, we do not know if there are other relative maxima or minima, and we know almost nothing about the concavity. Let us now use calculus to refine our results.

The first two derivatives,

$$y' = \frac{x}{(x-1)^{2/3}(x+2)^{1/3}}, \qquad y'' = \frac{-2}{(x-1)^{5/3}(x+2)^{4/3}},$$

give the following results:

$(0, -\sqrt[3]{4})$ is a relative minimum with horizontal tangent (since $y' = 0$ at $x = 0$),

$(1, 0)$ is neither relative maximum nor minimum but it has a vertical tangent (since y' is nonexistent at $x = 1$),

$(-2, 0)$ is a relative maximum with a vertical tangent (again because y' is nonexistent at $x = -2$),

$(1, 0)$ is a point of inflection, no slope.

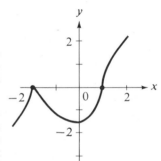

Note that, of the three critical values ($x = 0$, $x = 1$, and $x = -2$), only the first can be tested by the second derivative—there is no second derivative at $x = 1$ or $x = -2$.

This now tells us that the graph is as shown in Figure 5.39. We have located the relative minimum and have found that there are vertical tangents at both x intercepts. Furthermore, we know that there is no other maximum nor minimum; and there is no change in concavity except at $(1, 0)$.

Figure 5.39

Note the basic difference between the critical point $(0, -\sqrt[3]{4})$ and the other two critical points, $(1, 0)$ and $(-2, 0)$. At $x = 0$, $y' = 0$; but at $x = 1$ and $x = -2$, y' does not exist. Thus when $y' = 0$, the tangent is horizontal; when y' does not exist (the denominator of the expression for y' is 0, but the numerator is not), the tangent (if there is one) is vertical. This is summed up graphically in Figure 5.40

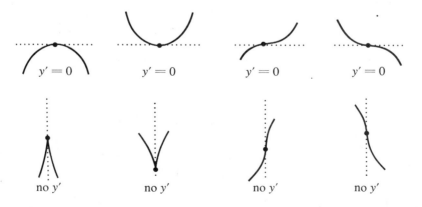

Figure 5.40

Example 4

Sketch $y^2 = \dfrac{27(x-1)^2}{x^3}$.

Suppose that we first sketch

$$z = \frac{27(x-1)^2}{x^3}.$$

We have intercept $(1, 0)$ even, vertical asymptote $x = 0$ odd, and horizontal asymptote $z = 0$ and no symmetry. The derivatives

$$z' = -\frac{27(x-1)(x-3)}{x^4} \quad \text{and} \quad z'' = \frac{54(x^2-6x+6)}{x^5}$$

give (for the graph of z)

$(1, 0)$ is a relative minimum,

$(3, 4)$ is a relative maximum,

$(3 + \sqrt{3}, 3(3 + \sqrt{3})/4)$ is a point of inflection,

$(3 - \sqrt{3}, 3(3 - \sqrt{3})/4)$ is a point of inflection.

In the resulting graph (Figure 5.41), z is represented by the dotted curve and y by the solid curve.

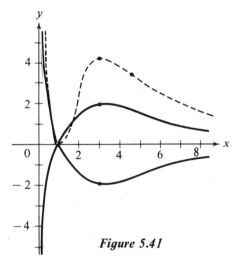

Figure 5.41

Note that the graphs of y and z have relative maxima at the same value of x. This of course, occurs only when $z > 0$. Furthermore, points of inflection are not preserved here; in other words, the fact that z has points of inflection at $x = 3 \pm \sqrt{3}$ does not imply that y has points of inflection at these values of x. In fact, a check on the second derivative of y shows that there is only one point of inflection, and it is at $x = 5$. *Do not try to carry over points of inflection from the auxiliary graph to the final one.* Similarly, slant asymptotes do not carry over; and if $z = a$ $(a \ge 0)$ is a horizontal asymptote of the auxiliary curve, then $y = \pm\sqrt{a}$ are horizontal asymptotes. Vertical asymptotes are preserved provided they are approached for $z > 0$.

Problems

A

In Problems 1–8, find all relative maxima and minima and all points of inflection, and sketch the curve.

1. $y = x^3 + 6x^2$.
2. $y = x^4 - 4x^3$.
3. $y = x^3 - 6x^2 + 9x - 3$.
4. $y = x^5 - 5x$.
5. $y = (x + 1)^2(x - 3)^3$.
6. $y = x^3(x - 5)^4$.
7. $y = (x^2 - 6)(2x - 3)$.
8. $y = x^2(x^2 - 6)$.

B

In Problems 9–16, find all relative maxima and minima and all points of inflection, and sketch the curve.

9. $y = (x - 1)^{1/3}(x + 3)^{2/3}$.

10. $y = x^{2/3}(x - 3)^{1/3}$.

11. $y = \sqrt{x}(x - 3)$.

12. $y = x^{2/5}(x + 5)^{3/5}$.

13. $y = \dfrac{x^2}{x + 1}$.

14. $y = \dfrac{4(x + 1)}{x^2}$.

15. $y = \dfrac{x^2 + 1}{x}$.

16. $y = \dfrac{1}{x^2 - 2x}$.

In Problems 17–24, find all relative maxima and minima, and sketch the curve.

17. $y^2 = x^4 - 4x^3$.

18. $y^2 = -2x^3 + 9x^2 - 12x + 54$.

19. $y^2 = 6x^2 - x^3$.

20. $y^2 = -\dfrac{1}{2}(x - 3)(x + 3)^2$.

21. $y^2 = \dfrac{x - 1}{x^3}$.

22. $y^2 = \dfrac{1}{x^2 - 1}$.

23. $y^2 = \dfrac{9x}{(x - 2)^3}$.

24. $y^2 = \dfrac{x^2(x - 4)}{x - 6}$.

C

25. The quadratic equation $Ax^2 + Bx + C = 0$ has complex roots if and only if $f(x) = Ax^2 + Bx + C$ has either a positive minimum or a negative maximum. Use this to show that $Ax^2 + Bx + C = 0$ has complex roots if and only if $B^2 - 4AC < 0$.

Review Problems

A

In Problems 1–4, sketch the graphs of the equations as rapidly as possible without using derivatives. Show all intercepts, asymptotes, and symmetry about either axis or the origin.

1. $y = \dfrac{x^2 - 1}{x^2 + 1}$.

2. $y = \dfrac{x^2(x - 3)}{(x^2 - 4)^2}$.

3. $y = \dfrac{x^2 - 9}{x^2}$.

4. $y = \dfrac{x^3 - 1}{x^2 + 3x}$.

In Problems 5 and 6, find all relative maxima and minima. Indicate which, if any, are absolute maxima or minima.

5. $y = 2x^3 - 3x^2 - 12x + 8$.

6. $y = \dfrac{x^2 + 1}{x}$.

In Problems 7 and 8, find all relative maxima and minima and all points of inflection. Indicate which, if any, are absolute maxima or minima.

7. $y = x^4 - 6x^2$.

8. $y = x^3(x - 4)$.

B

In Problems 9–16, sketch the graphs of the equations as rapidly as possible without using derivatives. Show all intercepts, asymptotes, and symmetry about either axis or the origin.

9. $y = (x - 1)^2(x + 3)^3(2x - 3)$.

10. $y = x + 1 - \dfrac{6}{x}$.

11. $y = x - \sqrt{x^2 - 4}$.

12. $y = \sqrt{\dfrac{(x - 1)(x + 2)}{x + 1}}$.

13. $y^2 = \dfrac{x(x - 2)}{(x + 1)^2}$.

14. $y^2 = \dfrac{2x - 1}{x + 2}$.

15. $y = \dfrac{x^2 - 3x - 4}{x^2 + 4x + 3}$.

16. $y = \dfrac{x^2}{x^2 - 2x}$.

In Problems 17–20, find all relative maxima and minima. Indicate which, if any, are absolute maxima or minima.

17. $y = x^3(3x - 1)^2$.

18. $y = \dfrac{(x^2 - 4)^2}{x^4 - 1}$.

19. $y = x - 3x^{1/3}$.

20. $y = (2x - 4)^{1/3}(x + 4)^{2/3}$.

In Problems 21 and 22, find all relative maxima and minima and all points of inflection. Indicate which, if any, are absolute maxima or minima.

21. $y = (x^2 - 9)^2$.

22. $y = \dfrac{x^2 + x + 9}{x + 1}$.

In Problems 23–26, find all relative maxima and minima and all points of inflection. Sketch the graph.

23. $y = x^3 - 6x^2 + 17$.

24. $y = (x - 4)(x + 2)^2$.

25. $y = \dfrac{3x^2}{3x^2 - 1}$.

26. $y = \dfrac{x^3}{x^3 + 2}$.

In Problems 27 and 28, find all relative maxima and minima. Sketch the graph.

27. $y^2 = x^3 - 12x + 16$.

28. $y^2 = \dfrac{(x^2 - 1)^2}{x^4 + 1}$.

29. Given the graph of $f(x)$ shown in Figure 5.42, indicate whether $f(x)$, $f'(x)$, and $f''(x)$ are positive, negative, zero, or undefined at each of the eight values of x indicated.

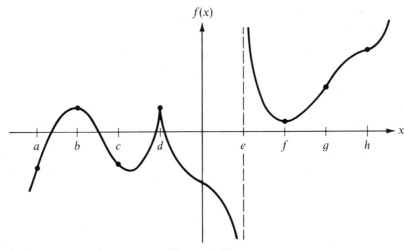

Figure 5.42

C

30. Suppose $f(x) = Ax^3 + Bx^2 + Cx + D$. Show that if this function has critical values $x = x_1$ and $x = x_2$, then it has a point of inflection at $x = (x_1 + x_2)/2$. Is this also true for the general fourth-degree polynomial function $f(x) = Ax^4 + Bx^3 + Cx^2 + Dx + E$?

31. Suppose $f(x) = (x - a)^m(x - b)^n$, where m and n are integers greater than 1. Two critical points are $(a, 0)$ and $(b, 0)$. Find the third and show that it is between these two. This is a special case of Rolle's theorem, which is stated and proved in Section 17.5.

6

Further Applications of the Derivative

6.1

Applications of Maxima and Minima

Maxima and minima have many practical applications. The engineer wants to maximize the strength of structures; the businessperson wants to minimize taxes, maximize profit, and minimize cost; and so on. Moreover, nature determines many maximum or minimum problems: light, when reaching an interface or a reflecting surface, takes a path that gives a minimum propagation time (see Problems 32 and 36); soap bubbles assume the shape with a minimum surface area. Thus the determination of maxima and minima can, become a very practical problem. We, of course, are interested in an *absolute* maximum or minimum; however, finding the relative maxima or minima is a first step in determining the absolute maximum or minimum. We proceed as follows.

1. From the description of the situation, decide what is to be made a maximum or minimum and what is allowed to vary (there may be several quantities that vary).
2. Write an equation expressing the quantity that is to be a maximum or minimum in terms of *one* of the variables. An expression in terms of two variables is often quite easy to obtain. In such a case, some condition will give a relationship between the two variables; thus, one variable can be expressed in terms of the other and substituted into the expression that is to be maximized or minimized. At this stage, we also want to determine the domain of our function, since end-point extrema may be significant.
3. Use the procedure developed in Chapter 5, remembering that the final objective is an absolute—not relative—maximum or minimum. Often, the physical situation will make it obvious whether or not a relative maximum or minimum exists.

Example 1

A farmer wishes to fence a field bordering a straight stream with 1000 yd of fencing material. It is not necessary to fence the side bordering the stream. What is the area of the largest rectangular field that can be fenced in this way?

Let us follow the three steps given above. First, we want to decide what is to be maximized and what is allowed to vary. Since we want the largest field, the area enclosed by the fence and stream is to be a maximum. Furthermore, the dimensions of the field—its length and width—are allowed to vary. If the situation is represented by Figure 6.1, we want to maximize the area A by varying x and y.

Our second step is to express A, which is to be maximized in terms of either x or y. It is a simple matter to express the area in terms of x and y:

$$A = xy.$$

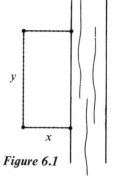

Figure 6.1

Now we must eliminate either x or y by finding a relationship between them.

Since there is 1000 yd of fencing material, we see that

$$2x + y = 1000, \quad \text{or} \quad y = 1000 - 2x.$$

Thus

$$A = x(1000 - 2x) = 1000x - 2x^2.$$

Before going on to the next step, we might note that there are restrictions on the values that x may take. Since x and y are lengths, neither one of them can be negative. Thus $x \geq 0$ and $y \geq 0$. Since $y = 1000 - 2x$, the condition $y \geq 0$ implies that $x \leq 500$. Thus

$$0 \leq x \leq 500.$$

Finally, we are ready to go on to step 3.

$$A = 1000x - 2x^2, \quad 0 \leq x \leq 500,$$
$$A' = 1000 - 4x = 0,$$
$$x = 250 \text{ yd},$$
$$A = 125{,}000 \text{ sq yd}.$$

Since $A'' = -4$ for all x, the graph of A as a function of x is concave downward. Thus $x = 250$ gives the relative and absolute maximum value for A. It might be noted here that we have the minimum value for A (namely, $A = 0$) at the end points $x = 0$ and $x = 500$. These end-point extrema make it clear that the maximum value must be assumed at some point between them.

Example 2

Find the volume of the largest right circular cylinder that can be inscribed in a sphere of radius 3.

Our first step is to note that we are trying to maximize the volume of the inscribed cylinder. The dimensions of this cylinder—its radius r and height h—are allowed to vary. The situation is illustrated in Figure 6.2(a).

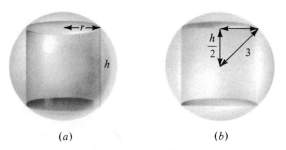

(a)   (b)

Figure 6.2

Our second step is to express the volume V of the cylinder in terms of either r or h. It is a simple matter to express V in terms of both r and h:

$$V = \pi r^2 h.$$

However, we see that r and h are not independent. For example, if h is fixed, then the fact that the cylinder is inscribed in a sphere of radius 3 determines the value of r. In order to see this relationship, we consider a cross section through the center of the sphere, as shown in Figure 6.2(b). From the right triangle formed, we see that

$$9 = r^2 + \frac{h^2}{4}, \quad \text{or} \quad r^2 = 9 - \frac{h^2}{4}.$$

Thus

$$V = \pi\left(9 - \frac{h^2}{4}\right)h$$

$$= \pi\left(9h - \frac{h^3}{4}\right).$$

We also note that both r and h are nonnegative because they are lengths. But $r \geq 0$ and $r^2 = 9 - (h^2/4)$ imply that $h \leq 6$. Thus

$$0 \leq h \leq 6.$$

Finally, we go on to step 3, which is finding the maximum.

$$V' = \pi\left(9 - \frac{3h^2}{4}\right) = 0,$$

$$h^2 = 12,$$

$$h = 2\sqrt{3}.$$

Thus $h = 2\sqrt{3}$ is the only critical value, and

$$V = 12\pi\sqrt{3}.$$

In addition, we have possible end-point extrema at $h = 0$ and $h = 6$. The inscribed "cylinder" collapses to give $V = 0$ at both of these values. This assures us that

$$V_{\text{max}} = 12\pi\sqrt{3}.$$

This can also be verified using the second derivative (which is negative for all positive values of h).

Example 3

Find the area of the largest rectangle that can be inscribed in $\dfrac{x^2}{a^2} + \dfrac{y^2}{b^2} = 1$.

Before we consider the three steps that give a solution, let us concentrate on sketching the graph of the given equation. The methods of the preceding chapter tell us three things about the graph of this equation: it has intercepts $(\pm a,\ 0)$ and $(0,\ \pm b)$; it is symmetric about both axes and about the origin; and it is restricted to $-a \leq x \leq a$, $-b \leq y \leq b$ (we assume $a, b > 0$). Implicit differentiation tells us that the graph also has horizontal tangents at $(0,\ \pm b)$ and vertical tangents at $(\pm a, 0)$. The graph, shown in Figure 6.3, is drawn with the assumption that $a > b$; however, the subsequent material makes no assumption concerning the relative sizes of a and b.

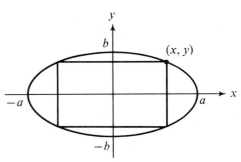

Figure 6.3

Now let us consider the first step of our solution. The inscribed rectangle must have all four corners on the curve; therefore, its sides are parallel to the axes. Suppose we label (x, y) the corner for which $0 \leq x \leq a$ and $0 \leq y \leq b$. It is now clear that we want to maximize the area of this rectangle by varying the position of the point (x, y). Since (x, y) is on the curve, the value of either x or y determines the other. Thus we may look upon either x or y as a variable.

Going on to step 2, we see that it is easy to express A in terms of both x and y.

$$A = 4xy.$$

Since (x, y) is on the ellipse, it satisfies the equation

$$\frac{x^2}{a^2} + \frac{y^2}{b^2} = 1;$$

and

$$y = \frac{b}{a}\sqrt{a^2 - x^2}.$$

We take only the nonnegative square root, since $0 \leq y \leq b$. Now

$$A = \frac{4b}{a}\, x\sqrt{a^2 - x^2}, \qquad 0 \leq x \leq a,$$

Finally, we carry out the maximization:

$$A' = \frac{4b}{a}\left(x\,\frac{-2x}{2\sqrt{a^2 - x^2}} + \sqrt{a^2 - x^2} \right) = \frac{4b}{a}\,\frac{a^2 - 2x^2}{\sqrt{a^2 - x^2}}.$$

Our function has two critical values within its domain, namely, $x = a/\sqrt{2}$ and $x = a$. Furthermore we have a possible end-point extremum at $x = 0$. If $x = a$ or $x = 0$, the inscribed rectangle collapses and $A = 0$. These give minimum values for A. Hence $x = a/\sqrt{2}$ must give the desired maximum. This can be verified using the first- (or second-) derivative test. At $x = a/\sqrt{2}$,

$$A_{\max} = 2ab.$$

Example 4

Find the point of the curve $y = x^2$ that is closest to $(4, -1/2)$.

Here we are trying to minimize the distance between the point $(4, -1/2)$ and a point (x, y) on the curve. The position of the point on the curve is allowed to vary. Again, because the point is on the curve, the value of x determines y uniquely (note that the value of y does not uniquely determine x).

The distance between the point $(4, -1/2)$ and an arbitrary point (x, y) on the curve (see Figure 6.4) is given by

$$d = \sqrt{(x - 4)^2 + (y + 1/2)^2}.$$

Since (x, y) is on the curve, its coordinates satisfy the equation $y = x^2$; and

$$d = \sqrt{(x - 4)^2 + (x^2 + 1/2)^2}.$$

Note here that there is no restriction on the values that x may take.

Finally, we maximize d:

$$d' = \frac{2(x - 4) + 2(x^2 + 1/2)2x}{2\sqrt{(x - 4)^2 + (x^2 + 1/2)^2}}$$

$$= \frac{2x^3 + 2x - 4}{\sqrt{(x - 4)^2 + (x^2 + 1/2)^2}}.$$

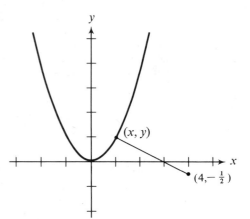

Figure 6.4

The only real value of x that makes the numerator zero is $x = 1$; the denominator cannot be zero. (If it were, then $x - 4 = 0$ and $x^2 + 1/2 = y + 1/2 = 0$, which would give $(x, y) = (4, -1/2)$, and $(4, -1/2)$ is not a point of $y = x^2$.) Thus the point we are seeking is $(1, 1)$.

A trick that could have been used to some advantage in Example 4 is to square d before taking the derivative.

$$D = d^2 = (x - 4)^2 + (x^2 + 1/2)^2.$$

Since d is always positive, both d and D must have a maximum or minimum at the same value of x. The advantage is that the derivative of D is easier to find than the derivative of d.

$$D' = 2(x - 4) + 2(x^2 + 1/2)2x = 4x^3 + 4x - 8.$$

Not only is it easier to take the derivative, but the result is far simpler. This could also have been used in Example 3 (see Problems 24 and 25).

Occasionally we encounter functions that are defined only for integers. Consider, for example, the function

$$f(n) = 20n - n^2, \quad n = 1, 2, 3, \ldots,$$

whose graph is given in Figure 6.5(a). Suppose we wish to find the maximum value of $f(n)$. It is clear that there is no tangent, and therefore no derivative, at any point of the graph. But we can differentiate if we consider the function

$$f(x) = 20x - x^2,$$

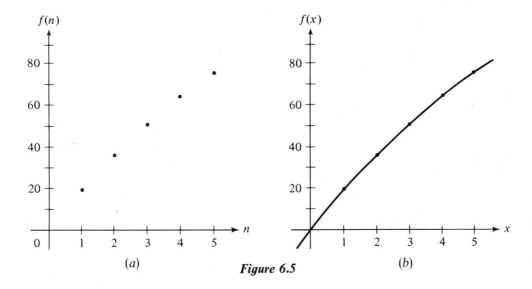

Figure 6.5

which is defined for all real numbers x. The graph is given in Figure 6.5(b).

$$f'(x) = 20 - 2x, \qquad f''(x) = -2.$$

The first derivative tells us that $x = 10$ is a critical value, and the second derivative tells us that it is a maximum. Thus the maximum value of $f(n)$ is 100 when $n = 10$.

Example 5

An efficiency expert determines that the time t, in hours, needed to perform a given job is

$$t = \frac{n^2 + 5n + 30}{n^2},$$

where n is the number of workers assigned to the job. Assuming that all workers are paid at the same rate, how many should be used to minimize the cost?

We want to minimize the cost C by varying the number n of workers. Suppose the workers are paid k dollars per hour. Then the cost is

$$C = knt = k\,\frac{n^2 + 5n + 30}{n}.$$

Of course, n must be a positive integer; but we ignore this restriction temporarily.

Now we minimize C:

$$\frac{dC}{dn} = k\,\frac{n^2 - 30}{n^2}.$$

This is zero if $n = \pm\sqrt{30} = \pm 5.477$. Of course, negative values of n are completely meaningless; and the second derivative tells us that $n = \sqrt{30}$ makes C a minimum. Thus, the graph of C is as shown in Figure 6.6. Since $n = \sqrt{30} = 5.477$ is not an integer, it is not our solution; but we can easily see from Figure 6.6

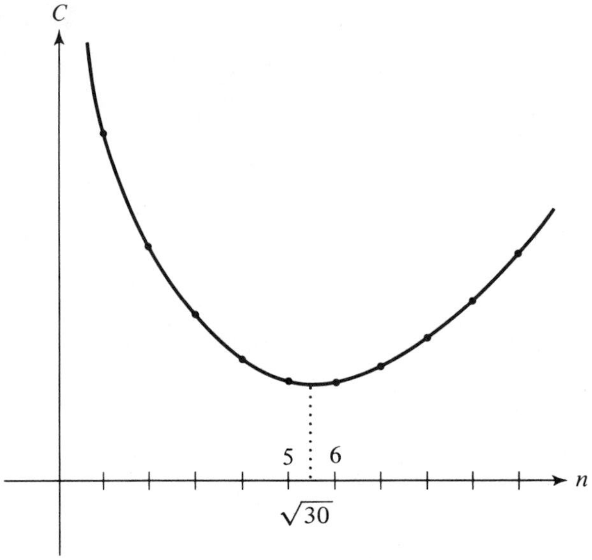

C

5 : 6

$\sqrt{30}$

n

Figure 6.6

that our solution is either $n = 5$ or $n = 6$. (Do not conclude that $n = 5$ gives the smaller value for C simply because $\sqrt{30}$ is closer to 5 than to 6.) Having narrowed the choice to two values of n, it is a simple matter to check: $C = 16k$ when $n = 5$; $C = 16k$ when $n = 6$. Thus the cost is a minimum when either 5 or 6 workers are used.

Problems

A

1. Find two numbers x and y whose sum is 48 and whose product is a maximum.
2. Find two numbers x and y whose sum is A such that the sum of their squares is a minimum.
3. A farmer wants to fence in 60,000 sq ft of land in a rectangular plot along a straight highway. The fence along the highway costs $1.00 per foot, while the fence for the other three sides costs $0.50 per foot. How much of each type of fence will have to be bought in order to keep expenses to a minimum? What is the minimum expense?
4. A farmer wants to fence in 60,000 sq ft of land in a rectangular plot and then divide it in half with a fence parallel to one pair of sides. What are the dimensions of the rectangular plot that will require the least amount of fence?
5. A farmer wants to fence in 180,000 sq ft of land in a rectangular plot and then divide it into three equal plots with a pair of fences both parallel to the same pair of sides (see Figure 6.7(a)). What is the least amount of fence needed to accomplish this?
6. The farmer in Problem 5 wonders if it would be cheaper to fence the same area (though not necessarily having the same dimensions) into three equal plots using the plan of Figure 6.7(b). Would this plan cost more than, less than, or the same amount as the earlier plan?

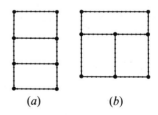

(*a*)   (*b*)

Figure 6.7

7. Suppose the farmer of Problem 5 wants to subdivide into four equal plots by either of the plans of Figure 6.8. Which plan (if either) is cheaper and what is the least amount of fence needed?

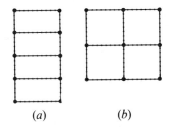

(a)   (b)

Figure 6.8

8. A square piece of cardboard with each side 12 in. long has a square cut out at each corner. The sides are then turned up to form an open box. Find the side of the cut-out square that will produce a box of maximum volume.

9. A rectangular box, open at the top, with a square base, is to have a volume of 4000 cu in. What must be its dimensions if the box is to require the least possible material?

10. If the box of Problem 9 is to be closed at the top, what must be its dimensions?

B

11. Find the dimensions of the rectangle of greatest area with its base on the x axis and its other two corners above the x axis and on $y = 9 - x^2$.

12. Find the dimensions of the trapezoid of greatest area inscribed in $y = 16 - x^2$ and having its longer base on the x axis.

13. Find the dimensions of the isosceles triangle of greatest area that can be inscribed in a circle of radius R.

14. Find the dimensions of the rectangle of greatest area that can be inscribed in a circle of radius $\sqrt{2}$.

15. Find the dimensions of the trapezoid of greatest area inscribed in a circle of radius R and having one base a diameter of the circle.

16. Find the dimensions of the isosceles triangle of least area that can be circumscribed about a circle of radius R.

17. Find the volume of the largest right circular cylinder that can be inscribed in a right circular cone of radius 3 and height 9.

18. Find the volume of the largest right circular cone that can be inscribed in a sphere of radius 3.

19. Find the dimensions of the right circular cylinder of greatest lateral surface that can be inscribed in a sphere of radius R.

20. Find the volume of the smallest right circular cone that can be circumscribed about a sphere of radius R.

21. A tin can is to be made with a capacity 2π cu in. What dimensions for it will require the smallest amount of tin?

22. Find the point on $y^2 = 4x$ closest to (3, 0).

23. Find the point in the first quadrant on $xy = 3$ closest to $(-8, 0)$.

24. Use the second-derivative test to show that the distance d of Example 4 is the minimum distance. Repeat using $D = d^2$ (see the paragraph following Example 4).

25. Solve the problem of Example 3 by maximizing A^2 rather than A.

26. A printed page has 1-in. margins at the top and bottom and 3/4-in. margins at the sides. If the area of the printed portion is to be 44 in.², what should the dimensions of the page be to use the least paper?

27. A 10-in. wire is cut into two pieces. One of them is bent into a square and the other into an equilateral triangle and the sum of the areas determined. What is the maximum such area?

28. Recall the traditional Christmas carol, *The Twelve Days of Christmas.*

> On the first day of Christmas
> My true love gave to me
> A partridge in a pear tree.
>
> On the second day of Christmas
> My true love gave to me
> Two turtle doves
> And a partridge in a pear tree.
>
> On the third day of Christmas
> My true love gave to me
> Three French hens,
> Two turtle doves,
> And a partridge in a pear tree.

It continues in the same vein through the twelve days of Christmas (in which a total of 364 gifts are presented by "true love"). The year after receiving these gifts the girl decided to reciprocate, since it was a leap year; but she continued throughout the entire year. Let us label the gifts $G_1, G_2, G_3, \ldots, G_{366}$, where n G_n's are presented on the n-th day of the year and every day thereafter. Which gift(s) did her "true love" receive most of, and how many? Which did he receive the least of, and how many?

29. From past experience, it is known that when a colony of bacteria of a certain type is sprayed with α-toxin, the number of bacteria present is related to the time by the formula

$$x = \frac{k(t + a)}{t^2 + b},$$

where x is the number of bacteria, t is the number of minutes after spraying, and a, b, and k are constants. It is observed that a colony of 20,000 bacteria reaches its maximum size of 25,000 one minute after spraying. Predict the size of the colony after 30 minutes.

30. The strength of a wooden beam of a given length is proportional to its width and the square of its height. Find the dimensions of the strongest beam that can be cut from a circular log of diameter 4 ft.

31. A ship is anchored 4 miles off a straight shore. Opposite a point 9 miles down the coast, another ship is anchored 8 miles from the shore. A boat from the first ship is to land a passenger on the shore and then proceed to the other ship to pick up another passenger before returning. At what point along the shore should the boat land the passenger in order to run the shortest course?

32. A ray of light from A is reflected to B in XY (see Figure 6.9). Find the value of x (in terms of a, b and c) such that the total time from A to B is a minimum. Show that, for this value of x, $\alpha = \beta$. Compare with Problem 31.

33. A man is in a boat 4 miles off a straight coast. He wants to reach a point 10 miles down the coast in the least possible time. If he can row 4 miles per hour and run 5 miles per hour, where should he land the boat?

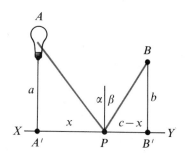

Figure 6.9

34. Suppose the point to be reached in Problem 33 is only 5 miles down the coast. Where should he land the boat?

35. Suppose the man of Problem 33 can both row and run at 5 miles per hour. Where should he land the boat?

36. A ray of light is to move from A to B (see Figure 6.10) in the least possible time. Its velocity above XY is v_1; its velocity below XY is v_2. Show that

$$\frac{\sin \alpha}{\sin \beta} = \frac{v_1}{v_2}.$$

When v_1 is the velocity of light in a vacuum and v_2 is the velocity of light in some other medium, v_1/v_2 is called the index of refraction of the medium.

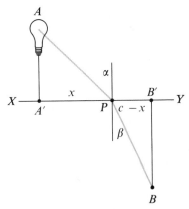

Figure 6.10

37. The current I in a voltaic cell is

$$I = \frac{E}{R + r},$$

where E is the electromotive force and R and r are the external and internal resistance, respectively. E and r are internal characteristics of the cell; they cannot be changed. The power developed is $P = RI^2$. Show that P is a maximum when $R = r$.

38. The efficiency of a screw is given by the formula

$$E = \frac{h(1 - h\mu)}{h + \mu},$$

where μ is the coefficient of friction and h is the tangent of the pitch angle of the screw. Find the value of h for which the efficiency is a maximum.

39. An impulse turbine consists of a high speed jet of water striking circularly mounted blades. The power P developed by such a turbine is directly proportional to the speed V of the jet, the speed U of the turbine and the speed of the jet relative to the turbine $V - U$. That is,

$$P = kVU(V - U).$$

For a given jet speed V, determine the turbine speed that will develop maximum power.

40. The cost of fuel used in propelling a ship varies as the cube of her speed, and is $12.80 per hour when the speed is 8 miles per hour. The other expenses are $50.00 per hour. Find the most economical speed and the minimum cost of a voyage of 1000 miles.

41. A widget manufacturer finds that the number of widgets that can be sold is inversely proportional to the square of the price charged for them. If it costs 50¢ to manufacture a widget, how should widgets be priced in order to maximize profits?

42. An automobile manufacturer finds that 50,000 cars can be sold if each is priced at $4000. However, the number sold increases by 30 for every $1 decrease in the price. The manufacturer has fixed costs of $20,000,000; in addition, it costs $2000 to produce each car. How should the cars be priced to maximize profits?

C

43. Find the shortest distance from the point (x_1, y_1) to the line $Ax + By + C = 0$.

44. The lower corner of a page of width k is folded over so as just to reach the inner edge of the page (see Figure 6.11). Find the width x of the part folded over when the length of the crease y is a minimum.

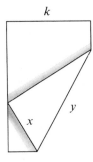

Figure 6.11

45. In Problem 44, find the width of the part folded over when the area of the triangle folded over is a minimum.
46. The intensity of light at a distance d from a source of intensity I is kI/d^2. If two light sources have intensities I_1 and I_2 and are at a distance d from each other, find the point between them where the intensity is the least.
47. A steel pipe 25 feet long is carried down a narrow corridor 5.4 feet wide. At the end of the corridor is a right-angle turn into a wider hall. How wide must the hall be in order to get the pipe around the corner? (Assume that the pipe bends just enough so that you may neglect its width.)
48. Two children want to meet at a point between their houses so that the total distance walked by them is a minimum. Where should they meet? Suppose there are three children with their houses on a line? Generalize to n children, assuming their houses are all on a line.

6.2

Related Rates

In many situations, two or more rates are related to each other in some way. For example, when a person is walking near a lamppost at night, the rate of walking and the rate at which the person's shadow is moving are related.

Problems involving related rates are solved by the following procedure.

1. List all the rates that are involved—those given and those we want to find.
2. If the rates are dx/dt, dy/dt, and dz/dt, then we must relate x, y, and z by an equation that holds true generally and not just at some particular time. The problem itself often suggests the equation, such as the area of a triangle or the volume of a sphere. Plane geometric problems often use the theorem of Pythagoras or relationships between the sides of similar triangles. The literals x, y, and z are essentially functions of time.

3. Remembering that the literals are functions of time, we differentiate both sides with respect to time. This results in dx/dt, dy/dt, dz/dt, and perhaps one or more of x, y, and z. We then substitute the known rates, as well as the particular values of x, y, and z that we have at the time in question, and solve for the unknown rate.

Example 1

A person 6 ft tall is walking away from a lamppost 15 ft high at the rate of 6 ft/sec. At what rate is the end of the person's shadow moving away from the lamppost?

We begin by drawing a diagram of the situation, as shown in Figure 6.12. Now we follow the procedure given above. First, we list all the rates in which we are interested. They are

$$\frac{dx}{dt} = 6 \text{ ft/sec}, \qquad \frac{dy}{dt} \quad \text{(unknown)}.$$

Next, we find an equation relating x and y. We can see from Figure 6.12 that triangles *ABC* and *DEC* are similar. Thus

$$\frac{y - x}{6} = \frac{y}{15}, \quad \text{or} \quad 3y = 5x.$$

Finally, we differentiate with respect to time, giving

$$3\frac{dy}{dt} = 5\frac{dx}{dt}$$

Substituting $dx/dt = 6$ gives

$$\frac{dy}{dt} = 10 \text{ ft/sec}.$$

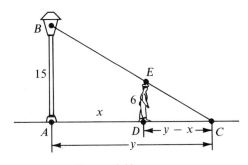

Figure 6.12

Example 2

The radius of a cylinder is decreasing at the rate of 4 ft/min, while the height is increasing at the rate of 2 ft/min. Find the rate of change of the volume when the radius is 2 ft and the height is 6 ft.

First, the rates in which we are interested are the two known rates,

$$\frac{dr}{dt} = -4 \text{ ft/min} \quad \text{and} \quad \frac{dh}{dt} = 2 \text{ ft/min},$$

and the unknown rate, dV/dt.

Now we want to relate r, h, and V. This is easily done since we have a formula for the volume of a right circular cylinder in terms of its radius and height:

$$V = \pi r^2 h.$$

Differentiating with respect to time (remember V, r, and h are functions of time), we have

$$\frac{dV}{dt} = \pi r^2 \frac{dh}{dt} + 2\pi r h \frac{dr}{dt}.$$

Now we substitute the values of r, h, dr/dt, and dh/dt. All four of these are given in the problem. Thus

$$\frac{dV}{dt} = \pi r^2 \frac{dh}{dt} + 2\pi r h \frac{dr}{dt}$$

$$= \pi(2)^2(2) + 2\pi(2)(6)(-4), \quad \text{when } r = 2 \text{ and } h = 6,$$

$$= -88\pi \ \text{ft}^3/\text{min}.$$

The minus here indicates that the volume is decreasing.

Example 3

Sand is poured onto the top of a conical pile at the rate of 10 ft^3/min. The coefficient of friction of the sand is such that the height and the radius are always the same. At what rate is the height increasing when the pile is 8 ft high?

We are given the rate of change of the volume of the pile, $dV/dt = 10$ ft^3/min; and we want to find the rate of increase of the height, dh/dt. Thus we want to relate V and h. Since we are dealing with a cone, we know that the volume is given by

$$V = \frac{1}{3}\pi r^2 h.$$

But we are also given that the height and radius are always equal. Thus

$$V = \frac{1}{3}\pi h^3.$$

Differentiation with respect to time gives

$$\frac{dV}{dt} = \pi h^2 \frac{dh}{dt}.$$

At the moment when $h = 8$,

$$10 = \pi(8)^2 \frac{dh}{dt},$$

$$\frac{dh}{dt} = \frac{10}{64\pi} \approx 0.05 \ \text{ft/min}.$$

(The symbol $\approx$ indicates approximate equality.)

Problems

A

1. The three dimensions of a box are increasing at the rates of 5 in./min, 7 in./min, and 2 in./min. At what rate is the volume increasing at the moment when the box is a cube with edge 10 in.? At what rate is the surface area increasing?

2. The base of a triangle is increasing at the rate of 3 in./min, while the altitude is decreasing at the same rate. At what rate is the area changing when (a) the base is 10 in. and the altitude is 6 in.?   (b) the base is 6 in. and the altitude is 10 in.?

3. The area of a circle is increasing at the rate of π in.^{2}/min. At what rate is the radius increasing when the area is 4π in.^{2}?

4. A 10-ft ladder is leaning against a vertical wall. If the bottom of the ladder is pulled away from the wall at the rate of 2 ft/sec, at what rate is the top of the ladder moving down the wall when the top is 6 ft from the ground?

B

5. A kite is 100 ft high. There are 260 ft of string out and it is being paid out at the rate of 5 ft/sec. If this results in the kite being carried along horizontally, what is the horizontal speed of the kite?

6. Helium is pumped into a spherical balloon at the rate of 3π ft^3/min. At what rate is the radius increasing (a) when the radius is 3 ft?   (b) when the volume is 36π ft^3?

7. In Problem 6, find the rate at which the surface area is increasing (a) when the radius is 2 ft and (b) when the volume is 36π ft^3.

8. A person 6 ft tall is walking away from a lamppost 18 ft high at the rate of 4 ft/sec. At what rate is the end of the person's shadow moving away from the lamppost? At what rate is the end of the shadow moving away from the person?

9. A light inside a garage is 10 ft above the floor and 6 ft behind the door opening. If the overhead door is descending at the rate of 1 ft/sec, with the bottom of the door remaining in the same vertical plane, at what rate is the door's shadow approaching the garage when the bottom of the door is 2 ft above the floor?

10. Ship A is steaming north at 10 miles per hour. Ship B, which is 8 miles west of ship A, is steaming east at 12 miles per hour. At what rate is the distance between them changing? At what rate will it be changing 1 hour from now?

11. A gas has a volume of 1 liter at 1 atmosphere pressure. If the pressure is increasing at the rate of 0.1 atmosphere/min, at what rate is the volume changing? Assume the gas to be ideal—that is, it satisfies the equation $Pv = k$, where k is a constant.

12. If the gas of Problem 11 is not ideal, but satisfies the equation $Pv = 22.41 + 0.01P$, at what rate is the volume changing?

13. When a gas is compressed adiabatically (with no gain or loss of heat), it satisfies the equation $Pv^{1.4} = k$, where k is a constant. At a given instant the pressure P is 40 lb/in.^{2} and the volume is 28 in.^{3} and decreasing at the rate of 2 in.^{3}/min. At what rate is the pressure changing?

14. The sides of an equilateral triangle are increasing at the rate of 3 in./min. At what rate is the area increasing when the side is 8 in.?

15. The area of an equilateral triangle is increasing at the rate of 4 in.^{2}/min. At what rate is one side increasing when the area is 16 in.^{2}?

16. A point is moving along the curve $y = x^2$ in such a way that its x coordinate is increasing at the rate of 3 units per minute. At what rate is y changing (a) when $x = 0$?   (b) when $x = 1$?   (c) when $x = 2$?

17. A point is moving along the curve $y = \sqrt{x}$ in such a way that its x coordinate is increasing at the rate of 3 units per minute. At what rate is y changing (a) when $x = 1$?   (b) when $x = 4$?

18. A point is moving along the curve $y = \sqrt{x}$ in such a way that its y coordinate is increasing at the rate of 3 units per minute. At what rate is its x coordinate changing (a) when $x = 1$?   (b) when $x = 4$?

19. A point is moving along the curve $y = \sqrt{x}$ in such a way that its x coordinate is increasing at the rate of 2 units per minute. At what rate is its slope changing (a) when $x = 1$?   (b) when $x = 4$?

20. A tank has the shape of an inverted cone with height 10 ft and radius 4 ft. Water is being pumped into it at the rate of 2π ft^3/min. How fast is the depth of the water increasing when it is 5 ft deep?

21. A trough that is 12 ft long and 2 ft high is 2 ft wide at the top and has triangular ends. If water is put in at the rate of 1 ft³/min, how fast is the depth increasing when it is 1.5 ft deep?

22. A trough with trapezoidal ends is 20 ft long, 2 ft high, 3 ft wide at the top, and 2 ft wide at the bottom. If water is being pumped in at the rate of 2 ft³/min, how fast is the depth increasing when the water is 1 ft deep?

C

23. If ship A in Problem 10 is steaming 30° west of north, at what rate is the distance between ships A and B changing? At what rate will it be changing 1 hour from now?

24. The walls of an A-frame cottage make an angle of 60° with the ground. A 12-ft ladder leans against the wall with the bottom of the ladder touching the wall. If the bottom of the ladder is pulled away from the cottage at 1 ft/sec, how fast is the top of the ladder moving down the wall when the ladder is just starting to move?

6.3

The Differential

In our first encounter with the so-called differential notation for the derivative, dy/dx, in Chapter 4 we did not discuss the separate meanings of dy and dx. We want to consider these now. First, however, it will help to learn some new notation in connection with the derivative.

By definition, the derivative, when it exists, is given by

$$f'(x) = \lim_{h \to 0} \frac{f(x+h) - f(x)}{h}.$$

The symbol h in this expression is sometimes called a dummy variable because it does not appear in the final result which, in this case, is the derivative. Any other symbol can be used instead of h to obtain the same result. A particular symbol that is sometimes used is Δx (delta x). This is not a product but is regarded as a single symbol. Thus

$$f'(x) = \lim_{\Delta x \to 0} \frac{f(x+\Delta x) - f(x)}{\Delta x}.$$

If we let $y = f(x)$, the right-hand expression is further simplified by the introduction of another symbol, Δy, given by

$$\Delta y = f(x + \Delta x) - f(x).$$

Thus

$$f'(x) = \lim_{\Delta x \to 0} \frac{\Delta y}{\Delta x}.$$

As mentioned above, Δx and Δy are not to be looked upon as products; in fact, the delta notation was used originally to represent a difference. Thus Δx and Δy may be thought of as differences of pairs of values of x and y, respectively. This is evident when we recall that $\Delta y/\Delta x$ is the slope of a line; and we have already seen that the slope is the difference between two y coordinates divided by the difference between the corresponding x coordinates. This is seen graphically in Figure 6.13. Note that the value of Δy depends upon the values of both x and Δx and that

$$y + \Delta y = f(x + \Delta x).$$

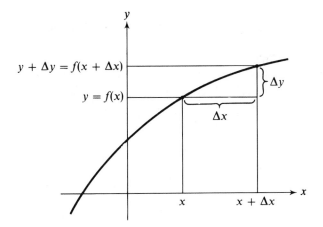

Figure 6.13

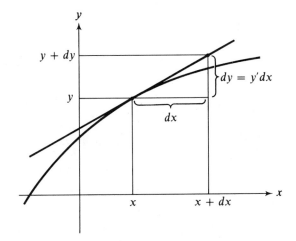

Figure 6.14

Now let us consider the differential. Again we assume that $y = f(x)$. We introduce another new variable, dx, called the differential of x. As in the case of Δx, dx is not to be viewed as a product, but rather as a single symbol. Corresponding to this is another new variable, dy, called the differential of y, which is dependent upon both x and dx in the following way:

$$dy = f'(x)\, dx.$$

In our discussion of the derivative, dy/dx was not considered as a quotient, since we had not given meanings to dy and dx individually. Now that we have meanings for them, we see that if $dx \neq 0$, then

$$\frac{dy}{dx} = f'(x).$$

If we compare Figure 6.14 with Figure 6.13, we see that dx and dy are related to the tangent line to $y = f(x)$ at (x, y) in the same way that Δx and Δy are related to the original curve.

Since Δy is defined to be $f(x + \Delta x) - f(x)$, Δy can be found only when both $f(x)$ and $f(x + \Delta x)$ are defined. Thus, for a given value of x in the domain of f, Δx is restricted to those numbers for which $x + \Delta x$ is also in the domain of f. Note that this restriction is not necessary for dx. Since dx and dy are related to each other by the tangent line (see Figure 6.14), which (provided $f'(x)$ exists) has an unrestricted domain, dx may be any real number. Of course, x must be in the domain of f' as well as in that of x. Actually, the applications of the differential we consider later in this section require that Δx and dx be "small."

Since $dy = y' \, dx$, finding the differential of a given function is simply a matter of finding the derivative and multiplying by dx.

Example 1

Given $y = x^3 - 2x^2$, find dy.

Since $y' = 3x^2 - 4x$, $dy = (3x^2 - 4x) \, dx$.

In the case of implicit functions, an alternate procedure may be followed.

Example 2

Given $y^2 - xy + 2x^2 = 5$, find dy.

First we find the differential on each side,

$$2y \, dy - x \, dy - y \, dx + 4x \, dx = 0.$$

We now solve for dy.

$$(2y - x) \, dy = (y - 4x) \, dx,$$

$$dy = \frac{y - 4x}{2y - x} \, dx.$$

Let us now consider the application of the differential to the problem of approximations and small errors. We note first of all that the terms "approximation" and "small errors" are not precise terms. For example we might ask: How small must an error be to qualify as a small error? Any answer must depend upon the size of the quantity we are measuring. But if a small error is defined as a certain percentage of the quantity measured, we cannot know the precise boundary between an error that is small and one that is not unless we know the exact value of the quantity in question. In that case it is unlikely that we shall care to talk about errors at all.

Thus we see that "approximation" and "small" are certainly vague terms. Nevertheless we have some idea—however vague it might be—of what we mean by an approximation and by small errors. So we shall call upon this vague idea and use it as if we knew exactly what is meant by approximations and small errors.

Let us see how the differential can help us compute small errors and make approximations. We have already seen that dx and dy are differences between pairs of values of x and y, respectively, determined by the tangent line to $y = f(x)$. Similarly, Δx and Δy are differences between pairs of values of x and of y, respectively, determined by $y = f(x)$ at x. We may select any value we want for dx; let us choose $dx = \Delta x$. The relationship between dy and Δy is given in Figure 6.15. Although dy and Δy are not necessarily the same, dy is a reasonable approximation of Δy provided $dx = \Delta x$ is "small" and $f(x)$ does not vary wildly for small values of x. Thus we can use differentials to compute dy as an approximation of Δy. This can be formalized in the following theorem.

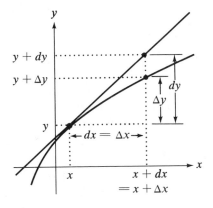

Figure 6.15

Theorem 6.1

If the function f is differentiable at $x = x_0$ and continuous on some interval (a, b) containing x_0, Δx is a number such that $a - x_0 < \Delta x < b - x_0$, and $dx = \Delta x$, then

$$\lim_{\Delta x \to 0} \frac{\Delta y - dy}{\Delta x} = 0.$$

Proof

Referring back to the meanings of Δy and dy, we have

$$\lim_{\Delta x \to 0} \frac{\Delta y - dy}{\Delta x} = \lim_{\Delta x \to 0} \frac{f(x_0 + \Delta x) - f(x_0) - f'(x_0)\, dx}{\Delta x}$$

$$= \lim_{\Delta x \to 0} \left[\frac{f(x_0 + \Delta x) - f(x_0)}{\Delta x} - \frac{f'(x_0)\, dx}{dx} \right]$$

$$= \lim_{\Delta x \to 0} \left[\frac{f(x_0 + \Delta x) - f(x_0)}{\Delta x} - f'(x_0) \right]$$

$$= f'(x_0) - f'(x_0) = 0.$$

This theorem not only says that $dy \to \Delta y$ as $\Delta x \to 0$, but that the difference $\Delta y - dy$ approaches zero more rapidly than Δx does.

Example 3

The edge of a cube was found to be ten inches. It is felt that this measurement is accurate to within 0.02 in. If the volume of the cube is computed using this measurement, what is the maximum error in the volume?

$$V = x^3.$$

We are given $x = 10.00$ with error $|\Delta x| \leq 0.02$. We want to find the maximum value of $|\Delta V|$ corresponding to Δx. Setting

$$|dx| = |\Delta x| \leq 0.02,$$

we compute dV.

$$\begin{aligned} |dV| &= |3x^2 \, dx| \\ &\leq 3(10.00)^2(0.02) \\ &= 6. \end{aligned}$$

Thus $|\Delta V| \approx |dV| \leq 6$ (we use the symbol $\approx$ for approximate equality).

How good is the approximation we just made? How does it compare with the actual value of $|\Delta V|$? We can easily compute $|\Delta V|$.

$$\begin{aligned} |\Delta V| &= |(x + \Delta x)^3 - x^3| \\ &\leq (10.02)^3 - (10.00)^3 \\ &= 6.012008. \end{aligned}$$

We see that $|dV| \leq 6$ is correct to two significant figures and it differs from $|\Delta V|$ by only 1 in the third figure. It might be noted that, while we computed $|\Delta V|$ exactly, one would hardly be interested in more than the first two or three figures.

We see that, for the small value of Δx given, this method is quite accurate for estimating ΔV. This method becomes less accurate as Δx increases. For instance, suppose you use no measuring device to find x but simply estimate it to be ten inches with an error of at most one inch. The values of $|dV|$ and $|\Delta V|$ are:

$$|dV| \leq 300, \qquad |\Delta V| \leq 331.$$

We see that $|dV|$ is not nearly so good an estimate of $|\Delta V|$ in this case.

Approximating numbers by differentials is very much like determining errors by differentials. We have seen that if Δx represents a small error in determining x, then Δy, the corresponding error in y, is found by letting $dx = \Delta x$. Then

$$dy \approx \Delta y.$$

Similarly we may use $y + dy$ to approximate $y + \Delta y$.

Example 4

Approximate $\sqrt{63}$ by differentials.

What we want here is a number y corresponding to $x = 63$ in the equation

$$y = \sqrt{x}.$$

If $x = 64$, then $y = 8$. Unfortunately, x is not 64; it is 63. We have the error $\Delta x = -1$, so that $x + \Delta x = 63$. Now we want to find $y + \Delta y$ corresponding to $x + \Delta x = 63$. Again let us take $dx = \Delta x = -1$ and compute $y + dy$, instead of $y + \Delta y$.

$$y = \sqrt{x} \qquad\qquad dy = \frac{dx}{2\sqrt{x}}$$

$$= \sqrt{64} \qquad\qquad = \frac{-1}{2\sqrt{64}}$$

$$= 8. \qquad\qquad = -0.0625.$$

$$y + dy = 7.938.$$

We have rounded off $y + dy$ to only three decimal places—remember this method gives only an approximation; do not try to push it too far. The actual value of $\sqrt{63}$ to three decimal places is 7.937.

An expression that is often more meaningful than the error is the relative error or the percentage error.

$$\text{relative error in } x = \frac{\text{error in } x}{x}.$$

$$\text{percentage error in } x = \frac{\text{error in } x}{x} \cdot 100.$$

Example 5

If the percentage error in measuring the edge of a cube is 3%, what is the percentage error in finding its volume?

$$V = x^3, \qquad dV = 3x^2\, dx,$$

$$\frac{dV}{V} = \frac{3x^2\, dx}{x^3} = 3 \cdot \frac{dx}{x},$$

$$\frac{dV}{V} \cdot 100 = 3 \cdot \frac{dx}{x} \cdot 100 = 3 \cdot 3\% = 9\%.$$

Problems

A

In Problems 1–10, find the differential dy.

1. $y = x^3 - x^2$.
2. $y = 2x^4 + 3x$.
3. $y = \dfrac{x + 4}{2x - 3}$.
4. $y = \dfrac{3x - 2}{x + 4}$.
5. $y = (x^2 + 1)^4$.
6. $y = x^3(3x - 4)^2$.
7. $x^2 + 4y^2 = 4$.
8. $y^2 + 2xy = 3$.
9. $x^3 - 3x^2y + 3xy^2 = 5$.
10. $\sqrt{x} + \sqrt{y} = 1$.

11. What is the error in determining the surface area in Example 3?
12. The radius of a circle was measured and found to be 8.19 ± 0.02 cm. Approximate the maximum error in the circumference. In the area.
13. The diameter of a circle was measured and found to be 7.38 ± 0.03 cm. Approximate the maximum error in the circumference. In the area.
14. The side of a square was measured and found to be 9.48 ± 0.01 in. What is the greatest error in computing the area? Approximate the error by differentials and find the exact value. Compare.
15. The relationship between centigrade, °C, and Fahrenheit, °F, temperatures is given by the formula

$$5(°F) = 9(°C) + 160.$$

If a Fahrenheit thermometer can be read to the nearest 1°, find the corresponding error in the centigrade temperature.

16. The radius of a sphere is measured and found to be 5.3 ± 0.2 cm. Approximate the error in the volume. In the surface area.

B

17. An unknown electrical resistance R is determined with a Wheatstone bridge by adjusting two things: a compensating resistance r read off a resistance box and the position x of a key along a slide wire of length l. Resistance R is then

$$R = r\frac{x}{l - x}.$$

Suppose $r = 30$ ohms, $l = 100$ cm, and $x = 50 \pm 0.1$ cm. Find R and approximate the error, ΔR.

18. Suppose the following figures are obtained from a Wheatstone bridge (see Problem 17): $r = 270$ ohms, $l = 100$ cm, and $x = 10 \pm 0.1$ cm. Find R and approximate the error ΔR. What do the results of this problem and Problem 17 suggest concerning the operation of a Wheatstone bridge for maximum precision?

In Problems 19–26, use differentials to approximate the numbers given.

19. $\sqrt{83}$.
20. $\sqrt{35}$.
21. $\sqrt[3]{25}$.
22. $\sqrt[3]{65}$.
23. $\sqrt[4]{17}$.
24. $\sqrt[5]{245}$.
25. $\dfrac{1}{\sqrt{27}}$.
26. $\dfrac{1}{\sqrt[3]{120}}$.

27. Show that if $y = mx$, the relative error in y is the same as the relative error in x.
28. Show that if $y = kx^n$, the relative error in y is n times the relative error in x.

29. If the volume of a cube is to be determined with a percentage error no greater than 3%, what is the greatest percentage error that can be tolerated in determining the edge?

30. If the volume of a sphere is to be determined with an error no greater than 3%, what is the greatest percentage error that can be tolerated in determining the radius?

31. The height and diameter of a right circular cone are known to be equal. If the volume is to be determined with an error no greater than 2%, what is the greatest percentage error that can be tolerated in determining the height?

32. If the lateral surface of the cone of Problem 31 is to be determined with an error no greater than 2%, what is the greatest percentage error that can be tolerated in determining the height?

C

33. Show that, for small values of h, $\sqrt[n]{1 + h}$ is approximated by $1 + (h/n)$.

34. Use the result of Problem 33 to approximate the numbers of Problems 19 and 21. Compare your results with those found in Problems 19 and 21.

6.4

Applications to Economics

Calculus has long been used in the physical sciences and engineering; in fact, physical applications prompted much of its early development. More recently calculus has been applied to business and the biological and social sciences. Let us consider some applications to economics.

If x units of a given commodity are produced, the cost of production, of course, depends upon the size of x. The *total cost function* gives the cost C of producing the x units as a function of x. A typical cost function is

$$C = 0.01x^2 + 20x + 400, \quad x \geq 0.$$

(We normally omit the proviso $x \geq 0$, because negative values of x are meaningless in the present context. Unless something is said to the contrary, it is understood that $x \geq 0$.) Notice that $C = 400$ even when no units of the commodity are produced. The constant term, 400, is called the *fixed cost*; it represents such cost items as rent, depreciation on machinery, and so on, which continue even when nothing is produced. The sum of the remaining terms, which depends upon x, is called the *variable cost*. The *average cost Q* (or the cost per unit) is simply C/x. For our typical cost function, the corresponding average cost function is

$$Q = \frac{C}{x} = 0.01x + 20 + \frac{400}{x}.$$

An obvious application of calculus presents itself at this point.

Example 1

If the cost function is $C = 0.01x^2 + 20x + 400$ (where C is in dollars), find the minimum cost per unit. How many units should be produced to attain this minimum?

$$Q = 0.01x + 20 + \frac{400}{x}, \qquad Q' = 0.01 - \frac{400}{x^2}, \qquad Q'' = \frac{800}{x^3}.$$

By setting $Q' = 0$, we find the critical value $x = 200$, for which $Q = 24$. Since $Q'' = 10^{-4} > 0$ when $x = 200$, this is a minimum. Thus $24 is the minimum cost per unit, and this is attained when 200 units are produced.

While it is reasonable to want to minimize the average cost, it is more likely that a businessperson will be interested in maximizing profit; and the value of x that gives the minimum average cost is often not the value that maximizes the profit. To determine the profit we need not only the cost function but also the revenue function. The *total revenue function* gives the revenue (or income) R as a function of the number of units x of the commodity sold. A typical revenue function is

$$R = 40x - 0.0525x^2.$$

Finally the profit P is given by

$$P = R - C.$$

Let us see the effect of considering both the cost and revenue functions.

Example 2

Determine the number of units of a commodity that should be produced to maximize the profit when the total cost and revenue functions are

$$C = 0.01x^2 + 20x + 400,$$
$$R = 40x - 0.0525x^2.$$

$$P = R - C = (40x - 0.0525x^2) - (0.01x^2 + 20x + 400)$$
$$= -0.0625x^2 + 20x - 400;$$
$$P' = -0.125x + 20;$$
$$P'' = -0.125.$$

The first derivative gives the critical value $x = 160$, and the second derivative assures us that it gives a maximum.

Let us compare the results of Examples 1 and 2. The total cost function was the same in both cases. In Example 1 we saw that producing 200 units of the commodity resulted in the lowest cost per unit, but Example 2 showed that the profit is a maximum when only 160 units are produced. Thus the maximum profit and the minimum cost per unit do not necessarily occur together at the same production level. This discrepancy is explained by demand equations. The *demand* for a commodity is the number of units of it that can be sold in a given period of time. The demand for a commodity and its price are normally related, with the demand increasing as the price decreases. Thus a *demand equation* gives the relationship between the price p and the number x of units of the commodity that can be sold at that price. The equation

$$p = 400 - 2x$$

is a typical demand equation. Because the demand goes up when the price goes down, dp/dx is negative for all values of x in any demand equation.

Example 3

The demand equation for a given commodity is $p = 100 - 0.005x$, while the total cost function is $C = 0.02x^2 + 50x + 1000$. Determine how many units must be produced in order to maximize the profit.

$$R = px = 100x - 0.005x^2,$$
$$P = R - C = (100x - 0.005x^2) - (0.02x^2 + 50x + 1000)$$
$$= -0.025x^2 + 50x - 1000;$$
$$P' = -0.05x + 50 = 0;$$
$$x = 1000.$$

Since $P'' = -0.05$, we are assured that 1000 units give the maximum profit.

It might be pointed out here that the last two examples make it appear that one would never have an occasion to minimize the average cost, as we did in Example 1. This is not the case. In some instances the price cannot be varied; for example, the price may be fixed by law, as in the case of some utilities and transportation, or the price may be determined by the total market conditions which are not affected by any one producer, as in the case of agricultural production. Moreover the cost function and demand equations are not easily determined. They require empirical data and market research, and they do not remain unchanged from one month to another. This is especially true in the case of the demand equation, which determines the total revenue function. Thus one might have to base decisions on a knowledge of the cost function alone.

Another economic tool is the marginal function. We have already seen that the total cost function gives the total cost C of production of x units of a commodity as a function of x. The *marginal total cost function* is, then, dC/dx. Similarly the *marginal revenue, marginal profit*, and *marginal average cost functions* are dR/dx, dP/dx, and dQ/dx, respectively. These give estimates of the changes in total cost, total revenue, profit, and average cost that result from a small change in production.

Example 4

The total cost function is $C = 0.05x^2 + 20x + 5000$. Find the average cost function and the marginal average cost function. If the level of production is currently 1000 units, will a small increase in production result in an increase or a decrease in the average cost?

$$Q = \frac{C}{x} = 0.05x + 20 + \frac{5000}{x}, \qquad \frac{dQ}{dx} = 0.05 - \frac{5000}{x^2}.$$

When $x = 1000$, $dQ/dx = 0.045$. Since this is positive, an increase in x results in an increase in Q.

It might be noted that the marginal functions may be looked upon in the same way that we considered the differential in the last section. We are using the tangent line rather than the curve itself to estimate small changes. For instance, in Example 4 the differential form gives $dQ = 0.045 \, dx$ when $x = 1000$. Thus a small increase in x implies that $dx > 0$; this in turn implies that $dQ > 0$. Q is increasing.

Since many new terms and symbols have been introduced in this section, we summarize them in the following table.

Symbol	Meaning
C	Cost of producing x units.
p	Price—the demand function relates price and the number of units x sold at that price.
R	Revenue from selling x units; $R = px$.
P	Profit; $P = R - C$.
Q	Average cost; $Q = C/x$.
dC/dx	Marginal cost.

Problems

A

1. A farmer determines that the total cost (in dollars) of producing x bushels of wheat is $C = 0.000008x^2 + 0.08x + 3200$. How many bushels should be produced to give the smallest cost per bushel?

2. The total cost and revenue functions for a certain commodity are
$$C = 0.05x^2 + 20x + 1000 \quad \text{and} \quad R = 50x - x^2.$$
For what value of x is the profit a maximum?

3. A theater manager determines that the number x of tickets sold for each performance is related to their dollar price p by the equation $p = 3 - 0.000004x^2$ What price should be charged to maximize the total revenue?

4. The total dollar cost of manufacturing x gizmos is $C = 25x + 4000$. The demand equation for gizmos is $p = 150 - 0.5x$. How many gizmos should be manufactured and what should be their selling price in order to maximize the profit?

5. A radio manufacturer determines that the total cost C (in dollars) of manufacturing x radios is $C = 0.004x^2 + 25x + 4000$ and the demand equation is $p = 60 - 0.01x$. What should be charged for the radios in order to maximize profit?

6. The total revenue function for a certain commodity is $R = 90x - 0.2x^2$. Find the marginal revenue (a) when $x = 200$ and (b) when $x = 400$.

7. The total cost function for a commodity is $C = 0.02x^2 + 16x + 4000$. Find the marginal cost and the marginal average cost when $x = 100$.

8. The total cost and revenue functions for a given commodity are
$$C = 0.03x^2 + 60x + 4000 \quad \text{and} \quad R = 90x - 0.15x^2.$$
Find the marginal cost, revenue, and profit when $x = 100$.

9. A given commodity can be manufactured at a total dollar cost of $C = 30x + 1000$, where x is the number produced. If the demand function is $p = 70 - 0.1x$, what price gives a maximum profit? What is the production level and profit at that price?

10. Suppose the fixed cost in Problem 9 increases by $50. What price gives a maximum profit? What is the production level and profit at that price?

B

11. Suppose that a tax of $1.00 per item is imposed on the manufacturer in Problem 9. What price gives a maximum profit? What is the production level and profit at that price?

12. Suppose that a tax of 10% of the sale price is imposed on the manufacturer in Problem 9. What price gives a maximum profit? What is the production level and profit at that price?

13. A toaster manufacturer has fixed costs of $100 per week and variable costs of $12 per toaster. The weekly demand for the toasters is related to the price p by the equation $x = 400 - 10p$. What is the maximum weekly profit?

14. Suppose the theater of Problem 3 seats only 450 people. What price should be charged for tickets?

15. Show that if the cost function is linear, there is no minimum average cost, but the average cost decreases as production increases.

16. If the total cost function is $C = ax + b$, where a and b are both positive, show that the marginal cost is always positive, while the marginal average cost is always negative.

17. Show that the marginal revenue and the marginal cost functions are equal when the profit is a maximum.

C

18. Show that the average cost and the marginal cost are equal when the average cost is a minimum.

19. Show that the average cost is a minimum when the average cost equals the marginal cost and C'' is positive.

6.5

Antiderivatives

We now consider the derivative problem in reverse; that is, given the derivative of a function, find the function. This is called *antidifferentiation*. In order to define this term, let us first define an antiderivative.

Definition

*If F is a function with domain D, then f is an **antiderivative** of F if $f'(x) = F(x)$ for all x in D.*

Note that we have defined *an* antiderivative, not *the* antiderivative of F. For example, if $F(x) = 2x$, then an antiderivative is $f(x) = x^2$ since $d/dx \, (x^2) = 2x$. But the derivative of $(x^2 + 1)$ is $2x$ and $d/dx \, (x^2 - 3) = 2x$; in fact, $d/dx \, (x^2 + C) = 2x$ no matter what number C represents. While it is certainly true that if $f(x) = x^2 + C$, then $f'(x) = 2x$, this fact does not necessarily imply that if $f'(x) = 2x$, then $f(x) = x^2 + C$ for some value of C. There still remains the possibility that there is some other expression not in the form $x^2 + C$ whose derivative is $2x$. This question is answered by the following theorem.

Theorem 6.2

If f and g are functions such that $f'(x) = g'(x)$ for all x, then $f(x) - g(x)$ is a constant.

The proof of this theorem (see Appendix B) requires the mean-value theorem, which we consider later (page 573). For now, we shall assume the theorem is true. Once we have found one antiderivative, we know that any other differs from it by a constant. Thus if $F(x) = 2x$,   then any antiderivative, $f(x)$, must be in the form $x^2 + C$. We now define antidifferentiation.

Definition

Antidifferentiation *of a function F is the process of finding all antiderivatives of F.*

There are still some unanswered questions about antidifferentiation. One question is, "Given a function F, must it have an antiderivative?" We are certainly in no position to answer such a question here. However, we shall restrict ourselves to functions that are known to have antiderivatives. To emphasize this, we shall use f' instead of F to indicate an antidifferentiable function.

Let us see if we can get a general formula for finding antiderivatives. Suppose $f'(x) = x^n$; what is $f(x)$? First of all

$$\frac{d}{dx}x^{n+1} = (n+1)x^n \quad \text{and} \quad \frac{d}{dx}\frac{x^{n+1}}{(n+1)} = x^n.$$

Theorem 6.3

If $f'(x) = x^n$ $(n \neq -1)$, then $f(x) = \dfrac{x^{n+1}}{n+1} + C.$

The case for which $n = -1$, giving $f'(x) = x^{-1} = 1/x$, will be considered in Chapter 12 (also see Problem 40).

Example 1

Given $f'(x) = \sqrt{x}$,   find $f(x)$.

$$f'(x) = x^{1/2}, \quad f(x) = \frac{x^{(1/2)+1}}{(1/2)+1} + C = \frac{2x^{3/2}}{3} + C.$$

This result is easily checked by differentiation, which gives the original expression $f'(x) = x^{1/2}$.

We can extend the use of this theorem by considering the results of Theorems 4.3 and 4.5. For example, Theorem 4.5 tells us that $c \cdot u(x)$ is a function whose derivative is $c \cdot u'(x)$; that is, $c \cdot u(x)$ is an antiderivative of $c \cdot u'(x)$. Thus all antiderivatives of $c \cdot u'(x)$ are of the form $c \cdot u(x) + C$. A similar argument can be used to state Theorem 4.3 in antiderivative form.

Theorem 6.4

If $f'(x) = c \cdot u'(x)$, then
$$f(x) = c \cdot u(x) + C;$$
if $f'(x) = u'(x) + v'(x)$, then
$$f(x) = u(x) + v(x) + C.$$

Of course, the statement concerning the antiderivative of a sum holds for differences as well as for the sum of three or more terms.

Example 2

Given $f'(x) = 3x$, find $f(x)$.

$$f(x) = 3 \cdot \frac{x^2}{2} + C = \frac{3x^2}{2} + C.$$

Example 3

Given $f'(x) = 4x + 2$, find $f(x)$.

$$f(x) = 4 \cdot \frac{x^2}{2} + 2x + C = 2x^2 + 2x + C.$$

Example 4

Given $f'(x) = \dfrac{x^4 - x^2 + 1}{x^2}$, find $f(x)$.

$$f'(x) = \frac{x^4 - x^2 + 1}{x^2} = x^2 - 1 + x^{-2};$$

$$f(x) = \frac{x^3}{3} - x + \frac{x^{-1}}{-1} + C$$

$$= \frac{x^3}{3} - x - \frac{1}{x} + C.$$

A question you may ask is, "When we have an antiderivative involving two or more terms, why is there not a constant for each term?" The answer is that the constants we get for each term can all be combined into a single constant. Thus, for Example 4, we have

$$f(x) = \frac{x^3}{3} + C_1 - x + C_2 - \frac{1}{x} + C_3$$

$$= \frac{x^3}{3} - x - \frac{1}{x} + (C_1 + C_2 + C_3)$$

$$= \frac{x^3}{3} - x - \frac{1}{x} + C.$$

If we are given nothing but $f'(x)$, we cannot find the original function $f(x)$, because we cannot evaluate C; the best we can do is find a family of functions all having the given derivative. But sometimes we are given additional information that allows us to choose the one member of the family we want.

Example 5

Given $f'(x) = 6x^2 + 6x - 4$ and $f(1) = 3$, find $f(x)$.

$$f(x) = 2x^3 + 3x^2 - 4x + C,$$
$$f(1) = 2 \cdot 1^3 + 3 \cdot 1^2 - 4 \cdot 1 + C.$$

Since $f(1) = 3$, we have

$$3 = 1 + C$$
$$C = 2.$$

Thus $f(x) = 2x^3 + 3x^2 - 4x + 2$.

Example 6

For $f''(x) = 32$, $f'(1) = 36$, and $f(1) = 16$, find $f(x)$.

$$f''(x) = 32, \qquad f'(x) = 32x + C_1.$$

Since $f'(1) = 36$, we have

$$36 = 32 + C_1$$
$$C_1 = 4,$$
$$f'(x) = 32x + 4,$$
$$f(x) = 16x^2 + 4x + C_2.$$

Since $f(1) = 16$, we have

$$16 = 16 + 4 + C_2$$
$$C_2 = -4.$$

Thus $f(x) = 16x^2 + 4x - 4$.

Example 7

For $f''(x) = 12$, $f(1) = 2$, and $f(2) = 15$, find $f(x)$.

$$f''(x) = 12, \qquad f'(x) = 12x + C_1, \qquad f(x) = 6x^2 + C_1x + C_2.$$

Using $f(1) = 2$ and $f(2) = 15$, we have

$$2 = 6 + C_1 + C_2 \qquad\qquad 15 = 24 + 2C_1 + C_2$$
$$C_1 + C_2 = -4, \qquad\qquad 2C_1 + C_2 = -9,$$
$$C_1 = -5,$$
$$C_2 = 1.$$

Thus $f(x) = 6x^2 - 5x + 1$.

The ability to find antiderivatives is a great aid when considering falling bodies. By Newton's law of universal gravitation, the force of attraction F of two objects is directly proportional to the product of their masses M_1 and M_2 and inversely proportional to the square of the distance r between them (actually between their centers of mass). In symbols,

$$F = G\frac{M_1 M_2}{r^2},$$

where G is a universal constant, called the gravitational constant. Furthermore, Newton's second law of motion says that the force acting upon a moving body is the product of the mass and acceleration of that body. If we take M to be the mass of the earth and m to be that of a falling object, then we have

$$F = G\frac{Mm}{r^2} = ma,$$

or

$$a = \frac{GM}{r^2}.$$

G and M are fixed; and, for problems in which the object falls only a short distance, r varies only a small amount. Thus for short falls, the acceleration due to gravity can be taken to be a constant (see Problem 39) which is independent of the mass of the falling object. At sea level this constant is approximately -32 ft/sec/sec (the minus indicates that acceleration is in the downward or negative direction).

Example 8

A ball is dropped from a 40-ft platform. Find its equation of motion.

From the foregoing discussion,

$$a = s'' = -32.$$

Furthermore at $t = 0$, $s = 40$ and $v = 0$ (since the ball is merely dropped rather than being thrown in one direction or the other). Therefore

$$v = s' = -32t + C_1.$$

Since $v = 0$ when $t = 0$, $C_1 = 0$.

$$v = -32t,$$
$$s = -16t^2 + C_2.$$

Since $s = 40$ when $t = 0$, $C_2 = 40$. Thus the equation of motion of the ball is

$$s = -16t^2 + 40, \quad 0 \le t \le ?.$$

The upper limit for t is the time at which the ball hits the ground. Setting $s = 0$, we have

$$-16t^2 + 40 = 0,$$
$$t^2 = 5/2,$$
$$t = \sqrt{10}/2.$$

The equation of motion is

$$s = -16t^2 + 40, \quad 0 \le t \le \sqrt{10}/2.$$

Example 9

A ball is thrown upward from ground level at a speed of 64 ft/sec. How high is it after 2 seconds? After 5 seconds?

First let us find its equation of motion, as in the previous example.

$$a = -32,$$
$$v = -32t + C_1.$$

Since $v = 64$ when $t = 0$, $C_1 = 64$.

$$v = -32t + 64,$$
$$s = -16t^2 + 64t + C_2.$$

Since the ball was thrown from ground level, $s = 0$ when $t = 0$. Therefore $C_2 = 0$, and the equation of motion is

$$s = -16t^2 + 64t, \quad 0 \leq t \leq ?.$$

Again the upper limit for t is the time at which the ball returns to ground level.

$$-16t^2 + 64t = 0,$$
$$16t(4 - t) = 0,$$
$$t = 0 \quad \text{or} \quad t = 4.$$

Thus the equation of motion is

$$s = -16t^2 + 64t, \quad 0 \leq t \leq 4.$$

At $t = 2$,

$$s = -16 \cdot 2^2 + 64 \cdot 2 = 64 \text{ ft.}$$

Since the ball hit the ground after 4 seconds, $s = 0$ when $t = 5$ (assuming that the ball does not bounce).

Problems

A

In Problems 1–8, find $f(x)$.

1. $f'(x) = 12x^3 - 6x^2 + 3$.
2. $f'(x) = 12x^3 + 3x^2 - 2x$.
3. $f'(x) = x^{3/2}$.
4. $f'(x) = x^{1/3}$.
5. $f'(x) = x^{-1/3}$, $f(1) = 2$.
6. $f'(x) = 3x^2$, $f(0) = 2$.
7. $f'(x) = 8x^3 + 4x - 3$, $f(2) = 19$.
8. $f'(x) = 4x^3 - 6x^2 + 4$, $f(0) \doteq 3$.

B

In Problems 9–22, find $f(x)$.

9. $f'(x) = \dfrac{1}{x^2}$.
10. $f'(x) = \dfrac{x^4 - 1}{x^3}$.

11. $f'(x) = \dfrac{2x^2 + 4x + 1}{\sqrt{x}}$.
12. $f'(x) = \dfrac{(x - 3)^2}{x^{1/3}}$.

13. $f'(x) = (x - 1)^2$.
14. $f'(x) = (x + 3)(x + 2)$.
15. $f'(x) = x^{-1/2}$, $f(4) = 3$.
16. $f'(x) = x^{1/2} + x^{1/3}$, $f(1) = 5/12$.
17. $f''(x) = 12$, $f'(1) = 3$, $f(1) = 4$.
18. $f''(x) = 6x + 2$, $f'(1) = 4$, $f(1) = 5$.
19. $f''(x) = -6x$, $f(1) = 3$, $f(2) = 5$.
20. $f''(x) = 12x - 4$, $f(2) = 8$, $f(-1) = 6$.
21. $f'''(x) = 6$, $f''(1) = 6$, $f'(1) = 4$, $f(1) = 0$.
22. $f'''(x) = -12$, $f(0) = 1$, $f(1) = 6$, $f(-1) = 4$.

In Problems 23–28, find the equation of motion.

23. $a = -32$; $s = 0$ and $v = 80$ when $t = 0$.
24. $a = -32$; $s = 0$ when $t = 0$ and again when $t = 12$.
25. $a = 4t$; $s = 4$ and $v = 16$ when $t = 0$.
26. $a = 6t - 1$; $s = 3$ and $v = 15$ when $t = 0$.
27. $v = 4t + 1$; $s = 5$ when $t = 1$.
28. $v = 6t^2 - t$; $s = 7$ when $t = 2$.

29. An object is thrown upward with initial speed v_0 from an initial position s_0. Give its equation of motion. What is its maximum height?

30. A person is standing on top of a building 256 ft high. How long will it take a stone to hit the ground   (a) if the person drops it?   (b) if the person throws it downward with a speed of 64 ft/sec?   (c) if the person throws it upward with a speed of 64 ft/sec?

31. A ball is dropped from a height of 8 ft. When it hits the ground it bounces back up with a speed that is three-fourths of its speed of impact. How high does it go after the first bounce? after the second bounce?

32. A car accelerates from 0 to 60 miles per hour (88 ft/sec) in 30 sec. Assuming the acceleration to be constant, what is that constant? Assuming that the car can continue the same rate of acceleration for another 30 sec, how fast will it be going then?

33. Suppose $a = kt$ in Problem 32. Find k. How fast will the car be going at the end of the first minute?

34. A ball is dropped from a tall building. Two seconds later another ball is dropped. Where is the first ball when the second is dropped? Do the two balls remain this distance apart? If not, give a formula for the distance between them as a function of the time t after the second ball is dropped.

35. A ball is thrown upward from ground level at a speed of 48 ft/sec. How high does it go? When does it hit the ground?

36. Work Problem 35 for the moon, where $a = -5.3$ ft/sec/sec.

37. Given the marginal cost function of $3x^2 - 4x + 5$ and a cost of $1000 for producing 10 items, find the cost function.

38. If the marginal revenue function is $(x - 10)^2$ and $100 is the revenue received for the sale of 16 items, what is the revenue function?

C

39. The mean radius of the earth is 3959 miles and the acceleration at sea level due to gravity is -32.15 ft/sec/sec. Find the acceleration due to gravity (a) at an elevation of 528 ft (0.1 mile) and (b) at an elevation of 10 miles.

40. Theorem 6.3 specifies that $n \neq -1$. If $f'(x) = 1/x$, we cannot find the antiderivative by Theorem 6.3. There is one, but it is not algebraic. Using the fact that $f'(x)$ is the slope of the graph and assuming that $f(1) = 0$, sketch the graph of f for $x > 0$.

Review Problems

A

1. A function f has derivative $f'(x) = x^2 + (1/x^2)$, and $f(1) = 1$. Find f.

2. A manufacturer's total cost and total revenue functions are

$$C = 20x + 10\sqrt{x + 1} + 4000 \quad \text{and} \quad R = 50x - x^2.$$

Find the marginal cost, revenue, and profit when $x = 15$. Should production be increased or decreased?

3. A manufacturer has fixed costs of $16,000 per month and can produce x widgets at a cost of $0.1x^2 + 20x$. How many widgets should be produced monthly to minimize the cost per widget?

4. The side of an equilateral triangle is measured and found to be 13.16 in., with a maximum error of 0.03 in. What are the area and the maximum error in the area?

5. If gas is pumped into a spherical balloon at the rate of 5 ft^3/min, at what rate is the radius increasing when $r = 3$ ft?

6. Estimate $\sqrt[3]{10}$ by differentials.

7. Estimate $1/\sqrt{23}$ by differentials.

B

8. A ball is thrown upward at 64 ft/sec from the top ledge of a building that is 128 ft high. How long afterward, and with what speed, will it hit the ground?

9. A line through (4, 5) forms, with the coordinate axes, a triangle in the first quadrant. Find the area of the smallest such triangle.

10. A point moves along the curve $y = x^3$ in such a way that the abscissa increases at the rate of 3 in./min. At what rate is the distance from the origin changing when $x = -2$? When $x = 3$?

11. A conical container with no top is to have a volume of 36π in.^{3}. What dimensions for the cone would minimize the amount of material used?

12. A fence is 4 ft high and 4 ft away from a brick wall. What is the shortest ladder that can reach the wall when the foot of the ladder is outside the fence?

13. A Norman window consists of a rectangle surmounted by a semicircle. If the perimeter is P, find the dimensions for the Norman window that will admit the most light.

14. Suppose the semicircular portion of the window of Problem 13 is tinted to admit only half as much light as the rest of the window. What should be the window's dimensions?

15. A car traveling at 40 ft/sec crosses a bridge over a canal at the same moment that a canal boat, traveling at 10 ft/sec, passes under the bridge. If the canal and the road intersect at right angles, how fast will the car and the canal boat be moving apart 10 sec later?

16. A soft drink vendor at an amusement park finds that the number x of drinks sold each day is inversely proportional to the square of the price charged and that 400 are sold when the price is 35¢ per drink. Overhead costs are $10 per day and the vendor pays a supplier 10¢ per drink. What should the vendor charge to maximize profits?

17. A pedometer determines the distance a person walks by counting the number of steps taken and multiplying by the length of one stride to get the distance in inches, which is then converted to miles. A man measures the length of his stride, which he finds to be 28 in. with a maximum error of 2 in. After a long hike he notes that the pedometer reading is 10.3 miles. Assuming that the pedometer counts steps accurately, find the maximum error in the distance.

18. A 10-ft plank (see Figure 6.16(a)) is suspended horizontally by a pair of cables attached to the ends of the plank. A 500-lb weight to be placed on the plank must be placed at least 1 ft away from the supporting cables, because of its size. Where should it be placed so that the product of the forces in the two supporting cables is a minimum? (*Hint:* The forces in the supporting cables are determined as shown in Figure 6.16(b).)

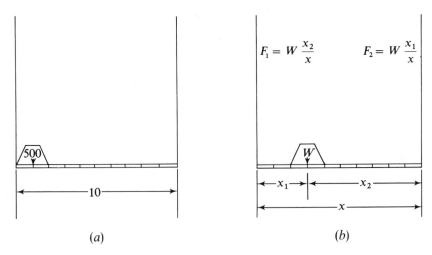

$$F_1 = W\,\frac{x_2}{x} \qquad F_2 = W\,\frac{x_1}{x}$$

(a)   (b)

Figure 6.16

19. A weight is attached to the end of a 20-ft rope that passes over a pulley 10 ft above the floor (see Figure 6.17). If the free end of this rope is moved along the floor away from the point under the weight at a rate of 2 ft/sec, how fast is the weight rising when it is 5 ft off the floor (neglect the diameter of the pulley)?

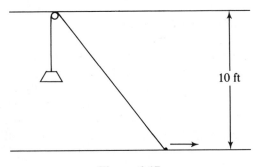

10 ft

Figure 6.17

C

20. Suppose two hallways of widths a and b intersect at right angles. Show that the longest rod that can be taken horizontally around the corner is of length $(a^{2/3} + b^{2/3})^{3/2}$. (Neglect the thickness of the rod.)

21. The equation $y = x^3 + C$ represents a family of curves; each value of C gives a member of the family. Find another family of curves orthogonal to this one—that is, every member of the new family intersects all the curves of the original family at right angles.

22. A line through the point (a, b) in the first quadrant forms, with the coordinate axes, a triangle in the first quadrant. Find the length of the shortest hypotenuse.

7

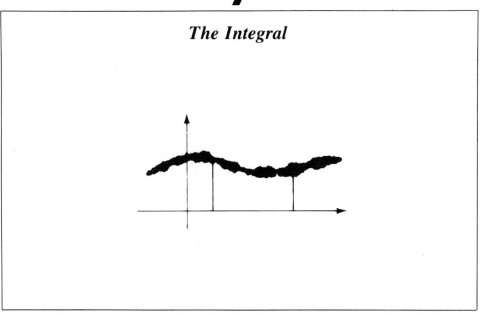

The Integral

7.1

Sigma Notation

Just as the problem of finding tangents to curves led us, by a limiting process, to the derivative, so we shall see that the problem of finding areas of regions will lead us, by another limiting process, to the integral. Differential and integral calculus are the two traditional subdivisions of the calculus; and, while they were inspired by seemingly unrelated problems, we shall see that there is really a very close connection between them. Let us, then, turn our attention to the problem of finding areas of regions. We shall attack it by subdividing the original region into many smaller ones and adding the areas of all of the small regions. Thus we shall consider sums with many terms, such as

$$1^2 + 3^2 + 5^2 + \cdots + 15^2 \quad \text{or} \quad 1 + 2 + 3 + \cdots + n.$$

In each case the dots are meant to indicate that the terms continue in the same form to the last term indicated. In the second example, the last term, of course, depends upon the value of n.

There are several disadvantages to this type of notation. First of all, the individual terms may not be clear from the first few terms. In addition, the second example looks a bit strange when n is one or two. Finally, this notation can be rather long and tedious. For these reasons the sigma notation is often used. In this notation, the sum

$$1^2 + 3^2 + 5^2 + \cdots + 15^2$$

is represented by

$$\sum_{i=1}^{8} (2i - 1)^2.$$

The Greek letter $\sum$ (uppercase sigma) is used to indicate that the given expression is a sum. The form of each term is $(2i - 1)^2$, where the i represents an integer. The subscript $(i = 1)$ and superscript (8) on $\sum$ indicate the values of i to be used for the first and last terms, respectively, of the sum. Thus the first term is

$$(2 \cdot 1 - 1)^2 = 1^2,$$

and the last term is

$$(2 \cdot 8 - 1)^2 = 15^2.$$

The intermediate terms are found by replacing i by consecutive integers between 1 and 8. The i in $(2i - 1)^2$ is a "dummy variable"; that is, it does not appear in the expanded form. Any other symbol can be used in its place with the same result. Thus

$$\sum_{i=1}^{8} (2i - 1)^2 = \sum_{j=1}^{8} (2j - 1)^2 = \sum_{k=1}^{8} (2k - 1)^2 = \text{etc.}$$

The sum $1 + 2 + 3 + \cdots + n$ can be represented by

$$\sum_{i=1}^{n} i \ .$$

Example 1

Give the expanded form of $\displaystyle\sum_{i=1}^{10} \frac{1}{i(i + 1)}$.

$$\sum_{i=1}^{10} \frac{1}{i(i + 1)} = \frac{1}{1(1 + 1)} + \frac{1}{2(2 + 1)} + \frac{1}{3(3 + 1)} + \cdots + \frac{1}{10(10 + 1)}$$

$$= \frac{1}{1 \cdot 2} + \frac{1}{2 \cdot 3} + \frac{1}{3 \cdot 4} + \cdots + \frac{1}{10 \cdot 11}.$$

Example 2

Give a sigma representation for $2 + 5 + 10 + \cdots + 122$.

We must find some expression $f(i)$ such that

$$f(1) = 2, \quad f(2) = 5, \quad \text{and} \quad f(3) = 10$$

(since we do not know the value of i corresponding to the last term, we cannot use it yet). Let us note that each term is one more than a perfect square: that is,

$$f(1) = 1 + 1 = 1^2 + 1,$$
$$f(2) = 4 + 1 = 2^2 + 1,$$
$$f(3) = 9 + 1 = 3^2 + 1.$$

Thus, all are in the form $i^2 + 1$. Furthermore, the last term is in the form $i^2 + 1$, with $i = 11$. Thus the above sum may be represented by

$$\sum_{i=1}^{11} (i^2 + 1).$$

It is often possible to express the same sum in more than one way, as the following example demonstrates.

Example 3

Show that

$$\sum_{i=1}^{12} (2i - 1)^2 \quad \text{and} \quad \sum_{i=-1}^{10} (2i + 3)^2$$

represent the same sum.

Let us write both in the expanded form:

$$\sum_{i=1}^{12} (2i - 1)^2 = (2 \cdot 1 - 1)^2 + (2 \cdot 2 - 1)^2 + (2 \cdot 3 - 1)^2 + \cdots + (2 \cdot 12 - 1)^2$$
$$= 1^2 + 3^2 + 5^2 + \cdots + 23^2.$$

$$\sum_{i=-1}^{10} (2i + 3)^2 = [2(-1) + 3]^2 + (2 \cdot 0 + 3)^2 + (2 \cdot 1 + 3)^2 + \cdots + (2 \cdot 10 + 3)^2$$
$$= 1^2 + 3^2 + 5^2 + \cdots + 23^2.$$

If all of the terms are known, it is a simple matter to find the sum. Thus,

$$\sum_{i=1}^{8} (2i - 1)^2 = 1^2 + 3^2 + 5^2 + \cdots + 15^2 = 680.$$

But what of the sum

$$\sum_{i=1}^{n} i = 1 + 2 + 3 + \cdots + n?$$

We cannot give the numerical value of this sum without knowing the value of n. But we can give a formula for the sum.

Theorem 7.1

If n is a positive integer, then $\displaystyle\sum_{i=1}^{n} i = \frac{n(n + 1)}{2}.$

Proof

Since we want to prove that the formula holds for any positive integer n, this theorem can be proved by mathematical induction. Let us recall that we must do two things. We must verify that the statement given by the formula is true when $n = 1$, and we must show that if it is true when $n = k$, then it is true when $n = k + 1$. If $n = 1$, then

$$\sum_{i=1}^{1} i = 1 \quad \text{and} \quad \frac{n(n + 1)}{2} = \frac{1 \cdot 2}{2} = 1.$$

Suppose it is true when $n = k$; that is,

$$\sum_{i=1}^{k} i = \frac{k(k+1)}{2}.$$

Now

$$\sum_{i=1}^{k+1} i = \sum_{i=1}^{k} i + (k+1)$$

$$= \frac{k(k+1)}{2} + \frac{2(k+1)}{2}$$

$$= \frac{(k+1)(k+2)}{2}.$$

Thus it is true when $n = k+1$. By mathematical induction, the formula holds for every positive integer n.

Just as there is a formula for the sum of the first n integers, there are formulas for the sums of the first n squares, the first n cubes, and so on. The following theorem gives several such formulas.

Theorem 7.2

If n is a positive integer, then

(a) $\displaystyle\sum_{i=1}^{n} 1 = n,$

(b) $\displaystyle\sum_{i=1}^{n} i = \frac{n(n+1)}{2},$

(c) $\displaystyle\sum_{i=1}^{n} i^2 = \frac{n(n+1)(2n+1)}{6},$

(d) $\displaystyle\sum_{i=1}^{n} i^3 = \frac{n^2(n+1)^2}{4},$

(e) $\displaystyle\sum_{i=1}^{n} i^4 = \frac{n(n+1)(6n^3 + 9n^2 + n - 1)}{30}.$

The formula of Theorem 7.1 is included in Theorem 7.2 for completeness. These formulas will be quite useful in the next section. All of them can be proved by mathematical induction as well as by other methods (see Problems 37–41).

Theorem 7.3

If f and g are functions, c is a real number, and m and n are positive integers, then

(a) $\displaystyle\sum_{i=1}^{n} cf(i) = c \sum_{i=1}^{n} f(i),$

(b) $\displaystyle\sum_{i=1}^{n} [f(i) + g(i)] = \sum_{i=1}^{n} f(i) + \sum_{i=1}^{n} g(i),$

(c) $\displaystyle\sum_{i=1}^{n} [f(i) - g(i)] = \sum_{i=1}^{n} f(i) - \sum_{i=1}^{n} g(i),$

(d) $\displaystyle\sum_{i=1}^{n} f(i) = \sum_{j=1}^{n} f(j) = \sum_{k=1}^{n} f(k),$

(e) $\displaystyle\sum_{i=1}^{m} f(i) + \sum_{i=m+1}^{n} f(i) = \sum_{i=1}^{n} f(i), \ (m < n).$

Since these formulas are easily proved by writing them in expanded form, the proof is left to the student.

Example 4

Simplify $\displaystyle\sum_{i=1}^{n} \left[\left(\frac{i}{n}\right)^2 + \frac{i}{n} \right] \frac{1}{n}.$

$$\sum_{i=1}^{n} \left[\left(\frac{i}{n}\right)^2 + \frac{i}{n} \right] \frac{1}{n} = \sum_{i=1}^{n} \left(\frac{i^2}{n^3} + \frac{i}{n^2} \right)$$

$$= \sum_{i=1}^{n} \frac{i^2}{n^3} + \sum_{i=1}^{n} \frac{i}{n^2}$$

$$= \frac{1}{n^3} \sum_{i=1}^{n} i^2 + \frac{1}{n^2} \sum_{i=1}^{n} i$$

$$= \frac{1}{n^3} \cdot \frac{n(n+1)(2n+1)}{6} + \frac{1}{n^2} \cdot \frac{n(n+1)}{2}$$

$$= \frac{(n+1)(2n+1)}{6n^2} + \frac{3n(n+1)}{6n^2}$$

$$= \frac{(n+1)(5n+1)}{6n^2}.$$

Problems

A

In Problems 1–10, express the sum in the expanded form.

1. $\displaystyle\sum_{i=1}^{7} i^3.$ 2. $\displaystyle\sum_{i=1}^{4} (i^2 + 3).$ 3. $\displaystyle\sum_{i=1}^{6} (2i - 5).$ 4. $\displaystyle\sum_{i=-2}^{4} 2i^2.$

5. $\displaystyle\sum_{i=0}^{7} (2i + 1).$ 6. $\displaystyle\sum_{i=1}^{n} (i^3 + 1).$ 7. $\displaystyle\sum_{i=1}^{n-1} (i^3 + 1).$ 8. $\displaystyle\sum_{i=1}^{n} (i^2 + 3).$

9. $\displaystyle\sum_{i=1}^{n} (2i + 1).$ 10. $\displaystyle\sum_{i=1}^{n-1} (i^2 - 2i).$

In Problems 11–20, express the sums in sigma notation.

11. $1 + 2 + 3 + \cdots + 15.$ 12. $1 + 3 + 5 + \cdots + 25.$

13. $2 + 4 + 6 + \cdots + 22.$ 14. $1 + 4 + 7 + \cdots + 28.$

15. $1 + 2 + 4 + 7 + \cdots + 56.$ 16. $1^3 + 2^3 + 3^3 + \cdots + (n - 1)^3.$

17. $1 + 3 + 5 + \cdots + (2n - 1).$ 18. $\dfrac{1}{2} + \dfrac{2}{3} + \dfrac{3}{4} + \cdots + \dfrac{n+1}{n+2}.$

19. $10 + 17 + 26 + \cdots + (n^2 + 1).$ 20. $\dfrac{2}{3} + \dfrac{3}{5} + \dfrac{4}{7} + \cdots + \dfrac{n-1}{2n-3}.$

In Problems 21–28, indicate whether or not the two given expressions are equal.

21. $\sum_{i=1}^{n}(2i-1)$, $\sum_{i=0}^{n-1}(2i+1)$.

22. $\sum_{i=1}^{n}(i^3+1)$, $\sum_{i=2}^{n+1}i^3$

23. $\sum_{i=1}^{n}(i-1)$, $\sum_{i=3}^{n+1}(i-2)$.

24. $\sum_{i=0}^{n}(i+1)^2$, $\sum_{i=1}^{n+1}i^2$.

25. $\sum_{i=0}^{n}(i^2+1)$, $\sum_{i=1}^{n+1}[(i-1)^2+1]$.

26. $\sum_{i=1}^{n}i^2(i+1)$, $\sum_{k=2}^{n+1}[(k-1)^3+(k-1)^2]$.

27. $\sum_{j=1}^{n-1}(2j+3)$, $\sum_{i=2}^{n}(2i+2)$.

28. $\sum_{i=1}^{n-1}i(i-1)$, $\sum_{j=1}^{n}(j^2-3j+2)$.

B

In Problems 29–36, simplify the given expression.

29. $\sum_{i=1}^{n}\left(\frac{i}{n}\right)^2\frac{1}{n}$.

30. $\sum_{i=1}^{n}\left(\frac{2i}{n}\right)^3\frac{2}{n}$.

31. $\sum_{i=1}^{n}\left(\frac{i}{n}+1\right)^2\frac{1}{n}$.

32. $\sum_{i=1}^{n}\left[\left(\frac{i}{n}\right)^2-\frac{3i}{n}\right]\frac{1}{n}$.

33. $\sum_{i=1}^{n}\left(\frac{i-1}{n}\right)^2\frac{1}{n}$.

34. $\sum_{i=1}^{2n}\left(\frac{i}{n}\right)^3\frac{1}{n}$.

35. $\sum_{i=1}^{n}\left(2+\frac{i}{n}\right)^2\frac{1}{n}$

36. $\sum_{i=1}^{n}\left[3\left(-1+\frac{i}{n}\right)^2-\left(-1+\frac{i}{n}\right)\right]\frac{1}{n}$.

C

37. Prove Theorem 7.2(a).
38. Prove Theorem 7.2(c).
39. Prove Theorem 7.2(d).
40. Prove Theorem 7.2(e).
41. Prove Theorem 7.2(b) (Theorem 7.1) without using induction. (*Hint:* Arrange the terms as indicated below and add columns.

$$1+\quad 2\quad +\quad 3\quad +\cdots+n,$$
$$n+(n-1)+(n-2)+\cdots+1.)$$

42. Prove Theorem 7.3.
43. Find formulas for $\sum_{i=1}^{n}2i$ and $\sum_{i=1}^{n}(2i-1)$. Show that their sum is $\sum_{i=1}^{2n}i$.

7.2

The Area under a Curve

Let us now consider the problem of finding the area of a given plane region. We have an intuitive idea of what we mean by the area of a plane region; yet when someone says that the area of some irregular region is 5 square units, exactly what is meant? In other words, can we give a general definition of the area of a plane region? Before attempting to do so, let us consider some properties of area.

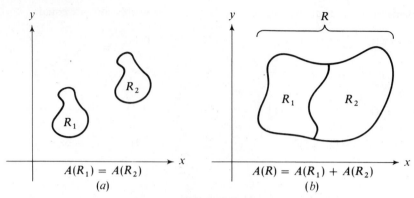

Figure 7.1

1. If R is any region, then the area of R, $A(R)$, is a real number and $A(R) \geq 0$.
2. If R_1 and R_2 are congruent regions, then $A(R_1) = A(R_2)$ (see Figure 7.1(a)).
3. If $R = R_1 \cup R_2$, where R_1 and R_2 have only boundary points in common, then $A(R) = A(R_1) + A(R_2)$ (see Figure 7.1(b)).

While these three properties tell us something about area, they give us no way of assigning a specific number as the area of a given region. In order to make such an assignment, we consider the area of a very simple region.

4. The area of a rectangle of length l and width w is $A = lw$.

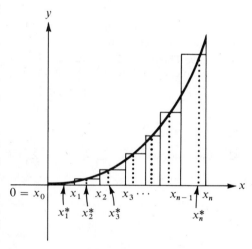

With these four properties it is possible to determine the area of any polygon (recall that a polygon has a finite number of *straight* sides). But suppose we want to find the area "under the curve" $y = x^2$ from $x = 0$ to $x = 1$, that is, the area of the region bounded by $y = x^2$, the x axis, and the vertical lines $x = 0$ and $x = 1$ (see Figure 7.2). This region has one curved side. How can we handle this case? Suppose we take a lesson from our discussion of the slope of a graph. There we approximated the tangent line with a secant line and noted what happened to its slope as it moved closer and closer to the tangent line

Figure 7.2

Since the area of a rectangle is known, let us approximate the area we want with areas of rectangles. This is done in the following way. We subdivide the interval $[0, 1]$ into n (not necessarily equal) subintervals (see Figure 7.2) by means of the numbers

$$x_0, x_1, x_2, \ldots, x_{n-1}, x_n, \quad \text{where} \quad x_0 = 0 \text{ and } x_n = 1.$$

Within each subinterval we select a number in any way we choose from the left- to the right-hand end point. Let us call these numbers

$$x_1^*, x_2^*, x_3^*, \ldots, x_n^*.$$

Now let us construct a rectangle for each subinterval, using the subinterval itself as the base and $f(x_i^*)$ as the altitude. Thus the sum of the areas of all of these rectangles gives an approximation (although perhaps a very poor one) of the area we seek. This sum is

$$f(x_1^*)(x_1 - x_0) + f(x_2^*)(x_2 - x_1) + f(x_3^*)(x_3 - x_2) + \cdots + f(x_n^*)(x_n - x_{n-1})$$
$$= \sum_{i=1}^{n} f(x_i^*)(x_i - x_{i-1}).$$

In order to simplify this expression, we reintroduce the delta notation that was used in Section 6.3. It was noted there that Δx represents a difference of two values of x. In particular, we define

$$\Delta x_i = x_i - x_{i-1}.$$

Thus Δx_i represents the width of the ith rectangle. With this notation our approximating sum is

$$\sum_{i=1}^{n} f(x_i^*)\, \Delta x_i.$$

Let us now consider the question of the error committed in making this approximation. There are two types of errors (see Figure 7.3), which we shall call positive and negative errors. A positive error occurs when a portion of a rectangle lies outside the original region R; errors of this type tend to make the sum greater than the area desired. A negative error occurs when a portion of R lies outside all of the rectangles; errors of this type tend to make the sum less than the area desired. In order to get a better approximation of the area of R, we must find a way to decrease both types of errors. Let us consider the positive errors first.

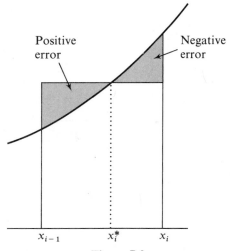

Figure 7.3

Suppose the interval $[0, 1]$ is subdivided as shown in Figure 7.4(a). If the x_i^*'s are chosen to be the right-hand end points, the positive error is the largest possible one for that subdivision. It is represented by the shaded portion of Figure 7.4(a). Now suppose that each interval of the first subdivision is itself subdivided to give the finer subdivision of Figure 7.4(b). The largest possible positive error is now smaller than it was for the first subdivision. This error is represented by the shaded portion of Figure 7.4(b). The difference between this error and the first is represented by the unshaded rectangles above the curve. By further subdivision, as in Figure 7.4(c), the largest possible positive error is reduced still further. Thus, as we take ever finer subdivisions, the maximum positive error (and thus any positive error) can be made as small as we choose. A similar analysis shows that we can make the negative error as small as we choose by taking the subdivision fine enough.

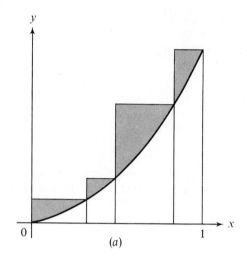

(a)

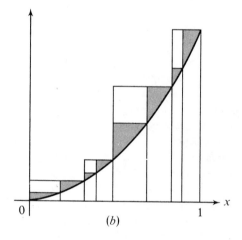

(b)

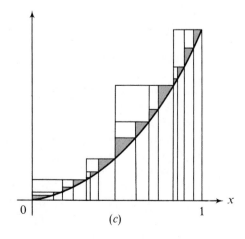

(c)

Figure 7.4

All of the foregoing analysis suggests a limiting process. Thus it appears that the area we want is the limit of the approximating sum as the lengths of the subintervals approach zero. The analysis also suggests that the result is independent of the way in which we subdivide (as long as the lengths of all of the subintervals approach zero) and the way in which the x_i^*'s are chosen. Let us now apply this method to our original problem.

Example 1

Find the area under the curve $y = x^2$ from $x = 0$ to $x = 1$.

First of all, we must subdivide the interval [0, 1] and then choose the x_i^*'s. In order to simplify the algebra involved, let us subdivide the interval into n *equal* subintervals and choose the x_i^*'s to be the right-hand end points of the subintervals (see Figure 7.5). Thus

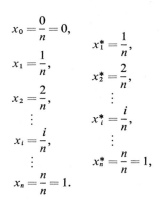

$$x_0 = \frac{0}{n} = 0,$$
$$x_1 = \frac{1}{n},$$
$$x_2 = \frac{2}{n},$$
$$\vdots$$
$$x_i = \frac{i}{n},$$
$$\vdots$$
$$x_n = \frac{n}{n} = 1.$$

$$x_1^* = \frac{1}{n},$$
$$x_2^* = \frac{2}{n},$$
$$\vdots$$
$$x_i^* = \frac{i}{n},$$
$$\vdots$$
$$x_n^* = \frac{n}{n} = 1,$$

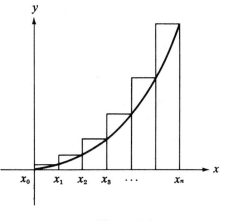

Figure 7.5

We see that the length of each of the n equal subintervals is $1/n$, and the lengths of all of them are approaching zero as n gets large and positive. This last observation allows us to simplify our notation for the limit.

$$A = \lim_{n \to +\infty} \sum_{i=1}^{n} f(x_i^*)\Delta x_i$$

$$= \lim_{n \to +\infty} \sum_{i=1}^{n} f\left(\frac{i}{n}\right)\left(\frac{1}{n}\right)$$

$$= \lim_{n \to +\infty} \sum_{i=1}^{n} \frac{i^2}{n^2} \cdot \frac{1}{n} \qquad \text{(See Note 1.)}$$

$$= \lim_{n \to +\infty} \frac{1}{n^3} \sum_{i=1}^{n} i^2 \qquad \text{(By Theorem 7.3(a))}$$

$$= \lim_{n \to +\infty} \frac{1}{n^3} \frac{n(n+1)(2n+1)}{6} \qquad \text{(By Theorem 7.2(c))}$$

$$= \lim_{n \to +\infty} \frac{\left(1 + \frac{1}{n}\right)\left(2 + \frac{1}{n}\right)}{6} \qquad \text{(See Note 2.)}$$

$$= \frac{1}{3}.$$

Note 1: Since our original function is in the form $f(x) = x^2$, it follows that

$$f(i/n) = (i/n)^2 = i^2/n^2.$$

Note 2: We have divided both numerator and denominator by n^3. In the numerator, each factor was divided by one of the n's.

Perhaps you feel critical of our answer in Example 1, because we used the right-hand end points throughout, so that our approximating sums are *all* bigger than the area we want. If the answer is incorrect, it must be too large. Let us now use the left-hand end points. Since the approximating sums are all smaller than the area we want (see Figure 7.6), we feel that the result will certainly not be greater than the area under the curve —either it will be the area we want, or it will be less than that area. In this case $x_i = i/n$ as before, but $x_i^* = (i-1)/n$. Thus

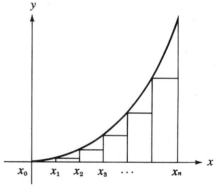

$$A = \lim_{n \to +\infty} \sum_{i=1}^{n} f(x_i^*)\Delta x_i$$

$$= \lim_{n \to +\infty} \sum_{i=1}^{n} f\left(\frac{i-1}{n}\right)\left(\frac{1}{n}\right)$$

$$= \lim_{n \to +\infty} \sum_{i=1}^{n} \frac{(i-1)^2}{n^2} \cdot \frac{1}{n}$$

$$= \lim_{n \to +\infty} \frac{1}{n^3} \sum_{i=1}^{n} (i-1)^2$$

Figure 7.6

$$= \lim_{n \to +\infty} \frac{1}{n^3} \frac{(n-1)n(2n-1)}{6} \quad \text{(see Problem 33, Section 7.1)}$$

$$= \lim_{n \to +\infty} \frac{\left(1 - \frac{1}{n}\right)\left(2 - \frac{1}{n}\right)}{6}$$

$$= \frac{1}{3}.$$

We see that we have exactly the same result as in the previous case. It seems reasonable to expect that, if we get 1/3 in both of these extreme cases, we should get 1/3 in any case.

The fact that the result is independent of the choice of the x_i^*'s is important. Similarly, it is important to note that the result is independent of the choice of the subdivision (see Problem 20). Unfortunately, these important facts are also difficult to prove, and we shall not attempt proofs here.

Example 2

Find the area under the curve $y = x^2$, from $x = 1$ to $x = 3$.

Since we have inferred that the result we get is independent of both the sub-division and the choice of the x_i^*'s, we shall subdivide the interval [1, 3] into n equal intervals and choose the x_i^*'s to be right-hand end points (see Figure 7.7).

Because we are subdividing an interval of length 2 into n equal intervals, the length of each is $2/n$. Thus

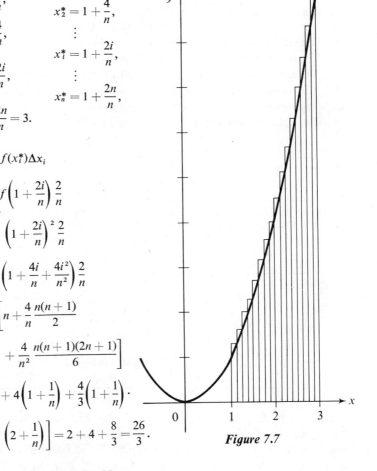

$$x_0 = 1, \qquad\qquad x_1^* = 1 + \frac{2}{n},$$

$$x_1 = 1 + \frac{2}{n}, \qquad x_2^* = 1 + \frac{4}{n},$$

$$x_2 = 1 + \frac{4}{n}, \qquad \vdots$$

$$\vdots \qquad\qquad x_i^* = 1 + \frac{2i}{n},$$

$$x_i = 1 + \frac{2i}{n}, \qquad \vdots$$

$$\vdots \qquad\qquad x_n^* = 1 + \frac{2n}{n},$$

$$x_n = 1 + \frac{2n}{n} = 3.$$

$$A = \lim_{n \to +\infty} \sum_{i=1}^{n} f(x_i^*)\Delta x_i$$

$$= \lim_{n \to +\infty} \sum_{i=1}^{n} f\left(1 + \frac{2i}{n}\right)\frac{2}{n}$$

$$= \lim_{n \to +\infty} \sum_{i=1}^{n} \left(1 + \frac{2i}{n}\right)^2 \frac{2}{n}$$

$$= \lim_{n \to +\infty} \sum_{i=1}^{n} \left(1 + \frac{4i}{n} + \frac{4i^2}{n^2}\right)\frac{2}{n}$$

$$= \lim_{n \to +\infty} \frac{2}{n}\left[n + \frac{4}{n}\frac{n(n+1)}{2}\right.$$

$$\left. + \frac{4}{n^2}\frac{n(n+1)(2n+1)}{6}\right]$$

$$= \lim_{n \to +\infty}\left[2 + 4\left(1 + \frac{1}{n}\right) + \frac{4}{3}\left(1 + \frac{1}{n}\right)\cdot\right.$$

$$\left.\left(2 + \frac{1}{n}\right)\right] = 2 + 4 + \frac{8}{3} = \frac{26}{3}.$$

Figure 7.7

Problems

B

1. Use the area of a rectangle and the properties of area on page 184 to find the area of a right triangle.

2. Use the area of a rectangle and the properties of area on page 184 to find the area of a triangle.

3. Use the area of a triangle and the properties of area on page 184 to find the area of a parallelogram.

4. Use the area of a triangle and the properties of area on page 184 to find the area of a trapezoid.

5. Use the limit method to find the area of the region bounded by $y = x$, the x axis, and $x = 1$. Since this region is triangular, it provides a method of checking the limit method against the method of Problem 1.

6. Find the area of the region bounded by $y = x^4$, the x axis, and $x = 1$, taking the x_i^*'s to be the right-hand end points of the subintervals.

7. Find the area of Problem 6, using the left-hand end points of the subintervals.

In Problems 8–18, find the area of the given region.

8. The region bounded by $y = x^3$, $y = 0$, and $x = 1$.
9. The region bounded by $y = x^2$ and $y = 0$ and between $x = 1$ and $x = 2$.
10. The region bounded by $y = x^2$, $y = 0$, and $x = 3$.
11. The region bounded by $y = x^3$ and $y = 0$ and between $x = 1$ and $x = 2$.
12. The region bounded by $y = x + x^2$ and $y = 0$ and between $x = 0$ and $x = 2$.
13. The region bounded by $y = 3x^2$, $y = 0$, and $x = 1$.
14. The region bounded by $y = 2x^2 + 3x$ and $y = 0$ and between $x = 2$ and $x = 3$.
15. The region bounded by $y = x^2 - x$ and $y = 0$ and between $x = 2$ and $x = 3$.
16. The region bounded by $y = x^3 + 2x^2 + 3$ and $y = 0$ and between $x = -1$ and $x = 0$.
17. The region bounded by $y = x - x^2$ and $y = 0$.
18. The region bounded by $y = (x^2 - 4)^2$ and $y = 0$.

C

19. Find the area of the region bounded by $y = x^2$, the x axis, and $x = 1$, taking the x_i^*'s to be the midpoints of the subintervals. (See Problem 43 of the previous section.)

20. Subdivide $[0, 1]$ into $2n$ subintervals with odd-numbered intervals of length $2/3n$ and even-numbered intervals of length $1/3n$. Choose the x_i^*'s in any convenient way. Show that the area of Example 1 is $1/3$ when using this subdivision.

7.3

The Definite Integral

Let us now formalize the material of the preceding section as we did with the derivative. First of all we restrict this discussion to functions that are bounded on the interval $[a, b]$. To say that the function f is *bounded* on $[a, b]$ means that there is a number M such that $|f(x)| < M$ for all x in $[a, b]$. Graphically this means that if $\mathcal{G}$ is the graph of $y = f(x), a \le x \le b$, there are two horizontal lines (namely, $y = \pm M$) with all of $\mathcal{G}$ between them. This rules out vertical asymptotes in the interval $[a, b]$. For example, $f(x) = 1/x$ is bounded on the interval $[1, 5]$, since $|f(x)| = |1/x| < 2$ for all x in $[1, 5]$ (see Figure 7.8(a)). On the other hand, $f(x) = 1/x$ is not bounded on $[-1, 1]$; no matter how large a number M we choose, there is a number x in $[-1, 1]$ with $|1/x| > M$ (see Figure 7.8(b)). We shall consider the unbounded case in Chapter 16.

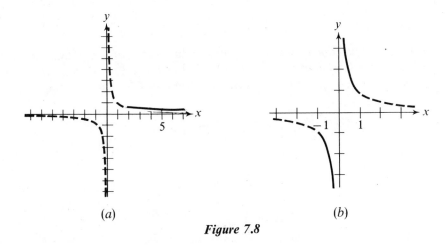

Figure 7.8

   Now let us recall what we did in the last section. In each case we subdivided an interval $[a, b]$ into n subintervals with the numbers

$$x_0 = a, x_1, x_2, x_3, \ldots, x_n = b$$

(although we used n *equal* subintervals in the examples, this was just an algebraic convenience—we could have subdivided in any way). Then we selected a number x_i^* in each subinterval,

$$x_{i-1} \leq x_i^* \leq x_i,$$

and found the sum,

$$\sum_{i=1}^{n} f(x_i^*) \, \Delta x_i.$$

   Finally, we took the limit of this sum as n increased indefinitely, in such a way that the lengths of all the subintervals approached zero.

   Some additional notation will simplify the situation a bit. For a given subdivision $S = \{x_0 = a, x_1, x_2, \ldots, x_n = b\}$, we shall represent the length of a longest subinterval by $\|S\|$, called the *norm* of the subdivision. As $\|S\| \to 0$, the lengths of all the subintervals must approach zero, and we can express the limit of the approximating sum as

$$\lim_{\|S\| \to 0} \sum_{i=1}^{n} f(x_i^*) \, \Delta x_i.$$

Just as the slope of a graph is far more important than it appeared to be at first glance, this limit is also far more important than it appears to be. Although it was inspired by the problem of finding an area, we shall see that the same type of expression (with appropriate changes in f) can be used to find many things that have the properties of area given on page 184; volume, arc length, work, force, and many others. Because of its importance and wide application, we give it a special name and a special symbol that does not suggest area.

Definition

*If f is a bounded function defined on the interval [a, b] (a < b), if S = {$x_0 = a, x_1, x_2,$ ..., $x_n = b$} is a subdivision of [a, b] with norm $\|S\|$ and if $x_{i-1} \leq x_i^* \leq x_i$, for i = 1, 2, 3, ..., n, and $\Delta x_i = x_i - x_{i-1}$, then the **definite integral** of f from a to b is*

$$\int_a^b f(x)\, dx = \lim_{\|S\| \to 0} \sum_{i=1}^n f(x_i^*)\, \Delta x_i,$$

provided this limit exists. If

$$\int_a^b f(x)\, dx$$

*exists, **f is integrable** on the interval [a, b].*

This limit is much more complex than it appears to be. The approximating sum

$$\sum_{i=1}^n f(x_i^*)\, \Delta x_i$$

is dependent upon the function f, the limits a and b, the subdivision S of $[a, b]$, and the choice of the x_i^*'s. Furthermore, the condition that $\|S\| \to 0$ requires that the length of every subinterval approach zero. If this condition is to be met, the interval $[a, b]$ must be further subdivided repeatedly. Of course, there is no limit to the number of ways this can be done.

If f is integrable on $[a, b]$, then by definition

$$\int_a^b f(x)\, dx$$

is independent of the choices of S and x_i^* and the way in which $\|S\| \to 0$. Thus we shall continue to subdivide $[a, b]$ into equal subintervals.

The numbers a and b are called the limits of integration; a is the lower limit and b the upper limit. The foregoing definition had the restriction $a < b$. We now extend this by the following definitions.

Definition

$$\int_a^a f(x)\, dx = 0.$$

Definition

If $a < b$ and f is integrable on $[a, b]$, then

$$\int_b^a f(x)\, dx = -\int_a^b f(x)\, dx.$$

The definite integral has the following properties.

Theorem 7.4

If $a, b,$ and c are real numbers and if the integrals in the following equations exist, then

(a) $\displaystyle\int_a^b kf(x)\, dx = k\int_a^b f(x)\, dx,$ where k is a constant,

(b) $\displaystyle\int_a^b [f(x) \pm g(x)]\, dx = \int_a^b f(x)\, dx \pm \int_a^b g(x)\, dx,$

(c) $\displaystyle\int_a^b f(x)\, dx = \int_a^c f(x)\, dx + \int_c^b f(x)\, dx.$

Note that part (a) is true only when k is a *constant*; do not attempt to bring a function of x outside the integral sign. Furthermore, part (b) is stated only for a sum or difference; it does not extend to products or quotients. These properties can be proved using the definition of the integral. The proofs are left to the student (see Problems 23–25).

Let us recall that this definition was motivated by a desire to find the area under a curve. But we used only intuitive ideas about area—we never actually defined area. Now that we have the integral (which is *not* defined in terms of area) we can use it to give a definition of the area under a curve. However, we must use some care. The integral does *not* give the area under the curve in every case—only when $f(x) \geq 0$ for all x in $[a, b]$ (note that this is the only case we considered in the preceding section). Thus we want to use $|f(x)|$ instead of $f(x)$.

Definition

The **area** of the region bounded by $y = f(x)$, the x axis, and the vertical lines $x = a$ and $x = b$ $(a < b)$ is

$$A = \int_a^b |f(x)|\, dx.$$

If $f(x) \geq 0$, then $|f(x)| = f(x)$; but if $f(x) \leq 0$, then $|f(x)| = -f(x)$. If a portion of the graph is below the x axis, we must use $|f(x)| = -f(x)$ for that portion. Of course, we might have defined an integral in such a way that it always gives the area under a curve without further change. While our present definition results in some inconvenience in this respect, it is more than compensated for by other applications of the integral, including areas of more generally defined regions.

Example 1

Find the area of the region bounded by $y = x^2 - 1$, the x axis, and the vertical lines $x = 0$ and $x = 2$.

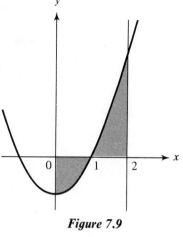

The region we want is indicated by the shaded portion of Figure 7.9. For $0 \le x \le 1$, $f(x) = y \le 0$; for $1 \le x \le 2$, $f(x) = y \ge 0$. Thus,

$$A = \int_0^2 |f(x)|\, dx$$

$$= \int_0^1 -f(x)\, dx + \int_1^2 f(x)\, dx$$

$$= \int_0^1 (1 - x^2)\, dx + \int_1^2 (x^2 - 1)\, dx.$$

Figure 7.9

Each of these integrals can be evaluated separately by the method of the preceding section. To evaluate $\int_0^1 (1 - x^2)\, dx$, we use the following subdivision and choice of the x_i^*'s.

$$x_0 = \frac{0}{n} = 0,$$
$$x_1^* = \frac{1}{n},$$
$$x_1 = \frac{1}{n},$$
$$x_2^* = \frac{2}{n}$$
$$x_2 = \frac{2}{n},$$
$$\vdots$$
$$x_i^* = \frac{i}{n},$$
$$x_i = \frac{i}{n},$$
$$\vdots$$
$$x_n^* = \frac{n}{n} = 1,$$
$$x_n = \frac{n}{n} = 1.$$

$$\int_0^1 (1 - x^2)\, dx = \lim_{n \to +\infty} \sum_{i=1}^n f(x_i^*)\Delta x_i$$

$$= \lim_{n \to +\infty} \sum_{i=1}^n f\left(\frac{i}{n}\right)\frac{1}{n}$$

$$= \lim_{n \to +\infty} \sum_{i=1}^n \left(\frac{n^2 - i^2}{n^2}\right)\frac{1}{n}$$

$$= \lim_{n \to +\infty} \frac{1}{n^3}\left[n^2 \sum_{i=1}^n 1 - \sum_{i=1}^n i^2\right]$$

$$= \lim_{n \to +\infty} \frac{1}{n^3}\left[n^3 - \frac{n(n+1)(2n+1)}{6}\right]$$

$$= \lim_{n \to +\infty} \left[1 - \frac{\left(1 + \frac{1}{n}\right)\left(2 + \frac{1}{n}\right)}{6}\right]$$

$$= 1 - \frac{1}{3} = \frac{2}{3}.$$

A similar argument is used on $\int_1^2 (x^2 - 1)\, dx$. Subdividing the interval $[1, 2]$ into n equal subintervals and choosing the right-hand end point, we have $\Delta x_i = 1/n$ and $x_i^* = 1 + (i/n)$.

$$
\begin{aligned}
\int_1^2 (x^2 - 1)\, dx &= \lim_{n \to +\infty} \sum_{i=1}^n f(x_i^*)\Delta x_i \\
&= \lim_{n \to +\infty} \sum_{i=1}^n f\left(1 + \frac{i}{n}\right)\frac{1}{n} \\
&= \lim_{n \to +\infty} \sum_{i=1}^n \left[\left(1 + \frac{i}{n}\right)^2 - 1\right]\frac{1}{n} \\
&= \lim_{n \to +\infty} \sum_{i=1}^n \left(\frac{i^2}{n^2} + \frac{2i}{n}\right)\frac{1}{n} \\
&= \lim_{n \to +\infty} \left[\frac{1}{n^3}\sum_{i=1}^n i^2 + \frac{2}{n^2}\sum_{i=1}^n i\right] \\
&= \lim_{n \to +\infty} \left[\frac{1}{n^3}\cdot\frac{n(n+1)(2n+1)}{6} + \frac{2}{n^2}\cdot\frac{n(n+1)}{2}\right] \\
&= \lim_{n \to +\infty} \left[\frac{\left(1 + \dfrac{1}{n}\right)\left(2 + \dfrac{1}{n}\right)}{6} + \left(1 + \frac{1}{n}\right)\right] \\
&= \frac{1}{3} + 1 = \frac{4}{3}.
\end{aligned}
$$

Finally,

$$
\begin{aligned}
A &= \int_0^1 (1 - x^2)\, dx + \int_1^2 (x^2 - 1)\, dx \\
&= \frac{2}{3} + \frac{4}{3} = 2.
\end{aligned}
$$

Compare this result with that of the next example.

Example 2

Evaluate $\int_0^2 (x^2 - 1)\, dx$.

Subdividing $[0, 2]$ and choosing x_i^*'s, we have (see Figure 7.10);

$x_0 = \dfrac{0}{n} = 0,$

$\qquad x_1^* = \dfrac{2}{n},$

$x_1 = \dfrac{2}{n},$

$\qquad x_2^* = \dfrac{4}{n},$

$x_2 = \dfrac{4}{n},$

$\qquad \vdots$

$\vdots$

$\qquad x_i^* = \dfrac{2i}{n},$

$x_i = \dfrac{2i}{n},$

$\qquad \vdots$

$\vdots$

$\qquad x_n^* = \dfrac{2n}{n} = 2,$

$x_n = \dfrac{2n}{n} = 2.$

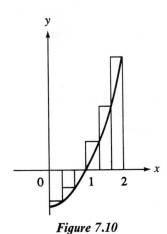

Figure 7.10

$$\int_0^2 (x^2 - 1)\, dx = \lim_{n \to +\infty} \sum_{i=1}^n f(x_i^*)\, \Delta x_i$$

$$= \lim_{n \to +\infty} \sum_{i=1}^n f\left(\frac{2i}{n}\right)\frac{2}{n} = \lim_{n \to +\infty} \sum_{i=1}^n \frac{4i^2 - n^2}{n^2} \cdot \frac{2}{n}$$

$$= \lim_{n \to +\infty} \frac{2}{n^3}\left[4\sum_{i=1}^n i^2 - n^2 \sum_{i=1}^n 1\right] = \lim_{n \to +\infty} \frac{2}{n^3}\left[\frac{4n(n+1)(2n+1)}{6} - n^3\right]$$

$$= \lim_{n \to +\infty}\left[\frac{4n(n+1)(2n+1)}{3n^3} - 2\right] = \lim_{n \to +\infty}\left[\frac{4\left(1+\frac{1}{n}\right)\left(2+\frac{1}{n}\right)}{3} - 2\right]$$

$$= \frac{8}{3} - 2 = \frac{2}{3}.$$

Note that the result is not the area we found in Example 1—it is really the area of the portion above the x axis minus the area of the portion below the x axis. This can easily be checked in Example 1, where these two areas were individually determined.

Although we used area to motivate our definition of an integral, it must be noted that the definition is independent of area. In fact, the area under a curve was defined in terms of an integral. The thing that makes the integral so important and useful is the many interpretations that can be given to it. Later in this chapter, we will interpret the integral in terms of work. In Chapter 16, there are other interpretations, such as volume, arc length, and surface area. Any one of these can be used to motivate the definition of the integral.

In addition, the integral may be introduced by means of upper and lower sums. The upper sum for a given subdivision is found by choosing the x_i^*'s that give the maximum value of f in the interval $[x_{i-1}, x_i]$. Similarly, the lower sum is found by choosing the x_i^*'s that give the minimum value of f in $[x_{i-1}, x_i]$. If the limits of these sums are both equal to some number I, then I is defined to be the value of the integral. This is essentially what we did in Example 1 of Section 7.2 and the discussion following it.

Problems

B

In Problems 1–10, evaluate the integrals.

1. $\displaystyle\int_0^3 (x^2 - 1)\, dx.$

2. $\displaystyle\int_{-1}^2 (x^2 - 4)\, dx.$

3. $\displaystyle\int_{-1}^1 x^4\, dx.$

4. $\displaystyle\int_{-1}^0 x^4\, dx.$

5. $\displaystyle\int_0^1 x^3\, dx.$

6. $\displaystyle\int_{-1}^1 x^3\, dx.$

7. $\displaystyle\int_{-1}^0 (x^2 + 1)\, dx.$

8. $\displaystyle\int_0^1 (x^2 - 1)\, dx.$

9. $\displaystyle\int_0^1 (x^2 - x)\, dx.$

10. $\displaystyle\int_3^4 (3x^2 - 9)\, dx.$

In Problems 11–20, find the area of the given region.

11.   The region bounded by $y = x^3$, $y = 0$, and $x = -2$.
12.   The region bounded by $y = x^3$ and $y = 0$ and between $x = -1$ and $x = 2$.
13.   The region bounded by $y = x^2 - x$ and $y = 0$ and between $x = 0$ and $x = 2$.
14.   The region bounded by $y = x^2 - x$ and $y = 0$ and between $x = -1$ and $x = 1$.
15.   The region bounded by $y = x^2 - 3x$ and $y = 0$.
16.   The region bounded by $y = x^3 - x^2$ and $y = 0$ and between $x = 0$ and $x = 3$.
17.   The region bounded by $y = x^3 - x^2$ and $y = 0$.
18.   The region bounded by $y = x^3 - 9x$ and $y = 0$ and between $x = 0$ and $x = 2$.
19.   The region bounded by $y = x^3 - 4x$ and $y = 0$.
20.   The region bounded by $y = x^4 - x$ and $y = 0$ and between $x = 0$ and $x = 2$.

C

21.   A function f is odd if $f(-x) = -f(x)$ and even if $f(-x) = f(x)$. What can be said about $\int_{-a}^{a} f(x)\,dx$ if f is odd?   if f is even?

22.   Show that the function f, where

$$f(x) = \begin{cases} 0 & \text{if } x \text{ is rational,} \\ 1 & \text{if } x \text{ is irrational,} \end{cases}$$

is not integrable on $[0, 1]$. (*Hint:* Evaluate the approximating sum in two different ways. First, use only rational values for all x_i^*'s; then, use only irrational values.)

23.   Prove Theorem 7.4(a).
24.   Prove Theorem 7.4(b).
25.   Prove Theorem 7.4(c).
26.   Suppose $f(x) \le g(x)$ on the interval $[a, b]$. Show that

$$\int_a^b f(x)\,dx \le \int_a^b g(x)\,dx.$$

7.4

The Fundamental Theorem of Integral Calculus

In the third century B.C., Archimedes developed a way of determining areas that is very similar to the method given in the last two sections. Nevertheless, the first systematic developments of calculus were published by Newton and Leibniz almost 2000 years later. Why was there such a long delay, and what happened in the meantime to make two simultaneous developments of calculus possible? The answer to the first question lies in the complexity of the method. Not only is integration by the definition of an integral very tedious, but the need for Theorem 7.2 restricts us to polynomials of degree four or less. Any really significant progress had to await a much simpler method of integration, which finally came about with the development of the derivative and the discovery of the relationship between the derivative and the integral. This discovery is given by the fundamental theorem of integral calculus, which served as the foundation of both Newton's and Leibniz's developments of calculus. The English mathematician Isaac Barrow, who was Newton's teacher, is generally looked upon as the first to recognize this relationship.

Theorem 7.5

(*Fundamental theorem of integral calculus*) *If the function f is continuous on the interval* [a, b], *then*

$$\int_a^b f(x)\, dx = F(b) - F(a),$$

where F is any function such that $F'(x) = f(x)$ *for all x in* [a, b].

Proof

For simplicity, let us assume that $f(x) \geq 0$ on $[a, b]$. (The general case can be handled by adding a sufficiently large constant. See Problem 31.) For each x_0, $a \leq x_0 \leq b$, define

$$G(x_0) = \int_a^{x_0} f(x)\, dx,$$

the area "under the curve" $y = f(x)$ from a to x_0 (see Figure 7.11). We next show that $G(x)$ is an antiderivative of $f(x)$:

$$G(x_0 + h) - G(x_0) = \int_a^{x_0+h} f(x)\, dx - \int_a^{x_0} f(x)\, dx$$

$$= \int_{x_0}^{x_0+h} f(x)\, dx, \qquad \text{by Theorem 7.4(c).}$$

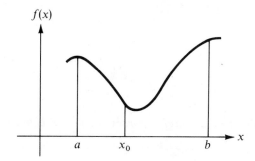

Figure 7.11

(If $h > 0$, then $G(x_0 + h) - G(x_0)$ is the area under $y = f(x)$ from x_0 to $x_0 + h$; see Figure 7.12(a).) Let m and M be the minimum and maximum values, respectively, of $f(x)$ in $[x_0, x_0 + h]$ (see Figure 7.12(b) and Note 1). Then

$$mh = \int_{x_0}^{x_0+h} m\, dx \leq \int_{x_0}^{x_0+h} f(x)\, dx \leq \int_{x_0}^{x_0+h} M\, dx = Mh$$

(see Problem 26 in Section 7.3), so

$$m \leq \frac{G(x_0 + h) - G(x_0)}{h} \leq M.$$

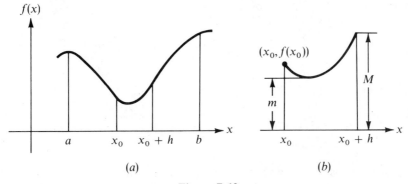

Figure 7.12

In Figure 7.12(b), we can see that m and M depend upon the value of h and that both approach $f(x_0)$ as h approaches zero (see Note 2).

$$\lim_{h \to 0} m = f(x_0), \qquad \lim_{h \to 0} M = f(x_0).$$

Since

$$\frac{G(x_0 + h) - G(x_0)}{h}$$

is squeezed between m and M, then

$$\lim_{h \to 0} m = \lim_{h \to 0} \frac{G(x_0 + h) - G(x_0)}{h} = \lim_{h \to 0} M = f(x_0),$$

so $G'(x_0)$ exists and equals $f(x_0)$. Thus, $G(x)$ is an antiderivative of $f(x)$. Furthermore, $G(a) = 0$ and

$$G(b) = \int_a^b f(x)\,dx.$$

Now let $F(x)$ be any antiderivative of $f(x)$. By Theorem 6.2, $G(x) = F(x) + C$ for some constant C. But $F(a) + C = G(a) = 0$, so $C = -F(a)$. Thus, $G(x) = F(x) - F(a)$, and

$$G(b) = \int_a^b f(x)\,dx = F(b) - F(a).$$

Note 1: A continuous function on a closed interval is bounded and assumes a maximum and a minimum on the interval (see Theorem 17.5).

Note 2: At this point we need the function f to be continuous. There are functions for which one or both of the statements $\lim_{h \to 0} m = f(x)$ and $\lim_{h \to 0} M = f(x)$ are false (see Problem 32).

A complete proof of this theorem, using the mean-value theorem (see page 573), is given in Appendix B. This theorem now tells us that we can often avoid the complicated limit in the definition of an integral and just substitute into a function F whose derivative is f. Our problem is to find F. This is a problem we have already encountered in Chapter 6—the antiderivative.

Example 1

Evaluate $\int_0^1 x^2\, dx$.

$$F'(x) = x^2, \qquad F(x) = \frac{x^3}{3} + C,$$

$$\int_0^1 x^2\, dx = F(1) - F(0) = \left(\frac{1}{3} + C\right) - C = \frac{1}{3}.$$

Note that the constant C cancels out in every case, since we are always taking the difference $F(b) - F(a)$. Since it always cancels out, we simply omit it. The notation is further shortened in the following way:

$$\int_0^1 x^2\, dx = \frac{x^3}{3}\bigg|_0^1 = \frac{1}{3} - 0 = \frac{1}{3}.$$

Example 2

Evaluate $\int_1^2 x^3\, dx$.

$$\int_1^2 x^3\, dx = \frac{x^4}{4}\bigg|_1^2 = 4 - \frac{1}{4} = \frac{15}{4}.$$

We see that we can now integrate much faster than before. Moreover, we can solve problems that we previously could not.

Example 3

Evaluate $\int_1^4 \sqrt{x}\, dx$.

$$\int_1^4 \sqrt{x}\, dx = \frac{2}{3} x^{3/2}\bigg|_1^4 = \frac{16}{3} - \frac{2}{3} = \frac{14}{3}.$$

You are warned that the fundamental theorem is stated for continuous functions— do not try to extend it to all integrable functions. There do exist functions that are integrable over an interval $[a, b]$ but have no antiderivative for that interval. One example is

$$f(x) = [\![x]\!]$$

on the interval $[0, 3]$. Actually this function does have an antiderivative in $[1/4, 3/4]$, since it is continuous on this smaller interval; however some functions that are integrable on any interval have antiderivatives on none.†

---

† See Bernard R. Gelbaum and John M. Olmsted, *Counterexamples in Analysis* (San Francisco: Holden-Day, 1964), p. 43.

Problems

A

In Problems 1–14, evaluate the given integrals.

1. $\int_0^2 x^2 \, dx.$

2. $\int_1^2 3x^2 \, dx.$

3. $\int_{-1}^2 (x^3 - 1) \, dx.$

4. $\int_{-1}^1 (x^4 + x) \, dx.$

5. $\int_1^2 (3x^2 + 2x) \, dx.$

6. $\int_0^1 (4x^3 - x) \, dx.$

7. $\int_0^1 x^2(x - 3) \, dx.$

8. $\int_{-2}^1 x(x^2 - 4) \, dx.$

9. $\int_1^2 x^2(x - 1) \, dx.$

10. $\int_1^2 x(x - 2)^2 \, dx.$

11. $\int_1^4 (x + 3\sqrt{x}) \, dx.$

12. $\int_0^4 (1 - \sqrt{x})^2 \, dx.$

13. $\int_1^8 \sqrt[3]{x} \, dx.$

14. $\int_1^8 5x^{2/3} \, dx.$

B

In Problems 15–20, evaluate the given integrals.

15. $\int_1^4 \frac{3}{\sqrt{x}} \, dx.$

16. $\int_4^9 (3x^{1/2} + x^{-1/2}) \, dx.$

17. $\int_1^2 \frac{1}{x^2} \, dx.$

18. $\int_1^3 \frac{2}{x^3} \, dx.$

19. $\int_1^4 \frac{2x - 3}{\sqrt{x}} \, dx.$

20. $\int_{-1}^2 \sqrt{|x|} \, dx.$

In Problems 21–30, find the area of the given region.

21. The region bounded by $y = x^2 + 1$, $y = 0$, $x = 0$, and $x = 1$.
22. The region bounded by $y = x^2 - 4$ and $y = 0$.
23. The region bounded by $y = x^2(x - 2)^2$ and $y = 0$.
24. The region bounded by $y = (x^2 + x)^2$ and $y = 0$.
25. The region bounded by $y = (x - 1)^2$, $y = 0$, and $x = 0$.
26. The region bounded by $y = x(x - 3)^2$ and $y = 0$.
27. The region bounded by $y = 1/\sqrt{x}$ and $y = 0$ and between $x = 1$ and $x = 4$.
28. The region bounded by $y = 1/x^2$ and $y = 0$ and between $x = 1/2$ and $x = 1$.
29. The region bounded by $y = \sqrt{x}$, $y = 0$, and $x = 4$.
30. The region bounded by $y = x - 2\sqrt{x}$ and $y = 0$.

C

31. How would the proof of the fundamental theorem have to be altered if $f(x) < 0$ for some x?
32. Give an example of a function f for which $\lim_{h \to 0} m \neq f(x)$ for some x. Give an example of a function f for which both $\lim_{h \to 0} m \neq f(x)$ and $\lim_{h \to 0} M \neq f(x)$ for the same x. (The symbols m, M, and h are given on page 198.)

33. For each x in $[a, b]$ the function G used in the proof of the fundamental theorem can be written as

$$G(x) = \int_a^x f(t)\, dt.$$

Since $G'(x) = f(x)$,

$$\frac{d}{dx} \int_a^x f(t)\, dt = f(x).$$

Given

$$f(x) = \int_1^x \frac{t}{\sqrt{t^4 + 1}}\, dt,$$

find $f'(3)$ and $f'(-2)$.

34. If

$$f(x) = \int_0^{x^2} \sqrt{1 + t^3}\, dt,$$

find $f'(x)$. (*Hint*: Use the chain rule.)

7.5

The Integral As an Antiderivative

The fundamental theorem of integral calculus indicates the importance of the antiderivative. Let us now give it a name that emphasizes its importance in integration.

Definition

The **indefinite integral** of the function f, represented by $\int f(x)\, dx$, is the set of all antiderivatives of f.

Rather than using set notation for the indefinite integral, we simply give the most general function. For example, $\int 3x^2\, dx$ *is expressed as* $x^3 + C$ rather than $\{x^3 + C \mid C$ real$\}$. The results of Section 6.5 concerning antiderivatives can now be put into integral notation.

Theorem 7.6

If $u(x)$ and $v(x)$ are integrable, then

(a) $\displaystyle \int x^n \, dx = \frac{x^{n+1}}{n+1} + C \quad (n \neq -1),$

(b) $\displaystyle \int cv(x) \, dx = c \int v(x) \, dx, \quad$ where c is a constant.

(c) $\displaystyle \int [u(x) + v(x)] \, dx = \int u(x) \, dx + \int v(x) \, dx.$

Example 1

Evaluate $\displaystyle \int (3x^2 - 4x) \, dx.$

$$\int (3x^2 - 4x) \, dx = \int 3x^2 \, dx - \int 4x \, dx$$

$$= 3 \int x^2 \, dx - 4 \int x \, dx$$

$$= 3 \cdot \frac{x^3}{3} - 4 \cdot \frac{x^2}{2} + C$$

$$= x^3 - 2x^2 + C.$$

With very little practice, you should be able to write the answer directly without the intermediate steps. They have been put in here to show how Theorem 7.6 was used.

We might make one observation about the notation. It is possible to think of an expression such as $\int x^2 \, dx$ in two ways: one way is to think of $\int \cdots dx$ as one symbol indicating that whatever is inside is to be integrated with respect to x; the other way is to think of $\int$ as the symbol of integration, where $\int x^2 \, dx$ means to integrate the differential $x^2 \, dx$. Each point of view has its advantages as well as some disadvantages. We shall generally follow the former point of view.

One other useful formula follows from the power rule stated in Theorem 4.11.

Theorem 7.7

If n is a rational number different from -1, then

$$\int [u(x)]^n \cdot u'(x) \, dx = \frac{[u(x)]^{n+1}}{n+1} + C,$$

or, abbreviated,

$$\int u^n \cdot u' \, dx = \frac{u^{n+1}}{n+1} + C.$$

Since $du = u' \, dx$, this formula is often written

$$\int u^n \, du = \frac{u^{n+1}}{n+1} + C.$$

Example 2

Evaluate $\int (x^2 + 1)^3 \cdot 2x\, dx.$

This integral is in the form $\int u^3 \cdot u'\, dx$, with $u = x^2 + 1$. Thus

$$\int (x^2 + 1)^3 \cdot 2x\, dx = \frac{(x^2 + 1)^4}{4} + C.$$

A question many ask at this point is, Where did the $2x$ go? The only answer is a reminder that integration is the inverse of differentiation and that

$$\frac{d}{dx}\left[\frac{(x^2 + 1)^4}{4} + C\right] = (x^2 + 1)^3 \cdot 2x.$$

Since $2x$ appears on differentiation, it must disappear on integration—wherever it comes from when we differentiate is where it must go when we integrate.

Some people prefer actually to carry out the substitution in the integral of Example 2. In that case, the formula

$$\int u^n\, du = \frac{u^{n+1}}{n+1} + C$$

is more convenient. Making the substitution $u = x^2 + 1$ and taking the differential, we have $du = 2x\, dx.$ Thus

$$\int (x^2 + 1)^3 \cdot 2x\, dx = \int u^3\, du$$

$$= \frac{u^4}{4} + C$$

$$= \frac{(x^2 + 1)^4}{4} + C.$$

The advantage of carrying out the substitution is that the need for the $2x$ in the integral becomes more obvious. On the other hand, the disadvantage is that it requires that we go through the tedious mechanics of carrying out the substitution to get the integral entirely in terms of u, and then substituting back after integration to get the result in terms of x. You might prefer to carry out the substitution on the first few problems and then go back to the method given above after you are more familiar with the situation.

Example 3

Evaluate $\int (x^2 - 1)^4 \cdot x\, dx.$

If we let $u = x^2 - 1$, then $u' = 2x$. Unfortunately we do not have $2x$; we have x. So, if we want something we do not have, we put it in and compensate with something else. Let us put in 2 and compensate with $1/2$.

$$\int (x^2 - 1)^4 \cdot x \, dx = \int \frac{1}{2}(x^2 - 1)^4 \cdot 2x \, dx$$

$$= \frac{1}{2}\int (x^2 - 1)^4 \cdot 2x \, dx \quad \text{(By Theorem 7.6(b))}$$

$$= \frac{1}{2}\frac{(x^2 - 1)^5}{5} + C \quad\quad \text{(By Theorem 7.7)}$$

$$= \frac{1}{10}(x^2 - 1)^5 + C.$$

If we make the substitution $u = x^2 - 1$, then $du = 2x \, dx$ or $x \, dx = 1/2 \, du$. Thus

$$\int (x^2 - 1)^4 \cdot x \, dx = \int u^4 \cdot \frac{1}{2} \, du$$

$$= \frac{1}{2}\int u^4 \, du \quad \text{(By Theorem 7.6(b))}$$

$$= \frac{1}{2}\frac{u^5}{5} + C \quad \text{(By Theorem 7.7)}$$

$$= \frac{1}{10}(x^2 - 1)^5 + C.$$

This method works only when what we have and what we want differ by a *constant* factor, since Theorem 7.6(b) allows us to take only constants outside the integral sign.

Example 4

Evaluate $\int (x^2 - 1)^2 \, dx.$

Letting $u = x^2 - 1$, we have $u' = 2x$. Let us try the method of the preceding example and see why it fails here.

$$\int (x^2 - 1)^2 \, dx = \int \frac{1}{2x}(x^2 - 1)^2 \cdot 2x \, dx$$

$$\neq \frac{1}{2x}\int (x^2 - 1)^2 \cdot 2x \, dx.$$

Theorem 7.6(b) cannot be used here, because $1/(2x)$ is not a constant. The integration must be carried out as follows:

$$\int (x^2 - 1)^2 \, dx = \int (x^4 - 2x^2 + 1) \, dx$$

$$= \frac{x^5}{5} - \frac{2x^3}{3} + x + C.$$

Example 5

Evaluate $\int (3x - 2)^2 \, dx$.

We can use the method of either Example 3 or Example 4. Let us try both methods.

$$\int (3x - 2)^2 \, dx \qquad\qquad \int (3x - 2)^2 \, dx$$

$$= \frac{1}{3} \int (3x - 2)^2 \cdot 3 \, dx \qquad = \int (9x^2 - 12x + 4) \, dx$$

$$= \frac{1}{9} (3x - 2)^3 + C_1. \qquad = 3x^3 - 6x^2 + 4x + C_2.$$

The results we get by the two different methods look quite different. Are they equivalent? Let us see.

$$\frac{1}{9}(3x - 2)^3 + C_1 = \frac{1}{9}(27x^3 - 54x^2 + 36x - 8) + C_1$$

$$= 3x^3 - 6x^2 + 4x - \frac{8}{9} + C_1$$

$$= 3x^3 - 6x^2 + 4x + C_2, \quad \text{where } C_2 = -\frac{8}{9} + C_1.$$

Although the two results look quite different, they are really equivalent with different constants of integration. Thus, one could carry out an integration in two different ways and arrive at two answers that look different but are both correct. One way to check is to differentiate the result. If differentiation gives the original expression, the integration is correct (assuming that the differentiation is correct).

Problems

A
Evaluate the integrals.

1. $\int 2(2x - 3)^2 \, dx$.

2. $\int 2(2x - 3)^{2/3} \, dx$.

3. $\int (x^3 - x)(3x^2 - 1) \, dx$.

4. $\int (x^3 - x)^2 \, dx$.

5. $\int \sqrt{4x - 3} \, dx$.

6. $\int \sqrt{3x + 1} \, dx$.

7. $\int (x^2 + 2)^2 x \, dx$.

8. $\int (x^2 + 3)^2 \, dx$.

9. $\int (x^2 + 4)^2 x^3 \, dx$.

10. $\int (x^3 + 1)^2 \, dx$.

11. $\int (x^3 + 3)^2 x^2 \, dx$.

12. $\int \frac{(\sqrt{x} + 1)^3}{\sqrt{x}} \, dx$.

B
Evaluate the integrals.

13. $\int \frac{(x^{1/3} - 1)^5}{x^{2/3}} \, dx$.

14. $\int (x^{-1} + 1)^4 x^{-2} \, dx$.

15. $\displaystyle\int (x^{-2} + 3x)^5(3 - 2x^{-3})\,dx.$

16. $\displaystyle\int x\sqrt{2x^2 - 3}\,dx.$

17. $\displaystyle\int x\sqrt[3]{3x^2 - 5}\,dx.$

18. $\displaystyle\int (x^{-2} - 3)^{2/5}x^{-3}\,dx.\cdot$

19. $\displaystyle\int \frac{x^2 + 4}{x^2}\,dx.$

20. $\displaystyle\int \frac{x^3 + 2x^2 - 7x - 1}{(x - 1)^2}\,dx.$

21. $\displaystyle\int \frac{x^3 + 8x^2 + 20x + 20}{(x + 2)^2}\,dx.$

22. $\displaystyle\int \frac{x^4 - 4x}{(x - 1)^2}\,dx.$

23. $\displaystyle\int_0^2 (x^2 - 3)^2 x\,dx.$

24. $\displaystyle\int_0^1 (4x + 3)^2\,dx.$

25. $\displaystyle\int_1^2 (x^3 - x)^2(3x^2 - 1)\,dx.$

26. $\displaystyle\int_{-1}^3 \sqrt{2x + 3}\,dx.$

27. $\displaystyle\int_1^{16} \sqrt{3x + 1}\,dx.$

28. $\displaystyle\int_0^1 (x^2 + 2)^3\,dx.$

29. $\displaystyle\int_1^2 (x^2 - 3)^3 x\,dx.$

30. $\displaystyle\int_0^1 4x\sqrt{x^2 + 3}\,dx.$

C

31. The polynomial P is positive for $x > 0$, and the area of the region bounded by $P(x)$, the x axis, and the vertical lines $x = 0$ and $x = k$ is $k^2(k + 3)/3$. Find $P(x)$.

7.6

The Area between Two Curves

By altering the $f(x)$ in $\int_a^b f(x)\,dx$, we can extend considerably what we are able to do with the integral. Figure 7.13(a) shows a representative rectangle used to set up the integral for the area. It is advisable to draw such rectangles in all cases, as an aid to setting up the integral—the integral $\int_a^b f(x)\,dx$ works only in the restricted case we have studied so far. Figure 7.13(b) shows a representative rectangle used to set up the integral for the area of the region bounded above by $y = f(x)$ and below by $y = g(x)$, and between the vertical lines $x = a$ and $x = b$. Basically, we have the same situation as before. The interval $[a, b]$ is subdivided and rectangles are formed. But this time, and this is the only difference, the top of the rectangle is on $y = f(x)$ and the bottom is on $y = g(x)$. Thus the height of the rectangle is the difference between the y coordinates: $f(x) - g(x)$. The relative position of the curves with respect to the x axis has no bearing on the situation (see page 3); the distance between two points on a vertical line is the larger y coordinate minus the smaller. Thus

$$A = \lim_{\|S\| \to 0} \sum_{i=1}^n [f(x_i^*) - g(x_i^*)]\,\Delta x_i$$

$$= \int_a^b [f(x) - g(x)]\,dx.$$

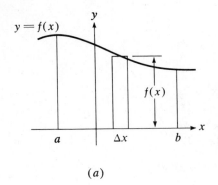

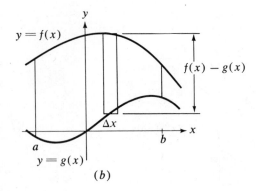

(a)   (b)

Figure 7.13

Example 1

Find the area of the region bounded by
$y = 4 - 4x^2$ and $y = 1 - x^2$.

The graphs of the two curves are given in Figure 7.14. We see that the points of intersection are $(-1, 0)$ and $(1, 0)$, representing the left- and right-hand extremes, respectively, of the region under consideration. Since the height of the rectangle is $y_1 - y_2$ and the width is Δx, we have

$$A = \int_{-1}^{1} (y_1 - y_2) \, dx$$

$$= \int_{-1}^{1} (3 - 3x^2) \, dx$$

$$= 3x - x^3 \Big|_{-1}^{1}$$

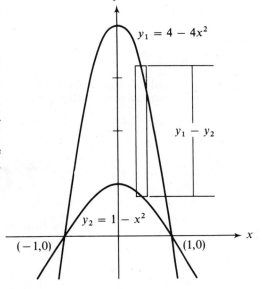

Figure 7.14

$$= (3 - 1) - (-3 + 1) = 4.$$

In this case we could have found the areas under each curve separately and subtracted. This would have given

$$A = \int_{-1}^{1} y_1 \, dx - \int_{-1}^{1} y_2 \, dx,$$

which is equivalent to

$$\int_{-1}^{1} (y_1 - y_2) \, dx.$$

Example 2

Find the area of the region bounded by $y^2 = x + 2$ and $y = x$.

The graphs are given in Figure 7.15. The points of intersection are found by solving the two equations simultaneously. In this case they do not represent the left- and right-hand extremes of the region. The extremes are $(-2, 0)$ and $(2, 2)$. Now we have a new problem: both ends of the vertical strip are on the curve $y^2 = x + 2$ for $-2 \le x \le -1$, while the top end is on $y^2 = x + 2$ and the bottom on the line $y = x$ for $-1 \le x \le 2$. We shall find the areas of the two parts separately, noting that the top and bottom halves of $y^2 = x + 2$ are $y = \sqrt{x + 2}$ and $y = -\sqrt{x + 2}$, respectively. Now

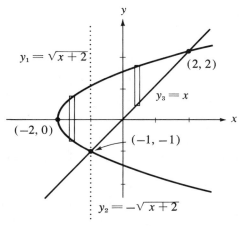

Figure 7.15

$$A = \int_{-2}^{-1} (y_1 - y_2) \, dx + \int_{-1}^{2} (y_1 - y_3) \, dx$$

$$= \int_{-2}^{-1} (\sqrt{x + 2} + \sqrt{x + 2}) \, dx + \int_{-1}^{2} (\sqrt{x + 2} - x) \, dx$$

$$= \frac{4}{3} (x + 2)^{3/2} \Big|_{-2}^{-1} + \left[\frac{2}{3} (x + 2)^{3/2} - \frac{1}{2} x^2 \right] \Big|_{-1}^{2}$$

$$= \frac{4}{3} + \frac{19}{6} = \frac{9}{2}.$$

The same result is found more simply by reversing the roles of x and y—that is, by using horizontal strips rather than vertical ones, noting that $(2, 2)$ and $(-1, -1)$ represent the top and bottom extremes of the region (see Figure 7.16). In this case, one end is always on $x = y^2 - 2$ and the other on $x = y$.

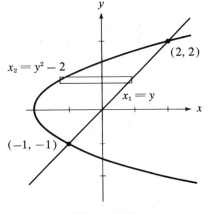

Figure 7.16

$$A = \int_{-1}^{2} (x_1 - x_2) \, dy$$

$$= \int_{-1}^{2} (y - y^2 + 2) \, dy$$

$$= \left(\frac{y^2}{2} - \frac{y^3}{3} + 2y \right) \Big|_{-1}^{2}$$

$$= \frac{9}{2}.$$

Example 3

Find the total area between $y = 4x - x^2$ and $y = x$ from $x = 0$ to $x = 4$.

The graphs are given in Figure 7.17.

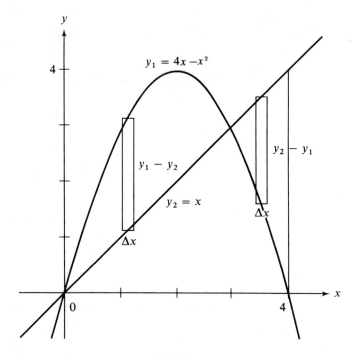

Figure 7.17

Notice that $y_1 = 4x - x^2$ is above $y_2 = x$ for $0 < x < 3$, while $y_2 = x$ is above $y_1 = 4x - x^2$ for $3 < x < 4$. Since the height of the vertical strip must be positive, we must always take it to be the larger y coordinate minus the smaller. Thus the height is $y_1 - y_2$ for $0 \le x \le 3$ and $y_2 - y_1$ for $3 \le x \le 4$. Thus two integrals must be used to find the area.

$$A = \int_0^3 (y_1 - y_2)\, dx + \int_3^4 (y_2 - y_1)\, dx$$

$$= \int_0^3 [(4x - x^2) - x]\, dx + \int_3^4 [x - (4x - x^2)]\, dx$$

$$= \int_0^3 (3x - x^2)\, dx + \int_3^4 (x^2 - 3x)\, dx$$

$$= \left(\frac{3x^2}{2} - \frac{x^3}{3} \right) \Big|_0^3 + \left(\frac{x^3}{3} - \frac{3x^2}{2} \right) \Big|_3^4$$

$$= \left(\frac{27}{2} - 9 \right) - 0 + \left(\frac{64}{3} - 24 \right) - \left(9 - \frac{27}{2} \right) = \frac{19}{3}.$$

Problems

A

In Problems 1–12, find the area of the given region.

1. The region bounded by $y = 2x - x^2$ and $y = 2x - 4$.
2. The region bounded by $y = 2 + x - x^2$ and $y = -4$.
3. The region bounded by $y = x^2 - 5x + 4$ and $y = -(x - 1)^2$.
4. The region bounded by $y = x^2 - x - 6$ and $y = x - 3$.
5. The region bounded by $x = y^2 - 4y$ and $x = 0$ and between $y = 2$ and $y = 5$.
6. The region bounded by $x = y^2 + 2y - 3$ and $x = 0$ and between $y = 0$ and $y = 2$.
7. The region bounded by $y = x^2$ and $x - y + 6 = 0$.
8. The region bounded by $y = x^2 - 1$ and $y = 11 - 2x^2$.
9. The region bounded by $y = 2x - x^2$ and $x + y = 0$.
10. The region bounded by $y = x^3$ and $y = \sqrt{x}$.
11. The region bounded by $x = y^2$ and $y = x^2$.
12. The region bounded by $x = y^2 - y - 2$ and $x + y = 7$.

B

In Problems 13–20, find the area of the given region.

13. The region bounded by $x^2 y = 4$ and $3x + y = 7$.
14. The region bounded by $x - y = 0$, $x + 3y = 0$, and $3x + y = 8$.
15. The region bounded by $y = x(x - 2)^2$ and $y = x$.
16. The region bounded by $x = y^3 - 3y^2$ and $x - y + 3 = 0$.
17. The region bounded by $y = x^2 + 2x$ and $y = 2 - x - x^2$.
18. The region bounded by $x = y^2 - 9$ and $x = 3 - 2y - y^2$.
19. The region bounded by $y = x^2 - 1$ and $y = x^3 - 1$.
20. The region bounded by $y = x^3 - x$ and $y = 3 - 3x^2$.

21. Find the area of the region bounded by $y = x^2$ and the line through $(0, 0)$ and $(2, 4)$.
22. Find the area of the region bounded by $y = x^2 - 2x$ and the line through $(-2, 8)$ and $(2, 0)$.
23. Find the area of the region bounded by $y = x^2 - x - 2$ and the line through $(-2, 4)$ and $(0, -2)$.
24. Find the area of the region bounded by $x = y^2 - 4$ and the line through $(5, 3)$ and $(0, -2)$.
25. Find the area of the triangle with sides $x + 2y - 7 = 0$, $x - y - 4 = 0$, and $5x + y - 8 = 0$.
26. Find the area of the triangle with vertices $(2, 3)$, $(-1, 0)$, and $(3, -2)$.

7.7

Work

Let us now consider another of the many applications of the integral—that of determining the amount of work done. Work is defined as the product of the force exerted on an object and the distance the object is moved by the force, assuming that the force is applied in the direction of the motion. For instance, if a 5-lb weight is lifted 10 ft, the work performed is the product of the force, 5 lb, and the distance, 10 ft, which is 50 ft-lb of work. Unfortunately, not all work problems are so simple.

Consider the work done in stretching a spring. At first, relatively little force is needed to stretch it; but as the spring is stretched more and more, more and more force is required. Since the force applied varies continuously as the spring is stretched, we cannot use any one number for the force. Similarly, if water is pumped out of a tank over the top, more work is required as the water level recedes, because the water must be lifted farther. In such instances as these, the integral can help us.

Example 1

A spring is stretched six inches by a force of 12 pounds. How much work is done in stretching the spring two feet?

First of all, let us assume that the spring obeys Hooke's law, which states that the force F required to stretch a spring a distance x is proportional to x. $F = kx$. Now k may be determined.

$$F = kx, \qquad 12 = k \cdot \frac{1}{2}, \qquad k = 24.$$

Thus, the force required to hold the spring stretched a distance x feet from the equilibrium position is

$F = 24x$ lb.

We can approximate the work done in the following way. Let us subdivide the interval $[0, 2]$ into n subintervals (see Figure 7.18). Let Δx_i be the length of one such subinterval and let x_i^* be a number in that subinterval. For the ith subinterval, the force is approximately $24x_i^*$ and the distance is Δx_i.

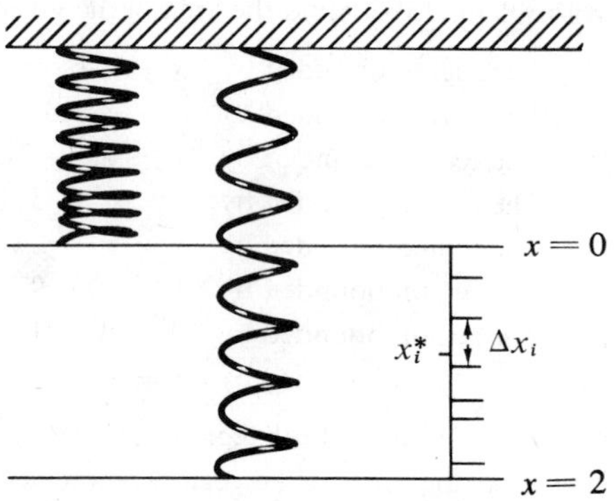

The work done throughout that subinterval is approximately

$24x_i^* \, \Delta x_i,$

Figure 7.18

and the total work from $x = 0$ to $x = 2$ is approximately

$$\sum_{i=1}^{n} 24x_i^* \, \Delta x_i.$$

But we do not want an approximation of the work done—we want the exact value. In order to find it, let us take the limit of our approximation as the lengths of the subintervals approach 0.

$$W = \lim_{\|S\| \to 0} \sum_{i=1}^{n} 24x_i^* \, \Delta x_i$$

$$= \int_{0}^{2} 24x \, dx$$

$$= 12x^2 \Big|_{0}^{2}$$

$$= 48 \text{ ft} \cdot \text{lb}.$$

We see here that defining an integral as a limit of a certain approximating sum is what allows us to extend its use from the computation of areas to other applications.

Example 2

A chain 50 feet long and weighing two pounds per foot is hanging from the top of a cliff. How much work is needed to pull 20 feet of it to the top?

The force needed to hold the chain is equal to the weight of the chain hanging down. Thus, if x feet of chain are hanging down,

$$F = 2x.$$

The work is then

$$W = \int_{30}^{50} 2x \, dx$$

$$= x^2 \Big|_{30}^{50}$$

$$= 1600 \text{ ft-lb.}$$

Example 3

A cylindrical tank ten feet in diameter and ten feet high is full of water. How much work is required to pump the water over the top?

We see (Figure 7.19) that different "layers" must be raised different distances to the top. If x represents the distance from the top of the tank, suppose we subdivide the interval $[0, 10]$ into n subintervals. If Δx_i is the length of the ith subinterval and x_i^* is a number in that interval, then the weight of the ith layer is its volume times the density (weight per unit volume) of water (62.4 lb/ft^3),

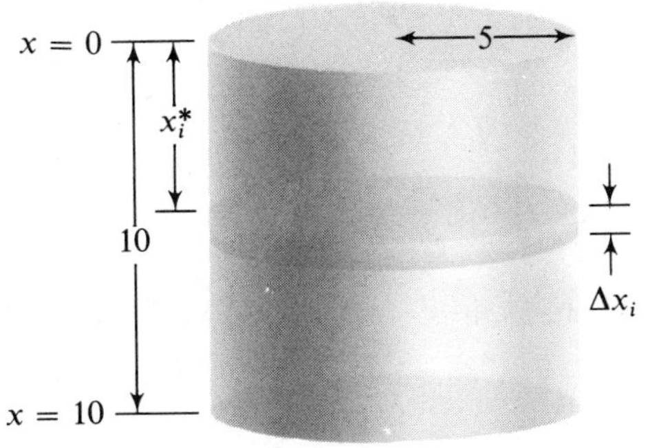

Figure 7.19

$$F = (62.4)\pi \cdot 5^2 \cdot \Delta x_i$$

$$= 1560\pi \cdot \Delta x_i,$$

and the approximate work done in raising it to the top is

$$W = 1560\pi \, \Delta x_i \cdot x_i^*.$$

Taking the limit of the sum of all such layers gives

$$W = \int_0^{10} 1560\pi x \, dx$$

$$= 780\pi x^2 \Big|_0^{10}$$

$$= 78{,}000\pi \text{ ft-lb.}$$

It might be noted that this same amount of work would be done *by* the water if it were allowed to run out at the bottom of the tank. This work could (theoretically) be recovered by having the water drive some machine (a waterwheel, for example). In practice, some of the work will be lost.

Suppose that we are filling a tank by pumping the water in through a hole in the bottom. In order to simplify this discussion, we assume that the tank is a cylinder with a horizontal cross section of area A. Now let us consider the work done in raising the level from x to $x + \Delta x$. We consider two ways of looking at the situation. In the first method we note that the force necessary to support the water in the tank is equal to the weight of the water. When the water is at the x level, that weight is $62.4Ax$. To increase the level to $x + \Delta x$ we need to increase the force to $62.4A(x + \Delta x)$. Thus the force is approximated by

$$F = 62.4Ax^*,$$

where $x \leq x^* \leq x + \Delta x$. The distance through which this force is acting is the increase in the level, which is

$$d = \Delta x,$$

giving the work

$$W = Fd = 62.4Ax^*\Delta x.$$

It is often simpler to think of the water already in the tank as remaining stationary, and the water necessary to increase the level by Δx as being elevated to the top layer. By this way of thinking, no work is done on the water already in the tank because it is not moved—the work is done only on the layer of water which is put on top. The force is equal to the weight of that layer; it is

$$F = 62.4A\Delta x.$$

The distance it is elevated is approximated by

$$d = x^*,$$

where again $x \leq x^* \leq x + \Delta x$. Thus the work is again seen to be

$$W = Fd = 62.4Ax^*\Delta x.$$

Example 4

Suppose the tank of Example 3 is placed on a ten-foot platform. How much work is done in filling the tank from a source of water at ground level if the water is pumped in through a hole in the bottom?

Let us approach the problem from the second point of view considered above. Suppose the vertical axis has 0 at ground level. Then the bottom of the tank is at 10 ft and the top at 20 ft. Thus we subdivide the interval $[10, 20]$ into n subintervals. This is shown on the right-hand side of Figure 7.20.

If Δx_i is the length of the ith subinterval and x_i^* is a number in that interval, then, as in Example 3, the force required to raise the layer of water in this interval is

$$F = (62.4)\pi \cdot 5^2 \cdot \Delta x_i = 1560\pi \cdot \Delta x_i$$

and the approximate work done in raising it to the top is

$$W = 1560\pi\Delta x_i \cdot x_i^*.$$

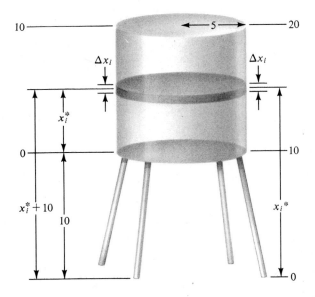

Figure 7.20

Taking the limit of the sum of all such layers from $x = 10$ to $x = 20$, we have

$$W = \int_{10}^{20} 1560\pi x \, dx = 780\pi x^2 \Big|_{10}^{20} = 234{,}000\pi \text{ ft-lb.}$$

A second way of labeling the vertical axis is shown on the left-hand side of Figure 7.20. In this case, we subdivide the interval $[0, 10]$ into n subintervals. The force required to raise the layer of water is still

$$F = 1560\pi \cdot \Delta x_i,$$

but the distance it is raised is

$$x_i^* + 10.$$

Thus the approximate work done in raising the layer to its proper position is

$$W = 1560\pi \Delta x_i (x_i^* + 10).$$

Now when we take the limit of the sum of all such layers from $x = 0$ to $x = 10$, we have

$$W = \int_0^{10} 1560\pi(x + 10) \, dx$$

$$= 1560\pi \left(\frac{x^2}{2} + 10x \right) \Big|_0^{10}$$

$$= 234{,}000\pi \text{ ft-lb.}$$

The foregoing example illustrates the fact that the placement of the axes is quite arbitrary.

Problems

A

1. A force of 3 lb stretches a spring 4 in. How much work is done in stretching the spring 1 ft?

2. How much work is done in stretching the spring of Problem 1 from 1 to 2 ft? From 2 to 3 ft?

3. A force of 6 lb stretches a spring 1 ft. How much work is done in stretching the spring the first 6 in.?

4. A force of 3 lb stretches a spring 9 in. How much work is done in stretching the spring 2 ft?

5. A force of 5 lb stretches a spring 8 in. How much work is done in stretching the spring from 1 ft to 3 ft?

6. If 135 ft-lb of work is expended in stretching the spring of Problem 5 from its equilibrium position, how far is it stretched?

7. An object moves on a line from 0 to 10 subject to force $F(x) = x^2 + 1$. Find the work done.

8. An object moves on a line from 1 to 7 subject to force $F(x) = x^2 + x - 1$. Find the work done.

B

9. A cylindrical tank 8 ft in diameter and 12 ft high is filled with water. How much work is done in pumping the water out over the top?

10. How much work is done in Problem 9 if only half the water is pumped out?

11. A conical tank (right circular cone) filled with water is 4 ft across the top and 4 ft high. How much work is done in pumping the water out over the top? How much work is done by the water if it runs out the bottom?

12. How much work is done in pumping the top 2.5 ft of water over the top of the tank of Problem 11? What portion of the entire volume of the tank does this represent?

13. A chain 50 ft long and weighing 1 lb/ft is hanging vertically. How much work is done in raising the chain to a horizontal position at the level of the top of the chain?

14. How much work is done on the chain of Problem 13 if only half is brought to the horizontal position and the other half is left hanging?

15. Suppose the bottom of the chain of Problem 13 is raised to the level of the top so that the chain is doubled but is still hanging vertically. How much work is expended in doing so?

16. A chain 12 ft long and weighing 0.5 lb/ft is lying on the floor. How much work is done if one end of the chain is raised to a level of 15 ft?

17. A 100-lb weight is suspended from a 20-ft cable weighing 0.25 lb/ft. The weighted end is raised 15 ft and the other end remains in its original position. How much work is done?

18. If the tank of Problem 9 is emptied through a hole 6 ft from the bottom, what is the net amount of work done in emptying it? (Note that the water above the hole is doing work for us as it runs out the hole. Assume that this work is recovered.) If the work done by the top 6 ft of water is lost, how much work must be done to empty the tank?

19. Suppose the tank of Problem 9 is elevated so that the bottom of the tank is 40 ft above the ground. How much work is done in filling the tank from a source of water at ground level if the water is pumped in over the top of the tank? What if it is pumped in through the bottom?

C

20. A bucket weighing 1 lb and holding 1 ft³ of water is suspended by a rope weighing 0.1 lb/ft in a well 15 ft deep. How much work is necessary to bring the bucket of water to the top?

21. Suppose the water in the bucket of Problem 20 is leaking out at the rate of 0.01 ft³/sec and the bucket is raised at the rate of 1 ft/sec. How much work is necessary to bring it to the top?

22. A ship has an anchor weighing 1000 lb out of water and 900 lb in water. The anchor chain weighs 5 lb/ft out of water and 4.5 lb/ft in water. The deck of the ship is 10 ft above the water, and the ship is directly above the anchor with no slack in the anchor chain. If the water is 50 ft deep, how much work is done to weigh anchor?

7.8

Approximate Integration

Integration is generally more difficult than differentiation. Although we can now differentiate any algebraic function, our present methods do not allow us to evaluate many relatively simple algebraic integrals. Of course we shall as time goes on extend the range of integrals that we can evaluate. Even so, it is easy to imagine extremely complicated integrals whose evaluation by any method is difficult or impossible. For such integrals as these, several methods of approximate integration have been devised. We consider two of them here. Both methods are based upon the interpretation of an integral as an area. We have already noted that we can determine areas by integrals. We now turn this idea around. Given an integral $\int_a^b f(x)\, dx$, which may come from a problem having nothing to do with area (for instance, a work problem), we can think of it as representing the "signed area" (with the portions below the x axis negative) under the curve $y = f(x)$ from $x = a$ to $x = b$. If we can approximate the area in some way, we shall have an approximation of the integral.

One way of approximating the area involves trapezoids (see Figure 7.21). The interval $[a, b]$ is subdivided into n equal subintervals, and the curve is approximated by straight-line segments forming n trapezoids. All have the common height

$$h = \frac{b - a}{n},$$

and the coordinates of the subdivisions are

$$x_0 = a,$$
$$x_1 = a + h,$$
$$x_2 = a + 2h,$$
$$\vdots$$
$$x_i = a + ih,$$
$$\vdots$$
$$x_n = a + nh = b.$$

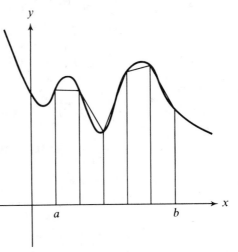

Figure 7.21

The area of the ith trapezoid is

$$A_i = \frac{h}{2}\, [f(x_{i-1}) + f(x_i)],$$

and the total area of all the approximating trapezoids is

$$A = \frac{h}{2}\, \{[f(x_0) + f(x_1)] + [f(x_1) + f(x_2)] + [f(x_2) + f(x_3)] + \cdots + [f(x_{n-1}) + f(x_n)]\}$$

$$= \frac{h}{2}\, [f(x_0) + 2f(x_1) + 2f(x_2) + \cdots + 2f(x_{n-1}) + f(x_n)]$$

$$= \frac{h}{2}\, (y_0 + 2y_1 + 2y_2 + \cdots + 2y_{n-1} + y_n).$$

This result is called the *trapezoidal rule*.

Another formula allows us to estimate the accuracy of our approximation. If A_T is the trapezoidal rule approximation and

$$\int_a^b f(x)\, dx = A_T + E_T,$$

then, assuming that f is continuous on $[a, b]$ and f'' exists on $[a, b]$, the error is given by

$$E_T = -\frac{h^2}{12}(b-a)f''(c),$$

where $a \leq c \leq b$. While this formula seldom allows us to determine the exact error, because we do not know the exact value of c, it does let us determine the maximum error. We shall not attempt to prove this error formula.

Example 1

Evaluate $\int_0^1 x^3\, dx$ by the trapezoidal rule, using $n = 4$. Find the maximum value of $|E_T|$ and the range of values of the integral.

$$h = \frac{1-0}{4} = 0.25;$$

$x_0 = 0,$	$y_0 = 0,$
$x_1 = 0.25,$	$y_1 = 0.015625,$
$x_2 = 0.5,$	$y_2 = 0.125,$
$x_3 = 0.75,$	$y_3 = 0.421875,$
$x_4 = 1.$	$y_4 = 1.$

$$\int_0^1 x^3\, dx \approx \frac{h}{2}(y_0 + 2y_1 + 2y_2 + 2y_3 + y_4)$$

$$= \frac{0.25}{2}(0 + 0.03125 + 0.25 + 0.84375 + 1) = 0.265625.$$

The error is given by

$$E_T = -\frac{h^2}{12}(b-a)f''(c) \quad (a \leq c \leq b)$$

$$= -\frac{(1/4)^2}{12}(1)(6c) \quad (0 \leq c \leq 1).$$

The maximum value of $|E_T|$ corresponds to the maximum value of c, namely, $c = 1$. Thus

$$|E_T| \leq \frac{(1/4)^2}{12}(1)(6) = 0.031.$$

Since E_T is negative,

$$0.235 \leq \int_0^1 x^3\, dx \leq 0.266.$$

In this case, integration in the usual way shows that the exact value is $1/4$.

The other method of approximate integration is called *Simpson's rule*. In this case, the interval $[a, b]$ is subdivided into n equal subintervals *where n is even*. Again,

$$h = \frac{b-a}{n} \quad \text{and} \quad x_i = a + ih.$$

This time the intervals are taken in pairs giving three points on the curve $y = f(x)$. For the first pair of intervals the points are (x_0, y_0), (x_1, y_1), and (x_2, y_2), where $y_i = f(x_i)$. See Figure 7.22. The curve $y = f(x)$ is approximated by

$$y = Ax^2 + Bx + C$$

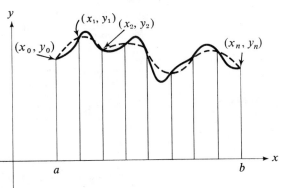

containing the three points (x_0, y_0), (x_1, y_1), and (x_2, y_2). The area under this curve from x_0 to x_2 is then found and taken as an approximation of

$$\int_{x_0}^{x_2} f(x)\, dx.$$

Figure 7.22

Let us find the approximating curve and the area under it. Since (x_0, y_0), (x_1, y_1), and (x_2, y_2) are on this curve,

$$y_0 = Ax_0^2 + Bx_0 + C,$$
$$y_1 = Ax_1^2 + Bx_1 + C,$$
$$y_2 = Ax_2^2 + Bx_2 + C.$$

Thus, by subtraction,

$$y_1 - y_0 = A(x_1^2 - x_0^2) + B(x_1 - x_0),$$
$$y_2 - y_1 = A(x_2^2 - x_1^2) + B(x_2 - x_1).$$

Since $x_1 - x_0 = x_2 - x_1 = h$, these simplify to

$$\frac{y_1 - y_0}{h} = A(x_1 + x_0) + B,$$

$$\frac{y_2 - y_1}{h} = A(x_2 + x_1) + B.$$

Again, by subtraction,

$$\frac{(y_2 - y_1) - (y_1 - y_0)}{h} = A(x_2 - x_0).$$

Furthermore, since $x_2 - x_0 = 2h$,

$$A = \frac{y_0 - 2y_1 + y_2}{2h^2}. \tag{1}$$

In addition, the second of our three original equations gives

$$Bx_1 + C = y_1 - Ax_1^2. \tag{2}$$

Now the area under the curve

$$y = Ax^2 + Bx + C$$

from $x = x_0$ to $x = x_2$ is

$$\int_{x_0}^{x_2} (Ax^2 + Bx + C)\, dx = \int_{x_1-h}^{x_1+h} (Ax^2 + Bx + C)\, dx$$

$$= \left(\frac{Ax^3}{3} + \frac{Bx^2}{2} + Cx \right)\Bigg|_{x_1-h}^{x_1+h}$$

$$= \frac{A}{3}[(x_1 + h)^3 - (x_1 - h)^3] + \frac{B}{2}[(x_1 + h)^2 - (x_1 - h)^2]$$

$$+ C[(x_1 + h) - (x_1 - h)]$$

$$= \frac{A}{3}(6hx_1^2 + 2h^3) + 2h(Bx_1 + C)$$

$$= \frac{A}{3}(6hx_1^2 + 2h^3) + 2h(y_1 - Ax_1^2), \quad \text{by (2)},$$

$$= 2h\left(\frac{Ah^2}{3} + y_1 \right)$$

$$= 2h\left(\frac{y_0 - 2y_1 + y_2}{6} + y_1 \right), \quad \text{by (1)},$$

$$= \frac{h}{3}(y_0 + 4y_1 + y_2).$$

Repeating this process for pairs of intervals from a to b and adding, we have

$$A = \frac{h}{3}[(y_0 + 4y_1 + y_2) + (y_2 + 4y_3 + y_4) + \cdots + (y_{n-2} + 4y_{n-1} + y_n)]$$

$$= \frac{h}{3}(y_0 + 4y_1 + 2y_2 + 4y_3 + 2y_4 + \cdots + 2y_{n-2} + 4y_{n-1} + y_n).$$

This is Simpson's rule. It generally gives a better approximation than the trapezoidal rule. In fact, if

$$\int_a^b f(x)\, dx = A_S + E_S,$$

then, assuming that f is continuous on $[a, b]$ and $f^{(4)}$ exists on $[a, b]$, the error is given by

$$E_S = -\frac{h^4}{180}(b - a)f^{(4)}(c),$$

where $a \le c \le b$. Again we shall not attempt to prove this error formula.

Example 2

Approximate $\int_0^1 x^3 \, dx$ by Simpson's rule, using $n = 4$. Find the maximum value of $|E_S|$ and the range of values of the integral.

$$h = \frac{1 - 0}{4} = 0.25;$$

$$
\begin{array}{ll}
x_0 = 0, & y_0 = 0, \\
x_1 = 0.25, & y_1 = 0.015625, \\
x_2 = 0.5, & y_2 = 0.125, \\
x_3 = 0.75 & y_3 = 0.421875, \\
x_4 = 1. & y_4 = 1.
\end{array}
$$

$$\int_0^1 x^3 \, dx \approx \frac{h}{3}(y_0 + 4y_1 + 2y_2 + 4y_3 + y_4)$$

$$= \frac{0.25}{3}(0 + 0.0625 + 0.25 + 1.6875 + 1) = 0.25.$$

The error is given by

$$E_S = -\frac{h^4}{180}(b - a)f^{(4)}(c) \quad (a \le c \le b).$$

But $E_S = 0$ because $f^{(4)}(x) = 0$, and the Simpson's rule approximation gives the exact value of the integral.

The trapezoidal rule and Simpson's rule are only two of several methods available for the numerical evaluation of integrals. The recent increased use of high speed computers has led to a renewed interest in this area. If you have access to a computer, you might consider writing a program to carry out the numerical integration given here. In any case, a hand calculator takes away much of the tediousness of these problems. For other methods of numerical integration, see Carl-Erik Fröberg, *Introduction to Numerical Analysis*, 2nd ed. (Reading, Mass.: Addison-Wesley, 1969), pp. 195–223.

Problems

A

In Problems 1–10, evaluate the given integral by both the trapezoidal rule and Simpson's rule and compare with the exact value.

1. $\int_1^2 (x^2 + 1) \, dx, \quad n = 4.$

2. $\int_1^2 (x^3 - 4) \, dx, \quad n = 4.$

3. $\int_0^1 (x + 1)^2 \, dx, \quad n = 4.$

4. $\int_0^1 x(x^2 + 1)^2 \, dx, \quad n = 6.$

5. $\int_0^1 \sqrt{x + 1} \, dx, \quad n = 4.$

6. $\int_0^1 \sqrt{x + 1} \, dx, \quad n = 6.$

7. $\int_0^1 2x\sqrt{x^2 + 3} \, dx, \quad n = 4.$

8. $\int_0^1 2x(x^2 + 3)^{3/2} \, dx, \quad n = 6.$

9. $\int_1^2 \frac{x^2 + 1}{x^2} \, dx, \quad n = 4.$

10. $\int_0^1 \frac{2x \, dx}{\sqrt{x^2 + 1}}, \quad n = 4.$

B

In Problems 11–16, evaluate the integral by the trapezoidal rule. Find the maximum value of $|E_T|$ and the range of values of the integral.

11. $\int_0^1 \sqrt{2x^2 + 1}\, dx, \quad n = 5.$

12. $\int_0^1 (x^2 + 3)^{3/2}\, dx, \quad n = 10.$

13. $\int_1^3 \frac{x + 1}{x}\, dx, \quad n = 5.$

14. $\int_1^3 \frac{dx}{\sqrt{x^2 + 1}}, \quad n = 4.$

15. $\int_0^4 \sqrt{16 - x^2}\, dx, \quad n = 4.$

16. $\int_0^7 \sqrt[3]{1 + 2x}\, dx, \quad n = 7.$

In Problems 17–22, evaluate the integral by Simpson's rule. Find the maximum value of $|E_S|$ and the range of values of the integral.

17. $\int_0^1 \sqrt{1 - x^2}\, dx, \quad n = 4.$

18. $\int_0^2 \sqrt{x^2 + x}\, dx, \quad n = 6.$

19. $\int_0^3 x\sqrt{4 + x}\, dx, \quad n = 6.$

20. $\int_0^3 \frac{x\, dx}{\sqrt{1 + 5x}}, \quad n = 6.$

21. $\int_0^1 \frac{x\, dx}{1 + x}, \quad n = 4.$

22. $\int_0^2 \sqrt[3]{1 + 2x^2}\, dx, \quad n = 4.$

In Problems 23–28, evaluate the integral by the trapezoidal rule and Simpson's rule.

23. $\int_2^4 x^3\sqrt{x^2 - 4}\, dx, \quad n = 4.$

24. $\int_1^3 (9 - x^2)^{2/3}\, dx, \quad n = 4.$

25. $\int_1^3 \frac{x^2 - 1}{x}\, dx, \quad n = 8.$

26. $\int_0^4 \frac{\sqrt{x}\, dx}{2 + 3x}, \quad n = 4.$

27. $\int_1^2 \sqrt{x^2 - x}\, dx, \quad n = 4.$

28. $\int_0^2 x^2\sqrt{4x + 1}\, dx, \quad n = 4.$

29. If $\int_0^1 \sqrt{x^2 + 4}\, dx$ is approximated by the trapezoidal rule, how big must n be so that $|E_T| < 0.001$?

30. If $\int_1^2 \frac{dx}{x}$ is approximated by Simpson's rule, how big must n be so that $|E_S| < 0.001$?

C

31. Show that the trapezoidal rule gives the exact value of $\int_a^b f(x)\, dx$ when $f(x)$ is a linear expression and that Simpson's rule gives the exact value when $f(x)$ is a third-degree polynomial.

32. Show that we can find the exact value of $\int_a^b f(x)\, dx$ when $f(x)$ is a second-degree polynomial by using the trapezoidal rule with error. Show that we can find the exact value when $f(x)$ is a fourth-degree polynomial by using Simpson's rule with error.

Review Problems

A

1. Simplify the expression $\displaystyle\sum_{i=1}^{n}\left[\left(\frac{2i}{n}\right)^2 - \frac{6i}{n}\right]\frac{2}{n}$.

In Problems 2–7, evaluate the given integrals.

2. $\displaystyle\int_{-1}^{2} (x^2 + 1)\,dx$.

3. $\displaystyle\int_{0}^{2} (x^2 + 1)^2\,dx$.

4. $\displaystyle\int_{0}^{2} x\sqrt{2x^2 + 1}\,dx$.

5. $\displaystyle\int_{1}^{2} \frac{(x^2 - 2)^2}{x^2}\,dx$.

6. $\displaystyle\int_{0}^{1} (4x + 1)^5\,dx$.

7. $\displaystyle\int_{-1}^{0} (2x + 1)(x^2 + x)^4\,dx$.

B

In Problems 8–16, find the areas of the given regions.

8. The region in the first quadrant bounded by the coordinate axes and $y = 6 + 5x - x^2$.
9. The region between $x = -3$ and $x = 2$ bounded by the x axis and $y = x^3 - 4x + 15$.
10. The region between $x = -1$ and $x = 2$ bounded by the x axis and $y = x^3 - 9x + 8$.
11. The region bounded by the y axis and $x = y^2 - 3y$.
12. The region bounded by $y = 5x - x^2$ and $y = 5 - x$.
13. The region bounded by $y = x^3$ and $y = x^4$.
14. The region bounded by $y = 5x - x^3$ and $y = x$.
15. The region between $x = 1/2$ and $x = 2$ bounded by $y = x$ and $y = 1/x^2$.
16. The region bounded by $y = x$, $y = 1/x^2$, and $y = 2$.

17. A spring is stretched 6 in. by a 3-lb weight. How much work is done in stretching it 1 ft? How much work is done in stretching it from 1 ft to 2 ft?

18. A 15-ft high cylindrical water tank with a diameter of 10 ft rests on a platform 20 ft above the ground. How much work is done in filling the tank with water through a hole in the bottom if it is pumped up from ground level?

19. A 20-ft by 50-ft swimming pool is 3 ft deep at one end and 8 ft deep at the other, and the bottom is a sloping plane (see Figure 7.23). How much work is done in pumping the water out over the top?

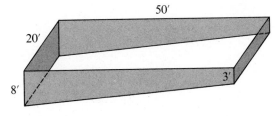

Figure 7.23

20. A 20-ft chain weighing 1/2 lb/ft hangs from the top of a 10-ft building (with 10 ft of chain on the ground). How much work is done in pulling the chain to the roof of the building?

21. Evaluate $\int_0^4 \sqrt{x^3 + 1}\, dx$ by both the trapezoidal rule and Simpson's rule using $n = 4$.

22. Evaluate $\int_1^2 \frac{x - 2}{x}\, dx$ by the trapezoidal rule using $n = 5$. Find the maximum value of $|E_T|$ and the range of values of the integral.

23. Evaluate $\int_1^4 \frac{x^2 + 1}{x}\, dx$ by Simpson's rule using $n = 6$. Find the maximum value of $|E_S|$ and the range of values of the integral.

24. Evaluate $\int_0^3 2x^3\, dx$ by the definition of the definite integral.

25. Evaluate $\int_1^2 (3x^2 - 4x + 1)\, dx$ by the definition of the definite integral.

C
26. Find a polynomial $P(x)$ such that $P'(x) - P(x) = 3x^2$.

8

Vectors in the Plane

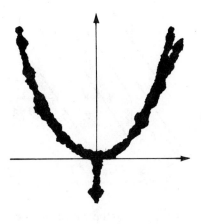

8.1

Introduction

Since quantities such as force, velocity, and acceleration have direction as well as magnitude, it is convenient to have a representation for them that reflects this dual nature. This is done by means of vectors, which have both magnitude and direction. Not only are vectors important in physics and engineering, but their use can simplify geometric problems considerably, especially in solid analytic geometry. One reason vectors are so useful is that they have such a wide range of interpretations.

Definition

*A **vector** v in the Cartesian plane is an ordered pair of real numbers, $v = (x, y)$, where x and y are called the **components** of the vector.*

Some people prefer the notation $v = <x, y>$ to distinguish a vector from a point. We do not want to make such a distinction. A point and a vector are basically identical; the difference is in the interpretation. In fact, the two viewpoints for a vector—namely, a vector as a point and a vector as an ordered pair of real numbers— are equally valid. The former point of view provides some geometric intuition, while the latter gives us a means for computation with vectors. To further enhance the geometry of vectors and bring out the notions of both magnitude and direction, we find it convenient to think of the vector (x, y) as an arrow or directed line segment

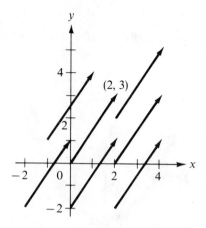

Figure 8.1

from the origin to the point (x, y). Furthermore, we do not distinguish between a directed line segment and a parallel displacement of it. Thus all the directed line segments shown in Figure 8.1 represent the vector $\mathbf{v} = (2, 3)$. We see that they all have the same length and direction, in keeping with our idea of a vector having both magnitude and direction. If we look at them in terms of their components, we see that the head of each arrow is 2 units to the right of and 3 units above the tail. This suggests the following theorem.

Theorem 8.1

The vector from the point $A = (x_1, y_1)$ *to* $B = (x_2, y_2)$ *is* $\mathbf{v} = (x_2 - x_1, y_2 - y_1)$. *We shall use the notation* $\overrightarrow{AB}$ *for the vector from A to B.*

The proof of this theorem is quite simple. It is left to the student (see Problem 36).

Example 1

Find the vector from $(3, -2)$ to $(-1, 5)$.

$$\begin{aligned}
\mathbf{v} &= (x_2 - x_1, y_2 - y_1) \\
&= (-1 - 3, 5 - (-2)) \\
&= (-4, 7).
\end{aligned}$$

When dealing with vectors, we use the term *scalar* for a real number. A scalar has magnitude but not direction. A special vector that is quite important is the *zero vector*, $\mathbf{0} = (0, 0)$. Its geometric representation is merely a point at the origin. As we shall see later, it behaves in many ways like the number zero; nevertheless, these two should not be confused: $\mathbf{0}$ is a vector; 0 is a scalar.

Let us now consider how vectors may be combined with each other and with scalars.

Definition

If $\mathbf{v}_1 = (x_1, y_1)$ and $\mathbf{v}_2 = (x_2, y_2)$ are vectors and k is a scalar, then

(a) the **sum** of $\mathbf{v}_1$ and $\mathbf{v}_2$ is

$$\mathbf{v}_1 + \mathbf{v}_2 = (x_1, y_1) + (x_2, y_2) = (x_1 + x_2, y_1 + y_2);$$

(b) the **difference**, $\mathbf{v}_1$ minus $\mathbf{v}_2$, is

$$\mathbf{v}_1 - \mathbf{v}_2 = (x_1, y_1) - (x_2, y_2) = (x_1 - x_2, y_1 - y_2);$$

(c) the **absolute value** or **length** of $\mathbf{v}_1$ is

$$|\mathbf{v}_1| = |(x_1, y_1)| = \sqrt{x_1^2 + y_1^2};$$

(d) the **scalar multiple** of $\mathbf{v}_1$ by k is

$$k\mathbf{v}_1 = k(x_1, y_1) = (kx_1, ky_1).$$

Example 2

If $\mathbf{u} = (3, -2)$ and $\mathbf{v} = (-1, 5)$, find $\mathbf{u} + \mathbf{v}, \mathbf{u} - \mathbf{v}, |\mathbf{v}|$, and $2\mathbf{v}$.

$$\mathbf{u} + \mathbf{v} = (3 - 1, -2 + 5) = (2, 3)$$
$$\mathbf{u} - \mathbf{v} = (3 - (-1), -2 - 5) = (4, -7)$$
$$|\mathbf{v}| = \sqrt{(-1)^2 + 5^2} = \sqrt{26}$$
$$2\mathbf{v} = 2(-1, 5) = (-2, 10)$$

The sum and difference of two vectors $\mathbf{v}_1$ and $\mathbf{v}_2$ are the diagonals of the parallelogram formed by $\mathbf{v}_1, \mathbf{v}_2$, and translations of $\mathbf{v}_1$ and $\mathbf{v}_2$, as shown in Figure 8.2. Figure 8.3 shows how the parallelogram rule for the addition of vectors follows from the definition. On the other hand, Theorem 8.1 tells us that the vector from the point (x_2, y_2) to the point (x_1, y_1) (see Figure 8.4) is $(x_1 - x_2, y_1 - y_2)$, which is equal to $\mathbf{v}_1 - \mathbf{v}_2$.

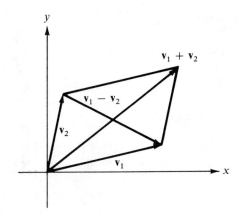

Figure 8.2

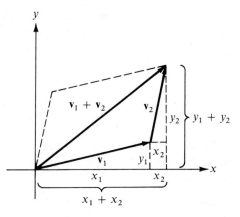

Figure 8.3

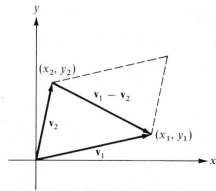

Figure 8.4

The absolute value or length of the vector $\mathbf{v}_1 = (x_1, y_1)$ is simply the length of the segment from the origin to (x_1, y_1).

Finally, the scalar multiple $k\mathbf{v}_1$ is a vector of length $|k|\,|\mathbf{v}_1|$ (see Problem 39) and with direction the same as that of $\mathbf{v}_1$ when $k > 0$ and opposite that of $\mathbf{v}_1$ when $k < 0$. Figure 8.5 gives some examples of scalar multiples of a vector $\mathbf{v}$.

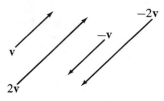

Figure 8.5

Let us take note of the fact that we are not adding and multiplying ordinary numbers; thus it is not obvious that the rules of ordinary arithmetic hold—they must be proved from the given definitions.

Theorem 8.2

The following properties hold for arbitrary vectors $\mathbf{u}$, $\mathbf{v}$, *and* $\mathbf{w}$ *and arbitrary scalars* a *and* b.

(a) $\mathbf{u} + \mathbf{v} = \mathbf{v} + \mathbf{u}$. (b) $\mathbf{u} + (\mathbf{v} + \mathbf{w}) = (\mathbf{u} + \mathbf{v}) + \mathbf{w}$.
(c) $(ab)\mathbf{v} = a(b\mathbf{v})$. (d) $(a + b)\mathbf{v} = a\mathbf{v} + b\mathbf{v}$.
(e) $\mathbf{v} + \mathbf{0} = \mathbf{v}$. (f) $0\mathbf{v} = \mathbf{0}$.
(g) $a\mathbf{0} = \mathbf{0}$. (h) $|a\mathbf{v}| = |a|\,|\mathbf{v}|$.
(i) $|\mathbf{u} + \mathbf{v}| \leq |\mathbf{u}| + |\mathbf{v}|$. (j) $a(\mathbf{u} + \mathbf{v}) = a\mathbf{u} + a\mathbf{v}$.

If we form the scalar multiple of the vector $\mathbf{v}$ and the scalar $1/|\mathbf{v}|$, the result is easily seen to be the unit vector (that is, the vector of length 1) in the direction of $\mathbf{v}$. It is usually written

$$\frac{\mathbf{v}}{|\mathbf{v}|}.$$

Of special interest are the unit vectors along the axes.

Definition

The vectors $\mathbf{i} = (1, 0)$ *and* $\mathbf{j} = (0, 1)$ *are called* **basis vectors.**

We now have an alternate way of expressing a vector in terms of its components, since

$$\mathbf{v} = (x, y) = x\mathbf{i} + y\mathbf{j}.$$

Similarly, Theorem 8.1 can be restated to give: *The vector from point* $A = (x_1, y_1)$ *to point* $B = (x_2, y_2)$ *is* $\mathbf{v} = (x_2 - x_1)\mathbf{i} + (y_2 - y_1)\mathbf{j}.$

To illustrate the use of the foregoing operations with vectors, let us extend a result we found in Section 1.1. There we showed that the midpoint of the segment AB, where $A = (x_1, y_1)$ and $B = (x_2, y_2)$, has coordinates $x = (x_1 + x_2)/2$ and $y = (y_1 + y_2)/2$. This is a special case of the more general point-of-division formulas. Suppose that, instead of finding the point halfway from A to B, we find the point one-third of the way from A to B or three-fourths of the way, or, more generally, some fraction r of the way from A to B. To do this, let us think of (x_1, y_1) and (x_2, y_2), not only as the points A and B, respectively, but also as the vectors $\mathbf{u}$ and $\mathbf{v}$, respectively. Then $\mathbf{v} - \mathbf{u}$ is the vector from A to B (see Figure 8.6). Now let (x, y) be the point P that is some fraction r of the way from A to B; or, alternately, (x, y) is the vector $\mathbf{w}$. It then follows that $\overrightarrow{AP}$ represents the vector $r(\mathbf{v} - \mathbf{u})$. It is easily seen from Figure 8.6 that

$$\mathbf{w} = \mathbf{u} + r(\mathbf{v} - \mathbf{u}).$$

Putting this result in component form, we have

$$x\mathbf{i} + y\mathbf{j} = x_1\mathbf{i} + y_1\mathbf{j} + r[(x_2 - x_1)\mathbf{i} + (y_2 - y_1)\mathbf{j}]$$
$$= [x_1 + r(x_2 - x_1)]\mathbf{i} + [y_1 + r(y_2 - y_1)]\mathbf{j}.$$

It follows from this (see Problem 38) that

$$x = x_1 + r(x_2 - x_1), \quad \text{and} \quad y = y_1 + r(y_2 - y_1).$$

These are the point-of-division formulas.

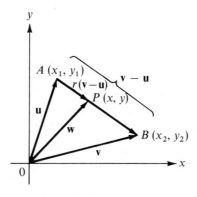

Figure 8.6

Theorem 8.3

If $A = (x_1, y_1)$, $B = (x_2, y_2)$, and P is a point such that $r = \overline{AP}/\overline{AB}$, then the co-ordinates of P are

$$x = x_1 + r(x_2 - x_1)$$

and

$$y = y_1 + r(y_2 - y_1).$$

Example 3

Find the point one-third of the way from $A = (2, 5)$ to $B = (8, -1)$.

$$r = \frac{\overline{AP}}{\overline{AB}} = \frac{1}{3}; \quad x = x_1 + r(x_2 - x_1) \quad y = y_1 + r(y_2 - y_1)$$

$$= 2 + \frac{1}{3}(8 - 2) \quad\quad = 5 + \frac{1}{3}(-1 - 5)$$

$$= 4; \quad\quad\quad\quad\quad = 3.$$

Thus the desired point is $(4, 3)$.

In deriving Theorem 8.3, we indicated that $0 < r < 1$ or that P is between A and B. But there is nothing in the argument that puts any restriction on the values that the scalar r may take. It is easily seen that $r = 0$ and $r = 1$ give $P = A$ and $P = B$, respectively. If $r > 1$, then P is on line AB and B separates A and P; if $r < 0$, then P is on line AB and A separates B and P.

Example 4

If the segment AB, where $A = (-3, 1)$ and $B = (2, 5)$, is extended beyond B to a point P twice as far from A as B is, find P.

$$r = \frac{\overline{AP}}{\overline{AB}} = 2; \quad x = x_1 + r(x_2 - x_1) \quad y = y_1 + r(y_2 - y_1)$$

$$= -3 + 2[2 - (-3)] \quad = 1 + 2(5 - 1)$$

$$= 7; \quad\quad\quad\quad\quad\quad = 9.$$

Thus $P = (7, 9)$.

Of course, the midpoint formulas given in Section 1.1 are easily seen to be a special case of the point-of-division formulas. It might be noted that the point-of-division formulas can be derived by use of similar triangles as we did in finding the midpoint formulas; however, the vector approach is somewhat shorter and illustrates the use of vectors in geometric problems.

Problems

A

In Problems 1–4, give in component form the vector **v** that is represented by $\overrightarrow{AB}$.

1. $A = (4, 3)$, $B = (-2, 1)$.
2. $A = (3, -2)$, $B = (4, 0)$.
3. $A = (-1, 5)$, $B = (3, 4)$.
4. $A = (-2, -3)$, $B = (-2, 5)$.

In Problems 5–8, give the unit vector in the direction of **v**.

5. $\mathbf{v} = (3, -1)$.
6. $\mathbf{v} = (5, 12)$.
7. $\mathbf{v} = 4\mathbf{i} - 3\mathbf{j}$.
8. $\mathbf{v} = 2\mathbf{j}$.

In Problems 9–12, find the point P such that $\overline{AP}/\overline{AB} = r$.

9. $A = (3, 4)$, $B = (7, 0)$; $r = 1/4$.
10. $A = (3, -5)$, $B = (0, 1)$; $r = 2/3$.
11. $A = (2, -1)$, $B = (3, 7)$; $r = 3$.
12. $A = (-5, 2)$, $B = (1, 7)$; $r = -2$.

In Problems 13–20, find the end points of the representative $\overrightarrow{AB}$ of **v** from the given information.

13. $\mathbf{v} = (3, -1)$, $A = (1, 4)$.
14. $\mathbf{v} = (1, 3)$, $A = (-1, 5)$.
15. $\mathbf{v} = (-1, 4)$, $B = (-2, 5)$.
16. $\mathbf{v} = (2, -3)$, $B = (5, 1)$.
17. $\mathbf{v} = 3\mathbf{i} - \mathbf{j}$, $A = (4, -2)$.
18. $\mathbf{v} = 2\mathbf{i} + 7\mathbf{j}$, $A = (3, 2)$.
19. $\mathbf{v} = 3\mathbf{i} - 4\mathbf{j}$, $B = (5, -2)$.
20. $\mathbf{v} = 2\mathbf{i} - 5\mathbf{j}$, $B = (-2, 8)$.

21. For $\mathbf{u} = (3, -1)$ and $\mathbf{v} = (1, 2)$, find $\mathbf{u} + \mathbf{v}$. Draw a diagram showing **u**, **v**, and $\mathbf{u} + \mathbf{v}$.
22. For $\mathbf{u} = (3, 2)$ and $\mathbf{v} = (2, -4)$, find $\mathbf{u} + \mathbf{v}$. Draw a diagram showing **u**, **v**, and $\mathbf{u} + \mathbf{v}$.
23. For $\mathbf{u} = \mathbf{i} - 2\mathbf{j}$ and $\mathbf{v} = 3\mathbf{i} + 5\mathbf{j}$, find $\mathbf{u} - \mathbf{v}$. Draw a diagram showing **u**, **v**, and $\mathbf{u} - \mathbf{v}$.
24. For $\mathbf{u} = \mathbf{i} - 5\mathbf{j}$ and $\mathbf{v} = 3\mathbf{i} + 2\mathbf{j}$, find $2\mathbf{u} + \mathbf{v}$. Draw a diagram showing **u**, **v**, and $2\mathbf{u} + \mathbf{v}$.

B

In Problems 25–28, find the end points of the representative $\overrightarrow{AB}$ of **v** from the given information.

25. $\mathbf{v} = (3, 5)$, $(4, 1)$ is the midpoint of AB.
26. $\mathbf{v} = (4, -6)$, $(3, -1)$ is the midpoint of AB.
27. $\mathbf{v} = 3\mathbf{i} - \mathbf{j}$, $(5, -2)$ is the midpoint of AB.
28. $\mathbf{v} = 2\mathbf{i} - 3\mathbf{j}$, $(4, -8)$ is the midpoint of AB.
29. If $P = (2, -5)$, $B = (4, -3)$, and $\overline{AP}/\overline{AB} = 1/2$, find A.
30. If $A = (4, 1)$, $P = (6, -3)$, and $\overline{AP}/\overline{AB} = 2/3$, find B.
31. If $P = (5, 3)$, $B = (1, 8)$, and $\overline{AP}/\overline{AB} = 2/5$, find A.
32. If $A = (5, 2)$, $P = (3, 3)$, and $\overline{AP}/\overline{AB} = 3/5$, find B.

C

33. Prove analytically that the medians of a triangle are concurrent at a point two-thirds of the way from each vertex to the midpoint of the opposite side.
34. Find the point of intersection of the medians of the triangle with vertices $(5, 2)$, $(0, 4)$, and $(-1, -1)$.
35. Show that if a triangle has vertices (x_1, y_1), (x_2, y_2), and (x_3, y_3), then the point of intersection of its medians is

$$\left(\frac{x_1 + x_2 + x_3}{3}, \frac{y_1 + y_2 + y_3}{3}\right).$$

36. Prove Theorem 8.1.
37. Prove Theorem 8.2.
38. Show that $a\mathbf{i} + b\mathbf{j} = c\mathbf{i} + d\mathbf{j}$ if and only if $a = c$ and $b = d$.
39. Show that $|k\mathbf{v}| = |k|\,|\mathbf{v}|$.

8.2

The Dot Product

We have considered the sums and differences of two vectors, but the only product considered so far is the scalar multiple—the product of a scalar and a vector. We now shall consider the product of two vectors. There are two different product operations for a pair of vectors, the dot product and the cross product. We shall take up the dot product in this section but defer a discussion of the cross product to Chapter 20 (since it requires three dimensions). First let us consider the angle between two vectors.

Definition

The angle between two nonzero vectors **u** *and* **v** *is the angle between representatives of* **u** *and* **v** *having a common tail. If* **u** $= k$**v** *for some scalar k, then the angle between them is* $0°$ *if* $k > 0$ *and* $180°$ *if* $k < 0$.

Note that the angle between two vectors is nondirected. That is, we do not consider the angle from one vector to another, which would imply a preferred direction, but rather the angle between two vectors. If θ is the angle between two vectors, then

$$0° \leq \theta \leq 180° \quad \text{or} \quad 0 \leq \theta \leq \pi.$$

Theorem 8.4

If **u** $= a_1\mathbf{i} + b_1\mathbf{j}$ *and* **v** $= a_2\mathbf{i} + b_2\mathbf{j}$ (**u** $\neq$ **0** *and* **v** $\neq$ **0**) *and if θ is the angle between them, then*

$$\cos \theta = \frac{a_1 a_2 + b_1 b_2}{|\mathbf{u}|\,|\mathbf{v}|}.$$

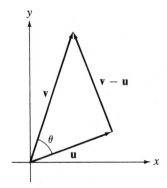

Figure 8.7

Proof

By the law of cosines (see Figure 8.7),

$$|\mathbf{v} - \mathbf{u}|^2 = |\mathbf{u}|^2 + |\mathbf{v}|^2 - 2|\mathbf{u}||\mathbf{v}| \cos \theta.$$

Since $\mathbf{v} - \mathbf{u} = (a_2 - a_1)\mathbf{i} + (b_2 - b_1)\mathbf{j}$, we have

$$(a_2 - a_1)^2 + (b_2 - b_1)^2 = a_1^2 + b_1^2 + a_2^2 + b_2^2 - 2|\mathbf{u}||\mathbf{v}| \cos \theta$$

$$|\mathbf{u}||\mathbf{v}| \cos \theta = a_1 a_2 + b_1 b_2$$

$$\cos \theta = \frac{a_1 a_2 + b_1 b_2}{|\mathbf{u}||\mathbf{v}|}$$

Example 1

Find the cosine of the angle between $\mathbf{u} = 3\mathbf{i} - 4\mathbf{j}$ and $\mathbf{v} = 5\mathbf{i} + 12\mathbf{j}$.

$$\cos \theta = \frac{a_1 a_2 + b_1 b_2}{|\mathbf{u}||\mathbf{v}|}$$

$$= \frac{(3)5 + (-4)12}{\sqrt{9 + 16}\sqrt{25 + 144}}$$

$$= -\frac{33}{65}.$$

If $\mathbf{u}$ and $\mathbf{v}$ are unit vectors, the denominator of the expression for $\cos \theta$ is one—it may be omitted. Thus $1/|\mathbf{u}||\mathbf{v}|$ is a normalizing factor, which is not needed if $\mathbf{u}$ and $\mathbf{v}$ are unit vectors. The dot product of two vectors is simply the expression for $\cos \theta$ without this normalizing factor. Thus, while the dot product of two vectors is generally not the same as the cosine of the angle between them, it does equal this cosine when either $|\mathbf{u}||\mathbf{v}| = 1$ or (more significantly) when $\cos \theta = 0$.

Definition

If $\mathbf{u} = a_1 \mathbf{i} + b_1 \mathbf{j}$ *and* $\mathbf{v} = a_2 \mathbf{i} + b_2 \mathbf{j}$, *then the **dot product (scalar product, inner product)** of* $\mathbf{u}$ *and* $\mathbf{v}$ *is*

$$\mathbf{u} \cdot \mathbf{v} = a_1 a_2 + b_1 b_2.$$

Note that the dot product of two vectors is *not* another vector; it is a scalar.

Example 2

Find the dot product of $\mathbf{u} = 3\mathbf{i} - 2\mathbf{j}$ and $\mathbf{v} = \mathbf{i} + \mathbf{j}$.

$$\mathbf{u} \cdot \mathbf{v} = (3)(1) + (-2)(1) = 1.$$

Let us now consider some applications of the dot product.

Theorem 8.5

The vectors $\mathbf{u}$ *and* $\mathbf{v}$ *(not both* $\mathbf{0}$*) are orthogonal (perpendicular) if and only if* $\mathbf{u} \cdot \mathbf{v} = 0$ *(the zero vector is taken to be orthogonal to every other vector).*

This follows directly from Theorem 8.4 and the definition of the dot product. Thus we have a simple test for the orthogonality (perpendicularity) of two vectors. As we shall see later, orthogonality of vectors is an important concept.

Example 3

Determine whether or not $\mathbf{u} = 2\mathbf{i} - \mathbf{j}$ and $\mathbf{v} = \mathbf{i} + 2\mathbf{j}$ are orthogonal.

$$\mathbf{u} \cdot \mathbf{v} = (2)(1) + (-1)(2) = 0.$$

They are orthogonal, since $\mathbf{u} \cdot \mathbf{v} = 0$.

Again there is the question of whether or not the dot product of vectors has the same properties as the product of numbers. The definition itself shows one difference, in that the dot product of two vectors is not itself a vector. While there are other differences, let us first note the similarities.

Theorem 8.6

If $\mathbf{u}$, $\mathbf{v}$, and $\mathbf{w}$ are vectors, then

$$\mathbf{u} \cdot \mathbf{v} = \mathbf{v} \cdot \mathbf{u},$$
$$(\mathbf{u} + \mathbf{v}) \cdot \mathbf{w} = \mathbf{u} \cdot \mathbf{w} + \mathbf{v} \cdot \mathbf{w}.$$

This is easily proved from the definition. The proof is left to the student. It might be noted that the dot product of three vectors $\mathbf{u} \cdot \mathbf{v} \cdot \mathbf{w}$ is meaningless, since the dot product of any pair of them is a scalar.

Theorem 8.7

If $\mathbf{u}$ and $\mathbf{v}$ are vectors and θ is the angle between them, then

$$\mathbf{u} \cdot \mathbf{v} = |\mathbf{u}|\,|\mathbf{v}|\,\cos\theta,$$
$$\mathbf{v} \cdot \mathbf{v} = |\mathbf{v}|^2.$$

The proof is left to the student. We might note some special cases of this theorem. If $\mathbf{u}$ and $\mathbf{v}$ are orthogonal, then $\theta = \pi/2$ and $\mathbf{u} \cdot \mathbf{v} = 0$, as we have seen before. If $\mathbf{u}$ and $\mathbf{v}$ are parallel, $\theta = 0$ or $\theta = \pi$ and $\mathbf{u} \cdot \mathbf{v} = \pm|\mathbf{u}|\,|\mathbf{v}|$.

Another use of the dot product of two vectors concerns the projection of one vector upon another.

Definition

*Suppose the vectors $\mathbf{u}$ and $\mathbf{v}$ are represented by the directed line segments $\overrightarrow{OA}$ and $\overrightarrow{OB}$, respectively. Then the **projection** of $\mathbf{u}$ on $\mathbf{v}$ is the vector $\mathbf{w}$ represented by $\overrightarrow{OC}$, where C is on the line OB and $AC \perp OB$ (see Figure 8.8).*

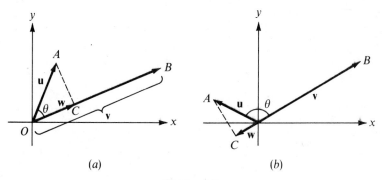

Figure 8.8

Since $\overrightarrow{OC}$ and $\overrightarrow{OB}$ have either the same or opposite directions, it follows that $\mathbf{w} = k\mathbf{v}$ for the proper choice of k. The value of k depends in part on the length of the projection $\mathbf{w}$. Thus finding this length is the first step in determining $\mathbf{w}$. The result is given in the following theorem.

Theorem 8.8

If $\mathbf{w}$ is the projection of $\mathbf{u}$ on $\mathbf{v}$ and θ is the angle between $\mathbf{u}$ and $\mathbf{v}$, then

$$|\mathbf{w}| = \frac{|\mathbf{u} \cdot \mathbf{v}|}{|\mathbf{v}|};$$

and

$$\mathbf{w} = \left(\frac{\mathbf{u} \cdot \mathbf{v}}{|\mathbf{v}|}\right) \frac{\mathbf{v}}{|\mathbf{v}|}.$$

Proof

We can see from Figure 8.8 that

$$|\mathbf{w}| = |\mathbf{u}| |\cos \theta|$$

$$= \frac{|\mathbf{u}| |\mathbf{v}| |\cos \theta|}{|\mathbf{v}|}$$

$$= \frac{|\mathbf{u} \cdot \mathbf{v}|}{|\mathbf{v}|}. \qquad \text{(By Theorem 8.7.)}$$

Now we can easily find $\mathbf{w}$. Its length is $|\mathbf{u} \cdot \mathbf{v}| / |\mathbf{v}|$, and its direction is determined by $\mathbf{v}$, since the projection is upon $\mathbf{v}$. Thus

$$\mathbf{w} = \pm \left(\frac{|\mathbf{u} \cdot \mathbf{v}|}{|\mathbf{v}|}\right) \frac{\mathbf{v}}{|\mathbf{v}|}.$$

Now if the angle θ between $\mathbf{u}$ and $\mathbf{v}$ is less than $\pi/2$, $\mathbf{u} \cdot \mathbf{v} > 0$. But $\mathbf{w}$ and $\mathbf{v}$ have the same direction in this case and

$$\mathbf{w} = + \left(\frac{|\mathbf{u} \cdot \mathbf{v}|}{|\mathbf{v}|}\right) \frac{\mathbf{v}}{|\mathbf{v}|} = \left(\frac{\mathbf{u} \cdot \mathbf{v}}{|\mathbf{v}|}\right) \frac{\mathbf{v}}{|\mathbf{v}|}.$$

If $\theta > \pi/2$, $\mathbf{u} \cdot \mathbf{v} < 0$ and $\mathbf{w}$ and $\mathbf{v}$ have opposite directions. This leads to

$$\mathbf{w} = - \left(\frac{|\mathbf{u} \cdot \mathbf{v}|}{|\mathbf{v}|}\right) \frac{\mathbf{v}}{|\mathbf{v}|} = \left(\frac{\mathbf{u} \cdot \mathbf{v}}{|\mathbf{v}|}\right) \frac{\mathbf{v}}{|\mathbf{v}|}.$$

Example 4

Find the projection **w** of **u** $= 3\mathbf{i} + \mathbf{j}$ on **v** $= 3\mathbf{i} + 4\mathbf{j}$.

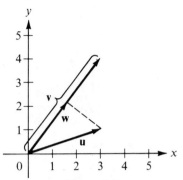

$$\mathbf{u} \cdot \mathbf{v} = (3)(3) + (1)(4) = 13,$$
$$|\mathbf{v}| = \sqrt{3^2 + 4^2} = 5.$$

Thus

$$\mathbf{w} = \left(\frac{\mathbf{u} \cdot \mathbf{v}}{|\mathbf{v}|}\right)\frac{\mathbf{v}}{|\mathbf{v}|}$$

$$= \frac{13}{5}\frac{3\mathbf{i} + 4\mathbf{j}}{5} = \frac{39}{25}\mathbf{i} + \frac{52}{25}\mathbf{j}.$$

Figure 8.9 gives a graphical representation of **u**, **v**, and **w**.

Figure 8.9

Problems

A

In Problems 1–8, find the angle θ between the given vectors.

1. $\mathbf{u} = 3\mathbf{i} - \mathbf{j}, \quad \mathbf{v} = \mathbf{i} + 2\mathbf{j}$.
2. $\mathbf{u} = 7\mathbf{i} + \mathbf{j}, \quad \mathbf{v} = \mathbf{i} - \mathbf{j}$.
3. $\mathbf{u} = -\mathbf{i} + 3\mathbf{j}, \quad \mathbf{v} = 3\mathbf{i} + \mathbf{j}$.
4. $\mathbf{u} = \mathbf{i} + \mathbf{j}, \quad \mathbf{v} = 3\mathbf{i} - \mathbf{j}$.
5. $\mathbf{u} = 2\mathbf{i} - \mathbf{j}, \quad \mathbf{v} = \mathbf{i} + 2\mathbf{j}$.
6. $\mathbf{u} = 4\mathbf{i} + 3\mathbf{j}, \quad \mathbf{v} = \mathbf{i} - \mathbf{j}$.
7. $\mathbf{u} = 2\mathbf{i} - 3\mathbf{j}, \quad \mathbf{v} = 8\mathbf{i} + \mathbf{j}$.
8. $\mathbf{u} = 13\mathbf{i} + \mathbf{j}, \quad \mathbf{v} = 2\mathbf{i} + 9\mathbf{j}$.

In Problems 9–16, find $\mathbf{u} \cdot \mathbf{v}$ and indicate whether or not **u** and **v** are orthogonal.

9. $\mathbf{u} = \mathbf{i} - \mathbf{j}, \quad \mathbf{v} = 2\mathbf{i} + \mathbf{j}$.
10. $\mathbf{u} = 3\mathbf{i} + \mathbf{j}, \quad \mathbf{v} = \mathbf{i} - 2\mathbf{j}$.
11. $\mathbf{u} = 6\mathbf{i} + 2\mathbf{j}, \quad \mathbf{v} = \mathbf{i} - 3\mathbf{j}$.
12. $\mathbf{u} = 2\mathbf{i} - 3\mathbf{j}, \quad \mathbf{v} = 2\mathbf{i} + \mathbf{j}$.
13. $\mathbf{u} = \mathbf{i} - \mathbf{j}, \quad \mathbf{v} = 3\mathbf{i} + 4\mathbf{j}$.
14. $\mathbf{u} = 6\mathbf{i} + 4\mathbf{j}, \quad \mathbf{v} = 2\mathbf{i} - 3\mathbf{j}$.
15. $\mathbf{u} = 2\mathbf{i} - 3\mathbf{j}, \quad \mathbf{v} = 3\mathbf{i} - \mathbf{j}$.
16. $\mathbf{u} = 4\mathbf{i}, \quad \mathbf{v} = 2\mathbf{j}$.

B

In Problems 17–24, find the projection of **u** on **v**.

17. $\mathbf{u} = 2\mathbf{i} - \mathbf{j}, \quad \mathbf{v} = \mathbf{i} + \mathbf{j}$.
18. $\mathbf{u} = \mathbf{i} + 3\mathbf{j}, \quad \mathbf{v} = 2\mathbf{i} + \mathbf{j}$.
19. $\mathbf{u} = \mathbf{i} + \mathbf{j}, \quad \mathbf{v} = \mathbf{i} - 7\mathbf{j}$.
20. $\mathbf{u} = 4\mathbf{i} + \mathbf{j}, \quad \mathbf{v} = 5\mathbf{i} + 3\mathbf{j}$.
21. $\mathbf{u} = \mathbf{i} - \mathbf{j}, \quad \mathbf{v} = 2\mathbf{i} + \mathbf{j}$.
22. $\mathbf{u} = 11\mathbf{i} - 3\mathbf{j}, \quad \mathbf{v} = 8\mathbf{i} + \mathbf{j}$.
23. $\mathbf{u} = 5\mathbf{i} + \mathbf{j}, \quad \mathbf{v} = 3\mathbf{i} - 2\mathbf{j}$.
24. $\mathbf{u} = 9\mathbf{i} - \mathbf{j}, \quad \mathbf{v} = 7\mathbf{i} + 4\mathbf{j}$.

In Problems 25–36, determine the value(s) of a so that the given conditions are satisfied.

25. $\mathbf{u} = 3\mathbf{i} - \mathbf{j}, \quad \mathbf{v} = \mathbf{i} + a\mathbf{j}; \quad$ **u** and **v** are perpendicular.
26. $\mathbf{u} = \mathbf{i} + 2\mathbf{j}, \quad \mathbf{v} = 4\mathbf{i} - a\mathbf{j}; \quad$ **u** and **v** are perpendicular.
27. $\mathbf{u} = 4\mathbf{i} + \mathbf{j}, \quad \mathbf{v} = 2\mathbf{i} + a\mathbf{j}; \quad$ **u** and **v** are perpendicular.
28. $\mathbf{u} = 2\mathbf{i} - \mathbf{j}, \quad \mathbf{v} = a\mathbf{i} + \mathbf{j}; \quad$ **u** and **v** are perpendicular.
29. $\mathbf{u} = \mathbf{i} + \mathbf{j}, \quad \mathbf{v} = a\mathbf{i} - \mathbf{j}; \quad$ **u** and **v** are parallel.
30. $\mathbf{u} = a\mathbf{i} + 5\mathbf{j}, \quad \mathbf{v} = 2\mathbf{i} + \mathbf{j}; \quad$ **u** and **v** are parallel.
31. $\mathbf{u} = \mathbf{i} - 3\mathbf{j}, \quad \mathbf{v} = a\mathbf{i} + \mathbf{j}; \quad$ **u** and **v** are parallel.
32. $\mathbf{u} = a\mathbf{i} - \mathbf{j}, \quad \mathbf{v} = 3\mathbf{i} + a\mathbf{j}; \quad$ **u** and **v** are parallel.
33. $\mathbf{u} = a\mathbf{i} + 2\mathbf{j}, \quad \mathbf{v} = \mathbf{i} - \mathbf{j}; \quad$ the angle between **u** and **v** is $\pi/3$.
34. $\mathbf{u} = 3\mathbf{i} - a\mathbf{j}, \quad \mathbf{v} = 2\mathbf{i} + 3\mathbf{j}; \quad$ the angle between **u** and **v** is $\pi/4$.
35. $\mathbf{u} = 2\mathbf{i} + 3\mathbf{j}, \quad \mathbf{v} = a\mathbf{i} - \mathbf{j}; \quad$ the angle between **u** and **v** is $2\pi/3$.
36. $\mathbf{u} = 3\mathbf{i} - \mathbf{j}, \quad \mathbf{v} = 4\mathbf{i} + a\mathbf{j}; \quad$ the angle between **u** and **v** is $\pi/6$.

In Problems 37–40, let **u** be represented by $\overrightarrow{AB}$, **v** by $\overrightarrow{AC}$, and **w** by $\overrightarrow{BC}$. Find the projections of **v** and **w** on **u**.

37. $A = (0, 0)$, $B = (1, 4)$, $C = (2, -1)$.
38. $A = (4, -1)$, $B = (7, 3)$, $C = (-1, 5)$.
39. $A = (3, 5)$, $B = (-2, 7)$, $C = (0, 2)$.
40. $A = (1, 0)$, $B = (4, -2)$, $C = (-3, 5)$.

C

41.   Prove Theorem 8.5.
42.   Prove Theorem 8.6.
43.   Prove Theorem 8.7.
44.   Show that $\mathbf{v} = A\mathbf{i} + B\mathbf{j}$ is perpendicular to $Ax + By + C = 0$.
45.   Show that $\mathbf{v} = B\mathbf{i} - A\mathbf{j}$ is parallel to $Ax + By + C = 0$.
46.   Show that $|\mathbf{u} \cdot \mathbf{v}| \le |\mathbf{u}|\,|\mathbf{v}|$.

8.3

Applications of Vectors

Let us now consider some applications of vectors. Vector methods can be used in some of the proofs of elementary geometry. Sometimes the resulting proof is much shorter than either an analytic or a synthetic argument.

Example 1

Using vector methods, prove that the line joining the midpoints of two sides of a triangle is parallel to and one-half the length of the third side.

Suppose D and E are the midpoints of AB and BC, respectively. Let us give directions to the line segments involved and consider them to be representatives of vectors as shown in Figure 8.10. Then

$$\mathbf{u} + \mathbf{v} = \mathbf{w},$$

$$\tfrac{1}{2}\mathbf{u} + \mathbf{q} + \tfrac{1}{2}\mathbf{v} = \mathbf{w},$$

$$\mathbf{q} = \mathbf{w} - \tfrac{1}{2}(\mathbf{u} + \mathbf{v}) = \mathbf{w} - \tfrac{1}{2}\mathbf{w} = \tfrac{1}{2}\mathbf{w}.$$

Since vectors have both magnitude and direction, we have proved that **q** has the same direction as **w** and that it is half the length of **w**.

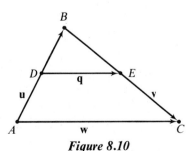

Figure 8.10

One reason for the brevity of the foregoing proof is that we were interested in both the direction and magnitude of $\overrightarrow{DE}$. Thus a representation by vectors was very efficient. When perpendicularity is involved, we consider the dot product. Since this product is defined in terms of the components, it is convenient to give component representations of the vectors.

Example 2

Prove the Pythagorean theorem by vector methods.

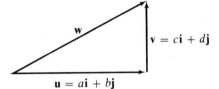

Figure 8.11

The triangle is given in Figure 8.11. Since **u** and **v** are perpendicular, $\mathbf{u} \cdot \mathbf{v} = 0$, which implies that

$$ac + bd = 0.$$

Then

$$\mathbf{w} = \mathbf{u} + \mathbf{v} = (a+c)\mathbf{i} + (b+d)\mathbf{j};$$
$$\begin{aligned} |\mathbf{w}| &= (a+c)^2 + (b+d)^2 \\ &= a^2 + 2ac + c^2 + b^2 + 2bd + d^2 \\ &= a^2 + b^2 + c^2 + d^2 \quad \text{(since } ac + bd = 0) \\ &= |\mathbf{u}|^2 + |\mathbf{v}|^2. \end{aligned}$$

On page 27, we derived the formula for the distance from the point (x_1, y_1) to the line $Ax + By + C = 0$. The proof given there, while not conceptually difficult, was tedious. Let us now consider a proof of this theorem using vector methods. Before considering the distance from a point to a line, let us note the result of Problem 44 of the previous section: the vector $\mathbf{v} = A\mathbf{i} + B\mathbf{j}$ is perpendicular to $Ax + By + C = 0$. This perpendicularity allows us to find the distance from any point to a given line.

Theorem 8.9

The distance from the point (x_1, y_1) to the line $Ax + By + C = 0$ is

$$d = \frac{|Ax_1 + By_1 + C|}{\sqrt{A^2 + B^2}}.$$

Proof

The distance we are considering here is the shortest, or perpendicular, distance. As noted above, the vector $\mathbf{v} = A\mathbf{i} + B\mathbf{j}$ is perpendicular to $Ax + By + C = 0$. Let (x, y) be a point on $Ax + By + C = 0$ and **u** be the vector represented by the segment from (x, y) to (x_1, y_1) (see Figure 8.12). Thus

$$\mathbf{u} = (x_1 - x)\mathbf{i} + (y_1 - y)\mathbf{j}.$$

The length we seek is the length of the projection **w** of **u** upon **v**. By Theorem 8.8,

$$d = |\mathbf{w}| = \frac{|\mathbf{v} \cdot \mathbf{u}|}{|\mathbf{v}|} = \frac{|A(x_1 - x) + B(y_1 - y)|}{\sqrt{A^2 + B^2}}$$

$$= \frac{|Ax_1 + By_1 - (Ax + By)|}{\sqrt{A^2 + B^2}}$$

$$= \frac{|Ax_1 + By_1 + C|}{\sqrt{A^2 + B^2}}.$$

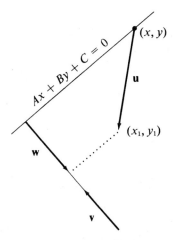

Figure 8.12

Still another way (though not a simple one) to find this formula is to work it as a minimization problem (see Problem 43, page 153).

One of the commonest uses of vectors is in analyzing the forces on an object. Let us consider a simple example of this.

Example 3

The following forces are exerted on an object: 5 lb to the right, 10 lb upward, 2 lb upward and to the right, inclined to the horizontal at an angle of 30°. What single force is equivalent to them?

The three given forces are first represented by vectors; the equivalent force is their sum.

$$\mathbf{v}_1 = 5\mathbf{i} \quad \text{and} \quad \mathbf{v}_2 = 10\mathbf{j}.$$

For the third vector we have an inclination of 30° and $|\mathbf{v}_3| = 2$ (see Figure 8.13). Thus

$$\mathbf{v}_3 = k \cos 30° \, \mathbf{i} + k \sin 30° \, \mathbf{j} = \frac{k\sqrt{3}}{2}\mathbf{i} + \frac{k}{2}\mathbf{j},$$

where k is chosen so that $|\mathbf{v}_3| = 2$ and k is positive.

$$|\mathbf{v}_3| = \sqrt{\frac{3k^2}{4} + \frac{k^2}{4}} = \sqrt{k^2} = k = 2.$$

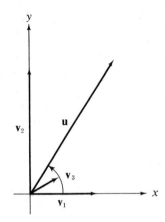

Figure 8.13

This gives $v_3 = \sqrt{3}\,i + j$.

$$u = v_1 + v_2 + v_3 = (5 + \sqrt{3})i + 11j,$$

$$|u| = \sqrt{(5 + \sqrt{3})^2 + 11^2} = \sqrt{149 + 10\sqrt{3}} \approx 12.9,$$

$$\cos \alpha = \frac{5 + \sqrt{3}}{\sqrt{149 + 10\sqrt{3}}} = 0.5220, \quad \alpha = 58°30'.$$

Thus the three given forces are equivalent to a single force of 12.9 lb directed upward to the right at an angle of 58°30′ with the (horizontal) x axis.

Let us now consider a more complex application of the same thing—the forces on the cables of a suspension bridge.

Suppose we consider a section OP (see Figure 8.14) of the cable, where O is the lowest point and $P(x, y)$ is another point. There are three forces acting upon OP: two forces of tension acting on the two ends, and the weight of that portion of the bridge supported by OP. At the point P the force exerted by the cable can be represented by a vector u of magnitude T (the tension in the cable at that point) and tangent to the curve of the cable. If the inclination of the tangent line is θ, then

$$u = T \cos \theta\, i + T \sin \theta\, j.$$

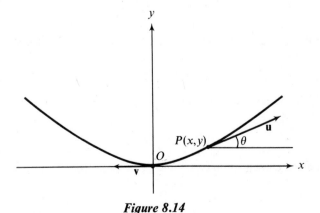

Figure 8.14

If the tension at O is T_0, then the vector $\mathbf{v}$ representing this force is

$$\mathbf{v} = -T_0\,\mathbf{i}.$$

Finally the weight of the roadway is uniformly distributed along the x axis, and we assume the weight of the cable to be negligible in comparison with the weight of the roadway. Thus if the roadway weighs w lb/ft, then the weight vector is

$$\mathbf{w} = -xw\mathbf{j}.$$

Now, since the forces must be in equilibrium,

$$\mathbf{u} + \mathbf{v} + \mathbf{w} = 0.$$

Thus

$$T\cos\theta = T_0 \quad \text{and} \quad T\sin\theta = xw.$$

Dividing the second equation by the first, we have

$$\tan\theta = \frac{xw}{T_0}.$$

Since θ is the inclination at P, $\tan\theta$ is the slope of the curve, or the derivative, at P. Thus, by integration,

$$y = \frac{x^2 w}{2T_0} + C.$$

Since $O = (0, 0)$ is on the curve, $C = 0$ and the cable defines the arc

$$y = \frac{x^2 w}{2T_0},$$

which is called a parabola. We shall say more about parabolas in Chapter 9.

While this equation tells us that the supporting cable is parabolic, T_0 cannot be determined directly. However it can be found from the dimensions of the bridge. If the span L is the distance between supporting piers (see Figure 8.15) and the sag H is the height from the highest to the lowest points of the cable, then $(L/2, H)$ is a point on the parabola. By simple substitution we see that

$$T_0 = \frac{L^2 w}{8H}.$$

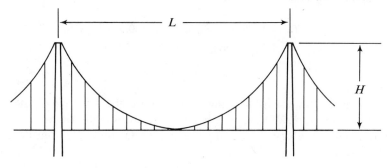

Figure 8.15

Thus a knowledge of H, L, and w allows us to determine the equation of the cable's curve. More important, it allows us to find the tension at any point of the cable. By squaring and adding the equations

$$T \cos \theta = T_0 \quad \text{and} \quad T \sin \theta = xw,$$

we find that the tension T at P is given by

$$T^2 = T_0^2 + x^2 w^2, \quad \left(-\frac{L}{2} \le T \le \frac{L}{2}\right).$$

Since T is a function of x, it is a simple matter to determine the maximum and minimum values of T. From the derivatives

$$\frac{d}{dx} T^2 = 2xw^2, \quad \frac{d^2}{dx^2} T^2 = 2w^2 > 0$$

we see that T^2 (and therefore T) is a minimum at $x = 0$, and the minimum tension is T_0. The maximum tension is attained at the end points, $x = \pm L/2$, from which we have

$$T_{max}^2 = T_0^2 + \frac{L^2 w^2}{4}, \quad T_{max} = \frac{1}{2}\sqrt{4T_0^2 + L^2 w^2}.$$

Thus the maximum tension occurs at the ends of the span, while the minimum occurs in the center.

Example 4

A suspension bridge with span 160 ft and sag 30 ft is to support a roadway weighing 3000 lb/ft. It is to be supported by two cables. Find the equation of the cables (with the placement of axes shown in Figure 8.14) and the maximum tension.

Since there are two cables, each one supports only half the weight of the roadway. Thus $w = 1500$ lb/ft.

$$T_0 = \frac{L^2 w}{8H} = \frac{(160)^2(1500)}{8(30)} = 160{,}000 \text{ lb.}$$

The equation is

$$y = \frac{x^2 w}{2T_0} = \frac{1500 x^2}{2(160{,}000)} = \frac{3}{640} x^2,$$

and the maximum tension is

$$T_{max} = \sqrt{T_0^2 + (L/2)^2 w^2} = \sqrt{(160{,}000)^2 + (80)^2(1500)^2}$$
$$= 200{,}000 \text{ lb.}$$

It might be noted that a wire hanging of its own weight, say a telephone wire, does not define a parabolic arc. When a wire hangs of its own weight, the weight it supports (itself) is not uniformly distributed along a horizontal line. We consider this more complicated situation on page 413.

Problems

A

In Problems 1–6, the given forces are acting on a body. What single force is equivalent to them?

1. $f_1 = 4i + 3j$, $f_2 = i - 2j$, $f_3 = i + j$.
2. $f_1 = 3i - j$, $f_2 = 2i + 5j$, $f_3 = -i + 5j$.
3. 4 lb to the right,   7 lb upward.
4. 3 lb downward,   8 lb to the right,   4 lb upward.
5. 3 lb downward,   4 lb to the right and inclined upward at an angle of 45° with the horizontal.
6. 3 lb to the left,   6 lb to the right and inclined upward at an angle of 60° with the horizontal.

B

In Problems 7–12, the given forces are acting on a body. What additional force will result in equilibrium? (A set of forces is in equilibrium if the sum of all of them is the zero vector.)

7. $f_1 = 5i + j$, $f_2 = i - 4j$, $f_3 = -2i + j$.
8. $f_1 = 3i - j$, $f_2 = 4i + 3j$, $f_3 = 2i - 5j$.
9. 2 lb to the left,   5 lb upward.
10. 3 lb to the right,   5 lb upward,   6 lb to the left.
11. 2 lb to the right,   6 lb to the right and inclined upward at an angle of 45° with the horizontal.
12. 4 lb upward,   3 lb to the right and inclined upward at an angle of 30° with the horizontal,   5 lb to the left and inclined downward at an angle of 60° with the horizontal.

In Problems 13–20, use vector methods to prove the given theorem.

13. The segment joining the midpoints of the nonparallel sides of a trapezoid is parallel to and one-half the sum of the lengths of the parallel sides.
14. The lines joining consecutive midpoints of a quadrilateral form a parallelogram.
15. If the diagonals of a parallelogram are perpendicular, then it is a rhombus.
16. The sum of the squares of the four sides of a parallelogram is equal to the sum of the squares of the two diagonals.
17. The diagonals of a rectangle are equal.
18. The base angles of an isosceles triangle are equal
19. If one of the parallel sides of a trapezoid is twice the length of the other, then the diagonals intersect at a point of trisection of both of them.
20. The medians of a triangle are concurrent at a point two-thirds of the way from each vertex to the midpoint of the opposite side. (*Hint:* Let $\overrightarrow{BP} = r\overrightarrow{BE}$ and $\overrightarrow{AP} = s\overrightarrow{AD}$ in Figure 8.16, and use the fact that $a\mathbf{u} + b\mathbf{v} = \mathbf{0}$, where neither $\mathbf{u}$ nor $\mathbf{v}$ is a scalar multiple of the other, implies $a = b = 0$.)

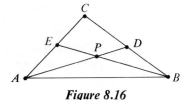

Figure 8.16

In Problems 21–24, data are given for a suspension bridge supported by two cables. Find the equation of the cables (with the placement of axes shown in Figure 8.14) and the maximum tension.

21. $L = 120$ ft, $H = 12.5$ ft, $w = 2000$ lb/ft.
22. $L = 1000$ ft, $H = 200$ ft, $w = 10{,}000$ lb/ft.
23. $L = 240$ ft, $H = 20$ ft, $w = 2000$ lb/ft.
24. $L = 240$ ft, $H = 30$ ft, $w = 2000$ lb/ft.
25. Given $T_0 = 500{,}000$ lb, $L = 400$ ft,   and $H = 40$ ft, find w.
26. Given $T_0 = 300{,}000$ lb, $L = 150$ ft, and $H = 15$ ft, find w.

27. A suspension bridge with a span of 320 ft is to support a roadway weighing 1500 lb/ft. It is to be supported by two cables, each of which can withstand a force of tension of 200,000 lb. What is the minimum sag that can be allowed?

28. If the span L and the weight load w of a suspension bridge are fixed, show that $T_{max} \to +\infty$ as $H \to 0$ and $T_{max} \to Lw/2$ as $H \to +\infty$.

Review Problems

A

1. Find the unit vector in the direction of $v = (2, -3)$.
2. Suppose $\overrightarrow{AB}$ is a representative of $v = 2i - 5j$. Find B when $A = (1, 3)$.
3. Suppose $\overrightarrow{AB}$ is a representative of $v = 5i - 3j$, and $(-2, 1)$ is the midpoint of AB. Find A and B.
4. If $u = (3, -2)$ and $v = (1, 6)$, find $u + v$, $u - v$, $3u$, and $2u + v$. Sketch all the vectors.
5. If $u = 2i - j$ and $v = 3i - 2j$, find $u \cdot v$.
6. Find the points of trisection of the segment joining $(2, -5)$ and $(-3, 7)$.
7. Find the angle θ between $u = 2i - 3j$ and $v = i + 5j$.

B

8. Find the projection w of $u = i + 4j$ upon $v = 2i - 9j$.
9. Find the projection w of $u = 4i + j$ upon $v = 2i - 2j$.
10. Determine a so that the angle between $u = 5i + j$ and $v = ai - j$ is $\pi/4$.
11. Determine a so that $u = i - 4j$ and $v = 3i + aj$ are orthogonal.
12. Use vectors to prove that the diagonals of an isosceles trapezoid are equal.
13. Use vectors to prove that if the lengths of the parallel sides of a trapezoid are in the ratio $1:n$, then the point of intersection of the diagonals divides both of them in the ratio $1:(n + 1)$.
14. The following forces are exerted on an object: 30 lb to the left, 15 lb downward, $10\sqrt{2}$ lb upward and to the right inclined to the horizontal at an angle of 45°. What single force is equivalent to them?
15. The handle of a lawnmower is inclined to the horizontal at an angle of 30°. If a man pushes forward and down in the direction of the handle with a force of 20 lb, with what force is the mower being pushed forward? With what force is it being pushed into the ground?
16. The data for a suspension bridge supported by two cables are: $L = 250$ ft, $H = 25$ ft, $w = 2000$ lb/ft. Find the maximum tension in the cables.

C

17. Given $u = 3i + 4j$ and $v = 12i - 5j$, find vectors v_1 and v_2 such that $v = v_1 + v_2$, $v_1 = ku$, and $v_2 \cdot u = 0$.

9

Conic Sections

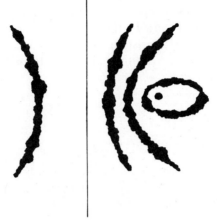

9.1

The Parabola

Up to this point the only second-degree equations that we have considered systematically have been equations of circles. We shall see that equations of the second degree represent (with two trivial exceptions) conic sections—that is, curves formed by the intersection of a plane with a right circular cone. Three general types of curves are formed in this way, the parabola, the ellipse, and the hyperbola.

Definition

*A **parabola** is the set of all points in a plane equidistant from a fixed point (focus) and a fixed line (directrix) not containing the focus.*

Suppose we choose the focus to be the point $(c, 0)$ and the directrix to be $x = -c$, $c \neq 0$ (see Figure 9.1). Let us choose a point (x, y) on the parabola and see what condition must be satisfied by x and y. From the definition, we have

$$\overline{PF} = \overline{PD},$$

$$\sqrt{(x - c)^2 + y^2} = |x + c|, \qquad \text{(See Note 1)}$$

$$(x - c)^2 + y^2 = (x + c)^2, \qquad \text{(See Note 2)}$$

$$x^2 - 2cx + c^2 + y^2 = x^2 + 2cx + c^2,$$

$$y^2 = 4cx.$$

Note 1: Since $\overline{PD}$ is a horizontal distance,

$$\overline{PD} = |x - (-c)| = |x + c|.$$

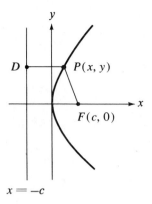

You might feel that we should drop the absolute-value signs, since it is clear from Figure 9.1 that $x + c$ must be positive. However, we did not insist that c be positive (although Figure 9.1 is given for a positive value of c). If c is negative, x is also negative and $x + c$ is negative.

Figure 9.1

Note 2: When we square both sides of an equation, there is a possibility of introducing extraneous roots. For instance, $(0, 1)$ is not a root of $x + y = x - y$, but it is a root of $(x + y)^2 = (x - y)^2$. The reason is that $x + y = 1$, while $x - y = -1$ for $(0, 1)$, and $1^2 = (-1)^2 = 1$. Whenever $(x + y)^2 = (x - y)^2$, and $x + y$ and $x - y$ are both positive, both negative, or both zero, extraneous roots cannot occur. Since $\sqrt{(x - c)^2 + y^2}$ and $|x + c|$ must both be positive in any case, we have introduced no extraneous roots; that is, any point satisfying

$$(x - c)^2 + y^2 = (x + c)^2$$

must also satisfy

$$\sqrt{(x - c)^2 + y^2} = |x + c|.$$

We see then that if a point is on the parabola with focus $(c, 0)$ and directrix $x = -c$, it must satisfy the equation $y^2 = 4cx$. Furthermore, since Note 2 indicates that all steps in the above argument are reversible, any point satisfying the equation $y^2 = 4cx$ is on the given parabola.

Theorem 9.1

A point (x, y) is on the parabola with focus $(c, 0)$ and directrix $x = -c$ if and only if it satisfies the equation

$$y^2 = 4cx.$$

Let us observe some properties of this parabola before considering others. First of all, it has a line of symmetry—in this case the x axis. This line is called the *axis* of the parabola. It is perpendicular to the directrix and contains the focus (see Figure 9.2). The point of intersection of the axis and the parabola is the *vertex*. The vertex of the parabola $y^2 = 4cx$ is the origin. Finally, the line segment through the focus perpendicular to the axis and having both ends on the parabola is the *latus rectum* (literally, straight side). Since the latus rectum of $y^2 = 4cx$ must be vertical and since it contains $(c, 0)$, the x coordinate of both ends is c. Substituting $x = c$ into $y^2 = 4cx$, we have

$$y^2 = 4c^2, \qquad y = \pm 2c.$$

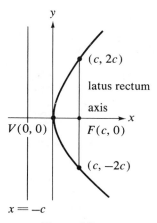

Figure 9.2

Thus one end of the latus rectum is $(c, 2c)$ and the other $(c, -2c)$. Its length is $4|c|$.

Finally the roles of the x and y may be reversed throughout, as the next theorem states.

Theorem 9.2

A point (x, y) is on the parabola with focus $(0, c)$ and directrix $y = -c$ if and only if it satisfies the equation

$$x^2 = 4cy.$$

Example 1

Sketch and discuss $y^2 = 8x$.

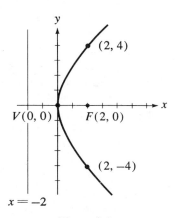

The equation is of the form

$$y^2 = 4cx,$$

with $c = 2$. Thus, it represents a parabola with vertex at the origin and axis on the x axis. The focus is at $(2, 0)$, and the directrix is $x = -2$. Finally, the length of the latus rectum is 8. This length may be used to determine the ends, $(2, \pm 4)$, of the latus rectum, which helps in sketching the curve (see Figure 9.3).

Figure 9.3

Example 2

Sketch and discuss $x^2 = -12y$.

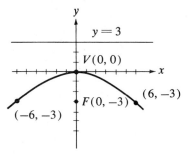

This equation is in the form

$$x^2 = 4cy,$$

with $c = -3$. Thus, it is a parabola with vertex at the origin and axis on the y axis. The focus is $(0, -3)$, the length of the latus rectum is 12, and the equation of the directrix is $y = 3$ (see Figure 9.4).

Figure 9.4

Example 3

Find an equation(s) of the parabola(s) with vertex at the origin and focus $(-4, 0)$.

Since the focus and vertex are on the x axis, the x axis is the axis of the parabola. Thus the equation is in the form $y^2 = 4cx$. Since the focus is $(-4, 0)$, $c = -4$ and the equation is $y^2 = -16x$.

Problems

A

In Problems 1–10, sketch and discuss the given parabola.

1. $y^2 = 16x$.
2. $y^2 = -8x$.
3. $x^2 = 8y$.
4. $x^2 = -4y$.
5. $y^2 = 10x$.
6. $x^2 = -6y$.
7. $x^2 = -3y$.
8. $y^2 = 5x$.
9. $x^2 = -2y$.
10. $y^2 = 7x$.

In Problems 11 and 12, find an equation(s) of the parabola(s) described.

11. Vertex: $(0, 0)$;   axis: x axis;   contains $(-3, 2)$.
12. Vertex: $(0, 0)$;   axis: y axis;   contains $(-3, 2)$.

B

In Problems 13–18, find an equation(s) of the parabola(s) described.

13. Vertex: $(0, 0)$;   axis: x axis;   length of latus rectum: 5.
14. Vertex: $(0, 0)$;   focus: $(0, -3)$.
15. Focus: $(-5, 0)$;   directrix: $x = 5$.
16. Focus: $(0, 7)$;   directrix: $y = -7$.
17. Vertex: $(0, 0)$;   contains $(2, 3)$ and $(-2, 3)$.
18. Vertex: $(0, 0)$;   contains $(-3, -5)$ and $(-3, 5)$.
19. Prove Theorem 9.2.
20. Find an equation of the line tangent to $y^2 = 5x$ at $(5, -5)$.
21. Find an equation of the line tangent to $x^2 = -5y$ at $(5, -5)$.
22. Find an equation of the line tangent to $y^2 = -16x$ and parallel to $x - y = 3$.
23. Find an equation of the line tangent to $x^2 = 6y$ and perpendicular to $x + 2y = 2$.
24. Find an equation(s) of the line(s) tangent to $y^2 = 4x$ and containing $(-2, -1)$.
25. Find an equation(s) of the line(s) tangent to $x^2 = -8y$ and containing $(4, 0)$.
26. Show that the line tangent to $y^2 = 4cx$ at (x_0, y_0) is $yy_0 = 2c(x + x_0)$.
27. Show that the point of a parabola closest to the focus is the vertex.
28. A comet has a parabolic orbit with the sun at the focus (see Figure 9.5). When the comet is 100,000,000 miles from the sun, the line joining the sun and the comet makes an angle of $60°$ with the axis of the parabola. How close to the sun will the comet get? (*Hint:* See Problem 27.)

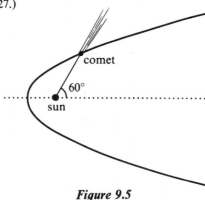

comet

$60°$

sun

Figure 9.5

C

29. Suppose the angle of Problem 28 increases from 60° to 90° in 36 hours. In how many more hours will the comet be at perihelion—the closest point to the sun? (*Hint:* Use the fact that the line joining the sun and the comet sweeps out equal areas within the orbit in equal times.)

30. Prove that the ordinate of any point P of the parabola $y^2 = 4cx$ is the mean proportional between the length of the latus rectum and the abscissa of P.

31. If a ray of light strikes a curved reflecting surface at the point P, it is reflected as if it had struck a flat surface tangent to the curve at P. That is, the rays make equal angles with the tangent at P (see Figure 9.6). Suppose we have a parabolic mirror with a light source at the focus. Show that no matter what point P on the parabola is struck with a ray of light, the light is reflected in a line parallel to the axis.

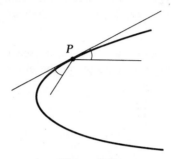

Figure 9.6

32. Suppose V is the vertex of the parabola $y^2 = 4cx$, F is the focus, P is a point of the parabola different from V, T is the point of intersection of the tangent at P and the x axis, N is the point of intersection of the normal at P and the x axis, X is the foot of the perpendicular from P to the x axis, and Q is the point of intersection of the tangent and the y axis (see Figure 9.7). Show that

(a) $\overline{TF} = \overline{FP}$,   (b) $\overline{TV} = \overline{VX}$,   (c) $\overline{XN} = \frac{1}{2}$ (latus rectum),   (d) $QF \perp TP$.

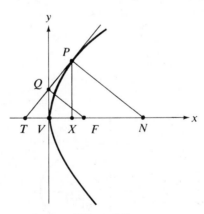

Figure 9.7

33. Suppose P_1 and P_2 are two points of a parabola, Q is the point of intersection of the tangents at P_1 and P_2, and F is the focus. Show that FQ bisects the angle P_1FP_2.

34. Suppose P_1 and P_2 are two points of a parabola and Q is the point of intersection of the tangents at P_1 and P_2. Show that the line through Q and parallel to the axis bisects P_1P_2.

35. Suppose P_1 and P_2 are two points of a parabola such that the line P_1P_2 contains the focus. Show that the point of intersection of the tangents at P_1 and P_2 is on the directrix.

9.2

The Ellipse

Definition

*An **ellipse** is the set of all points (x, y) such that the sum of the distances from (x, y) to a pair of distinct fixed points (foci) is a fixed constant.*

Let us choose the foci to be $(c, 0)$ and $(-c, 0)$ (see Figure 9.8), and let the fixed constant be $2a$. If (x, y) represents a point on the ellipse, we have

$$\sqrt{(x - c)^2 + y^2} + \sqrt{(x + c)^2 + y^2} = 2a$$

$$\sqrt{(x - c)^2 + y^2} = 2a - \sqrt{(x + c)^2 + y^2}$$

$$x^2 - 2cx + c^2 + y^2 = 4a^2 - 4a\sqrt{(x + c)^2 + y^2} + x^2 + 2cx + c^2 + y^2$$

$$4a\sqrt{(x + c)^2 + y^2} = 4a^2 + 4cx$$

$$\sqrt{(x + c)^2 + y^2} = a + \frac{cx}{a}$$

$$x^2 + 2cx + c^2 + y^2 = a^2 + 2cx + \frac{c^2 x^2}{a^2}$$

$$\frac{a^2 - c^2}{a^2} x^2 + y^2 = a^2 - c^2$$

$$\frac{x^2}{a^2} + \frac{y^2}{a^2 - c^2} = 1.$$

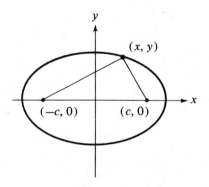

Figure 9.8

The triangle of Figure 9.8, with vertices $(c, 0)$, $(-c, 0)$, and (x, y), has one side of length $2c$. The sum of the lengths of the other two sides is $2a$. Thus

$$2a > 2c$$

$$a > c$$

$$a^2 > c^2$$

$$a^2 - c^2 > 0.$$

Since $a^2 - c^2$ is positive, we may replace it by another positive number, b^2. Thus

$$\frac{x^2}{a^2} + \frac{y^2}{b^2} = 1, \quad \text{where } b^2 = a^2 - c^2.$$

Observe that we squared both sides of the equation at two of the steps. In both cases, both sides of the equation are nonnegative. Thus we have introduced no extraneous roots, and the steps may be reversed.

Note that there are two axes of symmetry: the x axis and the y axis. Furthermore $(\pm a, 0)$ are the x intercepts and $(0, \pm b)$ are the y intercepts, where $a > b$ (since $b^2 = a^2 - c^2$). Thus the x axis is called the *major axis* and the y axis is the *minor axis*. The ends of the major axis $(\pm a, 0)$ are called the *vertices*; the ends of the minor axis $(0, \pm b)$ are the *covertices*; and the point of intersection $(0, 0)$ is called the *center* (see Figure 9.9). The *foci* $(\pm c, 0)$ are on the major axis.

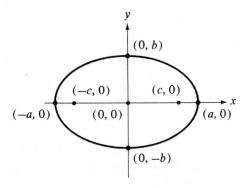

Figure 9.9

Theorem 9.3

A point (x, y) is on the ellipse with vertices $(\pm a, 0)$ and foci $(\pm c, 0)$ if and only if it satisfies the equation

$$\frac{x^2}{a^2} + \frac{y^2}{b^2} = 1,$$

where $b^2 = a^2 - c^2$.

An ellipse has two *latera recta* (plural of latus rectum), which are chords of the ellipse perpendicular to the major axis and containing the foci. If $x = \pm c$, then

$$\frac{c^2}{a^2} + \frac{y^2}{b^2} = 1$$

$$\frac{y^2}{b^2} = \frac{a^2 - c^2}{a^2} = \frac{b^2}{a^2}$$

$$y^2 = \frac{b^4}{a^2}$$

$$y = \pm \frac{b^2}{a}.$$

Thus, one latus rectum has end points $(c, \pm b^2/a)$, while the other has end points $(-c, \pm b^2/a)$. In both cases the length is $2b^2/a$. As with the parabola, this length may be used as an aid in sketching; however, the vertices and covertices allow one to make a reasonable sketch.

Again, the roles of the x and y may be reversed.

Theorem 9.4

A point (x, y) is on the ellipse with vertices $(0, \pm a)$ and foci $(0, \pm c)$ if and only if it satisfies the equation

$$\frac{y^2}{a^2} + \frac{x^2}{b^2} = 1,$$

where $b^2 = a^2 - c^2$.

One question that immediately arises is, How can we tell whether we have

$$\frac{x^2}{a^2} + \frac{y^2}{b^2} = 1 \quad \text{or} \quad \frac{y^2}{a^2} + \frac{x^2}{b^2} = 1?$$

The numbers in the denominators are not labeled a and b, so how do we know which is a and which is b? The answer is "size." In both cases $a > b$. Thus, the larger denominator is a^2, and the smaller is b^2.

Example 1

Sketch and discuss $9x^2 + 25y^2 = 225$.

First, we put the equation into standard form by dividing through by 225:

$$\frac{x^2}{25} + \frac{y^2}{9} = 1.$$

Now

$$a^2 = 25, \quad b^2 = 9, \quad \text{and} \quad c^2 = a^2 - b^2 = 16.$$

This ellipse has center $(0, 0)$, vertices $(\pm 5, 0)$, covertices $(0, \pm 3)$, and foci $(\pm 4, 0)$. The latera recta have length $2b^2/a = 2 \cdot 9/5 = 3.6$ (see Figure 9.10).

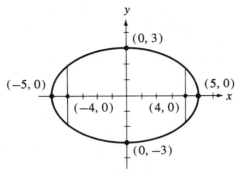

Figure 9.10

Example 2

Sketch and discuss $25x^2 + 16y^2 = 400$.

Putting the equation into standard form gives

$$\frac{x^2}{16} + \frac{y^2}{25} = 1.$$

Now

$$a^2 = 25, \quad b^2 = 16$$

and

$$c^2 = a^2 - b^2 = 9.$$

This ellipse has center $(0, 0)$, vertices $(0, \pm 5)$, covertices $(\pm 4, 0)$, and foci $(0, \pm 3)$. The latera recta have length

$$2b^2/a = 2 \cdot 16/5 = 6.4$$

(see Figure 9.11).

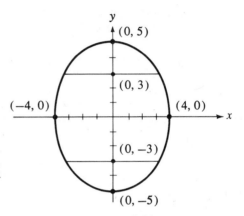

Figure 9.11

Example 3

Find an equation of the ellipse with vertices $(0, \pm 8)$ and foci $(0, \pm 5)$.

Since the vertices are on the y axis, we have the form

$$\frac{y^2}{a^2} + \frac{x^2}{b^2} = 1.$$

Furthermore, $a = 8$ and $c = 5$; thus $b^2 = a^2 - c^2 = 64 - 25 = 39$. The final result is

$$\frac{y^2}{64} + \frac{x^2}{39} = 1.$$

In addition to the quantities named, each ellipse is associated with a number, called the eccentricity. For any ellipse the *eccentricity* is

$$e = \frac{c}{a}.$$

The eccentricity of an ellipse satisfies the inequalities $0 < e < 1$. It gives a measure of the shape of the ellipse: the closer the eccentricity is to 0, the more nearly circular is the ellipse. For instance, in Example 1, $e = 4/5$, while in Example 2, $e = 3/5$. The ellipse of Example 2 is more nearly circular than the ellipse of Example 1, as can be easily seen by the sketches.

There is also a directrix associated with each focus of an ellipse. Associated with the focus $(c, 0)$ of the ellipse

$$\frac{x^2}{a^2} + \frac{y^2}{b^2} = 1$$

is the *directrix*

$$x = \frac{a}{e} = \frac{a^2}{c}.$$

If P is any point of the ellipse, the distance from P to the focus divided by the distance from P to the directrix is equal to the eccentricity. This is sometimes used as the definition of an ellipse.

Suppose we start with focus $(c, 0)$, directrix $x = a^2/c$, and eccentricity $e = c/a$. Now let us find the set of all points $P(x, y)$ such that the distance from P to the focus divided by the distance from P to the directrix equals the eccentricity.

$$\frac{\sqrt{(x - c)^2 + y^2}}{\dfrac{a^2}{c} - x} = \frac{c}{a}$$

$$\sqrt{(x - c)^2 + y^2} = a - \frac{cx}{a}$$

$$x^2 - 2cx + c^2 + y^2 = a^2 - 2cx + \frac{c^2 x^2}{a^2}$$

$$\frac{a^2 - c^2}{a^2} x^2 + y^2 = a^2 - c^2$$

$$\frac{x^2}{a^2} + \frac{y^2}{a^2 - c^2} = 1.$$

With $b^2 = a^2 - c^2$, this becomes

$$\frac{x^2}{a^2} + \frac{y^2}{b^2} = 1.$$

The same result can be obtained using focus $(-c, 0)$, directrix $x = -a^2/c$, and eccentricity $e = c/a$.

For a parabola, the distance from a point on the parabola to the focus divided by the distance of the point from the directrix is always 1. Thus we define $e = 1$ for every parabola.

Problems

A

In Problems 1–10, sketch and discuss the given ellipse.

1. $\dfrac{x^2}{169} + \dfrac{y^2}{25} = 1.$ 2. $\dfrac{x^2}{169} + \dfrac{y^2}{144} = 1.$

3. $\dfrac{x^2}{9} + \dfrac{y^2}{36} = 1.$ 4. $\dfrac{x^2}{81} + \dfrac{y^2}{4} = 1.$

5. $\dfrac{x^2}{25} + \dfrac{y^2}{49} = 1.$

6. $4x^2 + 49y^2 = 196.$

7. $9x^2 + 5y^2 = 45.$

8. $16x^2 + y^2 = 16.$

9. $16x^2 + 9y^2 = 144.$

10. $4x^2 + 25y^2 = 100.$

In Problems 11 and 12, find an equation(s) of the ellipse(s) described.

11. Center: $(0, 0)$;   vertex: $(0, -13)$;   focus: $(0, 12)$.
12. Center: $(0, 0)$;   covertex: $(0, -12)$;   focus: $(-5, 0)$.

B

In Problems 13–18, find an equation(s) of the ellipse(s) described.

13. Center: $(0, 0)$;   vertex: $(5, 0)$;   contains $(\sqrt{15}, 2)$.
14. Center: $(0, 0)$;   axes on the coordinate axes;   contains $(2, 2)$ and $(-4, 1)$.
15. Vertices: $(\pm 4, 0)$;   length of latus rectum: 6.
16. Covertices: $(\pm 3, 0)$;   length of latus rectum: 2.
17. Foci: $(\pm 6, 0)$; $e = 3/5$.
18. Foci: $(\pm 3, 0)$;   directrices: $x = \pm 12$.

19. Prove Theorem 9.4.
20. Find an equation of the line tangent to $x^2 + 4y^2 = 40$ at $(2, 3)$.
21. Find an equation of the line tangent to $2x^2 + 3y^2 = 11$ at $(2, 1)$.
22. Find an equation of the line containing $(3, -2)$ and tangent to $4x^2 + y^2 = 8$.
23. Find an equation of the line containing $(2, 4)$ and tangent to $3x^2 + 8y^2 = 84$.
24. Show that the line tangent to $(x^2/a^2) + (y^2/b^2) = 1$ at (x_0, y_0) is

$$\frac{xx_0}{a^2} + \frac{yy_0}{b^2} = 1.$$

C

25. Show that, given an elliptical mirror with a light source at one focus, no matter what point P on the ellipse is struck by a ray of light, the light is reflected to the other focus. (See Problem 31 of Section 9.1 for reflection by a curved surface.)
26. Find an expression for the length of a diameter (a chord through the center) of an ellipse. What is its maximum value? Its minimum value?
27. Suppose, in Figure 9.12, that A and B are fixed pins on the arm ABP and that AP and BP have lengths a and b, respectively. Show that if A is free to slide in channel XX' and B in channel YY', the point P traces an ellipse.

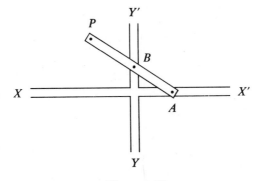

Figure 9.12

28. Given the focus $(-c, 0)$, directrix $x = -a^2/c$, and eccentricity $e = c/a$, show that these define the ellipse

$$\frac{x^2}{a^2} + \frac{y^2}{a^2 - c^2} = 1.$$

29. The earth moves in an elliptical orbit about the sun, with the sun at one focus. The least and greatest distances of the earth from the sun are 91,446,000 miles and 94,560,000 miles, respectively. What is the eccentricity of the ellipse?

9.3

The Hyperbola

Definition

A **hyperbola** *is the set of all points (x, y) in a plane such that the positive difference between the distances from (x, y) to a pair of distinct fixed points (foci) is a fixed constant.*

Again, let us choose the foci to be $(c, 0)$ and $(-c, 0)$ (see Figure 9.13) and choose the fixed constant to be $2a$. If (x, y) represents a point on the ellipse, we have

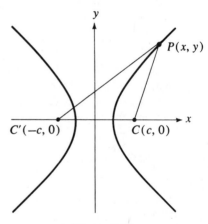

Figure 9.13

$$\sqrt{(x - c)^2 + y^2} - \sqrt{(x + c)^2 + y^2} = \pm 2a$$

$$\sqrt{(x - c)^2 + y^2} = \sqrt{(x + c)^2 + y^2} \pm 2a$$

$$x^2 - 2cx + c^2 + y^2 = x^2 + 2cx + c^2 + y^2 \pm 4a\sqrt{(x + c)^2 + y^2} + 4a^2$$

$$\mp 4a\sqrt{(x + c)^2 + y^2} = 4a^2 + 4cx$$

$$\mp \sqrt{(x + c)^2 + y^2} = a + \frac{cx}{a}$$

$$x^2 + 2cx + c^2 + y^2 = a^2 + 2cx + \frac{c^2 x^2}{a^2}$$

$$\frac{c^2 - a^2}{a^2} x^2 - y^2 = c^2 - a^2$$

$$\frac{x^2}{a^2} - \frac{y^2}{c^2 - a^2} = 1.$$

In the triangle PCC' of Figure 9.13,

$$\overline{PC'} < \overline{PC} + \overline{CC'}$$
$$\overline{PC'} - \overline{PC} < \overline{CC'}$$
$$2a < 2c$$
$$a < c$$
$$c^2 - a^2 > 0.$$

Since $c^2 - a^2$ is positive, we may replace it by another positive number, b^2. Thus

$$\frac{x^2}{a^2} - \frac{y^2}{b^2} = 1,$$

where $b^2 = c^2 - a^2$.

Again we squared both sides of the equation at two of the steps. The first time, both sides of the equation were positive; the second time, both were either positive or negative. Thus we have introduced no extraneous roots, and the steps may be reversed.

Again, both the x axis and the y axis are axes of symmetry and again $(\pm a, 0)$ are the x intercepts. However, there are no y intercepts; when $x = 0$, we have

$$-\frac{y^2}{b^2} = 1,$$

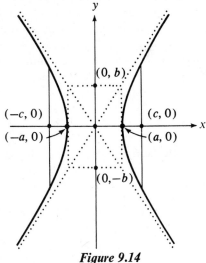

Figure 9.14

which is not satisfied by any real number y. The x axis (containing two points of the hyperbola) is called the *transverse axis*; the y axis is called the *conjugate axis*. The ends of the transverse axis, $(\pm a, 0)$, are called the *vertices*, and the point of intersection of the axes, $(0, 0)$, is called the *center* (see Figure 9.14).

Theorem 9.5

A point (x, y) is on the hyperbola with vertices $(\pm a, 0)$ and foci $(\pm c, 0)$ if and only if it satisfies the equation

$$\frac{x^2}{a^2} - \frac{y^2}{b^2} = 1,$$

where $b^2 = c^2 - a^2$.

A hyperbola has a pair of asymptotes—something no other conic section has. Thus a hyperbola is not—as might appear from inaccurate diagrams—a pair of parabolas. The hyperbola

$$\frac{x^2}{a^2} - \frac{y^2}{b^2} = 1, \quad \text{or} \quad y = \pm\frac{b}{a}\sqrt{x^2 - a^2},$$

has the slant asymptotes

$$y = \pm \frac{b}{a} x.$$

This can be proved by showing that

$$\lim_{x \to +\infty} \left[\left(\frac{b}{a} \sqrt{x^2 - a^2} \right) - \left(\frac{b}{a} x \right) \right] = 0,$$

$$\lim_{x \to +\infty} \left[\left(-\frac{b}{a} \sqrt{x^2 - a^2} \right) - \left(-\frac{b}{a} x \right) \right] = 0,$$

$$\lim_{x \to -\infty} \left[\left(\frac{b}{a} \sqrt{x^2 - a^2} \right) - \left(-\frac{b}{a} x \right) \right] = 0,$$

$$\lim_{x \to -\infty} \left[\left(-\frac{b}{a} \sqrt{x^2 - a^2} \right) - \left(\frac{b}{a} x \right) \right] = 0.$$

These are left to the student (see Problems 33 and 34). A convenient way of sketching the asymptotes is to plot both ($\pm a$, 0) and (0, $\pm b$) (even though the second pair of points is not on the hyperbola) and sketch the rectangle determined by them (see Figure 9.14). The diagonals of this rectangle are the asymptotes.

Again, two *latera recta* contain the foci and are perpendicular to the transverse axis. By using the same method as in the case of the parabola and ellipse, we can show their length to be

$$\frac{2b^2}{a}.$$

As with the parabola and the ellipse, the roles of x and y can be reversed.

Theorem 9.6

A point (x, y) is on the hyperbola with vertices $(0, \pm a)$ and foci $(0, \pm c)$ if and only if it satisfies the equation

$$\frac{y^2}{a^2} - \frac{x^2}{b^2} = 1,$$

where $b^2 = c^2 - a^2$.

It might be noted that a and b are determined by the sign of the term in which they appear; a^2 is always the denominator of the positive term and b^2 the denominator of the negative term. There is no requirement that a be greater than b, as there was for an ellipse.

The asymptotes of the hyperbola

$$\frac{y^2}{a^2} - \frac{x^2}{b^2} = 1$$

are

$$y = \pm \frac{a}{b} x.$$

Since the formulas for the asymptotes for the two cases are rather easy to confuse, a method that always works is to replace the 1 by 0 in the standard form and solve for y.

Example 1

Sketch and discuss $\dfrac{x^2}{9} - \dfrac{y^2}{16} = 1.$

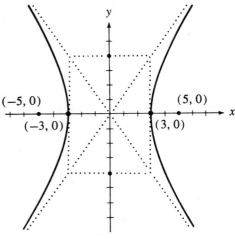

We see that $a^2 = 9$, $b^2 = 16$, and $c^2 = a^2 + b^2 = 25$. This hyperbola has center $(0, 0)$, vertices $(\pm 3, 0)$, and foci $(\pm 5, 0)$. The asymptotes are found by replacing the 1 of the standard form by 0 and solving for y.

$$\frac{x^2}{9} - \frac{y^2}{16} = 0$$

$$y^2 = \frac{16x^2}{9}$$

$$y = \pm \frac{4}{3} x.$$

Figure 9.15

The length of the latera recta is $2b^2/a = 32/3$ (see Figure 9.15).

Example 2

Sketch and discuss $16x^2 - 9y^2 + 144 = 0.$

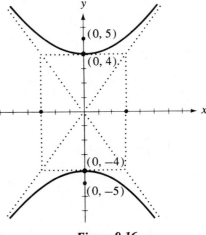

Putting this equation into standard form, we have

$$\frac{y^2}{16} - \frac{x^2}{9} = 1.$$

We see that $a^2 = 16$, $b^2 = 9$, and $c^2 = a^2 + b^2 = 25$. This hyperbola has center $(0, 0)$, vertices $(0, \pm 4)$, and foci $(0, \pm 5)$. Its asymptotes are $y = \pm 4x/3$ and the length of the latera recta is $2b^2/a = 9/2$ (see Figure 9.16).

Figure 9.16

Note the relationship between the equations of these two examples when in the standard forms; the left-hand sides are simply opposite in sign. Such hyperbolas are called *conjugate hyperbolas*.

Example 3

Find an equation of the hyperbola with foci $(\pm 4, 0)$ and vertex $(2, 0)$.

Since the foci are on the transverse axis and we are given that they are on the x axis, we must have the form

$$\frac{x^2}{a^2} - \frac{y^2}{b^2} = 1.$$

The foci tell us that $c = 4$, and the vertex gives $a = 2$; thus $b^2 = c^2 - a^2 = 12$. The resulting equation is

$$\frac{x^2}{4} - \frac{y^2}{12} = 1,$$

or

$$3x^2 - y^2 = 12.$$

Hyperbolas, as well as the other conic sections, can be determined by a single focus, a directrix, and an eccentricity. For the hyperbola

$$\frac{x^2}{a^2} - \frac{y^2}{b^2} = 1,$$

we have *eccentricity*

$$e = \frac{c}{a}$$

and *directrices*

$$x = \pm\frac{a}{e} = \pm\frac{a^2}{c},$$

where $x = a^2/c$ is used in conjunction with the focus $(c, 0)$ and $x = -a^2/c$ with the focus $(-c, 0)$. Since $c > a$, $e = c/a > 1$. Furthermore a single focus and directrix gives the entire hyperbola—not merely one branch. Either focus with its corresponding directrix generates a hyperbola.

Problems

A

In Problems 1–14, sketch and discuss the given hyperbola.

1. $\dfrac{x^2}{16} - \dfrac{y^2}{9} = 1.$

2. $\dfrac{x^2}{4} - \dfrac{y^2}{4} = 1.$

3. $\dfrac{y^2}{9} - \dfrac{x^2}{1} = 1.$

4. $\dfrac{y^2}{4} - \dfrac{x^2}{9} = 1.$

5. $\dfrac{x^2}{144} - \dfrac{y^2}{25} = 1.$

6. $\dfrac{y^2}{25} - \dfrac{x^2}{144} = 1.$

7. $\dfrac{y^2}{36} - \dfrac{x^2}{9} = 1.$

8. $x^2 - 9y^2 = 36.$

9. $4x^2 - y^2 = 4.$

10. $4x^2 - y^2 + 64 = 0.$

11. $x^2 - y^2 = 16.$

12. $16x^2 - 9y^2 = -25.$

13. $36y^2 - 100x^2 = 225.$

14. $9x^2 - 4y^2 - 16 = 0.$

In Problems 15 and 16, find an equation(s) of the hyperbola(s) described.

15. Vertices: $(\pm 2, 0)$; focus: $(-5, 0)$.

16. Foci: $(0, \pm 5)$; vertex: $(0, -3)$.

B

In Problems 17–26, find an equation(s) of the hyperbola(s) described.

17. Asymptotes: $y = \pm 2x/3$; vertex: $(6, 0)$.
18. Asymptotes: $y = \pm 3x/4$; focus: $(0, -9)$.
19. Asymptotes: $y = \pm 4x/3$; contains $(6, -4\sqrt{2})$.
20. Asymptotes: $y = \pm 3x/4$; length of latera recta: 9.

21. Vertices: $(\pm 5, 0)$;   contains $(9/5, -4)$.
22. Foci: $(\pm 13, 0)$;   contains $(-5\sqrt{2}, 12)$.
23. Vertices: $(0, \pm 5)$; $e = 13/5$.
24. Foci: $(\pm 6, 0)$; $e = 4/3$.
25. Directrices: $x = \pm 9/5$; $e = 5/3$.
26. Directrices: $y = \pm 5/3$;   focus: $(0, -6)$.

27. Find an equation of the line tangent to $16x^2 - y^2 = 144$ at $(5, -16)$.
28. Find an equation of the line tangent to $x^2 - y^2 = 9$ at $(-5, -4)$.
29. Find an equation(s) of the line(s) tangent to $x^2 - y^2 = 9$ and containing $(9, 9)$.
30. Find an equation(s) of the line(s) tangent to $4x^2 - 9y^2 = 7$ and containing $(-7, 7)$.
31. Show that the line tangent to $(x^2/a^2) - (y^2/b^2) = 1$ at the point (x_0, y_0) is

$$\frac{xx_0}{a^2} - \frac{yy_0}{b^2} = 1.$$

C

32. Show that there is a number k such that, if P is any point of a hyperbola, the product of the distances of P from the asymptotes of the hyperbola is k.

33. Show that $\lim\limits_{x \to +\infty} \left[\left(\frac{b}{a} \sqrt{x^2 - a^2} \right) - \left(\frac{b}{a} x \right) \right] = 0$.   (*Hint:* Rationalize the numerator.)

34. Show that $\lim\limits_{x \to -\infty} \left[\left(\frac{b}{a} \sqrt{x^2 - a^2} \right) - \left(-\frac{b}{a} x \right) \right] = 0$.

35. Show that if a ray of light is directed toward one focus of a hyperbolic mirror, it is reflected toward the other focus. This results in the path shown in Figure 9.17.

36. An airplane sends out an impulse that travels at the speed of sound (1100 ft/sec). Two receiving stations, whose positions are accurately known, record the times of reception of the impulse (they do not know the time the impulse was sent). How can this information be used to determine the position of the airplane, and to what extent can the position be determined? How many receiving stations are necessary to pinpoint the position of the airplane?

Figure 9.17

37. A man standing at a point $Q(x, y)$ hears the crack of a rifle at point $P_1(1000, 0)$ and the sound of the bullet hitting the target $P_2(-1000, 0)$ at the same time. If the bullet travels at 2000 ft/sec and sound travels at 1100 ft/sec, find an equation relating x and y.

9.4

Translation

The coordinate axes are something of an artificiality that we introduced on the plane in order to represent points and curves algebraically. Since the axes are of this nature, their placement is quite arbitrary. Thus we might prefer to move them in order to simplify some equation. Any change in the position of the axes may be represented by a combination of a translation and a rotation. A translation of the axes gives a new set of axes parallel to the old ones (see Figure 9.18(a)), while in a rotation, the axes are rotated about the origin (see Figure 9.18(b)).

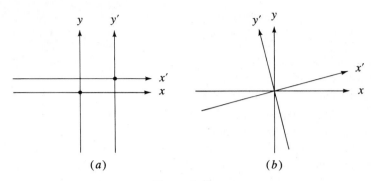

Figure 9.18

Let us consider translation first. If the axes are translated in such a way that the origin of the new coordinate system is the point (h, k) of the old system (see Figure 9.19), then every point has two representations: (x, y) in the old coordinate system, and (x', y') in the new. The relationship between the old and new coordinate system is easily seen from Figure 9.19 to be

$$x = x' + h \quad \text{or} \quad x' = x - h,$$
$$y = y' + k \quad \text{or} \quad y' = y - k.$$

These equations (either set) are called equations of translation.

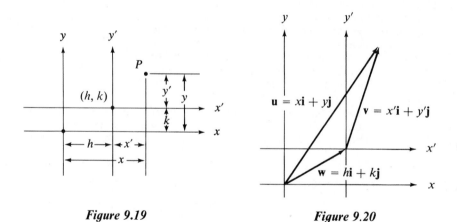

Figure 9.19   **Figure 9.20**

We can also consider translation from a vector point of view. The vector $x\mathbf{i} + y\mathbf{j}$ can be used to represent the point (x, y). A graphical representation of this vector is a directed line segment with its tail at the origin and its head at (x, y). It is easily seen from Figure 9.20 that $\mathbf{u} = \mathbf{v} + \mathbf{w}$ or

$$x\mathbf{i} + y\mathbf{j} = (x' + h)\mathbf{i} + (y' + k)\mathbf{j}.$$

Thus

$$x = x' + h,$$
$$y = y' + k.$$

Suppose we have a parabola with vertex at (h, k) and axis $y = k$ (see Figure 9.21).

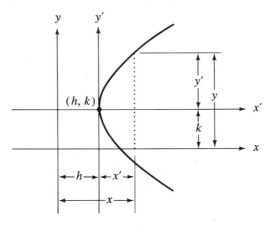

Figure 9.21

Let us put in a new pair of axes, the x' and y' axes, which are parallel to and in the same directions as the original axes and have their origin at the point (h, k) of the original system. Since the parabola's vertex is now at the origin of this coordinate system, its equation is

$$y'^2 = 4cx',$$

where $|c|$ is the distance from vertex to focus. Now the relationship between the old and new coordinates is given by the equations of translation

$$x' = x - h, \qquad y' = y - k.$$

Thus the equation of the parabola in the original coordinate system is

$$(y - k)^2 = 4c(x - h).$$

We can go through the same analysis for all three of our conics and obtain the following results.

Theorem 9.7

A point (x, y) is on the parabola with focus $(h + c, k)$ and directrix $x = h - c$ if and only if it satisfies the equation
$$(y - k)^2 = 4c(x - h).$$

A point (x, y) is on the parabola with focus $(h, k + c)$ and directrix $y = k - c$ if and only if it satisfies the equation
$$(x - h)^2 = 4c(y - k).$$

Theorem 9.8

A point (x, y) is on the ellipse with center (h, k), vertices $(h \pm a, k)$, and covertices $(h, k \pm b)$ if and only if it satisfies the equation

$$\frac{(x - h)^2}{a^2} + \frac{(y - k)^2}{b^2} = 1.$$

The foci are $(h \pm c, k)$, where $c^2 = a^2 - b^2$.

A point (x, y) *is on the ellipse with center* (h, k), *vertices* $(h, k \pm a)$, *and covertices* $(h \pm b, k)$ *if and only if it satisfies the equation*

$$\frac{(y - k)^2}{a^2} + \frac{(x - h)^2}{b^2} = 1.$$

The foci are $(h, k \pm c)$, *where* $c^2 = a^2 - b^2$.

Theorem 9.9

A point (x, y) *is on the hyperbola with center* (h, k), *vertices* $(h \pm a, k)$, *and foci* $(h \pm c, k)$ *if and only if it satisfies the equation*

$$\frac{(x - h)^2}{a^2} - \frac{(y - k)^2}{b^2} = 1,$$

where $b^2 = c^2 - a^2$.

A point (x, y) *is on the hyperbola with center* (h, k), *vertices* $(h, k \pm a)$, *and foci* $(h, k \pm c)$, *if and only if it satisfies the equation*

$$\frac{(y - k)^2}{a^2} - \frac{(x - h)^2}{b^2} = 1,$$

where $b^2 = c^2 - a^2$.

Suppose we now consider the equation

$$(y - k)^2 = 4c(x - h)$$

and carry out the indicated multiplications. The result is

$$y^2 - 2ky + k^2 = 4cx - 4ch$$
$$y^2 - 4cx - 2ky + (k^2 + 4ch) = 0.$$

This is in the form

$$y^2 + D'x + E'y + F' = 0,$$

where $D' = -4c$, $E' = -2k$, and $F' = k^2 + 4ch$. If we now multiply through by some number C $(C \neq 0)$, we have

$$Cy^2 + Dx + Ey + F = 0.$$

We can consider the other five standard forms of Theorems 9.7–9.9 in the same way. We find that in every case we get an equation of the form

$$Ax^2 + Cy^2 + Dx + Ey + F = 0.$$

If we have a parabola, either $A = 0$ or $C = 0$; if we have an ellipse, A and C are both positive or both negative; if we have a hyperbola, A and C have opposite signs. This is summarized in the following theorem.

Theorem 9.10

Every conic with axis (or axes) parallel to or on a coordinate axis (or axes) may be represented by an equation of the form

$$Ax^2 + Cy^2 + Dx + Ey + F = 0,$$

where A and C are not both zero. Furthermore, $AC = 0$ if the conic is a parabola, $AC > 0$ if it is an ellipse, and $AC < 0$ if it is a hyperbola.

The equations of Theorems 9.7–9.9 are called the standard forms for conics, while the equation of Theorem 9.10 is called the general form. It is a simple matter to go from the standard form to the general form; one merely carries out the indicated multiplications. We go from the general form to the standard form by completing the square, just as we did for a circle. Once we have the conic in standard form, it is a simple matter to carry out a translation.

Example 1

Sketch and discuss the parabola $9y^2 + 36x - 6y - 23 = 0$.

First we put the equation into standard form by completing the square on the y terms. To do this we first isolate the y terms on one side of the equation. Then we make the coefficient on y^2 one by dividing by 9. Finally we divide the resulting coefficient of y by 2 and square; this number is then added to both sides, making one side a perfect square.

$$9y^2 - 6y = -36x + 23$$

$$y^2 - \frac{2}{3}y = -4x + \frac{23}{9}$$

$$y^2 - \frac{2}{3}y + \frac{1}{9} = -4x + \frac{24}{9}$$

$$\left(y - \frac{1}{3}\right)^2 = -4\left(x - \frac{2}{3}\right).$$

The equations of translation,

$$x' = x - \frac{2}{3}, \qquad y' = y - \frac{1}{3},$$

now give

$$y'^2 = -4x'.$$

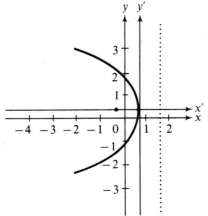

Figure 9.22

In the $x'y'$ system we have a parabola with vertex at the origin, axis the x' axis ($y' = 0$), focus at $(-1, 0)$, directrix $x' = 1$, and a latus rectum of length 4. By using the equations of translation, we express the vertex, axis, and so on, in the xy coordinate system. Thus we have a parabola with vertex at $(2/3, 1/3)$, axis $y = 1/3$, focus at $(-1/3, 1/3)$, directrix $x = 5/3$, and a latus rectum of length 4. The parabola with both sets of coordinate axes is given in Figure 9.22.

Example 2

Sketch and discuss the hyperbola $9x^2 - 4y^2 - 18x - 24y - 63 = 0$.

Here we must complete the square on both the x and the y terms. Thus we must make the coefficients of both x^2 and y^2 one. We do this by factoring.

$$9x^2 - 18x - 4y^2 - 24y = 63$$
$$9(x^2 - 2x) - 4(y^2 + 6y) = 63$$
$$9(x^2 - 2x + 1) - 4(y^2 + 6y + 9) = 63 + 9 \cdot 1 - 4 \cdot 9$$
$$9(x - 1)^2 - 4(y + 3)^2 = 36$$
$$\frac{(x - 1)^2}{4} - \frac{(y + 3)^2}{9} = 1.$$

Now the equations of translation,

$$x' = x - 1, \qquad y' = y + 3,$$

give

$$\frac{x'^2}{4} - \frac{y'^2}{9} = 1.$$

In the $x'y'$ system we have a hyperbola with center at $(0, 0)$, $a = 2$, $b = 3$, and $c^2 = a^2 + b^2 = 13$. Thus we have vertices $(\pm 2, 0)$, foci $(\pm\sqrt{13}, 0)$, asymptotes $y' = \pm 3x'/2$, and latera recta of length $2b^2/a = 2 \cdot 9/2 = 9$. Using the equations of translation, we see that in the xy system we have a hyperbola with center at $(1, -3)$, vertices $(3, -3)$ and $(-1, -3)$, foci $(1 \pm\sqrt{13}, -3)$, asymptotes $y + 3 = \pm\frac{3}{2}(x - 1)$, or $3x - 2y - 9 = 0$ and $3x + 2y + 3 = 0$. The latera recta are, of course, unchanged by the translation. The hyperbola with both coordinate systems is given in Figure 9.23.

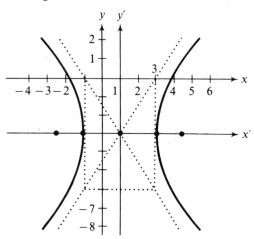

Figure 9.23

While Theorem 9.10 tells us that any conic with axis (or axes) on or parallel to the coordinate axes can be represented in the form

$$Ax^2 + Cy^2 + Dx + Ey + F = 0,$$

it does not follow that any equation in this form represents a parabola, ellipse, or hyperbola. This point is demonstrated by the following example.

Example 3

Sketch and discuss $4x^2 + 3y^2 - 16x + 18y + 43 = 0$.

Since $A = 4$ and $C = 3$, $AC = 12 > 0$. The equation appears to represent an ellipse. But completing the square gives

$$4(x^2 - 4x) + 3(y^2 + 6y) = -43$$
$$4(x^2 - 4x + 4) + 3(y^2 + 6y + 9) = -43 + 4 \cdot 4 + 3 \cdot 9$$
$$4(x - 2)^2 + 3(y + 3)^2 = 0.$$

Since neither term on the left is negative, the sum can be zero only if both terms are zero. Thus $x = 2$ and $y = -3$. This equation is satisfied only by the point $(2, -3)$.

The foregoing example is called a degenerate case of an ellipse, since the equation has the form of an ellipse but does not actually represent an ellipse. It is comparable to the degenerate cases of a circle, which we saw in Chapter 2. The degenerate cases of the three conics are given in the table below.

Conic	AC	Degenerate cases
Parabola	0	One line (two coincident lines) Two parallel lines No graph
Ellipse	+	Circle Point No graph
Hyperbola	−	Two intersecting lines

The method of completing the square is simple to use, but it is rather limited in scope. It can be used only on second-degree equations with no xy term. If there is an xy term or if the equation is not of the second degree, another method, illustrated by the following example, can be used.

Example 4

Translate axes so that the constant and the x term of $y = x^3 - 5x^2 + 7x - 5$ are eliminated.

Since we do not know what values of h and k to choose, we simply use the equations of translation,

$$x = x' + h \quad \text{and} \quad y = y' + k,$$

and see what values of h and k are needed to eliminate the terms specified.

$$y' + k = (x' + h)^3 - 5(x' + h)^2 + 7(x' + h) - 5$$
$$y' = x'^3 + (3h - 5)x'^2 + (3h^2 - 10h + 7)x' + (h^3 - 5h^2 + 7h - 5 - k).$$

Now we must choose h and k so that

$$3h^2 - 10h + 7 = 0$$
$$h^3 - 5h^2 + 7h - 5 - k = 0.$$

The first of these two equations gives

$$h = 1 \quad \text{or} \quad h = 7/3.$$

Substituting these values into the second, we have

$$k = -2 \quad \text{or} \quad k = -86/27.$$

Using $h = 1$ and $k = -2$, we get

$$y' = x'^3 - 2x'^2.$$

Using $h = 7/3$ and $k = -86/27$, we get

$$y' = x'^3 + 2x'^2.$$

The graphs of both cases are given in Figure 9.24. While there are two different translations giving two different equations, both graphs are the same when referred to the original xy system. As an added bonus, this method has located the relative maximum, $(1, -2)$, and minimum, $(7/3, -86/27)$. This method can also be used on second-degree equations with no xy term, but completing the square is so much simpler that most would prefer to use it.

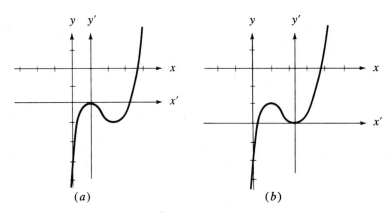

(a) (b)

Figure 9.24

Problems

A

In Problems 1–10, translate axes so as to have the center or vertex of the conic section at the origin of the new coordinate system. Sketch the curve showing both the old and new axes.

1. $y^2 - 4x - 2y + 9 = 0$.
2. $x^2 - 6x - 8y + 33 = 0$.
3. $4x^2 + y^2 - 16x + 6y + 21 = 0$.
4. $x^2 + 9y^2 - 4x + 72y + 139 = 0$.
5. $9x^2 - 4y^2 + 90x + 32y + 125 = 0$.
6. $9x^2 - 16y^2 + 36x + 32y - 124 = 0$.
7. $9x^2 + 4y^2 - 18x + 24y + 9 = 0$.
8. $4x^2 - y^2 - 40x - 2y + 95 = 0$.
9. $4x^2 - 4x - 4y - 5 = 0$.
10. $16x^2 + 36y^2 - 16x - 108y - 491 = 0$.

In Problems 11–18, sketch and discuss.

11. $x^2 - 6x - 6y + 3 = 0$.
12. $4x^2 + y^2 - 24x + 4y + 36 = 0$.
13. $16x^2 - 9y^2 + 54y - 225 = 0$.
14. $4x^2 - 9y^2 - 4x - 18y - 26 = 0$.
15. $8x^2 + 9y^2 + 48x - 18y + 9 = 0$.
16. $2x^2 - 5x - 3 = 0$.
17. $3x^2 - 3y^2 - 2x - 4y - 13 = 0$.
18. $25x^2 + 4y^2 + 50x - 12y - 66 = 0$.

B

In Problems 19–24, find an equation(s) of the given conic(s).

19. The ellipse with vertices (2, 6), (2, −4) and focus (2, −3).
20. The parabola with vertex (3, −1) and containing (7, 1).
21. The hyperbola with foci (−2, 5), (−2, −3) and vertex (−2, 2).
22. The parabola with vertex (2, −4) and directrix $x = 5$.
23. The ellipse with vertex (6, 1) and covertex (4, 2).
24. The hyperbola with asymptotes $3x - 4y - 13 = 0$, $3x + 4y - 5 = 0$ and focus (−2, −1).

In Problems 25–32, translate axes to eliminate the terms indicated.

25. $x^2 - 2xy + 4y^2 + 8x - 26y + 38 = 0$; first-degree terms.
26. $2x^2 - xy - y^2 - 2x + 5y - 5 = 0$; first-degree terms.
27. $xy - 4x + 3y - 11 = 0$; first-degree terms.
28. $y = x^3 + 9x^2 + 27x + 21$; constant, x^2 term.
29. $y = x^3 - 3x + 6$; constant, x term.
30. $y = x^4 - 3x^2 + 2x - 4$; constant, x term.
31. $y = x^5 - 10x^4 + 42x^3 - 92x^2 + 105x - 53$; constant, x^2 term.
32. $x^2y - 2x^2 + 2xy + 4x - 3y + 10 = 0$; first-degree terms.

C

33. Suppose that a translation changes the equation
$$Ax^2 + Bxy + Cy^2 + Dx + Ey + F = 0$$
into
$$A'x'^2 + B'x'y' + C'y'^2 + D'x' + E'y' + F' = 0.$$

Show that $A' = A$, $B' = B$, and $C' = C$ for any translation. A, B, and C are said to be invariant under translation.

34. It can easily be seen graphically that two conic sections have at most four points in common. But
$$2x^2 + xy - y^2 + 3y - 2 = 0,$$
$$2x^2 + 3xy + y^2 - 6x - 5y + 4 = 0$$

have in common the five points (1, 0), (−2, 3), (5, −4), (−6, 7), and (10, −9). Why?

9.5

Rotation

The second transformation of the axes that we wish to consider is a rotation of the axes about the origin (see Figure 9.25). If the axes are rotated through an angle θ, then every point of the plane has two representations: (x, y) in the original coordinate system and (x', y') in the new coordinate system. Alternatively, every vector **v** in the plane has two representations: $\mathbf{v} = x\mathbf{i} + y\mathbf{j}$ in the original coordinate system and

$\mathbf{v} = x'\mathbf{i}' + y'\mathbf{j}'$ in the new coordinate system (see Figure 9.25). In order to find the relationships between the x and y of one coordinate system and the x' and y' of the other let us consider the relationships of $\mathbf{i}$ and $\mathbf{j}$ with $\mathbf{i}'$ and $\mathbf{j}'$. Remembering that $\mathbf{i}$, $\mathbf{j}$, $\mathbf{i}'$ and $\mathbf{j}'$ are all unit vectors, we see from Figure 9.25 that

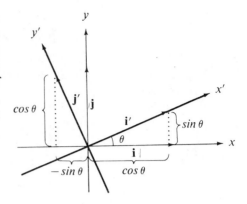

$$\mathbf{i}' = \cos\theta\mathbf{i} + \sin\theta\mathbf{j},$$
$$\mathbf{j}' = -\sin\theta\mathbf{i} + \cos\theta\mathbf{j}.$$

Thus

Figure 9.25

$$\mathbf{v} = x'\mathbf{i}' + y'\mathbf{j}'$$
$$= x'(\cos\theta\mathbf{i} + \sin\theta\mathbf{j}) + y'(-\sin\theta\mathbf{i} + \cos\theta\mathbf{j})$$
$$= (x'\cos\theta - y'\sin\theta)\mathbf{i} + (x'\sin\theta + y'\cos\theta)\mathbf{j}$$
$$= x\mathbf{i} + y\mathbf{j}$$

which gives the equations of rotation

$$x = x'\cos\theta - y'\sin\theta,$$
$$y = x'\sin\theta + y'\cos\theta.$$

Example 1

Find the new representation of

$$x^2 - xy + y^2 - 2 = 0$$

after rotating through an angle of 45°. Sketch the curve, showing both the old and new coordinate systems.

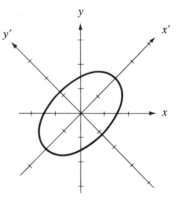

Since $\sin 45° = \cos 45° = 1/\sqrt{2}$, the equations of rotation are

$$x = \frac{x' - y'}{\sqrt{2}} \quad \text{and} \quad y = \frac{x' + y'}{\sqrt{2}}.$$

Substituting into the original equation, we have

Figure 9.26

$$\frac{(x'-y')^2}{2} - \frac{x'-y'}{\sqrt{2}} \cdot \frac{x'+y'}{\sqrt{2}} + \frac{(x'+y')^2}{2} - 2 = 0$$

$$\frac{x'^2 - 2x'y' + y'^2 - x'^2 + y'^2 + x'^2 + 2x'y' + y'^2}{2} = 2$$

$$x'^2 + 3y'^2 = 4.$$

Figure 9.26 shows the final result.

Example 2

Find a new representation of $x^2 + 4xy - 2y^2 - 6 = 0$ after rotating through an angle $\theta = \text{Arctan } 1/2$. Sketch the curve, showing both the old and new coordinate systems.

Figure 9.27 shows that

$$\sin \theta = \frac{1}{\sqrt{5}} \quad \text{and} \quad \cos \theta = \frac{2}{\sqrt{5}},$$

giving equations of rotation

$$x = \frac{2x' - y'}{\sqrt{5}} \quad \text{and} \quad y = \frac{x' + 2y'}{\sqrt{5}}.$$

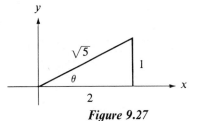

Figure 9.27

Substituting into the original equation, we have

$$\frac{(2x' - y')^2}{5} + 4 \frac{2x' - y'}{\sqrt{5}} \cdot \frac{x' + 2y'}{\sqrt{5}} - 2 \frac{(x' + 2y')^2}{5} - 6 = 0$$

$$\frac{4x'^2 - 4x'y' + y'^2 + 8x'^2 + 12x'y' - 8y'^2 - 2x'^2 - 8x'y' - 8y'^2}{5} = 6$$

$$2x'^2 - 3y'^2 = 6.$$

Figure 9.28 shows the final result. Note that Figure 9.27 can be used to determine the position of the new coordinate axes.

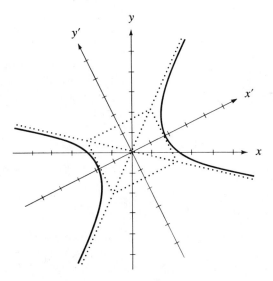

Figure 9.28

In both of these examples, we have seen that the given rotation has eliminated the xy term. Of course, not every rotation will do so—it must be specially chosen. Suppose we start with an equation in the form

$$Ax^2 + Bxy + Cy^2 + Dx + Ey + F = 0,$$

where $B \neq 0$, and rotate the axes through an angle θ, where $0° < \theta < 90°$. Then the equations of rotation are

$$x = x' \cos \theta - y' \sin \theta,$$
$$y = x' \sin \theta + y' \cos \theta.$$

Substituting these into our given equation, we get

$$A(x' \cos \theta - y' \sin \theta)^2 + B(x' \cos \theta - y' \sin \theta)(x' \sin \theta + y' \cos \theta)$$
$$+ C(x' \sin \theta + y' \cos \theta)^2 + D(x' \cos \theta - y' \sin \theta)$$
$$+ E(x' \sin \theta + y' \cos \theta) + F = 0.$$

After carrying out the multiplication and combining similar terms, we find that the coefficient of $x'y'$ is

$$(C - A)2 \sin \theta \cos \theta + B(\cos^2 \theta - \sin^2 \theta) = (C - A) \sin 2\theta + B \cos 2\theta.$$

We want this coefficient to be zero for the proper choice of θ. Let us set it equal to zero and see what θ should be.

$$(C - A) \sin 2\theta + B \cos 2\theta = 0$$

$$(A - C) \sin 2\theta = B \cos 2\theta$$

$$\frac{\sin 2\theta}{\cos 2\theta} = \frac{B}{A - C}, \quad A \neq C$$

$$\tan 2\theta = \frac{B}{A - C}, \quad A \neq C.$$

We can easily solve this equation for θ, but it would involve us in inverse trigonometric functions—let us try to get around them. To do this we shall try to find expressions for $\sin \theta$ and $\cos \theta$ that we can use in the equations of rotation.

First, we note that if $A \neq C$, then $\tan 2\theta$ exists, $2\theta \neq 90°$, and $\theta \neq 45°$. For any value of 2θ,

$$\sin^2 2\theta + \cos^2 2\theta = 1.$$

Dividing through by $\cos^2 2\theta$ (which is not zero since $\theta \neq 45°$), we get

$$\tan^2 2\theta + 1 = \frac{1}{\cos^2 2\theta}, \quad \cos^2 2\theta = \frac{1}{1 + \tan^2 2\theta}, \quad \cos 2\theta = \frac{\pm 1}{\sqrt{1 + \tan^2 2\theta}}.$$

The $\pm$ presents the question of which one to use. Since $0° < \theta < 90°$, $0° < 2\theta < 180°$. Both the tangent and cosine are positive for a first-quadrant angle and both are negative for a second-quadrant angle. Thus we choose the sign to agree with the sign of $\tan 2\theta$.

Now let us recall the half-angle identities

$$\sin \frac{A}{2} = \pm \sqrt{\frac{1 - \cos A}{2}} \quad \text{and} \quad \cos \frac{A}{2} = \pm \sqrt{\frac{1 + \cos A}{2}}.$$

Replacing A by 2θ and noting that both $\sin \theta$ and $\cos \theta$ must be positive, since $0° < \theta < 90°$, we have

$$\sin \theta = \sqrt{\frac{1 - \cos 2\theta}{2}}, \quad \cos \theta = \sqrt{\frac{1 + \cos 2\theta}{2}}.$$

Finally, if $A = C$, then

$$B \cos 2\theta = 0$$
$$\cos 2\theta = 0$$
$$2\theta = 90°$$
$$\theta = 45°.$$

Thus, in either case, we are able to rotate axes to eliminate the xy term. The resulting equation must then represent a conic or degenerate conic.

Let us sum up the results of the previous discussion. If $B \neq 0$, then the axes may be rotated to eliminate the xy term in the following way:

If $A = C$, then $\theta = 45°$. If $A \neq C$, then

$$\tan 2\theta = \frac{B}{A - C},$$

$$\cos 2\theta = \frac{\pm 1}{\sqrt{1 + \tan^2 2\theta}} \quad \text{(sign agrees with the sign of } \tan 2\theta),$$

$$\sin \theta = \sqrt{\frac{1 - \cos 2\theta}{2}},$$

$$\cos \theta = \sqrt{\frac{1 + \cos 2\theta}{2}}.$$

Example 3

Rotate axes to eliminate the xy term of $x^2 + 4xy - 2y^2 - 6 = 0$. Sketch, showing both sets of axes.

$$\tan 2\theta = \frac{B}{A - C} = \frac{4}{1 - (-2)} = \frac{4}{3},$$

$$\cos 2\theta = \frac{1}{\sqrt{1 + \tan^2 2\theta}} \quad \text{(since } \tan 2\theta \text{ is positive,} \\ \cos 2\theta \text{ is also positive)}$$

$$= \frac{1}{\sqrt{1 + \left(\frac{4}{3}\right)^2}} = \frac{3}{5},$$

$$\sin \theta = \sqrt{\frac{1 - \cos 2\theta}{2}} = \sqrt{\frac{1 - \frac{3}{5}}{2}} = \frac{1}{\sqrt{5}},$$

$$\cos \theta = \sqrt{\frac{1 + \cos 2\theta}{2}} = \sqrt{\frac{1 + \frac{3}{5}}{2}} = \frac{2}{\sqrt{5}},$$

$$\tan \theta = \frac{\sin \theta}{\cos \theta} = \frac{1}{2} \quad \text{(we shall use this for sketching)};$$

$$x = \frac{2x' - y'}{\sqrt{5}}, \quad y = \frac{x' + 2y'}{\sqrt{5}}.$$

Substituting these equations of rotation into the original equation (see Example 2), we have

$$2x'^2 - 3y'^2 = 6.$$

The sketch is given in Figure 9.28.

Example 4

Rotate axes to eliminate the xy term of

$$2x^2 - xy + 2y^2 - 2 = 0.$$

Sketch, showing both sets of axes.

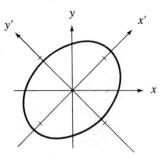

Since $A = C$, $\theta = 45°$ and the equations of rotation are

$$x = \frac{x' - y'}{\sqrt{2}}, \qquad y = \frac{x' + y'}{\sqrt{2}}.$$

Figure 9.29

Substituting these into the original equation, we have

$$2\frac{(x' - y')^2}{2} - \frac{x' - y'}{\sqrt{2}}\frac{x' + y'}{\sqrt{2}} + 2\frac{(x' + y')^2}{2} - 2 = 0$$

$$\frac{2x'^2 - 4x'y' + 2y'^2 - x'^2 + y'^2 + 2x'^2 + 4x'y' + 2y'^2}{2} = 2$$

$$3x'^2 + 5y'^2 = 4$$

$$\frac{x'^2}{4/3} + \frac{y'^2}{4/5} = 1.$$

The sketch is given in Figure 9.29.

The sketching of conic sections by rotating the axes is tedious at best. It is often possible to avoid this tediousness by using the methods of Chapter 5. This is especially true when we have a hyperbola, since the asymptotes are very helpful in providing a sketch of the graph. The following theorem can be used to determine which one of the three conic sections we have.

Theorem 9.11

Given the equation $Ax^2 + Bxy + Cy^2 + Dx + Ey + F = 0$, it represents a hyperbola, ellipse, or parabola (or a degenerate case of one of them) if $B^2 - 4AC$ is positive, negative, or zero, respectively.

The proof of this theorem is based upon the invariance of $B^2 - 4AC$. It is left to the student (see Problem 26).

Example 5

Sketch $x^2 - xy - 3y - 1 = 0$ without rotating axes.

First of all, $B^2 - 4AC = (-1)^2 - 4(1)(0) = 1$, indicating that the conic is a hyperbola or a degenerate case of one. Solving for y, we have

$$y = \frac{x^2 - 1}{x + 3}.$$

The methods of Chapter 5 give intercepts $(\pm 1, 0)$, $(0, -1/3)$ and vertical asymptote $x = -3$. There is no horizontal asymptote, but we know that there must be a second asymptote. To find it, we carry out the division.

$$y = \frac{x^2 - 1}{x + 3} = x - 3 + \frac{8}{x + 3}.$$

We now see that, if x is large in absolute value, $8/(x + 3)$ is almost zero and y is very near $x - 3$. Thus the slant asymptote is

$$y = x - 3.$$

With this we can easily sketch the hyperbola (see Figure 9.30).

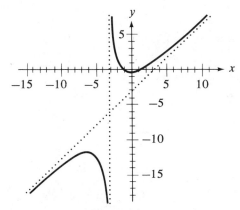

Figure 9.30

While we already have a relatively accurate sketch of the curve, we can use the derivative to determine the relative maximum and minimum if necessary.

$$y = \frac{x^2 - 1}{x + 3},$$

$$y' = \frac{(x + 3)2x - (x^2 - 1)}{(x + 3)^2} = \frac{x^2 + 6x + 1}{(x + 3)^2}.$$

Setting the numerator equal to zero and solving by the quadratic formula, we have $x = -3 \pm 2\sqrt{2}$. The y coordinates can then be found by substitution into the original equation. Thus we have a relative minimum at $(-3 + 2\sqrt{2}, 4\sqrt{2} - 6)$ and a relative maximum at $(-3 - 2\sqrt{2}, -4\sqrt{2} - 6)$. These points are *not* the vertices of the hyperbola.

Problems

A

In Problems 1–4, find a new representation of the given equation after rotating through the given angle. Sketch the curve, showing both the old and new coordinate systems.

1. $2x + 3y = 6$; $\theta = \text{Arctan } 3/2$.
2. $2x - y = 4$; $\theta = \text{Arctan } 2$.
3. $xy = 7$; $\theta = 45°$.
4. $x^2 - 6xy + y^2 - 8 = 0$; $\theta = 45°$.

B

In Problems 5–8, find a new representation of the given equation after rotating through the given angle. Sketch the curve, showing both the old and new coordinate systems.

5. $31x^2 + 10\sqrt{3}xy + 21y^2 - 144 = 0;$ $\theta = 30°.$
6. $13x^2 - 6\sqrt{3}xy + 7y^2 = 16;$ $\theta = 60°.$
7. $9x^2 - 24xy + 16y^2 - 320x - 240y = 0;$ $\theta = \text{Arctan } 3/4.$
8. $6x^2 - 5xy - 6y^2 + 26 = 0;$ $\theta = \text{Arctan } (-1/5).$

In Problems 9–18, rotate axes to eliminate the xy term. Sketch, showing both sets of axes.

9. $x^2 + xy + y^2 + 4\sqrt{2}x - 4\sqrt{2}y = 0.$ 10. $2x^2 + 5xy + 2y^2 - 9 = 0.$
11. $2x^2 + 12xy - 3y^2 - 42 = 0.$ 12. $2x^2 - 4xy - y^2 = 6.$
13. $5x^2 - 4xy + 8y^2 - 36 = 0.$ 14. $7x^2 - 20xy - 8y^2 + 52 = 0.$
15. $4x^2 + 12xy + 9y^2 + 8\sqrt{13}x + 12\sqrt{13}y - 65 = 0.$
16. $8x^2 - 12xy + 17y^2 + 20 = 0.$
17. $9x^2 - 6xy + y^2 - 12\sqrt{10}x - 36\sqrt{10}y = 0.$
18. $20x^2 + 15xy - 16y^2 = 7.$

In Problems 19–22, sketch without rotating axes.

19. $xy - x + y + 3 = 0$ 20. $2xy - x - y - 2 = 0.$
21. $x^2 - xy - y - 4 = 0.$ 22. $x^2 - xy + x + 2y = 0.$
23. Show that $x^2 + y^2 = 25$ is invariant under rotation through any angle.

C

24. Show that $2x^2 - 2xy + y^2 - 9 = 0$ is an equation of an ellipse. This equation is a quadratic equation in y. Rearranging the terms, we have $y^2 - 2xy + (2x^2 - 9) = 0,$ which can be solved by the quadratic formula to give $y = x \pm \sqrt{9 - x^2}.$ Sketch $y = x$ and $y = \pm \sqrt{9 - x^2}$ (square both sides first in the latter equation) on the same set of axes. For each value of x, add the y coordinates for these two curves to get points on the original curve. Use this method to find the graph of the given curve.

25. Given the equation $Ax^2 + Bxy + Cy^2 + Dx + Ey + F = 0$, which yields

$$A'x'^2 + B'x'y' + C'y'^2 + D'x' + E'y' + F' = 0$$

after rotation through the angle θ, show that $A' + C' = A + C$ for any value of θ (that is, $A + C$ is invariant under rotation).

26. Show that, in the general equation of second degree, $B^2 - 4AC$ is invariant under rotation. (See Problem 25 for definition of invariant.) Use the invariance of $\Delta = B^2 - 4AC$ to show that $\Delta > 0$ for a hyperbola, $\Delta < 0$ for an ellipse, and $\Delta = 0$ for a parabola.

27. Show that a second form for the equations of rotation is

$$x' = x \cos \theta + y \sin \theta,$$
$$y' = -x \sin \theta + y \cos \theta.$$

28. Suppose the axes have been rotated through an angle θ. Then the point P (see Figure 9.31) has representations (x, y) in the old coordinate system and (x', y') in the new; that is,

$$\overline{OQ} = x, \qquad \overline{OR} = x',$$
$$\overline{PQ} = y, \qquad \overline{PR} = y'.$$

Use Figure 9.31 to find the equations of rotation

$$x = x' \cos \theta - y' \sin \theta,$$
$$y = x' \sin \theta + y' \cos \theta.$$

(*Hint:* First find $\overline{RS}, \overline{OS}, \overline{TR},$ and $\overline{PT}$ in terms of $\theta, x',$ and y'.)

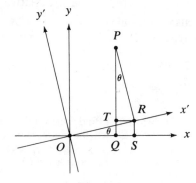

Figure 9.31

Review Problems

A

In Problems 1–12, sketch and discuss.

1. $y^2 = 16x$.

2. $4x^2 + 9y^2 = 36$.

3. $x^2 - y^2 + 9 = 0$.

4. $4x^2 + y^2 - 8x + 6y + 9 = 0$.

5. $y^2 - x + 2y + 4 = 0$.

6. $y^2 - 8x + 4y + 28 = 0$.

7. $9x^2 - 16y^2 + 36x - 128y - 364 = 0$.

8. $9x^2 + 25y^2 + 18x + 100y - 116 = 0$.

9. $9x^2 - 16y^2 + 18x - 16y - 139 = 0$.

10. $x^2 - 4x - 4y = 0$.

11. $3x^2 + 4y^2 + 30x - 16y + 91 = 0$.

12. $4x^2 - 9y^2 - 16x - 54y - 65 = 0$.

B

13. Find an equation(s) of the parabola(s) with vertex $(0, 0)$ and containing the points $(2, 4)$ and $(8, 8)$.

14. Find an equation(s) of the ellipse(s) with center $(0, 0)$, vertex $(10, 0)$, and focus $(-8, 0)$.

15. Find an equation(s) of the hyperbola(s) with vertices $(0, \pm 6)$ and focus $(0, 10)$.

16. Find an equation(s) of the ellipse(s) with center $(-4, 1)$, axes parallel to the coordinate axes, and tangent to both coordinate axes.

17. Find an equation(s) of the parabola(s) with focus $(3, 5)$ and directrix $x = -1$.

18. Find an equation(s) of the hyperbola(s) with asymptotes $5x - 4y + 22 = 0$ and $5x + 4y - 18 = 0$ and containing the point $(32/5, -7)$.

19. Find an equation(s) of the line(s) tangent to $2x^2 - y^2 + 4x + 3y - 6 = 0$ at $(1, 0)$.

20. Find an equation(s) of the line(s) tangent to $x^2 - 2y^2 - 3x + 4y = 0$ and containing $(-9/5, -27/20)$.

21. Translate axes to eliminate the constant and second-degree terms of

$$y = x^3 + 6x^2 + 3x - 14.$$

Sketch, showing both sets of axes.

22. Translate axes to eliminate the constant and third-degree terms of

$$y = x^4 - 16x^3 + 88x^2 - 192x + 140.$$

Sketch, showing both sets of axes.

23. Sketch $x^2 + xy - 2x + y = 0$ without rotating axes.

24. Rotate axes to eliminate the xy term of $3x^2 + 12xy - 2y^2 + 42 = 0$. Sketch, showing both sets of axes.

25. Rotate axes to eliminate the xy term of $2x^2 - \sqrt{3}\,xy + y^2 - 10 = 0$. Sketch, showing both sets of axes.

26. Rotate axes to eliminate the xy term of $4x^2 - 4xy + y^2 + \sqrt{5}\,x + 2\sqrt{5}\,y - 10 = 0$. Sketch, showing both sets of axes.

C

27. Sketch $5x^2 - 4xy + y^2 - 4x + 2y - 8 = 0$ without rotating axes.

10

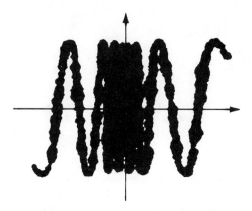

Limits and Continuity: A Geometric Approach

10.1

Limit Points of the Domain

In our earlier dealings with limits, the derivative and the integral were defined in terms of certain limits; and our determination of asymptotes was based on evaluating limits of the original function. But we had no definition of a limit of a function; we relied upon an intuitive notion of its meaning. In this chapter we shall approach a definition of the expression

$$\lim_{x \to a} f(x) = b$$

from a purely geometric point of view. We assume throughout that the graph of the function f is well known. Later (Chapter 17), we shall approach limits and continuity from an analytic standpoint.

Defining a limit has never been easy. One of the earliest attempts was made by the French mathematician Joseph Louis Lagrange (1736–1813). However, his reasoning was shown to be faulty. Further progress was made by the French mathematician Augustin-Louis Cauchy (1789–1857) and the Czech priest Bernhard Bolzano (1781–1848), who independently gave definitions of limit and continuity which are equivalent to present day definitions. Nevertheless they still had a degree of vagueness which is no longer acceptable. The present day definitions were first published by the German mathematician Heinrich Eduard Heine in 1872. But his publication was strongly influenced by the lectures at the University of Berlin of Karl Weierstrass, to whom some of the credit must go.

Before we consider a definition of limit, some points must be identified. In the expression

$$\lim_{x \to a} f(x) = b,$$

the point (a, b), where a and b are the numbers approached by x and $f(x)$ in the above limit statement, is represented by P. Similarly, A represents the point $(a, f(a))$ provided a is in the domain of f. We might note the following facts about A and P:

(1) for a given function f and the given limit statement $\lim_{x \to a} f(x) = b$, the point P must necessarily exist but A may or may not exist;
(2) even when A does exist, A and P are not necessarily the same point;
(3) A, when it exists, is a point of the graph of $y = f(x)$, while P may or may not be on the graph.

Suppose we consider a few examples:

Limit Statement	A	P
$\lim_{x \to a} f(x) = b$	$(a, f(a))$	(a, b)
$\lim_{x \to 0} x^2 = 0$	$(0, 0)$	$(0, 0)$
$\lim_{x \to 1} \dfrac{x^2 - 1}{x - 1} = 2$	does not exist	$(1, 2)$
$\lim_{x \to 2} (x^2 + 1) = 5$	$(2, 5)$	$(2, 5)$
$\lim_{x \to 2} (x^2 + 1) = 4$	$(2, 5)$	$(2, 4)$

You may object to the last limit statement because $\lim_{x \to 2}(x^2 + 1)$ is not 4. Nevertheless, if we were to prove that the statement $\lim_{x \to 2}(x^2 + 1) = 4$ is not true, we would have to identify A and P for this case just as we do for cases in which a given limit statement is true.

One other convention is needed here. We shall use $\mathcal{G}$ to represent the graph of $y = f(x)$ and $\mathcal{G} - A$ to represent the set of all points of $\mathcal{G}$ except A. If A does not exist, $\mathcal{G} - A$ will be taken to mean $\mathcal{G}$. A thorough knowledge of the meanings of P, A, $\mathcal{G}$ and $\mathcal{G} - A$ are necessary for the definitions of this chapter.

Before considering a definition of a limit, we shall discuss a related term—a limit point of the domain of a function. We define it here only for a real-valued function of a single real variable.

Definition

*To say that the number a is a **limit point** of the domain of the function f means that if h and k are two vertical lines with the line $x = a$ between them, then there is a point of $\mathcal{G} - A$ between them.*

Why do we refer to a *number* as a limit *point* of the domain? Why not a limit *number*? Remember that the class of functions we are dealing with is only one of many classes. In addition to functions of a single real number, there are functions of two or more real numbers, functions of one or more complex numbers, functions of vectors, and so forth, In order to have a single term for all types, we use the term "limit *point*." Thus, for our functions, a limit point is a real number. Other terms used in place of limit point are "point of accumulation" and "cluster point."

Now what does it mean? If the number a is a limit point of the domain of f, there are, roughly speaking, other numbers in the domain of f near a; in fact, there are other numbers in the domain "leading right up to" a. Let us consider some examples.

Example 1

Is the number 1 a limit point of the domain of $f(x) = x + 1$?

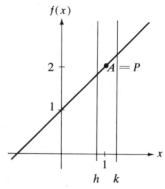

The graph $\mathscr{G}$ of $y = x + 1$ is given in Figure 10.1. The point A is $(1, f(1)) = (1, 2)$. If h and k are two vertical lines with $x = 1$ between them, then it is clear from Figure 10.1 that there is some point of $\mathscr{G} - A$ (that is, there is some point of the graph different from $(1, 2)$) that is also between h and k. Obviously this is true for *any* pair of vertical lines with $x = 1$ between them. Thus the number 1 is a limit point of the domain of f.

Figure 10.1

Perhaps you feel that it must be possible to find lines h and k close enough together so that A is the only point of $\mathscr{G}$ between them. If this were so, k would have to cross the x axis at the next point to the right of 1 and h would cross at the next point to the left of 1. But there is no "next point." If k crosses the x axis to the right of 1, then it has an equation of the form $x = c$, where $c > 1$. Now it is evident that there is a point on the x axis between 1 and c; one such point is the midpoint $(1 + c)/2$. Since the function $x + 1$ is defined for all x, there must be some point of $\mathscr{G}$ between $x = 1$ and k as well as a point of $\mathscr{G}$ between h and $x = 1$.

Example 2

Is the number 0 a limit point of the domain of $f(x) = \sqrt{x^2(x^2 - 1)}$?

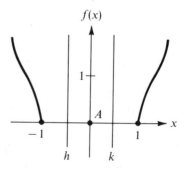

The graph $\mathscr{G}$ is given in Figure 10.2. The point A is $(0, f(0)) = (0, 0)$. We can see that A is an isolated point of $\mathscr{G}$. Let us choose h to be the line $x = -1/2$ and k to be the line $x = 1/2$ (see Figure 10.2). Now A is the only point of the graph between h and k; there is no point of $\mathscr{G} - A$ between h and k. Thus the number 0 is not a limit point of the domain.

Figure 10.2

It may seem that we twist things around to suit ourselves—that we could take h to be $x = -2$ and k to be $x = 2$ and thus find some point of $\mathscr{G} - A$ between h and k. Thus, you may think we can make zero a limit point of the domain or not as we choose. This is not really the case. In order to show that zero is a limit point of the

domain, we must show (as we did in Example 1) that, for *every* pair of vertical lines
h and k with $x = 0$ between them, there is a point of $\mathcal{G} - A$ between them. To prove
this statement false, we need find only one pair of vertical lines with $x = 0$ between
them but no point of $\mathcal{G} - A$ between them. This is what we did when we let h be
$x = -1/2$ and k be $x = 1/2$. It is very much like proving the statement "Every person
in this room is wearing a hat." In order to prove it false, we merely have to point
out *one* person in this room who is not wearing a hat—it is not necessary to show
that no one in the room is wearing a hat.

Example 3

Is the number 1 a limit point of the domain of

$$f(x) = \frac{x^2 - 1}{x - 1} \ ?$$

The graph is given in Figure 10.3. Since this
function is not defined for $x = 1$, there is no
point A. Thus $\mathcal{G} - A$ is $\mathcal{G}$. The function is
defined for all other values of x. Thus the
$\mathcal{G} - A$ here is the same as the $\mathcal{G} - A$ of
Example 1. By exactly the same argument we
used there, the number 1 is a limit point of the
domain of this function.

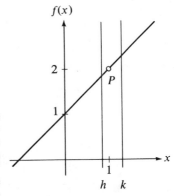

Figure 10.3

Note that the existence or nonexistence of A has nothing to do with the number
1 being a limit point of the domain of f. Since we consider only $\mathcal{G} - A$, we are interested only in what happens near A, not at it.

Problems

A

In Problems 1–10, identify the points P and A for the given limit statement.

1. $\lim\limits_{x \to 0} x^2 = 0.$

2. $\lim\limits_{x \to 1} x^2 = 2.$

3. $\lim\limits_{x \to 2} (x - 1) = 3.$

4. $\lim\limits_{x \to 2} (x - 1) = 1.$

5. $\lim\limits_{x \to 2} \dfrac{x^2 - 4}{x - 2} = 4.$

6. $\lim\limits_{x \to 2} \dfrac{x^2 - 4}{x - 2} = 0.$

7. $\lim\limits_{x \to 2} \dfrac{x^2 - 9}{x - 3} = 5.$

8. $\lim\limits_{x \to 1} \sqrt{x} = -1.$

9. $\lim\limits_{x \to 0} \sqrt{x} = 0.$

10. $\lim\limits_{x \to -1} \sqrt{x} = -1.$

B

In Problems 11–25, indicate whether or not the given number is a limit point of the domain of
the given function.

11. $4; \quad f(x) = \sqrt{x}.$

12. $2; \quad f(x) = \sqrt{x}.$

13. $-1; \quad f(x) = \sqrt{x}.$

14. $0; \quad f(x) = 1/x.$

15. $-1; \quad f(x) = 1/x.$

16. $0; \quad f(x) = |x|.$

17. 0; $f(x) = \begin{cases} 1 & \text{if } x > 0, \\ -1 & \text{if } x < 0. \end{cases}$

18. 1; $f(x) = \begin{cases} 3 & \text{if } x = 1, \\ 2 & \text{if } x = 2, \\ 1 & \text{if } x = 3. \end{cases}$

19. -1; $f(x) = \sqrt{x^2(x^2 - 1)}$.

20. $1/\sqrt{2}$; $f(x) = \sqrt{x^2(x^2 - 1)}$.

21. -1; $f(x) = \sqrt{x^2(x + 1)^2(x - 1)}$.

22. 0; $f(x) = \sqrt{x^2(x + 1)^2(x - 1)}$.

23. 1; $f(x) = \sqrt{x^2(x + 1)^2(x - 1)}$.

24. 0; $f(x) = \begin{cases} x - 1 & \text{if } x \geq 1, \\ -x - 1 & \text{if } x \leq -1. \end{cases}$

25. 1; $f(x) = \begin{cases} x - 1 & \text{if } x \geq 1, \\ -x - 1 & \text{if } x \leq -1. \end{cases}$

10.2

The Limit of a Function

We are now ready to consider the definition of a limit of a function. Bear in mind that the definition given here and the one in Chapter 17 are equivalent. They are merely in different dress.

Definition

The statement $\lim_{x \to a} f(x) = b$ *(where a and b are real numbers) means:*

(a) *If h and k are two vertical lines with P between them, then there is also a point of $\mathcal{G} - A$ between them;*

(b) *If α and β are two horizontal lines with P between them, then there also exists a pair of vertical lines l and m with P between them such that every point of $\mathcal{G} - A$ between l and m is also between α and β.*

The first part of this definition is easily seen to be a restatement of the condition that a is a limit point of the domain of f. Since P is (a, b), it is on the line $x = a$. Thus, two vertical lines with P between them also have the line $x = a$ between them. Recall that this assures us of points of the graph that are, roughly speaking, "close" to a in the x direction. Part (b), which primarily characterizes the limit, says that no matter what pair of horizontal lines we are given with P between them, there is a pair of vertical lines, with P between them, close enough together so that every point of $\mathcal{G} - A$ which is between the vertical lines is also between the horizontal lines (see Figure 10.4). It does *not* say that there is one pair of vertical lines that works for every pair of horizontal lines; the choice of l and m depends, not only upon the function f and the number a, but also upon the horizontal lines α and β. In general, the closer together α and β are, the closer together l and m have to be.

Note that both parts of the definition are concerned with $\mathcal{G} - A$ rather than $\mathcal{G}$. We are not concerned with the graph at $x = a$ but only with that portion near $x = a$ (near because we consider only the portion between the vertical lines). Let us look at some specific examples.

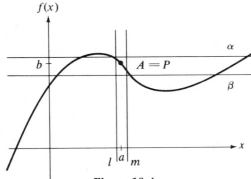

Figure 10.4

Example 1

Show that $\lim\limits_{x \to 0} \sqrt{x^2(x^2 - 1)}$ does not exist.

We must show that, no matter what number b represents,

$$\lim_{x \to 0} \sqrt{x^2(x^2 - 1)} = b$$

is false. The first part of the definition (which has nothing to do with the number b) says that, for the limit statement to be true, 0 must be a limit point of the domain of $f(x) = \sqrt{x^2(x^2 - 1)}$. We saw in Example 2 of the previous section that 0 is not a limit point of the domain of this function. Thus we need not consider the second part at all; the limit does not exist.

Example 2

Show that $\lim\limits_{x \to 1} (x + 1) = 2.$

We saw in Example 1 of the previous section that 1 is a limit point of the domain of $f(x) = x + 1$. Now let us consider the second part of the definition of a limit. First of all,

$$P = (1, 2) \quad \text{and} \quad A = (1, f(1)) = (1, 2).$$

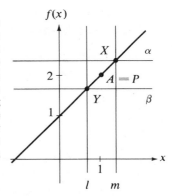

If α and β are two horizontal lines with α above P and β below it, then each of these lines has exactly one point in common with the graph of $y = x + 1$. There is a point X to the right of P that is common to α and $\mathscr{G}$, and there is a point Y to the left of P that is common to β and $\mathscr{G}$ (see Figure 10.5). Letting l be the vertical line through Y and m the vertical line through X, we see that every point of $\mathscr{G} - A$ between l and m is also between α and β. Thus the definition is satisfied and

$$\lim_{x \to 1} (x + 1) = 2.$$

Figure 10.5

While A itself is also between α and β, we are not interested in it—were it not between α and β, it would have no effect on the result (see Problem 7).

Example 3

Show that $\lim\limits_{x \to 1} \dfrac{x^2 - 1}{x - 1} = 2.$

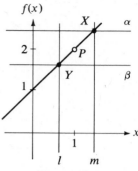

Note that $f(x) = (x^2 - 1)/(x - 1)$ has the same graph as $f(x) = x + 1$ except that it is not defined for $x = 1$. Thus, while P is still the point $(1, 2)$, there is no point A. The same arguments we used for both parts of the previous example can be used here—the absence of the point A has no effect on them whatsoever (see Figure 10.6). It must be remembered that $\mathscr{G} - A$ is the same as $\mathscr{G}$, since there is no A.

Figure 10.6

As this last example illustrates, the function f need not be defined at $x = a$ in order that $\lim_{x \to a} f(x)$ exist. On the other hand, the example using $\lim_{x \to 0} \sqrt{x^2(x^2 - 1)}$ illustrates that the mere existence of a functional value at $x = a$ is not enough to guarantee the existence of the limit.

Problems

A

In Problems 1–6, show whether or not part (a) of the definition of a limit is satisfied. The given limit statement is not necessarily true.

1. $\lim\limits_{x \to 3} f(x) = 1$, where $f(x) = \begin{cases} 1 & \text{if } x = 1, \\ 2 & \text{if } x = 2, \\ 1 & \text{if } x = 3. \end{cases}$

2. $\lim\limits_{x \to 0} \dfrac{\sqrt{x}}{x + 2} = 0.$

3. $\lim\limits_{x \to -2} \dfrac{\sqrt{x}}{x + 2} = \dfrac{\sqrt{2}}{4}.$

4. $\lim\limits_{x \to 0} \dfrac{1}{x^2} = 1.$

5. $\lim\limits_{h \to 0} |h| = 0.$

6. $\lim\limits_{h \to 0} f(h) = 0$, where $f(h) = \begin{cases} h & \text{if } h > 0, \\ 1 & \text{if } h < 0. \end{cases}$

B

In Problems 7–20, show that the given limit statement is true.

7. $\lim\limits_{x \to 1} f(x) = 2$, where $f(x) = \begin{cases} x + 1 & \text{if } x \neq 1, \\ 4 & \text{if } x = 1. \end{cases}$

8. $\lim\limits_{x \to 1} \dfrac{x^2 - 3x + 2}{x - 1} = -1.$

9. $\lim\limits_{x \to 0} \sqrt{x} = 0.$

10. $\lim\limits_{x \to 4} \dfrac{x - 4}{\sqrt{x} - 2} = 4.$

11. $\lim\limits_{x \to 1} f(x) = 1$, where $f(x) = \begin{cases} 2 - x & \text{if } x < 1, \\ x & \text{if } x \geq 1. \end{cases}$

12. $\lim\limits_{x \to 1} f(x) = 1$, where $f(x) = \begin{cases} 2 - x & \text{if } x < 1, \\ x & \text{if } x > 1. \end{cases}$

13. $\lim\limits_{x \to 1} f(x) = 1$,   where $f(x) = \begin{cases} 2 - x & \text{if } x < 1, \\ 2 & \text{if } x = 1, \\ x & \text{if } x > 1. \end{cases}$

14. $\lim\limits_{x \to -1} (|x| + x) = 0.$

15. $\lim\limits_{x \to 2} x^2 = 4.$

16. $\lim\limits_{x \to -2} x^2 = 4.$

17. $\lim\limits_{h \to 0} (3 + h) = 3.$

18. $\lim\limits_{h \to 0} (4x + 2h - 3) = 4x - 3.$

19. $\lim\limits_{u \to 2} \dfrac{u^2 + 1}{u} = \dfrac{5}{2}.$

20. $\lim\limits_{y \to -1} (y^3 + 1) = 0.$

C

21. Prove that $\lim\limits_{x \to a} k = k$, where k is a constant.

10.3

The Limit of a Function (Continued)

In considering the definition of a limit, we have only concerned ourselves with verifying that a given limit statement is true, except in those cases in which the limit statement is false because *a* is not a limit point of the domain of *f*—that is, it fails to satisfy the first part of the definition. We should like to turn our attention now to the case in which the first part of the definition is satisfied but the second is not.

Now the second part of the definition states that

> *for every pair of horizontal lines α and β with P between them, there exists a pair of vertical lines l and m with P between them satisfying the condition that every point of 𝒢 − A between l and m is also between α and β.*

Suppose we take it in smaller pieces. According to the definition, for every pair of horizontal lines α and β with P between them, there exist two vertical lines *l* and *m*, with P between them, that satisfy certain conditions. For this to be false, *there must exist a pair of horizontal lines α and β with P between them, such that* no such pair of vertical lines *l* and *m* exists. Another way of saying that no such pair of vertical lines exists is to say that *for every pair of vertical lines l and m with P between them*, the condition is not satisfied. Now the condition we are considering here is that every point of 𝒢 − A between *l* and *m* is also between α and β. If this condition is not satisfied, *there must be some point of 𝒢 − A between l and m which is not between α and β.* Now suppose we put all the pieces (the italicized portions above) together. If the second part of the definition is not true, then

> *there exists a pair of horizontal lines α and β with P between them such that, for every pair of vertical lines l and m with P between them, there is a point of 𝒢 − A between l and m which is not between α and β.*

Suppose we consider some specific examples.

Example 1

Show that if

$$f(x) = \begin{cases} 0 & \text{if } x \leq 0, \\ 1 & \text{if } x > 0, \end{cases}$$

then the limit statement $\lim\limits_{x \to 0} f(x) = 0$ is false.

First of all

$$A = P = (0, 0).$$

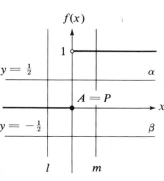

Since the function is defined for all x, the first part of the definition is easily seen to be satisfied. Thus if the limit is not 0, it must be because the second part of the definition is not satisfied. To show this, we must simply exhibit a pair of horizontal lines α and β with P between them such that, no matter how the vertical lines l and m are chosen with P between them, there is some point of $\mathcal{G} - A$ between l and m which is not between α and β. The lines $y = 1/2$ and $y = -1/2$ constitute one such pair (see Figure 10.7). No matter how close to P we take m, there is some point between m and the y axis (and thus between l and m) having a y coordinate 1 which is above both $y = 1/2$ and $y = -1/2$. Thus, the second part of the definition is not satisfied, and

Figure 10.7

$$\lim_{x \to 0} f(x) \neq 0.$$

Example 2

Show that if f is the function of Example 1, then the limit statement

$$\lim_{x \to 0} f(x) = 1$$

is false.

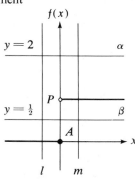

Now $P = (0, 1)$ and $A = (0, 0)$.

This time we choose α and β to be $y = 2$ and $y = 1/2$ (see Figure 10.8). Now no matter how close to P we take l, there is a point between l and the y axis (and thus between l and m) having a y coordinate 0, which is below both $y = 2$ and $y = 1/2$. Again, the second part of the definition is not satisfied and

$$\lim_{x \to 0} f(x) \neq 1.$$

Figure 10.8

We have seen that $\lim_{x \to 0} f(x)$ is neither zero nor one. This might lead you to suspect that $\lim_{x \to 0} f(x) = b$ is false no matter what number we choose for b. If this is the case, we say that this limit does not exist.

Definition

To say that the limit $\lim\limits_{x \to a} f(x)$ *does not exist means that*

$$\lim_{x \to a} f(x) = b$$

is false for every real number b.

Let us show that the limit of Examples 1 and 2 does not exist. At first this problem seems to be quite formidable—after all, there are infinitely many choices for *b*, and we cannot repeat this argument infinitely many times! But there is a way of overcoming this difficulty, and that is by making one case handle infinitely many values of *b*.

Example 3

Show that if *f* is the function of Example 1, then

$$\lim_{x \to 0} f(x)$$

does not exist; that is,

$$\lim_{x \to 0} f(x) = b$$

is false no matter what number *b* represents.

Let us consider two cases. First we assume that

$$\lim_{x \to 0} f(x) = b, \qquad \text{where} \quad b > 0.$$

Now $A = (0, 0)$, as always, and $P = (0, b)$. Let α be any horizontal line above P and β the line $y = b/2$ (see Figure 10.9). Note that $0 < b/2 < b$. No matter

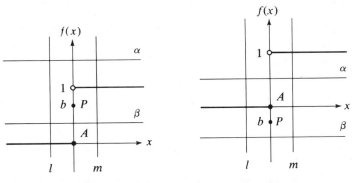

Figure 10.9   **Figure 10.10**

how we choose *l* and *m*, there is a point of $\mathcal{G} - A$ between them with negative abscissa and ordinate 0. Since this point is not between α and β,

$$\lim_{x \to 0} f(x) \neq b \qquad \text{where} \quad b > 0.$$

If this limit exists at all, it cannot be positive.

Now let us assume that

$$\lim_{x \to 0} f(x) = b \qquad \text{where} \quad b \leq 0.$$

As in the previous case, $A = (0, 0)$ and $P = (0, b)$. Choose α to be the line $y = 1/2$ and β to be any horizontal line below P (see Figure 10.10). Again, no matter how we choose *l* and *m*, there is a point of $\mathcal{G} - A$ between them with positive abscissa and ordinate 1; and again this point is not between α and β. Thus, the limit cannot be zero, nor can it be negative either—the limit simply does not exist.

One final word. We seem to have been pulling α's and β's out of a hat—how do we know where to choose them? The following procedure is helpful. Let us ask whether or not it is always possible to choose an l and m satisfying the definition if we take the α and β closer and closer to P. If it is not always possible to find such vertical lines, it will be detected when α and β are "close" to P. Note that in the case discussed our arguments would not hold if α and β were very far apart—that is, if α were above $y = 1$ and β below $y = 0$. But remember, in order to prove the falsity of a limit statement, we need only produce one offending pair α and β (or show that a is not a limit point of the domain of f).

Problems

B

In Problems 1–6, show that the given limit statement is false.

1. $\lim\limits_{x \to 0} f(x) = 1$,   where $f(x) = \begin{cases} 1 & \text{if } x \leq 0, \\ x & \text{if } x > 0. \end{cases}$

2. $\lim\limits_{x \to 0} f(x) = 0$,   where $f(x) = \begin{cases} 1 & \text{if } x \leq 0, \\ x & \text{if } x > 0. \end{cases}$

3. $\lim\limits_{x \to 2} f(x) = 0$,   where $f(x) = \begin{cases} x + 2 & \text{if } x \neq 2, \\ 0 & \text{if } x = 2. \end{cases}$

4. $\lim\limits_{x \to 0} (3x + 1) = 4.$ 5. $\lim\limits_{x \to 0} \dfrac{1}{x} = 0.$ 6. $\lim\limits_{x \to -2} \dfrac{1}{x + 2} = 1.$

In Problems 7–12, show whether the given limit statement is true or false.

7. $\lim\limits_{x \to -3} \dfrac{x^2 + 2x - 3}{x + 3} = -4.$ 8. $\lim\limits_{x \to 2} \dfrac{x^2 + 2x - 3}{x + 3} = 3.$ 9. $\lim\limits_{x \to 0} |x| = 0.$

10. $\lim\limits_{x \to 0} f(x) = 0$,   where $f(x) = \begin{cases} x & \text{if } x \geq 0, \\ 2 - x & \text{if } x < 0. \end{cases}$

11. $\lim\limits_{x \to 1} f(x) = 0$,   where $f(x) = \begin{cases} 0 & \text{if } x \neq 0, \\ 1 & \text{if } x = 0. \end{cases}$

12. $\lim\limits_{x \to 1} f(x) = 1$,   where $f(x) = \begin{cases} x & \text{if } x \geq 1, \\ 2 - x & \text{if } x < 1. \end{cases}$

C

In Problems 13–15, show whether the given limit statement is true or false.

13. $\lim\limits_{x \to 0} \sin \dfrac{1}{x} = 0.$ $\left(\textit{Hint: } \text{Sketch} \quad y = \sin \dfrac{1}{x} \text{ by plotting the points corresponding to} \right.$

$$x = \frac{1}{\pi/2}, \frac{1}{\pi}, \frac{1}{3\pi/2}, \frac{1}{2\pi}, \frac{1}{5\pi/2}, \text{etc.}\Bigg)$$

14. $\lim\limits_{x \to 0} x \sin \dfrac{1}{x} = 0.$ (See hint for Problem 13.)

15. $\lim\limits_{x \to 0} x^2 \sin \dfrac{1}{x} = 0.$ (See hint for Problem 13.)

In Problems 16–20, show that the given limit does not exist.

16. $\lim\limits_{x \to 0} f(x)$,   where $f(x) = \begin{cases} 3 & \text{if } x \leq 0, \\ 3x & \text{if } x > 0. \end{cases}$ 17. $\lim\limits_{x \to 0} \dfrac{1}{x}.$ 18. $\lim\limits_{x \to 0} \dfrac{1}{x^2}.$

19. $\lim\limits_{x \to 1} f(x)$,   where $f(x) = \begin{cases} 0 & \text{if } x < 1, \\ 1 & \text{if } x = 1, \\ x + 1 & \text{if } x > 1. \end{cases}$ 20. $\lim\limits_{x \to 0} \dfrac{1}{x} \sin \dfrac{1}{x}.$

10.4

One-Sided Limits

In the preceding section we considered in some detail the function

$$f(x) = \begin{cases} 0 & \text{if } x \le 0, \\ 1 & \text{if } x > 0 \end{cases}$$

and the limit $\lim_{x \to 0} f(x)$, which we showed does not exist. However, it might be noted that if a point moves along the curve in such a way that its x coordinate approaches zero, the y coordinate approaches zero or one, depending upon whether the point approaches from the left or right. This leads directly to the definitions of one-sided limits, in which the notations $x \to a^+$ and $x \to a^-$ are used to indicate the right-hand and left-hand limits, respectively, as x approaches a, or the limit as x approaches a from the right or left.

Definition

The statement $\lim_{x \to a^+} f(x) = b$ means $\lim_{x \to a} G(x) = b$, where $G(x)$ is the function that coincides with $f(x)$ for all $x \ge a$ and is undefined for $x < a$.

Definition

The statement $\lim_{x \to a^-} f(x) = b$ means $\lim_{x \to a} g(x) = b$, where $g(x)$ is the function that coincides with $f(x)$ for all $x \le a$ and is undefined for $x > a$.

Example 1

Show that if $f(x) = \begin{cases} 0 & \text{if } x \le 0, \\ 1 & \text{if } x > 0, \end{cases}$ then

$$\lim_{x \to 0^+} f(x) = 1.$$

In order to show that $\lim_{x \to 0^+} f(x) = 1$, we must show that $\lim_{x \to 0} G(x) = 1$, where

$$G(x) = \begin{cases} 0 & \text{if } x = 0, \\ 1 & \text{if } x > 0. \end{cases}$$

For this limit statement, $A = (0,\ 0)$ and $P = (0,\ 1)$. Since $G(x)$ is defined for $x \ge 0$, 0 is clearly a limit point of the domain of G. It is easily verified (see Figure 10.11), that if α and

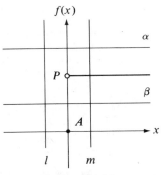

Figure 10.11

β are any pair of horizontal lines with P between them, then every point of $\mathcal{G} - A$ is between them no matter how l and m are selected.

Example 2

Show that if f is the function of Example 1, then

$$\lim_{x \to 0^-} f(x) = 0.$$

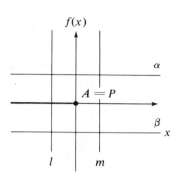

Figure 10.12

Now $\lim_{x \to 0^-} f(x) = 0$ becomes $\lim_{x \to 0} g(x) = 0$, where

$$g(x) = 0 \quad \text{if } x \leq 0.$$

For this limit statement, $A = P = (0, 0)$. Again, zero is clearly a limit point of the domain of $g(x)$. It is easy to verify (see Figure 10.12) that if α and β are any two horizontal lines with P between them, every point of $\mathscr{G} - A$ is between them no matter how l and m are selected.

Since right- and left-hand limits are nothing more than ordinary limits of new functions, their existence or nonexistence is verified in exactly the same way that we verified the existence or nonexistence of ordinary limits.

We might ask how the right- and left-hand limits of a given function are related to the limit of that function. We saw in the case of

$$f(x) = \begin{cases} 0 & \text{if } x \leq 0, \\ 1 & \text{if } x > 0 \end{cases}$$

that $\lim_{x \to 0^-} f(x) = 0$ and $\lim_{x \to 0^+} f(x) = 1$, but $\lim_{x \to 0} f(x)$ does not exist. This is a special case of the following theorem.

Theorem 10.1

If $\lim_{x \to a^-} f(x) = b_1$ and $\lim_{x \to a^+} f(x) = b_2$, where $b_1 \neq b_2$, then $\lim_{x \to a} f(x)$ does not exist.

Proof

Let us suppose that $b_1 < b_2$. Since we want to prove that $\lim_{x \to a} f(x)$ does not exist, we must show that $\lim_{x \to a} f(x) = b$ is not true no matter what number we choose for b. To do this, we are going to consider two cases: $b > b_1$ and $b \leq b_1$. But first, let us consider the points A and P for the three limits. Since all three are limits as x approaches a, $A = (a, f(a))$, for all three cases, provided $f(a)$ exists (if $f(a)$ does not exist, there is no point A in any of the three cases). Let $P_1 = (a, b_1)$, $P_2 = (a, b_2)$, and $P = (a, b)$; these are the P's for each of the three limit statements.

Assume that $\lim_{x \to a} f(x) = b$, where $b > b_1$. P and P_1 are on the same vertical line with P_1 below P (see Figure 10.13). Let α be any horizontal line above P, and let β be the line $y = (b + b_1)/2$ half-

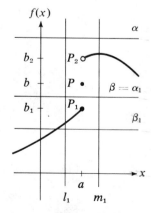

Figure 10.13

way between P and P_1. Now, since $\lim_{x \to a^-} f(x) = b_1$, it follows that no matter what pair of horizontal lines we choose with P_1 between them, there exists a pair of vertical lines l_1 and m_1 satisfying the conditions of the definition. Choose $\alpha_1 = \beta$ and let β_1 be any horizontal line below P_1. There is a pair of vertical lines l_1 and m_1 with P_1 between them such that every point of $\mathscr{G} - A$ between l_1 and m_1 and to the left of P_1 is between α_1 and β_1 (remember $\lim_{x \to a^-} f(x) = b_1$ is a left-hand limit). Now no matter how we choose l and m with P between them, some of these points are between them. But none of them is between α and β. Thus the assumption that $\lim_{x \to a} f(x) = b$, with $b > b_1$, is false. The case in which $\lim_{x \to a} f(x) = b$ with $b \leq b_1$ is left to the student.

This argument was made under the supposition that $b_1 < b_2$. A similar argument can be given for the case in which $b_1 > b_2$.

Theorem 10.2

If $\lim_{x \to a^+} f(x) = \lim_{x \to a^-} f(x) = b$,   then $\lim_{x \to a} f(x) = b$.

Proof

For all three limits, $A = (a, f(a))$ if it exists and $P = (a, b)$. The first part of the definition is easily seen to be satisfied, since it is satisfied for the points of $f(x)$ to the left (or right) of P. If α and β are two horizontal lines with P between them (see Figure 10.14), then, since $\lim_{x \to a^-} f(x) = b$, there is a pair of vertical lines l_1 and m_1 such that every point of $\mathscr{G} - A$ between l_1 and m_1 and to the left of P is also between α and β. Similarly, since $\lim_{x \to a^+} f(x) = b$, there is a pair of vertical lines l_2 and m_2 such that every point of $\mathscr{G} - A$ between l_2 and m_2 and to the right of P is also between α and β. Now, let $l = l_1$ and $m = m_2$. Then every point of $\mathscr{G} - A$ between l and m is also between α and β. Thus $\lim_{x \to a} f(x) = b$.

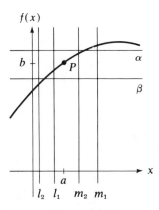

Figure 10.14

The converses of the last two theorems are not true. This is obvious for Theorem 10.1. We have already seen many examples in which $\lim_{x \to a} f(x)$ did not exist, but neither right- nor left-hand limits existed either. It is not nearly so obvious that the converse of Theorem 10.2 is false; however, a consideration of the limits of Problem 11 shows that this is the case.

Problems

B

In Problems 1–4, show that the given limit statements are true.

1. $\lim_{x \to 0^+} f(x) = 0$,   where $f(x) = \begin{cases} 1 & \text{if } x \leq 0, \\ x & \text{if } x > 0. \end{cases}$

2. $\lim_{x \to 1^+} f(x) = 2$,   where $f(x) = \begin{cases} 3 & \text{if } x \leq 1, \\ 2x & \text{if } x > 1. \end{cases}$

3. $\lim_{x \to 1^-} f(x) = -1$,   where $f(x) = \begin{cases} -x & \text{if } x < 1, \\ x & \text{if } x > 1. \end{cases}$

4. $\lim\limits_{x \to -2^+} f(x) = -3$,   where $f(x) = \begin{cases} 2x + 1 & \text{if } x \geq -2, \\ x - 2 & \text{if } x < -2. \end{cases}$

In Problems 5–8, show that the given limit statements are false.

5. $\lim\limits_{x \to 0^+} f(x) = 0$,   where $f(x) = \begin{cases} x & \text{if } x \leq 0, \\ 1 & \text{if } x > 0. \end{cases}$

6. $\lim\limits_{x \to 1^+} f(x) = 2$,   where $f(x) = \begin{cases} 2 + x & \text{if } x \leq 1, \\ 2 - x & \text{if } x > 1. \end{cases}$

7. $\lim\limits_{x \to 1^-} f(x) = 4$,   where $f(x) = \begin{cases} x - 2 & \text{if } x \neq 1, \\ 4 & \text{if } x = 1. \end{cases}$

8. $\lim\limits_{x \to 0^+} \sin \dfrac{1}{x} = 0.$

In Problems 9–14, find $\lim_{x \to a^-} f(x)$, $\lim_{x \to a^+} f(x)$, and $\lim_{x \to a} f(x)$ for the given function and the given value of a, provided the indicated limits exist.

9. $f(x) = \begin{cases} x + 1 & \text{if } x < 1, \\ x - 1 & \text{if } x > 1; \end{cases}$ $a = 1.$

10. $f(x) = \begin{cases} 1 + x & \text{if } x \geq 0, \\ 1 - x & \text{if } x < 0; \end{cases}$ $a = 0.$

11. $f(x) = \sqrt{x - 2};$ $a = 2.$

12. $f(x) = \dfrac{1}{x};$ $a = 0.$

13. $f(x) = \begin{cases} \dfrac{1}{x} & \text{if } x < 0, \\ 0 & \text{if } x \geq 0; \end{cases}$ $a = 0.$

14. $f(x) = \begin{cases} 2x + 1 & \text{if } x < 1, \\ 3x - 1 & \text{if } x > 1; \end{cases}$ $a = 0.$

C

In Problems 15–18, show that the given limits do not exist.

15. $\lim\limits_{x \to 0^+} \dfrac{1}{x}.$

16. $\lim\limits_{x \to 1^-} f(x)$,   where $f(x) = \begin{cases} \dfrac{1}{x - 1} & \text{if } x < 1, \\ x + 1 & \text{if } x > 1. \end{cases}$

17. $\lim\limits_{x \to 0^+} \log x.$

18. $\lim\limits_{x \to 0^+} \sin \dfrac{1}{x}.$

19.   Complete the proof of Theorem 10.1.

20.   Show that if $\lim_{x \to a^+} f(x) = b_1$ and $\lim_{x \to a^-} f(x) = b_2$ but $\lim_{x \to a} f(x)$ does not exist, then $b_1 \neq b_2$.

10.5

Infinite Limits

We have seen several ways of showing the nonexistence of a given limit. First of all, a may not be a limit point of the domain of the given function—that is, the first part of the definition might not be satisfied. When a is a limit point of the domain, there are basically three different situations in which the limit might fail to exist.

First is the situation in which two different numbers are approached from the two different sides. We discussed this case in some detail in the preceding section dealing with right- and left-hand limits. A simple example (see Figure 10.10) of this type of nonexistence is

$$\left.\begin{aligned}
\lim_{x \to 0^-} f(x) &= 0 \\[1ex]
\lim_{x \to 0^+} f(x) &= 1 \\[1ex]
\lim_{x \to 0} f(x) &\text{ does not exist}
\end{aligned}\right\} \text{where } f(x) = \begin{cases} 0 & \text{if } x \le 0, \\ 1 & \text{if } x > 0. \end{cases}$$

A second case of nonexistence occurs when no definite number is approached from either side but $f(x)$ gets large and positive as x approaches a (or $f(x)$ gets large and negative as x approaches a). This is exemplified (see Figure 10.15) by

$$\lim_{x \to 0} \frac{1}{x^2} \quad \text{does not exist.}$$

The third situation may be illustrated (see Figure 10.16) by

$$\lim_{x \to 0} \sin \frac{1}{x} \quad \text{does not exist.}$$

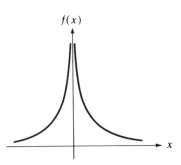

Figure 10.15

In this case, no definite number is approached from either side, but the function does not get larger as x approaches zero.

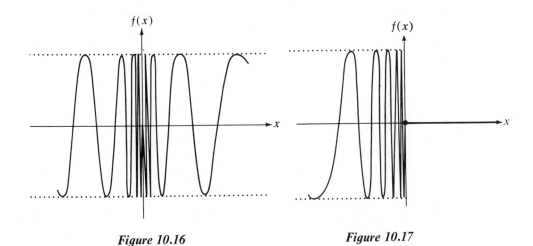

Figure 10.16   **Figure 10.17**

Of course there may be cases that combine two of these, such as (see Figure 10.17)

$$\lim_{x \to 0} f(x) \quad \text{where } f(x) = \begin{cases} \sin \dfrac{1}{x} & \text{if } x < 0, \\[2ex] 0 & \text{if } x \ge 0, \end{cases}$$

which combines the first and third types. Another is $\lim_{x\to 0} 1/x$ (see Figure 10.18), which combines the first two types; that is, $|f(x)|$ gets large as x approaches 0 from either side but $f(x)$ gets large and positive from one side and large and negative from the other.

Since the second type of nonexistence occurs relatively frequently, the special symbols $+\infty$ and $-\infty$ are used to identify it. Thus $\lim_{x\to a} f(x) = +\infty$ means (roughly speaking) that $f(x)$ gets large and positive as x approaches a. Of course, this is unsatisfactory as a definition, because the meaning of the word "large" is quite vague. We shall use the following definition.

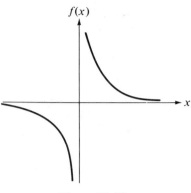

Figure 10.18

Definition

The statement $\lim_{x\to a} f(x) = +\infty \; (-\infty)$ *means that*

(a) *if h and k are two vertical lines with $x = a$ between them, then there is a point of $\mathcal{G} - A$ between them, and*

(b) *if α is a horizontal line, then there exists a pair of vertical lines l and m with $x = a$ between them such that every point of $\mathcal{G} - A$ between l and m is above (below) α.*

Note here that A has the same meaning as before, but there is no point P. The first part of the definition is basically the same as the first part of the definition of $\lim_{x\to a} f(x) = b$; but the second part says that, no matter how high (low) α may be, there is some point of $\mathcal{G} - A$ "close" to the vertical line $x = a$ and higher (lower) than α.

Example 1

Show that $\displaystyle\lim_{x\to 0} \frac{1}{x^2} = +\infty.$

The first part of the definition is easily seen to be satisfied, since $f(x) = 1/x^2$ has as its domain all real numbers different from zero. If α is a horizontal line above the x axis (see Figure 10.19), it intersects $\mathcal{G}$ at a pair of points X and Y. If l and m are vertical lines through X and Y, respectively, the second part of the definition is easily seen to be satisfied.

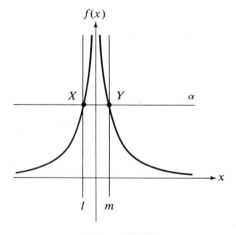

Figure 10.19

Remember here that the statement $\lim_{x\to a} f(x) = +\infty$ says that $\lim_{x\to a} f(x)$ does not exist (it is not equal to any number), but its nonexistence is of a special type.

The ways of defining

$$\lim_{x\to a^+} f(x) = +\infty, \quad \lim_{x\to a^+} f(x) = -\infty, \quad \lim_{x\to a^-} f(x) = +\infty, \quad \text{and} \quad \lim_{x\to a^-} f(x) = -\infty$$

are self-evident and are left to the student.

Up to this point we have considered limits in which x approaches a given real number a. Let us now consider the behavior of a function at its ends: that is, as x gets increasingly large in either the positive or the negative direction. Such limits are written

$$\lim_{x\to +\infty} f(x) \quad \text{and} \quad \lim_{x\to -\infty} f(x).$$

Since there is no numerical value for x to approach, there is neither an A nor a P.

Definition

The statement $\lim_{x\to +\infty} f(x) = b$ *means*

(a) *if h is a vertical line, then there is a point of $\mathcal{G}$ to the right of h, and*
(b) *if α and β are two horizontal lines with $y = b$ between them, then there is a vertical line l such that every point of $\mathcal{G}$ to the right of l is between α and β.*

The definition of $\lim_{x\to -\infty} f(x) = b$ is the same except that "right" is replaced by "left."

The two parts of the definition correspond to the two parts of the definition of $\lim_{x\to a} f(x) = b$. The first part assures us that, no matter how far to the right we go, there is still some graph there. Thus $\lim_{x\to +\infty} \sqrt{-x}$ does not exist, because $\sqrt{-x}$ is not defined for any x to the right of $x = 0$.

Example 2

Show that $\lim\limits_{x\to +\infty} \dfrac{x+1}{x} = 1$.

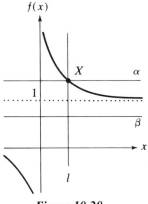

Figure 10.20

Since $y = (x + 1)/x$ is defined for all x except zero, the first part of the definition is easily seen to be satisfied. Suppose now that α and β are two horizontal lines with $y = 1$ between them (see Figure 10.20). If α is the higher one, α has a point X in common with the graph. Let l be the vertical line through X. Now every point of $\mathcal{G}$ to the right of l is below α and certainly above β. Thus the definition is seen to be satisfied.

In what ways might $\lim_{x\to +\infty} f(x)$ fail to exist? Of course, the first part of the definition might fail to be satisfied; we have already seen an example of this. However, instead of the basic three types of nonexistence we saw for $\lim_{x\to a} f(x)$, here we have only two. Obviously we cannot have the first type, in which two different numbers are approached from the two different sides, because you cannot approach $+\infty$ from both sides—you must approach it from the left. Thus the other two types of nonexistence are the only ones possible. They are illustrated by

$$\lim_{x \to +\infty} x^2 \quad \text{and} \quad \lim_{x \to +\infty} \sin x.$$

While neither limit exists, $\lim_{x \to +\infty} x^2$ is the special type of nonexistence given by $\lim_{x \to +\infty} x^2 = +\infty$ that we saw earlier. The actual definitions of such expressions as

$$\lim_{x \to +\infty} f(x) = +\infty, \qquad \lim_{x \to +\infty} f(x) = -\infty, \qquad \lim_{x \to -\infty} f(x) = +\infty, \quad \text{and} \qquad \lim_{x \to -\infty} f(x) = -\infty$$

are left to the student.

The symbol ∞ was introduced by the English mathematician John Wallis (1616–1703), a contemporary of Isaac Newton. Wallis was also the first to give a complete explanation of zero, negative, and fractional exponents as well as one of the first to look upon conics algebraically as curves represented by second-degree equations, rather than geometrically as sections of a cone.

Problems

B

In Problems 1–8, verify that the given limit statements are true.

1. $\displaystyle \lim_{x \to 1} \frac{1}{(x - 1)^2} = +\infty.$

2. $\displaystyle \lim_{x \to -3} \frac{4}{(x + 3)^2} = +\infty.$

3. $\displaystyle \lim_{x \to 1} \frac{x - 3}{(x - 1)^2} = -\infty.$

4. $\displaystyle \lim_{x \to +\infty} \frac{x + 3}{(x - 1)^2} = 0.$

5. $\displaystyle \lim_{x \to +\infty} \frac{1}{x - 1} = 0.$

6. $\displaystyle \lim_{x \to +\infty} \frac{x + 3}{x - 2} = 1.$

7. $\displaystyle \lim_{x \to -\infty} \frac{3x + 1}{2x - 1} = \frac{3}{2}.$

8. $\displaystyle \lim_{x \to -\infty} \frac{2x^2 + 1}{(x - 1)^2} = 2.$

9. Define $\displaystyle \lim_{x \to +\infty} f(x) = +\infty.$

10. Define $\displaystyle \lim_{x \to -\infty} f(x) = +\infty.$

11. Define $\displaystyle \lim_{x \to a^+} f(x) = -\infty.$

12. Define $\displaystyle \lim_{x \to a^-} f(x) = -\infty.$

13. Show that $\displaystyle \lim_{x \to 0^-} \frac{1}{x} = -\infty$ and $\displaystyle \lim_{x \to 0^+} \frac{1}{x} = +\infty$, but $\displaystyle \lim_{x \to 0} \frac{1}{x}$ is neither $+\infty$ nor $-\infty$.

In Problems 14–18, show whether the given limit statement is true or false.

14. $\displaystyle \lim_{x \to 0^+} \frac{x + 1}{x} = +\infty.$

15. $\displaystyle \lim_{x \to 0} \frac{2x + 1}{3x} = +\infty.$

16. $\displaystyle \lim_{x \to +\infty} \frac{2x + 1}{x} = 2.$

17. $\displaystyle \lim_{x \to +\infty} x^2 = +\infty.$

18. $\displaystyle \lim_{x \to -\infty} x^2 = -\infty.$

C

In Problems 19 and 20, show whether the given limit statement is true or false.

19. $\displaystyle \lim_{x \to +\infty} (x - \sqrt{x^2 + 1}) = 0.$

20. $\displaystyle \lim_{x \to -\infty} (x - \sqrt{x^2 + 1}) = 0.$

10.6

Continuity

Continuity, another very important concept in calculus, is closely related to limits.

Definition

*To say that the function f is **continuous at the point** A of 𝒢 means that if α and β are two horizontal lines with A between them, then there are two vertical lines l and m with A between them such that every point of 𝒢 between l and m is also between α and β.*

This definition is very much like that of a limit. There are, however, two important differences. First, this is only a one-part definition; it contains nothing corresponding to the first part of the definition of a limit. The second difference seems quite minor but, in reality, is far more significant: In considering the portion of the graph between the horizontal lines α and β and the vertical lines l and m, we are concerned with 𝒢 rather than 𝒢 − A.

Suppose we consider the effect of these two differences. The first difference implies that

$$f(x) = \sqrt{x^2(x^2 - 1)}$$

is continuous at the point (0, 0), which is an isolated point of its graph (see Figure 10.21). No matter what α and β are, l and m simply have to be taken close enough together so that the origin is the only point of 𝒢 between l and m. Recall that $\lim_{x \to 0} \sqrt{x^2(x^2 - 1)}$ does not exist, because $x = 0$ is not a limit point of the domain.

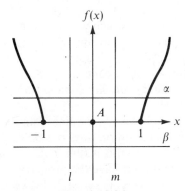

Figure 10.21

Some mathematicians use a different definition of continuity that implies that a function is not continuous at isolated points. This point of view has the advantage that it is more in keeping with out intuitive idea of continuity. On the other hand, it is not in step with other branches of mathematics, which define continuity so that discrete functions (functions defined at isolated values of x) are continuous on their entire domains.†

---

†See Lynn A. Steen and J. Arthur Seebach, Jr., *Counterexamples in Topology* (New York: Holt, Rinehart and Winston, 1970), p. 41.

The effect of the second difference is that, roughly speaking, the point A is where you expect it to be. That is, $f(x) = x^2$ is continuous at $(0, 0)$ (see Figure 10.22), while

$$f(x) = \begin{cases} x^2 & \text{if } x \neq 0, \\ 1 & \text{if } x = 0 \end{cases}$$

is not continuous at $(0, 1)$; in fact, $(0, 1)$ seems to be out of place (see Figure 10.23). Let us consider these differences in greater detail.

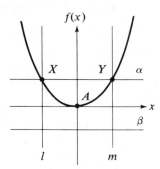

Figure 10.22

Example 1

Show that $f(x) = x^2$ is continuous at $(0, 0)$.

The argument is basically the same as the one showing that $\lim_{x \to 0} x^2 = 0$ (see Figure 10.22). The horizontal line α has two points, X and Y, in common with $\mathcal{G}$. If l and m are vertical lines through X and Y, it is evident that every point (including the origin) of $\mathcal{G}$ between l and m is also between α and β.

Before considering an example in which we show that a function is not continuous at a given point, let us consider what it means to say that a function is not continuous at the point A. If the function f is continuous at A, then, for two horizontal lines α and β with A between them, a certain condition holds. And if f is not continuous at A, then *there is some pair of horizontal lines α and β with A between them such that* the condition does not hold.

The condition is that there are two vertical lines l and m with A between them that have a certain property. If this condition does not hold, then two such vertical lines do not exist. Another way of saying this is that, *for every pair of vertical lines l and m with A between them,* l and m do not have the property.

Finally, the property is that every point of $\mathcal{G}$ between l and m is also between α and β. Then the property is absent if *there is some point of $\mathcal{G}$ between l and m but not between α and β.*

Putting together the foregoing statements in italics, we see that when we say f is not continuous at A, we mean that

> *there is some pair of horizontal lines α and β with A between them such that, for every pair of vertical lines l and m with A between them, there is some point of $\mathcal{G}$ between l and m but not between α and β.*

Let us now use this to verify a statement we made preceding Example 1.

Example 2

Show that

$$f(x) = \begin{cases} x^2 & \text{if } x \neq 0, \\ 1 & \text{if } x = 0 \end{cases}$$

is not continuous at (0, 1).

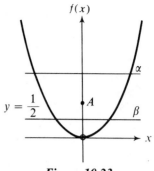

Figure 10.23

In order to show that this function is not continuous at $A = (0, 1)$ we must show that, for some pair of horizontal lines with A between them, no vertical lines can be found to satisfy the conditions of the definition (see Figure 10.23). Thus we select α to be any horizontal line above A and β to be the line $y = 1/2$. No matter how l and m are taken, there is some point of $\mathcal{G}$ between l and m but below both α and β. Thus this function is not continuous at A. Another way of saying this is to say that it is discontinuous at A.

Definition

*If f is a function with x_0 in its domain, then f is **discontinuous** at $(x_0, f(x_0))$ if and only if it is not continuous there.*

Note that we have defined continuity and discontinuity of a function at a single point or, equivalently, at a single real number x_0. Nevertheless, it is often convenient to talk about continuity on a set of numbers. This is easily defined.

Definition

A function f is continuous on a set E of real numbers if it is continuous at every number in E.

Example 3

If f is the function of Example 2, is f continuous on each of the following sets?

(a) $[0, 2]$ (b) $(0, 2)$ (c) $[-1, 1]$

We have already seen that f is discontinuous at the point (0, 1)—that is, when $x = 0$. It is easily verified that f is continuous at every other value of x. Thus f is not continuous on either $[0, 2]$ or $[-1, 1]$ because both of these sets include $x = 0$, where f is discontinuous; however, f is continuous on $(0, 2)$ because it is continuous at every number in this set.

In addition to talking about continuity on a set, some people refer to *continuity of a function* without reference to any particular point or set. Unfortunately, there is no standard definition for this term (although many mathematicians feel that the definition they use is the standard one). Some define a continuous function of a real variable to be one that is continuous at every real number. Others define it to be a function that is continuous at every number in its domain. There are still other variations. Thus the function defined by

$$f(x) = \frac{1}{x},$$

. with domain

$$\{x \mid x \text{ real and } x \neq 0\},$$

is continuous by the second definition because it is continuous at every point in its domain; however, f is not a continuous function by the first definition because it is not continuous at $x = 0$. Because of this difference in usage, you must be careful when a function is described as continuous to note which definition is being used.

We shall avoid the use of the ambiguous term *continuous function*. Instead, we shall refer to a function as being continuous at a single number or on a specific set of numbers.

Finally, how can we quickly observe discontinuities; that is, what does a curve look like at a discontinuity? Or at a point of continuity? A popular way of describing a function that is continuous on some *interval* (finite or infinite) is that it can be drawn without lifting a pencil from a piece of paper. Thus $f(x) = x^2$ is continuous on the set of real numbers, since we can draw its graph without picking up our pencil. But $f(x) = 1/x$ cannot be so drawn. Thus $f(x) = 1/x$ is not continuous on the set of real numbers—it is not continuous at $x = 0$. But this is no great surprise, because it is not defined at $x = 0$. It is only a useful criterion if the function is defined on the interval in question.

Another way of looking at the problem is from the opposite point of view; that is, we describe a function at a discontinuity. A function is discontinuous at one of its points if its graph has a vertical jump at that point. For example, the function of Example 2 is discontinuous at $x = 0$. Alternatively, we cannot draw the graph of this function over any interval containing 0 without picking up our pencil.

Problems

B

In Problems 1–12, indicate whether the given function is continuous at the point given and verify your result by means of the definition of continuity.

1. $f(x) = x^2$; at $(1, 1)$.

2. $f(x) = x^2 - 3x + 2$; at $(1, 0)$.

3. $f(x) = \begin{cases} 1 & \text{if } x \leq 0, \\ x & \text{if } x > 0; \end{cases}$ at $(0, 0)$.

4. $f(x) = \begin{cases} 1 & \text{if } x \leq 0, \\ x & \text{if } x > 0; \end{cases}$ at $(1, 1)$.

5. $f(x) = \begin{cases} 0 & \text{if } x \leq 0, \\ 1 & \text{if } x > 0; \end{cases}$ at $(-1, 0)$.

6. $f(x) = \dfrac{1}{x + 1}$; at $(1, 1/2)$.

7. $f(x) = \dfrac{1}{x}$; at $\left(\dfrac{1}{100}, 100\right)$

8. $f(x) = \dfrac{x^2 + 1}{x + 1}$; at $(-1, 2)$.

9. $f(x) = \begin{cases} x + 1 & \text{if } x \neq 1, \\ 2 & \text{if } x = 1; \end{cases}$ at $(1, 2)$.

10. $f(x) = \begin{cases} x + 1 & \text{if } x \neq 2, \\ 1 & \text{if } x = 2; \end{cases}$ at $(2, 1)$.

11. $f(x) = \begin{cases} 1 & \text{if } x = 1, \\ 3 & \text{if } x = 2, \\ 7 & \text{if } x = 3; \end{cases}$ at $(1, 1)$.

12. $f(x) = 2x + 1$ if $x = 0, 1, 2, \ldots$; at $(1, 3)$.

In Problems 13–18, indicate whether the given function is continuous on the set of real numbers and on the domain of the function. Where, in each case, does the function fail to be continuous?

13. $f(x) = \dfrac{1}{(x - 3)^2}$.

14. $f(x) = \dfrac{x^2 - 4}{x + 2}$.

15. $f(x) = \begin{cases} x + 1 & \text{if } x \neq 1, \\ 3 & \text{if } x = 1. \end{cases}$

16. $f(x) = x^2$ if $x = 1, 2, 3, \ldots$.

17. $f(x) = \begin{cases} 0 & \text{if } x \leq 0, \\ x^2 & \text{if } 0 < x < 1, \\ 0 & \text{if } 1 \leq x. \end{cases}$

18. $f(x) = \begin{cases} 0 & \text{if } x \leq 0, \\ x^2 & \text{if } 0 < x < 1, \\ 1 & \text{if } 1 \leq x. \end{cases}$

C

In Problems 19–24, indicate whether the given function is continuous at the point given and verify your result by means of the definition of continuity.

19. $f(x) = \begin{cases} \sin \dfrac{1}{x} & \text{if } \neq 0, \\ 0 & \text{if } x = 0; \end{cases}$ at $(0, 0)$.

20. $f(x) = \begin{cases} x \sin \dfrac{1}{x} & \text{if } x \neq 0, \\ 0 & \text{if } x = 0; \end{cases}$ at $(0, 0)$.

21. $f(x) = \begin{cases} x^2 \sin \dfrac{1}{x} & \text{if } x \neq 0, \\ 0 & \text{if } x = 0; \end{cases}$ at $(0, 0)$.

22. $f(x) = \begin{cases} \dfrac{1}{x} \sin \dfrac{1}{x} & \text{if } x \neq 0, \\ 0 & \text{if } x = 0; \end{cases}$ at $(0, 0)$.

23. $f(x) = \begin{cases} \sin \dfrac{1}{x} & \text{if } x \neq 0, \\ 0 & \text{if } x = 0; \end{cases}$ at $\left(\dfrac{1}{2\pi}, 0\right)$.

24. $f(x) = \begin{cases} x \sin \dfrac{1}{x} & \text{if } x \neq 0, \\ 0 & \text{if } x = 0; \end{cases}$ at $\left(\dfrac{1}{2\pi}, 0\right)$.

In Problems 25 and 26, indicate where the given function is continuous and discontinuous.

25. $f(x) = \begin{cases} 0 & \text{if } x \text{ is rational,} \\ 1 & \text{if } x \text{ is irrational.} \end{cases}$

26. $f(x) = \begin{cases} 0 & \text{if } x \text{ is rational,} \\ x & \text{if } x \text{ is irrational.} \end{cases}$

27. At what value(s) of x is f continuous if

$$f(x) = \begin{cases} x & \text{if } x \text{ is rational,} \\ 1 - x & \text{if } x \text{ is irrational?} \end{cases}$$

10.7

Limits, Continuity, and Derivatives

In the last section we were concerned with the difference between limits and continuity. We shall now consider the similarities and their implications.

Perhaps you wonder, why all the fuss about limits? We've been taking limits for some time now, with no trouble. Why suddenly make it hard? Actually we have had no trouble with limits, because we have considered only the very simplest of limits for "well-behaved" functions. When the functions get more complicated and the limits more difficult, we would have considerable difficulty were we to rely on an intuitive idea of limits. Mathematics has a way of playing tricks on people who rely entirely on intuitive ideas and rules-of-thumb.

In past discussion, the determination of limits has seemed to be a matter of substitution. When will substitution give us the limit we want? The answer is found in the similarity of the definitions of limits and continuity.

Theorem 10.3

$\lim_{x \to a} f(x) = f(a)$ *if and only if f is continuous at x = a and a is a limit point of the domain of f.*

Proof

Part I: Given that $\lim_{x \to a} f(x) = f(a)$, let us prove that $f(x)$ is continuous at $x = a$ and a is a limit point of the domain of f.

Since $\lim_{x \to a} f(x) = f(a)$, it follows that a is a limit point of the domain of f (by the mere existence of the limit). It also follows that if α and β are two horizontal lines with P between them, then there exists a pair of vertical lines l and m with P between them such that every point of $\mathcal{G} - A$ between l and m is also between α and β. Since $A = P = (a, f(a))$, A is also between α and β. Thus, the earlier statement can be changed to read: If α and β are two horizontal lines with A between them, then there exists a pair of vertical lines l and m with A between them such that every point of $\mathcal{G}$ between l and m is also between α and β. But this *means* that f is continuous at $x = a$.

Part II: Given that f is continuous at $x = a$ and a is a limit point of the domain of f, we now prove that $\lim_{x \to a} f(x) = f(a)$.

Since we are given that a is a limit point of the domain of f, the first part of the definition of the required limit statement is true. We merely have to verify that the second part is also true. Since f is given to be continuous at $x = a$, it must be defined at $x = a$ and $A = (a, f(a))$. By the definition of continuity, if α and β are two horizontal lines with A between them, then there exist two vertical lines l and m with A between them such that every point of $\mathcal{G}$ between l and m is also between α and β. Considering the limit statement that we want to prove, we have $P = A = (a, f(a))$. Since $P = A$, the above statement is true with A replaced by P. Furthermore, if every point of $\mathcal{G}$ between l and m is also between α and β, then every point of the smaller set $\mathcal{G} - P$ between l and m is also between α and β. Thus, the second part of the definition of limit is satisfied and $\lim_{x \to a} f(x) = f(a)$.

Many mathematicians use this property to define continuity. In fact, we used it in Section 3.4. Note that we required that f be defined on an interval $[c, d]$ that contains a. This implies that a is a limit point of the domain. Some authors omit the condition that a be a limit point of the domain. When this is done, the resulting definition of continuity is not equivalent to the one we have given. Such a definition implies that a function f is not continuous at isolated points. The pros and cons of these definitions were given in Section 10.6.

This theorem allows us to find limits such as $\lim_{x \to 2} x^2 = 4$ by a simple substitution after noting that $f(x) = x^2$ is a continuous function and that $x = 2$ is a limit point of its domain. However, most of the limits with which we have dealt have been somewhat more complicated. Consider, for example,

$$\lim_{x \to 1} \frac{x^2 - 1}{x - 1}.$$

While $f(x) = (x^2 - 1)/(x - 1)$ is continuous at every one of its points, it simply is not defined at $x = 1$. Of course, if it is not defined at $x = 1$, it is not continuous there and we cannot use Theorem 10.3. We handle this by factoring the numerator and cancelling factors to obtain

$$\lim_{x \to 1} \frac{x^2 - 1}{x - 1} = \lim_{x \to 1} (x + 1).$$

How do we justify this step? First, we note that the functions $f(x) = (x^2 - 1)/(x - 1)$ and $g(x) = x + 1$ are not the same; however, their graphs differ by only a single point. The point (1, 2) is on the graph of g but not on the graph of f. This slight difference is not enough to give different limits, as the following theorem states.

Theorem 10.4

If the functions f and g are identical except for one point and $\lim_{x \to a} f(x)$ exists, then $\lim_{x \to a} g(x)$ exists and

$$\lim_{x \to a} f(x) = \lim_{x \to a} g(x).$$

Proof

Let $A_f = (a, f(a))$ and $A_g = (a, g(a))$, provided these points exist. Since $\lim_{x \to a} f(x)$ exists, it is equal to some number b and $P = (a, b)$. This point P is also used for the other limit, since we want to verify that $\lim_{x \to a} g(x) = b$. Now let us split the argument into two cases: that is, $A_f = A_g$ and $A_f \neq A_g$.

Case I: $A_f = A_g$. Since these points are identical, the graphs must differ at some other point. Now choose l and m close enough together so that the point of difference is not between them; thus, whatever we say about $\mathcal{G} - A_f$ between l and m can also be said about $\mathcal{G} - A_g$ between l and m. Since this is the portion of the graph we are concerned about in the definition of a limit, it follows that $\lim_{x \to a} f(x) = \lim_{x \to a} g(x)$.

Case II: $A_f \neq A_g$. Since these points are different, $\mathcal{G} - A_f = \mathcal{G} - A_g$. Again, this is the only portion of the graph we are concerned about in the definition of a limit. Thus $\lim_{x \to a} f(x) = \lim_{x \to a} g(x)$.

It might be noted that this theorem would still be true if f and g differed at any finite number of points. The proof is quite similar. Both of these theorems are used to evaluate

$$\lim_{x \to a} \frac{x^2 - a^2}{x - a}.$$

Theorem 10.4 is used to replace the function $(x^2 - a^2)/(x - a)$ by $x + a$, which is defined and continuous at $x = a$. Theorem 10.3 then allows us to evaluate the limit by a simple substitution. Thus, we see that these two theorems form the whole foundation for taking limits by the simple methods we have been using. It must be remembered, however, that in order for this to work the functions with which we are dealing must be continuous at $x = a$. Fortunately, *most* of the functions encountered in elementary calculus are continuous.

Theorem 10.5

A function f is continuous at $x = a$ and a is a limit point of the domain of f if and only if $\lim_{h \to 0} f(a + h) = f(a)$.

The proof of this follows quite directly from Theorem 10.3 by substituting $a + h$ for the x of that theorem. The details are left to the student. It might be pointed out here that many authors take this as the definition of continuity.

Recall that we made some other assumptions about limits when we considered derivative formulas in Chapter 4. We assumed there that the limit of a sum, difference, product, or quotient is the sum, difference, product, or quotient of the limits. Let us now consider proofs of these.

Theorem 10.6

If $\lim_{x \to a} f(x) = b$ *and* $\lim_{x \to a} g(x) = c$, *then*

$$\lim_{x \to a} (f + g)(x) = \lim_{x \to a} [f(x) + g(x)] = b + c,$$

provided a is a limit point of the domain of f + g.

Proof

The graphs of f, g, and $f + g$ are given in Figure 10.24. For clarity they are graphed

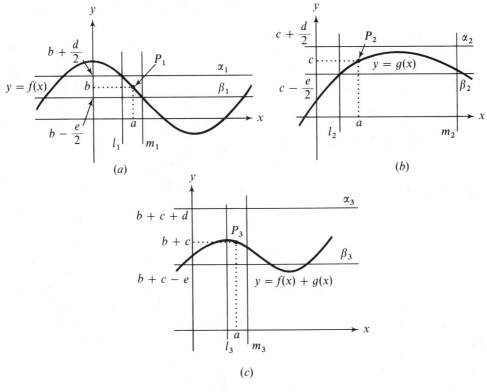

(a)

(b)

(c)

Figure 10.24

on three sets of axes. The graphs, points, and lines associated with f, g, and $f + g$ are identified by the subscripts 1, 2, and 3, respectively. Let

$$\alpha_3 : y = b + c + d \quad \text{and} \quad \beta_3 : y = b + c - e$$

be a pair of horizontal lines with P_3 between them. Then

$$\alpha_1 : y = b + \frac{d}{2} \quad \text{and} \quad \beta_1 : y = b - \frac{e}{2}$$

are two horizontal lines with P_1 between them. Thus there are two vertical lines l_1 and m_1 with P_1 between them such that every point of $\mathcal{G}_1 - A_1$ between l_1 and m_1 is also between α_1 and β_1. Similarly

$$\alpha_2: y = c + \frac{d}{2} \quad \text{and} \quad \beta_2: y = c - \frac{e}{2}$$

are two horizontal lines with P_2 between them. There are two vertical lines l_2 and m_2 with P_2 between them such that every point of $\mathcal{G}_2 - A_2$ between l_2 and m_2 is also between α_2 and β_2.

Now let l_3 be the rightmost of the two lines l_1 and l_2 and m_3 the leftmost of m_1 and m_2. Any point between l_3 and m_3 is also between l_1 and m_1 as well as between l_2 and m_2. Now let us consider a point $X_3\,(x, y_3)$ of $\mathcal{G}_3 - A_3$ between l_3 and m_3. The point $X_1\,(x, y_1)$ of $\mathcal{G}_1 - A_1$ is between l_1 and m_1 and therefore between α_1 and β_1. Thus

$$b - \frac{e}{2} < y_1 < b + \frac{d}{2}.$$

Similarly the point $X_2\,(x, y_2)$ of $\mathcal{G}_2 - A_2$ is between l_2 and m_2 and therefore between α_2 and β_2, giving

$$c - \frac{e}{2} < y_2 < c + \frac{d}{2}.$$

Since $y_1 + y_2 = y_3$,

$$b + c - e < y_3 < b + c + d,$$

or X_3 is between α_3 and β_3. Therefore any point of $\mathcal{G}_3 - A_3$ between l_3 and m_3 is between α_3 and β_3, and

$$\lim_{x \to a} (f + g)(x) = b + c.$$

Theorem 10.7

If $\lim_{x \to a} f(x) = b$ *and* $\lim_{x \to a} g(x) = c,$ *then*

$$\lim_{x \to a} (f - g)(x) = \lim_{x \to a} [f(x) - g(x)] = b - c,$$

provided a is a limit point of the domain of $f - g$.

The proof of this theorem is left to the student (see Problem 17).

Theorem 10.8

If $\lim_{x \to a} f(x) = b$ *and* $\lim_{x \to a} g(x) = c,$ *then*

$$\lim_{x \to a} fg(x) = \lim_{x \to a} f(x) \cdot g(x) = \lim_{x \to a} f(x) \cdot \lim_{x \to a} g(x) = bc,$$

provided a is a limit point of the domain of fg.

Proof

First we shall prove the theorem for a special case—namely, the case in which $c = 0$. We shall then use this result to prove the more general case. If $c = 0$, we break down the argument further to two cases: $b > 0$, $b \le 0$. We consider here only $b > 0$. The other case is left to the student (see Problem 18).

Again the graphs, points, and lines associated with f, g, and fg are identified by the subscripts 1, 2, and 3, respectively. Now let α_3 be the line $y = B$ ($B > 0$) and β_3 be the line $y = C$ ($C < 0$). We need to show that there are two vertical lines l_3 and m_3 such that every point of $\mathscr{G}_3 - A_3$ between l_3 and m_3 is also between α_3 and β_3.

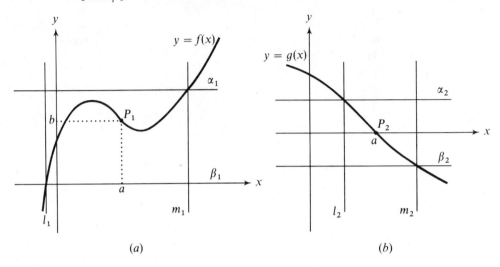

(a)   (b)

Figure 10.25

Let α_1 be the line $y = b + 1$ and β_1 be the line $y = 0$ (see Figure 10.25(a)). Then there are vertical lines l_1 and m_1 with P_1 between them such that every point of $\mathscr{G}_1 - A_1$ between l_1 and m_1 is also between α_1 and β_1. For such a point,

$$0 < f(x) < b + 1.$$

Let α_2 and β_2 be respectively

$$y = \frac{B}{b+1} > 0 \quad \text{and} \quad y = \frac{C}{b+1} < 0$$

(see Figure 10.25(b)). Then there are two vertical lines l_2 and m_2 with P_2 between them such that every point of $\mathscr{G}_2 - A_2$ between l_2 and m_2 is also between α_2 and β_2. For such a point,

$$\frac{C}{b+1} < g(x) < \frac{B}{b+1}.$$

Now let l_3 be the rightmost of the lines l_1 and l_2 and let m_3 be the leftmost of m_1 and m_2. Any point of $\mathscr{G}_1 - A_1$ between l_3 and m_3 is between α_1 and β_1, and any point of $\mathscr{G}_2 - A_2$ between l_3 and m_3 is between α_2 and β_2. Thus, for any point of $\mathscr{G}_3 - A_3$ between l_3 and m_3,

$$C = (b+1)\frac{C}{b+1} < f(x)\frac{C}{b+1} < f(x) \cdot g(x) < f(x)\frac{B}{b+1} < (b+1)\frac{B}{b+1} = B,$$

or such a point is between α_3 and β_3. Thus

$$\lim_{x \to a} f(x) \cdot g(x) = 0.$$

Now, for the general case. Since

$$f(x) \cdot g(x) = f(x)[g(x) - c] + c[f(x) - b] + bc$$

and

$$\lim_{x \to a} [f(x) - b] = \lim_{x \to a} f(x) - \lim_{x \to a} b = b - b = 0,$$

$$\lim_{x \to a} [g(x) - c] = \lim_{x \to a} g(x) - \lim_{x \to a} c = c - c = 0 \qquad \text{(by Theorem 10.7)},$$

it follows from our special case that

$$\lim_{x \to a} f(x) \cdot g(x) = \lim_{x \to a} f(x)[g(x) - c] + \lim_{x \to a} c[f(x) - b] + \lim_{x \to a} bc$$
$$= 0 + 0 + bc = bc.$$

Theorem 10.9

If $\lim_{x \to a} f(x) = b$ *and* $\lim_{x \to a} g(x) = c,$ *where* $c \neq 0,$ *then*

$$\lim_{x \to a} \frac{f}{g}(x) = \lim_{x \to a} \frac{f(x)}{g(x)} = \frac{b}{c},$$

provided a is a limit point of the domain of f/g.

Proof

We follow the convention of the previous theorems in representing f, g, and f/g with the subscripts 1, 2, and 3, respectively. Again we shall consider a special case first. Suppose $f(x) = 1$ for all x. Then we need to consider only the function g.

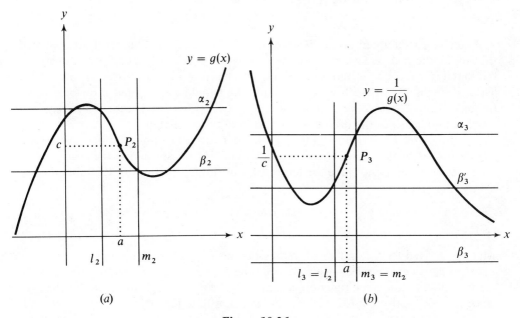

(a) $\qquad\qquad\qquad\qquad\qquad\qquad (b)$

Figure 10.26

We suppose here that $c > 0$. The case $c < 0$ is similar (see Problem 19). Let $\alpha_3 : y = C > 1/c$ and β_3 be two horizontal lines with P_3 between them (see Figure 10.26(b)). Let $\beta_3' : y = B < 1/c$ be a horizontal line which is below P_3 and above both β_3 and the x axis. It then follows that

$$0 < B < \frac{1}{c} < C.$$

Thus

$$\frac{1}{C} < c < \frac{1}{B}.$$

If α_2 is the line $y = 1/B$ and β_2 is $y = 1/C$ (see Figure 10.26(a)), then there are two vertical lines l_2 and m_2 with $P_2 (a, c)$ between them such that every point of $\mathcal{G}_2 - A_2$ between l_2 and m_2 is also between α_2 and β_2; that is,

$$\frac{1}{C} < g(x) < \frac{1}{B}.$$

Since this implies that

$$B < \frac{1}{g(x)} < C,$$

l_3 and m_3 can be taken to be l_2 and m_2 and every point of $\mathcal{G}_3 - A_3$ between l_3 and m_3 is between α_3 and β_3' and therefore between α_3 and β_3. Thus

$$\lim_{x \to a} \frac{1}{g(x)} = \frac{1}{c}.$$

Now, by the previous theorem,

$$\lim_{x \to a} \frac{f}{g}(x) = \lim_{x \to a} \frac{f(x)}{g(x)} = \lim_{x \to a} f(x) \cdot \frac{1}{g(x)} = \lim_{x \to a} f(x) \cdot \lim_{x \to a} \frac{1}{g(x)} = b \cdot \frac{1}{c} = \frac{b}{c}.$$

When we encountered $\lim_{h \to 0} f(x + h)$ in several proofs in Chapter 4, we assumed this limit to be $f(x)$. We noted there that this is not necessarily true for *all* functions, but that it is true for continuous functions. This has now been proved in Theorem 10.5. Thus, it might seem that Theorems 4.4 and 4.6 should specify that u (or v) be continuous. Actually, since we assume that $u'(x)$ exists, it is not necessary to say that u is continuous, since the derivative and continuity are related.

Let us look at the derivative. First of all, a derivative is a limit, but not every limit is a derivative. A derivative is a very special limit. If we are taking the derivative of the function f, we do not take the limit of f, but rather, the limit of the function

$$\frac{f(x + h) - f(x)}{h},$$

which is derived from f and which represents the slope of a certain line.

Let us consider the relationship between the derivative and continuity. First of all, the mere fact that a given function is continuous is not enough to guarantee the existence of a derivative. This is easily seen with the function defined by $y = |x|$. It is continuous at every one of its points, but it does not have a derivative at the origin (see Problem 12). But if a function has a derivative at a given point, it is necessarily continuous at that point.

Theorem 10.10

If f has a derivative at $x = a$, then it is continuous at $x = a$.

Proof

Since

$$\lim_{h \to 0} [f(a+h) - f(a)] = \lim_{h \to 0} \frac{f(a+h) - f(a)}{h} \cdot h$$

$$= \lim_{h \to 0} \frac{f(a+h) - f(a)}{h} \cdot \lim_{h \to 0} h \quad \text{(by Theorem 10.8)}$$

$$= f'(a) \cdot 0 = 0,$$

it follows that

$$0 = \lim_{h \to 0} [f(a+h) - f(a)] = \lim_{h \to 0} f(a+h) - \lim_{h \to 0} f(a) \quad \text{(by Theorem 10.7),}$$

and so

$$\lim_{h \to 0} f(a+h) = \lim_{h \to 0} f(a) = f(a).$$

By Theorem 10.5, f is continuous at $x = a$.

Problems

A

In Problems 1–6, evaluate the limits and indicate which theorem is used at each step.

1. $\lim\limits_{x \to 0} \dfrac{x^2 + x}{x^2 - x}$.

2. $\lim\limits_{x \to 2} \dfrac{x^2 - x - 2}{x^2 - 2x}$.

3. $\lim\limits_{x \to 1} \dfrac{x^3 + x^2 - x - 1}{x^2 - 1}$.

4. $\lim\limits_{x \to 0} \dfrac{x^3 - 3x}{x^3 + 2x^2 - x}$.

5. $\lim\limits_{h \to 0} \dfrac{(2 + h)^2 - 4}{h}$.

6. $\lim\limits_{h \to 0} \dfrac{2(1 + h)^2 + (1 + h) - 3}{h}$.

B

In Problems 7–10, evaluate the limits and indicate which theorem is used at each step.

7. $\lim\limits_{h \to 0} \dfrac{\dfrac{1}{(2 + h)^2} - \dfrac{1}{4}}{h}$.

8. $\lim\limits_{h \to 0} \dfrac{\sqrt{4 + h} - 2}{h}$.

9. $\lim\limits_{h \to 0} \dfrac{\dfrac{x + h}{x + h + 1} - \dfrac{x}{x + 1}}{h}$.

10. $\lim\limits_{h \to 0} \dfrac{\sqrt{x + h} - \sqrt{x}}{h}$.

11. Prove that Theorem 10.4 is still true if "one point" is replaced by "a finite number of points."

12. Show that $y = |x|$ is continuous at $(0, 0)$ but has no derivative there.

13. Give an example of a function that is continuous everywhere but has no derivative at two of its points.

14. Give an example of a function that is continuous everywhere but has no derivative at infinitely many of its points.

15. Supply the details for the proof of Theorem 10.5.

C

16. Give an example of two functions f and g such that 0 is a limit point of the domain of each but is not a limit point of the domain of $f + g$.

17. Prove Theorem 10.7. (*Hint:* First show that $\lim_{x \to a} [-g(x)] = -c$.)

18. Prove that if $\lim_{x \to a} g(x) = 0$ and $\lim_{x \to a} f(x) = b \leq 0$, then

$$\lim_{x \to a} f(x) \cdot g(x) = 0.$$

(*Hint:* Consider $\lim_{x \to a} [(1 + |b|) + f(x)]g(x)$.)

19. Prove that if $\lim_{x \to a} g(x) = c < 0$, then

$$\lim_{x \to a} \frac{1}{g(x)} = \frac{1}{c}.$$

Review Problems

A

In Problems 1 and 2, indicate where the function is continuous and where it is discontinuous.

1. $f(x) = \begin{cases} x & \text{if } x \leq -2, \\ x^2 & \text{if } -1 \leq x \leq 0, \\ x + 1 & \text{if } 0 < x < 1, \\ 1 - x & \text{if } x \geq 1. \end{cases}$

2. $f(x) = \begin{cases} 0 & \text{if } x \leq 0, \\ x & \text{if } 0 < x < 1, \\ 2 & \text{if } x = 1, \\ x + 2 & \text{if } x > 2. \end{cases}$

B

In Problems 3–16, use the proper definition to show that the given limit statement is either true or false.

3. $\lim\limits_{x \to 1} f(x) = 2$, where $f(x) = \begin{cases} x - 1 & \text{if } x \leq 0, \\ 2 & \text{if } x = 1, \\ x & \text{if } x \geq 2. \end{cases}$

4. $\lim\limits_{x \to 1} \dfrac{x^2 + 4x - 5}{x - 1} = 5.$

5. $\lim\limits_{x \to 2} f(x) = 5$, where $f(x) = \begin{cases} x^2 + 1 & \text{if } x \neq 0, \\ 0 & \text{if } x = 0. \end{cases}$

6. $\lim\limits_{x \to 0} f(x) = 0$, where $f(x) = \begin{cases} x^2 + 1 & \text{if } x \neq 0, \\ 0 & \text{if } x = 0. \end{cases}$

7. $\lim\limits_{x \to 0} f(x) = 1$, where $f(x) = \begin{cases} x^2 + 1 & \text{if } x \neq 0, \\ 0 & \text{if } x = 0. \end{cases}$

8. $\lim\limits_{x \to -1} \sqrt{-x^2(1 - x^2)^2} = 0.$

9. $\lim\limits_{x \to 3^+} \dfrac{1}{x - 3} = +\infty.$

10. $\lim\limits_{x \to 1} f(x) = 3$, where $f(x) = \begin{cases} 2x + 1 & \text{if } x < 1, \\ 2x - 1 & \text{if } x \geq 1. \end{cases}$

11. $\lim\limits_{x \to 1^-} f(x) = 3$, where $f(x) = \begin{cases} 2x + 1 & \text{if } x < 1, \\ 2x - 1 & \text{if } x \geq 1. \end{cases}$

12. $\lim\limits_{x \to +\infty} \dfrac{2x^2}{x^2 - 1} = 2.$

13. $\lim\limits_{x \to -\infty} \dfrac{x^2}{x - 1} = -\infty.$

14. $\lim\limits_{x \to +\infty} \dfrac{1}{x} \sin \dfrac{1}{x} = 0.$

15. $\lim\limits_{x \to 0} \dfrac{1}{x} \sin \dfrac{1}{x} = 0.$

16. $\lim\limits_{x \to 0^+} |x| = 0.$

In Problems 17 and 18, use the definition of continuity to show whether or not the function is continuous at the given point.

17. $f(x) = \begin{cases} x^2 & \text{if } x \le 1, \\ 2 - x & \text{if } x > 1; \end{cases}$ (1, 1).

18. $f(x) = \begin{cases} x^2 + 1 & \text{if } x \ne 1, \\ 1 & \text{if } x = 1; \end{cases}$ (1, 1).

C

In Problems 19 and 20, show that the given limit does not exist.

19. $\lim_{x \to 1} f(x)$, where $f(x) = \begin{cases} x^2 + 1 & \text{if } x < 1, \\ 2x - 1 & \text{if } x > 1. \end{cases}$

20. $\lim_{x \to 0} \dfrac{1}{x^2} \sin \dfrac{1}{x^2}$.

21. Show that if $f(x) = [\![x]\!] - [\![n - x]\!]$, then
$$\lim_{x \to n^+} f(x) - \lim_{x \to n^-} f(x) = 2.$$

22. Suppose $\lim_{x \to a} f(x) = \lim_{x \to a} g(x) = b$ and $f(x) \le h(x) \le g(x)$. Prove that
$$\lim_{x \to a} h(x) = b.$$

This is an important result that we shall use later.

11

Trigonometric Functions

11.1

Trigonometry Review

We now turn to some of the nonalgebraic (or transcendental) functions, beginning with the trigonometric functions and their inverses. The trigonometric functions are periodic, or repeating; they are very useful in the study of other periodic functions. A review of the trigonometric functions here will probably be useful.

An *angle* consists of a pair of rays (half-lines) OA and OB with a common end point O, called the *vertex* of the angle. An angle is in *standard position* if its vertex is at the origin and one of the rays, the *initial side*, is on the right half of the x axis. The other ray, the *terminal side*, is looked upon as a rotation of the initial side in either a clockwise or counterclockwise direction. The amount of this rotation determines the measure of the angle. A counterclockwise rotation has a positive measure and a clockwise rotation has a negative measure. The two most common ways of measuring angles are in *degrees* and *radians*. While measuring angles in degrees is convenient for many purposes, radian measure is more natural when we differentiate and integrate functions. If the vertex of the angle is at the center of a circle (see Figure 11.1), then the radian measure of the angle θ is the length s of the arc subtended by θ† divided by the radius of the circle:

$$\theta = \frac{s}{r}.$$

---

†While an angle (geometric) and its measure (arithmetic) are not the same, we follow the usual practice of using a single symbol θ to represent both. There should be no confusion between the two ideas because the context will make it clear which one we are referring to.

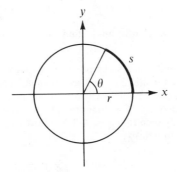

Figure 11.1

Since a complete rotation of 360° subtends an arc whose length is the circumference of the circle, it follows that 360° is equivalent to

$$\theta = \frac{s}{r} = \frac{2\pi r}{r} = 2\pi \text{ radians.}$$

With this result we can find a formula for the area of a sector (a pie-shaped region). If the central angle of a sector is 2π radians, the sector becomes a circle with area πr^2. Assuming that the area of a sector is proportional to the measure of the central angle, we compare the sector with central angle 2π and area πr^2 with one having central angle θ and area A. This gives the ratio

$$\frac{A}{\pi r^2} = \frac{\theta}{2\pi};$$

therefore the area of the sector is

$$A = \frac{1}{2} r^2 \theta.$$

Of course, θ is the *radian* measure of the angle. We shall use this result in the next section.

The six common trigonometric functions are defined in terms of the coordinates of a point other than the origin on the terminal side (see Figure 11.2). Thus we have:

$$\sin \theta = \frac{y}{r}, \qquad \tan \theta = \frac{y}{x}, \qquad \sec \theta = \frac{r}{x},$$

$$\cos \theta = \frac{x}{r}, \qquad \cot \theta = \frac{x}{y}, \qquad \csc \theta = \frac{r}{y}.$$

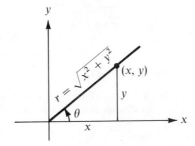

Figure 11.2

Example 1

If sin θ = 4/5 and θ is not a first-quadrant angle, find the values of the other five trigonometric functions.

Since sin θ = y/r and r is always positive, it follows that sin θ is positive when y is positive—that is, for the first- and second-quadrant angles. But we are given that θ is not a first-quadrant angle; then it must be in the second quadrant (see Figure 11.3). Using the theorem of Pythagoras on the triangle of Figure 11.3 with y = 4 and r = 5, it is easily seen that x = -3. We now use the definitions to find the other five trigonometric functions:

$$\tan \theta = \frac{y}{x} = -\frac{4}{3}, \qquad \sec \theta = \frac{r}{x} = -\frac{5}{3},$$

$$\cos \theta = \frac{x}{r} = -\frac{3}{5}, \qquad \cot \theta = \frac{x}{y} = -\frac{3}{4}, \qquad \csc \theta = \frac{r}{y} = \frac{5}{4}.$$

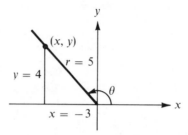

Figure 11.3

The definitions of the trigonometric functions together with the two familiar triangles of Figure 11.4 allow us to evaluate the trigonometric functions for any multiple of 30° = $\pi/6$ or 45° = $\pi/4$.

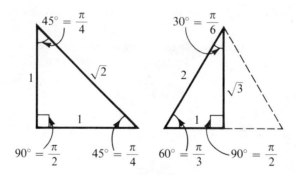

Figure 11.4

Example 2

Find the exact values of cos $(3\pi/4)$ and tan $(-\pi/3)$.

By placing the triangles of Figure 11.4 on the coordinate axes as shown in Figure 11.5 and noting which coordinates are negative, we may refer to the definitions to get the desired results:

$$\cos \frac{3\pi}{4} = \frac{x}{r} = \frac{-1}{\sqrt{2}} = -\frac{1}{\sqrt{2}}, \quad \text{and} \quad \tan -\frac{\pi}{3} = \frac{y}{x} = \frac{-\sqrt{3}}{1} = -\sqrt{3}.$$

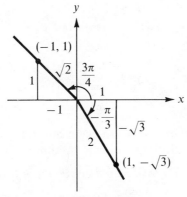

Figure 11.5

The definitions of the trigonometric functions dictate the following simple identities.

$$\cot \theta = \frac{1}{\tan \theta}, \quad \sec \theta = \frac{1}{\cos \theta}, \quad \csc \theta = \frac{1}{\sin \theta}, \quad \tan \theta = \frac{\sin \theta}{\cos \theta}, \quad \cot \theta = \frac{\cos \theta}{\sin \theta}.$$

Only slightly more difficult is the following derivation of a very important identity.
  By the Pythagorean theorem,

$$x^2 + y^2 = r^2.$$

Dividing by r^2 and using our definitions, we have

$$\frac{x^2}{r^2} + \frac{y^2}{r^2} = 1$$

$$\left(\frac{x}{r}\right)^2 + \left(\frac{y}{r}\right)^2 = 1$$

$$\cos^2 \theta + \sin^2 \theta = 1.$$

Once we have this identity, we may find two other useful ones by dividing through by $\cos^2 \theta$ and by $\sin^2 \theta$.

$$\frac{\cos^2 \theta}{\cos^2 \theta} + \frac{\sin^2 \theta}{\cos^2 \theta} = \frac{1}{\cos^2 \theta}$$

$$1 + \tan^2 \theta = \sec^2 \theta.$$

Similarly,

$$\cot^2 \theta + 1 = \csc^2 \theta.$$

Example 3

Show that $\tan^4 \theta = \sec^2 \theta \tan^2 \theta - \sec^2 \theta + 1.$

> Although an identity can be verified by working on both sides simultaneously, the problems we shall encounter later in this book require us to start with one side and work to the other. Thus we shall establish the identities in this section by starting with the left-hand side and working toward the right. In this case, we use
>
> $$1 + \tan^2 \theta = \sec^2 \theta$$
>
> in the form
>
> $$\tan^2 \theta = \sec^2 \theta - 1.$$

Thus

$$\tan^4 \theta = \tan^2 \theta \tan^2 \theta$$
$$= (\sec^2 \theta - 1) \tan^2 \theta$$
$$= \sec^2 \theta \tan^2 \theta - \tan^2 \theta$$
$$= \sec^2 \theta \tan^2 \theta - (\sec^2 \theta - 1)$$
$$= \sec^2 \theta \tan^2 \theta - \sec^2 \theta + 1.$$

Other important identities involve the sum and difference of two angles, double angles, and half-angles. Identities involving the sum and difference of two angles are

$$\sin (\theta \pm \phi) = \sin \theta \cos \phi \pm \cos \theta \sin \phi,$$

$$\cos (\theta \pm \phi) = \cos \theta \cos \phi \mp \sin \theta \sin \phi,$$

$$\tan (\theta \pm \phi) = \frac{\tan \theta \pm \tan \phi}{1 \mp \tan \theta \tan \phi}.$$

The identities for the difference of two angles may be derived from the corresponding sum identities by replacing ϕ by $-\phi$ throughout and simplifying. There are several ways of deriving the sum identities; one way is given in Problems 35–37.

By replacing ϕ by θ in the above sum identities and simplifying, we get the following double-angle identities ($\cos^2 \theta + \sin^2 \theta = 1$ is also used in the cos 2θ identity to get the last two forms).

$$\sin 2\theta = 2 \sin \theta \cos \theta,$$

$$\cos 2\theta = \cos^2 \theta - \sin^2 \theta = 1 - 2 \sin^2 \theta = 2 \cos^2 \theta - 1,$$

$$\tan 2\theta = \frac{2 \tan \theta}{1 - \tan^2 \theta}.$$

Finally, we may solve the last two cos 2θ identities for sin θ and cos θ, respectively, replacing 2θ by ϕ to get the half-angle identities for sine and cosine. These are then easily extended to the tangent. Thus we have the following.

$$\sin \frac{\phi}{2} = \pm \sqrt{\frac{1 - \cos \phi}{2}},$$

$$\cos \frac{\phi}{2} = \pm \sqrt{\frac{1 + \cos \phi}{2}},$$

$$\tan \frac{\phi}{2} = \pm \sqrt{\frac{1 - \cos \phi}{1 + \cos \phi}} = \frac{1 - \cos \phi}{\sin \phi} = \frac{\sin \phi}{1 + \cos \phi}.$$

Example 4

Express $\sin 3\theta$ as $\sin \theta \cdot f(\cos \theta)$.

We write $\sin 3\theta$ as $\sin (\theta + 2\theta)$, using the identity for the sum of two angles and the double-angle identities.

$$\sin 3\theta = \sin (\theta + 2\theta)$$
$$= \sin \theta \cos 2\theta + \cos \theta \sin 2\theta$$
$$= \sin \theta (2 \cos^2 \theta - 1) + \cos \theta \cdot 2 \sin \theta \cos \theta \quad \text{(See Note.)}$$
$$= \sin \theta (4 \cos^2 \theta - 1).$$

Note: We use the identity $\cos 2\theta = 2\cos^2\theta - 1$ rather than either of the other two because we want the result to be $\sin\theta$ times an expression involving only $\cos\theta$. Since we have a factor of $\sin\theta$ already in the term $\sin\theta\cos 2\theta$, we want $\cos 2\theta$ to be intirely in terms of $\cos\theta$.

Example 5

Find $\sin(\pi/12)$ and $\cos(5\pi/12)$.

Since $\pi/12 = (1/2)(\pi/6)$, we use a half-angle identity to find $\sin(\pi/12)$ (the positive root is used because $\sin(\pi/12)$ is known to be positive).

$$\sin\frac{\pi}{12} = \sqrt{\frac{1 - \cos(\pi/6)}{2}}$$

$$= \sqrt{\frac{1 - \sqrt{3/2}}{2}}$$

$$= \sqrt{\frac{2 - \sqrt{3}}{4}}$$

$$= \frac{\sqrt{2 - \sqrt{3}}}{2}.$$

Since

$$\frac{5\pi}{12} = \frac{3\pi}{12} + \frac{2\pi}{12} = \frac{\pi}{4} + \frac{\pi}{6},$$

we use a sum identity to find $\cos(5\pi/12)$.

$$\cos\frac{5\pi}{12} = \cos\left(\frac{\pi}{4} + \frac{\pi}{6}\right)$$

$$= \cos\frac{\pi}{4}\cos\frac{\pi}{6} - \sin\frac{\pi}{4}\sin\frac{\pi}{6}$$

$$= \frac{1}{\sqrt{2}}\frac{\sqrt{3}}{2} - \frac{1}{\sqrt{2}}\frac{1}{2}$$

$$= \frac{\sqrt{3} - 1}{2\sqrt{2}} = \frac{\sqrt{6} - \sqrt{2}}{4}.$$

The graphs of the six common trigonometric functions are given in Figure 11.6. Note that all of them are periodic (repeating). To say that a function f has period p means that $f(x + p) = f(x)$ for all x; in other words, the function repeats itself after every p units. The period of $y = \tan x$ and $y = \cot x$ is π; the other four functions have period 2π. Note also that four of them have vertical asymptotes, although there is no denominator to be zero. This is reasonable when we consider that all of them are defined as *ratios* of certain lengths.

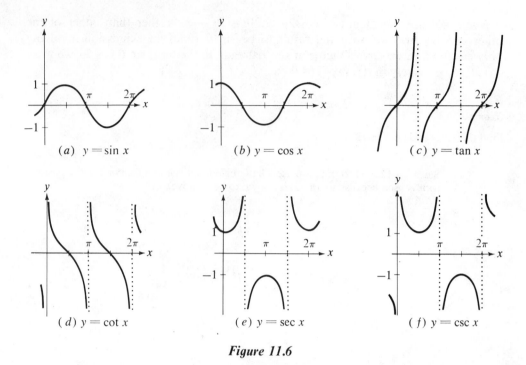

(a) $y = \sin x$ (b) $y = \cos x$ (c) $y = \tan x$

(d) $y = \cot x$ (e) $y = \sec x$ (f) $y = \csc x$

Figure 11.6

Since the graphs of $y = \sin x$ and $y = \cos x$ are waves, they are the most important of the six. Let us see how they are altered by changing certain constants. The equation

$$y = A \sin B(x - C)$$

has a graph of the form shown in Figure 11.7. The amplitude is $|A|$, with the curve inverted if A is negative. The period is $2\pi/|B|$, again with the curve inverted if B is negative. The displacement is C. If C is positive, the displacement is to the right; if it is negative, there is a negative (or left) displacement. Note, however, that we have $x - C$ in the given equation. Thus $y = \sin (x + 3)$ has a negative displacement; $C = -3$. We have the same results with the cosine curve except that a negative value of B does not invert the curve.

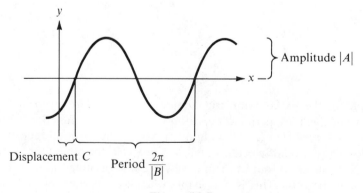

Displacement C Period $\dfrac{2\pi}{|B|}$ Amplitude $|A|$

Figure 11.7

Example 6

Sketch $y = 3 \sin 2x$.

First of all, note that $-1 \le \sin x \le 1$. Thus the factor of 3 in $3 \sin 2x$ alters this range by a factor of 3. The fact that we have $\sin 2x$ instead of $\sin x$ does not alter the range. Now it takes one complete cycle for whatever we are taking the sine of to go from 0 to 2π; that is, we have one complete cycle for

$$0 \le 2x \le 2\pi \quad \text{or} \quad 0 \le x \le \pi.$$

We have a sine curve with an amplitude of 3, a period of $2\pi/2 = \pi$, and no displacement. The result is shown in Figure 11.8.

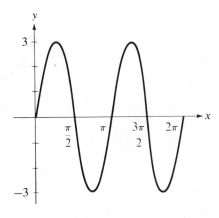

Figure 11.8

Example 7

Sketch $y = 4 \cos \left(2x + \dfrac{\pi}{2}\right)$.

First let us write the equation in the form

$$y = 4 \cos 2\left(x + \frac{\pi}{4}\right).$$

Now we see that we have a cosine curve with an amplitude of 4, a period of $2\pi/2 = \pi$, and a displacement of $-\pi/4$, as shown in Figure 11.9.

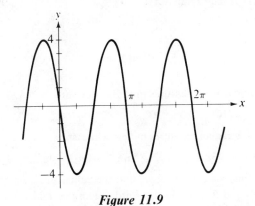

Figure 11.9

Example 8

Sketch $y = \cos x + \sin 2x$.

The method of addition of ordinates is useful for equations of the form $y = f(x) + g(x)$. We note that $y = y_1 + y_2$, where $y_1 = f(x)$ and $y_2 = g(x)$. Now we sketch both $y_1 = f(x)$ and $y_2 = g(x)$ on the same coordinate axes; and for each value of x we add the y values y_1 and y_2 from the two graphs to get the y of the original equation. Thus we sketch $y = \cos x$ and $y = \sin 2x$ and add the ordinates; the result is shown as the solid curve in Figure 11.10.

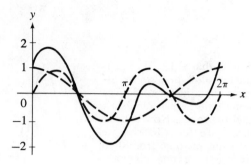

Figure 11.10

Example 9

Sketch $y = x + \sin x$.

Perhaps you wonder how we can add x and $\sin x$ when x is an angle and $\sin x$ is a number. Actually, both x and $\sin x$ are numbers. Although it is convenient to talk about the sine of an angle, we actually take trigonometric functions, not of angles, but of numbers. The numbers are simply the *measures* of angles. It is quite possible to consider trigonometric functions of numbers independently of any angular interpretations; but if we do want to impose such an interpretation, the value of x is the measure of an angle in radians. Again, addition of ordinates works very well and Figure 11.11 is self-explanatory.

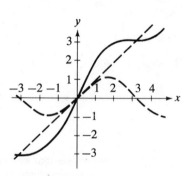

Figure 11.11

Problems

A

1. Express the following degree measures in radian measure.

$$45°, \quad -210°, \quad 270°, \quad 30°, \quad -180°, \quad -60°, \quad 135°, \quad 150°.$$

2. Express the following radian measures in degree measure.

$$\pi/3, \quad \pi, \quad 3\pi/4, \quad -\pi/2, \quad 5\pi/6, \quad -2\pi/3, \quad 3\pi/2, \quad 10\pi/6.$$

3. If $\tan \theta = 5/12$ and $\sin \theta$ is negative, find the values of all six trigonometric functions.
4. If $\sec \theta = -4/3$ and θ is not in the third quadrant, find the values of the remaining five trigonometric functions.

In Problems 5–8, find exact values of the given trigonometric functions.

5. $\tan 225°$.   6. $\sin -120°$.
7. $\cos (7\pi/6)$.   8. $\sec (-3\pi/4)$.

B

9. Express $(\sin u + \cos u)^2$ as $f(\sin 2u)$.
10. Prove that $2 \sin^2 (x/2) + \cos x = 1$.
11. Express $\sec^2 x \cos^5 x$ as $\cos x \cdot f(\sin x)$.

12. Express $\tan^6 \theta$ as $k + \sec^2 \theta \cdot f(\tan \theta)$.
13. Express $\sec^3 \theta \tan^3 \theta$ as $\sec \theta \tan \theta \cdot f(\sec \theta)$.
14. Prove $(\sec x + \tan x)^2 = 2 \sec^2 x + 2 \sec x \tan x - 1$ by working only with the left-hand side.
15. Prove $\cot^5 x = \cot^3 x \csc^2 x - \cot x \csc^2 x + \cot x$ by working only with the left-hand side.
16. Express $\sec^6 x$ as $\sec^2 x \cdot f(\tan x)$.
17. Express $\sec 2\theta$ as $f(\sec \theta)$.
18. If $\sin \theta = x - 1$ and θ is a first-quadrant angle, find $\frac{1}{2} \sin 2\theta$.
19. If $x = 3 \tan \theta$ and θ is a first-quadrant angle, find $\sec \theta + \tan \theta$.
20. Use a double-angle identity for cosines to show that $\cos^2 x = \frac{1}{2}(1 + \cos 2x)$. Use this result to express $\cos^4 x$ in terms of cosines to the first power only.
21. Express $\sin^2 x \cos^2 x$ in terms of cosines to the first power only (see Problem 20).
22. If $f(x) = \sin x$, show that

$$\frac{f(x+h) - f(x)}{h} = \sin x \left(\frac{\cos h - 1}{h} \right) + \cos x \left(\frac{\sin h}{h} \right).$$

In Problems 23–32, sketch one complete cycle.

23. $y = 3 \sin x$.
24. $y = 2 \cos 4x$.
25. $y = 4 \sin \pi x$.
26. $y = 2 \sin (2x + \pi)$.
27. $y = -\cos \left(x - \dfrac{\pi}{3} \right)$.
28. $y = 3 \cos \left(2\pi x + \dfrac{\pi}{2} \right)$.
29. $y = 2 \sin x + \sin 2x$.
30. $y = \cos x - \sin 2x$.
31. $y = 3 \cos x + \sin x$.
32. $y = 4 \sin x + 2 \sin 2x - \sin 4x$.

In Problems 33 and 34, sketch the graph.

33. $y = x - \sin x$.
34. $y = x^2 + \sin x$.

C

35. Show that $(x, y) = (\cos \theta, \sin \theta)$ in Figure 11.12. Use this result and the fact that $\overline{AC} = \overline{BD}$ in Figure 11.13 to show that $\cos (\theta + \phi) = \cos \theta \cos \phi - \sin \theta \sin \phi$.

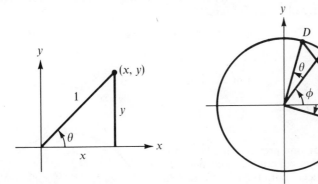

Figure 11.12   *Figure 11.13*

36. The graphs of $y = \sin x$ and $y = \cos x$ are identical except that they are 90° or $\pi/2$ radians out of phase. That is,

$$\sin (x - (\pi/2)) = -\cos x \quad \text{and} \quad \cos (x - (\pi/2)) = \sin x.$$

Use this and the identity derived in Problem 35 to prove that $\sin (\theta + \phi) = \sin \theta \cos \phi + \cos \theta \sin \phi$.

37.   Use the results of Problems 35 and 36 to show that

$$\tan(\theta + \phi) = \frac{\tan\theta + \tan\phi}{1 - \tan\theta\tan\phi}.$$

38.   Sketch

$$y = \frac{\pi}{2} + 2\sin x, \qquad y = \frac{\pi}{2} + 2\sin x + \frac{2}{3}\sin 3x,$$

and

$$y = \frac{\pi}{2} + 2\sin x + \frac{2}{3}\sin 3x + \frac{2}{5}\sin 5x$$

on the same coordinates. What do you think the graph of

$$y = \frac{\pi}{2} + 2\left(\sin x + \frac{1}{3}\sin 3x + \frac{1}{5}\sin 5x + \cdots\right)$$

looks like? This is an example of a Fourier series.

11.2

Derivatives of Trigonometric Functions

Before finding the derivatives of trigonometric functions, we shall first evaluate some difficult limits that will be useful later.

Theorem 11.1

$$\lim_{x\to 0} \frac{\sin x}{x} = 1.$$

Proof

Note first that our previous methods for evaluating limits cannot be used here. In the past, whenever both the numerator and denominator of a fraction approached zero, we (perhaps after some algebraic manipulation) canceled a factor to allow evaluation of the limit. This cannot be done here; we must try something entirely different.

Suppose we restrict the values of x to

$$0 < x < \pi/2.$$

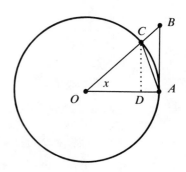

Figure 11.14

Now consider Figure 11.14, where the circle has radius one. Since $\overline{OA} = \overline{OC} = 1$, we have

$$\overline{AB} = \overline{OA} \tan x = \tan x,$$
$$\overline{CD} = \overline{OC} \sin x = \sin x.$$

Comparing areas gives us

$$\triangle OAC < \text{sector } OAC < \triangle OAB$$

$$\frac{1}{2} \cdot \overline{OA} \cdot \overline{CD} < \frac{1}{2} r^2 x < \frac{1}{2} \cdot \overline{OA} \cdot \overline{AB}$$

$$\frac{1}{2} \sin x < \frac{1}{2} x < \frac{1}{2} \tan x$$

$$1 < \frac{x}{\sin x} < \frac{1}{\cos x}$$

$$\cos x < \frac{\sin x}{x} < 1. \qquad \text{(See Note 1)}$$

Taking limits, we have

$$\lim_{x \to 0^+} 1 = 1 \quad \text{and} \quad \lim_{x \to 0^+} \cos x = 1 \qquad \text{(See Note 2)}$$

and, for $0 < x < \pi/2$, $(\sin x)/x$ is between 1 and $\cos x$. Thus $(\sin x)/x$ must also approach 1 (see Problem 22, page 311), or

$$\lim_{x \to 0^+} \frac{\sin x}{x} = 1.$$

Finally

$$\lim_{x \to 0^-} \frac{\sin x}{x} = \lim_{x \to 0^+} \frac{\sin (-x)}{-x} \qquad \text{(See Note 3)}$$

$$= \lim_{x \to 0^+} \frac{-\sin x}{-x}$$

$$= \lim_{x \to 0^+} \frac{\sin x}{x}$$

$$= 1.$$

Since

$$\lim_{x \to 0^+} \frac{\sin x}{x} = \lim_{x \to 0^-} \frac{\sin x}{x} = 1,$$

it follows from Theorem 10.2 that

$$\lim_{x \to 0} \frac{\sin x}{x} = 1.$$

Note 1: If three positive numbers are replaced by their reciprocals, the inequalities are reversed. For example $2 < 3 < 4$, but $\frac{1}{2} > \frac{1}{3} > \frac{1}{4}$, or $\frac{1}{4} < \frac{1}{3} < \frac{1}{2}$ (see Problem 25, Appendix A.1).

Note 2: These must be right-hand limits, because of the restriction imposed on x, namely, $0 < x < \pi/2$.

Note 3: $\lim_{x \to 0^-} (\sin x)/x$ means the limit of $(\sin x)/x$ as x approaches zero, with x remaining negative throughout. The result is the same no matter what symbol is used in place of x. In particular, if x is replaced by $-x$, we have the limit of $\sin(-x)/(-x)$ as $-x$ approaches zero with $-x$ remaining negative throughout. But this means that x approaches zero with x remaining positive throughout, or

$$\lim_{x \to 0^+} \frac{\sin(-x)}{-x}.$$

Theorem 11.2

$$\lim_{x \to 0} \frac{\cos x - 1}{x} = 0.$$

Proof

Again, both the numerator and denominator are approaching zero. Let us use the result of Theorem 11.1 to evaluate this limit.

$$\lim_{x \to 0} \frac{\cos x - 1}{x} = \lim_{x \to 0} \frac{\cos^2 x - 1}{x(\cos x + 1)}$$

$$= \lim_{x \to 0} \frac{-\sin^2 x}{x(\cos x + 1)}$$

$$= \lim_{x \to 0} \frac{\sin x}{x} \cdot \frac{-\sin x}{\cos x + 1}$$

$$= \left(\lim_{x \to 0} \frac{\sin x}{x}\right)\left(\lim_{x \to 0} \frac{-\sin x}{\cos x + 1}\right)$$

$$= 1 \cdot \frac{0}{2} = 0.$$

Before using these two theorems, let us recall once more that they are true whether we use the symbol x or any other. In particular, if x is replaced by h, we have

$$\lim_{h \to 0} \frac{\sin h}{h} = 1 \quad \text{and} \quad \lim_{h \to 0} \frac{\cos h - 1}{h} = 0.$$

Now let us make use of these limits.

Theorem 11.3

If $f(x) = \sin x$, then $f'(x) = \cos x$.

Proof

$$f'(x) = \lim_{h \to 0} \frac{f(x+h) - f(x)}{h}$$

$$= \lim_{h \to 0} \frac{\sin (x+h) - \sin x}{h}$$

$$= \lim_{h \to 0} \frac{\sin x \cos h + \cos x \sin h - \sin x}{h}$$

$$= \lim_{h \to 0} \left(\sin x \frac{\cos h - 1}{h} + \cos x \frac{\sin h}{h} \right)$$

$$= \left(\lim_{h \to 0} \sin x \right) \left(\lim_{h \to 0} \frac{\cos h - 1}{h} \right) + \left(\lim_{h \to 0} \cos x \right) \left(\lim_{h \to 0} \frac{\sin h}{h} \right)$$

$$= \sin x \cdot 0 + \cos x \cdot 1 = \cos x.$$

Theorem 11.4

If $f(x) = \sin u(x)$, then $f'(x) = \cos u(x) \cdot u'(x)$.

Proof

This follows directly from Theorem 11.3 and the chain rule (Theorem 4.9). Letting $y = f(x)$ and abbreviating $u(x)$ to u, we get

$$\frac{dy}{dx} = \frac{dy}{du} \cdot \frac{du}{dx} = \cos u \cdot \frac{du}{dx}.$$

The last step uses Theorem 11.3 with the x replaced by u.

Theorem 11.5

If $f(x) = \cos u(x)$, then $f'(x) = -\sin u(x) \cdot u'(x)$.

Proof

Again abbreviating $u(x)$ to u, we have

$$f(x) = \cos u = \sin (u + \pi/2),$$
$$f'(x) = \cos (u + \pi/2) \cdot u'$$
$$= -\sin u \cdot u'.$$

Theorem 11.6

If $f(x) = \tan u(x)$, then $f'(x) = \sec^2 u(x) \cdot u'(x)$.

Proof

$f(x) = \tan u = \sin u/\cos u$. By the quotient rule,

$$f'(x) = \frac{(\cos u)(\cos u \cdot u') - (\sin u)(-\sin u \cdot u')}{\cos^2 u}$$

$$= \frac{(\sin^2 u + \cos^2 u)u'}{\cos^2 u}$$

$$= \frac{1}{\cos^2 u} \cdot u' = \sec^2 u \cdot u'.$$

Theorem 11.7

If $f(x) = \cot u(x)$, then $f'(x) = -\csc^2 u(x) \cdot u'(x)$.

Theorem 11.8

If $f(x) = \sec u(x)$, then $f'(x) = \sec u(x) \tan u(x) \cdot u'(x)$.

Theorem 11.9

If $f(x) = \csc u(x)$, then $f'(x) = -\csc u(x) \cot u(x) \cdot u'(x)$.

The proofs of these are similar to the proof of Theorem 11.6 and are left to the student. Summarizing the results of these theorems, we have:

$$\frac{d}{dx} \sin u = \cos u \cdot \frac{du}{dx}, \qquad \frac{d}{dx} \cos u = -\sin u \cdot \frac{du}{dx},$$

$$\frac{d}{dx} \tan u = \sec^2 u \cdot \frac{du}{dx}, \qquad \frac{d}{dx} \cot u = -\csc^2 u \cdot \frac{du}{dx},$$

$$\frac{d}{dx} \sec u = \sec u \tan u \cdot \frac{du}{dx}, \qquad \frac{d}{dx} \csc u = -\csc u \cot u \cdot \frac{du}{dx}.$$

You are urged to note certain similarities and differences that will make these formulas easier to remember.

Example 1

Differentiate $y = \sin x^2$.

$y' = 2x \cos x^2.$

Example 2

Differentiate $y = \sec (2x + 1)$.

$y' = 2 \sec (2x + 1) \tan (2x + 1).$

Example 3

Differentiate $y = \sin^2 x$.

First of all, let us compare this with Example 1 and note the difference. The expression $\sin x^2$ indicates that we are to square x and take the sine of the result. On the other hand, $\sin^2 x$, which may also be written $(\sin x)^2$, indicates that we are to take the sine of x and square the result. In order to differentiate $\sin^2 x$, we must use the power rule.

$$y' = 2 \sin x \cdot \frac{d}{dx} \sin x$$

$$= 2 \sin x \cos x.$$

Example 4

Differentiate $y = \dfrac{\sin x}{1 + \cos x}$.

$$y' = \frac{(1 + \cos x)(\cos x) - (\sin x)(-\sin x)}{(1 + \cos x)^2}$$

$$= \frac{\cos^2 x + \cos x + \sin^2 x}{(1 + \cos x)^2}$$

$$= \frac{1 + \cos x}{(1 + \cos x)^2}$$

$$= \frac{1}{1 + \cos x}.$$

Example 5

Differentiate $y = \sqrt{\tan 2x}$.

$$y' = \frac{1}{2} (\tan 2x)^{-1/2} \sec^2 2x \cdot 2$$

$$= \frac{\sec^2 2x}{\sqrt{\tan 2x}}.$$

Example 6

Find an equation of the line tangent to $y = 3(1 - \sin 2x)$ at $x = \pi$.

Since $x = \pi$ yields $y = 3(1 - \sin 2\pi) = 3$, the point of tangency is $(\pi, 3)$. We differentiate to find the slope.

$$y' = -6 \cos 2x.$$

At $x = \pi$,

$$y' = -6 \cos 2\pi = -6.$$

Thus the tangent line is

$$y - 3 = -6(x - \pi)$$
$$6x + y = 3(1 + 2\pi).$$

Example 7

Given $y = \sin x + \cos x$, find all maxima, minima, and points of inflection; sketch the curve.

$$y' = \cos x - \sin x, \quad \text{and} \quad y'' = -\sin x - \cos x.$$

For critical values we set $y' = 0$.

$$\cos x - \sin x = 0,$$

$$\tan x = \frac{\sin x}{\cos x} = 1,$$

$$x = \frac{\pi}{4} + n\pi \quad (n \text{ an integer}).$$

When n is even,

$$y = \frac{1}{\sqrt{2}} + \frac{1}{\sqrt{2}} = \sqrt{2},$$

$$y'' = -\frac{1}{\sqrt{2}} - \frac{1}{\sqrt{2}} = -\sqrt{2} < 0.$$

Thus $(\pi/4 + 2n\pi, \sqrt{2})$ are relative maxima. When n is odd

$$y = -\frac{1}{\sqrt{2}} - \frac{1}{\sqrt{2}} = -\sqrt{2},$$

$$y'' = \frac{1}{\sqrt{2}} + \frac{1}{\sqrt{2}} = \sqrt{2} > 0.$$

Thus $(\pi/4 + (2n+1)\pi, -\sqrt{2})$ are relative minima. Since the given function is bounded and periodic, the relative maxima and minima are also absolute maxima and minima.

For points of inflection we set $y'' = 0$.

$$-\sin x - \cos x = 0,$$

$$\tan x = \frac{\sin x}{\cos x} = -1,$$

$$x = \frac{3\pi}{4} + n\pi \quad (n \text{ an integer}).$$

For n even,

$$y = \frac{1}{\sqrt{2}} - \frac{1}{\sqrt{2}} = 0;$$

for n odd,

$$y = -\frac{1}{\sqrt{2}} + \frac{1}{\sqrt{2}} = 0.$$

Furthermore, we found when testing for maxima and minima that the values of y'' were alternately negative and positive at $x = \pi/4$, $5\pi/4$, $9\pi/4$, and so on, assuring us that y'' does change sign at $x = 3\pi/4$, $7\pi/4$, and so forth. It follows that $(3\pi/4 + n\pi, 0)$ are points of inflection.

Sketching by addition of ordinates (with the maxima, minima, and points of inflection now accurately determined), we have the curve shown in Figure 11.15.

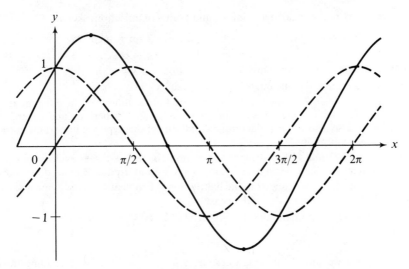

Figure 11.15

Problems

A

In Problems 1–18, differentiate and simplify.

1. $y = \sin 2x$.
2. $y = \tan x^2$.
3. $y = \cos 3x$.
4. $y = \sec (x^2 + 2)$.
5. $y = \sin^2 3x$.
6. $y = \cos^3 x^2$.
7. $y = (\sec 4x)^{3/2}$.
8. $y = \cot \sqrt{x}$.
9. $y = \sec x \tan x$.
10. $y = \sec^2 x - \tan^2 x$.
11. $y = (\sin x - \cos x)^2$.
12. $y = \dfrac{2 \cos x}{\sin 2x}$.
13. $y = \csc^2 x - \cot^2 x$.
14. $y = (\csc x + \cot x)^2$.
15. $y = x + \sec x$.
16. $y = \dfrac{\tan x}{1 + x^2}$.
17. $y = \dfrac{\tan 2x}{2x}$.
18. $y = x^2 \sin x$.

In Problems 19–22, find the derivative for the given value of x.

19. $y = \sin 2x + \cos x$, $x = \dfrac{\pi}{4}$.
20. $y = \sec^2 x + \tan^2 x$, $x = \dfrac{5\pi}{6}$.
21. $y = \sqrt{\sin 2x}$, $x = \pi$.
22. $y = \dfrac{\sin x}{1 + \cos x}$, $x = \dfrac{3\pi}{4}$.

B

In Problems 23–26, differentiate and simplify.

23. $y = \dfrac{\sin^2 x}{(1 - \cos x)^2}$.
24. $y = \dfrac{\sin x + 2 \cos x}{\sin x - 2 \cos x}$.
25. $y = \cos (\cos x)$.
26. $y = \sec (\sin x)$.

In Problems 27–32, find dy/dx.

27. $\sin x = \tan y$.
28. $(\sin x + \cos y)^2 = 1$.
29. $\tan (x + y) = y$.
30. $\sec (x + y) = \tan x - \tan y$.
31. $x + y = \cot (x - y)$.
32. $\sin (x + y) = (x - y)^2$.

In Problems 33–36, find all maxima, minima, and points of inflection; sketch.

33. $y = 2\sqrt{3}\sin x + \cos 2x$.

34. $y = 2\sin x + \cos 2x$.

35. $y = \sin x + \sqrt{2}\cos x$.

36. $y = x - \cos x$.

37. Find an equation of the line tangent to $y = \sin x$ at $x = \pi/3$.

38. Find an equation of the line tangent to $y = \tan^3 x$ at $x = \pi/4$.

39. Verify that $y = \sin 2x$ and $y = \cos 2x$ are both solutions of the differential equation $y'' + 4y = 0$. Verify that $y = c_1 \sin 2x + c_2 \cos 2x$, where c_1 and c_2 are arbitrary constants, is also a solution. (Differential equations of this type are important in the theory of undamped vibrations.)

40. A lighthouse is 2 miles off a straight shore. Its light makes three revolutions per minute. How fast does the light beam move along a sea wall at a point 2 miles down the coast?

41. How fast does the light beam from the lighthouse of Problem 40 move along a sea wall at a point 3.464 $(= 2\sqrt{3})$ miles down the coast?

42. Use differentials to estimate $\sin (11\pi/60)$. (*Hint:* $11\pi/60 = \pi/6 + \pi/60$.)

C

43. In Example 7 we solved the equation $\cos x - \sin x = 0$ by first dividing through by $\cos x$. We sometimes lose roots when dividing through by an expression that is zero for some value of x. Why is that not a problem here?

44. Sketch $y = \sin (1/x)$. (*Hint:* Find y for $x = 1/(\pi/2),\ 1/\pi,\ 1/(3\pi/2),\ 1/2\pi,\ \ldots$.)   Find y'. What happens to y and y' as $x \to 0$?

45. Repeat Problem 44 for $y = x \sin (1/x)$.

46. Repeat Problem 44 for $y = x^2 \sin (1/x)$.

11.3

Integrals Involving Trigonometric Functions

Since an indefinite integral is an antiderivative, there is an integral formula corresponding to every derivative formula. Corresponding to Theorems 11.4–11.9, we have the formulas of the following theorem.

Theorem 11.10

If u is a function of x, then

$$\int \sin u \cdot u'\, dx = \int \sin u\, du = -\cos u + C,$$

$$\int \cos u \cdot u'\, dx = \int \cos u\, du = \sin u + C,$$

$$\int \sec^2 u \cdot u'\, dx = \int \sec^2 u\, du = \tan u + C,$$

$$\int \csc^2 u \cdot u'\, dx = \int \csc^2 u\, du = -\cot u + C,$$

$$\int \sec u \tan u \cdot u' \, dx = \int \sec u \tan u \, du = \sec u + C,$$

$$\int \csc u \cot u \cdot u' \, dx = \int \csc u \cot u \, du = -\csc u + C.$$

For example, since

$$\frac{d}{dx} \cot u(x) = -\csc^2 u(x) \cdot u'(x),$$

it follows that

$$\int -\csc^2 u(x) \cdot u'(x) \, dx = \cot u(x) + C$$

or

$$\int \csc^2 u(x) \cdot u'(x) \, dx = -\cot u(x) + C.$$

Again, let us recall that $du = u' \, dx$. This substitution may be made in all of these formulas if desired.

Example 1

Evaluate $\int 2 \sin 2x \, dx$.

With $u = 2x$, $u' = 2$ and the integral is in the form $\int \sin u \cdot u' \, dx$. Thus

$$\int 2 \sin 2x \, dx = -\cos 2x + C.$$

Example 2

Evaluate $\int x \cos x^2 \, dx$.

In this case $u = x^2$ and $u' = 2x$. Since we do not have the factor 2, we must adjust the constant in order to get it.

$$\int x \cos x^2 \, dx = \frac{1}{2} \int 2x \cos x^2 \, dx$$

$$= \frac{1}{2} \sin x^2 + C.$$

Example 3

Evaluate $\int \sec (2x - 1) \tan (2x - 1) \, dx$.

$$\int \sec (2x - 1) \tan (2x - 1) \, dx = \frac{1}{2} \int \sec (2x - 1) \tan (2x - 1) \cdot 2 \, dx$$

$$= \frac{1}{2} \sec (2x - 1) + C.$$

Example 4

Evaluate $\int \tan^2 x \, dx$.

None of our formulas allows us to evaluate this integral directly. However, we can put it into a form we can evaluate by using the identity $\tan^2 x = \sec^2 x - 1$.

$$\int \tan^2 x \, dx = \int (\sec^2 x - 1) \, dx = \tan x - x + C.$$

Example 5

Evaluate $\int \sin^2 x \cos x \, dx$.

Again, this does not fit any of the forms of Theorem 11.10. However, it is in the form $\int u^2 \cdot u' \, dx$, where $u = \sin x$. Thus,

$$\int \sin^2 x \cos x \, dx = \frac{\sin^3 x}{3} + C.$$

Example 6

Evaluate $\int \frac{\sin x}{\cos^2 x} \, dx$.

This integral may be evaluated either by the use of identities, as in Example 4, or by use of the power rule, as in Example 5. Let us use both methods.

$$\int \frac{\sin x}{\cos^2 x} \, dx = \int \frac{1}{\cos x} \cdot \frac{\sin x}{\cos x} \, dx$$

$$= \int \sec x \tan x \, dx$$

$$= \sec x + C.$$

$$\int \frac{\sin x}{\cos^2 x} \, dx = -\int (\cos x)^{-2}(-\sin x) \, dx$$

$$= -\frac{-1}{\cos x} + C$$

$$= \sec x + C.$$

The identities

$$\sin^2 x = \frac{1}{2}(1 - \cos 2x),$$

$$\cos^2 x = \frac{1}{2}(1 + \cos 2x)$$

are very useful in integrating even powers of $\sin x$ and $\cos x$. We shall consider them in more detail in Section 15.3. The following example illustrates their use.

Example 7

Evaluate $\int \sin^2 x\, dx$.

$$\int \sin^2 x\, dx = \frac{1}{2} \int (1 - \cos 2x)\, dx$$

$$= \frac{1}{2} \int \left(1 - \frac{1}{2} \cdot 2 \cos 2x\right) dx$$

$$= \frac{1}{2} \left(x - \frac{1}{2} \sin 2x\right) + C$$

$$= \frac{1}{4} (2x - \sin 2x) + C.$$

Example 8

Find the area of the region bounded by $y = \sin x$ and $y = \cos x$ between $x = 0$ and $x = \pi/2$.

The two curves are given in Figure 11.16. By symmetry, the portions to the left and right of $x = \pi/4$ have equal area. The area we want is twice the area from $x = 0$ to $x = \pi/4$.

$$A = 2 \int_0^{\pi/4} (y_1 - y_2)\, dx$$

$$= 2 \int_0^{\pi/4} (\cos x - \sin x)\, dx$$

$$= 2(\sin x + \cos x) \Big|_0^{\pi/4}$$

$$= 2 \left[\left(\frac{1}{\sqrt{2}} + \frac{1}{\sqrt{2}}\right) - (0 + 1) \right] = 2(\sqrt{2} - 1).$$

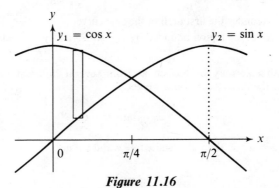

Figure 11.16

Problems

A

Evaluate the integrals in Problems 1–10.

1. $\int \sin 2\theta\, d\theta$.

2. $\int \cos (3x + 2)\, dx$.

3. $\int \csc 4x \cot 4x\, dx$.

4. $\int \sec^2 3u\, du$.

5. $\displaystyle\int x^2 \sec x^3 \tan x^3 \, dx.$ 6. $\displaystyle\int x \sec^2 3x^2 \, dx.$

7. $\displaystyle\int_0^{\pi/2} \sin \frac{x}{2} \, dx.$ 8. $\displaystyle\int_0^{\pi/6} \sec x \tan x \, dx.$

9. $\displaystyle\int_{-\pi/4}^{\pi/4} \tan^2 x \, dx.$ 10. $\displaystyle\int_{-\pi}^{\pi} \sin \theta \cos \theta \, d\theta.$

B

Evaluate the integrals in Problems 11–28.

11. $\displaystyle\int (\sin x - 2 \cos x)^2 \, dx.$ 12. $\displaystyle\int \tan^2 5x \, dx.$

13. $\displaystyle\int (\sin^2 x - \cos^2 x) \, dx.$ 14. $\displaystyle\int \cot^2 3x \, dx.$

15. $\displaystyle\int \sin^4 3\theta \cos 3\theta \, d\theta.$ 16. $\displaystyle\int \frac{\sec^2 2x}{\tan^3 2x} \, dx.$

17. $\displaystyle\int \frac{\tan x}{\sec^3 x} \, dx.$ 18. $\displaystyle\int \csc^3 2x \cot 2x \, dx.$

19. $\displaystyle\int (\sec x + 2 \tan x)^2 \, dx.$ 20. $\displaystyle\int (\csc x - \cot x)^2 \, dx.$

21. $\displaystyle\int \cos^2 \theta \, d\theta.$ 22. $\displaystyle\int \sin^2 4x \, dx.$

23. $\displaystyle\int (3 - \csc x \cot x)^2 \, dx.$ 24. $\displaystyle\int (\sin x)^{3/5} \cos x \, dx.$

25. $\displaystyle\int \sqrt{\cot x} \csc^2 x \, dx.$ 26. $\displaystyle\int \sec \theta \, (\sec \theta - \tan \theta) \, d\theta.$

27. $\displaystyle\int_0^{2\pi} (1 - \sin x)^2 \, dx.$ 28. $\displaystyle\int_{\pi/4}^{\pi/2} \frac{dx}{\sin^2 x}.$

29. Find the area under the first arch of the sine curve.
30. Find the area of the region bounded by $y = \sin 2x$ and $y = 0$ between $x = 0$ and $x = \pi/2$.

C

31. $\int \sec^2 x \tan x \, dx$ may be looked at from several different points of view. If we let $u = \tan x$, then $u' = \sec^2 x$ and we get

$$\frac{1}{2} \tan^2 x + C.$$

If we let $u = \sec x$, then $u' = \sec x \tan x$ and we get

$$\frac{1}{2} \sec^2 x + C.$$

Can both answers be correct? Explain.

32. Show that

$$\int_{-\pi}^{\pi} \sin x \, dx = \int_0^{2\pi} \sin x \, dx = \int_{\pi}^{3\pi} \sin x \, dx.$$

Prove that

$$\int_a^{a+2\pi} \sin x \, dx = \int_0^{2\pi} \sin x \, dx$$

for any real number a. Can this result be generalized?

11.4

Inverse Trigonometric Functions

Suppose we have the equation $y = \sin x$ and want to express x in terms of y. To do so, we introduce a new notation for the solution,

$$x = \arcsin y \quad \text{or} \quad x = \sin^{-1} y.$$

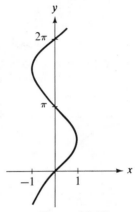

This is read: x is an inverse of sine y, or x is an angle whose sine is y. Since we take the sine of a number rather than of an angle, the former is the preferable way of reading it. Thus $x = \arcsin y$ is equivalent to $y = \sin x$, or $y = \arcsin x$ is equivalent to $x = \sin y$. To graph $y = \arcsin x$, we merely graph $x = \sin y$ (see Figure 11.17). This looks exactly like the graph of $y = \sin x$ with the x and y reversed. Note that arcsin x is *not* a function, since one value of x gives many values of arcsin x. The remaining five trigonometric functions have inverses that are defined analogously.

Figure 11.17

Example 1

Sketch $y = 2 \arcsin 3x$.

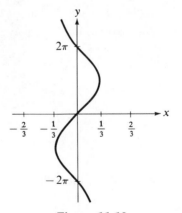

We first convert this to the equivalent equation involving the sine.

$$\frac{y}{2} = \arcsin 3x$$

$$3x = \sin \frac{y}{2}$$

$$x = \frac{1}{3} \sin \frac{y}{2}.$$

The sketch is shown in Figure 11.18.

Figure 11.18

To make the inverses functions, we make them single-valued by restricting their ranges. Figure 11.19 gives the graphs of the principal inverse trigonometric functions together with their domains and ranges. The reason for restricting the ranges as we have will be apparent when we consider derivatives of the functions.

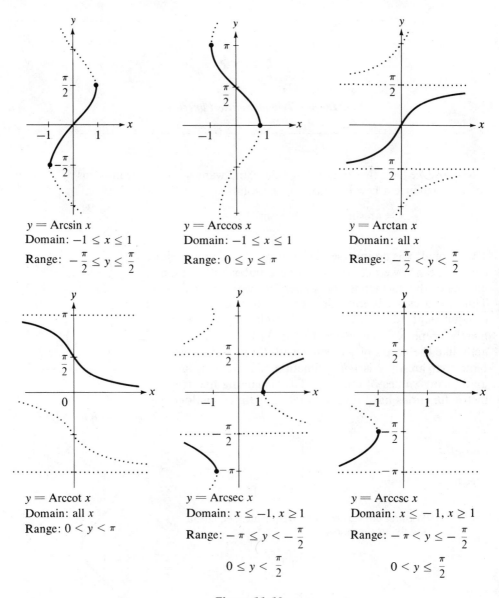

$y = \text{Arcsin } x$
Domain: $-1 \leq x \leq 1$
Range: $-\dfrac{\pi}{2} \leq y \leq \dfrac{\pi}{2}$

$y = \text{Arccos } x$
Domain: $-1 \leq x \leq 1$
Range: $0 \leq y \leq \pi$

$y = \text{Arctan } x$
Domain: all x
Range: $-\dfrac{\pi}{2} < y < \dfrac{\pi}{2}$

$y = \text{Arccot } x$
Domain: all x
Range: $0 < y < \pi$

$y = \text{Arcsec } x$
Domain: $x \leq -1, x \geq 1$
Range: $-\pi \leq y < -\dfrac{\pi}{2}$
$0 \leq y < \dfrac{\pi}{2}$

$y = \text{Arccsc } x$
Domain: $x \leq -1, x \geq 1$
Range: $-\pi < y \leq -\dfrac{\pi}{2}$
$0 < y \leq \dfrac{\pi}{2}$

Figure 11.19

We shall use the following notation.

$$y = \arcsin x, \quad \text{or} \quad y = \sin^{-1} x$$

represents the inverse sine *relation* (not function) with domain $-1 \leq x \leq 1$ and unrestricted range.

$$y = \text{Arcsin } x, \quad \text{or} \quad y = \text{Sin}^{-1} x$$

represents the principal inverse sine *function* with domain $-1 \leq x \leq 1$ and range $-\pi/2 \leq y \leq \pi/2$. This same type of notation is used for the other five inverse trigonometric functions. Now that we are dealing with functions, we can proceed to find the derivatives.

Theorem 11.11

If $f(x) = \text{Arcsin } u(x),$ *then* $f'(x) = \dfrac{u'(x)}{\sqrt{1 - [u(x)]^2}}.$

Proof

Let us abbreviate $u(x)$ to u and replace $f(x)$ by y, remembering that u and y are both functions of x. Now we have

$$y = \text{Arcsin } u,$$

or

$$u = \sin y, \quad \text{where} \quad -\pi/2 \le y \le \pi/2.$$

Differentiating this implicit function, we have

$$u' = \cos y \cdot y', \qquad y' = \frac{u'}{\cos y}.$$

Now since $\sin^2 y + \cos^2 y = 1,$

$$\cos y = \pm\sqrt{1 - \sin^2 y}$$
$$= \pm\sqrt{1 - u^2}.$$

Remembering that $-\pi/2 \le y \le \pi/2$, we see that $\cos y$ must be positive. Thus we may drop the $\pm$. Substituting this value of $\cos y$ back into the derivative, we have

$$y' = \frac{u'}{\sqrt{1 - u^2}}.$$

Theorem 11.12

If $f(x) = \text{Arccos } u(x),$ *then* $f'(x) = \dfrac{-u'(x)}{\sqrt{1 - [u(x)]^2}}.$

The proof, which is similar to that of Theorem 11.11, is left to the student.

Theorem 11.13

If $f(x) = \text{Arctan } u(x),$ *then* $f'(x) = \dfrac{u'(x)}{1 + [u(x)]^2}.$

Proof

Again, using u and y for $u(x)$ and $f(x)$, respectively, we have

$$y = \text{Arctan } u,$$
$$u = \tan y, \quad \text{where} \quad -\pi/2 < y < \pi/2,$$
$$u' = \sec^2 y \cdot y',$$
$$y' = \frac{u'}{\sec^2 y}$$
$$= \frac{u'}{1 + \tan^2 y}$$
$$= \frac{u'}{1 + u^2}.$$

Theorem 11.14

If $f(x) = \text{Arccot } u(x)$, then $f'(x) = \dfrac{-u'(x)}{1 + [u(x)]^2}$.

The proof is similar to that of Theorem 11.13.

Theorem 11.15

If $f(x) = \text{Arcsec } u(x)$, then $f'(x) = \dfrac{u'(x)}{u(x)\sqrt{[u(x)]^2 - 1}}$.

Proof

Again, using u and y for $u(x)$ and $f(x)$, we have $y = \text{Arcsec } u$.

$$u = \sec y, \quad \text{where } -\pi \le y < -\pi/2 \text{ or } 0 \le y < \pi/2,$$

$$u' = \sec y \tan y \cdot y',$$

$$y' = \frac{u'}{\sec y \tan y}.$$

Now $\sec y = u$ and, from the identity $1 + \tan^2 y = \sec^2 y$, we get

$$\tan y = \pm\sqrt{\sec^2 y - 1}$$

$$= \pm\sqrt{u^2 - 1}.$$

Since y is either in the first or third quadrant, $\tan y$ is positive and we may again drop the $\pm$, obtaining

$$y' = \frac{u'}{u\sqrt{u^2 - 1}}.$$

Theorem 11.16

If $f(x) = \text{Arccsc } u(x)$, then $f'(x) = \dfrac{-u'(x)}{u(x)\sqrt{[u(x)]^2 - 1}}$.

The proof, similar to that of Theorem 11.15, is left to the student. Summing up the results of Theorems 11.11–11.16, we have

$$\frac{d}{dx}\text{Arcsin } u = \frac{u'}{\sqrt{1 - u^2}}, \qquad \frac{d}{dx}\text{Arccos } u = \frac{-u'}{\sqrt{1 - u^2}},$$

$$\frac{d}{dx}\text{Arctan } u = \frac{u'}{1 + u^2}, \qquad \frac{d}{dx}\text{Arccot } u = \frac{-u'}{1 + u^2},$$

$$\frac{d}{dx}\text{Arcsec } u = \frac{u'}{u\sqrt{u^2 - 1}}, \qquad \frac{d}{dx}\text{Arccsc } u = \frac{-u'}{u\sqrt{u^2 - 1}}.$$

Example 2

Differentiate $y = \text{Arcsin } x^2$.

$$y' = \frac{2x}{\sqrt{1 - x^4}}.$$

Example 3

Differentiate $y = \text{Arctan } \dfrac{x}{a}$.

$$y' = \frac{\dfrac{1}{a}}{1 + \dfrac{x^2}{a^2}} = \frac{a}{a^2 + x^2}.$$

Example 4

Differentiate $y = x \text{ Arcsec } x$.

$$y' = x \frac{1}{x\sqrt{x^2 - 1}} + \text{Arcsec } x = \frac{1}{\sqrt{x^2 - 1}} + \text{Arcsec } x.$$

Expressions involving trigonometric and inverse trigonometric functions can sometimes be simplified. Suppose, for example, we have

$$\sin (\text{Arcsin } x).$$

Let us make the substitution $\theta = \text{Arcsin } x$. Then we have

$$\sin (\text{Arcsin } x) = \sin \theta.$$

But $\theta = \text{Arcsin } x$ is equivalent to $\sin \theta = x$. Thus

$$\sin (\text{Arcsin } x) = x.$$

This is self-evident if we read the left-hand side, "the sine of the principal angle whose sine is x."

Similarly,

$$\cos (\text{Arcos } x) = x, \qquad \tan (\text{Arctan } x) = x,$$

and so forth. However,

$$\text{Arcsin } (\sin x) \neq x,$$

because we have the *principal* inverse sine, whose value is restricted to $[-\pi/2, \pi/2]$ (see Figure 11.19). If $x = \pi$, then

$$\text{Arcsin } (\sin \pi) = \text{Arcsin } 0 = 0 \neq \pi.$$

But if x is restricted to $-\pi/2 \le x \le \pi/2$, then Arcsin (sin x) = x. Similar arguments hold for the other functions, giving

$$\text{Arcsin (sin } x) = x \quad \text{if } -\pi/2 \le x \le \pi/2,$$
$$\text{Arccos (cos } x) = x \quad \text{if } 0 \le x \le \pi,$$
$$\text{Arctan (tan } x) = x \quad \text{if } -\pi/2 < x < \pi/2,$$

etc.

Finally, let us consider

$$\sin \text{ (Arccos } x).$$

Proceeding as we did with sin (Arcsin x), we have

$$\sin \text{ (Arccos } x) = \sin \theta, \quad \text{where } \cos \theta = x \quad (0 \le \theta \le \pi).$$

Since $\sin^2 \theta + \cos^2 \theta = 1$,

$$\sin \theta = \pm\sqrt{1 - \cos^2 \theta} = \pm\sqrt{1 - x^2}.$$

But $\sin \theta \ge 0$, since $0 \le \theta \le \pi$. Thus we may drop the $\pm$, and

$$\sin \text{ (Arccos } x) = \sqrt{1 - x^2}.$$

Example 5

Differentiate $y = \sin$ (Arccos x).

We can either simplify this expression and then take the derivative or take the derivative immediately. Differentiating the given expression, we have

$$y' = \cos \text{ (Arccos } x) \frac{-1}{\sqrt{1 - x^2}}$$

$$= x \frac{-1}{\sqrt{1 - x^2}}$$

$$= \frac{-x}{\sqrt{1 - x^2}}.$$

Simplifying first, we have (from above)

$$y = \sqrt{1 - x^2}.$$

Differentiating this expression, we get the same result as before.

Example 6

Find an equation of the line tangent to $y = $ Arcsin x at $x = 1/2$.

Since Arcsin $1/2 = \pi/6$, the point on the curve is $(1/2, \pi/6)$. Now

$$y' = \frac{1}{\sqrt{1 - x^2}}.$$

At $x = 1/2$,

$$y' = \frac{1}{\sqrt{1 - 1/4}} = \frac{2}{\sqrt{3}}.$$

The tangent line is

$$y - \frac{\pi}{6} = \frac{2}{\sqrt{3}} \left(x - \frac{1}{2} \right),$$

$$12x - 6\sqrt{3}y = 6 - \sqrt{3}\pi.$$

Problems

A

In Problems 1–4, give the values.

1. Arcsin 0.
2. Arctan (-1).
3. Arccos (cos π).
4. Arcsin (sin π).

In Problems 5–16, differentiate and simplify.

5. $y = $ Arcsin $2x$.
6. $y = $ Arctan x^3.
7. $y = $ Arccsc $4x$.
8. $y = $ Arccos $(3x + 2)$.
9. $y = $ Arccot $(-2x)$.
10. $y = $ Arcsec $\sqrt{x}$.
11. $y = \sqrt{1 - x^2} + $ Arcsin x.
12. $y = \sqrt{1 - x^2} + $ Arccos x.
13. $y = x - $ Arctan x.
14. $y = \sqrt{x^2 - 1} + $ Arccsc x.
15. $y = $ Arcsin^2 x.
16. $y = x$ Arcsin x^2.

In Problems 17–20, find the derivative at the given value of x.

17. $y = $ Arcsin $2x$, $x = 1/2$.
18. $y = $ Arcsec x, $x = \sqrt{3}$.
19. $y = $ Arccos x^2, $x = 1/\sqrt{2}$.
20. $y = x^2$ Arctan x, $x = 1$.

B

In Problems 21–28, differentiate and simplify.

21. $y = $ Arcsin $x - x\sqrt{1 - x^2}$.

22. $y = -\frac{1}{8}[x\sqrt{1 - x^2}(2x^2 - 5) + 3$ Arccos $x]$.

23. $y = $ cot (Arcsin x).
24. $y = $ cos (Arctan x).

25. $y = $ Arcsin $\frac{1}{x}$.
26. $y = $ Arctan $\frac{1}{x}$.

27. $y = \dfrac{x}{\sqrt{1 - x^2}} - $ Arcsin x.
28. $y = x$ Arcsin $x + \sqrt{1 - x^2}$.

In Problems 29–32, find an equation of the line tangent to the given curve at the given point.

29. $y = $ Arctan x, $(-1, -\pi/4)$.
30. $y = x + $ Arccos x, $(1, 1)$.

31. $y = \dfrac{\text{Arctan } x}{x}$, $x = 1$.
32. $y = $ Arcsin^2 x, $x = 1/2$.

In Problems 33–36, sketch.

33. $y = 4$ arcsin x.
34. $y = 2$ arccos $\dfrac{x}{3}$.

35. $y = \dfrac{1}{4}$ arcsec $\dfrac{x}{2}$.
36. $y = -$arcsin $2x$.

In Problems 37 and 38, find all relative maxima and minima, and sketch.

37. $y = \text{Arctan } x - x$.   38. $y = \text{Arccos } x^2$.

39. Differentiate $y = \text{Arcsin } x - \text{Arcsec } x$. Watch out! (*Hint*: What is the domain of the function?)

40. Differentiate $y = \text{Arcsin } x + \text{Arccos } x$ and integrate the result. Use this result to show that $y = \pi/2$.

41. Differentiate $y = \text{Arctan } x + \text{Arccot } x$ and integrate the result. Use this result to show that $y = \pi/2$.

42. Use differentials to estimate Arctan 1.1.

43. A picture 3 ft high is hung with the bottom 2 ft above eye level. How far back should a person stand so that the picture appears to be the largest? (*Hint*: The picture appears to be the largest when the angle θ of Figure 11.20 is a maximum.)

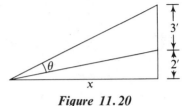

Figure 11.20

C

44. Solve for x: $\text{Arctan } x - \text{Arccot } x = \pi/4$.

11.5

Integrals Involving Inverse Trigonometric Functions

Again the derivative formulas of Theorems 11.11–11.16 may be stated in integral form.

Theorem 11.17

If u is a function of x, then

$$\int \frac{u' \, dx}{\sqrt{1 - u^2}} = \int \frac{du}{\sqrt{1 - u^2}} = \text{Arcsin } u + C,$$

$$\int \frac{u' \, dx}{1 + u^2} = \int \frac{du}{1 + u^2} = \text{Arctan } u + C,$$

$$\int \frac{u' \, dx}{u\sqrt{u^2 - 1}} = \int \frac{du}{u\sqrt{u^2 - 1}} = \text{Arcsec } u + C.$$

Of course, this theorem gives the integral form of only three of the six derivative formulas. Since the derivatives of the other three principal inverse functions are simply the negatives of the three we have here and since constants can be adjusted, there is no need to consider the other three. For instance,

$$\int \frac{-u' \, dx}{\sqrt{1 - u^2}} = \text{Arccos } u + C$$

or

$$\int \frac{-u'\, dx}{\sqrt{1-u^2}} = -\int \frac{u'\, dx}{\sqrt{1-u^2}} = -\text{Arcsin } u + C.$$

In Chapter 15 we shall consider a general method of integration whose use will make it unnecessary to memorize these integration formulas. However you might find it convenient to remember the first two of these, especially the second.

Example 1

Evaluate $\displaystyle\int \frac{dx}{\sqrt{1-2x^2}}.$

This is almost in the form

$$\int \frac{u'\, dx}{\sqrt{1-u^2}},$$

with $u = \sqrt{2}\,x$. Since $u' = \sqrt{2}$, we need to adjust the constant factor to put the expression into the above form. Thus

$$\int \frac{dx}{\sqrt{1-2x^2}} = \frac{1}{\sqrt{2}}\int \frac{\sqrt{2}\, dx}{\sqrt{1-(\sqrt{2}\,x)^2}} = \frac{1}{\sqrt{2}}\text{Arcsin }\sqrt{2}\,x + C.$$

Example 2

Evaluate $\displaystyle\int \frac{x\, dx}{\sqrt{1-x^2}}.$

This is *not* in any of the forms given in Theorem 11.17. But, since the derivative of $1-x^2$ is $-2x$, we can adjust the constant and use the power rule.

$$\int \frac{x\, dx}{\sqrt{1-x^2}} = -\frac{1}{2}\int (1-x^2)^{-1/2}(-2x)\, dx$$

$$= -\frac{1}{2}\frac{(1-x^2)^{1/2}}{1/2} + C$$

$$= -\sqrt{1-x^2} + C.$$

Example 3

Evaluate $\displaystyle\int \frac{dx}{4+x^2}.$

This again is not in any of the forms of Theorem 11.17, but this time it can be put into the form of one of them. The constant terms in all of the integrals of Theorem 11.17 are 1; thus, we need to get a 1 in place of the 4 we have. Let us divide both numerator and denominator by 4.

$$\int \frac{dx}{4+x^2} = \int \frac{\frac{1}{4}\, dx}{1+\frac{x^2}{4}}$$

$$= \frac{1}{2} \int \frac{\frac{1}{2} \, dx}{1 + \left(\frac{x}{2}\right)^2}$$

$$= \frac{1}{2} \operatorname{Arctan} \frac{x}{2} + C.$$

Example 4

Evaluate $\displaystyle\int \frac{dx}{x\sqrt{x^2 - 9}}$.

$$\int \frac{dx}{x\sqrt{x^2 - 9}} = \frac{1}{3} \int \frac{\frac{1}{3} \, dx}{\frac{x}{3} \sqrt{\left(\frac{x}{3}\right)^2 - 1}}$$

$$= \frac{1}{3} \operatorname{Arcsec} \frac{x}{3} + C.$$

Example 5

Evaluate $\displaystyle\int \frac{dx}{\sqrt{2x - x^2}}$.

There are two ways of evaluating this integral. The first is by completing the square under the radical.

$$\sqrt{2x - x^2} = \sqrt{-(x^2 - 2x)}$$

$$= \sqrt{-(x^2 - 2x + 1) + 1}$$

$$= \sqrt{1 - (x - 1)^2};$$

$$\int \frac{dx}{\sqrt{2x - x^2}} = \int \frac{dx}{\sqrt{1 - (x - 1)^2}}$$

$$= \operatorname{Arcsin}(x - 1) + C_1.$$

The other method is somewhat like that used in Example 3.

$$\int \frac{dx}{\sqrt{2x - x^2}} = \int \frac{\frac{1}{\sqrt{2x}} \, dx}{\sqrt{1 - \frac{x}{2}}}$$

$$= 2 \int \frac{\frac{1}{2\sqrt{2x}} \, dx}{\sqrt{1 - \left(\sqrt{\frac{x}{2}}\right)^2}}$$

$$= 2 \operatorname{Arcsin} \sqrt{\frac{x}{2}} + C_2.$$

Although the two results look quite different, we can see by differentiating that both are correct.

Example 6

Find the area of the region bounded by $y = 1/x\sqrt{4x^2 - 1}$ and the x axis between $x = 1$ and $x = 2$.

The graph is given in Figure 11.21. From this we see that the area is

$$A = \int_1^2 y\,dx = \int_1^2 \frac{dx}{x\sqrt{4x^2 - 1}} = \int_1^2 \frac{2\,dx}{2x\sqrt{(2x)^2 - 1}}$$

$$= \operatorname{Arcsec} 2x \,\Big|_1^2 = \operatorname{Arcsec} 4 - \operatorname{Arcsec} 2 = \operatorname{Arcsec} 4 - \pi/3.$$

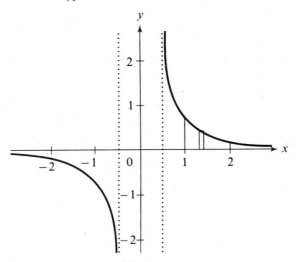

Figure 11.21

Problems

A

In Problems 1–14, evaluate the integral.

1. $\displaystyle\int \frac{-x\,dx}{\sqrt{1 - x^2}}.$

2. $\displaystyle\int \frac{-x\,dx}{\sqrt{1 - x^4}}.$

3. $\displaystyle\int \frac{x\,dx}{1 + x^4}.$

4. $\displaystyle\int \frac{dx}{x\sqrt{x^2 - 9}}.$

5. $\displaystyle\int \frac{dx}{\sqrt{1 - 9x^2}}.$

6. $\displaystyle\int \frac{x\,dx}{\sqrt{1 - 9x^2}}.$

7. $\displaystyle\int \frac{dx}{1 + 4x^2}.$

8. $\displaystyle\int \frac{dx}{1 + (x - 3)^2}.$

9. $\displaystyle\int \frac{dx}{\sqrt{1 - (2x + 1)^2}}.$

10. $\displaystyle\int \frac{dx}{(x + 4)\sqrt{(x + 4)^2 - 1}}.$

11. $\displaystyle\int_0^{1/2} \frac{dx}{\sqrt{1 - x^2}}.$

12. $\displaystyle\int_{-\sqrt{3}/2}^0 \frac{dx}{\sqrt{1 - x^2}}.$

13. $\displaystyle\int_0^1 \frac{dx}{1 + x^2}.$

14. $\displaystyle\int_0^{1/\sqrt{2}} \frac{x\,dx}{\sqrt{1 - x^4}}.$

B

In Problems 15–28, evaluate the integral.

15. $\displaystyle \int \frac{dx}{x^2 - 2x + 5}.$

16. $\displaystyle \int \frac{dx}{\sqrt{-x^2 + 8x - 15}}.$

17. $\displaystyle \int \frac{dx}{(x-2)\sqrt{x^2 - 4x + 3}}.$

18. $\displaystyle \int \frac{dx}{\sqrt{x - x^2}}.$

19. $\displaystyle \int \frac{dx}{2x^2 + 6x + 5}.$

20. $\displaystyle \int \frac{dx}{\sqrt{-3x^2 + 4x - 1}}.$

21. $\displaystyle \int \frac{\cos x \, dx}{1 + \sin^2 x}.$

22. $\displaystyle \int \frac{(1 + \tan^2 x) \, dx}{\sqrt{1 - \tan^2 x}}.$

23. $\displaystyle \int \frac{\sec^2 x \tan x \, dx}{\sqrt{1 - \tan^2 x}}.$

24. $\displaystyle \int \frac{\sec^2 x \, dx}{\tan x \sqrt{\tan^4 x - 1}}.$

25. $\displaystyle \int \frac{\text{Arctan } x}{1 + x^2} \, dx.$

26. $\displaystyle \int \frac{\text{Arcsin } x}{\sqrt{1 - x^2}} \, dx.$

27. $\displaystyle \int_{-3/2}^{-1} \frac{dx}{\sqrt{-x^2 - 2x}}.$

28. $\displaystyle \int_{0}^{\pi/4} \frac{\sin x \, dx}{1 + \cos^2 x}.$

29. Evaluate

$$\int \frac{u' \, dx}{a^2 + u^2} = \int \frac{du}{a^2 + u^2},$$

where a is a constant and u is a function of x. Use the result to evaluate the integral of Example 3.

30. Evaluate

$$\int \frac{u' \, dx}{\sqrt{a^2 - u^2}} = \int \frac{du}{\sqrt{a^2 - u^2}},$$

where a is a constant and u is a function of x.

31. Evaluate

$$\int \frac{u' \, dx}{u\sqrt{u^2 - a^2}} = \int \frac{du}{u\sqrt{u^2 - a^2}},$$

where a is a constant and u is a function of x.

In Problems 32–35, find the area of the given region.

32. The region bounded by $y = 1/(1 + x^2)$, $y = 0$, $x = -1$, and $x = 1$.
33. The region bounded by $y = 1/\sqrt{1 - x^2}$ and $y = 2$.
34. The region bounded by $y = 1/(4 + x^2)$ and $y = 1/8$.
35. The region bounded by $y = x/(1 + x^4)$, $y = 0$, and $x = 1$.

C

36. Show that the two results of Example 5 are equivalent by showing that

$$\text{Arcsin } (x - 1) - 2 \text{ Arcsin } \sqrt{\frac{x}{2}}$$

is a constant. (*Hint:* Show that the sine of this expression is a constant.)

37. Find the area of the region bounded by $y = 1/(1 + x^2)$, $y = 0$, $x = 0$, and $x = k$. What number does the area approach as $k \to +\infty$? What does this number represent?

Review Problems

A

In Problems 1–9, differentiate and simplify.

1. $y = \dfrac{1 + \sin x}{\cos x}$.

2. $y = \dfrac{\tan^5 x}{5} - \dfrac{\tan^3 x}{3}$.

3. $y = \tan^4 (x^2 + 1)$.

4. $y = \dfrac{3}{8} x - \dfrac{1}{4} \sin 2x + \dfrac{1}{32} \sin 4x$.

(Express the result in terms of $\sin x$.)

5. $y = 9x + 3 \cot 3x - \cot^3 3x$.

6. $y = \dfrac{1}{3} \operatorname{Arctan} \dfrac{x + 1}{3}$.

7. $y = \operatorname{Arcsin} (x - 1) + (x - 1)\sqrt{2x - x^2}$.

8. $y = \operatorname{Arcsec} x^2$.

9. $y = \operatorname{Arccos}^2 2x$.

In Problems 10 and 11, evaluate the integral.

10. $\displaystyle\int \sec 3x \tan 3x \, dx$.

11. $\displaystyle\int \sin^2 x \cos x \, dx$.

B

In Problems 12–17, evaluate the integral.

12. $\displaystyle\int \tan^2 4x \, dx$.

13. $\displaystyle\int \cos^{5/2} 3x \tan 3x \, dx$.

14. $\displaystyle\int \dfrac{x \, dx}{1 + 9x^4}$.

15. $\displaystyle\int \dfrac{x \, dx}{\sqrt{16 - x^2}}$.

16. $\displaystyle\int \dfrac{\operatorname{Arcsec} 2x}{x\sqrt{4x^2 - 1}} \, dx$.

17. $\displaystyle\int \dfrac{dx}{(x + 3)\sqrt{x^2 + 6x + 8}}$.

18. Find an equation of the line tangent to $y = \sin^2 x$ at $x = \pi/3$.
19. Find an equation of the line tangent to $y = \operatorname{Arccsc} \sqrt{5}x$ at $x = -2/\sqrt{10}$.
20. Find the area of the region bounded by $y = \sin 2x$ and $y = \tan^2 x$ between $x = 0$ and $x = \pi/4$.
21. Find the area of the region bounded by $y = 1/(1 + x^2)$, the x axis, and the vertical lines $x = 0$ and $x = 2$.
22. Find the maximum and minimum values of $y = a \sin x + b \cos x$. Show that it has a point of inflection whenever it crosses the x axis.
23. Use differentials to estimate Arcsin 0.51.

In Problems 24 and 25, find all relative maxima and minima and points of inflection; sketch.

24. $y = \sqrt{3} \cos x + \sin x$.

25. $y = \dfrac{x}{2} + \cos x$.

C

26. Use multiplication of ordinates to sketch $y = x \sin x$. Use this result and addition of ordinates to sketch $y = x \sin x + \cos x$ and find all relative maxima and minima.
27. At what rate is the angle between the minute and hour hands of a clock decreasing at 3 o'clock? If the minute hand is 4 in. long and the hour hand is 3 in. long, at what rate is the distance between their ends decreasing at 3 o'clock? (*Hint*: Use the law of cosines.)
28. A weight is dropped from the top of a building 500 ft high. The path of the weight is observed through a telescope atop another building that is 400 ft high and 200 ft away from the first building. Where is the weight when the rate of turning of the telescope is a maximum?
29. Use the methods of proof of Theorems 11.3 and 11.4 to prove Theorem 11.5.

12

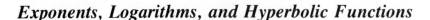

Exponents, Logarithms, and Hyperbolic Functions

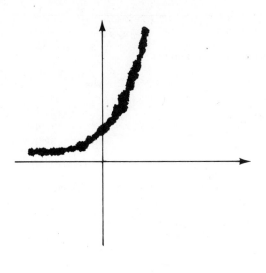

12.1

Exponential and Logarithmic Functions

Let us consider exponents and trace the development of a^n for various values of n. In your first encounter with exponents you learned that, if n is a positive integer, then

$$a^n = a \cdot a \cdot a \cdots a \quad (n \text{ factors}).$$

Later your knowledge of exponents was extended to include zero and negative integers:

$$a^0 = 1 \quad (a \neq 0),$$

$$a^{-n} = \frac{1}{a^n} \quad (a \neq 0, n \text{ a positive integer}),$$

Finally, this was extended to all rational numbers.

$$a^{p/q} = \sqrt[q]{a^p} = (\sqrt[q]{a})^p \quad (p, q \text{ integers}, q > 0, a \geq 0).$$

In the last extension, negative values of the base a were ruled out, because they lead to such problems as

$$\sqrt{(-1)^2} = \sqrt{1} = 1, \qquad \sqrt{(-1)^3} = \sqrt{-1} = i,$$

$$(\sqrt{-1})^2 = i^2 = -1, \qquad (\sqrt{-1})^3 = i^3 = -i,$$

$$(-1)^{2/2} = ? \qquad\qquad (-1)^{3/2} = ?$$

We see that the order in which we carry out the operations makes a difference if the base is negative. To avoid this difficulty, the base was restricted to positive numbers, for which the order of the operations is irrelevant.

All of the extensions from the original definition were made in such a way that the following properties of exponents, which were first derived for positive-integer exponents, remained true when exponents were extended to negative and, finally, to rational exponents.

$$\text{I} \quad a^n \cdot a^m = a^{n+m}.$$

$$\text{II} \quad \frac{a^n}{a^m} = a^{n-m} \quad (a \neq 0).$$

$$\text{III} \quad (a^n)^m = a^{nm}.$$

$$\text{IV} \quad (ab)^n = a^n b^n.$$

Let us now make a final extension, to irrational exponents.

Definition

For any fixed positive number a,

$$a^x = \lim_{r \to x} a^r \quad (r \text{ rational}).$$

Although we shall not attempt to do so here (proofs of the following statements can be found in Franklin†), the above limit can be shown to exist for all real values of x; and, for rational values of x, the value of a^x given by the limit is the same as that previously defined for rational exponents. Furthermore properties I-IV are satisfied for real exponents, and the function

$$F(x) = a^x,$$

where a is a fixed positive constant, is continuous over the set of all real numbers.

The graphs of $y = a^x$ for various values of a are given in Figure 12.1. The graph of $y = 1^x$ is the graph of $y = 1$, since $1^x = 1$ for all x. For all other positive values of a the graphs have the x axis as a horizontal asymptote. For $a > 1$, the graph rises steeply on the right and approaches the x axis on the left. The bigger the base, the more rapidly the graph increases on the right and the more rapidly approaches zero on the left. For $0 < a < 1$, the graph rises steeply on the left and approaches the x axis on the right. The smaller the base the more rapidly the graph increases on the left and the more rapidly it approaches zero on the right. All graphs contain the point $(0, 1)$ and are continuous.

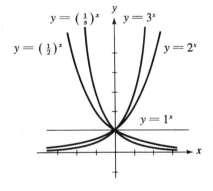

Figure 12.1

† Philip Franklin, *Treatise on Advanced Calculus* (New York, John Wiley, 1940), pp. 58–63.

Let us note at this time the distinction between $y = 2^x$ and $y = x^2$. The first, with a constant base and variable exponent, is an exponential function, which we are discussing here. The other, with a variable base and constant exponent, is a power function, which we have already discussed.

With this, we are in a position to define a logarithm in the customary way.

Definition

*The **logarithm**, base a $(a > 0, a \neq 1)$, of the number x $(x > 0)$ is the number y such that $a^y = x$. Thus,*

$$y = \log_a x \quad means \quad x = a^y.$$

We see that the logarithm and the exponential function are inverses of each other; that is, $y = \log_a x$, which is equivalent to $x = a^y$, is simply the exponential function $y = a^x$ with the x and y reversed. This can also be expressed by the following laws of logarithms, which follow directly from the foregoing definition.

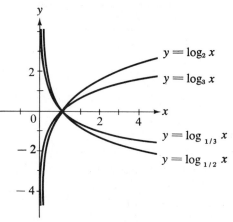

Figure 12.2

$$\text{I} \quad a^{\log_a n} = n,$$

$$\text{II} \quad \log_a a^n = n.$$

Thus, since the range of every exponential function with positive base other than one is the set of all positive numbers, we are assured that every positive number has a logarithm. The graphs of $y = \log_a x$ for various values of a are given in Figure 12.2. For all, the set of positive numbers is the domain and the set of real numbers is the range. The y axis is a vertical asymptote and $y = \log_a x$ increases slowly for $a > 1$ and decreases slowly for $0 < a < 1$. All contain the point $(1, 0)$ and are continuous.

The laws of logarithms, which are easily derived from the laws of exponents, are:

$$\text{III} \quad \log_a mn = \log_a m + \log_a n.$$

$$\text{IV} \quad \log_a \frac{m}{n} = \log_a m - \log_a n.$$

$$\text{V} \quad \log_a n^m = m \log_a n.$$

One other useful law is the equation for change of base.

$$\text{VI} \quad \log_a n = \frac{\log_b n}{\log_b a}.$$

All of the laws of exponents and logarithms hold for any choice of the base a (provided $a > 0$ and $a \neq 1$). When working with logarithms in the past, you probably worked almost exclusively with base 10. This is the most convenient base for computational work, since it coincides with the base of our system for representing numbers. But it is not very convenient in calculus. As we shall see in the next two sections, the most convenient base is the number e, which is defined as follows.

Definition

$$e = \lim_{x \to 0} (1 + x)^{1/x}.$$

This is a difficult limit to evaluate. Although it appears at first glance that the limit must be 1 (since the base approaches 1 and 1 to any power is still 1), actually this is not the case. The graph of

$$y = (1 + x)^{1/x}$$

is given in Figure 12.3. It has domain

$$\{x \mid x > -1, \; x \neq 0\}$$

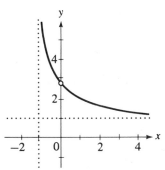

(because we ruled out negative or zero bases and the exponent is undefined when $x = 0$) and asymptotes $x = -1$ and $y = 1$. Although it is not defined at $x = 0$, the limit as x approaches zero exists and is a number between 2 and 3. This is the number we call e. The value of e to six significant figures is $2.71828 \ldots$. Since the base 10 and the base e are used so frequently, logarithms with these bases are abbreviated. Throughout this book we shall use the conventions

Figure 12.3

$$\log_{10} x = \log x \quad \text{and} \quad \log_e x = \ln x.$$

These are called the common logarithm and the natural logarithm, respectively. One word of warning: the abbreviations we are using are common in elementary mathematics texts as well as science and engineering texts; but most advanced mathematics books use $\log x$ to represent $\log_e x$. Thus, some confusion is inevitable, and you must check to see whether the base is 10 or e when confronted with $\log x$.

It is sometimes convenient to make the substitution $x = 1/n$ in the defintion of e. Thus

$$e = \lim_{x \to 0} (1 + x)^{1/x} = \lim_{n \to +\infty} \left(1 + \frac{1}{n}\right)^n.$$

Furthermore, restricting n to integer values has no effect on the limit (see Problem 40).

Problems

A

Sketch the graphs of the equations in Problems 1–8.

1. $y = 2^{x+1}$.
2. $y = 2^{x-2}$.
3. $y = 3^{-x}$.
4. $y = e^x$.
5. $y = \log_2 (x + 1)$.
6. $y = \log_3 (x - 1)$.
7. $y = \log_2 (-x)$
8. $y = \ln x$.

In Problems 9–12, solve for x. Do not use tables.

9. $y = e^x$.
10. $y = \log_3 x$.
11. $y = \log_5 (x - 1)$.
12. $y = 3^{\log_3 x}$.

In Problems 13 and 14, express the given logarithm in terms of simpler ones by the use of laws III–V of logarithms.

13. $\log x^2(x + 2)$.

14. $\log_3 \dfrac{x + 1}{x - 2}$.

B

Sketch the graphs of the equations in Problems 15–22.

15. $y = 2^{2x}$.

16. $y = 2^{-2x}$.

17. $y = 2^{|x|}$.

18. $y = 3^{x^2}$.

19. $y = \log_4 4x$.

20. $y = \log_2 x^3$.

21. $y = \log_2 |x|$.

22. $y = \log_3 (-9x)$.

In Problems 23–32, solve for x. Do not use tables.

23. $y = 4^{x-2}$.

24. $y = 3^{3x-5}$.

25. $y = \log_3 3^x$.

26. $y = \ln x^3$.

27. $\log x = 3 \log 2 - \log 4$.

28. $\log_2 x = \log_2 6 - \dfrac{1}{2} \log_2 9$.

29. $\log_3 x = \dfrac{1}{2} \log_3 5 + \log_3 4 - 2 \log_3 3$.

30. $\ln x = 3 \ln 2 - 2 \ln 3 - \ln 6$.

31. $\ln x^2 - \ln 2x = 3 \ln 3 - \ln 6$.

32. $\log x^2 = \dfrac{1}{2} \log 81 + \log 4 - 2 \log 5$.

In Problems 33–36, express the given logarithm in terms of simpler ones by the use of laws III–V of logarithms.

33. $\log_5 \dfrac{x^2(x + 3)}{x - 1}$.

34. $\log_2 \dfrac{\sqrt[3]{x(x + 4)}}{(x - 1)^2}$.

35. $\ln \dfrac{(x - 1)^{3/5}(x + 2)^{1/2}}{(x + 3)^{1/3}}$.

36. $\log \sqrt{\dfrac{(x - 1)x}{(x + 2)}}$.

C

37.   Prove laws I and II of logarithms, page 350.

38.   Prove laws III–V of logarithms.

39.   Prove the change-of-base formula. (*Hint:* Let $\log_a n = x$. Then $a^x = n$. Now take the logarithm (base b) of both sides.)

40.   Use the definition of e to find a decimal approximation of e correct to two decimal places. (*Hint:* Use the substititution $x = 1/n$ and expand the result by the binomial theorem, assuming n to be an integer.)

41.   If a principal P is invested at a rate r of interest, which is compounded n times per year, the final amount F (principal plus interest) at the end of one year is

$$F = P\left(1 + \frac{r}{n}\right)^n.$$

(This formula is found by repeated application of the simple-interest formula, $I = Prt$.) Find a formula for F if interest is compounded continuously ($n \to +\infty$). Suppose $1000 is invested at 6%. Find the final amount at the end of one year if interest is compounded (a) annually, (b) semiannually, (c) continuously.

12.2

Derivatives of Logarithmic Functions

Before finding derivatives of logarithmic functions, let us recall the definition of e.

$$e = \lim_{x \to 0} (1 + x)^{1/x}.$$

Although x appears in the limit expression, the number e has nothing to do with the choice of x. Any symbol may be used in place of x. In particular,

$$e = \lim_{h/x \to 0} \left(1 + \frac{h}{x}\right)^{x/h}.$$

Theorem 12.1

If $f(x) = \log_a x$, then

$$f'(x) = \frac{1}{x} \log_a e.$$

Proof

$$f'(x) = \lim_{h \to 0} \frac{f(x + h) - f(x)}{h}$$

$$= \lim_{h \to 0} \frac{\log_a (x + h) - \log_a x}{h}$$

$$= \lim_{h \to 0} \frac{\log_a \dfrac{x + h}{x}}{h} \qquad \text{(by law IV)}$$

$$= \lim_{h \to 0} \frac{1}{h} \log_a \left(1 + \frac{h}{x}\right)$$

$$= \lim_{h \to 0} \frac{1}{x} \cdot \frac{x}{h} \log_a \left(1 + \frac{h}{x}\right)$$

$$= \lim_{h \to 0} \frac{1}{x} \log_a \left(1 + \frac{h}{x}\right)^{x/h} \qquad \text{(by law V)}$$

$$= \frac{1}{x} \log_a e. \qquad\qquad \text{(See Note)}$$

Note: $\displaystyle \lim_{h \to 0} \frac{1}{x} \log_a \left(1 + \frac{h}{x}\right)^{x/h} = \lim_{h \to 0} \frac{1}{x} \lim_{h \to 0} \log_a \left(1 + \frac{h}{x}\right)^{x/h}$ (by Theorem 10.8)

$$= \frac{1}{x} \log_a \lim_{h \to 0} \left(1 + \frac{h}{x}\right)^{x/h},$$

since $\lim_{h \to 0} 1/x = 1/x$ by Problem 21, page 285; the second limit is altered by use of Problem 31, page 568, together with the fact that the logarithmic function is continuous.

Theorem 12.2

If $f(x) = \log_a u(x)$,   then

$$f'(x) = \frac{u'(x)}{u(x)} \log_a e.$$

This follows directly from Theorem 12.1 and the chain rule. The factor $\log_a e$ is rather bothersome; but we can eliminate it by choosing $a = e$, since $\ln e = 1$.

Theorem 12.3

If $f(x) = \ln u(x)$,   then

$$f'(x) = \frac{u'(x)}{u(x)}.$$

Example 1

Differentiate $y = \log_2 (x^2 + 1)$.

$$y' = \frac{2x}{x^2 + 1} \log_2 e.$$

Example 2

Differentiate $y = \ln |x|$.

$$y = \ln |x| = \begin{cases} \ln x & x > 0, \\ \ln(-x) & x < 0. \end{cases}$$

If $y = \ln x$, then $y' = 1/x$. If $y = \ln (-x)$, then $y' = (-1)/(-x) = 1/x$; thus $y' = 1/x$ in either case.

Example 3

Differentiate $y = \ln \sqrt{x}$.

Let us first differentiate this expression just as it stands:

$$y' = \frac{\dfrac{1}{2\sqrt{x}}}{\sqrt{x}} = \frac{1}{2x}.$$

We can get the same result more easily by simplifying the original expression:

$$y = \ln \sqrt{x} = \ln (x)^{1/2} = \frac{1}{2} \ln x.$$

Now the differentiation is simpler:

$$y' = \frac{1}{2} \cdot \frac{1}{x} = \frac{1}{2x}.$$

The second method was not much simpler than the first in this case, but let us consider another example.

Example 4

Differentiate $y = \ln \dfrac{x^2(x+1)}{(x-2)^3}$.

$$y = \ln x^2 + \ln (x+1) - \ln (x-2)^3$$
$$= 2 \ln x + \ln (x+1) - 3 \ln (x-2).$$
$$y' = \frac{2}{x} + \frac{1}{x+1} - \frac{3}{x-2}$$
$$= -\frac{7x+4}{x(x+1)(x-2)}.$$

Although there was some algebraic manipulation in simplifying the derivative, this method is still much simpler than differentiating the original expression. The use of logarithms can sometimes simplify complicated algebraic expressions.

Example 5

Differentiate $y = \dfrac{x^{2/3}(x-1)^{1/3}}{x+2}$.

Let us first take the natural logarithm of both sides of the equation.

$$\ln y = \ln \frac{x^{2/3}(x-1)^{1/3}}{x+2}$$
$$= \frac{2}{3} \ln x + \frac{1}{3} \ln (x-1) - \ln (x+2).$$

Now we have an implicit function. Let us take the derivatives of both sides.

$$\frac{y'}{y} = \frac{2}{3x} + \frac{1}{3(x-1)} - \frac{1}{x+2},$$
$$y' = y\left(\frac{2}{3x} + \frac{1}{3(x-1)} - \frac{1}{x+2}\right).$$

If the result can be left in this form, we have saved a great deal of time. Unfortunately, the most useful form requires that we combine the three terms into a single one.

The foregoing method of differentiation is sometimes called *logarithmic differentiation*.

Example 6

Differentiate $y = \sin x^2 \cdot \ln (x^2 + 1)$.

$$y' = \sin x^2 \cdot \frac{2x}{x^2+1} + \ln (x^2+1)(\cos x^2)2x.$$
$$= 2x\left[\frac{\sin x^2}{x^2+1} + \ln (x^2+1) \cos x^2\right].$$

Example 7

Given $y = (4 \ln^2 x)/x$,   find all relative maxima and minima and all points of inflection. Sketch.

$$y' = 4\,\frac{x(2 \ln x \cdot 1/x) - \ln^2 x}{x^2}$$

$$= 4\,\frac{2 \ln x - \ln^2 x}{x^2}$$

$$= \frac{4 \ln x(2 - \ln x)}{x^2}\,.$$

$$y'' = 4\,\frac{x^2(2/x - 2 \ln x \cdot 1/x) - 2x(2 \ln x - \ln^2 x)}{x^4}$$

$$= \frac{8(\ln^2 x - 3 \ln x + 1)}{x^3}\,.$$

Setting the first derivative equal to 0 to find critical points, we have

$$\ln x = 0, \qquad \ln x = 2,$$
$$x = 1; \qquad\quad x = e^2.$$

Thus the critical points are $(1, 0)$ and $(e^2, 16/e^2)$. Since $y'' = 8$ when $x = 1$ and $y'' = -8/e^6$ when $x = e^2$, it follows that $(1, 0)$ is a relative minimum and $(e^2, 16/e^2)$ is a relative maximum.

Setting the second derivative equal to 0 to find possible points of inflection, we have

$$\ln^2 x - 3 \ln x + 1 = 0.$$

Solving for $\ln x$ by the quadratic formula, we have

$$\ln x = \frac{3 \pm \sqrt{9 - 4}}{2} = \frac{3 \pm \sqrt{5}}{2} = \begin{cases} 2.618, \\ 0.382. \end{cases}$$

$$\ln x = 0.382, \qquad \ln x = 2.618,$$
$$x = 1.465, \qquad\quad x = 13.71,$$
$$(1.465, 0.398). \qquad (13.71, 2.000).$$

The values of y'' at $x = 1$ and $x = e^2$ assure us that the first of these points is a point of inflection. To see that the second is also a point of inflection, we note first that $y \geq 0$ for all values of x and is decreasing for $x > e^2$. Thus the curve cannot continue to be concave downward beyond $x = 13.71$, and $(13.71, 2.000)$ must be a point of inflection.

We can easily see that there is a vertical asymptote at $x = 0$, and the domain is the set of all positive values of x. Because $y \geq 0$ and there is no critical point to the right of $x = e^2$, there must be a horizontal asymptote. Furthermore, it must be below $y = 2.000$ (the ordinate of the rightmost point of inflection) but no lower than the x axis (since $y \geq 0$ for all x). However, we cannot identify it more exactly, since we have no method of evaluating the limit

$$\lim_{x \to +\infty} \frac{4 \ln^2 x}{x}$$

at this time. (We shall consider limits of this type in Chapter 18.) The graph is given in Figure 12.4.

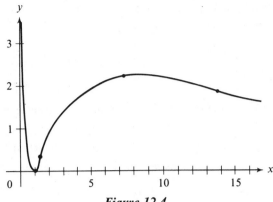

Figure 12.4

Problems

A

In Problems 1–14, differentiate and simplify.

1. $y = \log 4x$.

2. $y = \log_3 x^2$.

3. $y = \ln (x^2 + 4)$.

4. $y = \ln \sqrt{x^2 + 4}$.

5. $y = \ln \dfrac{1}{x}$.

6. $y = \ln \dfrac{2x + 1}{3x + 1}$.

7. $y = \ln \sqrt{\dfrac{2x + 1}{2x - 3}}$.

8. $y = \ln \dfrac{x^2 - 2}{x^2 + 2}$.

9. $y = \log_5 \left(\dfrac{3x + 1}{3x - 1}\right)^{2/3}$.

10. $y = \log \dfrac{x(x - 3)}{x + 5}$.

11. $y = x^2 \ln x$.

12. $y = \sqrt{x} \ln \sqrt{x}$.

13. $y = \dfrac{\ln x}{x}$.

14. $y = \dfrac{\ln x}{x^3}$.

In Problems 15–18, find the derivative at the value of x indicated.

15. $y = \ln x^2$, $x = 2$.

16. $y = \ln \sqrt{x}$, $x = 4$.

17. $y = x \ln x$, $x = 1$.

18. $y = \dfrac{\ln x}{x^2}$, $x = 2$.

B

In Problems 19–28, differentiate and simplify.

19. $y = \ln (\cos x)$.

20. $y = \ln (\sec x)$.

21. $y = \sin (\ln x)$.

22. $y = \ln (\cos^2 x + 1)$.

23. $y = \ln (x + \sqrt{x^2 - 1})$.

24. $y = \ln (x - \sqrt{x^2 - 1})$.

25. $y = \ln (\ln x)$.

26. $y = \ln (\ln x^3)$.

27. $y = \ln (x^2 \ln x)$.

28. $y = \ln (\ln (\ln x))$.

In Problems 29–34, find dy/dx.

29. $\ln y = \sin x$.

30. $\ln xy = x + y$.

31. $\ln y = \cos (x + y)$.

32. $\ln (\cos y) = x + y$.

33. $x^2 + y^2 = \ln (x + y)$.

34. $xy = \ln (\sin (x + y))$.

In Problems 35–38, differentiate by logarithmic differentiation (see Example 5).

35. $y = \sqrt{\dfrac{x(x + 4)}{x - 2}}.$

36. $y = \dfrac{x^2(x - 2)^5}{(x - 4)^3}.$

37. $y = \dfrac{x^{2/3}(x + 1)^{4/3}}{(x - 5)^{1/3}}.$

38. $y = \dfrac{(x + 3)^{3/2}(x - 1)^{1/2}}{x^{5/2}(x - 2)^{1/2}}.$

In Problems 39–42, find an equation of the line tangent to the given curve at the point indicated.

39. $y = \ln x,$ $(1, 0).$

40. $y = \ln x^3,$ $(1, 0).$

41. $y = \ln (\cos x),$ $(\pi/4, - (\ln 2)/2).$

42. $y = \sin (\ln x),$ $(1, 0).$

In Problems 43 and 44, find all relative maxima and minima, and sketch.

43. $y = x \ln x.$

44. $y = \dfrac{\ln x}{x}.$

45. A submarine telegraph cable consists of a conducting circular core surrounded by a circular layer of insulation. If x is the ratio of the radius of the core to the thickness of the insulation, then the speed of the signal is proportional to $x^2 \ln (1/x)$. For what value of x is the speed of the signal a maximum?

C

46. Find an equation of the line through the origin and tangent to $y = \ln x$. Use your result to determine the values of k for which $(\ln x)/x = k$ has   (a) no solution,   (b) exactly one solution,   (c) two solutions. Use a hand calculator or tables to find the value(s) of x for which $(\ln x)/x = 0.1$.

12.3

Derivatives of Exponential Functions

Since logarithmic and exponential functions are inverses of each other, it is a simple matter to find the derivative of one if we know the derivative of the other.

Theorem 12.4

If $f(x) = a^{u(x)},$ then $f'(x) = a^{u(x)} \cdot u'(x) \cdot \ln a.$

Proof

Again let us abbreviate $u(x)$ to u and replace $f(x)$ by y. By logarithmic differentiation,

$$y = a^u,$$

$$\ln y = \ln a^u = u \ln a,$$

$$\frac{y'}{y} = u' \ln a,$$

$$y' = y \cdot u' \ln a$$

$$= a^u \cdot u' \ln a.$$

Example 1

Differentiate $y = 2^x$.

$$y' = 2^x \cdot 1 \cdot \ln 2 = 2^x \ln 2.$$

Theorem 12.5

If $f(x) = e^{u(x)}$,   then $f'(x) = e^{u(x)} \cdot u'(x)$.

This follows directly from Theorem 12.4.

Example 2

Differentiate $y = e^x$.

$$y' = e^x \cdot 1 = e^x.$$

Perhaps you find this result a bit disappointing. All it means is that for any point on the graph of $y = e^x$, the y coordinate and the slope at that point are the same.

Example 3

Differentiate $y = e^{\sqrt{x}}$.

$$y' = e^{\sqrt{x}} \cdot \frac{1}{2\sqrt{x}} = \frac{e^{\sqrt{x}}}{2\sqrt{x}}.$$

Example 4

Differentiate $y = xe^x$.

Using the product rule, we have $y' = xe^x + e^x \cdot 1 = e^x(x + 1)$.

Example 5

Differentiate $y = x^2 \cdot 2^x$.

This illustrates the basic difference between a power function x^2, in which a variable base is raised to a constant power, and an exponential function, in which a constant base is raised to a variable power. Thus, the power rule is used on the first factor, while the derivative of an exponential function is used on the second. Of course, we must use the product rule, too.

$$y' = x^2(2^x \cdot 1 \cdot \ln 2) + 2^x(2x)$$
$$= x \cdot 2^x(x \ln 2 + 2).$$

Let us note that there are four possible combinations of the base and exponent. The forms, with an example of each, are given in the table.

Form	Example
(Constant)$^{\text{(Constant)}}$	e^2
(Variable)$^{\text{(Constant)}}$	x^3
(Constant)$^{\text{(Variable)}}$	2^x
(Variable)$^{\text{(Variable)}}$	x^x

We can now differentiate three of these four forms. The first is quite trivial, since a constant base to a constant power is still constant and its derivative is zero. The next two were illustrated in Example 5. Let us now consider the case in which both the base and the exponent are variable.

Example 6

Differentiate $y = x^x$.

We cannot differentiate by the power rule, since that requires a constant exponent. We cannot use the exponential formulas (Theorems 12.4 or 12.5), since they require a constant base. In the equation's present form, we cannot use any formula we have. Let us change its form by taking the logarithm of both sides.

$$\ln y = \ln x^x,$$
$$\ln y = x \ln x. \qquad \text{(by property V)}$$

Now we can take the derivative of this implicit function.

$$\frac{y'}{y} = x \cdot \frac{1}{x} + \ln x \cdot 1,$$

$$y' = y(1 + \ln x)$$
$$= x^x(1 + \ln x).$$

We may use any constant base we choose when taking the logarithm—we chose e because it is the simplest (its derivative does not have the bothersome constant factor $\log_a e$). This method of taking the logarithm of both sides can be used in any case involving exponents, but is unnecessary in those cases in which either the base or the exponent is constant, since we have formulas to handle these. Actually we have proved the power rule only for the case in which the exponent is rational (Theorem 4.10). We may use this method to extend that theorem to any n.

Theorem 12.6

If $f(x) = x^n$, where n is any real number, then $f'(x) = nx^{n-1}$.

The proof is left to the student.

Example 7

For $y = e^x/x$ find all maxima, minima, and points of inflection. Sketch.

$$y' = \frac{xe^x - e^x}{x^2} = \frac{e^x(x-1)}{x^2},$$

$$y'' = \frac{x^2(xe^x + e^x - e^x) - 2x(xe^x - e^x)}{x^4} = \frac{e^x(x^2 - 2x + 2)}{x^3}.$$

There is a vertical asymptote at $x = 0$ and a critical point at $(1, e)$. Since $y'' = e$ when $x = 1$, $(1, e)$ is a relative minimum. Since $x^2 - 2x + 2 = (x-1)^2 + 1 > 0$, there is no point of inflection.

We have already indicated that $x = 0$ is a vertical asymptote. Furthermore $y = 0$ is a horizontal asymptote, since

$$\lim_{x \to -\infty} \frac{e^x}{x} = 0.$$

But we have no way of evaluating the limit

$$\lim_{x \to +\infty} \frac{e^x}{x} ;$$

both the numerator and denominator increase indefinitely as x approaches $+\infty$. Nevertheless we see that this limit cannot exist, since the curve is concave upward everywhere to the right of the minimum $(1, e)$. The graph is shown in Figure 12.5.

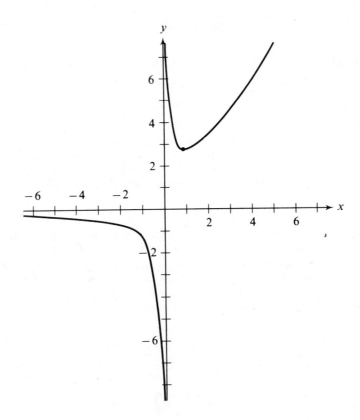

Figure 12.5

Problems

A

In Problems 1–12, differentiate and simplify.

1. $y = 3^x$.

2. $y = e^{x^3}$.

3. $y = 3^{\sqrt{x}}$.

4. $y = 4^{2x+1}$.

5. $y = e^{2x+2}$.

6. $y = e^{x^2-4}$.

7. $y = 2^{x^2+3}$.

8. $y = 3^{x^2-x}$.

9. $y = x + e^x$.

10. $y = \dfrac{e^x}{x^2}$.

11. $y = x^3 e^{x^3}$.

12. $y = \dfrac{1 + e^x}{x^2}$.

B

In Problems 13–22, differentiate and simplify.

13. $y = 3^x(x^3 - 1)$.   14. $y = \ln(x - e^x)$.   15. $y = e^{x^2 - \ln x}$.   16. $y = e^{\cos x}$.
17. $y = e^{\tan x}$.   18. $y = \sin e^{x^2}$.   19. $y = x^{x^3}$.   20. $y = x^{\sin x}$.
21. $y = (\sin x)^x$.   22. $y = x^{x^x}$.

In Problems 23–26, find dy/dx.

23. $\ln y = e^{x^2}$.   24. $\sin(x + y) = e^y$.
25. $\ln(\ln x) = e^y$.   26. $\ln(\sin^2 y) = xe^x$.

In Problems 27–30, find the derivative at the value of x indicated.

27. $y = e^{x^2}$, $x = 2$.   28. $y = e^x/x$, $x = 1$.
29. $y = e^{\sin x}$, $x = \pi/2$.   30. $y = x^3 e^{x^3}$, $x = 2$.

In Problems 31–34, find an equation of the line tangent to the curve at the point indicated.

31. $y = e^x$, $(2, e^2)$.   32. $y = 2^{x^2}$, $(1, 2)$.
33. $y = e^x/x$, $(1, e)$.   34. $y = e^{\sin x}$, $(\pi/6, \sqrt{e})$.

In Problems 35–38, find all relative maxima and minima, and sketch.

35. $y = xe^x$.   36. $y = x^2 e^x$.
37. $y = \cos e^x$.   38. $y = e^{-x} \sin x$.

39. Show that $y = e^x$ is a solution of the differential equation $y'' - 2y' + y = 0$. Show that $y = xe^x$ is also a solution. Show that $y = c_1 e^x + c_2 xe^x$, where c_1 and c_2 are arbitrary constants, is a solution. Differential equations of this type are useful in the theory of damped vibrations—that is, vibrations with continuously decreasing amplitude.

C

40. Find a formula for the derivative of $y = u^v$, where u and v are functions of x.
41. Prove Theorem 12.6.
42. Find an equation of the line through the origin and tangent to $y = e^x$. Use your result to determine the values of k for which $e^x/x = k$ has (a) no solution, (b) exactly one solution, (c) two solutions. Use a hand calculator or tables to find the value(s) of x for which $e^x/x = 3$. Compare with Problem 46 of the previous section.

12.4

Integrals Involving Logarithms and Exponentials

The theorems of the previous sections can be expressed in integral form. It would seem that the integral form of Theorem 12.3 is

$$\int \frac{u'}{u}\, dx = \ln u + C.$$

But this is unnecessarily restrictive; ln u exists only when $u > 0$. If $u < 0$, we alter the integral to

$$\int \frac{u'}{u}\,dx = \int \frac{-u'}{-u}\,dx.$$

Since $-u$ is positive, we may use the integration formula given above with the u of the formula replaced by $-u$ to get

$$\int \frac{u'}{u}\,dx = \int \frac{-u'}{-u}\,dx = \ln(-u) + C.$$

The two integrals

$$\int \frac{u'}{u}\,dx = \ln u + C \qquad \text{if } u > 0$$

$$\int \frac{u'}{u}\,dx = \ln(-u) + C \quad \text{if } u < 0$$

are combined to give the following result.

Theorem 12.7

If u is a function of x, then

$$\int \frac{u'}{u}\,dx = \int \frac{du}{u} = \ln |u| + C.$$

It might be noted that this theorem fills the gap left by the power rule. The power rule gives

$$\int u^n u'\,dx = \int u^n\,du = \frac{u^{n+1}}{n+1} + C \quad (n \neq -1).$$

If $n = -1$, Theorem 12.7 is used. We do not give the integral form of Theorem 12.2, because none is needed.

$$\int \frac{u'}{u} \log_a e\,dx = \log_a e \int \frac{u'}{u}\,dx$$

$$= \log_a e \ln |u| + C$$

$$= \log_a e \frac{\log_a |u|}{\log_a e} + C$$

$$= \log_a |u| + C.$$

The integral forms of Theorems 12.4 and 12.5 are given in Theorem 12.8, where we have changed the base of the logarithm in order to have the factor in the numerator.

Theorem 12.8

If u is a function of x, then

$$\int e^u \cdot u' \, dx = \int e^u \, du = e^u + C,$$

$$\int a^u \cdot u' \, dx = \int a^u \, du = \frac{a^u}{\ln a} + C = a^u \log_a e + C.$$

Example 1

Evaluate $\int \dfrac{x}{x^2 + 1} \, dx.$

If we take $u = x^2 + 1$, then $u' = 2x$. By adjusting the constant, we can put the integral into the form $\int \dfrac{u'}{u} \, dx.$

$$\int \frac{x}{x^2 + 1} \, dx = \frac{1}{2} \int \frac{2x}{x^2 + 1} \, dx$$

$$= \frac{1}{2} \ln |x^2 + 1| + C$$

$$= \frac{1}{2} \ln (x^2 + 1) + C.$$

The absolute value was dropped, because $x^2 + 1$ cannot be negative.

Example 2

Evaluate $\int \dfrac{x^3 + 3x^2 - x + 3}{x^2 + 1} \, dx.$

$$\int \frac{x^3 + 3x^2 - x + 3}{x^2 + 1} \, dx = \int \left(x + 3 - \frac{2x}{x^2 + 1} \right) dx$$

$$= \frac{x^2}{2} + 3x - \ln (x^2 + 1) + C.$$

Example 3

Evaluate $\int x e^{x^2} \, dx.$

$$\int x e^{x^2} \, dx = \frac{1}{2} \int 2x e^{x^2} \, dx = \frac{1}{2} e^{x^2} + C.$$

Note that we could not have used Theorem 12.8 if the factor x were not present. In fact, no one can evaluate the integral $\int e^{x^2} \, dx$ in terms of elementary functions.

Example 4

Evaluate $\displaystyle\int \frac{dx}{x \ln x}$.

If $u = \ln x$, then $u' = 1/x$. Thus

$$\int \frac{dx}{x \ln x} = \int \frac{1/x}{\ln x}\, dx$$
$$= \ln |\ln x| + C.$$

Example 5

Find the area of the region bounded by $y = e^x$ and the line joining $(1, e)$ and $(-1, 1/e)$.

The graphs are given in Figure 12.6. First let us find an equation for the line. Finding the slope and using the point-slope form of the line, we have

$$m = \frac{e - 1/e}{1 + 1} = \frac{e^2 - 1}{2e},$$

$$y - e = \frac{e^2 - 1}{2e} (x - 1)$$

$$y = \frac{e^2 - 1}{2e} x + \frac{e^2 + 1}{2e}.$$

From Figure 12.6, the area is

$$A = \int_{-1}^{1} (y_1 - y_2)\, dx$$

$$= \int_{-1}^{1} \left(\frac{e^2 - 1}{2e} x + \frac{e^2 + 1}{2e} - e^x \right) dx$$

$$= \left(\frac{e^2 - 1}{4e} x^2 + \frac{e^2 + 1}{2e} x - e^x \right) \Big|_{-1}^{1}$$

$$= \left(\frac{e^2 - 1}{4e} + \frac{e^2 + 1}{2e} - e \right) - \left(\frac{e^2 - 1}{4e} - \frac{e^2 + 1}{2e} - \frac{1}{e} \right)$$

$$= \frac{2}{e}.$$

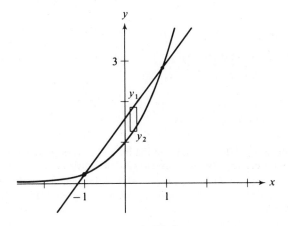

Figure 12.6

Problems

A

Evaluate the integrals in Problems 1–8.

1. $\displaystyle\int \frac{dx}{2x + 1}$

2. $\displaystyle\int \frac{dx}{4x - 3}.$

3. $\displaystyle\int e^{-2x}\, dx.$

4. $\displaystyle\int e^{3x}\, dx.$

5. $\displaystyle\int \frac{\sin x}{\cos x}\, dx.$

6. $\displaystyle\int \frac{2x + 3}{x^2 + 3x - 1}\, dx.$

7. $\displaystyle\int_0^2 \frac{dx}{x + 4}.$

8. $\displaystyle\int_0^1 \frac{dx}{e^{2x}}.$

B

Evaluate the integrals in Problems 9–30.

9. $\displaystyle\int \frac{x^2 + 3x + 2}{x - 1}\, dx.$

10. $\displaystyle\int \frac{3\sec^2 x}{\tan x}\, dx.$

11. $\displaystyle\int \frac{2e^x}{e^x + 1}\, dx.$

12. $\displaystyle\int \frac{e^{2x}}{e^x + 1}\, dx.$

13. $\displaystyle\int \frac{2x + 1}{\sqrt{x^2 + x}}\, dx.$

14. $\displaystyle\int \frac{2x + 1}{x^2 + 1}\, dx.$

15. $\displaystyle\int 3^x\, dx.$

16. $\displaystyle\int (2e^x + e^{-x})\, dx.$

17. $\displaystyle\int (e^x - e^{-x})\, dx.$

18. $\displaystyle\int (2e^x + e^{-x})^2\, dx.$

19. $\displaystyle\int \frac{4x^3 - 3x^2 + x - 5}{x}\, dx.$

20. $\displaystyle\int \frac{2e^x + e^{-x}}{2e^x - e^{-x}}\, dx.$

21. $\displaystyle\int \sec^2 x\, e^{\tan x}\, dx.$

22. $\displaystyle\int (\sin x + \cos x)e^{\sin x - \cos x}\, dx.$

23. $\displaystyle\int \frac{e^{2x} + 3e^x + 2}{e^x}\, dx.$

24. $\displaystyle\int \frac{dx}{x \ln \sqrt{x}}.$

25. $\displaystyle\int \frac{x^3 + 3x^2 + 5x + 3}{(x + 1)^2}\, dx.$

26. $\displaystyle\int \frac{dx}{(x^2 + 1)\operatorname{Arctan} x}.$

27. $\displaystyle\int_3^4 \frac{x^2 + 5}{x - 2}\, dx.$

28. $\displaystyle\int_1^3 \frac{x\, dx}{x^2 + 1}.$

29. $\displaystyle\int_{\pi/6}^{\pi/2} \frac{\cos x}{\sin x}\, dx.$

30. $\displaystyle\int_0^1 x^3\, e^{x^4}\, dx.$

In Problems 31–34, find the area of the region bounded by the given curves.

31. $y = e^x,\quad y = 0,\quad x = 0,\quad x = -2.$

32. $y = e^{-2x},\quad y = 0,\quad x = 0,\quad x = 2.$

33. $y = \ln x,\quad y = 0,\quad x = 2.$
    (*Hint:* Use horizontal strips.)

34. $y = \dfrac{e^x}{1 + e^x},\quad y = 0,\quad x = 0,\quad x = 2.$

C

35. Find the area of the region bounded by $y = e^{-x}$, both axes, and $x = k$, where $k > 0$. What is the limit of this area as k increases indefinitely? What does this limit represent?

36. It has been noted that $y' = y$ if $y = e^x$. In addition, $y' = y$ if $y = 0$. Is there any other function for which $y' = y$? If so, give one; if not, why not? (*Hint:* If $y' = y$, then $y'/y = 1$. Integrate both sides.)

37. Evaluate $\displaystyle\int \frac{1}{e^x + 1}\, dx.$

12.5

Differential Equations and Rates of Growth and Decay

Certain types of problems—some of which we shall consider shortly—lead to what are called *differential equations*. A differential equation is nothing more than an equation involving one or more derivatives or differentials. For example,

$$\frac{dy}{dx} + xy = x,$$

$$\frac{d^2y}{dx^2} - \frac{dy}{dx} = 4,$$

$$(y')^2 + xy' = y,$$

$$(x + y)\,dx + (2x - 5)\,dy = 0,$$

$$2x\,dx + (x - y)\,dy = 0,$$

are all examples of differential equations. Actually we have already seen some very simple differential equations, which we solved when reconstructing a function from its derivative. The differential equation

$$y' = f(x)$$

has as its solution

$$y = \int f(x)\,dx + C.$$

In fact, every time we find an indefinite integral we are solving a differential equation.

We consider now a differential equation that is only slightly more difficult. Sometimes the variables can be separated from each other by purely algebraic manipulation; that is, we can express the equation in the form

$$y'u(y) = v(x)$$

and solve by integrating both sides.

Example 1

Solve the differential equation $y' + xy = x.$

$$y' = x(1 - y), \qquad \frac{y'}{1 - y} = x.$$

We now integrate both sides with respect to x.

$$\int \frac{y'}{1 - y}\,dx = \int x\,dx.$$

$$-\ln|1 - y| = \frac{x^2}{2} + C.$$

While this form is sometimes suitable, we often prefer to solve for y. In that case

$$\ln|1 - y| = -\frac{x^2}{2} - C$$

$$|1 - y| = e^{-x^2/2 - c}$$

$$1 - y = \pm e^{-x^2/2} e^{-c}$$

$$1 - y = \pm Ke^{-x^2/2} \quad \text{(letting } K = e^{-c})$$

$$y = 1 \pm Ke^{-x^2/2}.$$

By allowing K to take on both positive and negative values, we may drop the $\pm$. Thus we have the solution

$$y = 1 + Ke^{-x^2/2}.$$

Note that the foregoing solution involves a constant of integration. This solution, called the *general solution*, represents a family of curves (one curve for each value of the constant). If, as sometimes happens, additional information allows us to determine the constant, the resulting solution is called a *particular solution*.

Example 2

Find the particular solution of $y' = 2xy^2 + y^2$ satisfying the condition that $y = 1$ when $x = 2$.

First we find the general solution. Dividing by y^2 and integrating, we have

$$\frac{y'}{y^2} = 2x + 1,$$

$$-\frac{1}{y} = x^2 + x + C.$$

Now we find the particular value of C from the condition that $y = 1$ when $x = 2$.

$$-1 = 4 + 2 + C$$
$$C = -7.$$

Hence the particular solution is

$$-\frac{1}{y} = x^2 + x - 7,$$

or

$$y = \frac{-1}{x^2 + x - 7}.$$

Now let us look at some physical situations that lead to differential equations of this type. The rate at which a population grows is directly proportional to the size of the population, assuming that the growing conditions remain fixed and that migration has no influence. This leads to a differential equation with separable variables. Moreover, the solution shows a rate of growth that follows an exponential curve.

Example 3

The rate of growth of a colony of bacteria is directly proportional to the size of the colony. If there are 5000 bacteria at noon and 10,000 at 2 P.M., how many will there be at 5 P.M.?

Let x be the number of bacteria at time t, the number of hours after noon. Then dx/dt is the rate of growth. Since the rate of growth is proportional to x, there is a constant k of proportionality such that

$$\frac{dx}{dt} = kx.$$

Dividing both sides by x and integrating with respect to t, we have

$$\frac{dx/dt}{x} = k$$

$$\int \frac{dx/dt}{x}\, dt = \int k\, dt$$

$$\ln|x| = kt + c$$

$$|x| = e^{kt+c} = e^{kt}e^{c}$$

$$x = \pm\, e^{c}e^{kt} = Ce^{kt}.$$

Now we must evaluate the constants C and k. Since $x = 5000$ when $t = 0$, and $x = 10,000$ when $t = 2$,

$$5000 = Ce^{0} \quad \text{and} \quad 10,000 = Ce^{2k}.$$

This yields $C = 5000$ and

$$10,000 = 5000e^{2k}$$

$$e^{2k} = 2$$

$$2k = \ln 2$$

$$k = \frac{1}{2}\ln 2 = 0.3466.$$

We now have

$$x = 5000e^{0.3466t}.$$

When $t = 5$,

$$x = 5000e^{(0.3466)5} = 28,300.$$

Radioactive elements, such as radium, uranium, and plutonium, spontaneously decompose into other elements and subatomic particles. The rate at which the decomposition takes place is directly proportional to the amount of the radioactive substance present. This leads to the differential equation

$$\frac{dx}{dt} = kx.$$

This is exactly the same differential equation that we encountered in Example 3 on rates of population growth. Of course, we have the same general solution,

$$x = Ce^{kt}.$$

However there is one very important difference. In Example 3 the population was increasing, while in this case the amount of radioactive substance is decreasing. In the first case k is positive; in the second it is negative. Thus exponential rates of growth and decay are two special cases of the general exponential curve. Their difference is illustrated graphically in Figure 12.7.

The half-life of a radioactive element is the time required for half of it to decompose. The half-life is a characteristic that is independent of the amount of the element present. Let us see why. The differential equation

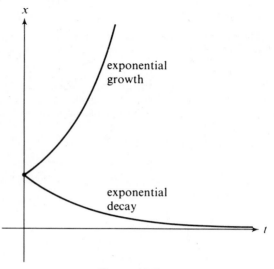

exponential growth

exponential decay

Figure 12.7

$$\frac{dx}{dt} = kx$$

has a general solution

$$x = Ce^{kt}.$$

Suppose x_0 is the amount present when $t = 0$. Then

$$x_0 = Ce^{k \cdot 0} = C,$$

and we have

$$x = x_0 e^{kt}.$$

The half-life is the value of t corresponding to $x = x_0/2$. Thus

$$\frac{x_0}{2} = x_0 e^{kt}$$

$$e^{kt} = \frac{1}{2}$$

$$kt = \ln \frac{1}{2} = -\ln 2$$

$$t = -\frac{1}{k} \ln 2.$$

We see here that the half-life is not dependent upon x_0; it depends only upon k, which is a characteristic of the radioactive element.

Example 4

If 0.025 g of a radioactive element decays to 0.020 g in 48 hours, find the half-life of the element.

Again the differential equation

$$\frac{dx}{dt} = kx$$

has solution

$$x = Ce^{kt}.$$

When $t = 0$, $x = 0.025$ and when $t = 48$, $x = 0.020$. From the first of these we get $C = 0.025$. Let us now use the second condition to determine k.

$$x = 0.025e^{kt}$$

$$0.020 = 0.025e^{48k}$$

$$e^{48k} = \frac{0.020}{0.025} = 0.8$$

$$48k = \ln 0.8 = 9.7769 - 10 = -0.2231$$

$$k = -0.00465.$$

Thus the relationship between x and t is

$$x = 0.025e^{-0.00465t}.$$

To find the half-life, we find the value of t corresponding to $x = 0.025/2 = 0.0125$. We have already seen that this gives

$$t = -\frac{1}{k}\ln 2 = -\frac{0.6931}{-0.00465} = 149 \text{ hr.}$$

The rate of a chemical reaction leads to similar differential equations. An nth order reaction proceeds at a rate that is proportional to the nth power of the amount of the reacting compounds. Of course this leads to an exponential solution only when $n = 1$. Note that n need not be an integer.

Newton's law of cooling states that the rate of change of temperature of an object is proportional to the difference of the temperatures of the object and its surroundings. This again leads to differential equations with variables that are separable and solutions that are exponential.

Example 5

A thermometer reading 75°F is placed in a refrigerator in which the temperature is 35°F. After 10 min the thermometer reads 50°F. What will it read after 20 min?

By Newton's law of cooling, the rate of change of the temperature T of the thermometer is proportional to the difference of the temperatures of the refrigerator and the thermometer. While the thermometer's temperature is changing, that of the refrigerator remains at 35°F. Thus we have the differential equation

$$\frac{dT}{dt} = k(T - 35).$$

We now separate the variables and integrate.

$$\frac{dT/dt}{T - 35} = k$$

$$\ln|T - 35| = kt + c.$$

We may drop the absolute values, since T cannot go below 35.

$$T - 35 = e^{kt+c} = Ce^{kt}$$

$$T = 35 + Ce^{kt}.$$

Now $T = 75$ when $t = 0$, and $T = 50$ when $t = 10$. From the first condition we have

$$75 = 35 + Ce^0$$

$$C = 75 - 35 = 40$$

$$T = 35 + 40e^{kt}.$$

For the second condition we have

$$50 = 35 + 40e^{10k}$$

$$40e^{10k} = 15$$

$$e^{10k} = \frac{15}{40} = 0.375$$

$$10k = \ln 0.375 = -0.9808$$

$$k = -0.0981,$$

$$T = 35 + 40e^{-0.0981t}.$$

Finally, when $t = 20$,

$$T = 35 + 40e^{-0.0981(20)}$$

$$= 35 + 40e^{-1.962}$$

$$= 35 + 40(0.1406)$$

$$= 40.6°F.$$

Problems

A

In Problems 1–4, find a general solution of the given differential equation.

1. $y' = xy + x.$
2. $y' = 2xy^2 - y^2.$
3. $y' = \sin^2 y \cos x.$
4. $x^2 + yy' = 0.$

B

In Problems 5 and 6, find a general solution of the given differential equation.

5. $y' - e^{x+y} = 0.$
6. $y' \ln y = 3xy + y.$

In Problems 7–10, find a particular solution satisfying the given differential equation and the given condition.

7. $(2 + x)y' = y^2;$ $y = 1$ when $x = -1.$
8. $2xy' = y;$ $y = 2$ when $x = 1.$
9. $(1 - x^2)y' = x;$ $y = 2$ when $x = 0.$
10. $y' = ye^x;$ $y = 1$ when $x = 0.$

11. If, under certain conditions, the number of bacteria in a quart of milk triples in 1 hour, in how many hours will it be 100 times the original number?

12. The following table† gives estimates of world populations.

Year	Population	Year	Population
1650	508,000,000	1920	1,811,000,000
1750	711,000,000	1930	2,015,000,000
1800	912,000,000	1940	2,249,000,000
1850	1,131,000,000	1950	2,509,000,000
1900	1,590,000,000	1960	3,008,000,000

Let us suppose that the rate of growth of the population is directly proportional to its size. Use the years 1650 and 1750 to find the population for the years 1850 and 1950. Compare your results with the values given in the table. Explain the discrepancy.

† Data taken from "Population," *Encyclopedia Americana International Edition*, (1972) XXII, 366. Reprinted with permission of the The Encyclopedia Americana, copyright 1972, The Americana Corporation.

13. Use the rate of growth indicated by world population figures (see Problem 12) for the years 1940 and 1960 to predict the world population in 2000.

14. If the wolf population of a certain isolated region decreased from 100,000 in 1950 to 75,000 in 1970, when can it be expected to be down to 10,000, given the same rate of decrease?

15. A radioactive element has a half-life of 50 hours. How long will it take for 95% of it to decompose?

16. Radium-226 has a half-life of 1622 years. How much of a 0.1 g sample remains after 50 years?

17. It is determined that 0.0001% of the hydrogen atoms in a water sample are H-3. One year later only 0.0000947% are H-3. Find the half-life of H-3.

18. It has been found that the bombardment of the upper atmosphere by cosmic rays converts nitrogen to radioactive carbon-14 with a half-life of 5760 years. This C-14 is then assimilated by all plants and animals and is known to exist in them in certain small percentages. When the plant or animal dies, it cannot assimilate new C-14; and the amount present at death decreases over the years by radioactivity. This is the basis of the carbon-14 method of dating organic remains. A bone was dated by burning a small sample of it and testing the resulting carbon dioxide with a Geiger counter. It was found that 90% of its C-14 had decomposed. How old is the bone?

19. The decomposition of nitrogen pentoxide into nitrogen tetroxide and oxygen is a first-order reaction; that is, the rate of decomposition of nitrogen tetroxide is directly proportional to its concentration. A 5-mole solution of nitrogen pentoxide decomposes to give a 3-mole solution in 10 min. How long does it take for 90% of the nitrogen pentoxide to decompose?

20. The rate of decomposition of a certain gas is directly proportional to the amount of the gas present. If 250 ml of the gas decompose to leave only 200 ml in 30 min, how long will it take for 100 ml of the original 250 ml to decompose?

21. Work Problem 20 assuming that the decomposition is a second-order reaction; that is, reaction is directly proportional to the square of the amount of gas present.

22. An outdoor thermometer registering 20°F is brought into a room in which the temperature is 75°F. After 5 min, the thermometer reading is 50°F. How long does it take for the temperature of the thermometer to come within 1/2°F of room temperature?

23. A thermometer reading 20°F is brought into a room in which the temperature is 75°F. After 2 min, the thermometer reads 38°F. When will it read 70°F?

C

24. A thermometer reading 10°F is brought indoors where the temperature is 75°F. After 1 min, the thermometer reads 30°F. After 5 min, the thermometer is taken back outside where the temperature is 10°F. What does the thermometer read 5 min after it is taken outside?

25. Suppose a nuclear reactor leaks a radioactive element x into the surrounding environment at a constant rate p. It decays at a rate that is proportional to the amount x that is present. If $x = 0$ when $t = 0$ (t is time), find the amount of the radioactive element in the environment as a function of time. Find the limiting value of x as $t \to +\infty$.

12.6

Utilization of Natural Resources

Some of the methods developed in the last section can be used to consider the ways in which our natural resources are being used. Both animal and mineral resources can be subjected to mathematical analysis; we consider first the use of mineral resources, since it represents the simpler problem. The minerals of the earth represent a natural resource whose supply is fixed. Once they are consumed, we must either find substitutes or learn to live without them. It is then natural to ask how long our supply can be expected to last.

We noted in the foregoing section that the growth of populations tends to follow an exponential curve. There, the data of Problem 12 bear out the contention that the world population follows an exponential curve. Thus the use of iron, coal, petroleum, and so forth, which is directly related to population size, can also be expected to follow an exponential curve. If x represents the yearly consumption of a given mineral at time t and x_0 is the yearly consumption when $t = 0$, then

$$x = x_0 e^{kt},$$

as we saw earlier. Now we are interested in the amount of the mineral used over a period of time from $t = 0$ to $t = t_1$. If we subdivide the time interval $[0, t_1]$ into n subintervals, then the ith subinterval is of length Δt_i and it corresponds to a rate of consumption x_i^*. This gives a consumption over the ith subinterval approximated by $x_i^* \Delta t_i$, and an approximate total consumption from $t = 0$ to $t = t_1$ of

$$\sum_{i=1}^{n} x_i^* \, \Delta t_i.$$

The limit of this approximating sum as the lengths of the subintervals approach 0 gives the total amount A of the mineral used in the time interval $[0, t_1]$. Thus

$$A = \lim_{\|S\| \to 0} \sum_{i=1}^{n} x_i^* \, \Delta t_i = \int_0^{t_1} x \, dt$$

$$= \int_0^{t_1} x_0 e^{kt} \, dt = \frac{x_0}{k} e^{kt} \bigg|_0^{t_1}$$

$$= \frac{x_0}{k} (e^{kt_1} - 1).$$

From this result we can determine the depletion times of various minerals using the data in Table 12.1.

Table 12.1†

Mineral	World Reserves	1972 Rate of Consumption	Rate of Increase of Consumption
Bauxite	15,000,000	61,900	6.5%
Coal (bituminous)	1,993,200,000	2,100,000	(4%)
Copper	340,000	7,100	4%
Feldspar	913,000	2,500	4.1–6.7%
Iron ore	250,000,000	763,000	(4%)
Natural gas	1,755,000	42,900	5.7%
Petroleum	631,900	18,500	(17,600 in 1971)
Potash	25,700,000	22,500	3.9%

† Data taken from *Commodity Data Summaries* (Bureau of Mines, United States Department of the Interior, January 1973). Percentages in parentheses are taken from earlier editions of the same publication. The figures (except for natural gas and petroleum) are given in thousands of tons. The figures for natural gas are given in billions of cubic feet; those for petroleum are in millions of barrels.

Example 1

Use the data in Table 12.1 to determine the depletion date for feldspar (used in the manu-facture of glass, pottery, and enamel) if it is used at the 1972 rate. Determine the depletion date if the consumption of feldspar increases at the lowest rate given.

The time required to deplete the reserves of feldspar if it is used at a constant rate is found by division.

$$t = \frac{913,000}{2,500} = 365.2 \text{ years.}$$

Thus the depletion date is

$$1972 + 365 = 2337.$$

Now let us use the lowest rate of increase given for feldspar, 4.1%. This is the value of k in the formula

$$A = \frac{x_0}{k}(e^{kt_1} - 1).$$

To determine the time needed to deplete the reserves, we let A equal the amount of the reserves and solve for t_1. Thus

$$913,000 = \frac{2,500}{0.041}(e^{0.041t_1} - 1)$$

$$15.0 = e^{0.041t_1} - 1$$

$$e^{0.041t_1} = 16.0$$

$$0.041t_1 = \ln 16.0 = 2.77$$

$$t_1 = 67.6 \text{ years.}$$

Thus the depletion date with the use of feldspar increasing at the rate of 4.1% per year is

$$1972 + 68 = 2040.$$

Let us now consider some problems connected with the use of animal resources. As we saw in the last section, a population tends to grow exponentially because the number of offspring is directly related to the size of the population. But this holds only up to a certain point. For example, the number of fish in a given lake cannot continue to increase exponentially; eventually a point is reached at which the lake is incapable of supporting more fish. Even sooner, however, other factors tend to limit the growth of the population. For example, adults of some species eat the young of other species. Thus, as the number of adults increases, the population is liable to be replenished at a slower rate because of the increased incidence of predation. Effects such as this tend to bring about self-regulation of a system.

Biologists who study this situation have found that the growth curve, which they call the *reproduction curve*, tends to be S-shaped, as in Figure 12.8. Thus if the population of a given species in a certain ecosystem is x_1 at the beginning of one year, it is $x_2 = f(x_1)$ at the beginning of the second, $x_3 = f(x_2)$ at the start of the third, and so on. Note that the shape of the curve gives values $x_1, x_2, x_3, \ldots$, which tend to level off at a stable population level s. If the population level is greater than s at the beginning of one year, as is x_1' of Figure 12.8, then $x_2' = f(x_1') < x_1'$. That is, the population level decreases, again tending to the stable population level s.

Now suppose we have a commercially marketable species in a (more or less) closed ecosystem and have determined that its reproduction curve is given by $y = f(x)$. We want to catch and kill a certain number of them each year. Our problem is to determine the *maximum sustainable yield*—that is, the maximum annual "harvest" that can be continued indefinitely. Now if the animal reaches the stable population level, s, then, no matter what the harvest, the population will not be back to s in only one year. An examination of Figure 12.8 shows that s is approached only after several years with no yield. But we can lower the population level to a point such that, with the exception of the first

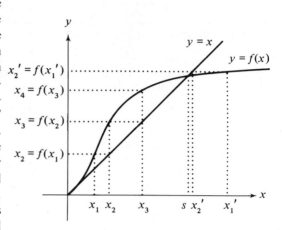

Figure 12.8

harvest, all annual harvests are equal and do not decrease the resulting population from year to year. If the population level after harvesting is x_1, then (see Figure 12.8) at the time of the next annual harvest, it will be up to $x_2 = f(x_1)$. We then harvest enough animals to lower the population to x_1. Thus the population is sustainable, as is the yield Y, where

$$Y = x_2 - x_1 = f(x_1) - x_1.$$

The only problem that remains is to determine x_1 so that the sustainable yield, Y, is a maximum. Differentiation with respect to x_1 gives

$$Y' = f'(x_1) - 1.$$

Setting this equal to 0, we see that

$$f'(x_1) = 1.$$

Thus the maximum sustainable yield is attained if we harvest to the level x_1 at which $f'(x_1) = 1$.

Example 2

It is determined that the reproduction curve for lake trout in Blue Lake is

$$y = x + 3.75 \times 10^{-5} x^2 - 1.25 \times 10^{-9} x^3.$$

Determine the stable population level, s. To what level should the lake trout population be reduced to give the maximum sustainable yield? What is that yield?

Since the reproduction curve crosses the line $y = x$ at the stable population level, we simply solve those two equations simultaneously.

$$x = x + 3.75 \times 10^{-5} x^2 - 1.25 \times 10^{-9} x^3$$
$$3.75 \times 10^{-5} x^2 - 1.25 \times 10^{-9} x^3 = 0$$
$$1.25 \times 10^{-9} x^2(30,000 - x) = 0$$
$$x = 0 \quad \text{or} \quad x = 30,000.$$

While a zero population is stable, that is hardly the number we want; the stable population level is $s = 30,000$.

The population level giving the maximum sustainable yield also gives $y' = 1$. Thus

$$y' = 1 + 7.5 \times 10^{-5}\, x - 3.75 \times 10^{-9}\, x^2 = 1$$
$$3.75 \times 10^{-9}\, x(20,000 - x) = 0$$
$$x = 0 \quad \text{or} \quad x = 20,000.$$

Clearly $x = 0$ does not give a maximum; it must be $x = 20,000$. The yield is

$f(20,000) - 20,000$

$= 20,000 + 3.75 \times 10^{-5}(20,000)^2 - 1.25 \times 10^{-9}(20,000)^3 - 20,000$

$= 20,000 + 15,000 - 10,000 - 20,000$

$= 5000.$

The foregoing discussion gives us a means of determining the maximum sustainable yield in the harvesting of animal resources. But the maximum yield does not necessarily correspond to the maximum profit. To see why, we must first consider the difference between money in hand and money due at some time in the future. It is clear that money in hand is worth more than money due at some time in the future; the only question is, How much more? The answer depends upon the rate at which we can invest money. For example, suppose we have $100 that we can invest at 6% compounded annually. After one year, we shall have our original $100 together with the interest of $6. Thus $100 today is worth $106 one year from today. Since the interest is compounded annually, after the first year, interest is paid, not only on the original principal, but also on all interest accumulated in the first year. We see then that, at the end of the second year, we have $106 together with the interest on $106, which is $6.36. This means that $100 today is worth $112.36 two years from today. We may continue in this way to find the value of $100 at any time in the future. More generally the above methods show that if a principal, P, is invested at an effective rate (that is, an interest rate compounded annually), i, then the final amount, A, to which P accumulates after n years is

$$A = P(1 + i)^n.$$

Now let us reverse the problem. If we are to receive an amount A in n years, what is the present value (that is, the equivalent at the present time) of A? If we can invest money at an effective rate, i, then we simply solve our accumulation formula for P, obtaining

$$P = A(1 + i)^{-n}.$$

Thus if money can be invested at 6% effective, then $100 due in 2 years has a present value of

$$P = 100(1 + 0.06)^{-2} = \$89.00.$$

We see, then, that money that we receive at some time in the future is worth a smaller amount at the present time.

Now let us take this into consideration in the harvesting of wildlife. Suppose that an animal species with reproduction curve $y = f(x)$ has a population level x_1 and is harvested to the level x each year. Then the first year's harvest is $x_1 - x$; thereafter there is the sustainable annual yield $Y = f(x) - x$. If we assume that the price per animal is p and that this price can be expected to remain at that level indefinitely, then the value of each year's harvest *at the time of harvesting* is $p(x_1 - x)$ the first year and $p[f(x) - x]$ for all other years. Now let us consider the present value, not only of this year's harvest, but of all future harvests. If money can be invested at an effective rate i, then the present value of that animal resource is

$$P = p(x_1 - x) + \frac{p}{1 + i}[f(x) - x] + \frac{p}{(1 + i)^2}[f(x) - x] + \frac{p}{(1 + i)^3}[f(x) - x] + \cdots$$

$$= p(x_1 - x) + p[f(x) - x]\left[\frac{1}{1 + i} + \frac{1}{(1 + i)^2} + \frac{1}{(1 + i)^3} + \cdots\right].$$

Now the infinite sum in brackets can be shown (see Theorem 19.2, page 605) to have the value $1/i$. Thus the present value is

$$P = p(x_1 - x) + \frac{p}{i}[f(x) - x].$$

Now let us differentiate to find the value of x that maximizes P.

$$P' = -p + \frac{p}{i}[f'(x) - 1] = 0,$$

$$f'(x) - 1 = i$$
$$f'(x) = 1 + i.$$

If this is compared with our former result, we see that harvesting to give the maximum sustainable yield reduces the population to the value x for which

$$f'(x) = 1,$$

while harvesting to give the maximum profit reduces the population to the level x at which

$$f'(x) = 1 + i.$$

The difference is illustrated graphically in Figure 12.9. It is seen there that the maximum yield occurs at that value of x at which the slope of $y = f(x)$ is 1, while the maximum profit occurs when the slope is $1 + i$. This larger slope normally corresponds to a smaller population level. Furthermore, the larger the value of i, the smaller the population level that should be maintained for a maximum profit. If $y = f(x)$ and i are such that $f'(x) = 1 + i$ has no solution, then the most profitable use of the given animal resource is immediate annihilation! As we can see, good business practices and good conservation practices do not necessarily go hand in hand.

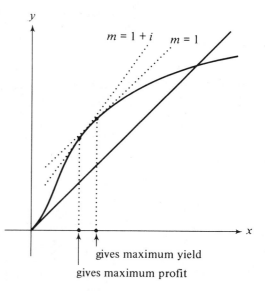

gives maximum yield

gives maximum profit

Figure 12.9

Example 3

Using the data of Example 2 and assuming that money can be invested at an effective 7.5%, determine the population level to which the lake trout must be reduced to achieve the maximum profit.

Noting that the maximum profit occurs when $f'(x) = 1 + i$, we have

$$1 + 7.5 \times 10^{-5} x - 3.75 \times 10^{-9} x^2 = 1.075$$
$$7.5 \times 10^{-5} x - 3.75 \times 10^{-9} x^2 = 0.075$$
$$20{,}000x - x^2 = 20{,}000{,}000$$
$$x^2 - 20{,}000x + 20{,}000{,}000 = 0.$$

By the quadratic formula,

$$x = \frac{20{,}000 \pm \sqrt{400 \times 10^6 - 80 \times 10^6}}{2}$$

$$= 10{,}000 \pm 4000\sqrt{5}$$

$$= \begin{cases} 18{,}900, \\ 1100. \end{cases}$$

The population level must be one of these two numbers. Although the smaller one is absurdly low, we can be sure of the correct figure by putting both back into the equation for the profit.

$$P = p\left(x_1 - x + \frac{f(x) - x}{i}\right).$$

For $x = 18{,}900$, $P = p(x_1 + 47{,}800)$; for $x = 1100$, $P = p(x_1 - 570)$. It is clear that the profit is a maximum if the lake trout in Blue Lake are harvested to a level of 18,900.

Problems

A

In Problems 1–4, use the data of Table 12.1 to determine the depletion date for the given mineral

(a) if it is used at the 1972 rate,
(b) if the consumption increases at the rate given in the table.

1. Bauxite.
2. Bituminous coal.
3. Natural gas.
4. Potash.

5. Determine the depletion date for feldspar if the consumption increases at the highest rate shown in Table 12.1

6. Determine the depletion date for petroleum if the consumption increases at the same rate that it did between 1971 and 1972.

B

7. Figure 12.10(a) gives a reproduction curve that is typical for many species. The population level c is called the critical level. What happens when the population level falls below c? It is feared that several species have already fallen below this critical level.

8. A few species (lemming and Arctic hare, for example) have a reproduction curve with a maximum as shown in Figure 12.10(b). Trace the annual population levels from the level x_1 given in Figure 12.10(b).

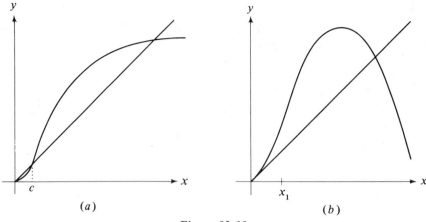

(a) (b)

Figure 12.10

In Problems 9–12, use the reproduction curve given to determine the stable population level, s. Also determine the level to which that population should be harvested to obtain the maximum sustainable yield and calculate that yield.

9. $y = 1.4x - 4 \times 10^{-5} x^2$.
10. $y = x + 9.375 \times 10^{-4} x^2 - 78.125 \times 10^{-10} x^3$.
11. $y = x + 10^{-11} x^4 - 2 \times 10^{-15} x^5$.
12. $y = x + 4 \times 10^{-9} x^3 - 2.5 \times 10^{-13} x^4$.

In Problems 13–16, use the given reproduction curve and effective rate of interest to determine the level to which the population should be harvested to obtain the maximum profit.

13. $y = 1.4x - 4 \times 10^{-5} x^2$; $i = 8\%$.
14. $y = 1.4x - 4 \times 10^{-5} x^2$; $i = 12\%$.
15. $y = x + 3.75 \times 10^{-5} x^2 - 1.25 \times 10^{-9} x^3$; $i = 15\%$.
16. $y = x + 1.5 \times 10^{-4} x^2 - 2.5 \times 10^{-8} x^3$; $i = 10.8\%$.

17. What is the lowest effective rate of interest at which money can be invested that would make annihilation of the population of Example 2 the most profitable? (*Hint:* Consider Figure 12.9. At what point is the slope the greatest?)

18. What is the lowest effective rate of interest that would make annihilation of the populations of Problems 13 and 14 the most profitable? (See Problem 17.)

12.7

Hyperbolic Functions

Certain combinations of functions occur frequently enough in applications of mathematics that it is convenient to set them apart by giving them special names. In this section we consider the hyperbolic functions, which are defined in terms of exponential functions but are like the trigonometric functions in many ways. They are called the hyperbolic sine, hyperbolic cosine, and so on, and are abbreviated sinh, cosh, and so forth, respectively.

Definition

(a) $\sinh x = \dfrac{e^x - e^{-x}}{2}.$

(b) $\cosh x = \dfrac{e^x + e^{-x}}{2}.$

(c) $\tanh x = \dfrac{\sinh x}{\cosh x}.$

(d) $\coth x = \dfrac{\cosh x}{\sinh x}.$

(e) $\operatorname{sech} x = \dfrac{1}{\cosh x}.$

(f) $\operatorname{csch} x = \dfrac{1}{\sinh x}.$

The first two hyperbolic functions are easily graphed by addition of ordinates, and the remaining four by division of ordinates. For example, since

$$\sinh x = \frac{e^x}{2} - \frac{e^{-x}}{2},$$

we graph $y = e^x/2$ and $y = -e^{-x}/2$ and add the ordinates. This is given in (a) of Figure 12.11. The graph of $y = \tanh x$ is found by noting that

$$\tanh x = \frac{\sinh x}{\cosh x}.$$

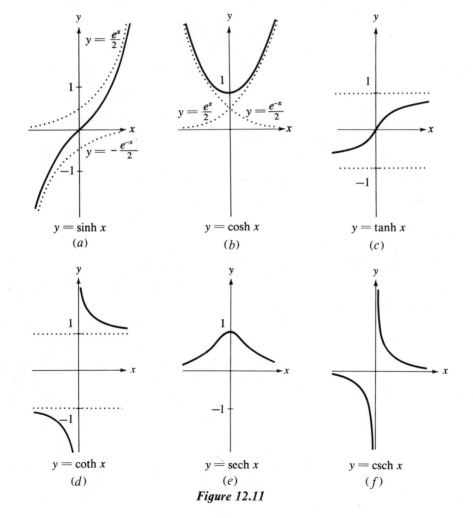

$y = \sinh x$
(a)

$y = \cosh x$
(b)

$y = \tanh x$
(c)

$y = \coth x$
(d)

$y = \operatorname{sech} x$
(e)

$y = \operatorname{csch} x$
(f)

Figure 12.11

For each value of x, we divide $y_1 = \sinh x$ by $y_2 = \cosh x$ to find $\tanh x$. The result is given in Figure 12.11(c). The graphs of all six hyperbolic functions are given in the figure. They are not periodic functions as are the trigonometric functions; but many identities involving hyperbolic functions are quite similar to those for the corresponding trigonometric functions. Some of the more important ones are given below.

$$\cosh^2 x - \sinh^2 x = 1.$$

$$1 - \tanh^2 x = \operatorname{sech}^2 x.$$

$$\coth^2 x - 1 = \operatorname{csch}^2 x.$$

$$\sinh (x \pm y) = \sinh x \cosh y \pm \cosh x \sinh y.$$

$$\cosh (x \pm y) = \cosh x \cosh y \pm \sinh x \sinh y.$$

$$\tanh (x \pm y) = \frac{\tanh x \pm \tanh y}{1 \pm \tanh x \tanh y}.$$

$$\sinh 2x = 2 \sinh x \cosh x.$$

$$\cosh 2x = \cosh^2 x + \sinh^2 x = 2 \cosh^2 x - 1 = 2 \sinh^2 x + 1.$$

$$\tanh 2x = \frac{2 \tanh x}{1 + \tanh^2 x}.$$

Example 1

Prove that $\cosh^2 x - \sinh^2 x = 1.$

$$\cosh^2 x - \sinh^2 x = \left(\frac{e^x + e^{-x}}{2}\right)^2 - \left(\frac{e^x - e^{-x}}{2}\right)^2$$

$$= \frac{e^{2x} + 2 + e^{-2x}}{4} - \frac{e^{2x} - 2 + e^{-2x}}{4}$$

$$= 1.$$

Example 2

Prove that $\sinh (x + y) = \sinh x \cosh y + \cosh x \sinh y.$

$\sinh x \cosh y + \cosh x \sinh y$

$$= \frac{e^x - e^{-x}}{2} \cdot \frac{e^y + e^{-y}}{2} + \frac{e^x + e^{-x}}{2} \cdot \frac{e^y - e^{-y}}{2}$$

$$= \frac{e^{x+y} + e^{x-y} - e^{-x+y} - e^{-x-y} + e^{x+y} - e^{x-y} + e^{-x+y} - e^{-x-y}}{4}$$

$$= \frac{2e^{x+y} - 2e^{-x-y}}{4}$$

$$= \frac{e^{x+y} - e^{-(x+y)}}{2}$$

$$= \sinh (x + y).$$

Example 3

Prove that $\sinh 2x = 2 \sinh x \cosh x$.

From Example 2,
$$\sinh (x + y) = \sinh x \cosh y + \cosh x \sinh y.$$
Replacing y by x, we have
$$\sinh 2x = 2 \sinh x \cosh x.$$

Example 4

Given $\tanh x = 3/5$, find the values of the other five hyperbolic functions.

$$\coth x = \frac{1}{\tanh x} = 5/3.$$

$$\operatorname{sech}^2 x = 1 - \tanh^2 x = 1 - \left(\frac{3}{5}\right)^2 = \frac{16}{25}.$$

$$\operatorname{sech} x = \frac{4}{5} \quad \text{(it cannot be negative)}.$$

$$\cosh x = \frac{1}{\operatorname{sech} x} = \frac{5}{4}.$$

$$\sinh^2 x = \cosh^2 x - 1 = \frac{25}{16} - 1 = \frac{9}{16}.$$

$$\sinh x = \frac{3}{4} \quad \text{(tanh } x \text{ and sinh } x \text{ agree in sign)}.$$

$$\operatorname{csch} x = \frac{1}{\sinh x} = \frac{4}{3}.$$

Theorem 12.9

If u is a function of x, then

(1) $\dfrac{d}{dx} \sinh u = \cosh u \cdot u'$,

(2) $\dfrac{d}{dx} \cosh u = \sinh u \cdot u'$,

(3) $\dfrac{d}{dx} \tanh u = \operatorname{sech}^2 u \cdot u'$,

(4) $\dfrac{d}{dx} \coth u = -\operatorname{csch}^2 u \cdot u'$,

(5) $\dfrac{d}{dx} \operatorname{sech} u = -\operatorname{sech} u \tanh u \cdot u'$,

(6) $\dfrac{d}{dx} \operatorname{csch} u = -\operatorname{csch} u \coth u \cdot u'$.

Since the hyperbolic functions are defined in terms of exponential functions, it is a simple matter to prove the theorem. The proof is left to the student.

Example 5

Differentiate $y = \cosh(3x + 1)$.

$$y' = 3 \sinh(3x + 1).$$

Example 6

Differentiate $y = x \sinh x$.

$$y' = x \cosh x + \sinh x.$$

Hyperbolic functions occur as solutions of certain types of differential equations (see Problems 39 and 40). Furthermore a wire hanging of its own weight is represented by a hyperbolic cosine; however a full analysis of this situation must be delayed until we consider the length of an arc (see page 413).

Problems

A

In Problems 1–6, find the values of the other hyperbolic functions.

1. $\sinh x = 4/3$.
2. $\cosh x = 5/3, \quad x < 0$.
3. $\sinh x = -12/5$.
4. $\cosh x = 13/12, \quad x > 0$.
5. $\tanh x = -5/13$.
6. $\operatorname{sech} x = 15/17, \quad x > 0$.

In Problems 7–10, differentiate and simplify.

7. $y = \tanh(3x - 2)$.
8. $y = \sinh^2 3x$.
9. $y = \sinh x^2$.
10. $y = x^2 \cosh x$.

B

In Problems 11–20, differentiate and simplify.

11. $y = \operatorname{csch}^2 x + \coth^2 x$.
12. $y = (2 \sinh x + \cosh x)^2$.
13. $y = \operatorname{sech} \sqrt{x}$.
14. $y = \operatorname{sech} x \tanh x$.
15. $y = e^{\tanh x}$.
16. $y = \cos x \sinh x$.
17. $y = \sinh^2 x + \cosh^2 x$.
18. $y = \ln(\tanh x)$.
19. $y = \sin(\cosh x)$.
20. $y = \cosh(\sin x)$.

In Problems 21–24, find an equation of the line tangent to the curve at the point indicated.

21. $y = \sinh x, \quad (0, 0)$.
22. $y = x \cosh x, \quad (0, 0)$.
23. $y = \tanh 3x, \quad (0, 0)$.
24. $y = \cosh x, \quad (1, (e^2 + 1)/2e)$.

In Problems 25–34, prove the identities.

25. $1 - \tanh^2 x = \operatorname{sech}^2 x$.
26. $\coth^2 x - 1 = \operatorname{csch}^2 x$.
27. $\cosh(x + y) = \cosh x \cosh y + \sinh x \sinh y$.
28. $\tanh(x + y) = \dfrac{\tanh x + \tanh y}{1 + \tanh x \tanh y}$.
29. $\sinh(x - y) = \sinh x \cosh y - \cosh x \sinh y$.
30. $\cosh 2x = \cosh^2 x + \sinh^2 x = 2 \sinh^2 x + 1 = 2 \cosh^2 x - 1$.
31. $\tanh 2x = \dfrac{2 \tanh x}{1 + \tanh^2 x}$.
32. $\sinh \dfrac{x}{2} = \pm \sqrt{\dfrac{\cosh x - 1}{2}}$.
33. $\cosh \dfrac{x}{2} = \sqrt{\dfrac{\cosh x + 1}{2}}$.
34. $\sinh 3x = 3 \sinh x + 4 \sinh^3 x$.

In Problems 35–38, prove the given statement.

35. $\dfrac{d}{dx}\sinh u = \cosh u \cdot u'$.

36. $\dfrac{d}{dx}\cosh u = \sinh u \cdot u'$.

37. $\dfrac{d}{dx}\tanh u = \text{sech}^2 u \cdot u'$.

38. $\dfrac{d}{dx}\text{sech } u = -\text{sech } u \tanh u \cdot u'$.

39. Verify that $y = \sinh 2x$ and $y = \cosh 2x$ are both solutions of the differential equation $y'' - 4y = 0$. Verify that $y = c_1 \sinh 2x + c_2 \cosh 2x$, where c_1 and c_2 are arbitrary constants, is also a solution. Compare with Problem 39, page 330.

40. Verify that $y = k \cosh (x/k) + C$ is a solution of the differential equation $(ky'')^2 - (y')^2 = 1$.

C

41. The Gudermannian (named for Christoph Gudermann, who first investigated this function) is defined by
$$\text{gd } x = \text{Arctan (sinh } x).$$
Show that if $y = \text{gd } x$, then $y' = \text{sech } x$. Also show that $y = \pm \pi/2$ are horizontal asymptotes. Sketch the graph.

42. If e^{ix}, where $i = \sqrt{-1}$, is defined to be $\cos x + i \sin x$, show that $e^{ix} = \cosh ix + \sinh ix$.

43. We see in Problem 39 that $y = \sinh 2x$ and $y = \cosh 2x$ are solutions of the differential equation $y'' = 4y$. Similarly, we saw in Problem 39, page 330, that $y = \sin 2x$ and $y = \cos 2x$ are solutions of $y'' = -4y$.   Use these results to find a single differential equation having all four of the above functions as solutions. (*Hint*: $y'' = \pm 4y$ are two–not one–differential equations; but they are a starting point for finding the desired equation.)

12.8

Integrals of Hyperbolic Functions

The formulas for the derivatives of hyperbolic functions can be put into integral form.

Theorem 12.10

If u is a function of x, then

(1) $\displaystyle\int \sinh u \cdot u' \, dx = \int \sinh u \, du = \cosh u + C,$

(2) $\displaystyle\int \cosh u \cdot u' \, dx = \int \cosh u \, du = \sinh u + C,$

(3) $\displaystyle\int \text{sech}^2 u \cdot u' \, dx = \int \text{sech}^2 u \, du = \tanh u + C,$

(4) $\displaystyle\int \text{csch}^2 u \cdot u' \, dx = \int \text{csch}^2 u \, du = -\coth u + C,$

(5) $\displaystyle\int \text{sech } u \tanh u \cdot u' \, dx = \int \text{sech } u \tanh u \, du = -\text{sech } u + C,$

(6) $\displaystyle\int \text{csch } u \coth u \cdot u' \, dx = \int \text{csch } u \coth u \, du = -\text{csch } u + C.$

Example 1

Evaluate $\int x \sinh x^2 \, dx$.

$$\int x \sinh x^2 \, dx = \frac{1}{2} \int 2x \sinh x^2 \, dx = \frac{1}{2} \cosh x^2 + C.$$

Example 2

Evaluate $\int \dfrac{\operatorname{sech} \sqrt{x} \tanh \sqrt{x}}{\sqrt{x}} \, dx$.

$$\int \frac{\operatorname{sech} \sqrt{x} \tanh \sqrt{x}}{\sqrt{x}} \, dx = 2 \int \frac{\operatorname{sech} \sqrt{x} \tanh \sqrt{x}}{2\sqrt{x}} \, dx$$

$$= -2 \operatorname{sech} \sqrt{x} + C.$$

Example 3

Evaluate $\int \tanh^2 x \, dx$.

We must first use an identity to put this into a form we can integrate.

$$\int \tanh^2 x \, dx = \int (1 - \operatorname{sech}^2 x) \, dx$$

$$= x - \tanh x + C.$$

Example 4

Evaluate $\int \sinh^2 x \cosh x \, dx$.

While this expression does not fit the form of any part of Theorem 12.10, we can handle it in the same way that we handled trigonometric expressions of this type. Since

$$\frac{d}{dx} \sinh x = \cosh x,$$

the above integral is in the form

$$\int u^2 \, u' \, dx.$$

Thus

$$\int \sinh^2 x \cosh x \, dx = \frac{1}{3} \sinh^3 x + C.$$

Problems

A

In Problems 1 and 2, evaluate the integrals.

1. $\displaystyle\int \cosh (2x + 1) \, dx$.

2. $\displaystyle\int_0^3 \sinh x \, dx$.

B

In Problems 3–12, evaluate the integrals.

3. $\displaystyle\int \frac{\text{sech}^2\,(1/x)}{x^2}\,dx.$

4. $\displaystyle\int 2\coth^2 2x\,dx.$

5. $\displaystyle\int \sinh^3 x \cosh x\,dx.$

6. $\displaystyle\int \frac{\text{sech}^2\,2x}{\tanh 2x}\,dx.$

7. $\displaystyle\int \text{csch}^5 x \coth x\,dx.$

8. $\displaystyle\int \frac{\sinh 3x}{\cosh 3x}\,dx.$

9. $\displaystyle\int \frac{\sinh x\,dx}{1 + \cosh^2 x}.$

10. $\displaystyle\int \frac{\cosh x\,dx}{\sqrt{1 - \sinh^2 x}}.$

11. $\displaystyle\int \frac{\text{sech}^2\,\sqrt{x}}{\sqrt{x}}\,dx$

12. $\displaystyle\int_{-1/2}^{2} \text{sech}^2\,(2x + 1)\,dx.$

In Problems 13 and 14, find the area of the region bounded by the given curves.

13. $y = \cosh x, \quad y = 0, \quad x = 0, \quad x = 1.$

14. $y = \sinh x, \quad y = 0, \quad x = \ln 3.$

C

15. Evaluate $\displaystyle\int \sinh^2 x\,dx.$

12.9

Inverse Hyperbolic Functions

We can consider the inverses of the hyperbolic functions just as we did those of the trigonometric functions. They are called the inverse hyperbolic sine, etc., and are represented by $\sinh^{-1}$, etc.

Definition

$y = \sinh^{-1} x$ *means* $x = \sinh y$. *The other five inverse hyperbolic functions are defined similarly except for* $\cosh^{-1} x$ *and* $\text{sech}^{-1} x$, *for which y must be nonnegative.*

The graphs of these six functions, together with their domains and ranges, are given in Figure 12.12. Since the hyperbolic functions are defined in terms of exponential functions, it is reasonable to expect that the inverse hyperbolic functions can be expressed in terms of the inverses of the exponential functions—that is, in terms of the logarithmic functions. In fact, this is the case.

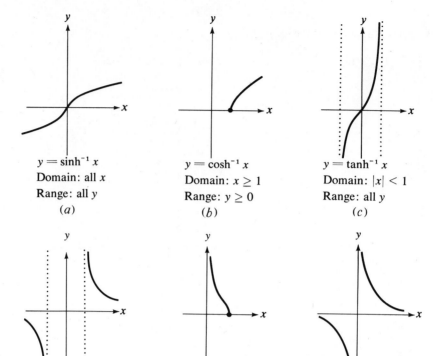

$y = \sinh^{-1} x$
Domain: all x
Range: all y
(a)

$y = \cosh^{-1} x$
Domain: $x \geq 1$
Range: $y \geq 0$
(b)

$y = \tanh^{-1} x$
Domain: $|x| < 1$
Range: all y
(c)

$y = \coth^{-1} x$
Domain: $|x| > 1$
Range: $|y| > 0$
(d)

$y = \operatorname{sech}^{-1} x$
Domain: $0 < x \leq 1$
Range: $y \geq 0$
(e)

$y = \operatorname{csch}^{-1} x$
Domain: $|x| > 0$
Range: $|y| > 0$
(f)

Figure 12.12

Theorem 12.11

(1) $\sinh^{-1} x = \ln (x + \sqrt{x^2 + 1})|;$

(2) $\cosh^{-1} x = \ln (x + \sqrt{x^2 - 1}), \quad x \geq 1;$

(3) $\tanh^{-1} x = \dfrac{1}{2} \ln \dfrac{1 + x}{1 - x}, \quad |x| < 1;$

(4) $\coth^{-1} x = \dfrac{1}{2} \ln \dfrac{x + 1}{x - 1}, \quad |x| > 1;$

(5) $\operatorname{sech}^{-1} x = \ln \dfrac{1 + \sqrt{1 - x^2}}{x}, \quad 0 < x \leq 1;$

(6) $\operatorname{csch}^{-1} x = \begin{cases} \ln \dfrac{1 + \sqrt{1 + x^2}}{x} & x > 0, \\[3mm] -\ln \dfrac{1 + \sqrt{1 + x^2}}{-x} & x < 0. \end{cases}$

Proof of (1)

$$y = \sinh^{-1} x,$$

$$x = \sinh y = \frac{e^y - e^{-y}}{2},$$

$$e^y - 2x - e^{-y} = 0$$

$$e^{2y} - 2xe^y - 1 = 0.$$

This is now a quadratic equation in e^y. By the quadratic formula,

$$e^y = x \pm \sqrt{x^2 + 1}.$$

Since $\sqrt{x^2 + 1} > x$,

$$x - \sqrt{x^2 + 1} < 0.$$

But $e^y > 0$. Thus the only possibility is

$$e^y = x + \sqrt{x^2 + 1}$$

$$y = \ln (x + \sqrt{x^2 + 1}).$$

Proof of (4)

$$y = \coth^{-1} x,$$

$$x = \coth y = \frac{\cosh y}{\sinh y} = \frac{e^y + e^{-y}}{e^y - e^{-y}},$$

$$e^y(x - 1) = e^{-y}(x + 1)$$

$$e^{2y} = \frac{x + 1}{x - 1}$$

$$e^y = \sqrt{\frac{x + 1}{x - 1}}$$

$$y = \ln \sqrt{\frac{x + 1}{x - 1}} = \frac{1}{2} \ln \frac{x + 1}{x - 1}.$$

The proofs of the other four parts of the theorem, which are similar, are left to the student. These formulas give a convenient way of evaluating the inverse hyperbolic functions for particular values of x.

Example 1

Evaluate $\sinh^{-1} 2$.

$$\sinh^{-1} 2 = \ln (2 + \sqrt{2^2 + 1})$$

$$= \ln (2 + \sqrt{5}).$$

This can now be approximated. We can either use a table of natural logarithms or change the base and use common logarithms, to get

$$\sinh^{-1} 2 \approx 1.444.$$

We can find the derivatives of the inverse hyperbolic functions by one of two methods: we may use the same method that was used to find the derivatives of the inverse trigonometric functions or we may use the results of Theorem 12.11.

Theorem 12.12

If u is a function of x, then

(1) $\dfrac{d}{dx} \sinh^{-1} u = \dfrac{u'}{\sqrt{1 + u^2}}$;

(2) $\dfrac{d}{dx} \cosh^{-1} u = \dfrac{u'}{\sqrt{u^2 - 1}}$;

(3) $\dfrac{d}{dx} \tanh^{-1} u = \dfrac{u'}{1 - u^2}$;

(4) $\dfrac{d}{dx} \coth^{-1} u = \dfrac{u'}{1 - u^2}$;

(5) $\dfrac{d}{dx} \operatorname{sech}^{-1} u = \dfrac{-u'}{u\sqrt{1 - u^2}}$;

(6) $\dfrac{d}{dx} \operatorname{csch}^{-1} u = \begin{cases} \dfrac{-u'}{u\sqrt{1 + u^2}}, & u > 0, \\[3mm] \dfrac{u'}{u\sqrt{1 + u^2}}, & u < 0. \end{cases}$

Proof

We shall prove the first derivative formula by both methods.
By the first method:

$$y = \sinh^{-1} u,$$
$$u = \sinh y,$$
$$u' = \cosh y \cdot y',$$
$$y' = \frac{u'}{\cosh y}$$
$$= \frac{u'}{\sqrt{1 + \sinh^2 y}}$$
$$= \frac{u'}{\sqrt{1 + u^2}} .$$

By Theorem 12.11,

$$y = \sinh^{-1} u = \ln (u + \sqrt{u^2 + 1}),$$

$$y' = \frac{u' + \dfrac{uu'}{\sqrt{u^2 + 1}}}{u + \sqrt{u^2 + 1}} = \frac{u'(\sqrt{u^2 + 1} + u)}{(u + \sqrt{u^2 + 1})\sqrt{u^2 + 1}}$$

$$= \frac{u'}{\sqrt{u^2 + 1}} .$$

The other formulas can be proved similarly. Their proofs are left to the student.

Example 2

Differentiate $y = \sinh^{-1} x^2$.

$$y' = \frac{2x}{\sqrt{1 + x^4}}.$$

Example 3

Differentiate $y = x \tanh^{-1} x$.

$$y' = x \frac{1}{1 - x^2} + \tanh^{-1} x = \frac{x}{1 - x^2} + \tanh^{-1} x.$$

Of course, we can use Theorem 12.12 to find the corresponding integral formulas. However we shall not consider these here, because we develop a general method of handling all of these (and many others) in Chapter 15.

Problems

A

In Problems 1–6, give the value to three decimal places.

1. $\sinh^{-1} 3$.
2. $\tanh^{-1} 1/2$.
3. $\operatorname{sech}^{-1} 1/\sqrt{2}$.
4. $\cosh^{-1} 2$.
5. $\coth^{-1} 3$.
6. $\operatorname{csch}^{-1} 1$.

In Problems 7–10, differentiate and simplify.

7. $y = \cosh^{-1} (3x + 1)$.
8. $y = \tanh^{-1} (2x - 5)$.
9. $y = \sinh^{-1} \sqrt{x}$.
10. $y = x \sinh^{-1} x$.

B

In Problems 11–20, differentiate and simplify.

11. $y = \dfrac{\coth^{-1} x}{x}$.
12. $y = \ln \cosh^{-1} x$.
13. $y = \tanh^{-1} e^x$.
14. $y = \operatorname{sech}^{-1} \sin x$.
15. $y = \sqrt{1 + x^2} + \sinh^{-1} x$.
16. $y = \sqrt{1 + x} + \operatorname{csch}^{-1} \sqrt{x}$.
17. $y = (\sinh^{-1} x)^2$.
18. $y = \sqrt{\cosh^{-1} x}$.
19. $y = (1 + \tanh^{-1} x)^2$.
20. $y = (\sinh^{-1} x + \sin^{-1} x)^2$.

21. Prove Theorem 12.11 (2).
22. Prove Theorem 12.11 (3).
23. Prove Theorem 12.11 (5).
24. Prove Theorem 12.11 (6).
25. Prove Theorem 12.12 (2).
26. Prove Theorem 12.12 (3).
27. Prove Theorem 12.12 (4).
28. Prove Theorem 12.12 (5).
29. Prove Theorem 12.12 (6).

C

30. The tractrix has equation $x = a \operatorname{sech}^{-1} y/a - \sqrt{a^2 - y^2}$. Find the distance from a point (x, y) on the tractrix and the x intercept of the line tangent to it at the point (x, y). Does this suggest a way of sketching the tractrix?

31.   A box of supplies weighing 200 lb is dropped from an airplane. If gravity were the only force acting on the box, then $F = ma = 200(-32) = -6400$. But air resistance presents an opposing force equal to the square of the velocity. Thus the total force acting on the box is $F = v^2 - 6400$, which leads to the differential equation $200 \, dv/dt = v^2 - 6400$. Solve this equation to find the velocity as a function of time. (*Hint:* Use Theorem 12.12 to carry out the integration.) Show that the velocity approaches a limit as t approaches $+\infty$. What is that limit? If the box is dropped by parachute, the force of air resistance is 16 times the square of the velocity. What is the limiting velocity in this case?

32.   Evaluate $\lim_{x \to +\infty} (\sinh^{-1} x - \ln x)$.

33.   Evaluate $\displaystyle\int \frac{\tanh^{-1} x}{1 - x^2} \, dx$.

Review Problems

A

In Problems 1 and 2, differentiate and simplify.

1. $y = \dfrac{e^x}{\sin x}$.

2. $y = \ln (\csc x)$.

In Problems 3 and 4, evaluate the integral.

3. $\displaystyle\int xe^{2x^2 + 1} \, dx$

4. $\displaystyle\int \frac{(x^2 + 4x - 3) \, dx}{x - 2}$.

B

In Problems 5–12, differentiate and simplify.

5. $y = \ln \dfrac{x^2(x - 1)}{\sqrt{x + 1}}$.

6. $y = x \cdot 3^x$.

7. $y = x \sinh^2 3x$.

8. $y = 2^e \cdot e^2$.

9. $y = \operatorname{sech} e^x$.

10. $y = x^3 \ln^2 x$.

11. $y = (\ln x)^{\sin x}$.

12. $y = \tanh^{-1} (3x + 2)$.

In Problems 13–16, evaluate the integral.

13. $\displaystyle\int \frac{(x - 2)^2}{x^2 + 4} \, dx$.

14. $\displaystyle\int \frac{(e^x - 3)^2}{e^x} \, dx$.

15. $\displaystyle\int \frac{2(x + 4) \, dx}{(x + 4)^2 + 4}$.

16. $\displaystyle\int \operatorname{sech} 3x \tanh 3x \, dx$.

In Problems 17–20, find an equation of the line tangent to the given curve at the point indicated.

17. $y = \ln^2 x$; $(e, 1)$.

18. $y = xe^{x^2}$; $(1, e)$.

19. $y = x^2 \cosh x$; $(0, 0)$.

20. $y = \ln (\cos x)$; $(\pi/3, -\ln 2)$.

In Problems 21–24, find all relative maxima and minima, all points of inflection, and sketch.

21. $y = \dfrac{\ln x}{x^2}$.

22. $y = x^2 e^x$.

23. $y = \dfrac{e^x}{x + 2}$.

24. $y = x^3 \ln |x|$.

In Problems 25–28, find the area of the indicated region.

25.　The region bounded by　$y = e^x$　and　$y = e^{2x}$, between $x = 0$ and $x = 1$.

26.　The region bounded by　$y = e^x$　and the line joining $(0, 1)$ and $(2, e^2)$.

27.　The region bounded by　$y = \ln x$,　$x = 0$,　and between $y = -\ln 2$ and $y = 0$.

28.　The region bounded by　$y = \cosh x$ and　$y = (e^4 + 1)/2e^2$.

29.　Show that for any real number t, $(\cosh t, \sinh t)$ satisfies the equation　$x^2 - y^2 = 1$,　which represents a hyperbola. (Hence the name "hyperbolic functions.")

30.　Prove that

$$\tanh^2 \frac{x}{2} = \frac{\cosh x - 1}{\cosh x + 1} \quad \text{and} \quad \tanh \frac{x}{2} = \frac{\cosh x - 1}{\sinh x}.$$

31.　Find a general solution of　$e^x y' = y$.

32.　Find a particular solution of　$y' \ln y = xy$　if $y = 1$ when $x = 2$.

33.　Use the rate of growth indicated by the population figures for 1800 and 1900 (see Problem 12, page 372) to predict the world population in 2000 and 2500.

34.　It is found that a 0.001-g sample of an unknown radioactive element decomposes to 0.0009 g after 24 hours. Find the half-life of the element.

35.　A thermometer reading 70°F is placed in a freezer in which the temperature is 0°F. After 5 min, the thermometer reading is 40°F. How long will it take for the thermometer to be within 2°F of the freezer temperature?

36.　Using the data of Table 12.1, determine the depletion date for iron ore
(a) if it is used at the 1972 rate,
(b) if the consumption increases at the rate given.

37.　Repeat Problem 36 for copper.

38.　The reproduction curve for a certain species is

$$y = 1.8x - 8 \times 10^{-6} x^2.$$

To what level should that species be harvested to give the maximum sustainable yield? What is that yield?

39.　The reproduction curve for a certain species of whale is

$$y = x + 9.375 \times 10^{-5} x^2 - 15.625 \times 10^{-9} x^3.$$

To what level should that species be harvested to give the maximum sustainable yield? What is that yield?

40.　If money can be invested at 8% effective interest, to what level should the population of Problem 38 be harvested to give the maximum profit? What is the annual sustainable yield at this level?

41.　If money can be invested at 9% effective interest, to what level should the whale population of Problem 39 be harvested to give the maximum profit? What is the annual sustainable yield at this level?

C

42.　A cylindrical tank has a cross-sectional area of 6 ft² and a height of 10 ft. Water runs out the bottom at a rate that is proportional to the height of the water in the tank; the rate is 3 ft³/min when the tank is full. Suppose water is also being added to the tank at the rate of 2 ft³/min. If we start with a full tank, how long will it take for the water level to drop to 8 ft?　7 ft?　6 ft?

13

Parametric Equations

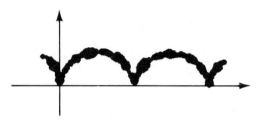

13.1

Parametric Equations

Up to now all of the equations we have dealt with have been in the form

$$y = f(x) \quad \text{or} \quad F(x, y) = 0.$$

In either case, a direct relationship between x and y is given. Sometimes, however, it is more convenient to express both x and y in terms of a third variable, called a parameter; that is,

$$x = f(t), \qquad y = g(t).$$

Each value of the parameter t gives a value of x and a value of y.

For instance, if $t = 0$ in the parametric equations

$$x = \sin t, \qquad y = \cos t,$$

we see that $x = 0$ and $y = 1$. Thus the point $(0, 1)$ is a point of the graph. Note that we still have just the x and y axes; t does not appear on the graph. Let us continue with this process. The resulting graph is given in Figure 13.1. Of course, we could continue with values of t beyond 360°, but we would simply go over the same points again. Although the value of t need not appear anywhere on the graph, we have labeled several points with their corresponding values of t. Once the points are plotted, they are joined in the order of increasing (or decreasing) values of t.

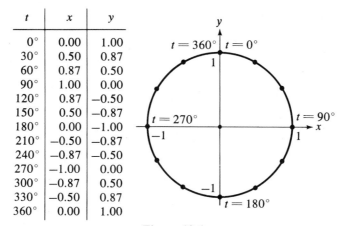

t	x	y
0°	0.00	1.00
30°	0.50	0.87
60°	0.87	0.50
90°	1.00	0.00
120°	0.87	−0.50
150°	0.50	−0.87
180°	0.00	−1.00
210°	−0.50	−0.87
240°	−0.87	−0.50
270°	−1.00	0.00
300°	−0.87	0.50
330°	−0.50	0.87
360°	0.00	1.00

Figure 13.1

The result seems to resemble a circle. How can we be *sure* it is a circle? If we had a single equation in x and y, we could easily tell by the form of the equation. Let us try to eliminate the parameter t between the equations $x = \sin t$ and $y = \cos t$:

$$\sin^2 t + \cos^2 t = 1,$$
$$x^2 + y^2 = 1.$$

We now see that we have a circle with center at the origin and radius 1.

Elimination of the parameter not only assures us that this particular curve is a circle; it also gives us a basis for sketching more rapidly than can be done by point-by-point plotting. However, we must be careful with the domain of the resulting equation. Let us illustrate this with some examples and see how the domain of $F(x, y) = 0$ plays an important role in sketching the graph.

Example 1

Graph the following two pairs of parametric equations by eliminating the parameter.

$$x = t, \qquad x = t^2,$$
$$y = t; \qquad y = t^2.$$

Elimination of the parameter gives $y = x$ in both cases. But the graphs are not the same, as the domains in the two cases will show. The domain can be determined from the first of the two parametric equations in each case. In the first

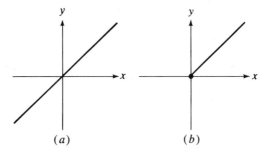

(a)   (b)

Figure 13.2

case, $x = t$ and, since there is no restriction on t, there is none on x; the domain is the set of all real numbers. In the second case, $x = t^2$. The domain of $y = x$ is the range of $x = t^2$, which is $\{x \mid x \geq 0\}$. Thus we have a restricted domain here that we did not have in the first case. The graphs are given in Figure 13.2.

Example 2

Eliminate the parameter between $x = t + 1$ and $y = t^2 + 3t + 2$ and sketch.

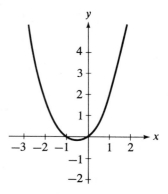

Solving $x = t + 1$ for t, we have

$$t = x - 1.$$

If this is substituted into $y = t^2 + 3t + 2$, then

$$y = (x - 1)^2 + 3(x - 1) + 2$$
$$= x^2 + x.$$

Note that there is no restriction on x; the domain of $y = x^2 + x$ is the set of all real numbers. It is now a simple matter to sketch the curve; it is given in Figure 13.3.

Figure 13.3

Of course, eliminating the parameter is not always so simple as in these examples. Occasionally it is difficult or impossible. In such cases the curve must be plotted point by point as we did with $x = \sin t$, $y = \cos t$.

Closely related to parametric equations are vector-valued functions. They are represented in the following way:

$$\mathbf{f}(t) = f_1(t)\mathbf{i} + f_2(t)\mathbf{j}.$$

Thus, when a value of t is substituted into the equation, the function takes on a vector value. For example, if

$$\mathbf{f}(t) = t\mathbf{i} + t^2\mathbf{j},$$

then

$$\mathbf{f}(1) = \mathbf{i} + \mathbf{j} \quad \text{and} \quad \mathbf{f}(2) = 2\mathbf{i} + 4\mathbf{j}.$$

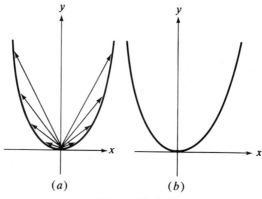

(a)   (b)

Figure 13.4

Recall that, in graphing vectors, we graph only representatives. Thus, in graphing vector functions, let us graph representatives of the vectors, each having its tail at the origin. Thus,

$$\mathbf{f}(t) = t\mathbf{i} + t^2\mathbf{j}$$

has the graphical representation shown in Figure 13.4(a). Normally we shall omit the directed line segments and show only their heads, as in part (b) of the figure. The result is equivalent to graphing the curve represented parametrically by

$$x = t, \qquad y = t^2.$$

Example 3

Sketch the curve $\mathbf{f}(t) = (t+2)\mathbf{i} + (t^2 + 7t + 12)\mathbf{j}.$

This is equivalent to the parametric equations

$$x = t + 2, \qquad y = t^2 + 7t + 12.$$

Eliminating the parameter, we have

$$y = x^2 + 3x + 2.$$

Its graph is given in Figure 13.5.

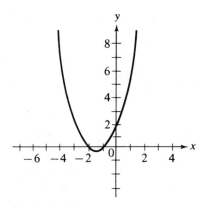

Figure 13.5

Let us consider a line determined by two of its points, $A(x_1, y_1)$ and $B(x_2, y_2)$. In Chapter 8 we derived the point-of-division formulas $x = x_1 + r(x_2 - x_1)$ and $y = y_1 + r(y_2 - y_1)$,   where $P(x, y)$ is a point on the line such that $r = \overline{AP}/\overline{AB}$. (Recall that r is negative if P and B are on opposite sides of A.) These equations are parametric equations of the line containing A and B with parameter r. This use of the point-of-division formulas as a parametric representation for a line is convenient; it is one of the principal methods we shall use to represent lines in three-dimensional space (see page 670).

Example 4

Find a parametric representation for the line through $(1, 5)$ and $(-2, 3)$.

Letting $(1, 5)$ and $(-2, 3)$ be the first and second points, respectively, of

$$x = x_1 + r(x_2 - x_1),$$
$$y = y_1 + r(y_2 - y_1),$$

we then have

$$x = 1 - 3r,$$
$$y = 5 - 2r.$$

The choice of $(1, 5)$ and $(-2, 3)$ as the first and second points is quite arbitrary; we could have reversed the designation. In that case the parametric representation would be

$$x = -2 + 3s,$$
$$y = 3 + 2s.$$

While the two representations appear to have little in common, it is easily seen that they represent the same line. We get the point $(1, 5)$ when $r = 0$ or when $s = 1$; we get $(-2, 3)$ when $r = 1$ or $s = 0$. In fact, whatever point we get for a given value of r, we get the same point for $s = 1 - r$. By using other points on the line, we can get still more parametric representations. Thus a line does not have a unique parametric representation.

Problems

A

In Problems 1–16, eliminate the parameter and sketch the curve.

1. $x = t^2 - 2, \quad y = t + 1.$
2. $x = t^2 + t - 2, \quad y = t - 1.$
3. $x = t - 1, \quad y = t^2 - 2t.$
4. $x = 2t^2 + t - 3, \quad y = t + 1.$
5. $x = t^2 + t, \quad y = t^2 - t.$
6. $x = t^2 - 1, \quad y = t^2 - 1.$
7. $x = t^3, \quad y = t^2.$
8. $x = a \cos \theta, \quad y = b \sin \theta.$
9. $x = 1 + \cos \theta, \quad y = -2 + \sin \theta.$
10. $x = 2 - \cos \theta, \quad y = 3 + 4 \sin \theta.$
11. $x = 2 + \sinh \theta, \quad y = 1 + \cosh \theta.$
12. $x = 4 + 2 \cosh \theta, \quad y = 1 - 4 \sinh \theta.$
13. $\mathbf{f}(t) = (t + 2)\mathbf{i} + t^2 \mathbf{j}.$
14. $\mathbf{f}(t) = t^2 \mathbf{i} + (t^2 + 1)\mathbf{j}.$
15. $\mathbf{f}(t) = t^2 \mathbf{i} + t^3 \mathbf{j}.$
16. $\mathbf{f}(t) = 2 \cos t \mathbf{i} + 2 \sin t \mathbf{j}.$

In Problems 17–22, give equations in parametric form for the line through the given pair of points.

17. $(1, 5), \quad (7, 3).$
18. $(4, 2), \quad (-1, 3).$
19. $(4, 3), \quad (-2, 1).$
20. $(4, 1), \quad (-8, 3).$
21. $(-1, 3), \quad (2, 3).$
22. $(-3, 2), \quad (-3, 5).$

B

In Problems 23 and 24, eliminate the parameter and sketch the curve.

23. $\mathbf{f}(t) = (t + 2)\mathbf{i} + (t^3 + 1)\mathbf{j}.$
24. $\mathbf{f}(t) = t^2 \mathbf{i} + e^t \mathbf{j}.$

In Problems 25–29, sketch the curve.

25. $x = \theta - \sin \theta, \quad y = 1 - \cos \theta.$
26. $x = \cos \theta + \theta \sin \theta, \quad y = \sin \theta - \theta \cos \theta.$
27. $x = a \cos^3 \theta, \quad y = a \sin^3 \theta.$
28. $x = 2a \cos \theta - a \cos 2\theta, \quad y = 2a \sin \theta - a \sin 2\theta.$

29. $x = t - a \tanh \dfrac{t}{a}, \quad y = a \operatorname{sech} \dfrac{t}{a}.$

C

30. Sketch the curve $x = e^t, \quad y = \sin t.$
31. Show that the parametric representation

$$x = x_1 + r(x_2 - x_1),$$
$$y = y_1 + r(y_2 - y_1)$$

of the line through (x_1, y_1) and (x_2, y_2) is equivalent to the two-point form of the line given on page 19.
32. Sketch each of the following parametric equations and note the similarities and differences.

   (a) $x = t, \quad y = t^2;$ (b) $x = \sqrt{t}, \quad y = t;$
   (c) $x = e^t, \quad y = e^{2t};$ (d) $x = \sin t, \quad y = 1 - \cos^2 t.$
33. Sketch each of the following parametric equations and note the similarities and differences.

   (a) $x = t, \quad y = \sqrt{t^2 - 1};$ (b) $x = \sqrt{t}, \quad y = \sqrt{t - 1};$
   (c) $x = \sec t, \quad y = \tan t;$ (d) $x = \cosh t, \quad y = \sinh t.$

13.2

Parametric Equations of a Locus

   The principal advantage of parametric equations is in the determination of equations of a locus. It is frequently simpler to relate x and y to some third variable than to relate them to each other directly. One example of this is the determination of the path of a projectile. We have already considered the path of an object thrown vertically upward or downward. The position is measured by the distance above or below a fixed reference point. But if the object is thrown in any nonvertical direction, then its position must have both a horizontal and a vertical component. Of course these can be determined by reference to a pair of axes. However, it is easier to relate the horizontal and vertical components of the position to time than to relate them directly to each other.
   As soon as a projectile is released, it becomes a falling object. Thus the vertical motion is governed by the laws of falling bodies, and the y coordinate is easily related to time by previous methods. The x coordinate is even easier to deal with. Because the force of gravity is the *only* force acting upon the projectile (we neglect the resistance of the air), there is no force tending to change the horizontal velocity—it remains constant. This can be integrated to give the x coordinate as a function of time.

Example 1

A gun is inclined to the horizontal at an angle of 30° and fired from ground level with an initial speed of 1500 ft/sec. Determine the path of the bullet.

Let us place the axes so that the gun is at the origin. Let $v(t)$ be the velocity of the bullet at time $t \geq 0$ and resolve the velocity vector into its horizontal and vertical components:

$$v(t) = v_x i + v_y j.$$

Similarly, the acceleration vector $a(t) = a_x i + a_y j$. The initial velocity $v(0)$ has components

$$v_x = |v(0)| \cos 30° = 1500 \cdot \frac{\sqrt{3}}{2} = 750\sqrt{3},$$

$$v_y = |v(0)| \sin 30° = 1500 \cdot \frac{1}{2} = 750,$$

where $|v(0)|$ is the initial speed and 30° gives the initial direction of the velocity vector (see Figure 13.6).

Now we consider the vertical component, which is a falling-body problem.

$$a_y = -32,$$

$$v_y = -32t + C_1.$$

Since $v_y = 750$ when $t = 0$, $C_1 = 750$.

$$v_y = -32t + 750,$$

$$y = -16t^2 + 750t + C_2.$$

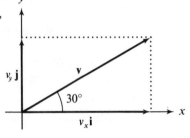

Figure 13.6

By the placement of the axes, $y = 0$ when $t = 0$. Thus $C_2 = 0$ and

$$y = -16t^2 + 750t.$$

For the horizontal component, the velocity is a constant.

$$v_x = 750\sqrt{3},$$

$$x = 750\sqrt{3}\, t + C_3.$$

Again by the placement of the axes, $x = 0$ when $t = 0$, which gives $C_3 = 0$ and

$$x = 750\sqrt{3}\, t.$$

Thus, in parametric form, the path of the projectile is given by

$$x = 750\sqrt{3}\, t, \quad y = -16t^2 + 750t.$$

By eliminating the parameter, we see that the path is parabolic.

$$y = \frac{-4x^2}{421,875} + \frac{x}{\sqrt{3}}.$$

In the past we have found equations of curves from a geometric description. Again this can often be accomplished by relating the x and y coordinates of a point on the curve to some third variable.

Example 2

Find parametric equations for the set of all points P which are determined as illustrated in Figure 13.7.

Since the ray OA is determined by the angle θ, θ is a convenient parameter. The x coordinate of P is the x coordinate of A; thus

$$x = a \cos \theta.$$

Similarly, the y coordinate of P is the y coordinate of B.

$$y = b \sin \theta.$$

Thus we have the curve in parametric form. It might be noted here that the parameter can easily be eliminated to give

$$\frac{x^2}{a^2} + \frac{y^2}{b^2} = 1.$$

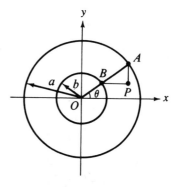

Figure 13.7

Example 3

A wheel of radius a is rolling along a line. Find the path traced by a point on the circumference if the line is the x axis and the point starts at the origin.

Suppose we use the angle θ (see Figure 13.8) as the parameter. Since the wheel is rolling along the x axis,

$$\overline{OT} = \overparen{PT} = a\theta.$$

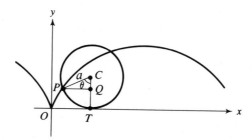

Figure 13.8

Thus C is the point $(a\theta, a)$. Furthermore, from triangle CPQ,

$$\overline{PQ} = a \sin \theta \quad \text{and} \quad \overline{CQ} = a \cos \theta.$$

Thus,

$$
\begin{aligned}
x &= \overline{OT} - \overline{PQ} & y &= \overline{CT} - \overline{CQ} \\
&= a\theta - a \sin \theta & &= a - a \cos \theta \\
&= a(\theta - \sin \theta); & &= a(1 - \cos \theta).
\end{aligned}
$$

This curve is called a cycloid.

This last problem can also be solved by using vectors. Suppose $\overrightarrow{OP}$, $\overrightarrow{OC}$, and $\overrightarrow{CP}$ represent the vectors **u**, **v**, and **w**, respectively (see Figure 13.9(a)). Clearly **u** = **v** + **w**. Since $\overrightarrow{OC} = \overrightarrow{OT} + \overrightarrow{TC}$,

$$\mathbf{v} = a\theta\mathbf{i} + a\mathbf{j}.$$

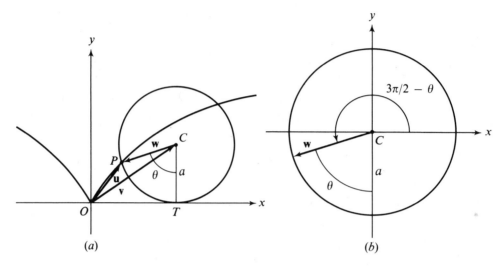

Figure 13.9

In order to express **w** in terms of θ, let us translate the axes so that the new origin is at C (see Figure 13.9(b)). We can now see that

$$\mathbf{w} = a \cos\left(\frac{3\pi}{2} - \theta\right)\mathbf{i} + a \sin\left(\frac{3\pi}{2} - \theta\right)\mathbf{j}$$

$$= -a \sin \theta\, \mathbf{i} - a \cos \theta\, \mathbf{j}.$$

Thus

$$\mathbf{u} = \mathbf{v} + \mathbf{w} = (a\theta - a \sin \theta)\mathbf{i} + (a - a \cos \theta)\,\mathbf{j}.$$

Problems

A

1. A gun that is inclined to the horizontal at an angle of 60° is fired from ground level with an initial speed of 2500 ft/sec. Determine the path of the bullet.

2. A gun is clamped in a horizontal position 8 ft above the ground and fired with an initial speed of 1600 ft/sec. Determine the path of the bullet. Where does it hit the ground?

3. A ball is thrown upward from ground level at a 30° angle of inclination. Given that it is thrown at 32 ft/sec, find the path of the ball. Where does it hit the ground?

4. A ball is thrown downward and inclined to the horizontal at an angle of 60° from a building 297 ft high. Find the path of the ball when it is thrown at 16 ft/sec. How far from the base of the building does it hit the ground?

5. A cannon, inclined at an angle θ to the horizontal, is fired from ground level with speed v_0. Determine the path of the projectile.

6. Use the result of Problem 5 to find the maximum range and the angle of inclination necessary to realize this maximum. (The range is the distance between the point of firing and the point at which the projectile hits the ground.)

B

7. Find the parametric equations for the set of all points P determined as shown in Figure 13.10. Eliminate the parameter.

8. Suppose, in Example 3, the point starts at $(0, 2a)$. Find the parametric equations for the curve.

9. Find the path traced by a point P a distance b from the center of the circle of Example 3 if P starts at $(0, a - b)$.

10. Sketch the curve of Problem 9 given that
    (a) $a = 2$ and $b = 1$; (b) $a = 1$ and $b = 2$.

11. In Figure 13.11, $\overline{BA} = \overline{BP} = \overline{BP'}$. Find parametric equations for the set of all points P and P' determined as shown. Sketch. This curve is called a strophoid.

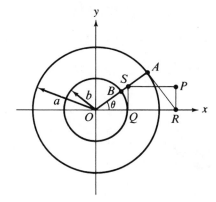

Figure 13.10

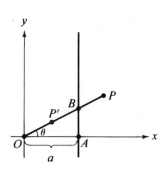

Figure 13.11

12. Find parametric equations for the set of all points P and P' determined as shown in Figure 13.12. This curve is called a conchoid.

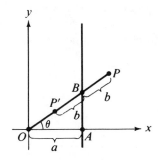

Figure 13.12

13. Sketch the conchoid (see Problem 12) for
    (a) $a = b = 1$; (b) $a = 1$ and $b = 2$; (c) $a = 2$ and $b = 1$.

C

14. If a string that is wound on a spool is unwound while the string is kept taut (see Figure 13.13), the curve traced by the end of the string is called the involute of the circle. Find parametric equations for the involute of a circle of radius a.

15. Find the path traced by a point P on the circumference of a circle of radius a that rolls inside a circle of radius $4a$ (see Figure 13.14). This curve is called a hypocycloid.

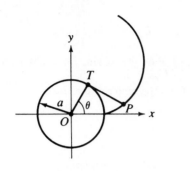

Figure 13.13

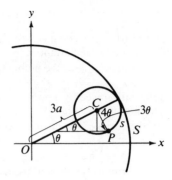

Figure 13.14

16. Find the path of the point P of Problem 15 when the smaller circle rolls outside the larger one. This curve is called an epicycloid.

17. What is the result if both of the circles of Problem 16 have radius a? This curve is called a cardioid.

13.3

Derivatives of Parametric Equations

Suppose we are interested in finding the slope of a graph that is represented in parametric form. This means finding dy/dx. If we can eliminate the parameter to get an equation in x and y, we can use our previous methods to find the derivative. Even when eliminating the parameter is impossible or extremely complicated, we can still find the derivative dy/dx.

Theorem 13.1

If $\ \ x = f(t), \ \ y = g(t), \ \ $ and $\ \ dx/dt \neq 0, \ \ $ then

$$\frac{dy}{dx} = \frac{dy/dt}{dx/dt}.$$

Proof

Since our definition of dy/dx requires that y be a function of x and we cannot be sure that y is a function of x, we cannot use this definition directly. However, we can alter it to fit our purposes. The derivative was defined as a certain limit of the slope of a secant line. Keeping this in mind and considering Figure 13.15, we see that

$$\frac{dy}{dx} = \lim_{h \to 0} \frac{g(t+h) - g(t)}{f(t+h) - f(t)}.$$

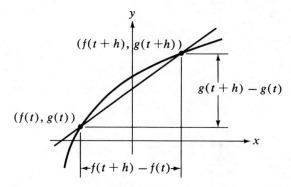

Figure 13.15

Dividing both numerator and denominator by h, we have

$$\frac{dy}{dx} = \lim_{h \to 0} \frac{\dfrac{g(t+h) - g(t)}{h}}{\dfrac{f(t+h) - f(t)}{h}} = \frac{\lim\limits_{h \to 0} \dfrac{g(t+h) - g(t)}{h}}{\lim\limits_{h \to 0} \dfrac{f(t+h) - f(t)}{h}} = \frac{dy/dt}{dx/dt}.$$

Example 1

Find dy/dx if $x = \sin t$, $y = \cos t$.

$$\frac{dy}{dx} = \frac{\dfrac{dy}{dt}}{\dfrac{dx}{dt}} = \frac{-\sin t}{\cos t} = -\tan t.$$

Suppose we want to take a second derivative, which is, of course, nothing more than the derivative of the first derivative. The first derivative is a function of the parameter t, while its derivative is taken with respect to x. By the chain rule,

$$\frac{d}{dt}\left(\frac{dy}{dx}\right) = \frac{d}{dx}\left(\frac{dy}{dx}\right) \cdot \frac{dx}{dt} = \frac{d^2y}{dx^2} \cdot \frac{dx}{dt},$$

so

$$\frac{d^2y}{dx^2} = \frac{\dfrac{d}{dt}\left(\dfrac{dy}{dx}\right)}{\dfrac{dx}{dt}};$$

and similarly,

$$\frac{d^3y}{dx^3} = \frac{\dfrac{d}{dt}\left(\dfrac{d^2y}{dx^2}\right)}{\dfrac{dx}{dt}}, \quad \text{etc.}$$

Example 2

Find dy/dx and d^2y/dx^2 for $x = \sin t$, $y = 1 + \cos t$.

$$\frac{dy}{dx} = \frac{\dfrac{dy}{dt}}{\dfrac{dx}{dt}} = \frac{-\sin t}{\cos t} = -\tan t.$$

$$\frac{d^2y}{dx^2} = \frac{\dfrac{d}{dt}\left(\dfrac{dy}{dx}\right)}{\dfrac{dx}{dt}} = \frac{-\sec^2 t}{\cos t} = -\sec^3 t.$$

Example 3

Find dy/dx and d^2y/dx^2 at $t = 0$ for $x = e^t$, $y = e^{-t}$.

$$\frac{dy}{dx} = \frac{-e^{-t}}{e^t} = -e^{-2t}, \qquad \frac{d^2y}{dx^2} = \frac{2e^{-2t}}{e^t} = 2e^{-3t}.$$

At $t = 0$,

$$\frac{dy}{dx} = -1, \qquad \frac{d^2y}{dx^2} = 2.$$

Problems

A

In Problems 1–10, find dy/dx and d^2y/dx^2.

1. $x = t^2 + t$, $y = t + 1$.
2. $x = t + 3$, $y = t^2 + 3t$.
3. $x = t^2 - 5t + 10$, $y = t + 3$.
4. $x = t^2 + 5t + 9$, $y = t + 3$.
5. $x = t^2 + t$, $y = t^2 - t$.
6. $x = 2t + 5$, $y = 4t^2$.
7. $x = 2 + \sin t$, $y = -1 + \cos t$.
8. $x = 3 + 2\cos t$, $y = 1 - \sin t$.
9. $x = 2 + \cosh t$, $y = -1 + \sinh t$.
10. $x = -2 + 4\cosh t$, $y = 3 + 2\sinh t$.

In Problems 11–16, find dy/dx and d^2y/dx^2 at the given value of t.

11. $x = t + 4$, $y = t^2 + 2t - 4$; $t = 1$.
12. $x = t^3 - 3t^2 + 5$, $y = 2t - 7$; $t = 2$.
13. $x = t^3$, $y = t^2$; $t = -3$.
14. $x = 2 + \cos t$, $y = 2 - \sin t$; $t = \pi/2$.
15. $x = e^t + 1$, $y = e^t + e^{-t}$; $t = 0$.
16. $x = 2 + \cosh t$, $y = 3 + \sinh t$; $t = 0$.

In Problems 17–20, find an equation of the line tangent to the given curve at the given point.

17. $x = 2t - 1$, $y = 4t^2 - 2t$; $t = 1$.
18. $x = t - 3$, $y = t^3 - 2t^2 + 5t + 2$; $t = 0$.
19. $x = 2\cos\theta$, $y = 3\sin\theta$; $\theta = \pi/4$.
20. $x = 2\cosh\theta$, $y = 3\sinh\theta$; $\theta = 1$.

In Problems 21–24, find all points at which the curve has a horizontal or vertical tangent.

21. $x = t + 3$, $y = t^3 - 3t^2$.
22. $x = t + 2$, $y = (t^2 + 4t)^2$.
23. $x = 3\cos\theta$, $y = 5\sin\theta$.
24. $x = 4\cosh\theta$, $y = 2\sinh\theta$.

B

In Problems 25–29, find all points at which the curve has a horizontal or vertical tangent.

25. $x = \theta - \sin\theta$, $y = 1 - \cos\theta$.   26. $x = \theta + \sin\theta$, $y = 1 - \cos\theta$.

27. $x = \cos^3\theta$, $y = \sin^3\theta$.   28. $x = \cos\theta + \theta\sin\theta$, $y = \sin\theta - \theta\cos\theta$.

29. $x = t - \tanh t$, $y = \operatorname{sech} t$.

30. Derive a general form for d^2y/dx^2 in terms of dx/dt and dy/dt.

13.4

Arc Length

Suppose we have a curve represented in parametric form,

$$x = f(t), \qquad y = g(t),$$

and want to find the length of the arc from some point (x_1, y_1) corresponding to $t = a$ to the point (x_2, y_2) corresponding to $t = b$ ($a < b$). Since we can find straight-line distances very easily, let us begin by approximating the length by a series of straight-line distances. Suppose the interval $[a, b]$ is subdivided by

$$S : a = t_0, t_1, t_2, \ldots, t_n = b.$$

To each t_i there corresponds a point $(x_i, y_i) = (f(t_i), g(t_i))$; these points are then used to approximate the arc length by a series of straight-line distances (see Figure 13.16). The length of the ith subarc is approximated by the distance s_i between (x_{i-1}, y_{i-1}) and (x_i, y_i),

$$s_i = \sqrt{(x_i - x_{i-1})^2 + (y_i - y_{i-1})^2}.$$

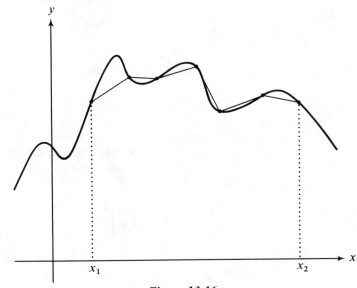

Figure 13.16

We now multiply and divide by $t_i - t_{i-1}$, noting that $t_i - t_{i-1} = \sqrt{(t_i - t_{i-1})^2}$, since $t_i - t_{i-1}$ is positive. Thus,

$$s_i = \sqrt{\left(\frac{x_i - x_{i-1}}{t_i - t_{i-1}}\right)^2 + \left(\frac{y_i - y_{i-1}}{t_i - t_{i-1}}\right)^2} \, (t_i - t_{i-1})$$

$$= \sqrt{\left(\frac{f(t_i) - f(t_{i-1})}{t_i - t_{i-1}}\right)^2 + \left(\frac{g(t_i) - g(t_{i-1})}{t_i - t_{i-1}}\right)^2} \, \Delta t_i,$$

where $\Delta t_i = t_i - t_{i-1}$. Now the mean-value theorem (see page 573) asserts the existence of a number t_i^* $(t_{i-1} < t_i^* < t_i)$ such that

$$\frac{f(t_i) - f(t_{i-1})}{t_i - t_{i-1}} = f'(t_i^*).$$

Similarly, there is a number t_i^{**} $(t_{i-1} < t_i^{**} < t_i)$ such that

$$\frac{g(t_i) - g(t_{i-1})}{t_i - t_{i-1}} = g'(t_i^{**}).$$

Thus

$$s_i = \sqrt{[f'(t_i^*)]^2 + [g'(t_i^{**})]^2} \, \Delta t_i.$$

The sum of the n approximating lengths gives an approximation for the entire arc, and the limit of this sum as the norm of the subdivision S approaches zero gives the exact value. If $t_i^* = t_i^{**}$, it is easily seen that

$$\lim_{\|S\| \to 0} \sum_{i=1}^{n} s_i = \int_a^b \sqrt{[f'(t)]^2 + [g'(t)]^2} \, dt;$$

however t_i^* and t_i^{**} are not necessarily equal. Nevertheless the conclusion is still valid. A proof of this fact is beyond the scope of this book,† but it follows intuitively if we note that, by taking $\|S\|$ small enough, we can make $|t_i^* - t_i^{**}|$ as small as we please—that is, smaller than any preassigned positive number. Thus

$$s = \int_a^b \sqrt{\left(\frac{dx}{dt}\right)^2 + \left(\frac{dy}{dt}\right)^2} \, dt.$$

Before using this formula, note that the preceding discussion is *not* a proof of the formula. We have not *defined* what is meant by the length of an arc. Our position is the same as when we first began to consider area; although we used our intuitive idea of area to determine what we meant by an integral, the actual definition of an integral was independent of area; in fact, area was defined in terms of the integral. Similarly we may now *define* arc length as

$$s = \int_{t_1}^{t_2} \sqrt{\left(\frac{dx}{dt}\right)^2 + \left(\frac{dy}{dt}\right)^2} \, dt.$$

---

† See Duhamel's principle, which is proved in Angus E. Taylor and Robert Mann, *Advanced Calculus.* 2nd ed. (Lexington, Mass.: Xerox, 1972), pp. 575–577.

We label the limits of integration t_1 and t_2 rather than a and b to emphasize that these numbers are values of t.

Example 1

Find the circumference of the circle $x = a \cos t, y = a \sin t$.

$$\frac{dx}{dt} = -a \sin t \quad \text{and} \quad \frac{dy}{dt} = a \cos t,$$

$$s = \int_0^{2\pi} \sqrt{(-a \sin t)^2 + (a \cos t)^2} \, dt$$

$$= \int_0^{2\pi} \sqrt{a^2 \sin^2 t + a^2 \cos^2 t} \, dt$$

$$= \int_0^{2\pi} a \, dt$$

$$= at \Big|_0^{2\pi}$$

$$= 2\pi a.$$

If the curve is represented by $y = f(x)$ rather than by a pair of parametric equations, we can proceed as follows.

$$s = \int_{t_1}^{t_2} \sqrt{\left(\frac{dx}{dt}\right)^2 + \left(\frac{dy}{dt}\right)^2} \, dt$$

$$= \int_{t_1}^{t_2} \sqrt{1 + \frac{(dy/dt)^2}{(dx/dt)^2}} \frac{dx}{dt} \, dt$$

$$= \int_{x_1}^{x_2} \sqrt{1 + \left(\frac{dy/dt}{dx/dt}\right)^2} \, dx$$

$$= \int_{x_1}^{x_2} \sqrt{1 + \left(\frac{dy}{dx}\right)^2} \, dx.$$

Similarly, if we have $x = g(y)$, then

$$s = \int_{y_1}^{y_2} \sqrt{1 + \left(\frac{dx}{dy}\right)^2} \, dy.$$

Note that in each case the limits of integration must be changed to the corresponding values of x or y when we integrate with respect to x or y, respectively.

Example 2

Find the length of $y = x^{2/3}$ from $(1, 1)$ to $(8, 4)$.

$$y' = \frac{2}{3} x^{-1/3} = \frac{2}{3x^{1/3}},$$

$$s = \int_1^8 \sqrt{1 + \left(\frac{2}{3x^{1/3}}\right)^2} \, dx$$

$$= \int_1^8 \sqrt{1 + \frac{4}{9x^{2/3}}} \, dx$$

$$= \int_1^8 \sqrt{\frac{9x^{2/3} + 4}{9x^{2/3}}} \, dx$$

$$= \int_1^8 \frac{\sqrt{9x^{2/3} + 4}}{3x^{1/3}} \, dx$$

$$= \frac{1}{18} \int_1^8 6x^{-1/3} \sqrt{9x^{2/3} + 4} \, dx$$

$$= \frac{1}{27} (9x^{2/3} + 4)^{3/2} \Big|_1^8$$

$$= \frac{80\sqrt{10} - 13\sqrt{13}}{27}.$$

Example 3

Find the length of $y = \frac{x^2}{4} - \frac{\ln x}{2}$ from $x = 1$ to $x = 2$.

$$y' = \frac{x}{2} - \frac{1}{2x},$$

$$s = \int_1^2 \sqrt{1 + \left(\frac{x}{2} - \frac{1}{2x}\right)^2} \, dx$$

$$= \int_1^2 \sqrt{1 + \frac{x^2}{4} - \frac{1}{2} + \frac{1}{4x^2}} \, dx$$

$$= \int_1^2 \sqrt{\frac{x^2}{4} + \frac{1}{2} + \frac{1}{4x^2}} \, dx$$

$$= \int_1^2 \sqrt{\left(\frac{x}{2} + \frac{1}{2x}\right)^2} \, dx$$

$$= \int_1^2 \left(\frac{x}{2} + \frac{1}{2x}\right) dx$$

$$= \frac{x^2}{4} + \frac{\ln x}{2} \Big|_1^2$$

$$= \frac{3}{4} + \frac{1}{2} \ln 2.$$

Example 4

Find the length of $x = y^{3/2}$ from $(1, 1)$ to $(8, 4)$.

This is really the same problem as Example 2, the only difference being the form of the equation.

$$x' = \frac{3}{2} y^{1/2},$$

$$s = \int_1^4 \sqrt{1 + \frac{9}{4} y}\, dy$$

$$= \frac{4}{9} \int_1^4 \frac{9}{4} \sqrt{1 + \frac{9}{4} y}\, dy$$

$$= \frac{4}{9} \frac{\left(1 + \frac{9}{4} y\right)^{3/2}}{3/2} \Bigg|_1^4$$

$$= \frac{8}{27} \left(1 + \frac{9}{4} y\right)^{3/2} \Bigg|_1^4$$

$$= \frac{80\sqrt{10} - 13\sqrt{13}}{27}.$$

Of course there are other forms for arc length. We shall consider one such in the next chapter (see Problem 21, page 444), but for the present we content ourselves with the three forms

$$s = \int_{x_1}^{x_2} \sqrt{1 + (dy/dx)^2}\, dx \quad \text{if } y = f(x),$$

$$s = \int_{y_1}^{y_2} \sqrt{1 + (dx/dy)^2}\, dy \quad \text{if } x = g(y),$$

$$s = \int_{t_1}^{t_2} \sqrt{(dx/dt)^2 + (dy/dt)^2}\, dt \quad \text{if } x = f(t) \text{ and } y = g(t).$$

Problems

A

In Problems 1–10, find the length of the arc described.

1. $x = \cos^3 \theta$, $y = \sin^3 \theta$ from $\theta = 0$ to $\theta = \pi/2$.
2. $x = e^t \sin t$, $y = e^t \cos t$ from $t = 0$ to $t = \pi$.
3. $x = 4t^3$, $y = 3t^2$ from $t = 0$ to $t = 1$.
4. $x = \cos \theta + \theta \sin \theta$, $y = \sin \theta - \theta \cos \theta$ from $\theta = 0$ to $\theta = \pi$.
5. $y = x^{3/2}$ from $x = 0$ to $x = 4/3$.　　　6. $y^3 = x^2$ from $x = 1$ to $x = 27$.

7. $y = \dfrac{x^4}{4} + \dfrac{1}{8x^2}$ from $x = 1$ to $x = 2$.　　8. $y = \dfrac{x^3}{6} + \dfrac{1}{2x}$ from $x = 1$ to $x = 3$.

9. $x = \cosh y$ from $y = 0$ to $y = 1$.　　10. $x = \dfrac{y^3}{3} + \dfrac{1}{4y}$ from $y = 1$ to $y = 3$.

B

In Problems 11–14, find the length of the arc described.

11. $y = \ln \dfrac{e^x + 1}{e^x - 1}$ from $x = 1$ to $x = 2$.   12. $y = \ln(1 - x^2)$ from $x = 0$ to $x = 1/2$.

(*Hint*: $\dfrac{2}{1 - x^2} = \dfrac{1}{1 + x} + \dfrac{1}{1 - x}$.)

13. $x = \text{Arccos}(1 - y) + \sqrt{2y - y^2}$ from $y = 1/2$ to $y = 1$.
14. $x = \text{sech}^{-1} y - \sqrt{1 - y^2}$ from $y = 1/2$ to $y = 1$.
15. Find the length of the loop of $9y^2 = x(x - 3)^2$.
16. Find the total length of $x^{2/3} + y^{2/3} = a^{2/3}$.
17. Find the distance between $(-1, 5)$ and $(2, 1)$ by using the distance formula. Now find an equation of the line through the given points and use an arc-length formula to find the same distance.
18. Use an arc-length formula to find the distance between (x_1, y_1) and (x_2, y_2) (see Problem 17). This reassures us that we have not been inconsistent in defining arc length as we have.

C

19. Let a curve have parametric equations $x = f(t)$, $y = g(t)$ and let α be the angle of inclination of the tangent line to the curve. The *curvature* κ at a point on the curve is the rate of change of α with respect to arc length s along the curve:

$$\kappa = \frac{d\alpha}{ds}.$$

Since $\tan \alpha = dy/dx$, $\alpha = \arctan(dy/dx)$.   Use the chain rule and the result of Problem 30, page 407, to show that

$$\kappa = \frac{d\alpha}{ds} = \frac{d}{dt}\left(\arctan \frac{dy}{dx}\right) \cdot \frac{dt}{ds}$$

$$= \frac{f'g'' - g'f''}{[(f')^2 + (g')^2]^{3/2}}.$$

20. If a curve is given by $y = f(x)$, show that

$$\kappa = \frac{y''}{[1 + (y')^2]^{3/2}}.$$

Show that $y = f(x)$ is concave up if $\kappa > 0$ and concave down if $\kappa < 0$.

21. Show that the curvature of the circle $x = R \cos \theta$, $y = R \sin \theta$, $0 \le \theta \le 2\pi$,   is $\kappa = -1/R$.
22. Find the curvature at a general point of $y = x^3/3$.
23. Find the curvature of $x = t^2 - 1$, $y = \frac{1}{3}t^3 - 1$ at $t = 2$.
24. Let l be the tangent line to $y = f(x)$ at $P(x, y)$. The *circle of curvature* is the circle with radius $R = 1/|\kappa|$ that lies on the same side of l as the curve $y = f(x)$ and is tangent to l at P. The radius R is called the *radius of curvature*. The center C of the circle of curvature is the *center of curvature*. Show that the coordinates (h, k) of the center of curvature are

$$h = x - R \sin \alpha, \qquad k = y + R \cos \alpha.$$

Show that

$$\sin \alpha = \frac{y'}{\sqrt{1 + (y')^2}}, \qquad \cos \alpha = \frac{1}{\sqrt{1 + (y')^2}},$$

and thus

$$h = x - \frac{y'[1 + (y')^2]}{y''}, \qquad k = y + \frac{1 + (y')^2}{y''}.$$

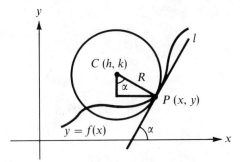

Figure 13.17

25.   Find the radius of curvature of $x = a \cos^3 t, \quad y = a \sin^3 t$.
26.   Find the radius of curvature of $x^{2/3} + y^{2/3} = a^{2/3}$.
27.   Find the center of curvature of $y = x/(x + 1)$ at $(0, 0)$.
28.   Find the center of curvature of $x^3 + y^3 = 4xy$ at $(2, 2)$.
29.   If point P moves along a curve C_1, the center of curvature corresponding to P describes a curve C_2, called the *evolute* of C_1; conversely, C_1 is the *involute* of C_2. Show that parametric equations of the evolute of $y = \frac{1}{2}x^2$ are $h = -x^3, \quad k = \frac{3}{2}x^2 + 1$. Eliminate the parameter x to obtain

$$h^2 = \frac{8}{27}(k - 1)^3.$$

30.   Find parametric equations and a rectangular equation for the evolute of $x = a \cos \theta$, $y = b \sin \theta$.

13.5

The Catenary

We saw in Section 8.3 that the supporting cables of a suspension bridge (in which the weight of the roadway is uniformly distributed horizontally) form a parabolic arc. We also noted there that a wire that simply hangs of its own weight does not hang in a parabolic arc, because of the different weight distribution. Now that we have considered arc length we are in a position to look into this more complicated problem.

First, however, let us look at the relationship between the definite integral and the derivative. Suppose

$$F'(x) = f(x).$$

Then

$$I = \int_a^x f(t)\, dt = F(t)\Big|_a^x = F(x) - F(a),$$

and

$$I' = F'(x) = f(x).$$

Thus

$$\frac{d}{dx} \int_a^x f(t)\, dt = f(x).$$

(see Problem 33, page 202). For an arc-length integral, if

$$s = \int_a^x \sqrt{1 + [f'(t)]^2}\ dt,$$

then

$$\frac{ds}{dx} = \sqrt{1 + [f'(x)]^2}.$$

If $y = f(x)$, then

$$\frac{ds}{dx} = \sqrt{1 + \left(\frac{dy}{dx}\right)^2}.$$

Now let us consider a wire hanging of its own weight. Suppose the axes are placed as shown in Figure 13.18, so that the minimum point O is on the y axis and the x axis is parallel to the tangent line at O. This placement of the axes implies that any equation representing the curve of the cable satisfies the conditions that

$$\frac{dy}{dx} = 0 \quad \text{and} \quad y = k$$

when $x = 0$.

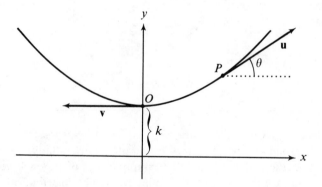

Figure 13.18

Again we consider a section OP of the cable, where O is the lowest point and $P(x, y)$ is another point. Three forces are acting upon it: the two forces of tension acting upon the two ends, and the weight of OP. At the point P the force exerted by the cable can be represented by a vector $\mathbf{u}$ of magnitude T (the tension in the cable at that point) and tangent to the curve representing the cable. Taking θ to be the inclination at P, we have

$$\mathbf{u} = T \cos \theta \mathbf{i} + T \sin \theta \mathbf{j}.$$

Letting T_0 be the tension at O, we represent the force at O by

$$\mathbf{v} = -T_0 \mathbf{i}.$$

Now the weight of *OP* is simply the product of the weight *w* per unit length and the length *s* of *OP*,

$$\mathbf{w} = -ws\mathbf{j}.$$

Since the three forces must be in equilibrium,

$$\mathbf{u} + \mathbf{v} + \mathbf{w} = 0,$$

$$T \cos \theta \mathbf{i} + T \sin \theta \mathbf{j} - T_0 \mathbf{i} - ws\mathbf{j} = 0.$$

Thus

$$T \cos \theta = T_0 \quad \text{and} \quad T \sin \theta = ws.$$

Dividing the second equation by the first yields

$$\tan \theta = \frac{ws}{T_0}.$$

Since θ is the inclination at *P*, $\tan \theta$ is the slope.

$$y' = \frac{w}{T_0} s,$$

$$y'' = \frac{w}{T_0} s' = \frac{w}{T_0} \sqrt{1 + (y')^2}.$$

In order to simplify we make the substitution

$$p = \frac{dy}{dx}.$$

Thus we have

$$\frac{dp}{dx} = \frac{w}{T_0} \sqrt{1 + p^2}$$

$$\frac{dp/dx}{\sqrt{1 + p^2}} = \frac{w}{T_0}.$$

The integral form of Theorem 12.12(1) is

$$\int \frac{u' \, dx}{\sqrt{1 + u^2}} = \sinh^{-1} u + C.$$

Using this, we have

$$\sinh^{-1} p = \frac{w}{T_0} x + C_1.$$

Since by our placement of axes $p = dy/dx = 0$ when $x = 0$, we have $C_1 = 0$. Therefore

$$\sinh^{-1} p = \frac{w}{T_0} x,$$

$$\frac{dy}{dx} = p = \sinh \frac{w}{T_0} x.$$

Integrating once more, we have

$$y = \frac{T_0}{w} \cosh \frac{w}{T_0} x + C_2.$$

Since $y = k$ when $x = 0$,

$$k = \frac{T_0}{w} + C_2, \quad \text{or} \quad C_2 = k - \frac{T_0}{w}.$$

In order to make the representation as simple as possible, let us take the constant k to be T_0/w. In that case, $C_2 = 0$ and the resulting equation for the cable is

$$y = \frac{T_0}{w} \cosh \frac{w}{T_0} x.$$

The curve represented by this equation is called a catenary (from the Latin word for chain). The wires hanging between telephone poles form arcs of catenaries.

Instead of using the span and the sag to determine T_0 as we did with the suspension bridge, it is more convenient to use the total length of the wire and the sag H (see Figure 8.15 for the sag and the span L). Suppose we use the integral for arc length to find the length S of the wire from the equation of a catenary that we derived above.

$$y = \frac{T_0}{w} \cosh \frac{wx}{T_0},$$

$$y' = \sinh \frac{wx}{T_0};$$

$$S = 2 \int_0^{L/2} \sqrt{1 + \sinh^2 \frac{wx}{T_0}} \, dx$$

$$= 2 \int_0^{L/2} \cosh \frac{wx}{T_0} \, dx$$

$$= \frac{2T_0}{w} \sinh \frac{wx}{T_0} \Big|_0^{L/2}$$

$$= \frac{2T_0}{w} \sinh \frac{wL}{2T_0}.$$

Thus

$$\sinh \frac{wL}{2T_0} = \frac{wS}{2T_0}.$$

Since $y = H + T_0/w$ when $x = L/2$, the original equation of the catenary yields

$$\cosh \frac{wL}{2T_0} = \frac{wH}{T_0} + 1.$$

Using the hyperbolic identity

$$\cosh^2 x - \sinh^2 x = 1,$$

we have

$$T_0 = \frac{w(S^2 - 4H^2)}{8H}.$$

To find the tension T at any point P, we note that $\tan \theta = ws/T_0$, so

$$\cos \theta = \frac{T_0}{\sqrt{T_0^2 + w^2 s^2}},$$

where s is the arc length from O to P. But $T \cos \theta = T_0$, so

$$T^2 = T_0^2 + w^2 s^2.$$

Again it is easily seen that the minimum tension occurs at the center and the maximum at the ends.

Example 1

A 400-ft length of wire weighing 2 lb/ft hangs with a 50-ft sag. Find the maximum and minimum tension in the wire and its equation (with the placement of axes shown in Figure 13.18, where $k = T_0/w$).

$$
\begin{aligned}
T_0 &= \frac{w(S^2 - 4H^2)}{8H} \\
&= \frac{2(400^2 - 4 \cdot 50^2)}{8 \cdot 50} \\
&= 750 \text{ lb.}
\end{aligned}
$$

This is the minimum tension; the maximum tension is

$$
\begin{aligned}
T &= \sqrt{T_0^2 + w^2 S^2} \\
&= \sqrt{750^2 + 2^2 \cdot 200^2} \\
&= \sqrt{722,500} = 850 \text{ lb.}
\end{aligned}
$$

The equation for the wire is

$$y = \frac{T_0}{w} \cosh \frac{wx}{T_0}$$

$$y = 375 \cosh \frac{x}{375} .$$

Problems

A

In Problems 1–4, find the minimum and maximum tension in the wire and its equation (with the placement of axes shown in Figure 13.18, where $k = T_0/w$).

1. $S = 400$ ft, $w = 2$ lb/ft, $H = 25$ ft. 2. $S = 500$ ft, $w = 3$ lb/ft, $H = 50$ ft.
3. $S = 100$ ft, $w = 1$ lb/ft, $H = 10$ ft. 4. $S = 1000$ ft, $w = 1$ lb/ft, $H = 20$ ft.

B

5. A 500-ft wire weighing 2 lb/ft can withstand a tension of 10,000 lb. What is the minimum sag that can be allowed?

6. A 300-ft wire weighing 2 lb/ft can withstand a tension of 10,000 lb. What is the minimum sag that can be allowed?

7. What happens to T_0 as $H \to 0$ where S and w are fixed? What does this tell us about the minimum value of the sag H?

Review Problems

A

In Problems 1–5, eliminate the parameter and sketch the curve.

1. $x = t + 1, \quad y = t^2 + 4t - 2$.   2. $x = 1 - 3 \cos \theta, \quad y = 2 + 2 \sin \theta$.
3. $\mathbf{f}(t) = (t + 1)^2 \mathbf{i} + t^2 \mathbf{j}$.   4. $\mathbf{f}(t) = e^t \mathbf{i} + \ln t \, \mathbf{j}$.
5. $x = \sin \theta, \quad y = \sin 2\theta$.

6. Sketch $\mathbf{f}(\theta) = \cos \theta \, (1 + \cos \theta) \mathbf{i} + \sin \theta \, (1 + \cos \theta) \mathbf{j}$.
7. Give a parametric representation of the line containing (2, 3) and (4, −5).
8. Give a parametric representation of the line containing (−1, 4) and (3, −3).
9. Find dy/dx and d^2y/dx^2 when $\quad x = 2t^2 + 3, \quad y = t^3 - 4$.
10. Find dy/dx and d^2y/dx^2 when $\theta = \pi/3$, given $\quad x = \sin \theta, \quad y = \sin 2\theta$.
11. Find an equation of the line tangent to $\quad x = t^2 - t, \quad y = t^2 + t, \quad$ at $t = 1$.
12. Find an equation of the line tangent to $\quad x = 3 + 4 \cos \theta, \quad y = 1 + \sqrt{3} \sin \theta, \quad$ at $\theta = \pi/3$.

B

13. A stone is thrown upward at an angle of 60° with the horizontal from a position 64 ft above the ground. If it is thrown with an initial speed of 32 ft/sec, find its path. Where and when does it hit the ground?

14. Find all points at which $\quad x = t^2 + 1, \quad y = 2t^3 + 3t^2 - 12t \quad$ has horizontal or vertical tangents.

15. Find all points at which $\quad x = \theta + \cos \theta, \quad y = 1 - \sin \theta \quad$ has horizontal or vertical tangents.

16. Find the length of $\quad x = 3(t - 1)^2, \quad y = 8t^{3/2} \quad$ from $t = 0$ to $t = 1$.

17. Find the length of $\quad y = x^2/8 - \ln x \quad$ from $(1, \frac{1}{8})$ to $(2, \frac{1}{2} - \ln 2)$.

18. A 400-ft length of wire weighing 2 lb/ft hangs with a 40-ft sag. Find the maximum and minimum tension in the wire and its equation (with the placement of axes shown in Figure 13.18, where $k = T_0/w$).

C

19. A cannon is fired from ground level with initial speed v_0 and angle of inclination θ. Show that the range R of the projectile (see Problem 6, page 402) is given by

$$R = \frac{v_0^2 \sin 2\theta}{32}.$$

Find the range if the cannon is fired from a hill of height H above a plain and the projectile hits on the plain.

20. Find parametric equations for the set of all points P determined as shown in Figure 13.19. (*Hint:* Show that $\overline{OB} = 2a \sin \theta$.) Sketch. This curve is called the witch of Agnesi.

21. In Figure 13.20, $\overline{OP} = \overline{AB}$. Find parametric equations for the set of all points P determined as shown. (*Hint:* Show that $\overline{OB} = 2a \sin \theta$.) Sketch. This curve is called the cissoid of Diocles.

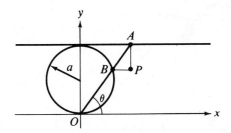

Figure 13.19

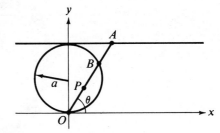

Figure 13.20

14

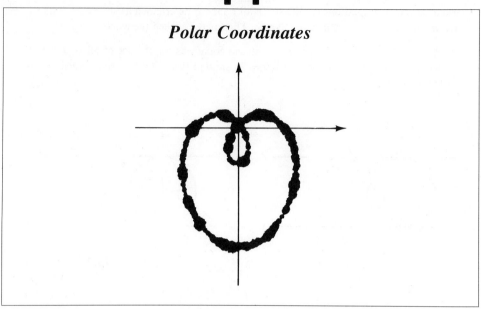

Polar Coordinates

14.1

Polar Coordinates

Up to now, a point in the plane has been represented by a pair of numbers, (x, y), which represent (for perpendicular axes) the distances of the point from the y and x axes, respectively. Another way of representing points is by *polar coordinates*. In this case, we need only one axis (the *polar axis*) and a point on it (the *pole*). These correspond to the positive x axis and the origin of the rectangular coordinate system, respectively. Normally we shall include the y axis, even though it is not necessary to do so.

Before considering points in polar coordinates, let us recall that an angle in the standard position has its vertex at the origin (or pole) and its initial side on the positive end of the x axis (or polar axis). The terminal side is another ray (or half-line) with the origin as its end point. The other ray with the same end point and on the same line as the terminal side is called the ray opposite the terminal side. For example, the terminal side of a $90°$ angle in standard position is the positive end of the y axis together with the origin; the ray opposite the terminal side is the negative end of the y axis together with the origin.

A point P is represented, in polar coordinates, by an ordered pair of numbers (r, θ). (See Figure 14.1.) It is determined in the following way: first find the terminal side of the angle θ in standard position; if $r \geq 0$, then P is on this terminal side and at a distance r from the pole; if $r < 0$, then P is on the ray opposite the terminal

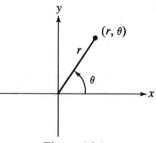

Figure 14.1

420

side and at a distance $|r|$ from the pole. A few points are given with their polar coordinates in Figure 14.2.

It might be noted that while the terminal side of the angle $-\pi/3$ is in the fourth quadrant, $(-1, -\pi/3)$ is in the second quadrant. The quadrant that a point is in is *not* determined by the signs of the two polar coordinates, as it is with rectangular coordinates. It is determined by the size of θ and the sign of r. If r is positive, the point is in whatever quadrant θ is in; if r is negative the point is in the opposite quadrant.

Polar coordinates present only one problem that we did not have with rectangular coordinates—a point has more than one representation in polar coordinates. For example, $(2, \pi/4)$ and $(-2, 5\pi/4)$ represent

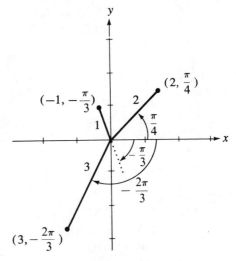

Figure 14.2

the same point. In fact, if (r, θ) is one representation of a point, then $(r, \theta + \pi n)$, where n is an even integer, and $(-r, \theta + \pi n)$, where n is an odd integer, are representations of the same point. Furthermore, $(0, \theta)$ is the pole for any choice of θ.

14.2

Graphs in Polar Coordinates

Equations in polar coordinates can be graphed by point-by-point plotting, as we graphed rectangular coordinates.

Example 1

Graph $r = \sin \theta$.

Note in Figure 14.3 that we have the entire graph for $0° \leq \theta < 180°$. The remaining values of θ simply repeat the graph a second time, since $(0, 0°) = (0, 180°)$, $(.5, 30°) = (-.5, 210°)$, and so forth. Of course, values of θ outside the range $0° \leq \theta \leq 360°$ would give no new points.

This method of point-by-point plotting is quite cumbersome here, as it was in the case of rectangular coordinates. One way to simplify the proceedings is to represent the table of values of r and θ by means of a graph. This may sound as if we are going in circles: we can get the graph from a table of values of r and θ that is represented by a graph. Actually, this is not so bad as it sounds. We shall represent the table by a graph in *rectangular coordinates*.

θ	r
0°	0.00
30°	0.50
60°	0.87
90°	1.00
120°	0.87
150°	0.50
180°	0.00
210°	−0.50
240°	−0.87
270°	−1.00
300°	−0.87
330°	−0.50
360°	0.00

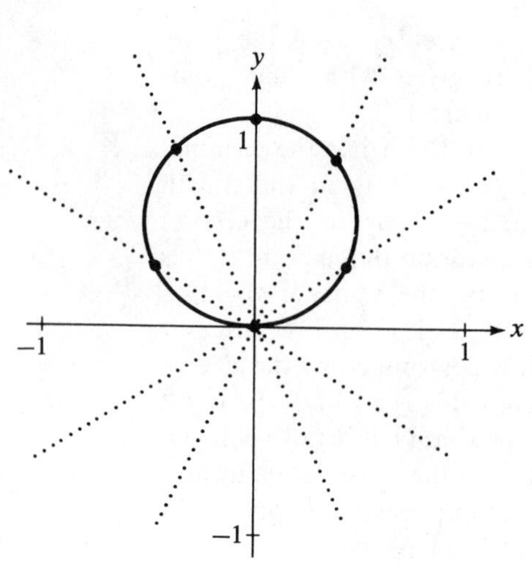

Figure 14.3

For example, the table of values of r and θ used in Example 1 can be represented by the graph shown in Figure 11.6(a). Of course, Figure 11.6(a) represents the graph of $y = \sin x$; we merely replace the symbols x and y by θ and r.

Example 2

Graph $r = 1 + \cos \theta$.

We can easily graph this equation in rectangular coordinates by using addition of ordinates. The result is given in Figure 14.4. Now we can read off values of r and θ just as we would from a table. As θ increases from 0° to 90°, r goes from 2 to 1. This gives the portion of the curve shown in (a) of Figure 14.5. As θ goes from 90° to 180°, r goes from 1 down to 0 (shown in (b)). As θ goes from 180° to 270°, r goes from 0 back up to 1 (as in (c)); and finally, as θ goes from 270° to 360°, we see in (d) that r goes from 1 to 2. The same path is traced for values of θ beyond 360° or less than 0°. Putting all of this together, we have the desired graph, shown in (e).

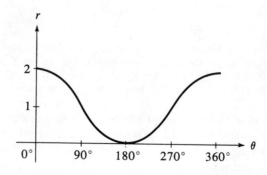

Figure 14.4

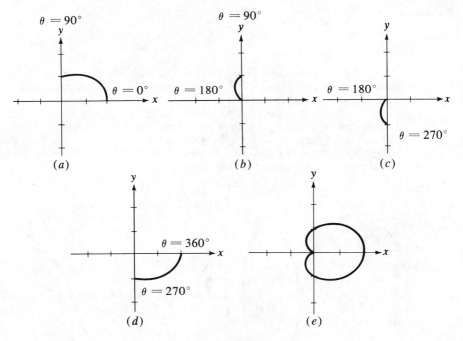

Figure 14.5

Example 3

Graph $r = \sin 2\theta$.

The graph is given in rectangular coordinates in (a) of Figure 14.6. This is then put on the polar graph shown in (b). Note that for θ in the range $90° < \theta < 180°$, r is negative. Thus instead of giving the loop in the second quadrant, it gives the one in the fourth quadrant. Similarly, r is negative for θ in the range $270° < \theta < 360°$. This gives the loop in the second quadrant.

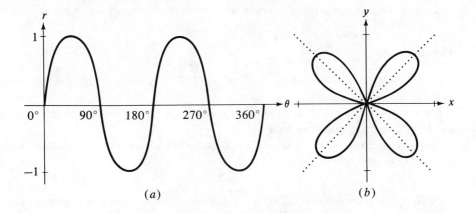

Figure 14.6

Example 4

Graph $r^2 = \sin 2\theta$.

Graphing in rectangular coordinates by the methods of Section 5.3, we have the result given in Figure 14.7(a). There are a couple of things of interest here. First of all, $r^2 = \sin 2\theta$ has two values of r for each θ in the ranges $0 < \theta < \pi/2$ and $\pi < \theta < 3\pi/2$, while it has no value at all for $\pi/2 < \theta < \pi$ and $3\pi/2 < \theta < 2\pi$. Since it has two values in the range $0 < \theta < \pi/2$, we get both loops for $0 \le \theta \le \pi/2$, shown in (b). Similarly we get both loops a second time for $\pi \le \theta \le 3\pi/2$. Because there is no value of r for $\pi/2 < \theta < \pi$ and $3\pi/2 < \theta < 2\pi$, there are no points of the graph in the second or fourth quadrants.

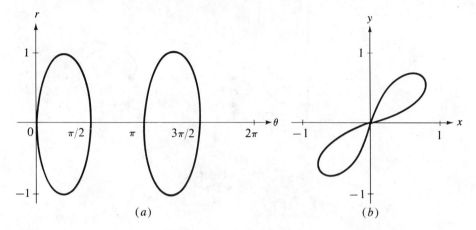

Figure 14.7

There are tests for symmetry in polar coordinates which are somewhat like those in rectangular coordinates. Suppose, for example, that for each point (r, θ) on a given curve there corresponds another point $(r, -\theta)$ on the same curve. Then (see Figure 14.8(a)) the curve is symmetric about the x axis. Thus if θ is replaced by $-\theta$ and the result is equivalent† to the original equation, then the graph is symmetric about the x axis. Figures 14.8(b) and (c) illustrate conditions leading to symmetry about the y axis and the pole, respectively. These tests are summarized in the following theorem.

Theorem 14.1

If θ is replaced by $-\theta$ and the result is equivalent to the original equation, then the graph is symmetric about the x axis.

If θ is replaced by $\pi - \theta$ and the result is equivalent to the original equation, then the graph is symmetric about the y axis.

If r is replaced by $-r$ and the result is equivalent to the original equation, then the graph is symmetric about the pole.

---

† Two equations are equivalent if any point that satisfies one of them also satisfies the other. To determine equivalence we use algebraic or trigonometric identities, add any expression to both sides of one equation, or multiply both sides of an equation by a nonzero constant in order to make that equation identical to the other.

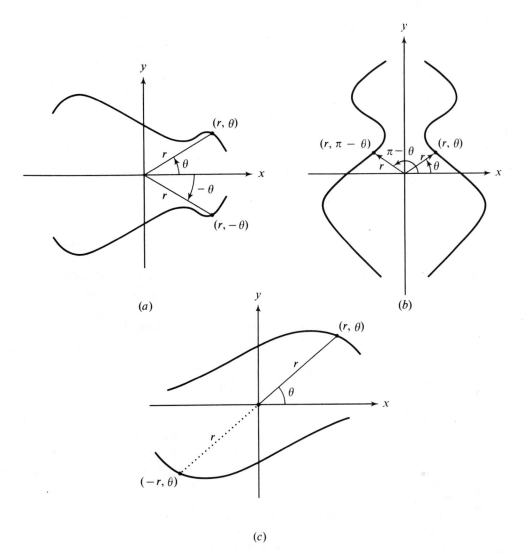

Figure 14.8

Example 5

Test $r = 1 + \cos\theta$ for symmetry.

The three tests are represented schematically below (the arrow is used to represent " is replaced by "):

(a) $\begin{cases} r \to r, \\ \theta \to -\theta. \end{cases}$

(b) $\begin{cases} r \to r, \\ \theta \to \pi - \theta. \end{cases}$

(c) $\begin{cases} r \to -r, \\ \theta \to \theta. \end{cases}$

    $r = 1 + \cos(-\theta)$
          $r = 1 + \cos(\pi - \theta)$
          $-r = 1 + \cos\theta.$

    $= 1 + \cos\theta.$
            $= 1 - \cos\theta.$

Since $r = 1 + \cos(-\theta)$ is equivalent to $r = 1 + \cos\theta$, we have symmetry about the x axis. The other two tests are negative; however, *this in itself is not enough to say that we do not have the other two types of symmetry* (we shall consider this in more detail in the next example). Nevertheless the graph of $r = 1 + \cos\theta$ (given in Figure 14.5(e)) indicates that we do have symmetry only about the x axis.

We see from this example that, while symmetry is sometimes an aid in graphing an equation, the graph can also be an aid in determining the presence or absence of symmetry. This is especially true in polar coordinates, because negative results in the tests of Theorem 14.1 do not necessarily imply a lack of symmetry, as the next example makes evident.

Example 6

Test $r = \sin 2\theta$ for symmetry.

(a) $\begin{cases} r \to r, \\ \theta \to -\theta. \end{cases}$ (b) $\begin{cases} r \to r, \\ \theta \to \pi - \theta. \end{cases}$ (c) $\begin{cases} r \to -r, \\ \theta \to \theta. \end{cases}$

$r = \sin(-2\theta)$ $r = \sin(2\pi - 2\theta)$ $-r = \sin 2\theta.$
$= -\sin 2\theta.$ $= -\sin 2\theta.$

All three tests are negative. Yet we can see from Figure 14.6(b) that all three types of symmetry are present.

The reason for the rather strange behavior of the last example can be traced directly to the fact that one point has many different representations in polar coordinates. For example, $(r, -\theta) = (-r, \pi - \theta) = (r, 2\pi - \theta)$, etc. Thus there are many tests for symmetry about the x axis. The equalities above lead to the three tests:

$$\begin{cases} r \to r \\ \theta \to -\theta \end{cases} \qquad \begin{cases} r \to -r \\ \theta \to \pi - \theta \end{cases} \qquad \begin{cases} r \to r \\ \theta \to 2\pi - \theta \end{cases}$$

If any one of these gives an equation that is equivalent to the original, there is symmetry about the x axis. Notice that while the first and third of these three tests give negative results in the equation of Example 6, the second gives a positive result. This is sufficient to assure us that there is symmetry about the x axis. If they all give negative results, nothing can be concluded, since there are still other possible tests. The student can easily devise other tests for all three types of symmetry. The result of all of this is that we must be content with tests for symmetry which do not guarantee a lack of symmetry when the test is negative.

Because of the multiplicity of the foregoing tests and the indecisiveness of negative results, you might prefer to rely upon the graphs of an equation. For example, Figure 14.6(b) suggests that we have all three types of symmetry for $r = \sin 2\theta$. Then the symmetry of Figure 14.6(a) assures us that we really do have the suspected symmetry. Note, however, that a particular type of symmetry in the rectangular coordinate graph does not necessarily imply the same type of symmetry in polar coordinates. Although the rectangular coordinate graph of $r^2 = \sin 2\theta$ (i.e. of $y^2 = \sin 2x$) shows symmetry about the x axis, the polar graph does not (see Figure 14.7). Furthermore the polar graph exhibits symmetry about the pole (or origin), while the rectangular coordinate graph does not. But the symmetry of the loops in rectangular coordinates does imply symmetry of the loops in polar coordinates and thus symmetry about the pole.

Problems

A

1. Plot the following points: $(1, \pi/6), (2, 45°), (0, 60°), (-2, 90°), (-1, 3\pi/4), (2, 300°)$.
2. Give an alternate polar representation with $0° \le \theta \le 180°$: $(4, 330°), (-2, 420°), (1, 210°), (0, 283°), (-3, 270°), (2, 240°)$.
3. Give an alternate polar representation with $r \ge 0$ and $0 \le \theta \le 2\pi$: $(-4, 2\pi/3), (3, -\pi/3), (0, 53\pi/18), (-1, 11\pi/6), (-2, 13\pi/6), (-2, 3\pi/4)$,
4. Give an alternate polar representation: $(2, 60°), (-2, \pi), (3, 210°), (0, \pi/3), (-1, 30°), (3, \pi/2)$.

In Problems 5–22, sketch the graph of the given equation and indicate any symmetry about either axis or the pole.

5. $r = \cos\theta$.

6. $r = 2\sin\theta$.

7. $r = 1 + \cos\theta$.

8. $r = 1 - \cos\theta$.

9. $r = 1 - \sin\theta$.

10. $r = \sin\theta - 1$.

11. $r = \cos 2\theta$.

12. $r = \cos 4\theta$.

13. $r = \sin 3\theta$.

14. $r = \cos 3\theta$.

15. $r = \sin 5\theta$.

16. $r = \cos 6\theta$.

17. $r = 1 + 2\sin\theta$.

18. $r = 1 - 2\cos\theta$.

19. $r = 2 + \cos\theta$.

20. $r = 3 + 4\sin\theta$.

21. $r = \tan\theta$.

22. $r = 2\sec\theta$.

B

In Problems 23–34, sketch the graph of the given equation and indicate any symmetry about either axis or the pole.

23. $r^2 = \sin\theta$.

24. $r^2 = \cos 3\theta$.

25. $r^2 = \cos 4\theta$.

26. $r^2 = \sin^2\theta$.

27. $r^2 = 1 + \sin\theta$.

28. $r^2 = 1 - \cos\theta$.

29. $r = \theta$.

30. $r = |\theta|$.

31. $r^2 = \theta^2$.

32. $r = \dfrac{3}{1 - \cos\theta}$.

33. $r = \dfrac{2}{1 - 2\cos\theta}$.

34. $r = \dfrac{3}{2 - \cos\theta}$.

C

35. Find two tests for symmetry about the y axis that are different from the one given in this section.

36. Find two tests for symmetry about the pole that are different from the one given in this section.

14.3

Points of Intersection

Suppose we have a pair of equations in polar form that we solve simultaneously to obtain pairs of numbers satisfying both equations—that is, points of intersection of the two curves.

Example 1

Find the points of intersection of $r = 1$ and $r = 2\sin\theta$.

Eliminating r from this pair of equations, we get

$$\sin\theta = \frac{1}{2}, \quad \text{or} \quad \theta = \pi/6,\ 5\pi/6,$$

giving the points $(1,\ \pi/6)$ and $(1,\ 5\pi/6)$. The graphs of these two curves, showing the two points of intersection, are given in Figure 14.9.

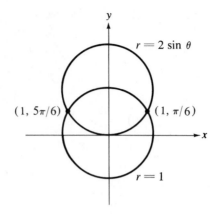

Figure 14.9

While the solutions of a pair of simultaneous equations must be points of inter-section of the curves represented by the equations, some points of intersection cannot be found in this way. The reason is that they have different representations on the two curves. Thus *we must graph both curves* to be sure that we have found all points of intersection.

Example 2

Find the points of intersection of $r = \sin \theta$ and $r = \cos \theta$.

Eliminating r between the two equations, we have

$$\sin \theta = \cos \theta.$$

If we divide by $\cos \theta$, then

$$\tan \theta = 1 \quad \text{and} \quad \theta = \pi/4 + \pi n.$$

In the range $0 \leq \theta < 2\pi$, we have $(1/\sqrt{2}, \pi/4)$ and $(-1/\sqrt{2}, 5\pi/4)$. But these are different representa-tions for the same point. Thus we have found only one point of intersection. As we can see from Figure 14.10, there are really two points of intersection—the one we found and the pole.

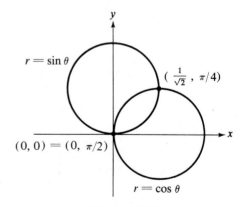

Figure 14.10

The pole has many different repre-sentations. On the curve $r = \sin \theta$ it is represented by $(0, \pi n)$; on $r = \cos \theta$ it is represented by $(0, \pi/2 + \pi n)$. Thus, while the pole is common to both curves, it does not have a common representation that satisfies both equations. So we cannot find this point of intersection by finding simultaneous solutions of the two equations. We might represent this point by $(0, 0) = (0, \pi/2)$.

One convenient way to think of the phenomenon of this last example is to imagine the two curves as paths traced by points as θ increases uniformly with time. From this point of view, the curves both go through the origin; but they do so at different times.

Example 3

Find all points of intersection of $r = \cos 2\theta$ and $r = \sin \theta$.

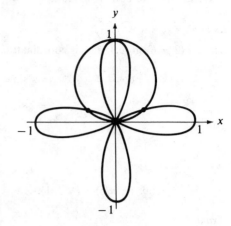

$$\cos 2\theta = \sin \theta.$$
$$1 - 2\sin^2 \theta = \sin \theta$$
$$2\sin^2 \theta + \sin \theta - 1 = 0$$
$$(2\sin \theta - 1)(\sin \theta + 1) = 0;$$
$$\sin \theta = \frac{1}{2}, \quad \sin \theta = -1;$$
$$\theta = \pi/6, \ 5\pi/6, \ 3\pi/2.$$

Thus, we have the points $(1/2, \pi/6)$, $(1/2, 5\pi/6)$, and $(-1, 3\pi/2)$. In addition we can see from Figure 14.11 that the pole is a point of intersection; it may be represented by $(0, \pi/4) = (0, 0)$. It might be also noted that the point $(-1, 3\pi/2)$ can also be written $(1, \pi/2)$, but this form satisfies only $r = \sin \theta$.

Figure 14.11

Problems

A

Find all points of intersection of the given curves.

1. $r = \sqrt{2}, \quad r = 2\cos \theta$.
2. $r = 1, \quad r = 2\sin \theta$.
3. $r = 2, \quad r = \cos \theta + 2$.
4. $r = 1, \quad r = 2\sin 2\theta$.
5. $r = \cos \theta, \quad r = 1 - \cos \theta$.
6. $r = \cos \theta, \quad r = 1 + \sin \theta$.
7. $r = \sin 2\theta, \quad r = \sin \theta$.
8. $r = \sin 2\theta, \quad r = \sqrt{2}\cos \theta$.
9. $r = \sec \theta, \quad r = \csc \theta$.
10. $r = \sec \theta, \quad r = \tan \theta$.
11. $r = 4\cos \theta + 5, \quad r = 3$.
12. $r = 2(1 + \cos \theta), \quad r(1 - \cos \theta) = 1$.
13. $r = 1 - \sin \theta, \quad r(1 - \sin \theta) = 1$.
14. $r^2 = \cos \theta, \quad r = \sec \theta$.
15. $r^2 = \sin \theta, \quad r = \sin \theta$.
16. $r = \sin 2\theta, \quad r = \cos 2\theta$.

B

17. $r = 1 - \sin \theta, \quad r = 1 - \cos \theta$.
18. $r^2 = \sin \theta, \quad r^2 = \cos \theta$.
19. $r = 2\cos \theta + 1, \quad r = 2\cos \theta - 1$.
20. $r^2 = \sin \theta, \quad r = \cos \theta$.

14.4

Relationships between Rectangular and Polar Coordinates

There are some simple relationships between rectangular and polar coordinates. These can be found easily by a consideration of Figure 14.12.

$$x = r\cos \theta,$$
$$y = r\sin \theta,$$
$$r^2 = x^2 + y^2,$$
$$\tan \theta = \frac{y}{x}.$$

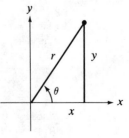

Figure 14.12

The last two, which may be solved for r and θ, would give us expressions involving $\pm$ and arctan. Thus, we prefer to leave them in their present form.

With these we can now change from one coordinate system to the other.

Example 1

Express $(2, \pi/6)$ in rectangular coordinates.

$$x = r \cos \theta \qquad\qquad y = r \sin \theta$$

$$= 2 \cos \frac{\pi}{6} \qquad\qquad = 2 \sin \frac{\pi}{6}$$

$$= 2 \cdot \frac{\sqrt{3}}{2} \qquad\qquad = 2 \cdot \frac{1}{2}$$

$$= \sqrt{3}; \qquad\qquad = 1.$$

Thus $(2, \pi/6) = (\sqrt{3}, 1)$.

Example 2

Express $(4, -4)$ in polar coordinates.

$$r^2 = x^2 + y^2 \qquad\qquad \tan \theta = \frac{y}{x}$$

$$= 16 + 16 \qquad\qquad = \frac{-4}{4}$$

$$= 32 \qquad\qquad = -1$$

$$r = \pm 4\sqrt{2}; \qquad\qquad \theta = \frac{3\pi}{4} + \pi n.$$

We have a choice for both r and θ. The values of r and θ cannot be selected independently; the value we choose for one will limit the available choices for the other. In this case, the point $(4, -4)$ is in the fourth quadrant. Thus, we may choose either a fourth-quadrant angle and a positive r or a second-quadrant angle and a negative r. Thus

$$(4, -4) = (4\sqrt{2}, 7\pi/4) = (-4\sqrt{2}, 3\pi/4) = (4\sqrt{2}, -\pi/4), \text{ etc.}$$

Of course we can use these equations to find a polar equation corresponding to one in rectangular coordinates, and vice versa.

Example 3

Express $y = x^2$ in polar coordinates.

$$y = x^2,$$

$$r \sin \theta = r^2 \cos^2 \theta$$

$$\sin \theta = r \cos^2 \theta \quad \text{or} \quad r = 0$$

$$r = \frac{\sin \theta}{\cos^2 \theta}$$

$$r = \sec \theta \tan \theta.$$

Since $r = 0$ represents only the pole and it is included in $r = \sec \theta \tan \theta$, we may drop $r = 0$. The result is

$$r = \sec \theta \tan \theta.$$

Example 4

Express $r = 1 - \cos \theta$ in rectangular coordinates.

First, multiply through by r.

$$r^2 = r - r \cos \theta.$$

At this point we could make the substitutions $r^2 = x^2 + y^2$, $r \cos \theta = x$, and $r = \pm\sqrt{x^2 + y^2}$. The last is rather bothersome, since it involves a $\pm$. In order to avoid this, let us isolate r on one side of the equation and square.

$$r = r^2 + r \cos \theta,$$
$$r^2 = (r^2 + r \cos \theta)^2,$$
$$x^2 + y^2 = (x^2 + y^2 + x)^2.$$

We have done two things that might introduce extraneous roots: (1) Multiplying by r may introduce only a single point, the pole, to the graph. Since the pole is already a point of the graph of $r = 1 - \cos \theta$, no new point is introduced here. (2) Squaring may introduce several new points. The equation

$$r^2 = (r^2 + r \cos \theta)^2$$

is equivalent to

$$r = \pm(r^2 + r \cos \theta).$$

Now $r = r^2 + r \cos \theta$ is equivalent to our original equation, $r = 1 - \cos \theta$, while $r = -(r^2 + r \cos \theta)$ is equivalent to $r = -1 - \cos \theta$. Thus

$$x^2 + y^2 = (x^2 + y^2 + x)^2$$

is equivalent to $r = 1 - \cos \theta$ together with $r = -1 - \cos \theta$. But $r = 1 - \cos \theta$ and $r = -1 - \cos \theta$ have the same graph. Thus we have introduced no new points by squaring.

Problems

A

1. The following points are given in polar coordinates. Give the rectangular coordinate representation of each.

$$(1, \pi/2), \quad (\sqrt{3}, \pi/6), \quad (-2, 3\pi), \quad (\sqrt{2}, 3\pi/4), \quad (3\sqrt{3}, 5\pi/3),$$
$$(-5, 7\pi/6), \quad (0, 5\pi/4), \quad (6, 0), \quad (-4, 7\pi/4).$$

2. The following points are given in rectangular coordinates. Give a polar coordinate representation of each.

$$(\sqrt{2}, -\sqrt{2}), \quad (-1, \sqrt{3}), \quad (4, 0), \quad (-1, -1), \quad (0, -2),$$
$$(0, 0), \quad (-2\sqrt{3}, 2), \quad (-3, 1), \quad (4, 3), \quad (-2, 4).$$

In Problems 3–18, express the given equation in polar coordinates.

3. $x = 3$.
4. $y = 2$.
5. $x^2 + y^2 = 4$.
6. $x^2 - y^2 = 2$.
7. $y^2 = 3x$.
8. $y = x^3$.
9. $(x + y)^2 = x - y$.
10. $x = y$.
11. $y = 4x$.
12. $y^2 = x^3$.
13. $2x - y + 3 = 0$.
14. $x^2 + y^2 - 6x = 0$.
15. $x^2 + y^2 + 4x - 6y + 1 = 0$.
16. $4x^2 + 16y^2 = 16$.
17. $xy = 1$.
18. $y = \dfrac{x}{x + 1}$.

In Problems 19–24, express the given equation in rectangular coordinates.

19. $r = a$. 20. $\theta = \pi/4$.
21. $\theta = \pi/6$. 22. $r = 3 \cos \theta$.
23. $r = 4 \sin \theta$. 24. $r = \sin 2\theta$.

B

In Problems 25–34, express the given equation in rectangular coordinates.

25. $r = \cos 2\theta$. 26. $r = 1 - \cos \theta$.
27. $r = 4 + 3 \sin \theta$. 28. $r^2 = \sin \theta$.
29. $r^2 = 1 + \sin \theta$. 30. $r^2 = \sin 2\theta$.

31. $r = \dfrac{1}{1 - \cos \theta}$. 32. $r = \dfrac{1}{1 - \sin \theta}$.

33. $r = 3 \sin \theta + 2 \cos \theta$. 34. $r = \sec \theta$.

14.5

Conics in Polar Coordinates

We found earlier that the equations of conic sections (in rectangular coordinates) have very simple forms if the center or vertex is at the origin and the axes are the coordinate axes. There are, however, three different forms corresponding to the three different types of conics. We find that conics can be easily represented in polar coordinates if a focus is at the origin and one axis is a coordinate axis. Furthermore, the same type of equation represents all three types of conics if we use the unifying concept of eccentricity.

Recall that any conic can be determined by a single focus, the corresponding directrix, and the eccentricity. If P is a point on the conic, then the distance from P to the focus divided by the distance from P to the directrix equals the eccentricity. The particular conic we get depends upon the eccentricity; the eccentricity is a positive number and

if $e < 1$, the conic is an ellipse,

if $e = 1$, the conic is a parabola,

if $e > 1$, the conic is a hyperbola.

If $P\ (r, \theta)$ is a point on a conic with focus O, directrix $x = p$ (p positive), and eccentricity e (see Figure 14.13), then

$$\frac{\overline{OP}}{\overline{PD}} = e, \quad \text{or} \quad \frac{|r|}{|p - r \cos \theta|} = e.$$

There are now two cases to consider:

$$\frac{r}{p - r \cos \theta} = e \quad \text{and} \quad \frac{r}{p - r \cos \theta} = -e.$$

Either of these yields an equation of the desired conic (see Problems 21 and 22); however, the first yields the commonly used form. Solving for r in this equation, we have

$$r = \frac{ep}{1 + e \cos \theta}.$$

If the directrix is $x = -p$ (p positive), then the equation is

$$r = \frac{ep}{1 - e \cos \theta}.$$

If the directrix is $y = \pm p$ (p positive), then the equation is

$$r = \frac{ep}{1 \pm e \sin \theta}.$$

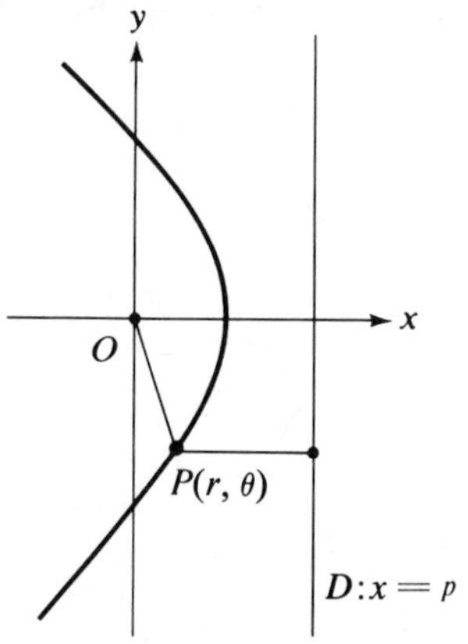

Figure 14.13

Theorem 14.2

The conic section with focus at the origin, directrix $x = \pm p$ (p positive), and eccentricity e has polar equation

$$r = \frac{ep}{1 \pm e \cos \theta};$$

if the directrix is $y = \pm p$ (p positive), it has equation

$$r = \frac{ep}{1 \pm e \sin \theta}.$$

Example 1

Describe $r = \dfrac{6}{4 + 3 \cos \theta}$.

Dividing numerator and denominator by 4, we have

$$r = \frac{\dfrac{3}{2}}{1 + \dfrac{3}{4} \cos \theta} = \frac{\dfrac{3}{4} \cdot 2}{1 + \dfrac{3}{4} \cos \theta}.$$

Thus the eccentricity is $3/4$ and the directrix is $x = 2$. The conic is an ellipse with focus at the origin, directrix $x = 2$, and eccentricity $3/4$.

Example 2

Sketch $r = \dfrac{15}{2 - 3 \cos \theta}$.

Dividing by 2, we have

$$r = \frac{\dfrac{15}{2}}{1 - \dfrac{3}{2}\cos\theta} = \frac{\dfrac{3}{2}\cdot 5}{1 - \dfrac{3}{2}\cos\theta}.$$

Thus we have a hyperbola with focus at the origin, eccentricity 3/2, and directrix $x = -5$. The vertices are on the x axis, one between the focus and directrix and the other to the left of the direc-
trix. When $\theta = 0$, $r = -15$; when $\theta = \pi$, $r = 3$. Thus the vertices are $(-15, 0)$ and $(3, \pi)$. When $\theta = \pi/2$ or $3\pi/2$, $r = 15/2$. Thus, the ends of one of the latera recta are $(15/2, \pi/2)$ and $(15/2, 3\pi/2)$. This information is enough to give a reason-
ably accurate picture of the hyper-
bola. If the asymptotes are desired, they can best be found by con-
sidering some of the above points in rectangular coordinates. Thus the vertices are $(-3, 0)$ and $(-15, 0)$, and the center is $(-9, 0)$, giving $a = 6$ and $c = 9$. We can now use the equation

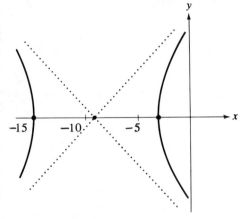

Figure 14.14

$$b^2 = c^2 - a^2$$

to find $b^2 = 45$ or $b = 3\sqrt{5}$. Once we have this, the asymptotes are easily found (see Figure 14.14).

Example 3

Find a polar equation of the parabola with focus at the origin and directrix $y = -4$.

The equation is in the form

$$r = \frac{ep}{1 - e\sin\theta},$$

since the directrix is a horizontal line below the focus. Furthermore, $e = 1$, since the conic is a parabola, and directrix $y = -4$ gives $p = 4$. Thus the equation is

$$r = \frac{4}{1 - \sin\theta}.$$

Problems

A

In Problems 1–8, state the type of conic and give a focus and its corresponding directrix and the eccentricity.

1. $r = \dfrac{6}{1 + 3\cos\theta}$.

2. $r = \dfrac{12}{1 - 2\sin\theta}$.

3. $r = \dfrac{4}{3 + \sin\theta}$.

4. $r = \dfrac{5}{4 - 4\cos\theta}$.

5. $r = \dfrac{4}{1 + \sin \theta}$

6. $r = \dfrac{10}{5 - 2 \cos \theta}.$

7. $r(3 + 2 \sin \theta) = 6.$

8. $r(3 - 6 \cos \theta) = 7.$

In Problems 9–14, find a polar equation of the conic with focus at the origin and the given eccentricity and directrix.

9. Directrix $x = 4$; $e = 2/3$.

10. Directrix $y = -2$; $e = 3$.

11. Directrix $y = 3$; $e = 1$.

12. Directrix $x = -3$; $e = 1$.

13. Directrix $x = 7$; $e = 7/4$.

14. Directrix $y = 3$; $e = 3/4$.

B

In Problems 15–20, sketch the given conic.

15. $r = \dfrac{2}{1 + \cos \theta}.$

16. $r = \dfrac{16}{5 - 3 \cos \theta}.$

17. $r = \dfrac{12}{4 - 5 \sin \theta}.$

18. $r(3 - 5 \cos \theta) = 9.$

19. $r(13 + 12 \sin \theta) = 26.$

20. $r(3 + 3 \sin \theta) = 4.$

21. Sketch $r = \dfrac{-2}{1 - \cos \theta}.$ Compare with the conic of Problem 15 (see Problem 22 also.)

22. Show that the conic section with focus at the origin, directrix $x = p$ (p positive), and eccentricity e has polar equation

$$r = \frac{-ep}{1 - e \cos \theta}.$$

Show that this equation is equivalent to the first equation of Theorem 14.2. (*Hint:* Note that $(r, \theta) = (-r, \theta + \pi)$.)

C

23. Suppose, in the equation $r = \dfrac{ep}{1 + e \cos \theta}$, $e \to 0$ and $p \to +\infty$ in such a way that ep remains constant. What happens to the shape of the conic? What happens to the equation of the conic?

24. A comet has a parabolic orbit with the sun at the focus. When the comet is 100,000,000 miles from the sun, the line joining the sun and the comet makes an angle of 60° with the axis of the parabola (see Figure 9.5, page 248). How close to the sun will the comet get?

25. A satellite has an elliptical orbit with the earth at one focus. At its closest point it is 100 miles above the surface of the earth; at its farthest point, 500 miles. Find a polar equation of its path. (Take the radius of the earth to be 4000 miles.)

26. Find a polar equation of a circle with center (k, α) and radius a by using the law of cosines (see Figure 14.15).

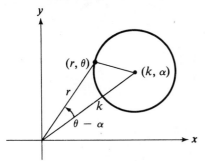

Figure 14.15

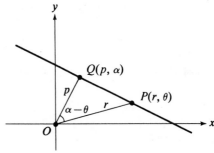

Figure 14.16

27. By using the trigonometry of right triangles, show that the line PQ (Figure 14.16) can be represented by the equation

$$x \cos \alpha + y \sin \alpha - p = 0.$$

This is called the *normal form* of the line, since it is expressed in terms of the polar coordinates of the point Q, which is the intersection of the original line and another perpendicular (or normal) to it and through the origin (see Problem 36, Section 2.3).

28. By using the identity

$$\sin^2 \alpha + \cos^2 \alpha = 1,$$

show that $Ax + By + C = 0$ can be put into the normal form by dividing through by $\pm\sqrt{A^2 + B^2}$ (see Problem 27 for the normal form).

29. Show that the distance from the point (x_1, y_1) to the line $Ax + By + C = 0$ is

$$d = \frac{|Ax_1 + By_1 + C|}{\sqrt{A^2 + B^2}}.$$

This result was obtained without the use of polar coordinates in Theorem 2.6, page 27. (*Hint:* Put the original line and the one parallel to it and through (x_1, y_1) into the normal form (see Problems 27 and 28).)

14.6

Derivatives in Polar Coordinates

If we are given the polar equation

$$r = f(\theta),$$

it is a simple matter to find the derivative $dr/d\theta$. The only trouble is that this does not represent the same thing as the derivative dy/dx, where $y = f(x)$. The latter represents the slope of the graph of $y = f(x)$. If we go back to the definition of the derivative, we see that

$$\frac{dr}{d\theta} = \lim_{h \to 0} \frac{f(\theta + h) - f(\theta)}{h}.$$

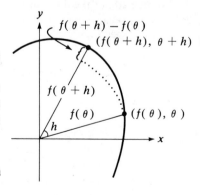

It is obvious from Figure 14.17 that this is not the slope.

Let us see if we can get the slope from some other expression. Since the slope of the graph is dy/dx, we are interested in expressing the equation

$$r = f(\theta)$$

in rectangular coordinates. Suppose we start with the relations

$$x = r \cos \theta \quad \text{and} \quad y = r \sin \theta.$$

Figure 14.17

By substituting $r = f(\theta)$, we get the desired equation in parametric form,

$$x = f(\theta) \cos \theta \quad \text{and} \quad y = f(\theta) \sin \theta.$$

Now we are in a position to find the desired slope.

$$\frac{dy}{dx} = \frac{\dfrac{dy}{d\theta}}{\dfrac{dx}{d\theta}} = \frac{f'(\theta) \sin \theta + f(\theta) \cos \theta}{f'(\theta) \cos \theta - f(\theta) \sin \theta}$$

$$= \frac{r' \sin \theta + r \cos \theta}{r' \cos \theta - r \sin \theta}.$$

You are encouraged to recall and use parametric differentiation to obtain this formula rather than memorizing it.

Example 1

Find the slope of the graph $r = 1 - \cos \theta$ at $\theta = \pi/2$.

$$\frac{dy}{dx} = \frac{r' \sin \theta + r \cos \theta}{r' \cos \theta - r \sin \theta} = \frac{\sin \theta \cdot \sin \theta + (1 - \cos \theta) \cos \theta}{\sin \theta \cos \theta - (1 - \cos \theta) \sin \theta}$$

$$= \frac{\sin^2 \theta + \cos \theta - \cos^2 \theta}{2 \sin \theta \cos \theta - \sin \theta}.$$

At $\theta = \pi/2$,

$$\frac{dy}{dx} = \frac{1^2 + 0 - 0}{2 \cdot 1 \cdot 0 - 1} = -1.$$

Example 2

Find an equation (in rectangular coordinates) of the line tangent to $r = \sin 2\theta$ at $(1, \pi/4)$.

$$\frac{dy}{dx} = \frac{r' \sin \theta + r \cos \theta}{r' \cos \theta - r \sin \theta} = \frac{2 \cos 2\theta \sin \theta + \sin 2\theta \cos \theta}{2 \cos 2\theta \cos \theta - \sin 2\theta \sin \theta}.$$

At $\theta = \pi/4$,

$$\frac{dy}{dx} = \frac{2 \cos \pi/2 \sin \pi/4 + \sin \pi/2 \cos \pi/4}{2 \cos \pi/2 \cos \pi/4 - \sin \pi/2 \sin \pi/4}$$

$$= \frac{2 \cdot 0 \cdot 1/\sqrt{2} + 1 \cdot 1/\sqrt{2}}{2 \cdot 0 \cdot 1/\sqrt{2} - 1 \cdot 1/\sqrt{2}} = -1.$$

Transforming the point $(1, \pi/4)$ into rectangular coordinates, we have

$$x = r \cos \theta = 1 \cos \pi/4 = 1/\sqrt{2},$$
$$y = r \sin \theta = 1 \sin \pi/4 = 1/\sqrt{2}.$$

Thus the desired equation is

$$y - 1/\sqrt{2} = -1 \, (x - 1/\sqrt{2}) \quad \text{or} \quad x + y - \sqrt{2} = 0.$$

Example 3

Find the relative maxima and minima of $r = 1 + \sin \theta$.

$$\frac{dy}{dx} = \frac{r' \sin \theta + r \cos \theta}{r' \cos \theta - r \sin \theta} = \frac{\cos \theta \sin \theta + (1 + \sin \theta) \cos \theta}{\cos^2 \theta - (1 + \sin \theta) \sin \theta}$$

$$= \frac{\cos \theta (2 \sin \theta + 1)}{\cos^2 \theta - \sin \theta - \sin^2 \theta}.$$

The numerator is 0 if either

$$\cos \theta = 0 \quad \text{or} \quad \sin \theta = -\frac{1}{2}.$$

Thus

$$\theta = \frac{\pi}{2}, \frac{3\pi}{2} \quad \text{and} \quad \theta = \frac{7\pi}{6}, \frac{11\pi}{6}$$

are critical values. Similarly the denominator is 0 if

$$\cos^2 \theta - \sin \theta - \sin^2 \theta = 0$$

$$1 - \sin^2 \theta - \sin \theta - \sin^2 \theta = 0$$

$$2 \sin^2 \theta + \sin \theta - 1 = 0$$

$$(2 \sin \theta - 1)(\sin \theta + 1) = 0,$$

$$\sin \theta = \frac{1}{2}, \qquad \sin \theta = -1;$$

$$\theta = \frac{\pi}{6}, \frac{5\pi}{6}, \frac{3\pi}{2}.$$

Since $\theta = 3\pi/2$ makes both numerator and denominator 0, one might expect both a horizontal and a vertical tangent. Actually there is a vertical tangent, as may be seen from Figure 14.18, although this cannot be determined from the derivative alone. We can also see from the figure that $(3/2, \pi/6)$ and $(3/2, 5\pi/6)$ are neither maxima nor minima. Thus, the relative maxima are $(2, \pi/2)$ and $(0, 3\pi/2)$, and the relative minima are $(1/2, 7\pi/6)$ and $(1/2, 11\pi/6)$.

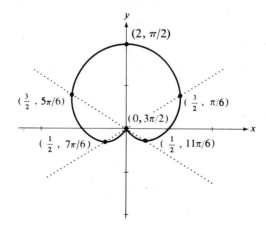

Figure 14.18

Problems

A

In Problems 1–8, find dy/dx.

1. $r = 1 + \cos \theta$.

2. $r = \dfrac{2 - \cos \theta}{2}$.

3. $r = 3 \csc \theta$.

4. $r = \cot \theta$.

5. $r = \cos 3\theta$.

6. $r = \sin 2\theta$.

7. $r = \theta$.

8. $r = \sin \theta$.

In Problems 9–14, find dy/dx for the given value of θ.

9. $r = 3 \sin \theta, \quad \theta = \pi/3$.

10. $r = 3 \cos \theta, \quad \theta = \pi/6$.

11. $r = 1 + \sin \theta, \quad \theta = \pi/4$.

12. $r = 3 - 2 \sin \theta, \quad \theta = \pi$.

13. $r = \sin 3\theta, \quad \theta = \pi/6$

14. $r = \tan \theta, \quad \theta = \pi/4$.

In Problems 15–20, find an equation (in rectangular coordinates) of the line tangent to the given curve at the given point.

15. $r = 1 + \cos \theta$ at $(1, \pi/2)$.

16. $r = 4 \cos \theta$ at $(2, \pi/3)$.

17. $r = \cos 2\theta$ at $(-1, \pi/2)$.

18. $r = 3 \sin \theta$ at $(3/2, \pi/6)$.

19. $r = 8 \sin^2 \theta$ at $(2, 5\pi/6)$.

20. $r = \tan \theta$ at $(1, \pi/4)$.

B

In Problems 21 and 22, find dy/dx.

21. $r = \dfrac{3}{1 + \sin \theta}$.

22. $r = \dfrac{\sin \theta}{1 + \cos \theta}$.

In Problems 23–28, find the relative maxima and minima.

23. $r = 1 + \cos \theta$.

24. $r = 1 - \sin \theta$.

25. $r = \sin 2\theta$.

26. $r = \cos^2 \theta$.

27. $r = 1 + 2 \cos \theta$.

28. $r = 2 - 3 \sin \theta$.

C

29. The slope is not most convenient for finding angles of intersection of two curves in polar coordinates. More convenient is tan ψ, where ψ is the smallest nonnegative angle from the radius vector (joining the origin to the point P) to the tangent line, where the counter-

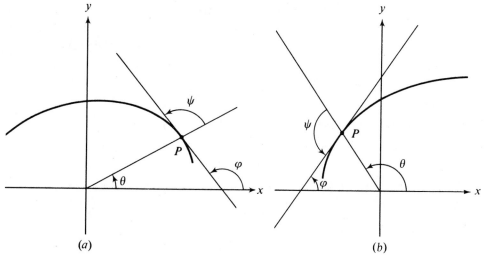

(a)   (b)

Figure 14.19

clockwise direction is taken to be positive as usual (see Figure 14.19). Show that $\psi = n\pi + (\varphi - \theta)$, where n is an integer. Use this result to show that

$$\tan \psi = \frac{f(\theta)}{f'(\theta)}.$$

30. Suppose the graphs of $r = f_1(\theta)$ and $r = f_2(\theta)$ intersect at (r, θ) with (r, θ) satisfying both equations. Let φ_1 be the inclination of the tangent line to $r = f_1(\theta)$ at (r, θ), and let ψ_1 be defined as in Problem 29 for $r = f_1(\theta)$ at (r, θ). Let φ_2 and ψ_2 be the corresponding quantities for $r = f_2(\theta)$ at (r, θ). Show that $\varphi_2 - \varphi_1$ is one of the following angles: $\psi_2 - \psi_1, \psi_2 - \psi_1 + \pi, \psi_2 - \psi_1 - \pi$. In any case

$$\tan (\varphi_2 - \varphi_1) = \tan (\psi_2 - \psi_1) = \frac{\tan \psi_2 - \tan \psi_1}{1 + \tan \psi_1 \tan \psi_2}.$$

If you take the angle of intersection of two curves to be the smallest nonnegative angle between their tangent lines, you may use the simpler $\tan \psi$ in place of dy/dx.

31. The curves $r = \cos 2\theta$ and $r = \sin \theta$ intersect at $(1/2, \pi/6)$ (see Example 3, page 429). Use dy/dx to find the angle of intersection. Repeat using $\tan \psi$.

32. Find the angle of intersection of $r = \cos \theta$ and $r = 1 - \cos \theta$ at each point of intersection (see Problem 5, page 429).

33. Find the angle of intersection of $r = 1$ and $r = 2 \sin \theta$ at each point of intersection.

14.7

Areas in Polar Coordinates

The problem of finding areas in polar coordinates is solved in basically the same way that it was in rectangular coordinates. Suppose we have a region bounded by $r = f(\theta)$ and the terminal sides of the angles α and β, where $\alpha < \beta$ (see Figure 14.20). Let us subdivide the (angular) interval $[\alpha, \beta]$ into n subintervals

$$\alpha = \theta_0, \theta_1, \theta_2, \ldots, \theta_n = \beta.$$

Now, for each value of i from 1 to n, let us choose θ_i^* such that

$$\theta_{i-1} \leq \theta_i^* \leq \theta_i.$$

A typical such interval might look like the one in Figure 14.21. We can approximate the area within the ith subinterval by the area of a sector. Let us recall that the area of a sector of angle θ (in radians) and radius r is

$$\frac{1}{2} \theta r^2.$$

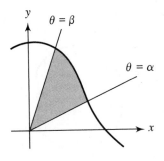

Figure 14.20

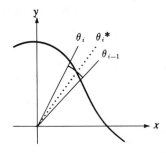

Figure 14.21

The angle of our sector is $\theta_i - \theta_{i-1}$, and we shall take the radius to be $f(\theta_i^*)$. Thus the area of the ith sector is

$$\frac{1}{2}(\theta_i - \theta_{i-1})[f(\theta_i^*)]^2,$$

and the sum of all of these areas is

$$\sum_{i=1}^{n} \frac{1}{2}(\theta_i - \theta_{i-1})[f(\theta_i^*)]^2, \quad \text{or} \quad \sum_{i=1}^{n} \frac{1}{2}[f(\theta_i^*)]^2 \, \Delta\theta_i,$$

where $\Delta\theta_i = \theta_i - \theta_{i-1}$. By taking the limit, we arrive at the desired area, which, by the definition of an integral, is

$$A = \int_\alpha^\beta \frac{1}{2}[f(\theta)]^2 \, d\theta, \quad \text{or} \quad A = \int_\alpha^\beta \frac{1}{2} r^2 \, d\theta$$

(provided this integral exists). Of course it can be shown (although we shall not attempt do to so) that the area found in this way is the same as the area as defined in Chapter 7.

Before considering examples of areas in polar coordinates, you would do well to review Example 7 on page 333. For convenience, we restate the identities used there to integrate even powers of sines and cosines.

$$\sin^2 x = \frac{1}{2}(1 - \cos 2x)$$

$$\cos^2 x = \frac{1}{2}(1 + \cos 2x)$$

Example 1

Find the area in the first quadrant within $r = 1 - \cos \theta$.

The interval from $\theta = 0$ to $\theta = \pi/2$ is subdivided into intervals, and the area within each is approximated by the area of a sector (see Figure 14.22). The limit of the sum gives the integral

$$\int_0^{\pi/2} \frac{1}{2} r^2 \, d\theta$$

$$= \int_0^{\pi/2} \frac{1}{2} (1 - \cos \theta)^2 \, d\theta$$

$$= \frac{1}{2} \int_0^{\pi/2} (1 - 2 \cos \theta + \cos^2 \theta) \, d\theta$$

$$= \frac{1}{2} \int_0^{\pi/2} \left(1 - 2 \cos \theta + \frac{1 + \cos 2\theta}{2} \right) d\theta$$

$$= \frac{1}{2} \int_0^{\pi/2} \left(\frac{3}{2} - 2 \cos \theta + \frac{1}{4} \cdot 2 \cos 2\theta \right) d\theta$$

$$= \frac{1}{2} \left(\frac{3}{2} \theta - 2 \sin \theta + \frac{1}{4} \sin 2\theta \right) \Big|_0^{\pi/2}$$

$$= \frac{3\pi - 8}{8}.$$

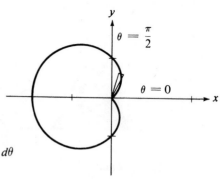

Figure 14.22

Example 2

Find the area within the inner loop of

$$r = 1 - 2 \sin \theta.$$

The graph of $r = 1 - 2 \sin \theta$ is given in Figure 14.23. The inner loop is determined by the interval $[\pi/6, 5\pi/6]$. The end points of this interval are found by noting that $r = 0$ at the ends of the loop. Thus

$$0 = 1 - 2 \sin \theta$$

$$\sin \theta = \frac{1}{2},$$

$$\theta = \pi/6, 5\pi/6.$$

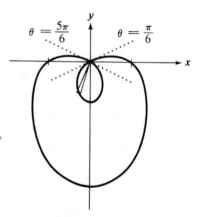

Figure 14.23

Although r is negative for all values of θ in the interval $(\pi/6, 5\pi/6)$, we need not change the sign, since we use r^2 in the integral and r^2 is never negative. Thus

$$A = \int_{\pi/6}^{5\pi/6} \frac{1}{2} r^2 \, d\theta$$

$$= \frac{1}{2} \int_{\pi/6}^{5\pi/6} (1 - 2 \sin \theta)^2 \, d\theta$$

$$= \frac{1}{2} \int_{\pi/6}^{5\pi/6} (1 - 4 \sin \theta + 4 \sin^2 \theta) \, d\theta$$

$$= \frac{1}{2} \int_{\pi/6}^{5\pi/6} \left(1 - 4 \sin \theta + 4 \cdot \frac{1 - \cos 2\theta}{2} \right) d\theta$$

$$= \frac{1}{2} \int_{\pi/6}^{5\pi/6} (3 - 4 \sin \theta - 2 \cos 2\theta) \, d\theta$$

$$= \frac{1}{2} (3\theta + 4 \cos \theta - \sin 2\theta) \Big|_{\pi/6}^{5\pi/6}$$

$$= \frac{2\pi - 3\sqrt{3}}{2}$$

Since the loop is symmetric about the y axis, we could have found the area of half of the loop (by integrating from $\theta = \pi/6$ to $\theta = \pi/2$) and then doubled the result.

Example 3

Find the area of the region inside both $r = \sin\theta$ and $r = 1 - \sin\theta$.

The graphs of these two equations are given in (a) of Figure 14.24. Since they are symmetric about the y axis, let us find the area in the first quadrant and

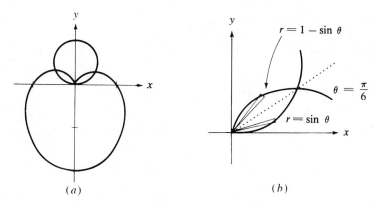

(a) (b)

Figure 14.24

double. We can see from (b) that the approximating sectors have their ends on either $r = \sin\theta$ or $r = 1 - \sin\theta$, depending upon the value of θ. For $0 \le \theta \le \pi/6$, the ends are on $r = \sin\theta$; for $\pi/6 \le \theta \le \pi/2$, the ends are on $r = 1 - \sin\theta$. Thus the area is

$$A = 2\left[\int_0^{\pi/6} \frac{1}{2}\sin^2\theta\,d\theta + \int_{\pi/6}^{\pi/2} \frac{1}{2}(1-\sin\theta)^2\,d\theta\right]$$

$$= \int_0^{\pi/6} \frac{1-\cos 2\theta}{2}\,d\theta + \int_{\pi/6}^{\pi/2}(1 - 2\sin\theta + \sin^2\theta)\,d\theta$$

$$= \int_0^{\pi/6} \frac{1-\cos 2\theta}{2}\,d\theta + \int_{\pi/6}^{\pi/2}\left(1 - 2\sin\theta + \frac{1-\cos 2\theta}{2}\right)d\theta$$

$$= \int_0^{\pi/6}\left(\frac{1}{2} - \frac{1}{2}\cos 2\theta\right)d\theta + \int_{\pi/6}^{\pi/2}\left(\frac{3}{2} - 2\sin\theta - \frac{1}{2}\cos 2\theta\right)d\theta$$

$$= \left(\frac{1}{2}\theta - \frac{1}{4}\sin 2\theta\right)\Bigg|_0^{\pi/6} + \left(\frac{3}{2}\theta + 2\cos\theta - \frac{1}{4}\sin 2\theta\right)\Bigg|_{\pi/6}^{\pi/2}$$

$$= \frac{2\pi - 3\sqrt{3}}{24} + \frac{4\pi - 7\sqrt{3}}{8}$$

$$= \frac{7\pi}{12} - \sqrt{3}.$$

Problems

A

In Problems 1–18, find the area of the region described.

1. Inside $r = 4 \sin \theta$.
2. Inside $r = \sin 2\theta$.
3. Inside $r = 1 + \sin \theta$.
4. Inside $r = 2 + \cos \theta$.
5. Inside $r = \sqrt{\sin \theta}$.
6. Inside $r^2 = \sin \theta$.
7. Inside $r = 3 - \sin \theta$.
8. Inside the inner loop of $r = 1 + 3 \sin \theta$.
9. Inside $r^2 = 9 \sin 2\theta$.
10. Inside $r^2 = 4 \cos 3\theta$.
11. Inside $r = 1$ and $r = 1 + \cos \theta$.
12. Inside $r = 1$ and $r = 2 \sin 2\theta$.
13. Inside $r = 1 + \cos \theta$ and $r = 1 + \sin \theta$.
14. Inside $r = 1 + \cos \theta$ and $r = 3 \cos \theta$.
15. Inside $r = 1 + \sin \theta$ and outside $r = 1$.
16. Inside $r = 1 - \sin \theta$ and outside $r = 2 \cos \theta$.
17. Inside $r = \sin \theta$ and $r = \sin 2\theta$.
18. Inside $r = 1$ and $r = 2 \sin \theta$.

B

In Problems 19 and 20, find the area of the region described.

19. Inside $r = \sin \theta + \cos \theta$.
20. Inside $r = \sin \theta - \cos \theta$.

21. Show that if $r = f(\theta)$, then the length of the arc from $\theta = \theta_1$ to $\theta = \theta_2$ is

$$s = \int_{\theta_1}^{\theta_2} \sqrt{r^2 + \left(\frac{dr}{d\theta}\right)^2}\, d\theta.$$

In Problems 22–25, use the result of Problem 21 to find the length of the given arc.

22. $r = \sin \theta$, from $\theta = 0$ to $\theta = \pi$.
23. $r = e^\theta$, from $\theta = 0$ to $\theta = 2$.

24. $r = 1 + \cos \theta$, from $\theta = 0$ to $\theta = 2\pi$.
25. $r = 2 \sin^3 \dfrac{\theta}{3}$, from $\theta = 0$ to $\theta = \pi/2$.

C

26. A comet has a parabolic orbit with the sun at the focus. When the comet is 100,000,000 miles from the sun, the line joining the sun and the comet makes an angle of 60° with the axis of the parabola (see Figure 9.5, page 248). If this angle increases from 60° to 90° in 36 hours, in how many more hours will the comet be at perihelion—the closest point to the sun? (*Hint:* Use the fact that the line joining the sun and the comet sweeps out equal areas within the orbit in equal times. Also use the result

$$\int \frac{d\theta}{(1 - \cos \theta)^2} = -\frac{1 + 3 \tan^2 \theta/2}{6 \tan^3 \theta/2} + C.)$$

Review Problems

A

In Problems 1–6, sketch the graph of the given equation and indicate any symmetry about either axis or the pole.

1. $r = 1 + 3 \cos \theta$.
2. $r = \cos 4\theta$.
3. $r = 1 + \sin 2\theta$.
4. $r = e^\theta$.
5. $r^2 = \sin 2\theta$.
6. $r^2 = 1 + \cos \theta$.

In Problems 7–12, find all points of intersection.

7. $r = 1$, $r = 1 + \cos\theta$.
8. $r = 1 + \sin\theta$, $r = 1 - \cos\theta$.
9. $r = \sin\theta$, $r = 1 + 2\sin\theta$.
10. $r = 2\sin 2\theta$, $r = 1$.
11. $r = \cos\theta$, $r = \cos 3\theta$.
12. $r = 1 + \sin\theta$, $r = \cos\theta - 1$.
13. (a) The following points are given in polar coordinates; give the rectangular coordinate representation of each: $(-1, \pi/2)$, $(2\sqrt{2}, 3\pi/4)$, $(6, 7\pi/6)$, $(-2\sqrt{3}, 2\pi/3)$.
   (b) The following points are given in rectangular coordinates; give a polar coordinate representation of each: $(-2, 2)$, $(-5, 0)$, $(1, -\sqrt{3})$, $(10, 4)$.
14. Express $x^2 + y^2 + 3y = 0$ in polar coordinates.
15. Give a polar equation for the ellipse with focus at the pole, the corresponding directrix $y = 3$, and eccentricity 2/3.
16. Give a polar equation for the parabola with focus at the pole and directrix $x = -6$.

In Problems 17 and 18, find the area of the region described.

17. Inside $r = 2 + \sin\theta$.
18. Inside $r = \sin 3\theta$.

B
19. Express $r = 1 + \sin\theta$ in rectangular coordinates.
20. Express $r(3\cos\theta - 5) = 16$ in rectangular coordinates.
21. Sketch the conic $r = 2/(1 - \sin\theta)$. Identify the focus, directrix, and eccentricity.
22. Sketch the conic $r = 2/(1 - 2\cos\theta)$. Identify a focus, the corresponding directrix, and the eccentricity.
23. Find an equation (in rectangular coordinates) of the line tangent to $r = 1/(1 - \cos\theta)$ at $\theta = \pi/2$.
24. Find an equation (in rectangular coordinates) of the line tangent to $r = 2 + \cos\theta$ at $\theta = \pi/3$.
25. Find all relative maxima and minima of $r = \sin\theta + \cos\theta$.
26. Find all relative maxima and minima of $r = (1 + \sin\theta)^2$.

In Problems 27 and 28, find the area of the region described.

27. Inside the outer loop but outside the inner loop of $r = 1 + 2\cos\theta$.
28. Inside $r = 1$ and $r = \sin\theta - 1$.

15

Methods of Integration

15.1

Fundamental Formulas

We had many integration formulas in Chapters 7, 11, and 12. Nevertheless, we still do not have integration formulas for integrals of four of the six trigonometric functions. Let us consider them now.

Theorem 15.1

$$\int \tan u(x) \cdot u'(x)\, dx = -\ln |\cos u(x)| + C.$$

Proof

Abbreviating $u(x)$ to u and $u'(x)$ to u', we have

$$\int \tan u \cdot u'\, dx = -\int \frac{-\sin u \cdot u'}{\cos u}\, dx.$$

The derivative of $\cos u$ is $-\sin u \cdot u'$. Our integral is in the form $\int \frac{v'}{v}\, dx$, where $v = \cos u$. Thus,

$$\int \tan u \cdot u'\, dx = \int \tan u\, du = -\ln |\cos u| + C.$$

Theorem 15.2

$\int \cot u(x) \cdot u'(x) \, dx = \ln |\sin u(x)| + C.$

The proof of this theorem, similar to that of Theorem 15.1, is left to the student. Similar formulas can be derived for the hyperbolic functions. Since these two integral formulas are so easy to derive, you might prefer to remember the derivations rather than the formulas themselves.

Theorem 15.3

$\int \sec u(x) \cdot u'(x) \, dx = \ln |\sec u(x) + \tan u(x)| + C.$

Proof

Again, abbreviating $u(x)$ to u and $u'(x)$ to u', we have

$$\int \sec u \cdot u' \, dx = \int \frac{\sec^2 u + \sec u \tan u}{\sec u + \tan u} \, u' \, dx.$$

Again, this integral is in the form $\int \frac{v'}{v} \, dx$, where $v = \sec u + \tan u$. Thus

$$\int \sec u \cdot u' \, dx = \int \sec u \, du = \ln |\sec u + \tan u| + C.$$

Theorem 15.4

$\int \csc u(x) \cdot u'(x) \, dx = \ln |\csc u - \cot u| + C.$

The proof, similar to that of Theorem 15.3, is left to the student. We now have the following integration formulas.

1. $\displaystyle\int x^n \, dx = \frac{x^{n+1}}{n+1} + C, \quad n \neq -1.$

2. $\displaystyle\int u^n \cdot u' \, dx = \int u^n \, du = \frac{u^{n+1}}{n+1} + C, \quad n \neq -1.$

3. $\displaystyle\int e^u \cdot u' \, dx = \int e^u \, du = e^u + C.$

4. $\displaystyle\int a^u \cdot u' \, dx = \int a^u \, du = a^u \cdot \log_a e + C.$

5. $\displaystyle\int \frac{u'}{u} \, dx = \int \frac{du}{u} = \ln |u| + C.$

6. $\displaystyle\int \sin u \cdot u' \, dx = \int \sin u \, du = -\cos u + C.$

7. $\displaystyle\int \cos u \cdot u' \, dx = \int \cos u \, du = \sin u + C.$

8. $\displaystyle\int \sec^2 u \cdot u' \, dx = \int \sec^2 u \, du = \tan u + C.$

9. $\displaystyle\int \csc^2 u \cdot u' \, dx = \int \csc^2 u \, du = -\cot u + C.$

10. $\displaystyle\int \sec u \tan u \cdot u' \, dx = \int \sec u \tan u \, du = \sec u + C.$

11. $\displaystyle\int \csc u \cot u \cdot u' \, dx = \int \csc u \cot u \, du = -\csc u + C.$

12. $\displaystyle\int \tan u \cdot u' \, dx = \int \tan u \, du = -\ln |\cos u| + C.$

13. $\displaystyle\int \cot u \cdot u' \, dx = \int \cot u \, du = \ln |\sin u| + C.$

14. $\displaystyle\int \sec u \cdot u' \, dx = \int \sec u \, du = \ln |\sec u + \tan u| + C.$

15. $\displaystyle\int \csc u \cdot u' \, dx = \int \csc u \, du = \ln |\csc u - \cot u| + C.$

16. $\displaystyle\int \sinh u \cdot u' \, dx = \int \sinh u \, du = \cosh u + C.$

17. $\displaystyle\int \cosh u \cdot u' \, dx = \int \cosh u \, du = \sinh u + C.$

18. $\displaystyle\int \operatorname{sech}^2 u \cdot u' \, dx = \int \operatorname{sech}^2 u \, du = \tanh u + C.$

19. $\displaystyle\int \operatorname{csch}^2 u \cdot u' \, dx = \int \operatorname{csch}^2 u \, du = -\coth u + C.$

20. $\displaystyle\int \operatorname{sech} u \tanh u \cdot u' \, dx = \int \operatorname{sech} u \tanh u \, du = -\operatorname{sech} u + C.$

21. $\displaystyle\int \operatorname{csch} u \coth u \cdot u' \, dx = \int \operatorname{csch} u \coth u \, du = -\operatorname{csch} u + C.$

22. $\displaystyle\int \frac{u'}{1 + u^2} \, dx = \int \frac{du}{1 + u^2} = \operatorname{Arctan} u + C.$

23. $\displaystyle\int \frac{u'}{\sqrt{1 - u^2}} \, dx = \int \frac{du}{\sqrt{1 - u^2}} = \operatorname{Arcsin} u + C.$

24. $\displaystyle\int \frac{u'}{u\sqrt{u^2 - 1}} \, dx = \int \frac{du}{u\sqrt{u^2 - 1}} = \operatorname{Arcsec} u + C.$

This is quite a long list of formulas to remember and it is a very unusual student who can remember them all for more than a week. However, things are not so bad as they seem. Only the first fifteen formulas are a must for you to memorize. The hyperbolic functions are defined in terms of exponentials. Thus if you forget an integral of a hyperbolic function, you can integrate by expressing it in terms of exponentials. Furthermore, hyperbolic functions are encountered rather infrequently. In addition, we shall consider methods of integration in this chapter that allow one to integrate any of the expressions of formulas 22–24. Thus, while you may find it convenient to memorize more than fifteen of the formulas, you will be relieved of the necessity of memorizing all of them.

Example 1

Evaluate $\displaystyle\int \frac{x^3 - 1}{x + 1}\, dx$.

Since the degree of the numerator is greater than that of the denominator, we shall divide until the remainder is of lower degree than that of the denominator.

$$\int \frac{x^3 - 1}{x + 1}\, dx = \int \left(x^2 - x + 1 - \frac{2}{x + 1} \right) dx.$$

Now we can integrate term by term.

$$\int \frac{x^3 - 1}{x + 1}\, dx = \frac{x^3}{3} - \frac{x^2}{2} + x - 2 \ln |x + 1| + C.$$

Example 2

Evaluate $\displaystyle\int e^x (e^x + 1)^2\, dx$.

This is in the form $\int u^2 \cdot u'\, dx$, where $u = e^x + 1$.

$$\int e^x (e^x + 1)^2\, dx = \frac{(e^x + 1)^3}{3} + C.$$

Example 3

Evaluate $\displaystyle\int \frac{\sin x}{1 + \cos x}\, dx$.

Since the derivative of the denominator is $-\sin x$, we adjust the constant factor and integrate.

$$\int \frac{\sin x}{1 + \cos x}\, dx = -\int \frac{-\sin x}{1 + \cos x}\, dx = -\ln (1 + \cos x) + C.$$

We dropped the absolute value signs, since $1 + \cos x$ cannot be negative.

Example 4

Evaluate $\displaystyle\int \tan 3x\, dx$.

$$\int \tan 3x\, dx = \frac{1}{3} \int 3 \tan 3x\, dx = -\frac{1}{3} \ln |\cos 3x| + C.$$

Example 5

Evaluate $\displaystyle\int \frac{dx}{x^2 + 4}$.

This is almost in the form of formula 22. The only difference is that we have a 4 where we want a 1. Let us get a 1 by dividing numerator and denominator by 4.

$$\int \frac{dx}{x^2 + 4} = \int \frac{\frac{1}{4} dx}{\frac{x^2}{4} + 1} = \int \frac{\frac{1}{4} dx}{\left(\frac{x}{2}\right)^2 + 1}$$

$$= \frac{1}{2} \int \frac{\frac{1}{2} dx}{\left(\frac{x}{2}\right)^2 + 1} = \frac{1}{2} \text{Arctan} \frac{x}{2} + C.$$

Problems

A

Evaluate the integrals.

1. $\int \frac{x^2 + 2x - 6}{x} \, dx.$

2. $\int \frac{x^2 - 5x + 2}{x - 1} \, dx.$

3. $\int \frac{4x^2 - 2x + 3}{2x + 1} \, dx.$

4. $\int \frac{y^3 + 1}{y - 1} \, dy.$

5. $\int \frac{u^3 + u}{u - 1} \, du.$

6. $\int \frac{4x^3 + 4x^2 + 3x + 8}{2x + 3} \, dx.$

7. $\int \sqrt{3x + 1} \, dx.$

8. $\int x(3x^2 - 5)^4 \, dx.$

9. $\int (x + 1)(x^2 + 2x)^{2/3} \, dx.$

10. $\int (x^2 - 2)^3 \, dx.$

11. $\int xe^{x^2} \, dx.$

12. $\int \sin x e^{\cos x} \, dx.$

13. $\int \frac{\cos x}{1 - \sin x} \, dx.$

14. $\int \frac{du}{(1 - 3u)^4}.$

15. $\int \frac{e^x + 1}{e^x} \, dx.$

16. $\int \frac{e^x}{e^x + 1} \, dx.$

17. $\int \frac{\sec^2 u}{1 + \tan u} \, du.$

18. $\int \sin x \cos x \, dx.$

19. $\int \sin 3\theta (\cos 3\theta + 1) \, d\theta.$

20. $\int \frac{\cosh x}{1 - \sinh x} \, dx.$

21. $\int \frac{\text{csch}^2 x}{1 - \coth x} \, dx.$

22. $\int \frac{dx}{x^2 + 16}.$

23. $\int \frac{dx}{x\sqrt{x^2 - 4}}.$

24. $\int \frac{x \, dx}{x^4 + 1}.$

25. $\int \frac{dx}{\sqrt{2 - x^2}}.$

26. $\int (\sec \theta + \tan \theta)^2 \, d\theta.$

27. $\int (1 + \tan \theta)^2 \, d\theta.$

28. $\int (1 + \csc \theta)^2 \, d\theta.$

B

Evaluate the integrals.

29. $\int (1 - \sqrt{x})^2 \, dx.$

30. $\int \frac{\ln x}{x} \, dx.$

31. $\int \frac{4 - \ln x}{x} \, dx.$

32. $\int \frac{1 + \sin x}{\sin x} \, dx.$

33. $\int \frac{e^x}{e^{2x} + 1} \, dx.$

34. $\int \sin 2x (\sin^2 x + 1) \, dx.$

35. $\int \frac{\sin \theta \cos \theta \, d\theta}{\cos^2 \theta + 1}.$

36. $\int \frac{\cos \theta \, d\theta}{\sin^2 \theta + 1}.$

37. $\int \frac{e^{1/x}}{x^2} \, dx.$

38. $\int \left(\frac{\sec \theta}{1 - \tan \theta} \right)^2 \, d\theta.$

39. $\int \frac{(1 - \sqrt{x})^2}{\sqrt{x}} \, dx.$

40. $\int \frac{dx}{x\sqrt{x^4 - 1}}.$

C

41. Evaluate $\int \frac{e^x - 1}{e^x + 1} \, dx.$

42. (a) Evaluate $\int \frac{dx}{e^x + 1}$ by dividing $1 + e^x$ into 1.

    (b) Evaluate by multiplying numerator and denominator by e^{-x}.

    (c) Show that the results above are equivalent.

43. Prove Theorem 15.2.

44. Prove Theorem 15.4.

45. Evaluate $\int \csc u \cdot u' \, dx$ by multiplying the numerator and denominator by $\csc u + \cot u$. Show that your result is equivalent to that given in Theorem 15.4.

46. Evaluate $\int \tanh u \cdot u' \, dx.$ 47.   Evaluate $\int \operatorname{sech} u \cdot u' \, dx.$

15.2

Integration by Substitution

In the method of integration by substitution, many different substitutions can be made, depending upon the particular expression we have to integrate. We shall consider only one type of substitution in this section—others will come later.

First, let us see what is involved. Given the integral $\int f(x) \, dx$ to evaluate, we want to find an $F(x)$ whose derivative with respect to x is $f(x)$:

$$F'(x) = f(x), \quad \text{so} \quad \int f(x) \, dx = F(x) + C.$$

Suppose u is a function of x, $u = g(x)$, such that $f(x)$ can be written in the form

$$f(x) = h(g(x))g'(x)$$
$$= h(u) \cdot u'.$$

If $H(u)$ is an antiderivative of $h(u)$, then $dH/du = h(u)$, and by the chain rule,

$$\frac{d}{dx} H(u) = \frac{d}{du} H(u) \cdot \frac{du}{dx} = h(u) \cdot u' = f(x).$$

Thus $F(x) = H(g(x)) = H(u)$ is the desired antiderivative of $f(x) = h(u) \cdot u'$, and

$$\int f(x)\,dx = \int h(u) \cdot u'\,dx = \int h(u)\,du = H(u) + C = H(g(x)) + C.$$

To summarize, if $f(x) = h(u)u'$, where $u = g(x)$, and $\int h(u)\,du = H(u) + C$, then $\int f(x)\,dx = H(g(x)) + C$.

To transform the expression $f(x)\,dx$ to the form $h(u)\,du$ by a substitution $u = g(x)$, we often let $u = g(x)$ be a complicated part of $f(x)$. Then, if $x = \phi(u)$ is the inverse function of $u = g(x)$, we replace x elsewhere in the integrand $f(x)$ by $\phi(u)$ and dx by $(d\phi/du)\,du$. Again, the chain rule justifies this procedure (see Problems 33 and 34). The result is that *when making the substitution $u = g(x)$, we must substitute not only into the integrand $f(x)$ but also into the differential dx.*

Now let us consider a particular substitution. We have seen that we can easily integrate expressions of the form $(ax + b)^n$ for any value of n by using formula 2 or 5. When n is a positive integer, we can also expand $(ax + b)^n$ by the binomial theorem and integrate term by term. But when we have an expression of the form $x^m(ax + b)^n$, where $m \neq 0$, we cannot use formula 2 or 5. If n is a positive integer, we can still expand and integrate term by term, but we cannot integrate this expression for other values of n.

In particular, we cannot integrate expressions of the form $x^m \sqrt{ax + b}$. Because the radical prevents us from expanding, we would like to get rid of it. This is where a substitution is useful. If we make the substitution

$$u = \sqrt{ax + b},$$

we are rid of the radical. Solving for x, we have

$$x = \frac{u^2 - b}{a} \quad \text{and} \quad dx = \frac{2u\,du}{a}.$$

Thus

$$\int x^m \sqrt{ax + b}\,dx = \int \left(\frac{u^2 - b}{a}\right)^m u \cdot \frac{2u}{a}\,du.$$

Now if m is a positive integer, we can expand and integrate term by term.

Let us consider some examples of this type of substitution.

Example 1

Evaluate $\int x\sqrt{x+1}\ dx$.

Since the derivative of $x+1$ (the expression under the radical) is 1 and we have a factor of x, we cannot use the power rule. Because we have a radical, we cannot multiply it out. Our only alternative is to make a substitution. Let

$$u = \sqrt{x+1}.$$

Solving for x, we have

$$x = u^2 - 1 \quad \text{and} \quad dx = 2u\ du.$$

Thus

$$\int x\sqrt{x+1}\ dx = \int (u^2 - 1)u \cdot 2u\ du$$

$$= \int (2u^4 - 2u^2)\ du$$

$$= \frac{2u^5}{5} - \frac{2u^3}{3} + C$$

$$= \frac{2}{15}\ u^3(3u^2 - 5) + C.$$

Finally, now that the original expression has been integrated, it must be put back in terms of x, since it was given in terms of x. Substituting

$$u = \sqrt{x+1},$$

we have

$$\int x\sqrt{x+1}\ dx = \frac{2}{15}\ (x+1)^{3/2}[3(x+1) - 5] + C$$

$$= \frac{2}{15}\ (x+1)^{3/2}(3x - 2) + C.$$

In this example, we let $u = \sqrt{x+1} = g(x)$, $x = u^2 - 1 = \phi(u)$,   and $dx = 2u\ du = (d\phi/du)\ du$. It might be noted here that the substitution $u = \sqrt{x+1}$ is not the only one that allows us to carry out the above integration. Another substitution that works just as well is $u = x + 1$ (see Problems 31 and 32).

Example 2

Evaluate $\int x^2\sqrt{x-2}\ dx$.

Substituting

$$u = \sqrt{x-2}$$

and solving for x, we have

$$x = u^2 + 2 \quad \text{and} \quad dx = 2u \, du.$$

$$\int x^2 \sqrt{x - 2} \, dx = \int (u^2 + 2)^2 \cdot u \cdot 2u \, du$$

$$= \int (2u^6 + 8u^4 + 8u^2) \, du$$

$$= \frac{2u^7}{7} + \frac{8u^5}{5} + \frac{8u^3}{3} + C$$

$$= \frac{2u^3}{105} (15u^4 + 84u^2 + 140) + C.$$

Substituting $u = \sqrt{x - 2}$, we have

$$\int x^2 \sqrt{x - 2} \, dx = \frac{2}{105} (x - 2)^{3/2} [15(x - 2)^2 + 84(x - 2) + 140] + C$$

$$= \frac{2}{105} (x - 2)^{3/2} (15x^2 + 24x + 32) + C.$$

Example 3

Evaluate $\displaystyle\int_0^5 x\sqrt{x + 4} \, dx.$

First, note that the limits of integration are values of x, since we have an integral with respect to x. If we substitute to give an integral with respect to u, it is assumed then that the limits of integration are values of u unless something is said to the contrary.

There are two ways of handling the limits of integration when using a substitution. One way is to substitute, integrate, substitute back to get the result in terms of the original x, and then put in the limits of integration. The second method is to change the limits of integration to the corresponding values of u when the original substitution is made, thus eliminating the need to get the result back in terms of x. Both methods are illustrated here. In either case, we use the same substitution,

$$u = \sqrt{x + 4}, \quad \text{when } x = 0, u = \sqrt{4} = 2;$$

$$x = u^2 - 4, \quad \text{when } x = 5, u = \sqrt{9} = 3;$$

$$dx = 2u \, du.$$

$$\int x\sqrt{x + 4} \, dx \qquad\qquad\qquad \int_0^5 x\sqrt{x + 4} \, dx$$

$$= \int (u^2 - 4)u \cdot 2u \, du \qquad\qquad = \int_2^3 (u^2 - 4)u \cdot 2u \, du$$

$$= \int (2u^4 - 8u^2) \, du \qquad\qquad = \int_2^3 (2u^4 - 8u^2) \, du$$

$$= \frac{2u^5}{5} - \frac{8u^3}{3} + C \qquad\qquad = \frac{2u^5}{5} - \frac{8u^3}{3} \Big|_2^3$$

$$= \frac{2}{15} u^3(3u^2 - 20) + C \qquad\qquad = \frac{2}{15} u^3(3u^2 - 20) \Big|_2^3$$

$$= \frac{2}{15} (x+4)^{3/2}[3(x+4) - 20] + C \qquad = \frac{2}{15} [3^3 \cdot 7 - 2^3(-8)]$$

$$= \frac{2}{15} (x+4)^{3/2}(3x - 8) + C; \qquad\qquad = \frac{506}{15}.$$

$$\int_0^5 x\sqrt{x+4}\, dx$$

$$= \frac{2}{15} (x+4)^{3/2}(3x - 8)\, \Big|_0^5$$

$$= \frac{2}{15} [9^{3/2} \cdot 7 - 4^{3/2} \cdot (-8)]$$

$$= \frac{506}{15}.$$

The method of substitution is not restricted to integrals of the form $\int x^m \sqrt{ax + b}\, dx$, as can be seen by the following example.

Example 4

Evaluate $\displaystyle\int \frac{x}{\sqrt{x+1}}\, dx.$

Let

$$u = \sqrt{x+1}, \qquad x = u^2 - 1, \qquad dx = 2u\, du.$$

Then

$$\int \frac{x}{\sqrt{x+1}}\, dx = \int \frac{u^2 - 1}{u}\, 2u\, du$$

$$= \int (2u^2 - 2)\, du$$

$$= \frac{2u^3}{3} - 2u + C$$

$$= \frac{2}{3} u(u^2 - 3) + C$$

$$= \frac{2}{3} \sqrt{x+1}\, (x - 2) + C.$$

Problems

A

Evaluate the integrals.

1. $\displaystyle\int x\sqrt{x+3}\,dx.$

2. $\displaystyle\int x\sqrt{x-2}\,dx.$

3. $\displaystyle\int x^2\sqrt{x+2}\,dx.$

4. $\displaystyle\int x^2\sqrt{1-x}\,dx.$

5. $\displaystyle\int x\sqrt{2x+1}\,dx.$

6. $\displaystyle\int x\sqrt{2x-3}\,dx.$

7. $\displaystyle\int \frac{x}{\sqrt{x-3}}\,dx.$

8. $\displaystyle\int \frac{x}{\sqrt{x+3}}\,dx.$

9. $\displaystyle\int \frac{x^2}{\sqrt{x-2}}\,dx.$

10. $\displaystyle\int \frac{x^2}{\sqrt{2x-1}}\,dx.$

11. $\displaystyle\int_1^5 x\sqrt{x-1}\,dx.$

12. $\displaystyle\int_0^4 x\sqrt{2x+1}\,dx.$

13. $\displaystyle\int_5^7 \frac{x}{\sqrt{x-4}}\,dx.$

14. $\displaystyle\int_0^4 \frac{x^2}{\sqrt{2x+1}}\,dx.$

15. $\displaystyle\int_1^2 x^3\sqrt{x-1}\,dx.$

16. $\displaystyle\int_2^5 \frac{x+2}{\sqrt{x-1}}\,dx.$

17. $\displaystyle\int_0^4 \frac{3x-2}{\sqrt{2x+1}}\,dx.$

18. $\displaystyle\int_0^7 x\sqrt[3]{x+1}\,dx.$

B

Evaluate the integrals.

19. $\displaystyle\int \frac{x^3}{\sqrt{x-2}}\,dx.$

20. $\displaystyle\int \frac{x^3}{\sqrt{2x+3}}\,dx.$

21. $\displaystyle\int x\sqrt[3]{x+1}\,dx.$

22. $\displaystyle\int x\sqrt[3]{2x+1}\,dx.$

23. $\displaystyle\int x\sqrt[4]{x-1}\,dx.$

24. $\displaystyle\int x\sqrt[5]{x-1}\,dx.$

25. $\displaystyle\int (2x+2)\sqrt{2x+1}\,dx.$

26. $\displaystyle\int (x-2)\sqrt{x+1}\,dx.$

27. $\displaystyle\int \frac{2x+1}{\sqrt{x-2}}\,dx.$

28. $\displaystyle\int \frac{x^2+x}{\sqrt{x+1}}\,dx.$

29. $\displaystyle\int_2^4 x^2\sqrt{2x-3}\,dx.$

30. $\displaystyle\int \sqrt{e^x-1}\,dx.$

C

31. Use the substitution $u = x + 1$ to evaluate the integral of Example 1.

32. Show that any integral of the form $\int x^m\sqrt{ax+b}\,dx,$ where m is a positive integer, can be evaluated by using the substitution $u = ax + b.$

33. Using the chain rule, show that if $x = \phi(u),$ then $\int f(x)\,dx = \int [f(\phi(u))\,d\phi/du]\,du.$

34. Let $u = x^2,$ $x = \sqrt{u},$ and $dx = du/2\sqrt{u}$ in $\displaystyle\int_{-1}^1 x^2\,dx$ to obtain

$$\frac{2}{3} = \int_{-1}^1 x^2\,dx = \frac{1}{2}\int_1^1 \sqrt{u}\,du = 0!$$

The explanation of this paradox lies in the fact that $u = g(x) = x^2$ does not have a single-valued inverse: $x = -\sqrt{u},$ $-1 \le x < 0,$ while $x = \sqrt{u},$ $0 \le x \le 1.$ Show that under the substitution $u = x^2,$

$$\int_{-1}^1 x^2\,dx = -\frac{1}{2}\int_1^0 \sqrt{u}\,du + \frac{1}{2}\int_0^1 \sqrt{u}\,du = \frac{2}{3}.$$

15.3

Trigonometric Integrals

A very powerful method of integration uses trigonometric substitutions. Before considering this method, let us first consider integration of trigonometric functions.

Case I:

$$\int \sin^m u \cos^n u \cdot u' \, dx = \int \sin^m u \cos^n u \, du.$$

We can use the identity

$$\sin^2 u + \cos^2 u = 1$$

to carry out the integration if either m or n is an odd positive integer. It is illustrated here for the case in which n is an odd positive integer.

$$\sin^m u \cos^n u = \sin^m u \cos^{n-1} u \cos u \quad (n - 1 \text{ even})$$
$$= \sin^m u \ (\cos^2 u)^{(n-1)/2} \cos u$$
$$= \sin^m u \ (1 - \sin^2 u)^{(n-1)/2} \cos u$$
$$= f(\sin u) \cos u,$$

where f is an ordinary polynomial. Thus,

$$\int \sin^m u \cos^n u \cdot u' \, dx = \int f(\sin u) \cos u \cdot u' \, dx = \int f(\sin u) \cos u \, du.$$

Since $\cos u \cdot u'$ is the derivative of $\sin u$, this can easily be integrated term by term. A similar method can be used if m is an odd positive integer.

Example 1

Evaluate $\int \sin^2 x \cos^3 x \, dx$.

$$\int \sin^2 x \cos^3 x \, dx = \int \sin^2 x \cos^2 x \cos x \, dx$$
$$= \int \sin^2 x \ (1 - \sin^2 x) \cos x \, dx$$
$$= \int (\sin^2 x - \sin^4 x) \cos x \, dx$$
$$= \frac{\sin^3 x}{3} - \frac{\sin^5 x}{5} + C.$$

Example 2

Evaluate $\int \sin^5 x \, dx$.

$$\int \sin^5 x \, dx = \int \sin^4 x \sin x \, dx$$

$$= \int (1 - \cos^2 x)^2 \sin x \, dx$$

$$= - \int (1 - 2\cos^2 x + \cos^4 x)(-\sin x) \, dx$$

$$= - \left(\cos x - \frac{2\cos^3 x}{3} + \frac{\cos^5 x}{5} \right) + C$$

$$= -\cos x + \frac{2\cos^3 x}{3} - \frac{\cos^5 x}{5} + C.$$

If m and n are both positive even integers, the above method does not work (see Problem 29). In this case we can use the identities

$$\cos 2x = 1 - 2\sin^2 x, \qquad \cos 2x = 2\cos^2 x - 1.$$

When these are solved for $\sin^2 x$ and $\cos^2 x$, respectively, we have

$$\sin^2 x = \frac{1 - \cos 2x}{2}, \qquad \cos^2 x = \frac{1 + \cos 2x}{2}.$$

In each case, we are replacing the second power of $\sin x$ or $\cos x$ by the first power of $\cos 2x$. Repeated applications of these identities will eventually lead to odd powers, which can be handled by the previous method.

Example 3

Evaluate $\int \sin^4 x \, dx$.

$$\int \sin^4 x \, dx = \int (\sin^2 x)^2 \, dx$$

$$= \int \left(\frac{1 - \cos 2x}{2} \right)^2 dx$$

$$= \frac{1}{4} \int (1 - 2\cos 2x + \cos^2 2x) \, dx$$

$$= \frac{1}{4} \int \left(1 - 2\cos 2x + \frac{1 + \cos 4x}{2} \right) dx$$

$$= \frac{1}{8} \int (3 - 4\cos 2x + \cos 4x) \, dx$$

$$= \frac{1}{8} \left(3x - 2\sin 2x + \frac{1}{4}\sin 4x \right) + C$$

$$= \frac{3}{8} x - \frac{1}{4} \sin 2x + \frac{1}{32} \sin 4x + C.$$

Example 4

Evaluate $\int \sin^2 2x \cos^2 2x\, dx.$

$$\int \sin^2 2x \cos^2 2x\, dx = \int \frac{1 - \cos 4x}{2}\, \frac{1 + \cos 4x}{2}\, dx$$

$$= \frac{1}{4} \int (1 - \cos^2 4x)\, dx$$

$$= \frac{1}{4} \int \left(1 - \frac{1 + \cos 8x}{2}\right) dx$$

$$= \frac{1}{8} \int (1 - \cos 8x)\, dx$$

$$= \frac{1}{8} \left(x - \frac{1}{8} \sin 8x\right) + C$$

$$= \frac{1}{8} x - \frac{1}{64} \sin 8x + C.$$

Sometimes other trigonometric functions can be changed to sines and cosines and the foregoing methods used.

Example 5

Evaluate $\int \frac{\tan^3 x}{\sec^4 x}\, dx.$

$$\int \frac{\tan^3 x}{\sec^4 x}\, dx = \int \frac{(\sin^3 x)/(\cos^3 x)}{1/(\cos^4 x)}\, dx$$

$$= \int \sin^3 x \cos x\, dx$$

$$= \frac{1}{4} \sin^4 x + C.$$

Problems

A

Evaluate the integrals.

1. $\int \sin^2 x \cos x\, dx.$

2. $\int \sin \theta \cos^2 \theta\, d\theta.$

3. $\int \cos \theta \sin^3 \theta\, d\theta.$

4. $\int \frac{\cos x}{\sin^3 x}\, dx.$

5. $\int \sin^3 3x \cos^2 3x\, dx.$

6. $\int \cos 4x \sin^4 4x\, dx.$

7. $\int_0^\pi \cos^2 \theta \sin^5 \theta\, d\theta.$

8. $\int \sin^2 \theta \cos^4 \theta\, d\theta.$

9. $\displaystyle\int \sin^2 2x \cos^2 2x \, dx.$

10. $\displaystyle\int \sin^4 \theta \cos^2 \theta \, d\theta.$

11. $\displaystyle\int \cos^6 \theta \, d\theta.$

B

Evaluate the integrals.

12. $\displaystyle\int_{\pi/12}^{\pi/6} \frac{\cos^3 2\theta}{\sqrt{\sin 2\theta}} \, d\theta.$

13. $\displaystyle\int_{-\pi/4}^{0} \sin^4 2\theta \, d\theta.$

14. $\displaystyle\int_{-\pi/2}^{-\pi/4} \sin^5 2x \cos^2 2x \, dx.$

15. $\displaystyle\int \frac{\cos^3 x}{\sin^2 x} \, dx.$

16. $\displaystyle\int \frac{dx}{\cos^2 \pi x}.$

17. $\displaystyle\int \cos^{3/2} x \sin^3 x \, dx.$

18. $\displaystyle\int x \sin^2 x^2 \cos^2 x^2 \, dx.$

19. $\displaystyle\int x \cos^3 x^2 \sin^2 x^2 \, dx.$

20. $\displaystyle\int \frac{\sec x}{\tan^2 x} \, dx.$

21. $\displaystyle\int \frac{dx}{\csc^2 x \cot x}.$

22. $\displaystyle\int \frac{\cot^2 x}{\csc^2 x} \, dx.$

23. $\displaystyle\int_{\pi/6}^{\pi/3} \sin^4 \theta \cot^2 \theta \, d\theta.$

24. $\displaystyle\int_{\pi/6}^{5\pi/6} \cos^4 \theta \tan^3 \theta \, d\theta.$

25. $\displaystyle\int \cos^3 t \sqrt{\sin t} \, dt.$

26. $\displaystyle\int \sqrt{\sec x} \tan x \, dx.$

C

27. $\int \sin x \cos x \, dx$ can be evaluated by three different methods: by noting that $\cos x$ is the derivative of $\sin x$, by noting that $-\sin x$ is the derivative of $\cos x$, and by using the identity $\sin 2x = 2 \sin x \cos x$. Carry out the integration by all three methods and show that the three answers are equivalent.

28. Evaluate $\int \sin \theta \cos^3 \theta \, d\theta$ by two different methods and show that the answers are equivalent.

29. Show why the use of the identity $\sin^2 u + \cos^2 u = 1$ does not allow us to evaluate $\int \sin^m u \cos^n u \cdot u' \, dx$ when both m and n are positive even integers.

15.4

Trigonometric Integrals (Continued)

Case II:

$$\int \sec^m u \tan^n u \cdot u' \, dx = \int \sec^m u \tan^n u \, du,$$

$$\int \csc^m u \cot^n u \cdot u' \, dx = \int \csc^m u \cot^n u \, du.$$

These integrals can be evaluated if either m is a positive even integer or n is a positive odd integer. Both depend upon the identities

$$\sec^2 u = 1 + \tan^2 u \quad \text{and} \quad \csc^2 u = 1 + \cot^2 u.$$

If m is a positive even integer, then

$$\int \sec^m u \tan^n u \cdot u' \, dx = \int \sec^{m-2} u \tan^n u \sec^2 u \cdot u' \, dx$$

$$= \int (\sec^2 u)^{(m-2)/2} \tan^n u \sec^2 u \cdot u' \, dx$$

$$= \int (1 + \tan^2 u)^{(m-2)/2} \tan^n u \sec^2 u \cdot u' \, dx$$

$$= \int f(\tan u) \sec^2 u \cdot u' \, dx = \int f(\tan u) \sec^2 u \, du.$$

Since $\sec^2 u \cdot u'$ is the derivative of $\tan u$, the result is easily evaluated. A similar method can be used for $\int \csc^m u \, \cot^n u \cdot u' \, dx$.

Example 1

Evaluate $\displaystyle\int \sec^4 x \tan^2 x \, dx$.

$$\int \sec^4 x \tan^2 x \, dx = \int \sec^2 x \tan^2 x \sec^2 x \, dx$$

$$= \int (1 + \tan^2 x) \tan^2 x \sec^2 x \, dx$$

$$= \int (\tan^2 x + \tan^4 x) \sec^2 x \, dx$$

$$= \frac{\tan^3 x}{3} + \frac{\tan^5 x}{5} + C.$$

Example 2

Evaluate $\displaystyle\int \sec^4 \theta \, d\theta$.

$$\int \sec^4 \theta \, d\theta = \int \sec^2 \theta \sec^2 \theta \, d\theta$$

$$= \int (1 + \tan^2 \theta) \sec^2 \theta \, d\theta$$

$$= \tan \theta + \frac{\tan^3 \theta}{3} + C.$$

If n is a positive odd integer, then

$$\int \sec^m u \tan^n u \cdot u' \, dx = \int \sec^{m-1} u \tan^{n-1} u \sec u \tan u \cdot u' \, dx$$

$$= \int \sec^{m-1} u \, (\tan^2 u)^{(n-1)/2} \sec u \tan u \cdot u' \, dx$$

$$= \int \sec^{m-1} u \, (\sec^2 u - 1)^{(n-1)/2} \sec u \tan u \cdot u' \, dx$$

$$= \int f(\sec u) \sec u \tan u \cdot u' \, dx = \int f(\sec u) \sec u \tan u \, du.$$

Since $\sec u \tan u \cdot u'$ is the derivative of $\sec u$, the result is easily evaluated. Again the same method can be used for $\int \csc^m u \cot^n u \cdot u' \, dx$.

Example 3

Evaluate $\displaystyle\int \sec^3 x \tan^3 x \, dx$.

$$\int \sec^3 x \tan^3 x \, dx = \int \sec^2 x \tan^2 x \sec x \tan x \, dx$$

$$= \int \sec^2 x \, (\sec^2 x - 1) \sec x \tan x \, dx$$

$$= \int (\sec^4 x - \sec^2 x) \sec x \tan x \, dx$$

$$= \frac{\sec^5 x}{5} - \frac{\sec^3 x}{3} + C.$$

Example 4

Evaluate $\displaystyle\int \cot^3 x \, dx$.

$$\int \cot^3 x \, dx = \int \frac{\cot^2 x}{\csc x} \csc x \cot x \, dx$$

$$= \int \frac{\csc^2 x - 1}{\csc x} \csc x \cot x \, dx$$

$$= \int \left(\frac{1}{\csc x} - \csc x\right) (-\csc x \cot x) \, dx$$

$$= \ln |\csc x| - \frac{\csc^2 x}{2} + C.$$

Case III:

$$\int \tan^n u \cdot u' \, dx = \int \tan^n u \, du \quad \text{and} \quad \int \cot^n u \cdot u' \, dx = \int \cot^n u \, du.$$

The method of Case II can be used here only if n is odd. The following method works for any positive integer n—odd or even. Of course, $\int \tan u \cdot u' \, dx$ and $\int \tan^2 u \cdot u' \, dx$ can be evaluated quite easily. If $n > 2$, then

$$\int \tan^n u \cdot u' \, dx = \int \tan^{n-2} u \tan^2 u \cdot u' \, dx$$

$$= \int \tan^{n-2} u \, (\sec^2 u - 1)u' \, dx$$

$$= \int (\tan^{n-2} u \sec^2 u - \tan^{n-2} u)u' \, dx.$$

$$= \int (\tan^{n-2} u \sec^2 u - \tan^{n-2} u) \, du.$$

The first term can easily be integrated, and the exponent on the second has been reduced by 2. If $n - 2 > 2$, this can be repeated until the exponent is down to 1 or 2.

Example 5

Evaluate $\int \tan^6 x \, dx$.

$$\int \tan^6 x \, dx = \int \tan^4 x \tan^2 x \, dx$$

$$= \int \tan^4 x \, (\sec^2 x - 1) \, dx$$

$$= \int (\tan^4 x \sec^2 x - \tan^4 x) \, dx$$

$$= \int (\tan^4 x \sec^2 x - \tan^2 x \tan^2 x) \, dx$$

$$= \int [\tan^4 x \sec^2 x - \tan^2 x(\sec^2 x - 1)] \, dx$$

$$= \int (\tan^4 x \sec^2 x - \tan^2 x \sec^2 x + \tan^2 x) \, dx$$

$$= \int (\tan^4 x \sec^2 x - \tan^2 x \sec^2 x + \sec^2 x - 1) \, dx$$

$$= \frac{\tan^5 x}{5} - \frac{\tan^3 x}{3} + \tan x - x + C.$$

Example 6

Evaluate $\int \cot^3 x \, dx$.

This is the same problem as Example 4. Let us evaluate the integral by this new method.

$$\int \cot^3 x \, dx = \int \cot x \cot^2 x \, dx$$

$$= \int \cot x \, (\csc^2 x - 1) \, dx$$

$$= \int (\cot x \csc^2 x - \cot x) \, dx$$

$$= -\frac{\cot^2 x}{2} - \ln |\sin x| + C.$$

(See Problem 32.)

Problems

A

Evaluate the integrals.

1. $\int \sec^2 x \tan^2 x \, dx.$

2. $\int \sec^4 \theta \tan^3 \theta \, d\theta.$

3. $\int \csc^4 x \cot^4 x \, dx.$

4. $\int \csc^6 2x \cot^2 2x \, dx.$

5. $\int \csc \theta \cot^3 \theta \, d\theta.$

6. $\int \sec \theta \tan^5 \theta \, d\theta.$

7. $\int \sec^3 x \tan x \, dx.$

8. $\int \sec^5 x \tan^3 x \, dx.$

9. $\int \sec^6 x \, dx.$

10. $\int \csc^4 2x \, dx.$

11. $\int \tan^3 \theta \, d\theta.$

12. $\int \tan^5 x \, dx.$

13. $\int_0^{\pi/3} \tan^2 x \, dx.$

B

Evaluate the integrals.

14. $\int \cot^5 2\theta \, d\theta.$

15. $\int \frac{\cos^2 \theta}{\sin^4 \theta} \, d\theta.$

16. $\int_0^{\pi/4} \frac{\sin^3 \theta}{\cos^4 \theta} \, d\theta.$

17. $\int \tan^4 3x \, dx.$

18. $\int x \csc^4 x^2 \cot^2 x^2 \, dx.$

19. $\int \cot^4 (2x + 1) \, dx.$

20. $\int \sec^3 4x \tan^3 4x \, dx.$

21. $\int \frac{\tan \theta}{1 - \tan^2 \theta} \, d\theta.$

22. $\int (\cos^2 x - \sin^2 x) \, dx.$

23. $\int \frac{\sin^2 \theta}{1 - \cos \theta} \, d\theta.$

24. $\int \frac{1 - \cos \theta}{\sin^2 \theta} \, d\theta.$

25. $\int \frac{\sin \theta \cos \theta}{\sin^2 \theta - \cos^2 \theta} \, d\theta.$

26. $\int \frac{\cot \theta}{\csc^3 \theta} \, d\theta.$

27. $\int \frac{\sec^4 \theta}{\tan^2 \theta} \, d\theta.$

28. $\int \frac{\cot^3 \theta}{\csc^2 \theta} \, d\theta.$

29. $\int \frac{\tan^2 x}{\sec^3 x} \, dx.$

30. $\int \frac{\sec x + \tan x}{\sec^2 x} \, dx.$

C

31. $\int \sec^2 x \tan x \, dx$ can be integrated by three different methods—using the fact that the exponent on $\sec x$ is even, using the fact that the exponent on $\tan x$ is odd, and changing to $\sin x$ and $\cos x$. Carry out the integration by all three methods and show that the three results are equivalent.

32. Show that the results of Examples 4 and 6 are equivalent.

15.5

Trigonometric Substitutions

In the past two sections we have widened the range of trigonometric functions that we can integrate. But many relatively simple trigonometric functions remain that we still cannot integrate. For example, $\int \sec^3 x \, dx$ cannot be integrated by any method we have had. We shall see more of this integral in Section 15.7.

As noted earlier, our attention to trigonometric expressions so far is in preparation for integration by trigonometric substitutions. We have seen that integrals such as $\int x\sqrt{x+1}\,dx$, in which the expression under the radical is linear, can be evaluated by means of the substitution $u = \sqrt{x+1}$. Unfortunately, this type of substitution is usually of no avail if the expression under the radical is quadratic. When dealing with an integral involving the square root of a quadratic expression, we use a trigonometric substitution.

Just as in the previous substitution, the trigonometric substitution is used to eliminate the radical. There are three substitutions to handle the three possible situations:

If the integral involves	use the substitution	and the identity
$\sqrt{a^2 - u^2}$,	$u = a \sin \theta$	$\cos^2 \theta = 1 - \sin^2 \theta$
$\sqrt{a^2 + u^2}$,	$u = a \tan \theta$	$\sec^2 \theta = 1 + \tan^2 \theta$
$\sqrt{u^2 - a^2}$,	$u = a \sec \theta$	$\tan^2 \theta = \sec^2 \theta - 1$

Example 1

Evaluate $\displaystyle\int \frac{dx}{\sqrt{1-x^2}}$.

Although this integral can be evaluated by formula 23 on page 448, you were promised there that we would provide a method of integration that would relieve you of the necessity of memorizing formulas 22–24. Trigonometric substitution is that method. Let us use the substitution

$$x = \sin \theta,$$

which gives

$$dx = \cos \theta \, d\theta;$$

then

$$\int \frac{dx}{\sqrt{1-x^2}} = \int \frac{\cos \theta \, d\theta}{\sqrt{1 - \sin^2 \theta}}$$

$$= \int \frac{\cos \theta \, d\theta}{\sqrt{\cos^2 \theta}}$$

$$= \int d\theta$$

$$= \theta + C.$$

Having carried out the integration, we want to express the result in terms of x. The substitution $x = \sin \theta$ gives $\theta = \text{Arcsin } x$ and

$$\int \frac{dx}{\sqrt{1-x^2}} = \text{Arcsin } x + C.$$

Example 2

Evaluate $\int \frac{x^3}{\sqrt{x^2+4}}\,dx.$

Let us use the substitution

$$x = 2\tan\theta,$$

which gives

$$dx = 2\sec^2\theta\,d\theta;$$

then

$$\begin{aligned}\int \frac{x^3}{\sqrt{x^2+4}}\,dx &= \int \frac{8\tan^3\theta\cdot 2\sec^2\theta}{\sqrt{4\tan^2\theta+4}}\,d\theta\\ &= \int \frac{16\sec^2\theta\tan^3\theta}{\sqrt{4\sec^2\theta}}\,d\theta\\ &= \int 8\sec\theta\tan^3\theta\,d\theta\\ &= \int 8\tan^2\theta\sec\theta\tan\theta\,d\theta\\ &= \int 8(\sec^2\theta-1)\sec\theta\tan\theta\,d\theta\\ &= 8\left(\frac{\sec^3\theta}{3}-\sec\theta\right)+C\\ &= \frac{8}{3}\sec^3\theta - 8\sec\theta + C.\end{aligned}$$

In order to express this result in terms of x, we see that the original substitution gives

$$\tan\theta = \frac{x}{2}.$$

Figure 15.1

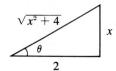

We can use either Figure 15.1 or the identity $\sec^2\theta = 1+\tan^2\theta$ to get

$$\sec\theta = \frac{\sqrt{x^2+4}}{2}.$$

Thus

$$\begin{aligned}\int \frac{x^3}{\sqrt{x^2+4}}\,dx &= \frac{8}{3}\,\frac{(x^2+4)^{3/2}}{8} - 8\,\frac{\sqrt{x^2+4}}{2} + C\\ &= \frac{1}{3}\sqrt{x^2+4}\,[(x^2+4)-12]+C\\ &= \frac{1}{3}\sqrt{x^2+4}\,(x^2-8)+C.\end{aligned}$$

Example 3

Evaluate $\displaystyle\int \frac{dx}{(4x^2 - 9)^{3/2}}$.

Since the 3/2 power gives a square root, this is handled in the same way as the other examples. Here we have $u^2 - a^2$, where $u = 2x$ and $a = 3$. Thus we want the substitution

$$2x = 3 \sec \theta, \quad \text{or} \quad x = \frac{3}{2} \sec \theta;$$

$$dx = \frac{3}{2} \sec \theta \tan \theta \, d\theta.$$

$$\int \frac{dx}{(4x^2 - 9)^{3/2}} = \int \frac{\frac{3}{2} \sec \theta \tan \theta \, d\theta}{(9 \sec^2 \theta - 9)^{3/2}}$$

$$= \int \frac{\frac{3}{2} \sec \theta \tan \theta \, d\theta}{(9 \tan^2 \theta)^{3/2}}$$

$$= \int \frac{\frac{3}{2} \sec \theta \tan \theta \, d\theta}{3^3 \tan^3 \theta}$$

$$= \frac{1}{18} \int \frac{\sec \theta}{\tan^2 \theta} \, d\theta$$

$$= \frac{1}{18} \int \frac{\dfrac{1}{\cos \theta}}{\dfrac{\sin^2 \theta}{\cos^2 \theta}} \, d\theta$$

$$= \frac{1}{18} \int \frac{\cos \theta}{\sin^2 \theta} \, d\theta$$

$$= \frac{1}{18} \cdot \frac{-1}{\sin \theta} + C$$

$$= -\frac{1}{18} \csc \theta + C.$$

Using our original substitution, we have $\sec \theta = 2x/3$. From Figure 15.2, we have

$$\csc \theta = \frac{2x}{\sqrt{4x^2 - 9}}.$$

Thus

$$\int \frac{dx}{(4x^2 - 9)^{3/2}} = -\frac{1}{18} \frac{2x}{\sqrt{4x^2 - 9}} + C$$

$$= \frac{-x}{9\sqrt{4x^2 - 9}} + C.$$

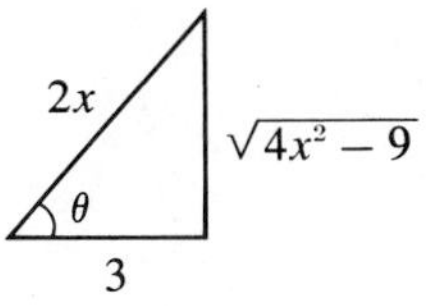

Figure 15.2

Although trigonometric substitution is particularly useful when radicals of quadratic expressions are present, the integral need not involve a radical for trigonometric substitution to be useful.

Example 4

Evaluate $\displaystyle \int \frac{dx}{(1+x^2)^2}$.

Since we do not have the derivative of $1 + x^2$, we cannot use the power rule. But a trigonometric substitution will enable us to combine the two terms, 1 and x^2, into a single term; thus we may carry out the division. Hence we use the substitution

$$x = \tan \theta,$$

which gives

$$dx = \sec^2 \theta \, d\theta.$$

This leads to

$$\int \frac{dx}{(1+x^2)^2} = \int \frac{\sec^2 \theta \, d\theta}{(1 + \tan^2 \theta)^2}$$

$$= \int \frac{\sec^2 \theta \, d\theta}{\sec^4 \theta}$$

$$= \int \frac{d\theta}{\sec^2 \theta}$$

$$= \int \cos^2 \theta \, d\theta$$

$$= \frac{1}{2} \int (1 + \cos 2\theta) \, d\theta$$

$$= \frac{1}{2} \theta + \frac{1}{4} \sin 2\theta + C.$$

Now, from the original substitution, we have

$$\theta = \text{Arctan } x;$$

and, from Figure 15.3 or trigonometric identities,

$$\sin \theta = \frac{x}{\sqrt{1+x^2}} \quad \text{and} \quad \cos \theta = \frac{1}{\sqrt{1+x^2}}.$$

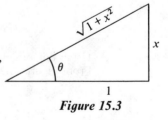

Figure 15.3

Thus

$$\sin 2\theta = 2 \sin \theta \cos \theta = \frac{2x}{1+x^2}.$$

This gives

$$\int \frac{dx}{(1+x^2)^2} = \frac{1}{2} \theta + \frac{1}{4} \sin 2\theta + C$$

$$= \frac{1}{2} \text{Arctan } x + \frac{x}{2(1+x^2)} + C.$$

Example 5

Evaluate $\displaystyle\int_0^3 \sqrt{9-x^2}\,dx$.

By making the substitution

$$x = 3\sin\theta,$$

we get

$$dx = 3\cos\theta\,d\theta.$$

Now when $x = 0$, $\sin\theta = 0$ and $\theta = 0$; when $x = 3$, $\sin\theta = 1$ and $\theta = \pi/2$. Thus

$$\int_0^3 \sqrt{9-x^2}\,dx = \int_0^{\pi/2} \sqrt{9 - 9\sin^2\theta}\ 3\cos\theta\,d\theta$$

$$= \int_0^{\pi/2} 9\cos^2\theta\,d\theta$$

$$= \frac{9}{2}\int_0^{\pi/2} (1 + \cos 2\theta)\,d\theta$$

$$= \frac{9}{2}\left(\theta + \frac{1}{2}\sin 2\theta\right)\Bigg|_0^{\pi/2}$$

$$= \frac{9}{2}\left[\left(\frac{\pi}{2} + \frac{1}{2}\sin\pi\right) - \left(0 + \frac{1}{2}\sin 0\right)\right]$$

$$= \frac{9\pi}{4}.$$

If we had not changed the limits of integration we would have had to express

$$\frac{9}{2}\left(\theta + \frac{1}{2}\sin 2\theta\right)$$

in terms of x before putting in the limits of integration. In order to do so we would have had to use the identity $\sin 2\theta = 2\sin\theta\cos\theta$.

$$\frac{9}{2}\left(\theta + \frac{1}{2}\sin 2\theta\right) = \frac{9}{2}(\theta + \sin\theta\cos\theta).$$

Now, since

$$\sin\theta = \frac{x}{3},$$

we have

$$\cos\theta = \frac{\sqrt{9-x^2}}{3} \quad\text{and}\quad \theta = \text{Arcsin}\,\frac{x}{3}.$$

$$\int_0^3 \sqrt{9-x^2}\,dx = \frac{9}{2}\left(\text{Arcsin}\,\frac{x}{3} + \frac{x\sqrt{9-x^2}}{9}\right)\Bigg|_0^3$$

$$= \frac{9}{2}\left[(\text{Arcsin}\,1 + 0) - (\text{Arcsin}\,0 + 0)\right]$$

$$= \frac{9}{2}\frac{\pi}{2}$$

$$= \frac{9\pi}{4}.$$

It might be noted that the substitutions, as we have stated them, do not define θ as a function of x. The substitution $u = a \sin \theta$ is a more convenient way of saying $\theta = \text{Arcsin} (u/a)$. It is this restriction on θ that allows us to say $\sqrt{\cos^2 \theta} = \cos \theta$. The same holds for the other two substitutions. This is also why we were able to use the positive square roots in Figures 15.1–15.3.

We now have the promised general method of integration that makes it unnecessary to memorize formulas 22–24 on page 448. Since all three formulas involve $1 \pm u^2$ or $u^2 - 1$ either with or without a radical, a trigonometric substitution can be used on any one of them. It might be noted that hyperbolic substitutions can be used to evaluate any of these integrals, and occasionally they lead to much simpler integrals than the trigonometric substitutions (see Problem 22); however, since both trigonometric and hyperbolic substitutions often lead to relatively complicated expressions, we normally prefer to remain with the more familiar trigonometric functions.

Problems

A

Evaluate the integrals without using formulas 22–24 on page 448.

1. $\displaystyle\int \frac{\sqrt{1 - x^2}}{x^2}\, dx.$ 2. $\displaystyle\int \frac{x^2}{\sqrt{9 - x^2}}\, dx.$ 3. $\displaystyle\int \frac{dx}{\sqrt{x^2 + 16}}.$

4. $\displaystyle\int \frac{dx}{(16 + x^2)^{3/2}}.$ 5. $\displaystyle\int \frac{x^3\, dx}{\sqrt{9 - x^2}}.$ 6. $\displaystyle\int \frac{x^3\, dx}{\sqrt{4 + x^2}}.$

7. $\displaystyle\int \frac{dx}{(x^2 - 5)^{3/2}}.$ 8. $\displaystyle\int \sqrt{9 - 4x^2}\, dx.$ 9. $\displaystyle\int \frac{\sqrt{4 - x^2}}{x}\, dx.$

10. $\displaystyle\int \frac{dx}{\sqrt{1 + x^2}}.$ 11. $\displaystyle\int \frac{dx}{x\sqrt{1 - x^2}}.$ 12. $\displaystyle\int_0^1 \frac{dx}{(x^2 + 1)^{3/2}}.$

B

Evaluate the integrals without using formulas 22–24 on page 448.

13. $\displaystyle\int (9 - x^2)^{3/2}\, dx.$ 14. $\displaystyle\int \frac{\sqrt{4 - 3x^2}}{x^4}\, dx.$ 15. $\displaystyle\int \frac{\sqrt{4x^2 - 9}}{x^3}\, dx.$

16. $\displaystyle\int \frac{x^3\, dx}{\sqrt{9x^2 + 4}}.$ 17. $\displaystyle\int \frac{dx}{(3x^2 + 2)^{3/2}}.$ 18. $\displaystyle\int_0^{\sqrt{2}} \frac{x^3\, dx}{\sqrt{4 - x^2}}.$

19. $\displaystyle\int_0^{3/2} x^2\sqrt{9 - 4x^2}\, dx.$ 20. $\displaystyle\int_{\sqrt{3}}^2 \frac{\sqrt{x^2 - 3}}{x}\, dx.$

C

21. Find the length of $x = \text{Arcsin } e^y$ from $y = -1$ to $y = -1/2$.
22. Evaluate the integral of Problem 3 by using the substitution $x = 4 \sinh t$. Compare with the integral form of Theorem 12.12(1) on page 390.

15.6

Integrals Involving $ax^2 + bx + c$

When the integral involves expressions of the form $ax^2 + bx + c$, it is often necessary to complete the square in order to get the expression into the form $au^2 + d$.

Example 1

Evaluate $\displaystyle\int \frac{dx}{x^2 + 2x + 10}$.

Let us complete the square on the first two terms of $x^2 + 2x + 10$.

$$x^2 + 2x + 10 = x^2 + 2x + 1 - 1 + 10$$
$$= (x + 1)^2 + 9.$$

At this point we may complete the integration either by a trigonometric substitution or by dividing both numerator and denominator by 9 and using formula 22 on page 448. Both methods are illustrated here.

$$x + 1 = 3 \tan \theta,$$
$$dx = 3 \sec^2 \theta \, d\theta.$$

$$\int \frac{dx}{x^2 + 2x + 10} = \int \frac{dx}{(x + 1)^2 + 9} \qquad\qquad \int \frac{dx}{x^2 + 2x + 10} = \int \frac{dx}{(x + 1)^2 + 9}$$

$$= \int \frac{3 \sec^2 \theta \, d\theta}{9 \tan^2 \theta + 9} \qquad\qquad\qquad = \int \frac{1/9}{\dfrac{(x + 1)^2}{9} + 1} \, dx$$

$$= \int \frac{3 \sec^2 \theta \, d\theta}{9 \sec^2 \theta} \qquad\qquad\qquad = \frac{1}{3} \int \frac{1/3}{\left(\dfrac{x + 1}{3}\right)^2 + 1} \, dx$$

$$= \frac{1}{3} \int d\theta = \frac{1}{3} \theta + C \qquad\qquad\qquad = \frac{1}{3} \operatorname{Arctan} \frac{x + 1}{3} + C.$$

$$= \frac{1}{3} \operatorname{Arctan} \frac{x + 1}{3} + C.$$

Example 2

Evaluate $\displaystyle\int \frac{dx}{(4x^2 + 16x + 15)^{3/2}}$.

Completing the square inside the parentheses, we have

$$4x^2 + 16x + 15 = 4(x^2 + 4x) + 15$$
$$= 4(x^2 + 4x + 4) - 16 + 15$$
$$= 4(x + 2)^2 - 1.$$

We now complete the integration by the substitution

$$2(x + 2) = \sec\theta,$$
$$dx = \frac{1}{2}\sec\theta\tan\theta\,d\theta;$$

$$\int \frac{dx}{(4x^2 + 16x + 15)^{3/2}} = \int \frac{dx}{[4(x+2)^2 - 1]^{3/2}}$$

$$= \int \frac{\frac{1}{2}\sec\theta\tan\theta\,d\theta}{(\sec^2\theta - 1)^{3/2}}$$

$$= \int \frac{\sec\theta\tan\theta\,d\theta}{2\tan^3\theta}$$

$$= \frac{1}{2}\int \frac{\sec\theta}{\tan^2\theta}\,d\theta$$

$$= \frac{1}{2}\int \frac{\frac{1}{\cos\theta}}{\frac{\sin^2\theta}{\cos^2\theta}}\,d\theta$$

$$= \frac{1}{2}\int \frac{\cos\theta}{\sin^2\theta}\,d\theta$$

$$= \frac{1}{2}\frac{-1}{\sin\theta} + C$$

$$= -\frac{1}{2}\csc\theta + C.$$

Our original substitution leads to Figure 15.4, which gives

$$\csc\theta = \frac{2(x+2)}{\sqrt{4x^2 + 16x + 15}}.$$

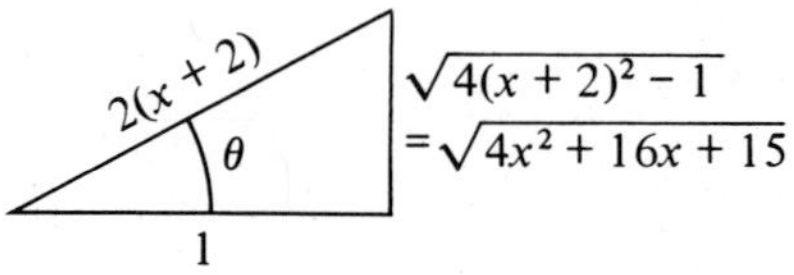

Figure with right triangle: hypotenuse $2(x + 2)$, vertical side $\sqrt{4(x+2)^2 - 1} = \sqrt{4x^2 + 16x + 15}$, base 1, angle θ.

Thus

$$\int \frac{dx}{(4x^2 + 16x + 15)^{3/2}}$$

$$= -\frac{1}{2}\csc\theta + C = -\frac{x+2}{\sqrt{4x^2 + 16x + 15}} + C.$$

Figure 15.4

Example 3

Evaluate $\displaystyle\int \frac{x-4}{\sqrt{-9x^2 + 36x - 32}}\, dx.$

Completing the square under the radical gives

$$
\begin{aligned}
-9x^2 + 36x - 32 &= -9(x^2 - 4x) - 32 \\
&= -9(x^2 - 4x + 4) + 36 - 32 \\
&= 4 - 9(x-2)^2.
\end{aligned}
$$

Thus we use the substitution

$$3(x-2) = 2\sin\theta$$

$$x = \frac{2}{3}\sin\theta + 2,$$

$$dx = \frac{2}{3}\cos\theta\, d\theta;$$

$$
\begin{aligned}
\int \frac{x-4}{\sqrt{-9x^2 + 36x - 32}}\, dx &= \int \frac{x-4}{\sqrt{4 - 9(x-2)^2}}\, dx \\[2mm]
&= \int \frac{\left(\dfrac{2}{3}\sin\theta + 2 - 4\right)\dfrac{2}{3}\cos\theta\, d\theta}{\sqrt{4 - 4\sin^2\theta}} \\[2mm]
&= \int \frac{\left(\dfrac{2}{3}\sin\theta - 2\right)\dfrac{2}{3}\cos\theta\, d\theta}{2\cos\theta} \\[2mm]
&= \int \left(\frac{2}{9}\sin\theta - \frac{2}{3}\right) d\theta \\[2mm]
&= -\frac{2}{9}\cos\theta - \frac{2}{3}\theta + C.
\end{aligned}
$$

From the original substitution we have

$$\sin\theta = \frac{3(x-2)}{2}.$$

Thus, from Figure 15.5 or trigonometric identities, we have

$$\cos\theta = \frac{\sqrt{-9x^2 + 36x - 32}}{2},$$

$$\theta = \mathrm{Arcsin}\, \frac{3x - 6}{2}.$$

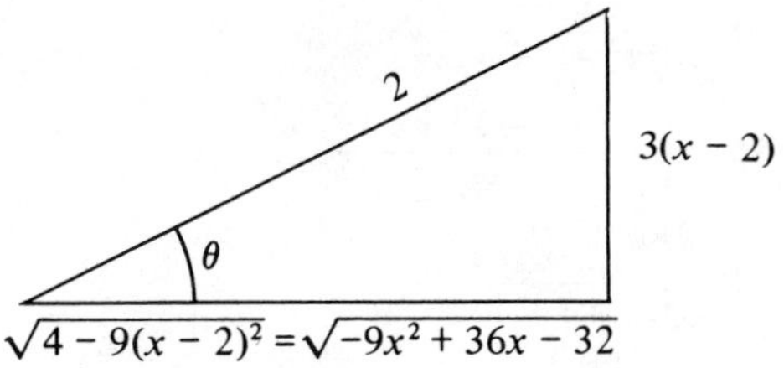

Figure 15.5

$$\int \frac{x-4}{\sqrt{-9x^2 + 36x - 32}} \, dx$$

$$= -\frac{2}{9} \cos \theta - \frac{2}{3} \theta + C$$

$$= -\frac{2}{9} \frac{\sqrt{-9x^2 + 36x - 32}}{2} - \frac{2}{3} \text{Arcsin} \frac{3x-6}{2} + C$$

$$= -\frac{1}{9} \sqrt{-9x^2 + 36x - 32} - \frac{2}{3} \text{Arcsin} \frac{3x-6}{2} + C.$$

Problems

A

Evaluate the integrals.

1. $\displaystyle\int \frac{dx}{x^2 + 2x + 2}.$

2. $\displaystyle\int \frac{dx}{\sqrt{x^2 - 6x + 8}}.$

3. $\displaystyle\int \frac{4\,dx}{\sqrt{-16x^2 - 32x - 15}}.$

4. $\displaystyle\int \frac{3\,dx}{(2x^2 - 8x - 10)^{3/2}}.$

5. $\displaystyle\int \sqrt{8x - 4x^2}\, dx.$

6. $\displaystyle\int (8 + 2x - x^2)^{3/2}\, dx.$

7. $\displaystyle\int \frac{2x + 3}{x^2 - 10x + 41}\, dx.$

8. $\displaystyle\int \frac{(x - 2)\,dx}{\sqrt{9x^2 + 12x + 8}}.$

9. $\displaystyle\int \frac{\sqrt{-x^2 + 10x - 21}}{x - 5}\, dx.$

10. $\displaystyle\int \frac{(4x - 1)\,dx}{(16x^2 - 8x + 17)^{3/2}}.$

11. $\displaystyle\int \frac{(2x + 3)\,dx}{\sqrt{-x^2 + 10x - 21}}.$

12. $\displaystyle\int (x - 2)\sqrt{-5x^2 + 20x - 13}\, dx.$

B

Evaluate the integrals.

13. $\displaystyle\int \frac{x^3 + 1}{x^2 + 2x + 2}\, dx.$

14. $\displaystyle\int \frac{8x^2\,dx}{4x^2 - 8x + 29}.$

15. $\displaystyle\int \frac{(x^2 - 5)\,dx}{(-x^2 - 6x - 8)^{3/2}}.$

16. $\displaystyle\int \frac{dx}{\sqrt{6x - x^2}}.$

17. $\displaystyle\int_{-1}^{2} \frac{dx}{x^2 + 2x + 10}.$

18. $\displaystyle\int_{1}^{5/4} \frac{dx}{-4x^2 + 8x - 3}.$

19. $\displaystyle\int_{-2}^{-1} \frac{dx}{(x^2 + 4x + 5)^{3/2}}.$

20. $\displaystyle\int_{-1/2}^{0} \frac{dx}{\sqrt{3 - 4x - 4x^2}}.$

15.7

Integration by Parts

Perhaps the most powerful method of integration is integration by parts. It is based upon the formula for the derivative of a product,

$$\frac{d}{dx}(uv) = u\frac{dv}{dx} + v\frac{du}{dx},$$

where u and v are functions of x. In the integral form, it is written

$$uv = \int uv' \, dx + \int vu' \, dx$$

or

$$\int uv' \, dx = uv - \int vu' \, dx \quad \text{or} \quad \int u \, dv = uv - \int v \, du.$$

With this formula we are able to trade old integrals for new (and, hopefully, simpler) ones.

Example 1

Evaluate $\int x \sin x \, dx$.

Suppose we let $u = x$ and $v' = \sin x$;
then $u' = 1$ and $v = -\cos x + C$.

$$\int x \sin x \, dx = x(-\cos x + C) - \int (-\cos x + C) \, dx$$
$$= -x \cos x + Cx + \sin x - Cx + K$$
$$= -x \cos x + \sin x + K.$$

Note that the constant of integration, C, does not appear in the final answer above. This is not merely the case for this particular example—it always happens, because we always have

$$u \cdot C - \int u' \cdot C \, dx = u \cdot C - u \cdot C + K = K.$$

Thus we may use any value of C we choose. Usually it is simplest to use $C = 0$, which we normally do, but occasionally it is more convenient to use some other value of C.†

Perhaps you wonder how one knows which factor to call u and which v'. To a large extent, it is a matter of trial and error, but the following guide is helpful in many (but not all) cases: *Let v' be the most complicated part of the expression that can be easily integrated.* Suppose we try using this rule on other examples.

---

† For an example in which it is more convenient to use a value of C other than 0, see Borman, J. L., "A Remark on Integration by Parts," *American Mathematical Monthly*, Vol. 51 (1944): pp. 32–33.

Example 2

Evaluate $\int x^3 e^{x^2} dx$.

Certainly e^{x^2} is a more "complicated" function than x^3, but we cannot integrate it. Let us then include a factor x, making it xe^{x^2}. This is even more complicated, and it can be integrated easily.

$$u = x^2, \qquad v' = xe^{x^2},$$

$$u' = 2x, \qquad v = \frac{1}{2} e^{x^2};$$

$$\int x^3 e^{x^2} dx = \frac{1}{2} x^2 e^{x^2} - \int xe^{x^2} dx$$

$$= \frac{1}{2} x^2 e^{x^2} - \frac{1}{2} e^{x^2} + C$$

$$= \frac{1}{2} e^{x^2}(x^2 - 1) + C.$$

Example 3

Evaluate $\int x \ln x \, dx$.

Again, $\ln x$ is the more complicated of the two factors, but it is not easily integrated. Thus x is the most complicated part that can still be integrated easily.

$$u = \ln x, \qquad v' = x,$$

$$u' = \frac{1}{x}, \qquad v = \frac{x^2}{2}.$$

$$\int x \ln x \, dx = \frac{x^2}{2} \ln x - \int \frac{x}{2} dx$$

$$= \frac{x^2}{2} \ln x - \frac{x^2}{4} + C$$

$$= \frac{x^2}{4} (2 \ln x - 1) + C.$$

Example 4

Evaluate $\int \ln x \, dx$.

We have only one factor, but we can easily have two, by using 1 as one of them. Since we cannot integrate $\ln x$ easily, we must choose

$$u = \ln x, \qquad v' = 1,$$

$$u' = \frac{1}{x}, \qquad v = x;$$

$$\int \ln x \, dx = x \ln x - \int dx$$

$$= x \ln x - x + C$$

$$= x(\ln x - 1) + C.$$

Sometimes more than one application of integration by parts is required, as the following example shows.

Example 5

Evaluate $\int x^2 \cos x \, dx$.

$$u = x^2, \qquad v' = \cos x,$$
$$u' = 2x, \qquad v = \sin x;$$

$$\int x^2 \cos x \, dx = x^2 \sin x - \int 2x \sin x \, dx.$$

Now the new integral, while it is better than the original, still cannot be evaluated by our integration formulas. Let us integrate by parts once more.

$$u = 2x, \qquad v' = \sin x,$$
$$u' = 2, \qquad v = -\cos x;$$

$$\int x^2 \cos x \, dx = x^2 \sin x - \left(-2x \cos x + \int 2 \cos x \, dx \right)$$
$$= x^2 \sin x + 2x \cos x - 2 \sin x + C.$$

When we can see from the start that more than one application of parts is needed, the process can be carried out more expeditiously by tabular integration. Let us note that the u and v' for the second application of parts corresponds to the u' and v from the first application. By telescoping the several applications of parts, we can make it into a table, as shown below. First we break up the integral into parts in the normal way. Then, instead of differentiating u only once, we repeat until we get 0. Similarly we integrate v' repeatedly. Finally, we take products diagonally, appending $+$, $-$, $+$, $-$ as shown. The integral equals the sum of these products.

$$\int x^2 \cos x \, dx = x^2 \sin x + 2x \cos x - 2 \sin x + C.$$

Example 6

Evaluate $\int x^3 e^x \, dx$.

$$\int x^3 e^x \, dx = x^3 e^x - 3x^2 e^x + 6xe^x - 6e^x + C.$$

Sometimes a repeated application of integration by parts leads back to an integral similar to our original one. If so, this expression can be combined with the original integral, as the next example illustrates.

Example 7

Evaluate $\int e^x \sin x \, dx.$

$$u = e^x, \qquad v' = \sin x,$$
$$u' = e^x, \qquad v = -\cos x;$$

$$\int e^x \sin x \, dx = -e^x \cos x + \int e^x \cos x \, dx.$$

$$u = e^x, \qquad v' = \cos x,$$
$$u' = e^x, \qquad v = \sin x;$$

$$\int e^x \sin x \, dx = -e^x \cos x + e^x \sin x - \int e^x \sin x \, dx.$$

Now let us add $\int e^x \sin x \, dx$ to both sides.

$$2 \int e^x \sin x \, dx = e^x(\sin x - \cos x)$$

$$\int e^x \sin x \, dx = \frac{1}{2} e^x(\sin x - \cos x) + C.$$

Example 8

Evaluate $\int \sec^3 \theta \, d\theta.$

$$u = \sec \theta, \qquad v' = \sec^2 \theta,$$
$$u' = \sec \theta \tan \theta, \quad v = \tan \theta;$$

$$\int \sec^3 \theta \, d\theta = \sec \theta \tan \theta - \int \sec \theta \tan^2 \theta \, d\theta$$

$$= \sec \theta \tan \theta - \int \sec \theta \, (\sec^2 \theta - 1) \, d\theta$$

$$= \sec \theta \tan \theta + \int \sec \theta \, d\theta - \int \sec^3 \theta \, d\theta,$$

$$2 \int \sec^3 \theta \, d\theta = \sec \theta \tan \theta + \ln \, |\sec \theta + \tan \theta|$$

$$\int \sec^3 \theta \, d\theta = \frac{1}{2} \, (\sec \theta \tan \theta + \ln \, |\sec \theta + \tan \theta|) + C.$$

Problems

A

Evaluate the integrals.

1. $\int x \cos x \, dx.$

2. $\int \theta \sin 5\theta \, d\theta.$

3. $\int x e^x \, dx.$

4. $\int 3x e^{-x} \, dx.$

5. $\int \ln x^2 \, dx.$

6. $\int \ln 5x \, dx.$

7. $\displaystyle\int \frac{\ln x}{x^2}\, dx.$ 8. $\displaystyle\int \frac{\ln (x + 2)}{\sqrt{x + 2}}\, dx.$ 9. $\displaystyle\int \text{Arcsin } x\, dx.$

10. $\displaystyle\int \text{Arccos } 3x\, dx.$ 11. $\displaystyle\int \text{Arctan } 2x\, dx.$ 12. $\displaystyle\int \cos \theta \ln \sin \theta\, d\theta.$

13. $\displaystyle\int \sinh^{-1} x\, dx.$ 14. $\displaystyle\int \cosh^{-1} x\, dx.$ 15. $\displaystyle\int x^3 \sqrt{9 - x^2}\, dx.$

16. $\displaystyle\int \frac{x^3}{\sqrt{9 - x^2}}\, dx.$ 17. $\displaystyle\int x^5 \sqrt{x^3 + 3}\, dx.$ 18. $\displaystyle\int x \sec^2 x\, dx.$

B

Evaluate the integrals.

19. $\displaystyle\int \text{Arccot } \sqrt{x}\, dx.$ 20. $\displaystyle\int x \text{ Arcsin } x\, dx.$ 21. $\displaystyle\int x^{2n-1} \sqrt{x^n + 3}\, dx.$

22. $\displaystyle\int \sin \ln x\, dx.$ 23. $\displaystyle\int \frac{\ln x\, dx}{(x - 1)^2}.$ 24. $\displaystyle\int x^2 \sin x\, dx.$

25. $\displaystyle\int x^4 \sin x\, dx.$ 26. $\displaystyle\int x^3 \cos 2x\, dx.$ 27. $\displaystyle\int x^3 e^{3x}\, dx.$

28. $\displaystyle\int x^4 e^{-x}\, dx.$ 29. $\displaystyle\int x^3 \ln^2 x\, dx.$ 30. $\displaystyle\int x^n e^x\, dx$ (n a positive integer).

31. $\displaystyle\int e^x \cos x\, dx.$ 32. $\displaystyle\int \csc^3 \theta\, d\theta.$ 33. $\displaystyle\int \sec^5 \theta\, d\theta.$

34. $\displaystyle\int \sec^n \theta\, d\theta.$ 35. $\displaystyle\int e^x \sin 3x\, dx.$ 36. $\displaystyle\int \sin x \cos 2x\, dx.$

37. $\displaystyle\int \sin x \sin 5x\, dx.$ 38. $\displaystyle\int \cos x \cos 3x\, dx.$

C

39. Use integration by parts to establish the recursion formula

$$\int \sin^n x\, dx = -\frac{\sin^{n-1} x \cos x}{n} + \frac{n - 1}{n} \int \sin^{n-2} x\, dx.$$

   Use this formula to evaluate $\displaystyle\int \sin^4 x\, dx.$

40. Use integration by parts to establish the recursion formula

$$\int \ln^n x\, dx = x \ln^n x - n \int \ln^{n-1} x\, dx.$$

   Use this formula to evaluate $\displaystyle\int \ln^3 x\, dx.$ Why would tabular integration not work here?

41. Evaluate $\int \sin^2 \theta\, d\theta$ by parts. Compare your answer with the one obtained in Section 11.3 using a double-angle formula and show that they are equivalent.

42. Evaluate $\int \cos^2 \theta\, d\theta$ by parts. Compare your answer with the one obtained in Section 11.3 using a double-angle formula and show that they are equivalent.

43. Evaluate the integral of Problem 37 by using the identity

$$\sin A \sin B = \frac{1}{2} \left[\cos (A - B) - \cos (A + B) \right].$$

44. Evaluate the integral of Problem 38 by using the identity

$$\cos A \cos B = \frac{1}{2} \left[\cos (A + B) + \cos (A - B) \right].$$

15.8

Partial Fractions: Linear Factors

This section and the next are devoted to a method of integrating rational fractions. In the past we have seen many problems in which we were to add two or more fractions,

$$\frac{1}{x + 1} + \frac{2}{x - 2} - \frac{4}{x},$$

to give a single fraction

$$\frac{- x^2 + 4x + 8}{x(x + 1)(x - 2)}.$$

Since the first expression above is easier to integrate than the single fraction, we should like to work this problem backward; that is, starting with

$$\frac{- x^2 + 4x + 8}{x(x + 1)(x - 2)},$$

we would like to break it down into the sum of the partial fractions,

$$\frac{1}{x + 1} + \frac{2}{x - 2} - \frac{4}{x}.$$

Once this is done, we can easily integrate term by term. We shall not attempt to go into the theory of partial fractions—this is basically an algebraic problem.† We shall merely summarize the results.

First of all, the numerator must be of lower degree than the denominator. If this is not the case, we divide until the remainder term is in the proper form. Next, the denominator is factored so that every factor is either a linear or a quadratic factor with real coefficients (this can always be done—but not always easily). Finally, this fraction can be broken down into partial fractions in a way that is dependent upon the factors of the denominator. There are four cases:

---

†See, for example, Garrett Birkhoff and Saunders MacLane, *A Survey of Modern Algebra*, 4th ed. (New York: Macmillan, 1977), pp. 90–93.

I. nonrepeated linear factors,
II. repeated linear factors,
III. nonrepeated quadratic factors,
IV. repeated quadratic factors.

We shall consider the first two cases in this section and the last two in the next section.

Case I Nonrepeated linear factors:

$$\frac{P(x)}{(a_1x + b_1)(a_2x + b_2) \cdots (a_nx + b_n)} = \frac{A_1}{a_1x + b_1} + \frac{A_2}{a_2x + b_2} + \cdots + \frac{A_n}{a_nx + b_n},$$

where $A_1, A_2, \ldots, A_n$ are constants.

Example 1

Evaluate $\displaystyle\int \frac{dx}{(x + 1)(x - 2)}$.

Let us break the fraction into partial fractions by the above rule.

$$\frac{1}{(x + 1)(x - 2)} = \frac{A}{x + 1} + \frac{B}{x - 2}.$$

Bear in mind that this equation (for the proper choices of A and B) is an identity; that is, it is true for all x for which both sides of the equation have meaning. This is an important consideration in determining A and B, which we shall now do. Multiplying both sides by $(x + 1)(x - 2)$, we have

$$1 = A(x - 2) + B(x + 1)$$
$$= (A + B)x + (-2A + B).$$

Since this is also an identity, the coefficients must be the same on both sides of the equation; that is, the coefficient of x on the right, $A + B$, must equal the coefficient of x on the left, 0; and the constant term on the right, $-2A + B$, must equal the constant term on the left, 1.

$$A + B = 0,$$
$$-2A + B = 1.$$

Solving simultaneously, we have

$$A = -\frac{1}{3} \quad \text{and} \quad B = \frac{1}{3}.$$

Thus

$$\frac{1}{(x + 1)(x - 2)} = \frac{-1/3}{x + 1} + \frac{1/3}{x - 2}$$

and

$$\int \frac{dx}{(x + 1)(x - 2)} = \int \left(-\frac{1}{3} \frac{1}{x + 1} + \frac{1}{3} \frac{1}{x - 2} \right) dx$$

$$= -\frac{1}{3} \ln |x + 1| + \frac{1}{3} \ln |x - 2| + C$$

$$= \frac{1}{3} \ln \left| \frac{x - 2}{x + 1} \right| + C.$$

We might note here that there is another method of determining the constants A and B. The equation

$$\frac{1}{(x+1)(x-2)} = \frac{A}{x+1} + \frac{B}{x-2}$$

is an identity for those values of A and B that make $1 = A(x-2) + B(x+1)$ an identity. Since this polynomial is an identity provided it is true for all values of x, we choose two particular values of x: namely, $x = 2$ and $x = -1$. For $x = 2$ we have

$$1 = 3B, \qquad B = \frac{1}{3};$$

for $x = -1$

$$1 = -3A, \qquad A = -\frac{1}{3}.$$

Moreover the computation of A and B in this way is simple enough to do in one's head, making this a very short method indeed. But while it is a very short method when we have only nonrepeated linear factors, it is not so useful when we have quadratic factors. In fact, it is often longer in those cases. For this reason we shall continue to use the method of equating coefficients on the two sides of an equation.

Example 2

Evaluate $\int \frac{x^3 + 1}{x^2 - x}\, dx.$

Since the degree of the numerator exceeds the degree of the denominator, we must divide before breaking it down into partial fractions.

$$\int \frac{x^3 + 1}{x^2 - x}\, dx = \int \left(x + 1 + \frac{x+1}{x^2 - x} \right) dx$$

$$= \int \left(x + 1 + \frac{x+1}{x(x-1)} \right) dx.$$

Using partial fractions, we have

$$\frac{x+1}{x(x-1)} = \frac{A}{x} + \frac{B}{x-1},$$

$$x + 1 = A(x-1) + Bx$$
$$= (A+B)x - A.$$

This gives

$$A + B = 1,$$
$$-A = 1.$$

Solving, we have

$$A = -1 \quad \text{and} \quad B = 2.$$

Thus

$$\int \frac{x^3+1}{x^2-x}\, dx = \int \left(x+1-\frac{1}{x}+\frac{2}{x-1}\right) dx$$

$$= \frac{x^2}{2}+x-\ln|x|+2\ln|x-1|+C$$

$$= \frac{x^2}{2}+x+\ln\left|\frac{(x-1)^2}{x}\right|+C.$$

Case II   Repeated linear factors:

$$\frac{P(x)}{(ax+b)^n} = \frac{A_1}{ax+b}+\frac{A_2}{(ax+b)^2}+\cdots+\frac{A_n}{(ax+b)^n},$$

where $A_1, A_2, \ldots, A_n$ are constants.

Example 3

Evaluate $\displaystyle\int \frac{x\,dx}{(x-2)^3}.$

$$\frac{x}{(x-2)^3} = \frac{A}{x-2}+\frac{B}{(x-2)^2}+\frac{C}{(x-2)^3}.$$

Multiplying by $(x-2)^3$, we have

$$x = A(x-2)^2+B(x-2)+C$$
$$= Ax^2+(-4A+B)x+(4A-2B+C).$$

Equating the coefficients, we have

$$A=0,$$
$$-4A+B=1,$$
$$4A-2B+C=0,$$

or

$$A=0, \qquad B=1, \qquad C=2.$$

Thus

$$\int \frac{x\,dx}{(x-2)^3} = \int \left(\frac{1}{(x-2)^2}+\frac{2}{(x-2)^3}\right) dx$$

$$= \frac{-1}{x-2}-\frac{1}{(x-2)^2}+K$$

$$= \frac{1-x}{(x-2)^2}+K,$$

where K is the constant of integration.

Example 4

Evaluate $\int \dfrac{x-2}{x^2(x-1)^2}\,dx$.

$$\frac{x-2}{x^2(x-1)^2} = \frac{A}{x} + \frac{B}{x^2} + \frac{C}{x-1} + \frac{D}{(x-1)^2},$$

$$x-2 = Ax(x-1)^2 + B(x-1)^2 + Cx^2(x-1) + Dx^2$$
$$= (A+C)x^3 + (-2A+B-C+D)x^2 + (A-2B)x + B;$$

$$A + C = 0,$$
$$-2A + B - C + D = 0,$$
$$A - 2B = 1,$$
$$B = -2.$$

Solving simultaneously, we have

$$A = -3, \qquad B = -2, \qquad C = 3, \qquad D = -1.$$

Thus

$$\int \frac{x-2}{x^2(x-1)^2}\,dx = \int \left(-\frac{3}{x} - \frac{2}{x^2} + \frac{3}{x-1} - \frac{1}{(x-1)^2}\right)dx$$

$$= -3\ln|x| + \frac{2}{x} + 3\ln|x-1| + \frac{1}{x-1} + K$$

$$= 3\ln\left|\frac{x-1}{x}\right| + \frac{3x-2}{x(x-1)} + K.$$

Problems

A
Evaluate the integrals.

1. $\displaystyle\int \frac{dx}{x(x+1)}.$

2. $\displaystyle\int \frac{dx}{x(x-3)}.$

3. $\displaystyle\int \frac{x\,dx}{(x+2)(x-1)}.$

4. $\displaystyle\int \frac{dx}{x^2-4x-5}.$

5. $\displaystyle\int \frac{x^2+3}{x^2+3x}\,dx.$

6. $\displaystyle\int \frac{x^3-4}{x^2-4x}\,dx.$

7. $\displaystyle\int \frac{dx}{x^3-x}.$

8. $\displaystyle\int \frac{dx}{(x+1)(x+2)(x+3)}.$

9. $\displaystyle\int \frac{(x^3+8)\,dx}{x^3-3x^2+2x}.$

10. $\displaystyle\int \frac{x^3+9}{9x^3-4x}\,dx.$

11. $\displaystyle\int \frac{x\,dx}{(x+2)^2}.$

12. $\displaystyle\int \frac{x\,dx}{(x+1)^2}.$

13. $\displaystyle\int \frac{x^2\,dx}{(x-1)^2}.$

14. $\displaystyle\int \frac{x+4}{x(x-2)^2}\,dx.$

15. $\displaystyle\int \frac{dx}{x^3-x^2}.$

B
Evaluate the integrals.

16. $\displaystyle\int \frac{dx}{x^2(x-1)^2}.$

17. $\displaystyle\int \frac{(x+2)\,dx}{x^4+2x^3-3x^2}.$

18. $\displaystyle\int \frac{dx}{(x-1)^2(x-2)^3}.$

19. $\displaystyle\int \frac{dx}{x(x-1)^2(x+1)^2}.$ 20. $\displaystyle\int \frac{dx}{(2x+1)^2(x-2)^2}.$ 21. $\displaystyle\int \frac{\cos x \, dx}{\sin^3 x + \sin^2 x}.$
$\qquad\qquad\qquad\qquad\qquad\qquad\qquad\qquad\qquad\qquad\qquad\qquad$ (*Hint:* Let $u = \sin x$.)

22. $\displaystyle\int \frac{\cos x \, dx}{\sin^2 x - 2\sin x - 3}.$

15.9

Partial Fractions: Quadratic Factors

Case III *Nonrepeated quadratic factors:*

$$\frac{P(x)}{(a_1 x^2 + b_1 x + c_1)\cdots(a_n x^2 + b_n x + c_n)}$$

$$= \frac{A_1 x + B_1}{a_1 x^2 + b_1 x + c_1} + \frac{A_2 x + B_2}{a_2 x^2 + b_2 x + c_2} + \cdots + \frac{A_n x + B_n}{a_n x^2 + b_n x + c_n}.$$

Example 1

Evaluate $\displaystyle\int \frac{dx}{x^3 + x}.$

The denominator can be factored to

$$x(x^2 + 1),$$

but the quadratic factor cannot be factored further without using complex coefficients. Thus,

$$\frac{1}{x(x^2+1)} = \frac{A}{x} + \frac{Bx+C}{x^2+1}.$$

The constants are now evaluated exactly as in the previous section.

$$1 = A(x^2 + 1) + (Bx + C)x$$
$$= (A + B)x^2 + Cx + A;$$

$$A + B = 0,$$

$$C = 0,$$

$$A = 1.$$

This gives

$$A = 1, \quad B = -1, \quad \text{and} \quad C = 0.$$

Thus

$$\int \frac{dx}{x^3 + x} = \int \left(\frac{1}{x} - \frac{x}{x^2 + 1}\right) dx$$

$$= \ln |x| - \frac{1}{2} \ln (x^2 + 1) + K$$

$$= \ln \frac{|x|}{\sqrt{x^2 + 1}} + K.$$

Example 2

Evaluate $\int \dfrac{dx}{x^4 + 5x^2 + 4}.$

Again the denominator can be factored to

$$(x^2 + 1)(x^2 + 4),$$

but neither factor can be factored further without using complex coefficients. Thus

$$\frac{1}{(x^2 + 1)(x^2 + 4)} = \frac{Ax + B}{x^2 + 1} + \frac{Cx + D}{x^2 + 4},$$

$$1 = (Ax + B)(x^2 + 4) + (Cx + D)(x^2 + 1)$$
$$= (A + C)x^3 + (B + D)x^2 + (4A + C)x + (4B + D);$$

$$A + C = 0,$$

$$B + D = 0,$$

$$4A + C = 0,$$

$$4B + D = 1.$$

Solving simultaneously, we have

$$A = 0, \quad B = 1/3, \quad C = 0, \quad \text{and} \quad D = -1/3.$$

Thus

$$\int \frac{dx}{x^4 + 5x^2 + 4} = \int \left(\frac{1/3}{x^2 + 1} - \frac{1/3}{x^2 + 4}\right) dx$$

$$= \int \left(\frac{1}{3}\frac{1}{x^2 + 1} - \frac{1}{6}\frac{1/2}{(x/2)^2 + 1}\right) dx$$

$$= \frac{1}{3} \text{Arctan } x - \frac{1}{6} \text{Arctan } \frac{x}{2} + K.$$

Case IV   Repeated quadratic factors:

$$\frac{P(x)}{(ax^2 + bx + c)^n} = \frac{A_1 x + B_1}{ax^2 + bx + c} + \frac{A_2 x + B_2}{(ax^2 + bx + c)^2} + \cdots + \frac{A_n x + B_n}{(ax^2 + bx + c)^n}.$$

Example 3

Evaluate $\int \dfrac{(x-1)^2}{(x^2+1)^2}\, dx$.

$$\frac{(x-1)^2}{(x^2+1)^2} = \frac{Ax+B}{x^2+1} + \frac{Cx+D}{(x^2+1)^2},$$

$$
\begin{aligned}
x^2 - 2x + 1 &= Ax^3 + Bx^2 + Ax + B + Cx + D \\
&= Ax^3 + Bx^2 + (A+C)x + (B+D).
\end{aligned}
$$

Equating coefficients, we have

$$
\begin{aligned}
A &= 0, \\
B &= 1, \\
A + C &= -2, \\
B + D &= 1,
\end{aligned}
$$

or

$$A = 0, \quad B = 1, \quad C = -2, \quad \text{and} \quad D = 0.$$

Thus

$$\int \frac{(x-1)^2}{(x^2+1)^2}\, dx = \int \left(\frac{1}{x^2+1} - \frac{2x}{(x^2+1)^2} \right) dx$$

$$= \text{Arctan}\, x + \frac{1}{x^2+1} + K.$$

Actually the use of partial fractions could have been avoided in this case by the following observation.

$$\frac{(x-1)^2}{(x^2+1)^2} = \frac{x^2-2x+1}{(x^2+1)^2} = \frac{x^2+1}{(x^2+1)^2} - \frac{2x}{(x^2+1)^2} = \frac{1}{x^2+1} - \frac{2x}{(x^2+1)^2}$$

While tricks like this can sometimes shorten our work, they cannot be depended upon to handle more than a few special cases. Nevertheless, it may be advantageous to take a moment to look for such situations.

Example 4

Evaluate $\int \dfrac{x^4+x^3+8x^2+16}{x(x^2+4)^2}\, dx$.

$$\frac{x^4+x^3+8x^2+16}{x(x^2+4)^2} = \frac{A}{x} + \frac{Bx+C}{x^2+4} + \frac{Dx+E}{(x^2+4)^2},$$

$$
\begin{aligned}
x^4+x^3+8x^2+16 &= A(x^2+4)^2 + (Bx+C)x(x^2+4) + (Dx+E)x \\
&= (A+B)x^4 + Cx^3 + (8A+4B+D)x^2 + (4C+E)x + 16A.
\end{aligned}
$$

Equating coefficients, we have

$$
\begin{aligned}
A + B &= 1, \\
C &= 1, \\
8A + 4B + D &= 8, \\
4C + E &= 0, \\
16A &= 16,
\end{aligned}
$$

or

$$A = 1, \quad B = 0, \quad C = 1, \quad D = 0, \quad \text{and} \quad E = -4.$$

Thus

$$\int \frac{x^4 + x^3 + 8x^2 + 16}{x(x^2 + 4)^2} \, dx = \int \left(\frac{1}{x} + \frac{1}{x^2 + 4} - \frac{4}{(x^2 + 4)^2} \right) dx.$$

Integrating term by term, we have

$$\int \frac{1}{x} \, dx = \ln |x| + K_1,$$

$$\int \frac{dx}{x^2 + 4} = \int \frac{1/4}{x^2/4 + 1} \, dx = \frac{1}{2} \int \frac{1/2}{(x/2)^2 + 1} \, dx = \frac{1}{2} \text{Arctan} \frac{x}{2} + K_2.$$

The third term requires a substitution.

$$x = 2 \tan \theta, \qquad dx = 2 \sec^2 \theta \, d\theta.$$

$$\int \frac{-4}{(x^2 + 4)^2} \, dx = \int \frac{-4 \cdot 2 \sec^2 \theta \, d\theta}{(4 \tan^2 \theta + 4)^2}$$

$$= \int \frac{-8 \sec^2 \theta \, d\theta}{16 \sec^4 \theta}$$

$$= -\frac{1}{2} \int \frac{d\theta}{\sec^2 \theta}$$

$$= -\frac{1}{2} \int \cos^2 \theta \, d\theta$$

$$= -\frac{1}{4} \int (1 + \cos 2\theta) \, d\theta$$

$$= -\frac{1}{4} \left(\theta + \frac{1}{2} \sin 2\theta \right) + K_3$$

$$= -\frac{1}{4} (\theta + \sin \theta \cos \theta) + K_3$$

$$= -\frac{1}{4} \text{Arctan} \frac{x}{2} - \frac{1}{4} \frac{x}{\sqrt{x^2 + 4}} \frac{2}{\sqrt{x^2 + 4}} + K_3$$

$$= -\frac{1}{4} \text{Arctan} \frac{x}{2} - \frac{x}{2(x^2 + 4)} + K_3.$$

Putting it all together, we have

$$\int \frac{x^4 + x^3 + 8x^2 + 16}{x(x^2 + 4)^2} \, dx = \ln |x| + \frac{1}{2} \text{Arctan} \frac{x}{2}$$

$$- \frac{1}{4} \text{Arctan} \frac{x}{2} - \frac{x}{2(x^2 + 4)} + K$$

$$= \ln |x| + \frac{1}{4} \text{Arctan} \frac{x}{2} - \frac{x}{2(x^2 + 4)} + K.$$

Problems

A

Evaluate the integrals.

1. $\displaystyle\int \frac{dx}{x^2(x^2 + 1)}.$

2. $\displaystyle\int \frac{8\,dx}{(x + 2)(x^2 + 4)}.$

3. $\displaystyle\int \frac{(3x + 3)\,dx}{x^3 - 1}.$

4. $\displaystyle\int \frac{4x\,dx}{x^4 - 1}.$

5. $\displaystyle\int \frac{3\,dx}{x^3 + x^2 + x + 1}.$

6. $\displaystyle\int \frac{(x - 1)\,dx}{x^4 + x^2}.$

7. $\displaystyle\int \frac{dx}{(x^2 + 4)(x + 2)^2}.$

8. $\displaystyle\int \frac{(x^2 + 1)\,dx}{x^3 + x^2 + x}.$

9. $\displaystyle\int \frac{2x\,dx}{(x + 1)(x^2 + 1)}.$

10. $\displaystyle\int \frac{x^3 + 8}{x(x^2 + 4)}\,dx.$

B

Evaluate the integrals.

11. $\displaystyle\int \frac{(x^2 - 2x + 2)\,dx}{(x^2 + 1)^2}.$

12. $\displaystyle\int \frac{x^3\,dx}{(x^2 + 4)^3}.$

13. $\displaystyle\int \frac{16\,dx}{x(x^2 + 4)^2}.$

14. $\displaystyle\int \frac{9(x^2 + 9x + 9)}{x^2(x^2 + 9)^2}.$

15. $\displaystyle\int \frac{x^4 + 4x^2 + 16}{(x^3 - 8)^2}\,dx.$

16. $\displaystyle\int \frac{2(x^2 + 2x + 2)}{x^2(x^2 + 2)^2}.$

17. $\displaystyle\int \frac{3x^2 + 9x + 12}{(x^2 + 1)(x^2 + 4)^2}\,dx.$

18. $\displaystyle\int \frac{50x\,dx}{(x^2 - 1)(x^2 + 4)^2}.$

19. $\displaystyle\int \frac{dx}{x(x^2 + 2x + 2)}.$

20. $\displaystyle\int \frac{16\,dx}{(x^4 - 1)^2}.$

15.10

Miscellaneous Substitutions

If none of the previous methods works, there are still a few substitutions that may be tried. One of these is the rationalizing substitution. We have already seen this type of substitution, on simple problems, in Section 15.2. Let us consider some other examples of it.

Example 1

Evaluate $\displaystyle\int \frac{dx}{\sqrt{x} + \sqrt[3]{x}}.$

We can eliminate the square root by the substitution $u = \sqrt{x}$, or $x = u^2$, but then we still have the cube root to deal with. Similarly, we can eliminate the cube root by the substitution $u = \sqrt[3]{x}$ or $x = u^3$, but then we still have the square root to deal with. Instead of substituting $x = u^2$, so we can find its square root, or $x = u^3$, so we can find its cube root, we want a substitution that will allow us to find both the square root and the cube root without getting fractional exponents. Thus we want a substitution of the form $x = u^k$, where k is a multiple of both 2 and 3. Let us use the least common multiple, 6.

$$x = u^6, \qquad dx = 6u^5 \, du.$$

$$\int \frac{dx}{\sqrt{x} + \sqrt[3]{x}} = \int \frac{6u^5 \, du}{u^3 + u^2}$$

$$= \int \frac{6u^3 \, du}{u + 1}$$

$$= \int \left(6u^2 - 6u + 6 - \frac{6}{u+1} \right) du$$

$$= 2u^3 - 3u^2 + 6u - 6 \ln |u + 1| + C$$

$$= 2\sqrt{x} - 3\sqrt[3]{x} + 6\sqrt[6]{x} - 6 \ln (\sqrt[6]{x} + 1) + C.$$

Example 2

Evaluate $\displaystyle \int \frac{dx}{\sqrt{x-1} + (x-1)^{3/2}}.$

Let us use the substitution $u = \sqrt{x-1}$ or $x = u^2 + 1$.

$$dx = 2u \, du.$$

Then

$$\int \frac{dx}{\sqrt{x-1} + (x-1)^{3/2}} = \int \frac{2u \, du}{u + u^3}$$

$$= \int \frac{2 \, du}{u^2 + 1}$$

$$= 2 \operatorname{Arctan} u + C$$

$$= 2 \operatorname{Arctan} \sqrt{x-1} + C.$$

Another substitution helpful in evaluating trigonometric integrals is

$$u = \tan \frac{\theta}{2}.$$

This substitution allows us to evaluate certain difficult trigonometric integrals by rational fractions. In particular it allows us to reduce a rational fraction in $\sin \theta$ and $\cos \theta$ to one involving u. By a trigonometric identity,

$$\tan^2 \frac{\theta}{2} = \frac{1 - \cos \theta}{1 + \cos \theta}.$$

Replacing $\tan \theta/2$ by u and solving for $\cos \theta$, we have

$$\cos \theta = \frac{1 - u^2}{1 + u^2}.$$

Using the identity

$$\tan \frac{\theta}{2} = \frac{\sin \theta}{1 + \cos \theta}$$

and replacing $\tan \theta/2$ by u, and $\cos \theta$ by $(1 - u^2)/(1 + u^2)$ gives

$$\sin \theta = \left(1 + \frac{1 - u^2}{1 + u^2}\right)u, \quad \text{or} \quad \sin \theta = \frac{2u}{1 + u^2}.$$

Finally let us consider the relation between the differentials du and $d\theta$. From the original substitution, $u = \tan(\theta/2)$, we have

$$\theta = 2 \operatorname{Arctan} u.$$

Thus

$$d\theta = \frac{2\, du}{1 + u^2}.$$

Summing up, the substitution $u = \tan \theta/2$ gives

$$\sin \theta = \frac{2u}{1 + u^2}, \quad \cos \theta = \frac{1 - u^2}{1 + u^2}, \quad \text{and} \quad d\theta = \frac{2\, du}{1 + u^2}.$$

Now any trigonometric expression can be expressed in terms of u.

Example 3

Evaluate $\displaystyle \int \frac{d\theta}{3 - 5 \sin \theta}$.

Using the substitution $u = \tan(\theta/2)$, we have

$$\int \frac{d\theta}{3 - 5 \sin \theta} = \int \frac{\dfrac{2\, du}{1 + u^2}}{3 - 5 \dfrac{2u}{1 + u^2}}$$

$$= \int \frac{2\, du}{3u^2 - 10u + 3}$$

$$= \int \frac{2\, du}{(3u - 1)(u - 3)}.$$

By use of partial fractions, we get

$$\int \frac{d\theta}{3 - 5 \sin \theta} = \int \left(-\frac{3}{4}\frac{1}{3u - 1} + \frac{1}{4}\frac{1}{u - 3}\right) du$$

$$= -\frac{1}{4} \ln|3u - 1| + \frac{1}{4} \ln|u - 3| + C$$

$$= \frac{1}{4} \ln \left|\frac{u - 3}{3u - 1}\right| + C$$

$$= \frac{1}{4} \ln \left|\frac{\tan \dfrac{\theta}{2} - 3}{3 \tan \dfrac{\theta}{2} - 1}\right| + C.$$

Example 4

Evaluate $\displaystyle\int\frac{d\theta}{1+\sin\theta+\cos\theta}$.

Again using the substitution $u = \tan(\theta/2)$, we have

$$\int\frac{d\theta}{1+\sin\theta+\cos\theta} = \int\frac{\dfrac{2\,du}{1+u^2}}{1+\dfrac{2u}{1+u^2}+\dfrac{1-u^2}{1+u^2}}$$

$$= \int\frac{2\,du}{2u+2}$$

$$= \int\frac{du}{u+1}$$

$$= \ln|u+1| + C$$

$$= \ln\left|\tan\frac{\theta}{2}+1\right| + C.$$

Problems

A

Evaluate the integrals.

1. $\displaystyle\int\frac{x\,dx}{x+2+\sqrt{x+2}}$.

2. $\displaystyle\int\frac{dx}{x+2\sqrt{x-2}}$.

3. $\displaystyle\int\frac{\sqrt{x}\,dx}{2x+\sqrt{x}}$.

4. $\displaystyle\int\frac{dx}{\sqrt{x}-\sqrt[3]{x}}$.

5. $\displaystyle\int\frac{dx}{\sqrt{x}-\sqrt[4]{x}}$.

6. $\displaystyle\int\frac{dx}{x^{1/2}-x^{2/3}}$.

7. $\displaystyle\int\frac{dx}{x^{2/3}-x^{3/4}}$.

8. $\displaystyle\int\frac{\sqrt{x}-1}{\sqrt{x}+1}\,dx$.

9. $\displaystyle\int\frac{\sqrt[3]{x}+1}{\sqrt[3]{x}-1}\,dx$.

10. $\displaystyle\int\frac{dx}{\sin x+\cos x+2}$.

11. $\displaystyle\int\frac{d\theta}{\sin\theta-\cos\theta+1}$.

12. $\displaystyle\int\frac{x^3}{\sqrt{x^2+1}}\,dx$.

13. $\displaystyle\int\frac{dx}{(x+2)^2+\sqrt{x+2}}$.

B

Evaluate the integrals.

14. $\displaystyle\int\frac{\sqrt{2x+1}}{x-2}\,dx$.

15. $\displaystyle\int\frac{\sqrt[3]{2x+1}}{x}\,dx$.

16. $\displaystyle\int\frac{d\theta}{\sin\theta-\cos\theta+2}$.

17. $\displaystyle\int\frac{d\theta}{3\cos\theta-2\sin\theta+1}$.

18. $\displaystyle\int\frac{dx}{\cos x-3\sin x-9}$.

19. $\displaystyle\int\frac{d\theta}{\sin\theta+\cos\theta}$.

20. $\displaystyle\int\frac{\sqrt{1-x^2}}{x}\,dx$.

21. $\displaystyle\int x^3\sqrt{x^2-4}\,dx$.

C

22.   Evaluate $\displaystyle\int \frac{\sqrt{x^2 + 2x + 2}}{x + 1}\, dx.$

23.   Show that

$$\int \frac{d\theta}{(1 - \cos\theta)^2} = -\frac{1 + 3\tan^2(\theta/2)}{6\tan^3(\theta/2)} + C.$$

This result was used in Problem 26, page 444.

15.11

Tables of Integrals

Extensive tables of integrals are available. For example, *Standard Mathematical Tables*, 22nd ed., by Samuel M. Selby (Cleveland: Chemical Rubber Co., 1974) lists almost 700 integrals. You will find a short table of integrals inside the cover of this book. A table of integrals is often useful, in conjunction with the techniques developed in this chapter, in evaluating a given integral.

Example 1

Evaluate $\displaystyle\int \frac{\sqrt{4x^2 - 9}}{x}\, dx.$

$$\int \frac{\sqrt{4x^2 - 9}}{x}\, dx = \int \frac{\sqrt{(2x)^2 - 3^2}}{2x} \cdot 2\, dx.$$

With $u = 2x$ and $a = 3$, this is of the form

$$\int \frac{\sqrt{u^2 - a^2}}{u}\, du,$$

so by formula 47,

$$\int \frac{\sqrt{4x^2 - 9}}{x}\, dx = \sqrt{4x^2 - 9} - 3\,\operatorname{Sec}^{-1}\frac{2x}{3} + C.$$

Example 2

Evaluate $\displaystyle\int \sin^4 x \cos^2 x\, dx.$

With $m = 4$ and $n = 2$ in formula 99,

$$\int \sin^4 x \cos^2 x\, dx = -\frac{\sin^3 x \cos^3 x}{4 + 2} + \frac{4 - 1}{4 + 2}\int \sin^2 x \cos^2 x\, dx$$

$$= -\frac{1}{6}\sin^3 x \cos^3 x + \frac{1}{2}\int \sin^2 x \cos^2 x\, dx.$$

Letting $m = 2$ and $n = 2$ in the right-hand integral, formula 99 gives

$$\int \sin^4 x \cos^2 x \, dx = -\frac{1}{6} \sin^3 x \cos^3 x + \frac{1}{2} \left(-\frac{\sin x \cos^3 x}{4} + \frac{1}{4} \int \cos^2 x \, dx \right)$$

$$= -\frac{1}{6} \sin^3 x \cos^3 x - \frac{1}{8} \sin x \cos^3 x$$

$$+ \frac{1}{8} \left(\frac{x}{2} + \frac{1}{4} \sin 2x \right) + C, \quad \text{by formula 79,}$$

$$= -\frac{1}{6} \sin^3 x \cos^3 x - \frac{1}{8} \sin x \cos^3 x + \frac{x}{16} + \frac{1}{32} \sin 2x + C.$$

This is an example of the use of a *reduction formula*; the exponent of the sine term was reduced by 2 each time.

Example 3

Evaluate $\displaystyle\int xe^x \cos x \, dx.$

Let us try an integration by parts with $u = x$, $u' = 1$, and $v' = e^x \cos x$. By formula 120,

$$v = \int e^x \cos x \, dx = \frac{e^x}{2} (\cos x + \sin x)$$

and

$$\int xe^x \cos x \, dx = \frac{xe^x}{2} (\cos x + \sin x) - \frac{1}{2} \int e^x (\cos x + \sin x) \, dx.$$

Making use of formulas 119 and 120,

$$\int xe^x \cos x \, dx = \frac{xe^x}{2} (\cos x + \sin x)$$

$$- \frac{1}{2} \left[\frac{e^x}{2} (\cos x + \sin x) + \frac{e^x}{2} (\sin x - \cos x) \right] + C$$

$$= \frac{e^x}{2} [x \cos x + (x - 1) \sin x] + C.$$

Example 4

Evaluate $\displaystyle\int \frac{x^4 + x^3 + 8x^2 + 16}{x(x^2 + 4)^2} \, dx.$

This is the same as Example 4, Section 15.9. Using partial fractions,

$$\int \frac{x^4 + x^3 + 8x^2 + 16}{x(x^2 + 4)^2} \, dx = \int \left[\frac{1}{x} + \frac{1}{x^2 + 4} - \frac{4}{(x^2 + 4)^2} \right] dx.$$

$$\int \frac{dx}{x} = \ln |x| + K_1;$$

$$\int \frac{dx}{x^2 + 4} = \frac{1}{2} \operatorname{Tan}^{-1} \frac{x}{2} + K_2, \quad \text{by formula 18;}$$

$$- 4 \int \frac{dx}{(x^2 + 4)^2} = -\frac{1}{4} \operatorname{Arctan} \frac{x}{2} - \frac{x}{2(x^2 + 4)} + K_3,$$

by formula 32 (with $n = 2$), followed by formula 18;

so

$$\int \frac{x^4 + x^3 + 8x^2 + 16}{x(x^2 + 4)^2} \, dx = \ln |x| + \frac{1}{4} \operatorname{Arctan} \frac{x}{2} - \frac{x}{2(x^2 + 4)} + K.$$

It is certainly clear by this time that integration is considerably more difficult than differentiation. Moreover the methods of this chapter will not allow us to integrate all functions. In fact, there are some functions that cannot be integrated in finite terms; that is, there is no expression involving any finite combination of the algebraic or the transcendental functions we have considered that has one of these functions for derivative. For instance, none of the following can be integrated in finite terms:

$$\int \sin x^2 \, dx, \qquad \int e^{-x^2} \, dx, \qquad \int \frac{dx}{\ln x},$$

$$\int x \tan x \, dx, \qquad \int \frac{\sin x}{x} \, dx, \qquad \int \frac{dx}{\sqrt{1 + x^3}}.$$

We shall consider integrals of this type in Chapter 19, which deals with infinite series. Of course, when appropriate limits of integration are given, these integrals can be approximated as closely as we wish by, say, Simpson's rule or the trapezoidal rule.

Problems

A

Evaluate the integrals. Use tables where convenient.

1. $\displaystyle \int \frac{x \, dx}{\sqrt{2x + 5}}.$

2. $\displaystyle \int \frac{dx}{x^2(3x - 2)}.$

3. $\displaystyle \int \frac{x \, dx}{x^4 \sqrt{9 + x^4}}.$

4. $\displaystyle \int \frac{\sqrt{4 + 9x^2}}{x^2} \, dx.$

5. $\displaystyle \int e^{-6x} \sin 8x \, dx.$

6. $\displaystyle \int e^{2x} \cos 5x \, dx.$

7. $\displaystyle \int \frac{x^2 \, dx}{\sqrt{6x - x^2}}.$

8. $\displaystyle \int \frac{x \, dx}{\sqrt{8x - 4x^2}}.$

B

Evaluate the integrals. Use tables where convenient.

9. $\displaystyle \int \cos^6 3x \, dx.$

10. $\displaystyle \int \sin^6 3x \, dx.$

11. $\displaystyle \int \frac{x^3 \, dx}{\sqrt{3 + 2x}}.$

12. $\displaystyle \int \frac{dx}{x^2\sqrt{2 + 3x}}.$

13. $\displaystyle \int x^2 \cos x \, dx.$

14. $\displaystyle \int \sin^2 x \cos^4 x \, dx.$

15. $\displaystyle \int x^2 \operatorname{Sin}^{-1} x \, dx.$

16. $\displaystyle \int x^2 \operatorname{Arctan} x \, dx.$

17. $\displaystyle \int \frac{dx}{\sqrt{4x^2 - 8x - 21}}.$

18. $\displaystyle \int (8 - 2x - x^2)^{3/2} \, dx.$

C

19. Example 3 is a special case of the formula

$$\int xv'(x) \, dx = xv(x) - \int v(x) \, dx.$$

Let $v_1(x) = \int v(x) \, dx,\ v_2(x) = \int v_1(x) \, dx, \ldots, v_{k+1}(x) = \int v_k(x) \, dx, \ldots, v_{n+1}(x) = \int v_n(x) \, dx,$ and let $P(x)$ be a polynomial of degree n with leading coefficient 1. Show that

$$\int P(x)v(x) \, dx = P(x)v_1(x) - P'(x)v_2(x) + P''(x)v_3(x) - + \ldots + (-1)^n n! v_{n+1}(x).$$

See Examples 5 and 6, Section 15.7.

20. Use the method of Problem 19 to show that

$$\int (x^2 - 2x + 6)e^{4x}\, dx = \frac{8x^2 - 20x + 53}{32} e^{4x} + C.$$

21. Use the method of Problem 19 to evaluate

$$\int (x^2 - 2x + 6) \sin 2x\, dx.$$

Review Problems

A
Evaluate the integrals.

1. $\displaystyle\int (1 + \tan 2\theta)^2\, d\theta.$

2. $\displaystyle\int (\sin \theta + \cos \theta)^3\, d\theta.$

3. $\displaystyle\int \sec^6 x \tan^3 x\, dx.$

4. $\displaystyle\int xe^{-5x}\, dx.$

5. $\displaystyle\int_{\pi/4}^{\pi/2} (\csc x + 1)(\cot x - 1)\, dx.$

6. $\displaystyle\int_0^{\pi/3} (\sin \theta + 2 \cos \theta)(2 \sin \theta - \cos \theta)\, d\theta.$

7. $\displaystyle\int \sec^4 \theta \tan^{5/2} \theta\, d\theta.$

8. $\displaystyle\int \csc^3 3x \cot^3 3x\, dx.$

9. $\displaystyle\int x^2 \operatorname{Arctan} x\, dx.$

10. $\displaystyle\int \frac{3x\, dx}{x^2 + x - 2}.$

11. $\displaystyle\int \frac{(\sec x + \tan x)^2}{\sec^2 x}\, dx.$

12. $\displaystyle\int \frac{-16\, dx}{x^3 - 5x^2}.$

13. $\displaystyle\int_1^2 \frac{dx}{2x^2 - 6x + 5}.$

14. $\displaystyle\int \frac{e^{-x} - 1}{e^{-x} + 1}\, dx.$

15. $\displaystyle\int \frac{x\, dx}{(x - 4)^{3/2}}.$

16. $\displaystyle\int \cot^4 3x\, dx.$

17. $\displaystyle\int_1^2 x\sqrt{3x - 2}\, dx.$

18. $\displaystyle\int \frac{dx}{\sqrt{4x^2 - 1}}.$

19. $\displaystyle\int \frac{\ln x}{x^3}\, dx.$

20. $\displaystyle\int \frac{x^2\, dx}{x^2 - 5x + 6}.$

21. $\displaystyle\int \frac{1 + \tan^2 x}{1 + \tan x}\, dx.$

22. $\displaystyle\int x\sqrt[3]{x - 6}\, dx.$

B
Evaluate the integrals.

23. $\displaystyle\int_1^{1 + \sqrt{3}} (2 + 2x - x^2)^{3/2}\, dx.$

24. $\displaystyle\int x^2\sqrt{4x + 1}\, dx.$

25. $\displaystyle\int \frac{dx}{2\sqrt{x} - 3\sqrt[3]{x}}.$

26. $\displaystyle\int_0^4 \frac{x + 2}{\sqrt{2x + 1}}\, dx.$

27. $\displaystyle\int_0^{2/3} \frac{x^3\,dx}{\sqrt{9x^2+4}}$.

28. $\displaystyle\int \frac{dx}{(x-2)\sqrt{x^2-4x+3}}$.

29. $\displaystyle\int \frac{-16\,dx}{x^2(x-2)^3}$.

30. $\displaystyle\int \frac{4\,dx}{x^{2/3}-x^{2/5}}$.

31. $\displaystyle\int \sin 2x \cos 3x\,dx$.

32. $\displaystyle\int \frac{(x^2+3x)\,dx}{x^3+x^2+x+1}$.

33. $\displaystyle\int \frac{(x+1)^2\,dx}{1-4x-x^2}$.

34. $\displaystyle\int \frac{(9x-81)\,dx}{x^5+9x^3}$.

35. $\displaystyle\int \frac{dx}{(5-x^2)^2}$.

36. $\displaystyle\int \frac{4\sin\theta\,d\theta}{3+\sin^2\theta}$.

37. $\displaystyle\int \frac{x^5+x^4+4x^3+8x^2+12x+16}{(x^2+4)^3}\,dx$.

38. $\displaystyle\int \frac{\sin\theta\,d\theta}{\sin^2\theta+4\cos^2\theta}$.

C

Evaluate the integrals.

39. $\displaystyle\int_0^1 (3x+x^3)e^x\,dx$.

40. $\displaystyle\int_{-3}^3 x^3\sqrt{25+x^2}\,dx$.

41. $\displaystyle\int_1^2 x^4 e^{2x}\,dx$.

42. Evaluate $\displaystyle\int e^{ax}\sin bx\,dx$, where a and b are nonzero constants.

43. Find the length of $y=3x^2$ from $x=0$ to $x=1$.

44. Show that $\displaystyle\int \tan^n x\,dx = \frac{\tan^{n-1}x}{n-1} - \int \tan^{n-2}x\,dx$, where $n \neq 1$.

16

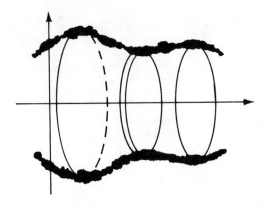

Further Applications of the Integral

16.1

Improper Integrals

We saw in Chapter 7 that the integral can be used to determine the amount of work done in a physical problem as well as to find the area of a plane region. Before going on to further applications, let us extend the definition of the integral to include infinite limits and infinite integrands.

Suppose we consider the integral

$$\int_1^k \frac{dx}{x^2},$$

where $k > 1$. This integral can be evaluated to give

$$\int_1^k \frac{dx}{x^2} = -\frac{1}{x} \Big|_1^k = 1 - \frac{1}{k}.$$

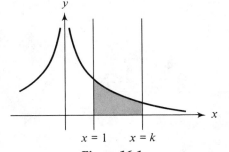

Figure 16.1

We now have the value of the integral for any value of k $(k > 1)$ we choose. Interpreting the integral geometrically, we see that it represents the area of the region bounded by $y = 1/x^2$, the x axis, and the vertical lines $x = 1$ and $x = k$ (see Figure 16.1). As k takes on larger and larger values, the area increases, getting closer and closer to the number 1; that is,

$$\lim_{k \to +\infty} \int_1^k \frac{dx}{x^2} = \lim_{k \to +\infty} \left(1 - \frac{1}{k}\right) = 1.$$

We shall represent this limit by

$$\int_{1}^{+\infty} \frac{dx}{x^2}$$

and take it to be the area of the region between $y = 1/x^2$ and $y = 0$ and to the right of $x = 1$. More generally, we make the following definition.

Definition

$$\int_{a}^{+\infty} f(x)\,dx = \lim_{k \to +\infty} \int_{a}^{k} f(x)\,dx$$

and

$$\int_{-\infty}^{a} f(x)\,dx = \lim_{k \to -\infty} \int_{k}^{a} f(x)\,dx$$

*whenever these limits exist. If an integral of either type exists, it is said to **converge** to the value of the integral; if it does not exist, it is said to **diverge**. Both such integrals are called **improper integrals**.*

Example 1

Evaluate $\int_{-\infty}^{0} e^x\,dx.$

$$\int_{-\infty}^{0} e^x\,dx = \lim_{k \to -\infty} \int_{k}^{0} e^x\,dx$$

$$= \lim_{k \to -\infty} e^x \Big|_{k}^{0}$$

$$= \lim_{k \to -\infty} (1 - e^k)$$

$$= 1.$$

Example 2

Evaluate $\int_{1}^{+\infty} \frac{dx}{x}.$

$$\int_{1}^{+\infty} \frac{dx}{x} = \lim_{k \to +\infty} \int_{1}^{k} \frac{dx}{x}$$

$$= \lim_{k \to +\infty} \ln x \Big|_{1}^{k}$$

$$= \lim_{k \to +\infty} \ln k$$

$$= +\infty.$$

The given integral diverges.

Although the graphs of $y = 1/x$ and $y = 1/x^2$ are quite similar for x positive, we have seen that

$$\int_{1}^{+\infty} \frac{dx}{x^2} = 1, \quad \text{but} \quad \int_{1}^{+\infty} \frac{dx}{x} \quad \text{diverges.}$$

Why should this be so when both have the x axis for a horizontal asymptote? We can see that the mere fact that $f(x)$ approaches 0 as x approaches $+\infty$ is not enough to guarantee the existence of $\int_a^{+\infty} f(x)\, dx$. The difference is that $1/x^2$ is approaching zero faster than $1/x$. In this case, the difference is enough to give a value of 1 for one of the integrals, while the other diverges.

If f is a function such that $\lim_{x\to+\infty} f(x) = \pm\infty$, then the improper integral $\int_a^{+\infty} f(x)\, dx$ diverges for any value of a (see Problem 32). However it is possible for $\int_a^{+\infty} f(x)\, dx$ to converge when $\lim_{x\to+\infty} f(x)$ does not exist but is neither $+\infty$ nor $-\infty$ (see Problem 33).

Let us now turn to the integral

$$\int_\epsilon^1 \frac{dx}{\sqrt{x}}, \quad \text{where } 0 < \epsilon < 1.$$

Again, we can evaluate this integral for any allowable value of ϵ.

$$\int_\epsilon^1 \frac{dx}{\sqrt{x}} = 2\sqrt{x}\,\Big|_\epsilon^1$$

$$= 2 - 2\sqrt{\epsilon}.$$

Interpreted geometrically, the result represents the area of the region bounded by $y = 1/\sqrt{x}$, the x axis, and the vertical lines $x = \epsilon$ and $x = 1$ (see Figure 16.2). Again the area increases as ϵ approaches 0. In fact,

$$\lim_{\epsilon\to 0^+} \int_\epsilon^1 \frac{dx}{\sqrt{x}} = \lim_{\epsilon\to 0^+} (2 - 2\sqrt{\epsilon})$$

$$= 2.$$

This limit can be interpreted as the area of the region bounded by $y = 1/\sqrt{x}$, the x axis, and the vertical lines $x = 0$ and $x = 1$, or as

$$\int_0^1 \frac{dx}{\sqrt{x}}.$$

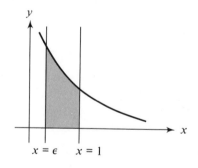

Figure 16.2

Definition

If f is continuous for $a < x \le b$ and if $x = a$ is a vertical asymptote, then

$$\int_a^b f(x)\, dx = \lim_{\epsilon\to 0^+} \int_{a+\epsilon}^b f(x)\, dx,$$

provided the limit exists. If f is continuous for $a \le x < b$ and $x = b$ is a vertical asymptote, then

$$\int_a^b f(x)\, dx = \lim_{\epsilon\to 0^+} \int_a^{b-\epsilon} f(x)\, dx,$$

provided the limit exists. If f is continuous for $a \leq x < c$ and $c < x \leq b$ and $x = c$ is a vertical asymptote, then

$$\int_a^b f(x)\, dx = \int_a^c f(x)\, dx + \int_c^b f(x)\, dx$$

$$= \lim_{\epsilon \to 0^+} \int_a^{c-\epsilon} f(x)\, dx + \lim_{\delta \to 0^+} \int_{c+\delta}^b f(x)\, dx,$$

provided both limits on the right exist. Again, an integral of one of the above types is called an **improper integral***; it is said to* **converge** *to its value when it exists and* **diverge** *when it does not.*

Similar definitions can be given when there are two or more (up to any finite number k) vertical asymptotes. Note that the definition of

$$\int_a^b f(x)\, dx$$

that was given in Chapter 7 (see page 192) considered bounded functions only.

Example 3

Evaluate $\int_0^1 \dfrac{dx}{x}$.

$$\int_0^1 \frac{dx}{x} = \lim_{\epsilon \to 0^+} \int_\epsilon^1 \frac{dx}{x}$$

$$= \lim_{\epsilon \to 0^+} \ln\, x \Big|_\epsilon^1$$

$$= \lim_{\epsilon \to 0^+} (-\ln\, \epsilon)$$

$$= +\infty.$$

Since the limit does not exist, the integral diverges.

It might be noted that we may have several vertical asymptotes or a vertical asymptote together with an infinite limit, as, for example,

$$\int_0^{+\infty} \frac{dx}{x}.$$

In such a case, it is convenient to break up the original integral into two new ones and handle the two cases separately. Thus

$$\int_0^{+\infty} \frac{dx}{x} = \int_0^1 \frac{dx}{x} + \int_1^{+\infty} \frac{dx}{x}.$$

We have already considered both of the integrals on the right-hand side (in Examples 2 and 3) and found that both diverge; thus $\int_0^{+\infty} dx/x$ diverges. The original integral converges provided *both* integrals on the right converge. It might be noted that, while we split the integrals at $x = 1$, this was quite an arbitrary choice—we could have used any positive number.

Example 4

Evaluate $\displaystyle\int_0^1 \frac{dx}{(x-1)^{2/3}}$.

$$\int_0^1 \frac{dx}{(x-1)^{2/3}} = \lim_{\epsilon \to 0^+} \int_0^{1-\epsilon} \frac{dx}{(x-1)^{2/3}}$$

$$= \lim_{\epsilon \to 0^+} 3(x-1)^{1/3} \Big|_0^{1-\epsilon}$$

$$= \lim_{\epsilon \to 0^+} (3 - 3\epsilon^{1/3})$$

$$= 3.$$

Note in this last example that we used the definition of an improper integral to give

$$\int_0^1 \frac{dx}{(x-1)^{2/3}} = \lim_{\epsilon \to 0^+} \int_0^{1-\epsilon} \frac{dx}{(x-1)^{2/3}}.$$

We could just as well have used

$$\int_0^1 \frac{dx}{(x-1)^{2/3}} = \lim_{\epsilon \to 0^-} \int_0^{1+\epsilon} \frac{dx}{(x-1)^{2/3}}.$$

In the first case, we have $1 - \epsilon < 1$, since $\epsilon > 0$; in the second, $1 + \epsilon < 1$, since $\epsilon < 0$.

Example 5

Evaluate $\displaystyle\int_0^3 \frac{dx}{(x-1)^2}$.

In this case the vertical asymptote does not correspond to either end point of the interval of integration but rather to some value of x ($x = 1$) between 0 and 3. Thus

$$\int_0^3 \frac{dx}{(x-1)^2} = \int_0^1 \frac{dx}{(x-1)^2} + \int_1^3 \frac{dx}{(x-1)^2}$$

$$= \lim_{\epsilon \to 0^+} \int_0^{1-\epsilon} \frac{dx}{(x-1)^2} + \lim_{\delta \to 0^+} \int_{1+\delta}^3 \frac{dx}{(x-1)^2}$$

$$= \lim_{\epsilon \to 0^+} \frac{-1}{x-1} \Big|_0^{1-\epsilon} + \lim_{\delta \to 0^+} \frac{-1}{x-1} \Big|_{1+\delta}^3$$

$$= \lim_{\epsilon \to 0^+} \left(\frac{1}{\epsilon} - 1 \right) + \lim_{\delta \to 0^+} \left(\frac{1}{\delta} - \frac{1}{2} \right).$$

Neither limit exists (they are both $+\infty$); therefore

$$\int_0^3 \frac{dx}{(x-1)^2}$$

does not exist.

Some care must be taken to find this vertical asymptote and to take it into consideration when evaluating the integral. If it had not been considered here, we would have had the following *erroneous* result.

$$\int_0^3 \frac{dx}{(x-1)^2} = \frac{-1}{x-1} \Big|_0^3 = -\frac{3}{2}, \quad \textit{which is false.}$$

Example 6

Evaluate $\int_{-1}^{1} \dfrac{dx}{x}$.

Again the vertical asymptote is at $x = 0$, which is between -1 and 1. Thus,

$$\int_{-1}^{1} \frac{dx}{x} = \int_{-1}^{0} \frac{dx}{x} + \int_{0}^{1} \frac{dx}{x}$$

$$= \lim_{\epsilon \to 0+} \int_{-1}^{-\epsilon} \frac{dx}{x} + \lim_{\delta \to 0+} \int_{\delta}^{1} \frac{dx}{x}$$

$$= \lim_{\epsilon \to 0+} \ln |x| \Big|_{-1}^{-\epsilon} + \lim_{\delta \to 0+} \ln |x| \Big|_{\delta}^{1}$$

$$= \lim_{\epsilon \to 0+} \ln \epsilon + \lim_{\delta \to 0+} (-\ln \delta).$$

Again, neither limit exists, since

$$\lim_{\epsilon \to 0+} \ln \epsilon = -\infty \quad \text{and} \quad \lim_{\delta \to 0+} (-\ln \delta) = +\infty.$$

Thus $\int_{-1}^{1} dx/x$ diverges.

One might be tempted to consider this last case in a somewhat different light than the previous example. Since one of the two limits is $+\infty$ and the other is $-\infty$, there is the temptation to say that the given integral converges to zero. In fact, one might argue further that since

$$\lim_{\delta \to 0+} \int_{\delta}^{1} \frac{dx}{x} = \lim_{\epsilon \to 0+} \int_{\epsilon}^{1} \frac{dx}{x},$$

it then follows that

$$\int_{-1}^{1} \frac{dx}{x} = \int_{-1}^{0} \frac{dx}{x} + \int_{0}^{1} \frac{dx}{x}$$

$$= \lim_{\epsilon \to 0+} \int_{-1}^{-\epsilon} \frac{dx}{x} + \lim_{\epsilon \to 0+} \int_{\epsilon}^{1} \frac{dx}{x}$$

$$= \lim_{\epsilon \to 0+} \left(\int_{-1}^{-\epsilon} \frac{dx}{x} + \int_{\epsilon}^{1} \frac{dx}{x} \right)$$

$$= \lim_{\epsilon \to 0+} [\ln \epsilon + (-\ln \epsilon)]$$

$$= \lim_{\epsilon \to 0+} 0$$

$$= 0.$$

This result is incorrect, because we are misapplying the statement

$$\lim_{x \to a} f(x) + \lim_{x \to a} g(x) = \lim_{x \to a} [f(x) + g(x)],$$

which is true, provided

$$\lim_{x \to a} f(x) \quad \text{and} \quad \lim_{x \to a} g(x)$$

both exist. In our case, neither

$$\lim_{\epsilon \to 0^+} \int_{-1}^{-\epsilon} \frac{dx}{x} \quad \text{nor} \quad \lim_{\epsilon \to 0^+} \int_{\epsilon}^{1} \frac{dx}{x}$$

exists; one is $+\infty$ and the other $-\infty$, both of which are special types of nonexistence.
   Remember that if

$$\int_a^b f(x)\, dx = \int_a^c f(x)\, dx + \int_c^b f(x)\, dx,$$

then, in order that $\int_a^b f(x)\, dx$ converge, *both integrals on the right must converge.* In this connection, it might also be noted that the results in Problem 21, page 197, hold if f is *continuous* on the interval $[-a, a]$; that is, $\int_{-a}^a f(x)\, dx = 0$ if f is odd and *continuous* on $[-a, a]$.
   However do not confuse the foregoing situation with

$$\int_a^{+\infty} [f(x) + g(x)]\, dx = \int_a^{+\infty} f(x)\, dx + \int_a^{+\infty} g(x)\, dx.$$

While it is true that the integral on the left converges whenever both integrals on the right converge, it is also possible for the integral on the left to converge when both of those on the right diverge. See Problem 31 in this connection.

Problems

A

In Problems 1–18, evaluate the given integrals.

1. $\displaystyle\int_2^{+\infty} \frac{dx}{x^3}.$

2. $\displaystyle\int_{-\infty}^{-1} \frac{dx}{x}.$

3. $\displaystyle\int_1^{+\infty} \frac{dx}{(x+2)^2}.$

4. $\displaystyle\int_3^{+\infty} \frac{dx}{(x-1)^3}.$

5. $\displaystyle\int_1^{+\infty} \frac{dx}{3x+1}.$

6. $\displaystyle\int_{-\infty}^0 \frac{dx}{(3x-2)^2}.$

7. $\displaystyle\int_1^{+\infty} \frac{dx}{\sqrt{3x-1}}.$

8. $\displaystyle\int_1^{+\infty} e^{-x}\, dx.$

9. $\displaystyle\int_2^3 \frac{dx}{\sqrt{x-2}}.$

10. $\displaystyle\int_{-2}^0 \frac{dx}{(x+2)^3}.$

11. $\displaystyle\int_{-1}^0 \frac{dx}{\sqrt[3]{2x+1}}.$

12. $\displaystyle\int_0^2 \frac{dx}{(3x-1)^{2/3}}.$

13. $\displaystyle\int_{-1}^{27} \frac{dx}{\sqrt[3]{x}}.$

14. $\displaystyle\int_0^{+\infty} \frac{dx}{\sqrt{x}}.$

15. $\displaystyle\int_0^{+\infty} \frac{dx}{x^2}.$

16. $\displaystyle\int_{-2}^2 \frac{dx}{x^2-4}.$

17. $\displaystyle\int_{-3}^6 \frac{dx}{(x+1)^{4/3}}.$

18. $\displaystyle\int_1^3 \frac{dx}{(2x-3)^3}.$

In Problems 19–24, find the area of the region bounded by the given curves.

19. $y = e^{-x}, \quad y = 0, \quad$ right of $x = 0$.

20. $y = \dfrac{1}{\sqrt{x}}, \quad y = 0, \quad$ between $x = 0$ and $x = 4$.

21. $y = \dfrac{1}{x^2}, \quad y = 0, \quad$ right of $x = 2$.

22. $y = \dfrac{1}{x^2}, \quad y = 0, \quad$ between $x = 0$ and $x = 2$.

23. $y = \dfrac{1}{(x-1)^{1/3}}$, $y = 0$,   between $x = 0$ and $x = 2$.

24. $y = \dfrac{1}{\sqrt{x-1}}$, $y = 0$,   right of $x = 5$.

B

In Problems 25–28, evaluate the given integrals.

25. $\displaystyle\int_{-1}^{0} \dfrac{dx}{x}$.   26. $\displaystyle\int_{0}^{1} \dfrac{\ln x}{x}\, dx$.   27. $\displaystyle\int_{0}^{2\pi/3} \tan x\, dx$.   28. $\displaystyle\int_{-\infty}^{+\infty} \dfrac{dx}{x^2+1}$.

29.   Show that $\displaystyle\int_{1}^{+\infty} \dfrac{dx}{x^n}$ converges if and only if $n > 1$.

30.   Show that $\displaystyle\int_{0}^{1} \dfrac{dx}{x^n}$ converges if and only if $n < 1$.

C

31.   Show that $\dfrac{1}{x(x+1)} = \dfrac{1}{x} - \dfrac{1}{x+1}$.

Show that
$$\int_{1}^{+\infty} \frac{dx}{x(x+1)}$$
converges, but
$$\int_{1}^{+\infty} \frac{dx}{x} \quad \text{and} \quad \int_{1}^{+\infty} \frac{dx}{x+1}$$
both diverge.

32.   Show that if $\displaystyle\lim_{x \to +\infty} f(x) = +\infty$,   then
$$\int_{a}^{+\infty} f(x)\, dx$$
diverges.

33.   Give an example of a function f such that
(i) f is nonnegative and continuous,
(ii) $\displaystyle\lim_{x \to +\infty} f(x)$ does not exist, and

(iii) $\displaystyle\int_{0}^{+\infty} f(x)\, dx$ converges.

(*Hint:* Interpret an integral as an area and use the fact that
$$0.3 + 0.03 + 0.003 + \cdots = 0.333\ldots = 1/3.)$$

16.2

Volumes of Solids of Revolution: Disc Method

Suppose the region bounded by $y = f(x)$, the x axis, and the vertical lines $x = a$ and $x = b$ is revolved about the x axis (see Figure 16.3). What is the volume of the resulting solid? In order to answer this question, let us cut up the original plane region into vertical strips and revolve them along with the region. The result is a series of discs approximating the solid region.

Let us consider the ith disc. The original interval $[a, b]$ has been cut into subintervals by the subdivision

$$a = x_0, x_1, x_2, \ldots, x_n = b.$$

The ith subinterval is $[x_{i-1}, x_i]$, and a number x_i^* has been selected in this subinterval. Thus we have a rectangle of width $\Delta x_i = x_i - x_{i-1}$ and height $f(x_i^*)$. After revolution about the x axis, the rectangle sweeps out a cylinder with radius $f(x_i^*)$ and height Δx_i. Thus its volume is

$$\pi[f(x_i^*)]^2 \, \Delta x_i,$$

and the total volume for all such discs is

$$\sum_{i=1}^{n} \pi[f(x_i^*)]^2 \, \Delta x_i.$$

Figure 16.3

This is only an approximation of the original volume—the actual volume may be found by taking the limit of this expression as the norm of the subdivision, $\|S\|$, approaches zero. This limit is an integral.

$$V = \lim_{\|S\| \to 0} \sum_{i=1}^{n} \pi[f(x_i^*)]^2 \, \Delta x_i$$

$$= \int_a^b \pi[f(x)]^2 \, dx.$$

The foregoing discussion does *not* constitute a proof that the formula gives the desired volume. In fact, we have not even *defined* "volume." If we were to try doing so we should find ourselves in the same position as when we tried to define area—we have to define it in terms of an integral. We shall not attempt a formal definition here, but it is clear that we could do so (at least for solids of revolution).

What has been noted here holds throughout this chapter: namely, the discussions given are not intended as *proofs*—but simply to make the given applications of integration seem plausible.

Example 1

The region bounded by $y = x^2$, the x axis, and $x = 1$ is revolved about the x axis. Find the volume of the resulting solid.

A vertical strip in the original region would give the disc shown in Figure 16.4. Thus

$$V = \int_0^1 \pi y^2 \, dx$$

$$= \pi \int_0^1 x^4 \, dx$$

$$= \frac{\pi x^5}{5} \bigg|_0^1$$

$$= \frac{\pi}{5}.$$

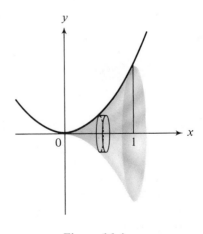

Figure 16.4

Although this example fits the given integration formula perfectly, you are *not* encouraged to memorize the formula. The only thing you need to memorize is the formula for the volume of a cylinder. Furthermore, it is best for you to draw a figure, including the disc, because the plane region can be revolved about *any* line—not merely about the *x* axis. It would be too difficult to try to catalog all possibilities and memorize a formula for each.

Example 2

The region bounded by $y = x^2$, the *y* axis, and $y = 1$ is revolved about the *y* axis. Find the volume of the resulting solid.

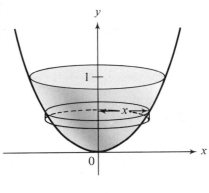

If the plane region is subdivided into horizontal strips and these are revolved about the *y* axis, we have discs with radius *x* and height Δy (see Figure 16.5). Thus

$$V = \int_0^1 \pi x^2 \, dy$$

$$= \int_0^1 \pi y \, dy$$

$$= \frac{\pi y^2}{2} \Big|_0^1$$

$$= \frac{\pi}{2}.$$

Figure 16.5

Example 3

Suppose the region of Example 1 is revolved about the line $x = 1$. Find the volume of the resulting solid.

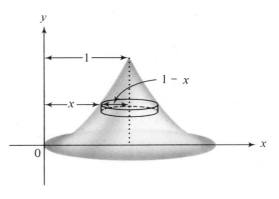

A horizontal strip, when revolved (see Figure 16.6), gives a cylinder with radius $1 - x$ and height Δy. Thus

$$V = \int_0^1 \pi (1 - x)^2 \, dy$$

$$= \int_0^1 \pi (1 - \sqrt{y})^2 \, dy$$

$$= \pi \int_0^1 (1 - 2\sqrt{y} + y) \, dy$$

$$= \pi \left(y - \frac{4}{3} y^{3/2} + \frac{1}{2} y^2 \right) \Big|_0^1$$

$$= \frac{\pi}{6}.$$

Figure 16.6

Example 4

Suppose the region of Example 1 is revolved about the y axis. Find the volume of the resulting solid.

If a horizontal strip is revolved about the y axis, the result is not a solid disc but a disc with a hole in it (see Figure 16.7). In order to find the volume of the disc with a hole in it, we simply find the volume of a solid disc and subtract the volume of the hole. If r and R are the smaller and larger radii, respectively, and h is the thickness of the disc, then

$$V = \pi R^2 h - \pi r^2 h$$
$$= \pi (R^2 - r^2)h.$$

In our case the larger radius is 1, the smaller is x, and the thickness is Δy. Thus the volume of the disc is

$$\pi(1 - x^2)\,\Delta y,$$

and the volume of the solid is

$$V = \int_0^1 \pi(1 - x^2)\,dy$$
$$= \pi \int_0^1 (1 - y)\,dy$$
$$= \pi\left(y - \frac{y^2}{2}\right)\Big|_0^1$$
$$= \frac{\pi}{2}.$$

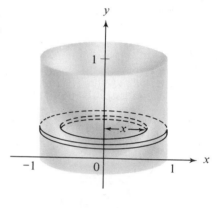

Figure 16.7

Problems

A

In Problems 1–9, the region bounded by the given curves is revolved about the x axis. Find the volume of the resulting solid.

1. $y = x^3$, x axis, $x = 1$.
2. $y = x^4$, x axis, $x = 1$.
3. $y = x^2 - x$, x axis.
4. $y = x^3 - x$, x axis, between $x = -1$ and $x = 1$.
5. $y = \sin x$, x axis, between $x = 0$ and $x = \pi$.
6. $y = e^x$, x axis, left of the y axis.
7. $y = 1/x$, x axis, between $x = 1$ and $x = e$.
8. $y = x$, $y = x^2$.
9. $y = x$, $y = x^3$ (first quadrant).

In Problems 10–15, each region bounded by the given curves is revolved about the y axis. Find the volume of the resulting solid.

10. $y = x^3$, y axis, $y = 8$.
11. $y = x^4$, y axis, $y = 1$ (first quadrant).
12. $y = \ln x$, x axis, between $x = 0$ and $x = 1$.
13. $y = x^3$, x axis, $x = 1$.
14. $y = x$, $y = x^2$.
15. $y = x$, $y = x^3$ (first quadrant).

B

16. The region bounded by $y = \ln x$ and the x axis between $x = 1$ and $x = e$ is revolved about the x axis. Find the volume of the resulting solid.
17. The region bounded by $y = \cos x$ and the x axis between $x = 0$ and $x = \pi/2$ is revolved about the y axis. Find the volume of the resulting solid.

In Problems 18–23, the region bounded by the given curves is revolved about the line indicated. Find the volume of the resulting solid.

18. $y = x^3$, x axis, $x = 1$; about $x = 1$. 19. $y = x^3$, y axis, $y = 1$; about $x = 1$.
20. $y = x^3$, x axis, $x = 1$; about $y = 1$.
21. $y = \cos x$, x axis, between $x = 0$ and $x = \pi/2$; about $y = 1$.
22. $y = 1 - x$, both axes; about $x = 2$. 23. $y = x$, $y = x^2$; about $x = 1$.

C

In Problems 24–26, the region bounded by the given curves is revolved about the y axis. Find the volume of the resulting solid.

24. $y = \sin x$, x axis, between $x = 0$ and $x = \pi$.
25. $y = x^2 - x$, x axis.
26. $y = e^x$, x axis, left of the y axis. (*Hint*: $\lim_{\epsilon \to 0} \epsilon \ln \epsilon = 0$ and $\lim_{\epsilon \to 0} \epsilon \ln^2 \epsilon = 0$.)

In Problems 27–30, the region bounded by the given curves is revolved about the line indicated. Find the volume of the resulting solid.

27. $y = \cos x$, x axis, between $x = 0$ and $x = \pi/2$; about $x = \pi$.
28. $x^2 + y^2 - 2x = 0$; about the y axis.
29. $x^2 + y^2 - 4x + 3 = 0$; about the y axis.
30. $y = x$, $y = x^2$; about $y = x$.
31. Find the volume of a sphere of radius r.
32. Find the volume of a cone of radius r and height h.
33. We saw in the last section that the area of the region between $y = 1/x$ and the x axis and to the right of $x = 1$ is infinite. Show that revolving this region about the x axis gives a solid with a finite volume.

16.3

Volumes of Solids of Revolution: Shell Method

In Example 4 of the previous section the region bounded by $y = x^2$, the x axis, and $x = 1$ was revolved about the y axis. In that case we noted that a horizontal strip, when rotated about the y axis, generates a disc with a hole in it. Let us now consider what happens if a vertical strip is used. The result is a hollow shell (see Figure 16.8) much like a tin can with both ends removed. This shell is a cylinder with a cylindrical hole in it— its volume is easily found.

Let R_1 and R_2 be the radii of the inner and outer cylinders, respectively, and let h be the height. Then $R = (R_1 + R_2)/2$ is the average radius and $t = R_2 - R_1$ is the thickness of the shell. Thus, the volume is

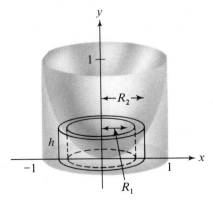

$$V = \pi R_2^2 h - \pi R_1^2 h$$
$$= \pi h(R_2^2 - R_1^2)$$
$$= \pi h(R_2 + R_1)(R_2 - R_1)$$
$$= 2\pi h \frac{R_1 + R_2}{2}(R_2 - R_1)$$
$$= 2\pi R h t.$$

Figure 16.8

Another way of considering this volume is to imagine that the shell has been cut on one side from top to bottom and rolled flat. Then we have a flat plate with length h and width $2\pi R$ (the circumference of the cylinder) and thickness t. Its volume is

$$V = 2\pi Rht.$$

Example 1

Use the shell method to find the volume of the solid of Example 4 of the previous section.

From Figure 16.8 we have $R = x$, $h = y$, and $t = \Delta x$. Thus

$$V = \int_0^1 2\pi xy \, dx = \int_0^1 2\pi x^3 \, dx = \frac{\pi x^4}{2} \bigg|_0^1 = \frac{\pi}{2}.$$

One advantage this method has over the disc method is that the same formula may be used for the volume in any case—whether the original solid has a hole in it or not. We have seen a case in which the solid has a hole; let us consider the other case.

Example 2

Use the shell method to find the volume of the solid of Example 2 of the previous section.

From Figure 16.9 we see that a vertical strip, when revolved, gives a shell with radius x, height $1 - y$, and thickness Δx. Thus

$$V = \int_0^1 2\pi x(1 - y) \, dx$$

$$= 2\pi \int_0^1 x(1 - x^2) \, dx$$

$$= 2\pi \int_0^1 (x - x^3) \, dx$$

$$= 2\pi \left(\frac{x^2}{2} - \frac{x^4}{4} \right) \bigg|_0^1$$

$$= \frac{\pi}{2}.$$

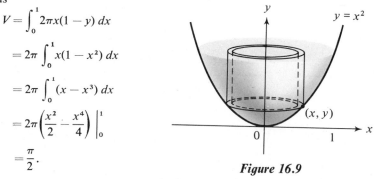

Figure 16.9

In many cases it is difficult, or impossible, to find a volume using the disc method, while the shell method works quite well.

Example 3

Find the volume of the solid generated by revolving about the y axis the region bounded by $y = x^2 - x^3$ and the x axis.

The graph of $y = x^2 - x^3$ is given in Figure 16.10(a). The relative maximum is at $(2/3, 4/27)$ (the scale on the y axis is enlarged for convenience). Let us first see what happens when we try to use the disc method. Figure 16.10(b) gives the solid and a representative disc. From this figure we get

$$V = \int_0^{4/27} \pi(x_2^2 - x_1^2) \, dy.$$

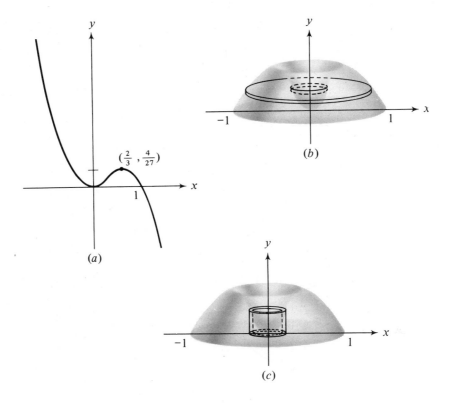

Figure 16.10

Now we are presented with the chore of solving $y = x^2 - x^3$ for x in terms of y. Not only that, but (since there are three values of x for a given value of y between 0 and 4/27) we must find the two positive values and assign x_2 as the larger and x_1 as the smaller. This is prohibitively difficult—we shall not attempt it.

Let us now consider the shell method. From Figure 16.10(c) we have

$$V = \int_0^1 2\pi x y \, dx$$

$$= 2\pi \int_0^1 x(x^2 - x^3) \, dx$$

$$= 2\pi \int_0^1 (x^3 - x^4) \, dx$$

$$= 2\pi \left(\frac{x^4}{4} - \frac{x^5}{5} \right) \Big|_0^1$$

$$= \frac{\pi}{10}.$$

We see that this method gives the volume with very little work—not only do we avoid solving $y = x^2 - x^3$ for x, but we do not need to know the coordinates of the relative maximum. Of course the shell method is not always the easier of the two.

Example 4

Find the volume of the solid generated by revolving about the x axis the region bounded by $x = y^3 - y$ and the y axis, between $y = 0$ and $y = 1$.

The plane region is given in Figure 16.11(a). Notice that x is negative and y positive for the portion of the curve in which we are interested. Figure 16.11(b) gives the solid and a representative shell in it. From this figure we have

$$V = \int_0^1 2\pi y (-x)\, dy.$$

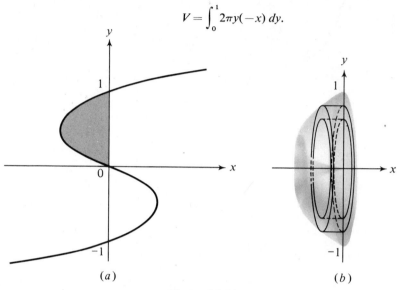

(a) (b)

Figure 16.11

Note that the height of the shell is $-x$ rather than x. Since x is negative (or zero) for all points on the portion of the curve in which we are interested, the height is the corresponding positive number, that is, $-x$.

$$V = 2\pi \int_0^1 y(y - y^3)\, dy$$

$$= 2\pi \int_0^1 (y^2 - y^4)\, dy$$

$$= 2\pi \left(\frac{y^3}{3} - \frac{y^5}{5} \right) \Big|_0^1$$

$$= \frac{4\pi}{15}.$$

It might be noted again that this problem would be more difficult by the disc method.

Problems

A

In Problems 1–9, each region bounded by the given curves is revolved about the line indicated. Use the shell method to find the volume.

1. $y = x^2$, y axis, $y = 1$ (first quadrant); about x axis.
2. $y = x^2$, x axis, $x = 1$; about x axis.
3. $y = x^3$, x axis, $x = 2$; about y axis.

4. $y = x^3$, x axis, $x = 2$;   about x axis.
5. $y = x^2$, x axis, $x = 1$;   about $x = 1$.
6. $y = e^x$, x axis,   between $x = 0$ and $x = 1$;   about y axis.
7. $y = x^2 - x$, x axis;   about y axis.
8. $y = x^2 - 3x + 2$, x axis;   about y axis.
9. $x = y^2 - 3y$, y axis;   about x axis.

In Problems 10–14, each region bounded by the given curves is revolved about the line indicated. Find the volume by any convenient method.

10. $y = x^2 + 2x - 3$, x axis;   about x axis.
11. $y = x^2 - 5x + 6$, x axis;   about y axis.
12. $y = x(x - 2)^2$, x axis;   about y axis.
13. $y = x(x - 2)^2$, x axis;   about x axis.
14. $y = \sin x$, x axis,   between $x = 0$ and $x = \pi$;   about y axis.

B

In Problems 15–19, each region bounded by the given curves is revolved about the line indicated. Use the shell method to find the volume.

15. $y = \ln x$, x axis,   between $x = 1$ and $x = e$;   about y axis.
16. $x = y^2 - 3y$, y axis;   about y axis.
17. $y = 1/x$, x axis,   right of $x = 1$;   about x axis.
18. $y = 1/x$, x axis,   right of $x = 1$;   about y axis.
19. $y = 1/x$, x axis,   right of $x = 1$;   about $x = 1$.

In Problems 20–26, each region bounded by the given curves is revolved about the line indicated. Find the volume by any convenient method.

20. $y = x(x - 2)^2$, x axis;   about $x = 2$.
21. $y = x(x - 2)^2$, x axis;   about $y = 2$.
22. $y = e^x$, x axis,   left of the y axis;   about y axis.
23. $y = xe^x$, x axis,   between $x = -2$ and $x = 0$;   about x axis.
24. $y = xe^x$, x axis,   between $x = -2$ and $x = 0$;   about y axis.
25. $x = (y^2 - 1)^2$, y axis;   about x axis.
26. $x = y^2(2 - y)$, y axis;   about x axis.

16.4

Volumes of Other Solids

In many cases we are interested in the volumes of solids that are not solids of revolution. These can be found by using double or triple integrals (see Chapter 22). But certain special cases can be handled by the single integrals we have studied—in particular, whenever parallel cross sections all have the same simple shape (all squares, all triangles, and so on). It is based upon the volume of a disc (not necessarily circular), which is the product of the cross-sectional area and the thickness.

$$V = At.$$

Example 1

A solid has a circular base of radius 1. Parallel cross sections perpendicular to the base are squares. Find the volume of the solid.

The solid is given in Figure 16.12(a) (although it is not necessary to know what the solid looks like in order to find the volume). One of the square cross sections $ABCD$ is given. Suppose we represent the base (shown in (b)) by the equation

$$x^2 + y^2 = 1.$$

Then $\overline{AB}$, the length of one side of the square cross section, is twice the y coordinate of B, or

$$2\sqrt{1 - x^2}.$$

Thus the cross-sectional area is

$$4(1 - x^2);$$

and, with thickness Δx, the volume of a cross-sectional square disc is

$$4(1 - x^2)\, \Delta x.$$

Thus,

$$V = \int_{-1}^{1} 4(1 - x^2)\, dx$$

$$= 4\left(x - \frac{x^3}{3}\right)\Bigg|_{-1}^{1}$$

$$= \frac{16}{3}.$$

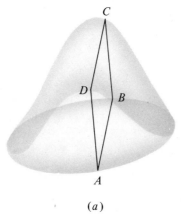

(a)   (b)

Figure 16.12

Example 2

A solid has a circular base of radius 1. Parallel cross sections perpendicular to the base are equilateral triangles. Find the volume of the solid.

Figure 16.13(a) shows the resulting solid with the triangular cross section ABC. Again, if we represent the base by the equation

$$x^2 + y^2 = 1,$$

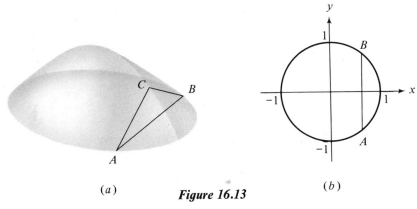

(a)   *Figure 16.13*   (b)

AB is of length

$$2\sqrt{1-x^2},$$

and the altitude is $\sqrt{3}/2$ times the base, or

$$\sqrt{3}\sqrt{1-x^2}.$$

Thus,

$$V = \int_{-1}^{1} \frac{1}{2} \cdot 2\sqrt{1-x^2}\sqrt{3}\sqrt{1-x^2}\, dx$$

$$= \int_{-1}^{1} \sqrt{3}(1-x^2)\, dx$$

$$= \sqrt{3}\left(x - \frac{x^3}{3}\right)\Big|_{-1}^{1}$$

$$= \frac{4\sqrt{3}}{3}.$$

Example 3

A circular cylinder of radius 1 is cut by two planes. One is perpendicular to the axis of the cylinder, and the other plane is inclined to the first at an angle of 45°, intersecting the first in a line that is a diameter of the cylinder. Find the volume of one of the wedges so formed.

There are two convenient ways of finding the volume. Both are illustrated here.

The wedge is illustrated in (a) of Figure 16.14. As can be seen there, cross sections perpendicular to the line of intersection of the two planes are right isosceles

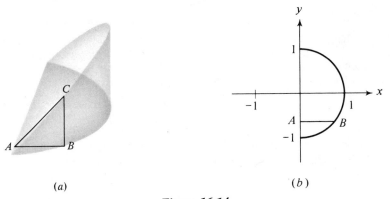

(a)   (b)

Figure 16.14

triangles. Let us again represent the base by the right half of the circle (shown in (b))

$$x^2 + y^2 = 1.$$

The length of AB is the x coordinate of B, or

$$\sqrt{1 - y^2}.$$

Since $\overline{BC} = \overline{AB}$, the area of the triangular cross section is

$$\frac{1}{2} \overline{AB} \cdot \overline{BC} = \frac{1}{2} \sqrt{1 - y^2} \sqrt{1 - y^2} = \frac{1}{2} (1 - y^2).$$

Thus

$$V = \int_{-1}^{1} \frac{1}{2} (1 - y^2) \, dy$$

$$= \frac{1}{2} \left(y - \frac{y^3}{3} \right) \Big|_{-1}^{1}$$

$$= \frac{2}{3}.$$

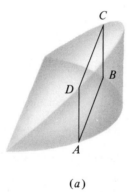

(a)

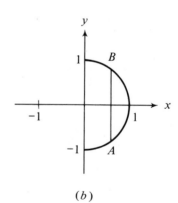

(b)

Figure 16.15

If the cross sections are taken perpendicular to the base but parallel to the line of intersection of the two planes (see Figure 16.15(a)) then the cross section is a rectangle $ABCD$. Again representing the base as before, we see that the length of AB is twice the y coordinate of B, or

$$2\sqrt{1 - x^2}.$$

Since the planes intersect at an angle of 45°, the length of AD is the same as the distance of AB from the y axis, or the x coordinate of B. Thus the area of $ABCD$ is

$$2x\sqrt{1 - x^2},$$

and

$$V = \int_0^1 2x\sqrt{1 - x^2}\, dx$$

$$= -\frac{2}{3}(1 - x^2)^{3/2}\Big|_0^1$$

$$= \frac{2}{3}.$$

Problems

A

In each of Problems 1–6, we have a solid whose base is a circle of radius 1. The parallel cross sections taken perpendicular to each base are described. Find each volume.

1. Rectangles of height 1.
2. Isosceles triangles of height 1.
3. Isosceles right triangles with one leg as base.
4. Isosceles right triangles with the hypotenuse as base.
5. Semicircles.
6. Semi-ellipses of height 2. (*Hint:* The area of an ellipse is $A = \pi ab$.)

Each of Problems 7–14 involves a solid whose base is bounded by $y = 4 - x^2$ and the x axis. Cross sections perpendicular to the base and parallel to the x axis are described. Find each volume.

7. Squares.
8. Equilateral triangles.
9. Rectangles of height 2.
10. Isosceles triangles of height 1.
11. Isosceles right triangles with one leg as base.
12. Isosceles right triangles with the hypotenuse as base.
13. Semicircles.
14. Semi-ellipses of height 6. (See Problem 6.)

B

Each of Problems 15–22 concerns a solid whose base is an ellipse with major axis 8 and minor axis 6. Cross sections perpendicular to the base and parallel to the minor axis are described. Find each volume.

15. Squares.
16. Equilateral triangles.
17. Rectangles of height 4.
18. Isosceles triangles of height 2.
19. Isosceles right triangles with one leg as base.
20. Isosceles right triangles with hypotenuse as base.
21. Semicircles.
22. Semi-ellipses of height 5. (See Problem 6.)

In Problems 23 and 24, find the volume of the wedge described in Example 3, given that the second plane is inclined to the first at the given angle.

23. $\pi/6$.   24. $\pi/3$.

C

25. Two circular cylinders of radius 2 intersect each other in such a way that their axes intersect at right angles. Find the volume of the portion inside both cylinders.
26. Repeat Problem 25 with both cylinders of radius r.

16.5

Surfaces of Revolution

The area of a surface of revolution is based upon the area of the lateral surface of a frustum of a right circular cone. This area (see Figure 16.16) is

$$A = 2\pi \frac{r_1 + r_2}{2} s,$$

where $(r_1 + r_2)/2$ is the average radius and s is the slant height. This formula still holds if $r_1 = r_2$. In this case we have a cylinder for which the average radius is the radius of the cylinder and the slant height is the height of the cylinder.

Figure 16.16

Now suppose we have a curve represented parametrically by $x = f(t)$, $y = g(t)$ and we revolve about the x axis the portion between $t = a$ and $t = b$ (see Figure 16.17). Suppose further that the straight-line segments used in Section 13.4 to approximate the arc length are revolved along with the curve. Each rotated segment gives a frustum of a cone (or a cylinder if it is parallel to the x axis). If the interval $[a, b]$ is subdivided to give

$$a = t_0, t_1, t_2, \ldots, t_n = b,$$

then the surface area for the ith interval is

$$\pi |g(t_{i-1}) + g(t_i)| \sqrt{[f(t_i) - f(t_{i-1})]^2 + [g(t_i) - g(t_{i-1})]^2}.$$

Again multiplying and dividing by $t_i - t_{i-1} = \Delta t_i$, we have

$$\pi |g(t_{i-1}) + g(t_i)| \sqrt{\left(\frac{f(t_i) - f(t_{i-1})}{t_i - t_{i-1}}\right)^2 + \left(\frac{g(t_i) - g(t_{i-1})}{t_i - t_{i-1}}\right)^2} (t_i - t_{i-1})$$

$$= \pi |g(t_{i-1}) + g(t_i)| \sqrt{\left(\frac{f(t_i) - f(t_{i-1})}{t_i - t_{i-1}}\right)^2 + \left(\frac{g(t_i) - g(t_{i-1})}{t_i - t_{i-1}}\right)^2} \Delta t_i.$$

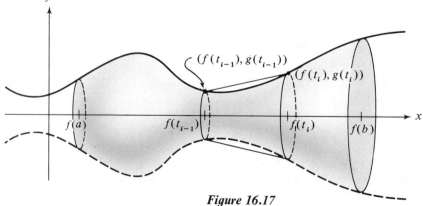

Figure 16.17

Now the surface area we want is the limit of the sum of the above areas as the norm of the subdivision approaches zero. We saw in Section 13.4 that

$$\lim_{\|S\| \to 0} \sum_{i=1}^{n} \sqrt{\left(\frac{f(t_i) - f(t_{i-1})}{t_i - t_{i-1}}\right)^2 + \left(\frac{g(t_i) - g(t_{i-1})}{t_i - t_{i-1}}\right)^2} \, \Delta t_i = \int_{a}^{b} \sqrt{\left(\frac{dx}{dt}\right)^2 + \left(\frac{dy}{dt}\right)^2} \, dt.$$

Furthermore, as the norm of the subdivision approaches zero, t_{i-1} and t_i approach a common number t. Thus it is reasonable to assume that the limit we are seeking is

$$S_x = \int_{a}^{b} \pi |2g(t)| \sqrt{\left(\frac{dx}{dt}\right)^2 + \left(\frac{dy}{dt}\right)^2} \, dt$$

$$= \int_{a}^{b} 2\pi |y| \sqrt{\left(\frac{dx}{dt}\right)^2 + \left(\frac{dy}{dt}\right)^2} \, dt.$$

Of course, this argument does not constitute a proof that the given formula represents the surface area desired;[†] it merely points up the reasonableness of our result. As in Section 13.4, the form of the integral can be changed to fit the form of the given equation. Thus we have the following three forms for the surface area, corresponding to the three forms for arc length on page 411.

$$S_x = \int_{t_1}^{t_2} 2\pi |y| \sqrt{\left(\frac{dx}{dt}\right)^2 + \left(\frac{dy}{dt}\right)^2} \, dt \quad \text{if } x = f(t) \quad \text{and} \quad y = g(t),$$

$$S_x = \int_{x_1}^{x_2} 2\pi |y| \sqrt{1 + \left(\frac{dy}{dx}\right)^2} \, dx \quad \text{if } y = f(x),$$

$$S_x = \int_{y_1}^{y_2} 2\pi |y| \sqrt{1 + \left(\frac{dx}{dy}\right)^2} \, dy \quad \text{if } x = g(y).$$

(As in Section 13.4, we have changed the limits of integration on the first integral from a and b to t_1 and t_2 to emphasize the fact that these limits are values of t.)

This can be summarized by

$$S_x = \int_{a}^{b} 2\pi |y| \, ds,$$

where ds (sometimes called the differential of arc) is given by

$$ds = \sqrt{\left(\frac{dx}{dt}\right)^2 + \left(\frac{dy}{dt}\right)^2} \, dt \quad \text{if } x = f(t) \text{ and } y = g(t),$$

$$ds = \sqrt{1 + \left(\frac{dy}{dx}\right)^2} \, dx \quad \text{if } y = f(x),$$

$$ds = \sqrt{1 + \left(\frac{dx}{dy}\right)^2} \, dy \quad \text{if } x = g(y).$$

Of course the limits of integration are values of t, x, or y in the first, second, or third cases, respectively.

---

† In general, surface area is a much more difficult problem than plane area or arc length. For a discussion of surface areas see Tibor Rado, "What Is the Area of a Surface?," *American Mathematical Monthly*, Vol. 50 (1943), pp. 139–141.

Example 1

Find the area of the surface formed by rotating about the x axis the arc $x = 4 - t^2$, $y = t$ from $t = 0$ to $t = \sqrt{2}$.

By eliminating the parameter we have the equation $x = 4 - y^2$, whose graph is given in Figure 16.18. When $t = 0$, we have the point $(4, 0)$; when $t = \sqrt{2}$, we have $(2, \sqrt{2})$. This gives the surface indicated. Now

$$\frac{dx}{dt} = -2t, \qquad \frac{dy}{dt} = 1,$$

and $|y| = y$ for all values of t in $[0, \sqrt{2}]$.

$$
\begin{aligned}
S_x &= \int_0^{\sqrt{2}} 2\pi y \sqrt{\left(\frac{dx}{dt}\right)^2 + \left(\frac{dy}{dt}\right)^2}\, dt \\
&= \int_0^{\sqrt{2}} 2\pi t \sqrt{4t^2 + 1}\, dt \\
&= \frac{\pi}{4} \int_0^{\sqrt{2}} 8t\sqrt{4t^2 + 1}\, dt \\
&= \frac{\pi}{4} \frac{(4t^2 + 1)^{3/2}}{3/2} \Bigg|_0^{\sqrt{2}} \\
&= \frac{\pi}{6} (4t^2 + 1)^{3/2} \Bigg|_0^{\sqrt{2}} \\
&= \frac{\pi}{6} (27 - 1) \\
&= \frac{13\pi}{3}.
\end{aligned}
$$

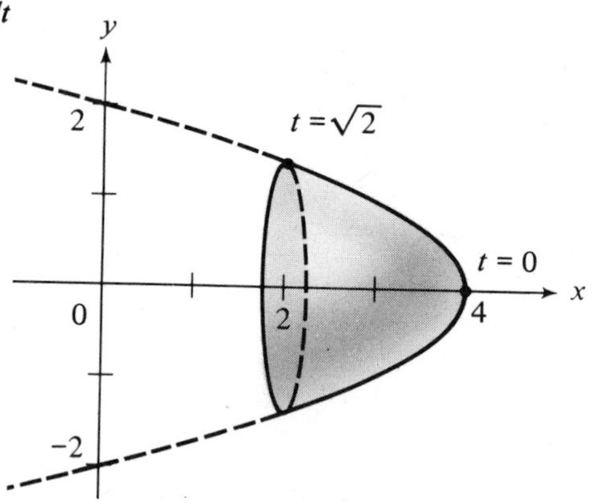

Figure 16.18

Because we have a formula with which we can set up the integral for a surface area, it is usually not necessary to draw a figure. We omit it in the remaining examples.

Example 2

Find the area of the surface formed by rotating about the x axis the arc $y = x^3/3$ from $x = 0$ to $x = 2$.

$$y' = x^2.$$

Furthermore, $y \geq 0$ for all x in the interval $[0, 2]$. Thus $|y| = y$.

$$
\begin{aligned}
S_x &= \int_0^2 2\pi y \sqrt{1 + (y')^2}\, dx \\
&= \int_0^2 \frac{2\pi x^3}{3} \sqrt{1 + x^4}\, dx \\
&= \frac{\pi}{6} \int_0^2 4x^3 \sqrt{1 + x^4}\, dx \\
&= \frac{\pi}{6} \frac{(1 + x^4)^{3/2}}{3/2} \Bigg|_0^2 \\
&= \frac{\pi}{9} (1 + x^4)^{3/2} \Bigg|_0^2 \\
&= \frac{\pi}{9} (17^{3/2} - 1).
\end{aligned}
$$

By a process similar to the one used to derive the formulas for S_x, it can be shown that if the arc is rotated about the y axis, the surface area is

$$S_y = \int_a^b 2\pi |x| \, ds,$$

where we again have three choices for ds, depending upon the form of the given equation.

Example 3

Find the area of the surface formed by rotating about the y axis the arc

$$x = \frac{y^3}{6} + \frac{1}{2y} \quad \text{from } (2/3, 1) \text{ to } (14/3, 3).$$

$$\frac{dx}{dy} = \frac{y^2}{2} - \frac{1}{2y^2}.$$

Clearly, x is positive throughout the given interval. Thus $|x| = x$.

$$S_y = \int_1^3 2\pi x \sqrt{1 + \left(\frac{dx}{dy}\right)^2} \, dy$$

$$= \int_1^3 2\pi \left(\frac{y^3}{6} + \frac{1}{2y}\right) \sqrt{1 + \left(\frac{y^2}{2} - \frac{1}{2y^2}\right)^2} \, dy$$

$$= \int_1^3 2\pi \left(\frac{y^3}{6} + \frac{1}{2y}\right) \left(\frac{y^2}{2} + \frac{1}{2y^2}\right) \, dy$$

$$= 2\pi \int_1^3 \left(\frac{y^5}{12} + \frac{y}{3} + \frac{1}{4y^3}\right) \, dy$$

$$= 2\pi \left(\frac{y^6}{72} + \frac{y^2}{6} - \frac{1}{8y^2}\right) \Big|_1^3 = \frac{208\pi}{9}$$

Problems

A

In Problems 1–8, the given arc is rotated about the x axis. Find the area of the resulting surface.

1. $x = a \cos^3 \theta$, $y = a \sin^3 \theta$; from $\theta = 0$ to $\theta = \pi/2$.
2. $x = t + 2$, $y = t^3$; from $t = 0$ to $t = 2$.

3. $y = \frac{x^3}{3} + \frac{1}{4x}$; from $x = 1$ to $x = 3$.   4. $y = \frac{x^4}{4} + \frac{1}{8x^2}$; from $x = 1$ to $x = 2$.
5. $y = \cosh x$; from $x = -1$ to $x = 1$.   6. $y^2 = 8x$; from $(0, 0)$ to $(2, 4)$.
7. $9y^2 = x(3 - x)^2$; from $(0, 0)$ to $(3, 0)$.   8. $8y^2 = x^2 - x^4$; from $(0, 0)$ to $(1, 0)$.

In Problems 9–13, the given arc is rotated about the y axis. Find the area of the resulting surface.

9. $x = \sqrt{y}$; from $y = 0$ to $y = 4$.   10. $y = \frac{x^2}{4} - \frac{\ln x}{2}$; from $x = 1$ to $x = e$.
11. $y = \cosh x$; from $x = 0$ to $x = 1$.   12. $y = \ln x$; from $(1, 0)$ to $(e, 1)$.

13. $y = \frac{x^3}{3}$; from $(0, 0)$ to $(3, 9)$.

14. Find the surface area of a sphere of radius R.
15. The portion of the circle $x^2 + y^2 = 4$ to the right of $x = 1$ is rotated about the x axis, forming a zone of the sphere. Find its surface area.

16. A zone of the sphere of Problem 15 is formed by rotating about the x axis the portion of the circle between $x = 0$ and $x = 1$. Find the surface area. Compare with the result of Problem 15.

17. A zone of the sphere of Problem 15 is formed by rotating about the x axis the portion of the circle between $x = -1/2$ and $x = 1/2$. Find the surface area. Compare with the results of Problems 15 and 16.

18. Find the surface area of a zone of height h of a sphere of radius r.

B

In Problems 19–24, the given arc is rotated about the x axis. Find the area of the resulting surface.

19. $x = t^3, \quad y = t^2; \quad$ from $t = 0$ to $t = 2$.
20. $x^{2/3} + y^{2/3} = a^{2/3}; \quad$ from $(0, a)$ to $(a, 0)$.
21. $y = \sin x; \quad$ from $x = 0$ to $x = \pi$.
22. $y = e^x; \quad$ left of $x = 0$.

23. $y = \dfrac{x^2}{4} - \dfrac{\ln x}{2}; \quad$ from $x = 1$ to $x = 4$.

24. $x = a\theta - a \sin \theta, \quad y = a - a \cos \theta; \quad$ from $\theta = 0$ to $\theta = \pi$.

25. Find the lateral surface area of a cone of height h and radius r.

In Problems 26–28, the given arc is rotated about the y axis. Find the area of the resulting surface.

26. $x = y^{2/3}; \quad$ from $y = 0$ to $y = 1$.
27. $x = e^t \cos t, \quad y = e^t \sin t; \quad$ from $t = 0$ to $t = \pi$.
28. $x = t + 1, \quad y = t^2 - 1; \quad$ from $t = -1$ to $t = 1$.

C

29. We have seen (Section 16.1) that the region bounded by $y = 1/x$ and the x axis and to the right of $x = 1$ has infinite area, but the solid formed by revolving this region about the x axis has finite volume (see Problem 33, page 509). Is the area of the surface of this solid finite or infinite?

30. The ellipse $x^2/a^2 + y^2/b^2 = 1 \quad (a > b)$ is rotated about the x axis. Find the area of the resulting surface.

31. The ellipse of Problem 30 is rotated about the y axis. Find the area of the resulting surface.

16.6

Center of Mass and Moments

The center of mass of an object (or system of objects) is the point on which that object (or system) balances. For example, if a board is weighted at each end, the point at which it balances is the center of mass of the given system.

In the preceding sections we used integrals to determine areas, volumes, arc lengths, and so forth, because they are all additive; for example, the area of a large region can be found by cutting it up into smaller regions and adding all of their areas. Unfortunately, we cannot use the additive process with centers of mass.

For example, suppose we have a barbell four feet long which we place on the x axis with one end at -2 and the other at 2. Suppose furthermore that we put a 100-pound mass at $x = -2$ and a ten-pound mass at $x = 2$. (When we say, "put a 100-pound mass at $x = -2$," we mean that the center of mass—or balancing point—

is at $x = -2$.) Since one center of mass is at $x = -2$ and the other at $x = 2$, we find that the sum is $x = 0$, which is not the center of mass of the barbell. Obviously the barbell would balance at a point nearer to the 100-pound mass than to the 10-pound mass.

Since we cannot add centers of mass, we must find something that is additive before we can even consider using the integral. We need to consider not only the locations of masses, but also their relative sizes. For this purpose, we consider the *moment* about a certain axis. The moment is the product of the mass and the distance from the axis.

The following problem illustrates the use of moments. We have a board lying on the x axis and balanced at the origin. A 100-pound mass is placed at $x = -2$. Where shall we put a 50-pound mass in order to have the board balance again? The moments should be the same on both sides. On the left, there is a mass of 100 pounds at a distance 2 from the origin; the moment is 200. On the right, there is a mass of 50 pounds at an unknown distance x from the origin; the moment is $50x$. Equating moments, we have

$$50x = 200, \quad \text{or} \quad x = 4.$$

Thus a 100-pound mass at $x = -2$ and a 50-pound mass at $x = 4$ balance the board at the origin; that is, its center of mass is at the origin.

To use a more convenient method, we assign a $+$ or $-$ to the distance, depending upon the direction from the point of reference. Thus, if the point of reference is the origin, we can use the x coordinate of the point at which a mass is located, rather than the distance. In using this method in the last example, we should equate the sum of all moments to zero, rather than equating the moments on the two sides of the point of reference. Thus

$$-2 \cdot 100 + 50x = 0$$
$$x = 4.$$

This method also makes finding the center of mass much simpler. In the last example we asked where to put a certain mass so that the system's center of mass would be at the origin. If the center of mass of the system is to be determined, the new method is easier.

Example 1

Find the center of mass of a system consisting of a 100-pound mass at $x = -2$ and a 10-pound mass at $x = 2$.

Taking the origin as the point of reference, we have

$$M_0 = 100(-2) + 10 \cdot 2$$
$$= -200 + 20$$
$$= -180.$$

Since the sum of the moments of the individual masses equals the moment of the entire system, -180 is the moment about the origin of the entire system. If $\bar{x}$ is the coordinate of the center of mass of the system, then

$$M_0 = 110\bar{x}.$$

Thus

$$110\bar{x} = -180,$$

$$\bar{x} = -\frac{18}{11}.$$

We see that the center of mass can be determined by finding (a) the moment of the system about a given point and (b) the total mass. The center of mass is then the moment divided by the mass.

Suppose the masses are distributed about a plane rather than a line. If we find the moment about the y axis for one of the masses, we see that this is dependent only upon the mass and its x coordinate; the y coordinate is irrelevant. Thus we may find the total moment about the y axis and the x coordinate of the center of mass by using the given masses and their x coordinates only. Similarly the y coordinate of the center of mass is determined only by the masses and their y coordinates.

Example 2

Find the center of mass of the system consisting of a ten-pound mass at $(1, 3)$, a twenty-pound mass at $(-2, 2)$, and a four-pound mass at $(-1, 8)$.

The moment about the x axis, M_x, is the product of the mass and the (directed) distance from the x axis, which is the y *coordinate*. Similarly M_y is the product of the mass and the distance from the y axis, which is the x coordinate. Thus

$$M_x = 10 \cdot 3 + 20 \cdot 2 + 4 \cdot 8$$
$$= 102,$$
$$M_y = 10 \cdot 1 + 20(-2) + 4(-1)$$
$$= -34,$$
$$m = 10 + 20 + 4$$
$$= 34,$$
$$\bar{x} = \frac{M_y}{m} = \frac{-34}{34} = -1,$$
$$\bar{y} = \frac{M_x}{m} = \frac{102}{34} = 3.$$

Example 3

Find the center of mass of a thin plate of uniform density and having the shape shown in (a) of Figure 16.19.

First of all let us choose a pair of coordinate axes. This choice is quite arbitrary—any convenient one will do. Let us put the axes in the position shown in (b) of the figure. Now the region can be subdivided into three rectangles. Because of their symmetry, the centers of mass of the rectangles are at their geometric centers. Thus we have a rectangle of area 8 with center at $(1, 2)$, another of area 3 with center at $(7/2, 1/2)$, and a third of area 4 with center at $(6, 1)$. Because the region is of uniform density, its mass is proportional to its area; thus the mass is the product of the area and the density (which is the mass per unit area). Furthermore, the density is a factor of both the total mass and any moment. Since the

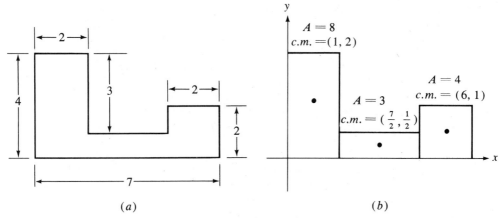

(a)   *(b)*

Figure 16.19

coordinates of the center of mass are found by dividing the moments by the total mass, the density cancels out—thus we shall neglect this factor. We now have

$$A = 8 + 3 + 4 = 15,$$

$$M_x = 8 \cdot 2 + 3 \cdot \frac{1}{2} + 4 \cdot 1 = \frac{43}{2},$$

$$M_y = 8 \cdot 1 + 3 \cdot \frac{7}{2} + 4 \cdot 6 = \frac{85}{2};$$

and

$$\bar{x} = \frac{M_y}{A} = \frac{85/2}{15} = \frac{17}{6},$$

$$\bar{y} = \frac{M_x}{A} = \frac{43/2}{15} = \frac{43}{30}.$$

Of course these are the coordinates of the center of mass with respect to the chosen set of axes. If the axes had been in a different position, the result would give the same point but with a different representation. Note here that the center of mass of the plate is not within the plate.

Problems

A

In Problems 1–10, find the center of mass of each system described.

1. 2 lb at $(4, 0)$,   10 lb at $(2, 0)$,   4 lb at $(-6, 0)$.
2. 3 lb at $(4, 0)$,   5 lb at $(-5, 0)$.
3. 3 lb at $(2, 0)$,   5 lb at $(6, 0)$.
4. 4 lb at $(2, 0)$,   2 lb at $(5, 0)$,   2 lb at $(-2, 0)$,   2 lb at $(-3, 0)$.
5. 2 lb at $(2, 5)$,   5 lb at $(-2, 2)$,   3 lb at $(4, -2)$.
6. 3 lb at $(5, 5)$,   4 lb at $(2, 1)$,   4 lb at $(1, -1)$.
7. 5 lb at $(1, 0)$,   3 lb at $(3, 5)$,   2 lb at $(-4, 1)$.
8. 4 lb at $(0, 0)$,   4 lb at $(3, 4)$,   3 lb at $(-2, 2)$.
9. 10 lb at $(2, 2)$,   3 lb at $(6, -2)$,   7 lb at $(0, -4)$.
10. 4 lb at $(3, -1)$,   3 lb at $(5, 0)$,   2 lb at $(-3, 1)$,   1 lb at $(0, -4)$.

11. There is a 5-lb mass at $(6, 0)$ and a 4-lb mass at $(-1, 0)$. Where should a 3-lb mass be placed in order to have the system balance at the origin?

12. There is a 3-lb mass at $(1, 0)$, a 2-lb mass at $(-3, 0)$, and a 2-lb mass at $(4, 0)$. Where should a 2-lb mass be placed in order to have the system balance at the origin?

13. There is a 5-lb mass at $(3, 0)$ and a 2-lb mass at $(-2, 0)$. Where should a 3-lb mass be placed in order to have the system balance at $(2, 0)$?

14. There is a 10-lb mass at $(3, 0)$, a 5-lb mass at $(-5, 0)$, and a 20-lb mass at $(5, 0)$. Where should a 10-lb mass be placed in order to have the system balance at $(1, 0)$?

15. There is a 2-lb mass at $(5, 2)$ and a 4-lb mass at $(-2, 4)$. Where should a 4-lb mass be placed in order to have the system balance at the origin?

16. There is a 3-lb mass at $(4, -2)$, a 4-lb mass at $(-1, -1)$, and a 2-lb mass at $(-2, 4)$. Where should a 1-lb mass be placed in order to have the system balance at the origin?

17. There is a 4-lb mass at $(2, 2)$ and a 2-lb mass at $(-3, 4)$. Where should a 2-lb mass be placed in order to have the system balance at $(-1, 1)$?

18. There is a 5-lb mass at $(-5, -1)$, a 10-lb mass at $(4, 2)$, and a 15-lb mass at $(5, 1)$. Where should a 20-lb mass be placed in order to have the system balance at $(2, 2)$?

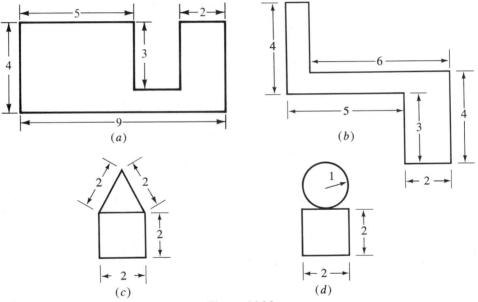

Figure 16.20

In Problems 19–22, find the center of mass of a plate of uniform density with the shape of the indicated figure.

19. Figure 16.20(a).   20. Figure 16.20(b).
21. Figure 16.20(c).   22. Figure 16.20(d).

B

23. Suppose that the triangular portion of the plate of Figure 16.20(c) has twice the mass per unit area as the square. Find the center of mass of the entire plate.

24. Suppose that the square portion of the plate of Figure 16.20(c) has twice the mass per unit area as the triangle. Find the center of mass of the entire plate.

25. Suppose the circular portion of the plate of Figure 16.20(d) has twice the mass per unit area as the square. Find the center of mass of the entire plate.

26. Suppose the square portion of the plate of Figure 16.20(d) has three times the mass per unit area as the circle. Find the center of mass of the entire plate.

16.7

Centroids of Plane Regions

In the preceding section we found that the center of mass of a plate of uniform density is independent of the actual value of the density; it depends only upon the shape of the plate. Thus we can consider the center of mass of a region quite apart from any actual physical interpretation; when we do so, we use the term *centroid* rather than center of mass to distinguish it from the physical problem. A similar situation holds for solids.

Let us now consider the centroid of a region bounded by $y = f(x)$ $(y \geq 0)$, the x axis, and the vertical lines $x = a$ and $x = b$ (see Figure 16.21). Let us cut the region into vertical strips. We use the subdivision

$$S : x_0, x_1, x_2, \ldots, x_n,$$

where $x_0 = a$ and $x_n = b$; and choose x_i^* to be

$$x_i^* = \frac{x_{i-1} + x_i}{2} \quad (i = 1, 2, 3, \ldots, n).$$

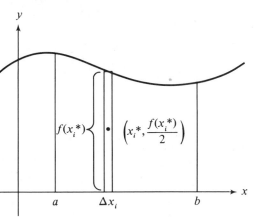

Figure 16.21

Then $f(x_i^*)$ is the height of each rectangle, and $\Delta x_i = x_i - x_{i-1}$ is its width. The centroid of each rectangle (see Figure 16.21) is its geometric center, $(x_i^*, f(x_i^*)/2)$. Thus, the moments about the x and y axes for this strip are

$$M_x = \frac{f(x_i^*)}{2} \cdot f(x_i^*) \, \Delta x_i \quad \text{and} \quad M_y = x_i^* f(x_i^*) \, \Delta x_i.$$

Since moments can be added, we can approximate the moments for the desired region by adding the moments for all of the strips. Finally we get the exact values by taking the limit of this sum as the widths of all of them approach zero. Of course this gives integrals:

$$M_x = \lim_{\|S\| \to 0} \sum_{i=1}^{n} \frac{1}{2} [f(x_i^*)]^2 \, \Delta x_i = \int_a^b \frac{y^2}{2} \, dx.$$

$$M_y = \lim_{\|S\| \to 0} \sum_{i=1}^{n} x_i^* f(x_i^*) \, \Delta x_i = \int_a^b xy \, dx.$$

If these are compared with the integral for the area of this region,

$$A = \int_a^b y \, dx,$$

and if the centroid of the vertical strip is represented by $(x, y/2)$, we see that the moment about the x axis, M_x, is given by the integral for area with $y/2$ (the distance of the centroid of the strip from the x axis) as an additional factor. Similarly, the moment about the y axis is given by the integral for area, with x (the distance from the y axis) as an additional factor.

You are advised *not* to memorize these formulas. As in the case of area or volume, there are too many different possibilities to try to catalog all of them. *The important thing to remember is the method* leading to the formulas rather than the formulas themselves.

Example 1

Find the centroid of the region bounded by $y = x^2$, $y = 0$, and $x = 1$.

The situation is illustrated graphically in Figure 16.22. This is exactly the same as the case illustrated previously. Thus

$$A = \int_0^1 y\, dx = \int_0^1 x^2\, dx = \frac{x^3}{3}\Big|_0^1 = \frac{1}{3},$$

$$\cdot M_x = \int_0^1 \frac{y^2}{2}\, dx = \int_0^1 \frac{x^4}{2}\, dx = \frac{x^5}{10}\Big|_0^1 = \frac{1}{10},$$

$$M_y = \int_0^1 xy\, dx = \int_0^1 x^3\, dx = \frac{x^4}{4}\Big|_0^1 = \frac{1}{4};$$

and

$$\bar{x} = \frac{M_y}{A} = \frac{1/4}{1/3} = \frac{3}{4},$$

$$\bar{y} = \frac{M_x}{A} = \frac{1/10}{1/3} = \frac{3}{10}.$$

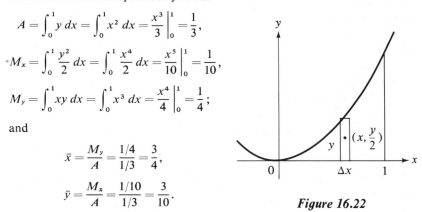

Figure 16.22

Example 2

Find the centroid of the region in the first quadrant bounded by $x = -y^2 + 2y + 3$.

The graph is given in Figure 16.23. In this case it is much easier to cut the region into horizontal strips than into vertical ones. Thus

$$A = \int_0^3 x\, dy = \int_0^3 (-y^2 + 2y + 3)\, dy = -\frac{y^3}{3} + y^2 + 3y\Big|_0^3 = 9,$$

$$M_x = \int_0^3 xy\, dy = \int_0^3 (-y^3 + 2y^2 + 3y)\, dy = -\frac{y^4}{4} + \frac{2y^3}{3} + \frac{3y^2}{2}\Big|_0^3 = \frac{45}{4},$$

$$M_y = \int_0^3 \frac{x^2}{2}\, dy = \frac{1}{2}\int_0^3 (-y^2 + 2y + 3)^2\, dy$$

$$= \frac{1}{2}\int_0^3 (y^4 - 4y^3 - 2y^2 + 12y + 9)\, dy$$

$$= \frac{1}{2}\left(\frac{y^5}{5} - y^4 - \frac{2y^3}{3} + 6y^2 + 9y\right)\Big|_0^3 = \frac{153}{10};$$

and

$$\bar{x} = \frac{M_y}{A} = \frac{153/10}{9} = \frac{17}{10},$$

$$\bar{y} = \frac{M_x}{A} = \frac{45/4}{9} = \frac{5}{4}.$$

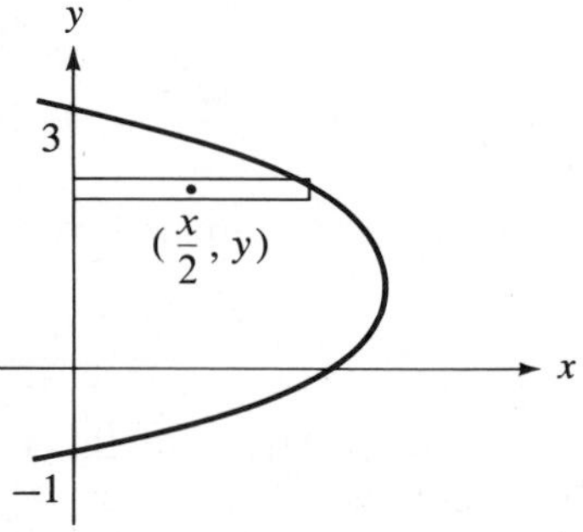

Figure 16.23

Example 3

Find the centroid of the region bounded by $y = x$ and $y = x^2$.

The graph is given in Figure 16.24. Let us note that the y coordinate of the centroid of the vertical strip is half-way between the top and bottom, which is the average of the y coordinates. Thus

$$A = \int_0^1 (y_1 - y_2)\, dx = \int_0^1 (x - x^2)\, dx$$

$$= \frac{x^2}{2} - \frac{x^3}{3} \Big|_0^1 = \frac{1}{6},$$

$$M_x = \int_0^1 (y_1 - y_2) \frac{y_1 + y_2}{2}\, dx$$

$$= \frac{1}{2} \int_0^1 (y_1^2 - y_2^2)\, dx = \frac{1}{2} \int_0^1 (x^2 - x^4)\, dx$$

$$= \frac{1}{2} \left(\frac{x^3}{3} - \frac{x^5}{5} \right) \Big|_0^1 = \frac{1}{15},$$

$$M_y = \int_0^1 (y_1 - y_2) x\, dx = \int_0^1 (x^2 - x^3)\, dx$$

$$= \frac{x^3}{3} - \frac{x^4}{4} \Big|_0^1 = \frac{1}{12};$$

and

$$\bar{x} = \frac{M_y}{A} = \frac{1/12}{1/6} = \frac{1}{2},$$

$$\bar{y} = \frac{M_x}{A} = \frac{1/15}{1/6} = \frac{2}{5}.$$

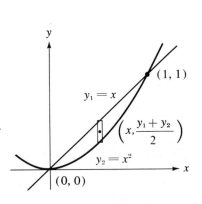

Figure 16.24

Problems

A

In Problems 1–13, find the centroids of the regions bounded by the given curves.

1. $y = x^2$, $y = 0$, $x = 2$.
2. $y = x^3$, $y = 0$, $x = 1$.
3. $y = x^4$, $y = 0$, $x = 1$.
4. $y = \sqrt{x}$, $y = 0$, $x = 1$.
5. $y = x^2$, $x = 0$, $y = 1$ (first quadrant).
6. $y = x^3$, $x = 0$, $y = 1$.
7. $y = x^3 - x^2$, $y = 0$.
8. $y = x(x - 1)^2$, $y = 0$.
9. $x = y^4 + 1$, $x = 0$, $y = 0$, $y = 1$.
10. $x = y^2 - 1$, $x = 0$.
11. $y = \sin x$, $y = 0$ between $x = 0$ and $x = \pi$.
12. $y = x^3$, $y = x$ (first quadrant).
13. $y = x^4$, $y = x$.

B

In Problems 14–24, find the centroids of the regions bounded by the given curves.

14. $y = e^x$, $y = 0$, left of $x = 0$.
    (*Hint*: $\lim_{k \to -\infty} (k - 1)e^k = 0$.)
15. $y = \dfrac{1}{x^2 + 1}$, $y = 0$.
16. $y = \dfrac{1}{x^2}$, $y = 0$, right of $x = 1$.
17. $y = 9 - x^2$, $y = \dfrac{x^2}{4} - 1$.
18. $x^2/16 - y^2/9 = 1$, $x = 5$.
19. $\dfrac{x^2}{a^2} + \dfrac{y^2}{b^2} = 1$, $x = 0$, $y = 0$ (first quadrant).
20. $x^{2/3} + y^{2/3} = a^{2/3}$, $x = 0$, $y = 0$ (first quadrant).
21. $\sqrt{x} + \sqrt{y} = \sqrt{a}$, $x = 0$, $y = 0$.
22. $y = x^2$, $x - y + 2 = 0$.
23. $x^2 + y^2 = 9$, above $y = 0$.
24. $x^2 + y^2 = 9$, above $y = 1$.

C

25. The First Theorem of Pappus states that if a region R is entirely on one side of a line, then the volume of the solid generated by revolving R about that line is the product of the area of R and the length of the path of the centroid of R. Verify this for the region bounded by $y = x^2$, the y axis, and $y = 1$ revolved about the x axis.

26. Verify the First Theorem of Pappus (see Problem 25) for the region bounded by $y = x^2$, the x axis, and $x = 1$ revolved about the x axis.

16.8

Centroids of Solids of Revolution

In considering the centroid of a solid, we need to find moments about planes, rather than about axes. Just as we considered only moments about two axes in the plane, we shall now consider only moments about two planes. One of these contains the x axis and is perpendicular to the y axis; we call this the xz plane. The other contains the y axis and is perpendicular to the x axis; we call it the yz plane.

By an analysis similar to the one of the previous section, we can find moments for solids of revolution in much the same way that we did for plane regions: that is, the moment about the xz plane is given by the integral for volume with an additional factor, which is the distance (with appropriate sign) from the centroid of the disc or shell to the xz plane. Since the axis of revolution is also an axis of symmetry, the centroid is on it. Thus, only one moment is needed.

Example 1

Find the centroid of the solid formed by revolving about the x axis the region bounded by $y = x^2$, $y = 0$, and $x = 1$.

The graph is given in Figure 16.25. Since the x axis is an axis of symmetry, the centroid is on it. All we need to find is its x coordinate. From Figure 16.25 we have

$$V = \int_0^1 \pi y^2 \, dx = \pi \int_0^1 x^4 \, dx$$

$$= \frac{\pi x^5}{5} \bigg|_0^1 = \frac{\pi}{5}.$$

Since the centroid of the disc is $(x, 0)$, it is a distance x from the yz plane. Thus, M_{yz} is the integral for volume above with the additional factor x.

$$M_{yz} = \int_0^1 \pi y^2 x \, dx = \pi \int_0^1 x^5 \, dx$$

$$= \frac{\pi x^6}{6} \bigg|_0^1 = \frac{\pi}{6},$$

and

$$\bar{x} = \frac{M_{yz}}{V} = \frac{\pi/6}{\pi/5} = \frac{5}{6}.$$

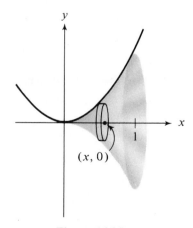

Figure 16.25

Example 2

The region of Example 1 is revolved about the y axis. Find its centroid.

By symmetry, the centroid is on the y axis—we need only find the y coordinate. Let us consider both the disc method and the shell method for finding the volume and moment. Figure 16.26 shows a representative disc. The centroid of the disc is in the center of the hole at $(0, y)$. From the figure, we have

$$V = \int_0^1 \pi(1 - x^2)\, dy = \pi \int_0^1 (1 - y)\, dy = \pi \left(y - \frac{y^2}{2} \right) \bigg|_0^1 = \frac{\pi}{2}.$$

Since the centroid of the disc is $(0, y)$, it is a distance y from the xz plane. Thus we include the factor y to get M_{xz}.

$$M_{xz} = \int_0^1 \pi(1 - x^2)y\, dy = \pi \int_0^1 (y - y^2)\, dy = \pi \left(\frac{y^2}{2} - \frac{y^3}{3} \right) \bigg|_0^1 = \frac{\pi}{6},$$

and

$$\bar{y} = \frac{M_{xz}}{V} = \frac{\pi/6}{\pi/2} = \frac{1}{3}.$$

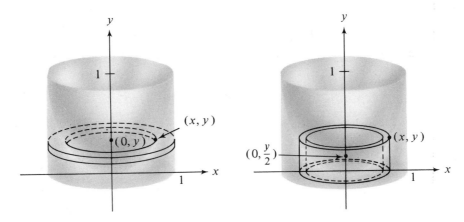

Figure 16.26   **Figure 16.27**

Figure 16.27 shows the same solid with a representative shell. Again the centroid of the shell is in the center of the hole at $(0, y/2)$. Thus we have

$$V = \int_0^1 2\pi xy\, dx = 2\pi \int_0^1 x^3\, dx = \frac{\pi x^4}{2} \bigg|_0^1 = \frac{\pi}{2}.$$

This time we get M_{xz} by including the factor $y/2$, since the centroid of the shell is a distance $y/2$ from the xz plane.

$$M_{xz} = \int_0^1 2\pi xy \cdot \frac{y}{2}\, dx = \pi \int_0^1 x^5\, dx = \frac{\pi x^6}{6} \bigg|_0^1 = \frac{\pi}{6},$$

and

$$\bar{y} = \frac{M_{xz}}{V} = \frac{\pi/6}{\pi/2} = \frac{1}{3}.$$

Although both the volume and moment were computed by both methods in the foregoing example, there is no need to use the same method for both of them. If it is easier to find the volume by one method and the moment by the other, by all means do so.

Example 3

Find the centroid of the solid formed by revolving about the y axis the region bounded by $y = \cos x$ and $y = 0$, between $x = 0$ and $x = \pi/2$.

The solid is given in Figure 16.28 with a representative cylindrical shell (the use of circular discs is considerably more difficult). Thus we have

$$V = \int_0^{\pi/2} 2\pi xy \, dx$$

$$= 2\pi \int_0^{\pi/2} x \cos x \, dx \qquad \begin{cases} u = x, \ v' = \cos x \\ u' = 1, \ v = \sin x \end{cases}$$

$$= 2\pi(x \sin x + \cos x) \Big|_0^{\pi/2}$$

$$= \pi(\pi - 2).$$

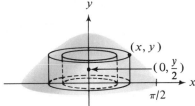

Figure 16.28

$$M_{xz} = \int_0^{\pi/2} 2\pi xy \cdot \frac{y}{2} \, dx = \int_0^{\pi/2} \pi xy^2 \, dx$$

$$= \pi \int_0^{\pi/2} x \cos^2 x \, dx \qquad \begin{cases} u = x, \ v' = \cos^2 x = \dfrac{1 + \cos 2x}{2} \\ u' = 1, \ v = \dfrac{x}{2} + \dfrac{1}{4}\sin 2x \end{cases}$$

$$= \pi\left(\frac{x^2}{4} + \frac{x}{4}\sin 2x + \frac{1}{8}\cos 2x\right)\Big|_0^{\pi/2}$$

$$= \frac{\pi(\pi^2 - 4)}{16};$$

and

$$\bar{y} = \frac{M_{xz}}{V} = \frac{\pi(\pi^2 - 4)/16}{\pi(\pi - 2)} = \frac{\pi + 2}{16}.$$

Of course, the centroid is on the y axis.

Problems

A

Find the centroid of the solid formed by revolving about the given axis the region bounded by the given curves.

1. $y = x^3$, $y = 0$, $x = 1$; about the x axis.
2. $y = x^3$, $y = 0$, $x = 1$; about the y axis.
3. $y = (x - 1)^2$, $y = 0$, $x = 0$; about the y axis.
4. $y = 2 - x$, $x = 0$, $y = 0$; about the y axis.
5. $y = 2 - 2x$, $x = 0$, $y = 0$; about the y axis.
6. $y = 2x - x^2$, $x = 0$; about the y axis.

7. $x^2 + y^2 = 1$, right of $x = 0$; about the x axis.
8. $x^2 + y^2 = 9$, right of $x = 1$; about the x axis.
9. $x^2/a^2 + y^2/b^2 = 1$, right of $x = 0$; about the x axis.
10. $x^2/a^2 + y^2/b^2 = 1$, above $y = 0$; about the y axis.
11. $x^2 - y^2 = 1$, $x = 3$; about the x axis.
12. $x^2 - y^2 = 1$, $y = 0$, $y = 2$; about the y axis.
13. $y = x^2 - x$, x axis; about the y axis.

B

Find the centroid of the solid formed by revolving about the given axis the region bounded by the given curves.

14. $y = 1/(x^2 + 1)$, $y = 0$, right of $x = 0$; about the x axis.
15. $y = \sin x$, $y = 0$, between $x = 0$ and $x = \pi/2$; about the x axis.
16. $y = \cos x^2$, $y = 0$, between $x = 0$ and $x = \sqrt{\pi/2}$; about the y axis.
17. $y = e^{-x}$, $y = 0$, right of $x = 0$; about the x axis. (*Hint*: $\lim_{k \to +\infty} (2k + 1)e^{-2k} = 0$.)
18. $y = \ln x$, $y = 0$, between $x = 1$ and $x = e$; about the y axis.
19. $x = y^2 - 4y$, y axis; about the x axis.
20. $y = 1/x^2$, $y = 0$, right of $x = 1$; about the x axis.

16.9

Centroids of Arcs and Surfaces

By an analysis similar to the one used for centroids of plane regions in Section 16.7, we can find the centroid of an arc or a surface of revolution. For an arc we have

$$s = \int_a^b ds, \qquad M_x = \int_a^b y \, ds, \qquad M_y = \int_a^b x \, ds;$$

and

$$\bar{x} = \frac{M_y}{s}, \qquad \bar{y} = \frac{M_x}{s},$$

where ds is the differential of arc given in Section 16.5 (page 519).

If an arc is rotated about the x axis to give a surface of revolution, then

$$S_x = \int_a^b 2\pi |y| \, ds, \qquad M_{yz} = \int_a^b 2\pi x |y| \, ds, \qquad \text{and} \quad \bar{x} = \frac{M_{yz}}{S_x}.$$

Of course, the centroid is on the x axis by symmetry.

Similarly, if the arc is rotated about the y axis, then

$$S_y = \int_a^b 2\pi |x| \, ds, \qquad M_{xz} = \int_a^b 2\pi |x| \, y \, ds, \qquad \text{and} \quad \bar{y} = \frac{M_{xz}}{S_y}.$$

Example 1

Find the centroid of the arc $y = \sqrt{1 - x^2}$.

The curve is a semicircle—its graph is given in Figure 16.29. We see by symmetry $\bar{x} = 0$; we need only find $\bar{y}$.

$$y' = \frac{-x}{\sqrt{1 - x^2}},$$

$$s = \int_{-1}^{1} \sqrt{1 + (y')^2} \, dx$$

$$= 2 \int_{0}^{1} \sqrt{1 + \frac{x^2}{1 - x^2}} \, dx$$

$$= 2 \lim_{\epsilon \to 0+} \int_{0}^{1-\epsilon} \frac{dx}{\sqrt{1 - x^2}}$$

$$= 2 \lim_{\epsilon \to 0+} \text{Arcsin } x \Big|_{0}^{1-\epsilon}$$

$$= 2 \lim_{\epsilon \to 0+} \text{Arcsin } (1 - \epsilon) = \pi;$$

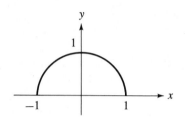

Figure 16.29

$$M_x = \int_{-1}^{1} y\sqrt{1 + (y')^2} \, dx$$

$$= \int_{-1}^{1} \sqrt{1 - x^2} \, \frac{1}{\sqrt{1 - x^2}} \, dx$$

$$= \int_{-1}^{1} dx$$

$$= x \Big|_{-1}^{1} = 2.$$

Thus $\bar{y} = \dfrac{M_x}{s} = \dfrac{2}{\pi}$.

Example 2

Find the centroid of the first quadrant arc of the circle $x = \cos \theta, \ y = \sin \theta$.

This is again a circle of radius one with center at the origin. The first quadrant corresponds to values of θ from $\theta = 0$ to $\theta = \pi/2$.

$$\frac{dx}{d\theta} = -\sin \theta, \qquad \frac{dy}{d\theta} = \cos \theta;$$

$$s = \int_{0}^{\pi/2} \sqrt{\left(\frac{dx}{d\theta}\right)^2 + \left(\frac{dy}{d\theta}\right)^2} \, d\theta$$

$$= \int_{0}^{\pi/2} \sqrt{\sin^2 \theta + \cos^2 \theta} \, d\theta$$

$$= \int_{0}^{\pi/2} d\theta$$

$$= \theta \Big|_{0}^{\pi/2} = \pi/2;$$

$$M_x = \int_0^{\pi/2} y \sqrt{\left(\frac{dx}{d\theta}\right)^2 + \left(\frac{dy}{d\theta}\right)^2} \, d\theta$$

$$= \int_0^{\pi/2} \sin \theta \, d\theta$$

$$= -\cos \theta \Big|_0^{\pi/2} = 1.$$

$$M_y = \int_0^{\pi/2} x \sqrt{\left(\frac{dx}{d\theta}\right)^2 + \left(\frac{dy}{d\theta}\right)^2} \, d\theta$$

$$= \int_0^{\pi/2} \cos \theta \, d\theta$$

$$= \sin \theta \Big|_0^{\pi/2} = 1.$$

Thus

$$\bar{x} = \frac{M_y}{s} = \frac{1}{\pi/2} = \frac{2}{\pi}.$$

From symmetry, $\bar{y} = \bar{x} = 2/\pi$.

Example 3

Find the centroid of the surface formed by revolving about the x axis the arc of Example 2.

$$S_x = \int_0^{\pi/2} 2\pi |y| \sqrt{\left(\frac{dx}{d\theta}\right)^2 + \left(\frac{dy}{d\theta}\right)^2} \, d\theta$$

$$= \int_0^{\pi/2} 2\pi \sin \theta \, d\theta$$

$$= -2\pi \cos \theta \Big|_0^{\pi/2} = 2\pi;$$

$$M_{yz} = \int_0^{\pi/2} 2\pi x |y| \sqrt{\left(\frac{dx}{d\theta}\right)^2 + \left(\frac{dy}{d\theta}\right)^2} \, d\theta$$

$$= \int_0^{\pi/2} 2\pi \cos \theta \sin \theta \, d\theta$$

$$= \pi \sin^2 \theta \Big|_0^{\pi/2} = \pi.$$

Thus the centroid is on the x axis, with

$$\bar{x} = \frac{M_{yz}}{S_x} = \frac{\pi}{2\pi} = \frac{1}{2}.$$

Example 4

Find the centroid of the surface formed by revolving about the y axis the portion of $x^2 + y^2 = 1$ in the first quadrant.

This is basically the same problem as Example 3, except this time it is in rectangular coordinates and the arc is revolved about the y axis.

$$y = \sqrt{1 - x^2}, \qquad y' = \frac{-x}{\sqrt{1 - x^2}};$$

$$S_y = \int_0^1 2\pi |x| \sqrt{1 + (y')^2}\, dx$$

$$= \lim_{\epsilon \to 0^+} \int_0^{1-\epsilon} 2\pi \frac{x}{\sqrt{1 - x^2}}\, dx$$

$$= \lim_{\epsilon \to 0^+} -2\pi \sqrt{1 - x^2}\ \Big|_0^{1-\epsilon}$$

$$= \lim_{\epsilon \to 0^+} [2\pi - 2\pi \sqrt{1 - (1 - \epsilon)^2}] = 2\pi;$$

$$M_{xz} = \int_0^1 2\pi y |x| \sqrt{1 + (y')^2}\, dx$$

$$= \int_0^1 2\pi x\, dx$$

$$= \pi x^2 \ \Big|_0^1 = \pi.$$

Thus the centroid is on the y axis, with

$$\bar{y} = \frac{M_{xz}}{S_y} = \frac{\pi}{2\pi} = \frac{1}{2}.$$

Problems

A

In Problems 1–6, find the centroid of the given arc.

1. The portion of $x^{2/3} + y^{2/3}$ in the first quadrant.

2. $y = \cosh x$, from $x = -1$ to $x = 1$.   3. $y = \dfrac{x^3}{6} + \dfrac{1}{2x}$, from $x = 1$ to $x = 3$.

4. $x = \dfrac{y^3}{3} + \dfrac{1}{4y}$, from $y = 1$ to $y = 3$.

5. The portion of $x^2 + 4y = 4$ in the first quadrant.
6. $x = \cos^3 \theta$, $y = \sin^3 \theta$, from $\theta = 0$ to $\theta = \pi/2$.

In Problems 7–11, the given arc is revolved about the x axis. Find the centroid of the resulting surface.

7. All of $x = \sqrt{1 - y^2}$.
8. $x^2 + y^2 = 9$, from $(3, 0)$ to $(1, 2\sqrt{2})$ (first quadrant).
9. $y = \cosh x$, from $x = -1$ to $x = 1$.   10. $y = 2x$, from $(0, 0)$ to $(1, 2)$.

11. $y = \dfrac{x^3}{3} + \dfrac{1}{4x}$, from $x = 1$ to $x = 3$.

In Problems 12 and 13, the given arc is revolved about the y axis. Find the centroid of the resulting surface.

12. $y = hx/r$, from $(0, 0)$ to (r, h).   13. Top half of $x^2 + y^2 - 2x = 0$.

B

In Problems 14–17, find the centroid of the given arc.

14. $x^2 = 8y$, from $(-4, 2)$ to $(4, 2)$.
15. $x = \theta - \sin \theta$, $y = 1 - \cos \theta$, from $\theta = 0$ to $\theta = 2\pi$.
    (The curve is symmetric about $x = \pi$.)
16. $x = a(\theta - \sin \theta)$, $y = a(1 - \cos \theta)$, from $\theta = 0$ to $\theta = \pi$.
17. $y^2 = 4x$, from $(1, -2)$ to $(1, 2)$.

In Problems 18 and 19, the given arc is revolved about the y axis. Find the centroid of the resulting surface.

18. $x = \sqrt{y}$, from $y = 0$ to $y = 4$.   19. $y = \cosh x$, from $x = 0$ to $x = 1$.

C

20. The arc $x = \theta - \sin \theta$, $y = 1 - \cos \theta$, from $\theta = 0$ to $\theta = \pi/2$, is revolved about the x axis. Find the centroid of the resulting surface.
21. The Second Theorem of Pappus states that if a plane arc lies entirely on one side of a line, then the area of the surface formed by revolving the arc about that line is the product of the length of the arc and the length of the path of the centroid of the arc. Verify this for the arc $y = \sqrt{1 - x^2}$ rotated about the x axis.
22. Verify the Second Theorem of Pappus (see Problem 21) for the arc $y = x^2$, from $(0, 0)$ to $(1, 1)$, rotated about the y axis.

16.10

Moments of Inertia

In previous sections we have dealt with moments in the determination of centroids. In each case the moment was the product of the mass and a distance. This is called the first moment.

Let us consider another moment. If a mass m is concentrated at a distance x from a given line, then the moment of inertia of the mass about the line is defined to be

$$I = mx^2.$$

This is called the second moment. Inertia is the property of matter to resist motion when at rest and to resist a change in speed or direction when in motion. The moment of inertia is a measure of this tendency—the greater the moment of inertia, the greater the tendency to resist changes in motion.

The moment of inertia is also of great interest to engineers, since it is related to the strength of materials. For example, the stiffness of a beam is directly proportional to the moment of inertia of its cross section about the axis of the beam (see Problem 23).

Example 1

Find the moment of inertia about the y axis of a five-pound mass at $x = 3$.

$$I_y = mx^2$$
$$= 5 \cdot 3^2$$
$$= 45.$$

Example 2

Find the moment of inertia about the y axis of a system consisting of a five-pound mass at $x = 3$ and a five-pound mass at $x = -3$.

$$I_y = m_1 x_1^2 + m_2 x_2^2$$
$$= 5 \cdot 3^2 + 5(-3)^2$$
$$= 90.$$

We see by these examples that the symmetry of the system does not bring about a cancellation of the two moments as it did with first moments. Since the distance is squared, the individual moments are never negative; and, instead of canceling, the individual moments add up to give a bigger moment. Note also that here we are not using the moment to find something else as we did with first moments. Thus, the density (mass per unit area) will not cancel out as it did when finding centers of mass. We must take density into consideration when finding a moment of inertia.

Example 3

Find the moment of inertia about the y axis of the region bounded by $y = x^2$, the x axis, and $x = 1$ where the region has a density of 2 g/cm². (The units for all of the equations are cm.)

$$I_y = \lim_{\|S\| \to 0} \sum_{i=1}^{n} x_i^2 m_i$$

$$= \lim_{\|S\| \to 0} \sum_{i=1}^{n} x_i^2 \cdot 2y_i \, \Delta x_i$$

$$= \int_0^1 x^2 \cdot 2y \, dx$$

$$= 2 \int_0^1 x^4 \, dx$$

$$= \frac{2x^5}{5} \Big|_0^1$$

$$= \frac{2}{5} \text{ g cm}^2.$$

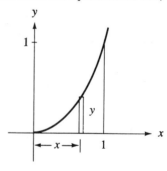

Figure 16.30

Example 4

Find the moment of inertia about the x axis of the region of Example 3.

We cannot use vertical strips as we did in Example 3, because the points on them are not all the same distance from the x axis (nor will they approach that in the limit). We must use horizontal strips (see Figure 16.31).

$$I_x = \int_0^1 y^2 \cdot 2(1 - x) \, dy$$

$$= 2 \int_0^1 y^2 (1 - \sqrt{y}) \, dy$$

$$= 2 \int_0^1 (y^2 - y^{5/2}) \, dy$$

$$= 2 \left(\frac{y^3}{3} - \frac{2}{7} y^{7/2} \right) \Big|_0^1$$

$$= \frac{2}{21} \text{ g cm}^2.$$

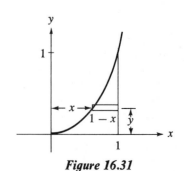

Figure 16.31

A quantity related to the moment of inertia is the radius of gyration. If the entire mass m of a region is located at a distance R from the axis of reference, then

$$I = mR^2, \quad \text{or} \quad R = \sqrt{I/m}.$$

Thus, for any region, the radius of gyration is given by $R = \sqrt{I/m}$. When a region has a radius of gyration R about a given line, it means that with respect to rotation about that line, it behaves as if the entire mass were concentrated at a distance R from the line.

Example 5

Find the radius of gyration about the y axis of the region of Example 3.

From Example 3, $I_y = 2/5$ g cm².

$$
\begin{aligned}
m &= \int_0^1 2y \, dx \\
&= 2 \int_0^1 x^2 \, dx \\
&= \frac{2x^3}{3} \bigg|_0^1 \\
&= \frac{2}{3} \text{ g.}
\end{aligned}
$$

Thus

$$R = \sqrt{I/m} = \sqrt{\frac{2/5}{2/3}} = \sqrt{3/5} \text{ cm.}$$

This indicates that, with respect to motion about the y axis, the region resists changes in motion as if all of the mass were concentrated $\sqrt{3/5}$ cm from the y axis. It might be noted that the density is not needed for finding the radius of gyration, since it must always cancel out.

Example 6

Find the radius of gyration about the x axis of the region of Example 4.

From Example 4, $I_x = 2/21$. From Example 5, $m = 2/3$; thus

$$R = \sqrt{I/m} = \sqrt{\frac{2/21}{2/3}} = \frac{1}{\sqrt{7}} \text{ cm.}$$

Problems

A

Find the moment of inertia and radius of gyration about the given axis.

1. 4-lb mass at $x = 2$,   2-lb mass at $x = -3$;   about y axis.
2. 10-lb mass at $x = 1$,   10-lb mass at $x = 5$;   about y axis.
3. 3-lb mass at $x = -2$,   6-lb mass at $x = -3$;   about y axis.
4. 2-lb mass at $x = 5$,   3-lb mass at $x = -2$;   about y axis.

5. The region bounded by $y = x^2$, the y axis, $y = 1$ (first quadrant); density 6 g/cm^2; about y axis.

6. The region of Problem 5; about x axis.

7. The region bounded by $y = x^3$, the x axis, $x = 1$; density 8 g/cm^2; about x axis.

8. The region of Problem 7; about y axis.

9. The region bounded by $y = 2x - x^2$ and the x axis; density 1 g/cm^2; about y axis.

10. The region of Example 3; about $x = 1$.

11. The region of Example 3; about $y = 1$.

12. The region bounded by $y = 1/x$, the x axis and between $x = 1$ and $x = 2$; density 6 g/cm^2; about y axis.

13. The region of Problem 12; about x axis.

B

Find the moment of inertia and radius of gyration about the given axis.

14. The region of Problem 9; about x axis.

15. The region bounded by $y = \sin x$ and the x axis between $x = 0$ and $x = \pi$; density 2 g/cm^2; about y axis.

16. The region of Problem 15; about $x = \pi/2$.

17. The region bounded by $y = e^x$, the x axis and between $x = -1$ and $x = 0$; density 7 g/cm^2; about y axis.

18. The region of Problem 17; about $x = -1$.

19. Inside $x^2 + y^2 = 1$; density 2 g/cm^2; about y axis.

20. Inside $4x^2 + y^2 = 4$; density 2 g/cm^2; about y axis.

21. The region of Problem 20; about x axis.

22. The region bounded by $\sqrt{x} + \sqrt{y} = 1$ and both axes; density 5 g/cm^2; about x axis.

C

23. The stiffness of a beam is proportional to the moment of inertia of its cross section about the axis of the beam. Show that a pipe with inside diameter 4 in. and outside diameter 5 in. and a rod of diameter 3 in. have the same cross-sectional area. Which is stiffer?

16.11

Fluid Force

Suppose a swimming pool is 15 feet wide and 30 feet long, has a horizontal bottom, and is 8 feet deep. Suppose further that we want to find the force of the water on the bottom of the pool. This is a relatively easy problem. The pressure (force per unit area) of the water on the bottom is the weight of the water above a region of area 1; thus the pressure is the product of the depth and the density of water.

$$P = 8(62.4)$$
$$= 499.2 \text{ lb/ft}^2.$$

The total force is then the product of the pressure and the area.

$$F = P \cdot A$$
$$= (499.2)(15 \cdot 30)$$
$$= 224{,}640 \text{ lb.}$$

Now let us consider the more difficult problem of finding the force on one end of the pool. Since different parts of the wall are at different depths, the pressure is not a constant; it varies from 0 at the top to $8(62.4) = 499.2$ lb/ft^2 at the bottom. This continuous variation suggests the use of the integral. Let us introduce coordinate axes, as in Figure 16.32. Now we subdivide the interval [0, 8] on the y axis by

$$y_0 = 0, y_1, y_2, y_3, \ldots, y_{n-1}, y_n = 8,$$

and select a number y^* in each subinterval such that

$$y_{i-1} \leq y_i^* \leq y_i.$$

Let us consider the force on a horizontal strip corresponding to the ith subinterval (see Figure 16.32). We can approximate the depth below the surface by $8 - y_i^*$.

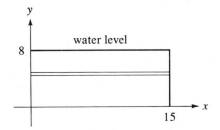

Figure 16.32

Although the points of the strip are not all at this depth, none differs by more than the width of the interval $y_i - y_{i-1}$. Thus the pressure is $62.4(8 - y_i^*)$, and the force on the horizontal strip is the product of this pressure and the area of the strip,

$$62.4(8 - y_i^*)15(y_i - y_{i-1}) = 936(8 - y_i^*)\, \Delta y_i,$$

where $\Delta y_i = y_i - y_{i-1}$. We can approximate the total force on the wall by adding together the forces for all of the horizontal strips. The exact value is found by taking the limit of this approximating sum as the lengths of all the subintervals approach 0.

$$F = \lim_{\|s\| \to 0} \sum_{i=1}^{n} 936(8 - y_i^*)\, \Delta y_i$$
$$= \int_0^8 936(8 - y)\, dy$$
$$= 936\left(8y - \frac{y^2}{2}\right)\Big|_0^8$$
$$= 29{,}952 \text{ lb.}$$

While this is a simple case, it illustrates how the integral may be used to determine fluid forces.

Example 1

A swimming pool is ten feet wide and twenty feet long. The bottom is flat (but not horizontal) and the sides are vertical. The water is three feet deep at one end and ten feet deep at the other. Find the force of the water on one twenty-foot side.

Coordinate axes are put in the position shown in Figure 16.33. Notice that the downward direction is taken to be positive so that we can deal with positive numbers throughout. The equation of the line through (20, 3) and (0, 10) is $7x + 20y = 200$. We see from Figure 16.33 that, while the horizontal strips

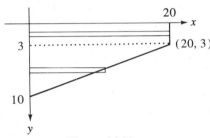

Figure 16.33

always have their left ends on the y axis, the right ends are sometimes on the vertical line $x = 20$ and sometimes on $7x + 20y = 200$. Thus we need two integrals.

$$F = \int_0^3 (62.4)y(20)\, dy + \int_3^{10} (62.4)y\left(\frac{200 - 20y}{7}\right) dy$$

$$= 1248 \int_0^3 y\, dy + \frac{1248}{7} \int_3^{10} (10y - y^2)\, dy$$

$$= 624y^2 \Big|_0^3 + \frac{1248}{7}\left(5y^2 - \frac{y^3}{3}\right)\Big|_3^{10}$$

$$= 28{,}912 \text{ lb.}$$

Of course, the placement of the axes has no effect on the final result. You are encouraged to see this for yourself by repeating this problem with the axes in another position.

Example 2

Find the force on a circular gate of diameter four feet in a vertical dam where the center of the gate is twenty feet below the surface of the water.

First let us put the coordinate axes in the position indicated in Figure 16.34. The circular gate is then represented by the equation $x^2 + y^2 = 4$. Thus we have

$$F = \int_{-2}^2 62.4(20 - y)2x\, dy$$

$$= 124.8 \int_{-2}^2 (20 - y)\sqrt{4 - y^2}\, dy.$$

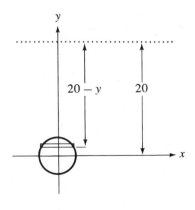

Figure 16.34

Using the substitution $y = 2 \sin \theta$, we have

$$F = 124.8 \int_{-\pi/2}^{\pi/2} (20 - 2 \sin \theta)\sqrt{4 - 4 \sin^2 \theta}\, 2 \cos \theta\, d\theta$$

$$= 124.8 \int_{-\pi/2}^{\pi/2} 8(10 - \sin \theta) \cos^2 \theta\, d\theta$$

$$= 998.4 \int_{-\pi/2}^{\pi/2} [5(1 + \cos 2\theta) - \sin \theta \cos^2 \theta]\, d\theta$$

$$= 998.4 \left[5\theta + \frac{5}{2} \sin 2\theta + \frac{1}{3} \cos^3 \theta \right] \Big|_{-\pi/2}^{\pi/2}$$

$$= 4992\pi \text{ lb.}$$

Problems

A

1. A dam contains a vertical rectangular gate 10 ft wide and 5 ft high. The top of the gate is horizontal and 10 ft below the surface of the water. Find the force on the gate.

2. A vertical wall of a swimming pool is 10 ft wide and 6 ft high. Find the force of water on the wall when the pool is half full.

3. A hollow metal cube has a 1-ft edge. It is suspended underwater with the top face horizontal and 8 ft below the surface. Find the total force on all six faces.

4. A dam has a square gate with side 2 ft long and one diagonal vertical. The highest point of the gate is 8 ft below the water surface. Find the force on the gate.

5. A trough is 6 ft long and 1 ft high. Vertical cross sections are equilateral triangles with the top side horizontal. Find the force on one end if the trough is filled with water.

6. A trough is 8 ft long and 1 ft high. Vertical cross sections are isosceles right triangles with the hypotenuse horizontal. Find the force on one end if the trough is filled with water.

7. Find the force on one side of the trough of Problem 5.

8. Find the force on one side of the trough of Problem 6.

9. Find the force on the bottom half of a vertical ellipse with major axis 10 and minor axis 6 if the major axis is on the surface of the water.

10. Find the force on the bottom half of the ellipse of Problem 9 if the minor axis is on the surface of the water.

11. A dam has a vertical gate that is an isosceles trapezoid with upper base 6 ft, lower base 8 ft, and height 3 ft. Find the force on the gate if the upper base is 10 ft below the surface.

12. Suppose the gate of Problem 11 were inverted. Find the force on it.

13. A dam has a vertical gate that is 10 ft wide and 3 ft high. Its top is horizontal and it can withstand a force of 35,000 lb. What is the highest level to which the dam can be filled?

B

14. An irrigation ditch has a vertical head gate that is a circle of diameter 2 ft. Find the force on the head gate if the water is level with the top of the gate.

15. Find the force on the head gate of Problem 14 if the level of the water is half-way up from the bottom.

16. Find the force on the head gate of Problem 14 if the level of the water is h ft $(0 \le h \le 2)$ from the bottom.

17. Find the force on the head gate of Problem 14 if the level of the water is h ft $(h \ge 0)$ above the top of the gate.

18. A dam is in the shape of a parabola 10 ft high and 8 ft across the top. Find the force on it when filled to the top.

19. Find the force on the dam of Problem 18 when filled half-way to the top.

20. A dam has a vertical, circular gate of radius 2 ft. It can withstand a force of 18,000 lb. What is the highest level to which the dam can be filled?

21. A swimming pool is in the form of a circular cylinder of radius 10 ft. What is the force on the wall of the pool if it is filled to a level of 4 ft?

22. Suppose that the gate of Problem 1 is inclined at an angle of 60° with the horizontal. What is the force on it?

23. Suppose that the gate of Problem 1 is inclined at an angle of 45° with the horizontal. What is the force on it?

24. Suppose that the dam and gate of Example 2 are inclined at an angle of 60° with the horizontal. What is the force on the gate?

C

25. Show that the force on the vertical face of a dam is the product of the density of the water behind the dam, the area of the face, and the depth of the centroid of the face.

26. Use the result of Problem 25 to solve   (a) Problem 1;   (b) Problem 5;   (c) Problem 8;   (d) Problem 15.

Review Problems

A

In Problems 1–3, evaluate the given integral.

1. $\displaystyle\int_{-\infty}^{+\infty} \frac{x\,dx}{x^4 + 1}$.

2. $\displaystyle\int_{-2}^{0} \frac{dx}{\sqrt[3]{x + 2}}$.

3. $\displaystyle\int_{0}^{3} \frac{x\,dx}{x - 2}$.

4. The region bounded by $y = 2x - x^2$ and the x axis is revolved about the x axis. Find the volume of the resulting solid. Find the volume of the solid formed by revolving this region about the y axis.

5. The region bounded by $y = e^x$, the y axis, and the line $y = e$ is revolved about the x axis. Find the volume of the resulting solid.

6. The region bounded by $xy = 5$ and $x + y = 6$ is revolved about the x axis. Find the volume of the resulting solid.

7. The region bounded by $y = 4x^2$, the x axis, and the line $x = 2$ is revolved about $x = 2$. Find the volume of the resulting solid.

8. Find the volume of the solid formed by revolving the region of Problem 7 about $x = 3$.

9. The region bounded by $y = x(x - 2)^2$ and the x axis is revolved about the y axis. Find the volume of the solid formed.

10. A solid has a base that is an isosceles right triangle with hypotenuse 4; cross sections perpendicular to the base and parallel to the hypotenuse are semicircles. Find the volume of the solid.

11. The arc $y = a \cosh (x/a)$ from $x = -a$ to $x = a$ is revolved about the x axis. Find the area of the resulting surface.

12. Find the area of the surface formed by revolving about the y axis the arc $y = \frac{2}{3}(x^2 - 1)^{3/2}$ from $x = 1$ to $x = 2$.

13. Find the center of mass of the system consisting of a 3-lb mass at $(3, 1)$, a 2-lb mass at $(-2, 4)$, and a 5-lb mass at $(0, -3)$.

14. A thin metal plate of uniform density has the shape given in Figure 16.35. Find its center of mass.

15. Suppose that the metals of the plate of Problem 14 have the following densities: the large center square has density 1 oz/in.^{2}, the rectangle on the right has density 2 oz/in.^{2}, and the small square on the left has density 3 oz/in.^{2}. Find the center of mass of the plate.

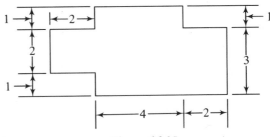

Figure 16.35

16. Find the centroid of the region bounded by $y = x^2 - 4x + 3$ and the x axis.

17. Find the centroid of the region bounded by $y = 6x - x^2$ and $y = 3x$.

18. Find the centroid of the region bounded by $y = 4(4 - x^2)$ and $y = 4 - x^2$.

19. The region bounded by $y = x^2(x - 2)$ and the x axis is revolved about the x axis. Find the centroid of the resulting solid.

20. Find the centroid of the solid formed by revolving the region of Problem 18 about the y axis.

21. Find the centroid of the solid formed by revolving about the x axis the region bounded by $y = 4x - x^2$ and $y = x$.

22. Find the moment of inertia and radius of gyration about the y axis of the region bounded by $y = 3x - x^2$ and the x axis, if the region has a density of 2 g/cm^2.

23. A region bounded by $y = x$ and $y = x^2$ has density 4 g/cm^2. Find its moment of inertia and radius of gyration about the x axis.

24. A vertical dam contains a rectangular gate 3 ft wide and 2 ft high. The top of the gate is horizontal and 10 ft below the water surface. Find the total force on the gate.

B

25. Evaluate $\int_0^{+\infty} e^{-x} \sin x \, dx$.

26. Find the volume of the solid formed by revolving about the line $y = 1$ the region bounded by $y = 1/x^2$, the x axis, and to the right of $x = 1$.

27. A solid has a base consisting of an equilateral triangle with side 4; cross sections perpendicular to the base and parallel to one side of the triangle are semi-ellipses of height 4. Find the volume of the solid.

28. A pyramid of height h has a rectangular base with dimensions $x \times y$. Show that the volume of the pyramid is $V = xyh/3$.

29. Find the area of the surface formed by revolving about the y axis the arc $x = \sqrt{t}/2$, $y = (t^3 + 128)/128t$, from $t = 1$ to $t = 4$.

30. The region to the right of $x = 1$ and between $y = 1/x^3$ and the x axis is revolved about the x axis. Find the centroid of the resulting solid.

31. Find the centroid of the arc $y = x^2 - (\ln x)/8$ from $x = 1$ to $x = 2$.

32. Suppose the arc of Problem 31 is revolved about the y axis. Find the centroid of the resulting surface.

33. Suppose the arc $x = \cos^3 \theta$, $y = \sin^3 \theta$ from $\theta = 0$ to $\theta = \pi/2$ is revolved about the x axis. Find the centroid of the resulting surface.

34. Find the moment of inertia and radius of gyration about the x axis of the region bounded by $y = 2x - x^2$ and the x axis if the region has a density of 1 g/cm^2.

35. A dam has the shape of a parabola 40 ft high and 40 ft across the top. Find the force on the dam when it is filled to the top and when it is filled to a level 10 ft below the top.

C

36. Suppose that $f(x) \geq 0$ for $x \geq 0$ and that the region bounded by $y = f(x)$, the x axis, $x = 0$, and $x = k$ is revolved about the x axis. The volume of the resulting solid is $3k^3 + 3k^2 + 4k$. Find f.

37. Find the centroid of the region in the first quadrant bounded by $y = x$ and $y = x^n$, where $n > 1$. What is the limiting position of the centroid as $n \to +\infty$?

38. Find the centroid of the region of Figure 16.36. What is the limiting position of the centroid as $r \to R$? Compare this with the centroid of a semicircular arc of radius R.

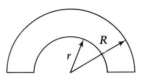

Figure 16.36

39. A rectangular tank is 10 ft high with a 10-ft × 10-ft base. It is filled with a 50–50 mixture of oil and water. If the oil has a density of 40 lb/ft^3, find the force on a vertical wall of the tank.

17

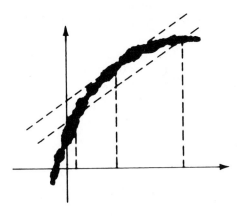

Limits and Continuity: The Epsilon-Delta Approach

17.1

The Limit

Since absolute values and inequalities are used quite extensively throughout this chapter, a review of these topics is given in Appendix A for those who need it. In Chapter 10 we considered limits and continuity by means of a geometric definition. The main disadvantage of that definition is that it depends upon knowledge of an accurate graph of the function in question. Furthermore, you are not likely to encounter that definition for limit in other books. We now consider a more standard definition of limit.

Definition

The statement $\lim\limits_{x \to a} f(x) = b$ *means*

(a) if δ is a positive number, then there is a number x in the domain of f such that

$$0 < |x - a| < \delta, \quad and$$

(b) if ϵ is a positive number, then there is a positive number δ such that

$$|f(x) - b| < \epsilon$$

for every number x in the domain of f for which

$$0 < |x - a| < \delta.$$

547

The two parts of this definition correspond exactly to the two parts of the definition in Chapter 10. In order to see the similarity, note first that $|x - a|$ can be interpreted as the horizontal distance between points with x coordinates x and a; similarly, $|f(x) - b|$ can be interpreted as the vertical distance between points with y coordinates $f(x)$ and b. The first part of this definition states that no matter how small a number δ we choose, there is some number x in the domain of f such that $x \neq a$ (since $0 < |x - a|$), and the distance between x and a is less than δ. In other words, there are numbers in the domain of f as close as we please to the number a. This is exactly what was stated by the first part of the definition in Chapter 10. Let us recall that the number a is a limit point of the domain of f if it satisfies the first part of the definition.

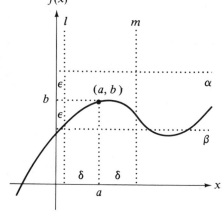

Figure 17.1

Definition

The number a is a limit point of the domain of the function f if for each $\delta > 0$ there is a number x in the domain of f (necessarily distinct from a) such that

$$0 < |x - a| < \delta.$$

Given a positive number ϵ, we see that it determines a pair of horizontal lines α and β (see Figure 17.1), which are a distance ϵ above and below (a, b). Thus, having a positive number ϵ is equivalent to having a pair of horizontal lines α and β with (a, b) between them. Similarly, the positive number δ determines a pair of vertical lines l and m, which are a distance δ to the right and left of the point (a, b). Now the set of all numbers x belonging to the domain of f and satisfying

$$0 < |x - a| < \delta$$

corresponds to the set of all points of the graph between l and m except the one with x coordinate a. Similarly the set of all numbers $f(x)$ in the range of f and satisfying

$$|f(x) - b| < \epsilon$$

corresponds to the set of all points of the graph between α and β. This explanation shows that the new definition is equivalent to the one given in Chapter 10. Let us now see how we can work with this definition.

Example 1

Show, by means of the definition, that $\lim\limits_{x \to 1} (x + 1) = 2$.

Since the function $f(x) = x + 1$ is defined for all x, the first part of the definition is satisfied. Suppose we are given a positive number ϵ. We must now find a positive number δ, such that $|f(x) - 2| < \epsilon$ whenever $0 < |x - 1| < \delta$. Let us suppose that $0 < |x - 1| < \delta$ for some choice of δ. Then

$$|f(x) - 2| = |(x + 1) - 2|$$
$$= |x - 1| < \delta.$$

We want $|f(x) - 2|$ to be less than ϵ. We get it if we choose $\delta = \epsilon$. Since we are able to find a δ in terms of the given ϵ, we see that, for any value of ϵ, there is a value of δ satisfying the condition that $|f(x) - 2| < \epsilon$ for any x in the domain such that $0 < |x - 1| < \delta$. Thus the given limit statement is true.

Example 2

Show that $\lim\limits_{x \to 1} x^2 = 1$.

Again $f(x) = x^2$ is defined for all x, and the first part of the definition holds. Given a positive number ϵ, we now want to find a positive number δ such that

$$|x^2 - 1| < \epsilon$$

whenever

$$0 < |x - 1| < \delta.$$

Again, suppose that $0 < |x - 1| < \delta$ for some choice of δ. The situation is not so simple here as it was in the previous example. But we do know this much: no matter what value of δ we choose, $|x^2 - 1|$ can be broken down into two factors, $|x + 1|$ and $|x - 1|$, one of which is always less than δ. All we need to do is limit the size of the other factor. We can do that by restricting the size of δ. Let us arbitrarily choose $\delta \leq 1$; that is, we shall never choose a value of δ greater than one, although we may choose a smaller value. Now

$$|x - 1| < \delta \leq 1$$
$$|x - 1| < 1$$
$$-1 < x - 1 < 1$$
$$1 < x + 1 < 3$$
$$|x + 1| < 3.$$

Thus, if $|x - 1| < \delta$ and $\delta \leq 1$, then

$$|x^2 - 1| = |x + 1| \cdot |x - 1| < 3\delta.$$

Now if we let $3\delta = \epsilon$ or $\delta = \epsilon/3$, then $|x^2 - 1| < \epsilon$. But this is also based on the assumption that $\delta \leq 1$. If $\epsilon > 3$, then $\epsilon/3 > 1$. Thus we must choose δ to be the smaller of the two numbers $\epsilon/3$ and 1, and we write it

$$\delta = \min \{\epsilon/3, 1\}.$$

In this last example, how did we know to choose $\delta \leq 1$? Why not $\delta \leq 3$ or $\delta \leq 1/2$? Actually we could have started with any of these restrictions. The choice of $\delta \leq 1$ was quite arbitrary; any other choice would work as well. This is not always the case, as the next example shows.

Example 3

Show that $\lim\limits_{x \to 1} (1/x) = 1$.

Again, the first part of the definition is obviously satisfied. Now we must show that if ϵ is a positive number, then there is a positive number δ such that

$$\left| \frac{1}{x} - 1 \right| < \epsilon$$

whenever $x \neq 0$ and $|x - 1| < \delta$. Suppose $0 < |x - 1| < \delta$. Then

$$\left| \frac{1}{x} - 1 \right| = \left| \frac{1 - x}{x} \right|$$

$$= \frac{|1 - x|}{|x|}$$

$$= \frac{1}{|x|} |x - 1|.$$

Again one of the two factors is less than δ, but the value of the other must be restricted. The only way we can do this is to restrict the value of δ. Let us try $\delta \leq 1$. Then

$$|x - 1| < \delta \leq 1$$
$$|x - 1| < 1$$
$$-1 < x - 1 < 1$$
$$0 < x < 2$$
$$\frac{1}{x} > \frac{1}{2}$$
$$\frac{1}{|x|} > \frac{1}{2}.$$

Unfortunately this does not get us anywhere. We want to be able to say that $1/|x|$ is less than something. But because $0 < x < 1$, there is no limit on how big $1/x$ can be. Our problem is that, by choosing $\delta \leq 1$, we did not restrict x enough. This restriction merely requires that $|x - 1| < 1$. Figure 17.2(a) shows

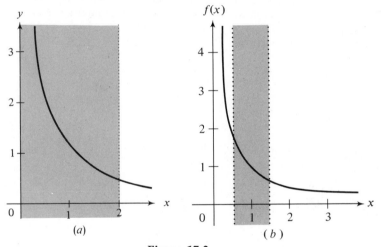

Figure 17.2

the points satisfying this inequality. The points of the graph that satisfy this inequality have y coordinates ($=1/x$) without an upper limit. We can see from Figure 17.2(b) that if the width of the band were narrowed, there would be an upper limit to $1/x$. With this in mind, let us choose $\delta \leq 1/2$. Now

$$|x - 1| < \delta \leq \frac{1}{2}$$
$$|x - 1| < \frac{1}{2}$$
$$-\frac{1}{2} < x - 1 < \frac{1}{2}$$
$$\frac{1}{2} < x < \frac{3}{2}$$
$$\frac{2}{3} < \frac{1}{x} < 2$$
$$\frac{2}{3} < \frac{1}{|x|} < 2.$$

Actually the only inequality we want is $1/|x| < 2$. Now we have

$$\left|\frac{1}{x} - 1\right| = \frac{1}{|x|}|x - 1| < 2\delta.$$

Thus we choose $2\delta = \epsilon$ or $\delta = \epsilon/2$, provided $\epsilon/2 \leq 1/2$, or

$$\delta = \min\{\epsilon/2, 1/2\}.$$

Problems

A

In Problems 1–10, determine whether a is a limit point of the domain of the given function.

1. $f(x) = x + 2, \quad a = 2.$

2. $f(x) = \dfrac{x^2 - 1}{x - 1}, \quad a = 1.$

3. $f(x) = \dfrac{x^2 - 1}{x - 1}, \quad a = 0.$

4. $f(x) = \sqrt{x}, \quad a = 0.$

5. $f(x) = \sqrt{x}, \quad a = -1.$

6. $f(x) = \sqrt{x^4 - 9x^2}, \quad a = 0.$

7. $f(x) = \sqrt{x^4 - 9x^2}, \quad a = 3.$

8. $f(x) = \sqrt{x^4 - 9x^2}, \quad a = 1.$

9. $f(x) = \dfrac{1}{x}, \quad a = 0.$

10. $f(x) = \text{Arcsin } x, \quad a = 2.$

11. Show that $\lim_{x \to 1} x^2 = 1$ by using $\delta \leq 3$.
12. Show that $\lim_{x \to 1} x^2 = 1$ by using $\delta \leq 1/2$.

In Problems 13–23, use the definition of limit stated in this section to show that the given limit statements are true.

13. $\lim\limits_{x \to 1} (x - 3) = -2.$

14. $\lim\limits_{x \to 1} (2x + 3) = 5.$

15. $\lim\limits_{x \to 2} \dfrac{x^2 - 4}{x - 2} = 4.$

16. $\lim\limits_{x \to 1} \dfrac{x^2 - 4}{x - 2} = 3.$

17. $\lim\limits_{x \to 0} x^2 = 0.$

18. $\lim\limits_{x \to 3} x^2 = 9.$

19. $\lim\limits_{x \to -1} x^2 = 1.$

20. $\lim\limits_{x \to 1} (x^2 + x) = 2.$

21. $\lim\limits_{x \to 1} (x^2 - x) = 0.$

22. $\lim\limits_{x \to 1} (x^2 + 3x) = 4.$

23. $\lim\limits_{x \to 0} \sqrt{x} = 0.$

B

In Problems 24–30, use the definition of limit stated in this section to show that the given limit statements are true.

24. $\lim\limits_{x \to 1} x^3 = 1.$

25. $\lim\limits_{x \to 2} x^3 = 8.$

26. $\lim\limits_{x \to 0} x^3 = 0.$

27. $\lim\limits_{x \to 0} \dfrac{1}{x + 1} = 1.$

28. $\lim\limits_{x \to 3} \dfrac{1}{x} = \dfrac{1}{3}.$

29. $\lim\limits_{x \to 1} \dfrac{1}{x^2} = 1.$

30. $\lim\limits_{x \to 0} \dfrac{1}{x - 1} = -1.$

17.2

The Limit (Continued)

The examples of the previous section required us only to verify that a given limit statement is true. We now consider the problem of showing that a given limit statement is false. Again there are *two* ways for a given limit statement to be false: it can fail to satisfy the first part of the definition *or* it can fail to satisfy the second part. It is easy to show whether or not the first part of the definition is satisfied; the second part is more difficult.

When considering the geometric definition of limit, we noted that if the second condition fails to hold, it fails for horizontal lines that are "close" together. Since these are related to the ϵ of our definition, we see that if the second condition of this definition fails to hold, it fails for "small" values of ϵ. More precisely, if the second part of the definition fails for one value of ϵ, it fails for any smaller value. In order to show that the second part of the definition fails, we must find *one* value of ϵ for which there is *no* δ satisfying the given conditions. Let us consider some examples.

Example 1

Show that $\lim_{x \to 0} f(x) \neq 1$ if

$$f(x) = \begin{cases} 0 & x \leq 0, \\ 1 & x > 0. \end{cases}$$

Since $f(x)$ is defined for all real numbers, the first part of the definition is clearly satisfied—zero is a limit point of the domain of f. The graph of this function is given in Figure 17.3. If we were working with horizontal and vertical lines, we would choose β high enough so that the portion of the graph on the x axis is not between α and β. Thus no matter how we choose l and m, there is some point of the graph between l and m that is not between α and β. This corresponds to a choice of ϵ less than or equal to 1. Let us choose $\epsilon = 1/2$. We now want to show that we cannot find a value of δ that satisfies the definition. Let us assume that there is a value of δ satisfying the definition. Then if x satisfies the inequality

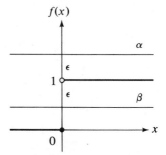

Figure 17.3

$$0 < |x - 0| < \delta,$$

we have $0 < |x| < \delta$, or

$$-\delta < x < \delta \quad \text{and} \quad x \neq 0.$$

For those values of x for which $-\delta < x < 0$, we see that $f(x) = 0$ and

$$|f(x) - 1| = |0 - 1| = 1 > \epsilon.$$

This contradicts the definition of limit. Thus, for $\epsilon = 1/2$, there is no value of δ that satisfies the definition of a limit,

$$\lim_{x \to 0} f(x) \neq 1.$$

Of course there are some values of x satisfying $0 < |x - 0| < \delta$ which also satisfy $|f(x) - 1| < 1/2$. We are not interested in them. We have shown that it is not true that *every* x satisfying $0 < |x - 0| < \delta$ also satisfies $|f(x) - 1| < 1/2$.

What happens here if we try to proceed as we did in the previous section? Suppose $0 < |x - 0| < \delta$. Then

$$|f(x) - 1| = \begin{cases} |0 - 1| = 1 & \text{if } x \leq 0, \\ |1 - 1| = 0 & \text{if } x > 0. \end{cases}$$

Now, can we choose δ in such a way that $|f(x) - 1| < \epsilon$, no matter what positive number ϵ represents? If $\epsilon \leq 1$, we cannot. No matter what δ we choose, there is a number x such that $-\delta < x < 0$ and $|f(x) - 1| = 1 \not< \epsilon$.

Example 2

Show that if $\quad f(x) = \begin{cases} 1 & \text{if } x \text{ is rational,} \\ -1 & \text{if } x \text{ is irrational,} \end{cases} \quad$ then $\lim\limits_{x \to 0} f(x) \neq 1$.

Recall that if a and b are any two rational numbers, however close, there are infinitely many irrationals between them. Likewise there are infinitely many rational numbers between any two irrationals. Because of this, the graph of f (see Figure 17.4) consists of infinitely many points tightly packed on the line $y = 1$ together with a tight packing of points on $y = -1$. Since f is defined for all real numbers, it is clear that 0 is a limit point of the domain of f.

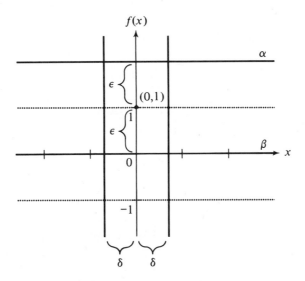

Figure 17.4

If we were to use the definition of Chapter 10 to prove that the limit is not 1, we should choose horizontal lines α and β so that $y = -1$ is not between them. Thus we want to choose $\epsilon \leq 2$. Let us choose $\epsilon = 1$. Now we want to show that no matter what positive number δ might be, there is a value of x satisfying

$$0 < |x - 0| < \delta$$

but not

$$|f(x) - 1| < 1.$$

But it is clear that there must be such an x for any value of δ; we merely choose an irrational number x between 0 and δ. Clearly the first inequality above is satisfied. Since x is irrational, $f(x) = -1$ and

$$|f(x) - 1| = |-1 - 1| = 2 \not< 1.$$

Thus

$$\lim_{x \to 0} f(x) \neq 1.$$

As we indicated in Chapter 10, we say that a given limit does not exist if it is not equal to any real number.

Example 3

Show that $\lim_{x \to 0} 1/x$ does not exist.

In order to show that $\lim_{x \to 0} 1/x$ does not exist we must show that no matter what number we choose for b, $\lim_{x \to 0} 1/x = b$ is not true. Again 0 is a limit point of the domain of the given function. Thus the first part of the definition is satisfied no matter what number b represents. Let us now arbitrarily choose $\epsilon = 1$. We must now show that for this choice of ϵ and for any choice of b, we cannot find a positive number δ satisfying the conditions of the definition. In order to simplify the problem, let us break it down into two cases depending on the value of b.

Case I $b \geq 0$: Now we want to find a value of x satisfying $0 < |x - 0| < \delta$ but not satisfying $|f(x) - b| < 1$. If

$$|f(x) - b| < 1,$$

then

$$\left| \frac{1}{x} - b \right| < 1$$

$$-1 < \frac{1}{x} - b < 1$$

$$b - 1 < \frac{1}{x} < b + 1.$$

Since $b \geq 0$, $b + 1$ is positive; if x is also positive, then

$$x > \frac{1}{b + 1} > 0.$$

Let us choose

$$0 < x < \min\left\{ \delta, \frac{1}{b + 1} \right\}.$$

Clearly $0 < |x| < \delta$. Since $0 < x < 1/(b + 1)$,

$$f(x) = \frac{1}{x} \geq b + 1.$$

Thus $|f(x) - b| \not< 1$.

Case II $b < 0$: Again we want to find a value of x satisfying $0 < |x - 0| < \delta$ but not $|f(x) - b| < 1$. Again,

$$|f(x) - b| < 1$$

leads to

$$b - 1 < \frac{1}{x} < b + 1.$$

Since $b < 0$, $b - 1$ is negative; if x is also negative, then

$$x < \frac{1}{b - 1} < 0.$$

Let us choose

$$\max \left\{ -\delta, \frac{1}{b - 1} \right\} < x < 0.$$

Since $-\delta < x < 0$, $0 < |x| = -x < \delta$.   Since $1/(b - 1) < x < 0$,

$$\frac{1}{x} < b - 1.$$

Thus $|f(x) - b| \not< 1$. The definition of limit fails to hold for any choice of b and $\lim_{x \to 0} 1/x$ does not exist.

As we have seen before, this is a special type of nonexistence. We shall consider it in more detail in the next section.

Problems

A

In Problems 1–9, show by means of the definition of limit that the given limit statement is false.

1. $\lim_{x \to 1} (x + 1) = 4.$

2. $\lim_{x \to 1} (2x + 3) = 4.$

3. $\lim_{x \to 1} f(x) = 4$,   where $f(x) = \begin{cases} x + 2 & \text{if } x \neq 1, \\ 4 & \text{if } x = 1. \end{cases}$

4. $\lim_{x \to 2} f(x) = 1$,   where $f(x) = \begin{cases} x - 3 & \text{if } x \neq 2, \\ 1 & \text{if } x = 2. \end{cases}$

5. $\lim_{x \to 0} f(x) = 0$,   where $f(x) = \begin{cases} 0 & \text{if } x \leq 0, \\ 1 & \text{if } x > 0. \end{cases}$

6. $\lim_{x \to 0} f(x) = 1$,   where $f(x) = \begin{cases} 0 & \text{if } x \leq 0, \\ 1 & \text{if } x > 0. \end{cases}$

7. $\lim_{x \to 0} f(x) = 0$,   where $f(x) = \begin{cases} x & \text{if } x < 0, \\ 1 & \text{if } x > 0. \end{cases}$

8. $\lim_{x \to 0} f(x) = 1$,   where $f(x) = \begin{cases} x & \text{if } x < 0, \\ 1 & \text{if } x > 0. \end{cases}$

9. $\lim_{x \to 0} \frac{1}{x^2} = 0.$

In Problems 10–14, indicate whether the given limit statement is true or false and show by means of the definition of limit that your answer is correct.

10. $\lim\limits_{x \to 1} \dfrac{x^2 - 4}{x - 2} = 3.$

11. $\lim\limits_{x \to 1} f(x) = 0,$ where $f(x) = \begin{cases} x - 1 & \text{if } x \neq 1, \\ 3 & \text{if } x = 1. \end{cases}$

12. $\lim\limits_{x \to 1} f(x) = 5,$ where $f(x) = \begin{cases} x + 1 & \text{if } x \neq 1, \\ 5 & \text{if } x = 1. \end{cases}$

13. $\lim\limits_{x \to 0} \dfrac{1}{x} = 0.$

14. $\lim\limits_{x \to 0} f(x) = 0,$ where $f(x) = \begin{cases} 1/x & \text{if } x \neq 0, \\ 0 & \text{if } x = 0. \end{cases}$

B

In Problems 15–17, show by means of the definition of limit that the given limit statement is false.

15. $\lim\limits_{x \to 0} \dfrac{1}{x^2} = 1.$ 16. $\lim\limits_{x \to 1} \dfrac{1}{x - 1} = 0.$

17. $\lim\limits_{x \to 2} f(x) = 1,$ where $f(x) = \begin{cases} 1 & \text{if } x \text{ is rational,} \\ -1 & \text{if } x \text{ is irrational.} \end{cases}$

18. Show that $\lim\limits_{x \to 1} \dfrac{1}{x} = 1.$

C

In Problems 19–24, show that the given limit does not exist.

19. $\lim\limits_{x \to 0} f(x),$ where $f(x) = \begin{cases} 0 & \text{if } x \leq 0, \\ 1 & \text{if } x > 0. \end{cases}$

20. $\lim\limits_{x \to 0} f(x),$ where $f(x) = \begin{cases} 1 & \text{if } x \leq 0, \\ x & \text{if } x > 0. \end{cases}$

21. $\lim\limits_{x \to 0} f(x),$ where $f(x) = \begin{cases} x + 1 & \text{if } x \leq 0, \\ x^2 & \text{if } x > 0. \end{cases}$

22. $\lim\limits_{x \to 0} f(x),$ where $f(x) = \begin{cases} 0 & \text{if } x \leq 0, \\ 1/x & \text{if } x > 0. \end{cases}$

23. $\lim\limits_{x \to 1} \dfrac{1}{x - 1}.$ 24. $\lim\limits_{x \to 0} \dfrac{1}{x^2}.$

17.3

One-Sided and Infinite Limits

In Chapter 10 we defined the right-hand limit statement

$$\lim_{x \to a^+} f(x) = b$$

by defining a new function G, which coincides with f when $x \geq a$ and is undefined when $x < a$. In effect we threw away the part of the graph left of $x = a$. Another approach that could have been used there is to restrict the vertical lines l and m (and h and k) so that no point of the graph to the left of $x = a$ is between them. Thus we could take l to be $x = a$ and m to be to the right of $x = a$. If m is taken to be $x = a + \delta$, then points between $x = a$ and $x = a + \delta$ satisfy the inequality

$$0 < x - a < \delta.$$

These lead us to the following definitions of one-sided limits.

Definition

The statement $\quad \lim_{x \to a^+} f(x) = b \quad$ *means*

(a) if δ is a positive number, then there is a number x in the domain of f such that

$$0 < x - a < \delta, \quad and$$

(b) if ϵ is a positive number, then there is a positive number δ such that

$$|f(x) - b| < \epsilon$$

for every number x in the domain of f for which

$$0 < x - a < \delta.$$

Definition

The statement $\quad \lim_{x \to a^-} f(x) = b \quad$ *means*

(a) if δ is a positive number, then there is a number x in the domain of f such that

$$-\delta < x - a < 0, \quad and$$

(b) if ϵ is a positive number, then there is a positive number δ such that

$$|f(x) - b| < \epsilon$$

for every number x in the domain of f for which

$$-\delta < x - a < 0.$$

Example 1

Show that if $\quad f(x) = \begin{cases} 0 & \text{if } x \le 0, \\ 1 & \text{if } x > 0, \end{cases} \quad$ then $\quad \lim_{x \to 0^+} f(x) = 1, \quad$ and $\quad \lim_{x \to 0^-} f(x) = 0.$

We have already seen in Problem 19 of the previous section that the ordinary two-sided limit does not exist. The graph of this function is given in Figure 17.3.

Looking at the right-hand limit, we see that since f is defined for all real values of x, it is defined for some (in fact, all) x satisfying the inequalities $0 < x < \delta$ no matter what positive number δ represents. Now suppose ϵ is a positive number. We want to find a positive number δ such that $|f(x) - 1| < \epsilon$ whenever $0 < x < \delta$. If, for some choice of δ, x satisfies the inequalities $0 < x < \delta$, then $f(x) = 1$ and

$$|f(x) - 1| = |1 - 1| = 0 < \epsilon.$$

This puts no restriction on δ. No matter what positive number δ we choose, the definition is satisfied; and

$$\lim_{x \to 0+} f(x) = 1.$$

A similar argument holds for the left-hand limit. The function f is defined for all real x; therefore it is defined for some x satisfying $-\delta < x < 0$, no matter what positive number δ represents. Suppose now that ϵ is a positive number; and suppose that, for some positive number δ, x satisfies

$$-\delta < x < 0.$$

Then $f(x) = 0$ and

$$|f(x) - 0| = |0 - 0| = 0 < \epsilon.$$

Again there is no restriction on δ; it may be chosen in any way, giving

$$\lim_{x \to 0-} f(x) = 0.$$

By Problem 24 of the last section, we see that $\lim_{x \to 0} 1/x^2$ does not exist. But this is an example of a special type of nonexistence which we want to distinguish from other types. We had a geometric definition of $\lim_{x \to a} f(x) = +\infty$ in Chapter 10; we now consider an algebraic definition.

Definition

The statement $\lim_{x \to a} f(x) = +\infty$ *means*

(a) *if δ is a positive number, then there is a number x in the domain of f such that $0 < |x - a| < \delta$, and*

(b) *if N is a number, then there is a positive number δ such that*

$$f(x) > N$$

for every number x in the domain of f for which $0 < |x - a| < \delta$.

The first part of this definition is exactly the same as the first part of the definition of $\lim_{x \to a} f(x) = b$. In the second part, ϵ has been replaced by N, which may be thought of as a large number. Then, roughly speaking, the second part of the definition says that no matter how big N is, all values of x in the domain of f which are sufficiently near a have functional values that are even bigger.

Example 2

Show that $\lim_{x \to 0} \dfrac{1}{x^2} = +\infty.$

Since the given function is defined for all x except 0, the first part of the definition is satisfied. If $N \le 0$, the entire graph is above $y = N$ (see Figure 17.5). Thus, for any choice of δ, $f(x) > N$ for every number in the domain for which $0 < |x - 0| < \delta$. If N is positive, then $y = N$ has two points in common with the graph, namely $(\pm 1/\sqrt{N}, N)$. Now if $|x| < 1/\sqrt{N}$, then

$$f(x) = \frac{1}{x^2} = \frac{1}{|x|^2} > (\sqrt{N})^2 = N.$$

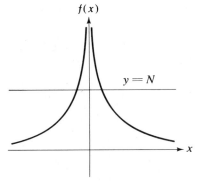

Figure 17.5

Let us choose $\delta = 1/\sqrt{N}$. It now follows that, if N is any number, there is a positive number δ such that $f(x) > N$ for every number x in the domain of f for which $0 < |x - 0| < \delta$. Thus

$$\lim_{x \to 0} \frac{1}{x^2} = +\infty.$$

We can define $\lim_{x \to a} f(x) = -\infty$ similarly.

Definition

The statement $\lim_{x \to a} f(x) = -\infty$ *means*

(a) *if δ is a positive number, then there is a number x in the domain of f such that*
$0 < |x - a| < \delta$,   *and*

(b) *if N is a number, then there is a positive number δ such that*

$$f(x) < N$$

for every number x in the domain of f for which $0 < |x - a| < \delta$.

Example 3

Show that $\lim_{x \to 0} \dfrac{1}{x}$ is neither $+\infty$ nor $-\infty$ but

$$\lim_{x \to 0^-} \frac{1}{x} = -\infty \quad \text{and} \quad \lim_{x \to 0^+} \frac{1}{x} = +\infty.$$

Let us first show that $\lim_{x \to 0} 1/x$ is neither $+\infty$ nor $-\infty$. In each of the two cases, we must find a particular value of N for which there is no value of δ satisfying the given conditions. Let us choose $N = 0$. There is a value of x ($x = -\delta/2$) in the domain of f satisfying $0 < |x - 0| < \delta$ but not $f(x) > 0$. Similarly there is a value of x ($x = \delta/2$) in the domain of f satisfying $0 < |x - 0| < \delta$ but not $f(x) < 0$. Thus, $\lim_{x \to 0} 1/x$ is neither $+\infty$ nor $-\infty$.

Let us show that $\lim_{x \to 0^-} 1/x = -\infty$. Since f is defined for all negative values of x, the first part of the definition is satisfied.

Now if $N \geq 0$, then, no matter what positive number δ represents, $f(x) < N$ if $-\delta < x < 0$. Suppose $N < 0$. If $f(x) = 1/x$ is to be less than N, then we must choose x such that

$$x > \frac{1}{N}.$$

Thus if $N < 0$, choose $\delta = -1/N$. Now if

$$-\delta < x < 0,$$

then

$$\frac{1}{N} < x < 0$$

and $f(x) = 1/x < N$. Thus $\lim_{x \to 0-} 1/x = -\infty$.

A similar argument shows that $\lim_{x \to 0+} 1/x = +\infty$. The first part of the definition is easily seen to be satisfied, since f is defined for all $x > 0$. If $N \leq 0$, then, for any positive δ, $f(x) > N$ if $0 < x < \delta$. Suppose $N > 0$. If $f(x) = 1/x > N$, then x must be chosen such that

$$x < \frac{1}{N}.$$

Therefore we choose δ to be $1/N$ if $N > 0$. If

$$0 < x < \delta = \frac{1}{N},$$

then

$$f(x) = \frac{1}{x} > N.$$

Hence $\lim_{x \to 0+} 1/x = +\infty$.

Let us now consider limits in which $x \to +\infty$ or $x \to -\infty$.

Definition

The statement $\quad \lim_{x \to +\infty} f(x) = b \quad$ *means*

(a) *if M is a number, then there is a number x in the domain of f such that $x > M$, and*

(b) *if ϵ is a positive number, then there is a number N such that*

$$|f(x) - b| < \epsilon$$

for every number x in the domain of f for which $x > N$.

Again, the two parts of this definition correspond exactly to the two parts of the definition given in Chapter 10. The first part says that no matter how big M is, there is a number x in the domain that is still bigger. In other words, no matter what vertical line we choose, there is some point of the graph to the right of it. The second part says that for large enough values of x, $f(x)$ is near (within a distance ϵ of) b. Of course, the same type of definition is used if $x \to -\infty$.

Definition

The statement $\lim\limits_{x \to -\infty} f(x) = b$ *means*

(a) *if M is a number, then there is a number x in the domain of f such that x < M, and*

(b) *if ϵ is a positive number, then there is a number N such that*

$$|f(x) - b| < \epsilon$$

for every number x in the domain of f for which x < N.

Example 4

Show that $\lim\limits_{x \to +\infty} \dfrac{1}{x} = 0.$

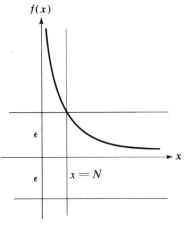

The first part of the definition is easily seen to be satisfied, since the domain of the function is the set of all real numbers different from 0. Suppose ϵ is a positive number. We want to show that there is a number N big enough so that, if $x > N$, then

$$|f(x) - 0| < \epsilon.$$

Let us find those values of x for which the last inequality is true.

$$\left|\frac{1}{x} - 0\right| < \epsilon$$

$$-\epsilon < \frac{1}{x} < \epsilon,$$

$$x < -\frac{1}{\epsilon} \quad \text{or} \quad x > \frac{1}{\epsilon}.$$

Figure 17.6

We are concerned with a limit as $x \to +\infty$; thus we are only interested in "large" values of x (those bigger than some number N). So we are interested in the second of the last two inequalities ($x > 1/\epsilon$). Let us then choose $N = 1/\epsilon$. Thus if

$$x > N = \frac{1}{\epsilon},$$

then

$$f(x) = \frac{1}{x} < \epsilon \quad \text{and} \quad |f(x) - 0| < \epsilon.$$

The given limit statement is therefore true.

Problems

A

In Problems 1–11, show that the given limit statement is true.

1. $\lim\limits_{x \to 0^+} f(x) = 0,$ where $f(x) = \begin{cases} -2 & \text{if } x \le 0, \\ x & \text{if } x > 0. \end{cases}$

2. $\lim\limits_{x \to 0^-} f(x) = -2,$ where $f(x) = \begin{cases} -2 & \text{if } x \le 0, \\ x & \text{if } x > 0. \end{cases}$

3. $\lim\limits_{x \to 0} \dfrac{1}{x^4} = +\infty.$

4. $\lim\limits_{x \to 1} \dfrac{1}{(x - 1)^2} = +\infty.$

5. $\lim\limits_{x\to 0} \dfrac{-1}{x^2} = -\infty.$

6. $\lim\limits_{x\to 1} \dfrac{-1}{(x-1)^2} = -\infty.$

7. $\lim\limits_{x\to -2^+} \dfrac{1}{x+2} = +\infty.$

8. $\lim\limits_{x\to -2^-} \dfrac{1}{x+2} = -\infty.$

9. $\lim\limits_{x\to +\infty} \dfrac{1}{x^2} = 0.$

10. $\lim\limits_{x\to -\infty} \dfrac{1}{x^2} = 0.$

11. $\lim\limits_{x\to +\infty} \dfrac{x+1}{x} = 1.$

B

In Problems 12–14, show that the given limit statement is true.

12. $\lim\limits_{x\to -\infty} \dfrac{x}{x+1} = 1.$
13. $\lim\limits_{x\to -\infty} e^x = 0.$
14. $\lim\limits_{x\to 0} \ln x = -\infty.$

In Problems 15–20, show that the given limit statement is false.

15. $\lim\limits_{x\to 0^-} f(x) = 0,$ where $f(x) = \begin{cases} 1/x & \text{if } x < 0, \\ x & \text{if } x \geq 0. \end{cases}$

16. $\lim\limits_{x\to 1^+} f(x) = 3,$ where $f(x) = \begin{cases} x & \text{if } x \neq 1, \\ 3 & \text{if } x = 1. \end{cases}$

17. $\lim\limits_{x\to 0} \dfrac{1}{x^3} = +\infty.$

18. $\lim\limits_{x\to 0} \dfrac{1}{x^3} = -\infty.$

19. $\lim\limits_{x\to 0} f(x) = +\infty,$ where $f(x) = \begin{cases} 0 & \text{if } x < 0, \\ 1/x & \text{if } x > 0. \end{cases}$

20. $\lim\limits_{x\to 0} f(x) = +\infty,$ where $f(x) = \begin{cases} 0 & \text{if } x < 0, \\ 2 & \text{if } x > 0. \end{cases}$

C

In Problems 21 and 22, show that the given limit statement is false.

21. $\lim\limits_{x\to +\infty} \dfrac{x^2-1}{x-1} = 1.$

22. $\lim\limits_{x\to +\infty} \sqrt{\dfrac{1-x^2}{1+x^2}} = 1.$

23. Define $\lim_{x\to +\infty} f(x) = +\infty.$
24. Define $\lim_{x\to -\infty} f(x) = +\infty.$
25. Show that $\lim_{x\to +\infty} x^2 = +\infty$ (see Problem 23).
26. Show that $\lim_{x\to -\infty} x^2 = +\infty$ (see Problem 24).

17.4

Continuity

Let us now consider an ϵ-δ definition of continuity.

Definition

*The function f is **continuous** at the point $(a, f(a))$ means that, if ϵ is a positive number, then there is a positive number δ such that*

$$|f(x) - f(a)| < \epsilon$$

for all x in the domain of f for which

$$|x - a| < \delta.$$

This concept of continuity at a point can be extended to continuity of the whole function and to discontinuity at a point in exactly the same way as was done in Chapter 10.

Definition

*The function f is **discontinuous** at the point $(a, f(a))$ if it is not continuous at $(a, f(a))$.*

Definition

*A function is **continuous** on a set E of real numbers if it is continuous at every number in E.*

Example 1

Show that $f(x) = x^2$ is continuous at $(1, 1)$.

The method of snowing continuity here is similar to that of howing that $\lim_{x \to 1} x^2 = 1$, as we did in Example 2 of Section 17.1. You might compare them. We need to show that if ϵ is a positive number, then there i a positive number δ such that

$$|x^2 - 1| < \epsilon$$

whenever

$$|x - 1| < \delta.$$

Again, note that

$$|x^2 - 1| = |x + 1| |x - 1|,$$

and we are restricting $|x - 1|$ to values less than δ. In order to restrict the value of $|x + 1|$, let us choose δ so that

$$\delta \leq 1.$$

Then

$$|x - 1| < \delta \leq 1$$
$$|x - 1| < 1$$
$$-1 < x - 1 < 1$$
$$1 < x + 1 < 3$$
$$|x + 1| < 3.$$

Thus

$$|x^2 - 1| = |x + 1| |x - 1| < 3\delta.$$

Now let us choose $\delta = \epsilon/3$ but subject to the condition that $\delta \leq 1$. Thus, if we choose

$$\delta = \min \left\{ 1, \frac{\epsilon}{3} \right\},$$

it follows that if

$$|x - 1| < \delta,$$

then

$$|x^2 - 1| < \epsilon.$$

Example 2

Show that $f(x) = \begin{cases} 0 & x \leq 0, \\ 1 & x > 0 \end{cases}$ is discontinuous at (0, 0).

We need to exhibit an ϵ for which no δ can be found that satisfies the conditions for continuity. Since the graph jumps from 0 to 1 at $x = 0$, let us choose $\epsilon = 1/2$ (any number ≤ 1 will do). Now we want to show that, for this choice of ϵ, there is no number δ such that

$$|f(x) - 0| < \epsilon$$

whenever

$$|x - 0| < \delta.$$

Assume there is such a number δ. Then let x be a number such that

$$0 < x < \delta.$$

For this choice of x,

$$|x - 0| < \delta,$$

but

$$|f(x) - 0| = |1 - 0| \not< \epsilon.$$

Thus, no such δ can be found, and the given function is discontinuous at (0, 0).

Let us now consider some of the properties of continuous functions that make them so important. First we repeat here some results that were stated and proved in Section 10.7.

Theorem 17.1

If $\lim_{x \to a} f(x) = b$ and $\lim_{x \to a} g(x) = c$, where b and c are real numbers (not $\pm \infty$), then

(a) $\lim_{x \to a} [f(x) + g(x)] = b + c,$

(b) $\lim_{x \to a} [f(x) - g(x)] = b - c,$

(c) $\lim_{x \to a} f(x)g(x) = bc,$

(d) $\lim_{x \to a} \dfrac{f(x)}{g(x)} = \dfrac{b}{c} \quad (c \neq 0),$

provided a is a limit point of the domain of each combination of functions.

Proof

(a) By hypothesis, a is a limit point of $f + g$. Given $\epsilon > 0$, we must find a $\delta > 0$ so that $|[f(x) + g(x)] - (b + c)| < \epsilon$ for each x in the domain of $f + g$ for which $0 < |x - a| < \delta$. Now,

$$|[f(x) + g(x)] - (b + c)| = |[f(x) - b] + [g(x) - c]| \leq |f(x) - b| + |g(x) + c|.$$

If we make both

$$|f(x) - b| < \frac{\epsilon}{2} \quad \text{and} \quad |g(x) - c| < \frac{\epsilon}{2},$$

then

$$\left|[f(x) + g(x)] - (b + c)\right| < \frac{\epsilon}{2} + \frac{\epsilon}{2} = \epsilon.$$

If $\epsilon > 0$, then $\epsilon/2 > 0$ also; thus, since $\lim_{x \to a} f(x) = b$, there is a $\delta_1 > 0$ so that $|f(x) - b| < \epsilon/2$ for all x in the domain of f for which $0 < |x - a| < \delta_1$. Likewise, there is a $\delta_2 > 0$ so that $|g(x) - c| < \epsilon/2$ for all x in the domain of g for which $0 < |x - a| < \delta_2$. Let $\delta = \min(\delta_1, \delta_2)$. If $0 < |x - a| < \delta$ and x is in the domain of $f + g$, then both $0 < |x - a| < \delta \le \delta_1$ and $0 < |x - a| < \delta \le \delta_2$; so

$$\left|[f(x) + g(x)] - (b + c)\right| \le |f(x) - b| + |g(x) - c| < \frac{\epsilon}{2} + \frac{\epsilon}{2} = \epsilon.$$

Thus
$$\lim_{x \to a}[f(x) + g(x)] = b + c.$$

(b)   The proof of part (b) is similar to the proof of part (a). (See Problem 32.)

(c)   The choice of δ depends on rearranging $f(x)g(x) - bc$ and then getting an upper bound $|g(x)|$ near $x = a$. To obtain a bound on $|g(x)|$, let $\epsilon = 1$. Since $\lim_{x \to a} g(x) = c$, there is a $\delta_1 > 0$ so that for x in the domain of g and $0 < |x - a| < \delta_1$, $|g(x) - c| < 1$. Thus

$$c - 1 < g(x) < c + 1 \quad \text{and} \quad |g(x)| < |c| + 1.$$

Now let $\epsilon > 0$ be given. There is a $\delta_2 > 0$ so that

$$|f(x) - b| < \frac{\epsilon}{2(|c| + 1)}$$

if $0 < |x - a| < \delta_2$, x in the domain of f; there is a $\delta_3 > 0$ so that

$$|g(x) - c| < \frac{\epsilon}{2(|b| + 1)}$$

if $0 < |x - a| < \delta_3$, x in the domain of g. Let $\delta = \min(\delta_1, \delta_2, \delta_3)$ and let $0 < |x - a| < \delta$, x in the domain of fg. Then

$$\begin{aligned}
|f(x)g(x) - bc| &= |f(x)g(x) - bg(x) + bg(x) - bc| \\
&\le |g(x)||f(x) - b| + |b||g(x) - c| \\
&< (|c| + 1)\frac{\epsilon}{2(|c| + 1)} + |b| \cdot \frac{\epsilon}{2(|b| + 1)} \\
&\le \frac{\epsilon}{2} + \frac{\epsilon}{2} = \epsilon.
\end{aligned}$$

Thus
$$\lim_{x \to a} f(x)g(x) = bc.$$

(d)   Since we divide by $g(x)$, we first restrict x in the domain of g sufficiently close to a so that $|g(x)| > |c|/2$. By hypothesis, $c \ne 0$. Let $\epsilon = |c|/2$. There is a $\delta_1 > 0$ so that $|g(x) - c| < |c|/2$ for $0 < |x - a| < \delta_1$, x in the domain of g. Thus

$$c - \frac{|c|}{2} < g(x) < c + \frac{|c|}{2}.$$

If $c > 0$, then

$$g(x) > c - \frac{c}{2} = \frac{c}{2}.$$

If $c < 0$, then $|c| = -c$ and

$$g(x) < c - \frac{c}{2} = \frac{c}{2} < 0.$$

In either case,

$$|g(x)| > \frac{|c|}{2} \quad \text{and} \quad \frac{1}{|g(x)|} < \frac{2}{|c|}.$$

Let $\epsilon > 0$ be given. There is a $\delta_2 > 0$ so that $|f(x) - b| < |c|\epsilon/4$ for $0 < |x - a| < \delta_2$, x in the domain of f. There is a $\delta_3 > 0$ so that $|g(x) - c| < |c|^2 \epsilon/4(|b| + 1)$ for $0 < |x - a| < \delta_3$, x in the domain of g. Let $\delta = \min(\delta_1, \delta_2, \delta_3)$. Then

$$\left| \frac{f(x)}{g(x)} - \frac{b}{c} \right| = \left| \frac{f(x) c - g(x) b}{cg(x)} \right|$$

$$= \frac{1}{|c| |g(x)|} \cdot |f(x)c - bc + bc - g(x)b|$$

$$\leq \frac{2}{|c|^2} (|c| |f(x) - b| + |b| |g(x) - c|), \quad \text{since} \frac{1}{|g(x)|} < \frac{2}{|c|},$$

$$< \frac{2}{|c|^2} \cdot \frac{\epsilon |c|^2}{4} + |b| \cdot \frac{\epsilon}{2(|b| + 1)}$$

$$< \frac{\epsilon}{2} + \frac{\epsilon}{2} = \epsilon.$$

Thus

$$\lim_{x \to a} \frac{f(x)}{g(x)} = \frac{b}{c}.$$

The choice of δ_2 and δ_3 clearly took some hindsight!

Theorem 17.2

$\lim_{x \to a} f(x) = f(a)$ *if and only if f is continuous at $x = a$ and a is a limit point of the domain of f.*

Proof

If f is continuous at $x = a$, then $f(a)$ is defined. Given $\epsilon > 0$, there is a $\delta > 0$ so that $|f(x) - f(a)| < \epsilon$ for all x in the domain of f for which $|x - a| < \delta$, and hence also for all x in the domain for which $0 < |x - a| < \delta$. Thus $\lim_{x \to a} f(x) = f(a)$, since a is a limit point of the domain by hypothesis.

Suppose $\lim_{x \to a} f(x) = f(a)$ and a is a limit point of the domain of f. Then $f(a)$ is defined, and for $\epsilon > 0$ there is a $\delta > 0$ so that $|f(x) - f(a)| < \epsilon$ for all x in the domain for which $0 < |x - a| < \delta$. Since $|f(a) - f(a)| = 0 < \epsilon$, then $|f(x) - f(a)| < \epsilon$ for $x = a$ also. Hence, f is continuous at a.

The following theorems are a direct consequence of the last two. We shall use these results in a later section. Their proofs are left to the student.

Theorem 17.3

If $\lim_{x \to a} f(x) = b$ and g is continuous at $x = b$, then

$$\lim_{x \to a} g(f(x)) = g(b).$$

Corollary

If f is continuous at $x = a$, $b = f(a)$, and g is continuous at b, then the composite function $g \circ f$ is continuous at a.

Theorem 17.4

If f and g are continuous at x = a, then f + g, f − g, and fg are continuous at x = a, and f/g is continuous at x = a, provided g(a) ≠ 0.

Other important properties of continuous functions are given in the following two theorems. Their proofs are beyond the scope of this book; however proofs may be found in Taylor and Mann.†

Theorem 17.5

If a function f is continuous on an interval [a, b], then there is a number x_0 in [a, b] such that $f(x_0) \geq f(x)$ for all x in [a, b]; similarly, there is a number x_1 in [a, b] such that $f(x_1) \leq f(x)$ for all x in [a, b].

Theorem 17.6

(Intermediate-value theorem) If f is continuous on [a, b] and N is a number between f(a) and f(b), then f(x) = N for some number x between a and b.

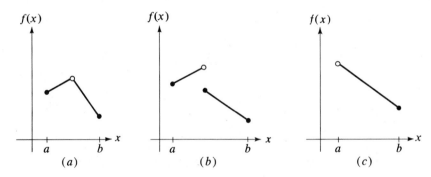

Figure 17.7

Theorem 17.5 says that a continuous function on a closed interval attains its maximum and its minimum within that interval. Since this theorem is used in the next section, let us show that the conditions are all needed for the result. First of all, note that f is defined on [a, b] because it is given that f is continuous on [a, b]; if f is continuous at x, it is also defined at x. We can see by Figure 17.7(a) that if f is not continuous, because it is not defined at some number between a and b, then the number x_0 does not necessarily exist. Figure 17.7(b) shows that if f is discontinuous for some number between a and b, then x_0 need not exist. Finally, (c) shows that the interval must be closed, since there is no number x_0 of (a, b) satisfying the conditions of the theorem. Similar examples can be used to show the necessity of the conditions of Theorem 17.6 (see Problem 28).

The intermediate-value theorem is very important in the theory of functions. Some applications are given in Problems 25, 26, and 29.

† See Angus E. Taylor and Robert Mann, *Advanced Calculus*. 2nd ed. (Lexington, Mass.: Xerox, 1972), pp. 95–98.

Problems

A

In Problems 1–10, use the definition of continuity to show that each function is continuous at the given point.

1. $f(x) = x$, at $(0, 0)$.
2. $f(x) = x$, at $(2, 2)$.
3. $f(x) = 2x + 2$, at $(1, 4)$.
4. $f(x) = x - 3$, at $(1, -2)$.
5. $f(x) = \dfrac{x^2 - 1}{x - 1}$, at $(0, 1)$.
6. $f(x) = x^2$, at $(0, 0)$.
7. $f(x) = x^2$, at $(3, 9)$.
8. $f(x) = x^3$, at $(0, 0)$.
9. $f(x) = \begin{cases} 0 & \text{if } x \le 0, \\ 1 & \text{if } x > 0; \end{cases}$ at $(1, 1)$.
10. $f(x) = \begin{cases} 0 & \text{if } x \le 0, \\ x & \text{if } x > 0; \end{cases}$ at $(1, 1)$.

11. Use the definition of continuity to show that $f(x) = \begin{cases} x^2 & \text{if } x \le 0, \\ x + 1 & \text{if } x > 0, \end{cases}$ is discontinuous at $(0, 0)$.

B

In Problems 12–15, use the definition of continuity to show that each function is continuous at the given point.

12. $f(x) = x^3$, at $(1, 1)$.
13. $f(x) = 1/x$, at $(1, 1)$.
14. $f(x) = 1/x$, at $(3, 1/3)$.
15. $f(x) = 1/x^2$, at $(1, 1)$.

In Problems 16–20, use the definition of continuity to show that each function is discontinuous at the given point.

16. $f(x) = \begin{cases} x + 1 & \text{if } x \ne 1, \\ 3 & \text{if } x = 1; \end{cases}$ at $(1, 3)$.
17. $f(x) = \begin{cases} 1 & \text{if } x \le 0, \\ x & \text{if } x > 0; \end{cases}$ at $(0, 1)$.
18. $f(x) = \begin{cases} x - 2 & \text{if } x \ne 2, \\ 1 & \text{if } x = 2; \end{cases}$ at $(2, 1)$.
19. $f(x) = \begin{cases} x^2 & \text{if } x \ne 0, \\ 1 & \text{if } x = 0; \end{cases}$ at $(0, 1)$.
20. $f(x) = \begin{cases} x + 4 & \text{if } x < 1, \\ x & \text{if } x \ge 1; \end{cases}$ at $(1, 1)$.

21. If $f(x) = (x^2 + x - 2)/(x - 1)$ for $x \ne 1$, how should we define $f(1)$ to make f continuous?

22. Show that every polynomial is continuous for all x. (*Hint:* Show that $f(x) = k$ and $f(x) = x$ are continuous and use Theorem 17.4.)

23. Does $f(x) = x^2$, $0 < x < 2$, have a maximum value? A minimum value? What does Theorem 17.5 tell us about this function?

24. Repeat Problem 23 for $f(x) = x^2$, $-1 < x < 2$.

25. Use the result of Problem 22 to show that if a polynomial is positive for one value of x and negative for another, it is zero somewhere between them.

26. Show that $\sqrt{2}$ exists. (*Hint:* Consider $f(x) = x^2$, $1 \le x \le 2$, and use the intermediate-value theorem.)

C

27. Use the definition of continuity to show that $f(x) = \begin{cases} \sin (1/x) & \text{if } x \ne 0, \\ 0 & \text{if } x = 0 \end{cases}$ is discontinuous at $(0, 0)$.

28. Give an example to show that Theorem 17.6 is not true if f is not continuous.

29. Show that a moving object with an average velocity v_{av} for the time interval $a \le t \le b$ moves with velocity v_{av} at some time in this interval.

30. Prove Theorem 17.4.

31. Prove Theorem 17.3. (*Hint:* Use the continuity of g at $x = b$ together with a given ϵ to determine a δ. Then use this δ in place of the ϵ in the definition of the limit.)

32. Prove Theorem 17.1 (b).

17.5

Rolle's Theorem

You will recall our mentioning earlier that several theorems that were stated but not proved might be proved by using the mean-value theorem. It should be clear by now that the mean-value theorem is very important and powerful. In this section we consider Rolle's theorem, which is a special case of the mean-value theorem as well as an aid in proving it. The mean-value theorem itself is stated in the next section.

Theorem 17.7

(*Rolle's theorem*) *If a and b are numbers* $(a < b)$ *and f is a function such that*

(a) *f is continuous on* $[a, b]$,

(b) $f'(x)$ *exists for all x in* (a, b), *and*

(c) $f(a) = f(b) = 0$,

then there is a number x_0 *between a and b such that* $f'(x_0) = 0$.

The theorem says, roughly speaking, that if the graph of f has no horizontal or vertical jumps and no sharp points and if the ends are on the x axis, then there is at least one point between a and b at which the graph has a horizontal tangent. All three of the conditions are needed, although the third condition can be weakened somewhat (see Problem 22). Figure 17.8 shows that there need not be such a number x_0 if (a) f is not continuous on $[a, b]$, (b) $f'(x)$ does not exist for some x in (a, b), and (c) $f(a)$ and $f(b)$ are not both 0, while Figure 17.9 shows situations in which all of the conditions of Rolle's theorem are satisfied. Note that all cases in Figure 17.9 have at least one horizontal tangent—there may be just one as in (a), two as in (b), three, a hundred, or even infinitely many as in (c). Note that if one or more of the conditions of Rolle's theorem is not satisfied, this does *not* mean that no such number x_0 exists. The theorem merely states what follows if the three conditions *are* satisfied; if they are not all satisfied, the theorem says absolutely nothing concerning the existence of the number x_0.

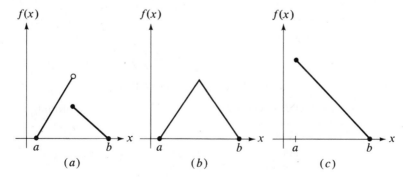

Figure 17.8

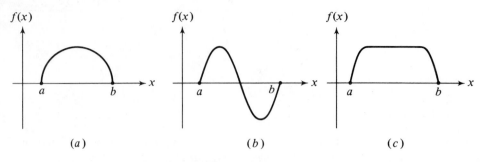

Figure 17.9

Proof

The proof of Rolle's theorem consists of three separate cases, which cover all possible functions f: Case I, $f(x) > 0$ for some x between a and b; Case II, $f(x) < 0$ for some x between a and b; Case III, $f(x) = 0$ for all x between a and b.

Case I $f(x) > 0$ for some x between a and b: By Theorem 17.5, there is a number x_0 in $[a, b]$ such that $f(x_0) \geq f(x)$ for every x in $[a, b]$. Since $f(x) > 0$ for some x between a and b and $f(x_0) \geq f(x)$, it follows that $f(x_0) > 0$. Thus, x_0 is neither a nor b, since $f(a) = f(b) = 0$, and x_0 is between a and b. Now define a function F by

$$F(x) = \frac{f(x_0) - f(x)}{x_0 - x}$$

for all x except x_0 in $[a, b]$. Since $f(x_0) \geq f(x)$ for all x in $[a, b]$,

$$f(x_0) - f(x) \geq 0$$

for all such values of x. If $x < x_0$, then $x_0 - x > 0$ and $F(x) \geq 0$. If $x > x_0$, then $x_0 - x < 0$ and $F(x) \leq 0$. By definition,

$$\lim_{x \to x_0} F(x) = \lim_{x \to x_0} \frac{f(x_0) - f(x)}{x_0 - x} = f'(x_0).$$

Since we are given that $f'(x)$ exists for all x in (a, b), we know that $\lim_{x \to x_0} F(x)$ is some number L. Let us assume L to be positive. If $\epsilon = L/2$, then, by the definition of limit, there is a positive number δ such that

$$|F(x) - L| < \frac{L}{2} \quad \text{whenever} \quad 0 < |x - x_0| < \delta.$$

Suppose $x < b$ and $x_0 < x < x_0 + \delta$. For this choice of x, $|x - x_0| < \delta$. But, since $x > x_0$,

$$F(x) \leq 0 \quad \text{and} \quad F(x) - L \leq -L.$$

Thus

$$|F(x) - L| \geq L > \frac{L}{2},$$

which is a contradiction. Thus, the assumption that L is positive is wrong.

Assume now that L is negative. If $\epsilon = -L/2$, then, by the definition of limit, there is a positive number δ such that

$$|F(x) - L| < -\frac{L}{2} \quad \text{whenever} \quad 0 < |x - x_0| < \delta.$$

Suppose $x > a$ and $x_0 - \delta < x < x_0$. For this choice of x, $|x - x_0| < \delta$. But, since $x < x_0$,

$$F(x) \geq 0 \quad \text{and} \quad F(x) - L \geq -L > 0.$$

Thus

$$|F(x) - L| \geq -L > -\frac{L}{2},$$

which is a contradiction. The assumption that L is negative is also wrong. Since L is neither positive nor negative, L must be 0, which means $f'(x_0) = 0$.

Case II $f(x) < 0$ for some x between a and b: Again by Theorem 17.5, there is a number x_0 in $[a, b]$ such that $f(x_0) \leq f(x)$ for every x in $[a, b]$. We can, by an argument similar to that used for Case I, show that x_0 is between a and b and $f'(x_0) = 0$.

Case III $f(x) = 0$ for all x between a and b: In this case, $f'(x) = 0$ for all x between a and b and, in particular, for some x_0 between a and b.

Thus, in every case, there is a number x_0 between a and b such that $f'(x_0) = 0$.

Example 1

Verify Rolle's theorem for the function $f(x) = x^2 - x$.

First of all $f(x) = 0$ if $x = 0$ or $x = 1$. These must be the values of a and b. This function is continuous for all x, certainly for all x in $[0, 1]$. $f'(x) = 2x - 1$ exists for all x and thus for all x in $(0, 1)$. Thus the three conditions of Rolle's theorem are satisfied; the theorem tells us that there must be a number x_0 between 0 and 1 such that $f'(x_0) = 0$. Let us verify that this is the case. $f'(x) = 0$ if $2x - 1 = 0$, or $x = 1/2$. Thus the number x_0 is $1/2$, which is between 0 and 1. Rolle's theorem is verified for this function (see Figure 17.10).

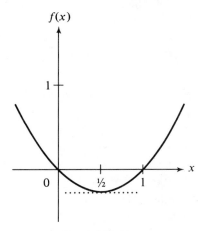

Figure 17.10

Example 2

Does Rolle's theorem tell us anything about the function $f(x) = x^{2/3} - 1$? If so, what? If not, why not?

Again $f(x) = 0$ if $x = \pm 1$. Thus $a = -1$ and $b = 1$. This function is continuous for all x and thus for all x in $[-1, 1]$. But

$$f'(x) = \frac{2}{3}x^{-1/3} = \frac{2}{3x^{1/3}}.$$

There is no derivative when $x = 0$. We *cannot* say that $f'(x)$ exists for all x in $(-1, 1)$—the conditions of Rolle's theorem are not satisfied. Thus, Rolle's theorem tells us nothing about this function (see Figure 17.11).

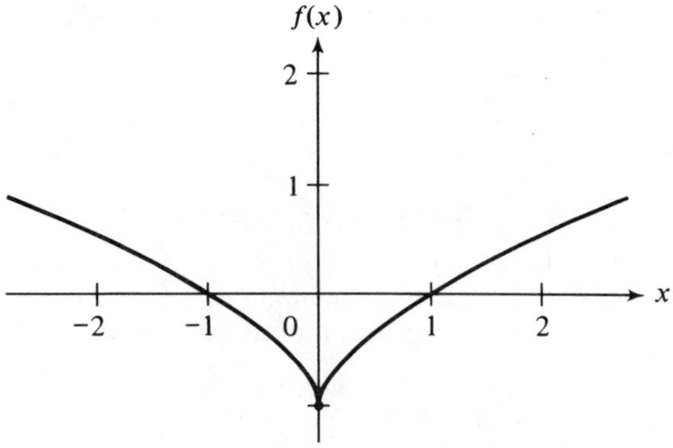

Figure 17.11

Note that Rolle's theorem does *not* say that there is no number between -1 and 1 for which the derivative is 0. It is possible to find functions for which the conditions of Rolle's theorem are not satisfied but for which there does exist a number x_0 between a and b for which $f'(x_0) = 0$. If the three conditions of Rolle's theorem are not satisfied, the theorem tells us absolutely nothing about the function; it tells us neither that there is nor that there is not such a number x_0.

Problems

A

In Problems 1–10, verify Rolle's theorem for the given function. Sketch the graph.

1. $f(x) = x^2 - 3x$.
2. $f(x) = x^2 + x$.
3. $f(x) = x^2 - x - 2$.
4. $f(x) = x^2 + x - 6$.
5. $f(x) = x^3 + x^2$.
6. $f(x) = x(x + 3)^2$.
7. $f(x) = x^3 - x$; $a = -1$, $b = 0$.
8. $f(x) = x^3 - x$; $a = 0$, $b = 1$.
9. $f(x) = \sin x$; $a = 0$, $b = \pi$.
10. $f(x) = \sin x$; $a = \pi$, $b = 3\pi$.

In Problems 11–15, does Rolle's theorem tell us anything about the given function? If so, what? If not, why not?

11. $f(x) = x - x^{1/3}$; $a = 0$, $b = 1$.
12. $f(x) = x - x^{1/3}$; $a = -1$, $b = 1$.
13. $f(x) = x^{4/5} - 1$.
14. $f(x) = \tan x$; $a = 0$, $b = \pi$.
15. $f(x) = x^{1/3} - 1$.

B

In Problems 16–19, does Rolle's theorem tell us anything about the given function? If so, what? If not, why not?

16. $f(x) = \begin{cases} x & \text{if } x \le 1, \\ 2 - x & \text{if } x > 1. \end{cases}$

17. $f(x) = \begin{cases} x & \text{if } x < 1, \\ 1 & \text{if } 1 \le x \le 2, \\ 3 - x & \text{if } x > 2. \end{cases}$

18. $f(x) = \begin{cases} x^2 - 1 & \text{if } x \le 0, \\ x^3 - 1 & \text{if } x > 0. \end{cases}$

19. $f(x) = \ln |x|$.

C

20. Does Rolle's theorem tell us anything about the function $f(x) = x \ln x$? If so, what? If not, why not?

21. Complete the proof for Case II of Rolle's theorem.

22. Prove that if the third condition of Rolle's theorem were changed to $f(a) = f(b)$, the resulting statement would be true.
23. Show that $x^3 + ax + b = 0$, where $a > 0$, cannot have two real roots.
24. Show that $x^3 - ax + b = 0$, where $a > 0$, cannot have three positive roots. Show that it cannot have three negative roots.
25. Show that if a, b, and c are real numbers $(a < b < c)$ and f is a function such that

>   (a) f is continuous on $[a, c]$,
>   (b) $f'(x)$ and $f''(x)$ exist for all x in (a, c), and
>   (c) $f(a) = f(b) = f(c) = 0$,

then there is a number d between a and c such that $f''(d) = 0$.

17.6

The Mean-Value Theorem

Now that we have proved Rolle's theorem let us consider the mean-value theorem, to which we have been leading.

Theorem 17.8

(*Mean-value theorem*)   *If a and b are numbers $(a < b)$ and f is a function such that*

(a) f is continuous on $[a, b]$, and
(b) $f'(x)$ exists for all x in (a, b),

   then there is a number x_0 between a and b such that

$$f'(x_0) = \frac{f(b) - f(a)}{b - a}.$$

Note that the conditions on f stated here are two of the three conditions of Rolle's theorem. Also note that

$$\frac{f(b) - f(a)}{b - a}$$

is the slope of the line joining $(a, f(a))$ and $(b, f(b))$ (see Figure 17.12). Thus, roughly speaking, the mean-value theorem says that if the graph of f has no horizontal or vertical jumps and no sharp points, then there is at least one point between a and b at which the tangent to the graph is parallel to the line joining the ends. Of course, there may be more than one such point; there may be two, three, a hundred, even infinitely many.

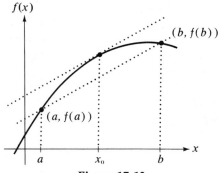

Figure 17.12

Proof

In order to prove the mean-value theorem, we would like to find a new function F in terms of f such that F still satisfies the two conditions of continuity and differentiability as well as the third condition of Rolle's theorem. Thus, we shall take the new function F to be the difference between the y coordinates of the given function and the line joining the end points. First, let us find an equation for that line. The slope of the line is

$$\frac{f(b) - f(a)}{b - a},$$

and it contains the point $(a, f(a))$. Using the point-slope formula, we have

$$y - f(a) = \frac{f(b) - f(a)}{b - a}(x - a),$$

or

$$y = f(a) + \frac{f(b) - f(a)}{b - a}(x - a).$$

Thus the new function F is defined by

$$F(x) = f(x) - f(a) - \frac{f(b) - f(a)}{b - a}(x - a).$$

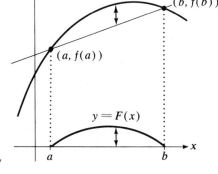

Figure 17.13

We now check that F satisfies the three hypotheses of Rolle's theorem. First of all, F is the difference of two functions—one is f and the other represents a line. Since these two are both continuous, F is continuous by Theorem 17.4. Finding the derivative, we have

$$F'(x) = f'(x) - \frac{f(b) - f(a)}{b - a}. \tag{1}$$

Since $f'(x)$ exists for every x in (a, b), $F'(x)$ exists for every x in (a, b). Finally

$$F(a) = f(a) - f(a) - \frac{f(b) - f(a)}{b - a}(a - a) = 0 - 0 = 0,$$

$$F(b) = f(b) - f(a) - \frac{f(b) - f(a)}{b - a}(b - a) = [f(b) - f(a)] - [f(b) - f(a)] = 0.$$

Thus, all the conditions of Rolle's theorem are satisfied. Then, by Rolle's theorem, there is a number x_0 between a and b such that

$$F'(x_0) = 0. \tag{2}$$

But now (1) and (2) imply

$$f'(x_0) - \frac{f(b) - f(a)}{b - a} = 0$$

$$f'(x_0) = \frac{f(b) - f(a)}{b - a}.$$

Example 1

Verify the mean-value theorem for $f(x) = x^2$, where $a = 0$ and $b = 1$.

Clearly f is continuous and differentiable everywhere; so f satisfies the conditions of the mean-value theorem. Thus the mean-value theorem asserts that there is a number x_0 between 0 and 1 such that

$$f'(x_0) = \frac{f(1) - f(0)}{1 - 0}.$$

Let us now verify that there is such a number.

$$\frac{f(1) - f(0)}{1 - 0} = \frac{1^2 - 0^2}{1 - 0} = 1,$$

$$f'(x) = 2x.$$

We now see that if $x_0 = 1/2$, then

$$f'(x_0) = \frac{f(1) - f(0)}{1 - 0}.$$

Thus the theorem is verified.

The following useful theorem is related to the mean-value theorem.

Theorem 17.9

(Extended mean-value theorem)   If a and b are numbers $(a < b)$ and f and g are functions such that

(a) *f and g are continuous on $[a, b]$,*
(b) *$f'(x)$ and $g'(x)$ exist for all x in (a, b), and*
(c) *$g'(x) \neq 0$ for each x in (a, b),*

then there is a number x_0 between a and b such that

$$\frac{f'(x_0)}{g'(x_0)} = \frac{f(b) - f(a)}{g(b) - g(a)}.$$

The choice $g(x) = x$ reduces the extended mean-value theorem to the mean-value theorem. This gives us a hint as to how the generalized version might be proved. In proving the mean-value theorem, we used the function

$$F(x) = f(x) - f(a) - \frac{f(b) - f(a)}{b - a}(x - a).$$

If $g(x) = x$, then $g(a) = a$ and $g(b) = b$, so it might be reasonable to replace x, a, and b by $g(x)$, $g(a)$, and $g(b)$, respectively. Thus, in order to prove the extended mean-value theorem, we shall consider the new function

$$F(x) = f(x) - f(a) - \frac{f(b) - f(a)}{g(b) - g(a)} [g(x) - g(a)].$$

By use of this new function, the theorem can be proved in exactly the same way as the mean-value theorem. The details of the proof are left to the student (see Problem 22). Actually, there is one additional problem here that was not encountered in the proof of the mean-value theorem. We cannot be sure that the expression for $F(x)$ is meaningful unless we can be sure that $g(b) - g(a) \neq 0$, for this expression appears as a denominator in $F(x)$. This can be shown by use of Rolle's theorem (see Problem 21).

Example 2

Verify the extended mean-value theorem for $f(x) = x^2 - 1$, $g(x) = x - 1$, $a = 0$, and $b = 3$.

Clearly, both functions are defined and continuous, as well as differentiable, everywhere. Furthermore, $g'(x) = 1$ for all x and so it is not 0 for any x. Thus the extended mean-value theorem asserts the existence of a number x_0 between 0 and 3 such that

$$\frac{f'(x_0)}{g'(x_0)} = \frac{f(3) - f(0)}{g(3) - g(0)}.$$

Let us verify that this is the case. If there is such a number x_0, then

$$\frac{2x_0}{1} = \frac{8 - (-1)}{2 - (-1)} \quad \text{or} \quad x_0 = \frac{3}{2}.$$

Since $0 < 3/2 < 3$, the theorem is verified for this case.

The proofs of several theorems were omitted from earlier chapters because they required the use of the mean-value theorem. The omitted proofs are given in Appendix B. A review of these theorems and their proofs will demonstrate the importance of the mean-value theorem in theoretical work. The extended mean-value theorem also has important applications, one of which is considered in the next chapter.

Problems

A

In Problems 1–9, verify the mean-value theorem for the given function and the given values of a and b.

1. $f(x) = x^2$; $a = -1$, $b = 3$.
2. $f(x) = x^2 + 2$; $a = -2$, $b = 3$.
3. $f(x) = x^3$; $a = 0$, $b = 3$.
4. $f(x) = x^3$; $a = -1$, $b = 1$.
5. $f(x) = \dfrac{1}{x}$; $a = 1$, $b = 3$.
6. $f(x) = \dfrac{1}{x^2}$; $a = 1$, $b = 2$.
7. $f(x) = \dfrac{1}{x + 1}$; $a = 0$, $b = 2$.
8. $f(x) = \dfrac{x}{x + 1}$; $a = 0$, $b = 3$.
9. $f(x) = e^x$; $a = 0$, $b = 1$.

In Problems 10–13, does the mean-value theorem tell you anything about the given function? If so, what? If not, why not?

10. $f(x) = \dfrac{1}{x}$; $a = -1$, $b = 1$.

11. $f(x) = |x|$; $a = 0$, $b = 1$.

12. $f(x) = |x|$; $a = -1$, $b = 2$.

13. $f(x) = \begin{cases} x & \text{if } x \leq 0, \\ 0 & \text{if } x > 0; \end{cases}$ $a = -1$, $b = 1$.

In Problems 14–17, verify the extended mean-value theorem for the given functions.

14. $f(x) = 2x + 1$, $g(x) = x^2$; $a = 0$, $b = 1$.

15. $f(x) = x^2 + 2$, $g(x) = \dfrac{1}{x}$; $a = 1$, $b = 2$.

16. $f(x) = x^3$, $g(x) = \dfrac{1}{x + 2}$; $a = 0$, $b = 1$.

17. $f(x) = (x + 1)^2$, $g(x) = 2x - 1$; $a = 1$, $b = 3$.

B

18. Verify the mean-value theorem for $f(x) = \tan x$; $a = 0$, $b = \pi/4$.

In Problems 19 and 20, does the mean-value theorem tell you anything about the given function? If so, what? If not, why not?

19. $f(x) = \ln |x|$; $a = -1$, $b = 1$.

20. $f(x) = |\ln x|$; $a = 1/2$, $b = 2$.

C

21. Show that if g is continuous on $[a, b]$ and if $g'(x) \neq 0$ for all x in (a, b), then $g(a) \neq g(b)$ (see Problem 22 of the previous section).

22. Complete the proof of the extended mean-value theorem.

23. Show that if f is a differentiable function and $f'(x) \leq 2$ for all x, then there is at most one real number a such that $a > 1$ and $f(a) = a^2$.

24. Show that $\sin x \leq x$ for all $x \geq 0$.

Review Problems

A

In Problems 1–6, indicate whether the given limit statement is true or false and show by means of the proper definition that your answer is correct.

1. $\lim\limits_{x \to 2} (2x + 1) = 5$.

2. $\lim\limits_{x \to 0} \sqrt{x^2 - 1} = 1$.

3. $\lim\limits_{x \to 0^-} f(x) = 1$,   where $f(x) = \begin{cases} -x & \text{if } x < 0, \\ 1 & \text{if } x \geq 0. \end{cases}$

4. $\lim\limits_{x \to 0} f(x) = -\infty$,   where $f(x) = \begin{cases} 1/x & \text{if } x < 0, \\ x^2 & \text{if } x \geq 0. \end{cases}$

5. $\lim\limits_{x \to 3^+} \dfrac{1}{x - 3} = +\infty$.

6. $\lim\limits_{x \to -\infty} x^2 = +\infty$.

B

In Problems 7–12, indicate whether the given limit statement is true or false and show by means of the proper definition that your answer is correct.

7. $\lim\limits_{x \to 2} [\![x]\!] = 2.$

8. $\lim\limits_{x \to 2^+} [\![x]\!] = 2.$

9. $\lim\limits_{x \to -2} x^2 = 4.$

10. $\lim\limits_{x \to -1} \dfrac{1}{x+3} = \tfrac{1}{2}.$

11. $\lim\limits_{x \to +\infty} \dfrac{x^2+1}{x+1} = 1.$

12. $\lim\limits_{x \to -\infty} \dfrac{2x}{x-3} = 2.$

13. Show that $\lim\limits_{x \to 0} \sin(1/x)$ does not exist.

14. Show that $\lim\limits_{x \to 1} [1/(x-1)]$ is neither $+\infty$ nor $-\infty$, but

$$\lim\limits_{x \to 1^+} \frac{1}{x-1} = +\infty \quad \text{and} \quad \lim\limits_{x \to 1^-} \frac{1}{x-1} = -\infty.$$

15. Show by means of the definition of continuity that $f(x) = x^2 + 3$ is continuous at $(1, 4)$.

16. Show by means of the definition of continuity that

$$f(x) = \begin{cases} 2x + 1 & \text{if } x < 0, \\ x + 2 & \text{if } x \geq 0 \end{cases}$$

is discontinuous at $(0, 2)$.

17. If $f(x) = (x^2 + 2x - 8)/(x + 4)$ for $x \neq -4$, what value should be given $f(-4)$ in order that f be continuous?

18. Does Rolle's theorem say anything about $f(x) = \sqrt{1 - x^2}$? If so, what? If not, why not?

19. Does Rolle's theorem say anything about $f(x) = (x^2 - 4)/x$? If so, what? If not, why not?

20. Does the mean-value theorem tell us anything about $f(x) = x^{2/3}$, where $a = -1$ and $b = 8$? If so, what? If not, why not?

21. Does the mean-value theorem tell us anything about $f(x) = x^{2/3}$, where $a = 0$ and $b = 8$? If so, what? If not, why not?

22. A function f is known to be positive when $x = 0$ and negative when $x = 1$. Is this sufficient to assure us that $f(a) = 0$ for some a between 0 and 1? Explain.

C

23. Show that if $f(x) = \begin{cases} 1 & \text{if } x \text{ is rational,} \\ 0 & \text{if } x \text{ is irrational,} \end{cases}$ then $\lim_{x \to a} f(x)$ does not exist for any real value of a or for $a = \pm\infty$.

24. Show that if $f(x) = \begin{cases} x & \text{if } x \text{ is rational,} \\ 0 & \text{if } x \text{ is irrational,} \end{cases}$ then $\lim_{x \to 0} f(x) = 0$ but $\lim_{x \to a} f(x)$ does not exist if $a \neq 0$.

25. Show that if a and b are numbers $(a < b)$ and f is a function such that
    (a) f is continuous on $[a, b]$,
    (b) $f'(x), f''(x)$, and $f'''(x)$ exist for all x in (a, b), and
    (c) $f(a) = f(b) = f'(a) = f'(b) = 0$,
    then there is a number c between a and b such that $f'''(c) = 0$.

26. Show that if f is continuous at $x = a$ and $f(a) > 0$, then there are numbers b and c with a between them such that $f(x) > 0$ for all x in (b, c).

18

18.1

The Forms $0/0$ *and* ∞/∞

Some of the very simple limits we have dealt with, such as

$$\lim_{x \to 1} \frac{x}{x+1} = \frac{1}{2},$$

can be evaluated simply by substitution. On the other hand, there have been limits such as

$$\lim_{x \to 1} \frac{x^2 - 1}{x - 1}$$

that cannot be evaluated by substitution, because both the numerator and denominator approach 0 as x approaches 1. When we first encountered limits of this type, we observed that the mere fact that both numerator and denominator approach 0 tells us nothing about the limit of the quotient. This particular form of limit is called an indeterminate form. The two types we have encountered most frequently in the past have been the forms $0/0$ and ∞/∞, by which is meant both the numerator and denominator approach 0 or else the numerator approaches $+\infty$ or $-\infty$ and the denominator approaches $+\infty$ or $-\infty$ (numerator and denominator not necessarily agreeing in sign).

Of course, we had little trouble evaluating limits like

$$\lim_{x \to 1} \frac{x^2 - 1}{x - 1}.$$

We simply factored the numerator and canceled the factors $x - 1$. Unfortunately we could not evaluate

$$\lim_{x \to 0} \frac{\sin x}{x}$$

so simply. In that case we went through a rather elaborate geometric argument in order to find the limit. Because the method of canceling factors has only a very limited application, we should like to find a more general method of handling indeterminate forms. A method which works for a very wide range of functions is called l'Hôpital's (pronounced low'-pee-tahls) rule.

In order to simplify the statements of l'Hôpital's rule, let us introduce the concept of a deleted neighborhood. If a is a real number, then a deleted neighborhood of a is the set of all numbers except a in an open interval (b, c) that contains a. For example: a deleted neighborhood of 3 is

$$\{x \,|\, x \text{ is in } (0, 4) \text{ and } x \neq 3\} = \{x \,|\, 0 < x < 3 \text{ or } 3 < x < 4\}.$$

Of course, there are many other deleted neighborhoods of 3. A deleted neighborhood of $+\infty$ is the set of all numbers greater than some real number N. Likewise, a deleted neighborhood of $-\infty$ is the set of all numbers less than some real number N.

Now let us consider l'Hôpital's rule. It is stated here as two separate theorems.

Theorem 18.1

(*l'Hôpital's rule*—0/0)   *If a is a real number, $+\infty$, or $-\infty$ and f and g are functions such that*

   (a)   *$f'(x)$ and $g'(x)$ exist for all x in some deleted neighborhood of a and $g'(x) \neq 0$ for all x in that neighborhood,*

   (b)   *$\lim_{x \to a} f(x) = \lim_{x \to a} g(x) = 0$, and*

   (c)   *$\lim_{x \to a} f'(x)/g'(x) = L$, where L is a real number, $+\infty$, or $-\infty$,*

then

$$\lim_{x \to a} \frac{f(x)}{g(x)} = L.$$

Proof

Let us consider the case in which a is a real number and the deleted neighborhood is $\{x \,|\, b < x < a \text{ or } a < x < c\}$. Let us further restrict ourselves to the right-hand limit.

The statement of the theorem gives no assurance that either $f(a)$ or $g(a)$ exists. Thus, let us consider two new functions F and G such that

$$F(x) = \begin{cases} f(x) & x \neq a, \\ 0 & x = a, \end{cases} \qquad G(x) = \begin{cases} g(x) & x \neq a, \\ 0 & x = a. \end{cases}$$

Then F and G are continuous at $x = a$. If $a < x < c$, then F and G are defined and continuous on $[a, x]$. (We have already noted that they are continuous at $x = a$; since they are differentiable, they are continuous at every other number in $[a, x]$.) Also, $F' = f'$ and $G' = g'$ exist at every number in (a, x) and $G' \neq 0$

everywhere in (a, x). By the extended mean-value theorem, there is a number y between a and x such that

$$\frac{F(x) - F(a)}{G(x) - G(a)} = \frac{F'(y)}{G'(y)}.$$

But $F(a) = G(a) = 0$. Thus

$$\frac{F(x)}{G(x)} = \frac{F'(y)}{G'(y)}.$$

As $x \to a^+$, we also have $y \to a^+$, because $a < y < x$. Therefore,

$$\lim_{x \to a^+} \frac{f(x)}{g(x)} = \lim_{x \to a^+} \frac{F(x)}{G(x)} = \lim_{y \to a^+} \frac{F'(y)}{G'(y)} = \lim_{y \to a^+} \frac{f'(y)}{g'(y)} = L.$$

A similar argument (see Problem 34) shows that

$$\lim_{x \to a^-} \frac{f(x)}{g(x)} = L,$$

which gives

$$\lim_{x \to a} \frac{f(x)}{g(x)} = L$$

for the case in which a is a real number.

Suppose $a = +\infty$. Let us introduce a new variable t such that

$$x = \frac{1}{t}.$$

Then

$$\lim_{x \to +\infty} \frac{f(x)}{g(x)} = \lim_{t \to 0^+} \frac{f\left(\frac{1}{t}\right)}{g\left(\frac{1}{t}\right)}$$

$$= \lim_{t \to 0^+} \frac{\dfrac{d}{dt} f\left(\frac{1}{t}\right)}{\dfrac{d}{dt} g\left(\frac{1}{t}\right)}$$

$$= \lim_{t \to 0^+} \frac{-t^2 \dfrac{d}{dt} f\left(\frac{1}{t}\right)}{-t^2 \dfrac{d}{dt} g\left(\frac{1}{t}\right)}$$

$$= \lim_{x \to +\infty} \frac{\dfrac{d}{dx} f(x)}{\dfrac{d}{dx} g(x)} \qquad \text{(See Note)}$$

$$= \lim_{x \to +\infty} \frac{f'(x)}{g'(x)} = L.$$

A similar argument handles the case in which $a = -\infty$ (see Problem 35).

Note: Since $x = 1/t$, $f(x) = f(1/t)$. By the chain rule,

$$\frac{d}{dt}f\left(\frac{1}{t}\right) = \frac{d}{dx}f(x) \cdot \frac{d}{dt}\left(\frac{1}{t}\right)$$

$$= \frac{d}{dx}f(x)\left(-\frac{1}{t^2}\right),$$

$$-t^2\frac{d}{dt}f\left(\frac{1}{t}\right) = \frac{d}{dx}f(x).$$

Theorem 18.2

(l' Hôpital's rule—∞/∞) If a is a real number, $+\infty$, or $-\infty$ and f and g are functions such that

(a) *$f'(x)$ and $g'(x)$ exist for all x in some deleted neighborhood of a and $g'(x) \neq 0$ for each x in that neighborhood,*

(b) *$\lim_{x\to a}|f(x)| = \lim_{x\to a}|g(x)| = +\infty$, and*

(c) *$\lim_{x\to a}f'(x)/g'(x) = L$, where L is a real number, $+\infty$, or $-\infty$,*

then

$$\lim_{x\to a}\frac{f(x)}{g(x)} = L.$$

The proof for this case is more difficult. We omit it.†

The conclusions of Theorem 18.1 and Theorem 18.2 remain valid if a is an end point of a finite interval and appropriate one-sided limits are taken as x approaches a. See Example 5.

It might appear at first glance that the conditions on the use of l'Hôpital's rule are quite stringent and that only a few functions satisfy all of them. Actually, quite the opposite is true. Differentiability is a condition that almost all of the functions we have encountered satisfy at all but a finite number of points. Furthermore, by taking the interval small enough, we can assure ourselves that the derivative of the denominator is not 0 for all values of x "near" $x = a$. The only thing we need to do for our simple functions (and it must always be done) is make sure we have one of the indeterminate forms 0/0 or ∞/∞.

Example 1

Evaluate $\lim_{x\to 0}\dfrac{\sin x}{x}$.

$$\lim_{x\to 0}\frac{\sin x}{x}\quad\begin{array}{c}\to 0\\\to 0\end{array}\quad = \lim_{x\to 0}\frac{\cos x}{1} = \frac{1}{1} = 1.$$

This, of course, agrees with the result we obtained in Chapter 11 (see pages 322–323). That is not to say that we could have used l'Hôpital's rule at that time. In order to use it, we need to know the derivative of sin x. We did not know it then; in fact, we used the value of this limit to find the derivative of sin x.

It might be noted that we did *not* take the derivative of (sin x)/x; we took the derivatives of the numerator and denominator individually.

---

† For a proof of this theorem see Angus E. Taylor and Robert Mann, *Advanced Calculus.* 2nd ed. (Lexington, Mass.: Xerox, 1972), pp. 115–118.

Example 2

Evaluate $\lim\limits_{x \to +\infty} \dfrac{e^x}{x}$.

$$\lim_{x \to +\infty} \frac{e^x}{x} \begin{matrix} \to +\infty \\ \to +\infty \end{matrix} = \lim_{x \to +\infty} \frac{e^x}{1} = +\infty.$$

Example 3

Evaluate $\lim\limits_{x \to 0} \dfrac{1 - \cos x}{x^2}$.

$$\lim_{x \to 0} \frac{1 - \cos x}{x^2} \begin{matrix} \to 0 \\ \to 0 \end{matrix} = \lim_{x \to 0} \frac{\sin x}{2x} \begin{matrix} \to 0 \\ \to 0 \end{matrix}.$$

At this point you might feel that l'Hôpital's rule has failed us, since we are still left with the indeterminate form 0/0. But there is nothing to prevent us from using l'Hôpital's rule a second time. The result is

$$\lim_{x \to 0} \frac{1 - \cos x}{x^2} = \lim_{x \to 0} \frac{\sin x}{2x} = \lim_{x \to 0} \frac{\cos x}{2} = \frac{1}{2}.$$

Example 4

Evaluate $\lim\limits_{x \to 0} \dfrac{x - \sin x}{2 + 2x + x^2 - 2e^x}$.

$$\lim_{x \to 0} \frac{x - \sin x}{2 + 2x + x^2 - 2e^x} \begin{matrix} \to 0 \\ \to 0 \end{matrix} = \lim_{x \to 0} \frac{1 - \cos x}{2 + 2x - 2e^x} \begin{matrix} \to 0 \\ \to 0 \end{matrix}$$

$$= \lim_{x \to 0} \frac{\sin x}{2 - 2e^x} \begin{matrix} \to 0 \\ \to 0 \end{matrix}$$

$$= \lim_{x \to 0} \frac{\cos x}{-2e^x} = -\frac{1}{2}.$$

Example 5

Evaluate $\lim\limits_{x \to 0^+} \dfrac{\cot x}{\cot 2x}$.

$$\lim_{x \to 0^+} \frac{\cot x}{\cot 2x} \begin{matrix} \to +\infty \\ \to +\infty \end{matrix} = \lim_{x \to 0^+} \frac{-\csc^2 x}{-2\csc^2 2x} \begin{matrix} \to -\infty \\ \to -\infty \end{matrix}$$

$$= \lim_{x \to 0^+} \frac{2\csc^2 x \cot x}{4\csc^2 2x \cot 2x} \begin{matrix} \to +\infty \\ \to +\infty \end{matrix}.$$

At this point let us stop and take stock. The derivative of $\cot x$ is $-\csc^2 x$; the derivative of $\csc x$ is $-\csc x \cot x$. Furthermore,

$$\lim_{x \to 0^+} \cot x = +\infty \quad \text{and} \quad \lim_{x \to 0^+} \csc x = +\infty.$$

Thus, by taking derivatives of cot x and csc x, we can only get a new expression involving cot x and csc x, and the limit in either case is $+\infty$. Thus, continued application of l'Hôpital's rule will get us nowhere. Let us return to the beginning and try something different.

$$\lim_{x\to 0^+} \frac{\cot x}{\cot 2x} = \lim_{x\to 0^+} \frac{\tan 2x}{\tan x} \quad \begin{matrix} \to 0 \\ \to 0 \end{matrix}$$

$$= \lim_{x\to 0^+} \frac{2\sec^2 2x}{\sec^2 x} = 2.$$

Example 6

Evaluate $\displaystyle\lim_{x\to 0^+} \frac{e^x}{x}$.

$$\lim_{x\to 0^+} \frac{e^x}{x} \quad \begin{matrix} \to 1 \\ \to 0^+ \end{matrix} \quad = +\infty.$$

We see that l'Hôpital's rule is not necessary here; in fact, it cannot be used, since we do not have one of the indeterminate forms. A blind application of the rule gives

$$\lim_{x\to 0^+} \frac{e^x}{x} = \lim_{x\to 0^+} \frac{e^x}{1} = 1,$$

which is incorrect. Before using l'Hôpital's rule *you must make sure you have one of the indeterminate forms* $0/0$ *or* ∞/∞.

L'Hôpital's rule is so named because it first appeared in print in 1696 in a calculus textbook (the first such text) by the Marquis de l'Hôpital (1661–1704). Actually the text consisted of the lecture notes of his teacher, Johann Bernoulli (1667–1748). Thus Theorems 18.1 and 18.2 might be more properly called Bernoulli's rule.

Problems

A

In Problems 1–16, evaluate the given limit.

1. $\displaystyle\lim_{x\to 1} \frac{x^2-1}{x-1}$.

2. $\displaystyle\lim_{x\to 0} \frac{e^x-1}{\sin x}$.

3. $\displaystyle\lim_{x\to -2} \frac{x^2+x-2}{x+2}$.

4. $\displaystyle\lim_{x\to a} \frac{\sqrt{x}-\sqrt{a}}{x-a}$.

5. $\displaystyle\lim_{x\to a} \frac{\sqrt[3]{x}-\sqrt[3]{a}}{x-a}$.

6. $\displaystyle\lim_{x\to +\infty} \frac{x^2+4x}{x^2-2}$.

7. $\displaystyle\lim_{x\to -\infty} \frac{x-5}{x^2+4}$.

8. $\displaystyle\lim_{x\to 0} \frac{\tan x}{x}$.

9. $\displaystyle\lim_{x\to 1^-} \frac{\ln x}{\sqrt{1-x}}$.

10. $\displaystyle\lim_{x\to 0} \frac{\sin x}{e^x}$.

11. $\displaystyle\lim_{x\to \pi/2} \frac{\sec x}{\sec 2x}$.

12. $\displaystyle\lim_{x\to 0^+} \frac{\ln x}{x}$.

13. $\displaystyle\lim_{x\to +\infty} \frac{\ln x}{x}$.

14. $\displaystyle\lim_{x\to \pi/2} \frac{\cos x + \sin x - 1}{\cos x - \sin x + 1}$.

15. $\displaystyle\lim_{x\to 0} \frac{\ln x}{1/x}$.

16. $\displaystyle\lim_{x\to 0} \frac{\sin x}{e^x - e^{-x}}$.

B

In Problems 17–26, evaluate the given limit.

17. $\displaystyle\lim_{x\to 0} \frac{\tan x - \sin x}{x^3}$.

18. $\displaystyle\lim_{x\to +\infty} \frac{\ln x}{x^n}$.
(n a positive integer).

19. $\displaystyle\lim_{x\to 0} \frac{a^x - b^x}{x}$.

20. $\lim\limits_{x\to 0} \dfrac{e^x - e^{-x} - 2x}{x - \sin x}$.

21. $\lim\limits_{x\to 0} \dfrac{\ln (\tan x)}{\ln (\tan 2x)}$.

22. $\lim\limits_{x\to 0^+} \dfrac{1 - \ln x}{e^{1/x}}$.

23. $\lim\limits_{x\to \pi/2^+} \dfrac{\ln [x - (\pi/2)]}{\tan x}$.

24. $\lim\limits_{x\to 0^+} \dfrac{\text{Arcsin} (1 - x)}{\sqrt{2x - x^2}}$.

25. $\lim\limits_{x\to 0} \dfrac{\sin x - x \cos x}{x - \sin x}$.

26. $\lim\limits_{x\to \pi/4} \dfrac{\sec^2 x - 2 \tan x}{1 + \cos 4x}$.

C

In Problems 27–29, evaluate the given limit.

27. $\lim\limits_{x\to 0} \dfrac{x - \text{Arcsin} x}{\sin^3 x}$.

28. $\lim\limits_{x\to 1} \dfrac{x^x - x}{1 - x + \ln x}$.

29. $\lim\limits_{x\to 0^+} \dfrac{e^{-1/x}}{x}$. (*Hint:* Use the substitution $u = 1/x$.)

30. Find real numbers a and b such that
$$\lim_{x\to 0} \frac{\sin x - ax + bx^2}{x^2} = 2.$$

31. Find real numbers a and b such that
$$\lim_{x\to 0} \frac{\cos ax + b}{x^2} = -2.$$

32. The sum of the first n terms of a geometric series is given by
$$S = a + ar + ar^2 + \cdots + ar^{n-1} = \frac{a(r^n - 1)}{r - 1} \quad (r \neq 1).$$
Use l'Hôpital's rule to evaluate
$$\lim_{r\to 1} \frac{a(r^n - 1)}{r - 1}.$$
Compare this result with the geometric series with $r = 1$.

33. Evaluate $\displaystyle\int_1^{+\infty} \dfrac{\ln x}{x^2}\, dx$.

34. Prove that $\lim\limits_{x\to a^-} \dfrac{f(x)}{g(x)} = L$ under the conditions of Theorem 18.1.

35. Prove that $\lim\limits_{x\to -\infty} \dfrac{f(x)}{g(x)} = L$ under the conditions of Theorem 18.1.

18.2

The Forms $0 \cdot \infty$ *and* $\infty - \infty$

Let us emphasize that l'Hôpital's rule handles only the two indeterminate forms $0/0$ and ∞/∞. However, some other indeterminate forms can be put into one of these two forms. We consider two of them here. The first is the form $0 \cdot \infty$: that is,
$$\lim_{x\to a} f(x) \cdot g(x),$$
where
$$\lim_{x\to a} f(x) = 0 \quad \text{and} \quad \lim_{x\to a} |g(x)| = +\infty.$$

One factor tends to make the product large in absolute value, while the other tends to make it small. Thus, the outcome is in doubt, and $0 \cdot \infty$ is an indeterminate form. It can be put into the form $0/0$ or ∞/∞ by one of the following methods.

$$\lim_{x \to a} f(x) \cdot g(x) = \lim_{x \to a} \frac{f(x)}{1/g(x)} \quad \begin{matrix} \to 0 \\ \to 0. \end{matrix}$$

$$\lim_{x \to a} f(x) \cdot g(x) = \lim_{x \to a} \frac{g(x)}{1/f(x)} \quad \begin{matrix} \to \pm\infty \\ \to \pm\infty. \end{matrix}$$

Either way gives a form that l'Hôpital's rule can handle; the choice depends upon which of the two is easier.

Example 1

Evaluate $\lim\limits_{x \to 0} x \ln x$.

$$\begin{aligned} \lim_{x \to 0} x \ln x \to 0(-\infty) &= \lim_{x \to 0} \frac{\ln x}{1/x} \quad \begin{matrix} \to -\infty \\ \to +\infty \end{matrix} \quad \text{(Since } \ln x \text{ is defined only for} \\ & \qquad\qquad\qquad\qquad\qquad\quad \text{positive values of } x, x \to 0^+\text{)} \\ &= \lim_{x \to 0} \frac{1/x}{-1/x^2} \quad \begin{matrix} \to +\infty \\ \to -\infty \end{matrix} \\ &= \lim_{x \to 0} (-x) = 0. \end{aligned}$$

In this example we had a choice of changing $x \ln x$ to

$$\frac{\ln x}{1/x} \quad \text{or} \quad \frac{x}{1/\ln x}.$$

Since the derivative of $1/\ln x$ is relatively complicated, we chose the other, simpler form. After using l'Hôpital's rule once, we were still left with the indeterminate form ∞/∞. Continued use of l'Hôpital's rule would have gotten us nowhere; but simplifying the expression algebraically led to an answer directly. Remember that l'Hôpital's rule is not a cure for all ills; sometimes another method works where it fails.

Example 2

Evaluate $\lim\limits_{x \to \pi/2^-} \sec 3x \cos 5x$.

$$\begin{aligned} \lim_{x \to \pi/2^-} \sec 3x \cos 5x \to (-\infty)0 &= \lim_{x \to \pi/2^-} \frac{\cos 5x}{\cos 3x} \quad \begin{matrix} \to 0 \\ \to 0 \end{matrix} \\ &= \lim_{x \to \pi/2^-} \frac{-5 \sin 5x}{-3 \sin 3x} = \frac{-5}{3}. \end{aligned}$$

Another form that leads to either $0/0$ or ∞/∞ is generally referred to as "$\infty - \infty$." In the past we have used ∞ without a $+$ or $-$ (as in ∞/∞ or $0 \cdot \infty$), when it made no difference if we had the $+$ or the $-$. It does make a difference in this case; in fact, *the signs must be watched carefully.* The form $\infty - \infty$ is a convenient notation for all of the following expressions: $(+\infty) - (+\infty,) (-\infty) - (-\infty)$ and $(+\infty) + (-\infty)$. Again, one term tends to make the expression large while the other tends to make it small, giving an indeterminate form.

Unfortunately there is no generally applicable, simple procedure for converting to one of the forms $0/0$ or ∞/∞. All that can be said is: combine the two terms into a single fraction and try to get one of the previous forms from that.

Example 3

Evaluate $\displaystyle\lim_{x \to \pi/2 -} (\sec x - \tan x)$.

$$\lim_{x \to \pi/2 -} (\sec x - \tan x) \to (+\infty) - (+\infty) = \lim_{x \to \pi/2 -} \left(\frac{1}{\cos x} - \frac{\sin x}{\cos x} \right)$$

$$= \lim_{x \to \pi/2 -} \frac{1 - \sin x}{\cos x} \quad \begin{array}{l} \to 0 \\ \to 0 \end{array}$$

$$= \lim_{x \to \pi/2 -} \frac{-\cos x}{-\sin x} = \frac{0}{-1} = 0.$$

Example 4

Evaluate $\displaystyle\lim_{x \to 1+} \left(\frac{1}{x-1} - \frac{2}{x^2-1} \right)$.

$$\lim_{x \to 1+} \left(\frac{1}{x-1} - \frac{2}{x^2-1} \right) \to (+\infty) - (+\infty) = \lim_{x \to 1+} \left(\frac{x+1}{x^2-1} - \frac{2}{x^2-1} \right)$$

$$= \lim_{x \to 1+} \frac{x-1}{x^2-1} \quad \begin{array}{l} \to 0 \\ \to 0 \end{array}$$

$$= \lim_{x \to 1+} \frac{1}{2x} = \frac{1}{2}.$$

While we used l'Hôpital's rule here, we had a choice of that or canceling factors.

Example 5

Evaluate $\displaystyle\lim_{x \to 0+} \left(\frac{1}{x} - \ln x \right)$.

$$\lim_{x \to 0+} \frac{1}{x} = +\infty, \qquad \lim_{x \to 0+} \ln x = -\infty.$$

Since the second term is subtracted from the first, the two terms are pulling in the same direction. This is *not* an indeterminate form and l'Hôpital's rule is not needed.

$$\lim_{x \to 0+} \left(\frac{1}{x} - \ln x \right) = +\infty.$$

Problems

A

In Problems 1–15, evaluate the limits.

1. $\displaystyle\lim_{x \to 0+} x^2 \ln x$.

2. $\displaystyle\lim_{x \to 0+} x^n \ln x$ (n a positive integer).

3. $\displaystyle\lim_{x \to 0+} \sin x \ln x$.

4. $\displaystyle\lim_{x \to 0+} \tan x \ln x$.

5. $\displaystyle\lim_{x \to \pi/2 -} \sec x \cos 5x$.

6. $\displaystyle\lim_{x \to \pi/4 -} (1 - \tan x) \sec 2x$.

7. $\displaystyle\lim_{x \to \pi/2 -} \sec 3x \cos 7x$.

8. $\displaystyle\lim_{x \to 1-} (1 - x) \tan \frac{\pi x}{2}$.

9. $\displaystyle\lim_{x\to+\infty} e^{-x}\ln x.$

10. $\displaystyle\lim_{x\to0^+} e^{-x}\ln x.$

11. $\displaystyle\lim_{x\to\pi/2^-} \sec x\left(x\sin x - \frac{\pi}{2}\right).$

12. $\displaystyle\lim_{x\to a^+} \ln(x-a)\tan(x-a).$

13. $\displaystyle\lim_{x\to+\infty} x\ln\left(1+\frac{a}{x}\right)\ (a>0).$

14. $\displaystyle\lim_{x\to1}\left(\frac{1}{\ln x}-\frac{x}{\ln x}\right).$

15. $\displaystyle\lim_{x\to0}\left(\frac{1}{2x}-\frac{1}{x(e^x+1)}\right).$

B

In Problems 16–24, evaluate the limits.

16. $\displaystyle\lim_{x\to0^+}\ln x\,\ln(1+x).$

17. $\displaystyle\lim_{x\to1^+}\left(\frac{1}{\ln x}-\frac{1}{x-1}\right).$

18. $\displaystyle\lim_{x\to\pi/2^-}(2x\tan x-\pi\sec x).$

19. $\displaystyle\lim_{x\to1}\left(\frac{x}{x-1}-\frac{1}{\ln x}\right).$

20. $\displaystyle\lim_{x\to0}\left(\frac{1}{x^2}-\frac{1}{x\tan x}\right).$

21. $\displaystyle\lim_{x\to0}\left(\frac{1}{x(1+x)}-\frac{\ln(1+x)}{x^2}\right).$

22. $\displaystyle\lim_{x\to0}\left(\frac{x}{\sin^3 x}-\cot^2 x\right).$

23. $\displaystyle\lim_{x\to1^+}\left(\frac{1}{1-x}-\frac{1}{\ln x}\right).$

24. $\displaystyle\lim_{x\to\pi^-}(\cot x-\csc x).$

25. Evaluate the limit of Example 1 by using the substitution $u=1/x.$

C

In Problems 26 and 27, evaluate the limits.

26. $\displaystyle\lim_{x\to0}\left(\frac{2}{\sin^2 x}-\frac{1}{1-\cos x}\right).$

27. $\displaystyle\lim_{x\to+\infty}\left[x-x^2\ln\left(1+\frac{1}{x}\right)\right].$

28. Evaluate $\displaystyle\int_0^1 \ln x\,dx.$

29. Evaluate $\displaystyle\int_0^{+\infty}\frac{dx}{x(x+1)}.$

30. Sketch $y=x\ln x.$

18.3

The Forms 0^0, ∞^0, *and* 1^∞

Let us now consider the limit

$$\lim_{x\to a}[f(x)]^{g(x)},$$

where

(a) $\displaystyle\lim_{x\to a}f(x)=0$ and $\displaystyle\lim_{x\to a}g(x)=0$ $(0^0),$

(b) $\displaystyle\lim_{x\to a}f(x)=+\infty$ and $\displaystyle\lim_{x\to a}g(x)=0$ $(\infty^0),$

(c) $\displaystyle\lim_{x\to a}f(x)=1$ and $\displaystyle\lim_{x\to a}|g(x)|=+\infty$ $(1^\infty).$

All three of these cases are handled in the same way. This method depends upon the fact (see Theorem 17.3, page **566**) that if G is a continuous function, then

$$G(\lim_{x\to a} F(x)) = \lim_{x\to a} G(F(x)).$$

In particular, if $G(x) = \ln x$, we have

$$\ln(\lim_{x\to a} F(x)) = \lim_{x\to a} \ln F(x).$$

Let us consider some examples.

Example 1

Evaluate $\lim\limits_{x\to 0^+} x^x$.

We see that this is the indeterminate form 0^0. Assuming that the given limit either exists or is $+\infty$, we represent it by y and take the logarithm of both sides.

$$y = \lim_{x\to 0^+} x^x,$$

$$\ln y = \ln \lim_{x\to 0^+} x^x = \lim_{x\to 0^+} \ln x^x$$

$$= \lim_{x\to 0^+} x \ln x \qquad \to 0(-\infty)$$

$$= \lim_{x\to 0^+} \frac{\ln x}{1/x} \qquad \begin{array}{l} \to -\infty \\ \to +\infty \end{array}$$

$$= \lim_{x\to 0^+} \frac{1/x}{-1/x^2} \qquad \begin{array}{l} \to +\infty \\ \to -\infty \end{array}$$

$$= \lim_{x\to 0^+} (-x) = 0.$$

We now have, not the limit we want, but the logarithm of that limit. Since $\ln y = 0$, $y = e^0 = 1$, or

$$\lim_{x\to 0^+} x^x = 1.$$

Example 2

Evaluate $\lim\limits_{x\to +\infty} x^{1/x}$.

This has the form ∞^0. Thus, if

$$y = \lim_{x\to +\infty} x^{1/x},$$

then

$$\ln y = \ln \lim_{x\to +\infty} x^{1/x} = \lim_{x\to +\infty} \ln x^{1/x}$$

$$= \lim_{x\to +\infty} \frac{1}{x} \ln x \qquad \to 0(+\infty)$$

$$= \lim_{x\to +\infty} \frac{\ln x}{x} \qquad \begin{array}{l} \to +\infty \\ \to +\infty \end{array}$$

$$= \lim_{x\to +\infty} \frac{1/x}{1} = \frac{0}{1} = 0.$$

Thus

$$\lim_{x\to +\infty} x^{1/x} = e^0 = 1.$$

Example 3

Evaluate $\lim\limits_{x \to 0} (1 + x)^{1/x}$.

This has the form 1^{∞}. Thus if

$$y = \lim_{x \to 0} (1 + x)^{1/x},$$

then

$$\ln y = \ln \lim_{x \to 0} (1 + x)^{1/x} = \lim_{x \to 0} \ln (1 + x)^{1/x}$$

$$= \lim_{x \to 0} \frac{1}{x} \ln (1 + x) \quad \to \infty \cdot 0$$

$$= \lim_{x \to 0} \frac{\ln (1 + x)}{x} \quad \begin{matrix} \to 0 \\ \to 0 \end{matrix}$$

$$= \lim_{x \to 0} \frac{\dfrac{1}{1 + x}}{1} = 1.$$

Thus

$$\lim_{x \to 0} (1 + x)^{1/x} = e^1 = e.$$

Let us recall that e was defined to be this limit (see page 351). Thus, our computations do not prove that this limit exists.

It might appear that 0^{∞} is also an indeterminate form. Actually a limit of the form $0^{(+\infty)}$ is always 0, while one of the form $0^{(-\infty)}$ is always $+\infty$. However, they can be handled in the same way as the indeterminate forms above. Let us consider

$$\lim_{x \to a} [f(x)]^{g(x)},$$

where

$$\lim_{x \to a} f(x) = 0 \quad \text{and} \quad \lim_{x \to a} g(x) = +\infty.$$

If we let

$$y = \lim_{x \to a} [f(x)]^{g(x)},$$

then

$$\ln y = \ln \lim_{x \to a} [f(x)]^{g(x)} = \lim_{x \to a} \ln [f(x)]^{g(x)}$$

$$= \lim_{x \to a} g(x) \ln f(x) \quad \to (+\infty)(-\infty)$$

$$= -\infty$$

and

$$y = \lim_{x \to a} [f(x)]^{g(x)} = \lim_{x \to a} e^{\ln[f(x)]^{g(x)}} \quad \to e^{-\infty}$$

$$= 0.$$

The form $0^{(-\infty)}$ can be shown to give $+\infty$ in the same way.

Problems

A

Evaluate the limits in Problems 1–18.

1. $\lim\limits_{x\to 0^+} x^{x^2}$.

2. $\lim\limits_{x\to 0^+} x^{\sin x}$.

3. $\lim\limits_{x\to 0^+} x^{\tan x}$.

4. $\lim\limits_{x\to 0} (1 + ax)^{1/x}$.

5. $\lim\limits_{x\to +\infty} \left(1 + \dfrac{1}{x}\right)^x$.

6. $\lim\limits_{x\to +\infty} \left(1 + \dfrac{1}{x^2}\right)^x$.

7. $\lim\limits_{x\to 0^+} (\sin x)^{\tan x}$.

8. $\lim\limits_{x\to \pi/2^-} (\sin x)^{\tan x}$.

9. $\lim\limits_{x\to \pi/2^-} (\tan x)^{\cos x}$.

10. $\lim\limits_{x\to +\infty} (1 + e^{-x})^{1/x}$.

11. $\lim\limits_{x\to 0} (e^x + x)^{1/x}$.

12. $\lim\limits_{x\to 0^+} x^{\ln x}$.

13. $\lim\limits_{x\to 1} x^{1/(1-x)}$.

14. $\lim\limits_{x\to 0^+} (\cot x)^{\sin x}$.

15. $\lim\limits_{x\to +\infty} (1 + x)^{1/x}$.

16. $\lim\limits_{x\to 0^+} x^{1/\ln(\sin x)}$.

17. $\lim\limits_{x\to 0} (1 - x)^{1/x}$.

18. $\lim\limits_{x\to 0^+} x^{1/\ln x}$.

B

Evaluate the limits in Problems 19–24.

19. $\lim\limits_{x\to +\infty} (\ln x)^{1/x}$.

20. $\lim\limits_{x\to 1^+} (\ln x)^{x-1}$.

21. $\lim\limits_{x\to 0^+} (-\ln x)^x$.

22. $\lim\limits_{x\to 0^+} \left(\dfrac{1}{x}\right)^{\tan x}$.

23. $\lim\limits_{x\to 0} (\cos ax)^{b/x}$.

24. $\lim\limits_{x\to 0} (\cos ax)^{b/x^2}$.

C

25. Evaluate $\lim\limits_{x\to 0} (1 + ax)^{b/x}$.

26. Evaluate $\lim\limits_{x\to 0} \left(1 + \dfrac{a}{x}\right)^{bx}$.

27. Show that a limit of the form $0^{(-\infty)}$ always gives $+\infty$.

28. Determine a such that $\lim_{x\to 0} (\cos ax)^{1/x^2} = 1/e^2$.

29. Sketch $y = x^x$.

Review Problems

A

In Problems 1–12, evaluate the limits.

1. $\lim\limits_{x\to 0} \dfrac{\sin x}{1 + \cos x}$.

2. $\lim\limits_{x\to 0} \dfrac{\sqrt{9 + x} - \sqrt{9 - x}}{\sqrt{4 + x} - \sqrt{4 - x}}$.

3. $\lim\limits_{x\to \pi/2} \dfrac{\sec x}{1 + \tan x}$.

4. $\lim\limits_{x\to +\infty} \dfrac{3^x}{9^x}$.

5. $\lim\limits_{x\to 0} (1 - e^x)\cot 3x$.

6. $\lim\limits_{x\to 0} \dfrac{\cos x - 1 + x^2/2}{x^3}$.

7. $\lim\limits_{x\to +\infty} \dfrac{x^3}{e^x - x}$.

8. $\lim\limits_{x\to \pi^-} (\csc x + \cot x)$.

9. $\lim\limits_{x\to 0} \dfrac{e^x - e^{-x}}{\cot 2x}$.

10. $\lim\limits_{x\to 0} (1 + \sin 3x)^{1/x}$.

11. $\displaystyle\lim_{x \to +\infty} (e^x + e^{-x})^{2/x}.$

12. $\displaystyle\lim_{x \to 1} \frac{x^4 + 2x^3 - 3x^2 - 4x + 4}{x^4 - 6x^3 + 13x^2 - 12x + 4}.$

B

In Problems 13–19, evaluate the limits.

13. $\displaystyle\lim_{x \to 0} (e^x - e^{-x})^x.$

14. $\displaystyle\lim_{x \to +\infty} (e^x - x^2)$

(*Hint:* Consider $\displaystyle\lim_{x \to +\infty} e^x/x^2$.)

15. $\displaystyle\lim_{x \to 0^+} (\sin x + x^2 - x) \ln x.$

16. $\displaystyle\lim_{x \to \pi^-} (\csc x - \cot x).$

17. $\displaystyle\lim_{x \to 0} \sin ax \ln ax.$

18. $\displaystyle\lim_{x \to +\infty} \frac{\cos x}{x^2}.$

19. $\displaystyle\lim_{x \to +\infty} (1 + e^{-x})^{x^2}.$

C

20.   Evaluate $\displaystyle\int_1^{+\infty} \frac{x \, dx}{(x + 1)(x + 2)}.$

21.   Evaluate $\displaystyle\lim_{x \to 1^+} \left(\frac{1}{x - 1} \right)^{\sin \pi x}$

22.   Evaluate $\displaystyle\int_1^{+\infty} \frac{x^2 \, dx}{e^x}.$

23.   Determine a, b, c, d, and e such that

$$\lim_{x \to 0} \frac{\cos ax + bx^3 + cx^2 + dx + e}{x^4} = \frac{2}{3}.$$

19

Infinite Series

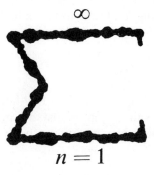

$$\sum_{n=1}^{\infty}$$

19.1

Sequences and Series

An infinite series is simply an indicated sum of infinitely many terms. But, with this seemingly simple statement, come many questions. What is meant by the sum of infinitely many numbers? Can we always add infinitely many numbers to get a sum? Can we ever add infinitely many numbers?

Actually you have already seen sums of infinitely many numbers. For example, when we write

$$\frac{1}{3} = 0.333\ldots,$$

we have an infinite sum. In our decimal notation, the symbol $0.333\ldots$ means

$$\frac{3}{10} + \frac{3}{100} + \frac{3}{1000} + \cdots.$$

Thus we maintain that there is a sum of this infinite set of numbers and that sum is $1/3$.

The ancient Greeks thought that no infinite set of numbers could possibly have a finite sum. Because of this feeling, they were caught in some logical paradoxes. Perhaps the most famous of these was given by Zeno of Elea (c. 500 B.C.). He pointed out that it is logically impossible to walk from one place to another. He reasoned that, before a person could go the entire distance, d, he first has to walk half of d. Then, of the distance $d/2$ that remained, he had to go half of that, leaving a distance $d/4$ yet to be covered. But he would have to go half of that distance. Continuing in this way, he could never walk the entire distance d!

You will notice that, except for the very last sentence, there is no mention of time in this argument. It was simply assumed that, since each segment walked requires some time and there are infinitely many segments, infinite time is required to cover all of them. But let us suppose that a man is walking at a constant rate, and that he walks half of the desired distance in a half-hour. Then half of the remaining distance requires half of the first time, or a quarter-hour. Similarly the next segment requires an eighth of an hour, and so on. Thus the total time in hours to walk the given distance is the infinite sum

$$\frac{1}{2} + \frac{1}{4} + \frac{1}{8} + \cdots.$$

Now surely, if he walked half the distance in a half-hour and he continued walking at a constant speed, he must walk the entire distance in one hour. This would imply that the infinite sum given above has the finite value 1:

$$\frac{1}{2} + \frac{1}{4} + \frac{1}{8} + \cdots = 1.$$

On the other hand, if the man were to walk progressively slower, so that each segment walked required the same time as any other segment, say a half-hour, then the total time is

$$\frac{1}{2} + \frac{1}{2} + \frac{1}{2} + \cdots.$$

This certainly cannot have a finite sum.

When mathematicians realized that we can sometimes find the sum of infinitely many numbers, there was a tendency to expect infinite sums to have the same properties as finite sums. This again led to difficulties. For example, let us consider

$$1 - 1 + 1 - 1 + 1 - 1 + \cdots.$$

Suppose we group the terms as $(1 - 1) + (1 - 1) + (1 - 1) + \cdots$. We then have

$$(1 - 1) + (1 - 1) + (1 - 1) + \cdots = 0 + 0 + 0 + \cdots = 0.$$

But if we group them in another way, we obtain

$$1 + (-1 + 1) + (-1 + 1) + (-1 + 1) + \cdots = 1 + 0 + 0 + 0 + \cdots = 1.$$

Clearly something is wrong. The practice of grouping terms, which is always valid for a finite sum, cannot always be used for sums of infinitely many terms. Problems such as this demand that we define more carefully what we are talking about.

Let us start with an idea that logically precedes that of a series, namely, a sequence.

Definition

*A **sequence** is a function whose domain is the set of all integers equal to or greater than a particular integer N. The functional values are called the **terms** of the sequence.*

In most cases the domain is the set of all positive integers; however, it is often convenient to have some other domain. One that is frequently used is the set of all nonnegative integers; in this case, the first term corresponds to $n = 0$, the second to $n = 1$, and so on. Occasionally it is convenient to have as domain all integers equal to or greater than 2; in this case, the first term corresponds to $n = 2$, the second to $n = 3$, and so on.

Suppose we consider the function

$$f(n) = 2n - 1, \quad n = 1, 2, 3, \ldots .$$

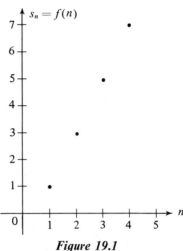

Since the domain of this function is defined to be the set of all positive integers (all integers ≥ 1), it is a sequence. The terms of this sequence are the numbers obtained when we replace n by $1, 2, 3$, and so forth, in turn. Thus, the terms are

$$1, 3, 5, 7, 9, \ldots .$$

The graph of this function is given in Figure 19.1.

Instead of using the notation $f(n)$ for the value of the nth term, we shall normally use the notation s_n. In this notation, the sequence is $\{s_n\}$, where

$$s_n = 2n - 1, \quad n = 1, 2, 3, \ldots .$$

Figure 19.1

Occasionally it is convenient to represent the sequence simply by listing the first few terms, as we did above. The two forms

$$s_n = 2n - 1, \quad n = 1, 2, 3, \ldots$$

and

$$1, 3, 5, 7, 9, \ldots$$

are called the *generator form* and the *expanded form*, respectively. It is often convenient to go from one form to the other. To go from the generator form to the expanded form is simply a matter of substitution.

Example 1

Find the first four terms of the sequence $\{s_n\}$, where

$$s_n = (n - 1)^2, \quad n = 2, 3, 4, \ldots .$$

The first four terms are s_2, s_3, s_4, and s_5, which are found by substituting $n = 2, 3, 4, 5$. The result is 1, 4, 9, and 16.

In going from the expanded form to the generator form we have the problem of finding the general term. This is exactly the same problem we had in Section 7.1.

Example 2

Find a generator form for the sequence $\dfrac{1}{2}, \dfrac{3}{4}, \dfrac{7}{8}, \dfrac{15}{16}, \ldots .$

We have some degree of ambiguity right from the start. Since we have a sequence, we have a function whose domain is the set of all integers equal to or greater than some integer N. But we do not know the value of N. Thus we arbitrarily choose $N = 1$. With that choice we have:

n	1	2	3	4	$\cdots$
s_n	$\dfrac{1}{2}$	$\dfrac{3}{4}$	$\dfrac{7}{8}$	$\dfrac{15}{16}$	$\cdots$

Note that the denominator of s_n is always a power of 2: in fact, it is 2^n. Furthermore, the numerator of s_n is one less than the denominator. The result is

$$s_n = \frac{2^n - 1}{2^n} = 1 - \frac{1}{2^n}.$$

Hence the generator form of the given sequence is $\{s_n\}$, where

$$s_n = 1 - \frac{1}{2^n}, \quad n = 1, 2, 3, \ldots .$$

Example 3

Express $1, -1, 1, -1, 1, -1, \ldots$ in generator form.

Again taking $N = 1$, we see that

$$s_n = \begin{cases} 1 & \text{if } n \text{ is odd,} \\ -1 & \text{if } n \text{ is even.} \end{cases}$$

However this can be expressed in a simpler form if we recall that all integer powers of -1 are either 1 or -1. Since -1 to an even power is 1 and -1 to an odd power is still -1, we merely need the exponent on -1 to be even when n is odd and odd when n is even. This can be accomplished by choosing the exponent to be $n + 1$ (actually we may use n plus or minus any odd number). Thus we have

$$s_n = (-1)^{n+1}, \quad n = 1, 2, 3, \ldots .$$

It might be noted here that the expression for s_n can be simplified somewhat by choosing $N = 0$. In that case we have

$$s_n = (-1)^n, \quad n = 0, 1, 2, \ldots .$$

While this representation is sometimes useful, we shall normally consider the domain of a sequence to be the set of positive integers. Henceforth, unless something is said to the contrary, it is assumed that the domain of a sequence $\{s_n\}$ is the set of all positive integers. The graphs of the sequences of Examples 1, 2, and 3 are given in Figure 19.2. Notice that Figure 19.2(b) differs from the other two in that the points of that sequence tend to level off to the number 1 as n gets larger. On the other hand, the points of Figure 19.2(a) simply go higher and higher as we move to the right, while those of Figure 19.2(c) oscillate between 1 and -1. This leads us to a consideration of limits of sequences. Because the domain of a sequence is a set of integers, no number is a limit point of the domain. Thus we cannot have limits of the form

$$\lim_{n \to a} s_n,$$

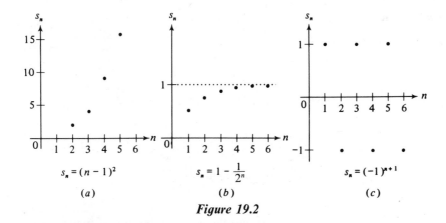

$s_n = (n-1)^2$

(a)

$s_n = 1 - \dfrac{1}{2^n}$

(b)

$s_n = (-1)^{n+1}$

(c)

Figure 19.2

where a is a real number. Furthermore, there is a smallest integer in the domain; so we cannot have limits of the form

$$\lim_{n \to -\infty} s_n.$$

The only limit that can possibly exist is of the form

$$\lim_{n \to +\infty} s_n.$$

Thus, when we talk of the limit of a sequence, we mean the limit as $n \to +\infty$.

Since a sequence is a function, we may use the definition of a limit as $n \to +\infty$ that was stated on page 560. The first part of the definition must be satisfied for any sequence; we shall not consider it here. Thus for a sequence $\{s_n\}$,

$$\lim_{n \to +\infty} s_n = b \quad (b \text{ finite})$$

means that if ϵ is a positive number, then there is a number N such that $|s_n - b| < \epsilon$ for every integer $n > N$. In other words, no matter how small ϵ is, there is an integer N large enough so that every term of the sequence beyond the Nth term is within a distance ϵ of b.

Definition

*To say that the sequence $\{s_n\}$ **converges** means that*

$$\lim_{n \to +\infty} s_n$$

*exists. The sequence **diverges** if it does not converge, that is, if*

$$\lim_{n \to +\infty} s_n$$

does not exist. (Remember that

$$\lim_{n \to +\infty} s_n = +\infty \quad and \quad \lim_{n \to +\infty} s_n = -\infty$$

are special cases of nonexistence.)

Let us consider several examples illustrating this definition.

Example 4

Does the sequence $\{s_n\}$ given by

$$S_n = 1 - \frac{1}{2^n}, \quad n = 1, 2, 3, \ldots,$$

converge or diverge? If it converges, to what number does it converge?

$$\lim_{n \to +\infty} \left(1 - \frac{1}{2^n}\right) = 1 - \lim_{n \to +\infty} \frac{1}{2^n} = 1 - 0 = 1.$$

Thus the sequence converges to the number 1.

Note here that it is not necessary to use the actual definition of limit; our previous methods of evaluating limits as $x \to +\infty$ can be used just as well with sequences as with any other function. However, Figure 19.2(b) does show that no matter how small a number ϵ we choose, there is an integer N large enough so that every term beyond the Nth term is within ϵ of the number 1.

Example 5

Does the sequence $\{s_n\}$ given by

$$s_n = \frac{2^n}{n^2}, \quad n = 1, 2, 3, \ldots,$$

converge or diverge? If it converges, to what number does it converge?

By l'Hôpital's rule, $\quad \lim_{n \to +\infty} \frac{2^n}{n^2} = \lim_{n \to +\infty} \frac{2^n \ln 2}{2n} = \lim_{n \to +\infty} \frac{2^n \ln^2 2}{2} = +\infty.$

Since this is a special case of nonexistence, the sequence diverges.

Example 6

Does the sequence $\{s_n\}$ given by

$$s_n = \frac{(-1)^{n+1} n^2}{n^2 + 1}, \quad n = 1, 2, 3, \ldots,$$

converge or diverge? If it converges, to what number does it converge?

$$\lim_{n \to +\infty} \frac{(-1)^{n+1} n^2}{n^2 + 1} = \lim_{n \to +\infty} \frac{(-1)^{n+1}}{1 + 1/n^2}.$$

Now

$$\frac{1}{1 + 1/n^2} \to 1 \quad \text{as} \quad n \to +\infty,$$

but $(-1)^{n+1}$ oscillates between 1 and -1. Thus the product oscillates, with the odd-numbered terms approaching 1 and the even-numbered terms approaching -1 as $n \to +\infty$. Thus $\lim_{n \to +\infty} s_n$ does not exist—the sequence diverges.

Note that the divergence in Example 5 is basically different from that in Example 6. In Example 5 the sequence diverges because there is no limit on how big the terms get as $n \to +\infty$; that is, the sequence is unbounded. On the other hand, the sequence of Example 6 is bounded—no term is greater than 1 or less than -1. Nevertheless, it diverges because the terms do not level off to any one number as $n \to +\infty$; they continue to oscillate. While Example 6 shows that boundedness is not enough to assure us that a sequence converges, it is enough if the sequence is increasing.

Definition

*The sequence $\{s_n\}$ is **increasing** if $s_{n+1} > s_n$ for all n in the domain of the sequence; it is **decreasing** if $s_{n+1} < s_n$ for all n in the domain.*

The sequences of Examples 1 and 2 are increasing, while that of Example 3 is neither increasing nor decreasing.

Theorem 19.1

Any bounded, increasing sequence $\{s_n\}$ converges.

Proof

Since $\{s_n\}$ is bounded, there is a number k such that $s_n \leq k$ for every integer n. Let $S = \{x \mid x \geq s_n$ for every integer $n\}$. This set is not empty, since k is in it. Let u be the smallest number in S (the existence of such a number u is a characteristic of the real numbers). Now let us show that

$$\lim_{n \to +\infty} s_n = u.$$

Let ϵ be a positive number. Assume that

$$s_n \leq u - \epsilon$$

for every positive integer n. Then $u - \epsilon$ is in S. But $u - \epsilon < u$, contradicting the statement that u is the smallest number in S. Thus the assumption is wrong, and there is a number N such that $s_N > u - \epsilon$. If $n > N$, then $s_n > s_N > u - \epsilon$ and $s_n < u$. Thus there is a number N such that, if $n > N$, $|s_n - u| < \epsilon$; and $\{s_n\}$ converges to u.

Note: This theorem remains true if the terms of the sequence are *nondecreasing*; that is, $s_{n+1} \geq s_n$ and some or all s_n may be equal.

Of course, we can prove similarly that any bounded, decreasing sequence converges. This theorem will be useful when we consider series. However one must be careful not to read into this theorem something that is not there. The theorem tells us only about sequences which are both bounded and increasing (or decreasing). If a sequence is bounded but neither increasing nor decreasing, we may not conclude from that information alone that the sequence converges or diverges. We have seen an example (Example 6) of a divergent sequence that is bounded and neither increasing nor decreasing. On the other hand, $\{s_n\}$, where

$$s_n = \frac{(-1)^{n+1}}{n}, \quad n = 1, 2, 3, \ldots,$$

is a convergent sequence which is bounded and neither increasing nor decreasing.

Example 7

Does the sequence $\{s_n\}$, where

$$s_n = 1 + \frac{1}{2!} + \frac{1}{3!} + \cdots + \frac{1}{n!}, \quad n = 1, 2, 3, \ldots,$$

converge or diverge?

We see here that s_n is much too complicated to allow us to find $\lim_{n \to +\infty} s_n$ without a great deal of difficulty. Furthermore, there appears to be little hope of simplifying s_n. But

$$s_{n+1} = 1 + \frac{1}{2!} + \frac{1}{3!} + \cdots + \frac{1}{n!} + \frac{1}{(n+1)!}$$

$$= s_n + \frac{1}{(n+1)!}.$$

Thus

$$s_{n+1} > s_n,$$

and our sequence is increasing.

Now let us see if it is bounded. To do so we compare s_n with a somewhat simpler expression. Since

$$n! = 1 \cdot 2 \cdot 3 \cdots n$$

and

$$2^{n-1} = 1 \cdot 2 \cdot 2 \cdots 2 \quad (n - 1 \text{ factors of } 2),$$

we see that

$$2^{n-1} \leq n!, \quad n = 1, 2, 3, \ldots,$$

or

$$\frac{1}{2^{n-1}} \geq \frac{1}{n!}, \quad n = 1, 2, 3, \ldots.$$

Thus

$$s_n = 1 + \frac{1}{2!} + \frac{1}{3!} + \cdots + \frac{1}{n!}$$

$$\leq 1 + \frac{1}{2} + \frac{1}{2^2} + \cdots + \frac{1}{2^{n-1}}.$$

Suppose we let

$$t_n = 1 + \frac{1}{2} + \frac{1}{2^2} + \cdots + \frac{1}{2^{n-1}};$$

then

$$\frac{1}{2} t_n = \frac{1}{2} + \frac{1}{2^2} + \frac{1}{2^3} + \cdots + \frac{1}{2^n}.$$

Subtracting, we have

$$\frac{1}{2} t_n = 1 - \frac{1}{2^n}$$

$$t_n = 2 - \frac{1}{2^{n-1}} < 2.$$

Hence

$$s_n \leq t_n < 2;$$

the given sequence is bounded. By Theorem 19.1, it converges.

Although Theorem 19.1 tells us that the foregoing sequence converges, it does not tell us the number to which it converges. Nevertheless the proof indicates that it converges to the smallest number u in the set $S = \{x|x \geq s_n$ for every integer $n\}$. Thus u is the smallest number that is at least as large as every term of the sequence; such a number is called the *least upper bound* of the sequence. Unfortunately, the determination of the least upper bound of this sequence is no simple matter. But since the least upper bound is smaller than any other upper bound, we can at least give limits for it. Since $s_n < 2$ for all n, 2 is an upper bound for the sequence. Therefore the sequence converges to some positive number that is no larger than 2.

Problems

A

In Problems 1–8, give first four terms of the sequence $\{s_n\}$ for the s_n given.

1. $s_n = n^2$, $n = 1, 2, 3, \ldots$

2. $s_n = \dfrac{1}{5^n}$, $n = 1, 2, 3, \ldots$

3. $s_n = 1 + (-1)^{n+1}$, $n = 1, 2, 3, \ldots$

4. $s_n = \dfrac{(-1)^{n+1}}{3^n}$, $n = 1, 2, 3, \ldots$

5. $s_n = \dfrac{(-1)^n}{\ln n}$, $n = 2, 3, 4, \ldots$

6. $s_n = \dfrac{1}{3^n}$, $n = 0, 1, 2, \ldots$

7. $s_n = 1 + \dfrac{1}{2} + \dfrac{1}{2^2} + \cdots + \dfrac{1}{2^n}$, $n = 0, 1, 2, \ldots$

8. $s_n = 1 + \dfrac{1}{2!} + \dfrac{1}{3!} + \cdots + \dfrac{1}{n!}$, $n = 1, 2, 3, \ldots$

In Problems 9–15, give the sequence in generator form.

9. $\dfrac{1}{2}, \dfrac{2}{3}, \dfrac{3}{4}, \dfrac{4}{5}, \ldots$

10. $1, \dfrac{1}{3}, \dfrac{1}{9}, \dfrac{1}{27}, \ldots$

11. $-\dfrac{1}{2}, \dfrac{3}{4}, -\dfrac{7}{8}, \dfrac{15}{16}, \ldots$

12. $\dfrac{3}{2}, -\dfrac{9}{4}, \dfrac{27}{8}, -\dfrac{81}{16}, \ldots$

13. $1 \cdot 3, 3 \cdot 5, 5 \cdot 7, 7 \cdot 9, \ldots$

14. $\dfrac{2}{1}, -\dfrac{3}{2}, \dfrac{4}{3}, -\dfrac{5}{4}, \ldots$

15. $1 \cdot 2, -2 \cdot 4, 3 \cdot 8, -4 \cdot 16, \ldots$

In Problems 16–22, indicate whether the sequence converges or diverges. If it converges, to what number does it converge?

16. $1, -\dfrac{1}{2}, \dfrac{1}{3}, -\dfrac{1}{4}, \ldots$

17. $1, \dfrac{1}{2}, \dfrac{1}{3}, \dfrac{1}{4}, \ldots$

18. $\dfrac{1}{2}, \dfrac{1}{4}, \dfrac{1}{8}, \dfrac{1}{16}, \ldots$

19. $s_n = \dfrac{n}{n^2 + 1}$.

20. $s_n = \dfrac{n}{n + 3}$.

21. $s_n = \dfrac{n^2 - 1}{n^2}$.

22. $s_n = \dfrac{(-1)^n n}{n + 3}$.

B

In Problems 23–25, indicate whether the sequence converges or diverges. If it converges, to what number does it converge?

23. $\dfrac{1}{2}, -\dfrac{3}{4}, \dfrac{7}{8}, -\dfrac{15}{16}, \ldots$

24. $s_n = \dfrac{4n(4n + 3)}{(4n + 1)(4n + 2)}$.

25. $s_n = \dfrac{n}{3^n}$.

C

26. Give the sequence in generator form: $\dfrac{3}{2}, \dfrac{2}{3}, \dfrac{5}{4}, \dfrac{4}{5}, \dfrac{7}{6}, \dfrac{6}{7}, \ldots$

In Problems 27 and 28, indicate whether the sequence converges or diverges. If it converges, to what number does it converge?

27. $s_n = \dfrac{\ln n}{n}$.

28. $s_n = \dfrac{n^2}{n!}$.

29. Prove that any bounded, decreasing sequence converges.

30. Use the binomial theorem to show that, for any positive integer n,

$$\left(1+\frac{1}{n}\right)^n = 1 + 1 + \frac{1}{2!}\left(1-\frac{1}{n}\right) + \frac{1}{3!}\left(1-\frac{1}{n}\right)\left(1-\frac{2}{n}\right) +$$

$$\cdots + \frac{1}{n!}\left(1-\frac{1}{n}\right)\left(1-\frac{2}{n}\right)\cdots\left(1-\frac{n-1}{n}\right).$$

Use this expansion and a similar one for $\left(1+\dfrac{1}{n+1}\right)^{n+1}$ to show that, for any positive integer n,

$$\left(1+\frac{1}{n}\right)^n < \left(1+\frac{1}{n+1}\right)^{n+1}.$$

Use the above expansion and the result of Example 7 to show that

$$\left(1+\frac{1}{n}\right)^n < 3$$

for any positive integer n. Show that the foregoing results imply that $\{s_n\}$, where

$$s_n = \left(1+\frac{1}{n}\right)^n, \quad n=1, 2, 3, \ldots,$$

is a convergent sequence.

31. On page 351 we defined the number e to be

$$e = \lim_{x \to 0} (1+x)^{1/x},$$

but we gave no proof that this limit exists. If we make the substitution $x = 1/n$, we have

$$e = \lim_{n \to +\infty} \left(1+\frac{1}{n}\right)^n.$$

Problem 30 shows that this limit (with n restricted to positive integers) exists. Suppose now that

$$\frac{1}{n+1} < x < \frac{1}{n}.$$

Show that

$$\left(1+\frac{1}{n+1}\right)^n < (1+x)^{1/x} < \left(1+\frac{1}{n}\right)^{n+1}.$$

Use this result to show that

$$\lim_{x \to 0^+} (1+x)^{1/x}$$

exists and equals

$$\lim_{n \to +\infty} \left(1+\frac{1}{n}\right)^n.$$

19.2

Series

Let us return to our primary interest—the series. First a word on notation. We have already used the sigma notation to represent sums. Thus

$$\sum_{n=1}^{5} \frac{1}{n} = \frac{1}{1} + \frac{1}{2} + \frac{1}{3} + \frac{1}{4} + \frac{1}{5}.$$

Let us extend this to an infinite sum.

$$\sum_{n=1}^{\infty} \frac{1}{n} = \frac{1}{1} + \frac{1}{2} + \frac{1}{3} + \cdots \quad \text{(without end)},$$

or, more generally,

$$\sum_{n=1}^{\infty} a_n = a_1 + a_2 + a_3 + \cdots \quad \text{(without end)}.$$

Of course, the values of n need not start with 1—they can start with 0 or 2 or any other integer.

Definition

A series is an expression of the form

$$\sum_{n=N}^{\infty} a_n,$$

where N is a fixed integer.

Thus the example

$$\sum_{n=1}^{\infty} \frac{1}{n} = 1 + \frac{1}{2} + \frac{1}{3} + \cdots$$

is a series. Again we have two types of notation: the *generator form* and the *expanded form*, given by

$$\sum_{n=1}^{\infty} \frac{1}{n} \quad \text{and} \quad 1 + \frac{1}{2} + \frac{1}{3} + \cdots,$$

respectively. The problem of changing from one form to the other is solved here in the same way as in Sections 7.1 and 19.1.

Note the difference between a sequence and a series. A sequence is a function; there is no question of adding the terms of a sequence—they are simply listed as functional values. When written in the expanded form, a sequence has commas between the terms. On the other hand, a series is a sum; we are very definitely interested in adding the terms of a series. When written in the expanded form, the series has pluses between the terms (or minuses if the terms are negative). Thus

$$1, \frac{1}{2}, \frac{1}{3}, \frac{1}{4}, \cdots$$

is a sequence, but

$$1 + \frac{1}{2} + \frac{1}{3} + \frac{1}{4} + \cdots$$

is a series, in accord with our idea of a series as an infinite sum. Let us now consider what it means to add infinitely many numbers. For simplicity we shall consider a series of the form

$$\sum_{n=1}^{\infty} a_n,$$

where n begins with 1. This is often abbreviated to $\sum a_n$. Other cases can be considered in exactly the same way.

Definition

If we have the series

$$\sum_{n=1}^{\infty} a_n = a_1 + a_2 + a_3 + \cdots,$$

and if we write

$$s_1 = a_1$$
$$s_2 = a_1 + a_2$$
$$s_3 = a_1 + a_2 + a_3$$
$$\vdots$$
$$s_n = a_1 + a_2 + a_3 + \cdots + a_n$$
$$\vdots$$

*then the sequence $\{s_n\}$ is called the **sequence of partial sums**. If $\{s_n\}$ converges to some number S, then $\sum a_n$ is said to **converge** to S and S is called the **sum** of the series. If $\{s_n\}$ diverges, then $\sum a_n$ is said to **diverge**.*

Example 1

Does the series

$$\sum_{n=1}^{\infty} \frac{1}{2^n} = \frac{1}{2} + \frac{1}{4} + \frac{1}{8} + \cdots$$

converge or diverge? If it converges, to what number does it converge?

We consider the sequence $\{s_n\}$ of partial sums

$$s_1 = \frac{1}{2}, \quad s_2 = \frac{3}{4}, \quad s_3 = \frac{7}{8}, \ldots, \quad s_n = 1 - \frac{1}{2^n}, \ldots.$$

Since

$$\lim_{n \to +\infty} s_n = \lim_{n \to +\infty} \left(1 - \frac{1}{2^n}\right) = 1,$$

the original series converges to 1.

Example 2

Does the series

$$1 - 1 + 1 - 1 + 1 - 1 + \cdots$$

converge or diverge? If it converges, to what number does it converge?

The sequence $\{s_n\}$ of partial sums is

$$s_1 = 1, \; s_2 = 0, \; s_3 = 1, \; s_4 = 0, \ldots, \; s_n = \begin{cases} 1 & \text{if } n \text{ is odd,} \\ 0 & \text{if } n \text{ is even.} \end{cases}$$

So $\lim_{n \to +\infty} s_n$ does not exist. Thus the original series diverges.

These two examples are special cases of the general geometric series. A geometric series is one of the form

$$a + ar + ar^2 + \cdots + ar^{n-1} + \cdots \quad (a \neq 0),$$

in which each term is found by multiplying the previous one by a constant r. The nth partial sum is

$$s_n = a + ar + ar^2 + \cdots + ar^{n-1},$$

and

$$rs_n = ar + ar^2 + \cdots + ar^{n-1} + ar^n.$$

Subtracting the second from the first, we have

$$s_n - rs_n = a - ar^n$$

$$s_n(1 - r) = a(1 - r^n)$$

$$s_n = \frac{a(1 - r^n)}{1 - r} \quad \text{(provided } r \neq 1).$$

If $|r| < 1$, then

$$\lim_{n \to +\infty} r^n = 0 \quad \text{and} \quad \lim_{n \to +\infty} s_n = \frac{a}{1 - r}.$$

If $|r| > 1$, then, neither $\lim_{n \to +\infty} r^n$ nor $\lim_{n \to +\infty} s_n$ exists. If $|r| = 1$, we cannot use the formula for s_n; but it is clear from a consideration of the original series that if $a \neq 0$, the series diverges.

Theorem 19.2

The geometric series given by

$$a + ar + ar^2 + \cdots + ar^{n-1} + \cdots, \quad a \neq 0,$$

converges to $a/(1 - r)$ if $|r| < 1$ and diverges if $|r| \geq 1$.

Using this result, we see that the series of Example 1 is a geometric series with $a = 1/2$ and $r = 1/2$. Thus, by Theorem 19.2, it converges to

$$\frac{a}{1 - r} = \frac{1/2}{1 - 1/2} = 1.$$

We mentioned on page 593 that an infinite decimal is a series. An infinite repeating decimal is a geometric series. Hence we can use Theorem 19.2 to show that it is a rational number and to express it in fractional form.

Example 3

Give $0.232323\ldots$ as a quotient of two integers.

$$0.232323\ldots = 0.23 + 0.0023 + 0.000023 + \cdots.$$

This is a geometric series with $a = 0.23$ and $r = 0.01$. It converges to

$$\frac{a}{1-r} = \frac{0.23}{1-0.01} = \frac{0.23}{0.99} = \frac{23}{99}.$$

This can easily be checked by division.

Another type of series in which the determination of the nth partial sum is relatively easy is the telescoping series. A telescoping series is one in which the nth term can be expressed in the form

$$a_n = b_n - b_{n+1}.$$

For example, the series

$$\frac{1}{1\cdot 2} + \frac{1}{2\cdot 3} + \frac{1}{3\cdot 4} + \cdots + \frac{1}{n(n+1)} + \cdots$$

is a telescoping series, since

$$\frac{1}{n(n+1)} = \frac{1}{n} - \frac{1}{n+1}.$$

Note that the two terms on the right have the same form, the only difference being that the second has $n + 1$ where the first has n.

Because each term can be represented in this way, the partial sum s_n takes a convenient form.

$$\begin{aligned}
s_n &= a_1 + a_2 + a_3 + \cdots + a_n \\
&= (b_1 - b_2) + (b_2 - b_3) + (b_3 - b_4) + \cdots + (b_n - b_{n+1}) \\
&= b_1 - b_{n+1}.
\end{aligned}$$

Now the convergence of the series depends upon the existence of the limit

$$\lim_{n\to +\infty} s_n = \lim_{n\to +\infty} (b_1 - b_{n+1}).$$

Clearly this limit exists if and only if

$$\lim_{n\to +\infty} b_{n+1} = \lim_{n\to +\infty} b_n = L$$

exists; and in that case

$$\lim_{n\to +\infty} s_n = b_1 - L.$$

Thus we have proved the following theorem.

Theorem 19.3

If $\sum a_n$ is a telescoping series with $a_n = b_n - b_{n+1}$, then $\sum a_n$ converges if and only if $\{b_n\}$ converges. Furthermore, if $\{b_n\}$ converges to L, then $\sum a_n$ converges to $b_1 - L$.

Example 4

Does the series

$$\sum_{n=1}^{\infty} \frac{1}{n(n+1)}$$

converge or diverge? If it converges, to what number does it converge?

As we have already indicated, this is a telescoping series, since

$$\frac{1}{n(n+1)} = \frac{1}{n} - \frac{1}{n+1}$$

(the expression on the right having been determined by partial fractions). Thus $b_n = 1/n$ and

$$L = \lim_{n \to +\infty} b_n = \lim_{n \to +\infty} \frac{1}{n} = 0.$$

By Theorem 19.3, the given series converges to $b_1 - L = 1 - 0 = 1$.

Returning to a consideration of general series, let us look at a relationship between the partial sums and the individual terms.

Theorem 19.4

If the series $\sum a_n$ converges, then

$$\lim_{n \to +\infty} a_n = 0.$$

Before giving a proof of this theorem, let us emphasize that the limit we are considering is the limit of the nth term of the series and *not* the nth term of the sequence of partial sums. Thus, the theorem says that if

$$s_1, s_2, s_3, \ldots \to S \quad (S \text{ a real number}),$$

then

$$a_1, a_2, a_3, \ldots \to 0.$$

Proof

$a_n = s_n - s_{n-1}.$

Since $\sum a_n$ converges

$$\lim_{n \to +\infty} s_n = S \quad \text{and} \quad \lim_{n \to +\infty} s_{n-1} = S.$$

Thus

$$\lim_{n \to +\infty} a_n = \lim_{n \to +\infty} (s_n - s_{n-1})$$
$$= \lim_{n \to +\infty} s_n - \lim_{n \to +\infty} s_{n-1}$$
$$= S - S = 0.$$

The following theorem is a direct consequence of Theorem 19.4.

Theorem 19.5

> *If* $\lim\limits_{n \to +\infty} a_n \neq 0,$ *then the series* $\sum a_n$ *diverges.*

The proof of this theorem, sometimes called the *nth term test*, is left to the student. The application of Theorem 19.5 to Example 2 shows that the series

$$1 - 1 + 1 - 1 + 1 - 1 + \cdots$$

diverges, since the sequence

$$1, \, -1, \, 1, \, -1, \, 1, \, -1, \ldots$$

clearly does not converge to 0.

   A word of warning. The converses of the last two theorems are not true. If the limit of the *n*th term is zero, that is no guarantee that $\sum a_n$ converges. To put it another way, the condition that the limit of the *n*th term be zero is necessary for convergence (without it the series must diverge), but it is not sufficient (more information is needed to assure us of convergence). The following example shows that $\lim_{n \to +\infty} a_n = 0$ is not a sufficient condition for convergence.

Example 5

Show that the series

$$\sum_{n=1}^{\infty} \frac{1}{n} = 1 + \frac{1}{2} + \frac{1}{3} + \frac{1}{4} + \cdots,$$

called the *harmonic series*, diverges even though the limit of the *n*th term is 0.

   It is a simple matter to verify that the limit of the *n*th term is 0.

$$\lim_{n \to +\infty} a_n = \lim_{n \to +\infty} \frac{1}{n} = 0.$$

In order to show that the series diverges, we consider the following partial sums.

$$s_1 = 1 > 1 \cdot \frac{1}{2},$$

$$s_2 = 1 + \frac{1}{2} > \frac{1}{2} + \frac{1}{2} = 2 \cdot \frac{1}{2},$$

$$s_4 = 1 + \frac{1}{2} + \left(\frac{1}{3} + \frac{1}{4}\right) > \frac{1}{2} + \frac{1}{2} + \left(\frac{1}{4} + \frac{1}{4}\right) = 3 \cdot \frac{1}{2},$$

$$s_8 = 1 + \frac{1}{2} + \left(\frac{1}{3} + \frac{1}{4}\right) + \left(\frac{1}{5} + \frac{1}{6} + \frac{1}{7} + \frac{1}{8}\right)$$

$$> \frac{1}{2} + \frac{1}{2} + \left(\frac{1}{4} + \frac{1}{4}\right) + \left(\frac{1}{8} + \frac{1}{8} + \frac{1}{8} + \frac{1}{8}\right) = 4 \cdot \frac{1}{2},$$

$$\vdots$$
$$s_{2^{n-1}} > n \cdot \frac{1}{2},$$
$$\vdots$$

Since s_n is increasing,

$$\lim_{n \to +\infty} s_n = \lim_{n \to +\infty} s_{2^{n-1}} = +\infty,$$

and the series diverges.

The preceding argument indicates that no matter how large a number we choose, there is a partial sum of this series that is greater than that number. For instance, if we choose the number $100 = 200 \cdot (1/2)$, then the 2^{199}-th partial sum is greater than 100. If we choose $1000 = 2000 \cdot (1/2)$, then the 2^{1999}-th partial sum exceeds it. This seems rather incredible when considering that the terms of the series are small to begin with and getting smaller as we go out. Things aren't always what they seem when we are dealing with infinite series.

Let us conclude this section with a simple theorem that is more self-evident.

Theorem 19.6

If $\sum a_n$ and $\sum b_n$ are convergent series and c is a number, then

$$\sum ca_n, \quad \sum (a_n + b_n), \quad and \quad \sum (a_n - b_n)$$

are convergent and

$$\sum ca_n = c \sum a_n, \quad \sum (a_n + b_n) = \sum a_n + \sum b_n, \quad and \quad \sum (a_n - b_n) = \sum a_n - \sum b_n.$$

This follows directly from a consideration of partial sums. The details are left to the student.

Problems

A

In Problems 1–6, give the first four terms of the series.

1. $\displaystyle\sum_{n=1}^{\infty} \frac{1}{n^2}$.

2. $\displaystyle\sum_{n=1}^{\infty} \frac{1}{3^n}$.

3. $\displaystyle\sum_{n=1}^{\infty} \frac{n+2}{n(n+1)}$.

4. $\displaystyle\sum_{n=1}^{\infty} \frac{1}{n(n+1)}$.

5. $\displaystyle\sum_{n=1}^{\infty} \frac{(-1)^n}{n!}$.

6. $\displaystyle\sum_{n=1}^{\infty} \frac{(-1)^n}{n^2+3}$.

In Problems 7–12, give the series in generator form.

7. $\dfrac{1}{10} + \dfrac{1}{100} + \dfrac{1}{1000} + \cdots$.

8. $1 + 4 + 7 + 10 + \cdots$.

9. $1 + 2 + 4 + 8 + \cdots$.

10. $1 \cdot 2 + 3 \cdot 4 + 5 \cdot 6 + 7 \cdot 8 + \cdots$.

11. $1 \cdot 3 - 2 \cdot 4 + 3 \cdot 5 - \cdots$.

12. $1 \cdot 2 - 2 \cdot 2^2 + 3 \cdot 2^3 - \cdots$.

In Problems 13–20, express the given repeating decimal as a quotient of two integers.

13. $0.2222\ldots$.

14. $0.535353\ldots$.

15. $0.134134134\ldots$.

16. $0.141414\ldots$.

17. $0.9999\ldots$.

18. $0.328732873287\ldots$.

19. $1.353535\ldots$.

20. $2.23121212\ldots$.

In Problems 21–25, indicate whether the series converges or diverges. If it converges, to what number does it converge?

21. $\dfrac{1}{4} + \dfrac{1}{16} + \dfrac{1}{64} + \cdots + \dfrac{1}{4^n} + \cdots$.

22. $\dfrac{1}{3} + \dfrac{1}{9} + \dfrac{1}{27} + \cdots + \dfrac{1}{3^n} + \cdots$.

23. $1 - 2 + 3 - 4 + \cdots + (-1)^{n+1}n + \cdots$.

24. $1 - \dfrac{1}{3} + \dfrac{1}{9} - \dfrac{1}{27} + \cdots + \dfrac{(-1)^{n-1}}{3^{n-1}} + \cdots$.

25. $\dfrac{1}{2} + \dfrac{2}{3} + \dfrac{3}{4} + \cdots + \dfrac{n}{n+1} + \cdots$.

B

In Problems 26–31, indicate whether the series converges or diverges. If it converges, to what number does it converge?

26. $1 + \dfrac{3}{4} + \dfrac{9}{16} + \cdots + \dfrac{3^{n-1}}{4^{n-1}} + \cdots$.

27. $\dfrac{1}{2} + \dfrac{1}{4} + \dfrac{1}{6} + \dfrac{1}{8} + \cdots + \dfrac{1}{2n} + \cdots$.

28. $\dfrac{1}{1 \cdot 2} - \dfrac{1}{2 \cdot 3} - \dfrac{1}{3 \cdot 4} - \cdots - \dfrac{1}{n(n+1)} - \cdots$.

29. $\dfrac{1}{1 \cdot 2} + \dfrac{3}{2 \cdot 5} + \dfrac{5}{5 \cdot 10} + \dfrac{7}{10 \cdot 17} + \cdots + \dfrac{2n-1}{[(n-1)^2+1](n^2+1)} + \cdots$.

30. $\dfrac{5}{1 \cdot 4} + \dfrac{11}{4 \cdot 9} + \dfrac{19}{9 \cdot 16} + \cdots + \dfrac{n^2+3n+1}{n^2(n+1)^2} + \cdots$.

31. $\dfrac{1 \cdot 2}{2 \cdot 3} + \dfrac{2 \cdot 4}{3 \cdot 4} + \dfrac{3 \cdot 8}{4 \cdot 5} + \cdots + \dfrac{n \cdot 2^n}{(n+1)(n+2)} + \cdots$.

C

In Problems 32 and 33, indicate whether the series converges or diverges. If it converges, to what number does it converge?

32. $\dfrac{1}{2} + \dfrac{1}{3} + \dfrac{1}{2^2} + \dfrac{1}{3^2} + \dfrac{1}{2^3} + \dfrac{1}{3^3} + \cdots$.   33. $\dfrac{1}{2} - \dfrac{1}{3} + \dfrac{1}{2^2} - \dfrac{1}{3^2} + \dfrac{1}{2^3} - \dfrac{1}{3^3} + \cdots$.

34. Prove Theorem 19.5.   35. Prove Theorem 19.6.

36. Show that if $\sum_{n=1}^{\infty} a_n$ converges, then $\sum_{n=N}^{\infty} a_n$, where $N \geq 1$, converges.

37. Show that if $\sum_{n=N}^{\infty} a_n$, where $N \geq 1$, converges, then $\sum_{n=1}^{\infty} a_n$ converges.

38. Show that if for some $N \geq 1$, $a_n = b_n$ for each $n \geq N$, then $\sum_{n=1}^{\infty} b_n$ converges if and only if $\sum_{n=1}^{\infty} a_n$ converges. Thus, changing any finite number of terms does not affect the convergence or divergence of the series.

19.3

The Comparison and Limit Comparison Tests

In the last section we used the definition of convergence to determine whether a given series converges or diverges. By finding the limit of the nth partial sum we were able to determine not merely whether or not a given series converges; but, if it did converge, we could determine the number to which it converged. However, in many cases the determination of the nth partial sum is extremely difficult if not impossible. Nevertheless, if we are content to know whether a given series converges or diverges, without, in the former case, determining the number to which it converges, then there are several tests for convergence available to us. We consider such tests in this section and the next two.

At first we shall restrict the type of series we consider. If all of the terms of a series are positive, then the sequence of partial sums is an increasing sequence. Thus all we need to do to show convergence is to show that the sequence of partial sums is bounded. Furthermore if the limit of the nth partial sum fails to exist, it can do so in just one way: namely,

$$\lim_{n \to +\infty} s_n = +\infty.$$

Hence in this section and the next we consider only series of positive terms.

In this section we consider two kinds of comparison tests. That is, we determine whether or not a given series converges by comparing it with another which is known to converge or to diverge. Therefore we need a supply of series whose convergence properties are known. We already have two such types of series: namely, geometric series and telescoping series. Let us consider one other type before we take up the comparison test—the p-series.

A p-series is of the form

$$\sum_{n=1}^{\infty} \frac{1}{n^p} = \frac{1}{1^p} + \frac{1}{2^p} + \frac{1}{3^p} + \cdots .$$

We have already seen (Example 5 of the previous section) an example of the p-series. That is the harmonic series

$$1 + \frac{1}{2} + \frac{1}{3} + \cdots ,$$

for which $p = 1$. We have seen that this series diverges.

Theorem 19.7

If p is a real number, then the p-series

$$\sum_{n=1}^{\infty} \frac{1}{n^p} = \frac{1}{1^p} + \frac{1}{2^p} + \frac{1}{3^p} + \cdots$$

converges if $p > 1$ and diverges if $p \le 1$.

Proof

Let us first consider the case in which $p > 1$. First we note that

$$\frac{1}{2^p} + \frac{1}{3^p} < 2 \cdot \frac{1}{2^p} = \frac{1}{2^{p-1}},$$

$$\frac{1}{4^p} + \frac{1}{5^p} + \frac{1}{6^p} + \frac{1}{7^p} < 4 \cdot \frac{1}{4^p} = \frac{1}{4^{p-1}},$$

$$\frac{1}{8^p} + \frac{1}{9^p} + \cdots + \frac{1}{15^p} < 8 \cdot \frac{1}{8^p} = \frac{1}{8^{p-1}},$$

and so forth. This gives

$$s_1 = 1,$$

$$s_3 = 1 + \frac{1}{2^p} + \frac{1}{3^p} < 1 + \frac{1}{2^{p-1}},$$

$$s_7 = 1 + \frac{1}{2^p} + \cdots + \frac{1}{7^p} < 1 + \frac{1}{2^{p-1}} + \frac{1}{4^{p-1}},$$

$$s_{15} = 1 + \frac{1}{2^p} + \cdots + \frac{1}{15^p} < 1 + \frac{1}{2^{p-1}} + \frac{1}{4^{p-1}} + \frac{1}{8^{p-1}},$$

and so on. We see that the expressions on the right are partial sums of the series

$$1 + \frac{1}{2^{p-1}} + \frac{1}{4^{p-1}} + \frac{1}{8^{p-1}} + \cdots = 1 + \frac{1}{2^{p-1}} + \frac{1}{(2^{p-1})^2} + \frac{1}{(2^{p-1})^3} + \cdots.$$

But the last series is a geometric series with

$$r = \frac{1}{2^{p-1}} < 1;$$

it converges to

$$S = \frac{1}{1 - 1/2^{p-1}} = \frac{2^{p-1}}{2^{p-1} - 1}.$$

Hence the partial sums of this geometric series—and therefore of the *p*-series—are bounded by S. By Theorem 19.1, the *p*-series converges.

We have already proved in the previous section that the *p*-series diverges if $p = 1$. Suppose now that $p < 1$. Then

$$n^p \leq n \quad \text{and} \quad \frac{1}{n^p} \geq \frac{1}{n}.$$

Thus the partial sums of the *p*-series with $p < 1$ are all equal to or greater than the corresponding partial sums of the harmonic series (*p*-series with $p = 1$). Since

$$\lim_{n \to +\infty} s_n = +\infty,$$

where s_n is the *n*th partial sum of the harmonic series, the same is true of the partial sums of the *p*-series. Thus the *p*-series diverges.

This theorem can also be proved by means of the integral test, which we consider in the next section (see Example 3, page 622).

Example 1

Determine whether $1 + \dfrac{1}{\sqrt{2}} + \dfrac{1}{\sqrt{3}} + \cdots$ converges or diverges.

This is a *p*-series with $p = 1/2$. It diverges.

Now that we know the convergence properties of geometric series, telescoping series, and *p*-series, let us consider the comparison test.

Theorem 19.8

(*Comparison test*) *If $\sum a_n$ is a series of positive terms, $\sum b_n$ is a convergent series of positive terms, and for some integer $N \geq 1$, $a_n \leq b_n$ for all $n \geq N$, then $\sum a_n$ converges. If $\sum b_n$ is a divergent series of positive terms and for some integer $N \geq 1$, $a_n \geq b_n$ for all $n \geq N$, then $\sum a_n$ diverges.*

Proof

Let us consider the first case. Let $\{s_n\}$ and $\{t_n\}$ be the sequences of partial sums of $\sum a_n$ and $\sum b_n$, respectively. Suppose $a_n \leq b_n$ for every integer $n \geq 1$;

then $s_n \le t_n$ for every n. But

$$\lim_{n \to +\infty} t_n = t.$$

Since $\sum b_n$ is a series of positive terms converging to t, $t_n \le t$ for every n. Thus $s_n \le t_n \le t$ for every n and, by Theorem 19.1, $\sum a_n$ converges. If $a_n \le b_n$ only for $n \ge N$,   where $N > 1$,   then the result follows from Problem 38 in Section 19.2 by replacing a_n by b_n, $n = 1, 2, \ldots, N - 1$,   in $\sum a_n$.

Suppose now that $\sum b_n$ diverges and $a_n \ge b_n$.   Assume that $\sum a_n$ converges. By the first part, $\sum b_n$ converges; but this is a contradiction. Thus $\sum a_n$ diverges.

It must be emphasized that $\sum a_n$ and $\sum b_n$ are series of *positive* terms. Neither part of Theorem 19.8 would hold without that restriction (see Problems 29 and 30).

Let us also note that the terms of the series being tested must be less than or equal to those of a convergent series or greater than or equal to those of a divergent series. Getting the terms less than or equal to those of a divergent series or greater than or equal to those of a convergent series tells us nothing about the series.

Example 2

Test for convergence

$$\frac{1}{1 \cdot 1} + \frac{1}{2 \cdot 3} + \frac{1}{3 \cdot 5} + \cdots + \frac{1}{n(2n - 1)} + \cdots.$$

Suppose we try to simplify the nth term to give something that we recognize.

$$\frac{1}{n(2n - 1)} = \frac{1}{2n^2 - n} > \frac{1}{2n^2}.$$

But the series

$$\sum_{n=1}^{\infty} \frac{1}{2n^2} = \frac{1}{2} \sum_{n=1}^{\infty} \frac{1}{n^2}$$

is convergent by Theorem 19.7 ($p = 2$). The combination of greater than and convergence tells us nothing. Let us try to reverse the inequality. In the situation above, we simplified

$$\frac{1}{2n^2 - n}$$

by throwing away the $-n$. This made the new denominator bigger and the new fraction smaller. In order to reverse this situation, let us replace $2n^2 - n$ by an expression that is smaller (or, at least, never bigger).

$$\frac{1}{n(2n - 1)} = \frac{1}{2n^2 - n} \le \frac{1}{2n^2 - n^2} = \frac{1}{n^2}.$$

Now the situation is much better.

$$\sum_{n=1}^{\infty} \frac{1}{n^2}$$

converges and

$$\frac{1}{2n^2 - n} \le \frac{1}{n^2}$$

for every positive integer n. By the comparison test, the original series converges.

A test which is generally easier to apply is the limit comparison test. First, a definition.

Definition

*Let $\sum a_n$ and $\sum b_n$ be two series of positive terms. $\sum a_n$ and $\sum b_n$ are of the **same order of magnitude** if*

$$0 < \lim_{n \to +\infty} \frac{a_n}{b_n} < +\infty;$$

$\sum a_n$ *is of a **lesser order of magnitude** than $\sum b_n$ if*

$$\lim_{n \to +\infty} \frac{a_n}{b_n} = 0;$$

$\sum a_n$ *is of a **greater order of magnitude** than $\sum b_n$ if*

$$\lim_{n \to +\infty} \frac{a_n}{b_n} = +\infty.$$

For example, if we compare $\sum n$ and $\sum (n + 1)$, we see that for large values of n the 1 in $\sum (n + 1)$ appears to be relatively insignificant. This is expressed by saying that these two series are of the same order of magnitude. This is verified by the limit

$$\lim_{n \to +\infty} \frac{n}{n + 1} = \lim_{n \to +\infty} \frac{1}{1 + 1/n} = 1.$$

Similarly, $\sum 1/n$ and $\sum 1/3n$ are of the same order of magnitude, since

$$\lim_{n \to +\infty} \frac{1/n}{1/3n} = \lim_{n \to +\infty} \frac{3n}{n} = 3.$$

But $\sum 1/n$ and $\sum 1/n^2$ are not of the same order of magnitude, since

$$\lim_{n \to +\infty} \frac{1/n}{1/n^2} = \lim_{n \to +\infty} n = +\infty;$$

in this case $\sum 1/n$ is of a greater order of magnitude than $\sum 1/n^2$. We are now in a position to state the limit comparison test.

Theorem 19.9

(Limit comparison test) Suppose $\sum a_n$ and $\sum b_n$ are two series of positive terms. If $\sum a_n$ and $\sum b_n$ are of the same order of magnitude, then either both series converge or both diverge.

If $\sum a_n$ is of a lesser order of magnitude than $\sum b_n$ and $\sum b_n$ converges, then $\sum a_n$ also converges.

If $\sum a_n$ is of a greater order of magnitude than $\sum b_n$ and $\sum b_n$ diverges, then $\sum a_n$ also diverges.

Proof

Suppose $\sum a_n$ and $\sum b_n$ are of the same order of magnitude. Then

$$L = \lim_{n \to +\infty} \frac{a_n}{b_n}$$

is a positive real number. Let $\epsilon = L/2$. Then there is a positive integer N such that

$$\left| \frac{a_n}{b_n} - L \right| < \frac{L}{2}$$

for $n > N$. Thus, for $n > N$,

$$-\frac{L}{2} < \frac{a_n}{b_n} - L < \frac{L}{2}$$

$$\frac{L}{2} < \frac{a_n}{b_n} < \frac{3L}{2}$$

$$\frac{L}{2} b_n < a_n < \frac{3L}{2} b_n.$$

Now let us suppose that $\sum b_n$ converges. By Theorem 19.6, $\sum 3Lb_n/2$ converges; and from the above, $a_n < 3\,Lb_n/2$. Thus, by the comparison test, $\sum a_n$ converges.

Suppose $\sum b_n$ diverges. Then $\sum Lb_n/2$ also diverges (for if $\sum Lb_n/2$ converges, then

$$\sum \frac{2}{L} \left(\frac{L}{2} b_n \right) = \sum b_n$$

converges). Again, from the above, $a_n > Lb_n/2$. By the comparison test, $\sum a_n$ diverges.

The proofs of the remaining two parts are similar and are left to the student (see Problems 32 and 33).

Example 3

Use the limit comparison test to test

$$\sum_{n=1}^{\infty} \frac{1}{n(2n-1)}$$

for convergence.

This is the series of Example 2; we have already seen that it converges. Now let us use the limit comparison test. First we show that

$$\frac{1}{n(2n-1)} \quad \text{and} \quad \frac{1}{n^2}$$

are of the same order of magnitude.

$$\lim_{n \to +\infty} \frac{1/[n(2n-1)]}{1/n^2} = \lim_{n \to +\infty} \frac{n^2}{n(2n-1)} = \lim_{n \to +\infty} \frac{1}{2 - 1/n} = \frac{1}{2}.$$

Now since $\sum 1/n^2$ converges, it follows from Theorem 19.9 that the given series converges.

Example 4

Test for convergence $\sum_{n=1}^{\infty} \dfrac{n+1}{n \cdot 2^n}$.

Since

$$\lim_{n \to +\infty} \frac{(n+1)/(n \cdot 2^n)}{1/2^n} = \lim_{n \to +\infty} \frac{n+1}{n} = 1,$$

it follows that the given series and $\sum 1/2^n$ are of the same order of magnitude. But $\sum 1/2^n$ is a geometric series with $r = 1/2$; it converges. Therefore by Theorem 19.9, the given series converges.

Example 5

Test for convergence $\sum_{n=1}^{\infty} \dfrac{\ln (n+1)}{n^2}$.

First of all, it is apparent that $\sum \ln (n+1)$ and $\sum \ln n$ are of the same order of magnitude. But, since we do not know the convergence properties of $\sum (\ln n)/n^2$, this is not much help. We need to simplify even more. Since, for $k > 0$,

$$\lim_{n \to +\infty} \frac{\ln n}{n^k} = \lim_{n \to +\infty} \frac{1/n}{kn^{k-1}} = \lim_{n \to +\infty} \frac{1}{kn^k} = 0,$$

it follows that $\sum \ln n$ is of a lesser order of magnitude than $\sum n^k$ for any $k > 0$. If we replace $\ln (n+1)$ by n, we expect $\sum [\ln (n+1)]/n^2$ to be of a lesser order of magnitude than $\sum n/n^2 = \sum 1/n$. But $\sum 1/n$ diverges. Theorem 19.9 tells us nothing in that case. Let us choose a smaller value of k, say $k = 1/2$. Then

$$\lim_{n \to +\infty} \frac{[\ln (n+1)]/n^2}{1/n^{3/2}} = \lim_{n \to +\infty} \frac{\ln (n+1)}{\sqrt{n}} = \lim_{n \to +\infty} \frac{1/(n+1)}{1/(2\sqrt{n})}$$

$$= \lim_{n \to +\infty} \frac{2\sqrt{n}}{n+1} = \lim_{n \to +\infty} \frac{2}{\sqrt{n} + 1/\sqrt{n}} = 0.$$

We know that $\sum [\ln (n+1)]/n^2$ is of a lesser order of magnitude than $\sum 1/n^{3/2}$. Furthermore, $\sum 1/n^{3/2}$ is a p-series with $p = 3/2 > 1$; it converges. Thus, by the second part of the limit comparison test, the original series converges.

Remember that the second and third parts of the limit comparison test require that $\sum a_n$ be of lesser magnitude than a convergent series or that $\sum a_n$ be of greater order of magnitude than a divergent series. If $\sum a_n$ is of greater order of magnitude than a convergent series or of lesser order of magnitude than a divergent series, the limit comparison test tells us nothing.

Problems

A

In Problems 1–17, test for convergence.

1. $1 + \dfrac{1}{4} + \dfrac{1}{9} + \cdots + \dfrac{1}{n^2} + \cdots$.

2. $\dfrac{1}{3} + \dfrac{1}{6} + \dfrac{1}{11} + \cdots + \dfrac{1}{n^2 + 2} + \cdots$.

3. $1 + \dfrac{1}{\sqrt[3]{2}} + \dfrac{1}{\sqrt[3]{3}} + \cdots + \dfrac{1}{\sqrt[3]{n}} + \cdots$.

4. $1 + \dfrac{1}{4} + \dfrac{1}{16} + \cdots + \dfrac{1}{4^{n-1}} + \cdots$.

5. $\dfrac{\ln 3}{3} + \dfrac{\ln 4}{4} + \dfrac{\ln 5}{5} + \cdots + \dfrac{\ln (n+2)}{n+2} + \cdots$.

6. $\dfrac{1}{3} + \dfrac{2}{4} + \dfrac{3}{5} + \cdots + \dfrac{n}{n+2} + \cdots$.

7. $\dfrac{1}{1 \cdot 3} + \dfrac{1}{2 \cdot 3^2} + \dfrac{1}{3 \cdot 3^3} + \cdots + \dfrac{1}{n \cdot 3^n} + \cdots$.

8. $\dfrac{1}{3} + \dfrac{1}{3^2} + \dfrac{1}{3^3} + \cdots + \dfrac{1}{3^n} + \cdots$.

9. $\dfrac{1}{2 \cdot 2} + \dfrac{2}{3 \cdot 2^2} + \dfrac{3}{4 \cdot 2^3} + \cdots + \dfrac{n}{(n+1)2^n} + \cdots$.

10. $\dfrac{1}{1 \cdot 3} + \dfrac{1}{2 \cdot 4} + \dfrac{1}{3 \cdot 5} + \cdots + \dfrac{1}{n(n+2)} + \cdots$.

11. $\displaystyle\sum_{n=1}^{\infty} \dfrac{2n}{2n+1}$.

12. $\displaystyle\sum_{n=1}^{\infty} \dfrac{1}{(n+1)(n+3)}$.

13. $\displaystyle\sum_{n=1}^{\infty} \frac{2n}{(n+1)(n+2)}$.

14. $\displaystyle\sum_{n=1}^{\infty} \frac{1}{(n+1)(n+4)}$.

15. $\displaystyle\sum_{n=1}^{\infty} \frac{n+3}{n(n+1)(n+2)}$.

16. $\displaystyle\sum_{n=1}^{\infty} \frac{n}{(n+1)(n+2)(n+3)}$.

17. $1 + \dfrac{2!}{2^2} + \dfrac{3!}{3^2} + \cdots + \dfrac{n!}{n^2} + \cdots$.

B

In Problems 18–24, test for convergence.

18. $1 + \dfrac{1}{2!} + \dfrac{1}{3!} + \cdots + \dfrac{1}{n!} + \cdots$.

19. $\displaystyle\sum_{n=1}^{\infty} \frac{1}{e^n}$.

20. $\displaystyle\sum_{n=2}^{\infty} \frac{n}{(n+1)\ln n}$.

21. $\displaystyle\sum_{n=1}^{\infty} \frac{1}{n^n}$.

22. $\displaystyle\sum_{n=1}^{\infty} \frac{1}{\sqrt{n(n+1)}}$.

23. $\displaystyle\sum_{n=1}^{\infty} \frac{1}{(2n-1)(3n-1)}$.

24. $\displaystyle\sum_{n=1}^{\infty} \frac{2n-1}{n(n+1)}$.

C

In Problems 25–28, test for convergence.

25. $\displaystyle\sum_{n=1}^{\infty} \frac{(n+1)2^{n+1}}{3^n}$.

26. $\displaystyle\sum_{n=1}^{\infty} \frac{\ln(2n+1)}{n(n+2)}$.

27. $\displaystyle\sum_{n=1}^{\infty} \frac{\ln\sqrt{3n-1}}{\sqrt{n}\sqrt{n^3+5n+2}}$.

28. $\displaystyle\sum_{n=1}^{\infty} \frac{\ln(2n+1)}{n+3}$.

29. Give an example of a series $\sum a_n$ and a convergent series $\sum b_n$ of positive terms such that $a_n \le b_n$ for all n but $\sum a_n$ is divergent.

30. Give an example of a series $\sum a_n$ of positive terms and a divergent series $\sum b_n$ such that $a_n \ge b_n$ for all b but $\sum a_n$ is convergent.

31. Find the maximum value of $y = \dfrac{\ln x}{x}$. Use this result to show that $\ln n < n$, $n = 1, 2, 3, \ldots$. Show that

$$\sum_{n=2}^{\infty} \frac{1}{\ln n}$$

diverges.

32. Prove the second part of Theorem 19.9.

33. Prove the third part of Theorem 19.9.

19.4

The Ratio and Integral Tests

While the comparison test of the last section can determine convergence or divergence for many series, it is a difficult, or impossible test for many others. Let us consider some other methods of testing series.

Theorem 19.10

(*Ratio test*)   *If $\sum a_n$ is a series of positive terms and*

$$r = \lim_{n \to +\infty} \frac{a_{n+1}}{a_n},$$

and if

(1) $r < 1$, *then the series converges,*
(2) $r > 1$ (*including $r = +\infty$*), *then the series diverges,*
(3) $r = 1$, *the test fails.*

Proof

Suppose $r < 1$; let v be a number such that $r < v < 1$ and let $\epsilon = v - r$. By the definition of a limit, there is a number N such that, if $n > N$, then

$$\left| \frac{a_{n+1}}{a_n} - r \right| < \epsilon$$

$$-\epsilon < \frac{a_{n+1}}{a_n} - r < \epsilon$$

$$r - \epsilon < \frac{a_{n+1}}{a_n} < r + \epsilon = v.$$

Suppose $M > N$. Then

$$\frac{a_{M+1}}{a_M} < v, \quad a_{M+1} < va_M,$$

$$\frac{a_{M+2}}{a_{M+1}} < v, \quad a_{M+2} < va_{M+1} < v^2 a_M,$$

$$\frac{a_{M+3}}{a_{M+2}} < v, \quad a_{M+3} < va_{M+2} < v^3 a_M,$$

and so forth. Let us consider the series

$$\sum_{n=1}^{\infty} v^n a_M = a_M \sum_{n=1}^{\infty} v^n.$$

Since $v < 1$, this geometric series converges. By the comparison test, the series

$$a_{M+1} + a_{M+2} + a_{M+3} + \cdots$$

also converges, and thus $\sum a_n$ converges (see Problem 37, page 610).

Suppose now that $r > 1$. Let $\epsilon = r - 1$. Again, by the definition of a limit, there is a number N such that, if $n > N$, then

$$\left| \frac{a_{n+1}}{a_n} - r \right| < \epsilon.$$

Then

$$-\epsilon < \frac{a_{n+1}}{a_n} - r < \epsilon$$

$$1 = r - \epsilon < \frac{a_{n+1}}{a_n} < r + \epsilon.$$

Thus, $a_{n+1} > a_n$ for all $n > N$. Therefore,

$$\lim_{n \to +\infty} a_n \neq 0,$$

and the series diverges.

To show that the test fails when $r = 1$, we need merely note that $r = 1$ for

$$\sum_{n=1}^{\infty} \frac{1}{n} \quad \text{and} \quad \sum_{n=1}^{\infty} \frac{1}{n^2},$$

while the first series diverges and the second converges.

Example 1

Test for convergence $\sum_{n=1}^{\infty} \frac{n}{3^n}$.

$$\lim_{n \to +\infty} \frac{a_{n+1}}{a_n} = \lim_{n \to +\infty} \frac{(n+1)/3^{n+1}}{n/3^n}$$

$$= \lim_{n \to +\infty} \frac{n+1}{3n}$$

$$= \frac{1}{3} < 1.$$

The series converges.

The ratio test is especially useful in handling expressions involving factorials.

Example 2

Test for convergence $\sum_{n=1}^{\infty} \frac{2^n}{n!}$.

$$\lim_{n \to +\infty} \frac{a_{n+1}}{a_n} = \lim_{n \to +\infty} \frac{2^{n+1}/(n+1)!}{2^n/n!}$$

$$= \lim_{n \to +\infty} \frac{2}{n+1}$$

$$= 0 < 1.$$

The series converges.

Recall that if $r > 1$, then the limit of the nth term of the series is not zero. Thus the ratio test indicates divergence only when the terms do not tend to zero—it is not a sensitive test for divergence. Actually, as we shall see in Section 19.6, this insensitivity is an advantage.

The ratio test is very simple—when it works. However, there are many series for which the test fails. If it does, we simply have to fall back on a comparison test or the integral test, which follows.

Theorem 19.11

(*Integral test*) *If* $\sum_{n=1}^{\infty} a_n$ *is a series of positive terms, f is a continuous function for $x \geq 1$ such that $f(n) = a_n$ for all positive integers n, and there is a positive integer N such that f is decreasing for all $x > N$, then either*

$$\sum_{n=1}^{\infty} a_n \quad and \quad \int_{1}^{+\infty} f(x) \, dx$$

both converge or both diverge.

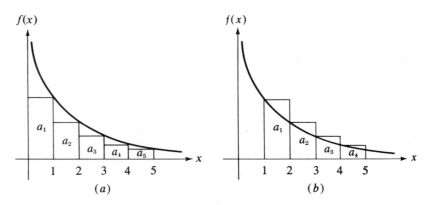

Figure 19.3

Proof

Let us first consider the case for which $N = 1$; that is, f is decreasing for all $x > 1$. Suppose the integral exists. In Figure 19.3(a) the first rectangle has height a_1 and width 1. Thus its area is a_1. Similarly the areas of the other rectangles are $a_2, a_3, a_4, \ldots$. Thus, from the figure, we see that

$$\int_{1}^{n} f(x) \, dx > a_2 + a_3 + \cdots + a_n,$$

or

$$a_1 + \int_{1}^{n} f(x) \, dx > a_1 + a_2 + a_3 + \cdots + a_n = s_n.$$

Thus,

$$\lim_{n \to +\infty} \left[a_1 + \int_{1}^{n} f(x) \, dx \right] = a_1 + \lim_{n \to +\infty} \int_{1}^{n} f(x) \, dx$$

$$= a_1 + \int_{1}^{+\infty} f(x) \, dx$$

$$\geq \lim_{n \to +\infty} s_n.$$

Since the first limit exists, the last one must also, and the series converges.

Now suppose that the integral does not exist. A similar consideration of Figure 19.3(b) gives

$$\int_{1}^{n} f(x) \, dx < a_1 + a_2 + \cdots + a_{n-1} = s_{n-1}$$

and

$$\lim_{n \to +\infty} \int_1^n f(x)\, dx \le \lim_{n \to +\infty} s_{n-1} = \lim_{n \to +\infty} s_n.$$

If we assume that the series converges, then the last limit exists and the first must also. But the first limit is the integral which does not exist, giving a contradiction. The series must diverge.

Now if $N > 1$ we may use the above argument to show that

$$\sum_{n=N}^{\infty} a_n \quad \text{and} \quad \int_N^{+\infty} f(x)\, dx$$

either both converge or both diverge. But

$$\sum_{n=1}^{\infty} a_n = \sum_{n=1}^{N-1} a_n + \sum_{n=N}^{\infty} a_n;$$

and since

$$\sum_{n=1}^{N-1} a_n$$

is finite, it has no effect upon the convergence of the original series—that is determined entirely by

$$\sum_{n=N}^{\infty} a_n$$

(see Problem 37, page 610). Similarly the convergence of

$$\int_1^{+\infty} f(x)\, dx = \int_1^{N-1} f(x)\, dx + \int_N^{+\infty} f(x)\, dx$$

is determined only by

$$\int_N^{+\infty} f(x)\, dx.$$

Thus

$$\sum_{n=1}^{\infty} a_n \quad \text{and} \quad \int_1^{+\infty} f(x)\, dx$$

either both converge or both diverge.

It might be noted that if the series were given by

$$\sum_{n=2}^{\infty} a_n,$$

then the integral would be

$$\int_2^{+\infty} f(x)\, dx.$$

Note also that, while this theorem states that the series and the integral converge together, it does *not* say that they converge to the same number.

The integral test gives us an alternate method of determining the conditions for convergence of a *p*-series.

Example 3

For what values of p does the p-series converge?

We consider only the p-series

$$\sum_{n=1}^{\infty} \frac{1}{n^p}$$

for $p > 0$; the nth term test shows that it diverges for $p \leq 0$. For $p > 0$, $f(x) = 1/x^p$ is easily seen to be continuous and decreasing for $x > 0$ (since $f'(x) = -p/x^{p+1} < 0$). Then if $p \neq 1$,

$$\int_1^{+\infty} \frac{dx}{x^p} = \lim_{k \to +\infty} \int_1^k x^{-p}\,dx$$

$$= \lim_{k \to +\infty} \left. \frac{x^{1-p}}{1-p} \right|_1^k$$

$$= \lim_{k \to +\infty} \left(\frac{k^{1-p}}{1-p} - \frac{1}{1-p} \right)$$

$$= \begin{cases} \dfrac{1}{p-1} & \text{if } p > 1, \\ +\infty & \text{if } p < 1. \end{cases}$$

Finally if $p = 1$, we have

$$\int_1^{+\infty} \frac{dx}{x} = \lim_{k \to +\infty} \int_1^k \frac{dx}{x}$$

$$= \lim_{k \to +\infty} \left. \ln x \right|_1^k$$

$$= \lim_{k \to +\infty} \ln k = +\infty.$$

Thus the p-series converges only when $p > 1$; it diverges if $p \leq 1$.

Example 4

Test for convergence $\displaystyle \sum_{n=2}^{\infty} \frac{1}{n\sqrt{\ln n}}$.

Let

$$f(x) = \frac{1}{x\sqrt{\ln x}}.$$

Note that the given series starts with $n = 2$ (since $\ln 1 = 0$); therefore we are concerned with this function for $x \geq 2$. Clearly f is continuous for $x \geq 2$; and since x and $\ln x$ are both increasing functions, $x\sqrt{\ln x}$ is increasing and $1/x\sqrt{\ln x}$ is decreasing for $x \geq 2$. Thus the conditions of Theorem 19.11 are satisfied.

$$\int_2^{+\infty} \frac{dx}{x\sqrt{\ln x}} = \lim_{k \to +\infty} \int_2^k (\ln x)^{-1/2} \frac{1}{x} \, dx$$

$$= \lim_{k \to +\infty} 2\sqrt{\ln x} \,\Big|_2^k$$

$$= \lim_{k \to +\infty} (2\sqrt{\ln k} - 2\sqrt{\ln 2}) = +\infty,$$

showing that the given series diverges.

Problems

A

In Problems 1–7, test for convergence by the ratio test. Use another test only if the ratio test fails.

1. $\displaystyle\sum_{n=1}^{\infty} \frac{n+1}{n \cdot 2^n}$.
2. $\displaystyle\sum_{n=1}^{\infty} \frac{n+1}{n!}$.
3. $\displaystyle\sum_{n=1}^{\infty} \frac{2^{n+1}}{3^n}$.
4. $\displaystyle\sum_{n=1}^{\infty} \frac{2^n}{3^{n+1}}$.

5. $\displaystyle\sum_{n=1}^{\infty} \frac{3^n}{n!}$.
6. $\displaystyle\sum_{n=1}^{\infty} \frac{3^n}{n \cdot 2^n}$.
7. $\displaystyle\sum_{n=1}^{\infty} \frac{3^n}{n^2 \cdot 2^n}$.

In Problems 8–12, test for convergence by any test.

8. $\displaystyle\sum_{n=1}^{\infty} \frac{1}{2^n + 1}$.
9. $\displaystyle\sum_{n=1}^{\infty} \frac{n(n+2)}{(n+1)(n+3)}$.
10. $\displaystyle\sum_{n=1}^{\infty} \frac{\sqrt{n}}{(n+1)^2}$.

11. $\displaystyle\sum_{n=1}^{\infty} \frac{1}{2^n - 1}$.
12. $\displaystyle\sum_{n=1}^{\infty} \frac{2n-1}{2^n}$.

13. Test for convergence by the integral test $\displaystyle\sum_{n=1}^{\infty} \frac{1}{2n+3}$.

B

In Problems 14–16, test for convergence by the ratio test.

14. $\displaystyle\sum_{n=1}^{\infty} \frac{2 \cdot 4 \cdot 6 \cdots 2n}{1 \cdot 3 \cdot 5 \cdots (2n-1)}$.
15. $\displaystyle\sum_{n=1}^{\infty} \frac{n!}{1 \cdot 3 \cdot 5 \cdots (2n-1)}$.

16. $\displaystyle\sum_{n=1}^{\infty} \frac{n!}{(2n)!}$.

In Problems 17–24, test for convergence by the integral test.

17. $\displaystyle\sum_{n=2}^{\infty} \frac{1}{n \ln n}$.
18. $\displaystyle\sum_{n=1}^{\infty} \frac{\ln n}{n}$.
19. $\displaystyle\sum_{n=3}^{\infty} \frac{1}{n \ln n \ln \ln n}$.

20. $\displaystyle\sum_{n=2}^{\infty} \frac{1}{n \ln^2 n}$.
21. $\displaystyle\sum_{n=1}^{\infty} \ln \frac{n+2}{n}$.
22. $\displaystyle\sum_{n=1}^{\infty} \frac{n}{e^n}$.

23. $\displaystyle\sum_{n=1}^{\infty} \frac{n^2}{e^n}$.
24. $\displaystyle\sum_{n=1}^{\infty} \frac{1}{\cosh^2 n}$.

In Problems 25 and 26, test for convergence by any test.

25. $\displaystyle\sum_{n=1}^{\infty} \frac{1 \cdot 3 \cdot 5 \cdots (2n-1)}{2 \cdot 5 \cdot 8 \cdots (3n-1)}$.
26. $\displaystyle\sum_{n=1}^{\infty} \frac{2^n}{n^2 + 1}$.

C

In Problems 27–32, test for convergence by any test.

27. $\sum\limits_{n=1}^{\infty}$ Arccot n.

28. $\sum\limits_{n=1}^{\infty}$ Arccsc n.

29. $\sum\limits_{n=1}^{\infty} (\sqrt{n^2 + 1} - n)$.

30. $\sum\limits_{n=1}^{\infty} \dfrac{n^n}{n!}$.

31. $\sum\limits_{n=1}^{\infty} \dfrac{2^n + n}{n! + 1}$.

32. $\sum\limits_{n=2}^{\infty} \dfrac{1}{n \ln^4 n}$.

33. Show that if $\sum a_n$ is a convergent series of positive terms and $a_{n+1}/a_n \le u < 1$ for all $n \ge N$, then $S - s_n \le a_N u/(1 - u)$, where S is the sum of the series and s_n is the nth partial sum.

34. Use the result of Problem 33 to approximate $\sum_{n=0}^{\infty} 1/n!$ to two decimal places.

35. Suppose we delete from the harmonic series all terms having the digit 9 in the denominator. Is the resulting series convergent or divergent? (*Hint:* Consider first all terms having a one-digit denominator, then all those having a two-digit denominator, and so on.)

19.5

Alternating Series and Conditional Convergence

The past two sections have been concerned exclusively with series of positive terms. Of course, series in which the terms are all negative are handled in exactly the same way. We now turn to series of both positive and negative terms. By far the most frequently encountered series of this type is the alternating series, in which the terms are alternately positive and negative. Testing for convergence of an alternating series is relatively easy in certain cases.

Theorem 19.12

(*Alternating series test*)   *The alternating series*

$$\sum_{n=1}^{\infty} (-1)^{n+1} b_n = b_1 - b_2 + b_3 - b_4 + \cdots \quad (b_n > 0)$$

converges if $b_{n+1} < b_n$ *for all n and* $\lim_{n \to +\infty} b_n = 0$; *furthermore, if S denotes the sum of the series and* s_n *the nth partial sum, then*

$$|S - s_n| < b_{n+1}.$$

Proof

We shall prove only the first part of this theorem; the second part is left to the student. Let us consider an even-numbered partial sum.

$$S_n = (b_1 - b_2) + (b_3 - b_4) + \cdots + (b_{n-1} - b_n)$$
$$= b_1 - (b_2 - b_3) - (b_4 - b_5) - \cdots - (b_{n-2} - b_{n-1}) - b_n.$$

Since $b_{n+1} < b_n$ for all n, the expressions in parentheses are all positive. Thus the first equation implies that $S_n > 0$, and the second implies that $S_n < b_1$. Let us consider the sequence

$$s_2, s_4, s_6, \ldots$$

of even-numbered partial sums. Since all sums are positive and less than b_1 and since they form an increasing sequence, it follows from Theorem 19.1 that they have a limit S; that is,

$$\lim_{\substack{n \text{ even} \\ n \to +\infty}} s_n = S.$$

Now let us consider the odd-numbered partial sums:

$$\lim_{\substack{n \text{ even} \\ n \to +\infty}} s_{n+1} = \lim_{\substack{n \text{ even} \\ n \to +\infty}} (s_n + b_{n+1})$$

$$= \lim_{\substack{n \text{ even} \\ n \to +\infty}} s_n + \lim_{\substack{n \text{ even} \\ n \to +\infty}} b_{n+1}$$

$$= S + 0 = S.$$

Thus

$$\lim_{n \to +\infty} s_n = S,$$

and the series converges. A similar argument can be given for an alternating series with a first term that is negative.

Example 1

Test for convergence $\displaystyle \sum_{n=1}^{\infty} \frac{(-1)^{n+1}}{n} = 1 - \frac{1}{2} + \frac{1}{3} - \frac{1}{4} + \cdots .$

This is an alternating series in which

$$b_{n+1} = \frac{1}{n+1} < \frac{1}{n} = b_n$$

and

$$\lim_{n \to +\infty} b_n = \lim_{n \to +\infty} \frac{1}{n} = 0.$$

The series converges.

Of course, the series

$$\sum_{n=1}^{\infty} \frac{1}{n} = 1 + \frac{1}{2} + \frac{1}{3} + \frac{1}{4} + \cdots$$

has already been shown to diverge, even though

$$a_{n+1} < a_n \quad \text{and} \quad \lim_{n \to +\infty} a_n = 0.$$

Example 2

Test for convergence $1 - \frac{1}{2} + \frac{1}{2} - \frac{1}{4} + \frac{1}{3} - \frac{1}{8} + \frac{1}{4} - \frac{1}{16} + \cdots .$

Although this is an alternating series with

$$\lim_{n \to +\infty} b_n = 0,$$

we cannot conclude from Theorem 19.12 that the series converges, because

$$b_{n+1} < b_n$$

is *not* true for all n. If we consider the positive and negative terms separately, we see that the series

$$-\frac{1}{2} - \frac{1}{4} - \frac{1}{8} - \frac{1}{16} - \cdots$$

of negative terms converges to -1 (geometric series with $a = -1/2$ and $r = 1/2$) while the series

$$1 + \frac{1}{2} + \frac{1}{3} + \frac{1}{4} + \cdots$$

of positive terms diverges to $+\infty$ (harmonic series). This means that no matter how large a number we choose, there is some partial sum of the harmonic series that exceeds it. But the partial sums $(-1/2, -3/4, -7/8, \ldots)$ of the geometric series of negative terms are all greater than -1. Thus if M is a number, there is a partial sum of the series of positive terms exceeding $M + 1$. Since the corresponding partial sum of the negative terms is greater than -1, the partial sum of both positive and negative terms together is greater than M. Thus, the series diverges.

Sometimes we find not only that a given alternating series converges, but also that the corresponding series of positive terms converges. Since series of this type have special properties, we want to single them out.

Definition

*The series $\sum a_n$ of positive and negative terms **converges absolutely** if $\sum |a_n|$ converges. If $\sum a_n$ converges but $\sum |a_n|$ diverges, $\sum a_n$ is said to **converge conditionally**.*

The series

$$\sum_{n=1}^{\infty} \frac{(-1)^{n+1}}{n}$$

of Example 1 converges conditionally, since it converges but the series of absolute values diverges. On the other hand,

$$\sum_{n=1}^{\infty} \frac{(-1)^{n+1}}{2^n}$$

converges absolutely, since the series of absolute values converges.

The following theorems give the relations between convergence and absolute convergence.

Theorem 19.13

If the series $\sum a_n$ converges absolutely, then it converges.

Proof

Let $c_n = (a_n + |a_n|)/2$. If $a_n \geq 0$, then $|a_n| = a_n$ and $c_n = (a_n + a_n)/2 = a_n$. If $a_n < 0$, then $|a_n| = -a_n$ and $c_n = (a_n - a_n)/2 = 0$. Thus $0 \leq c_n \leq |a_n|$

for all n. Let $\{s_n\}$ denote the sequence of partial sums of $\sum a_n$, $\{t_n\}$ denote the sequence of partial sums of $\sum |a_n|$, and $\{u_n\}$ denote the sequence of partial sums of $\sum c_n$. Since $2c_n = a_n + |a_n|$, $2u_n = s_n + t_n$. Also, $s_n \leq t_n$, so $0 \leq u_n \leq t_n$. Since it is given that $\sum |a_n|$ is convergent, $\lim_{t \to +\infty} t_n = t < \infty$. Thus $\{u_n\}$ also converges to a number $u \leq t$. But $s_n = 2u_n - t_n$, so $\{s_n\}$ converges to $2u - t$.

Theorem 19.14

If $\sum a_n$ diverges, then $\sum |a_n|$ diverges.

Proof

Assume that $\sum |a_n|$ converges. Then, by Theorem 19.13, $\sum a_n$ converges, contradicting our hypothesis. Thus $\sum |a_n|$ diverges.

There is no such term as *absolute divergence*. Theorem 19.14 tells us that if a series is divergent, the corresponding series of absolute values is also divergent. We do not need a special term to indicate that both diverge—simply saying that the series diverges is enough.

Let us consider some properties of absolutely convergent series.

Theorem 19.15

If $\sum a_n$ converges absolutely, then the series of all the positive terms converges, as does the series of all the negative terms.

The proof of this theorem is contained within the proof of Theorem 19.13, since $\sum c_n$ represents the series of nonnegative terms in $\sum a_n$ with sum u; and the series of negative terms has sum $u - t$. Note that this conclusion is not true for all convergent series. The series

$$\sum_{n=1}^{\infty} \frac{(-1)^{n+1}}{n} = 1 - \frac{1}{2} + \frac{1}{3} - \frac{1}{4} + \cdots$$

converges (conditionally), but neither

$$1 + \frac{1}{3} + \frac{1}{5} + \cdots = \sum_{n=1}^{\infty} \frac{1}{2n-1}$$

nor

$$-\frac{1}{2} - \frac{1}{4} - \frac{1}{6} - \cdots = \sum_{n=1}^{\infty} \frac{-1}{2n}$$

converges. The converse of this theorem is also true.

Theorem 19.16

If $\sum a_n$ is a series for which the series of positive terms and the series of negative terms both converge, then $\sum a_n$ converges absolutely.

The proof is left to the student. This theorem is useful in testing alternating series for which Theorem 19.12 cannot be used. Of course, the use of this theorem is not restricted to alternating series—it can be used on any series of both positive and negative terms.

Example 3

Test for convergence $\dfrac{1}{2} - \dfrac{1}{3} + \dfrac{1}{4} - \dfrac{1}{9} + \dfrac{1}{8} - \dfrac{1}{27} + \cdots .$

Theorem 19.12 cannot be used here, since the condition $b_{n+1} < b_n$ for all n is not satisfied; but the two series

$$\sum_{n=1}^{\infty} \frac{1}{2^n} \quad \text{and} \quad \sum_{n=1}^{\infty} \frac{-1}{3^n}$$

converge. By Theorem 19.16, the original series converges absolutely and thus converges.

Theorem 19.17

If $\sum a_n$ is an absolutely convergent series converging to S, then any series $\sum c_n$ formed by rearranging the terms of $\sum a_n$ also converges to S.

By a rearrangement we mean changing the order in which the terms appear. No terms are added or deleted—only the position of terms in the series is changed. Do not confuse a rearrangement with a regrouping; in a regrouping there is a replacement of several terms by a single term which is their sum.

The proof of this theorem is left to the student (see Problem 31). Perhaps you think a proof unnecessary, since the conclusion appears obvious. Furthermore, you may wonder why we have the condition that $\sum a_n$ is absolutely convergent—surely *any* convergent series can be rearranged in *any* way and converge to the same number. If that is what you are thinking, you are wrong. Incredible as it may seem, it is sometimes possible to take an infinite set of numbers, add them in one order to get one sum, then add them in a second order and get a different sum! Let us see how.

We have already seen that the series

$$\sum_{n=1}^{\infty} \frac{(-1)^{n+1}}{n} = 1 - \frac{1}{2} + \frac{1}{3} - \frac{1}{4} + \frac{1}{5} - \frac{1}{6} + \cdots$$

converges conditionally. More importantly, neither

$$1 + \frac{1}{3} + \frac{1}{5} + \frac{1}{7} + \cdots \quad \text{nor} \quad -\frac{1}{2} - \frac{1}{4} - \frac{1}{6} - \frac{1}{8} - \cdots$$

converges. In fact, it is the divergence of both of these series that makes it possible for different rearrangements of the original series to converge to different numbers. No matter how large a number we name, there is some partial sum of

$$1 + \frac{1}{3} + \frac{1}{5} + \frac{1}{7} + \cdots$$

which exceeds that number. Furthermore, if we delete the first n terms, for *any* number n, the same can be said for the resulting series. A similar statement can be made for the series of negative terms.

By grouping the partial sums as we did in the proof of Theorem 19.12; namely

$$s_n = \left(1 - \frac{1}{2}\right) + \left(\frac{1}{3} - \frac{1}{4}\right) + \cdots + \left(\frac{1}{n-1} - \frac{1}{n}\right)$$

$$= 1 - \left(\frac{1}{2} - \frac{1}{3}\right) - \left(\frac{1}{4} - \frac{1}{5}\right) - \cdots - \left(\frac{1}{n-2} - \frac{1}{n-1}\right) - \frac{1}{n} \quad (n \text{ even}),$$

we see that the sum S of the original series is positive but less than 1 (actually $S = \ln 2 = 0.693\ldots$). Now, let us rearrange these same terms to converge to 1. Suppose we take positive terms (beginning with 1 and going in descending order) until the partial sum is greater than 1:

$$1 + \frac{1}{3}.$$

Now let us take negative terms until the partial sum is less than 1:

$$1 + \frac{1}{3} - \frac{1}{2}.$$

Let us continue with the positive terms until the partial sum is again greater than 1:

$$1 + \frac{1}{3} - \frac{1}{2} + \frac{1}{5}.$$

Again with negative terms:

$$1 + \frac{1}{3} - \frac{1}{2} + \frac{1}{5} - \frac{1}{4}.$$

Continuing, we have

$$1 + \frac{1}{3} - \frac{1}{2} + \frac{1}{5} - \frac{1}{4} + \frac{1}{7} + \frac{1}{9} - \frac{1}{6} + \frac{1}{11} + \frac{1}{13} - \frac{1}{8} + \cdots.$$

Because of the divergence of the series of positive terms and the series of negative terms, we know, by our previous remarks, that this process can be carried on indefinitely.

Finally we observe that, as this process is continued, the terms we use tend toward 0. This implies that the difference between 1 and the partial sums must tend toward 0. Let us illustrate this last point with the first eleven terms of the rearrangement as given above. The tenth terms is $1/13$ and it makes $s_{10} > 1$. But since we used positive terms only when the previous partial sum was less than 1, we know that $s_{10} < 1 + 1/13$. Furthermore no partial sum beyond the tenth can be as large as s_{10}, since we shall be using smaller positive terms—and then only if the previous partial sum is less than 1. Similarly the eleventh term, $-1/8$, makes $s_{11} < 1$; but, since we know that $s_{10} > 1$, it follows that $s_{11} > 1 - 1/8$. Again no later partial sum will differ from 1 by more than $1/8$. As we continue, these differences from 1 must tend toward 0. Hence, the rearranged series converges to 1.

Perhaps you feel that, because we seem to be using up the positive terms faster than the negative ones, we shall deplete them while some negative ones remain. But there is no danger in that happening, since we have an infinite supply and we use only finitely many at each step. No matter what number of the original series we consider, it must appear somewhere in the rearrangement.

The same argument can be used to find a rearrangement converging to any pre-assigned number we want. In fact, the series can be rearranged to converge to any number, positive or negative, or to diverge in either direction!

Problems

A

In Problems 1–10, test for convergence and absolute convergence.

1. $\displaystyle\sum_{n=1}^{\infty} \frac{(-1)^{n+1}}{2n}$.

2. $\displaystyle\sum_{n=1}^{\infty} \frac{(-1)^{n+1}}{2n-1}$.

3. $\displaystyle\sum_{n=1}^{\infty} \frac{(-1)^{n-1}}{n^2}$.

4. $\displaystyle\sum_{n=1}^{\infty} \frac{(-1)^{n+1}}{n^2+1}$.

5. $\displaystyle\sum_{n=1}^{\infty} \frac{(-1)^{n-1}}{\sqrt{n}}$.

6. $\displaystyle\sum_{n=1}^{\infty} \frac{(-1)^n}{n\sqrt{n}}$.

7. $\displaystyle\sum_{n=2}^{\infty} \frac{(-1)^{n+3}}{\ln n}$.

8. $\displaystyle\sum_{n=1}^{\infty} \frac{(-1)^n}{3^n}$.

9. $\displaystyle\sum_{n=1}^{\infty} \frac{(-1)^n}{n!}$.

10. $\displaystyle\sum_{n=2}^{\infty} \frac{(-1)^{n+1}n}{\ln n}$.

B

In Problems 11–20, test for convergence and absolute convergence.

11. $\displaystyle\sum_{n=1}^{\infty} \frac{(-1)^{n-1}n^2}{2^n}$.

12. $\displaystyle\sum_{n=1}^{\infty} \frac{(-1)^n(2n+1)}{n^2}$.

13. $\displaystyle\sum_{n=1}^{\infty} \frac{(-1)^n n!}{n^2 \cdot 2^n}$.

14. $\displaystyle\sum_{n=1}^{\infty} \frac{(-1)^{n+1}n!}{2^n}$.

15. $\displaystyle\sum_{n=1}^{\infty} \frac{(-1)^n(n+1)}{n\sqrt{n}}$.

16. $\displaystyle\sum_{n=1}^{\infty} \frac{(-1)^{n+1}n}{e^n}$.

17. $\displaystyle\sum_{n=2}^{\infty} \frac{(-1)^{n+2}}{n \ln n}$.

18. $\displaystyle\sum_{n=1}^{\infty} (-1)^{n-1} \frac{1 \cdot 3 \cdot 5 \cdots (2n-1)}{1 \cdot 4 \cdot 7 \cdots (3n-2)}$.

19. $\displaystyle\sum_{n=1}^{\infty} \frac{\sin n}{n^2}$;

20. $\displaystyle\sum_{n=1}^{\infty} \frac{\cos \pi n}{n}$.

In Problems 21–24, use the second part of Theorem 19.12 to approximate the sum of the series so that the error does not exceed 0.1.

21. $\displaystyle\sum_{n=1}^{\infty} \frac{(-1)^{n-1}}{n^2}$.

22. $\displaystyle\sum_{n=1}^{\infty} \frac{(-1)^{n+1}}{n}$.

23. $\displaystyle\sum_{n=1}^{\infty} \frac{(-1)^n}{n^2+1}$.

24. $\displaystyle\sum_{n=1}^{\infty} \frac{(-1)^{n+1}n}{(n+1)^2}$.

C

25. Prove the second part of Theorem 19.12.
26. Prove the first part of Theorem 19.12 for the series $\sum_{n=1}^{\infty} (-1)^n b_n$.
27. Show that the terms of the series

$$\sum_{n=1}^{\infty} \frac{(-1)^{n+1}}{n}$$

can be rearranged so that the resulting series converges to 0.

28. Show that the terms of the series

$$\sum_{n=1}^{\infty} \frac{(-1)^{n+1}}{n}$$

can be rearranged so that the resulting series converges to 2.

29. Show that the terms of the series

$$\sum_{n=1}^{\infty} \frac{(-1)^{n+1}}{n}$$

can be rearranged so that the resulting series diverges to $+\infty$.

30. Prove Theorem 19.16.

31. Prove Theorem 19.17. (*Hint:* Prove it first for a series of nonnegative terms; then use Theorem 19.15.)

19.6

Series of Functions

Let us now consider series in which the terms, instead of being constants, are functions of x—that is, $\sum f_n(x)$. A special case of this is the power series

$$\sum_{n=0}^{\infty} a_n(x-a)^n,$$

where the a_n's are constants. There is a notational difficulty here when $x = a$. In that case, we have

$$\sum_{n=0}^{\infty} a_n \cdot 0^n = a_0 \cdot 0^0 + a_1 \cdot 0^1 + a_2 \cdot 0^2 + \cdots.$$

Of course all terms beyond the first are 0. But what is the first term? We have already seen that 0^0 is undefined, and a limit in that form is indeterminate. Since the first term is a_0 for any other value of x, we shall also take it to be a_0 when $x = a$. A more accurate, but more cumbersome, notation is

$$a_0 + \sum_{n=1}^{\infty} a_n(x-a)^n.$$

For each value of x we have a series of constants, which either converges or diverges. Our problem is to find all values of x for which the series converges. The problem is simple enough for a given value of x; we simply need to substitute for x to find the series of constants and test it for convergence by the tests of the previous sections. Unfortunately, we cannot rely upon this method to find all values of x for which the series converges since the test would have to be repeated infinitely many times. To complicate matters further, some values of x give series of positive terms, while others give alternating series. This problem can be overcome by taking absolute values and determining the values of x for which the series converges absolutely. The ratio test (when it works) is generally the simplest test to use on series of functions. Let us consider it in some detail.

Suppose

$$\lim_{n \to +\infty} \left| \frac{f_{n+1}(x)}{f_n(x)} \right| = r(x).$$

The values of x for which $r(x) < 1$ are values for which the series converges absolutely. There is no problem here, for, if the series converges absolutely, it converges. Now the values of x for which $r(x) > 1$ are those for which the series of absolute values diverges. But what is to prevent the series from converging conditionally? The answer is "the insensitivity of the ratio test." We have already seen (page 619) that, if $r > 1$, not only does the series (in this case the series of absolute values) diverge; the terms of the series do not even tend to zero. Now, if the absolute values of the terms do not tend to zero, the terms themselves cannot. Thus, the original series diverges. The only problem we have is with those values of x for which $r = 1$. But there are usually only finitely many such values (often just two), and they can be checked individually.

As we shall see in the following examples, a power series always converges on an interval (which, in the extreme cases, may consist of a single point or the entire x axis). This interval is called the *interval of convergence* of the power series. The interval of convergence may include neither, one, or both end points. Furthermore, the interval of convergence of the power series

$$\sum_{n=0}^{\infty} a_n(x - a)^n$$

is always centered at $x = a$. Thus the distance from $x = a$ to an end point of the interval (that is, half the length of the interval) is called the *radius of convergence* of the series.

Example 1

Find the interval of convergence of $\sum_{n=0}^{\infty} x^n$.

Using the ratio test, we have

$$\lim_{n \to +\infty} \frac{|x^{n+1}|}{|x^n|} = \lim_{n \to +\infty} |x| = |x|.$$

Now the original series converges absolutely for all x for which

$$|x| < 1$$

and diverges for all x for which

$$|x| > 1.$$

The only values of x in question are $x = \pm 1$. When $x = 1$, the series is

$$1 + 1 + 1 + 1 + \cdots,$$

which obviously diverges. When $x = -1$, the series is

$$1 - 1 + 1 - 1 + \cdots,$$

which also diverges. Thus the original series converges for

$$-1 < x < 1.$$

Example 2

Find the interval of convergence of $\displaystyle\sum_{n=1}^{\infty} \frac{x^n}{n}$.

$$\lim_{n \to +\infty} \left| \frac{x^{n+1}/(n+1)}{x^n/n} \right| = \lim_{n \to +\infty} \left| \frac{n}{n+1} x \right| = |x|.$$

Again the series converges absolutely for $|x| < 1$ or $-1 < x < 1$. Let us check the end points. When $x = 1$, we have the harmonic series

$$\sum_{n=1}^{\infty} \frac{1}{n},$$

which diverges. When $x = -1$, we have the alternating harmonic series

$$\sum_{n=1}^{\infty} \frac{(-1)^n}{n},$$

which converges. Thus the series converges for

$$-1 \le x < 1.$$

Example 3

Find the interval of convergence of $\displaystyle\sum_{n=0}^{\infty} \left(\frac{x}{2}\right)^n$.

$$\lim_{n \to +\infty} \left| \frac{(x/2)^{n+1}}{(x/2)^n} \right| = \lim_{n \to +\infty} \left| \frac{x}{2} \right| = \left| \frac{x}{2} \right|.$$

The series converges absolutely when $|x/2| < 1$ or $|x| < 2$, or $-2 < x < 2$. When $x = 2$, we have $1 + 1 + 1 + \cdots$, which diverges. When $x = -2$, we have $1 - 1 + 1 - \cdots$, which also diverges. Thus the series converges if and only if

$$-2 < x < 2.$$

We can also get this result by noting that the given series is a geometric series which converges if and only if

$$|r| = \left| \frac{x}{2} \right| < 1.$$

Example 4

Find the interval of convergence of $\displaystyle\sum_{n=0}^{\infty} \frac{x^n}{n!}$.

$$\lim_{n \to +\infty} \left| \frac{x^{n+1}/(n+1)!}{x^n/n!} \right| = \lim_{n \to +\infty} \left| \frac{x}{n+1} \right| = 0.$$

Since $r < 1$ for *any* choice of x, this series converges for every real number x.

Example 5

Find the interval of convergence of $\displaystyle\sum_{n=0}^{\infty} n! \, x^n$.

$$\lim_{n \to +\infty} \left| \frac{(n+1)! \, x^{n+1}}{n! \, x^n} \right| = \lim_{n \to +\infty} |(n+1)x| = \begin{cases} 0 & \text{if } x = 0, \\ +\infty & \text{if } x \ne 0. \end{cases}$$

Thus the series converges if and only if $x = 0$.

Examples 4 and 5 give the extremes of convergence ranges. We always have the trivial case of convergence when all the terms (or all but one) are zero, which occurs in any of the foregoing cases when $x = 0$. There is no other value that gives convergence in Example 5.

Example 6

Find the interval of convergence of $\sum_{n=1}^{\infty} \dfrac{n}{2^n} (x - 1)^n$.

$$\lim_{n \to +\infty} \left| \frac{(n+1)(x-1)^{n+1}/2^{n+1}}{n(x-1)^n/2^n} \right| = \lim_{n \to +\infty} \left| \frac{(n+1)(x-1)}{2n} \right| = \left| \frac{x-1}{2} \right|.$$

The series converges absolutely if

$$\left| \frac{x-1}{2} \right| < 1, \quad |x - 1| < 2, \quad -2 < x - 1 < 2, \quad \text{or} \quad -1 < x < 3.$$

When $x = 3$, we have

$$\sum_{n=1}^{\infty} \frac{n \cdot 2^n}{2^n} = \sum_{n=1}^{\infty} n,$$

which diverges. When $x = -1$, we have

$$\sum_{n=1}^{\infty} \frac{n(-2)^n}{2^n} = \sum_{n=1}^{\infty} (-1)^n n,$$

which also diverges. Thus the series converges if and only if $-1 < x < 3$.

It might be noted that in checking for convergence at the end points of the interval, there is one test that cannot work—the ratio test. The ratio test was used to determine the interval and it was found to fail at the end points.

Examples 1–6 are all power series. However, the same methods can sometimes be applied to more general series of functions.

Example 7

Find all values of x for which $\sum_{n=0}^{\infty} 2^n \cos^n x$ converges.

$$\lim_{n \to +\infty} \left| \frac{2^{n+1} \cos^{n+1} x}{2^n \cos^n x} \right| = \lim_{n \to +\infty} |2 \cos x| = |2 \cos x|.$$

The series converges absolutely when $|2 \cos x| < 1$ or $|\cos x| < 1/2$. This gives (see Figure 19.4)

$$\frac{\pi}{3} + n\pi < x < \frac{2\pi}{3} + n\pi, \quad n = 0, \pm 1, \pm 2, \ldots .$$

Note here that the convergence set is not a single interval and that we have infinitely many end points to check. However the task of checking is simplified by the fact that $\cos x = \pm 1/2$ at any end point. Thus if $\cos x = 1/2$, the original series becomes

$$\sum_{n=0}^{\infty} 2^n \frac{1}{2^n} = \sum_{n=0}^{\infty} 1,$$

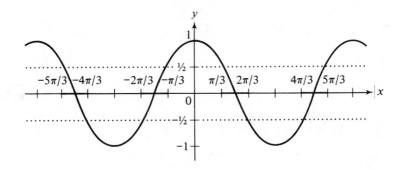

Figure 19.4

which diverges; if $\cos x = -1/2$, the series becomes

$$\sum_{n=0}^{\infty} 2^n \frac{(-1)^n}{2^n} = \sum_{n=0}^{\infty} (-1)^n,$$

which also diverges. Thus the series converges only for

$$\frac{\pi}{3} + n\pi < x < \frac{2\pi}{3} + n\pi, \quad n = 0, \pm 1, \pm 2, \dots .$$

Problems

A

In Problems 1–10, find the interval of convergence. Test end points for convergence.

1. $\displaystyle\sum_{n=1}^{\infty} nx^n$.

2. $\displaystyle\sum_{n=0}^{\infty} \frac{x^n}{n+1}$.

3. $\displaystyle\sum_{n=1}^{\infty} \frac{nx^n}{(n+1)^2}$.

4. $\displaystyle\sum_{n=1}^{\infty} \frac{nx^n}{n+1}$.

5. $\displaystyle\sum_{n=1}^{\infty} \frac{n^2 x^n}{2^n}$.

6. $\displaystyle\sum_{n=1}^{\infty} \frac{nx^n}{2^n}$.

7. $\displaystyle\sum_{n=0}^{\infty} \frac{2^n x^n}{3^n}$.

8. $\displaystyle\sum_{n=1}^{\infty} \frac{(-1)^n x^n}{n \cdot 2^n}$.

9. $\displaystyle\sum_{n=1}^{\infty} n^2 2^n x^n$.

10. $\displaystyle\sum_{n=1}^{\infty} \frac{n(n+1)x^n}{5^n}$.

B

In Problems 11–22, find the interval of convergence. Test end points for convergence.

11. $\displaystyle\sum_{n=1}^{\infty} (-2)^n (x+1)^n$.

12. $\displaystyle\sum_{n=1}^{\infty} n^2 (x-1)^n$.

13. $\displaystyle\sum_{n=0}^{\infty} \frac{(x+3)^n}{2^n}$.

14. $\displaystyle\sum_{n=1}^{\infty} \frac{(x+3)^n}{n \cdot 2^n}$.

15. $\displaystyle\sum_{n=1}^{\infty} \frac{(x+3)^n}{n^2 \cdot 2^n}$.

16. $\displaystyle\sum_{n=0}^{\infty} n!(x-4)^n$.

17. $\displaystyle\sum_{n=1}^{\infty} \frac{n(2x-3)^n}{(n+1)^2}$.

18. $\displaystyle\sum_{n=1}^{\infty} \frac{n \cdot 2^n (x-1)^n}{n+1}$.

19. $\displaystyle\sum_{n=1}^{\infty} \frac{(x+2)^n}{n^n}$.

20. $\displaystyle\sum_{n=1}^{\infty} \frac{(x-3)^n}{2 \cdot 5 \cdot 8 \cdots (3n-1)}$.

21. $\displaystyle\sum_{n=1}^{\infty} \frac{(x+3)^n}{n(n+1)}$.

22. $\displaystyle\sum_{n=1}^{\infty} \frac{n!}{1 \cdot 3 \cdot 5 \cdots (2n-1)} x^n$.

In Problems 23 and 24, find all values of x for which the series converges.

23. $\displaystyle\sum_{n=1}^{\infty} \frac{1}{nx^n}$.

24. $\displaystyle\sum_{n=1}^{\infty} \frac{2^n}{nx^n}$.

C

In Problems 25–27, find all values of x for which the series converges.

25. $\displaystyle\sum_{n=0}^{\infty} 2^n \sin^n x.$ 26. $\displaystyle\sum_{n=0}^{\infty} \tan^n x.$ 27. $\displaystyle\sum_{n=1}^{\infty} \frac{\sin n\pi x}{n^2}.$

(*Hint:* Use a comparison test.)

28.   Give an example of a sequence $\{f_n\}$ of nonnegative continuous functions defined on the closed interval $[0, 1]$ such that $\lim_{n\to +\infty} f_n(x) = 0$ for all x in $[0, 1]$, but

$$\lim_{n\to +\infty} \int_0^1 f_n(x)\, dx = +\infty.$$

(*Hint:* For a nonnegative function, the integral may be interpreted as the area under the curve.)

29.   Give an example of a sequence $\{f_n\}$ of nonnegative piecewise continuous functions defined on the closed interval $[0, 1]$ such that $\lim_{n\to +\infty} f_n(x)$ does not exist for any x in $[0, 1]$, but

$$\lim_{n\to +\infty} \int_0^1 f_n(x)\, dx = 0.$$

(A function f is piecewise continuous on $[0, 1]$ if it is continuous at all but a finite number of points of $[0, 1]$, and if a is a point of discontinuity in $[0, 1]$, then $\lim_{x\to a^-} f(x)$ and $\lim_{x\to a^+} f(x)$ both exist.)

19.7

Taylor's Series

A power series represents a function whose domain is the interval of convergence of the series. For example, as we shall see shortly,

$$e^x = 1 + x + \frac{x^2}{2!} + \frac{x^3}{3!} + \cdots$$

for all values of x. Now there are several advantages to having a function expressed in terms of a power series. One of these is that, by neglecting all but the first few terms of the series, we have an approximation to the given function. Furthermore, the function is approximated by a polynomial, which is perhaps the simplest of all expressions we have encountered. We can easily differentiate or integrate any polynomial term by term. Assuming that this termwise differentiation and integration is valid for the entire series (of course, this requires proof), we then have a method for differentiating and integrating functions that are difficult or impossible to handle by our former methods. For example, we have already noted (see page 495) that

$$\int e^{-x^2}\, dx$$

cannot be evaluated in finite terms. But if $f(x) = e^{-x^2}$ can be expressed as a series, we can (assuming the validity of term-by-term integration) evaluate the integral as a series.

As a first step then, we want to consider the problem of expanding a function to give a power series,

$$f(x) = \sum_{n=0}^{\infty} a_n(x - a)^n.$$

In order to determine the coefficients a_n, let us first *assume* that $f(x)$ has a power-series expansion for a given value of a. Let us further assume that all derivatives of f exist in some interval containing a and that term-by-term differentiation is valid. Thus

$$f(x) = a_0 + a_1(x - a) + a_2(x - a)^2 + a_3(x - a)^3 + \cdots,$$
$$f'(x) = a_1 + 2a_2(x - a) + 3a_3(x - a)^2 + 4a_4(x - a)^3 + \cdots,$$
$$f''(x) = 1 \cdot 2a_2 + 2 \cdot 3a_3(x - a) + 3 \cdot 4a_4(x - a)^2 + \cdots,$$
$$f'''(x) = 1 \cdot 2 \cdot 3a_3 + 2 \cdot 3 \cdot 4a_4(x - a) + 3 \cdot 4 \cdot 5a_5(x - a)^2 + \cdots,$$
$$f^{(4)}(x) = 1 \cdot 2 \cdot 3 \cdot 4a_4 + 2 \cdot 3 \cdot 4 \cdot 5a_5(x - a) + 3 \cdot 4 \cdot 5 \cdot 6a_6(x - a)^2 + \cdots.$$

Now when $x = a$, we have

$$f(a) = a_0,$$
$$f'(a) = a_1,$$
$$f''(a) = 2!a_2,$$
$$f'''(a) = 3!a_3,$$
$$f^{(4)}(a) = 4!a_4,$$

and so forth, or

$$a_0 = f(a)/0!,$$
$$a_1 = f'(a)/1!,$$
$$a_2 = f''(a)/2!,$$
$$a_3 = f'''(a)/3!,$$
$$a_4 = f^{(4)}(a)/4!,$$
$$\vdots$$
$$a_n = f^{(n)}(a)/n!,$$
$$\vdots$$

Theorem 19.18

If f has a power-series expansion in powers of $x - a$ and all derivatives of f exist in some interval containing a and if term-by-term differentiation is valid, then

$$f(x) = \sum_{n=0}^{\infty} \frac{f^{(n)}(a)}{n!} (x - a)^n$$

*for all x in the interval. This series is called the **Taylor's series** for f about a.*

This series is named for the English mathematician Brook Taylor (1685–1731), who first published this result in 1715. A Taylor's series expansion about 0 is often called a *Maclaurin's series* expansion, after Colin Maclaurin (1698–1746) of Scotland.

Example 1

Expand e^x in a Maclaurin's series about 0. Determine all values of x for which it converges.

$$f(x) = e^x, \qquad f(0) = e^0 = 1;$$
$$f'(x) = e^x, \qquad f'(0) = e^0 = 1;$$
$$f''(x) = e^x, \qquad f''(0) = e^0 = 1;$$
$$\vdots \qquad\qquad \vdots$$

$$e^x = \frac{1}{0!} x^0 + \frac{1}{1!} x^1 + \frac{1}{2!} x^2 + \frac{1}{3!} x^3 + \cdots$$

$$= 1 + x + \frac{x^2}{2!} + \frac{x^3}{3!} + \cdots = \sum_{n=0}^{\infty} \frac{x^n}{n!}.$$

By Example 4 of the previous section, this series converges for all real numbers x.

Example 2

Expand $\sin x$ in a Taylor's series about $\pi/2$. Determine all values of x for which it converges.

$$f(x) = \sin x, \qquad f(\pi/2) = 1;$$
$$f'(x) = \cos x, \qquad f'(\pi/2) = 0;$$
$$f''(x) = -\sin x, \qquad f''(\pi/2) = -1;$$
$$f'''(x) = -\cos x, \qquad f'''(\pi/2) = 0;$$
$$f^{(4)}(x) = \sin x, \qquad f^{(4)}(\pi/2) = 1;$$
$$\vdots \qquad\qquad \vdots$$

$$\sin x = \frac{1}{0!} (x - \pi/2)^0 + \frac{0}{1!} (x - \pi/2)^1 - \frac{1}{2!} (x - \pi/2)^2$$

$$+ \frac{0}{3!} (x - \pi/2)^3 + \frac{1}{4!} (x - \pi/2)^4 + \cdots$$

$$= 1 - \frac{(x - \pi/2)^2}{2!} + \frac{(x - \pi/2)^4}{4!} - \frac{(x - \pi/2)^6}{6!} + \cdots$$

$$= \sum_{n=0}^{\infty} (-1)^n \frac{(x - \pi/2)^{2n}}{(2n)!}.$$

$$\lim_{n \to +\infty} \left| \frac{(x - \pi/2)^{2n+2}/(2n+2)!}{(x - \pi/2)^{2n}/(2n)!} \right| = \lim_{n \to +\infty} \left| \frac{(x - \pi/2)^2}{(2n+1)(2n+2)} \right| = 0.$$

This series converges for all x.

Example 3

Expand $\ln x$ in a Taylor's series about 1. Determine all values of x for which it converges.

$$f(x) = \ln x, \qquad f(1) = 0;$$
$$f'(x) = 1/x, \qquad f'(1) = 1;$$
$$f''(x) = -1/x^2, \qquad f''(1) = -1;$$
$$f'''(x) = 2!/x^3, \qquad f'''(1) = 2!;$$
$$f^{(4)}(x) = -3!/x^4, \qquad f^{(4)}(1) = -3!;$$
$$\vdots \qquad\qquad \vdots$$

$$\ln x = \frac{0}{0!}(x-1)^0 + \frac{1}{1!}(x-1)^1 - \frac{1}{2!}(x-1)^2 + \frac{2!}{3!}(x-1)^3 - \frac{3!}{4!}(x-1)^4 + \cdots$$

$$= (x-1) - \frac{(x-1)^2}{2} + \frac{(x-1)^3}{3} - \frac{(x-1)^4}{4} + \cdots$$

$$= \sum_{n=1}^{\infty} (-1)^{n+1} \frac{(x-1)^n}{n}.$$

$$\lim_{n \to +\infty} \left| \frac{(x-1)^{n+1}/(n+1)}{(x-1)^n/n} \right| = \lim_{n \to +\infty} \left| \frac{n}{n+1}(x-1) \right| = |x-1|.$$

Thus, it converges if

$$|x-1| < 1$$
$$-1 < x - 1 < 1$$
$$0 < x < 2.$$

When $x = 0$, the series is

$$-1 - \frac{1}{2} - \frac{1}{3} - \frac{1}{4} - \cdots,$$

which diverges. When $x = 2$, the series is

$$1 - \frac{1}{2} + \frac{1}{3} - \frac{1}{4} + \cdots,$$

which converges. Thus, the series converges if

$$0 < x \leq 2.$$

We can often avoid the task of repeated differentiation when dealing with algebraic expressions. But first we need the following result.

Theorem 19.19

The power-series representation of a function about a is unique.

Proof

We again assume that term-by-term differentiation is valid. Suppose that $f(x)$ has two power-series representations.

$$f(x) = b_0 + b_1(x-a) + b_2(x-a)^2 + b_3(x-a)^3 + \cdots,$$
$$f(x) = c_0 + c_1(x-a) + c_2(x-a)^2 + c_3(x-a)^3 + \cdots.$$

Then

$$0 = (b_0 - c_0) + (b_1 - c_1)(x-a) + (b_2 - c_2)(x-a)^2 + (b_3 - c_3)(x-a)^3 + \cdots.$$

When $x = a$, this gives $b_0 = c_0$. Term-by-term differentiation gives

$$0 = (b_1 - c_1) + 2(b_2 - c_2)(x-a) + 3(b_3 - c_3)(x-a)^2 + \cdots.$$

Again setting $x = a$, we have $b_1 = c_1$. Proceeding in this way, we see that $b_i = c_i$, $i = 0, 1, 2, 3, \cdots$.

This theorem tells us that when we find a power-series expansion about a for a given function f, no matter how we found it, we must have the Taylor's series for f about a, since there can be only one.

Example 4

Expand $1/(x + 1)$ in a Maclaurin's series.

Instead of repeatedly differentiating $f(x) = \dfrac{1}{x + 1}$,   we carry out the division.

$$
\begin{array}{r}
1 - x + x^2 - x^3 + \cdots \\
1 + x \,\overline{\big|\,1 } \\
\underline{1 + x} \\
- x \\
\underline{-x - x^2} \\
x^2 \\
\underline{x^2 + x^3} \\
- x^3
\end{array}
$$

Thus

$$
\frac{1}{x + 1} = 1 - x + x^2 - x^3 + \cdots = \sum_{n=0}^{\infty} (-1)^n x^n.
$$

It appears that this short cut is only possible when we want a Maclaurin's series. Actually, it is possible to use this result to find a Taylor's series for any value of a that we choose, as the next example illustrates.

Example 5

Expand $1/(x + 1)$ in a Taylor's series about 3.

$$
\frac{1}{1 + x} = \frac{1}{4 + (x - 3)} = \frac{1}{4\left(1 + \dfrac{x - 3}{4}\right)} = \frac{1}{4}\,\frac{1}{1 + u},
$$

where $u = (x - 3)/4$. Using the result of Example 4, we have

$$
\frac{1}{4}\,\frac{1}{1 + u} = \frac{1}{4} \sum_{n=0}^{\infty} (-1)^n u^n.
$$

Thus

$$
\frac{1}{1 + x} = \frac{1}{4} \sum_{n=0}^{\infty} (-1)^n \left(\frac{x - 3}{4}\right)^n = \sum_{n=0}^{\infty} (-1)^n \frac{(x - 3)^n}{4^{n+1}}.
$$

Another short cut to finding a Taylor's series is the method of substitution.

Example 6

Expand $1/(1 - x^2)$ in a Maclaurin's series.

By long division,

$$
\frac{1}{1 - u} = 1 + u + u^2 + u^3 + \cdots = \sum_{n=0}^{\infty} u^n.
$$

Substituting x^2 for u,

$$\frac{1}{1 - x^2} = \sum_{n=0}^{\infty} (x^2)^n = \sum_{n=0}^{\infty} x^{2n}.$$

The series converges for $-1 < x < 1$.

Problems

A

In Problems 1–12, expand the function in a Taylor's series about the given value of a. Determine the interval of convergence.

1. $f(x) = e^x$, $a = 1$.
2. $f(x) = \cos x$, $a = 0$.
3. $f(x) = \sin x$, $a = 0$.
4. $f(x) = \cos x$, $a = \pi/6$.
5. $f(x) = \sin x$, $a = \pi/4$.
6. $f(x) = 1/x$, $a = 1$.
7. $f(x) = \dfrac{1}{1 - x}$, $a = 0$.
8. $f(x) = \sinh x$, $a = 0$.
9. $f(x) = \cosh x$, $a = 0$.
10. $f(x) = \dfrac{1}{(x + 1)^2}$, $a = 0$.
11. $f(x) = x^4 - 3x^2 - 2$, $a = 1$.
12. $f(x) = x^3 + x^2 - 2x + 1$, $a = 1$.

B

In Problems 13 and 14, expand the function in a Taylor's series about the given value of a. Determine the interval of convergence.

13. $f(x) = \operatorname{Arctan} x$, $a = 0$.
14. $f(x) = \dfrac{1}{1 + 2x}$, $a = 1$.

In Problems 15–19, expand in a Taylor's series about the given value of a.

15. $f(x) = \sin 2x$, $a = 0$.
16. $f(x) = \sqrt{x}$, $a = 1$.
17. $f(x) = \dfrac{1}{\sqrt{x}}$, $a = 1$.
18. $f(x) = \dfrac{1}{x^2 + 1}$, $a = 1$.
19. $f(x) = e^{-x^2}$, $a = 0$.

C

20. Expand $f(x) = \tan x$ in a Taylor's series about $a = 0$.
21. Explain why $f(x) = \ln x$ does not have a Maclaurin's series expansion.
22. Expand $\sin u$ in powers of u. If $u = x^2$, what is the result? Expand $\sin x^2$ in powers of x. Compare.
23. Expand $\sin x$ in powers of x. Differentiate the resulting series term by term. Compare with the expansion of $\cos x$.
24. Expand e^u in powers of u. If $u = x^2$, what is the result? Expand e^{x^2} in powers of x. Compare.
25. Expand $1/(1 + x^2)$ in powers of x. Integrate the resulting series term by term to obtain the Maclaurin's series for $\operatorname{Arctan} x$.
26. Expand $(1 + x)^n$ (n not an integer) in powers of x. Use the result to express $(a + b)^n$ as an infinite series. This series is the *binomial series*; it converges for $|b/a| < 1$.

In Problems 27–29, use the result of Problem 26 to expand the given function in a Maclaurin's series.

27. $\sqrt{1 - x^2}$.
28. $\dfrac{1}{\sqrt{1 - x^2}}$.
29. $(8 + x)^{1/3}$.

30. If $i = \sqrt{-1}$, show, from the Maclaurin expansion of e^x (Example 1), that

$$e^{i\theta} = \cos \theta + i \sin \theta \quad \text{(Euler's theorem)}$$

and thus deduce the important results

$$\sin \theta = \frac{e^{i\theta} - e^{-i\theta}}{2i} \quad \text{and} \quad \cos \theta = \frac{e^{i\theta} + e^{-i\theta}}{2}.$$

Compare with the definitions of sinh x and cosh x, and show that sinh $ix = i \sin x$ and cosh $ix = \cos x$.

31. Expand $f(x) = 3x/(x^2 - x - 2)$ in a Taylor's series about $a = 1$ and find the interval of convergence. (*Hint:* Write $f(x) = \dfrac{2}{x - 2} + \dfrac{1}{x + 1}$ and use Theorem 19.6.)

32. Evaluate $\displaystyle\sum_{n=0}^{\infty} \frac{(\ln 2)^n}{n!}$.

33. Evaluate $\displaystyle\sum_{n=1}^{\infty} \frac{(-1)^{n+1} (e^{-2} + 1)^n}{n}$.

19.8

Remainder Theorems

In the previous section, we showed how to find a Taylor's series from a given function. By previous methods, we can determine the values of x for which a given Taylor's series converges. But even if we have a Taylor's series determined by a given function f and we know that the series converges for some number x, we have no guarantee that the series converges to $f(x)$. This is illustrated by the following example.

Example 1

Find the Maclaurin's series expansion for

$$f(x) = \begin{cases} e^{-1/x^2} & \text{if } x \neq 0, \\ 0 & \text{if } x = 0. \end{cases}$$

Find the values of x for which it converges. Find the values of x for which it converges to $f(x)$.

In order to find the Maclaurin's series expansion we need to find $f(0)$, $f'(0)$, $f''(0)$, and so forth. While we may use derivative formulas to find $f'(x)$ for $x \neq 0$, we need to resort to the definition of a derivative to find $f'(0)$. Thus

$$f'(0) = \lim_{h \to 0} \frac{f(0 + h) - f(0)}{h} = \lim_{h \to 0} \frac{e^{-1/h^2} - 0}{h}.$$

We may use l'Hôpital's rule to evaluate this limit; however the use of l'Hôpital's rule at this point leads us nowhere (the student is invited to try it and see why). We must alter the form of the limit first.

$$f'(0) = \lim_{h \to 0} \frac{1/h}{e^{1/h^2}} = \lim_{h \to 0} \frac{-1/h^2}{-\dfrac{2}{h^3} e^{1/h^2}} = \lim_{h \to 0} \frac{h}{2e^{1/h^2}} = 0.$$

Thus

$$f'(x) = \begin{cases} \dfrac{2}{x^3} e^{-1/x^2} & \text{if } x \neq 0, \\ \quad 0 & \text{if } x = 0. \end{cases}$$

Similarly,

$$f''(x) = \begin{cases} (2 - 3x^2) \dfrac{2}{x^6} e^{-1/x^2} & \text{if } x \neq 0, \\ \quad 0 & \text{if } x = 0. \end{cases}$$

Because $f^{(n)}(0)$ is always a combination of limits of the form

$$\lim_{h \to 0} \frac{e^{-1/h^2}}{h^m}, \quad \text{where } m \text{ is a positive integer,}$$

which, by the above method, are all 0, $f^{(n)}(0) = 0$ for all positive integers n. Thus the Taylor's series expansion is

$$f(0) + f'(0)x + f''(0) \frac{x^2}{2!} + \cdots = 0 + 0 + 0 + \cdots.$$

Obviously this Maclaurin's series converges for all x, but it does not converge to $f(x)$ except when $x = 0$.

This illustrates a problem we have with Taylor's series. In order to settle this problem, we now consider finite series with remainders.

Theorem 19.20

If f is a function such that $f^{(n+1)}(x)$ exists for all x in an interval containing the number a, then, for all x in that interval,

$$f(x) = f(a) + \frac{f'(a)}{1!} (x - a) + \frac{f''(a)}{2!} (x - a)^2 + \cdots + \frac{f^{(n)}(a)}{n!} (x - a)^n + R_n,$$

where

$$R_n = \frac{f^{(n+1)}(c)}{(n + 1)!} (x - a)^{n+1}$$

for some number c between a and x.

Proof

For fixed x, let the number K be defined by

$$f(x) = f(a) + \frac{f'(a)}{1!}(x-a) + \frac{f''(a)}{2!}(x-a)^2 + \cdots + \frac{f^{(n)}(a)}{n!}(x-a)^n$$

$$+ \frac{K}{(n+1)!}(x-a)^{n+1}. \qquad (1)$$

Let

$$\varphi(t) = f(x) - f(t) - \frac{f'(t)}{1!}(x-t) - \frac{f''(t)}{2!}(x-t)^2 - \cdots - \frac{f^{(n)}(t)}{n!}(x-t)^n$$

$$- \frac{K}{(n+1)!}(x-t)^{n+1}.$$

Note here that φ is a function of t; x is to be looked upon as a constant. Since differentiability implies continuity (Theorem 10.10), $f(t), f'(t), f''(t), \ldots, f^{(n)}(t)$ are all continuous in the interval $[a, x]$. Thus, by Theorem 17.4, $\varphi(t)$ is continuous in $[a, x]$. Differentiating (again recall that φ is a function of t—we are differentiating with respect to t and x is a constant), we have

$$\varphi'(t) = -f'(t) + \left[f'(t) - \frac{f''(t)}{1!}(x-t) \right] + \left[\frac{f''(t)}{1!}(x-t) - \frac{f'''(t)}{2!}(x-t)^2 \right]$$

$$+ \cdots + \left[\frac{f^{(n)}(t)}{(n-1)!}(x-t)^{n-1} - \frac{f^{(n+1)}(t)}{n!}(x-t)^n \right] + \frac{K}{n!}(x-t)^n$$

$$= -\frac{f^{(n+1)}(t)}{n!}(x-t)^n + \frac{K}{n!}(x-t)^n,$$

which exists for all t in (a, x). Finally,

$$\varphi(a) = f(x) - f(a) - \frac{f'(a)}{1!}(x-a) - \frac{f''(a)}{2!}(x-a)^2 - \cdots$$

$$- \frac{f^{(n)}(a)}{n!}(x-a)^n - \frac{K}{(n+1)!}(x-a)^{n+1}$$

$$= 0 \quad \text{by equation (1)},$$

and

$$\varphi(x) = f(x) - f(x) = 0.$$

Thus φ satisfies the conditions of Rolle's theorem (Theorem 17.7, page 569); there is a number c between a and x such that

$$\varphi'(c) = -\frac{f^{(n+1)}(c)}{n!}(x-c)^n + \frac{K}{n!}(x-c)^n = 0.$$

This implies that

$$K = f^{(n+1)}(c).$$

Substituting this result back into equation (1), we have the desired result.

Note that the $(n + 1)$th partial sum s_{n+1} of a Taylor's series is the nth-degree polynomial

$$s_{n+1} = f(a) + \frac{f'(a)}{1!}(x - a) + \frac{f''(a)}{2!}(x - a)^2 + \cdots + \frac{f^{(n)}(a)}{n!}(x - a)^n.$$

It is called the nth-degree *Taylor polynomial* and is represented by $T_n(x)$.

Example 2

Find the Taylor's series expansion for $f(x) = e^x$ in powers of x. Find the values of x for which it converges. Find the values of x for which it converges to $f(x)$.

From Example 1 of the last section, the Taylor's series expansion is

$$\sum_{n=0}^{\infty} \frac{x^n}{n!},$$

and it converges for all x. By Theorem 19.20

$$R_n = \frac{x^{n+1}}{(n+1)!} e^c, \quad \text{where } 0 < c < x.$$

Since

$$\lim_{n \to +\infty} R_n = \lim_{n \to +\infty} \frac{x^{n+1}}{(n+1)!} e^c = 0,$$

the difference between e^x and the nth Taylor polynomial approaches zero as n approaches $+\infty$, no matter what number x represents. Thus the series converges to $f(x)$ for all values of x.

In addition to the theoretical applications mentioned above, this remainder theorem has some very practical applications. If we are interested in a numerical value such as $e^{0.1}$ or $\sin 2°$, we can approximate it by a series. Why approximate? The series gives us the exact value, doesn't it? The reason becomes apparent when one attempts to carry it out. From a practical point of view, we cannot add infinitely many numbers. We know that the infinite sum does exist and that it is the number we want, but we must approximate it with some partial sum. As soon as we begin doing this we must consider the error committed in discarding many of the terms. We would be rather unhappy to find, for instance, that we were throwing away more than we were keeping. Thus, we now consider the remainder after the nth Taylor polynomial.

Example 3

Approximate $e^{0.1}$ by $T_4(0.1)$; what is a maximum value of the error involved?

$$e^x = 1 + x + \frac{x^2}{2!} + \frac{x^3}{3!} + \cdots,$$

$$T_4(0.1) = 1 + 0.1 + \frac{(0.1)^2}{2!} + \frac{(0.1)^3}{3!} + \frac{(0.1)^4}{4!}$$

$$= 1.1051708333\ldots,$$

$$R_4 = \frac{(0.1 - 0)^5}{5!} e^c, \quad \text{where } 0 < c < 0.1,$$

$$< \frac{(0.1)^5}{5!} 2 \quad (\text{since } e^c < 2 \text{ if } c < 0.1)$$

$$< 0.0000002.$$

Thus, an approximation of $e^{0.1}$, accurate to six decimal places, is 1.105171.

Example 4

Approximate $\sin 2°$ accurate to four decimal places.

$$\sin x = x - \frac{x^3}{3!} + \frac{x^5}{5!} - \cdots, \qquad x = 2° = 0.0349 \ldots (\text{radians}).$$

Let us first see how many terms are needed for the desired accuracy.

$$R_n = \frac{x^{n+1}}{(n+1)!} f^{(n+1)}(c), \quad \text{where } 0 < c < 0.0349 \ldots.$$

Since $f(x) = \sin x$, the derivatives are either $\pm \sin x$ or $\pm \cos x$. In any case, $|f^{(n+1)}(c)| < 1$. Thus,

$$|R_n| < \frac{x^{n+1}}{(n+1)!} \approx \frac{(0.0349)^{n+1}}{(n+1)!}.$$

We want to choose n big enough that $|R_n| < 0.00005$. At this point, it is a matter of trial and error to see how big n should be. We see that

$$|R_2| < \frac{(0.035)^3}{3!} \approx 0.000007.$$

Thus we want T_2 which is $0 + x + 0 \cdot x^2 = x$. Hence

$$\sin 2° \approx 0.0349 \quad \text{to four decimal places.}$$

Example 5

Find $\sin 65°$ accurate to four decimal places.

Let us proceed as in Example 4.

$$\sin x = x - \frac{x^3}{3!} + \frac{x^5}{5!} \cdots, \qquad x = 65° = 1.13446 \ldots,$$

$$|R_n| = \frac{x^{n+1}}{(n+1)!} f^{(n+1)}(c), \quad \text{where } 0 < c < 1.13446,$$

$$< \frac{(1.14)^{n+1}}{(n+1)!}.$$

A few trials show that $n = 8$ is sufficiently large. Thus, $\sin 65°$ is approximated by the eighth-degree Taylor polynomial.

$$T_8(1.13446) = 1.13446 - \frac{1.13446^3}{3!} + \frac{1.13446^5}{5!} - \frac{1.13446^7}{7!}$$

$$\approx 0.9063.$$

This problem is considerably more tedious than the other two, since we have to raise 1.13446 to the third, fifth, and seventh powers. Let us alter the procedure slightly. Since 65° is rather near 60° $= \pi/3$, let us expand $\sin x$ in powers of $x - \pi/3$.

$$\sin x = \frac{\sqrt{3}}{2} + \frac{1}{2}(x - \pi/3) - \frac{\sqrt{3}}{2 \cdot 2!}(x - \pi/3)^2 - \frac{1}{2 \cdot 3!}(x - \pi/3)^3 + \cdots,$$

$$x = 65° \approx 1.13446,$$

$$x - \pi/3 \approx 1.13446 - 1.04720 = 0.08726,$$

$$R_n = \frac{(x - \pi/3)^{n+1}}{(n+1)!} f^{(n+1)}(c), \quad \text{where } \pi/3 < c < 1.13446,$$

$$< \frac{(0.09)^{n+1}}{(n+1)!}.$$

A few trials show that $n = 3$ is sufficiently large. Thus

$$\sin 65° \approx \sin 1.13446$$

$$\approx \frac{\sqrt{3}}{2} + \frac{1}{2}(0.08726) - \frac{\sqrt{3}}{2 \cdot 2!}(0.08726)^2 - \frac{1}{2 \cdot 3!}(0.08726)^3$$

$$\approx 0.9063.$$

In general, we want to choose a as a convenient number as near x as possible.

It might be noted that the method used here for determining the values of $e^{0.1}$, $\sin 2°$, and $\sin 65°$ is similar to the methods used in constructing the tables given in the back of this book.

Problems

A

In Problems 1–10, approximate the given number by the nth-degree Taylor polynomial of the Maclaurin's series expansion. What is a maximum value of the error involved?

1. e, $n = 2$.
2. $e^{0.2}$, $n = 2$.
3. $\sin 6°$, $n = 4$.
4. $\sin 8°$, $n = 4$.
5. $\cos 6°$, $n = 4$.
6. $\cos 8°$, $n = 4$.
7. $\sinh 1$, $n = 4$.
8. $\cosh 1$, $n = 4$.
9. $\tan 6°$, $n = 4$.
10. Arctan 0.4, $n = 4$.

In Problems 11–17, give the value of the number by using a Taylor's series about the given value of a and with the given error.

11. e; $a = 0$, $R_n < 0.0005$.
12. $e^{0.2}$; $a = 0$, $R_n < 0.00001$.
13. $\sin 10°$; $a = 0$, $R_n < 0.00005$.
14. $\sin 5°$; $a = 0$, $R_n < 0.00005$.
15. $\cos 35°$; $a = \pi/6$, $R_n < 0.0005$.
16. $\cos 35°$; $a = 0$, $R_n < 0.0005$.
17. $\sin 35°$; $a = \pi/6$, $R_n < 0.00005$.

B

In Problems 18–24, give the value of the number by using a Taylor's series about the given value of a and with the given error.

18. $\tan 10°$; $a = 0$, $R_n < 0.0001$.
19. Arctan 0.1; $a = 0$, $R_n < 0.0001$.
20. $\tan 40°$; $a = \pi/4$, $R_n < 0.0001$.
21. $\ln 0.3$; $a = 1$, $R_n < 0.1$.
22. $\ln 2$; $a = 1$, $R_n < 0.1$.
23. $\sqrt{17}$; $a = 16$, $R_n < 0.01$.
24. $\sqrt[3]{26}$; $a = 27$, $R_n < 0.01$.

19.9

Differentiation and Integration of Series

In the last section we saw one application of series—numerical approximation of transcendental functions. A second important application has to do with the differentiation and integration of power series. It is based upon the following two theorems.

Theorem 19.21

If the series $f(x) = \sum_{n=0}^{\infty} a_n(x-a)^n$ converges for $|x-a| < R$, then

$$\sum_{n=1}^{\infty} na_n(x-a)^{n-1}$$

also converges for $|x-a| < R$ and

$$f'(x) = \sum_{n=1}^{\infty} na_n(x-a)^{n-1}.$$

Theorem 19.22

If the series $f(x) = \sum_{n=0}^{\infty} a_n(x-a)^n$ converges for $|x-a| < R$, then

$$\sum_{n=0}^{\infty} \frac{a_n}{n+1}(x-a)^{n+1}$$

converges for $|x-a| < R$ and

$$\int f(x)\, dx = C + \sum_{n=0}^{\infty} \frac{a_n}{n+1}(x-a)^{n+1}$$

within its interval of convergence.

In other words, these two theorems say that a power series can be differentiated and integrated term by term within its interval of convergence to give a series that is the derivative or integral, respectively, of the function represented by the original series. We shall not prove either of these theorems, as the proofs† are beyond the scope of this book.

It might be noted that these theorems say nothing about the end points of the intervals of convergence; they say only that the series for a function, its derivative, and its integral all converge within the same open interval. In fact, it sometimes happens that, using termwise differentiation of series, we lose convergence at one or both end points; and we sometimes gain convergence at the end points when we integrate. This is illustrated in our first example.

---

† For proofs of these theorems see Angus E. Taylor and Robert Mann, *Advanced Calculus.* 2nd ed. (Lexington, Mass.: Xerox, 1972), pp. 665–669.

Example 1

Find the interval of convergence of

$$f(x) = \frac{x^2}{1 \cdot 2} + \frac{x^3}{2 \cdot 3} + \frac{x^4}{3 \cdot 4} + \cdots + \frac{x^{n+1}}{n(n+1)} + \cdots.$$

Use term-by-term differentiation to find $f'(x)$ and $f''(x)$, and determine their intervals of convergence.

$$\lim_{n \to +\infty} \left| \frac{x^{n+2}/(n+1)(n+2)}{x^{n+1}/n(n+1)} \right| = \lim_{n \to +\infty} \left| \frac{nx}{n+2} \right| = |x|.$$

Thus the series for $f(x)$ converges if $|x| < 1$. It is easily seen by the limit comparison test (comparing with $\sum 1/n^2$) that we have convergence at both end points. Thus the interval of convergence is

$$-1 \le x \le 1.$$

By term-by-term differentiation,

$$f'(x) = x + \frac{x^2}{2} + \frac{x^3}{3} + \cdots + \frac{x^n}{n} + \cdots.$$

$$\lim_{n \to +\infty} \left| \frac{x^{n+1}/(n+1)}{x^n/n} \right| = \lim_{n \to +\infty} \left| \frac{nx}{n+1} \right| = |x|.$$

We see that this series still converges if $|x| < 1$. But while it converges when $x = -1$ (alternating harmonic series), it does not when $x = 1$ (harmonic series). The interval of convergence of the series for $f'(x)$ is

$$-1 \le x < 1.$$

Finally

$$f''(x) = 1 + x + x^2 + \cdots + x^{n-1} + \cdots.$$

$$\lim_{n \to +\infty} \left| \frac{x^n}{x^{n-1}} \right| = \lim_{n \to +\infty} |x| = |x|.$$

It is easily seen that we have convergence at neither end point. The series for $f''(x)$ has the interval of convergence

$$-1 < x < 1.$$

Example 2

Use the result of Example 1 to identify $f(x)$ in finite form.

Since, as we can determine by division,

$$\frac{1}{1-x} = 1 + x + x^2 + x^3 + \cdots = f''(x),$$

we integrate to get

$$f'(x) = -\ln(1-x) + C = x + \frac{x^2}{2} + \frac{x^3}{3} + \cdots$$

(we have dropped the absolute values on $\ln |1 - x|$ because $1 - x$ is never negative in the interval of convergence). By letting $x = 0$, we find that $C = 0$. Thus

$$f'(x) = -\ln(1 - x) = x + \frac{x^2}{2} + \frac{x^3}{3} + \cdots.$$

Integrating once more, using parts on the left, we get

$$f(x) = x + (1 - x)\ln(1 - x) + C$$

$$= \frac{x^2}{1 \cdot 2} + \frac{x^3}{2 \cdot 3} + \frac{x^4}{3 \cdot 4} + \cdots.$$

Again by letting $x = 0$, we have $C = 0$ and

$$f(x) = x + (1 - x)\ln(1 - x).$$

Example 3

Use the Maclaurin's series expansion for $1/(1 + x)$ to find the series for $\ln(1 + x)$.

By dividing or by using the formula for the sum of a geometric series, we have

$$\frac{1}{1 + x} = 1 - x + x^2 - x^3 + \cdots,$$

for $|x| < 1$. Integrating, we get

$$\ln(1 + x) = C + x - \frac{x^2}{2} + \frac{x^3}{3} - \frac{x^4}{4} + \cdots.$$

When $x = 0$, we get $\ln 1 = C$, or $C = 0$. Thus

$$\ln(1 + x) = x - \frac{x^2}{2} + \frac{x^3}{3} - \frac{x^4}{4} + \cdots,$$

for $|x| < 1$. Note however that we have gained convergence at the end point $x = 1$.

We now consider one of the most important uses of series—evaluation of integrals by series methods.

Example 4

Evaluate $\int \sin x^2 \, dx$.

As noted on page 495, we have as yet had no method of integrating this expression. Let us see how it can be done by series.

$$\sin u = u - \frac{u^3}{3!} + \frac{u^5}{5!} - \frac{u^7}{7!} + \cdots.$$

By letting $u = x^2$ (see Problem 22, page 641), we have

$$\sin x^2 = x^2 - \frac{x^6}{3!} + \frac{x^{10}}{5!} - \frac{x^{14}}{7!} + \cdots.$$

Now

$$\int \sin x^2 \, dx = C + \frac{x^3}{3} - \frac{x^7}{7 \cdot 3!} + \frac{x^{11}}{11 \cdot 5!} - \frac{x^{15}}{15 \cdot 7!} + \cdots.$$

A problem that was impossible by former methods is quite simple now. Perhaps you are a little disappointed by the answer—you want the series expressed in finite terms. But, as noted on page **495**, it cannot be given in finite terms—the answer can *only* be given as a series. This is not an unusual situation. Many functions can be given only in terms of infinite series. Actually, the functions we have studied in previous chapters represent only a small class of special functions.

Recall that Theorems 19.21 and 19.22 are stated only for power series. Term-by-term differentiation and integration of other types of series can result in large variations in the convergence set. For example, it can be shown by a comparison test (compare with $\sum 1/n^2$) that

$$\sum_{n=1}^{\infty} \frac{\sin nx}{n^2}$$

converges absolutely for all x. However the series of derivatives,

$$\sum_{n=1}^{\infty} \frac{\cos nx}{n},$$

diverges when x is an integer multiple of 2π; and the series,

$$-\sum_{n=1}^{\infty} \sin nx,$$

of second derivatives converges only when x is a multiple of π. We can see that the situation here is not nearly so simple as it was for power series. Thus we confine our attention to power series.

Problems

A

In Problems 1–7, use the Maclaurin's series expansion of $f(x)$ to get the one for $g(x)$. Examine the behavior at the end points.

1. $f(x) = \dfrac{1}{1-x}$, $g(x) = \dfrac{1}{(1-x)^2}$.

2. $f(x) = \dfrac{-1}{1+x}$, $g(x) = \dfrac{1}{(1+x)^2}$.

3. $f(x) = \cos x$, $g(x) = \sin x$.

4. $f(x) = \sinh x$, $g(x) = \cosh x$.

5. $f(x) = \dfrac{1}{1+x^2}$, $g(x) = \text{Arctan } x$.

6. $f(x) = \dfrac{1}{1-x^2}$, $g(x) = \tanh^{-1} x$.

7. $f(x) = \dfrac{1}{1-x^2}$, $g(x) = \ln \dfrac{1+x}{1-x}$.

8. Use the identity $\sin^2 x = \dfrac{1 - \cos 2x}{2}$ to find the Maclaurin's series expansion of $f(x) = \sin^2 x$.

9. Use the identity $\cos^2 x = \dfrac{1 + \cos 2x}{2}$ to find the Maclaurin's series expansion of $f(x) = \cos^2 x$.

In Problems 10–13, use series to carry out the integration.

10. $\displaystyle\int \cos x^2 \, dx.$

11. $\displaystyle\int \frac{\sin x}{x} \, dx.$

12. $\displaystyle\int \frac{\cos x}{x} \, dx.$

13. $\displaystyle\int e^{-x^2} \, dx.$

In Problems 14 and 15, use series to approximate the given integral to four decimal places.

14. $\displaystyle\int_0^{0.2} \sin x^2 \, dx.$

15. $\displaystyle\int_0^{1/2} \cos x^3 \, dx.$

In Problems 16 and 17, use series to evaluate the limit.

16. $\displaystyle\lim_{x \to 0} \frac{\sin x}{x}.$

17. $\displaystyle\lim_{x \to 0} \frac{1 - \cos x}{\sin x}.$

B

In Problems 18–20, use the Maclaurin's series for $f(x)$ to get the one for $g(x)$.

18. $f(x) = \dfrac{1}{\sqrt{1 - x^2}}, \quad g(x) = \text{Arcsin } x.$

19. $f(x) = \dfrac{1}{\sqrt{1 + x^2}}, \quad g(x) = \sinh^{-1} x.$

20. $f(x) = \tan x, \quad g(x) = \ln \cos x.$

In Problems 21 and 22, use series to carry out the integration.

21. $\displaystyle\int \sqrt{1 + x^3} \, dx.$

22. $\displaystyle\int \frac{dx}{\sqrt{1 - x^3}}.$

In Problems 23 and 24, use series to approximate the given integral to three decimal places.

23. $\displaystyle\int_{1/2}^1 \frac{\cos x}{x} \, dx.$

24. $\displaystyle\int_{0.1}^{0.2} \frac{\ln (x + 1)}{x} \, dx.$

In Problems 25 and 26, use series to evaluate the limit.

25. $\displaystyle\lim_{x \to 0} \frac{e^x - 1}{\sin x}.$

26. $\displaystyle\lim_{x \to 0} \frac{e^x - e^{-x} - 2x}{x - \sin x}.$

C

27. The value of π may be determined by noting that $\text{Arctan } 1 = \pi/4$ and using the Maclaurin's series expansion for Arctan x. How many terms are needed to determine π to three decimal places? If we let $\alpha = \text{Arctan }(1/2)$ and $\beta = \text{Arctan }(1/3)$, then

$$\tan(\alpha + \beta) = \frac{\tan \alpha + \tan \beta}{1 - \tan \alpha \tan \beta} = \frac{\dfrac{1}{2} + \dfrac{1}{3}}{1 - \dfrac{1}{2} \cdot \dfrac{1}{3}} = 1,$$

or $\alpha + \beta = \pi/4$. How many terms are needed to determine α to four decimal places? β?

28. Expand $\ln \dfrac{1 + x}{1 - x}$ in a Maclaurin's series, converging for $-1 < x < 1$. Use the series to approximate $\ln 2$ to five decimal places.

29. Use series to approximate $\displaystyle\int_0^{0.1} \sin (\cos x) \, dx$ to two decimal places.

30. Show that $\dfrac{1}{1 - x} = -\displaystyle\sum_{n=1}^{\infty} x^{-n}, \quad |x| > 1.$ (*Hint:* Divide 1 by $x - 1$.)

31. Show that $1 - (1/x) = \displaystyle\sum_{n=0}^{\infty} (1 - x)^{-n}, \quad x < 0 \quad \text{or} \quad x > 2.$

Review Problems

A

1. Give the generator form of the sequence $1 \cdot 1, 2 \cdot 2, 3 \cdot 4, 4 \cdot 8, \ldots$.
2. Give the generator form of the sequence $2, -3/2, 4/6, -5/24, 6/120, -7/720, \ldots$.

In Problems 3–6, indicate whether the sequence converges or diverges. If it converges, to what number does it converge?

3. $\dfrac{3}{2}, \dfrac{5}{5}, \dfrac{7}{8}, \dfrac{9}{11}, \ldots$

4. $\dfrac{2}{3}, \dfrac{3}{7}, -\dfrac{4}{11}, \dfrac{5}{15}, \dfrac{6}{19}, -\dfrac{7}{23}, \ldots$

5. $s_n = \dfrac{n^2}{4n+5}$.

6. $s_n = (-1)^{n+1} \dfrac{2^n - 1}{3^n + 1}$.

In Problems 7–9, indicate whether the given series converges or diverges. If it converges, to what number does it converge?

7. $\dfrac{1}{4} - \dfrac{1}{16} + \dfrac{1}{64} - \cdots + \dfrac{(-1)^{n+1}}{4^n} + \cdots$.

8. $\dfrac{2}{3} + \dfrac{2}{9} + \dfrac{2}{27} + \cdots + \dfrac{2}{3^n} + \cdots$.

9. $\dfrac{2}{3} + \dfrac{2}{15} + \dfrac{2}{35} + \cdots + \dfrac{2}{4n^2 - 1} + \cdots$.

10. Express $3.521521521\ldots$ as a fraction.

In Problems 11–16, test for convergence.

11. $\displaystyle\sum_{n=1}^{\infty} \dfrac{n(n+1)}{(2n+1)(2n-1)}$.

12. $\displaystyle\sum_{n=1}^{\infty} \dfrac{1}{n+4}$.

13. $\displaystyle\sum_{n=1}^{\infty} \dfrac{n^2}{(4n^2-1)^2}$.

14. $\displaystyle\sum_{n=1}^{\infty} \dfrac{n!}{2^n}$.

15. $\displaystyle\sum_{n=2}^{\infty} \dfrac{\sqrt{n}}{n^2 - 1}$.

16. $\displaystyle\sum_{n=1}^{\infty} \dfrac{n+1}{3^n}$.

In Problems 17–19, test for convergence and absolute convergence.

17. $\displaystyle\sum_{n=1}^{\infty} \dfrac{(-1)^n n^3}{2^n}$.

18. $\displaystyle\sum_{n=1}^{\infty} \dfrac{(-1)^{n+1} n}{n+1}$.

19. $\displaystyle\sum_{n=1}^{\infty} \dfrac{(-1)^n}{n\sqrt{n+1}}$.

20. Find a Taylor's series for $f(x) = \sin x$, $a = \pi/3$. Determine the interval of convergence.

B

In Problems 21–29, test for convergence and absolute convergence.

21. $\dfrac{1}{2} + \dfrac{5}{6} + \dfrac{11}{12} + \cdots + \left(\dfrac{(n+1)^2 + 1}{n+1} - \dfrac{n^2 + 1}{n} \right) + \cdots$.

22. $\displaystyle\sum_{n=1}^{\infty} \dfrac{2^{2n} + 1}{3^n - 1}$.

23. $\displaystyle\sum_{n=1}^{\infty} \dfrac{n!}{2 \cdot 5 \cdot 8 \cdots (3n-1)}$.

24. $\displaystyle\sum_{n=2}^{\infty} \dfrac{1}{n \ln^4 n}$.

25. $\displaystyle\sum_{n=2}^{\infty} \dfrac{1}{n \ln n^4}$.

26. $\displaystyle\sum_{n=1}^{\infty} \dfrac{n}{2^n - 1}$.

27. $\displaystyle\sum_{n=1}^{\infty} \dfrac{(-1)^{n+1}(2n-1)}{n^2 + 1}$.

28. $\dfrac{1}{2} + \dfrac{1}{2^2} - 1 + \dfrac{1}{2^3} + \dfrac{1}{2^4} - \dfrac{1}{2^2} + \dfrac{1}{2^5} + \dfrac{1}{2^6} - \dfrac{1}{3^2} + \cdots$.

29. $\dfrac{1}{2} - \dfrac{2}{3} + \dfrac{2}{3} - \dfrac{4}{9} + \dfrac{3}{4} - \dfrac{8}{27} + \cdots$.

In Problems 30–33, find the interval of convergence.

30. $\displaystyle\sum_{n=1}^{\infty} \frac{3^n x^n}{\sqrt{n}}$.

31. $\displaystyle\sum_{n=0}^{\infty} \frac{(n^2 + 1)x^n}{3^n}$.

32. $\displaystyle\sum_{n=0}^{\infty} \frac{x^n}{\ln (n + 3)}$.

33. $\displaystyle\sum_{n=1}^{\infty} \frac{(-1)^n n! \, x^n}{n^4 4^n}$.

In Problems 34 and 35, find all values of x for which the series converges.

34. $\displaystyle\sum_{n=1}^{\infty} \frac{n^2 + n}{x^n}$.

35. $\displaystyle\sum_{n=1}^{\infty} (-1)^{n+1} \frac{\cos^n x}{n \cdot 2^n}$.

In Problems 36–38, expand the given function in a Taylor's series about the given value of a. Determine the interval of convergence.

36. $f(x) = \dfrac{1}{1 + x^3}, \quad a = 0.$

37. $f(x) = \dfrac{1}{1 - x}, \quad a = 2.$

38. $f(x) = \ln x, \quad a = e.$

In Problems 39–42, approximate the given number by the nth-degree Taylor polynomial of the Maclaurin's series expansion. What is the maximum value of the error involved?

39. $\sin 10°, \quad n = 5.$

40. $\sqrt{e}, \quad n = 3.$

41. $\cos 12°, \quad n = 4.$

42. $\sinh (1/2), \quad n = 5.$

43. Use the Maclaurin's series expansion for $f(x) = 1/(1 + x^3)$ to get one for

$$g(x) = \frac{x^2}{(1 + x^3)^2}.$$

44. Use the Maclaurin's series expansion for $f(x) = 1/(1 - x)$ to get one for

$$g(x) = \frac{1}{(1 - x)^3}.$$

45. Use series to evaluate $\displaystyle\int \frac{dx}{1 + x^4}.$

46. Use series to approximate $\displaystyle\int_0^1 e^{-x^3} \, dx$ to four decimal places.

47. Use series to evaluate $\displaystyle\lim_{x \to 0} \frac{\sin x - x + x^3/6}{x^5}.$

C

48. When a certain ball is dropped from a height h, it bounces to a height $h/2$. If the ball is dropped from a height of 12 ft and allowed to continue bouncing, what is the total distance traveled by the ball?

49. Test for convergence $\sum_{n=1}^{\infty} \text{Arccot } n$.

50. Test for convergence $\displaystyle\sum_{n=1}^{\infty} \frac{1}{\sqrt{n} \ln^4 n}.$

51. A spider wants to cross a bottomless gorge that is 10 ft across. He has at his disposal a very large supply of 1×2-in. cards, but he has no means of anchoring them or fastening them to each other. All he can do is place the first one on the ground extending a short distance over the gorge, place the second one on top of the first extending a short distance farther over the gorge, and so on. Is it possible for the spider to span the gorge in this way? Explain your answer. (*Hint:* Imagine that the card bridge is built from the top down.)

20

Solid Analytic Geometry

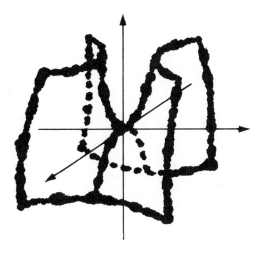

20.1

Introduction: Coordinates and Vectors in Space

So far we have dealt almost exclusively with plane figures. Let us now consider the geometry of solid figures. Forming the bridge between the algebra and geometry is the assignment of numbers to points in space, similar to the assignment we made to points in a plane; a point in space, however, is represented by a set of three numbers rather than two. We begin with a set of three lines, called *axes*, concurrent at a point (the *origin*). The only requirement is that these three lines not be *coplanar*—that is, that they not all lie in the same plane. However, we shall consider only the case in which the axes are mutually perpendicular. The three axes, labeled x, y, and z, with a scale on each, determine a set of three numbers, called *coordinates*, associated uniquely with any point in space. Since any pair of intersecting lines determines a plane, the three pairs of axes determine three *coordinate planes*, which we shall call the *xy plane*, the *xz plane*, and the *yz plane* (see Figure 20.1(a)). The *x coordinate* of a point P in space is the number associated with the point on the x axis that is the intersection of the x axis and the plane through P parallel to the yz plane. The y and z *coordinates* of P are defined in a similar fashion by considering the points of intersection of the y and z axes with planes through P parallel to the xz and xy coordinate planes, respectively (see Figure 20.1(b)).

The coordinate planes separate space into eight *octants*. Although we shall not number all of them, the one in which all three coordinates are positive is called the *first octant*. Note that points of the xy plane have z coordinate 0, points of the xz plane have y coordinate 0, and points of the yz plane have x coordinate 0. Similarly, points of the x axis have y and z coordinates 0, and so on. Of course, the origin has all of these coordinates 0.

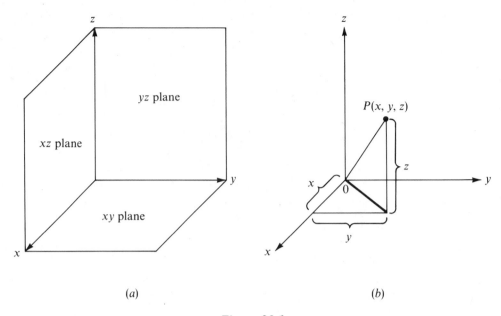

Figure 20.1

The two basic geometric representations of the axes are given in Figure 20.2; part (a) shows a *right-hand system*, while part (b) shows a *left-hand system*. Graphs of equations in the two systems are mirror images of each other. Since we shall normally represent space by a right-hand system, the axes will usually appear in the positions indicated in part (a). This is sometimes represented by the right-hand rule illustrated in Figure 20.2(c). If the index and second fingers of the right hand point in the direction of the x and y axes, respectively, then the thumb points in the direction of the z axis.

Many of the formulas of solid analytic geometry are simple extensions of plane analytic geometry. The one that follows is an example.

Theorem 20.1

The distance between two points (x_1, y_1, z_1) and (x_2, y_2, z_2) is

$$d = \sqrt{(x_1 - x_2)^2 + (y_1 - y_2)^2 + (z_1 - z_2)^2}.$$

The proof, which requires a double application of the Theorem of Pythagoras, is left to the student. Of course, if the line joining the two points is on or parallel to one of the coordinate planes, at least one of the three terms of this formula is zero, and it reduces to the plane case. Similarly, if the line joining the points is on or parallel to one of the axes, at least two terms are zero and the distance is the absolute value of the difference between the coordinates of the remaining pair.

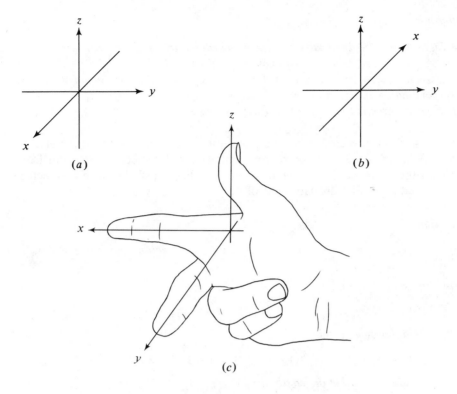

(a)

(b)

(c)

Figure 20.2

Example 1

Find the distance between $(1, -2, 5)$ and $(-3, 6, 4)$.

$$d = \sqrt{(x_1 - x_2)^2 + (y_1 - y_2)^2 + (z_1 - z_2)^2}$$
$$= \sqrt{(1 + 3)^2 + (-2 - 6)^2 + (5 - 4)^2}$$
$$= \sqrt{16 + 64 + 1}$$
$$= \sqrt{81}$$
$$= 9.$$

Vectors in three-dimensional space may be handled in much the same way as vectors in the plane. Vectors themselves, the sum and difference of two vectors, the absolute value of a vector, and scalar multiple of a vector are defined in the same way as they were in Chapter 8; and Theorem 8.2 (see page 228) holds for vectors in space as well as for vectors in the plane. The following theorems and definitions are the three-dimensional analogs of theorems and definitions of Chapter 8. The proofs of the theorems are simple extensions of the corresponding theorems in two dimensions.

Definition

A **vector** **v** in three-dimensional space is an ordered triple of real numbers, $\mathbf{v} = (x, y, z)$, where x, y, and z are called the **components** of the vector.

The vectors $\mathbf{v}_1 = (x_1, y_1, z_1)$ and $\mathbf{v}_2 = (x_2, y_2, z_2)$ are **equal** if and only if corresponding components are equal: $x_1 = x_2$, $y_1 = y_2$, and $z_1 = z_2$.

The **zero vector** is the vector $\mathbf{0} = (0, 0, 0)$.

As in the plane, a vector (x, y, z) can be thought of as an ordered triple of numbers or as a point in space, and can be represented as a directed line segment from the origin to the point (x, y, z). Again, we do not distinguish between a directed line segment and a parallel displacement of it.

Definition

If $\mathbf{v}_1 = (x_1, y_1, z_1)$ and $\mathbf{v}_2 = (x_2, y_2, z_2)$ are vectors and k is a scalar, then

(a) the sum of $\mathbf{v}_1$ and $\mathbf{v}_2$ is

$$\mathbf{v}_1 + \mathbf{v}_2 = (x_1, y_1, z_1) + (x_2, y_2, z_2) = (x_1 + x_2, y_1 + y_2, z_1 + z_2);$$

(b) the difference, $\mathbf{v}_1$ minus $\mathbf{v}_2$, is

$$\mathbf{v}_1 - \mathbf{v}_2 = (x_1, y_1, z_1) - (x_2, y_2, z_2) = (x_1 - x_2, y_1 - y_2, z_1 - z_2);$$

(c) the **absolute value** or **length** of $\mathbf{v}_1$ is

$$|\mathbf{v}_1| = |(x_1, y_1, z_1)| = \sqrt{x_1^2 + y_1^2 + z_1^2}; \quad \text{and}$$

(d) the scalar multiple of $\mathbf{v}_1$ by k is

$$k\mathbf{v}_1 = k(x_1, y_1, z_1) = (kx_1, ky_1, kz_1).$$

The geometric interpretations of the sum and difference of vectors, absolute value, and scalar multiple are the same as in the plane. The absolute value of $\mathbf{v}_1 = (x_1, y_1, z_1)$ is the length of the segment from the origin to (x_1, y_1, z_1). The scalar multiple $k\mathbf{v}_1$ is a vector of length $|k||\mathbf{v}_1|$ and with direction the same as that of $\mathbf{v}_1$ when $k > 0$ and opposite that of $\mathbf{v}_1$ when $k < 0$ ($\mathbf{v}_1 \neq \mathbf{0}$). The sum and difference of two vectors $\mathbf{v}_1$ and $\mathbf{v}_2$ are the diagonals of the parallelogram formed by $\mathbf{v}_1$, $\mathbf{v}_2$, and translates of $\mathbf{v}_1$ and $\mathbf{v}_2$, as shown in Figure 20.3.

Definition

The vectors $\mathbf{i} = (1, 0, 0)$, $\mathbf{j} = (0, 1, 0)$, and $\mathbf{k} = (0, 0, 1)$ are called **basis** vectors.

Theorem 20.2

Each vector in space $\mathbf{v} = (a, b, c)$ can be written uniquely in the form

$$\mathbf{v} = a\mathbf{i} + b\mathbf{j} + c\mathbf{k}.$$

The numbers a, b, and c are called the first, second, and third component, respectively, of $\mathbf{v}$.

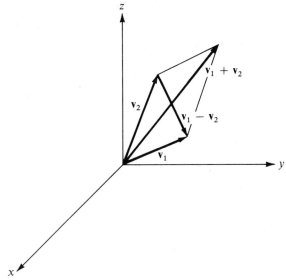

Figure 20.3

Theorem 20.3

*If $\overrightarrow{AB}$, where $A = (x_1, y_1, z_1)$ and $B = (x_2, y_2, z_2)$, represents a vector **v** in space, then*

$$\mathbf{v} = (x_2 - x_1, y_2 - y_1, z_2 - z_1).$$

Example 2

Give in component form the vector **v** that is represented by $\overrightarrow{AB}$ where $A = (4, 3, -1)$ and $B = (-1, 2, -3)$.

By Theorem 20.3,

$$\mathbf{v} = (-1 - 4)\mathbf{i} + (2 - 3)\mathbf{j} + (-3 + 1)\mathbf{k} = -5\mathbf{i} - \mathbf{j} - 2\mathbf{k}.$$

Another easy extension from two dimensions is the point-of-division formula.

Theorem 20.4

If $P_1 = (x_1, y_1, z_1)$ and $P_2 = (x_2, y_2, z_2)$ and P is a point such that $r = \overline{P_1P}/\overline{P_1P_2}$, then the coordinates of P are

$$x = x_1 + r(x_2 - x_1),$$
$$y = y_1 + r(y_2 - y_1),$$
$$z = z_1 + r(z_2 - z_1).$$

Again the proof is similar to the one for the two-dimensional case (see Theorem 8.3), and it is left to the student.

Example 3

Find the point $1/3$ of the way from $(-2, 4, 1)$ to $(4, 1, 7)$.

$$x = x_1 + r(x_2 - x_1) = -2 + \frac{1}{3}(4 + 2) = 0,$$

$$y = y_1 + r(y_2 - y_1) = 4 + \frac{1}{3}(1 - 4) = 3,$$

$$z = z_1 + r(z_2 - z_1) = 1 + \frac{1}{3}(7 - 1) = 3.$$

The desired point is $(0, 3, 3)$.

The following theorem is a direct consequence of the point-of-division formulas.

Theorem 20.5

If $P_1 = (x_1, y_1, z_1)$ and $P_2 = (x_2, y_2, z_2)$, then the coordinates of the midpoint of the segment P_1P_2 are

$$x = \frac{x_1 + x_2}{2}, \quad y = \frac{y_1 + y_2}{2}, \quad z = \frac{z_1 + z_2}{2}.$$

Example 4

Find the end points of the representative $\overrightarrow{AB}$ of $\mathbf{v}$ if $\mathbf{v} = 2\mathbf{i} - 4\mathbf{j} + \mathbf{k}$ and $(2, -3, 5)$ is the midpoint of AB.

If $A = (x_1, y_1, z_1)$ and $B = (x_2, y_2, z_2)$, then, by Theorem 20.3,

$$x_2 - x_1 = 2, \qquad y_2 - y_1 = -4, \qquad z_2 - z_1 = 1.$$

By Theorem 20.5,

$$\frac{x_1 + x_2}{2} = 2, \qquad \frac{y_1 + y_2}{2} = -3, \qquad \frac{z_1 + z_2}{2} = 5.$$

Solving simultaneously, we have

$$A = \left(1, -1, \frac{9}{2}\right), \qquad B = \left(3, -5, \frac{11}{2}\right).$$

Definition

The **angle** between two nonzero vectors $\mathbf{u}$ and $\mathbf{v}$ is the smaller angle between the representatives of $\mathbf{u}$ and $\mathbf{v}$ having their tails at the origin. If the representatives lie along a straight line, then $\mathbf{v}$ is a nonzero scalar multiple of $\mathbf{u}$,

$$\mathbf{v} = k\mathbf{u}, \quad k \neq 0.$$

In this case, the angle between $\mathbf{u}$ and $\mathbf{v}$ is $0°$ if $k > 0$ and $180°$ if $k < 0$ (see Figure 20.4).

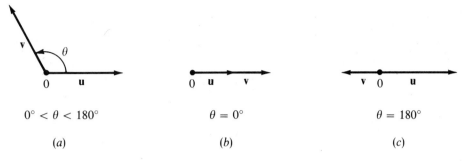

$0° < \theta < 180°$ $\theta = 0°$ $\theta = 180°$

(a)   *(b)*   *(c)*

Figure 20.4

Theorem 20.6

If $\mathbf{u} = a_1\mathbf{i} + b_1\mathbf{j} + c_1\mathbf{k}$ *and* $\mathbf{v} = a_2\mathbf{i} + b_2\mathbf{j} + c_2\mathbf{k}$ $(\mathbf{u} \neq 0$ *and* $\mathbf{v} \neq 0)$ *and if* θ *is the angle between them, then*

$$\cos \theta = \frac{a_1 a_2 + b_1 b_2 + c_1 c_2}{|\mathbf{u}|\,|\mathbf{v}|}.$$

Proof

By the law of cosines (see Figure 20.5),

$$|\mathbf{v} - \mathbf{u}|^2 = |\mathbf{u}|^2 + |\mathbf{v}|^2 - 2|\mathbf{u}|\,|\mathbf{v}| \cos \theta.$$

Since

$$\mathbf{v} - \mathbf{u} = (a_2 - a_1)\mathbf{i} + (b_2 - b_1)\mathbf{j}$$
$$+ (c_2 - c_1)\mathbf{k},$$

we have

$$(a_2 - a_1)^2 + (b_2 - b_1)^2 + (c_2 - c_1)^2$$
$$= a_1^2 + b_1^2 + c_1^2 + a_2^2 + b_2^2 + c_2^2$$
$$- 2|\mathbf{u}|\,|\mathbf{v}| \cos \theta.$$

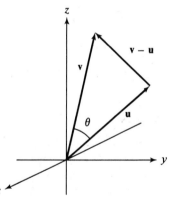

Figure 20.5

It follows that

$$|\mathbf{u}|\,|\mathbf{v}| \cos \theta = a_1 a_2 + b_1 b_2 + c_1 c_2,$$

and

$$\cos \theta = \frac{a_1 a_2 + b_1 b_2 + c_1 c_2}{|\mathbf{u}|\,|\mathbf{v}|}.$$

Compare this proof with the proof of Theorem 8.4 on page 232.

Definition

If $\mathbf{u} = a_1\mathbf{i} + b_1\mathbf{j} + c_1\mathbf{k}$ *and* $\mathbf{v} = a_2\mathbf{i} + b_2\mathbf{j} + c_2\mathbf{k},$ *then the* **dot product** *(scalar product, inner product) of* $\mathbf{u}$ *and* $\mathbf{v}$ *is*

$$\mathbf{u} \cdot \mathbf{v} = a_1 a_2 + b_1 b_2 + c_1 c_2.$$

Theorems 8.5–8.7 still hold for three-dimensional vectors. They are restated here for convenience.

Theorem 20.7

If **u** *and* **v** *are vectors and* θ *is the angle between them, then*

$$\mathbf{u} \cdot \mathbf{v} = |\mathbf{u}||\mathbf{v}| \cos \theta, \quad and \quad \mathbf{v} \cdot \mathbf{v} = |\mathbf{v}|^2.$$

Theorem 20.8

The vectors **u** *and* **v** *(not both* **0***) are orthogonal (perpendicular) if and only if* $\mathbf{u} \cdot \mathbf{v} = 0$ *(the zero vector is taken to be orthogonal to every vector).*

Theorem 20.9

If **u**, **v**, *and* **w** *are vectors, then*

$$\mathbf{u} \cdot \mathbf{v} = \mathbf{v} \cdot \mathbf{u}, \quad and \quad (\mathbf{u} + \mathbf{v}) \cdot \mathbf{w} = \mathbf{u} \cdot \mathbf{w} + \mathbf{v} \cdot \mathbf{w}.$$

Example 5

Given $\mathbf{u} = 2\mathbf{i} + \mathbf{j} - 3\mathbf{k}$ and $\mathbf{v} = \mathbf{i} - 2\mathbf{j} - \mathbf{k}$, find $\mathbf{u} + \mathbf{v}, \mathbf{u} - \mathbf{v}$, and $\mathbf{u} \cdot \mathbf{v}$.

$$\mathbf{u} + \mathbf{v} = (2 + 1)\mathbf{i} + (1 - 2)\mathbf{j} + (-3 - 1)\mathbf{k} = 3\mathbf{i} - \mathbf{j} - 4\mathbf{k},$$
$$\mathbf{u} - \mathbf{v} = (2 - 1)\mathbf{i} + (1 + 2)\mathbf{j} + (-3 + 1)\mathbf{k} = \mathbf{i} + 3\mathbf{j} - 2\mathbf{k},$$
$$\mathbf{u} \cdot \mathbf{v} = (2)(1) + (1)(-2) + (-3)(-1) = 3.$$

Example 6

Determine whether the vectors $\mathbf{u} = 3\mathbf{i} + \mathbf{j} - 2\mathbf{k}$ and $\mathbf{v} = 2\mathbf{i} - 4\mathbf{j} + \mathbf{k}$ are orthogonal.

$$\mathbf{u} \cdot \mathbf{v} = (3)(2) + (1)(-4) + (-2)(1) = 0.$$

Thus **u** and **v** are orthogonal.

Problems

A

In Problems 1–4, find the distance between the pair of points given.

1. $(4, -2, 1)$, $(2, 2, -3)$.
2. $(2, 6, 0)$, $(-3, 1, 3)$.
3. $(-2, 5, 1)$, $(-2, 8, 4)$.
4. $(2, 0, 5)$, $(4, 0, 4)$.

In Problems 5–8, give in component form the vector **v** that is represented by $\overrightarrow{AB}$. Sketch $\overrightarrow{AB}$.

5. $A = (3, -2, 4)$, $B = (5, 4, -1)$.
6. $A = (1, 3, -4)$, $B = (-3, 2, 3)$.
7. $A = (2, -3, 8)$, $B = (2, 5, 2)$.
8. $A = (6, 0, -1)$, $B = (2, -3, 2)$.

In Problems 9–12, give the unit vector in the direction of **v**.

9. $\mathbf{v} = (1, -2, 2)$.
10. $\mathbf{v} = (2, -1, 4)$.
11. $\mathbf{v} = 3\mathbf{i} - 4\mathbf{k}$.
12. $\mathbf{v} = 3\mathbf{i} - 2\mathbf{j} + \mathbf{k}$.

In Problems 13–16, find the midpoint of the segment AB.

13. $A = (4, 3, 5),\quad B = (-2, -1, 2)$.
15. $A = (4, 3, -1),\quad B = (4, 8, -3)$.
14. $A = (6, -2, 5),\quad B = (-3, 6, 7)$.
16. $A = (-3, 2, 0),\quad B = (3, 6, 5)$.

In Problems 17–20, find the point P such that $\overline{AP}/\overline{AB} = r$.

17. $A = (5, 2, 3),\quad B = (-5, 7, -2),\quad r = 2/5$.
18. $A = (6, 6, 3),\quad B = (3, -3, 0),\quad r = 1/3$.
19. $A = (3, 1, 5),\quad B = (-3, 4, 2),\quad r = 2$.
20. $A = (-2, 5, 1),\quad B = (4, -1, 2),\quad r = 3/2$.

In Problems 21–24, find $\mathbf{u} + \mathbf{v}$, $\mathbf{u} - \mathbf{v}$, and $\mathbf{u} \cdot \mathbf{v}$. Indicate whether $\mathbf{u}$ and $\mathbf{v}$ are orthogonal.

21. $\mathbf{u} = (3, 1, -4),\quad \mathbf{v} = (2, 6, 3)$.
23. $\mathbf{u} = 4\mathbf{i} + 3\mathbf{j} - \mathbf{k},\quad \mathbf{v} = \mathbf{i} + 2\mathbf{j} + 3\mathbf{k}$.
22. $\mathbf{u} = (2, -1, 3),\quad \mathbf{v} = (2, 4, 1)$.
24. $\mathbf{u} = 2\mathbf{i} - \mathbf{j} + 6\mathbf{k},\quad \mathbf{v} = 2\mathbf{i} - 2\mathbf{j} - \mathbf{k}$.

In Problems 25–28, find the angle θ between the vectors.

25. $\mathbf{u} = (1, 1, 2),\quad \mathbf{v} = (2, -1, 1)$.
27. $\mathbf{u} = 2\mathbf{i} + 4\mathbf{j} + 4\mathbf{k},\quad \mathbf{v} = 4\mathbf{i} - 3\mathbf{k}$.
26. $\mathbf{u} = (2, -2, 1),\quad \mathbf{v} = (-1, 4, -2)$.
28. $\mathbf{u} = 5\mathbf{i} - \mathbf{j} + 3\mathbf{k},\quad \mathbf{v} = 4\mathbf{i} + 5\mathbf{j} - 2\mathbf{k}$.

B

29. Given $A = (5, -2, 3)$, $P = (6, 0, 0)$, and $\overline{AP}/\overline{AB} = 1/3$, find B.
30. Given $B = (-4, 14, 4)$, $P = (-1, 8, -4)$, and $\overline{AP}/\overline{AB} = 2/5$, find A.
31. Given $B = (6, 0, 9)$, $P = (4, 1, 6)$, and $\overline{AP}/\overline{AB} = 3/4$, find A.
32. Given $A = (5, 3, -2)$, $P = (1, 5, 2)$, and $\overline{AP}/\overline{AB} = 2/3$, find B.

In Problems 33–36, find the unknown quantity.

33. $A = (5, 1, 0),\quad B = (1, y, 2),\quad \overline{AB} = 6$.
34. $A = (-2, 4, 3),\quad B = (x, -4, 2),\quad \overline{AB} = 9$.
35. $A = (x, 4, -2),\quad B = (-x, -6, 3),\quad \overline{AB} = 15$.
36. $A = (x, x, 5),\quad B = (-1, -2, 0),\quad \overline{AB} = 5\sqrt{2}$.

37. The point $(-1, 5, 2)$ is a distance 6 from the midpoint of the segment joining $(1, 3, 2)$ and $(x, -1, 6)$. Find x.
38. The point $(1, -2, 9)$ is a distance $5\sqrt{5}$ from the midpoint of the segment joining $(1, y, 2)$ and $(5, -1, 6)$. Find y.

In Problems 39 and 40, find the end points of the representative of $\mathbf{v}$ from the given information.

39. $\mathbf{v} = 4\mathbf{i} - 2\mathbf{j} + \mathbf{k}$; $(2, 5, -1)$ is the midpoint of AB.
40. $\mathbf{v} = 6\mathbf{i} + \mathbf{j} - 4\mathbf{k}$; $(3, 2, -5)$ is the midpoint of AB.

C

41. Prove Theorem 20.1.   42. Prove Theorem 20.2.   43. Prove Theorem 20.3.
44. Prove Theorem 20.4.   45. Prove Theorem 20.5.   46. Prove Theorem 20.9.
47. Prove that $\mathbf{u} \cdot \mathbf{u} = |\mathbf{u}|^2$.
48. Prove that $(\mathbf{u} + \mathbf{v}) \cdot (\mathbf{u} + \mathbf{v}) = |\mathbf{u}|^2 + 2\mathbf{u} \cdot \mathbf{v} + |\mathbf{v}|^2$.
49. Prove the triangle inequality $|\mathbf{u} + \mathbf{v}| \le |\mathbf{u}| + |\mathbf{v}|$.
50. Prove $|\mathbf{u} - \mathbf{v}| \ge |\,|\mathbf{u}| - |\mathbf{v}|\,|$.

20.2

Direction Angles, Cosines, and Numbers; Projections

The vectors in the same (or opposite) direction as a nonzero vector **v** are the nonzero scalar multiples of **v**. Thus, a direction for **v** is determined by any triple of numbers proportional to **v**.

Definition

If **v** *is a nonzero vector, then the components of any nonzero scalar multiple of* **v** *form a set of* **direction numbers** *for* **v**.
The components of **v**/|**v**| *are the* **direction cosines** *of* **v**.
No direction is assigned to the zero vector.

Note from Figure 20.6(b) that, since **v**/|**v**| is a vector of unit length, if **v**/|**v**| = (l, m, n), then

$$l = \cos \alpha, \qquad m = \cos \beta, \qquad n = \cos \gamma.$$

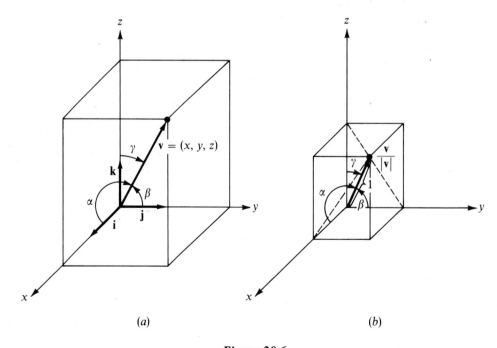

(a)   (b)

Figure 20.6

Definition

*The set $\{\alpha, \beta, \gamma\}$ is the set of **direction angles** for* **v**, *where α is the angle between* **v** *and* **i**, *β is the angle between* **v** *and* **j**, *and γ is the angle between* **v** *and* **k**.

The angle between two vectors was defined in the previous section. Note again that this is not a directed angle; in fact, we have given no convention for positive and negative angles in space. Thus, direction angles are never negative and never greater than 180°.

Since the length of $\mathbf{v}/|\mathbf{v}|$ is 1, we have the following theorem.

Theorem 20.10

If $\{l, m, n\}$ is a set of direction cosines for a vector, then

$$l^2 + m^2 + n^2 = 1.$$

Example 1

Find direction numbers, direction cosines, and direction angles for $\mathbf{v} = (1, 2, -2)$.

Since $1 \cdot \mathbf{v} = \mathbf{v}$,   then $\{1, 2, -2\}$ is a set of direction numbers for $\mathbf{v}$.

$$\frac{\mathbf{v}}{|\mathbf{v}|} = \left(\frac{1}{3}, \frac{2}{3}, -\frac{2}{3}\right), \quad \text{so} \quad l = \frac{1}{3}, \quad m = \frac{2}{3}, \quad n = -\frac{2}{3};$$

$\{1/3, 2/3, -2/3\}$ is the set of direction cosines.

$$\cos \alpha = \frac{1}{3} \qquad\qquad \cos \beta = \frac{2}{3} \qquad\qquad \cos \gamma = -\frac{2}{3}$$

$$\alpha = \text{Arccos}\,\frac{1}{3} \qquad \beta = \text{Arccos}\,\frac{2}{3} \qquad \gamma = \text{Arccos}\left(-\frac{2}{3}\right)$$

$\left\{\text{Arccos}\,\dfrac{1}{3}, \text{Arccos}\,\dfrac{2}{3}, \text{Arccos}\left(-\dfrac{2}{3}\right)\right\}$ is the set of direction angles for **v** (see Figure 20.7).

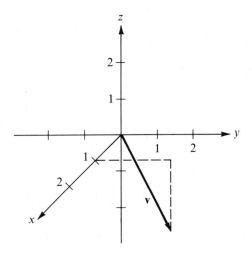

Figure 20.7

Example 2

Given that a vector **v** has direction numbers $\{4, 1, -2\}$ and is directed upward, find its direction cosines.

The vector $(4, 1, -2)$ is a scalar multiple of $\mathbf{v}/|\mathbf{v}| = (l, m, n)$, so there is a constant k such that

$$kl = 4, \qquad km = 1, \qquad \text{and} \qquad kn = -2.$$
$$k^2 l^2 + k^2 m^2 + k^2 n^2 = 16 + 1 + 4$$
$$k^2 (l^2 + m^2 + n^2) = 21.$$

Since $l^2 + m^2 + n^2 = 1$, then $k^2 = 21$ and $k = \pm\sqrt{21}$. But **v** is directed upward, so $\gamma < 90°$ and $n > 0$. Thus $k = -\sqrt{21}$.

$$l = -\frac{4}{\sqrt{21}}, \qquad m = -\frac{1}{\sqrt{21}}, \qquad n = \frac{2}{\sqrt{21}}.$$

The methods of solution in Examples 1 and 2 can be used to prove the following theorems.

Theorem 20.11

The vector $\mathbf{v} = (a, b, c)$ has direction numbers $\{a, b, c\}$.

Theorem 20.12

A vector with direction numbers $\{a, b, c\}$ has direction cosines

$$l = \frac{a}{\pm\sqrt{a^2 + b^2 + c^2}}, \qquad m = \frac{b}{\pm\sqrt{a^2 + b^2 + c^2}}, \qquad n = \frac{c}{\pm\sqrt{a^2 + b^2 + c^2}}.$$

(The ambiguity in sign can be resolved if it is known, for example, whether the vector points up or down, left or right, or forward or backward.)

Definition

*The statement "A vector **v** is directed along a line l," means that some parallel translate of **v** lies on l.*

Definition

*A set of direction angles, cosines, or numbers for a line l is any set of direction angles, cosines, or numbers, respectively, for any vector **v** directed along l.*

Note that every line has two sets of direction angles and two sets of direction cosines, corresponding to the two possible directions on the line. It is easily seen that if $\{l, m, n\}$ is one set of direction cosines for a line, then the other is $\{-l, -m, -n\}$.

Example 3

Find the two sets of direction cosines and direction angles for a line that has $\{1, 2, 2\}$ as a set of direction numbers.

By Theorem 20.12, $l = \pm 1/3$, $m = \pm 2/3$, $n = \pm 2/3$. Thus, the two possible sets of direction cosines are $\{1/3, 2/3, 2/3\}$ and $\{-1/3, -2/3, -2/3\}$, and they give approximate direction angles $\{71°, 48°, 48°\}$ and $\{109°, 132°, 132°\}$, respectively.

Example 4

Suppose a line has direction numbers $\{2, -4, 1\}$ and contains the point $(1, 3, 4)$. Find another point on the line.

Since the line has direction numbers $\{2, -4, 1\}$, the vector $(2, -4, 1)$ has a representative $\overrightarrow{AB}$ on the line with tail at $A = (1, 3, 4)$. Let $B = (x, y, z)$. By Theorem 20.3,

$$x - 1 = 2, \qquad y - 3 = -4, \qquad z - 4 = 1,$$

so $B = (3, -1, 5)$ is another point in the line.

Note that if $A = (x_1, y_1, z_1)$ and $B = (x_2, y_2, z_2)$ are points on a line l, then $\overrightarrow{AB}$ lies on l and $\{x_2 - x_1, y_2 - y_1, z_2 - z_1\}$ is a set of direction numbers for l.

It is clear that if two lines are parallel and directed the same way, they must have the same set of direction angles and, thus, the same set of direction cosines. If they are parallel and have opposite directions, their direction angles are supplementary, and corresponding direction cosines are opposite in sign. Thus, any set of direction numbers for one line is proportional to a set of direction numbers for the other. Furthermore, this reasoning can be reversed to show that if two distinct lines have proportional sets of direction numbers, they are parallel.

Theorem 20.13

Two lines are parallel if and only if sets of direction numbers for the two lines are proportional.

In general, two lines in space are *skew*; that is, they neither intersect nor are parallel. For such lines we make the following definition.

Definition

*Two skew lines are **perpendicular** if and only if there are parallel translates of the lines that intersect and are perpendicular.*

Suppose that lines l_1 and l_2 have direction numbers $\{a_1, b_1, c_1\}$ and (a_2, b_2, c_2), respectively. Then vectors $\mathbf{v}_1$ and $\mathbf{v}_2$ directed along lines l_1 and l_2, respectively, may be represented by

$$\mathbf{v}_1 = a_1\mathbf{i} + b_1\mathbf{j} + c_1\mathbf{k}, \qquad \mathbf{v}_2 = a_2\mathbf{i} + b_2\mathbf{j} + c_2\mathbf{k}.$$

By Theorem 20.8, $\mathbf{v}_1$ and $\mathbf{v}_2$ are orthogonal if and only if

$$\mathbf{v}_1 \cdot \mathbf{v}_2 = a_1 a_2 + b_1 b_2 + c_1 c_2 = 0.$$

This gives the following theorem for perpendicularity of lines.

Theorem 20.14

Two lines with direction numbers $\{a_1, b_1, c_1\}$ and $\{a_2, b_2, c_2\}$ are perpendicular if and only if

$$a_1 a_2 + b_1 b_2 + c_1 c_2 = 0.$$

In later sections of this book we will need the *projection* of one vector upon another. The definition and properties are formally the same as in the plane.

Definition

Suppose the nonzero vectors **u** *and* **v** *are represented by the directed line segments* $\overrightarrow{OA}$ *and* $\overrightarrow{OB}$, *respectively. Then the* **projection** *of* **u** *on* **v** *is the vector* **w** *represented by* $\overrightarrow{OC}$, *where C is on the line OB and AC $\perp$ OB (see Figure 20.8).*

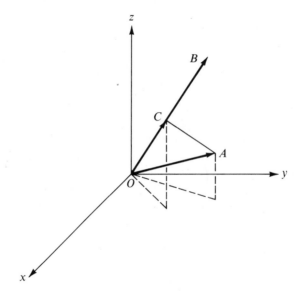

Figure 20.8

Theorem 20.15

If **w** *is the projection of* **u** *on* **v**, *then*

$$|\mathbf{w}| = \frac{|\mathbf{u} \cdot \mathbf{v}|}{|\mathbf{v}|} \quad and \quad \mathbf{w} = \left(\frac{\mathbf{u} \cdot \mathbf{v}}{|\mathbf{v}|} \right) \frac{\mathbf{v}}{|\mathbf{v}|}.$$

The proof of Theorem 20.15 is left to the student (see Problem 32).

Example 5

Find the projection **w** of $\mathbf{u} = 2\mathbf{i} - \mathbf{j} + \mathbf{k}$ upon $\mathbf{v} = 3\mathbf{i} + \mathbf{j} - 4\mathbf{k}$.

$$\mathbf{u} \cdot \mathbf{v} = (2)(3) + (-1)(1) + (1)(-4) = 1 \quad and \quad |\mathbf{v}| = \sqrt{3^2 + 1^2 + (-4)^2} = \sqrt{26}.$$

Thus

$$\mathbf{w} = \left(\frac{\mathbf{u} \cdot \mathbf{v}}{|\mathbf{v}|}\right) \frac{\mathbf{v}}{|\mathbf{v}|}$$

$$= \frac{1}{\sqrt{26}} \frac{3\mathbf{i} + \mathbf{j} - 4\mathbf{k}}{\sqrt{26}}$$

$$= \frac{3}{26}\mathbf{i} + \frac{1}{26}\mathbf{j} - \frac{2}{13}\mathbf{k}.$$

Problems

A

In Problems 1–4, find the projection of $\mathbf{u}$ upon $\mathbf{v}$.

1. $\mathbf{u} = \mathbf{i} - 2\mathbf{j} + 4\mathbf{k}$, $\mathbf{v} = 2\mathbf{j} + 3\mathbf{k}$.   2. $\mathbf{u} = 4\mathbf{i} - \mathbf{j} - \mathbf{k}$, $\mathbf{v} = \mathbf{i} + \mathbf{j} + \mathbf{k}$.
3. $\mathbf{u} = 2\mathbf{i} + \mathbf{j}$, $\mathbf{v} = \mathbf{j} - 2\mathbf{k}$.   4. $\mathbf{u} = 4\mathbf{i} + \mathbf{j} - 2\mathbf{k}$, $\mathbf{v} = \mathbf{i} - 2\mathbf{j} + \mathbf{k}$.

In Problems 5–10, find the set of direction angles for the vector described.

5. Direction numbers $\{4, -2, 4\}$;   directed to the right of the xz plane.
6. Direction numbers $\{4, 1, 8\}$;   directed to the right of the xz plane.
7. Direction numbers $\{2, -1, -3\}$;   directed above the xy plane.
8. Direction numbers $\{1, 4, -2\}$;   directed behind the yz plane.
9. Direction numbers $\{1, -1, 0\}$;   directed to the right of the xz plane.
10. Direction numbers $\{1, 1, 1\}$;   directed behind the yz plane.

In Problems 11–16, find a set of direction numbers for the lines containing the two given points.

11. $(1, 4, 3)$ and $(5, 2, -1)$.   12. $(2, 0, -4)$ and $(-1, 2, 3)$.
13. $(2, 2, 1)$ and $(0, 0, 3)$.   14. $(3, 6, -2)$ and $(-1, 3, 4)$.
15. $(0, 0, 0)$ and $(5, 2, -3)$.   16. $(-1, 4, 5)$ and $(3, -4, 0)$.

In Problems 17–26, two lines are described by a pair of points on each. Indicate whether the lines are parallel, perpendicular, coincident, or none of these.

17. $(2, 1, 5), (3, 3, -1)$; $(4, 2, 10), (1, -4, 5)$.
18. $(3, 4, 1), (4, 8, -1)$; $(2, 3, -5), (0, -5, -1)$.
19. $(4, 2, -1), (7, 6, 2)$; $(5, 10, 3), (-4, -2, -6)$.
20. $(4, 1, -4), (3, 2, 1)$; $(4, 1, -4), (11, 3, -3)$.
21. $(4, 5, 1), (3, 2, -4)$; $(4, 1, 2), (5, -1, 3)$.
22. $(2, 1, 4), (4, -3, 12)$; $(1, 3, 0), (6, -7, 20)$.
23. $(3, 1, 4), (4, 3, 3)$; $(5, 5, 2), (0, -5, 7)$.
24. $(2, 3, 1), (4, -2, 2)$; $(1, 0, 3), (3, -3, 1)$.
25. $(4, 4, -3), (1, 3, -1)$; $(2, 1, 5), (8, 3, 1)$.
26. $(2, 1, 3), (5, -1, 1)$; $(3, 4, -1), (5, 3, 3)$.

27. Give the direction angles and direction cosines for the coordinate axes with their usual directions.

B

28. Prove that the projection of $\mathbf{u} + \mathbf{v}$ on $\mathbf{w}$ is the sum of the projection of $\mathbf{u}$ on $\mathbf{w}$ and the projection of $\mathbf{v}$ on $\mathbf{w}$.
29. Let $\mathbf{u}$, $\mathbf{v}$, and $\mathbf{w}$ be noncoplanar unit vectors and let $\mathbf{v}_1$ be the projection of $\mathbf{v}$ on $\mathbf{u}$. Show that $\mathbf{u}$ is orthogonal to $\mathbf{v} - \mathbf{v}_1$.

30. (Problem 29 continued)   Let

$$\mathbf{v}_2 = \frac{(\mathbf{v} - \mathbf{v}_1)}{|\mathbf{v} - \mathbf{v}_1|},$$

and set

$$\mathbf{w}_1 = \mathbf{w} - (\mathbf{u} \cdot \mathbf{w})\mathbf{u} - (\mathbf{v}_2 \cdot \mathbf{w})\mathbf{v}_2.$$

Show that $\mathbf{u}$, $\mathbf{v}_2$, and $\mathbf{w}_2 = \mathbf{w}_1/|\mathbf{w}_1|$ are mutually orthogonal unit vectors.

31. Given $\mathbf{u} = (1, 0, 0)$, $\mathbf{v} = (1/3, 2/3, 2/3)$,   and $\mathbf{w} = (2/3, -1/3, 2/3)$, use the procedure of Problems 29 and 30 to obtain three mutually orthogonal unit vectors $\mathbf{u}$, $\mathbf{v}_2$, and $\mathbf{w}_2$.

C

32. Prove Theorem 20.15.

20.3

The Line

Recall that in Chapter 13 (pages 397–398) we derived a parametric representation for a line containing two given points. The result was basically the point-of-division formulas, with r as the parameter, that were obtained by a vector argument in Section 8.1. The same argument holds in three dimensions; but instead of using two points, let us consider the line l with direction numbers $\{a, b, c\}$ and containing the point P_0 (x_0, y_0, z_0). Let P (x, y, z) be any other point on l. If $\mathbf{u}$ and $\mathbf{w}$ are the vectors represented by $\overrightarrow{OP_0}$ and $\overrightarrow{OP}$, respectively (see Figure 20.9), it is seen that

$$\mathbf{u} = x_0\mathbf{i} + y_0\mathbf{j} + z_0\mathbf{k}, \qquad \mathbf{w} = x\mathbf{i} + y\mathbf{j} + z\mathbf{k}.$$

Since $\{a, b, c\}$ is a set of direction numbers for l,

$$\mathbf{v} = a\mathbf{i} + b\mathbf{j} + c\mathbf{k}$$

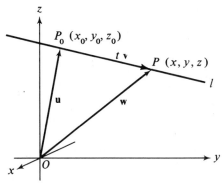

Figure 20.9

is a vector lying along *l*. But $\overrightarrow{P_0 P}$ represents another vector lying along *l*; it must be a scalar multiple, *t*v, of **v**. Since

$$\mathbf{w} = \mathbf{u} + t\mathbf{v},$$

it follows that

$$x = x_0 + ta,$$
$$y = y_0 + tb,$$
$$z = z_0 + tc.$$

The point P_0 on *l* is obtained when $t = 0$ in these equations.

Theorem 20.16

A parametric representation of the line containing (x_0, y_0, z_0) and having direction numbers $\{a, b, c\}$ is

$$x = x_0 + at, \quad y = y_0 + bt, \quad z = z_0 + ct.$$

Example 1

Find a parametric representation for the line containing $(1, 3, -2)$ and having direction numbers $\{3, 2, -1\}$.

$$x = 1 + 3t, \qquad y = 3 + 2t, \qquad z = -2 - t.$$

Example 2

Find a parametric representation of the line containing $(4, 2, -1)$ and $(0, 2, 3)$.

A set of direction numbers is $\{4 - 0, 2 - 2, -1 - 3\} = \{4, 0, -4\}$. Thus the line is

$$x = 4 + 4t, \qquad y = 2, \qquad z = -1 - 4t.$$

Once we have the direction numbers we may use them with either of the two given points. Thus, another representation is

$$x = 4s, \qquad y = 2, \qquad z = 3 - 4s.$$

Although this does not look much like the first representation, it is easily seen that they are equivalent. For instance, $t = 0$ gives the point $(4, 2, -1)$, as does $s = 1$; $t = -1$ gives the point $(0, 2, 3)$, as does $s = 0$, and so forth.

In fact,

$$
\begin{aligned}
x &= 4 + 4t & y &= 2, & z &= -1 - 4t \\
&= 4(t + 1) & & & &= 3 - 4 - 4t \\
&= 4s, & & & &= 3 - 4(t + 1) \\
& & & & &= 3 - 4s,
\end{aligned}
$$

where $s = t + 1$. Thus, whatever point we get using a value of t can be found by choosing $s = t + 1$.

A simpler set of direction numbers can also be found. Since the ones we have are all multiples of 4, we can multiply through by $1/4$ to get another set of direction numbers, $\{1, 0, -1\}$. Using these with the first point gives

$$x = 4 + u, \qquad y = 2, \qquad z = -1 - u.$$

Again, we see that $4t = u$, so the two representations are equivalent.

Perhaps you wonder what is needed to be able to say that two parametric representations are equivalent. If a value of t and another of s both give the same point, then, for those values of t and s, the three coordinates must be equal. Eliminating x, y, and z between the two parametric representations gives three equations in t and s (in some of these, the parameters may both be absent, as they are in the representation of y here). If all give the same result when they are solved for one parameter in terms of the other, and if the domain and range are the same, then the representations are equivalent.

Suppose we eliminate the parameter in the representation given by Theorem 20.16. If none of the direction numbers is zero, we can solve each equation for t and set them equal to each other. This gives

$$\frac{x - x_0}{a} = \frac{y - y_0}{b} = \frac{z - z_0}{c}.$$

Actually, this is just a shorter way of writing the three equations

$$\frac{x - x_0}{a} = \frac{y - y_0}{b},$$

$$\frac{y - y_0}{b} = \frac{z - z_0}{c},$$

$$\frac{x - x_0}{a} = \frac{z - z_0}{c}.$$

But these three equations are not independent—the last can be found from the first two. Let us discard it and consider only the first two, which, as we shall see in Section 20.5, represent planes. Any point that satisfies both equations is on both planes and therefore on the intersection of the two planes, which is a line. Thus this representation of a line gives it as the intersection of two planes. It might be noted that the equation we discarded is also a plane containing the same line.

What, now, if one of the direction numbers is zero. Let us suppose that $a = 0$. Then the line in parametric form is

$$x = x_0, \qquad y = y_0 + bt, \qquad z = z_0 + ct.$$

We do not have to eliminate the parameter from the first equation—it is already gone. By eliminating t between the last two equations as before, we have

$$\frac{y - y_0}{b} = \frac{z - z_0}{c}.$$

This, together with $x = x_0$ (or $x - x_0 = 0$) gives the line as the intersection of two planes.

If two of the direction numbers are zero, we have two equations in which the parameter is missing. The parameter in the third equation cannot be eliminated, because there is no second equation with which to combine it. But it is not necessary to eliminate it! The two equations without the parameter already give us the necessary two planes.

Theorem 20.17

If a line contains the point (x_0, y_0, z_0) *and has direction numbers* $\{a, b, c\}$, *then it can be represented by*

$$\text{(i)} \quad \frac{x - x_0}{a} = \frac{y - y_0}{b} = \frac{z - z_0}{c}$$

if none of the direction numbers is zero;

$$\text{(ii)} \quad x - x_0 = 0 \quad and \quad \frac{y - y_0}{b} = \frac{z - z_0}{c}$$

if $a = 0$ *and neither b nor c is zero (similar results follow if* $b = 0$ *or* $c = 0$);

$$\text{(iii)} \quad x - x_0 = 0 \quad and \quad y - y_0 = 0$$

if $a = 0$ *and* $b = 0$ *(again, similar results follow for some other pair of direction numbers equaling zero). These are called* **symmetric equations** *of the line.*

Example 3

Find symmetric equations of the line containing $(4, 1, -2)$ and having direction numbers $\{1, 3, -2\}$.

$$\frac{x - 4}{1} = \frac{y - 1}{3} = \frac{z + 2}{-2}.$$

Example 4

Find symmetric equations for the line containing $(4, 1, 3)$ and $(2, 1, -2)$.

A set of direction numbers is $\{4 - 2, 1 - 1, 3 + 2\} = \{2, 0, 5\}$. Since $b = 0$, we have (using the first point)

$$\frac{x - 4}{2} = \frac{z - 3}{5} \quad and \quad y - 1 = 0.$$

Example 5

Find the point of intersection (if any) of the lines

$$x = 3 + 2t, \qquad y = 2 - t, \qquad z = 5 + t$$

and

$$x = -3 - s, \qquad y = 7 + s, \qquad z = 16 + 3s.$$

Let us assume that there is a point of intersection. Then there is a value of t and a value of s which yield the same values of x, y, and z. For these particular values of t and s, we have

$$x = 3 + 2t = -3 - s,$$
$$y = 2 - t = 7 + s,$$
$$z = 5 + t = 16 + 3s,$$

or

$$2t + s = -6, \qquad t + s = -5, \qquad t - 3s = 11.$$

If we solve the first pair simultaneously, we get

$$t = -1 \quad \text{and} \quad s = -4.$$

We see that they also satisfy the third equation. Thus there is a point of intersection which corresponds to $t = -1$ (or $s = -4$). It is $(1, 3, 4)$.

It might be noted that there are three possibilities. One is the situation in which there is a value of t and a value of s satisfying all three of the equations in t and s, as above. This results in a single point of intersection. In a second possibility, there is no value for t or s satisfying all three of the equations; that is, the values of t and s that satisfy the first two equations fail to satisfy the third. Thus, there is no point of intersection. The third possibility is that any two of the three equations in t and s are dependent; that is, any pair of values for t and s that satisfies one of them, satisfies all three. In this case we have two different representations for the same line (see the discussion following Example 2).

Problems

A

In Problems 1–16, represent the given line in parametric form and symmetric form.

1. Containing $(2, -4, 2)$;   direction numbers $\{2, 3, 1\}$.
2. Containing $(2, 1, 5)$;   direction numbers $\{2, -1, 4\}$.
3. Containing $(2, 0, 3)$;   direction numbers $\{4, -1, 3\}$.
4. Containing $(4, -3, 2)$;   direction numbers $\{4, 2, -3\}$.
5. Containing $(1, 0, 5)$;   direction numbers $\{3, 1, 0\}$.
6. Containing $(1, 1, 1)$;   direction numbers $\{3, 0, 2\}$.
7. Containing $(3, 1, 2)$;   direction numbers $\{1, 0, 0\}$.
8. Containing $(3, 3, 2)$;   direction numbers $\{0, 0, 1\}$.
9. Containing $(3, 3, 1)$ and $(4, 0, 2)$. 　　10. Containing $(4, 0, 5)$ and $(1, 2, 2)$.
11. Containing $(-4, 2, 0)$ and $(3, 1, 2)$. 　　12. Containing $(-8, 4, -1)$ and $(-2, 0, 4)$.
13. Containing $(2, 2, 4)$ and $(1, 2, 7)$. 　　14. Containing $(5, 1, 3)$ and $(5, 3, 5)$.
15. Containing $(2, 4, -5)$ and $(5, 4, -5)$. 　　16. Containing $(1, -3, 3)$ and $(1, 5, 3)$.

In Problems 17–20, find the point of intersection (if any) of the given lines.

17. $x = 2 - t,$ $y = 3 + 2t,$ $z = 4 + t;$ $x = 1 + t,$ $y = -2 + t,$ $z = 5 - 4t.$
18. $x = 4 + t,$ $y = -8 - 2t,$ $z = 12t;$ $x = 3 + 2s,$ $y = -1 + s,$ $z = -3 - 3s.$
19. $x = 3 - t,$ $y = 5 + 3t,$ $z = -1 - 4t;$ $x = 8 + 2s,$ $y = -6 - 4s,$ $z = 5 + s.$
20. $x = 3 + t,$ $y = 4 - 2t,$ $z = 1 + 5t;$ $x = 5 - t,$ $y = 3 + 2t,$ $z = 8 + 4t.$

B

In Problems 21–24, find the point of intersection (if any) of the given lines.

21. $\dfrac{x-5}{1} = \dfrac{y+2}{-2} = \dfrac{z-3}{5}$; $\dfrac{x-4}{-2} = \dfrac{y-2}{1} = \dfrac{z-4}{3}$.

22. $\dfrac{x-2}{1} = \dfrac{y-3}{-2} = \dfrac{z+1}{1}$; $\dfrac{x-3}{2} = \dfrac{y-1}{-4} = \dfrac{z}{2}$.

23. $\dfrac{x-2}{1} = \dfrac{y-3}{-2} = \dfrac{z}{4}$; $x-4=0$, $\dfrac{y-2}{1} = \dfrac{z-3}{-1}$.

24. $\dfrac{x-3}{1} = \dfrac{y+3}{-4}$, $z+1=0$; $\dfrac{x}{-2} = \dfrac{y-2}{1} = \dfrac{z-3}{4}$.

In Problems 25–30, indicate whether the two given lines are parallel, perpendicular, coincident, or none of these.

25. $x = 4 - t$, $y = 3 + 2t$, $z = 1 + t$: $x = 1 + 2t$, $y = 4 - 4t$, $z = 3 - 2t$.
26. $x = 3 + 5t$, $y = -1 - 2t$, $z = 4 + t$; $x = 3$, $y = 4 + 2s$, $z = -2 + 4s$.

27. $\dfrac{x-2}{1} = \dfrac{y-5}{-3} = \dfrac{z+1}{2}$; $\dfrac{x-4}{-3} = \dfrac{y+1}{9} = \dfrac{z-3}{-6}$.

28. $x = 2 + t$, $y = 5 - 3t$, $z = 1 + 4t$; $x = 4 - t$, $y = 2 + 2t$, $z = 3t$.

29. $\dfrac{x-1}{2} = \dfrac{z+3}{4}$, $y - 5 = 0$; $\dfrac{x+2}{6} = \dfrac{y-5}{3} = \dfrac{z}{2}$.

30. $\dfrac{x+3}{1} = \dfrac{y-4}{3} = \dfrac{z+2}{-2}$; $\dfrac{x-5}{-3} = \dfrac{y+3}{-9} = \dfrac{z-1}{6}$.

31. Give equations for each of the coordinates axes.

C

32. Find the coordinates of the points in which the line $x = 3 + 3t$, $y = 2 - 2t$, $z = 3t$ intersects the coordinate planes. Sketch the line.

33. Show that if a line lies in the xy plane, then $n = \cos \gamma = 0$, and the line has equations

$$\dfrac{x - x_0}{\cos \alpha} = \dfrac{y - y_0}{\sin \alpha},\quad z = 0.$$

If θ is the angle of inclination of the line in the xy plane, then $y - y_0 = (\tan \theta)(x - x_0)$, $\theta \neq 90°$.   Compare Theorem 2.1.

20.4

The Cross Product

Let us now look at the other product of two vectors—the cross product.

Definition

If $\mathbf{u} = a_1\mathbf{i} + b_1\mathbf{j} + c_1\mathbf{k}$ *and* $\mathbf{v} = a_2\mathbf{i} + b_2\mathbf{j} + c_2\mathbf{k}$, *then the* **cross product (vector product, outer product)** *of* $\mathbf{u}$ *and* $\mathbf{v}$ *is*

$$\mathbf{u} \times \mathbf{v} = (b_1 c_2 - c_1 b_2)\mathbf{i} + (c_1 a_2 - a_1 c_2)\mathbf{j} + (a_1 b_2 - b_1 a_2)\mathbf{k}.$$

Some obvious questions arise. Why do we want to define a cross product this way? What is it good for? What are its properties? In some ways, all answers are the same. We define the cross product in this way to establish some interesting properties that are useful for certain applications. In a way, this is approaching the problem backward. It would be more logical to define the cross product of two vectors as that one having the desired properties and then show that such a vector must take the form given. Our way of doing it is by far the simpler approach. Before looking at some properties, let us consider a simpler form for the cross product.

Theorem 20.18

If $\mathbf{u} = a_1\mathbf{i} + b_1\mathbf{j} + c_1\mathbf{k}$ *and* $\mathbf{v} = a_2\mathbf{i} + b_2\mathbf{j} + c_2\mathbf{k}$, *then*

$$\mathbf{u} \times \mathbf{v} = \begin{vmatrix} \mathbf{i} & \mathbf{j} & \mathbf{k} \\ a_1 & b_1 & c_1 \\ a_2 & b_2 & c_2 \end{vmatrix}.$$

This theorem follows directly from the definition if we expand the symbolic determinant by minors along the first row.

Example 1

Given $\mathbf{u} = 3\mathbf{i} + \mathbf{j} - 2\mathbf{k}$ and $\mathbf{v} = \mathbf{i} + 2\mathbf{j} + \mathbf{k}$, find $\mathbf{u} \times \mathbf{v}$ and $\mathbf{v} \times \mathbf{u}$.

$$\mathbf{u} \times \mathbf{v} = \begin{vmatrix} \mathbf{i} & \mathbf{j} & \mathbf{k} \\ 3 & 1 & -2 \\ 1 & 2 & 1 \end{vmatrix} = 5\mathbf{i} - 5\mathbf{j} + 5\mathbf{k},$$

$$\mathbf{v} \times \mathbf{u} = \begin{vmatrix} \mathbf{i} & \mathbf{j} & \mathbf{k} \\ 1 & 2 & 1 \\ 3 & 1 & -2 \end{vmatrix} = -5\mathbf{i} + 5\mathbf{j} - 5\mathbf{k}.$$

Note that $\mathbf{u} \times \mathbf{v} \neq \mathbf{v} \times \mathbf{u}$!

Again we are not multiplying numbers; there is no reason to assume that the cross product of two vectors has the same properties as the product of two numbers. We have already seen one difference in Example 1. The cross product has the following properties.

Theorem 20.19

If $\mathbf{u}$, $\mathbf{v}$, *and* $\mathbf{w}$ *are vectors and a is a scalar, then*

(a) $\mathbf{u} \times \mathbf{v} = -(\mathbf{v} \times \mathbf{u})$;
(b) $\mathbf{u} \times (\mathbf{v} + \mathbf{w}) = (\mathbf{u} \times \mathbf{v}) + (\mathbf{u} \times \mathbf{w})$;
(c) $\mathbf{u} \times \mathbf{0} = \mathbf{0} \times \mathbf{u} = \mathbf{0}$;
(d) *if* $\mathbf{u} = a\mathbf{v}$, *then* $\mathbf{u} \times \mathbf{v} = \mathbf{0}$ (*that is, the cross product of parallel vectors is* $\mathbf{0}$);
(e) $(\mathbf{u} \times \mathbf{v}) \cdot \mathbf{w} = \mathbf{u} \cdot (\mathbf{v} \times \mathbf{w})$;
(f) $\mathbf{u} \times (a\mathbf{v}) = (a\mathbf{u}) \times \mathbf{v} = a(\mathbf{u} \times \mathbf{v})$.

Proof

(a) Suppose $\mathbf{u} = a_1\mathbf{i} + b_1\mathbf{j} + c_1\mathbf{k}$, $\mathbf{v} = a_2\mathbf{i} + b_2\mathbf{j} + c_2\mathbf{k}$, and $\mathbf{w} = a_3\mathbf{i} + b_3\mathbf{j} + c_3\mathbf{k}$. Since, by Theorem 20.18, $\mathbf{u} \times \mathbf{v}$ and $\mathbf{v} \times \mathbf{u}$ are given by determinants that are identical except for the reversal of the second and third rows, it follows that

$$\mathbf{u} \times \mathbf{v} = -(\mathbf{v} \times \mathbf{u}).$$

(b) Since $\mathbf{v} + \mathbf{w} = (a_2 + a_3)\mathbf{i} + (b_2 + b_3)\mathbf{j} + (c_2 + c_3)\mathbf{k}$,

$$\mathbf{u} \times (\mathbf{v} + \mathbf{w}) = \begin{vmatrix} \mathbf{i} & \mathbf{j} & \mathbf{k} \\ a_1 & b_1 & c_1 \\ a_2 + a_3 & b_2 + b_3 & c_2 + c_3 \end{vmatrix}$$

$$= \begin{vmatrix} \mathbf{i} & \mathbf{j} & \mathbf{k} \\ a_1 & b_1 & c_1 \\ a_2 & b_2 & c_2 \end{vmatrix} + \begin{vmatrix} \mathbf{i} & \mathbf{j} & \mathbf{k} \\ a_1 & b_1 & c_1 \\ a_3 & b_3 & c_3 \end{vmatrix}$$

$$= (\mathbf{u} \times \mathbf{v}) + (\mathbf{u} \times \mathbf{w}).$$

The proofs of the remaining parts are left to the student (see Problems 30 and 31).

It might be noted that the definition of the cross product was stated in terms of three-dimensional vectors. In fact, we must have a three-dimensional vector space, for $\mathbf{u} \times \mathbf{v}$ is not in the plane determined by $\mathbf{u}$ and $\mathbf{v}$, as shown in the next theorem.

Theorem 20.20

If $\mathbf{u}$ and $\mathbf{v}$ are nonzero vectors, then $\mathbf{u} \times \mathbf{v}$ is perpendicular to both $\mathbf{u}$ and $\mathbf{v}$.

Proof

$$\mathbf{u} \cdot (\mathbf{u} \times \mathbf{v}) = (\mathbf{u} \times \mathbf{u}) \cdot \mathbf{v} \quad \text{(why?)}$$
$$= \mathbf{0} \cdot \mathbf{v} \quad \text{(why?)}$$
$$= 0. \quad \text{(why?)}$$

Thus $\mathbf{u}$ and $\mathbf{u} \times \mathbf{v}$ are perpendicular. A similar argument shows that $\mathbf{u} \times \mathbf{v}$ and $\mathbf{v}$ are perpendicular.

This property of the cross product gives us its principal use. Certain problems in three-dimensional analytic geometry that were relatively difficult without the use of the cross product are easier now.

Example 2

Find a set of direction numbers for a line perpendicular to the plane containing

$$x = 1, \quad y = 3 + 2t, \quad z = 4 + t$$

and

$$x = 1 + 4s, \quad y = 3 + 2s, \quad z = 4 + 2s.$$

Any line perpendicular to a given plane is perpendicular to any line in that plane. This suggests the use of the cross product. Vectors directed along the given lines are

$$\mathbf{u} = 2\mathbf{j} + \mathbf{k} \quad \text{and} \quad \mathbf{v} = 4\mathbf{i} + 2\mathbf{j} + 2\mathbf{k}.$$

Since $\mathbf{u} \times \mathbf{v} = 2\mathbf{i} + 4\mathbf{j} - 8\mathbf{k}$, we have $\{2, 4, -8\}$ as one set of direction numbers for the desired line; $\{1, 2, -4\}$ is a simpler set.

Example 3

Find equations for the line containing $(1, 4, 3)$ and perpendicular to

$$\frac{x-1}{2} = \frac{y+3}{1} = \frac{z-2}{4} \quad \text{and} \quad \frac{x+2}{3} = \frac{y-4}{2} = \frac{z+1}{-2}.$$

Again, vectors along the two given lines are

$$\mathbf{u} = 2\mathbf{i} + \mathbf{j} + 4\mathbf{k} \quad \text{and} \quad \mathbf{v} = 3\mathbf{i} + 2\mathbf{j} - 2\mathbf{k};$$

and $\mathbf{u} \times \mathbf{v} = -10\mathbf{i} + 16\mathbf{j} + \mathbf{k}$ is perpendicular to both of them. The desired line is, therefore,

$$\frac{x-1}{-10} = \frac{y-4}{16} = \frac{z-3}{1}.$$

Example 4

Find the distance between the lines

$$x = 1 - 4t, \qquad y = 2 + t, \qquad z = 3 + 2t$$

and

$$x = 1 + s, \qquad y = 4 - 2s, \qquad z = -1 + s.$$

The desired distance is to be measured along a line perpendicular to both of the given lines. Again, vectors along the given lines are $\mathbf{u} = -4\mathbf{i} + \mathbf{j} + 2\mathbf{k}$ and $\mathbf{v} = \mathbf{i} - 2\mathbf{j} + \mathbf{k}$ (see Figure 20.10). Thus the distance is to be measured along

$$\mathbf{u} \times \mathbf{v} = 5\mathbf{i} + 6\mathbf{j} + 7\mathbf{k}.$$

The point A $(1, 2, 3)$ is on the first line, and B $(1, 4, -1)$ is on the second. The vector represented by $\overrightarrow{AB}$ is

$$\mathbf{w} = 2\mathbf{j} - 4\mathbf{k}.$$

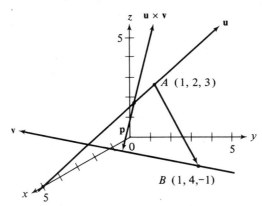

Figure 20.10

We want a vector whose representatives are all perpendicular to both of the given lines and with one representative having its head on one line and its tail on the other. All of the representatives of $\mathbf{u} \times \mathbf{v}$ are perpendicular to both lines and one of the representatives of $\mathbf{w}$ has its end points on the given lines. Thus, the projection $\mathbf{p}$ of $\mathbf{w}$ on $\mathbf{u} \times \mathbf{v}$ has the desired properties and its length is the distance between the given lines.

$$|\mathbf{p}| = \frac{|\mathbf{w} \cdot (\mathbf{u} \times \mathbf{v})|}{|\mathbf{u} \times \mathbf{v}|} = \frac{|0 \cdot 5 + 2 \cdot 6 - 4 \cdot 7|}{\sqrt{25 + 36 + 49}} = \frac{16}{\sqrt{110}}.$$

Up to this point we have been dealing exclusively with the direction of $\mathbf{u} \times \mathbf{v}$. Its length also has some interesting properties.

Theorem 20.21

If $\mathbf{u}$ *and* $\mathbf{v}$ *are vectors and* θ *is the angle between them, then*

$$|\mathbf{u} \times \mathbf{v}| = |\mathbf{u}| \, |\mathbf{v}| \sin \theta.$$

Proof

Since $\cos \theta = (\mathbf{u} \cdot \mathbf{v})/(|\mathbf{u}| \, |\mathbf{v}|)$ by Theorem 20.7,

$$|\mathbf{u}| \, |\mathbf{v}| \, \sin \theta = |\mathbf{u}| \, |\mathbf{v}| \sqrt{1 - \cos^2 \theta} \qquad \text{(See Note 1.)}$$

$$= |\mathbf{u}| \, |\mathbf{v}| \sqrt{1 - \frac{(\mathbf{u} \cdot \mathbf{v})^2}{|\mathbf{u}|^2 |\mathbf{v}|^2}} \quad .$$

$$= \sqrt{|\mathbf{u}|^2 |\mathbf{v}|^2 - (\mathbf{u} \cdot \mathbf{v})^2}.$$

If we let $\mathbf{u} = a_1 \mathbf{i} + b_1 \mathbf{j} + c_1 \mathbf{k}$ and $\mathbf{v} = a_2 \mathbf{i} + b_2 \mathbf{j} + c_2 \mathbf{k}$, then

$$|\mathbf{u}| \, |\mathbf{v}| \, \sin \theta = \sqrt{(a_1^2 + b_1^2 + c_1^2)(a_2^2 + b_2^2 + c_2^2) - (a_1 a_2 + b_1 b_2 + c_1 c_2)^2}$$

$$= \sqrt{(b_1 c_2 - c_1 b_2)^2 + (c_1 a_2 - a_1 c_2)^2 + (a_1 b_2 - b_1 a_2)^2}$$

(See Note 2.)

$$= |\mathbf{u} \times \mathbf{v}|.$$

Note 1: By the definition of the angle between two vectors, $0° \le \theta \le 180°$; and $\sin \theta \ge 0$.

Note 2: The algebra here is routine but tedious. It is left to the student.

Note the similarity between this theorem and the first part of Theorem 20.7. One consequence of this theorem is given in Problem 25.

It appears that Theorems 20.20 and 20.21 give a geometric description of $\mathbf{u} \times \mathbf{v}$, the first giving its length and the second, its direction. Actually this is not quite true. There are two vectors of a given length which are perpendicular to both $\mathbf{u}$ and $\mathbf{v}$; they have opposite orientations (that is, one is the negative of the other). It can be shown that $\mathbf{u} \times \mathbf{v}$ is the one that gives the system $\{\mathbf{u}, \mathbf{v}, \mathbf{u} \times \mathbf{v}\}$ a right-hand orientation; that is, if the index and second fingers of the right hand point in the directions of $\mathbf{u}$ and $\mathbf{v}$, respectively, then the thumb points in the direction of $\mathbf{u} \times \mathbf{v}$ (see Figure 20.11). This is summarized in the next theorem.

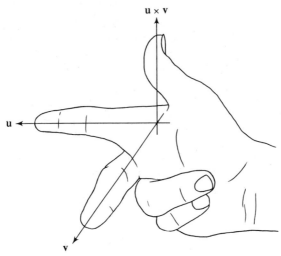

Figure 20.11

Theorem 20.22

*If **u** and **v** are vectors, θ is the angle between them, and **n** is the unit vector perpendicular to both **u** and **v** such that $\{\mathbf{u}, \mathbf{v}, \mathbf{n}\}$ forms a right-hand system, then*

$$\mathbf{u} \times \mathbf{v} = |\mathbf{u}||\mathbf{v}|\sin\theta\,\mathbf{n}.$$

Some authors take this as the definition of $\mathbf{u} \times \mathbf{v}$. Another direct result from Theorem 20.21 follows.

Theorem 20.23

*If **u** and **v** are two nonzero vectors, then $\mathbf{u} \times \mathbf{v} = \mathbf{0}$ if and only if $\mathbf{u} = k\mathbf{v}$ for some scalar k.*

This extends Theorem 20.19(d), which gives the "if" part of this theorem. The proof is left to the student (see **Problem 35**).

Problems

A

In Problems 1–6, find $\mathbf{u} \times \mathbf{v}$. Use the dot product to verify that your result is perpendicular to both **u** and **v**.

1. $\mathbf{u} = \mathbf{i} + \mathbf{j} + \mathbf{k}, \quad \mathbf{v} = 2\mathbf{i} - \mathbf{j} - 4\mathbf{k}$.
2. $\mathbf{u} = 3\mathbf{i} - \mathbf{j} + 2\mathbf{k}, \quad \mathbf{v} = 2\mathbf{i} + \mathbf{j} + \mathbf{k}$.
3. $\mathbf{u} = 4\mathbf{i} + 2\mathbf{j}, \quad \mathbf{v} = 3\mathbf{i} - \mathbf{j}$.
4. $\mathbf{u} = 2\mathbf{i} + 3\mathbf{j} - 2\mathbf{k}, \quad \mathbf{v} = -2\mathbf{i} + \mathbf{j}$.
5. $\mathbf{u} = 2\mathbf{i} + \mathbf{j} - \mathbf{k}, \quad \mathbf{v} = -\mathbf{i} - \mathbf{j} + 3\mathbf{k}$.
6. $\mathbf{u} = 3\mathbf{i} + 2\mathbf{k}, \quad \mathbf{v} = -2\mathbf{i} + \mathbf{j}$.

In Problems 7–12, find direction numbers for the line described.

7. Perpendicular to the plane containing $(2, 2, 3)$, $(-1, 4, 1)$, and $(0, 1, 2)$.
8. Perpendicular to the plane containing $(3, 1, 2)$, $(1, -1, 1)$, and $(2, 0, 4)$.

9. Perpendicular to the plane containing

   $$x = 2 + 4t, \quad y = 2 - t, \quad z = 4 \quad \text{and} \quad x = 2 - 2s, \quad y = 2 + 2s, \quad z = 4 - s.$$

10. Perpendicular to the plane containing

   $$x = 2 + t, \quad y = 3 - 2t, \quad z = -t \quad \text{and} \quad x = 2 - 2s, \quad y = 3 + s, \quad z = -s.$$

11. Perpendicular to the plane containing

   $$x = 2 + 2t, \quad y = 3 - t, \quad z = -1 + t \quad \text{and} \quad (4, 2, 4).$$

12. Perpendicular to the plane containing

   $$x = 4 + t, \quad y = -1 + 2t, \quad z = 2t \quad \text{and} \quad (2, 4, 1).$$

In Problems 13 and 14, find equations for the line described.

13. Containing $(4, -1, 0)$ and perpendicular to

   $$x = 3 + t, \quad y = 2 - t, \quad z = 2t \quad \text{and} \quad x = 4, \quad y = 2 + s, \quad z = -1 + s.$$

14. Containing $(3, 2, 1)$ and perpendicular to

   $$x = 1 - 2t, \quad y = 3 + t, \quad z = 4 - t \quad \text{and} \quad x = 2 + s, \quad y = -1 + 2s, \quad z = 3 - s.$$

In Problems 15–18, find the distance between the given lines.

15. $x = 2 + t, \quad y = 1 - t, \quad z = 4t \quad$ and $\quad x = 2 + s, \quad y = 4 - 2s, \quad z = 1 + 3s.$
16. $x = 1 + t, \quad y = -2 + 3t, \quad z = 4 + t \quad$ and $\quad x = 2 - s, \quad y = 3 + 2s, \quad z = 1 + s.$
17. $x = 2 + t, \quad y = -4 + t, \quad z = 1 - 3t \quad$ and $\quad x = 3 - s, \quad y = 4 + 2s, \quad z = 2 + s.$
18. $x = 1 + t, \quad y = 1 - 5t, \quad z = 2 + t \quad$ and $\quad x = 4 + s, \quad y = 5 + 2s, \quad z = -3 + 4s.$

B

In Problems 19 and 20, find the distance between the given lines.

19. $x = 4t, \quad y = 1 + t, \quad z = -2 - t \quad$ and $\quad x = 9 + 4s, \quad y = 1 + s, \quad z = -2 - s.$
20. $x = 2 + 3t, \quad y = 5 + t, \quad z = -1 - 2t \quad$ and $\quad x = 2 + 3s, \quad y = 3 + s, \quad z = 5 - 2s.$

In Problems 21–24, find equations for the line described.

21. Containing $(0, 4, -2)$ and perpendicular to the plane determined by $(0, 4, -2)$ and the line $x = -2 + 2t, \quad y = 8t, \quad z = -1 + t.$
22. Containing $(2, 3, 1)$ and perpendicular to the plane determined by $(2, 3, 1)$ and the line $x = 0, \quad y = 2t, \quad z = t.$
23. Containing $(1, 1, 2)$ and perpendicular to and containing a point of $x = 1 - t, \quad y = 2 + 2t, \quad z = 4t.$
24. Containing $(2, 0, 5)$ and perpendicular to and containing a point of $x = 4 + t, \quad y = 3 - 2t, \quad z = 1 + t.$
25. Suppose the vectors $\mathbf{u}$ and $\mathbf{v}$ are represented by $\overrightarrow{AB}$ and $\overrightarrow{AC}$, respectively. Show that the area of $\triangle ABC$ is $(1/2)|\mathbf{u} \times \mathbf{v}|$. (Equivalently, the parallelogram determined by AB and AC has area $|\mathbf{u} \times \mathbf{v}|$.) (*Hint:* Use Theorem 20.21.)

In Problems 26–29, use the result of Problem 25 to find the area of the triangles with the given vertices.

26. $(1, 0, 4), \quad (2, -1, 2), \quad (4, 4, 1).$
27. $(3, -2, 1), \quad (-1, 2, 0), \quad (4, 4, 2).$
28. $(2, 4, 3), \quad (1, 0, 1), \quad (-2, 2, 4).$
29. $(4, 2), \quad (3, -1), \quad (-1, 0).$

C

30. Prove parts (c), (d), and (f) of Theorem 20.19.

31. Show that if $\mathbf{u} = a_1\mathbf{i} + b_1\mathbf{j} + c_1\mathbf{k}$, $\mathbf{v} = a_2\mathbf{i} + b_2\mathbf{j} + c_2\mathbf{k}$, and $\mathbf{w} = a_3\mathbf{i} + b_3\mathbf{j} + c_3\mathbf{k}$, then

$$\mathbf{u} \cdot (\mathbf{v} \times \mathbf{w}) = \begin{vmatrix} a_1 & b_1 & c_1 \\ a_2 & b_2 & c_2 \\ a_3 & b_3 & c_3 \end{vmatrix}.$$

Use this result to prove Theorem 20.19(e).

32. Given that $\overrightarrow{AB}$, $\overrightarrow{AC}$, and $\overrightarrow{AD}$ represent the vectors $\mathbf{u}$, $\mathbf{v}$, and $\mathbf{w}$, respectively, show that the parallelepiped determined by AB, AC, and AD (see Figure 20.12) has volume $|\mathbf{u} \cdot (\mathbf{v} \times \mathbf{w})|$. (*Hint*: By Problem 25, the area of the base is $|\mathbf{v} \times \mathbf{w}|$.)

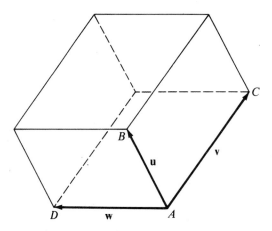

Figure 20.12

In Problems 33 and 34, use the result of Problem 32 to find the volume of the parallelepiped determined by AB, AC, and AD.

33. $A = (0, 0, 0)$, $B = (2, 1, 3)$, $C = (5, 3, 1)$, $D = (2, -1, 4)$.

34. $A = (1, 3, 2)$, $B = (4, 1, 5)$, $C = (1, 5, 2)$, $D = (0, 5, -1)$.

35. Prove Theorem 20.23

36. Let P, Q, and R be distinct points, and let $\mathbf{u}$, $\mathbf{v}$, and $\mathbf{w}$ be represented by $\overrightarrow{OP}$, $\overrightarrow{OQ}$, and $\overrightarrow{OR}$, respectively. Show that $\mathbf{u} \times \mathbf{v} + \mathbf{v} \times \mathbf{w} + \mathbf{w} \times \mathbf{u}$ is perpendicular to the plane containing P, Q, and R.

37. Show that symmetric equations of the line containing the points (x_1, y_1, z_1) and (x_2, y_2, z_2) are

$$\frac{x - x_1}{x_2 - x_1} = \frac{y - y_1}{y_2 - y_1} = \frac{z - z_1}{z_2 - z_1} \qquad (x_1 \neq x_2, \ y_1 \neq y_2, \ z_1 \neq z_2).$$

38. If $\mathbf{v} = (a, b, c)$, show that $a = \mathbf{v} \cdot \mathbf{i}$, $b = \mathbf{v} \cdot \mathbf{j}$, and $c = \mathbf{v} \cdot \mathbf{k}$. Use this result to prove that $\mathbf{u} \times (\mathbf{v} \times \mathbf{w}) = (\mathbf{u} \cdot \mathbf{w})\mathbf{v} - (\mathbf{u} \cdot \mathbf{v})\mathbf{w}$.

39. Show that $(\mathbf{u} \times \mathbf{v}) \cdot (\mathbf{w} \times \mathbf{t}) = (\mathbf{u} \cdot \mathbf{w})(\mathbf{v} \cdot \mathbf{t}) - (\mathbf{u} \cdot \mathbf{t})(\mathbf{v} \cdot \mathbf{w})$.

40. Show that $|\mathbf{u} \times \mathbf{v}|^2 = \begin{vmatrix} \mathbf{u} \cdot \mathbf{u} & \mathbf{u} \cdot \mathbf{v} \\ \mathbf{u} \cdot \mathbf{v} & \mathbf{v} \cdot \mathbf{v} \end{vmatrix}$.

20.5

The Plane

Let us now consider the plane. Perhaps the simplest way of determining a plane is by three noncollinear points. But, for the purpose of determining its equation, it is better to describe it by a single point and a line perpendicular to it.

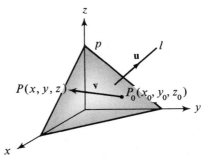

Let p be a plane in space containing the point $P_0(x_0, y_0, z_0)$ (see Figure 20.13); and let l, with direction numbers $\{a, b, c\}$, be a line perpendicular to p. To determine an equation of p, we consider any point $P(x, y, z)$ lying in the plane. The directed line segment $\overrightarrow{P_0 P}$ represents a vector

Figure 20.13

$$\mathbf{v} = (x - x_0)\mathbf{i} + (y - y_0)\mathbf{j} + (z - z_0)\mathbf{k}$$

in the plane p. Since l is perpendicular to p, it is perpendicular to any line in this plane; in particular, it is perpendicular to $P_0 P$. Since

$$\mathbf{u} = a\mathbf{i} + b\mathbf{j} + c\mathbf{k}$$

is a vector lying along l, we have

$$\mathbf{u} \cdot \mathbf{v} = 0,$$
$$[a\mathbf{i} + b\mathbf{j} + c\mathbf{k}] \cdot [(x - x_0)\mathbf{i} + (y - y_0)\mathbf{j} + (z - z_0)\mathbf{k}] = 0$$
$$a(x - x_0) + b(y - y_0) + c(z - z_0) = 0.$$

Furthermore, the argument may be traced backward to show that any point (x, y, z) that satisfies the last equation must lie in the plane p. Thus we have proved the following theorem.

Theorem 20.24

A point is on a plane containing (x_1, y_1, z_1) and perpendicular to a line with direction numbers $\{A, B, C\}$ if and only if it satisfies the equation

$$A(x - x_1) + B(y - y_1) + C(z - z_1) = 0.$$

Example 1

Find the equation of the plane containing $(1, 3, -2)$ and perpendicular to the line through $(2, 5, 1)$ and $(0, 1, -3)$.

A set of direction numbers for the given line is $\{2, 4, 4\}$ or $\{1, 2, 2\}$. Thus the desired plane is

$$1(x - 1) + 2(y - 3) + 2(z + 2) = 0$$
$$x + 2y + 2z - 3 = 0.$$

Example 2

Find an equation of the plane containing the two lines

$$x = 1, \, y = 3 + 2t, \, z = 4 + t \quad \text{and} \quad x = 1 + 4s, \, y = 3 + 2s, \, x = 4 + 2s.$$

These lines clearly intersect at (1, 3, 4). All we need, then, is a set of direction numbers for a line perpendicular to the desired plane. This was done in Example 2 of the previous section by using the cross product of two vectors. One such set is {1, 2, −4}. By Theorem 20.24, the corresponding plane is

$$1(x - 1) + 2(y - 3) - 4(z - 4) = 0$$
$$x + 2y - 4z + 9 = 0.$$

The following theorem is a direct consequence of Theorem 20.24.

Theorem 20.25

Any plane can be represented by an equation of the form

$$Ax + By + Cz + D = 0,$$

where {A, B, C} is a set of direction numbers for a line normal to (that is, perpendicular to) the plane. Conversely, an equation of the above form (where A, B, and C are not all zero) represents a plane with {A, B, C} a set of direction numbers for a normal line.

Example 3

Find an equation of the plane containing the points $P_1(1, 0, 1)$, $P_2(-1, -4, 1)$, and $P_3(-2, -2, 2)$.

This problem may be solved using either Theorem 20.25 or Theorem 20.24. Let us do it both ways.

By Theorem 20.25, the equation we seek is, for the proper choices of A, B, C, and D,

$$Ax + By + Cz + D = 0.$$

We get an equation in A, B, C, and D from each of the three given points.

$$
\begin{aligned}
P_1(1, 0, 1){:} &\quad A \phantom{{}-4B} + C + D = 0, \\
P_2(-1, -4, 1){:} &\quad -A - 4B + C + D = 0, \\
P_3(-2, -2, 2){:} &\quad -2A - 2B + 2C + D = 0.
\end{aligned}
$$

Although we cannot solve for A, B, C, and D directly, since we have only three equations in four unknowns, we can solve for three of them in terms of the other one. If we take A to be fixed and solve for the other three, we have $B = -A/2$, $C = 2A$, and $D = -3A$. We may give A any nonzero value we want; let us choose $A = 2$. Then $B = -1$, $C = 4$, and $D = -6$; the resulting equation is

$$2x - y + 4z - 6 = 0.$$

We now solve the same problem using Theorem 20.24. We let $\overrightarrow{P_1 P_2}$ and $\overrightarrow{P_1 P_3}$ represent the vectors **u** and **v**, respectively. Then

$$\mathbf{u} = (-1 - 1)\mathbf{i} + (-4 - 0)\mathbf{j} + (1 - 1)\mathbf{k} = -2\mathbf{i} - 4\mathbf{j},$$
$$\mathbf{v} = (-2 - 1)\mathbf{i} + (-2 - 0)\mathbf{j} + (2 - 1)\mathbf{k} = -3\mathbf{i} - 2\mathbf{j} + \mathbf{k}.$$

Since **u** and **v** lie in the desired plane, their cross product is perpendicular to it.

$$\mathbf{u} \times \mathbf{v} = \begin{vmatrix} \mathbf{i} & \mathbf{j} & \mathbf{k} \\ -2 & -4 & 0 \\ -3 & -2 & 1 \end{vmatrix} = -4\mathbf{i} + 2\mathbf{j} - 8\mathbf{k}.$$

Thus $\{-4, 2, -8\}$ is a set of direction numbers for a line perpendicular to the desired plane. A simpler set is $\{2, -1, 4\}$. Using this, together with the point $(1, 0, 1)$ in Theorem 20.24, we have

$$2(x - 1) - y + 4(z - 1) = 0$$
$$2x - y + 4z - 6 = 0.$$

Example 4

Sketch $x + 2y + 3z = 6.$

By Theorem 20.25, we know that this equation represents a plane. Knowing this, we merely need to find three points to determine the plane. The simplest points to find are the intercepts (the points where the plane crosses the coordinate axes), which are $(6, 0, 0)$, $(0, 3, 0)$, and $(0, 0, 2)$. Thus we have the plane shown in Figure 20.14. The lines connecting the intercepts are the *traces* of the plane in the coordinate planes.

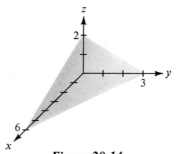

Figure 20.14

Example 5

Sketch $x + 2y = 4.$

This equation represents a line if we are considering only the xy plane (for which $z = 0$). But we get the same line when $z = 1$ or $z = 2$, and so forth. Thus the result is a plane that is parallel to the z axis (see Figure 20.15).

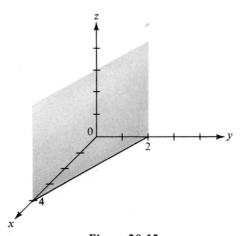

Figure 20.15

Of course the two planes

$$A_1x + B_1y + C_1z + D_1 = 0,$$
$$A_2x + B_2y + C_2z + D_2 = 0$$

are parallel if and only if their normal lines are parallel. Similarly the planes are perpendicular if and only if their normal lines are perpendicular. Thus from Theorems 20.13 and 20.14, we have the following theorem for planes.

Theorem 20.26

The planes

$$A_1x + B_1y + C_1z + D_1 = 0,$$
$$A_2x + B_2y + C_2z + D_2 = 0$$

are parallel (or coincident) if and only if there is a number k such that

$$A_2 = kA_1, \qquad B_2 = kB_1, \qquad C_2 = kC_1;$$

they are perpendicular if and only if

$$A_1A_2 + B_1B_2 + C_1C_2 = 0.$$

Example 6

Show that $3x + y - 4z = 2$ and $6x + 2y - 8z = 3$ are parallel planes and that $3x - y + 2z = 5$ is perpendicular to both of them.

The coefficients of x, y, and z in the three equations are

$$
\begin{array}{lll}
A_1 = 3, & A_2 = 6, & A_3 = 3, \\
B_1 = 1, & B_2 = 2, & B_3 = -1, \\
C_1 = -4, & C_2 = -8, & C_3 = 2.
\end{array}
$$

Since $A_2 = 2A_1$, $B_2 = 2B_1$, and $C_2 = 2C_1$, the first and second planes are either parallel or coincident. But since $D_2 = -3 \neq 2D_1 = 2(-2)$, they are not equivalent equations—the planes are parallel. Since

$$A_1A_3 + B_1B_3 + C_1C_3 = 3 \cdot 3 + 1(-1) - 4 \cdot 2 = 0,$$

the first and third planes are perpendicular. Of course, the third plane must then be perpendicular to the second as well; but this may also be checked using Theorem 20.26.

$$A_2A_3 + B_2B_3 + C_2C_3 = 6 \cdot 3 + 2(-1) - 8 \cdot 2 = 0.$$

Problems

A

In Problems 1–6, sketch the plane.

1. $3x - y + z = 9.$
2. $3x + 2y + z = 6.$
3. $x - y - 4z = 8.$
4. $x + 2y - 4z + 4 = 0.$
5. $y - 5 = 0.$
6. $2x + y = 3.$

In Problems 7–20, find an equation(s) of the plane(s) satisfying the given conditions.

7. Containing $(4, 2, 3)$ and perpendicular to a line with direction numbers $\{-2, 5, 1\}$.
8. Containing $(2, 3, -5)$ and perpendicular to a line with direction numbers $\{4, -3, 1\}$.
9. Containing $(3, 2, 5)$ and perpendicular to the line $x = 1 + t$, $y = 3t$, $z = 4 + t$.
10. Containing $(3, 4, -3)$ and perpendicular to the line $x = 1 - 3t$, $y = 2 + t$, $z = 3 - 2t$.

11. Containing $(4, -1, 2)$ and parallel to $x + y - 2z = 4$.
12. Containing $(5, 3, 1)$ and parallel to $4x - 3y + 2z = 3$.
13. Containing $(2, -2, -2)$, $(1, -3, 5)$, and $(-1, 4, 1)$.
14. Containing $(1, 1, 0)$, $(1, 4, 2)$, and $(2, -2, 1)$.
15. Containing $(3, 1, -4)$, $(2, 3, 1)$, and $(7, 4, -2)$.
16. Containing $(1, 4, 2)$, $(3, 2, -1)$, and $(5, 0, 2)$.
17. Containing $x = 2 + 2t$, $y = -1 + t$, $z = 4 - t$ and $x = 2 - s$, $y = -1 - 2s$, $z = 4 + 3s$.
18. Containing $x = 4 + 2t$, $y = 2 - t$, $z = 1 + t$ and $x = 4 - 2s$, $y = 2 + 2s$, $z = 1 + s$.
19. Containing $(-2, 3, -4)$ and $x = 1 + t$, $y = 3 - 2t$, $z = -2 + t$.
20. Containing $(4, 1, 2)$ and $x = 4 - t$, $y = 1 + 2t$, $z = 3 - t$.

In Problems 21 and 22, find equations of the given line.

21. Containing $(4, -2, 3)$ and perpendicular to $3x + 2y - z + 6 = 0$.
22. Containing $(2, 5, -1)$ and perpendicular to $2x - y + 3z + 2 = 0$.

In Problems 23–28, indicate whether the given planes are parallel, perpendicular, coincident, or none of these.

23. $x + 3y - z = 4$, $2x - y + z = 3$.   24. $3x + y - 5z = 2$, $x + 2y + z = 4$.
25. $4x - 2y + z = 1$, $x + y - 2z = 0$.   26. $4x + y - z = 5$, $x - y + 2z = 2$.
27. $4x - 2y + 2z = 6$, $2x - y + z = 3$.   28. $2x - y + 3z = 4$, $6x - 3y + 9z = 5$.

B

In Problems 29–32, find an equation(s) of the plane(s) satisfying the given conditions.

29. Containing $x = 4 + t$, $y = 2t$, $z = 5$ and $x = 1 + s$, $y = 3 + 2s$, $z = -2$.
30. Containing $x = 3 + 2t$, $y = 4 - t$, $z = 1 + t$ and $x = -1 + 2s$, $y = 3 - s$, $z = 4 + s$.
31. Containing $(3, 0, -4)$ and perpendicular to $2x - 5y + z = 1$ and $x - 2y - z = 3$.
32. Containing $(2, 4, -3)$ and perpendicular to $3x + 2y - z + 1 = 0$ and $x - y + 2z = 0$.

In Problems 33 and 34, find equations of the given line.

33. Containing $(4, 0, 5)$ and parallel to $2x - 5y + z + 1 = 0$ and $x + 2y - z + 2 = 0$.
34. Containing $(2, -4, 5)$ and parallel to $x - y + 3z = 4$ and $3x - 3y + 2z = 5$.
35. Does the line $x = 1$, $y = 5 + 4t$, $z = 2 + t$ lie in the plane $2x - y + 4z = 5$?
36. Find direction cosines of the line $2x - 3y + z = 0$, $x + 2y - 3z = 5$.
37. Find direction cosines of the line $x - y - z = 2$, $4x - 3y + 2z + 7 = 0$.
38. Find the point of intersection of the line $x = 2 + t$, $y = -1 + 3t$, $z = 2 + t$ and the plane $2x + y - 3z + 7 = 0$.
39. Find the point of intersection of the line $\dfrac{x-1}{2} = \dfrac{y+1}{-1} = \dfrac{z}{3}$ and the plane $3x + 2y - z = 5$.

C

40. Show that the plane containing (x_1, y_1, z_1), (x_2, y_2, z_2), and (x_3, y_3, z_3) may be written

$$\begin{vmatrix} x & y & z & 1 \\ x_1 & y_1 & z_1 & 1 \\ x_2 & y_2 & z_2 & 1 \\ x_3 & y_3 & z_3 & 1 \end{vmatrix} = 0.$$

What happens if the three points are collinear?

41. Suppose a line moves along the line $x + y = 4,\quad z = 0$ in such a way that it is always parallel to the vector $\mathbf{i} + \mathbf{j} + \mathbf{k}$. Find an equation of the plane generated by the moving line.

42. Find an equation of the plane that consists of the points equidistant from the points $(1, 1, 3)$ and $(2, 6, 1)$.

43. Find an equation of the plane containing the origin, which is perpendicular to $x + y + z = 6$ and has a normal line that makes an angle of $45°$ with the z axis.

20.6

Distance between a Point and a Plane or Line; Angles between Lines or Planes

In Section 20.4 we used the cross product and the projection of one vector upon another to find the distance between a pair of lines in space. Similar methods can be used to find the distance between a point and a plane or between a point and a line.

Example 1

Find the distance between $(3, -4, 1)$ and $x - 2y + 2z + 4 = 0$.

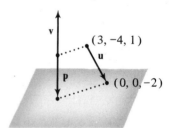

Figure 20.16

A vector perpendicular to the given plane is

$$\mathbf{v} = \mathbf{i} - 2\mathbf{j} + 2\mathbf{k}$$

(see Figure 20.16). We now choose an arbitrary point on $x - 2y + 2z + 4 = 0$, say $(0, 0, -2)$, and let $\mathbf{u}$ be the vector represented by the directed line segment from $(3, -4, 1)$ to $(0, 0, -2)$.

$$\mathbf{u} = (0 - 3)\mathbf{i} + (0 + 4)\mathbf{j} + (-2 - 1)\mathbf{k} = -3\mathbf{i} + 4\mathbf{j} - 3\mathbf{k}.$$

Now the distance we want is the length of the projection $\mathbf{p}$ of $\mathbf{u}$ upon $\mathbf{v}$.

$$d = \frac{|\mathbf{u} \cdot \mathbf{v}|}{|\mathbf{v}|} = \frac{|(-3)(1) + (4)(-2) + (-3)(2)|}{\sqrt{1 + 4 + 4}} = \frac{17}{3}.$$

Exactly the same method can be used to find the distance between the point (x_1, y_1, z_1) and the plane $Ax + By + Cz + D = 0$. We obtain the following result.

Theorem 20.27

The distance between the point (x_1, y_1, z_1) and the plane $Ax + By + Cz + D = 0$ is

$$d = \frac{|Ax_1 + By_1 + Cz_1 + D|}{\sqrt{A^2 + B^2 + C^2}}.$$

The proof is left to the student (see Problem 33). Notice that this formula is similar to the one on page 27 for the distance between a point and a line in two dimensions.

With this formula the distance of Example 1 is

$$d = \frac{|Ax_1 + By_1 + Cz_1 + D|}{\sqrt{A^2 + B^2 + C^2}}$$

$$= \frac{|1 \cdot 3 - 2(-4) + 2 \cdot 1 + 4|}{\sqrt{1^2 + (-2)^2 + 2^2}}$$

$$= \frac{17}{3}.$$

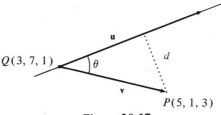

Figure 20.17

Example 2

Find the distance between $P(5, 1, 3)$ and the line $x = 3,\ y = 7 + t,\ z = 1 + t$.

A vector directed along the given line is

$$\mathbf{u} = \mathbf{j} + \mathbf{k}$$

(see Figure 20.17), and the point $Q(3, 7, 1)$ is on the line. Letting $\mathbf{v}$ be the vector represented by $\overrightarrow{QP}$,

$$\mathbf{v} = (5 - 3)\mathbf{i} + (1 - 7)\mathbf{j} + (3 - 1)\mathbf{k} = 2\mathbf{i} - 6\mathbf{j} + 2\mathbf{k}.$$

From Figure 20.17, the distance is

$$d = |\mathbf{v}|\ \sin\theta.$$

But by Theorem 20.21,

$$|\mathbf{u} \times \mathbf{v}| = |\mathbf{u}|\,|\mathbf{v}|\ \sin\theta.$$

Therefore

$$d = |\mathbf{v}|\ \sin\theta = \frac{|\mathbf{u} \times \mathbf{v}|}{|\mathbf{u}|} = \frac{|8\mathbf{i} + 2\mathbf{j} - 2\mathbf{k}|}{|\mathbf{j} + \mathbf{k}|} = \frac{\sqrt{64 + 4 + 4}}{\sqrt{2}}$$

$$= \frac{\sqrt{72}}{\sqrt{2}} = \sqrt{36} = 6.$$

The proof of the following theorem is left to the student (see Problem 34).

Theorem 20.28

The distance from the point (x_1, y_1, z_1) *to the line* $x = x_0 + at,\ \ y = y_0 + bt,$
$z = z_0 + ct$ *is*

$$d = \frac{|\mathbf{u} \times \mathbf{v}|}{|\mathbf{u}|},$$

where $\mathbf{u} = (a, b, c)$ *and* $\mathbf{v} = (x_1 - x_0, y_1 - y_0, z_1 - z_0)$.

We have already considered the angle between two vectors in Section 20.1 (see page 660). The relationship between the two vectors and the angle between them was given in Theorem 20.6. This is restated here for convenience.

If $\mathbf{u} = a_1\mathbf{i} + b_1\mathbf{j} + c_1\mathbf{k}$ *and* $\mathbf{v} = a_2\mathbf{i} + b_2\mathbf{j} + c_2\mathbf{k}$ ($\mathbf{u} \neq \mathbf{0}$ *and* $\mathbf{v} \neq \mathbf{0}$) *and if* θ *is the angle between them, then*

$$\cos\theta = \frac{a_1 a_2 + b_1 b_2 + c_1 c_2}{|\mathbf{u}|\,|\mathbf{v}|}.$$

An angle between two lines can be found in much the same way, since the direction numbers of a line are the components of some vector directed along that line. However, an angle between two lines is the angle between two vectors directed along these lines. Since vectors directed along a line can be oriented in one of two directions, there are two angles θ_1 and θ_2 between any pair of lines. Since

$$\theta_1 + \theta_2 = 180°$$

and

$$\cos\theta_2 = -\cos\theta_1,$$

we can determine the absolute value of $\cos\theta$, where θ is an angle between two lines; but we need additional information to determine the sign of $\cos\theta$.

Theorem 20.29

If θ *is an angle between two lines with direction numbers* $\{a_1, b_1, c_1\}$ *and* $\{a_2, b_2, c_2\}$, *then*

$$|\cos\theta| = \frac{|a_1 a_2 + b_1 b_2 + c_1 c_2|}{\sqrt{a_1^2 + b_1^2 + c_1^2}\sqrt{a_2^2 + b_2^2 + c_2^2}}.$$

Example 3

Find the acute angle between the lines

$$x = 2 - 3t, \qquad y = 4 + t, \qquad z = t$$

and

$$x = 3 + s, \qquad y = 1 - s, \qquad z = 2 + 2s.$$

Direction numbers for the lines are $\{-3, 1, 1\}$ and $\{1, -1, 2\}$. Thus

$$
\begin{aligned}
|\cos\theta| &= \frac{|a_1 a_2 + b_1 b_2 + c_1 c_2|}{\sqrt{a_1^2 + b_1^2 + c_1^2}\sqrt{a_2^2 + b_2^2 + c_2^2}} \\
&= \frac{|-3\cdot 1 + 1(-1) + 1\cdot 2|}{\sqrt{9 + 1 + 1}\sqrt{1 + 1 + 4}} \\
&= \frac{|-2|}{\sqrt{11}\sqrt{6}} = \frac{2}{\sqrt{66}} \approx 0.2462.
\end{aligned}
$$

Since the angle is acute, $\cos \theta$ is positive.

$$\cos \theta \approx 0.2462,$$

$$\theta \approx 76°.$$

If two lines are perpendicular, then $\cos \theta = 0$. This gives us the well-known test for perpendicularity

$$a_1 a_2 + b_1 b_2 + c_1 c_2 = 0,$$

where $\{a_1, b_1, c_1\}$ and $\{a_2, b_2, c_2\}$ are direction numbers for the lines.

Let us now consider an angle of intersection of two planes. This is defined to be an angle between their normal lines. With this definition, the problem is reduced to a familiar one. Note that we have again made no attempt to define *the* angle between two planes but only *an* angle. The particular one desired must be specified. The situation is illustrated in Figure 20.18.

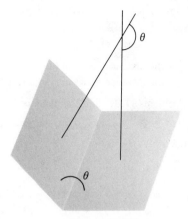

Figure 20.18

Theorem 20.30

If θ is an angle between the planes

$$A_1 x + B_1 y + C_1 z + D_1 = 0 \quad and \quad A_2 x + B_2 y + C_2 z + D_2 = 0,$$

then

$$|\cos \theta| = \frac{|A_1 A_2 + B_1 B_2 + C_1 C_2|}{\sqrt{A_1^2 + B_1^2 + C_1^2}\sqrt{A_2^2 + B_2^2 + C_2^2}}.$$

Example 4

Find the acute angle between the planes $2x + y - z + 3 = 0$ and $4x - y + z + 1 = 0$.

$$|\cos \theta| = \frac{|A_1 A_2 + B_1 B_2 + C_1 C_2|}{\sqrt{A_1^2 + B_1^2 + C_1^2}\sqrt{A_2^2 + B_2^2 + C_2^2}}$$

$$= \frac{|2 \cdot 4 + 1(-1) - 1 \cdot 1|}{\sqrt{4 + 1 + 1}\sqrt{16 + 1 + 1}}$$

$$= \frac{|6|}{\sqrt{6}\sqrt{18}} = \frac{6}{6\sqrt{3}} = \frac{\sqrt{3}}{3} \approx 0.5774.$$

Since we want the acute angle, $\cos \theta$ is positive.

$$\cos \theta \approx 0.5774,$$

$$\theta \approx 55°.$$

Problems

A

In Problems 1–8, find the distance between the plane and point given.

1. $4x + y - 8z + 1 = 0$; $(2, 0, 3)$.
2. $4x + 2y - 2z + 3 = 0$; $(1, 3, -2)$.
3. $2x - y + z + 5 = 0$; $(1, 0, 2)$.
4. $x + y - 3z - 3 = 0$; $(3, 3, 1)$.
5. $x - 2y + 4 = 0$; $(2, 2, 4)$.
6. $x + 3y + z - 3 = 0$; $(2, 1, -2)$.
7. $y + 7 = 0$; $(1, 3, 1)$.
8. $x + z - 5 = 0$; $(2, -4, 2)$.

9. If the distance between $(2, y, 3)$ and $4x - 4y + 2z - 5 = 0$ is 3/2, find y.
10. If the distance between $(1, 4, z)$ and $8x - y + 4z - 3 = 0$ is 1, find z.

In Problems 11–18, find the distance between the point and line given.

11. $(4, 3, 3)$; $x = 2 + 2t,\quad y = 5 - 5t,\quad z = -1 - t$.
12. $(1, 3, -2)$; $x = 4,\quad y = -1 + 4t,\quad z = 2 + t$.

13. $(-1, 0, 5)$; $\dfrac{x - 2}{2} = \dfrac{y - 1}{1} = \dfrac{z + 2}{3}$.

14. $(2, 4, -1)$; $x = 4 + t,\quad y = -2 + 3t,\quad z = 1 + t$.
15. $(2, 3, -1)$; $x = 4 + t,\quad y = 1 - t,\quad z = 3 + 2t$.
16. $(4, 1, -2)$; $x = 2 - t,\quad y = 3 - 2t,\quad z = 1 - t$.

17. $(3, -1, 4)$; $\dfrac{x + 2}{1} = \dfrac{y}{-2} = \dfrac{z + 4}{-2}$.

18. $(1, 3, 2)$; $\dfrac{x - 1}{3} = \dfrac{y + 2}{1} = \dfrac{z - 1}{-2}$.

In Problems 19–24, find the distance between the parallel planes.

19. $x - 4y - 2z = 5,\quad x - 4y - 2z = 10$.
20. $2x - y + 2z = 9,\quad 2x - y + 2z = -12$.
21. $x + y + 4z = 6,\quad 2x + 2y + 8z = 9$.
22. $x + 3y - 4z = 3,\quad 2x + 6y - 8z = -5$.
23. $x - y - z = 4,\quad 2x - 2y - 2z = -3$.
24. $x + 2y = 4,\quad x + 2y = -1$.

In Problems 25–32, find the angle described.

25. The angle between $\mathbf{u} = \mathbf{i} - 2\mathbf{j} + 6\mathbf{k}$ and $\mathbf{v} = 4\mathbf{i} + 5\mathbf{j}$.
26. The angle between $\mathbf{u} = 2\mathbf{i} + \mathbf{j} + 4\mathbf{k}$ and $\mathbf{v} = 4\mathbf{i} - \mathbf{j} + 2\mathbf{k}$.
27. The obtuse angle between $x = 1 - 2t,\quad y = 3 + t,\quad z = 4t$, and $x = 2 + t,\quad y = 3 - 2t,\quad z = 1 + 3t$.
28. The acute angle between $x = 3t,\quad y = 2t,\quad z = -t$ and $x = -t,\quad y = -4t,\quad z = 2t$.
29. The acute angle between $x = 3 - t,\quad y = 4 - 2t,\quad z = -1 - t$ and $x = t,\quad y = -4 + t,\quad z = 2 + t$.
30. The obtuse angle between $x = 1 - t,\quad y = 3 + 2t,\quad z = 5 - t$ and $x = 3 - t,\quad y = 4 - t,\quad z = 2t$.
31. The obtuse angle between $x - y + z - 4 = 0$ and $2x + y + z = 0$.
32. The acute angle between $x + 2y - z - 4 = 0$ and $x + y - 3z + 7 = 0$.

B

33. Prove Theorem 20.27.
34. Prove Theorem 20.28.

20.7

Cylinders and Spheres

We now turn our attention to more complex surfaces, beginning with the cylinder. A cylinder is formed by a line (*generatrix*) moving along a curve (directrix) while remaining parallel to a fixed line. If the generatrix is parallel to one of the coordinate axes, the equation of the cylinder is quite simple.

Theorem 20.31

A nonlinear equation of the form

$$f(x, y) = 0$$

is a cylinder with generatrix parallel to the z axis and directrix $f(x, y) = 0$ in the xy plane. Similar statements hold when one of the other variables is absent.

It is a simple matter to see why this is so. If $x = x_0$ and $y = y_0$ satisfies the equation $f(x, y) = 0$, then any point of the form (x_0, y_0, z), for *any* choice of z, is on the surface. But the set of all such points is a line parallel to the z axis. Thus any point on the curve $f(x, y) = 0$ in the xy plane determines a line parallel to the z axis in space. The result is then a cylinder.

Example 1

Sketch $x^2 + y^2 = 4$.

The surface is a cylinder with generatrix parallel to the z axis and directrix a circle in the xy plane. A portion of the cylinder is given in Figure 20.19.

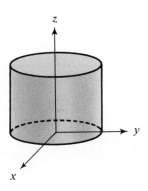

Figure 20.19

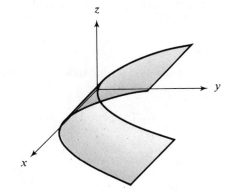

Figure 20.20

Example 2

Sketch $y = z^2$.

The surface is a cylinder with generatrix parallel to (or on) the x axis and directrix a parabola in the yz plane. A portion of the cylinder is given in Figure 20.20.

Another relatively simple surface is the sphere. The following theorems concerning
the sphere are analogous to those for a circle and are proved in much the same way.

Theorem 20.32

*A point (x, y, z) is on the sphere of radius r and center at (h, k, l) if and only if it
satisfies the equation*

$$(x - h)^2 + (y - k)^2 + (z - l)^2 = r^2.$$

Theorem 20.33

Any sphere can be represented by an equation of the form

$$Ax^2 + Ay^2 + Az^2 + Gx + Hy + Iz + J = 0,$$

where $A \neq 0$.

Theorem 20.34

An equation of the form

$$Ax^2 + Ay^2 + Az^2 + Gx + Hy + Iz + J = 0,$$

where $A \neq 0$, represents a sphere, a point, or no locus.

The proofs are left to the student.

Example 3

Give an equation for the sphere with center $(1, 3, -2)$ and radius 3.

By Theorem 20.32, the equation is

$$(x - 1)^2 + (y - 3)^2 + (z + 2)^2 = 9$$

or

$$x^2 + y^2 + z^2 - 2x - 6y + 4z + 5 = 0.$$

Example 4

Describe the locus of $x^2 + y^2 + z^2 + 2x - 4y - 8z + 5 = 0$.

Let us put the equation into the form of Theorem 20.32 by completing squares.

$$x^2 + 2x \qquad + y^2 - 4y \qquad + z^2 - 8z \qquad = -5$$
$$x^2 + 2x + 1 + y^2 - 4y + 4 + z^2 - 8z + 16 = -5 + 1 + 4 + 16$$
$$(x + 1)^2 + (y - 2)^2 + (z - 4)^2 = 16.$$

This represents a sphere with center $(-1, 2, 4)$ and radius 4.

Example 5

Describe the locus of $2x^2 + 2y^2 + 2z^2 - 2x + 6y - 4z + 7 = 0$.

$$x^2 + y^2 + z^2 - x + 3y - 2z + \frac{7}{2} = 0$$

$$x^2 - x \qquad + y^2 + 3y \qquad + z^2 - 2z \qquad = -\frac{7}{2}$$

$$x^2 - x + \frac{1}{4} + y^2 + 3y + \frac{9}{4} + z^2 - 2z + 1 = -\frac{7}{2} + \frac{1}{4} + \frac{9}{4} + 1$$

$$\left(x - \frac{1}{2}\right)^2 + \left(y + \frac{3}{2}\right)^2 + (z - 1)^2 = 0.$$

The equation represents the point $\left(\frac{1}{2}, -\frac{3}{2}, 1\right)$.

Problems

A

In Problems 1–10, sketch the given surface.

1. $x^2 + z^2 = 4$.
2. $y^2 + z^2 = 4$.
3. $x^2 - z^2 = 1$.
4. $y = 4x^2$.
5. $x^2 + z^2 + 2x = 0$.
6. $xy = 8$.
7. $x = \sin z$.
8. $z = 9 - y^2$.
9. $z = e^y$.
10. $x = \ln z$.

In Problems 11–20, identify the equation as representing a sphere, a point, or no locus. If it is a sphere, give its center and radius. If it is a point, give its coordinates.

11. $x^2 + y^2 + z^2 + 6x - 10y + 2z + 19 = 0$.
12. $x^2 + y^2 + z^2 - 2x + 4z - 4 = 0$.
13. $x^2 + y^2 + z^2 + 6x - 8y - 2z + 22 = 0$.
14. $x^2 + y^2 + z^2 - 8x + 4y - 8z + 37 = 0$.
15. $2x^2 + 2y^2 + 2z^2 - 2x + 2y - 10z + 13 = 0$.
16. $2x^2 + 2y^2 + 2z^2 + 2x - 6y + 4z - 1 = 0$.
17. $3x^2 + 3y^2 + 3z^2 + 4x - 2y - 8z + 7 = 0$.
18. $9x^2 + 9y^2 + 9z^2 - 12x + 6y + 6z - 2 = 0$.

19. $6x^2 + 6y^2 + 6z^2 - 6x - 4y - 3z = 0.$
20. $4x^2 + 4y^2 + 4z^2 - 4x - 8y + 16z + 21 = 0.$

In Problems 21 and 22, find an equation(s) in the general form of the sphere(s) described.

21. Center $(3, 1, 1)$ and containing the origin.
22. Center $(3, 2, 1)$ and radius 5.

B

In Problems 23–28, find an equation(s) in the general form of the sphere(s) described.

23. Center $(4, 1, -3)$ and tangent to $2x - y - 2z = 4.$
24. Center $(2, 4, 7)$ and tangent to $4x - 8y + z = 10.$
25. Tangent to $x + 2y + 2z - 17 = 0$ at $(1, 4, 4)$ with radius 3.
26. Tangent to $x - 3y + 4z + 23 = 0$ at $(1, 4, -3)$ with radius $\sqrt{26}$.
27. Containing $(4, 1, 0), (-2, -1, 0), (0, 2, 1),$ and $(1, 1, 1).$
28. Containing $(3, 1, -1), (2, 5, 2), (-3, 0, 1),$ and $(-1, 0, 0).$
29. Prove Theorem 20.32.   30. Prove Theorem 20.33.
31. Prove Theorem 20.34.
32. Write the equations of the coordinate planes.

C

In Problems 33–38, sketch the solid bounded by the given planes and surfaces.

33. $x^2 + y^2 = 9,\quad y = x,\quad x = 0,\quad z = 0,\quad z = 5;$ in the first octant.
34. $z = 4 - x^2,\quad y = 2x,\quad y = 4,\quad x = 0,\quad z = 0.$
35. $2z = 4 + x^2,\quad z = 0,\quad x = 2,\quad x = -2,\quad y = 4,\quad y = -4.$
36. $y^2 + z^2 = 4,\quad x = 2y,\quad x = 4,\quad y = 0,\quad z = 0.$
37. $x^2 + y^2 + z^2 = 25,\quad y = x,\quad x = 0,\quad z = 0;$ in the first octant.
38. $x^2 + y^2 = 1,\quad z = 1 - x,\quad z = 0.$

39. Find an equation of the set of points equidistant from the z axis and the plane $y = 3$. Sketch.

20.8

Quadric Surfaces

In the plane, a second-degree equation represents a parabola, ellipse, hyperbola, or a degenerate case of one of them. There are far more variations in space, where we have already seen that certain cylinders and the sphere are represented by second-degree equations. The traces in the coordinate planes of a given surface are simply the intersections of the surface with the coordinate planes. The traces in the coordinate planes of quadric surfaces, represented by second-degree equations, are conics or degenerate conics. We say that a quadric surface is in the standard position if its traces in the coordinate planes are in the standard position—that is, they have their center or vertex at the origin and axes along the coordinate axes. In the following discussion, all surfaces are assumed to be in the standard position.

The *ellipsoid* (see Figure 20.21) is represented by an equation of the form

$$\frac{x^2}{a^2} + \frac{y^2}{b^2} + \frac{z^2}{c^2} = 1.$$

Its traces in the coordinate planes are ellipses (or circles). We have already seen a special case of this, in which $a = b = c$. In that case, we have a sphere. There are two other special cases. One is the *prolate spheroid*. Here, two of the denominators are equal and both are less than the third. It has the shape of a football and may be generated by rotating an ellipse about its major axis. The other case is the *oblate spheroid*, in which two of the denominators are equal and both greater than the third. It has the shape of a doorknob and may be generated by rotating an ellipse about its minor axis.

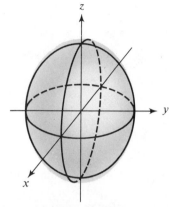

Figure 20.21

The *hyperboloid of one sheet* (see Figure 20.22) is represented by an equation of the form

$$\frac{x^2}{a^2} + \frac{y^2}{b^2} - \frac{z^2}{c^2} = 1.$$

Its traces in the xz plane and yz plane are hyperbolas; in the xy plane, the trace is an ellipse. If $a = b$, it may be generated by rotating a hyperbola about its conjugate axis.

The *hyperboloid of two sheets* (see Figure 20.23) is represented by an equation of the form

$$\frac{x^2}{a^2} - \frac{y^2}{b^2} - \frac{z^2}{c^2} = 1.$$

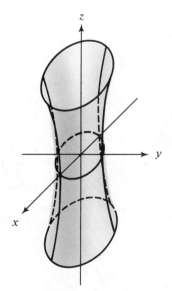

Figure 20.22

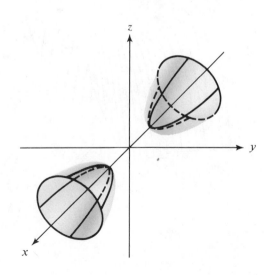

Figure 20.23

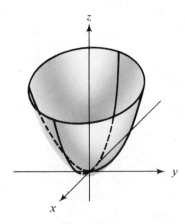

Figure 20.24

Its traces in the xy plane and xz plane are hyperbolas. It has no trace in the yz plane; however, if $|x| > a$, its intersection with a plane parallel to the yz plane is an ellipse. If $b = c$, it may be generated by rotating a hyperbola about its tranverse axis.

The *elliptic paraboloid* (see Figure 20.24) is represented by an equation of the form

$$\frac{x^2}{a^2} + \frac{y^2}{b^2} = \frac{z}{c}.$$

Its traces in the xz plane and yz plane are parabolas. Its trace in the xy plane is a single point. If $c > 0$, then its intersection with a plane parallel to and above the xy plane

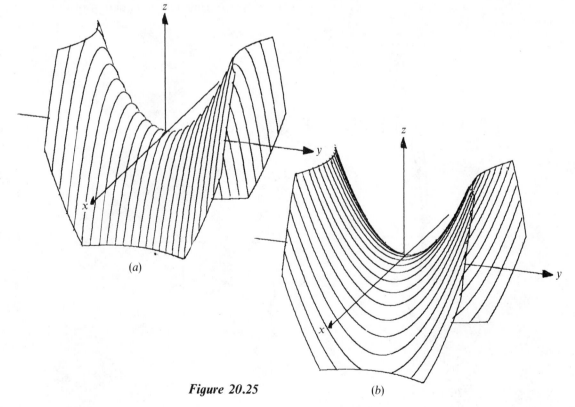

(a)

(b)

Figure 20.25

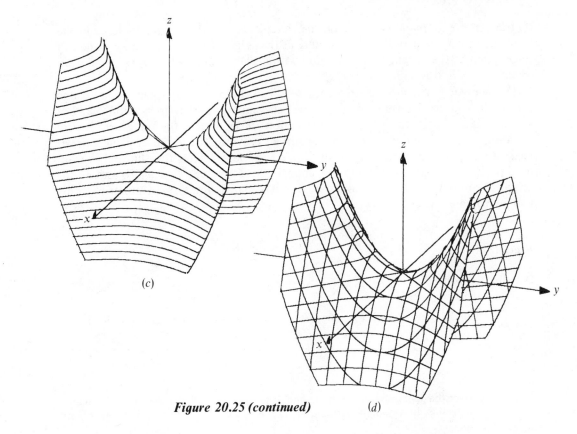

Figure 20.25 (continued) (d)

is an ellipse; below the xy plane there is no intersection. This situation is reversed if $c < 0$. In Figure 20.24, $c > 0$. If $a = b$, the elliptic paraboloid is generated by rotating a parabola about its axis.

The *hyperbolic paraboloid* (see Figure 20.25) is represented by an equation of the form

$$\frac{x^2}{a^2} - \frac{y^2}{b^2} = \frac{z}{c}.$$

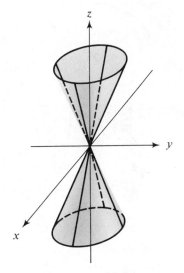

Its traces in the xz plane and yz plane are parabolas, one opening upward and the other down. Its trace in the xy plane is a pair of lines intersecting at the origin (a degenerate hyperbola). Its intersection with a plane parallel to the xy plane is a hyperbola. If $c > 0$, those hyperbolas above the xy plane have the transverse axis parallel to the x axis, while those below have it parallel to the y axis. If $c < 0$, this situation is reversed. In Figure 20.25, $c < 0$.

The *elliptic cone* (see Figure 20.26) is represented by an equation of the form

$$\frac{x^2}{a^2} + \frac{y^2}{b^2} - \frac{z^2}{c^2} = 0.$$

Figure 20.26

Its trace in the xz plane is a pair of lines intersecting at the origin. Its trace in the yz plane is also a pair of lines intersecting at the origin. Its trace in the xy plane is a single point at the origin. Its intersection with a plane parallel to the xy plane is an ellipse. If $a = b$, it is a circular cone.

The symmetry of a surface $F(x, y, z) = 0$ in the coordinate planes and around the origin can be determined by an analysis similar to that in Section 5.2. For example, if z is replaced by $-z$ in the equation $F(x, y, z) = 0$ and an equivalent equation results, then the corresponding surface is symmetric in the xy plane. Similar tests can be given for symmetry in the xz and yz planes. Thus the ellipsoid in standard position is symmetric in all three coordinate planes, and hence also around the origin, as are the hyperboloids and the cone. The paraboloids are symmetric in only two coordinate planes.

Example 1

Describe and sketch $9x^2 + 9y^2 - 4z^2 = 36$.

Dividing by 36, we have a hyperboloid of one sheet in the standard form:

$$\frac{x^2}{4} + \frac{y^2}{4} - \frac{z^2}{9} = 1.$$

Since the denominators of the x^2 and y^2 terms are equal, the trace in the xy plane, as well as in any plane parallel to it, is a circle. The surface is given in Figure 20.27.

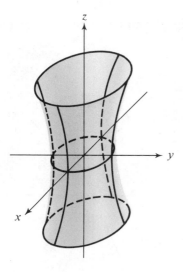

Figure 20.27

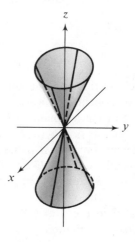

Figure 20.28

Example 2

Describe and sketch $9x^2 + 9y^2 - 4z^2 = 0$.

Again dividing by 36, we have

$$\frac{x^2}{4} + \frac{y^2}{4} - \frac{z^2}{9} = 0,$$

which is a circular cone with its axis the z axis. The cone is given in Figure 20.28.

If the given equation does not fit any of these forms, determine its traces in the coordinate planes and planes parallel to them in order to have some idea of the shape.

Problems

A

Describe and sketch the following quadric surfaces.

1. $4x^2 + 9y^2 + 9z^2 = 36$.
2. $36x^2 + 4y^2 + 9z^2 = 36$.
3. $4x^2 + 4y^2 - z^2 + 16 = 0$.
4. $x - y^2 - z^2 = 0$.
5. $x^2 - y - z^2 = 0$.
6. $x^2 - y^2 + z^2 = 0$.
7. $x^2 - 4y^2 - 4z^2 = 0$.
8. $x^2 + 4y + z^2 = 0$.
9. $16x^2 - 9y^2 - 9z^2 + 144 = 0$.
10. $4x^2 - y^2 + 4z^2 + 16 = 0$.
11. $25x^2 + 16y^2 + 25z^2 = 400$.
12. $x^2 + 4y - z^2 = 0$.
13. $16x^2 - 9y + 16z^2 = 0$.
14. $16x^2 - 9y^2 + 36z^2 + 144 = 0$.
15. $36x^2 - 4y - 9z^2 = 0$.
16. $16x^2 - 9y^2 - 9z^2 = 0$.
17. $x^2 + y^2 - 4z = 0$.
18. $16x^2 - 36y^2 + 9z^2 + 144 = 0$.
19. $x^2 + y^2 + 4z^2 = 4$.
20. $x^2 + y^2 - 4z^2 = 4$.
21. $9x^2 - y^2 - z^2 = 9$.
22. $x^2 - y^2 - 9z = 0$.
23. $x^2 + y^2 + 2z = 0$.
24. $9x^2 - y^2 - z^2 = 0$.
25. $25x^2 - 4y^2 + 25z^2 = 100$.

B

If the coordinate axes are translated with the origin moved to (h, k, l), the equations of translation are $x' = x - h$, $y' = y - k$, $z' = z - l$.   In many cases, translations may be carried out by completing the square. In Problems 26–30, translate and sketch.

26. $z = 4 - x^2 - 2y^2$.
27. $z = 1 + x^2 + y^2$.
28. $z = x^2 + y^2 + 2x + 4y + 7$.
29. $x^2 + 4y^2 + 9z^2 + 2x + 16y - 18z - 10 = 0$.
30. $x^2 + 4y^2 - z^2 - 2x - 24y - 8z + 17 = 0$.

C

In Problems 31–36, sketch the solid bounded by the given surfaces.

31. $z = x^2 + y^2$, $z = 4$, $y = x$, $y = 0$;   in the first octant.
32. $z = 9 - x^2 - y^2$, $y = x$, $y = 0$, $z = 0$;   in the first octant.
33. Above $x^2 + y^2 - z^2 = 0$,   below $x + y + z = 4$;   in the first octant.
34. $x^2 + y^2 - z^2 = 0$, $x + z = 4$.
35. $x^2 + y^2 = 9 - z$, $x^2 + y^2 = 9 + z$.

36. Inside $x^2 + y^2 - z^2 = 0$, and $x^2 + y^2 + z^2 = 16$.

37. Find equations for the set of points (x, y, z)
     (a) equidistant from $(0, 0, 4)$ and the plane $z = -4$;
     (b) twice as far from $(0, 0, 4)$ as from $z = -4$;
     (c) half as far from $(0, 0, 4)$ as from $z = -4$.

38. Find an equation of the set of points such that the square of the distance from the y axis equals the distance from the xz plane.

39. Show that through each point of the hyperboloid of one sheet

$$\frac{x^2}{a^2} + \frac{y^2}{b^2} - \frac{z^2}{c^2} = 1$$

there are two lines lying entirely on the surface; such a surface is called a *ruled* surface. (*Hint:* Write the equation as $\dfrac{x^2}{a^2} - \dfrac{z^2}{c^2} = 1 - \dfrac{y^2}{b^2}$ and factor both sides.)

20.9

Cylindrical and Spherical Coordinates

Here we look at two other coordinate systems in space that are useful. The first of these is a *cylindrical coordinate system*, which is convenient when the z axis is a line of symmetry. In this system, a point P with projection Q on the xy plane (see Figure 20.29) is represented by (r, θ, z), where (r, θ) is a polar representation of Q and z is the (directed) distance of P from the xy plane. The relations between rectangular and cylindrical coordinates are the same as the relations between rectangular and polar coordinates; that is,

$$x = r \cos \theta, \qquad y = r \sin \theta, \qquad z = z$$

or

$$r = \pm\sqrt{x^2 + y^2}, \qquad \theta = \arctan \frac{y}{x}, \qquad z = z.$$

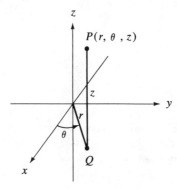

Figure 20.29

Example 1

Express $(4, 30°, -2)$ in rectangular coordinates.

$$x = r \cos \theta = 4 \cos 30° = 4 \cdot \frac{\sqrt{3}}{2} = 2\sqrt{3},$$

$$y = r \sin \theta = 4 \sin 30° = 4 \cdot \frac{1}{2} = 2,$$

$$z = -2.$$

Thus, the point is $(2\sqrt{3}, 2, -2)$.

Example 2

Express $(2, 2, 4)$ in cylindrical coordinates.

$$r = \pm\sqrt{x^2 + y^2} = \pm\sqrt{8} = \pm 2\sqrt{2},$$

$$\theta = \arctan \frac{y}{x} = \arctan 1 = \pi/4 + n\pi,$$

$$z = 4.$$

There are two choices for r and infinitely many for θ. The choices we make are not independent of each other—the choice of one puts restrictions on the other. If we choose θ to be $\pi/4$ (or any first-quadrant angle), we must choose $r = 2\sqrt{2}$; if we choose θ to be $5\pi/4$ (or any third-quadrant angle), we must choose $r = -2\sqrt{2}$. Thus, two possible representations are

$$(2\sqrt{2}, \pi/4, 4) = (-2\sqrt{2}, 5\pi/4, 4).$$

Example 3

Express $x + y + z = 1$ in cylindrical coordinates.

Substituting $x = r \cos \theta$ and $y = r \sin \theta$, we have

$$r \cos \theta + r \sin \theta + z = 1.$$

Example 4

Express $r = z \sin \theta$ in rectangular coordinates.

Multiplying both sides by r, we have

$$r^2 = z \cdot r \sin \theta, \qquad x^2 + y^2 = yz.$$

Another useful system for representing points in space uses *spherical coordinates*. This coordinate system is most convenient when there is symmetry about the origin. In this system, a point is represented by (ρ, θ, φ), where ρ is the distance of the point from the origin, θ has the same meaning as in cylindrical coordinates, and φ is the angle between the positive end of the z axis and the segment joining the origin to the given point (see Figure 20.30). Since φ is undirected, it is never negative.

Furthermore

$$0 \le \varphi \le \pi.$$

Likewise, we restrict ρ: $\rho \ge 0$.

From Figure 20.30, we see that $\overline{OQ} = \rho \sin \varphi$. Thus

$$x = \overline{OQ} \cos \theta = \rho \sin \varphi \cos \theta,$$

$$y = \overline{OQ} \sin \theta = \rho \sin \varphi \sin \theta,$$

$$z = \rho \cos \varphi.$$

We can easily see from these that

$$\rho^2 = x^2 + y^2 + z^2.$$

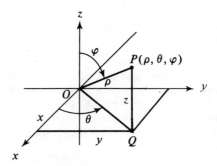

Figure 20.30

Example 5

Express $(2, \pi/6, 2\pi/3)$ in rectangular coordinates.

$$x = \rho \sin \varphi \cos \theta$$
$$= 2 \sin 2\pi/3 \cdot \cos \pi/6$$
$$= 2 \cdot \frac{\sqrt{3}}{2} \cdot \frac{\sqrt{3}}{2} = \frac{3}{2},$$
$$y = \rho \sin \varphi \sin \theta$$
$$= 2 \sin 2\pi/3 \sin \pi/6$$
$$= 2 \cdot \frac{\sqrt{3}}{2} \cdot \frac{1}{2} = \frac{\sqrt{3}}{2},$$
$$z = \rho \cos \varphi$$
$$= 2 \cos 2\pi/3$$
$$= 2\left(-\frac{1}{2}\right) = -1.$$

Thus the point is $(3/2, \sqrt{3}/2, -1)$.

Example 6

Express $(\sqrt{3}, \sqrt{3}, -\sqrt{2})$ in spherical coordinates.

$$\rho^2 = x^2 + y^2 + z^2$$
$$= 3 + 3 + 2 = 8,$$
$$\rho = 2\sqrt{2};$$
$$z = \rho \cos \varphi,$$
$$-\sqrt{2} = 2\sqrt{2} \cos \varphi,$$
$$\cos \varphi = -\frac{1}{2},$$
$$\varphi = 120°;$$
$$x = \rho \sin \varphi \cos \theta,$$
$$\sqrt{3} = 2\sqrt{2} \cdot \frac{\sqrt{3}}{2} \cos \theta,$$
$$\cos \theta = \frac{1}{\sqrt{2}},$$
$$\theta = 45°.$$

The point is $(2\sqrt{2}, 45°, 120°)$.

Example 7

Express $x^2 + y^2 - z^2 = 0$ in spherical coordinates.

$$\rho^2 \sin^2 \varphi \cos^2 \theta + \rho^2 \sin^2 \varphi \sin^2 \theta - \rho^2 \cos^2 \varphi = 0$$
$$\rho^2[\sin^2 \varphi(\cos^2 \theta + \sin^2 \theta) - \cos^2 \varphi] = 0$$
$$\rho^2(\sin^2 \varphi - \cos^2 \varphi) = 0$$
$$\sin^2 \varphi - \cos^2 \varphi = 0 \quad \text{or} \quad \rho = 0;$$
$$\cos 2\varphi = 0,$$
$$2\varphi = \pi/2, \ 3\pi/2$$
$$\varphi = \pi/4, \ 3\pi/4.$$

Thus, we have $\varphi = \pi/4$, $\varphi = 3\pi/4$, and $\rho = 0$. The value $\varphi = \pi/4$ gives the top half of the cone; $\varphi = 3\pi/4$ gives the bottom half; and $\rho = 0$ gives a single point—the origin. Since $\rho = 0$ is included in both of the others, we may drop it. The final result is

$$\varphi = \pi/4 \quad \text{and} \quad \varphi = 3\pi/4.$$

Example 8

Express $\rho^2 \sin \varphi \cos \varphi \cos \theta = 1$ in rectangular coordinates.

$$\rho^2 \sin \varphi \cos \varphi \cos \theta = 1$$
$$(\rho \sin \varphi \cos \theta)(\rho \cos \varphi) = 1,$$
$$xz = 1.$$

Problems

A

1. The following points are given in cylindrical coordinates. Express them in rectangular coordinates.

   (a) $(2, 45°, 2)$. (b) $(3, 2\pi/3, -2)$. (c) $(2, 0°, 1)$. (d) $(0, \pi/4, -3)$.

2. The following points are given in rectangular coordinates. Express them in cylindrical coordinates.

   (a) $(1, 1, 5)$. (b) $(0, 2, -2)$. (c) $(-1, \sqrt{3}, 3)$. (d) $(-2\sqrt{3}, -2, 3)$.

3. The following points are given in spherical coordinates. Express them in rectangular coordinates.

   (a) $(4, 45°, 30°)$. (b) $(1, \pi/6, 0)$. (c) $(2, 90°, 45°)$. (d) $(2, 5\pi/6, 3\pi/4)$.

4. The following points are given in rectangular coordinates. Express them in spherical coordinates.

   (a) $(2, 2, 0)$. (b) $(2, 1, -2)$. (c) $(4, -\sqrt{3}, 2)$. (d) $(1, 1, \sqrt{2})$.

5. The following points are given in cylindrical coordinates. Express them in spherical coordinates.

   (a) $(2, \pi/4, -2)$. (b) $(3, 30°, 4)$. (c) $(2, \pi/2, -4)$. (d) $(0, 45°, 3)$.

6. The following points are given in spherical coordinates. Express them in cylindrical coordinates.

   (a) $(4, 45°, 30°)$. (b) $(2, 2\pi/3, \pi/2)$. (c) $(2, 210°, 135°)$. (d) $(3, \pi/6, 2\pi/3)$.

In Problems 7–14, express the given equations in cylindrical and spherical coordinates.

7. $x^2 + y^2 + z^2 = 4$.   8. $x^2 + y^2 = 6$.
9. $x^2 - y^2 = z$.   10. $x^2 + y^2 = z$.
11. $x^2 - y^2 + z^2 = 1$.   12. $x^2 - y^2 - z^2 = 1$.
13. $4x^2 + 9y^2 + 9z^2 = 1$.   14. $x^2 + y^2 - z^2 = 1$.

B

In Problems 15–22, express the given equations in rectangular coordinates.

15. $z = r^2 \cos 2\theta$.   16. $z = r^2 \sin 2\theta$.
17. $z = r(1 + \sin \theta)$.   18. $z = r^2$.
19. $\rho \sin \varphi = 1$.   20. $\rho \sin \varphi \tan \varphi \sin 2\theta = 3$.
21. $\rho^2 \sin \varphi \cos \varphi = 1$.   22. $\rho = \sin \varphi \cos \theta$.

C

In Problems 23–29, describe the curves and regions determined by the given equations and inequalities.

23. $\theta = \pi/4$, $\phi = \pi/4$.   24. $\rho = 5$, $\theta = \pi/3$.
25. $r = 2\theta$, $z = 0$.   26. $\theta = z$.
27. $2 \le \rho \le 3$.   28. $\rho \le 4$, $\rho \sin \phi \ge 2$.
29. $0 \le \theta \le \pi/4$, $0 \le \phi \le \pi/4$, $0 \le \rho \le 4$.

20.10

Curves in Space; Velocity and Acceleration

Curves in space are most easily represented by parametric equations or, equivalently, vector-valued functions. We have already considered these in two dimensions, in Chapter 13.

Example 1

Graph $\mathbf{f}(t) = \cos t\,\mathbf{i} + \sin t\,\mathbf{j} + t\,\mathbf{k}.$

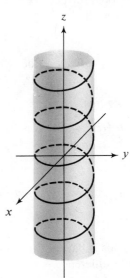

This has a three-dimensional graph which is equivalent to

$$x = \cos t, \qquad y = \sin t, \qquad z = t.$$

The result is the spiral shown in Figure 20.31.

Limits for vector functions can be defined in terms of limits for the components.

Definition

If $\mathbf{f}(t) = f_1(t)\mathbf{i} + f_2(t)\mathbf{j} + f_3(t)\mathbf{k},$ then

$$\lim_{t \to a} \mathbf{f}(t) = \left[\lim_{t \to a} f_1(t)\right]\mathbf{i} + \left[\lim_{t \to a} f_2(t)\right]\mathbf{j} + \left[\lim_{t \to a} f_3(t)\right]\mathbf{k}.$$

Figure 20.31

Now the derivative is defined as it was for real-valued functions.

Definition

$$\mathbf{f}'(t) = \lim_{h \to 0} \frac{\mathbf{f}(t + h) - \mathbf{f}(t)}{h}.$$

Theorem 20.35

If $\mathbf{f}(t) = f_1(t)\mathbf{i} + f_2(t)\mathbf{j} + f_3(t)\mathbf{k},$ then

$$\mathbf{f}'(t) = f_1'(t)\mathbf{i} + f_2'(t)\mathbf{j} + f_3'(t)\mathbf{k}.$$

It is a simple matter to show that this theorem follows from the definition of a derivative. The proof is left to the student.

Example 2

Find the derivative of $\mathbf{f}(t) = t^2\mathbf{i} + (2t + 1)\mathbf{j} + (2t - 1)\mathbf{k}.$

$$\mathbf{f}'(t) = 2t\,\mathbf{i} + 2\mathbf{j} + 2\mathbf{k}.$$

The derivative of a vector function at a particular value of t is another vector with representatives on or parallel to the tangent to the graph of the original function at the given value of t. This is seen by a graphical consideration of the definition of the derivative.

As noted earlier, velocity is a vector having both magnitude and direction. Up to now we have used the scalar speed and the vector velocity interchangeably, since we were dealing only with rectilinear motion. Let us now consider curvilinear motion. Suppose a particle is moving along a curve represented parametrically by

$$x = x(t), \qquad y = y(t), \qquad z = z(t).$$

Suppose further that at time t the particle is at $(x(t), y(t), z(t))$. This can be represented vectorially by

$$\mathbf{r}(t) = x(t)\mathbf{i} + y(t)\mathbf{j} + z(t)\mathbf{k}.$$

Now we define velocity and acceleration.

Definition

If $\mathbf{r}(t) = x(t)\mathbf{i} + y(t)\mathbf{j} + z(t)\mathbf{k}$ *represents the position of a particle at time t, then the* **velocity** *at time t is the vector*

$$\mathbf{v}(t) = \mathbf{r}'(t)$$

and the **acceleration** *at time t is the vector*

$$\mathbf{a}(t) = \mathbf{r}''(t).$$

This is basically the same thing that we encountered with rectilinear motion; the only difference is that we are dealing with vector functions here instead of the real-valued functions we have had in the past. The two situations are tied together by the following.

Definition

If $\mathbf{r}(t) = x(t)\mathbf{i} + y(t)\mathbf{j} + z(t)\mathbf{k}$ *represents the position of a particle at time t, then the* **speed** *at time t is*

$$\frac{ds}{dt} = |\mathbf{r}'(t)| = |\mathbf{v}(t)|$$

and the **rate of change of speed** *is*

$$\frac{d^2 s}{dt^2} = \frac{d}{dt}|\mathbf{r}'(t)| = \frac{d}{dt}|\mathbf{v}(t)|.$$

The speed, which is defined above as the absolute value of the velocity, is equivalent to the absolute value of the rectilinear velocity that we have already considered. Let us see how. Since the distance covered by a moving particle is the length of the arc traversed, we want to concern ourselves here with arc length. You will recall that in two dimensions, arc length is given by

$$s = \int_a^b \sqrt{\left(\frac{dx}{dt}\right)^2 + \left(\frac{dy}{dt}\right)^2} \, dt.$$

This can easily be extended to three dimensions, to get

$$s = \int_a^b \sqrt{\left(\frac{dx}{dt}\right)^2 + \left(\frac{dy}{dt}\right)^2 + \left(\frac{dz}{dt}\right)^2}\, dt.$$

Now if we take the limits of integration to be 0 and t_0, instead of a and b, we see that s (which is now a function of t_0) is the length of the arc traversed between time 0 and time t_0.

$$s(t_0) = \int_0^{t_0} \sqrt{\left(\frac{dx}{dt}\right)^2 + \left(\frac{dy}{dt}\right)^2 + \left(\frac{dz}{dt}\right)^2}\, dt.$$

Thus, $s(t_0)$ represents the distance traversed in a given time t_0. The speed is the rate of change of this distance, or ds/dt_0. Differentiating $s(t_0)$, we have

$$\frac{ds}{dt_0} = \sqrt{\left(\frac{dx}{dt_0}\right)^2 + \left(\frac{dy}{dt_0}\right)^2 + \left(\frac{dz}{dt_0}\right)^2}$$

$$= \left|\frac{d\mathbf{r}(t_0)}{dt}\right| = |\mathbf{r}'(t_0)|,$$

or

$$\frac{ds}{dt} = |\mathbf{r}'(t)|.$$

This is exactly the way the speed was defined. Thus, this definition is consistent with our former idea of speed.

Example 3

A particle moves along a path according to the equation $\mathbf{r}(t) = \cos t\,\mathbf{i} + \sin t\,\mathbf{j}$. Find the velocity, acceleration, speed, and rate of change of speed at $t = \pi/4$. Graph the path of the particle and give a graphical representation of $\mathbf{v}$ and $\mathbf{a}$ at $t = \pi/4$.

$$\mathbf{r}(t) = \cos t\,\mathbf{i} + \sin t\,\mathbf{j},$$
$$\mathbf{v}(t) = -\sin t\,\mathbf{i} + \cos t\,\mathbf{j},$$
$$\mathbf{a}(t) = -\cos t\,\mathbf{i} - \sin t\,\mathbf{j},$$

$$\frac{ds}{dt} = |\mathbf{v}(t)| = \sqrt{\sin^2 t + \cos^2 t} = 1,$$

$$\frac{d^2s}{dt^2} = 0.$$

At $t = \pi/4$, we have

$$\mathbf{r}\!\left(\frac{\pi}{4}\right) = \cos\frac{\pi}{4}\,\mathbf{i} + \sin\frac{\pi}{4}\,\mathbf{j} = \frac{1}{\sqrt{2}}\,\mathbf{i} + \frac{1}{\sqrt{2}}\,\mathbf{j},$$

$$\mathbf{v}\!\left(\frac{\pi}{4}\right) = -\sin\frac{\pi}{4}\,\mathbf{i} + \cos\frac{\pi}{4}\,\mathbf{j} = -\frac{1}{\sqrt{2}}\,\mathbf{i} + \frac{1}{\sqrt{2}}\,\mathbf{j},$$

$$\mathbf{a}\!\left(\frac{\pi}{4}\right) = -\cos\frac{\pi}{4}\,\mathbf{i} - \sin\frac{\pi}{4}\,\mathbf{j} = -\frac{1}{\sqrt{2}}\,\mathbf{i} - \frac{1}{\sqrt{2}}\,\mathbf{j},$$

$$\frac{ds}{dt} = 1,$$

$$\frac{d^2s}{dt^2} = 0.$$

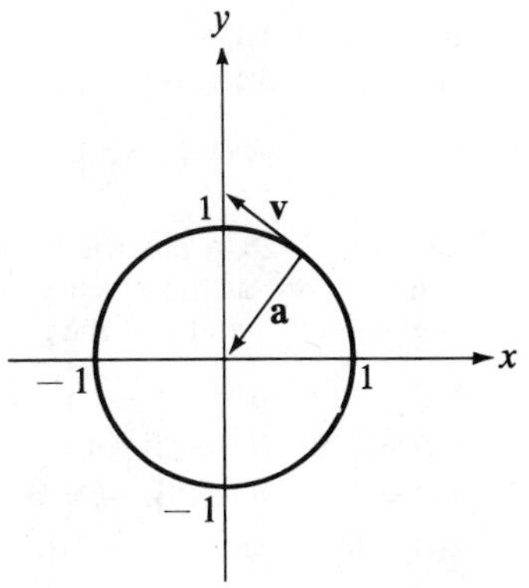

Figure 20.32

The path of the particle is a circle of radius 1 as shown in Figure 20.32. The velocity and acceleration are represented by directed line segments having their tails at $(1/\sqrt{2}, 1/\sqrt{2})$, which is the position of the particle at $t = \pi/4$.

Note that, while the speed is a constant, the velocity is not, since the particle is always changing direction. Similarly the rate of change of the speed is zero, but the acceleration is not. Note that the particle is always accelerating toward the center of the circle.

Example 4

A particle moves along a path according to the equation $r(t) = t\mathbf{i} + t^2\mathbf{j} + \dfrac{2}{3}t^3\mathbf{k}$. Find $\mathbf{v}$, $\mathbf{a}$, ds/dt and d^2s/dt^2.

$$r(t) = t\mathbf{i} + t^2\mathbf{j} + \frac{2}{3}t^3\mathbf{k},$$

$$\mathbf{v} = \mathbf{i} + 2t\mathbf{j} + 2t^2\mathbf{k},$$

$$\mathbf{a} = 2\mathbf{j} + 4t\mathbf{k},$$

$$\frac{ds}{dt} = |\mathbf{v}| = \sqrt{1 + 4t^2 + 4t^4} = 1 + 2t^2,$$

$$\frac{d^2s}{dt^2} = 4t.$$

It can be seen by both of these examples that, while the absolute value of the velocity is the speed, the absolute value of the acceleration is *not* the rate of change of speed.

Problems

A

In Problems 1–6, find $\mathbf{f}'(a)$. Sketch $\mathbf{f}(t)$ and $\mathbf{f}'(a)$.

1. $\mathbf{f}(t) = t^3\mathbf{i} + t^2\mathbf{j}, \quad a = 2.$
2. $\mathbf{f}(t) = 2t\mathbf{i} + t^3\mathbf{j}, \quad a = 1.$
3. $\mathbf{f}(t) = \cos t\mathbf{i} + \sin t\mathbf{j}, \quad a = 0.$
4. $\mathbf{f}(t) = (t - 1)\mathbf{i} + (t^2 - 1)\mathbf{j}, \quad a = 2.$
5. $\mathbf{f}(t) = \cos t\mathbf{i} + \sin t\mathbf{j} + e^t\mathbf{k}, \quad a = 0.$
6. $\mathbf{f}(t) = t\mathbf{i} + t^2\mathbf{j} + (t^2 + 2)\mathbf{k}, \quad a = 1.$

In Problems 7–12, find $\mathbf{f}'(t)$ and $\mathbf{f}''(t)$.

7. $\mathbf{f}(t) = (t^2 + 1)\mathbf{i} + (t^2 - 1)\mathbf{j} + t^2\mathbf{k}.$
8. $\mathbf{f}(t) = t^2\mathbf{i} + 2t\mathbf{j} - t^3\mathbf{k}.$
9. $\mathbf{f}(t) = (t^2 - 4)\mathbf{i} + (t^3 - 1)\mathbf{j} + (t + 2)\mathbf{k}.$
10. $\mathbf{f}(t) = (t^2 - 1)\mathbf{i} + (t^3 - 1)\mathbf{j} + t^4\mathbf{k}.$
11. $\mathbf{f}(t) = t\mathbf{i} + \sin t\mathbf{j} + \cos t\mathbf{k}.$
12. $\mathbf{f}(t) = \dfrac{1}{t}\mathbf{i} + \ln t\mathbf{j} - \dfrac{1}{t^2}\mathbf{k}.$

In Problems 13–26, a particle moves along a path according to the equation given. Find $\mathbf{v}$, $\mathbf{a}$, ds/dt, and d^2s/dt^2 at the given value of t. Sketch the path of the particle and give a graphical representation of $\mathbf{v}$ and $\mathbf{a}$ at the given value of t.

13. $r(t) = (t + 1)\mathbf{i} + (t^2 - 1)\mathbf{j}, \quad t = 0.$
14. $r(t) = 2t\mathbf{i} + t^2\mathbf{j}, \quad t = 1.$
15. $r(t) = t^2\mathbf{i} + (t^2 + t)\mathbf{j}, \quad t = 2.$
16. $r(t) = (t^2 - 1)\mathbf{i} + (t^2 + 1)\mathbf{j}, \quad t = 1.$
17. $r(t) = (t + 1)\mathbf{i} - t^3\mathbf{j}, \quad t = 0.$
18. $r(t) = t\mathbf{i} + t^3\mathbf{j}, \quad t = 1.$
19. $r(t) = (t^3 - 1)\mathbf{i} + (t^2 - 1)\mathbf{j}, \quad t = 1.$
20. $r(t) = (t^2 - 1)\mathbf{i} + (t^4 - 1)\mathbf{j}, \quad t = -1.$

21. $\mathbf{r}(t) = t^2\mathbf{i} + \dfrac{1}{t^2}\mathbf{j}, \quad t = 1.$

22. $\mathbf{r}(t) = t\mathbf{i} + \dfrac{2}{t}\mathbf{j}, \quad t = 1.$

23. $\mathbf{r}(t) = \cosh t\,\mathbf{i} + \sinh t\,\mathbf{j}, \quad t = 0.$

24. $\mathbf{r}(t) = \sin 2t\,\mathbf{i} + \cos 2t\,\mathbf{j}, \quad t = \pi/2.$

25. $\mathbf{r}(t) = t\mathbf{i} + \ln t\,\mathbf{j}, \quad t = 1.$

26. $\mathbf{r}(t) = t\mathbf{i} + e^t\mathbf{j}, \quad t = 1.$

B

In Problems 27–34, a particle moves along a path according to the equation given. Find $\mathbf{v}$, $\mathbf{a}$, ds/dt, and d^2s/dt^2. Sketch the path.

27. $\mathbf{r}(t) = \mathbf{i} + (t - 1)\mathbf{j} + (t^2 + 1)\mathbf{k}.$

28. $\mathbf{r}(t) = t\mathbf{i} - (t + 1)\mathbf{j} + t^2\mathbf{k}.$

29. $\mathbf{r}(t) = \sqrt{2}t\mathbf{i} + \sqrt{t}\mathbf{j} + \dfrac{4}{3}t^{3/2}\mathbf{k}.$

30. $\mathbf{r}(t) = t^2\mathbf{i} + (1 - t)\mathbf{j} + t\mathbf{k}.$

31. $\mathbf{r}(t) = \dfrac{1}{2}e^{2t}\mathbf{i} + \sqrt{2}e^t\mathbf{j} + t\mathbf{k}.$

32. $\mathbf{r}(t) = \cos t\,\mathbf{i} + \sin t\,\mathbf{j} + t\mathbf{k}.$

33. $\mathbf{r}(t) = \dfrac{1}{2}t^2\mathbf{i} + \sqrt{2}t\mathbf{j} + \ln t\,\mathbf{k}.$

34. $\mathbf{r}(t) = \dfrac{1}{t}\mathbf{i} + \sqrt{2}t\mathbf{j} + \dfrac{1}{3}t^3\mathbf{k}.$

20.11

Curves in Space: Tangent, Normal, and Curvature

Suppose that a particle moving in space traces out a path C given by the vector-valued function

$$\mathbf{r}(t) = x(t)\mathbf{i} + y(t)\mathbf{j} + z(t)\mathbf{k},$$

where $x'(t)$, $y'(t)$, and $z'(t)$ are continuous and never simultaneously zero for $a \le t \le b$. The arc length along C from the initial point $\mathbf{r}(a)$ to $\mathbf{r}(t_0)$ is

$$s(t_0) = \int_a^{t_0} s'(t)\, dt, \quad a \le t_0 \le b.$$

The position of the particle on C is uniquely determined by its distance $s(t)$ along the curve. Thus the curve C can also be described by a function

$$\mathbf{R}(s), \quad 0 \le s \le L,$$

with arc length as parameter, where

$$L = \int_a^b s'(t)\, dt$$

is the total length of C. Furthermore,

$$\mathbf{R}(s) = \mathbf{R}(s(t)) = \mathbf{r}(t), \quad a \le t \le b.$$

By the chain rule,

$$\frac{d\mathbf{R}}{dt} = \frac{d\mathbf{R}}{ds} \cdot \frac{ds}{dt} = \frac{d\mathbf{r}}{dt};$$

so

$$\frac{d\mathbf{R}}{ds} = \frac{d\mathbf{r}/dt}{ds/dt} = \frac{\mathbf{v}}{v},$$

where $v = |\mathbf{r}'(t)| = ds/dt$ is the speed of the particle and $\mathbf{v}(t) = \mathbf{r}'(t)$ is its velocity. Since $\mathbf{v}(t)$ is tangent to C, we see that $\mathbf{T} = d\mathbf{R}/ds$ is a *unit tangent vector* to C.

To investigate the motion of a particle further, we now state some results on derivatives of vector functions.

If $k(t)$ is a scalar function and $\mathbf{f}(t)$ and $\mathbf{g}(t)$ are vector functions, then

$$(\mathbf{f}(t) + \mathbf{g}(t))' = \mathbf{f}'(t) + \mathbf{g}'(t),$$

$$(k(t)\mathbf{f}(t))' = k'(t)\mathbf{f}(t) + k(t)\mathbf{f}'(t),$$

$$(\mathbf{f}(t) \cdot \mathbf{g}(t))' = \mathbf{f}'(t) \cdot \mathbf{g}(t) + \mathbf{f}(t) \cdot \mathbf{g}'(t),$$

$$(\mathbf{f}(t) \times \mathbf{g}(t))' = \mathbf{f}'(t) \times \mathbf{g}(t) + \mathbf{f}(t) \times \mathbf{g}'(t).$$

Theorem 20.36

If $\mathbf{f}(t)$ is a unit vector, then $\mathbf{f}(t)$ and $\mathbf{f}'(t)$ are orthogonal.

Proof

Since $|\mathbf{f}(t)| = 1$, then $|\mathbf{f}(t)|^2 = \mathbf{f}(t) \cdot \mathbf{f}(t) = 1$. Hence, $2\mathbf{f}(t) \cdot \mathbf{f}'(t) = 0$, and $\mathbf{f}(t)$ and $\mathbf{f}'(t)$ are orthogonal.

Since $\mathbf{T}$ is a unit vector, it follows from Theorem 20.36 that $d\mathbf{T}/ds$ is orthogonal to $\mathbf{T}$. This allows us to define a unique *principal normal vector* $\mathbf{N}$ to C by the equation

$$\mathbf{N} = \frac{1}{\kappa} \frac{d\mathbf{T}}{ds},$$

where

$$\kappa = \left| \frac{d\mathbf{T}}{ds} \right| = \frac{1}{v} \left| \frac{d\mathbf{T}}{dt} \right|$$

is the *curvature* of C.

In addition, we define the *binormal vector*

$$\mathbf{B} = \mathbf{T} \times \mathbf{N}.$$

Since $\mathbf{T}$ and $\mathbf{N}$ are orthogonal unit vectors, $\mathbf{B}$ is also a unit vector. The vector triple $\mathbf{T}, \mathbf{N}, \mathbf{B}$ forms the *moving trihedral* along the curve C.

Example 1

Find **T**, **N**, and **B** for $r(t) = t\mathbf{i} + \frac{1}{2}t^2\mathbf{j} + \mathbf{k}$.

$$\mathbf{r}'(t) = \mathbf{i} + t\mathbf{j}, \qquad \mathbf{v} = \sqrt{1 + t^2}, \qquad \mathbf{T} = \frac{1}{\sqrt{1 + t^2}}(\mathbf{i} + t\mathbf{j}).$$

$$\frac{d\mathbf{T}}{ds} = \frac{d\mathbf{T}/dt}{ds/dt} = \frac{1}{\mathbf{v}}\frac{d\mathbf{T}}{dt}$$

$$= \frac{1}{(1 + t^2)^2}(-t\mathbf{i} + \mathbf{j}).$$

$$\kappa = \left|\frac{d\mathbf{T}}{ds}\right| = \frac{1}{(1 + t^2)^{3/2}},$$

$$\mathbf{N} = \frac{1}{\kappa} \cdot \frac{d\mathbf{T}}{ds} = \frac{1}{\sqrt{1 + t^2}}(-t\mathbf{i} + \mathbf{j}).$$

$$\mathbf{B} = \mathbf{T} \times \mathbf{N} = \frac{1}{1 + t^2}[(\mathbf{i} + t\mathbf{j}) \times (-t\mathbf{i} + \mathbf{j})] = \mathbf{k}.$$

As a check, note that **T**, **N**, and **B** are mutually orthogonal unit vectors. Figure 20.33 shows the trihedral when $t = 2$.

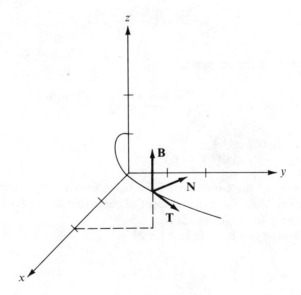

Figure 20.33

Let us use the abbreviation $a = s''(t) = v'$.   Recall from the last section that

$$\mathbf{v} = \mathbf{r}'(t) = \frac{d\mathbf{R}}{ds} \cdot \frac{ds}{dt} = v\frac{d\mathbf{R}}{ds}.$$

Thus

$$\mathbf{a} = \frac{d\mathbf{v}}{dt} = \frac{d}{dt}(v\mathbf{T}) = \frac{dv}{dt}\mathbf{T} + v\frac{d\mathbf{T}}{dt}$$

$$= \left(\frac{dv}{dt}\right)\mathbf{T} + v\frac{d\mathbf{T}}{ds}\cdot\frac{ds}{dt},$$

$$\mathbf{a} = a\mathbf{T} + \kappa v^2\mathbf{N}.$$

The coefficients a and κv^2 are the *tangential* and *normal components* of the acceleration $\mathbf{a}$.

To obtain expressions for $\mathbf{N}$, $\mathbf{B}$, and κ directly in terms of $\mathbf{r}'(t)$ and $\mathbf{r}''(t)$, let us cross multiply by $\mathbf{v}$ the equation for $\mathbf{a}$ above and note that $\mathbf{v} = v\mathbf{T}$.

$$\mathbf{v} \times \mathbf{a} = a(\mathbf{v} \times \mathbf{T}) + v^2\kappa(\mathbf{v} \times \mathbf{N})$$

$$= av(\mathbf{T} \times \mathbf{T}) + v^3\kappa(\mathbf{T} \times \mathbf{N})$$

$$= av(\mathbf{0}) + v^3\kappa\mathbf{B},$$

so

$$\mathbf{v} \times \mathbf{a} = v^3\kappa\mathbf{B},$$

where $\mathbf{B}$ is the unit binormal vector. Thus

$$\kappa = \frac{|\mathbf{v} \times \mathbf{a}|}{v^3} = \frac{|\mathbf{r}' \times \mathbf{r}''|}{|\mathbf{r}'|^3}.$$

Note: For vectors in the xy plane,

$$\kappa = \frac{|(x'\mathbf{i} + y'\mathbf{j} + 0\mathbf{k}) \times (x''\mathbf{i} + y''\mathbf{j} + 0\mathbf{k})|}{v^3}$$

$$= \frac{|x'y'' - x''y'|}{[(x')^2 + (y')^2]^{3/2}}$$

is the absolute value of the curvature of the plane curve $\mathbf{r}(t) = x(t)\mathbf{i} + y(t)\mathbf{j}$ (see Problem 19, Section 13.4). Thus the concept of curvature for a plane curve has been extended consistently to curves in space.

A similar expression can be obtained for $\mathbf{N}$.

$$(\mathbf{v} \times \mathbf{a}) \times \mathbf{v} = (\mathbf{v} \times a\mathbf{T}) \times \mathbf{v} + (\mathbf{v} \times \kappa v^2\mathbf{N}) \times \mathbf{v} \quad (\text{since } \mathbf{a} = a\mathbf{T} + \kappa v^2\mathbf{N})$$

$$= a\mathbf{0} \times \mathbf{v} + \kappa v^2(\mathbf{v} \times \mathbf{N}) \times \mathbf{v} \quad (\text{since } \mathbf{v} = c\mathbf{T} \text{ for some scalar } c)$$

$$= v^2\kappa[(\mathbf{v} \cdot \mathbf{v})\mathbf{N} - (\mathbf{v} \cdot \mathbf{N})\mathbf{v}] \quad (\text{by Problem 38, Section 20.4})$$

$$= v^4\kappa\mathbf{N} \quad (\text{since } \mathbf{v} \cdot \mathbf{N} = 0).$$

Thus

$$\mathbf{N} = \frac{(\mathbf{r}' \times \mathbf{r}'') \times \mathbf{r}'}{\kappa v^4}.$$

Example 2

Find **T**, **N**, κ, and the tangential and normal components of the acceleration for the motion $\mathbf{r}(t) = t\mathbf{i} + (1 - t)\mathbf{j} + \frac{1}{2}t^2\mathbf{k}$.

$$\mathbf{r}' = \mathbf{i} - \mathbf{j} + t\mathbf{k},$$

$$\mathbf{r}'' = 0\mathbf{i} + 0\mathbf{j} + \mathbf{k},$$

$$v = |\mathbf{r}'| = \sqrt{2 + t^2},$$

$$\kappa = \frac{|\mathbf{r}' \times \mathbf{r}''|}{v^3} = \frac{|-\mathbf{i} - \mathbf{j} + 0\mathbf{k}|}{(2 + t^2)^{3/2}} = \frac{\sqrt{2}}{(2 + t^2)^{3/2}}$$

$$\mathbf{T} = \frac{1}{\sqrt{2 + t^2}}(\mathbf{i} - \mathbf{j} + t\mathbf{k})$$

$$\mathbf{N} = \frac{(-\mathbf{i} - \mathbf{j} + 0\mathbf{k}) \times (\mathbf{i} - \mathbf{j} + t\mathbf{k})}{\sqrt{2}\,(2 + t^2)^{-3/2}\,(2 + t^2)^{4/2}}$$

$$= \frac{-t\mathbf{i} + t\mathbf{j} + 2\mathbf{k}}{\sqrt{2}(2 + t^2)^{1/2}}.$$

The components of **a** are

$$a = \frac{t}{\sqrt{2 + t^2}} \quad \text{and} \quad \kappa v^2 = \frac{\sqrt{2}}{\sqrt{2 + t^2}}.$$

Finally, let us investigate $d\mathbf{N}/ds$ and $d\mathbf{B}/ds$. Since $\mathbf{B} = \mathbf{T} \times \mathbf{N}$ is a unit vector, $d\mathbf{B}/ds \perp \mathbf{B}$, and hence $d\mathbf{B}/ds$ lies in the plane of **T** and **N**. Suppose

$$\frac{d\mathbf{B}}{ds} = \sigma\mathbf{T} - \tau\mathbf{N}$$

for some constants σ and τ. Differentiating $\mathbf{T} \cdot \mathbf{B} = 0$ with respect to s,

$$\mathbf{T} \cdot \frac{d\mathbf{B}}{ds} + \kappa\mathbf{N} \cdot \mathbf{B} = 0.$$

Since $\mathbf{N} \cdot \mathbf{B} = 0$, $\mathbf{T} \cdot d\mathbf{B}/ds = 0$ also, so $\sigma = 0$ and $d\mathbf{B}/ds = -\tau\mathbf{N}$, where the scalar function τ is called the *torsion*.

$$\frac{d\mathbf{N}}{ds} = \frac{d}{ds}(\mathbf{B} \times \mathbf{T}) = \mathbf{B} \times \kappa\mathbf{N} + (-\tau\mathbf{N}) \times \mathbf{T} = \tau\mathbf{B} - \kappa\mathbf{T}.$$

In summary,

$$\mathbf{T} = \frac{\mathbf{r}'}{|\mathbf{r}'|}, \quad \mathbf{N} = \frac{(\mathbf{r}' \times \mathbf{r}'') \times \mathbf{r}'}{\kappa|\mathbf{r}'|^4}, \quad \kappa = \frac{|\mathbf{r}' \times \mathbf{r}''|}{|\mathbf{r}'|^3}$$

and

$$\frac{d\mathbf{T}}{ds} = \kappa\mathbf{N}, \quad \frac{d\mathbf{N}}{ds} = \tau\mathbf{B} - \kappa\mathbf{T}, \quad \text{and} \quad \frac{d\mathbf{B}}{ds} = -\tau\mathbf{N}.$$

These are called the *Frenet-Serret formulas*.

Problems

A

In Problems 1–4, find **T**, **N**, and **B**.

1. $\mathbf{r}(t) = \cos t\,\mathbf{i} + \sin t\,\mathbf{j} + t\,\mathbf{k}$.
2. $\mathbf{r}(t) = t^2\mathbf{i} + (2t + 1)\mathbf{j} + (2t - 1)\mathbf{k}$.
3. $\mathbf{r}(t) = 2t\mathbf{i} + t^2\mathbf{j} + (1 - t)\mathbf{k}$.
4. $\mathbf{r}(t) = 2\sin t\,\mathbf{i} + 3\cos t\,\mathbf{j} + t\,\mathbf{k}$.

In Problems 5–8, find the curvature κ.

5. $\mathbf{r}(t) = \cos t\,\mathbf{i} + \sin t\,\mathbf{j}$.
6. $\mathbf{r}(t) = t\mathbf{i} + \frac{1}{2}t^2\mathbf{j}$.
7. $\mathbf{r}(t) = 2t\mathbf{i} + t^2\mathbf{j} + (1 - t)\mathbf{k}$.
8. $\mathbf{r}(t) = \cos t\,\mathbf{i} + \sin t\,\mathbf{j} + t\,\mathbf{k}$.

In Problems 9–12, find the tangential and normal components of the acceleration.

9. $\mathbf{r}(t) = t^2\mathbf{i} + t^3\mathbf{j}$.
10. $\mathbf{r}(t) = \cos t\,\mathbf{i} + \sin t\,\mathbf{j}$.
11. $\mathbf{r}(t) = (t^2 + 1)\mathbf{i} + (t^2 - 1)\mathbf{j} + t^2\mathbf{k}$.
12. $\mathbf{r}(t) = \mathbf{i} + (t - 1)\mathbf{j} + (t^2 + 1)\mathbf{k}$.

B

13. Suppose $\mathbf{r}'(t) = \mathbf{c} \times \mathbf{r}(t)$ for all t, where $\mathbf{c}$ is a constant vector. Show that $\mathbf{a}(t)$ is perpendicular to $\mathbf{c}$ and $v(t)$ is constant.
14. Suppose the curvature of the motion $\mathbf{r}(t)$ is identically zero. Show that $\mathbf{T}$ is constant and $\mathbf{T} \times \mathbf{r}(t)$ is constant, so the motion is along a straight line.
15. If the torsion τ is identically zero, show that the motion is restricted to a plane.

Review Problems

A

1. Find the point 1/3 of the way from $(-2, 1, 5)$ to $(1, 4, -4)$.
2. Find the projection of $\mathbf{u} = \mathbf{i} - 3\mathbf{j} + \mathbf{k}$ upon $\mathbf{v} = 4\mathbf{i} + 2\mathbf{j} + 3\mathbf{k}$.
3. Find the set of direction angles for the vector $\mathbf{v}$ with direction numbers $\{1, -1, \sqrt{2}\}$ if $\mathbf{v}$ is directed upward.
4. Suppose the line l_1 contains $(2, -2, 4)$ and $(5, 3, 0)$, while l_2 contains $(4, -3, 1)$ and $(3, -4, -1)$. Are l_1 and l_2 parallel, perpendicular, coincident, or none of these?
5. Suppose the line l_1 contains $(5, 1, -2)$ and $(2, -3, 1)$, while l_2 contains $(3, 8, 1)$ and $(-3, 0, 7)$. Are l_1 and l_2 parallel, perpendicular, coincident, or none of these?

In Problems 6 and 7, represent the given line in parametric form and in symmetric form.

6. Containing $(4, 0, 3)$ and $(5, 1, 3)$.
7. Containing $(2, -3, 5)$; direction numbers $\{3, -2, 1\}$.

In Problems 8 and 9, find the point of intersection (if any) of the given lines.

8. $\dfrac{x}{2} = \dfrac{y - 6}{-4} = \dfrac{z - 3}{1}$; $\dfrac{x - 2}{-1} = \dfrac{y - 4}{3} = \dfrac{z - 7}{1}$.
9. $x = 2 + t,\ y = 3 - 2t,\ z = 3 + 5t;\ x = -3 + 2s,\ y = -1 + 3s,\ z = -s$.

10. Find the distance between the lines

$$x = 4 - t,\ y = 3 + 2t,\ z = 4 + t \quad \text{and} \quad x = 3 + 2s,\ y = 5 - s,\ z = -1 + 4s.$$

In Problems 11 and 12, find an equation(s) of the plane(s) satisfying the given conditions.

11. Containing $(4, 0, -2)$ and perpendicular to $x = 4 - t$, $y = 3 + 2t$, $z = 1$.
12. Containing $(2, 5, 1)$, $(3, -2, 4)$, and $(1, 0, 2)$.

13. Find the distance between $(4, 3, 1)$ and $2x - 2y + z = 4$.
14. Find the distance between $x - 4y + z = 3$ and $x - 4y + z = 8$.
15. Find the distance between $(2, 5, -2)$ and the line $x = 3 + t$, $y = 4 - t$, $z = 2 + 2t$.
16. Find the distance between the lines

$$x = 1 + 2t, \quad y = 4 + 3t, \quad z = 2 - t \quad \text{and} \quad x = 4 + 2s, \quad y = 2 + 3s, \quad z = 5 - s.$$

17. Find the acute angle between the planes $2x - 5y + z = 3$ and $x + y - 2z = 1$.

In Problems 18 and 19, describe the locus of the given equation.

18. $x^2 + y^2 + z^2 - 2x - 4y + 8z + 5 = 0$.
19. $2x^2 + 2y^2 + 2z^2 - 2x - 4y + 6z + 7 = 0$.

In Problems 20–23, sketch and describe.

20. $9x^2 + 9y^2 - 4z^2 = 36$.   21. $z^2 - x^2 = 4$.
22. $9x^2 - 36y^2 + 4z^2 = 0$.   23. $x^2 + y - z^2 = 0$.

24. The following points are given in rectangular coordinates. Express them in cylindrical and spherical coordinates.

   (a) $(2, 2, 1)$,   (b) $(-1, 1, -2)$.

25. Express $z = r \cos \theta$ in rectangular coordinates.
26. Express $z = 4x^2 + 4y^2$ in cylindrical and spherical coordinates.
27. Express $\rho = \sin \phi (2 \cos \theta - \sin \theta)$ in rectangular coordinates.
28. For $\mathbf{f}(t) = t^3\mathbf{i} + (3t^2 + 1)\mathbf{j} + (6t + 1)\mathbf{k}$, find $\mathbf{f}'(t)$ and $\mathbf{f}''(t)$.

B

29. The point $(5, 3, -2)$ is a distance 3 from the midpoint of the segment joining $(5, 7, 2)$ and $(1, 1, z)$. Find z.

30. Find the end points of the representative $\overrightarrow{AB}$ of $\mathbf{v}$ if $\mathbf{v} = 3\mathbf{i} - \mathbf{j} + 4\mathbf{k}$ and the midpoint of AB is $(2, 4, -3)$.

31. Suppose $\mathbf{u} = 2\mathbf{i} - \mathbf{j} + 3\mathbf{k}$ and $\mathbf{v} = \mathbf{i} + 4\mathbf{j} - \mathbf{k}$. Express $\mathbf{u}$ in the form $\mathbf{u} = \mathbf{p} + \mathbf{q}$, where $\mathbf{p} = k\mathbf{v}$ and $\mathbf{q} \cdot \mathbf{v} = 0$. Interpret geometrically.

In Problems 32 and 33, find equations for the line described.

32. Containing $(3, 5, -2)$ and perpendicular to the plane containing

$$x = 5 - t, \quad y = 2 + 3t, \quad z = 4 + t \quad \text{and} \quad x = 3 - s, \quad y = 5 + 3s, \quad z = -1 + s.$$

33. Perpendicular at $(4, 2, 3)$ to the plane containing $(4, 2, 3)$, $(-1, 3, 1)$, and $(2, 5, -3)$.

34. Find an equation(s) of the plane(s) containing $(4, 2, -3)$ and perpendicular to

$$2x - y + 4z = 5 \quad \text{and} \quad x + 3y - z = 2.$$

35. Find, in parametric form, the line of intersection of

$$2x - y + 3z = 2 \quad \text{and} \quad x + 3y + 2z = 4.$$

36. Sketch and describe $x^2 + y^2 + 4z^2 - 2x + 4y + 1 = 0$.
37. Sketch the graph of $\mathbf{f}(t) = (t + 1)\mathbf{i} + (t^2 - 1)\mathbf{j} + (t + 1)\mathbf{k}$.
38. A particle moves along a path according to the equation $\mathbf{r}(t) = (t^2 - 2)\mathbf{i} + (t + 1)\mathbf{j}$. Find $\mathbf{v}$, $\mathbf{a}$, ds/dt, and d^2s/dt^2 when $t = 1$. Sketch the path of the particle and give a graphical representation of $\mathbf{v}$ and $\mathbf{a}$ at $t = 1$.

39.   A particle moves along a path according to $r(t) = (t^2 + 1)i + (t^2 - 1)j + 2tk$. Find v, a, ds/dt, and d^2s/dt^2 when $t = 2$.

40.   Find T, N, and B for $r(t) = (1 + t)i - t^2j + 2tk$.

41.   Find the tangential and normal components of the acceleration for $r(t) = (1 + t)i - t^2j + 2tk$.

42.   Find the curvature κ of $r(t) = 2ti + t^2j + (2 - t)k$.

C

43.   Prove or show to be false: $u \times (v \times w) = (u \times v) \times w$.

44.   Find an equation of the plane containing T and N at a general point of $r(t) = t^2i + (2 + t)j + \frac{1}{2}t^2k$.

45.   If $u = a_1i + b_1j$ and $v = a_2i + b_2j$ are noncollinear vectors, show that any vector $w = ci + dj$ can be written as a linear combination of u and v; that is, there exist scalars α and β so that $w = \alpha u + \beta v$.

21

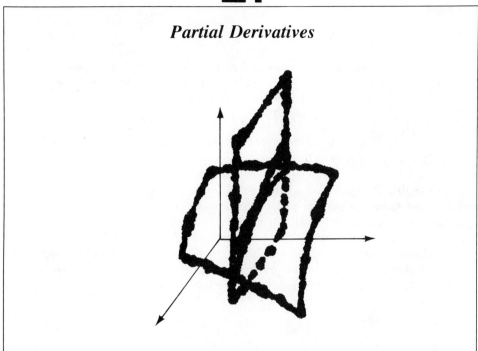

Partial Derivatives

21.1

Functions of Several Variables

In the last chapter our consideration of surfaces, represented by equations in three variables, x, y, and z, laid the ground-work for important concepts in calculus. Let us now consider some of these concepts. Just as we restricted ourselves to functions in the past, so we shall, for the most part, restrict ourselves to functions of several variables, $z = f(x, y)$.

In Section 3.1, we defined a function to be a set of ordered pairs such that no two ordered pairs have the same first object and different second ones. Thus a real-valued function of two variables, $z = f(x, y)$, is a set of ordered pairs

$$f = \{(z, (x, y)) \mid z = f(x, y)\},$$

where the pairs (x, y) belong to a subset of the plane, the domain of f. The variables x and y are independent variables and z is a dependent variable. The graph of a function of two variables is a surface in space, and any vertical line intersects the surface in at most one point.

Definition

$\lim\limits_{\substack{x \to a \\ y \to b}} f(x, y) = L$ *means that*

(a) *if δ is a positive number, then there is a number pair (x, y) of the domain of f such that $|x - a| < \delta$ and $|y - b| < \delta$, and $|x - a|$ and $|y - b|$ are not both 0, and*

(b) *if ϵ is a positive number, then there is a positive number δ such that*

$$|f(x, y) - L| < \epsilon$$

whenever (x, y) is in the domain of f, $|x - a| < \delta$ and $|y - b| < \delta$, and $|x - a|$ and $|y - b|$ are not both 0.

The geometric interpretation of this definition is given in Figure 21.1. The first part says that there is a number pair of the domain of f which is "close" to but not equal to (a, b)—"close" in this case meaning within a square of side 2δ, sides parallel to the x and y axes, and center at (a, b). The second part says that if we take a pair of horizontal planes with (a, b, L) between them, then there is a square of the type given above such that every point of the graph whose projection on the xy plane is inside the square, but not at its center, is between the horizontal planes.

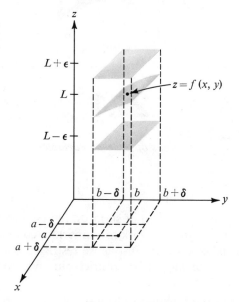

Figure 21.1

Example 1

Show, by means of the definition of a limit, that $\lim\limits_{\substack{x \to 1 \\ y \to 2}} (x + 2y) = 5$.

The first part of the definition is easily seen to be satisfied, since $f(x, y) = x + 2y$ is defined for all x and all y.

Let $\epsilon > 0$. Suppose that

$$|x - 1| < \delta \quad \text{and} \quad |y - 2| < \delta$$

for some choice of δ. Then

$$|f(x, y) - L| = |x + 2y - 5| = |(x - 1) + 2(y - 2)|$$
$$\leq |x - 1| + 2|y - 2| < 3\delta.$$

Then let us choose δ so that $3\delta = \epsilon$ (that is, we choose $\delta = \epsilon/3$). With that choice we see that $|f(x, y) - 5| < \epsilon$ whenever $|x - 1| < \delta$ and $|y - 2| < \delta$.

When dealing with a function of one variable, we found that $\lim_{x \to a} f(x)$ does not exist if we get different values for the limit as $x \to a$ from the two different sides—that is, if

$$\lim_{x \to a^-} f(x) \neq \lim_{x \to a^+} f(x).$$

When dealing with a function of two variables, we see that (x, y) can approach (a, b) along many different paths, rather than just two as before. Nevertheless, in order for the limit to exist, we must get the same value for the limit no matter what path is used in approaching (a, b). A convenient method of showing that a given limit does not exist is to show that we get different results when approaching (a, b) along different paths.

Example 2

Show that $\lim_{\substack{x \to 0 \\ y \to 0}} \dfrac{x^2 - y^2}{x^2 + y^2}$ does not exist.

Let us show that we get different limits as $(0, 0)$ is approached along two different paths. If $(0, 0)$ is approached along the line $y = 0$, then

$$\lim_{\substack{x \to 0 \\ y \to 0}} \frac{x^2 - y^2}{x^2 + y^2} = \lim_{x \to 0} \frac{x^2}{x^2} = 1.$$

But along the line $x = 0$, we have

$$\lim_{\substack{x \to 0 \\ y \to 0}} \frac{x^2 - y^2}{x^2 + y^2} = \lim_{y \to 0} \frac{-y^2}{y^2} = -1.$$

This implies that the given limit does not exist. Basically this says that there are points arbitrarily close to $(0, 0)$ for which $f(x, y)$ is near 1 and others for which $f(x, y)$ is near -1. This contradicts the definition of a limit.

Of course, a similar definition of limit holds for a function of three or more variables although there is no graphical representation. Continuity is defined analogously.

Definition

*If f is a real-valued function of two real variables, then to say that f is **continuous** at the point (a, b) of the domain of f means that if ϵ is a positive number, then there is a positive number δ such that*

$$|f(x, y) - f(a, b)| < \epsilon$$

whenever (x, y) is in the domain of f and $|x - a| < \delta$ and $|y - b| < \delta$.

We can see, by their similarity, the relationship between these two definitions; limits of continuous functions can be found by substitution provided the point in question is a limit point of the domain of the function.

Let us define a *neighborhood* of a point (a, b) to be a square centered at (a, b):

$$N(a, b) = \{(x, y) | |x - a| \le \delta, \quad |y - b| \le \delta\},$$

for some $\delta > 0$. Then $f(x, y)$ is continuous at (a, b) if for each $\epsilon > 0$, there is a neighborhood $N(a, b)$ such that $|f(x, y) - f(a, b)| < \epsilon$ for all points (x, y) in N that are also in the domain of f.

The theorems on limits in Sections 10.7 and 17.4 extend to functions of several variables, and the usual combinations and composition of continuous functions are continuous.

21.2

The Partial Derivative

Now we consider two derivatives, which we call *partial derivatives*, of a function of two variables.

Definition

*If $z = f(x, y)$, then the **partial derivative** of z with respect to x at (x, y) is*

$$\frac{\partial z}{\partial x} = \lim_{h \to 0} \frac{f(x + h, y) - f(x, y)}{h}$$

if this limit exists. The partial derivative of z with respect to y at (x, y) is

$$\frac{\partial z}{\partial y} = \lim_{h \to 0} \frac{f(x, y + h) - f(x, y)}{h}$$

if this limit exists.

Note that when the partial derivative of z is taken with respect to x, the y is unchanged in the two terms of the numerator; similarly the x is unchanged when the partial derivative is with respect to y. This suggests a very simple way of finding partial derivatives. When finding a partial derivative with respect to x, we simply differentiate, with y being regarded as a constant; similarly we find a partial derivative with respect to y by regarding x as a constant and differentiating with respect to the variable y.

Example 1

If $z = x^2 + xy$,   find $\partial z/\partial x$ and $\partial z/\partial y$.

Regarding y as a constant, we have $\dfrac{\partial z}{\partial x} = 2x + y$.

Regarding x as a constant, we have $\dfrac{\partial z}{\partial y} = x$.

Of course, if we want a partial derivative at a particular point of the graph, we merely substitute its coordinates into the result.

Let us consider a graphical interpretation of partial derivatives. Suppose z is a function of x and y and we are considering the partial derivative of z with respect to x at the point (x_1, y_1, z_1). Since y is taken to be a constant and an ordinary derivative is taken with respect to x, this is graphically equivalent to considering the intersection of the plane $y = y_1$ with the surface $z = f(x, y)$ (see Figure 21.2) and then finding the slope of the curve that is the intersection of this plane and the original surface.

Exactly the same type of definition holds for functions of n variables.

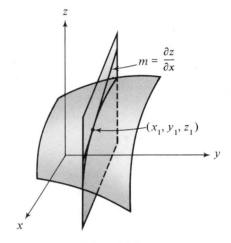

Figure 21.2

Definition

If $z = f(x_1, x_2, x_3, \ldots, x_n)$, then

$$\frac{\partial z}{\partial x_i} = \lim_{h \to 0} \frac{f(x_1, \ldots, x_i + h, \ldots, x_n) - f(x_1, \ldots, x_i, \ldots, x_n)}{h}$$

if this limit exists.

Again, $\partial z/\partial x_i$ is found by looking upon all variables but z and x_i as constants and finding an ordinary derivative. The only difference here is that if we go beyond a total of three variables, there is no easily visualized graph, since more than three dimensions would be required.

Example 2

Given $w = x^2 y + y^2 z - z^2 x$,   find $\partial w/\partial x$, $\partial w/\partial y$ and $\partial w/\partial z$.

$$\frac{\partial w}{\partial x} = 2xy - z^2,$$

$$\frac{\partial w}{\partial y} = x^2 + 2yz,$$

$$\frac{\partial w}{\partial z} = y^2 - 2zx.$$

There are several different types of notation used in connection with partial derivatives. If $z = f(x_1, x_2, \ldots, x_n)$, then the partial derivative of z (or f) with respect to x_i is denoted by

$$\frac{\partial z}{\partial x_i}, \quad z_{x_i}, \quad z_i, \quad D_{x_i}z, \quad D_i z,$$

$$\frac{\partial f}{\partial x_i}, \quad f_{x_i}, \quad f_i, \quad D_{x_i}f, \quad D_i f.$$

Each has its advantages and disadvantages, but all are used interchangeably; so you must learn all of them.

Just as with ordinary derivatives, it is possible to take a second, third, etc., partial derivative of a function. But there are more of them, since the second partial need not be with respect to the same variable as the first. For instance, if $z = f(x, y)$, we can take second partials both with respect to x, both with respect to y, first with respect to x and then y, or first with respect to y and then x. The notation is as follows.

Both with respect to x:

$$\frac{\partial^2 z}{\partial x^2} = z_{xx} = z_{11} = D_{11}z.$$

Both with respect to y:

$$\frac{\partial^2 z}{\partial y^2} = z_{yy} = z_{22} = D_{22} z.$$

First with respect to x and then y:

$$\frac{\partial^2 z}{\partial y \, \partial x} = z_{xy} = z_{12} = D_{12} z.$$

First with respect to y and then x:

$$\frac{\partial^2 z}{\partial x \, \partial y} = z_{yx} = z_{21} = D_{21}z.$$

Note that, with the exception of the differential notation, we read from left to right. For example, z_{xy} means the second partial derivative of z, taken first with respect to x and then with respect to y. In the differential notation this is written

$$\frac{\partial^2 z}{\partial y \, \partial x},$$

and is read from right to left. The reason for this order is apparent when it is written as follows:

$$\frac{\partial^2 z}{\partial y \, \partial x} = \frac{\partial}{\partial y}\left(\frac{\partial z}{\partial x}\right).$$

Example 3

Find all second partials of $z = x^2 - 2xy + 2y^2$.

$$\frac{\partial z}{\partial x} = 2x - 2y, \qquad\qquad \frac{\partial z}{\partial y} = -2x + 4y,$$

$$\frac{\partial^2 z}{\partial x^2} = 2, \qquad \frac{\partial^2 z}{\partial y\,\partial x} = -2, \qquad \frac{\partial^2 z}{\partial x\,\partial y} = -2, \qquad \frac{\partial^2 z}{\partial y^2} = 4.$$

It might be noted that, in this case,

$$\frac{\partial^2 z}{\partial y\,\partial x} = \frac{\partial^2 z}{\partial x\,\partial y}.$$

Actually there is a large class of functions for which this is true, and it includes most of the functions with which we shall be dealing.

Theorem 21.1†

If f is a function such that

$$\frac{\partial^2 z}{\partial y\,\partial x} \quad and \quad \frac{\partial^2 z}{\partial x\,\partial y}$$

are both continuous inside a neighborhood, then

$$\frac{\partial^2 z}{\partial y\,\partial x} = \frac{\partial^2 z}{\partial x\,\partial y}.$$

Similar theorems exist for higher mixed partial derivatives.

Problems

A

In Problems 1–10, find $\partial z/\partial x$ and $\partial z/\partial y$.

1. $z = x^2 + y^2$.
2. $z = x^2 - 3xy$.
3. $z = \dfrac{x}{y}$.

4. $z = \dfrac{x - y}{x + y}$.
5. $z = \sqrt{xy}$.
6. $z = (x^2 + y)^2$.

7. $z = (x - 2y)^3$.
8. $z = \ln (x^2 + y^2)$.
9. $z = \sin x \cos y$.
10. $z = \ln \sin xy$.

In Problems 11–16, find $\partial z/\partial x$ and $\partial z/\partial y$ at the point indicated.

11. $z = x + \dfrac{y^2}{x}$, $(1, 2, 5)$.
12. $z = x^2 - y^2$, $(1, 1, 0)$.

13. $z = xy - y^3$, $(2, -1, -1)$.
14. $z = e^{x/y}$, $(2, 1, e^2)$.

15. $z = x \sin y$, $(2, \pi/2, 2)$.
16. $z = e^{\sin xy}$, $(4, 0, 1)$.

---

† For a proof of this theorem, see Kenneth Rogers, *Advanced Calculus* (Columbus, Ohio: Charles E. Merrill, 1976), p. 102.

In Problems 17–22, find $\partial w/\partial x$, $\partial w/\partial y$, and $\partial w/\partial z$.

17. $w = x \sin yz$.

18. $w = xyz$.

19. $w = \dfrac{x^2 + y^2}{y^2 + z^2}$.

20. $w = x^2y^2 + y^3 + y^2z^2$.

21. $w = xe^{y/z}$.

22. $w = x^2 + y^2$.

In Problems 23–28, find $\partial^2 z/\partial x^2$, $\partial^2 z/\partial y\,\partial x$, $\partial^2 z/\partial x\,\partial y$, and $\partial^2 z/\partial y^2$.

23. $z = \dfrac{y}{x}$.

24. $z = x^3 - 2x^2y + y^3$.

25. $z = \sin x \cos y$.

26. $z = \dfrac{1}{(2x + 3y)^2}$.

27. $z = \ln(x^3 + y^3)$.

28. $z = e^x \tan y$.

29. Show that $\lim_{\substack{x \to 1 \\ y \to -2}} (4x - y + 2) = 8$.

30. Show that $\lim_{\substack{x \to 2 \\ y \to 3}} (2x + 3y + 1) = 14$.

B

31. Show that the limit of Example 2 in Section 21.1 is 0 if $(0, 0)$ is approached along the path $x = y$.

32. Show that the limit of Example 2 in Section 21.1 is 3/5 if $(0, 0)$ is approached along the path $x = 2y$.

33. Show that $\lim_{\substack{x \to 0 \\ y \to 0}} \dfrac{xy}{x^2 + y^2}$ does not exist.

34. Investigate $\lim_{\substack{x \to 0 \\ y \to 0}} \dfrac{x^2 - y^2}{x^2 + y^2}$ along each line $y = mx$, $-\infty < m < \infty$.

C

35. Show that $\lim_{\substack{x \to 0 \\ y \to 0}} \dfrac{x^2y}{x^4 + y^2} = 0$ along $y = mx$ for each m. What is $\lim_{\substack{x \to 0 \\ y \to 0}} \dfrac{x^2y}{x^4 + y^2}$ along $y = x^2$? Does $\lim_{\substack{x \to 0 \\ y \to 0}} \dfrac{x^2y}{x^4 + y^2}$ exist?

36. Let

$$f(x, y) = \begin{cases} \dfrac{x^2y}{x^4 + y^2} & \text{if } (x, y) \neq (0, 0), \\ 0 & \text{if } (x, y) = (0, 0). \end{cases}$$

Show that $\partial f/\partial x = 0$ and $\partial f/\partial y = 0$ at $(0, 0)$, but $f(x, y)$ is not continuous at $(0, 0)$. (See Problem 35.)

37. Formulate a definition of

$$\lim_{(x,\, y,\, z) \to (a,\, b,\, c)} f(x, y, z) = L.$$

38. Describe the domain of $f(x, y) = \sqrt{xy}$.
39. Describe the domain of $f(x, y) = \ln(x + y - 4)$.
40. Describe the domain of $f(x, y, z) = \sqrt{4 - x^2 - y^2 - z^2}$.
41. Given $\partial f/\partial x = y^2 \cos xy^2$ and $\partial f/\partial y = 3y^2 + 2xy \cos xy^2$, find $f(x, y)$.

21.3

The Chain Rule

Recall that in Section 4.4 we considered the chain rule for differentiating a function of a function. For $y = f(u)$ and $u = g(x)$, we had $y = f(g(x))$ and

$$\frac{dy}{dx} = \frac{dy}{du}\frac{du}{dx}.$$

This proved to be very useful, especially when deriving formulas for the derivatives of transcendental functions. In this section we extend the chain rule to functions of several variables.

Theorem 21.2

Suppose $z = f(x, y)$, *and* $x(t)$ *and* $y(t)$ *are differentiable functions in an open interval containing* t. *If* $\partial z/\partial x$ *and* $\partial z/\partial y$ *are continuous in a neighborhood of* $(x(t), y(t))$, *then*

$$\frac{dz}{dt} = \frac{\partial z}{\partial x}\cdot\frac{dx}{dt} + \frac{\partial z}{\partial y}\cdot\frac{dy}{dt}.$$

Proof

Before proving the chain rule, let us derive a useful increment formula for $f(x, y)$. Let Δx and Δy be increments in x and y, respectively. The corresponding increment in $z = f(x, y)$ is

$$\Delta z = f(x + \Delta x, y + \Delta y) - f(x, y).$$

Adding and subtracting $f(x, y + \Delta y)$ and applying the mean-value theorem to the resulting differences,

$$\Delta z = f(x + \Delta x, y + \Delta y) - f(x, y + \Delta y) + f(x, y + \Delta y) - f(x, y)$$
$$= f_x(x_1, y + \Delta y)\Delta x + f_y(x, y_1)\Delta y,$$

where x_1 is between x and $x + \Delta x$ and y_1 is between y and $y + \Delta y$. Since $\partial z/\partial x = f_x$ and $\partial z/\partial y = f_y$ are continuous,

$$f_x(x_1, y + \Delta y) = f_x(x, y) + \epsilon,$$
$$f_y(x, y_1) = f_y(x, y) + \eta,$$

where $\epsilon \to 0$ and $\eta \to 0$ as $(\Delta x, \Delta y) \to (0, 0)$. (The value 0 may be assumed by ϵ and η here.) Thus we have the *fundamental increment formula*,

$$\Delta z = f(x + \Delta x, y + \Delta y) - f(x, y) = f_x(x, y)\Delta x + f_y(x, y)\Delta y + \epsilon\Delta x + \eta\Delta y,$$

where $\epsilon \to 0$ and $\eta \to 0$ as $(\Delta x, \Delta y) \to (0, 0)$.

Suppose now that $z = f(x, y)$, $x = x(t)$, and $y = y(t)$. It then follows that z is a function of t, $z = f(x(t), y(t))$, and we can consider the derivative of z with respect to t.

$$\frac{dz}{dt} = \lim_{h \to 0} \frac{f(x(t + h), y(t + h)) - f(x(t), y(t))}{h}.$$

Corresponding to the increment h in t there are increments $\Delta x = x(t + h) - x(t)$ and $\Delta y = y(t + h) - y(t)$. Also, since dx/dt and dy/dt exist, $x(t)$ and $y(t)$ are continuous by Theorem 10.10, so both $\Delta x \to 0$ and $\Delta y \to 0$ as $h \to 0$. By the fundamental increment formula,

$$\frac{1}{h} \left[f(x(t + h), y(t + h)) - f(x(t), y(t)) \right]$$

$$= \frac{1}{h} \left[f(x(t) + \Delta x, y(t) + \Delta y) - f(x(t), y(t)) \right]$$

$$= \frac{1}{h} \left[f_x(x(t), y(t))\Delta x + f_y(x(t), y(t))\Delta y + \epsilon \Delta x + \eta \Delta y \right]$$

$$= f_x(x(t), y(t)) \cdot \frac{\Delta x}{h} + f_y(x(t), y(t)) \cdot \frac{\Delta y}{h} + \epsilon \cdot \frac{\Delta x}{h} + \eta \cdot \frac{\Delta y}{h}.$$

Thus

$$\frac{dz}{dt} = \lim_{h \to 0} \frac{f(x(t + h), y(t + h)) - f(x(t), y(t))}{h}$$

$$= f_x(x(t), y(t)) \cdot \lim_{h \to 0} \frac{x(t + h) - x(t)}{h}$$

$$+ f_y(x(t), y(t)) \cdot \lim_{h \to 0} \frac{y(t + h) - y(t)}{h}$$

$$+ \lim_{h \to 0} \epsilon \cdot \lim_{h \to 0} \frac{x(t + h) - x(t)}{h} + \lim_{h \to 0} \eta \cdot \lim_{h \to 0} \frac{y(t + h) - y(t)}{h}$$

$$= f_x(x(t), y(t)) \frac{dx}{dt} + f_y(x(t), y(t)) \frac{dy}{dt} + 0 \cdot \frac{dx}{dt} + 0 \cdot \frac{dy}{dt}$$

$$= \frac{\partial f}{\partial x} \cdot \frac{dx}{dt} + \frac{\partial f}{\partial y} \cdot \frac{dy}{dt}$$

Example 1

If $z = x^2 + y^2$, $x = \sin t$ and $y = e^t$, find dz/dt.

$$\frac{dz}{dt} = \frac{\partial z}{\partial x} \frac{dx}{dt} + \frac{\partial z}{\partial y} \frac{dy}{dt}$$

$$= 2x \cdot \cos t + 2y \cdot e^t$$

$$= 2(x \cos t + ye^t).$$

By substituting $x = \sin t$ and $y = e^t$, we can express the result entirely in terms of t.

$$\frac{dz}{dt} = 2(\sin t \cos t + e^{2t}).$$

Of course, this derivative could also be found by substituting first and then differentiating.

$$z = x^2 + y^2$$
$$= \sin^2 t + e^{2t};$$
$$\frac{dz}{dt} = 2 \sin t \cos t + 2e^{2t}.$$

Theorem 21.2 is especially useful when we do not know what all of the functions are. It can be extended in many ways. Perhaps the most obvious is the case in which z is a function of three or more variables, each of which is a function of t.

Theorem 21.3

If $z = f(x_1, x_2, \ldots, x_n)$ and $x_i = g_i(t)$, $i = 1, 2, \ldots, n$,   and if all partial derivatives of z are continuous and dx_i/dt exist, $i = 1, 2, \ldots, n$,   then

$$\frac{dz}{dt} = \sum_{i=1}^{n} \frac{\partial f}{\partial x_i} \frac{dx_i}{dt}.$$

A special case of Theorem 21.2 follows.

Theorem 21.4

If $z = f(x, y)$ and $y = g(x)$ and if $\partial z/\partial x$ and $\partial z/\partial y$ are continuous and dz/dx exists, then

$$\frac{dz}{dx} = \frac{\partial z}{\partial x} + \frac{\partial z}{\partial y} \cdot \frac{dy}{dx}.$$

Example 2

If $z = x^2 + xy + y^2$ and $y = \sin x$,   find dz/dx.

$$\frac{dz}{dx} = \frac{\partial z}{\partial x} + \frac{\partial z}{\partial y} \cdot \frac{dy}{dx} = (2x + y) + (x + 2y) \cos x.$$

Again, this expression can be put entirely in terms of x, or it could be found by substituting $y = \sin x$ first and differentiating in the ordinary way.

Note the distinction between $\partial z/\partial x$ and dz/dx. $\partial z/\partial x$ is determined by the original function $z = f(x, y)$, where x and y are assumed to be two independent variables. The fact that $y = g(x)$ puts an additional restriction upon x and y does not enter into consideration when one is finding $\partial z/\partial x$ or $\partial z/\partial y$. This restriction *is* taken into account when finding dz/dx.

Just as Theorem 21.2, in which z is a function of two variables, can be extended to give Theorem 21.3, in which z is a function of n variables, it can also be extended from the case in which x and y are functions of a single variable t to the case in which x and y are functions of m variables. Let us consider one special case here.

Theorem 21.5

If $z = f(x, y)$, $x = F(u, v)$, and $y = G(u, v)$ and if $\partial z/\partial x$ and $\partial z/\partial y$ are continuous and $\partial x/\partial u$, $\partial x/\partial v$, $\partial y/\partial u$, and $\partial y/\partial v$ exist, then

$$\frac{\partial z}{\partial u} = \frac{\partial z}{\partial x} \cdot \frac{\partial x}{\partial u} + \frac{\partial z}{\partial y} \cdot \frac{\partial y}{\partial u}$$

and

$$\frac{\partial z}{\partial v} = \frac{\partial z}{\partial x} \cdot \frac{\partial x}{\partial v} + \frac{\partial z}{\partial y} \cdot \frac{\partial y}{\partial v}.$$

The proof of this theorem is analogous to the proof of Theorem 21.2.

Example 3

Given $z = x^2 - y^3$, $x = u + v$, and $y = u - v$, find $\partial z/\partial u$ and $\partial z/\partial v$.

$$\frac{\partial z}{\partial u} = \frac{\partial z}{\partial x} \cdot \frac{\partial x}{\partial u} + \frac{\partial z}{\partial y} \cdot \frac{\partial y}{\partial u}$$

$$= 2x \cdot 1 - 3y^2 \cdot 1$$
$$= 2x - 3y^2.$$

$$\frac{\partial z}{\partial v} = \frac{\partial z}{\partial x} \cdot \frac{\partial x}{\partial v} + \frac{\partial z}{\partial y} \cdot \frac{\partial y}{\partial v}$$

$$= 2x \cdot 1 - 3y^2 \, (-1)$$
$$= 2x + 3y^2.$$

A question that arises is, "How do we know when to use a partial derivative and when to use an ordinary derivative?" It is simply a matter of noting whether the function in question is a function of one variable or more than one. For instance, in Theorem 21.5, z is a function of the two variables x and y. Thus we want the partial derivatives

$$\frac{\partial z}{\partial x} \quad \text{and} \quad \frac{\partial z}{\partial y}.$$

Again, both x and y are functions of the two variables u and v. We again want partial derivatives

$$\frac{\partial x}{\partial u}, \quad \frac{\partial x}{\partial v}, \quad \frac{\partial y}{\partial u}, \quad \frac{\partial y}{\partial v}.$$

Finally, if the x and y in $z = f(x, y)$ are replaced by $F(u, v)$ and $G(u, v)$, we have $z = f(F(u, v), G(u, v))$, which is still a function of two variables. Thus, we still want partial derivatives,

$$\frac{\partial z}{\partial u} \quad \text{and} \quad \frac{\partial z}{\partial v}.$$

The chain rule is useful in rate-of-change problems.

Example 4

Suppose the radius of a right circular cylinder is increasing at the rate of 2 in./min and the height is decreasing at 4 in./min. At what rate is the volume changing at the moment when the radius is 4 in. and the height is 10 in. ?

$V = \pi r^2 h$. Since both r and h are functions of the time t we use the chain rule to differentiate.

$$\frac{dV}{dt} = \frac{\partial V}{\partial r}\frac{dr}{dt} + \frac{\partial V}{\partial h}\frac{dh}{dt}$$

$$= 2\pi r h \frac{dr}{dt} + \pi r^2 \frac{dh}{dt}.$$

But $dr/dt = 2$, $dh/dt = -4$, $r = 4$, and $h = 10$. Therefore

$$\frac{dV}{dt} = 2\pi(4)(10)(2) + \pi(4)^2(-4)$$

$$= 96\pi \text{ in.}^3/\text{min.}$$

Problems

A

In Problems 1–24, use the theorems of this section to find the required derivatives.

1. $z = x^2 y$, $x = t^2 + 1$, $y = e^t$; find dz/dt.
2. $z = xy$, $x = \sin t$, $y = 2t + 1$; find dz/dt.
3. $z = x^3 - y$, $x = te^t$, $y = \sin t$; find dz/dt.
4. $z = x/y$, $x = t \sin t$, $y = \cos t$; find dz/dt.

5. $w = \dfrac{x^2 + y^2}{z}$, $x = t - 2$, $y = t^2 - 1$, $z = t^2 + 1$; find dw/dt.

6. $w = x^2 + y^2 + z^2$, $x = t - 1$, $y = t$, $z = t + 1$; find dw/dt.
7. $w = z(x^2 + y)$, $x = e^t$, $y = e^{-t}$, $z = t$; find dw/dt.
8. $w = x^2 y + y^2 z + z^2 x$, $x = \sin t$, $y = \cos t$, $z = \tan t$; find dw/dt.
9. $z = x/y$, $x = u \sin v$, $y = v \cos u$; find $\partial z/\partial u$ and $\partial z/\partial v$.
10. $z = x^2 + y^2$, $x = u^2 - v^2$, $y = u^2 + v^2$; find $\partial z/\partial u$ and $\partial z/\partial v$.
11. $z = x^2 y$, $x = 2u + v$, $y = 2v - u$; find $\partial z/\partial u$ and $\partial z/\partial v$.
12. $z = xy$, $x = u/v$, $y = u^2 + v^2$; find $\partial z/\partial u$ and $\partial z/\partial v$.
13. $w = xyz$, $x = u + v$, $y = u - v$, $z = 2u + 3v$; find $\partial w/\partial u$ and $\partial w/\partial v$.
14. $w = x^2 + y^2 + z^2$, $x = u^2 - v^2$, $y = u^2 + v^2$, $z = uv$; find $\partial w/\partial u$ and $\partial w/\partial v$.
15. $w = xy + yz + zx$, $x = u + v$, $y = 2u + v$, $z = u - 2v$; find $\partial w/\partial u$ and $\partial w/\partial v$.
16. $w = 3x + 2y - z$, $x = u \sin v$, $y = v \sin u$, $z = \sin u \sin v$; find $\partial w/\partial u$ and $\partial w/\partial v$.
17. $w = 2x + y - z$, $x = t^2 + u^2$, $y = u^2 + v^2$, $z = v^2 + t^2$; find $\partial w/\partial t$, $\partial w/\partial u$, and $\partial w/\partial v$.
18. $w = xyz$, $x = tuv$, $y = t/u$, $z = u/v$; find $\partial w/\partial t$, $\partial w/\partial u$, and $\partial w/\partial v$.

19. $v = 2x + y - z + 3w$, $x = t \sin t$, $y = \dfrac{\sin t}{t}$, $z = t \cos t$, $w = \dfrac{\cos t}{t}$; find dv/dt.

20. $v = xy + zw$, $x = st$, $y = s + t$, $z = s/t$, $w = s - t$; find $\partial v/\partial s$ and $\partial v/\partial t$.
21. $z = xy$, $y = e^x \sin x$; find dz/dx.
22. $z = x^2 + y^2$, $y = x \sin x$; find dz/dx.

23. $z = x/y$, $y = x + \sin x$; find dz/dx.
24. $z = x - 2y$, $y = xe^x$; find dz/dx.

B

25. For $z = f(x, y)$, $x = r \cos \theta$, and $y = r \sin \theta$, find $\partial z/\partial r$ and $\partial z/\partial \theta$. Show that

$$\left(\frac{\partial z}{\partial x}\right)^2 + \left(\frac{\partial z}{\partial y}\right)^2 = \left(\frac{\partial z}{\partial r}\right)^2 + \frac{1}{r^2}\left(\frac{\partial z}{\partial \theta}\right)^2.$$

26. For $z = f(r, \theta)$, $r = \sqrt{x^2 + y^2}$, and $\theta = \text{Arctan}\,(y/x)$, find $\partial z/\partial x$ and $\partial z/\partial y$.
27. For $w = f(x, y, z)$, $x = \rho \sin \varphi \cos \theta$, $y = \rho \sin \varphi \sin \theta$, and $z = \rho \cos \varphi$, find $\partial w/\partial \rho$, $\partial w/\partial \varphi$, and $\partial w/\partial \theta$.

28. If $z = f(x^2 + y^2)$, show that

$$x\frac{\partial z}{\partial y} - y\frac{\partial z}{\partial x} = 0.$$

29. For $z = f(ax + by)$, show that

$$b\frac{\partial z}{\partial x} = a\frac{\partial z}{\partial y}.$$

30. For $w = f(x - y, y - z, z - x)$, show that

$$\frac{\partial w}{\partial x} + \frac{\partial w}{\partial y} + \frac{\partial w}{\partial z} = 0.$$

31. If the radius of a right circular cone is increasing at the rate of 2 in./min and the height is increasing at the rate of 3 in./min, at what rate is the volume changing at the moment when the radius is 5 in. and the height is 6 in.?
32. Suppose a gas satisfies the ideal gas law, $PV = RT$, where P is the pressure in atmospheres (atm), V is the volume in liters, T is the temperature in degrees absolute, and R is the ideal gas constant, 0.082054. If the pressure is increasing at the rate of 0.02 atm/min and the volume is decreasing at the rate of 0.1 liter/min, find the rate of change of the temperature at the moment when $P = 2$ atm and $V = 3$ liters.
33. A vertical line is moving to the right at 2 in./min, and a horizontal line is moving upward at 3 in./min. Find the rate of change of the area of the rectangle formed by the coordinate axes and these two lines at the moment when the lines are $x = 5$ and $y = 4$.
34. A point moves on the surface $z = x^3 + 2xy^2$ with x increasing at the rate of 4 in./min and y increasing at 6 in./min. At what rate is $\partial z/\partial x$ changing at the moment when $x = 2$ and $y = 5$?
35. Prove Theorem 21.4.

C

36. Show that if $f(x, y)$ is continuous and has continuous first-order partial derivatives in a circle containing the points (x_1, y_1) and (x_2, y_2), then

$$f(x_2, y_2) - f(x_1, y_1) = f_x(x^*, y^*)(x_2 - x_1) + f_y(x^*, y^*)(y_2 - y_1)$$

where (x^*, y^*) is a point on the segment from (x_1, y_1) to (x_2, y_2). (*Hint:* Consider $F(t) = f(x_1 + t(x_2 - x_1), y_1 + t(y_2 - y_1))$, $0 \le t \le 1$.)
37. If $z = x^2 - y^2$, $x = t^2$, and $y = 1 - 2t$, find d^2z/dt^2.
38. If $z = x^2 - y^2$, $x = 2r + 3s$, and $y = 3r - s$, find $\partial^2 z/\partial r^2$, $\partial^2 z/\partial s \partial r$, and $\partial^2 z/\partial s^2$.
39. If $z = f(x + at) + g(x - at)$, show that

$$\frac{\partial^2 z}{\partial t^2} = a^2 \frac{\partial^2 z}{\partial x^2}.$$

40. If $z = f(y/x)$,   show that

$$x \frac{\partial z}{\partial x} + y \frac{\partial z}{\partial y} = 0.$$

41. If $z = f(x, y) = x^2 - 2xy - y^2$,   show that

$$\Delta z = f(x + \Delta x, y + \Delta y) - f(x, y) = f_x \Delta x + f_y \Delta y + \epsilon \Delta x + \eta \Delta y,$$

where $\epsilon = \Delta x - \Delta y$ and $\eta = -\Delta x - \Delta y$;   thus verify the fundamental increment formula for this function.

42. Verify the fundamental increment formula for the function $f(x, y) = xy + 3y^2 - 2x + y + 4$.

21.4

Tangent Plane to a Surface; Tangent Line to a Curve

Just as the derivative of $y = f(x)$ allowed us to find the equation of a line tangent to $y = f(x)$, the partial derivatives of $z = f(x, y)$ allow us to find the equation of a plane tangent to $z = f(x, y)$. Again we have the problem of determining what is meant by the plane tangent to a surface at a given point. Since we know what a tangent line to a curve is, let us define a tangent plane in terms of it. Suppose we consider a vertical plane containing (x_0, y_0, z_0). The intersection of this plane with the surface $z = f(x, y)$ is a curve (see Figure 21.3). It seems reasonable that the tangent line to this curve lies in the plane tangent to the surface at (x_0, y_0, z_0). As a tentative definition, we might say that the tangent plane to the surface $z = f(x, y)$ at (x_0, y_0, z_0) is the plane containing all such tangent lines. Of course, there is the question of whether or not all of these lines lie in a single plane. Let us assume for the time being that they do (we shall prove it later).

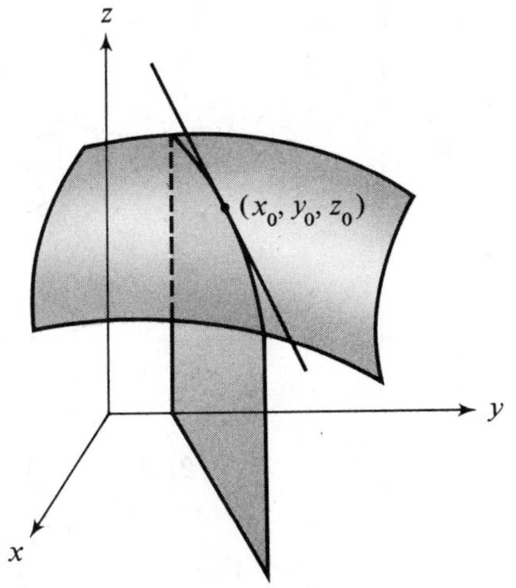

Let us now consider the two simplest such vertical planes—those parallel to the xz plane and the yz plane. Now $\partial z / \partial x$ evaluated at (x_0, y_0, z_0), which we represent by

$$\frac{\partial z}{\partial x}\bigg|_{(x_0, y_0, z_0)},$$

Figure 21.3

gives the slope of the line tangent to the curve determined by $z = f(x, y)$ and the plane $y = y_0$. Similarly,

$$\frac{\partial z}{\partial y}\bigg|_{(x_0, y_0, z_0)}$$

gives the slope of the line tangent to the curve determined by $z = f(x, y)$ and $x = x_0$.
Now we have the slopes of two lines in the desired plane; we want a set of direction
numbers for a line perpendicular to the plane. In order to do this, let us find vectors
along the two lines. For the line in the plane $y = y_0$, we have the vector

$$\mathbf{v} = 1 \cdot \mathbf{i} + 0 \cdot \mathbf{j} + \frac{\partial z}{\partial x} \mathbf{k}$$

(see Figure 21.4). We abbreviate the notation for the partial derivative evaluated at
(x_0, y_0, z_0) to simply $\partial z/\partial x$—it is understood that it is evaluated at (x_0, y_0, z_0).
Similarly, the vector along the line in the plane $x = x_0$ is

$$\mathbf{u} = 0 \cdot \mathbf{i} + 1 \cdot \mathbf{j} + \frac{\partial z}{\partial y} \mathbf{k}.$$

Now a vector perpendicular to both of these is the cross product $\mathbf{u} \times \mathbf{v}$.

$$\mathbf{u} \times \mathbf{v} = \begin{vmatrix} \mathbf{i} & \mathbf{j} & \mathbf{k} \\ 0 & 1 & \dfrac{\partial z}{\partial y} \\ 1 & 0 & \dfrac{\partial z}{\partial x} \end{vmatrix} = \frac{\partial z}{\partial x} \mathbf{i} + \frac{\partial z}{\partial y} \mathbf{j} - 1 \cdot \mathbf{k}.$$

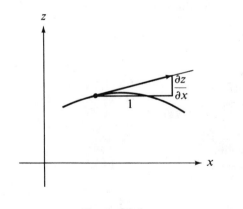

Figure 21.4

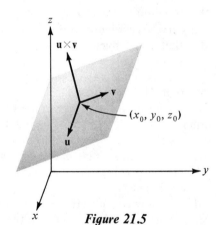

Figure 21.5

The components of this vector give a set of direction numbers for a line perpendicular
to the desired plane (see Figure 21.5). Using them, together with the point (x_0, y_0, z_0)
in the plane, we have the tangent plane.

$$A(x - x_0) + B(y - y_0) + C(z - z_0) = 0,$$

where

$$A = \frac{\partial z}{\partial x}\bigg|_{(x_0, y_0, z_0)}, \qquad B = \frac{\partial z}{\partial y}\bigg|_{(x_0, y_0, z_0)}, \qquad C = -1.$$

Theorem 21.6

If $z = f(x, y)$ *and the partial derivatives of z are continuous at* (x_0, y_0, z_0), *then the plane tangent at* (x_0, y_0, z_0) *to the surface represented by* $z = f(x, y)$ *is*

$$A(x - x_0) + B(y - y_0) + C(z - z_0) = 0$$

and the normal line is

$$\frac{x - x_0}{A} = \frac{y - y_0}{B} = \frac{z - z_0}{C},$$

where

$$A = \frac{\partial z}{\partial x}\bigg|_{(x_0, y_0, z_0)} , \qquad B = \frac{\partial z}{\partial y}\bigg|_{(x_0, y_0, z_0)} , \qquad C = -1.$$

Example 1

Find equations of the tangent plane and normal line to $z = x^2 + y^2$ at $(1, 1, 2)$.

$$\frac{\partial z}{\partial x}\bigg|_{(1,1,2)} = 2x \bigg|_{(1,1,2)} = 2,$$

$$\frac{\partial z}{\partial y}\bigg|_{(1,1,2)} = 2y \bigg|_{(1,1,2)} = 2.$$

The desired plane is

$$2(x - 1) + 2(y - 1) - (z - 2) = 0$$
$$2x + 2y - z - 2 = 0,$$

and the normal line is

$$\frac{x - 1}{2} = \frac{y - 1}{2} = \frac{z - 2}{-1}.$$

The tangent plane was found using only two tangent lines; there is still the question whether this plane contains all the tangent lines through (x_0, y_0, z_0). That this is indeed the case will be proved in Theorem 21.7, where the surface is given in the more general form $F(x, y, z) = 0$.

Theorem 21.7

If $F(x, y, z) = 0$ *and the partial derivatives of F are continuous and not all zero at* (x_0, y_0, z_0), *then the plane tangent at* (x_0, y_0, z_0) *to the surface determined by* $F(x, y, z) = 0$ *is*

$$A(x - x_0) + B(y - y_0) + C(z - z_0) = 0$$

and the normal line is

$$\frac{x - x_0}{A} = \frac{y - y_0}{B} = \frac{z - z_0}{C},$$

where

$$A = \frac{\partial F}{\partial x}\bigg|_{(x_0, y_0, z_0)} , \qquad B = \frac{\partial F}{\partial y}\bigg|_{(x_0, y_0, z_0)} , \qquad C = \frac{\partial F}{\partial z}\bigg|_{(x_0, y_0, z_0)} .$$

Proof

Let $x = x(t)$, $y = y(t)$, $z = z(t)$, $a < t < b$, be a parametric representation of a curve lying on the surface and containing the point (x_0, y_0, z_0) when $t = t_0$. Then $G(t) = F(x(t), y(t), z(t)) = 0$, $a < t < b$. Differentiating by the chain rule,

$$G'(t) = \frac{\partial F}{\partial x} \cdot x'(t) + \frac{\partial F}{\partial y} \cdot y'(t) + \frac{\partial F}{\partial z} \cdot z'(t) = 0, \quad a < t < b.$$

Thus the vector

$$\mathbf{n} = \frac{\partial F}{\partial x}\mathbf{i} + \frac{\partial F}{\partial y}\mathbf{j} + \frac{\partial F}{\partial z}\mathbf{k}$$

is orthogonal to the vector

$$\mathbf{w} = x'(t)\mathbf{i} + y'(t)\mathbf{j} + z'(t)\mathbf{k}$$

at each point on the curve. But $\mathbf{w}$ is a tangent vector to the curve, so at (x_0, y_0, z_0) the vector $\mathbf{n}$ is perpendicular to a plane containing the tangent line to each curve that lies on the surface and passes through (x_0, y_0, z_0). This plane is the tangent plane to the surface at (x_0, y_0, z_0).

Note that Theorem 21.6 is a special case of Theorem 21.7, with $F(x, y, z) = f(x, y) - z = 0$.

Example 2

Find equations of the tangent plane and normal line to

$$\frac{x^2}{4} + \frac{y^2}{4} + \frac{z^2}{9} = 1$$

at $(1, -1, 3/\sqrt{2})$.

$$A = \frac{\partial F}{\partial x}\bigg|_{(1,-1,3/\sqrt{2})} = \frac{x}{2}\bigg|_{(1,-1,3/\sqrt{2})} = \frac{1}{2},$$

$$B = \frac{\partial F}{\partial y}\bigg|_{(1,-1,3/\sqrt{2})} = \frac{y}{2}\bigg|_{(1,-1,3/\sqrt{2})} = -\frac{1}{2},$$

$$C = \frac{\partial F}{\partial z}\bigg|_{(1,-1,3/\sqrt{2})} = \frac{2z}{9}\bigg|_{(1,-1,3/\sqrt{2})} = \frac{\sqrt{2}}{3}.$$

Thus the tangent plane is

$$\frac{1}{2}(x - 1) - \frac{1}{2}(y + 1) + \frac{\sqrt{2}}{3}\left(z - \frac{3}{\sqrt{2}}\right) = 0$$

$$3x - 3y + 2\sqrt{2}\,z - 12 = 0,$$

and the normal line is

$$\frac{x - 1}{1/2} = \frac{y + 1}{-1/2} = \frac{x - 3/\sqrt{2}}{\sqrt{2}/3}.$$

The only convenient representation of a curve in space is the parametric form (or the vector form, which is equivalent to it). We saw an example of this in Section 20.10, where we graphed

$$\mathbf{f}(t) = \cos t\mathbf{i} + \sin t\mathbf{j} + t\mathbf{k},$$

which we noted is equivalent to

$$x = \cos t, \qquad y = \sin t, \qquad z = t.$$

We also noted there that the derivative of a vector function is another vector tangent to the curve representing the vector function. This gives us a simple way of finding an equation of the line tangent to a curve in space. If

$$\mathbf{f}(t) = x(t)\mathbf{i} + y(t)\mathbf{j} + z(t)\mathbf{k},$$

then

$$\mathbf{f}'(t) = x'(t)\mathbf{i} + y'(t)\mathbf{j} + z'(t)\mathbf{k}$$

is tangent to the curve represented by $f(t)$. Therefore, $\{x'(t_0), y'(t_0), z'(t_0)\}$ is a set of direction numbers for the tangent line at $(x(t_0), y(t_0), z(t_0))$.

The plane perpendicular to the tangent line to a curve at a point on the curve is called the *normal plane* to the curve at the point.

Theorem 21.8

If $x(t)$, $y(t)$, and $z(t)$ are differentiable at $t = t_0$ and $x_0 = x(t_0)$, $y_0 = y(t_0)$, and $z_0 = z(t_0)$, then the line tangent to

$$x = x(t), \quad y = y(t), \quad z = z(t)$$

at $t = t_0$ is

$$\frac{x - x_0}{x'(t_0)} = \frac{y - y_0}{y'(t_0)} = \frac{z - z_0}{z'(t_0)},$$

and the normal plane is

$$x'(t_0)(x - x_0) + y'(t_0)(y - y_0) + z'(t_0)(z - z_0) = 0.$$

Example 3

Find equations of the tangent line and normal plane to $x = \cos t$, $y = \sin t$, $z = t$ at $t = \pi$.

$$x_0 = \cos \pi = -1, \quad y_0 = \sin \pi = 0, \quad z_0 = \pi,$$
$$x' = -\sin t = -\sin \pi = 0,$$
$$y' = \cos t = \cos \pi = -1,$$
$$z' = 1.$$

The line in symmetric form is

$$x + 1 = 0, \quad \frac{y}{-1} = \frac{z - \pi}{1}.$$

In parametric form, it is

$$x = -1, \quad y = -t, \quad z = \pi + t.$$

The normal plane is

$$0(x + 1) - (y - 0) + 1(z - \pi) = 0$$
$$y - z + \pi = 0.$$

Problems

A

In Problems 1–16, find equations for the tangent plane and normal line to the given surface at the indicated point.

1. $z = xy$, at $(2, 1, 2)$.

2. $z = x^2 - y^2$, at $(3, 2, 5)$.

3. $z = x^2 - xy$, at $(2, 1, 2)$.

4. $z = x^2 + \dfrac{y^2}{4}$, at $(2, 2, 5)$.

5. $z = xy - x - 2y + 2$, at $(1, 2, -1)$.

6. $z = \dfrac{x + y}{xy}$, at $(2, 2, 1)$.

7. $z = xy^2$, at $(2, 1, 2)$.

8. $z = x^2 - y^3$, at $(1, 1, 0)$.

9. $z^2 = x^2 + y^2$, at $(3, 4, 5)$.

10. $x^2 + y^2 + z^2 = 9$, at $(2, 1, 2)$.

11. $x^2 - y^2 - z^2 = 9$, at $(9, 6, -6)$.

12. $x^2 + y^2 - z^2 = 4$, at $(3, 2, 3)$.

13. $xyz + x + y + z = -3$, at $(1, -2, 2)$.

14. $xy + yz = 4$, at $(2, 1, 2)$.

15. $y = x^2 - z^2$, at $(5, 9, 4)$.

16. $x = y^2 + z^2$, at $(5, 2, 1)$.

In Problems 17–24, find equations for the tangent line and normal plane to the curve at the given point.

17. $x = 4 \cos t$, $y = 3 \sin t$, $z = t$; at $t = \pi/2$.

18. $x = \cos t$, $y = \sin t$, $z = t$; at $t = 0$.

19. $x = t^2 + 1$, $y = t^2 - 1$, $z = t$; at $t = 2$.

20. $x = t^2$, $y = t - 1$, $z = t + 1$; at $t = 0$.

21. $x = t^2 - 1$, $y = t^3 + 1$, $z = 2t$; at $t = 3$.

22. $x = t^2$, $y = t^3$, $z = 1/t$; at $t = 1$.

23. $x = e^t$, $y = \ln t$, $z = t$; at $t = 1$.

24. $x = 3t - 1$, $y = \sin t$, $z = \cos t$; at $t = \pi$.

B

25. Show that the two surfaces $f(x, y, z) = 0$ and $g(x, y, z) = 0$ intersect orthogonally if and only if

$$\frac{\partial f}{\partial x} \cdot \frac{\partial g}{\partial x} + \frac{\partial f}{\partial y} \cdot \frac{\partial g}{\partial y} + \frac{\partial f}{\partial z} \cdot \frac{\partial g}{\partial z} = 0.$$

26. Use the result of Problem 25 to show that $xyz^2 = 1$ and $x^2 + y^2 - z^2 = 1$ intersect orthogonally.

27. Find the points on $x^2 + 4y^2 + 16z^2 - 2xy = 12$ where the tangent planes are parallel to the xz plane.

28. Find the tangent line to the curve of intersection of $x^2 + 2y^2 + z^2 = 7$ and $z - x^2 - y^2 = 0$ at the point $(1, 1, 2)$. (*Hint:* The tangent line is perpendicular to the normal to each surface.)

29. Find the tangent line to the curve of intersection of $x^2 + y^2 + z^2 = 12$ and $2x - 3y - 5z = 0$ at $(2, -2, 2)$.

30. Show that the plane tangent to

$$\frac{x^2}{a^2} + \frac{y^2}{b^2} + \frac{z^2}{c^2} = 1$$

at (x_0, y_0, z_0) is

$$\frac{xx_0}{a^2} + \frac{yy_0}{b^2} + \frac{zz_0}{c^2} = 1.$$

(Compare this with Problem 24, page 255.)

31. Show that the plane tangent to

$$\frac{x^2}{a^2} + \frac{y^2}{b^2} - \frac{z^2}{c^2} = 1$$

at (x_0, y_0, z_0) is

$$\frac{xx_0}{a^2} + \frac{yy_0}{b^2} - \frac{zz_0}{c^2} = 1.$$

(Compare this with Problem 31, page 261.)

21.5

Directional Derivatives

We have already noted that if $z = f(x, y)$, then $\partial z/\partial x$ represents the slope of the graph found by taking a vertical cross section parallel to the x axis (see Figure 21.2). Similarly $\partial z/\partial y$ is the slope of the graph found by taking a vertical cross section parallel to the y axis. These are just special cases of a more general type of derivative—the directional derivative.

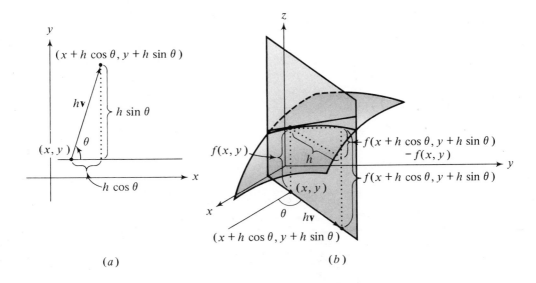

Figure 21.6

Suppose we consider the direction of a vertical plane to be represented by a unit vector $\mathbf{v}$ (see Figure 21.6(a)) in the xy plane, making an angle θ with the positive x axis. First of all, $\mathbf{v}$ can be expressed in the form

$$\mathbf{v} = \cos \theta \mathbf{i} + \sin \theta \mathbf{j}.$$

A vector of length h and having the same direction as $\mathbf{v}$ is

$$h\mathbf{v} = h \cos \theta \mathbf{i} + h \sin \theta \mathbf{j},$$

with tail at (x, y) and head at $(x + h \cos \theta, y + h \sin \theta)$.

Definition

If $z = f(x, y)$ and $\mathbf{v}$ is the unit vector $\cos \theta \mathbf{i} + \sin \theta \mathbf{j}$, then the directional derivative of z in the direction of $\mathbf{v}$ is

$$D_{\mathbf{v}} z = \lim_{h \to 0} \frac{f(x + h \cos \theta, y + h \sin \theta) - f(x, y)}{h},$$

if this limit exists.

Suppose we consider the vertical plane determined by the point (x, y) and the unit vector $\mathbf{v}$ in the xy plane (see Figure 21.6). The intersection of this plane with the surface $z = f(x, y)$ is some curve, and the slope of that curve at $(x, y, f(x, y))$ is given by the directional derivative. This is illustrated in Figure 21.6(b).

Note that the two partial derivatives are special cases of the directional derivative. When $\theta = 0$,

$$D_{\mathbf{v}} z = \lim_{h \to 0} \frac{f(x + h \cos 0, y + h \sin 0) - f(x, y)}{h}$$

$$= \lim_{h \to 0} \frac{f(x + h, y) - f(x, y)}{h}$$

$$= \frac{\partial z}{\partial x}.$$

Similarly, when $\theta = \pi/2$,

$$D_{\mathbf{v}} z = \lim_{h \to 0} \frac{f\left(x + h \cos \frac{\pi}{2}, y + h \sin \frac{\pi}{2}\right) - f(x, y)}{h}$$

$$= \lim_{h \to 0} \frac{f(x, y + h) - f(x, y)}{h}$$

$$= \frac{\partial z}{\partial y}.$$

Now let us consider the relationship between the general directional derivative and the two partial derivatives.

Theorem 21.9

If $z = f(x, y)$, *and if* $\partial z / \partial x$ *and* $\partial z / \partial y$ *are both continuous, and* $\mathbf{v} = \cos\theta\mathbf{i} + \sin\theta\mathbf{j}$, *then*

$$D_{\mathbf{v}}z = \frac{\partial z}{\partial x}\cos\theta + \frac{\partial z}{\partial y}\sin\theta.$$

Proof

From the fundamental increment formula (see the proof of Theorem 21.2),

$$D_{\mathbf{v}}f = \lim_{h\to 0}\frac{1}{h}\left[f(x + h\cos\theta, y + h\sin\theta) - f(x, y)\right]$$

$$= \lim_{h\to 0}\frac{1}{h}\left[f_x(x, y)(h\cos\theta) + f_y(x, y)(h\sin\theta) + \epsilon\,(h\cos\theta) + \eta(h\sin\theta)\right]$$

$$= f_x(x, y)\cos\theta + f_y(x, y)\sin\theta$$

$$= \frac{\partial z}{\partial x}\cos\theta + \frac{\partial z}{\partial y}\sin\theta.$$

Note that in the increment formula, we replaced Δx by $h\cos\theta$ and Δy by $h\sin\theta$, so $\epsilon \to 0$ and $\eta \to 0$ as $h \to 0$.

Example 1

Given that $z = x^2 + y^2$ and $\mathbf{v}$ is the unit vector making an angle $30°$ with the positive x axis, find $D_{\mathbf{v}}z$.

$$D_{\mathbf{v}}z = \frac{\partial z}{\partial x}\cos\theta + \frac{\partial z}{\partial y}\sin\theta$$

$$= 2x\cos 30° + 2y\sin 30°$$

$$= 2x \cdot \frac{\sqrt{3}}{2} + 2y \cdot \frac{1}{2}$$

$$= \sqrt{3}x + y.$$

Let us extend the preceding definition to functions of three variables. Recall that a unit vector in three dimensions is represented by its direction cosines:

$$\mathbf{v} = \cos\alpha\mathbf{i} + \cos\beta\mathbf{j} + \cos\gamma\mathbf{k}.$$

Definition

If $w = f(x, y, z)$ *and* $\mathbf{v}$ *is the unit vector*

$$\cos\alpha\mathbf{i} + \cos\beta\mathbf{j} + \cos\gamma\mathbf{k},$$

then the directional derivative in the direction of $\mathbf{v}$ *is*

$$D_{\mathbf{v}}w = \lim_{h\to 0}\frac{f(x + h\cos\alpha, y + h\cos\beta, z + h\cos\gamma) - f(x, y, z)}{h},$$

if this limit exists.

An argument similar to the one used for Theorem 21.9 proves the following theorem.

Theorem 21.10

If $w = f(x, y, z)$, and $\partial w/\partial x$, $\partial w/\partial y$, and $\partial w/\partial z$ are continuous, and

$$\mathbf{v} = \cos \alpha \mathbf{i} + \cos \beta \mathbf{j} + \cos \gamma \mathbf{k},$$

then

$$D_\mathbf{v} w = \frac{\partial w}{\partial x} \cos \alpha + \frac{\partial w}{\partial y} \cos \beta + \frac{\partial w}{\partial z} \cos \gamma.$$

The form of the conclusions of Theorem 21.9 and Theorem 21.10 suggests the dot product of two vectors. One of these vectors is the unit vector $\mathbf{v}$. The other is called the *gradient* of f and is represented by

$$\nabla f(x, y) = \frac{\partial f}{\partial x} \mathbf{i} + \frac{\partial f}{\partial y} \mathbf{j}$$

or

$$\nabla f(x, y, z) = \frac{\partial f}{\partial x} \mathbf{i} + \frac{\partial f}{\partial y} \mathbf{j} + \frac{\partial f}{\partial z} \mathbf{k}.$$

With this, we see that

$$D_\mathbf{v} f = \nabla f \cdot \mathbf{v}.$$

Furthermore, if $\mathbf{u}$ (not necessarily a unit vector) specifies any direction, then

$$D_\mathbf{u} f = \nabla f \cdot \frac{\mathbf{u}}{|\mathbf{u}|}.$$

The symbol ∇ is read "del." The gradient plays an important role in vector analysis and has some interesting properties.

Theorem 21.11

The maximum value of the directional derivative $D_\mathbf{v} f$ occurs when $\mathbf{v}$ has the same direction as ∇f. The value of this maximum is $|\nabla f|$.

Proof

By Theorem 8.7 (page 234) or Theorem 20.7,

$$D_\mathbf{v} f = \nabla f \cdot \mathbf{v} = |\nabla f| |\mathbf{v}| \cos \theta.$$

Since $|\mathbf{v}| = 1$,

$$D_\mathbf{v} f = |\nabla f| \cos \theta.$$

But ∇f is fixed for a given point. Therefore, $D_\mathbf{v} f$ is a maximum when $\cos \theta$ is a maximum. But the maximum value of $\cos \theta$ is 1 when $\theta = 0$. Thus the maximum value of $D_\mathbf{v} f$ is $|\nabla f|$. Since this occurs when $\theta = 0$ (that is, when the angle between ∇f and $\mathbf{v}$ is 0), $\mathbf{v}$ has the same direction as ∇f.

Example 2

A circular plate coincides with the region bounded by $x^2 + y^2 = 100$. If the temperature $T = x^2 + 2y^2 + 50$ at each point (x, y) in the plate, find the magnitude and direction of the maximum rate of change of temperature at the point $(2, 3)$.

$$\nabla T = \frac{\partial T}{\partial x}\mathbf{i} + \frac{\partial T}{\partial y}\mathbf{j} = 2x\mathbf{i} + 4y\mathbf{j}.$$

At $(2, 3)$, $\nabla T = (2 \cdot 2)\mathbf{i} + (4 \cdot 3)\mathbf{j} = 4\mathbf{i} + 12\mathbf{j}$.

The maximum rate of change is

$$|\nabla T(2, 3)| = |4\mathbf{i} + 12\mathbf{j}| = 4\sqrt{10}.$$

The direction is the same as that of ∇T:

$$\mathbf{v} = \frac{\nabla T}{|\nabla T|} = \frac{1}{\sqrt{10}}\mathbf{i} + \frac{3}{\sqrt{10}}\mathbf{j}$$

Another important property of the gradient is its directional relationship to the level curves or level surfaces of a function. Suppose that $z = f(x, y)$. Then for each number k in the range of f, the set of points (x, y) in the xy plane satisfying $f(x, y) = k$ is called a level curve. For example, if $z = 4x^2 + y^2$ (see Figure 21.7(a)), then several of its level curves are given in Figure 21.7(b).

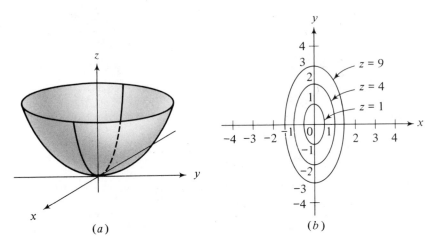

(a)

(b)

Figure 21.7

These level curves represent the intersection of $z = 4x^2 + y^2$ with the horizontal planes $z = 1, z = 4$, and $z = 9$. Now let us consider the directional derivatives at some point (x_0, y_0) of one of these level curves. If $\mathbf{v}$ is a unit vector tangent to the level curve, then $D_{\mathbf{v}}f(x_0, y_0) = 0$, because $z = f(x, y)$ is neither increasing nor decreasing in the direction given by $\mathbf{v}$. But, as we noted in the proof of Theorem 21.11,

$$D_{\mathbf{v}}f = \nabla f \cdot \mathbf{v} = |\nabla f||\mathbf{v}| \cos \theta = |\nabla f| \cos \theta,$$

where θ is the angle between the vectors $\mathbf{v}$ and ∇f. But $D_\mathbf{v} f = 0$ when $\cos \theta = 0$ and $\theta = 90°$ (remember that $0° \leq \theta \leq 180°$, since θ is the angle between two vectors). Thus the gradient of f at (x_0, y_0) is perpendicular (or normal) to the level curve at (x_0, y_0).

If we think of the family of level curves of a surface as a topographical map of the surface, then we see that the curves of "steepest descent" (or "steepest ascent") on the surface follow the direction of the gradient vector to the level curves at each point.

Note that in this analysis we started with $z = f(x, y)$ and considered level curves in the xy plane. We may also start with $w = f(x, y, z)$ and consider level surfaces in space. In the proof of Theorem 21.7 we showed that the vector $\mathbf{n} = (\partial F/\partial x)\mathbf{i} + (\partial F/\partial y)\mathbf{j} + (\partial F/\partial z)\mathbf{k}$ is perpendicular (or normal) to the surface $F(x, y, z) = f(x, y, z) - C = 0$. But this vector is exactly ∇f. Thus the gradient $\nabla f(x_0, y_0, z_0)$ is normal to the level surface $f(x, y, z) = f(x_0, y_0, z_0)$.

It is because of this normality of the gradient that the maximum directional derivative (which is $|\nabla f|$) at a given point is often called the *normal derivative* at that point. We use the notation df/dn for the normal derivative of the function f.

Example 3

Find the normal derivative of $z = x^2 + y^2$ and the vector $\mathbf{v}$ to which it corresponds.

$$\nabla z = \frac{\partial z}{\partial x}\mathbf{i} + \frac{\partial z}{\partial y}\mathbf{j} = 2x\mathbf{i} + 2y\mathbf{j},$$

$$\frac{df}{dn} = |\nabla z| = \sqrt{4x^2 + 4y^2} = 2\sqrt{x^2 + y^2},$$

$$\mathbf{v} = \frac{\nabla z}{|\nabla z|} = \frac{2x\mathbf{i} + 2y\mathbf{j}}{2\sqrt{x^2 + y^2}} = \frac{x\mathbf{i} + y\mathbf{j}}{\sqrt{x^2 + y^2}}.$$

Problems

A

In Problems 1–10, find the directional derivative in the direction of the given vector.

1. $z = x^3 + xy + y^3$, $\mathbf{v} = \frac{\sqrt{3}}{2}\mathbf{i} - \frac{1}{2}\mathbf{j}$.

2. $z = x^2 + 2xy$, $\mathbf{v} = \frac{1}{\sqrt{2}}\mathbf{i} + \frac{1}{\sqrt{2}}\mathbf{j}$.

3. $z = \frac{x + y}{xy}$, $\mathbf{v} = \frac{3}{\sqrt{10}}\mathbf{i} + \frac{1}{\sqrt{10}}\mathbf{j}$.

4. $z = \frac{x}{y}$, $\mathbf{v} = \frac{1}{\sqrt{2}}\mathbf{i} - \frac{1}{\sqrt{2}}\mathbf{j}$.

5. $z = e^{xy}$, $\mathbf{v} = \frac{1}{2}\mathbf{i} - \frac{\sqrt{3}}{2}\mathbf{j}$.

6. $z = x^3 - y + 1$, $\mathbf{v} = -\frac{\sqrt{2}}{\sqrt{3}}\mathbf{i} + \frac{1}{\sqrt{3}}\mathbf{j}$.

7. $w = xyz$, $\mathbf{v} = \mathbf{i} - 2\mathbf{j} + 2\mathbf{k}$.

8. $w = x^2 + y^2 + z^2$, $\mathbf{v} = \mathbf{i} + \mathbf{j} - \sqrt{2}\mathbf{k}$.

9. $w = \frac{x + 2y - z}{xy}$, $\mathbf{v} = \mathbf{i} + \sqrt{2}\mathbf{j} - \sqrt{5}\mathbf{k}$.

10. $w = x + y^2 - z^3$, $\mathbf{v} = \mathbf{i} - 2\mathbf{j} - \sqrt{5}\mathbf{k}$.

In Problems 11–20, find the directional derivative in the direction of the given vector at the given point.

11. $z = x^2 - 2y^2, \quad \mathbf{v} = -\dfrac{1}{\sqrt{2}}\mathbf{i} - \dfrac{1}{\sqrt{2}}\mathbf{j}; \quad (0, 1, -2).$

12. $z = x + y^2, \quad \mathbf{v} = \dfrac{1}{2}\mathbf{i} + \dfrac{\sqrt{3}}{2}\mathbf{j}; \quad (1, 2, 5).$

13. $z = (x + y)^2, \quad \mathbf{v} = \dfrac{1}{\sqrt{5}}\mathbf{i} - \dfrac{2}{\sqrt{5}}\mathbf{j}; \quad (1, -1, 0).$

14. $z = x^3 + xy^2, \quad \mathbf{v} = \dfrac{3}{\sqrt{13}}\mathbf{i} + \dfrac{2}{\sqrt{13}}\mathbf{j}; \quad (1, 1, 2).$

15. $z = xy^3, \quad \mathbf{v} = 4\mathbf{i} + \mathbf{j}; \quad (1, 1, 1).$

16. $z = (3x + 2y)^2, \quad \mathbf{v} = \dfrac{1}{2}\mathbf{i} + \dfrac{\sqrt{3}}{2}\mathbf{j}; \quad (-1, 1, 1).$

17. $w = x^2 - (y + z)^2, \quad \mathbf{v} = \mathbf{i} - 2\mathbf{j} + \mathbf{k}; \quad (2, 1, 1, 0).$
18. $w = xy + yz, \quad \mathbf{v} = \mathbf{i} + 3\mathbf{j} + 2\mathbf{k}; \quad (1, 2, 1, 4).$

19. $w = xy/z, \quad \mathbf{v} = 3\mathbf{i} - \mathbf{j} + \sqrt{6}\mathbf{k}; \quad (2, 1, -1, -2).$

20. $w = \dfrac{x + y}{y + z}, \quad \mathbf{v} = \mathbf{i} - \mathbf{j} + \mathbf{k}; \quad (1, 1, 1, 1).$

In Problems 21–30, find the normal derivative and the corresponding unit vector.

21. $z = x^3 y.$ 22. $z = 2x^2 - y.$
23. $z = x/y.$ 24. $z = x^2 - y^3.$
25. $w = xyz.$ 26. $w = x^2 + y^2 + z^2.$
27. $z = xy + 1$ at $(1, 1, 2).$ 28. $z = x^2 - 2y^2$ at $(2, 1, 2).$
29. $w = xy - yz$ at $(1, 1, -1, 2).$ 30. $w = x^2 y + yz$ at $(1, 0, 2, 0).$

31. The temperature of a metal plate is given by the formula

$$T = 50 - 5x - 10y + xy, \quad |x| \le 2, \quad |y| \le 2.$$

What is the temperature at $(1, 1)$? In what direction must one go from $(1, 1)$ in order to have the temperature increase fastest?

B

32. The electric potential on a plate is distributed according to the formula

$$E = 20 - 4x - 5y + xy, \quad |x| \le 5, \quad |y| \le 4.$$

A particle moves along the plate in such a way that the electric potential is always increasing as rapidly as possible. Assuming that it starts at $(5, 4)$, describe its path.

33. If the directional derivative is 7 in the direction from $P_0(2, -1)$ toward $P_1(0, 1)$ and 3 in the direction from P_0 toward $P_2(-1, 1)$, what is it in the direction from P_0 toward $P_3(6, 2)$?

34. The temperature in a solid is given by the formula

$$T = x^2 + 2y^2 + 3z^2 + 20.$$

What is the temperature at $(1, 2, 1)$? In what direction must one go from $(1, 2, 1)$ in order to have the temperature decrease fastest?

C

35. Prove Theorem 21.9 by utilizing the function $g(s) = f(x + s \cos \theta, y + s \sin \theta).$ (*Hint:* By the chain rule, show that $g'(0) = (\partial z/\partial x) \cdot \cos \theta + (\partial z/\partial y) \cdot \sin \theta,$ and from the definition of the derivative, show that $g'(0) = D_v z.$)

21.6

Implicit Functions

We first considered derivatives of implicit functions in Chapter 4. Now we shall reconsider them in the light of partial derivatives and go on to consider more complicated cases. Under suitable conditions,† an expression of the form

$$F(x, y) = 0$$

determines y as a differentiable function of x, $y = f(x)$, such that $F(x, y(x)) = 0$ on some x interval. Thus, looking upon y as a function of x, we take the derivative of both sides with respect to x. Assuming that the derivatives in question exist, then by Theorem 21.4,

$$\frac{\partial F}{\partial x} + \frac{\partial F}{\partial y} \cdot \frac{dy}{dx} = 0.$$

Solving for dy/dx, we have

$$\frac{dy}{dx} = -\frac{\partial F/\partial x}{\partial F/\partial y}.$$

Theorem 21.12

If $F(x, y) = 0$, $\partial F/\partial x$ and $\partial F/\partial y$ are continuous, and $\partial F/\partial y \neq 0$, then dy/dx exists and

$$\frac{dy}{dx} = -\frac{\partial F/\partial x}{\partial F/\partial y}.$$

Example 1

Given $x^2 + y^2 = 4$, find dy/dx.

Since the equation can be written in the form

$$F(x, y) = x^2 + y^2 - 4 = 0,$$

Theorem 21.12 gives·

$$\frac{dy}{dx} = -\frac{\partial F/\partial x}{\partial F/\partial y} = -\frac{x}{y}.$$

---

† If $f(a, b) = 0$, $\partial f/\partial x$ and $\partial f/\partial y$ are continuous in a neighborhood of (a, b), and $\partial f/\partial y \neq 0$ at (a, b), then there exists a differentiable function $y = y(x)$ such that $f(x, y(x)) = 0$ on some x interval containing $x = a$. For a proof of this result and its generalizations, see Kenneth Rogers, *Advanced Calculus* (Columbus, Ohio: Charles E. Merrill, 1976), pp. 111–119.

If we compare this example with what we did on pages 90–94, we see that the procedures are essentially the same. In effect, we were deriving this formula every time we needed it in Chapter 4. Let us consider a somewhat more difficult case. Suppose

$$F(x, y, z) = 0.$$

Again, under appropriate conditions, this equation determines one of the variables, say z, as a function of the other two,

$$z = f(x, y).$$

So we can find $\partial z/\partial x$ and $\partial z/\partial y$. Suppose we first take the derivative of both sides of $F(x, y, z) = 0$ with respect to x, remembering that $z = f(x, y)$ and x and y are independent.

$$\frac{\partial F}{\partial x} \cdot \frac{\partial x}{\partial x} + \frac{\partial F}{\partial y} \cdot \frac{\partial y}{\partial x} + \frac{\partial F}{\partial z} \cdot \frac{\partial z}{\partial x} = 0.$$

Of course $\partial x/\partial x = 1$. Since x and y are taken to be independent, $\partial y/\partial x = 0$, and the above equation can be simplified to

$$\frac{\partial F}{\partial x} + \frac{\partial F}{\partial z} \cdot \frac{\partial z}{\partial x} = 0.$$

Solving for $\partial z/\partial x$, we have

$$\frac{\partial z}{\partial x} = - \frac{\partial F/\partial x}{\partial F/\partial z}.$$

By a similar argument, we could show that

$$\frac{\partial z}{\partial y} = - \frac{\partial F/\partial y}{\partial F/\partial z}.$$

Theorem 21.13

If $F(x, y, z) = 0$, *the three partial derivatives of F are continuous, and* $\partial F/\partial z \neq 0$, *then* $\partial z/\partial x$ *and* $\partial z/\partial y$ *exist and*

$$\frac{\partial z}{\partial x} = - \frac{\partial F/\partial x}{\partial F/\partial z}, \qquad \frac{\partial z}{\partial y} = - \frac{\partial F/\partial y}{\partial F/\partial z}.$$

Example 2

Given $x^2 y + y^2 z + z^2 x = 0$, find $\partial z/\partial x$ and $\partial z/\partial y$.

$$\frac{\partial z}{\partial x} = - \frac{\partial F/\partial x}{\partial F/\partial z} = - \frac{2xy + z^2}{y^2 + 2zx},$$

$$\frac{\partial z}{\partial y} = - \frac{\partial F/\partial y}{\partial F/\partial z} = - \frac{x^2 + 2yz}{y^2 + 2zx}.$$

Actually, there is no particular reason to consider z as a function of x and y; we can think of any one of the three as a function of the other two. There are six possible derivatives we can consider:

$$\frac{\partial z}{\partial x} = -\frac{\partial F/\partial x}{\partial F/\partial z}, \quad \frac{\partial z}{\partial y} = -\frac{\partial F/\partial y}{\partial F/\partial z},$$

$$\frac{\partial y}{\partial x} = -\frac{\partial F/\partial x}{\partial F/\partial y}, \quad \frac{\partial y}{\partial z} = -\frac{\partial F/\partial z}{\partial F/\partial y},$$

$$\frac{\partial x}{\partial y} = -\frac{\partial F/\partial y}{\partial F/\partial x}, \quad \frac{\partial x}{\partial z} = -\frac{\partial F/\partial z}{\partial F/\partial x}.$$

As an aid to the memory, we might note that, except for the minus sign, the quotients behave as if the partial derivatives were fractions. There is another way of looking at the expressions which will carry over to other, more complicated situations. Suppose, for instance, that we want to find $\partial y/\partial x$ from the equation $F(x, y, z) = 0$. We must then consider y to be a function of x and z, $y = f(x, z)$. In effect, this says that x and z are two independent variables and that y is the dependent variable. In that case, $\partial F/\partial y$, the derivative with respect to this one dependent variable, must be in the denominator. The numerator is found by replacing the y in the denominator by x, since we are looking for $\partial y/\partial x$. This scheme works in all of the cases listed. However, you are warned not to draw any far-reaching conclusions from the above discussion— it is only a device to aid the memory. This result holds for an equation of any number of variables.

Let us now consider a somewhat more complicated situation. Suppose we have the two equations in x, y, and z

$$F(x, y, z) = 0,$$

$$G(x, y, z) = 0.$$

Suppose further that one of the variables, say z, can be eliminated to give a single equation, $H(x, y) = 0$ in x and y, and that we can solve for y as a function of x, that is, $y = f(x)$. Then we could consider dy/dx. Similarly, we can imagine y to be eliminated and solve the resulting equation for z as a function of x, $z = g(x)$. Then we could also consider dz/dx. To summarize, in general, two equations in three variables determine two of the variables as functions of the third variable; that is, there are two dependent variables and $3 - 2 = 1$ independent variable.

The geometric interpretation of this situation is of interest. The two equations $F(x, y, z) = 0$ and $G(x, y, z) = 0$ determine two surfaces that intersect in a curve. With x as parameter, the curve has parametric representation

$$x = x, \quad y = f(x), \quad z = g(x),$$

and the tangent line to the curve has direction numbers $\{1, f'(x), g'(x)\}$ (see Problem 35).

Suppose we now try to find dy/dx and dz/dx without first finding $y = f(x)$ and $z = g(x)$. We begin by taking the partial derivatives of both equations with respect to x. By Theorem 21.4 (extended to three variables), we have

$$\frac{\partial F}{\partial x} + \frac{\partial F}{\partial y} \cdot \frac{dy}{dx} + \frac{\partial F}{\partial z} \cdot \frac{dz}{dx} = 0,$$

$$\frac{\partial G}{\partial x} + \frac{\partial G}{\partial y} \cdot \frac{dy}{dx} + \frac{\partial G}{\partial z} \cdot \frac{dz}{dx} = 0$$

or

$$\frac{\partial F}{\partial y} \cdot \frac{dy}{dx} + \frac{\partial F}{\partial z} \cdot \frac{dz}{dx} = -\frac{\partial F}{\partial x},$$

$$\frac{\partial G}{\partial y} \cdot \frac{dy}{dx} + \frac{\partial G}{\partial z} \cdot \frac{dz}{dx} = -\frac{\partial G}{\partial x}.$$

We now have two simultaneous equations in dy/dx and dz/dx. Solving by determinants (Cramer's rule), we have

$$\frac{dy}{dx} = \frac{\begin{vmatrix} -\dfrac{\partial F}{\partial x} & \dfrac{\partial F}{\partial z} \\[2mm] -\dfrac{\partial G}{\partial x} & \dfrac{\partial G}{\partial z} \end{vmatrix}}{\begin{vmatrix} \dfrac{\partial F}{\partial y} & \dfrac{\partial F}{\partial z} \\[2mm] \dfrac{\partial G}{\partial y} & \dfrac{\partial G}{\partial z} \end{vmatrix}} = -\frac{\begin{vmatrix} \dfrac{\partial F}{\partial x} & \dfrac{\partial F}{\partial z} \\[2mm] \dfrac{\partial G}{\partial x} & \dfrac{\partial G}{\partial z} \end{vmatrix}}{\begin{vmatrix} \dfrac{\partial F}{\partial y} & \dfrac{\partial F}{\partial z} \\[2mm] \dfrac{\partial G}{\partial y} & \dfrac{\partial G}{\partial z} \end{vmatrix}}$$

and

$$\frac{dz}{dx} = \frac{\begin{vmatrix} \dfrac{\partial F}{\partial y} & -\dfrac{\partial F}{\partial x} \\[2mm] \dfrac{\partial G}{\partial y} & -\dfrac{\partial G}{\partial x} \end{vmatrix}}{\begin{vmatrix} \dfrac{\partial F}{\partial y} & \dfrac{\partial F}{\partial z} \\[2mm] \dfrac{\partial G}{\partial y} & \dfrac{\partial G}{\partial z} \end{vmatrix}} = -\frac{\begin{vmatrix} \dfrac{\partial F}{\partial y} & \dfrac{\partial F}{\partial x} \\[2mm] \dfrac{\partial G}{\partial y} & \dfrac{\partial G}{\partial x} \end{vmatrix}}{\begin{vmatrix} \dfrac{\partial F}{\partial y} & \dfrac{\partial F}{\partial z} \\[2mm] \dfrac{\partial G}{\partial y} & \dfrac{\partial G}{\partial z} \end{vmatrix}}$$

This result is often abbreviated to

$$\frac{dy}{dx} = -\frac{\dfrac{\partial(F, G)}{\partial(x, z)}}{\dfrac{\partial(F, G)}{\partial(y, z)}}, \qquad \frac{dz}{dx} = -\frac{\dfrac{\partial(F, G)}{\partial(y, x)}}{\dfrac{\partial(F, G)}{\partial(y, z)}}.$$

Expressions of the form

$$\frac{\partial(F, G)}{\partial(x, y)} = \begin{vmatrix} \dfrac{\partial F}{\partial x} & \dfrac{\partial F}{\partial y} \\ \dfrac{\partial G}{\partial x} & \dfrac{\partial G}{\partial y} \end{vmatrix}$$

are called *Jacobians* after the German mathematician Karl Gustav Jacob Jacobi (1804–1851).

Theorem 21.14

If

$$F(x, y, z) = 0 \quad and \quad G(x, y, z) = 0$$

and if all partial derivatives of both functions are continuous and

$$\frac{\partial(F, G)}{\partial(y, z)} \neq 0,$$

then dy/dx and dz/dx both exist and

$$\frac{dy}{dx} = -\frac{\dfrac{\partial(F, G)}{\partial(x, z)}}{\dfrac{\partial(F, G)}{\partial(y, z)}}, \qquad \frac{dz}{dx} = -\frac{\dfrac{\partial(F, G)}{\partial(y, x)}}{\dfrac{\partial(F, G)}{\partial(y, z)}},$$

where

$$\frac{\partial(F, G)}{\partial(y, z)} = \begin{vmatrix} \dfrac{\partial F}{\partial y} & \dfrac{\partial F}{\partial z} \\ \dfrac{\partial G}{\partial y} & \dfrac{\partial G}{\partial z} \end{vmatrix}.$$

Note that in the expressions for dy/dx and dz/dx, the denominator is a Jacobian involving the partial derivatives of F and G with respect to the dependent variables y and z. In the expression for dy/dx, y is replaced by the independent variable x in the Jacobian in the numerator; in the expression for dz/dx, z is replaced by x in the Jacobian in the numerator.

Now suppose we are asked to find dy/dx only. How do we know which are the dependent variables and which is the independent variable? When we see dy/dx, we know immediately that y must be a function of x. Thus y is a dependent variable and x an independent variable. Since two equations are given, there must be a second dependent variable—it must be z. So y and z are the two dependent variables, and the denominator must be

$$\frac{\partial(F, G)}{\partial(y, z)},$$

by the scheme we have already described. Since we want dy/dx, we replace the y of the denominator by x to get the proper expression for the numerator. Thus

$$\frac{dy}{dx} = -\frac{\dfrac{\partial(F, G)}{\partial(x, z)}}{\dfrac{\partial(F, G)}{\partial(y, z)}}.$$

While the order of the y and z of the denominator is quite arbitrary, we must simply substitute without changing the order if we are to get the proper expression for the numerator. If the order were reversed, we would not get the correct expression, but rather its negative.

Example 3

Given $x^2 + y^2 + z^2 = 4$ and $xyz = 1$,   find dy/dx.

The two equations are $F: x^2 + y^2 + z^2 - 4 = 0,$

$G: xyz - 1 = 0.$

Since we want dy/dx, we know that y is a dependent variable and x is independent. Thus z must be the other dependent variable. Therefore,

$$\frac{dy}{dx} = -\frac{\dfrac{\partial(F, G)}{\partial(x, z)}}{\dfrac{\partial(F, G)}{\partial(y, z)}} = -\frac{\begin{vmatrix} \dfrac{\partial F}{\partial x} & \dfrac{\partial F}{\partial z} \\[2mm] \dfrac{\partial G}{\partial x} & \dfrac{\partial G}{\partial z} \end{vmatrix}}{\begin{vmatrix} \dfrac{\partial F}{\partial y} & \dfrac{\partial F}{\partial z} \\[2mm] \dfrac{\partial G}{\partial y} & \dfrac{\partial G}{\partial z} \end{vmatrix}} = -\frac{\begin{vmatrix} 2x & 2z \\ yz & xy \end{vmatrix}}{\begin{vmatrix} 2y & 2z \\ xz & xy \end{vmatrix}}$$

$$= -\frac{2x^2 y - 2yz^2}{2xy^2 - 2xz^2} = -\frac{y(x^2 - z^2)}{x(y^2 - z^2)}.$$

This procedure can be extended to two equations in four, five, six, or more variables, as well as to three equations in four or more variables. In setting up the expression for the desired derivative, remember that the number of equations equals the number of dependent variables; the number of independent variables is then the total number of variables minus the number of dependent variables. An additional problem arises when the number of variables exceeds the number of equations by more than one. To see how, suppose we have the two equations

$$F(w, x, y, z) = 0,$$

$$G(w, x, y, z) = 0$$

and we are asked to find $\partial y/\partial x$. Now we know that, if we are to talk about $\partial y/\partial x$, then y must be a function of x and some other variable. Thus y is a dependent variable and

x is independent. But what of the other two variables? Which is the second dependent variable and which is independent? There is no way of determining it from the information given. The problem, as stated, is ambiguous—there are two possible answers, depending upon whether we take w or z to be the second dependent variable.

In order to avoid this ambiguity, we shall use the notation

$$\left(\frac{\partial y}{\partial x}\right)_w,$$

which is read: the partial derivative of y with respect to x, w held constant. It implies that w, as well as x, is an independent variable. Thus

$$\left(\frac{\partial y}{\partial x}\right)_w = -\frac{\dfrac{\partial(F, G)}{\partial(x, z)}}{\dfrac{\partial(F, G)}{\partial(y, z)}}.$$

Example 4

Given $w^2 + x^2 + y^2 + z^2 = 4$ and $wxyz = 1$,   find $\left(\dfrac{\partial z}{\partial y}\right)_x$.

Here x and y are the independent variables, and w and z are dependent variables.

$$F:\ w^2 + x^2 + y^2 + z^2 - 4 = 0,$$
$$G:\ wxyz - 1 = 0.$$

$$\left(\frac{\partial z}{\partial y}\right)_x = -\frac{\dfrac{\partial(F, G)}{\partial(w, y)}}{\dfrac{\partial(F, G)}{\partial(w, z)}} = -\frac{\begin{vmatrix} \dfrac{\partial F}{\partial w} & \dfrac{\partial F}{\partial y} \\[2mm] \dfrac{\partial G}{\partial w} & \dfrac{\partial G}{\partial y} \end{vmatrix}}{\begin{vmatrix} \dfrac{\partial F}{\partial w} & \dfrac{\partial F}{\partial z} \\[2mm] \dfrac{\partial G}{\partial w} & \dfrac{\partial G}{\partial z} \end{vmatrix}} = -\frac{\begin{vmatrix} 2w & 2y \\ xyz & wxz \end{vmatrix}}{\begin{vmatrix} 2w & 2z \\ xyz & wxy \end{vmatrix}}$$

$$= -\frac{2w^2xz - 2xy^2z}{2w^2xy - 2xyz^2} = -\frac{z(w^2 - y^2)}{y(w^2 - z^2)}.$$

Example 5

Given $u^2 + v^2 = 1$, $x^2 + y^2 = 1$,   and $u + v + x + y = 0$,   find dy/du.

$$F:\ u^2 + v^2 \qquad\qquad -1 = 0,$$
$$G:\ \qquad\quad x^2 + y^2 - 1 = 0,$$
$$H:\ u + v + \ x + y \qquad = 0.$$

Since there are three equations, there are three dependent variables. Since we want dy/du, u is the one independent variable. Thus,

$$\frac{dy}{du} = -\frac{\dfrac{\partial(F, G, H)}{\partial(v, x, u)}}{\dfrac{\partial(F, G, H)}{\partial(v, x, y)}} = -\frac{\begin{vmatrix} \dfrac{\partial F}{\partial v} & \dfrac{\partial F}{\partial x} & \dfrac{\partial F}{\partial u} \\[2mm] \dfrac{\partial G}{\partial v} & \dfrac{\partial G}{\partial x} & \dfrac{\partial G}{\partial u} \\[2mm] \dfrac{\partial H}{\partial v} & \dfrac{\partial H}{\partial x} & \dfrac{\partial H}{\partial u} \end{vmatrix}}{\begin{vmatrix} \dfrac{\partial F}{\partial v} & \dfrac{\partial F}{\partial x} & \dfrac{\partial F}{\partial y} \\[2mm] \dfrac{\partial G}{\partial v} & \dfrac{\partial G}{\partial x} & \dfrac{\partial G}{\partial y} \\[2mm] \dfrac{\partial H}{\partial v} & \dfrac{\partial H}{\partial x} & \dfrac{\partial H}{\partial y} \end{vmatrix}} = -\frac{\begin{vmatrix} 2v & 0 & 2u \\ 0 & 2x & 0 \\ 1 & 1 & 1 \end{vmatrix}}{\begin{vmatrix} 2v & 0 & 0 \\ 0 & 2x & 2y \\ 1 & 1 & 1 \end{vmatrix}}$$

$$= -\frac{4xv - 4xu}{4xv - 4yv} = -\frac{x(v - u)}{v(x - y)}.$$

Problems

A

In Problems 1–26, find the required derivative(s).

1. $x^2 - 2xy - y^2 = 1$; find dy/dx.
2. $x^3 - 2x^2y + y^3 = 2$; find dy/dx.
3. $(x^2 + y)^2 + y = 2$; find dy/dx.
4. $x + \sqrt{xy} + y = 4$; find dy/dx.
5. $e^{xy} + y^2 + 1 = 0$; find dy/dx.
6. $\sin xy + x = 2$; find dy/dx.
7. $x^2 - xyz + z^2 = 2$; find $\partial y/\partial x$ and $\partial y/\partial z$.
8. $x^2 + y^2 + z^2 = 9$; find $\partial y/\partial x$ and $\partial y/\partial z$.
9. $x^3 - y^3 - z^3 = 4$; find $\partial x/\partial y$ and $\partial x/\partial z$.
10. $xy + yz + zx = 3$; find $\partial x/\partial z$ and $\partial x/\partial y$.
11. $(x + y)^2 = (y - z)^3$; find $\partial y/\partial x$ and $\partial y/\partial z$.
12. $\sin xy + \sin yz = 1$; find $\partial x/\partial y$ and $\partial x/\partial z$.
13. $(x + y)^2 + (y - z)^2 = w$; find $\partial z/\partial y$.
14. $x^2 - 2y^2 + z^2 = w$; find $\partial z/\partial x$.
15. $(u + v)^2 - (x + y + z)^2 = 1$; find $\partial u/\partial x$ and $\partial u/\partial y$.
16. $u^2 - v^2 = x + y + z$; find $\partial x/\partial u$ and $\partial x/\partial v$.
17. $x^2 + y^2 = 1$, $y^2 + z^2 = 1$; find dy/dz.
18. $x^2 + y^2 + z^2 = 6$; $x + y + z = 2$; find dy/dx.
19. $x - y + z = 1$, $x^3 - y^3 = 4$; find dz/dx.
20. $x + 2y - z = 5$, $x^2 + y^2 - z^2 = 4$; find dx/dy.
21. $x + y + z = 1$, $y^2 + z^2 = w$; find $(\partial w/\partial x)_y$.
22. $x^2 = y$, $y^2 = z$; find dy/dx.
23. $x^2 - y^2 = 1$, $x + y + z + w = 1$; find $(\partial y/\partial x)_w$.
24. $u - 2v + x + 2y - z = 0$, $u^2 + v^2 + x^2 + y^2 + z^2 = 4$; find $(\partial u/\partial x)_{y,z}$.
25. $x = r \cos \theta$, $y = r \sin \theta$; find $(\partial r/\partial \theta)_x$ and $(\partial r/\partial \theta)_y$.
26. $x = r \cos \theta$, $y = r \sin \theta$, $z = t$; find $(\partial y/\partial t)_{r,\theta}$.

B

27. For $w + x + y + z = 1$, $w^2 + x^2 + y^2 + z^2 = 4$, $wxyz = 2$,   find dy/dx.
28. For $z = x/y$ and $u = 2x + y$ and $v = x - 2y$,   find $(\partial z/\partial v)_u$.
29. For $z = xy$ and $u = x + y$ and $v = x - y$,   find $(\partial z/\partial u)_v$.
30. For $z = f(x, y)$, $x = r \cos \theta$,   and $y = r \sin \theta$,   show that

$$\left(\frac{\partial z}{\partial r}\right)_\theta^2 + \frac{1}{r^2}\left(\frac{\partial z}{\partial \theta}\right)_r^2 = \left(\frac{\partial z}{\partial x}\right)_y^2 + \left(\frac{\partial z}{\partial y}\right)_x^2.$$

C

31. If $f(x, y) = 0$,   show that

$$\frac{d^2 y}{dx^2} = -\frac{f_y^2 f_{xx} - 2 f_x f_y f_{xy} + f_x^2 f_{yy}}{f_y^3}.$$

(*Hint:* Differentiate implicitly $f_x + f_y \dfrac{dy}{dx} = 0$.)

32. Find all first and second partial derivatives of z if $xy + yz + z^2 = 0$.
33. Find all first and second partial derivatives of z if $xy + 3yz - z^2 = 0$.

34. If $F(x, y, z) = 0$,   show that $\dfrac{\partial z}{\partial x} \cdot \dfrac{\partial x}{\partial y} \cdot \dfrac{\partial y}{\partial z} = -1$.

35. Show that the tangent line to the curve of intersection of $F(x, y, z) = 0$ and $G(x, y, z) = 0$ at the point (x_0, y_0, z_0) has equation

$$\frac{x - x_0}{A} = \frac{y - y_0}{B} = \frac{z - z_0}{C},$$

where

$$A = \frac{\partial(F, G)}{\partial(y, z)}, \qquad B = \frac{\partial(F, G)}{\partial(z, x)}, \qquad C = \frac{\partial(F, G)}{\partial(x, y)}.$$

If $A \neq 0$, show that direction numbers of the tangent line are $\{1, dy/dx, dz/dx\}$.
36. Find an equation of the tangent line to the circle $x^2 + y^2 + z^2 - 14 = 0$, $x^2 + y^2 - 5 = 0$ at $(1, 2, 3)$.

21.7

The Total Differential; Approximation by Differentials

In the proof of the chain rule (Theorem 21.2), we derived the fundamental increment formula

$$\Delta z = f(x + \Delta x, y + \Delta y) - f(x, y) = \frac{\partial f}{\partial x} \Delta x + \frac{\partial f}{\partial y} \Delta y + \epsilon \Delta x + \eta \Delta y.$$

In this formula, ϵ and η both tend to zero as Δx and Δy tend to zero. Thus when Δx and Δy (the increments or errors in x and y, respectively) are small, then $\epsilon \Delta x + \eta \Delta y$

is of a higher order of smallness, and we might expect that the first two terms,

$$\frac{\partial f}{\partial x}\,\Delta x + \frac{\partial f}{\partial y}\,\Delta y,$$

would be a close approximation to the exact increment Δz. Recall that when $y = f(x)$ (a function of one variable), we let dx be a new independent variable and we defined $dy = f'(x)\,dx$. We then used the differential dy to approximate Δy when the differential $dx = \Delta x$ was small. Now, in the expression for Δz, suppose we take $dx = \Delta x$ and $dy = \Delta y$. Then we can approximate the increment Δz by

$$dz = \frac{\partial f}{\partial x}\,dx + \frac{\partial f}{\partial y}\,dy.$$

Definition

If $z = f(x, y)$ and both partial derivatives exist, then the **total differential**, dz, of z is

$$dz = \frac{\partial f}{\partial x}\,dx + \frac{\partial f}{\partial y}\,dy.$$

Example 1

Find dz for $z = x^2 - xy$.

$$dz = \frac{\partial z}{\partial x}\,dx + \frac{\partial z}{\partial y}\,dy = (2x - y)\,dx - x\,dy.$$

Of course, the same type of definition holds for functions of three or more variables.

Example 2

Find dw for $w = x^3 - x^2 y + y^2 z$.

$$dw = \frac{\partial w}{\partial x}\,dx + \frac{\partial w}{\partial y}\,dy + \frac{\partial w}{\partial z}\,dz = (3x^2 - 2xy)\,dx + (2yz - x^2)\,dy + y^2\,dz.$$

The tangent plane to the surface $z = f(x, y)$ at the point (x_0, y_0, z_0) has the equation

$$\frac{\partial z}{\partial x}\,(x - x_0) + \frac{\partial z}{\partial y}\,(y - y_0) - (z - z_0) = 0,$$

where $\partial z/\partial x$ and $\partial z/\partial y$ are evaluated at (x_0, y_0, z_0). Let

$$dx = x - x_0, \quad dy = y - y_0, \quad \text{and} \quad dz = z - z_0 = \frac{\partial z}{\partial x}\,dx + \frac{\partial z}{\partial y}\,dy;$$

then the point $(x_0 + dx,\, y_0 + dy,\, z_0 + dz)$ lies on the plane tangent to the surface $z = f(x, y)$ at the point (x_0, y_0, z_0). If

$$\Delta z = f(x_0 + dx,\, y_0 + dy) - f(x_0, y_0) = \frac{\partial z}{\partial x}\,dx + \frac{\partial z}{\partial y}\,dy + \epsilon\,dx + \eta\,dy,$$

then $(x_0 + dx, y_0 + dy, z_0 + \Delta z)$ lies on the surface itself, and the difference in the z coordinates of these two points is

$$(z_0 + \Delta z) - (z_0 + dz) = \Delta z - dz = \epsilon\, dx + \eta\, dy.$$

Thus, approximating Δz by dz corresponds geometrically to approximating the surface by its tangent plane (see Figure 21.8).

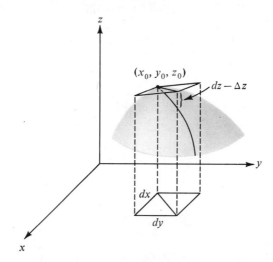

Figure 21.8

Example 3

The sides of a rectangle are found to be 4.00 in. and 6.50 in., with a maximum error of 0.02 in. in each case. What is the area of the rectangle and what is the maximum error in the area?

$$x = 4.00, \ |dx| \leq 0.02, \qquad |dA| = |y\, dx + x\, dy|$$
$$y = 6.50, \ |dy| \leq 0.02, \qquad \leq |y|\,|dx| + |x|\,|dy|$$
$$A = xy \qquad\qquad \leq (6.50)(0.02) + (4.00)(0.02)$$
$$= (4.00)(6.50) \qquad\qquad = 0.21 \text{ in.}^2$$
$$= 26 \text{ in.}^2$$

The value of $|\Delta A|$ is

$$|\Delta A| \leq (4.02)(6.52) - (4.00)(6.50)$$
$$= 0.2104 \text{ in.}^2$$

We see that $|dA|$ is very nearly $|\Delta A|$.

Example 4

The volume of a cone is found by measuring the diameter and height. Given a diameter of 4.12 ± 0.02 in. and a height of 6.21 ± 0.03 in., find the volume and the maximum error in volume.

$$r = 2.06 \text{ in.,} \qquad |dr| \leq 0.01 \text{ in.,}$$
$$h = 6.21 \text{ in.,} \qquad |dh| \leq 0.03 \text{ in.,}$$

$$V = \frac{1}{3}\pi r^2 h = \frac{1}{3}(3.14)(2.06)^2(6.21) = 27.6 \text{ in.}^3,$$

$$|dV| = \left|\frac{\partial V}{\partial r}\, dr + \frac{\partial V}{\partial h}\, dh\right| \leq \left|\frac{\partial V}{\partial r}\right| |dr| + \left|\frac{\partial V}{\partial h}\right| |dh|$$

$$= \left|\frac{2}{3}\pi r h\right| \left| dr \right| + \left|\frac{1}{3}\pi r^2\right| \left| dh \right|$$

$$= \frac{2}{3}(3.14)(2.06)(6.21)(0.01) + \frac{1}{3}(3.14)(2.06)^2(0.03)$$

$$= 0.40 \text{ in.}^3$$

Example 5

Use differentials to approximate $\sqrt{25.2}/2.13$.

We want the value of z corresponding to $x = 25.2$ and $y = 2.13$ in the equation

$$z = \frac{\sqrt{x}}{y}.$$

If $x = 25$ and $y = 2$, then $z = 2.5$ by an easy calculation. But we have made errors by letting $x = 25$ and $y = 2$. These errors are $dx = 0.2$ and $dy = 0.13$. An estimate of the corresponding error in z is

$$dz = \frac{\partial z}{\partial x}\, dx + \frac{\partial z}{\partial y}\, dy$$

$$= \frac{1}{2y\sqrt{x}}\, dx - \frac{\sqrt{x}}{y^2}\, dy$$

$$= \frac{0.2}{2(2)\sqrt{25}} - \frac{\sqrt{25}(0.13)}{2^2} = -0.15.$$

Thus

$$z + dz = 2.5 - 0.15 = 2.35,$$

and

$$\frac{\sqrt{25.2}}{2.13} \approx 2.35.$$

Problems

A

In Problems 1–10, find the total differential df.

1. $f(x, y) = x^2 y + xy^2.$

2. $f(x, y) = x \sin y.$

3. $f(x, y) = \dfrac{x}{y} + \ln y.$

4. $f(x, y) = \dfrac{x + y}{xy}.$

5. $f(x, y) = (x + \tan xy)^3$.

6. $f(x, y, z) = x^2 + y^2 + z^2$.

7. $f(x, y, z) = x^2 yz - y^2 z + z^2$.

8. $f(x, y, z) = e^{xy} + \sin yz$.

9. $f(x, y, z) = z \operatorname{Arcsin} y + x^2 y$.

10. $f(x, y, z, w) = x^2 + 2yz + w^2$.

In Problems 11–14, use differentials to approximate the numbers given.

11. $20.2\sqrt{50}$.

12. $\dfrac{(1.95)^3}{4.14}$.

13. $\dfrac{\sqrt[3]{8.02}}{4.11}$.

14. $\sqrt{50}\sqrt[3]{26}$.

15. The volume of a cylinder is determined by measuring the diameter and height. The diameter is 5.21 ± 0.04 in., and the height is 7.32 ± 0.05 in. Find the volume and the maximum error in the volume.

16. Find the total surface area and the maximum error in the area of the cylinder of Problem 15.

17. The sides of a box are found to be $4.00 \times 3.00 \times 6.00$ in. Find the maximum error in the volume if the maximum error in any of the three measurements is 0.01 in.

18. The density D of an object is given by $D = M/V$, where M is the mass and V is the volume of the object. A certain object is found to have mass 439.81 g and volume 25.34 ml. What is the density? If the maximum error in measuring the mass is ± 0.01 g and the maximum error in measuring the volume is ± 0.05 ml, find the maximum error in the density.

19. The two legs of a right triangle are found to be 4.00 ± 0.02 in. and 6.00 ± 0.02 in. Find the hypotenuse and the maximum error of the hypotenuse.

B

20. The angle of elevation of the top of a building is $23°00'00''$ when measured from a point 100 ft away. If the maximum error in measuring the angle is $30''$ and the maximum error in measuring the distance is 0.1 ft, what is the maximum error in the height of the building?

21. The angle of elevation of the top of a hill is $21°00'00''$ and the distance to the top of the hill is 500 ft. If the maximum error in measuring the angle is $30''$ and the maximum error in measuring the distance is 0.3 ft, find the maximum error in determining the height of the hill.

22. In determining the molecular weight of a low-boiling liquid by the Victor Meyer method, the ideal gas formula,

$$PV = nRT,$$

is used. The number of moles n of the liquid is represented by w/M, where w is the weight of the liquid and M is the molecular weight. Thus

$$M = \frac{wRT}{PV},$$

where w is the weight of the liquid in grams, T is the temperature of the vapor in degrees absolute, P is the pressure in atmospheres, V is the volume in liters, and R is the ideal gas constant 0.082054. The following data are taken:

$$T = 370.00 \pm 0.04°,$$
$$w = 0.05000 \pm 0.00001 \text{ g},$$
$$P = 0.9605 \pm 0.0001 \text{ atm},$$
$$V = 0.06070 \pm 0.00002 \text{ liter}.$$

Find M and the maximum error in M.

21.8

Maxima and Minima

One of the principal applications of ordinary derivatives is the determination of relative maxima and minima. This application can be extended to relative maxima and minima of functions of two variables by the use of partial derivatives. If $z = f(x, y)$ and $\partial z/\partial x$ and $\partial z/\partial y$ exist, it is clear that, in order to have a relative maximum or minimum, we must have a horizontal tangent plane. Thus both partial derivatives must be zero, which condition is enough for determining critical points.

Unfortunately, the use of second derivatives to determine whether we have a relative maximum, minimum, or neither is a complicated process. For instance, the fact that

$$\frac{\partial^2 z}{\partial x^2} < 0 \quad \text{and} \quad \frac{\partial^2 z}{\partial y^2} < 0$$

is not enough to guarantee that we have a relative maximum. An important consideration in the determination of relative maxima and minima is the determinant

$$D = \begin{vmatrix} \dfrac{\partial^2 z}{\partial x^2} & \dfrac{\partial^2 z}{\partial y\,\partial x} \\[2mm] \dfrac{\partial^2 z}{\partial x\,\partial y} & \dfrac{\partial^2 z}{\partial y^2} \end{vmatrix} = \frac{\partial^2 z}{\partial x^2} \cdot \frac{\partial^2 z}{\partial y^2} - \left(\frac{\partial^2 z}{\partial y\,\partial x}\right)^2 .$$

(Assuming that we are dealing with a function with continuous second partial derivatives, we can replace $\partial^2 z/\partial x\,\partial y$ by $\partial^2 z/\partial y\,\partial x$.)

Theorem 21.15

If $z = f(x, y)$, all second partial derivatives are continuous, $\partial z/\partial x = \partial z/\partial y = 0$ at (a, b), and

(a) if $D > 0$, $\partial^2 z/\partial x^2 > 0$ and $\partial^2 z/\partial y^2 > 0$ at (a, b), then (a, b) is a relative minimum;

(b) if $D > 0$, $\partial^2 z/\partial x^2 < 0$ and $\partial^2 z/\partial y^2 < 0$ at (a, b), then (a, b) is a relative maximum;

(c) if $D < 0$ at (a, b), then (a, b) is neither a relative maximum nor a relative minimum;

(d) if $D = 0$ at (a, b), the test fails.

We shall not give a proof of this theorem, since it involves some rather sophisticated ideas that we have not considered.† We can make a few observations, however. If $\partial^2 z/\partial x^2$ and $\partial^2 z/\partial y^2$ have opposite signs, their product is negative and $D < 0$. Thus if $D > 0$, then $\partial^2 z/\partial x^2$ and $\partial^2 z/\partial y^2$ are either both positive or both negative. However, it is possible for $\partial^2 z/\partial x^2$ and $\partial^2 z/\partial y^2$ to have the same signs and have $D < 0$; the implication cannot be reversed.

† For a proof see Angus E. Taylor and Robert Mann, *Advanced Calculus*, 2nd ed. (Lexington, Mass.: Xerox, 1972), pp. 227–230.

Example 1

Test $z = x^2 + y^2$ for relative maxima and minima.

$$\frac{\partial z}{\partial x} = 2x, \qquad \frac{\partial z}{\partial y} = 2y.$$

Setting both derivatives equal to zero, we get the critical point (0, 0).

$$\frac{\partial^2 z}{\partial x^2} = 2, \qquad \frac{\partial^2 z}{\partial y^2} = 2, \qquad \frac{\partial^2 z}{\partial y\,\partial x} = 0;$$

$$D = 2 \cdot 2 - 0^2 = 4.$$

At $x = 0$, $y = 0$ (or at any other point, since all three of the second partial derivatives are constant),

$$D > 0, \qquad \frac{\partial^2 z}{\partial x^2} > 0, \qquad \frac{\partial^2 z}{\partial y^2} > 0.$$

Thus we have a relative minimum at the origin. This result agrees with our previous notion of this surface (see Figure 20.24).

Example 2

Test $z = x^2 - y^2$ for relative maxima and minima.

$$\frac{\partial z}{\partial x} = 2x, \qquad \frac{\partial z}{\partial y} = -2y.$$

We again have the critical point (0, 0).

$$\frac{\partial^2 z}{\partial x^2} = 2, \qquad \frac{\partial^2 z}{\partial y^2} = -2, \qquad \frac{\partial^2 z}{\partial y\,\partial x} = 0;$$

$$D = 2(-2) - 0^2 = -4.$$

Since $D < 0$, we have neither a relative maximum nor a relative minimum. The given surface is a hyperbolic paraboloid (see Figure 21.9), which is saddle shaped. For this reason, critical points that are neither relative maxima nor minima are sometimes referred to as "saddle points."

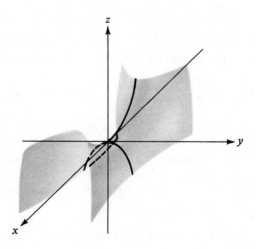

Figure 21.9

Example 3

Test $z = x^3 + y^3 - 3xy$ for relative maxima and minima.

$$\frac{\partial z}{\partial x} = 3x^2 - 3y, \qquad \frac{\partial z}{\partial y} = 3y^2 - 3x.$$

Setting both equal to zero and solving simultaneously, we have

$$3x^2 - 3y = 0 \quad \text{or} \quad y = x^2,$$
$$3y^2 - 3x = 0 \quad \text{or} \quad x = y^2.$$

Thus

$$x = x^4$$
$$x^4 - x = 0$$
$$x(x^3 - 1) = 0$$
$$x = 0 \quad \text{or} \quad x = 1.$$

Thus we have the two critical points (0, 0) and (1, 1).

$$\frac{\partial^2 z}{\partial x^2} = 6x, \qquad \frac{\partial^2 z}{\partial y^2} = 6y, \qquad \frac{\partial^2 z}{\partial y \, \partial x} = -3,$$

and

$$D = 6x \cdot 6y - (-3)^2 = 36xy - 9.$$

At (0, 0), $D = -9$ and (0, 0) is a saddle point. At (1, 1), $D = 27$, $\partial^2 z/\partial x^2 = 6$, and $\partial^2 z/\partial y^2 = 6$;   and (1, 1) is a relative minimum.

Example 4

Find three positive numbers x, y, and z such that $x + y + z = 12$ and x^2yz is a maximum.

$$M = x^2yz, \quad \text{where } x + y + z = 12,$$
$$= x^2y(12 - x - y)$$
$$= 12x^2y - x^3y - x^2y^2,$$

$$\frac{\partial M}{\partial x} = 24xy - 3x^2y - 2xy^2 = xy(24 - 3x - 2y),$$

$$\frac{\partial M}{\partial y} = 12x^2 - x^3 - 2x^2y = x^2(12 - x - 2y).$$

Neither $x = 0$ nor $y = 0$ gives a maximum value for M. Thus we have

$$24 - 3x - 2y = 0, \quad \text{or} \quad 3x + 2y = 24,$$
$$12 - x - 2y = 0, \quad \text{or} \quad x + 2y = 12.$$

Solving simultaneously, we have

$$x = 6, \quad y = 3, \quad z = 3.$$

It is clear from the given conditions that this must give a maximum although the second-derivative test may be used.

Problems

A

In Problems 1–14, find all critical points and test for relative maxima and minima.

1. $z = x^2 + 4y^2 + x + 8y + 1.$
2. $z = x^2 + y^2 - 2x + 4y - 2.$
3. $z = x^2 + 2xy - y^2.$
4. $z = x^2 - 2y^2 + 2x + 4y - 1.$
5. $z = 2x^2 + xy + y^2 + 2x - 3y + 2.$
6. $z = x^2 + 4xy + y^2 + 6x + 1.$

7. $z = x^3 + y^3 - 3x^2 - 6y^2 - 9x.$
8. $z = x^3 + y^3 + 3xy.$
9. $z = 2y^3 + x^2 - y^2 - 4x - 4y + 1.$
10. $z = x^3 - x^2 + y^2 - x + 2y + 2.$
11. $z = x^3 + x^2y + y^2 - 4y + 2.$
12. $z = x^2y + xy^2 - 3xy.$
13. $z = 3x^3 - xy^2 + y.$
14. $z = x^2 + xy + y^2 - x - 2y + 1.$

B

In Problems 15 and 16, find all critical points and test for relative maxima and minima.

15. $z = xy + \dfrac{1}{x} + \dfrac{8}{y}.$
16. $z = 3(x + y)^3 + (x - y)^2 - (x + y).$

17. Find three positive numbers x, y, and z such that $x + y + z = 25$ and x^2y^2z is a maximum.

18. Find three positive numbers x, y, and z such that $x + y + z = k$ and $x^ay^bz^c$ is a maximum.

19. Find the volume of the largest rectangular parallelepiped that can be inscribed in

$$\frac{x^2}{1} + \frac{y^2}{4} + \frac{z^2}{9} = 1.$$

20. Find the volume of the largest rectangular parallelepiped that can be inscribed in

$$\frac{x^2}{a^2} + \frac{y^2}{b^2} + \frac{z^2}{c^2} = 1.$$

C

21. Find the volume of the largest rectangular parallelepiped in the first octant with three faces in the coordinate planes and a vertex on $2x + y + z = 4$.

22. Find the volume of the largest rectangular parallelepiped in the first octant with three faces in the coordinate planes and a vertex on

$$\frac{x}{a} + \frac{y}{b} + \frac{z}{c} = 1 \quad (a, b, c > 0).$$

23. Find the point on the plane $2x - y + z = 1$ that is nearest $(1, 3, 4)$.
24. Find the shortest distance from (x_1, y_1, z_1) to $Ax + By + Cz + D = 0$.
25. The Ace Company manufactures two models of its product—a standard model and a deluxe model. It costs \$65 to manufacture the standard model and \$80 for the deluxe model. Market research has shown that if the standard and deluxe models are priced at \$x and \$y, respectively, then $1200 - 40x + 20y$ standard models are sold and $2650 + 10x - 30y$ deluxe models are sold. How should the prices be set to maximize the profit?

21.9

Constrained Maxima and Minima

Quite often we are called upon to maximize or minimize $w = f(x, y, z)$ subject to the condition that $g(x, y, z) = 0$. When presented with a problem of this type in the past, we solved $g(x, y, z)$ for one of the variables, say z, in terms of the others, $z = G(x, y)$, and substituted into the other equation, to get

$$w = f(x, y, G(x, y)) = F(x, y).$$

Thus the critical values were found by setting the two partial derivatives equal to zero:

$$\frac{\partial F}{\partial x} = 0, \qquad \frac{\partial F}{\partial y} = 0.$$

Unfortunately, this procedure is often tedious and sometimes impossible. In such cases, another method—called the method of Lagrangian multipliers, after the French mathematician Joseph Louis Lagrange (1736–1813)—can be used.

In geometric terms, we seek the points on the surface $g(x, y, z) = 0$ where $f(x, y, z)$ has an extremum (maximum or minimum). If $x = x(t)$, $y = y(t)$, $z = z(t)$ is a parametric representation of a curve on $g(x, y, z) = 0$, then $F(t) = f(x(t), y(t), z(t))$ should have an extremum at such a point. Thus

$$F'(t) = f_x x'(t) + f_y y'(t) + f_z z'(t) = 0,$$

and as in the proof of theorem 21.7, the gradient vector

$$\nabla f = f_x \mathbf{i} + f_y \mathbf{j} + f_z \mathbf{k}$$

is orthogonal to the tangent vector $\mathbf{v} = x'(t)\mathbf{i} + y'(t)\mathbf{j} + z'(t)\mathbf{k}$. But ∇g is orthogonal to $\mathbf{v}$, so ∇f and ∇g are parallel (if nonzero) and $\nabla f = -\lambda \nabla g$ for some scalar λ. Thus we have the equations

$$\frac{\partial f}{\partial x} + \lambda \frac{\partial g}{\partial x} = 0,$$

$$\frac{\partial f}{\partial y} + \lambda \frac{\partial g}{\partial y} = 0,$$

$$\frac{\partial f}{\partial z} + \lambda \frac{\partial g}{\partial z} = 0,$$

$$g(x, y, z) = 0.$$

This system is equivalent to $\nabla u = \mathbf{0}$, where

$$u(x, y, z, \lambda) = f(x, y, z) + \lambda g(x, y, z).$$

The constant λ in these equations is called a *Lagrange multiplier*. Solving these four equations for x, y, and z gives the desired critical values.

The Lagrangian method can be extended to a function with two or more constraining equations. It has the disadvantage of giving us critical points only. It does not tell us whether we have a relative maximum, minimum, or neither. Fortunately, this can often be determined from the physical situation for stated problems.

Example 1

Find the minimum value of $w = x^2 + y^2 + z^2$ subject to the condition $x + y + z = 1$.

From $u = x^2 + y^2 + z^2 + \lambda(x + y + z - 1)$, we get

$$\frac{\partial u}{\partial x} = 2x + \lambda = 0,$$

$$\frac{\partial u}{\partial y} = 2y + \lambda = 0,$$

$$\frac{\partial u}{\partial z} = 2z + \lambda = 0,$$

$$\frac{\partial u}{\partial \lambda} = x + y + z - 1 = 0.$$

The first three give us $x = y = z$. Substituting into the last equation, we have

$$x = y = z = \frac{1}{3} \quad \text{and} \quad w = \frac{1}{3},$$

which is clearly the minimum value.

Example 2

Find the maximum value of $w = xy + z$ subject to the condition

$$x^2 + y^2 + z^2 = 1.$$

$u = xy + z + \lambda(x^2 + y^2 + z^2 - 1),$

$\dfrac{\partial u}{\partial x} = y + 2\lambda x = 0,$

$\dfrac{\partial u}{\partial y} = x + 2\lambda y = 0,$

$\dfrac{\partial u}{\partial z} = 1 + 2\lambda z = 0,$

$\dfrac{\partial u}{\partial \lambda} = x^2 + y^2 + z^2 - 1 = 0.$

The first three equations give

$$y^2 z + 2\lambda xyz = 0,$$
$$x^2 z + 2\lambda xyz = 0,$$
$$xy + 2\lambda xyz = 0,$$

or

$$y^2 z = x^2 z = xy.$$

From the first equality, $y^2 z = x^2 z$, we get either $z = 0$ or $x^2 = y^2$. If $z = 0$, then

$$\frac{\partial u}{\partial z} = 1 + 2\lambda z = 1 \neq 0.$$

Thus $x^2 = y^2$,   or $x = \pm y$.

If $x = y$, then   If $x = -y$, then

$x^2 z = xy = x^2$ $x^2 z = xy = -x^2$

and   and

$x = 0$ or $z = 1$. $x = 0$ or $z = -1$.

Actually, both equalities hold in each case, because if we start with $x = 0$, we get $z = \pm 1$, while if we start with $z = 1$ or $z = -1$, we get $x = y = 0$. Thus, the critical values are $(0, 0, 1)$ and $(0, 0, -1)$. Clearly, the maximum corresponds to $(0, 0, 1)$, which gives

$$w = xy + z = 0 \cdot 0 + 1 = 1.$$

Example 3

Find the point of $x + 2y - z = 3$ which is nearest the origin.

The distance from a point (x, y, z) in the plane to the origin is

$$d = \sqrt{x^2 + y^2 + z^2}.$$

Since d cannot be negative, it is a minimum whenever $d^2 = u$ is a minimum. Thus we want to minimize

$$u = x^2 + y^2 + z^2$$

under the condition

$$x + 2y - z = 3.$$

Using Lagrangian multipliers, we have

$$w = x^2 + y^2 + z^2 + \lambda(x + 2y - z - 3).$$

$$\frac{\partial w}{\partial x} = 2x + \lambda = 0, \quad \text{or} \quad \lambda = -2x,$$

$$\frac{\partial w}{\partial y} = 2y + 2\lambda = 0, \quad \text{or} \quad \lambda = -y,$$

$$\frac{\partial w}{\partial z} = 2z - \lambda = 0, \quad \text{or} \quad \lambda = 2z,$$

$$\frac{\partial w}{\partial \lambda} = x + 2y - z - 3 = 0.$$

Thus

$$-2x = -y = 2z,$$

or

$$y = 2x \quad \text{and} \quad z = -x.$$

Substituting into $x + 2y - z - 3 = 0$, we have

$$6x = 3, \quad \text{or} \quad x = 1/2.$$

The desired point is $(1/2, 1, -1/2)$.

If there are two constraints $g(x, y, z) = 0$ and $h(x, y, z) = 0$,   then the function $f(x, y, z)$ is to have an extremum on the curve of intersection of the surfaces. In this case $\nabla f \cdot \nabla g \times \nabla h = 0$,   and $\nabla f, \nabla g$, and ∇h are coplanar. Thus there exist scalars λ and μ such that $\nabla f = -\lambda \nabla g - \mu \nabla h$ (see Problem 23).

Example 4

Find the minimum value of $w = x^2 + y^2 + z^2$ subject to the conditions

$$x + y + 2z = 12 \quad \text{and} \quad x - 3y - 2z = -16.$$

$$u = x^2 + y^2 + z^2 + \lambda(x + y + 2z - 12) + \mu(x - 3y - 2z + 16).$$

$$\frac{\partial u}{\partial x} = 2x + \lambda + \mu = 0,$$

$$\frac{\partial u}{\partial y} = 2y + \lambda - 3\mu = 0,$$

$$\frac{\partial u}{\partial z} = 2z + 2\lambda - 2\mu = 0,$$

$$\frac{\partial u}{\partial \lambda} = x + y + 2z - 12 = 0,$$

$$\frac{\partial u}{\partial \mu} = x - 3y - 2z + 16 = 0.$$

Eliminating λ and μ from the first three equations, we have $x + y - z = 0$. Solving this together with the last two equations we have $x = 1$, $y = 3$, $z = 4$. Thus the minimum value of w is

$$w = 1^2 + 3^2 + 4^2 = 26.$$

Problems

A

In Problems 1–10, use Lagrangian multipliers to find the desired critical points subject to the conditions given.

1. The minimum value of $w = x^2 + y^2 + z^2$; $xyz = 1$.
2. The maximum value of $w = x + y + z$; $x^2 + y^2 + z^2 = 4$.
3. The minimum value of $w = x^2 + y^2 + z^2$; $2x - y + 3z = 6$.
4. The maximum value of $w = xyz$; $x^2 + y^2 + z^2 = 1$.
5. The minimum value of $w = x^2 + 3y^2 + 2z^2$; $2x - 3y + 5z = 1$.
6. The minimum value of $w = 2x^2 + y^2 + z^2$; $x + 2y - 4z = 8$.
7. The minimum value of $w = x^4 + y^4 + z^4$; $2x - 3y + 6z = 6$.
8. The minimum value of $w = x^4 + y^4 + z^4$; $x + y + z = 1$.
9. The maximum value of $w = xyz$; $2x^2 + y^2 + 4z^2 = 4$.
10. The maximum value of $w = xyz$; $x^2 + 2y^2 + z^2 = 2$.

B

In Problems 11–16, use Lagrangian multipliers to find the desired critical points subject to the conditions given.

11. The maximum value of $w = xyz$; $x^3 + y^3 + z^3 = 24$.
12. The maximum value of $w = x^3 + yz^2$; $x^2 + y^2 + z^2 = 1$.
13. The minimum value of $w = x^2 + y^2 + z^2$; $x - y + z = 1$ and $2x + y + 3z = 6$.
14. The minimum value of $w = x^2 + y^2 + z^2$; $x + 2y - z = 2$ and $2x - y - z = 2$.
15. The maximum value of $w = xyz$; $x + y + z = 2$ and $x - y - z = 1$.
16. The minimum value of $w = x^2 + y^2 + z^2$; $x + 3y - z = 6$ and $2x + 2y + z = 2$.
17. Find the point of $-x^2 - z^2 + y = 2$ nearest the origin.
18. Find the point of $x^2 - y^2 - (z - 4)^2 = 1$ nearest the origin.
19. Find the volume of the largest rectangular parallelepiped that can be inscribed in $4x^2 + 9y^2 + 36z^2 = 36$.
20. Find the volume of the largest rectangular parallelepiped that can be inscribed in

$$\frac{x^2}{a^2} + \frac{y^2}{b^2} + \frac{z^2}{c^2} = 1.$$

21. Find the volume of the largest rectangular parallelepiped in the first octant that has three faces in the coordinate planes and a vertex on $x + 2y + z = 2$.
22. Find the volume of the largest rectangular parallelepiped in the first octant that has three faces in the coordinate planes and a vertex on

$$\frac{x}{a} + \frac{y}{b} + \frac{z}{c} = 1 \quad (a, b, c > 0).$$

Review Problems

A

1. For $w = x^2 + xy - yz$,   find $\partial w/\partial x$, $\partial w/\partial y$, and $\partial w/\partial z$.
2. For $z = x^4 + 4x^3 y - 2y^4$,   show that $\partial^2 z/\partial y \, \partial x = \partial^2 z/\partial x \, \partial y$.
3. For $z = 3x^3 - 5xy^2$ and $y = (x - 2)^2$,   find $\partial z/\partial x$ and dz/dx.
4. For $z = x^2 y - 2y^3$, $x = t^2 + 1$,   and $y = 1 - t^3$,   find dz/dt.
5. For $w = xy + 2yz + 3zx$, $x = u^2 + v^2$, $y = u^2 - v^2$,   and $z = 2u + 3v$,   find $\partial w/\partial u$ and $\partial w/\partial v$.
6. If the two legs of a right triangle are increasing at the rates of 3 in./min and 4 in./min, at what rate is the perimeter increasing at the moment when both legs are 8 in.?

In Problems 7–10, find the required derivative(s).

7. $3x^2 y - 2y^2 z + 5z^2 x = 4$;   find $\partial x/\partial y$ and $\partial x/\partial z$.
8. $3x^3 - 4x^2 y + xy^2 = 1$;   find dy/dx.
9. $4x - 3y^2 + z^2 = 1$, $w = 2x + y - z$;   find $(\partial x/\partial z)_y$.
10. $4x + 2y - 3z = 1$, $x^3 - y^3 - z^3 = 4$;   find dy/dx.
11. Find df for $f(x, y) = \sin xy + x \cos y$.
12. Find the directional derivative for $w = xy - 2yz + 4zx$ in the direction of $\mathbf{v} = 4\mathbf{i} + \mathbf{j} - \mathbf{k}$.
13. Find the directional derivative for $z = 2x^2 - 4xy + y^2$ in the direction of $\mathbf{v} = 4\mathbf{i} - 3\mathbf{j}$ at $(1, 2, -2)$.
14. Find the maximum value of the directional derivative of $z = 2xy + y^2$ at $(1, 2, 8)$. To what unit vector $\mathbf{v}$ does it correspond?

In Problems 15–17, find equations for the tangent plane and normal line to the given surface at the indicated point.

15. $z = 2x^3 - 4y^3$,   at $(1, 1, -2)$.   16. $2x^2 y - y^2 z + z^2 x = 4$,   at $(1, 1, 2)$.
17. $4x^2 - y^2 + 9z^2 = 36$,   at $(1, -2, 2)$.

In Problems 18 and 19, find equations for the line tangent to the curve at the given point.

18. $x = t^2 + 1$, $y = 2t - 3$, $z = t^3 - 1$;   at $t = 2$.
19. $x = \sin t$, $y = \cos 2t$, $z = t$;   at $t = \pi/6$.

20. Approximate $4.05\sqrt{25.1}$ by differentials.
21. The height of a building is determined by measuring a distance x out from the base of the building and the angle θ of elevation of the top. Given $x = 500 \pm 0.5$ ft   and $\theta = 30°00' \pm 5'$,   find the height of the building and the maximum error in determining it.

In Problems 22 and 23, find all critical points and test for maxima and minima.

22. $z = 4x^2 + y^2 + 8x - 8y + 18$.   23. $z = y^3 + xy^2 - 2xy$.

24. Find the maximum value of $w = 2x + y + 4z$ subject to the condition $x^2 + y^2 + z^2 = 9$.
25. Find the minimum value of $w = 2x^2 + y + z$ subject to the condition $x = y^2 + z^2$.

B

26. Find the minimum value of $w = x^2 + y^2 + z^2$ subject to the conditions $2x + y + 2z = 9$ and $5x + 5y + 7z = 29$.
27. Find the volume of the largest rectangular parallelepiped that has one face on the xy plane and all four vertices of the opposite face on $z = 4 - x^2 - 4y^2$.

C

28. If the directional derivative is 5 in the direction from $P_0(4, 2)$ toward $P_1(5, 4)$ and 2 in the direction from P_0 toward $P_2(-1, -1)$, what is it in the direction from P_0 toward $P_3(1, 6)$?

22

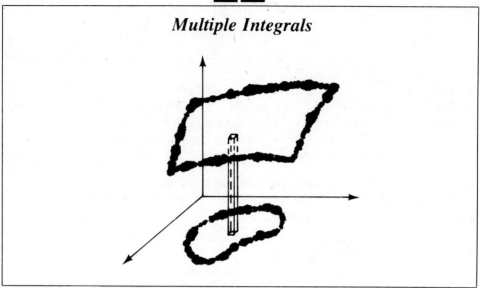

Multiple Integrals

22.1

The Double Integral

Let us recall that in defining the integral, we first considered a subdivision

$$a = x_0, x_1, x_2, x_3, \ldots, x_n = b$$

of the interval $[a, b]$; then for each subinterval $[x_{i-1}, x_i]$ we selected any number x_i^* such that

$$x_{i-1} \le x_i^* \le x_i \, ;$$

finally, we defined

$$\int_a^b f(x) \, dx = \lim_{\|s\| \to 0} \sum_{i=1}^n f(x_i^*)(x_i - x_{i-1}),$$

where $\|s\|$ is the norm of the subdivision—that is, the length of the longest subinterval. The result gave the area of the region "under the curve" (at least if $f(x) \ge 0$ in $[a, b]$).

Let us now increase the number of dimensions by one. We shall consider as our model the volume "under a surface." This, of course, brings up the question "What is volume?" While we might be able to answer this question for certain special types of solids that can be given in terms of a single integral, we cannot yet answer this question in general. We shall make the following assumptions about volume:

(1) the volume of a solid is a nonnegative number;
(2) if S_1 and S_2 are congruent solids, then their volumes are equal;
(3) if $S = S_1 \cup S_2$, where S_1 and S_2 have only boundary points in common, then the volume of S is the sum of the volumes of S_1 and S_2;
(4) the volume of a rectangular parallelepiped is the product of the length, width, and height.

768

Now suppose we have a rectangle R in the xy plane bounded by the lines $x = a$, $x = b$, $y = c$, and $y = d$, and suppose f is a function that is bounded, continuous, and nonnegative for all points in R. Let

$$x_0 = a < x_1 < \cdots < x_m = b$$

and

$$y_0 = c < y_1 < \cdots < y_n = d$$

be subdivisions of $[a, b]$ and $[c, d]$, respectively. Let

$$A_{ij} = (x_i - x_{i-1})(y_j - y_{j-1}) = \Delta x_i \Delta y_j$$

be the area of the corresponding rectangle R_{ij} (see Figure 22.1(a)), and for each i, j let (x_i^*, y_j^*) be chosen inside or on the rectangle R_{ij}. Now, for each rectangle R_{ij}, let us consider a rectangular parallelepiped with base R_{ij} of area A_{ij}, and height $f(x_i^*, y_j^*)$. Then $f(x_i^*, y_j^*)A_{ij}$ approximates the volume under $z = f(x, y)$ and above R_{ij} (see Figure 22.1(b)).

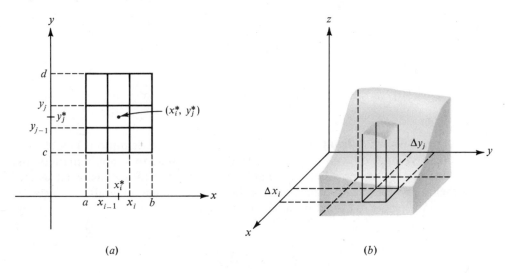

(a)   (b)

Figure 22.1

The sum of the volumes of all such parallelepipeds,

$$\sum_{i=1}^{m} \sum_{j=1}^{n} f(x_i^*, y_j^*)A_{ij},$$

gives an approximation of the volume under the surface. Such a sum is called a *Riemann sum*. As our subdivision gets finer (both dimensions of the rectangles approach zero), the approximations get better. The limit of these approximating sums is what we shall call the *volume under the surface*.

In order to say exactly what is meant by the subdivision getting finer, let us define the norm of the subdivision $\|S\|$ to be the length of the longest diagonal of all the rectangles R_{ij}.

Definition

*If f is bounded and continuous in a rectangle R and if S is a subdivision of R with (x_i^*, y_j^*) in R_{ij}, then the **double integral** of f over R is*

$$\iint\limits_R f(x, y) \, dA = \lim_{\|S\| \to 0} \sum_{i=1}^{n} f(x_i^*, y_j^*) A_{ij},$$

*provided this limit exists. If it exists, f is **integrable** over R.*

Of course, while the definition is inspired by volume, it gives the volume "under the surface" only when $f(x, y) \geq 0$ for all (x, y) in R. Since the following theorems are intuitively obvious from our idea of a double integral, we do not give proofs.

Theorem 22.1

If f is integrable over R and c is a real number, then

$$\iint\limits_R cf(x, y) \, dA = c \iint\limits_R f(x, y) \, dA.$$

Theorem 22.2

If f and g are both integrable over R, then

$$\iint\limits_R [f(x, y) + g(x, y)] \, dA = \iint\limits_R f(x, y) \, dA + \iint\limits_R g(x, y) \, dA.$$

Single integrals are difficult to evaluate by means of the definition. Double integrals are even more difficult. Let us see what can be done to simplify matters. In particular, since we can now evaluate many single integrals rather easily, let us try to put the double integral in terms of a combination of single integrals.

For each x, $a \leq x \leq b$, let us subdivide the interval $[c, d]$ and consider a sum of the form

$$\sum_{j=1}^{n} f(x, y_j^*) \Delta y_j.$$

The limit of this sum as the norm of the subdivision approaches zero is the integral

$$A(x) = \int_c^d f(x, y) \, dy.$$

If $f(x, y) \geq 0$, then $A(x)$ is the cross-sectional area at x. We see from Section 16.4 that the volume under the surface is also the integral of $A(x)$ from a to b,

$$\int_a^b \left[\int_c^d f(x, y) \, dy \right] dx$$

(see Figure 22.2(a)). Thus

$$\iint\limits_R f(x, y) \, dA = \int_a^b \left[\int_c^d f(x, y) \, dy \right] dx.$$

Usually the brackets are omitted:

$$\iint_R f(x, y)\, dA = \int_a^b \int_c^d f(x, y)\, dy\, dx.$$

The same thing can be done if we reverse the role of x and y, to get

$$\iint_R f(x, y)\, dA = \int_c^d \int_a^b f(x, y)\, dx\, dy$$

(see Figure 22.2(b)). These two forms are called *iterated integrals*. Thus we may use iterated integrals to evaluate double integrals.

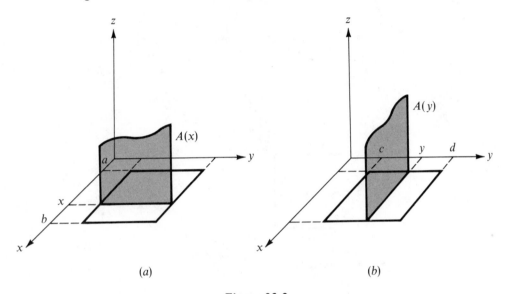

(a) (b)

Figure 22.2

Example 1

Evaluate $\displaystyle\int_0^2 \int_2^4 (x^3 + xy)\, dy\, dx.$ Describe the region R, reverse the order of integration, and evaluate.

$$\int_0^2 \int_2^4 (x^3 + xy)\, dy\, dx = \int_0^2 \left(x^3 y + \frac{xy^2}{2} \right) \Big|_{y=2}^{y=4} dx$$

$$= \int_0^2 [(4x^3 + 8x) - (2x^3 + 2x)]\, dx$$

$$= \frac{x^4}{2} + 3x^2 \Big|_0^2$$

$$= 8 + 12 = 20.$$

Note that in carrying out the integration with respect to y, we take x to be a constant—in keeping with the foregoing discussion. In this sense, multiple integration is like partial differentiation.

The rectangle R is bounded by $x = 0$, $x = 2$, $y = 2$, and $y = 4$ (see Figure 22.3). Reversing the order of integration,

$$\int_0^2 \int_2^4 (x^3 + xy)\, dy\, dx = \int_2^4 \int_0^2 (x^3 + xy)\, dx\, dy$$

$$= \int_2^4 \left(\frac{x^4}{4} + \frac{yx^2}{2} \right) \Bigg|_{x=0}^{x=2} dy$$

$$= \int_2^4 (4 + 2y)\, dy$$

$$= 4y + y^2 \Big|_2^4$$

$$= (16 + 16) - (8 + 4) = 20.$$

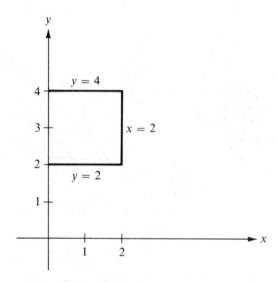

Figure 22.3

To define the double integral for a general region R, let us consider a rectangle with sides parallel to the coordinate axes that contains R in its interior. A subdivision of this rectangle into subrectangles R_{ij} induces a subdivision of R. If we consider only those R_{ij} that lie entirely inside R (see Figure 22.4), then the Riemann sum

$$\sum_{R_{ij} \text{ contained in } R} f(x_i^*, y_j^*) A_{ij}$$

approximates the volume under the surface $z = f(x, y)$ and above the region R. Again, as our subdivision gets finer, the approximations get better, and we define the volume under the surface as the limit of these approximating sums as the norm of the subdivision goes to zero, if the limit exists. The definition of the double integral of f over a rectangle and Theorems 22.1 and 22.2 hold for general bounded regions R. The double integral is also additive with respect to the region of integration, as described in Theorem 22.3 (the proof is again intuitively obvious).

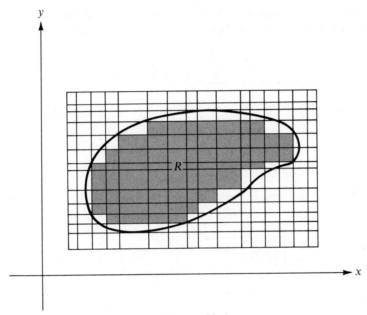

Figure 22.4

Theorem 22.3

If f is integrable over R and R = R₁ ∪ R₂, where R₁ and R₂ have only boundary points in common, then

$$\iint\limits_{R} f(x, y)\, dA = \iint\limits_{R_1} f(x, y)\, dA + \iint\limits_{R_2} f(x, y)\, dA.$$

To define iterated integrals, let us consider a region R bounded by curves $y = g_1(x)$, $y = g_2(x)$, $x = a$, and $x = b$ (see Figure 22.5(a)). A vertical line through

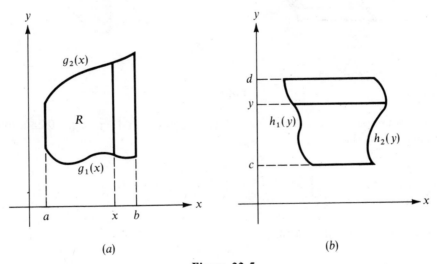

(a)   (b)

Figure 22.5

$(x, 0)$ intersects R in an interval from $(x, g_1(x))$ to $(x, g_2(x))$. As the norm of the subdivision of $[g_1(x), g_2(x)]$ approaches zero, a sum of the form

$$\sum_{j=1}^{n} f(x, y_j^*)\Delta y_j$$

has as a limit the integral

$$A(x) = \int_{g_1(x)}^{g_2(x)} f(x, y) \, dy.$$

Again, if $f(x, y) \geq 0$, this is the cross-sectional area of the volume at x, and the volume is

$$\iint_R f(x, y) \, dA = \int_a^b \int_{g_1(x)}^{g_2(x)} f(x, y) \, dy \, dx$$

(see Figure 22.6(a)).

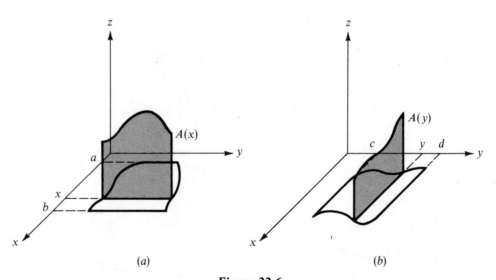

(a)   (b)

Figure 22.6

If R is bounded by $x = h_1(y)$, $x = h_2(y)$, $y = c$, and $y = d$, then a similar argument shows that

$$\iint_R f(x, y) \, dA = \int_c^d \int_{h_1(y)}^{h_2(y)} f(x, y) \, dx \, dy$$

(see Figures 22.5(b) and 22.6(b)).

Example 2

Evaluate $\displaystyle\int_0^1 \int_0^{x^2} (x^2 + xy - y^2)\, dy\, dx$ and describe R.

$$\int_0^1 \int_0^{x^2} (x^2 + xy - y^2)\, dy\, dx$$

$$= \int_0^1 \left(x^2 y + \frac{xy^2}{2} - \frac{y^3}{3} \right)\Bigg|_0^{x^2} dx$$

$$= \int_0^1 \left(x^4 + \frac{x^5}{2} - \frac{x^6}{3} \right) dx$$

$$= \frac{x^5}{5} + \frac{x^6}{12} - \frac{x^7}{21} \Bigg|_0^1$$

$$= \frac{1}{5} + \frac{1}{12} - \frac{1}{21} = \frac{33}{140}.$$

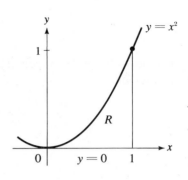

Figure 22.7

The region of integration is bounded below by $y = 0$, above by $y = x^2$, and on the left and right by $x = 0$ and $x = 1$. This is shown in Figure 22.7.

Note that the integrand, $x^2 + xy - y^2$, has nothing to do with the region R of integration, except that the integrand must be defined over R. The region of integration is determined only by the limits of integration. Conversely, the limits of integration are determined by R. The limits of Example 2 are determined by cutting R into vertical strips (see Figure 22.8(a)), to obtain the outer limits,

$$\int_0^1 --- dx;$$

then we subdivide each vertical strip, to obtain the inner limits,

$$\int_0^{x^2} --- dy.$$

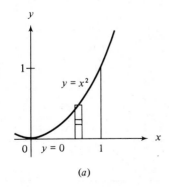

(a)

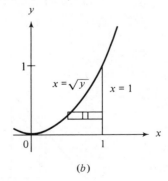

(b)

Figure 22.8

If we should start by using horizontal strips as in Figure 22.8(b), then the outer limits are the extreme values of y,

$$\int_0^1 ---dy.$$

By subdividing each horizontal strip, we see that $x = \sqrt{y}$ on the left and $x = 1$ on the right, giving

$$\int_{\sqrt{y}}^1 ---dx.$$

Thus

$$\int_0^1 \int_0^{x^2} (x^2 + xy - y^2)\, dy\, dx = \int_0^1 \int_{\sqrt{y}}^1 (x^2 + xy - y^2)\, dx\, dy.$$

You are encouraged to verify this by evaluating the integral on the right.

Example 3

Evaluate $\displaystyle\iint_R (x^2 + y^2)\, dA$,   where R is the triangle formed by $y = x$, $y = 0$, and $x = 1$.

There are two ways of expressing this double integral as an iterated integral. If R is first subdivided using vertical strips as in Figure 22.9(a), then the outer integral is an integral with respect to x and the limits of integration are the extreme values of x in R; namely, $x = 0$ and $x = 1$. This gives

$$\int_0^1 ---dx.$$

Now when each of these vertical strips is subdivided, the inner integral is one with respect to y and the limits of integration are the extreme values of y for each vertical strip. While the bottom of each vertical strip is at $y = 0$, no one constant describes the top. But we do know that the top is always on the line $y = x$. Thus the inner integral is

$$\int_0^x ---dy.$$

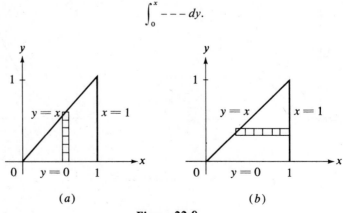

(a)   (b)

Figure 22.9

Putting these limits together with the integrand, $x^2 + y^2$, we have

$$\iint_R (x^2 + y^2) \, dA = \int_0^1 \int_0^x (x^2 + y^2) \, dy \, dx$$

$$= \int_0^1 \left(x^2 y + \frac{y^3}{3} \right) \Big|_0^x dx$$

$$= \int_0^1 \frac{4x^3}{3} \, dx$$

$$= \frac{x^4}{3} \Big|_0^1 = \frac{1}{3}.$$

If R is first cut into horizontal strips as in Figure 22.9(b), the outer integral is one with respect to y; and the limits are the extreme values of y,

$$\int_0^1 - - - dy.$$

When these horizontal strips are subdivided, we have the inner integral with respect to x, the limits being $x = y$ on the left and $x = 1$ on the right,

$$\int_y^1 - - - dx.$$

Again putting these together with the integrand, $x^2 + y^2$, we have

$$\iint_R (x^2 + y^2) \, dA = \int_0^1 \int_y^1 (x^2 + y^2) \, dx \, dy$$

$$= \int_0^1 \left(\frac{x^3}{3} + xy^2 \right) \Big|_y^1 dy$$

$$= \int_0^1 \left(\frac{1}{3} + y^2 - \frac{4y^3}{3} \right) dy$$

$$= \frac{y}{3} + \frac{y^3}{3} - \frac{y^4}{3} \Big|_0^1 = \frac{1}{3}.$$

Example 4

Evaluate $\displaystyle\iint_R (x^2 - xy) \, dA$,   where R is the region bounded by $y = x$ and $y = 3x - x^2$.

The simpler way of setting up the iterated integral is shown graphically in Figure 22.10(a); it gives

$$\iint_R (x^2 - xy) \, dA = \int_0^2 \int_x^{3x - x^2} (x^2 - xy) \, dy \, dx$$

$$= \int_0^2 \left(x^2 y - \frac{xy^2}{2} \right) \Big|_x^{3x - x^2} dx$$

$$= \int_0^2 \left[x^2(3x - x^2) - \frac{x(3x - x^2)^2}{2} - \left(x^3 - \frac{x^3}{2} \right) \right] dx$$

$$= \int_0^2 \left(-2x^3 + 2x^4 - \frac{x^5}{2} \right) dx$$

$$= \left(-\frac{x^4}{2} + \frac{2x^5}{5} - \frac{x^6}{12} \right) \Big|_0^2$$

$$= -\frac{8}{15}.$$

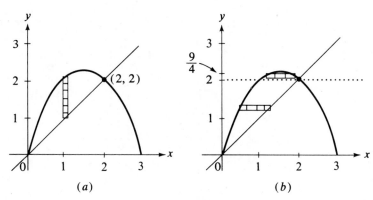

Figure 22.10

Since the result is negative, it cannot be a volume. The reason it is negative is that $x^2 - xy$ is negative (or zero) throughout R.

An alternate, though more difficult, way is illustrated in Figure 22.10(b). Two iterated integrals are needed here, since the horizontal strips sometimes have one end on the parabola and the other on the line and sometimes have both ends on the parabola. Solving the equation $y = 3x - x^2$ for x, we have

$$x = \frac{3 \pm \sqrt{9 - 4y}}{2}.$$

Thus,

$$\iint_R (x^2 - xy)\, dA = \int_0^2 \int_{\frac{3 - \sqrt{9-4y}}{2}}^{y} (x^2 - xy)\, dx\, dy + \int_2^{9/4} \int_{\frac{3 - \sqrt{9-4y}}{2}}^{\frac{3 + \sqrt{9-4y}}{2}} (x^2 - xy)\, dx\, dy.$$

We shall not evaluate these integrals—the work involved is considerable.

Problems

A

In Problems 1–8, evaluate the given integral and sketch R.

1. $\int_0^2 \int_0^3 (xy + x - y)\, dy\, dx.$

2. $\int_1^2 \int_0^3 (y^2 - 6xy)\, dx\, dy.$

3. $\int_2^4 \int_1^2 (2x + 2y - 1)\, dx\, dy.$

4. $\int_0^1 \int_0^x xy\, dy\, dx.$

5. $\int_0^2 \int_0^{x^2} (x^2 - y^2)\, dy\, dx.$

6. $\int_0^1 \int_0^{1-x} (x^2 y + xy^2)\, dy\, dx.$

7. $\int_0^1 \int_{y^2}^{y} (xy + 1)\, dx\, dy.$

8. $\int_1^2 \int_y^{y^2} \frac{x}{y}\, dx\, dy.$

In Problems 9–14, evaluate the given integral, sketch R, and give the integral with the order of integration reversed.

9. $\int_{-1}^1 \int_{-1}^1 dy\, dx.$

10. $\int_0^1 \int_0^{1-x} x^2\, dy\, dx.$

11. $\displaystyle\int_{-1}^{1}\int_{x^2}^{1}(x^2+y^2)\,dy\,dx.$

12. $\displaystyle\int_{0}^{\pi}\int_{0}^{1}e^{x}\sin y\,dx\,dy.$

13. $\displaystyle\int_{0}^{1}\int_{0}^{y}e^{x+y}\,dx\,dy.$

14. $\displaystyle\int_{0}^{\pi}\int_{0}^{\sin x}x\,dy\,dx.$

In Problems 15–20, evaluate the given integral.

15. $\displaystyle\iint_{R}(x^2-y^2)\,dA,$ where R is the region bounded by $x=0$, $y=1$, and $y=x$.

16. $\displaystyle\iint_{R}x^2y\,dA,$ where R is the region bounded by $y=x^2$, $y=0$, and $x=2$.

17. $\displaystyle\iint_{R}(x^3-y^3)\,dA,$ where R is the region bounded by $y=x^3$, $x=0$, and $y=1$.

18. $\displaystyle\iint_{R}xy\,dA,$ where R is the region bounded by $y=x$ and $y=x^2$.

19. $\displaystyle\iint_{R}(x+y)\,dA,$ where R is the region bounded by $xy=4$ and $x+y=5$.

20. $\displaystyle\iint_{R}(x^2+y^3)\,dA,$ where R is the region bounded by $x=1$, $x=3$, $y=0$, and $y=4$.

B

In Problems 21–25, evaluate the given integral.

21. $\displaystyle\iint_{R}(x+y)^2\,dA,$ where R is the region in the first quadrant bounded by $y=x^3$ and $y=x$.

22. $\displaystyle\iint_{R}xy\,dA,$ where R is the region bounded by $x=y^2$ and $x=y+2$.

23. $\displaystyle\iint_{R}xy\,dA,$ where R is the region bounded by $x=y^2-3y$ and $x=0$.

24. $\displaystyle\iint_{R}dA,$ where R is the region bounded by $y=x^2-4$ and $y=8+2x-x^2$.

25. $\displaystyle\iint_{R}\sqrt{xy}\,dA,$ where R is the triangle with vertices $(0,0)$, $(1,1)$, and $(9,1)$.

C

26. Evaluate $\displaystyle\iint_{R}y\sin x^3\,dA,$ where R is the triangle with vertices $(0,0)$, $(3,0)$, and $(3,2)$.

27.   Evaluate $\displaystyle\iint\limits_R e^{-x^2}\,dA,$ where R is the region bounded by $y = x,$ $y = 0,$ and $x = 2.$

28.   Evaluate $\displaystyle\int_1^2 \int_0^1 \frac{1}{y^2\sqrt{y^2 - x^2}}\,dx\,dy.$

29.   Show that $\displaystyle\int_a^b \int_a^y f(x)\,dx\,dy = \int_a^b (b - x) f(x)\,dx.$

22.2

Volume, Area, and Mass

The definition of the double integral was formulated with volume in mind. Let us consider some examples of this.

Example 1

Find the volume of the solid bounded by $y = x^2,$ $z = 0,$ and $y + z = 2.$

A graphical representation of this solid is given in Figure 22.11(a); its projection onto the xy plane is shown in part (b). First of all, by symmetry, we need only find the volume in the first octant and multiply by 2. Let us cut the solid by vertical slices parallel to the yz plane. These in turn are cut by slices parallel to the xz plane (see Figure 22.11). In setting up the integral, we determine the limits of integration by the region of Figure 22.11(b) (the right half only, since we double); the integrand is the height z of the vertical element of Figure 22.11(a). Finally, since the height

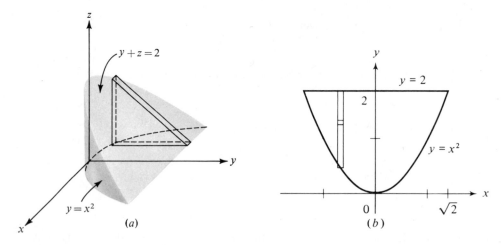

Figure 22.11

is represented by a z coordinate on the plane $y + z = 2$, we replace z by its equivalent in x and y (it happens to involve only y here), $z = 2 - y$. Thus we have

$$V = 2 \int_0^{\sqrt{2}} \int_{x^2}^{2} z \, dy \, dx = 2 \int_0^{\sqrt{2}} \int_{x^2}^{2} (2 - y) \, dy \, dx$$

$$= 2 \int_0^{\sqrt{2}} \left(2y - \frac{y^2}{2}\right)\bigg|_{x^2}^{2} dx = 2 \int_0^{\sqrt{2}} \left(2 - 2x^2 + \frac{x^4}{2}\right) dx$$

$$= 2\left(2x - \frac{2x^3}{3} + \frac{x^5}{10}\right)\bigg|_0^{\sqrt{2}} = 2\left(2\sqrt{2} - \frac{4\sqrt{2}}{3} + \frac{2\sqrt{2}}{5}\right) = \frac{32\sqrt{2}}{15}.$$

Example 2

Find the volume inside the paraboloid $z = x^2 + y^2$, between $z = 0$ and $z = 4$.

Again, by symmetry, we need consider only the portion in the first octant and multiply by 4. The solid is given in Figure 22.12(a) and its projection onto the xy plane is shown in part (b). Note that this projection corresponds to the widest portion existing when $z = 4$. Again let us take slices parallel to the yz plane and cut each of these slices parallel to the xz plane. Note from part (a) that the height of the resulting parallelepiped is not the z coordinate of a point on the paraboloid, but rather is $4 - z$. Thus

$$V = 4 \int_0^2 \int_0^{\sqrt{4 - x^2}} (4 - z) \, dy \, dx$$

$$= 4 \int_0^2 \int_0^{\sqrt{4 - x^2}} (4 - x^2 - y^2) \, dy \, dx$$

$$= 4 \int_0^2 \left(4y - x^2 y - \frac{y^3}{3}\right)\bigg|_0^{\sqrt{4 - x^2}} dx.$$

$$= 4 \int_0^2 \frac{2}{3}(4 - x^2)^{3/2} \, dx.$$

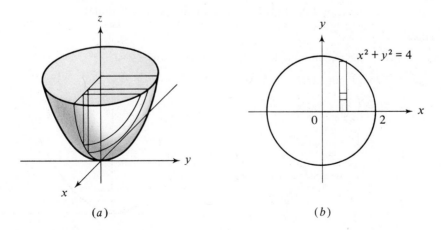

(a) $\qquad\qquad\qquad\qquad$ (b)

Figure 22.12

Substituting $x = 2 \sin \theta$, we have

$$V = \frac{8}{3} \int_0^{\pi/2} (4 - 4 \sin^2 \theta)^{3/2} \, 2 \cos \theta \, d\theta$$

$$= \frac{128}{3} \int_0^{\pi/2} \cos^4 \theta \, d\theta$$

$$= \frac{32}{3} \int_0^{\pi/2} (1 + \cos 2\theta)^2 \, d\theta$$

$$= \frac{32}{3} \int_0^{\pi/2} \left(1 + 2 \cos 2\theta + \frac{1 + \cos 4\theta}{2} \right) d\theta$$

$$= \frac{32}{3} \left. \left(\frac{3}{2} \theta + \sin 2\theta + \frac{1}{8} \sin 4\theta \right) \right|_0^{\pi/2}$$

$$= 8\pi.$$

While both of these examples were worked by projecting the solid onto the xy plane, we could make the projection onto any convenient coordinate plane. Both examples could have been done by the methods of Chapter 16, since Example 2 is a solid of revolution and parallel cross sections of Example 1 are all triangles or rectangles. However, where the methods of Chapter 16 can only be used for special cases, the use of the double integral is much more general.

The double integral giving the volume between two surfaces is analogous to the integral for the area between two curves. Suppose a volume is bounded below by $z = f_1(x, y)$, is bounded above by $z = f_2(x, y)$, and lies over the region R in the xy plane. Then the volume is

$$V = \iint\limits_R (z_2 - z_1) \, dA = \iint\limits_R [f_2(x, y) - f_1(x, y)] \, dA.$$

Thus, in Example 2, $z_1 = x^2 + y^2$, $z_2 = 4$, and R is the disc $x^2 + y^2 \le 4$.

Let us now consider the problem of finding area by a double integral. Since the double integral gives volume, it may be difficult to see how we can go back to a two-dimensional figure to find area. In effect, we don't. The area of a plane region is the same as the volume of a cylinder of height 1 and having that region as base. Thus, we can find an area by finding the volume of the proper cylinder.

Example 3

Find the area of the region bounded by $y = x^2$, $x = 1$, and $y = 0$.

From Figure 22.13, we have

$$A = \int_0^1 \int_0^{x^2} dy \, dx$$

$$= \int_0^1 \left. y \right|_0^{x^2} dx$$

$$= \int_0^1 x^2 \, dx$$

$$= \left. \frac{x^3}{3} \right|_0^1$$

$$= \frac{1}{3}.$$

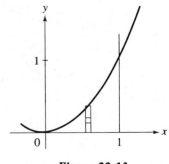

Figure 22.13

Perhaps you feel at this point that we are using a sledge hammer to swat a fly. Why not use a single integral in the example and get the same result with much less effort?

$$A = \int_0^1 y \, dx$$

$$= \int_0^1 x^2 \, dx$$

$$= \frac{x^3}{3} \Big|_0^1 = \frac{1}{3}.$$

You are quite right. We did more work than necessary—on this particular problem. Suppose, however, that we want to find the mass of a thin plate of varying density $\delta(x, y)$ that coincides with a region R. Let us consider a rectangular subdivision of R, and in each subrectangle R_{ij} let us evaluate the density at an arbitrary point (x_i^*, y_j^*). The mass of R_{ij} (density $\times$ area) is approximately $\delta(x_i^*, y_j^*)A_{ij}$, and

$$\sum_{i=1}^m \sum_{j=1}^n \delta(x_i^*, y_j^*)A_{ij}$$

is an approximation to the total mass of the region R. Taking the limit as the norm of the subdivision goes to zero, we get the mass

$$m = \iint_R \delta(x, y) \, dA.$$

Example 4

Find the mass of the region of Example 3 if the density at (x, y) is $1 + x + y$.

$$m = \int_0^1 \int_0^{x^2} (1 + x + y) \, dy \, dx$$

$$= \int_0^1 \left(y + xy + \frac{y^2}{2} \right) \Big|_0^{x^2} dx$$

$$= \int_0^1 \left(x^2 + x^3 + \frac{x^4}{2} \right) dx$$

$$= \frac{x^3}{3} + \frac{x^4}{4} + \frac{x^5}{10} \Big|_0^1 = \frac{41}{60}.$$

Problems

A

In Problems 1–7, use the double integral to find the volume of the given solid.

1. The solid bounded by the three coordinate planes and $x + y + z = 1$.
2. The solid bounded by the xz plane, the yz plane, $6x + 2y + 3z = 6$, and $4x + 2y - z = 4$.
3. The solid bounded by $x^2 + y^2 = 4$, $y = z$, and $z = 0$ (one side of the xz plane only).
4. The solid bounded by $x^2 + y^2 - 4x = 0$, $y = z$, and $z = 0$.

5. The solid in the first octant bounded by $x + y = 4$ and $z = xy$.
6. The solid bounded by $x^2 + y^2 = 4$ and $x^2 + z^2 = 4$.
7. The solid bounded by $x^2 + y^2 = 1$ and $x^2 + z^2 = 1$.

In Problems 8–13, find the area of the given region.

8. The region bounded by $y = x^3$, the x axis, and $x = 1$.
9. The region bounded by $y = x^2 - 2x$ and the x axis.
10. The region bounded by $y = x^2$ and $y = x$.
11. The region bounded by $y = x^2$ and $x - y + 2 = 0$.
12. The region bounded by $y = \sin x$ and the x axis between $x = 0$ and $x = \pi$.
13. The region between the x axis and $y = e^x$ to the left of the y axis.

In Problems 14–20, find the mass of the given plane region.

14. The region bounded by $y = x^3$, the x axis, and $x = 1$ with density xy.
15. The region bounded by $y = x^3$, the x axis, and $x = 1$ with density $2 - (x + y)$.
16. The region bounded by $y = 4 - x^2$ and the x axis with density $1 + y$.
17. The region bounded by $y = x^2$ and $y = x$ with density $x^2 + y^2$.
18. The region inside $x^2 + y^2 = 4$ with density $|x| + |y|$.
19. The region inside $x^2 + 4y^2 = 4$ with density $x^2 + y^2$.
20. The region bounded by $y = \sin x$ and the x axis between $x = 0$ and $x = \pi$ with density $|\cos x|$.

B

In Problems 21–25, use the double integral to find the volume of the given solid.

21. The solid in the first octant that is bounded by $z = y^2$ and $x^2 + y^2 = 4$.
22. The solid bounded by the three coordinate planes and $\sqrt{x} + \sqrt{y} + \sqrt{z} = 1$.
23. A sphere of radius R.
24. The portion of the sphere $x^2 + y^2 + z^2 = 9$ that is above $z = 0$.
25. The solid bounded by $z = x^2 + y^2$ and $z = 2$.

26. Find the mass of the region bounded by $y = \sin x$ and the x axis between $x = 0$ and $x = \pi$ with density $x + y$.

C

In Problems 27 and 28, use the double integral to find the volume of the given solid.

27. A cone of height h and radius r.
28. The ellipsoid $9x^2 + 4y^2 + 36z^2 = 36$.

22.3

Double Integrals in Polar Coordinates

Suppose the region R is given in polar coordinates. Then, instead of subdividing R by rectangles, we shall use portions of sectors as given in Figure 22.14. Now we want to express dA in terms of dr and $d\theta$. If the dimensions given are

$$\Delta\theta = \theta_i - \theta_{i-1} \quad \text{and} \quad \Delta r = r_j - r_{j-1}$$

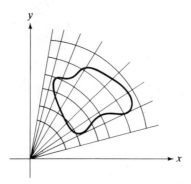

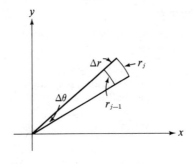

Figure 22.14   *Figure 22.15*

(see Figure 22.15), then, as we saw in Section 14.7, the area of the sector with radius r and central angle $\Delta\theta$ is

$$\frac{1}{2} r^2 \, \Delta\theta.$$

The portion of the sector between r_j and r_{j-1} is then

$$\frac{1}{2} r_j^2 \, \Delta\theta - \frac{1}{2} r_{j-1}^2 \, \Delta\theta = \frac{1}{2} (r_j^2 - r_{j-1}^2) \, \Delta\theta$$

$$= \frac{1}{2} (r_j + r_{j-1})(r_j - r_{j-1}) \, \Delta\theta$$

$$= \frac{r_j + r_{j-1}}{2} \, \Delta r \, \Delta\theta$$

$$= r_j^* \, \Delta r \, \Delta\theta,$$

where

$$r_{j-1} < r_j^* = \frac{r_{j-1} + r_j}{2} < r_j.$$

If we choose θ_i^* between θ_{i-1} and θ_i, then

$$\sum_{i=1}^{m} \sum_{j=1}^{n} f(r_j^* \cos \theta_i^*, r_j^* \sin \theta_i^*) r_j^* \, \Delta r_j \, \Delta\theta_i$$

is an approximation to the integral. Letting Δr_j and $\Delta\theta_i$ approach 0, we have

$$\iint\limits_{R} f(x, y) \, dA = \iint\limits_{R} F(r, \theta) r \, dr \, d\theta,$$

where $F(r, \theta) = f(r \cos \theta, r \sin \theta)$.

Note that we can think of the element of area in polar coordinates as a "curvilinear" rectangle with sides $r \, \Delta\theta$ and Δr, respectively, and with area $r \, \Delta r \, \Delta\theta$.

Example 1

Find the volume of the solid bounded above by $z = x^2 + y^2$, below by $z = 0$, and on the sides by $r = 1 - \cos\theta$.

Since the solid is symmetric about the xz plane (see Figure 22.16) we can consider just half of it and double the result.

$$V = 2 \int_0^\pi \int_0^{1-\cos\theta} zr\, dr\, d\theta$$

$$= 2 \int_0^\pi \int_0^{1-\cos\theta} (x^2 + y^2)r\, dr\, d\theta$$

$$= 2 \int_0^\pi \int_0^{1-\cos\theta} r^3\, dr\, d\theta$$

$$= 2 \int_0^\pi \frac{r^4}{4}\Big|_0^{1-\cos\theta} d\theta$$

$$= \frac{1}{2} \int_0^\pi (1 - \cos\theta)^4\, d\theta$$

$$= \frac{1}{2} \int_0^\pi (1 - 4\cos\theta + 6\cos^2\theta - 4\cos^3\theta + \cos^4\theta)\, d\theta$$

$$= \frac{1}{2} \int_0^\pi \left[1 - 4\cos\theta + 3(1 + \cos 2\theta) - 4(1 - \sin^2\theta)\cos\theta + \frac{(1 + \cos 2\theta)^2}{4}\right] d\theta$$

$$= \frac{1}{2} \int_0^\pi \left(\frac{17}{4} - 8\cos\theta + \frac{7}{2}\cos 2\theta + 4\sin^2\theta\cos\theta + \frac{1}{4}\cos^2 2\theta\right) d\theta$$

$$= \frac{1}{2} \int_0^\pi \left(\frac{17}{4} - 8\cos\theta + \frac{7}{2}\cos 2\theta + 4\sin^2\theta\cos\theta + \frac{1 + \cos 4\theta}{8}\right) d\theta$$

$$= \frac{1}{2} \int_0^\pi \left(\frac{35}{8} - 8\cos\theta + \frac{7}{2}\cos 2\theta + 4\sin^2\theta\cos\theta + \frac{1}{8}\cos 4\theta\right) d\theta$$

$$= \frac{1}{2}\left(\frac{35}{8}\theta - 8\sin\theta + \frac{7}{4}\sin 2\theta + \frac{4}{3}\sin^3\theta + \frac{1}{32}\sin 4\theta\right)\Big|_0^\pi$$

$$= \frac{35}{16}\pi.$$

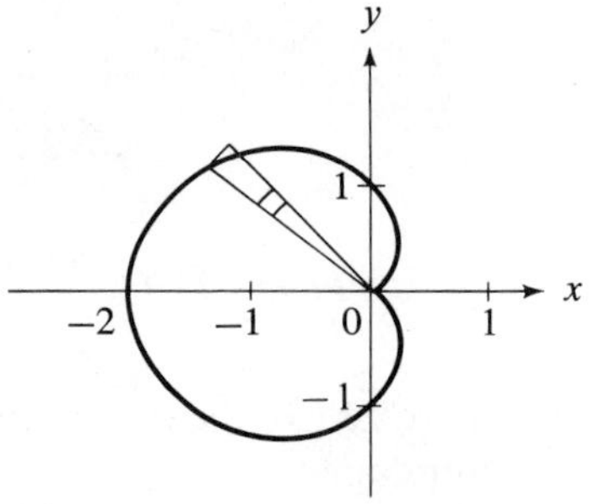

Figure 22.16

Example 2

Find the mass of the region inside $r = 1 + \sin\theta$ where the density at (r, θ) is r.

We see (Figure 22.17) that the region and the density are symmetric about the y axis. Again we may find the mass of half of the region and double the result.

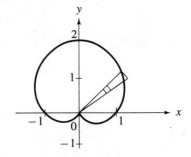

Figure 22.17

$$m = 2 \int_{-\pi/2}^{\pi/2} \int_0^{1+\sin\theta} r \cdot r \, dr \, d\theta$$

$$= 2 \int_{-\pi/2}^{\pi/2} \frac{r^3}{3} \bigg|_0^{1+\sin\theta} d\theta$$

$$= \frac{2}{3} \int_{-\pi/2}^{\pi/2} (1 + \sin\theta)^3 \, d\theta$$

$$= \frac{2}{3} \int_{-\pi/2}^{\pi/2} (1 + 3\sin\theta + 3\sin^2\theta + \sin^3\theta) \, d\theta$$

$$= \frac{2}{3} \int_{-\pi/2}^{\pi/2} \left[1 + 3\sin\theta + \frac{3}{2}(1 - \cos 2\theta) + (1 - \cos^2\theta)\sin\theta \right] d\theta$$

$$= \frac{2}{3} \int_{-\pi/2}^{\pi/2} \left(\frac{5}{2} + 4\sin\theta - \frac{3}{2}\cos 2\theta - \cos^2\theta\sin\theta \right) d\theta$$

$$= \frac{2}{3} \left(\frac{5}{2}\theta - 4\cos\theta - \frac{3}{4}\sin 2\theta + \frac{1}{3}\cos^3\theta \right) \bigg|_{-\pi/2}^{\pi/2}$$

$$= \frac{2}{3} \left[\frac{5\pi}{4} - \left(-\frac{5\pi}{4} \right) \right] = \frac{5\pi}{3}.$$

Even if the problem is given entirely in terms of rectangular coordinates, it is sometimes easier to integrate if polar coordinates are used.

Example 3

Find the volume of the solid inside $x^2 + y^2 + z^2 = 4$ and $x^2 + y^2 = 1$.

In cylindrical coordinates the surfaces are $z = \pm\sqrt{4 - r^2}$ and $r = 1$.

$$V = 8 \int_0^{\pi/2} \int_0^1 \sqrt{4 - r^2} \, r \, dr \, d\theta$$

$$= -4 \int_0^{\pi/2} \int_0^1 \sqrt{4 - r^2}(-2r) \, dr \, d\theta$$

$$= -4 \int_0^{\pi/2} \frac{2}{3}(4 - r^2)^{3/2} \bigg|_0^1 d\theta$$

$$= -4 \int_0^{\pi/2} \frac{2}{3}(3^{3/2} - 8) \, d\theta$$

$$= -\frac{8}{3}(3\sqrt{3} - 8) \, \theta \bigg|_0^{\pi/2}$$

$$= \frac{4\pi}{3}(8 - 3\sqrt{3}).$$

If the integral of this example had been set up in rectangular coordinates, it would be

$$V = 8 \int_0^1 \int_0^{\sqrt{1-x^2}} z \, dy \, dx = 8 \int_0^1 \int_0^{\sqrt{1-x^2}} \sqrt{4 - (x^2 + y^2)} \, dy \, dx.$$

Notice that if we make the substitutions

$$x = r\cos\theta \quad \text{and} \quad y = r\sin\theta$$

into the integrand, then $dy\,dx$ is replaced by $r\,dr\,d\theta$ and not $dr\,d\theta$. This is in keeping with the result

$$\iint\limits_{R} f(x, y)\, dA = \iint\limits_{R} F(r, \theta)\, r\, dr\, d\theta,$$

where $F(r, \theta) = f(r \cos \theta, r \sin \theta)$. See Problem 25 for more comments on substitutions.

Problems

A

In Problems 1–8, find the volume of the given solid.

1. The solid bounded above by $z = 3$, below by $z = 0$, and on the sides by $r = \sin \theta$.
2. The solid bounded above by $z = x$, below by $z = 0$, and on the sides by $r = \cos \theta$.
3. The solid bounded above by $z = 1 - (x^2 + y^2)$, below by $z = 0$, and on the sides by $r = \sin \theta$.
4. The solid bounded above by $z = \sqrt{x^2 + y^2}$, below by $z = 0$, and on the sides by $r = 1 + \sin \theta$.
5. The solid bounded above by $z = 1 + x^2 + y^2$, below by $z = 0$, and on the sides by $r = \sin 2\theta$.
6. The solid bounded above by $z = 1 + x^2 + y^2$, below by $z = 1 - x^2 - y^2$, and on the sides by $r = \cos \theta$.
7. The solid bounded above by $z = \sqrt{x^2 + y^2}$, below by $z = -\sqrt{x^2 + y^2}$, and on the sides by $r = \sin 3\theta$.
8. The solid bounded above by $z = x^2 + y^2$, below by $z = -\sqrt{x^2 + y^2}$, and on the sides by $r = \cos 4\theta$.

In Problems 9–15, find the mass of the given region.

9. Inside $r = 1 - \sin \theta$ with density $|r|$ at (r, θ).
10. Inside $r = 1 - \sin \theta$ with density $|\sin \theta|$ at (r, θ).
11. Inside $r = \sin \theta$ with density r^2 at (r, θ).
12. Inside $r = \cos \theta$ with density $\sin^2 \theta$ at (r, θ).
13. Inside $r = \cos 2\theta$ with density $|r|$ at (r, θ).
14. Inside $r = \sin 2\theta$ with density $|\sin 2\theta|$ at (r, θ).
15. Inside $r^2 = \sin \theta$ with density r^2 at (r, θ).

In Problems 16–22, use polar coordinates to find the given volume.

16. Inside $x^2 + y^2 = 1$ and bounded by $z = 0$ and $z = 4$.
17. Inside $x^2 + y^2 = 1$ and bounded by $z = 0$ and $z = x^2 + y^2$.
18. Inside $x^2 + y^2 = 1$ and $x^2 + y^2 + z^2 = 9$.
19. Inside $x^2 + y^2 + z^2 = 1$ and above $x^2 + y^2 = z^2$.
20. Inside $x^2 + y^2 - 2x = 0$ and bounded by $z = 0$ and $z = y$.
21. Inside $x^2 + y^2 - 2x = 0$ and bounded by $z^2 = x^2 + y^2$.
22. Inside $x^2 + y^2 - 2y = 0$ and bounded by $z = 0$ and $z = x^2 + y^2$.

B

23. Find the mass of the region inside $r = \sin^2 \theta$ with density $1 + |r|$ at (r, θ).
24. Find the volume inside $x^2 + y^2 = 1$ and $y^2 + z^2 = 1$.

25. The equations $x = \varphi(u, v)$ and $y = \psi(u, v)$ map a region R in the xy plane into another region S in the uv plane. It can be shown† that, with the proper conditions on φ, ψ, R, S, and f,

$$\iint_R f(x, y)\, dy\, dx = \iint_S f(\varphi(u, v), \psi(u, v)) \left| \frac{\partial(x, y)}{\partial(u, v)} \right| du\, dv,$$

where $\partial(x, y)/\partial(u, v)$ is the Jacobian defined on page 750. Show that, for $x = r \cos \theta$ and $y = r \sin \theta$,

$$\frac{\partial(x, y)}{\partial(r, \theta)} = r;$$

therefore

$$\iint_R f(x, y)\, dy\, dx = \iint_S f(r \cos \theta, r \sin \theta)\, r\, dr\, d\theta.$$

In Problems 26–28, use polar coordinates to evaluate the integral.

26. $\displaystyle \int_0^3 \int_0^{\sqrt{9-x^2}} e^{-(x^2+y^2)}\, dy\, dx.$

27. $\displaystyle \int_1^2 \int_0^x \frac{1}{\sqrt{x^2+y^2}}\, dy\, dx.$

28. $\displaystyle \int_0^1 \int_y^{\sqrt{2-y^2}} x^4\, dx\, dy.$

29. Find the mass of the smaller of the regions bounded by $r = 1/2$, $r = 1$, $\theta = 0$, and the part of the spiral $r\theta = 1$ from $\theta = 1$ to $\theta = 2$ if the density varies as the square of the distance from the origin. (*Hint:* Integrate first with respect to θ.)

30. Find the mass of the region inside the outer loop of $r = 1 + 2 \cos \theta$ if $\delta = k|y|$.

C

31. Find the mass of the region inside $r = 2 + \sin \theta$ if the density varies as the square of the distance from $(2, 0)$.

32. Find the mass of a circular disc of radius a if the density varies as the distance from a point on the circumference.

33. Of great interest in statistics is the area under the normal distribution curve, $y = e^{-x^2/2}$. Using the fact that

$$\int_{-\infty}^{+\infty} e^{-x^2/2}\, dx = \int_{-\infty}^{+\infty} e^{-y^2/2}\, dy$$

and

$$\int_{-\infty}^{+\infty} e^{-x^2/2}\, dx \int_{-\infty}^{+\infty} e^{-y^2/2}\, dy = \int_{-\infty}^{+\infty} \int_{-\infty}^{+\infty} e^{-x^2/2}\, e^{-y^2/2}\, dy\, dx,$$

show that

$$\int_{-\infty}^{+\infty} e^{-x^2/2}\, dx = \sqrt{2\pi}.$$

† See Angus E. Taylor and Robert Mann, *Advanced Calculus*, 2nd ed. (Lexington, Mass.: Xerox, 1972), pp. 490–493.

22.4

Centers of Mass and Moments of Inertia

Of course the double integral can be used in the determination of centers of mass and moments of inertia. We merely have to make the proper subdivision and recall the meaning of the first and second moments. To set up the integral for a first moment of a plane region, we set up the integral for the mass and include a factor for the (directed) distance from the element of area to the axis in question. If we are dealing with a second moment, that distance is squared (see Sections 16.6–16.10).

Example 1

Find the center of mass of the region bounded by $y = x^2$, the x axis, and $x = 1$ if the density at (x, y) is $x + y$.

From Figure 22.18, we have

$$m = \int_0^1 \int_0^{x^2} (x + y)\, dy\, dx$$

$$= \int_0^1 \left(xy + \frac{y^2}{2} \right) \Big|_0^{x^2} dx$$

$$= \int_0^1 \left(x^3 + \frac{x^4}{2} \right) dx$$

$$= \left(\frac{x^4}{4} + \frac{x^5}{10} \right) \Big|_0^1 = \frac{7}{20}.$$

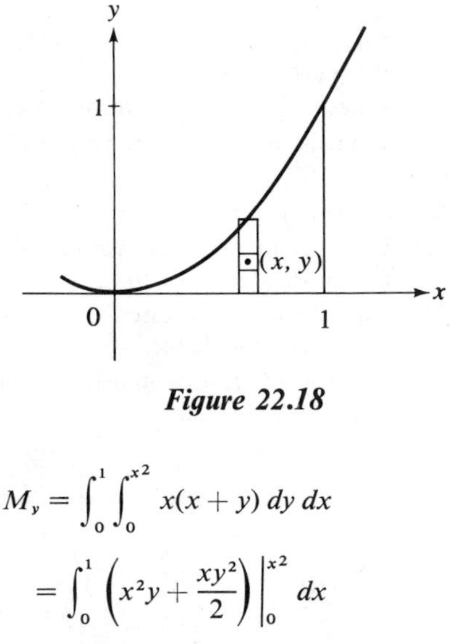

Figure 22.18

$$M_x = \int_0^1 \int_0^{x^2} y(x + y)\, dy\, dx \qquad M_y = \int_0^1 \int_0^{x^2} x(x + y)\, dy\, dx$$

$$= \int_0^1 \left(\frac{xy^2}{2} + \frac{y^3}{3} \right) \Big|_0^{x^2} dx \qquad = \int_0^1 \left(x^2 y + \frac{xy^2}{2} \right) \Big|_0^{x^2} dx$$

$$= \int_0^1 \left(\frac{x^5}{2} + \frac{x^6}{3} \right) dx \qquad = \int_0^1 \left(x^4 + \frac{x^5}{2} \right) dx$$

$$= \left(\frac{x^6}{12} + \frac{x^7}{21} \right) \Big|_0^1 \qquad = \left(\frac{x^5}{5} + \frac{x^6}{12} \right) \Big|_0^1$$

$$= \frac{11}{84}. \qquad = \frac{17}{60}.$$

$$\bar{x} = \frac{M_y}{m} = \frac{17/60}{7/20} = \frac{17}{21}, \quad \bar{y} = \frac{M_x}{m} = \frac{11/84}{7/20} = \frac{55}{147}.$$

Example 2

Find the moment of inertia and radius of gyration of the region of Example 1 about the y axis.

$$I_y = \int_0^1 \int_0^{x^2} x^2(x+y)\, dy\, dx = \int_0^1 \left(x^3 y + \frac{x^2 y^2}{2} \right) \Big|_0^{x^2} dx$$

$$= \int_0^1 \left(x^5 + \frac{x^6}{2} \right) dx = \left(\frac{x^6}{6} + \frac{x^7}{14} \right) \Big|_0^1 = \frac{5}{21},$$

$$I_y = m R_y^2,$$

$$R_y = \sqrt{\frac{I_y}{m}} = \sqrt{\frac{5/21}{7/20}} = \sqrt{\frac{100}{147}} = \frac{10}{7\sqrt{3}}.$$

Example 3

Find the moment of inertia and radius of gyration of the region of Example 1 about the z axis.

The distance from the point (x, y) to the z axis is $\sqrt{x^2 + y^2}$. Thus

$$I_z = \int_0^1 \int_0^{x^2} (x^2 + y^2)(x + y)\, dy\, dx$$

$$= \int_0^1 \left(x^3 y + \frac{x^2 y^2}{2} + \frac{xy^3}{3} + \frac{y^4}{4} \right) \Big|_0^{x^2} dx$$

$$= \int_0^1 \left(x^5 + \frac{x^6}{2} + \frac{x^7}{3} + \frac{x^8}{4} \right) dx$$

$$= \left(\frac{x^6}{6} + \frac{x^7}{14} + \frac{x^8}{24} + \frac{x^9}{36} \right) \Big|_0^1$$

$$= \frac{155}{504},$$

$$I_z = m R_z^2,$$

$$R_z = \sqrt{\frac{I_z}{m}} = \sqrt{\frac{155/504}{7/20}} = \frac{5\sqrt{31}}{21\sqrt{2}}.$$

Let us recall that first moments for solids are taken about planes rather than lines. Thus we set up integrals for first moments by setting up the integral for the mass and include a factor for the (directed) distance from the center of mass of the element of volume to the plane in question.

Example 4

Find the centroid of the solid inside $x^2 + y^2 = 1$ and between $z = 0$ and $z = 1 - x$.

The solid is given in (a) of Figure 22.19. We see by symmetry that $\bar{y} = 0$. Thus, we need to find only two of the three moments. The projection of the solid onto the xy plane is given in (b). From this we have

$$V = 2 \int_{-1}^1 \int_0^{\sqrt{1-x^2}} z\, dy\, dx$$

$$= 2 \int_{-1}^1 \int_0^{\sqrt{1-x^2}} (1 - x)\, dy\, dx$$

$$= 2 \int_{-1}^1 (1 - x)\sqrt{1 - x^2}\, dx.$$

Substituting $x = \sin\theta$, we have

$$V = 2\int_{-\pi/2}^{\pi/2} (1 - \sin\theta)\cos^2\theta\,d\theta$$

$$= \int_{-\pi/2}^{\pi/2} (1 + \cos 2\theta - 2\sin\theta\cos^2\theta)\,d\theta$$

$$= \theta + \frac{1}{2}\sin 2\theta + \frac{2}{3}\cos^3\theta\Big|_{-\pi/2}^{\pi/2}$$

$$= \pi.$$

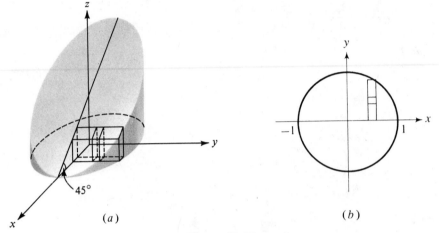

(a)   (b)

Figure 22.19

From (a) of the figure, we see that the element of volume is a vertical solid strip. Its center of mass is at the geometric center of this strip or at $(x, y, z/2)$ where (x, y, z) is a point of the plane $z = 1 - x$.

$$M_{yz} = 2\int_{-1}^{1}\int_{0}^{\sqrt{1-x^2}} x(1 - x)\,dy\,dx$$

$$= 2\int_{-1}^{1} (x - x^2)\sqrt{1 - x^2}\,dx.$$

Again substituting $x = \sin\theta$, we have

$$M_{yz} = 2\int_{-\pi/2}^{\pi/2} (\sin\theta - \sin^2\theta)\cos^2\theta\,d\theta$$

$$= \int_{-\pi/2}^{\pi/2}\left[2\sin\theta\cos^2\theta - \frac{1}{2}(\sin^2 2\theta)\right]d\theta$$

$$= \int_{-\pi/2}^{\pi/2}\left[2\sin\theta\cos^2\theta - \frac{1}{4}(1 - \cos 4\theta)\right]d\theta$$

$$= \int_{-\pi/2}^{\pi/2}\left[2\sin\theta\cos^2\theta - \frac{1}{4} + \frac{1}{4}\cos 4\theta\right]d\theta$$

$$= -\frac{2}{3}\cos^3\theta - \frac{1}{4}\theta + \frac{1}{16}\sin 4\theta\Big|_{-\pi/2}^{\pi/2}$$

$$= -\frac{\pi}{4};$$

$$M_{xy} = 2 \int_{-1}^{1} \int_{0}^{\sqrt{1-x^2}} \frac{z}{2}(1-x)\,dy\,dx$$

$$= \int_{-1}^{1} \int_{0}^{\sqrt{1-x^2}} (1-x)^2\,dy\,dx$$

$$= \int_{-1}^{1} (1-x)^2 \sqrt{1-x^2}\,dx.$$

Again substituting $x = \sin\theta$, we have

$$M_{xy} = \int_{-\pi/2}^{\pi/2} (1-\sin\theta)^2 \cos^2\theta\,d\theta$$

$$= \int_{-\pi/2}^{\pi/2} (\cos^2\theta - 2\sin\theta\cos^2\theta + \sin^2\theta\cos^2\theta)\,d\theta$$

$$= \int_{-\pi/2}^{\pi/2} \left[\frac{1}{2}(1+\cos2\theta) - 2\sin\theta\cos^2\theta + \frac{1}{4}\sin^2 2\theta \right] d\theta$$

$$= \int_{-\pi/2}^{\pi/2} \left[\frac{1}{2} + \frac{1}{2}\cos2\theta - 2\sin\theta\cos^2\theta + \frac{1}{8}(1-\cos4\theta) \right] d\theta$$

$$= \int_{-\pi/2}^{\pi/2} \left(\frac{5}{8} + \frac{1}{2}\cos2\theta - 2\sin\theta\cos^2\theta - \frac{1}{8}\cos4\theta \right) d\theta$$

$$= \frac{5}{8}\theta + \frac{1}{4}\sin2\theta + \frac{2}{3}\cos^3\theta - \frac{1}{32}\sin4\theta \Big|_{-\pi/2}^{\pi/2} = \frac{5}{8}\pi.$$

$$\bar{x} = \frac{M_{yz}}{V} = \frac{-\pi/4}{\pi} = -\frac{1}{4}, \quad \bar{y} = 0, \quad \bar{z} = \frac{M_{xy}}{V} = \frac{5\pi/8}{\pi} = \frac{5}{8}.$$

Second moments for solids are moments about lines. Moreover, since we are not dealing with centers of mass, the solid strip should be parallel to the line. Thus, if the moment of inertia is the moment about the z axis (or a line parallel to it), we must use solid strips parallel to the z axis—that is, vertical strips; if we want the moment about the y axis, we must use solid strips parallel to the y axis, and so on.

Example 5

Find the moment of inertia and radius of gyration of the solid of Example 4 about the z axis. The density is one.

First, we must cut up the solid in such a way that all points of the resulting slice are the same distance from the z axis. The vertical strips of Example 4 are the proper ones for the moment of inertia about the z axis or any line parallel to it. Thus we have

$$I_z = 2 \int_{-1}^{1} \int_{0}^{\sqrt{1-x^2}} (x^2+y^2)z\,dy\,dx$$

$$= 2 \int_{-1}^{1} \int_{0}^{\sqrt{1-x^2}} (x^2+y^2)(1-x)\,dy\,dx$$

$$= 2 \int_{-1}^{1} (1-x)\left(x^2 y + \frac{y^3}{3} \right)\Big|_{0}^{\sqrt{1-x^2}} dx$$

$$= \frac{2}{3} \int_{-1}^{1} (1-x)(2x^2+1)\sqrt{1-x^2}\,dx.$$

Substituting $x = \sin \theta$, we have

$$I_z = \frac{2}{3} \int_{-\pi/2}^{\pi/2} (1 - \sin \theta)(2 \sin^2 \theta + 1) \cos^2 \theta \, d\theta$$

$$= \frac{2}{3} \int_{-\pi/2}^{\pi/2} (-2 \sin^3 \theta \cos^2 \theta + 2 \sin^2 \theta \cos^2 \theta - \sin \theta \cos^2 \theta + \cos^2 \theta) \, d\theta$$

$$= \frac{2}{3} \int_{-\pi/2}^{\pi/2} \left[-2 \sin \theta \, (1 - \cos^2 \theta) \cos^2 \theta + \frac{1}{2} \sin^2 2\theta - \sin \theta \cos^2 \theta + \frac{1}{2} (1 + \cos 2\theta) \right] d\theta$$

$$= \frac{2}{3} \int_{-\pi/2}^{\pi/2} \left[-3 \sin \theta \cos^2 \theta + 2 \sin \theta \cos^4 \theta + \frac{1}{4} (1 - \cos 4\theta) + \frac{1}{2} + \frac{1}{2} \cos 2\theta \right] d\theta$$

$$= \frac{2}{3} \int_{-\pi/2}^{\pi/2} \left[-3 \sin \theta \cos^2 \theta + 2 \sin \theta \cos^4 \theta + \frac{3}{4} - \frac{1}{4} \cos 4\theta + \frac{1}{2} \cos 2\theta \right] d\theta$$

$$= \frac{2}{3} \left(\cos^3 \theta - \frac{2}{5} \cos^5 \theta + \frac{3}{4} \theta - \frac{1}{16} \sin 4\theta + \frac{1}{4} \sin 2\theta \right) \Big|_{-\pi/2}^{\pi/2}$$

$$= \frac{\pi}{2} \, ;$$

$$R_z = \sqrt{\frac{I_z}{V}} = \sqrt{\frac{\pi/2}{\pi}} = \frac{1}{\sqrt{2}}.$$

Problems

A

In Problems 1–8, find the center of mass of the given region.

1. The region in the first quadrant bounded by $y = x^2$, the y axis, and $y = 1$ with density xy at (x, y).
2. The region bounded by $y = x^2$, the x axis, and $x = 1$ with density xy at (x, y).
3. The region bounded by $y = x^3$, the x axis, and $x = 1$ with density $x + y$ at (x, y).
4. The region in the first quadrant bounded by $y = x^2$, the y axis, and $y = 1$ with density $x + y$ at (x, y).
5. The region bounded by $y = x^2$ and $y = x$ with density $x^2 y$ at (x, y).
6. The region bounded by $y = x^3$, the x axis, and $x = 1$ with density $2 - x - y$ at (x, y).
7. The region bounded by $y = x^2$ and $8y = 16 - x^2$ with density $1 + |x| + |y|$ at (x, y).
8. The region bounded by $y = x^2$ and $x + y = 2$ with density $|xy|$ at (x, y).

In Problems 9–14, find the moment of inertia and radius of gyration of the given region about the given line.

9. The region in the first quadrant bounded by $y = x^2$, the y axis, and $y = 1$ with density $x + y$; about the x axis.
10. The region bounded by $y = x^2$, the x axis, and $x = 1$ with density xy; about the y axis.
11. The region bounded by $y = x^3$, the x axis, and $x = 1$ with density xy; about the x axis.
12. The region in the first quadrant bounded by $y = x^2$, the y axis, and $y = 1$ with density $x + y$; about the z axis.
13. The region bounded by $y = x^3$, the x axis, and $x = 1$ with density $x + y$; about the y axis.
14. The region bounded by $y = x^2$ and $y = x$ with density $x + y$; about the z axis.

In Problems 15 and 16, find the centroid of the given solid.

15.   The solid bounded by $y = x^2$, $y = 1$, $z = 0$,   and $z = 1 + y$.
16.   The solid bounded by $y = x$, $y = x^2$, $z = 0$,   and $z = 1 + x + y$.

In Problems 17–19, find the moment of inertia and radius of gyration of the given solid about the given line. The density is one.

17.   The solid bounded by the coordinate planes and $x + y + z = 1$;   about the z axis.
18.   The solid bounded by the coordinate planes and $x + 2y + z = 4$;   about the x axis.
19.   The solid bounded by $z = 1 - y^2$, $x = 1 - y^2$, $z = 0$,   and $x = 0$;   about the x axis.

B

20.   Find the centroid of the solid bounded by $x^2 + y^2 = 1$, $z = 0$,   and $z = 1 - x^2$.
21.   Find the moment of inertia and radius of gyration about the y axis of the solid of Example 4.

C

22.   Find the centroid of the solid bounded by $x^2 + y^2 = 1$, $z = 0$,   and $x + y + z = 2$.
23.   Find the centroid of the region bounded by the x axis and one arch of $y = \sin x$.
24.   Find the centroid of the region bounded by $r = a(1 - \cos \theta)$.
25.   Find the center of mass of the region bounded by $r = 1 - \cos \theta$ if the density varies inversely as the distance from the pole. What happens if the density varies inversely as the square of the distance from the pole?
26.   Find the centroid of the quarter ring $0 < a \le r \le b$, $0 \le \theta \le \pi/4$.
27.   Find the centroid of the interior of one loop of $r = 4 \sin 2\theta$.
28.   Find the centroid of the interior of a loop of $r = \cos 2\theta$.
29.   Find the moments of inertia of the ring $0 < a \le r \le b$, $0 \le \theta \le 2\pi$,   about the x and y axes.
30.   Find the moment of inertia about its axis of a right circular cylinder of height h and radius of base a.
31.   Find the moment of inertia about its axis of a right circular cone of height h and radius of base a.
32.   Find the moment of inertia about the z axis of the region bounded by $r^2 = a^2 \cos 2\theta$.

22.5

Triple Integrals, Volume and Mass

Let us consider another method of determining the volume of a bounded solid. Instead of considering the projection of the solid onto a coordinate plane and subdividing that projection with two sets of lines, suppose we subdivide the solid itself with three sets of planes parallel to the three coordinate axes. This gives a set of rectangular parallelepipeds. Suppose we number from 1 to n all of the parallelepipeds lying entirely inside the given solid, S. The volume V_i of the ith parallelepiped is assumed to be the product of its three dimensions. The sum of the V_i's,

$$\sum_{i=1}^{n} V_i,$$

gives an approximation of the volume of the solid. The norm $\|s\|$ of the subdivision s is taken to be the length of the longest diagonal of the n parallelepipeds determined by the subdivision of the solid S. The volume of S is then taken to be the limit of the approximating sum as $\|s\|$ approaches zero. This gives the triple integral:

$$\iiint_{S} dV = \lim_{\|s\| \to 0} \sum_{i=1}^{n} V_i.$$

In most cases, we shall be interested, not in volume—which can be found by a double integral—but in something related to volume, like mass or moments. Therefore, we shall consider triple integrals of some real-valued function f which is bounded and continuous on S. We choose a point (x_i^*, y_i^*, z_i^*) on or inside the ith parallelepiped and consider the sum of the products,

$$\sum_{i=1}^{n} f(x_i^*, y_i^*, z_i^*) V_i.$$

The limit of this sum is the triple integral of f over S:

$$\iiint_{S} f(x, y, z)\, dV = \lim_{\|s\| \to 0} \sum_{i=1}^{n} f(x_i^*, y_i^*, z_i^*) V_i.$$

Triple integrals are very difficult to evaluate by this definition. But the process can be simplified by considering iterated integrals, as we did with the double integral. Suppose, for example, that S is bounded above by the surface $z = G(x, y)$ and below by $z = F(x, y)$, and that S projects onto the region in the xy plane between $y = f(x)$ and $y = g(x)$, $a \leq x \leq b$. Thus, S is determined by the inequalities

$$F(x, y) \leq z \leq G(x, y)$$

$$f(x) \leq y \leq g(x)$$

$$a \leq x \leq b$$

(see Figure 22.20). Then

$$\iiint_{S} f(x, y, z)\, dV = \int_{a}^{b} \int_{f(x)}^{g(x)} \int_{F(x, y)}^{G(x, y)} f(x, y, z)\, dz\, dy\, dx.$$

On the right-hand side, we first integrate with respect to z, holding both x and y fixed, then follow with a double iterated integral in y and x.

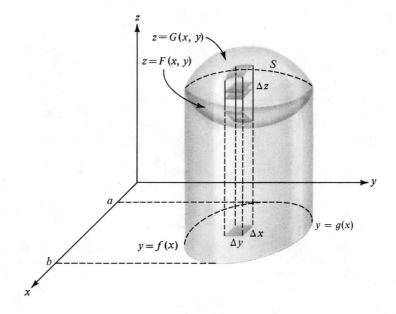

Figure 22.20

Example 1

Evaluate $\displaystyle\int_0^1 \int_0^{x^2} \int_{x^2+y^2}^{x+y} dz\,dy\,dx.$

$$\int_0^1 \int_0^{x^2} \int_{x^2+y^2}^{x+y} dz\,dy\,dx = \int_0^1 \int_0^{x^2} z \,\bigg|_{x^2+y^2}^{x+y} dy\,dx$$

$$= \int_0^1 \int_0^{x^2} (x + y - x^2 - y^2)\,dy\,dx$$

$$= \int_0^1 \left(xy + \frac{y^2}{2} - x^2 y - \frac{y^3}{3}\right)\bigg|_0^{x^2} dx$$

$$= \int_0^1 \left(x^3 + \frac{x^4}{2} - x^4 - \frac{x^6}{3}\right) dx$$

$$= \frac{x^4}{4} - \frac{x^5}{10} - \frac{x^7}{21}\bigg|_0^1 = \frac{43}{420}.$$

Note that a solid region can give rise to several iterated integrals. As we have seen, a triple iterated integral is like a double iterated integral with the process repeated a third time. Thus in setting up triple iterated integrals, we begin by considering the projection of the solid upon one of the coordinate planes. Since there are three co-ordinate planes upon which to project and there are two orders of iteration in any plane, a triple integral can be represented as an iterated integral in six different ways.

Example 2

Describe the solid S given by the iterated integral $\displaystyle\int_0^1 \int_{x^2}^1 \int_0^{1-y} dz\,dy\,dx,$ and set up the other five integrals determined by S.

The solid is determined by the limits on the integral. Since the two outer integrals are with respect to x and y, we consider the projection of S into the xy plane. From

$$\int_0^1 \int_{x^2}^1 ---\,dy\,dx,$$

we see that x has extreme values 0 and 1, with y between $y = x^2$ and $y = 1$ (see Figure 22.21(a)). Since the innermost integral shows that z is between $z = 0$ and $z = 1 - y$, it follows that S is bounded by the planes $z = 0$, $z = 1 - y$, $x = 0$, $y = 1$, and by the cylinder $y = x^2$. These are given by OAB, ABC, OAC, AB, and OBC, respectively, in Figure 22.21(b).

We can get a second iterated integral using the same projection simply by reversing the order of x and y (see Figure 22.21(c)). This has no effect on the innermost integral. Thus we have

$$\int_0^1 \int_0^{\sqrt{y}} \int_0^{1-y} dz\,dx\,dy.$$

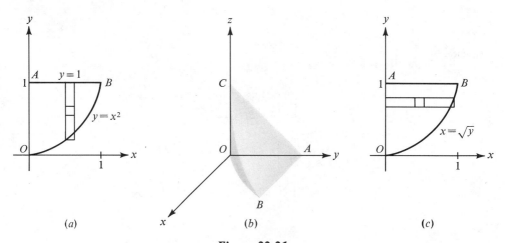

Figure 22.21

If S is projected upon the yz plane, we have the triangle determined by OAC (see Figure 22.22). Since BC is the intersection of $y = x^2$ and $z = 1 - y$, its projection upon the yz plane has equation $z = 1 - y$. Starting with either vertical strips as in Figure 22.22(a) or horizontal strips as in part (b), we have

$$\int_0^1 \int_0^{1-y} ---\,dz\,dy \quad \text{or} \quad \int_0^1 \int_0^{1-z} ---\,dy\,dz.$$

In either case, x goes from the plane OAC ($x = 0$) to the cylinder OBC ($x = \sqrt{y}$). Thus we have

$$\int_0^1 \int_0^{1-y} \int_0^{\sqrt{y}} dx\,dz\,dy \quad \text{or} \quad \int_0^1 \int_0^{1-z} \int_0^{\sqrt{y}} dx\,dy\,dz.$$

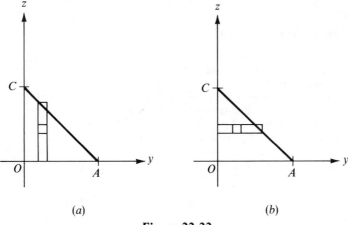

(a)   *(b)*

Figure 22.22

Finally, the projection of S upon the xz plane gives a region determined by ABC (see Figure 22.23). Since CB is the intersection of $y = x^2$ and $y = 1 - z$, its projection CD upon the xz plane has equation $x^2 = 1 - z$ or $z = 1 - x^2$. Again we may begin with either vertical strips (Figure 22.23(b)) or horizontal strips (Figure 22.23(c));

$$\int_0^1 \int_0^{1-x^2} --- dz\, dx \quad \text{or} \quad \int_0^1 \int_0^{\sqrt{1-z}} --- dx\, dz.$$

In either case, y goes from the cylinder $y = x^2$ (OBC) to the plane $y = 1 - z$ (ABC). Thus, the resulting integrals are

$$\int_0^1 \int_0^{1-x^2} \int_{x^2}^{1-z} dy\, dz\, dx \quad \text{or} \quad \int_0^1 \int_0^{\sqrt{1-z}} \int_{x^2}^{1-z} dy\, dx\, dz.$$

It is an easy matter to check that all six of these integrals have value 4/15.

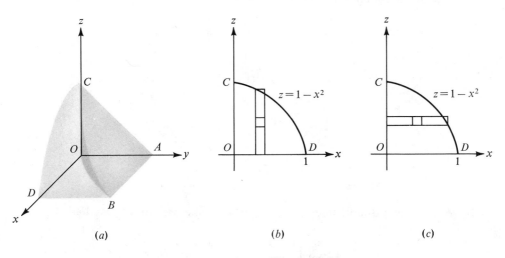

(a)   *(b)*   *(c)*

Figure 22.23

Example 3

Find the volume of the solid inside $x^2 + y^2 = 1$ and $x^2 + z^2 = 1$.

The solid is the intersection of the two cylinders shown in Figure 22.24(a). Since it is symmetric about all three coordinate axes, we need only find the volume in the first octant and multiply by eight. Let us consider the projection of the solid onto one of the coordinate planes—say, the xy plane. This is given in Figure 22.24(b). Cutting by planes, first parallel to the yz plane, then parallel to the xz plane, and finally to the xy plane, we have

$$V = 8 \int_0^1 \int_0^{\sqrt{1-x^2}} \int_0^{\sqrt{1-x^2}} dz\, dy\, dx.$$

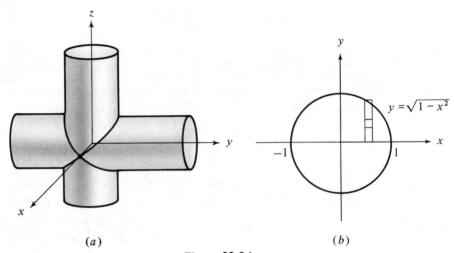

(a) (b)

Figure 22.24

The upper limits are found by solving $x^2 + y^2 = 1$ for y and $x^2 + z^2 = 1$ for z.

$$V = 8 \int_0^1 \int_0^{\sqrt{1-x^2}} \sqrt{1-x^2}\, dy\, dx$$

$$= 8 \int_0^1 (1 - x^2)\, dx$$

$$= 8 \left(x - \frac{x^3}{3} \right) \Big|_0^1$$

$$= \frac{16}{3}.$$

If we had used a double integral in the last example, we would have had

$$V = 8 \int_0^1 \int_0^{\sqrt{1-x^2}} \sqrt{1-x^2}\, dy\, dx,$$

which is what we had after the first integration. We see that in this instance there is no advantage in using a triple integral. Then what can a triple integral do that a double integral cannot do just as easily? A triple integral is especially useful in finding the mass of a solid that is not of uniform density. In that case, we simply have the same integral that we had for volume with the density as an additional factor in the integrand.

Example 4

Find the mass of the solid of Example 3 for a density of $|xyz|$.

We again have symmetry about all three axes, not only with respect to the solid itself, but also with respect to the density. Thus we have the same mass in all eight octants and we need to consider only the first. In that octant, xyz is always positive and $|xyz| = xyz$. Thus,

$$m = 8 \int_0^1 \int_0^{\sqrt{1-x^2}} \int_0^{\sqrt{1-x^2}} xyz \, dz \, dy \, dx$$

$$= 4 \int_0^1 \int_0^{\sqrt{1-x^2}} xyz^2 \Big|_0^{\sqrt{1-x^2}} dy \, dx$$

$$= 4 \int_0^1 \int_0^{\sqrt{1-x^2}} x(1-x^2)y \, dy \, dx$$

$$= 2 \int_0^1 x(1-x^2)y^2 \Big|_0^{\sqrt{1-x^2}} dx$$

$$= 2 \int_0^1 x(1-x^2)^2 \, dx$$

$$= -\frac{1}{3}(1-x^2)^3 \Big|_0^1 = \frac{1}{3}.$$

Problems

A

In Problems 1–4, evaluate the integral.

1. $\displaystyle\int_0^2 \int_0^x \int_0^{x+y} x \, dz \, dy \, dx.$

2. $\displaystyle\int_1^2 \int_0^{z^2} \int_0^{xz} z^2 \, dy \, dx \, dz.$

3. $\displaystyle\int_0^1 \int_{-x}^x \int_{-x}^x (x+z) \, dy \, dz \, dx.$

4. $\displaystyle\int_0^1 \int_0^{\sqrt{1-y^2}} \int_0^y xz \, dz \, dx \, dy.$

In Problems 5–8, describe the solid S given by the iterated integral. Set up the other five iterated integrals determined by S and find the volume of S.

5. $\displaystyle\int_1^4 \int_0^2 \int_0^3 dz \, dx \, dy.$

6. $\displaystyle\int_0^4 \int_0^3 \int_0^x dz \, dy \, dx.$

7. $\displaystyle\int_0^3 \int_0^y \int_0^x dz \, dx \, dy.$

8. $\displaystyle\int_0^1 \int_0^{x^2} \int_0^y dz \, dy \, dx.$

In Problems 9–11, use a triple integral to find the volume of the given solid.

9.   The solid bounded by the three coordinate planes and $2x - y + 3z = 6$.
10. The solid bounded by $y = x$, $y = x^2$, $z = 0$ and $z = x + y$.
11. The solid bounded by $y = x^2$, $y = 0$, $x = 1$, $z = 0$, and $z = x^2 + y^2$.

In Problems 12–18, find the mass of the given solid.

12. The unit cube lying in the first octant with three faces in the coordinate planes and with density xyz.
13. The cube of Problem 12 with density $x + y + z$.
14. The solid bounded by $x^2 + y^2 = 1$, $z = 0$, and $z = 1$ with density $|xyz|$.
15. The solid inside $x^2 + y^2 = 1$ and $y^2 + z^2 = 1$ with density $|y|$.
16. The solid bounded by $y = x^2$, $z = 0$, and $y + z = 1$ with density $|x|$.
17. The solid bounded by the coordinate planes and $x + y + z = 1$ with density $1 - x - y$.
18. The solid bounded by the coordinate planes and $x + 2y + 3z = 6$ with density $x + y + z$.

B

In Problems 19–22, use a triple integral to find the volume of the given solid.

19. The solid bounded by $x^2 + y^2 = 1$, $z = 0$, and $x + y + 4z = 4$.
20. The solid bounded by $x^2 + 4y^2 = 4$, $z = 0$, and $x + y + 4z = 4$.
21. The solid bounded by $x^2 + 4y^2 = 4$, $z = 0$, and $z = 1 - y^2$.
22. The solid inside $x^2 + y^2 + z^2 = 4$.

23. Find the mass of the solid in the first octant bounded by $a^2z = a^3 - xy^2$, $y^2 = ax$, and $y = a$ with density $x + y + z$.

C

24. Find the mass of the solid inside $x^2 + y^2 = 1$ and $x^2 + z^2 = 1$ with density $|z|$.

25. Use a triple integral to find the volume of the solid inside $\dfrac{x^2}{a^2} + \dfrac{y^2}{b^2} + \dfrac{z^2}{c^2} = 1$.

26. Show that $\displaystyle\int_a^b \int_a^z \int_a^y f(x)\, dx\, dy\, dz = \int_a^b \frac{(b - x)^2}{2} f(x)\, dx$.

22.6

Centers of Mass and Moments of Inertia

Triple integrals can also be used to determine centers of mass and moments of inertia. The triple integral's advantage over the double integral occurs when the solid is not of uniform density. To obtain the triple iterated integral for first and second moments, we set up the integral for mass and include a factor for the distance (raised to the first or second powers for first and second moments, respectively) from the element of mass to the line or plane in question.

Example 1

Find the center of mass of the solid bounded by $y = x^2$, $y = 0$, $x = 1$, $z = 0$, and $z = x$ with density y at (x, y, z).

The solid, shown in Figure 22.25(a), is first cut by planes parallel to the yz plane, then by planes parallel to the xz plane, and finally by planes parallel to the xy plane. (Note that we have interchanged the x and y axes in the figure for better visibility.) The projection onto the xy plane is given in Figure 22.25(b). From these two figures, we have

$$m = \int_0^1 \int_0^{x^2} \int_0^x y \, dz \, dy \, dx \qquad M_{yz} = \int_0^1 \int_0^{x^2} \int_0^x xy \, dz \, dy \, dx$$

$$= \int_0^1 \int_0^{x^2} yz \Big|_0^x dy \, dx \qquad = \int_0^1 \int_0^{x^2} xyz \Big|_0^x dy \, dx$$

$$= \int_0^1 \int_0^{x^2} xy \, dy \, dx \qquad = \int_0^1 \int_0^{x^2} x^2 y \, dy \, dx$$

$$= \int_0^1 \frac{1}{2} xy^2 \Big|_0^{x^2} dx \qquad = \frac{1}{2} \int_0^1 x^6 \, dx$$

$$= \int_0^1 \frac{x^5}{2} \, dx \qquad = \frac{x^7}{14} \Big|_0^1 = \frac{1}{14} \, ;$$

$$= \frac{x^6}{12} \Big|_0^1 = \frac{1}{12} \, ;$$

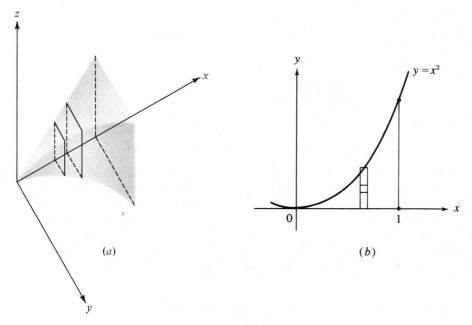

(a) (b)

Figure 22.25

$$M_{xz} = \int_0^1 \int_0^{x^2} \int_0^x y^2 \, dz \, dy \, dx \qquad\qquad M_{xy} = \int_0^1 \int_0^{x^2} \int_0^x yz \, dz \, dy \, dx$$

$$= \int_0^1 \int_0^{x^2} y^2 z \, \bigg|_0^x \, dy \, dx \qquad\qquad = \int_0^1 \int_0^{x^2} \frac{yz^2}{2} \, \bigg|_0^x \, dy \, dx$$

$$= \int_0^1 \int_0^{x^2} xy^2 \, dy \, dx \qquad\qquad = \int_0^1 \int_0^{x^2} \frac{x^2 y}{2} \, dy \, dx$$

$$= \int_0^1 \frac{xy^3}{3} \, \bigg|_0^{x^2} \, dx \qquad\qquad = \int_0^1 \frac{x^2 y^2}{4} \, \bigg|_0^{x^2} \, dx$$

$$= \int_0^1 \frac{x^7}{3} \, dx \qquad\qquad = \int_0^1 \frac{x^6}{4} \, dx$$

$$= \frac{x^8}{24} \, \bigg|_0^1 = \frac{1}{24} ; \qquad\qquad = \frac{x^7}{28} \, \bigg|_0^1 = \frac{1}{28} ;$$

$$\bar{x} = \frac{M_{yz}}{m} = \frac{1/14}{1/12} = \frac{6}{7} ;$$

$$\bar{y} = \frac{M_{xz}}{m} = \frac{1/24}{1/12} = \frac{1}{2} ;$$

$$\bar{z} = \frac{M_{xy}}{m} = \frac{1/28}{1/12} = \frac{3}{7} .$$

Example 2

Find the moment of inertia and radius of gyration of the solid of Example 1 about the z axis.

The distance of a point (x, y, z) from the z axis is $\sqrt{x^2 + y^2}$. Thus,

$$I_z = \int_0^1 \int_0^{x^2} \int_0^x y(x^2 + y^2) \, dz \, dy \, dx$$

$$= \int_0^1 \int_0^{x^2} xy(x^2 + y^2) \, dy \, dx = \int_0^1 \int_0^{x^2} (x^3 y + xy^3) \, dy \, dx$$

$$= \int_0^1 \left(\frac{x^3 y^2}{2} + \frac{xy^4}{4} \right) \bigg|_0^{x^2} \, dx = \int_0^1 \left(\frac{x^7}{2} + \frac{x^9}{4} \right) dx$$

$$= \left(\frac{x^8}{16} + \frac{x^{10}}{40} \right) \bigg|_0^1 = \frac{1}{16} + \frac{1}{40} = \frac{7}{80} ;$$

$$R_z = \sqrt{\frac{I_z}{m}} = \sqrt{\frac{7/80}{1/12}} = \frac{\sqrt{105}}{10} .$$

Sometimes a triple integral is simpler to evaluate than a double integral, even when the solid is of uniform density. Recall that when we used a double integral to obtain a moment of inertia, we found that the element of volume had to be parallel to the axis in question, so that all points of it would be approximately the same distance from that axis. Since the element of mass used to set up the triple iterated integral is a rectangular parallelepiped, rather than the solid strip used for the double iterated integral, the element becomes arbitrarily small. Thus, there is no need to consider the question of having the element of mass parallel to a certain axis.

Example 3

Find the moment of inertia and radius of gyration about the x axis of the solid of Example 4, Section 22.4. The density is one.

Since the density is 1, $m = V$, so from Figure 22.19,

$$m = 2 \int_{-1}^1 \int_0^{\sqrt{1-x^2}} \int_0^{1-x} dz \, dy \, dx = \pi.$$

The distance of the point (x, y, z) to the x axis is $\sqrt{y^2 + z^2}$, so we have

$$I_x = 2 \int_{-1}^{1} \int_{0}^{\sqrt{1-x^2}} \int_{0}^{1-x} (y^2 + z^2) \, dz \, dy \, dx$$

$$= 2 \int_{-1}^{1} \int_{0}^{\sqrt{1-x^2}} \frac{z}{3} (3y^2 + z^2) \Big|_{0}^{1-x} dy \, dx$$

$$= \frac{2}{3} \int_{-1}^{1} \int_{0}^{\sqrt{1-x^2}} (1-x)[3y^2 + (1-x)^2] \, dy \, dx$$

$$= \frac{2}{3} \int_{-1}^{1} (1-x)[y^3 + (1-x)^2 y] \Big|_{0}^{\sqrt{1-x^2}} dx$$

$$= \frac{4}{3} \int_{-1}^{1} (1-x)^2 \sqrt{1-x^2} \, dx.$$

Letting $x = \sin \theta$, we have

$$I_x = \frac{4}{3} \int_{-\pi/2}^{\pi/2} (1 - \sin \theta)^2 \cos^2 \theta \, d\theta$$

$$= \frac{4}{3} \int_{-\pi/2}^{\pi/2} (\cos^2 \theta - 2 \sin \theta \cos^2 \theta + \sin^2 \theta \cos^2 \theta) \, d\theta$$

$$= \frac{4}{3} \int_{-\pi/2}^{\pi/2} \left[\frac{1}{2}(1 + \cos 2\theta) - 2 \sin \theta \cos^2 \theta + \frac{1}{4}(1 - \cos^2 2\theta) \right] d\theta$$

$$= \frac{4}{3} \int_{-\pi/2}^{\pi/2} \left[\frac{3}{4} + \frac{1}{2} \cos 2\theta - 2 \sin \theta \cos^2 \theta - \frac{1}{8}(1 + \cos 4\theta) \right] d\theta$$

$$= \frac{4}{3} \int_{-\pi/2}^{\pi/2} \left(\frac{5}{8} + \frac{1}{2} \cos 2\theta - 2 \sin \theta \cos^2 \theta - \frac{1}{8} \cos 4\theta \right) d\theta$$

$$= \frac{4}{3} \left(\frac{5}{8} \theta + \frac{1}{4} \sin 2\theta + \frac{2}{3} \cos^3 \theta - \frac{1}{32} \sin 4\theta \right) \Big|_{-\pi/2}^{\pi/2}$$

$$= \frac{5\pi}{6};$$

$$R_x = \sqrt{\frac{I_x}{m}} = \sqrt{\frac{5\pi/6}{\pi}} = \sqrt{\frac{5}{6}}.$$

Problems

A

In Problems 1–8, find the center of mass of the given solid.

1. The unit cube in the first octant with three faces in the coordinate planes and density xyz.
2. The solid of Problem 1 with density $x + y + z$.
3. The solid bounded by $y = 0$, $y = x$, $x = 1$, $z = 0$, and $z = xy$ with density z.
4. The solid bounded by $x = 0$, $x = 1$, $y = 0$, $y = 1$, $z = 0$, and $z = xy$ with density xyz.
5. The solid inside $x^2 + y^2 = 1$ and between $z = 0$ and $z = 1$ with density $|xyz|$.
6. The portion of the solid of Problem 5 that lies in the first octant.
7. The solid inside $x^2 + y^2 = 1$ and between $z = 0$ and $z = x^2 + y^2$ with density $|y|$.
8. The solid bounded by $x^2 + y^2 = 1$, $z = 0$, and $z = 1$ with density $2 - z$.

In Problems 9–16, find the moment of inertia and radius of gyration of the given solid about the given axis.

9. The solid of Problem 1;   about the z axis.

10. The solid of Problem 1;   about the x axis.

11. The solid bounded by $y = x$, $y = x^2$, $z = 0$, and $z = xy$ with density 1; about the y axis.

12. The solid bounded by $x = 0$, $x = 1$, $y = 0$, $y = 1$, $z = 0$, and $z = x^2 y^2$ with density 2;   about the z axis.

13. The solid bounded by $y = 0$, $y = x^2$, $x = 1$, $z = 0$, and $z = x$ with density 4; about the x axis.

14. The solid inside $x^2 + y^2 = 4$, above $z = 0$, and below $z = y$ with density 1; about the x axis.

15. The solid of Problem 14 with density $|x|$;   about the x axis.

16. The solid in the first octant inside $x^2 + y^2 = 1$, above $z = 0$, and below $z = x^2 + y^2$ with density y;   about the z axis.

C

17. Prove that the moment of a body about any plane passing through the center of mass is zero. (*Hint:* The directed distance from a general point (x, y, z) to an arbitrary plane passing through $(\bar{x}, \bar{y}, \bar{z})$ is $r = l(x - \bar{x}) + m(y - \bar{y}) + n(z - \bar{z})$, where l, m, and n are direction cosines of the normal to the plane.)

18. Prove that if a body has zero moment with respect to a plane, then that plane must pass through the center of mass of the body. (*Hint:* First show that the moment about a general plane is md, where m is the mass of the body and d is the distance from the center of mass to the plane.)

19. Let l be an arbitrary line in space and let S be a solid with mass m. Show that the moment of inertia of S about l is $I_l = I_{l'} + md^2$, where l' is a line parallel to l through the center of mass of S and d is the distance between l and l'.

22.7

Triple Integrals in Cylindrical and Spherical Coordinates

Just as it is sometimes convenient to use polar coordinates when using single or double integrals, it is sometimes convenient to use cylindrical or spherical coordinates in triple integrals. Cylindrical coordinates are especially useful when there is a line of symmetry and the axes are placed in such a way that the line of symmetry corresponds to the z axis.

Recall that in cylindrical coordinates a point in space is represented by the triple (r, θ, z), where (r, θ) is a polar representation of the projection upon the xy plane and z is the same as the z coordinate in rectangular coordinates. A cylindrical representation of a point is related to the rectangular representation by

$$x = r \cos \theta, \qquad y = r \sin \theta, \qquad z = z.$$

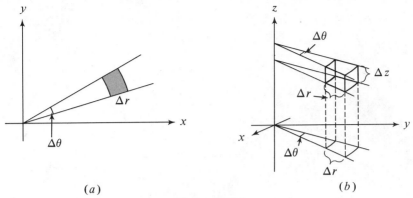

Figure 22.26

We saw in Section 22.3 that the element of area given in Figure 22.26(a) has area

$$r_{av} \, \Delta r \, \Delta \theta,$$

leading to the double iterated integral

$$\iint\limits_{R} F(r, \theta) r \, dr \, d\theta.$$

Similarly the element of volume of Figure 22.26(b) has volume

$$r_{av} \, \Delta r \, \Delta \theta \, \Delta z.$$

This gives us integrals of the form

$$\iiint\limits_{S} F(r, \theta, z) r \, dr \, d\theta \, dz.$$

Example 1

Find the mass of the solid inside the cylinder $x^2 + y^2 = 1$ and outside the cone $z^2 = x^2 + y^2$ when the density at (x, y, z) is $\sqrt{x^2 + y^2}$.

In cylindrical coordinates the cylinder is $r = 1$, the cone is $z^2 = r^2$ and the density at (r, θ, z) is $|r|$. The graph is given in Figure 22.27(a) and its projection upon the xy plane is in (b). By the symmetry of both the figure and the density, we consider only the first octant and multiply by 8. From Figure 22.27, we have

$$m = 8 \int_0^{\pi/2} \int_0^1 \int_0^r r^2 \, dz \, dr \, d\theta$$

$$= 8 \int_0^{\pi/2} \int_0^1 r^3 \, dr \, d\theta$$

$$= 8 \int_0^{\pi/2} \frac{1}{4} \, d\theta = \pi.$$

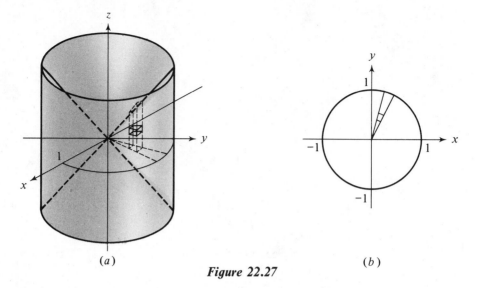

Figure 22.27

Note in this example that the integrand r^2 takes one r from the density at (r, θ, z) and the other from the element of volume

$$r_{av} \, \Delta r \, \Delta \theta \, \Delta z.$$

You might object to 0 as the lower limit for r because, for any level above the xy plane, r goes from the cone (not 0) to the cylinder. However, note that we are looking at the base in the projection of Figure 22.27(b). The fact that we go only to the cone is taken care of in the limits on z ($0 \le z \le r$). The situation is illustrated further in Figure 22.28(a). Looking only at the portion in the first octant, we first cut triangular wedges from $\theta = 0$ to $\theta = \pi/2$. Now each wedge (one of which is illustrated) is divided by cylindrical cuts, with the z axis being the axis of the cylinders. Here we see that r takes values from $r = 0$ to $r = 1$. Finally each vertical element is now cut by horizontal cuts, showing that z goes from the xy plane ($z = 0$) to the cone ($z = r$).

If we make the cuts in the order z, θ, r, as illustrated in Figure 22.28(b), the resulting integral is

$$m = 8 \int_0^1 \int_0^{\pi/2} \int_z^1 r^2 \, dr \, d\theta \, dz = \pi.$$

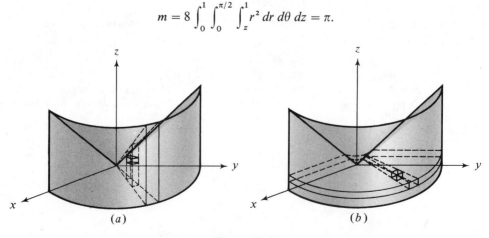

Figure 22.28

In this case we cannot consider a projection upon a coordinate plane—we must work directly with the three-dimensional figure. It is the role of the z coordinate in (r, θ, z) that allows us to project upon the xy plane. In such a case the order of integration is

$$\iiint\limits_{S} F(r, \theta, z) r \, dz \, dr \, d\theta \quad \text{or} \quad \iiint\limits_{S} F(r, \theta, z) r \, dz \, d\theta \, dr.$$

It is instructive to set up the integral for Example 1 in rectangular coordinates:

$$m = 8 \int_{0}^{1} \int_{0}^{\sqrt{1-x^2}} \int_{0}^{\sqrt{x^2+y^2}} \sqrt{x^2 + y^2} \, dz \, dy \, dx.$$

You are encouraged to carry out at least the integrations with respect to z and y to convince yourself that the integral in cylindrical coordinates is much easier to evaluate.

Next we shall consider triple integrals, which are represented in terms of spherical coordinates. Such coordinates are most useful when there is a point of symmetry and the axes are placed with the point of symmetry at the origin. Recall that in spherical coordinates a point is given by (ρ, θ, φ) (see Figure 22.29) with $\rho \geq 0$ and $0 \leq \varphi \leq \pi$.

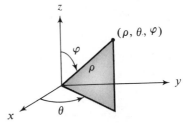

Figure 22.29

Spherical coordinates are related to rectangular coordinates by

$$x = \rho \sin \varphi \cos \theta, \qquad y = \rho \sin \varphi \sin \theta, \qquad z = \rho \cos \varphi.$$

The element of volume used for spherical coordinates is shown in Figure 22.30.

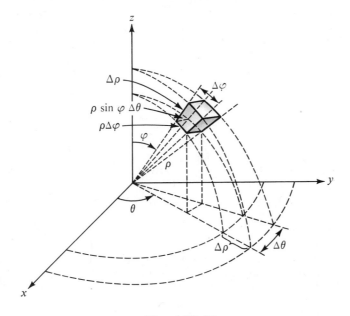

Figure 22.30

By using the formula for the length of a circular arc,

$$s = r\theta,$$

we find that the edges of this element of volume are $\Delta\rho$, $\rho\Delta\varphi$, and $\rho \sin \varphi \, \Delta\theta$. If we look upon the element as an approximation of a rectangular parallelepiped, we have the volume

$$\Delta\rho \cdot \rho \, \Delta\varphi \cdot \rho \sin \varphi \, \Delta\theta,$$

which leads to integrals of the form

$$\iiint_{S} f(\rho, \, \theta, \, \varphi)\rho^2 \sin \varphi \, d\rho \, d\theta \, d\varphi.$$

Example 2

Find the mass of the solid inside both the sphere $x^2 + y^2 + z^2 = 1$ and the top half of the cone $z^2 = x^2 + y^2$ when the density at (x, y, z) is $\sqrt{x^2 + y^2 + z^2}$.

In spherical coordinates the sphere is $\rho = 1$, the top half of the cone is $\varphi = \pi/4$ (see Example 7, page 705), and the density at $(\rho, \, \theta, \, \varphi)$ is ρ. By symmetry, we consider only the portion in the first octant and multiply by 4. The graph is given in Figure 22.31. Projections upon coordinate planes will not do here; we must consider this in three dimensions. First we cut S with planes containing the z axis (see Figure 22.31); this gives the outer integral

$$\int_{0}^{\pi/2} - - - \, d\theta.$$

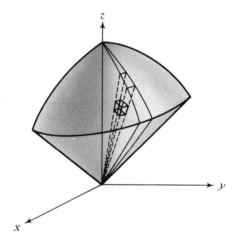

Figure 22.31

Next each slice is further cut by cones with apex at the origin and axis on the z axis from $\varphi = 0$ to the given cone, $\varphi = \pi/4$. This gives the middle integral

$$\int_{0}^{\pi/4} - - - \, d\varphi.$$

Finally each resulting element of volume is cut up by spheres with radius ranging from $\rho = 0$ to the given sphere $\rho = 1$. This now gives the inner integral,

$$\int_{0}^{1} - - - \, d\rho.$$

Putting all of this together with the density, we have

$$m = 4 \int_0^{\pi/2} \int_0^{\pi/4} \int_0^1 \rho^3 \sin \varphi \, d\rho \, d\varphi \, d\theta$$

$$= 4 \int_0^{\pi/2} \int_0^{\pi/4} \frac{1}{4} \sin \varphi \, d\varphi \, d\theta$$

$$= \int_0^{\pi/2} (-\cos \varphi) \Big|_0^{\pi/4} d\theta$$

$$= \int_0^{\pi/2} \left(-\frac{1}{\sqrt{2}} + 1 \right) d\theta$$

$$= \frac{\sqrt{2} - 1}{\sqrt{2}} \frac{\pi}{2} = \frac{(2 - \sqrt{2})\pi}{4} .$$

The integral for the mass in rectangular coordinates is

$$m = \int_0^1 \int_0^{\sqrt{1-x^2}} \int_{x^2+y^2}^{\sqrt{1-x^2-y^2}} \sqrt{x^2 + y^2 + z^2} \, dz \, dy \, dx,$$

so in this case, spherical coordinates simplify the problem.

Example 3

Find the center of mass of a hemisphere of radius 1 whose density at each point is proportional to that point's distance from the base.

In spherical coordinates, the sphere is $\rho = 1$, the base is $\varphi = \pi/2$, and the density is $kz = k\rho \cos \varphi$ (see Figure 22.32). Because of the symmetry, we know that the center of mass is on the z axis—we only need to find its z coordinate. Again because of the symmetry, we can determine the mass and the moment about the xy plane by restricting ourselves to the first octant and multiplying by 4. Cutting first with cones, then with planes, and then with spheres (see Figure 22.32) we have

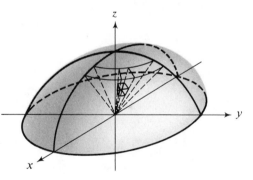

Figure 22.32

$$m = 4 \int_0^{\pi/2} \int_0^{\pi/2} \int_0^1 k\rho \cos \varphi \cdot \rho^2 \sin \varphi \, d\rho \, d\theta \, d\varphi$$

$$= 4 \int_0^{\pi/2} \int_0^{\pi/2} \int_0^1 k\rho^3 \sin \varphi \cos \varphi \, d\rho \, d\theta \, d\varphi$$

$$= 4 \int_0^{\pi/2} \int_0^{\pi/2} \frac{k}{4} \sin \varphi \cos \varphi \, d\theta \, d\varphi$$

$$= 4 \int_0^{\pi/2} \frac{\pi k}{8} \sin \varphi \cos \varphi \, d\varphi$$

$$= 4 \cdot \frac{\pi k}{8} \cdot \frac{\sin^2 \varphi}{2} \Big|_0^{\pi/2} = 4 \cdot \frac{\pi k}{8} \cdot \frac{1}{2} = \frac{\pi k}{4} .$$

In setting up the integral for M_{xy} we include a factor for the distance of our element of mass from the xy plane. This is $z = \rho \cos \varphi$.

$$M_{xy} = 4 \int_0^{\pi/2} \int_0^{\pi/2} \int_0^1 k(\rho \cos \varphi)^2 \, \rho^2 \sin \varphi \, d\rho \, d\theta \, d\varphi$$

$$= 4 \int_0^{\pi/2} \int_0^{\pi/2} \int_0^1 k\rho^4 \cos^2 \varphi \sin \varphi \, d\rho \, d\theta \, d\varphi$$

$$= 4 \int_0^{\pi/2} \int_0^{\pi/2} \frac{k}{5} \cos^2 \varphi \sin \varphi \, d\theta \, d\varphi$$

$$= 4 \int_0^{\pi/2} \frac{\pi k}{10} \cos^2 \varphi \sin \varphi \, d\varphi$$

$$= 4 \left(-\frac{\pi k}{10} \frac{\cos^3 \varphi}{3} \right) \Big|_0^{\pi/2}$$

$$= 4 \cdot \frac{\pi k}{10} \cdot \frac{1}{3} = \frac{2\pi k}{15} \, .$$

$$\bar{z} = \frac{M_{xy}}{m} = \frac{2\pi k/15}{\pi k/4} = \frac{8}{15} \, .$$

Thus the center of mass is $(0, 0, 8/15)$.

Again, the student should set up the integrals in rectangular coordinates.

Problems

A

In Problems 1–8, use cylindrical coordinates.

1. A right circular cylinder with radius 1 and height 4 has a density at each point that is proportional to its distance from the axis of the cylinder. Find the total mass.

2. A right circular cone has a base of radius 2 and height 5. The density at each point is proportional to the distance of the point from the base. Find the total mass.

3. Find the mass of the solid inside both the sphere $x^2 + y^2 + z^2 = 1$ and the top half of the cone $z^2 = x^2 + y^2$ given a density at (x, y, z) of z.

4. Find the mass of the solid inside the cylinder $x^2 + y^2 = 2y$ but outside the cone $z^2 = x^2 + y^2$ given a density at (x, y, z) of $\sqrt{x^2 + y^2}$.

5. Find the center of mass of the solid inside the cylinder $x^2 + y^2 = 1$, above $z = 0$, and outside the cone $z^2 = x^2 + y^2$ given the density at (x, y, z) of $|z|$.

6. Find the center of mass of the solid bounded by the two cones $z^2 = x^2 + y^2$ and $z^2 = 4x^2 + 4y^2$ and between $z = 0$ and $z = 4$ for a density at (x, y, z) of z.

7. Find the moment of inertia about the z axis of the solid inside the cylinder $x^2 + y^2 = 1$ but outside the cone $z^2 = 4x^2 + 4y^2$ given the density at (x, y, z) of $|z|$.

8. Find the moment of inertia about the z axis of the cylinder $x^2 + y^2 = 2x$ between $z = 0$ and $z = 4$ given the density at (x, y, z) of $\sqrt{x^2 + y^2}$.

In Problems 9–16, use spherical coordinates.

9. Find the mass of the sphere with radius 2 and density at each point equal to 4 minus the distance from the center to that point.

10. Find the mass of the solid inside the sphere $x^2 + y^2 + z^2 = 1$ and outside the cone $z^2 = x^2 + y^2$ given the density at (x, y, z) of $\sqrt{x^2 + y^2 + z^2}$.

11. Find the mass of the solid between the spheres $x^2 + y^2 + z^2 = r^2$ and $x^2 + y^2 + z^2 = R^2$ $(r < R)$ given the density at (x, y, z) of $\sqrt{x^2 + y^2 + z^2}$.

12. Find the mass of the solid inside the sphere $x^2 + y^2 + z^2 = 4$ but outside the cone $z^2 = 3x^2 + 3y^2$ for the density at (x, y, z) of $e^{\sqrt{x^2 + y^2 + z^2}}$.

13. Find the center of mass of the solid inside both the sphere $x^2 + y^2 + z^2 = 1$ and the top half of the cone $z^2 = x^2 + y^2$ for the density at (x, y, z) of $4 - 2\sqrt{x^2 + y^2 + z^2}$.

14. Find the center of mass of the solid inside the sphere $x^2 + y^2 + z^2 = 1$, outside the cone $z^2 = x^2 + y^2$, and above the xy plane given the density at (x, y, z) of $4 - \sqrt{x^2 + y^2 + z^2}$.

15. Find the center of mass of the solid between the spheres $x^2 + y^2 + z^2 = 1$ and $x^2 + y^2 + z^2 = 4$ and above the xy plane for the density at (x, y, z) of z.

16. Find the moment of inertia about the z axis of the solid inside both the sphere $x^2 + y^2 + z^2 = 1$ and the top half of the cone $z^2 = x^2 + y^2$ given the density at (x, y, z) of $2 - \sqrt{x^2 + y^2 + z^2}$.

B

17. Find the moment of inertia and radius of gyration about its axis of a right circular cone of height h and radius of base a when the density is constant.

18. Find the mass of a hemisphere of radius a given that the density varies as the nth power of the distance from the center.

19. Find the center of mass of the solid bounded by $z = 4 - x^2 - y^2$ and the xy plane for a density at (x, y, z) of $x^2 + y^2 + z^2$.

C

20. The equations $x = F(u, v, w)$, $y = G(u, v, w)$, and $z = H(u, v, w)$ map a solid region S in xyz space into another solid region S' in uvw space. It can be shown that, under the proper conditions,

$$\iiint_S f(x, y, z) \, dz \, dy \, dx = \iiint_{S'} f(F, G, H) \left| \frac{\partial(x, y, z)}{\partial(u, v, w)} \right| \, du \, dv \, dw.$$

Show that, for $x = \rho \sin \varphi \cos \theta$, $y = \rho \sin \varphi \sin \theta$, and $z = \rho \cos \varphi$,

$$\frac{\partial(x, y, z)}{\partial(\rho, \varphi, \theta)} = \rho^2 \sin \varphi;$$

and therefore

$$\iiint_S f(x, y, z) \, dz \, dy \, dx = \iiint_{S'} f(\rho \sin \varphi \cos \theta, \rho \sin \varphi \sin \theta, \rho \cos \varphi) \rho^2 \sin \varphi \, d\rho \, d\varphi \, d\theta.$$

Compare with Problem 25 in Section 22.3.

Review Problems

A

In Problems 1–4, evaluate and sketch R.

1. $\displaystyle\int_0^3 \int_0^{4x - x^2} (x + y) \, dy \, dx.$

2. $\displaystyle\int_0^4 \int_{(x - 2)^2}^4 (2x - y) \, dy \, dx.$

3. $\displaystyle\int_{-2}^1 \int_{y^2 - 1}^{1 - y} dx \, dy.$

4. $\displaystyle\int_0^{5/2} \int_{y^2 - 3y}^{2y - y^2} (2xy + 1) \, dx \, dy.$

In Problems 5 and 6, evaluate the given integral.

5. $\iint_R (x^2 - y)\, dA$,   where R is the region bounded by $y = x$, $x + y = 4$,   and $y = 0$.

6. $\iint_R (x - 4y)\, dA$,   where R is the region bounded by $y = x$ and $y = x^2 + x - 5$.

7. Find the volume of the solid bounded by the coordinate planes and $3x + y - 4z = 6$.

8. Find the mass of the solid bounded by $z = 1 - x^2 - y^2$ and the plane $y = z$,   given that the density at (x, y, z) is $|x|$.

9. Find the mass of the region bounded by $x = y$, $x + 2y = 4$,   and $y = 0$ if the density at (x, y) is $x + y$.

10. Find the mass of the region bounded by $y = 6x - x^2$ and $x = y$ if the density at (x, y) is $1 + xy$.

11. Find the volume of the cylinder $r = \cos 2\theta$ between $z = 0$ and $z = 2$.

12. Find the mass of the region inside $r = 2 + \cos \theta$ with density $|r|$ at (r, θ).

13. Use polar coordinates to find the mass of the region between the circles $x^2 + y^2 = a^2$ and $x^2 + y^2 = b^2$ $(a < b)$ with density $\sqrt{x^2 + y^2}$ at (x, y).

14. Find the center of mass of the region bounded by $y = x^2$ and $y = 12 - 3x^2$ if the density at (x, y) is $x^2 + 2y$.

15. Find the moment of inertia and radius of gyration about the x axis of the region bounded by $x = y^2$ and $x + y = 6$,   with density $2x$ at (x, y).

16. Find the centroid of the solid bounded by $y = x^2$, $y + z = 4$,   and $z = 0$.

17. Find the moment of inertia and radius of gyration about the z axis of the solid bounded by $y = x^2$, $y + z = 4$,   and $z = 0$.

In Problems 18–20, evaluate the integral.

18. $\int_0^1 \int_0^{x^2} \int_0^{x+z} (x - y)\, dy\, dz\, dx$.

19. $\int_0^2 \int_1^y \int_{x+y}^1 xy\, dz\, dx\, dy$.

20. $\int_0^1 \int_x^{x^2} \int_0^y (2x + 3y + z)\, dz\, dy\, dx$.

21. Find the mass of the solid bounded by the cylinder $y = x^2$ and the planes $y + 2z = 4$ and $y - 4z = 4$ for the density at (x, y, z) of $1 + y + z^2$.

22. Find the center of mass of the solid bounded by the coordinate planes and the plane $4x + 2y + z = 8$ for the density at (x, y, z) of $1 + x$.

23. Find the center of mass of the solid bounded by the coordinate planes, $y + z = 2$,   and $x = 4$ with density $1 + x + y + z$ at (x, y, z).

24. Find the moment of inertia and radius of gyration about the x axis of the solid bounded by $y = z^2$, $y = 4$, $x = 0$,   and $x = 4$ with density y at (x, y, z).

25. Find the mass of the solid inside $x^2 + y^2 + z^2 = 4$,   outside $z^2 = x^2 + y^2$,   and above $z = 0$ with density $1 + z$ at (x, y, z).

26. Find the center of mass of the hemisphere inside $x^2 + y^2 + z^2 = 4$ and above $z = 0$ with density z at (x, y, z).

B

27. Find the mass of the solid bounded by $y = x^2$ and $y = 4 - z^2$ given the density at (x, y, z) of $1 + y$.

28. Find the mass of a right circular cone of height h and radius of base a, given that the density varies jointly as the distance from the axis of the cone and the distance from the base of the cone.

29. Find the mass of the solid inside $x^2 + y^2 = 4$ but outside $z^2 = x^2 + y^2$ given the density at (x, y, z) of $|z|$.

30. Find the moment of inertia about the z axis of the solid inside $x^2 + y^2 = 1$ and $x^2 + y^2 + z^2 = 4$ with density $\sqrt{x^2 + y^2}$ at (x, y, z).

23

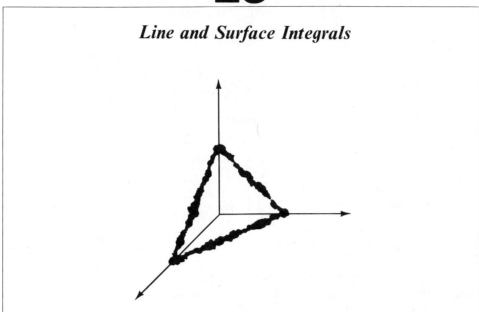

Line and Surface Integrals

23.1

Definition of the Line Integral

A natural extension of the integral of a function of a single variable $\int_a^b f(x)\,dx$ on an interval $[a, b]$ is the integral of a function of two variables $f(x, y)$ along a curve in the plane or the integral of a function $f(x, y, z)$ on a curve in space. Such curvilinear integrals are traditionally called *line integrals*. Again, an extension of the double integral over a region in the plane is the integral of a function of three variables on a surface in space; such an integral is called a *surface integral*. Line and surface integrals have important applications in the physical sciences, engineering, and mathematics, some of which we will consider in this chapter.

We begin by defining the line integral in the plane. Let C be a plane curve with parametric representation $\quad x = x(t), \quad y = y(t), \quad a \le t \le b, \quad$ where $x'(t)$ and $y'(t)$ are continuous and not simultaneously zero on $[a, b]$; such a curve is called a *smooth curve*. Suppose $f(x, y)$ is continuous at each point of C, and let $s(t)$ be arc length along C:

$$s(t) = \int_a^t \sqrt{[x'(u)]^2 + [y'(u)]^2}\,du$$

(see page 411). The *line integral* of $f(x, y)$ with respect to arc length along C is

$$\int_C f(x, y)\, ds = \int_a^b f(x(t), y(t))s'(t)\, dt,$$

where

$$s'(t) = \sqrt{\left(\frac{dx}{dt}\right)^2 + \left(\frac{dy}{dt}\right)^2}.$$

In addition to $\int_C f(x, y)\, ds$, we also define

$$\int_C P(x, y)\, dx = \int_a^b P(x(t), y(t))x'(t)\, dt$$

and

$$\int_C Q(x, y)\, dy = \int_a^b Q(x(t), y(t))y'(t)\, dt,$$

the line integral of P with respect to x along C and the line integral of Q with respect to y along C.

In applications there may be two functions $P(x, y)$ and $Q(x, y)$ defined on C—the x and y components of a force or velocity vector, for example—and these last two integrals are often combined:

$$\int_C P(x, y)\, dx + Q(x, y)\, dy = \int_a^b \left[P(x(t), y(t))x'(t) + Q(x(t), y(t))y'(t)\right] dt$$

$$= \int_C P(x, y)\, dx + \int_C Q(x, y)\, dy.$$

This last integral can be written in vector form. Let $\mathbf{r}(t) = x(t)\mathbf{i} + y(t)\mathbf{j}$, $a \le t \le b$, be a vector equation of the curve C and let $\mathbf{f}(x, y) = P(x, y)\mathbf{i} + Q(x, y)\mathbf{j}$. Then $\mathbf{r}'(t) = x'(t)\mathbf{i} + y'(t)\mathbf{j}$ and

$$\int_C P\, dx + Q\, dy = \int_a^b \left[P(x(t), y(t))x'(t) + Q(x(t), y(t))y'(t)\right] dt$$

$$= \int_a^b \mathbf{f}(x(t), y(t)) \cdot \mathbf{r}'(t)\, dt$$

$$= \int_C \mathbf{f} \cdot d\mathbf{r},$$

where we have defined

$$d\mathbf{r} = dx\mathbf{i} + dy\mathbf{j}.$$

Thus

$$\int_C P\, dx + Q\, dy = \int_C \mathbf{f} \cdot d\mathbf{r}.$$

Example 1

Evaluate $\int_C (x^2 - y^2)\, dx - 2xy\, dy$ where

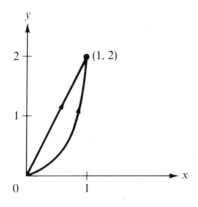

Figure 23.1

(a)   *C* is the line segment $x = t,$ $y = 2t,$ $0 \le t \le 1,$ and

(b)   *C* is the arc of the parabola $y = 2x^2$ from (0, 0) to (1, 2) (see Figure 23.1).

(a) $\displaystyle \int_C (x^2 - y^2)\, dx - 2xy\, dy = \int_0^1 \{[x^2(t) - y^2(t)]x'(t) - [2x(t)y(t)]y'(t)\}\, dt$

$$= \int_0^1 \{(t^2 - 4t^2) \cdot 1 - 2 \cdot t(2t) \cdot 2\}\, dt$$

$$= \int_0^1 (-3t^2 - 8t^2)\, dt = -\frac{11}{3} t^3 \bigg|_0^1 = -\frac{11}{3}.$$

(b)   Parametric equations of a curve given by $y = f(x),$ $a \le x \le b,$ can be obtained by setting $x = t,$ $y = f(t),$ $a \le t \le b,$ or, more simply, $x = x,$ $y = f(x).$ Thus, with $x = x,$ $y = 2x^2,$ $0 \le x \le 1,$

$$\int_C (x^2 - y^2)\, dx - 2xy\, dy = \int_0^1 [(x^2 - 4x^4) \cdot 1 - (2 \cdot x \cdot 2x^2) \cdot 4x]\, dx$$

$$= \int_0^1 (x^2 - 20x^4)\, dx = \frac{1}{3} x^3 - 4x^5 \bigg|_0^1 = -\frac{11}{3}.$$

Example 2

Evaluate $\int_C x\, ds$ where

$$C:\quad x = 2 \cos t,\quad y = 2 \sin t,\quad 0 \le t \le \pi.$$

See Figure 23.2.

$$s'(t) = \left[\left(\frac{dx}{dt}\right)^2 + \left(\frac{dy}{dt}\right)^2\right]^{1/2}$$

$$= [(-2 \sin t)^2 + (2 \cos t)^2]^{1/2} = 2,$$

so

$$\int_C x\, ds = \int_0^\pi x(t)s'(t)\, dt$$

$$= \int_0^\pi \cos t \cdot 2\, dt = 2 \sin t \bigg|_0^\pi = 0.$$

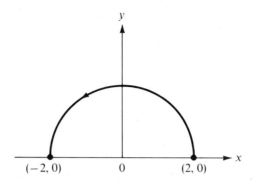

Figure 23.2

While we defined a line integral along a curve in terms of a given parametric representation, the value of a line integral is unchanged under a broad class of changes of parameter, so long as there is not a reversal of direction along the curve.

If we think of the independent variable t as time, then a change of parameter simply changes the speed with which a particle traverses the curve C. For example, suppose C is the circle $x = \cos t$, $y = \sin t$, $0 \le t \le 2\pi$, and we let $t = 2u$. Then C is given in terms of the parameter u by $x = \cos 2u$, $y = \sin 2u$, $0 \le u \le \pi$, and the circle is traversed in half the "time." The following theorem gives sufficient conditions for the value of a line integral to remain constant under a change of parameter.

Theorem 23.1

Let C: $x = x(t)$, $y = y(t)$, $a \le t \le b$, be a smooth curve and let $t = \phi(u)$, $c \le u \le d$, where $\phi'(u) > 0$ on (c, d) and $\phi(c) = a$, $\phi(d) = b$. Then a line integral along C has the same value regardless of whether t or u is used as the parameter for C.

Proof

Since $\phi'(u) > 0$, $t = \phi(u)$ is strictly increasing on $[c, d]$, the inverse function $u = \phi^{-1}(t) = \psi(t)$ exists and $du/dt = d\psi/dt = 1/(d\phi/du) = 1/(dt/du)$. Also, if we let $x_1(u) = x(\phi(u))$ and $y_1(u) = y(\phi(u))$, then $x = x_1(u)$, $y = y_1(u)$, $c \le u \le d$, is a parametric representation of C.

In $\int_C P(x, y)\, dx$, let us substitute $t = \phi(u)$, $dx/dt = (dx_1/du) \cdot (du/dt) = (dx_1/du)/(dt/du)$, and $dt = (d\varphi/du)\, du = (dt/du)\, du$ (see Section 15.2). Then

$$\int_C P(x, y)\, dx = \int_a^b P(x(t), y(t)) \frac{dx}{dt}\, dt$$

$$= \int_c^d P(x(\phi(u)), y(\phi(u))) \left(\frac{dx_1/du}{dt/du} \right) \frac{dt}{du}\, du$$

$$= \int_c^d P(x_1(u), y_1(u)) \frac{dx_1}{du}\, du.$$

The invariance of the other line integrals under a change of parameter can be shown in a similar fashion (see Problem 22).

Example 3

In part (b) of Example 1, make the change of parameter $t = \sin u$ and evaluate the resulting integral.

Let $t = \sin u$. Then $x = \sin u$, $y = 2 \sin^2 u$, $0 \le u \le \pi/2$, are parametric equations of the arc of $y = 2x^2$ from $(0, 0)$ to $(1, 2)$.

$$\int_C (x^2 - y^2)\, dx - 2xy\, dy$$

$$= \int_0^{\pi/2} [(\sin^2 u - 4 \sin^4 u) \cos u - 2 \sin u \cdot 2 \sin^2 u \cdot 4 \sin u \cos u]\, du$$

$$= \int_0^{\pi/2} (\sin^2 u - 20 \sin^4 u) \cos u\, du$$

$$= \frac{1}{3} \sin^3 u - 4 \sin^5 u \Big|_0^{\pi/2} = \frac{1}{3} - 4 = -\frac{11}{3},$$

which is the same value we obtained before.

The definition of the line integral can be extended to allow paths with corners, such as a broken-line path. Let C have parametric representation $x = x(t)$, $y = y(t)$, $a \le t \le b$, where $x(t)$ and $y(t)$ are continuous on $[a, b]$. The path C is said to be *sectionally smooth* if the interval $[a, b]$ can be subdivided into at most a finite number of subintervals on each of which $x'(t)$ and $y'(t)$ are continuous. Now suppose $C_1, C_2, \ldots, C_n$ are sectionally smooth curves, and the initial point of C_{i+1} is the terminal point of C_i, $i = 1, 2, \ldots, n - 1$. Then $C = \sum_{i=1}^n C_i$ is the curve obtained by first traversing C_1, then C_2, and continuing thus until C_n has been traversed. From the additive property of the integral,

$$\int_{\sum_{i=1}^n C_i} P\, dx + Q\, dy = \sum_{i=1}^n \int_{C_i} P\, dx + Q\, dy.$$

If C has parametric representation $x = x(t)$, $y = y(t)$, $a \le t \le b$, then $-C$ is defined to be the curve with parametric representation $x = x(a + b - t)$, $y = y(a + b - t)$, $a \le t \le b$. As t goes from a to b, $a + b - t$ goes from b to a, so the original curve C is traversed in the opposite direction. The initial point C is the terminal point of $-C$, and conversely. It is left to the student to show that

$$\int_{-C} P\, dx = -\int_C P\, dx,$$

$$\int_{-C} Q\, dy = -\int_C Q\, dy,$$

and thus

$$\int_{-C} \mathbf{f} \cdot d\mathbf{r} = -\int_C \mathbf{f} \cdot d\mathbf{r}.$$

Example 4

Evaluate $\int_C (x^2 - y^2)\, dx - 2xy\, dy$ along the path consisting of the line segment from $(0, 0)$ to $(0, 2)$ followed by the segment from $(0, 2)$ to $(1, 2)$ (see Figure 23.3).

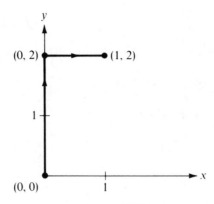

Figure 23.3

On the segment from $(0, 0)$ to $(0, 2)$, let $x = 0$, $y = t$, $0 \le t \le 2$. On the segment from $(0, 2)$ to $(1, 2)$, let $x = t$, $y = 2$, $0 \le t \le 1$. Then

$$\int_C (x^2 - y^2)\, dx - 2xy\, dy = \int_0^2 [(0^2 - t^2)0 - (2 \cdot 0 \cdot t)1]\, dt$$

$$+ \int_0^1 [(t^2 - 4)1 - (2 \cdot t \cdot 2)0]\, dt$$

$$= \int_0^2 0\, dt + \int_0^1 (t^2 - 4)\, dt$$

$$= \frac{t^3}{3} - 4t \Big|_0^1 = -\frac{11}{3}.$$

Example 5

Evaluate $\int_C \mathbf{f} \cdot d\mathbf{r}$, where $\mathbf{f} = x\mathbf{i} - y\mathbf{j}$ and C consists of the quarter circle C_1: $x = -2\cos t$, $y = 2\sin t$, $0 \le t \le \pi/2$, followed by the arc of the ellipse C_2: $x^2/9 + y^2/4 = 1$ from $(0, 2)$ to $(3, 0)$, followed by the line segment C_3 from $(3, 0)$ to $(0, -5)$ (see Figure 23.4).

Let us parametrize C_2 by

$$y = \frac{2}{3}\sqrt{9 - x^2}, \quad 0 \le x \le 3,$$

and $-C_3$ by

$$y = \frac{5}{3}x - 5, \quad 0 \le x \le 3.$$

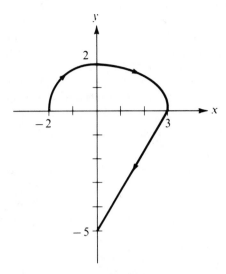

Figure 23.4

(Note that C_3 is traversed in the direction of decreasing x, from $x = 3$ to $x = 0$.)
Then

$$\int_C \mathbf{f} \cdot d\mathbf{r} = \int_{C_1} x\,dx - y\,dy + \int_{C_2} x\,dx - y\,dy - \int_{C_3} x\,dx - y\,dy$$

$$= \int_0^{\pi/2} \left[-2\cos t(2\sin t) - 2\sin t(2\cos t) \right] dt$$

$$+ \int_0^3 \left[x - \left(\frac{2}{3}\sqrt{9 - x^2} \right) \left(-\frac{2}{3}\frac{x}{\sqrt{9 - x^2}} \right) \right] dx$$

$$- \int_0^3 \left[x - \left(\frac{5}{3}x - 5 \right) \cdot \frac{5}{3} \right] dx$$

$$= 4\int_0^{\pi/2} \cos t(-\sin t)\,dt + \int_0^3 \frac{13}{9}x\,dx - \int_0^3 \left(-\frac{16}{9}x + \frac{25}{3} \right) dx$$

$$= 2\cos^2 t \Big|_0^{\pi/2} + \left(\frac{13}{18}x^2 \Big|_0^3 \right) + \left(\frac{8}{9}x^2 - \frac{25}{3}x \Big|_0^3 \right) = -\frac{29}{2}.$$

Problems

A

In Problems 1–12, sketch the path and evaluate the line integral.

1. $\displaystyle\int_C (2x - 3y)\,dx; \quad C:\quad x = 2t,\quad y = 3t,\quad 0 \le t \le 2.$

2. $\displaystyle\int_C (xy + y)\,dy; \quad C:\quad x = 2 - t,\quad y = t^2,\quad 0 \le t \le 2.$

3. $\displaystyle\int_C 2xy\,dy; \quad C:\quad x = \cos\theta,\quad y = \sin\theta,\quad 0 \le \theta \le \pi/2.$

4. $\displaystyle\int_C y^2\,dx + x^2\,dy; \quad C:\quad x = 2 - t,\quad y = 3t,\quad 0 \le t \le 1.$

5. $\int_C 2xy \, dy - x^2 \, dx; \quad C: \quad x = t^2, \quad y = 1 - t, \quad 0 \le t \le 1.$

6. $\int_C x^2 \, dx + y^2 \, dy; \quad C: \quad x = \sin 2t, \quad y = \cos 2t, \quad 0 \le t \le \pi.$

7. $\int_C (2x - 3y) \, dx; \quad C$ consists of the line segments from $(0, 0)$ to $(4, 0)$ and from $(4, 0)$ to $(4, 4)$.

8. $\int_C (x + 5y) \, dy; \quad C$ consists of the line segments from $(1, 0)$ to $(3, 0)$ and from $(3, 0)$ to $(3, 4)$.

9. $\int_C \mathbf{f} \cdot d\mathbf{r}, \mathbf{f} = 0\mathbf{i} + y(x + 1)\mathbf{j} \, ; C$ consists of the segment from $(1, 0)$ to $(1, 2)$ followed by the segment from $(1, 2)$ to $(2, 3)$.

10. $\int_C \mathbf{f} \cdot d\mathbf{r}, \quad \mathbf{f} = (x^2 + y^2)\mathbf{i} + 0\mathbf{j}; \quad C$ consists of the segment from $(0, 2)$ to $(2, 2)$ followed by the segment from $(2, 2)$ to $(2, 4)$.

11. $\int_C \mathbf{f} \cdot d\mathbf{r}, \quad \mathbf{f} = y\mathbf{i} + x\mathbf{j}; \quad C$ consists of the arc $y = x^2, \quad 0 \le x \le 2,$ followed by the segment from $(2, 4)$ to $(0, 0)$.

12. $\int_C \mathbf{f} \cdot d\mathbf{r}, \quad \mathbf{f} = y^2\mathbf{i} + x^2\mathbf{j}, \quad C$ consists of the segment $x = 2 - t, \quad y = 3t, \quad 0 \le t \le 1,$ followed by the segment $x = 1 - t, \quad y = 3 - 3t, \quad 0 \le t \le 1.$

B

In Problems 13–18, sketch the path and evaluate the integral.

13. $\int_C \mathbf{f} \cdot d\mathbf{r}, \quad \mathbf{f} = y^2\mathbf{i} - x^2\mathbf{j}; \quad C: \quad y = 1 - x^2, \quad -1 \le x \le 1.$

14. $\int_C \mathbf{f} \cdot d\mathbf{r}, \quad \mathbf{f} = y\mathbf{i} + x\mathbf{j}; \quad C: \quad y = x^2, \quad 0 \le x \le 2.$

15. $\int_C ds; \quad C: \quad x = 8t^3, \quad y = 4t^2, \quad 0 \le t \le 1.$

16. $\int_C (y - x) \, ds; \quad C: \quad x = 3 \sin \theta, \quad y = 3 \cos \theta, \quad 0 \le \theta \le 2\pi.$

17. $\int_C (x^2 - y^2) \, dx - xy \, dy$ around the rectangle with vertices $(1, 1), (-1, 1), (-1, 0), (1, 0)$.

18. $\int_C (x + y) \, ds; \quad C$ consists of the arc $x = e^t \sin t, \quad y = e^t \cos t, \quad 0 \le t \le \pi,$ followed by the segment from $(0, -e^\pi)$ to $(0, 1)$.

C

In Problems 19–21, sketch the path and evaluate the integral.

19. $\int_C \dfrac{y \, dx - x \, dy}{x^2 + y^2}; \quad C: \quad x = \cos^3 t, \quad y = \sin^3 t, \quad 0 \le t \le \pi/2.$ (*Hint:* Set $u = \tan^3 t.$)

20. $\int_C 2y \, dx - 3x \, dy$ around the circle $x^2 + y^2 = 1$ in the counterclockwise direction.

21. $\displaystyle\int_C (x^2 - y^2)\, ds$ around the circle $x^2 + y^2 = 9$ in the counterclockwise sense.

22. Show that $\displaystyle\int_C Q(x,\, y)\, dy$ and $\displaystyle\int_C f(x,\, y)\, ds$ are invariant under a strictly increasing differentiable change of parameter.

23. Let $\mathbf{T} = \mathbf{r}'(t)/|\mathbf{r}'(t)|$ be the unit tangent vector to the curve C: $\mathbf{r}(t) = x(t)\mathbf{i} + y(t)\mathbf{j}$, $a \leq t \leq b$,   and let $\mathbf{f} = P\mathbf{i} + Q\mathbf{j}$.   Show that $\displaystyle\int_C \mathbf{f} \cdot d\mathbf{r} = \int_C \mathbf{f} \cdot \mathbf{T}\, ds$.

23.2

Green's Theorem in the Plane

Green's theorem expresses a double integral over a region as a line integral along the boundary of the region. As we will see, it has many important consequences.

Before stating the result, it will be convenient to make a few definitions. Let C be a sectionally smooth curve with parametric representation $x = x(t)$, $y = y(t)$, $\alpha \leq t \leq \beta$.   If $(x(\alpha),\, y(\alpha)) = (x(\beta),\, y(\beta))$,   but $(x(t_1),\, y(t_1)) \neq (x(t_2),\, y(t_2))$ for any other values of t, then C is a *simple closed curve*. It is *closed* because the end points coincide; it is *simple* because it only intersects itself at the end points. We say that a simple closed curve is traversed in the *positive sense* if the inside of C is to the left as the curve is traversed in the direction of increasing t. Finally, we say that a function $f(x, y)$ is *smooth* on a set S if its first partial derivatives are continuous in S.

Theorem 23.2

(Green's theorem)   Let $P(x,\, y)$ and $Q(x,\, y)$ be smooth inside and on a sectionally smooth simple closed curve C. Then

$$\int_C P(x,\, y)\, dx + Q(x,\, y)\, dy = \iint\limits_R \left(\frac{\partial Q}{\partial x} - \frac{\partial P}{\partial y}\right) dA,$$

where R is the region bounded by C and C is traversed in the positive sense.

Proof

Suppose first that R can be represented both as

$$R = \{(x,\, y)\,|\,f_1(x) \leq y \leq f_2(x),\quad a \leq x \leq b\}$$

and

$$R = \{(x,\, y)\,|\,g_1(y) \leq x \leq g_2(y),\quad c \leq y \leq d\}$$

(see Figure 23.5). Such a region is said to be *simple*.

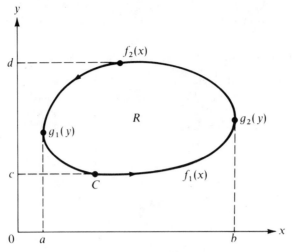

Figure 23.5

Let us consider the integral of $P(x, y)$ around C in the positive sense. As we go along the lower part of C, $y = f_1(x)$, so $P(x, y) = P(x, f_1(x))$, and x increases from a to b. Along the upper part of C, $y = f_2(x)$ and we go from right to left as C is traversed in the positive sense, so $P(x, y) = P(x, f_2(x))$ and x goes from b to a. Thus

$$\int_C P(x, y)\, dx = \int_a^b P(x, f_1(x))\, dx + \int_b^a P(x, f_2(x))\, dx$$

$$= \int_a^b P(x, f_1(x))\, dx - \int_a^b P(x, f_2(x))\, dx$$

$$= -\int_a^b \left[P(x, y) \Big|_{y=f_1(x)}^{y=f_2(x)} \right] dx$$

$$= -\int_a^b \left[\int_{f_1(x)}^{f_2(x)} \frac{\partial P(x, y)}{\partial y}\, dy \right] dx$$

$$= -\iint_R \frac{\partial P}{\partial y}\, dA.$$

Similarly, separating C into left- and right-hand parts, we have

$$\int_C Q(x, y)\, dy = \int_c^d Q(g_2(y), y)\, dy + \int_d^c Q(g_1(y), y)\, dy$$

$$= \int_c^d \left[Q(x, y) \Big|_{g_1(y)}^{g_2(y)} \right] dy$$

$$= \int_c^d \left[\int_{g_1(y)}^{g_2(y)} \frac{\partial Q(x, y)}{\partial x}\, dx \right] dy$$

$$= \iint_R \frac{\partial Q}{\partial x}\, dA.$$

Thus

$$\int_C P(x, y)\, dx + Q(x, y)\, dy = \iint_R \left(\frac{\partial Q}{\partial x} - \frac{\partial P}{\partial y} \right) dA.$$

Note: We can allow C to contain a vertical line segment σ, since x is constant along such a segment ($\int_\sigma P\,dx = 0$). Similarly, C can contain a horizontal line segment (C might be a rectangle, for example).

More complicated regions can often be broken up into several regions of the simple type we considered above. In Figure 23.6, we see an example of such a

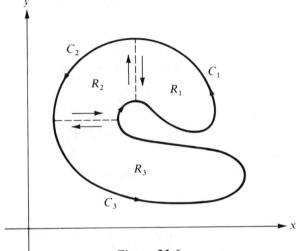

Figure 23.6

region. When we integrate in the positive sense around the boundary C_i of each subregion R_i ($i = 1, 2, 3$), the interior (dotted line) curves are traversed once in each direction, so the corresponding integrals along these interior curves cancel. Thus, in Figure 23.6,

$$\int_C P\,dx + Q\,dy = \int_{C_1 + C_2 + C_3} P\,dx + Q\,dy$$

$$= \iint_{R = R_1 \cup R_2 \cup R_3} (Q_x - P_y)\,dA,$$

where C is the boundary of R. The general case is proved in advanced texts.†

Example 1

Use Green's theorem to evaluate $\int_C (x^2 - y^2)\,dx - 2xy\,dy$, where C is the ellipse $x^2 + 4y^2 = 16$.

$$P(x, y) = x^2 - y^2, \qquad \frac{\partial P}{\partial y} = -2y,$$

$$Q(x, y) = -2xy, \qquad \frac{\partial Q}{\partial x} = -2y,$$

$$\int_C (x^2 - y^2)\,dx + 2xy\,dy = \iint_R [-2y - (-2y)]\,dA = 0.$$

---

† See, for example, Kenneth Rogers, *Advanced Calculus* (Columbus, Ohio: Charles E. Merrill, 1976), pp. 316–328.

Example 2

Evaluate $\int_C (\sin \sqrt{x} + y) \, dx + \ln (1 + y^2) \, dy$,   where C is the circle $x^2 + y^2 = 9$.

$$\int_C (\sin \sqrt{x} + y) \, dx + \ln (1 + y^2) \, dy = \iint\limits_{x^2 + y^2 \le 9} (0 - 1) \, dA = -9\pi,$$

where 9π is the area of the circle.

Example 3

If R is the region bounded by a simple closed curve C, show that the area of R is given by any one of the integrals $\int_C -y \, dx$, $\int_C x \, dy$, $1/2 \int_C x \, dy - y \, dx$,   where C is traversed in the positive sense.

$$\int_C -y \, dx = \iint\limits_R [0 - (-1)] \, dA = \iint\limits_R dA = \text{Area of } R.$$

The verification of the other two formulas is left to the student (see Problem 11).

Example 4

Calculate the area of the ellipse $x^2/a^2 + y^2/b^2 = 1$ by each of the formulas of Example 3.

$$A = \int_C -y \, dx = -4 \int_a^0 \left(\frac{b}{a}\right) \sqrt{a^2 - x^2} \, dx.$$

Let $x = a \sin \theta$, $dx = a \cos \theta \, d\theta$.   Then

$$A = -4 \int_a^0 \left(\frac{b}{a}\right) \sqrt{a^2 - x^2} \, dx = -4 \int_{\pi/2}^0 \left(\frac{b}{a}\right) a^2 \cos^2 \theta \, d\theta$$

$$= 4ab \int_0^{\pi/2} \left(\frac{1}{2} + \frac{\cos 2\theta}{2}\right) d\theta$$

$$= 4ab \left(\frac{\theta}{2} + \frac{\sin 2\theta}{4}\right)\Big|_0^{\pi/2} = \pi ab.$$

Next, we have

$$A = \int_C x \, dy = 4 \int_0^b \left(\frac{a}{b}\right) \sqrt{b^2 - y^2} \, dy = \pi ab.$$

Now let $x = a \cos \theta$, $y = b \sin \theta$, $0 \le \theta \le 2\pi$,   be parametric equations of the ellipse. Then

$$A = \frac{1}{2} \int_C x \, dy - y \, dx = \frac{1}{2} (4) \int_0^{\pi/2} [a \cos \theta \cdot b \cos \theta - b \sin \theta (-a \sin \theta)] \, d\theta$$

$$= 2 \int_0^{\pi/2} ab \, d\theta = \pi ab.$$

Green's theorem can be extended to multiply connected regions, that is, regions with "holes" in them. Suppose a region R is bounded by simple closed curves C_1, $C_2, \ldots, C_n$, where R lies inside C_1 and outside each of $C_2, C_3, \ldots, C_n$. Also suppose

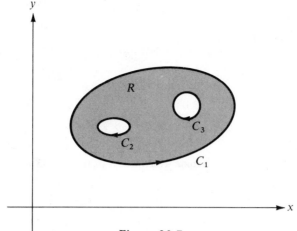

Figure 23.7

$P(x, y)$ and $Q(x, y)$ are smooth on R and its boundary. In Figure 23.7, the boundary of R consists of three curves: $C = C_1 + C_2 + C_3$. If C is traversed in the positive sense, keeping R to the left, then the outer curve, C_1, is traversed in the counterclockwise sense, while each interior boundary curve is traversed in the clockwise sense. If we divide R into simple regions $R_1, \ldots, R_N$ by disjoint arcs lying in R, as in the proof of Theorem 23.2, and apply Green's theorem to each of the resulting subregions R_i, then the line integrals along the interior arcs cancel in pairs, and

$$\int_C P\ dx + Q\ dy = \sum_{i=1}^{N} \iint_{R_i} \left(\frac{\partial Q}{\partial x} - \frac{\partial P}{\partial y} \right) dA$$

$$= \iint_R \left(\frac{\partial Q}{\partial x} - \frac{\partial P}{\partial y} \right) dA$$

(see Figure 23.8).

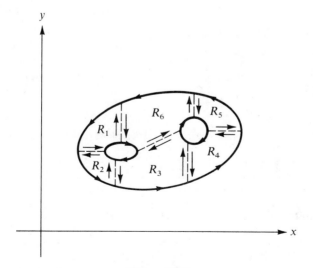

Figure 23.8

Problems

A

In Problems 1–6, use Green's theorem to evaluate the integral around the indicated curve in the positive sense.

1. $\displaystyle\int_C 2y\,dx - 3x\,dy$;   circle $x^2 + y^2 = 1$.

2. $\displaystyle\int_C x^2\,dx - y^2\,dy$;   circle $x^2 + y^2 = 9$.

3. $\displaystyle\int_C xy^2\,dx + (x^2 + y^2)\,dy$;   rectangle with vertices $(0, 0)$, $(3, 0)$, $(3, 4)$, and $(0, 4)$.

4. $\displaystyle\int_C (\cos x + xy^2)\,dx + (e^y + x^2)\,dy$;   semicircle $x^2 + y^2 = 4$, $x \geq 0$.

5. $\displaystyle\int_C (e^{x^2} + y^2)\,dx + (e^{y^2} + x^2)\,dy$;   triangle with vertices $(0, 0)$, $(4, 0)$, and $(0, 4)$.

6. $\displaystyle\int_C (\arctan x + y)\,dx + (\arcsin y - x)\,dy$;   circle $x^2 + y^2 = 16$.

B

7. Suppose $P(x, y) = -y/(x^2 + y^2)$, $Q(x, y) = x/(x^2 + y^2)$, and $C: x = \cos\theta$, $y = \sin\theta$, $0 \leq \theta \leq 2\pi$. Show that $\displaystyle\int_C P\,dx + Q\,dy = 2\pi$, while $\displaystyle\iint_R (Q_x - P_y)\,dA = 0$, where R is the inside of C. Does this contradict Green's theorem? Explain.

8. Verify Green's theorem when R is the annulus $0 < a^2 \leq x^2 + y^2 \leq 1$, $P(x, y) = -y/(x^2 + y^2)$,   and $Q = x/(x^2 + y^2)$.

9. Prove the *deformation of contours theorem*: If R is the region between two simple closed curves C_1 and C_2 and $\partial Q/\partial x = \partial P/\partial y$ in R, then

$$\int_{C_1} P\,dx + Q\,dy = \int_{C_2} P\,dx + Q\,dy.$$

10. Evaluate $\displaystyle\int_C [x/(x^2 + y^2)^{3/2}]\,dx + [y/(x^2 + y^2)^{3/2}]\,dy$ where C is the ellipse $4x^2 + 9y^2 = 36$ traversed in the positive sense.

11. Verify the last two formulas in Example 3.

In Problems 12 and 13, use the results of Example 3 to find the area enclosed by the given curves.

12. $x = a\cos^3\theta$, $y = a\sin^3\theta$, $0 \leq \theta \leq 2\pi$.
13. $y = x^2$ and $x - y + 2 = 0$.

14. Show that the coordinates of the centroid $(\bar{x}, \bar{y})$ of a plane region R with area A and boundary C satisfy $\displaystyle 2A\bar{x} = \int_C x^2\,dy$ and $\displaystyle 2A\bar{y} = -\int_C y^2\,dx$.

15. Show that the moments of inertia of the region in Problem 14 about the x and y axes, respectively, satisfy $\displaystyle 3I_x = -\int_C y^3\,dx$ and $\displaystyle 3I_y = \int_C x^3\,dy$.

16. Show that if A is the area of the region bounded by the polar coordinate curve C: $r = f(\theta)$, $0 \leq \theta \leq 2\pi$,   then

$$A = (1/2)\left(\int_C x\,dy - y\,dx\right) = \int_0^{2\pi} \frac{1}{2}r^2\,d\theta.$$

23.3

Independence of the Path; Exact Differentials

In Examples 1 and 4 of Section 23.1, we evaluated a line integral along three different paths from (0, 0) to (1, 2) and obtained the same value along each path. We will now show that this was not an accident. For a significant class of line integrals, the value of the integral depends only on the end points and is independent of the path between them. For convenience, let us first state a few definitions.

Definition

*A set S is **open** if for each point (x, y) in S there is a neighborhood of (x, y) contained in S.*

*A set S is **connected** if each two points in S can be connected by a broken-line path lying entirely in S. An open connected set is called a **region.***

*A set S is **simply connected** if the inside of each simple closed curve in S also lies in S; that is, S has no "holes" in it. It can be shown that the interior of a simple closed curve is itself simply connected.*

Definition

Suppose P(x, y) and Q(x, y) are defined in S. Then

$$\int_C P(x, y) \, dx + Q(x, y) \, dy$$

*is said to be **independent of the path** in S if for each pair of points A and B in S the values of the integrals along all paths in S from A to B are the same. Thus the value of the integral depends only on the end points and not on the path between the end points. In this case we can write*

$$\int_C P \, dx + Q \, dy = \int_A^B P \, dx + Q \, dy.$$

The fundamental theorem of calculus tells us that if $g'(x)$ is continuous on an interval $[a, b]$, then

$$\int_a^b g'(x) \, dx = g(b) - g(a).$$

Thus the value of the integral depends only on the values of the antiderivative $g(x)$ at the end points of $[a, b]$. Now suppose P and Q are smooth functions in a region R and suppose there is a scalar function $F(x, y)$ such that $\partial F/\partial x = P$ and $\partial F/\partial y = Q$. Then the gradient of $F = \nabla F = P\mathbf{i} + Q\mathbf{j} = \mathbf{f}$ and $F(x, y)$ is called a *potential* for the vector function **f.** The following theorem is the analog of the fundamental theorem of calculus for line integrals.

Theorem 23.3

Suppose $\nabla F = P\mathbf{i} + Q\mathbf{j}$ in a region R and let C: $x = x(t),   y = y(t),   a \le t \le b$, be a curve in R from $A = (x(a), y(a))$ to $B = (x(b), y(b))$.   Then

$$\int_C P\,dx + Q\,dy = \int_C \mathbf{f} \cdot d\mathbf{r} = F(B) - F(A).$$

That is, the integral is independent of the path in R and depends only on the values of a potential F at the end points.

Proof

$$\int_C P\,dx + Q\,dy = \int_a^b [P(x(t), y(t))x'(t) + Q(x(t), y(t))y'(t)]\,dt$$

$$= \int_a^b \left(\frac{\partial F}{\partial x}\frac{dx}{dt} + \frac{\partial F}{\partial y}\frac{dy}{dt}\right) dt$$

$$= \int_a^b F'(x(t), y(t))\,dt, \qquad \text{by the chain rule,}$$

$$= F(x(t), y(t))\Big|_a^b = F(B) - F(A).$$

Example 1

Evaluate $\int_C (3x^2 + y^2)\,dx + 2xy\,dy$ along the upper half of the ellipse $x^2/4 + y^2/9 = 1$ from $(-2, 0)$ to $(2, 0)$.

A potential for $(3x^2 + y^2)\mathbf{i} + 2xy\mathbf{j}$ is $F(x, y) = x^3 + xy^2$,   so

$$\int_C (3x^2 + y^2)\,dx + 2xy\,dy = F(2, 0) - F(-2, 0) = 8 - (-8) = 16.$$

The converse of Theorem 23.3 is also true.

Theorem 23.4

Suppose $P(x, y)$ and $Q(x, y)$ are smooth in a region R. If

$$\int_C P(x, y)\,dx + Q(x, y)\,dy$$

is independent of the path in R, then there exists a function $F(x, y)$ such that $\partial F/\partial x = P$ and $\partial F/\partial y = Q$ in R.

Proof

Let (x_0, y_0) be a fixed point in R, and for each point (x, y) in R define

$$F(x, y) = \int_{(x_0, y_0)}^{(x, y)} P\,dx + Q\,dy,$$

the integral taken along any convenient path in R from (x_0, y_0) to (x, y). Since the integral is assumed to be independent of the path, $F(x, y)$ is a single-valued function in R. Now fix (x_1, y_1) in R. Since R is open, for h sufficiently small, the segment from (x_1, y_1) to $(x_1 + h, y_1)$ lies in R. Then

$$\frac{1}{h} \left[F(x_1 + h, y_1) - F(x_1, y_1)\right]$$

$$= \frac{1}{h} \left[\int_{(x_0,\, y_0)}^{(x_1 + h,\, y_1)} P\, dx + Q\, dy - \int_{(x_0,\, y_0)}^{(x_1,\, y_1)} P\, dx + Q\, dy\right]$$

$$= \frac{1}{h} \int_{(x_1,\, y_1)}^{(x_1 + h,\, y_1)} P\, dx + Q\, dy.$$

Let us evaluate the integral along the segment from (x_1, y_1) to $(x_1 + h, y_1)$: $x = x$, $y = y_1$, $x_1 \le x \le x_1 + h$ (see Figure 23.9). Then

$$\frac{1}{h} \int_{(x_1,\, y_1)}^{(x_1 + h,\, y_1)} P\, dx + Q\, dy = \frac{1}{h} \int_{x_1}^{x_1 + h} \left[P(x, y_1) \cdot 1 + Q(x, y_1) \cdot 0\right] dx$$

$$= \frac{1}{h} \int_{x_1}^{x_1 + h} P(x, y_1)\, dx.$$

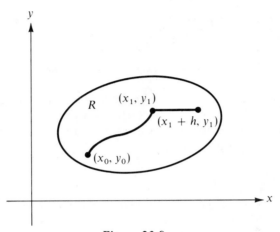

Figure 23.9

As in the proof of the fundamental theorem of calculus (Theorem 7.5),

$$\lim_{h \to 0} \frac{1}{h} \left[F(x_1 + h, y_1) - F(x_1, y_1)\right] = \lim_{h \to 0} \frac{1}{h} \int_{x_1}^{x_1 + h} P(x, y_1)\, dx$$

$$= P(x_1, y_1).$$

Thus $\partial F/\partial x = P(x, y)$ throughout R. By considering the segment from (x_1, y_1) to $(x_1, y_1 + k)$, a similar argument shows that

$$\frac{\partial F(x_1, y_1)}{\partial y} = \lim_{k \to 0} \frac{1}{k} \int_{y_1}^{y_1 + k} Q(x_1, y)\, dy = Q(x_1, y_1),$$

and $\partial F/\partial y = Q(x, y)$ in R.

We also have the following theorem (see Problem 18).

Theorem 23.5

The line integral $\int_C P\, dx + Q\, dy$ *is independent of the path in a region R if and only if* $\int_C P\, dx + Q\, dy = 0$ *for every simple closed curve that, together with its inside, lies in R.*

In Section 21.7 we considered the total differential of a function $f(x, y)$:

$$df = \frac{\partial f}{\partial x}\, dx + \frac{\partial f}{\partial y}\, dy.$$

We now see that if $\int_C P\, dx + Q\, dy$ is independent of the path, then $P\, dx + Q\, dy$ is the total differential of a function $F(x, y)$:

$$dF = \frac{\partial F}{\partial x}\, dx + \frac{\partial F}{\partial y}\, dy = P\, dx + Q\, dy.$$

Definition

The differential expression

$$P(x, y)\, dx + Q(x, y)\, dy$$

*is called an **exact differential** (or simply **exact**) if there is a function $F(x, y)$ such that*

$$dF = P\, dx + Q\, dy;$$

that is, if $P\, dx + Q\, dy$ is the total differential of some function $F(x, y)$.

The function $F(x, y)$ is a potential for the vector function $\mathbf{f} = P\mathbf{i} + Q\mathbf{j}.$ Thus $P\, dx + Q\, dy$ is exact if and only if $P\mathbf{i} + Q\mathbf{j}$ has a potential.

In the following theorem we give a simple test for determining when a differential expression is exact, or equivalently, when $P\mathbf{i} + Q\mathbf{j}$ possesses a potential.

Theorem 23.6

If $P(x, y)$ and $Q(x, y)$ have continuous first partial derivatives in a simply connected region R, then the differential expression

$$P(x, y)\, dx + Q(x, y)\, dy$$

is exact if and only if

$$\frac{\partial P}{\partial y} = \frac{\partial Q}{\partial x}$$

throughout R.

Proof

If $P\, dx + Q\, dy$ is exact, then there is a function $F(x, y)$ such that $\partial F/\partial x = P$ and $\partial F/\partial y = Q$. Since $\partial P/\partial y$ and $\partial Q/\partial x$ are continuous, the mixed partial derivatives of F are equal (by Theorem 21.1), and

$$\frac{\partial P}{\partial y} = \frac{\partial^2 F}{\partial y\, \partial x} = \frac{\partial^2 F}{\partial x\, \partial y} = \frac{\partial Q}{\partial x}.$$

Now suppose $\partial P/\partial y = \partial Q/\partial x$ in R. Let C_1 and C_2 be two paths in R from A to B. If C_1 and C_2 intersect only at A and B, then $C = C_1 - C_2$ is a simple closed curve in R. Since R is simply connected, the inside S of C also lies in R, and by Green's theorem,

$$\left| \int_C P\, dx + Q\, dy \right| = \left| \iint_S \left(\frac{\partial Q}{\partial x} - \frac{\partial P}{\partial y} \right) dA \right| = 0,$$

so

$$\int_{C_1} P\,dx + Q\,dy = \int_{C_2} P\,dx + Q\,dy.$$

If C_1 and C_2 intersect in a finite number of points, then this argument can be repeated for each simply connected subregion formed by C_1 and C_2. It can be shown that even if C_1 and C_2 intersect infinitely often, the integral has the same value along C_1 and C_2. Thus the integral is independent of the path, and by Theorem 23.4 the differential expression is exact.

Example 2

Show that $(x^2 - y^2)\,dx - 2xy\,dy$ is exact.

To show exactness, we have

$$P = x^2 - y^2, \qquad Q = -2xy;$$

$$\frac{\partial P}{\partial y} = -2y = \frac{\partial Q}{\partial x},$$

so the expression is exact (see Example 1, Section 23.1).

Example 3

Show that $(ye^{xy} + 3x^2)\,dx + (xe^{xy} - \cos y)\,dy$ is exact, and find the function $F(x, y)$ for which this is the differential.

To show exactness, we have

$$P = ye^{xy} + 3x^2, \qquad Q = xe^{xy} - \cos y;$$

$$\frac{\partial P}{\partial y} = xye^{xy} + e^{xy}, \qquad \frac{\partial Q}{\partial x} = xye^{xy} + e^{xy}.$$

Since

$$\frac{\partial P}{\partial y} = \frac{\partial Q}{\partial x},$$

the given expression is exact. Thus, there is a function $F(x, y)$ such that

$$\frac{\partial F}{\partial x} = P = ye^{xy} + 3x^2 \quad \text{and} \quad \frac{\partial F}{\partial y} = Q = xe^{xy} - \cos y.$$

Let us now integrate the first of these with respect to x, remembering that y is taken to be a constant.

$$F(x, y) = \int (ye^{xy} + 3x^2)\,dx$$

$$= e^{xy} + x^3 + C(y).$$

Note here that, since y is taken to be a constant, the constant of integration is a function of y. To evaluate $C(y)$ we differentiate with respect to y.

$$\frac{\partial F}{\partial y} = xe^{xy} + C'(y) = xe^{xy} - \cos y.$$

Thus

$$C'(y) = -\cos y,$$

$$C(y) = \int (-\cos y)\,dy = -\sin y + K.$$

This now gives

$$F(x, y) = e^{xy} + x^3 - \sin y + K.$$

Note that we have no way of finding K without additional information.

This gives us a method of solving certain types of differential equations. A differential equation of the form

$$P(x, y) \, dx + Q(x, y) \, dy = 0$$

is called *exact* if the left-hand side is an exact differential.

Example 4

Show that the following differential equation is exact and find its general solution (see page 368).

$$(\sin y + 2x) \, dx + (x \cos y - 3) \, dy = 0.$$

To test for exactness, we differentiate.

$$\frac{\partial P}{\partial y} = \cos y, \qquad \frac{\partial Q}{\partial x} = \cos y.$$

Since these are equal, this equation is exact. Thus there is a function $F(x, y)$ such that

$$\frac{\partial F}{\partial x} = \sin y + 2x \quad \text{and} \quad \frac{\partial F}{\partial y} = x \cos y - 3.$$

By integration with respect to x, we have

$$F = \int (\sin y + 2x) \, dx = x \sin y + x^2 + C(y).$$

Differentiating, we obtain

$$\frac{\partial F}{\partial y} = x \cos y + C'(y) = x \cos y - 3.$$

Thus

$$C'(y) = -3, \qquad C(y) = -3y + K.$$

This gives us the general solution

$$x \sin y + x^2 - 3y + K = 0.$$

Let us summarize the results of this section. Theorems 23.3 and 23.4 tell us that $P \, dx + Q \, dy$ is exact and $P\mathbf{i} + Q\mathbf{j}$ has a potential in a region R if and only if $\int_C P \, dx + Q \, dy$ is independent of the path in R. Theorem 23.5 tells us that this in turn is equivalent to $\int_C P \, dx + Q \, dy = 0$ for every simple closed curve that, together with its interior, lies in R. Finally, Theorem 23.6 gives a test for exactness in a simple connected region. Thus, if P and Q are smooth functions defined in a simply connected region R, then the following statements are equivalent.

(a) $P \, dx + Q \, dy$ is an exact differential in R.

(b) $P\mathbf{i} + Q\mathbf{j}$ possesses a potential in R.

(c) $\displaystyle\int_C P \, dx + Q \, dy$ is independent of the path in R.

(d) $\displaystyle\int_C P\,dx + Q\,dy = 0$ for every simple closed curve in R.

(e) $\dfrac{\partial P}{\partial y} = \dfrac{\partial Q}{\partial x}$ in R.

Problems

A

In Problems 1–6, test for exactness. If the differential expression is exact, find the function $F(x, y)$ for which it is the differential.

1. $(y^2 + 3x^2)\,dx + 2xy\,dy.$
2. $e^y\,dx + (xe^y - 2y)\,dy.$
3. $(x^3 \cos xy + 2x \sin xy)\,dx + x^2 y \cos xy\,dy.$
4. $(2xy^3 + e^x)\,dx + (3x^2y^2 - 2y)\,dy.$
5. $\left(\dfrac{1}{x} + 2xy\right)dx + \left(\dfrac{1}{y} + x^2 + 3y^2\right)dy.$
6. $(2x^3y + 3y^2)\,dx + (3x^2y^2 + 6x)\,dy.$

In Problems 7–10, test the given differential equation for exactness. If it is exact, find the general solution.

7. $(8xy + 3)\,dx + 4(x^2 + y)\,dy = 0.$
8. $(3x^2y + 2x)\,dx + (2 - x^3)\,dy = 0.$
9. $(xy^2e^{xy^2} + e^{xy^2} + 2y)\,dx + (2x^2ye^{xy^2} + 2x - 12y^2)\,dy = 0.$
10. $y \sec^2 xy - 12x^2 + (x \sec^2 xy)y' = 0.$

B

In Problems 11 and 12, test for exactness. If the differential expression is exact, find the function $F(x, y)$ for which it is the differential.

11. $(xye^{xy} + e^{xy} - 4x)\,dx + (x^2e^{xy} - \sin y)\,dy.$
12. $(x \cos y - 2 \cos x)\,dx + (\sin y + 2y \sin x)\,dy.$

13. Test $(ye^{xy} + 4xy + 2x - 2)\,dx + (xe^{xy} + 2x^2 + 1)\,dy = 0$ for exactness. If it is exact, find the general solution.

14. Show that $(3x + 2y)\,dx + x\,dy = 0$ is not an exact differential equation but, if we multiply through by x, the resulting differential equation is exact (x is called an *integrating factor* for the given differential equation). Find the general solution.

15. Show that $1/y^2$ is an integrating factor (see Problem 14) of

$$(2xy^2 + y)\,dx + (y^2 - x)\,dy = 0.$$

Find the general solution.

In Problems 16 and 17, find an integrating factor and solve.

16. $y\,dx - x\,dy = 0.$
17. $\sin y\,dx + x \sec y\,dy = 0.$

C

18. Show that if $\displaystyle\int_C P\,dx + Q\,dy$ is independent of the path in a region R, then $\displaystyle\int_C P\,dx + Q\,dy = 0$ for each simple closed curve C in R.

19. Evaluate $\displaystyle\int_C (x^2 + 2xy)\,dx + (x^2 + \cos y)\,dy,$ $C\colon\ y = \sin x^2,\ \ 0 \le x \le \sqrt{\pi}.$

20. Evaluate $\displaystyle\int_C e^x \sin y\,dx + (y^3 + e^x \cos y)\,dy$ around the rectangle with vertices $(1, 1)$, $(-1, 1), (-1, -1),$ and $(1, -1).$

23.4

Work; Vector Fields

Suppose a smooth curve C is given in vector form by $\mathbf{r}(t) = x(t)\mathbf{i} + y(t)\mathbf{j}$, $a \le t \le b$. A tangent vector to C is given by $\mathbf{r}'(t) = x'(t)\mathbf{i} + y'(t)\mathbf{j}$, and the unit tangent vector to C in the direction of increasing t is $\mathbf{T} = \mathbf{r}'(t)/|\mathbf{r}'(t)|$ (see Section 20.11). Then

$$\mathbf{T} = \frac{(dx/dt)\mathbf{i} + (dy/dt)\mathbf{j}}{ds/dt} = \frac{dx}{ds}\mathbf{i} + \frac{dy}{ds}\mathbf{j},$$

where the differential of arc length

$$ds = \sqrt{\left(\frac{dx}{dt}\right)^2 + \left(\frac{dy}{dt}\right)^2}\, dt.$$

Thus, the differential expression $\mathbf{T}\, ds = dx\mathbf{i} + dy\mathbf{j} = d\mathbf{r}$. If $\mathbf{f} = P\mathbf{i} + Q\mathbf{j}$ is a vector-valued function defined at each point of C, then

$$\int_C \mathbf{f} \cdot \mathbf{T}\, ds = \int_C \mathbf{f} \cdot d\mathbf{r} = \int_C P\, dx + Q\, dy.$$

As an application, suppose $\mathbf{f}$ is a variable force defined at each point of C. Then $\mathbf{f} \cdot \mathbf{T}$ is the tangential component of $\mathbf{f}$ in the direction of C, and the net work done by the force as a particle moves from $\mathbf{r}(a)$ to $\mathbf{r}(b)$ along C is $\int_C \mathbf{f} \cdot \mathbf{T}\, ds = \int_C \mathbf{f} \cdot d\mathbf{r}$. Thus

$$W = \int_C \mathbf{f} \cdot d\mathbf{r} = \int_C P\, dx + Q\, dy.$$

Example 1

Calculate the net work done in moving a particle from $(0, 0)$ to $(1, 1)$ along $y = x^2$ if the force at each point is $\mathbf{f}(x, y) = (2x + y)\mathbf{i} + (x - 3y)\mathbf{j}$. Assume force is measured in pounds and distance in feet.

Let C have vector representation

$$\mathbf{r}(x) = x\mathbf{i} + x^2\mathbf{j}, \quad 0 \le x \le 1.$$

Then

$$\begin{aligned}
W &= \int_0^1 [(2x + x^2) \cdot 1 + (x - 3x^2) \cdot 2x]\, dx \\
&= \int_0^1 (2x + 3x^2 - 6x^3)\, dx \\
&= x^2 + x^3 - \frac{3}{2}x^4 \Big|_0^1 = \frac{1}{2} \text{ ft-lb.}
\end{aligned}$$

If the unit tangent vector to a curve C is

$$\mathbf{T} = \frac{dx}{ds}\mathbf{i} + \frac{dy}{ds}\mathbf{j},$$

then the right-hand unit normal vector to C is

$$\mathbf{n} = \frac{dy}{ds}\mathbf{i} - \frac{dx}{ds}\mathbf{j}$$

(see Figure 23.10). Thus if $\mathbf{f} = P\mathbf{i} + Q\mathbf{j},$ then

$$\int_C \mathbf{f} \cdot \mathbf{n} \, ds = \int_C \left(P \frac{dy}{ds} - Q \frac{dx}{ds} \right) ds = \int_C P \, dy - Q \, dx.$$

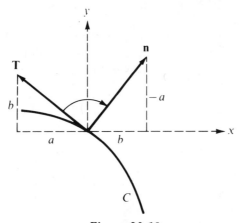

Figure 23.10

Example 2

Find the work done when a particle is moved along $y = 2 + x^2$ from $(0, 2)$ to $(1, 3)$ against a force that is directed toward the origin and is proportional to the distance from the origin, if μ is the coefficient of friction between the particle and the path.

The central force is $\mathbf{f} = -k\mathbf{r} = -kx\mathbf{i} - ky\mathbf{j}.$ The work done by the tangential component of the force is $\int_C \mathbf{f} \cdot \mathbf{T} \, ds = -k \int_C x \, dx + y \, dy.$ Since $d(x^2 + y^2) = 2x \, dx + 2y \, dy,$ the integrand is exact and depends only on the end points:

$$-k \int_C x \, dx + y \, dy = -\frac{k}{2} (x^2 + y^2) \Big|_{(0, 2)}^{(1, 3)} = -3k.$$

The frictional force $\mu \, \mathbf{f} \cdot \mathbf{n}$ is proportional to the normal component of the central force acting to press the particle against the path. Since the frictional force always opposes the motion, the work done by friction is

$$\int_C |\mu \mathbf{f} \cdot \mathbf{n}| \, ds = \mu k \int_C |y \, dx - x \, dy|$$

$$= \mu k \int_0^1 |-(2 + x^2) \cdot 1 + x(2x)| \, dx$$

$$= \mu k \int_0^1 (2 - x^2) \, dx = \frac{5}{3} \mu k.$$

The total work done is

$$W = -3k + \frac{5}{3}\,\mu k.$$

The term $-3k$ represents work stored as potential energy (hence recoverable); the term $5\mu k/3$ is work dissipated through friction (hence not recoverable).

A vector-valued function $\mathbf{f}$ defined in a region R associates with each point in the region a vector; in this case, $\mathbf{f}$ is often called a *vector field*. Examples of vector fields are gravitational and electromagnetic fields, where at each point there is defined a force of attraction or repulsion, and velocity fields, where at each point a velocity vector is defined. Suppose $\mathbf{f} = P(x,\,y)\mathbf{i} + Q(x,\,y)\mathbf{j}$, where P and Q are smooth in R. If $\mathbf{f}$ has a potential $F(x,\,y)$ so that the gradient of $F = \mathbf{f}$ (or $\nabla F = \mathbf{f}$), then the vector field is said to be *conservative*. The conditions for a vector field to be conservative are precisely the conditions that $P\,dx + Q\,dy$ be exact; for example, $\partial Q/\partial x = \partial P/\partial y$ in R.

Now suppose that $\mathbf{f}$ defines a conservative force field in a region R with potential $-F(x,\,y)$ (that is, $\nabla(-F) = \mathbf{f}$; the minus sign is introduced for physical reasons), and suppose a particle of constant mass m moves along a smooth curve $\mathbf{r}(t)$ in R. By Newton's second law,

$$m\mathbf{r}''(t) = \mathbf{f}(x(t),\,y(t)) = -\nabla F(x(t),\,y(t)).$$

Thus

$$m\mathbf{r}' \cdot \mathbf{r}'' + \nabla F \cdot \mathbf{r}' = 0.$$

Since $(v^2)' = (\mathbf{r}' \cdot \mathbf{r}')' = 2\mathbf{r}' \cdot \mathbf{r}''$ and $[F(x(t),\,y(t))]' = \nabla F \cdot \mathbf{r}'(t)$ by the chain rule, an integration gives

$$\frac{1}{2}\,mv^2 + F = \text{constant}.$$

This is the conservation law: The sum of the kinetic energy and the potential energy is a constant.

Example 3

Show that if a force at each point is inversely proportional to the square of the distance of that point to the origin, then the force field is conservative, and find a potential function for the field.

$$\mathbf{f}(x,\,y) = \frac{k}{r^2} \cdot \frac{\mathbf{r}}{r}, \quad \text{where} \quad \mathbf{r} = x\mathbf{i} + y\mathbf{j} \quad \text{and } r = |r|.$$

Thus

$$\mathbf{f}(x,\,y) = \frac{k}{r^3}\,\mathbf{r} = \frac{kx}{(x^2 + y^2)^{3/2}}\,\mathbf{i} + \frac{ky}{(x^2 + y^2)^{3/2}}\,\mathbf{j} = P\mathbf{i} + Q\mathbf{j}.$$

Since $P_y = -(3/2)\,kx \cdot 2y(x^2 + y^2)^{-5/2}$ and $Q_x = -(3/2)\,ky \cdot 2x(x^2 + y^2)^{-5/2}$, $P_y = Q_x$. Thus the field is conservative (here R is the plane minus the origin). A potential is $F(x,\,y) = -k/r = -k/(x^2 + y^2)^{1/2}$.

Suppose now that $\mathbf{f} = P\mathbf{i} + Q\mathbf{j}$ defines a velocity vector field of a fluid flow in a region R and C is a curve in R. At each point on C, $\mathbf{f} \cdot \mathbf{T}$ is the projection of the velocity $\mathbf{f}$ in the direction of the tangent vector and, if C is closed, then

$$\int_C \mathbf{f} \cdot \mathbf{T}\, ds = \int_C P\, dx + Q\, dy$$

is called the *circulation* of the velocity field f around C. On other hand, at each point on C, $\mathbf{f} \cdot \mathbf{n}$ is the projection of the velocity $\mathbf{f}$ in the direction of the normal, and

$$\int_C \mathbf{f} \cdot \mathbf{n}\, ds = \int_C P\, dy - Q\, dx$$

represents the rate of flow, or *flux*, of the fluid across C in the direction of $\mathbf{n}$ (see Figure 23.11).

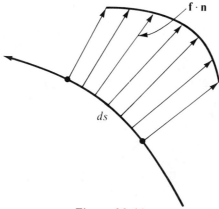

Figure 23.11

For the vector function $\mathbf{f} = P\mathbf{i} + Q\mathbf{j}$,   the *divergence* of $\mathbf{f}$ is defined to be

$$\operatorname{div} \mathbf{f} = \frac{\partial P}{\partial x} + \frac{\partial Q}{\partial y}.$$

The following theorem is a restatement of Green's theorem using vector language.

Theorem 23.7

(The divergence theorem in the plane)   Let R be a bounded region with complete boundary C oriented in the positive sense, and suppose $\mathbf{f}$ is a vector field defined on R and its boundary. Then

$$\int_C \mathbf{f} \cdot \mathbf{n}\, ds = \iint_R \operatorname{div} \mathbf{f}\, dA.$$

Example 4

Verify the divergence theorem for R the interior of $x^2 + y^2 = 1$ and $\mathbf{f} = 2x\mathbf{i} + 3y\mathbf{j}$.

$$\operatorname{div} \mathbf{f} = \frac{\partial P}{\partial x} + \frac{\partial Q}{\partial y} = 2 + 3 = 5.$$

$$\iint_R \operatorname{div} \mathbf{f}\, dA = 5 \cdot \text{area of circle} = 5\pi.$$

On $x^2 + y^2 = 1$ let $x = \cos t$, $y = \sin t$, $0 \le t \le 2\pi$. Then

$$\int_C P\, dy - Q\, dx = \int_0^{2\pi} [2\cos t \cdot \cos t - 3 \sin t(-\sin t)]\, dt$$

$$= \int_0^{2\pi} (2\cos^2 t + 3 \sin^2 t)\, dt$$

$$= 2\left(\frac{t}{2} + \frac{1}{4} \sin 2t\right) + 3\left(\frac{t}{2} - \frac{1}{4} \sin 2t\right)\Big|_0^{2\pi} = 5\pi.$$

Problems

A

In Problems 1–8, find the net work done in moving a particle along the given path under the given forces.

1. Along $y = 4 - x^2$, $0 \le x \le 2$; $\mathbf{f} = (4 - x)\mathbf{i} + (x + 2y)\mathbf{j}$.
2. Along $x = y^2$ from $(0, 0)$ to $(4, 2)$; $\mathbf{f} = (3x + 5y)\mathbf{i} + (2x - y)\mathbf{j}$.
3. Along $x = 3 \cos t$, $y = 3 \sin t$, $0 \le t \le \pi$; $\mathbf{f} = -\sin t\mathbf{i} + \cos t\mathbf{j}$.
4. Around $x^2 + y^2 = 4$ in the positive sense; $\mathbf{f} = xy\mathbf{i} - 2y\mathbf{j}$.
5. Along $y = e^x$, $0 \le x \le 2$; $\mathbf{f} = x^2\mathbf{i} + (e^{-x} + y)\mathbf{j}$.
6. Along $y = 2x^2$ from $(0, 0)$ to $(1, 2)$; $\mathbf{f} = 2xy\mathbf{i} - 4y\mathbf{j}$.
7. Along $x + y = 4$ from $(4, 0)$ to $(0, 4)$; $\mathbf{f} = x^2\mathbf{i} + xy\mathbf{j}$.
8. Along $2x - 3y = 0$ from $(0, 0)$ to $(6, 4)$; $\mathbf{f} = 2y\mathbf{i} - 3x\mathbf{j}$.

In Problems 9 and 10, use the divergence theorem to evaluate the integral.

9. $\displaystyle\int_C (x + y^2)\, dx + (2y - x^2)\, dy$; C the circle $x^2 + y^2 = 4$ in the positive sense.

10. $\displaystyle\int_C (4x + y)\, dx + (3x - 2y)\, dy$; C the complete boundary of the annulus $1 \le x^2 + y^2 \le 4$ in the positive sense.

In Problems 11 and 12, show that the vector field determined by $\mathbf{f}$ is conservative, and find a potential.

11. $\mathbf{f} = 3x\mathbf{i} + 2y\mathbf{j}$.   12. $\mathbf{f} = ye^x\mathbf{i} + e^x\mathbf{j}$.

B

In Problems 13–15, find the net work done in moving a particle along the given path under the given forces.

13. Along $x^2 + y^2 = 4$ from $(\sqrt{3}, 1)$ to $(\sqrt{2}, \sqrt{2})$; $\mathbf{f} = (1/x)\mathbf{i} + (1/y)\mathbf{j}$.
14. Along $y = 1 - (1/x)$ from $(1, 0)$ to $(4, 3/4)$; $\mathbf{f}$ directed toward $(0, 1)$ and proportional to the distance from $(0, 1)$, coefficient of friction μ.
15. Along $y = e^{-x}$, $0 \le x < \infty$; $\mathbf{f} = xy\mathbf{i} - y\mathbf{j}$, coefficient of friction $\mu = 0.1$.

In Problems 16–18, let $\mathbf{r} = x\mathbf{i} + y\mathbf{j}$ and $r = \sqrt{x^2 + y^2}$.

16. Show that

$$\nabla r^{-m} = -\frac{m}{r^{m+2}}\mathbf{r}, \quad m \text{ an integer} \ge 1.$$

17. Show that the vector field determined by a force that is always directed toward the origin and has magnitude inversely proportional to the kth power of the distance from the origin $(k \ge 1)$ is conservative, and find a potential function.

18. Suppose $g(r)$ has a continuous derivative for all $r \neq 0$. Show that

$$\mathbf{f} = \frac{g'(r)}{r}\mathbf{r}$$

defines a conservative force field in the plane minus the origin, and find a potential function.

C

19. Show from Newton's second law that the work done in moving a particle of mass m without friction along a curve $x = x(t)$, $y = y(t)$ from $t = a$ to $t = b$ represents the net change in kinetic energy of the mass.

23.5

Line Integrals in Space; Surface Integrals

The definitions and basic properties of line integrals extend readily to three dimensions.

A space curve C: $\mathbf{r}(t) = x(t)\mathbf{i} + y(t)\mathbf{j} + z(t)\mathbf{k}$, $a \leq t \leq b$, is *smooth* if $x'(t)$, $y'(t)$, and $z'(t)$ are continuous on $[a, b]$. If $\mathbf{f}(x, y, z) = P\mathbf{i} + Q\mathbf{j} + R\mathbf{k}$ and each of $P(x, y, z)$, $Q(x, y, z)$, and $R(x, y, z)$ is continuous on C, then we define the line integral

$$\int_C P \, dx + Q \, dy + R \, dz$$

$$= \int_a^b [P(x(t), y(t), z(t))x'(t) + Q(x(t), y(t), z(t))y'(t) + R(x(t), y(t), z(t))z'(t)] \, dt$$

$$= \int_a^b \mathbf{f} \cdot \mathbf{r}'(t) \, dt = \int_a^b \mathbf{f} \cdot d\mathbf{r}.$$

Suppose $\mathbf{f} = P\mathbf{i} + Q\mathbf{j} + R\mathbf{k}$ and P, Q, and R are smooth functions in a region V in space. The expression

$$\mathbf{f} \cdot d\mathbf{r} = P \, dx + Q \, dy + R \, dz$$

is an *exact differential* if there is a function $F(x, y, z)$ such that $\partial F / \partial x = P$, $\partial F / \partial y = Q$, and $\partial F / \partial z = R$; that is, $P \, dx + Q \, dy + R \, dz$ is the total differential of F. In this case, $F(x, y, z)$ is called a *potential* for the vector field defined by $\mathbf{f}$.

The following condition for exactness can be proved (see Problem 22).

Theorem 23.8

Suppose $P(x, y, z)$, $Q(x, y, z)$, and $R(x, y, z)$ are smooth in a region V of space. If $P \, dx + Q \, dy + R \, dz$ is exact, then

$$\frac{\partial R}{\partial y} = \frac{\partial Q}{\partial z}, \quad \frac{\partial P}{\partial z} = \frac{\partial R}{\partial x}, \quad and \quad \frac{\partial Q}{\partial x} = \frac{\partial P}{\partial y}.$$

These conditions are also sufficient for the exactness of $P\,dx + Q\,dy + R\,dz$ if V, for example, is the interior of a rectangular parallelepiped.

A modification of the proofs of Theorems 23.4 and 23.6 yields the following theorem.

Theorem 23.9

Let $\mathbf{f} = P\mathbf{i} + Q\mathbf{j} + R\mathbf{k}$, where $P(x, y, z)$, $Q(x, y, z)$, and $R(x, y, z)$ are smooth in a region V. Then $\int_C \mathbf{f} \cdot d\mathbf{r}$ is independent of the path in V if and only if $\mathbf{f} \cdot d\mathbf{r}$ is exact.

Example 1

Show that $(2xy + z^2)\,dx + (x^2 + 6yz^2)\,dy + (2xz + 6y^2z)\,dz$ is exact, and find a potential.

$$P = 2xy + z^2, \quad Q = x^2 + 6yz^2, \quad R = 2xz + 6y^2z;$$
$$P_y = 2x = Q_x, \quad Q_z = 12yz = R_y, \quad R_x = 2z = P_z;$$

so the differential expression is exact.

Let

$$F(x, y, z) = \int_0^x P(t, y, z)\,dt$$

$$= \int_0^x (2ty + z^2)\,dt = x^2y + xz^2 + g(y, z).$$

$$\frac{\partial F}{\partial y} = x^2 + \frac{\partial g}{\partial y} = Q = x^2 + 6yz^2,$$

$$\frac{\partial g}{\partial y} = 6yz^2,$$

$$g(y, z) = \int_0^y 6tz^2\,dt = 3y^2z^2 + h(z).$$

Thus

$$F(x, y, z) = x^2y + xz^2 + 3y^2z^2 + h(z).$$

$$\frac{\partial F}{\partial z} = 2xz + 6y^2z + h'(z) = R = 2xz + 6y^2z,$$

so $h'(z) = 0$ and $h(z) = K$. Then $F(x, y, z) = x^2y + xz^2 + 3y^2z^2 + K$.

To extend Green's theorem to space, we need to define a surface integral; and to do this, we need the element of surface area. If a plane rectangular area S makes an angle γ with the horizontal and projects down upon an area A, then $S = A \sec \gamma$ (see Figure 23.12). Now suppose $z = f(x, y)$ is a surface S lying over a region R in the xy plane. Let dA be an element of area of R and dS be the element of surface area that projects upon dA (see Figure 23.13). If we approximate dS by the part of the tangent plane to S at a point of dS that projects onto dA, then

$$dS \doteq \sec \gamma\, dA,$$

where γ is the angle between the upward pointing normal $\mathbf{n}$ to S and the positive z direction $\mathbf{k}$. But this normal has direction numbers $\{-f_x, -f_y, 1\}$ and $\mathbf{k} = (0, 0, 1)$,

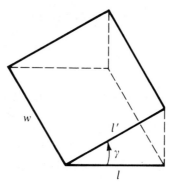

$$S = wl' = wl \sec \gamma = A \sec \gamma$$

Figure 23.12

so

$$\cos \gamma = \frac{1}{\sqrt{f_x^2 + f_y^2 + 1}}.$$

Thus

$$dS = \sec \gamma \, dA = \sqrt{f_x^2 + f_y^2 + 1} \, dA,$$

and the area of the surface is

$$S = \iint_R \sqrt{f_x^2 + f_y^2 + 1} \, dA.$$

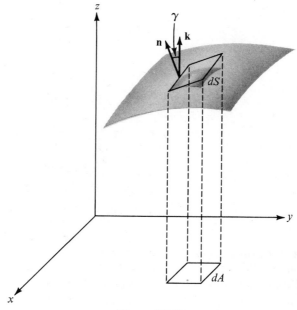

Figure 23.13

Similar formulas can be written for surfaces that project into the yz plane or the xz plane.

If $F(x, y, z)$ is continuous on the surface $z = f(x, y)$, then the surface integral of F is

$$\iint_S F(x, y, z) \, dS = \iint_R F(x, y, f(x, y)) \sec \gamma \, dA.$$

Example 2

Find the area of $z = 1 - y^2$ that is bounded by the planes $z = 0$, $x = 0$, $x = y$, and $y = 1$ (see Figure 23.14).

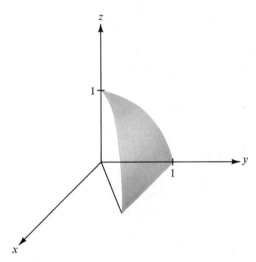

Figure 23.14

$$S = \int_0^1 \int_0^y \sqrt{0^2 + (2y)^2 + 1} \, dx \, dy$$

$$= \frac{1}{8} \int_0^1 8y \sqrt{1 + 4y^2} \, dy$$

$$= \frac{1}{12} (1 + 4y^2)^{3/2} \Big|_0^1$$

$$= \frac{1}{12} (5\sqrt{5} - 1).$$

An application of the surface integral is given by the flux across a surface immersed in a fluid with velocity field $\mathbf{f}(x, y, z) = P\mathbf{i} + Q\mathbf{j} + R\mathbf{k}$. For an element of surface area dS, $\mathbf{f} \cdot \mathbf{n} \, dS$ represents the rate of flow across the element dS, where $\mathbf{n}$ is the upward unit normal. Thus, $\iint_S \mathbf{f} \cdot \mathbf{n} \, dS$ is the flux across the surface S in the direction of the normal $\mathbf{n}$ (see Figure 23.15).

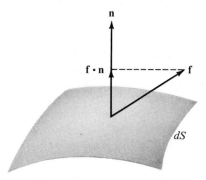

Figure 23.15

Example 3

Find the flux across the part of $z = 1 - y^2$ that is bounded by $x = 0$, $x = 1$, $y = 0$, and $y = 1$, if the velocity is $\mathbf{f} = x\mathbf{i} + \mathbf{j}$.

$$\text{Flux} = \iint_S \mathbf{f} \cdot \mathbf{n} \, dS$$

$$= \iint_R \frac{1}{\sqrt{1 + 4y^2}} (x\mathbf{i} + \mathbf{j} + 0\mathbf{k}) \cdot (0\mathbf{i} + 2y\mathbf{j} + \mathbf{k}) \sqrt{1 + 4y^2} \, dA$$

$$= \int_0^1 \int_0^1 2y \, dy \, dx = 1.$$

Note that

$$\mathbf{n} = \frac{1}{\sqrt{f_x^2 + f_y^2 + 1}} (-f_x\mathbf{i} - f_y\mathbf{j} + \mathbf{k}).$$

Problems

A

In Problems 1–4, evaluate the line integrals.

1. $\displaystyle\int_C x^2 \, dx + xz \, dy + yz \, dz;$ $C:$ $\mathbf{r}(t) = t\mathbf{i} + (2 - t)\mathbf{j} + t^2\mathbf{k},$ $0 \le t \le 1.$

2. $\displaystyle\int_C yz \, dx + xz \, dy + xy \, dz;$ $C:$ $\mathbf{r}(t) = t\mathbf{i} + t^2\mathbf{j} + t^3\mathbf{k},$ $0 \le t \le 1.$

3. $\displaystyle\int_C \mathbf{f} \cdot \mathbf{r}'(t) \, dt,$ where $\mathbf{f} = 2y\mathbf{i} + 3x\mathbf{j} + z\mathbf{k}$ and $\mathbf{r}(t) = 3t\mathbf{i} - 2t\mathbf{j} + 5t\mathbf{k},$ $0 \le t \le 1.$

4. $\displaystyle\int_C \mathbf{f} \cdot \mathbf{r}'(t) \, dt.$ where $\mathbf{f} = x\mathbf{i} + yz\mathbf{j} + xz\mathbf{k}$ and $\mathbf{r}(t) = t^2\mathbf{i} - \mathbf{j} + (2 - t)\mathbf{k},$ $0 \le t \le 2.$

In Problems 5–8, test for exactness; if the expression is exact, find a potential.

5. $(2xy + z^2) \, dx + (x^2 - 2) \, dy + (2xz + 1) \, dz.$
6. $yz \, dx + xz \, dy + xy \, dz.$
7. $(e^x + 2xy) \, dx + (x^2 - \sin y) \, dy + 4z^3 \, dz.$
8. $(y - 3z + 2z^2) \, dx + (x + 2z) \, dy + (2y - 3x + 4xz) \, dz.$

In Problems 9–14, use double integration to find the area of the given surface.

9. The part of $x + y + z = 4$ in the first octant.
10. The part of $y^2 + z^2 = 1$ in the first octant and between $x = 0$ and $x = 5$.
11. The part of $z = x^2 + y^2$ in the first octant bounded by $x = 0$, $y = 0$, and $x^2 + y^2 = 4$. (*Hint:* Use polar coordinates.)
12. The part of $z = x^2$ bounded by $x = 1$, $y = 0$, and $y = x$.
13. The part of $x^2 + z^2 = a^2$ in the first octant and bounded by $y = 2z$ and $4x^2 + y^2 = 4a^2$. (*Hint:* Project the surface upon the yz plane.)
14. The part of $x^2 = 4y$ in the first octant and bounded by $2x + 2z = 3$. (*Hint:* Project the surface upon the xz plane.)

In Problems 15 and 16, find the centroid and the moment of inertia about the z axis of the given surface using double integrals.

15. $z = x^2 + y^2$ between $z = 0$ and $z = 4$.
16. The conical surface $4x^2 + 4y^2 - z^2 = 0$ between $z = 0$ and $z = 4$.
17. Find the flux across the first octant part of $x + y + z = 4$ if $\mathbf{f} = 2x\mathbf{i} - 3y\mathbf{j} + z\mathbf{k}$.
18. Find the flux across the first octant part of $x + 2y + 6z = 12$ if $\mathbf{f} = x\mathbf{i} - 4z\mathbf{j} + 3y\mathbf{k}$.

B

19. Find the flux across the part of $z = xy$ bounded by $x = 0$, $x = 2$, $y = 0$, and $y = 1$ if $\mathbf{f} = -xy^2\mathbf{i} + z\mathbf{j}$.
20. Find the flux across $z = x^2 + y^2$ between $z = 0$ and $z = 4$ if $\mathbf{f} = x\mathbf{i} + y\mathbf{j} + z\mathbf{k}$.
21. Evaluate $\displaystyle\int_C z^2\, dx - 2xz\, dy + x^2\, dz$ around the triangle with vertices $(0, 0, 2)$, $(2, 0, 0)$, and $(1, 1, 1)$.
22. Show that if $P\, dx + Q\, dy + R\, dz$ is exact, then

$$\frac{\partial Q}{\partial z} = \frac{\partial R}{\partial y}, \quad \frac{\partial R}{\partial x} = \frac{\partial P}{\partial z}, \quad \frac{\partial P}{\partial y} = \frac{\partial Q}{\partial x}.$$

23.6

Divergence and Curl; Stokes' Theorem

The integral theorems in space involve the *divergence* and *curl* of a vector-valued function, and these in turn can be expressed in terms of an operator associated with the gradient. Recall that the gradient of a function is

$$\nabla F(x, y, z) = \frac{\partial F}{\partial x}\mathbf{i} + \frac{\partial F}{\partial y}\mathbf{j} + \frac{\partial F}{\partial z}\mathbf{k}.$$

This is often written in the operator form

$$\nabla F = \left(\mathbf{i}\frac{\partial}{\partial x} + \mathbf{j}\frac{\partial}{\partial y} + \mathbf{k}\frac{\partial}{\partial z}\right)F(x, y, z) = \frac{\partial F}{\partial x}\mathbf{i} + \frac{\partial F}{\partial y}\mathbf{j} + \frac{\partial F}{\partial z}\mathbf{k},$$

where

$$\nabla = \mathbf{i}\frac{\partial}{\partial x} + \mathbf{j}\frac{\partial}{\partial y} + \mathbf{k}\frac{\partial}{\partial z}.$$

If we take the dot product of this operator with a vector function $f(x, y, z) = P\mathbf{i} + Q\mathbf{j} + R\mathbf{k}$, we obtain the *divergence* of $\mathbf{f}$,

$$\text{div } \mathbf{f} = \nabla \cdot \mathbf{f}$$

$$= \left(\mathbf{i}\frac{\partial}{\partial x} + \mathbf{j}\frac{\partial}{\partial y} + \mathbf{k}\frac{\partial}{\partial z}\right) \cdot (P\mathbf{i} + Q\mathbf{j} + R\mathbf{k})$$

$$= \frac{\partial P}{\partial x} + \frac{\partial Q}{\partial y} + \frac{\partial R}{\partial z}.$$

If we take the cross product of ∇ with $\mathbf{f}$, we get a vector called the *curl* of $\mathbf{f}$:

$$\text{curl } \mathbf{f} = \nabla \times \mathbf{f} = \left(\mathbf{i}\frac{\partial}{\partial x} + \mathbf{j}\frac{\partial}{\partial y} + \mathbf{k}\frac{\partial}{\partial z}\right) \times (P\mathbf{i} + Q\mathbf{j} + R\mathbf{k})$$

$$= \begin{vmatrix} \mathbf{i} & \mathbf{j} & \mathbf{k} \\ \dfrac{\partial}{\partial x} & \dfrac{\partial}{\partial y} & \dfrac{\partial}{\partial z} \\ P & Q & R \end{vmatrix}$$

$$= \left(\frac{\partial R}{\partial y} - \frac{\partial Q}{\partial z}\right)\mathbf{i} + \left(\frac{\partial P}{\partial z} - \frac{\partial R}{\partial x}\right)\mathbf{j} + \left(\frac{\partial Q}{\partial x} - \frac{\partial P}{\partial y}\right)\mathbf{k}.$$

Example 1

Find the divergence and curl of $\mathbf{f} = 3x^2\mathbf{i} + yz\mathbf{j} + xz^2\mathbf{k}$.

$$P = 3x^2, \quad Q = yz, \quad R = xz^2.$$

$$\text{div } \mathbf{f} = \frac{\partial P}{\partial x} + \frac{\partial Q}{\partial y} + \frac{\partial R}{\partial z} = 6x + z + 2xz.$$

$$\nabla x \mathbf{f} = (0 - y)\mathbf{i} + (0 - z^2)\mathbf{j} + (0 - 0)\mathbf{k} = -y\mathbf{i} - z^2\mathbf{j}.$$

If $\mathbf{f} = P(x, y)\mathbf{i} + Q(x, y)\mathbf{j} + 0\mathbf{k}$ is defined in a region R of the xy plane with boundary C, then Green's theorem may be written

$$\int_C \mathbf{f} \cdot \mathbf{T} \, ds = \iint_R (\nabla \times \mathbf{f}) \cdot \mathbf{k} \, dA.$$

Here, $\mathbf{k}$ is the upward unit normal vector to the region R. A generalization of this to space is Stokes' theorem.

Theorem 23.10

(Stokes' theorem)   Let $\mathbf{f} = P\mathbf{i} + Q\mathbf{j} + R\mathbf{k}$,   *where P, Q, and R are smooth in a bounded region that contains a surface S:* $z = f(x, y)$ *with boundary C traversed in the positive sense. Then*

$$\int_C \mathbf{f} \cdot \mathbf{T} \, ds = \iint_S \text{curl } \mathbf{f} \cdot \mathbf{n} \, dS,$$

where $\mathbf{T}$ *is the unit tangent vector to* C *and* $\mathbf{n}$ *is the upward normal to* S. *In component form,*

$$\int_C P \, dx + Q \, dy + R \, dz$$

$$= \iint_S [(R_y - Q_z) \cos \alpha + (P_z - R_x) \cos \beta + (Q_x - P_y) \cos \gamma] \, dS$$

$$= \iint_S [(R_y - Q_z) \, dy \, dz + (P_z - R_x) \, dx \, dz + (Q_x - P_y) \, dx \, dy.$$

The proof of this theorem and its extension to more general orientable ("two-sided") surfaces is found in advanced texts.

Note from Theorem 23.10 that if $\mathbf{f}$ has a potential, or equivalently, if $P \, dx + Q \, dy + R \, dz$ is exact, then $\nabla \times \mathbf{f} = \mathbf{0}$ and $\int_C \mathbf{f} \cdot \mathbf{T} \, ds = 0$.

Example 2

Verify Stokes' theorem for the surface $z = 1 - y^2$, $0 \le x \le 1$, $0 \le y \le 1$, if $\mathbf{f} = (x - y)\mathbf{i} + (y - x)\mathbf{j} + 2z\mathbf{k}$.

$$\nabla \times \mathbf{f} = \begin{vmatrix} \mathbf{i} & \mathbf{j} & \mathbf{k} \\ \dfrac{\partial}{\partial x} & \dfrac{\partial}{\partial y} & \dfrac{\partial}{\partial z} \\ x - y & y - x & 2z \end{vmatrix} = \mathbf{0}.$$

Thus $\iint_S \nabla \times \mathbf{f} \cdot \mathbf{n} \, dS = 0$.

On C_1 (see Figure 23.16), $x = t$, $y = 0$, $z = 1$, $0 \le t \le 1$, so $\mathbf{T} \, ds = d\mathbf{r} = \mathbf{i} \, dt$ and

$$\int_{C_1} \mathbf{f} \cdot \mathbf{T} \, ds = \int_{C_1} \mathbf{f} \cdot d\mathbf{r} = \int_0^1 t \, dt = \frac{1}{2}.$$

On C_2, $x = 1$, $y = t$, $z = 1 - t^2$, $0 \le t \le 1$, so $d\mathbf{r} = (0\mathbf{i} + 1\mathbf{j} - 2t\mathbf{k}) \, dt$ and

$$\int_{C_2} \mathbf{f} \cdot \mathbf{T} \, ds = \int_0^1 [(t - 1) + 2(1 - t^2)(-2t)] \, dt$$

$$= \int_0^1 (-3t - 1 + 4t^3) \, dt = -\frac{3}{2}.$$

On C_3, $x = 1 - t$, $y = 1$, $z = 0$, $0 \le t \le 1$ (note the orientation of C_3), so $d\mathbf{r} = (-1\mathbf{i} + 0\mathbf{j} + 0\mathbf{k}) \, dt$ and

$$\int_{C_3} \mathbf{f} \cdot \mathbf{T} \, ds = \int_0^1 (1 - t - 1)(-1) \, dt = \frac{1}{2}.$$

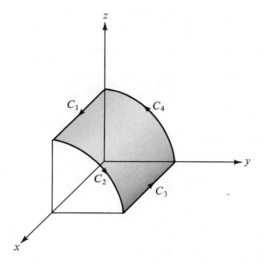

Figure 23.16

On C_4, $x = 0$, $y = 1 - t$, $z = 2t - t^2$, $0 \le t \le 1$, so $d\mathbf{r} = (0\mathbf{i} - \mathbf{j} + 2(1 - t)\mathbf{k}) \, dt$
and

$$\int_{C_4} \mathbf{f} \cdot \mathbf{T} \, ds = \int_0^1 \left[(t - 1) + 2(2t - t^2)2(1 - t) \right] dt$$

$$= \int_0^1 (-1 + 9t - 12t^2 + 4t^3) \, dt = -t + \frac{9}{2}t^2 - 4t^3 + t^4 \Big|_0^1 = \frac{1}{2}.$$

Thus

$$\int_C \mathbf{f} \cdot \mathbf{T} \, ds = \int_{C_1 + C_2 + C_3 + C_4} \mathbf{f} \cdot \mathbf{T} \, ds = 0.$$

Example 3

Evaluate the line integral

$$\int_C y^2 z^2 \, dx + 2xyz^2 \, dy + 2xy^2z \, dz$$

along the curve C: $\mathbf{r}(t) = \cos t\mathbf{i} + \sin t\mathbf{j} + \sin t\mathbf{k}$, $0 \le t \le 2\pi$.

Let $\mathbf{f} = y^2 z^2 \mathbf{i} + 2xyz^2 \mathbf{j} + 2xy^2z \mathbf{k}$. Then

curl $f = \nabla \times f = (4xyz - 4xyz)\mathbf{i} + (2y^2z - 2y^2z)\mathbf{j} + (2yz^2 - 2yz^2)\mathbf{k} = \mathbf{0}$.

Thus by Stokes' theorem, for any surface S with boundary C,

$$\int_C \mathbf{f} \cdot \mathbf{T} \, ds = \iint_S \nabla \times \mathbf{f} \cdot \mathbf{n} \, dS = 0.$$

Another generalization of Green's theorem is the divergence theorem. Let us call a surface S *simple* if each line parallel to a coordinate axis that intersects S does so in a single point, in two points, or in a line segment.

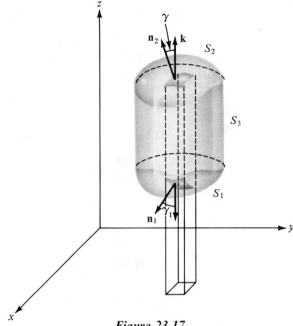

Figure 23.17

Theorem 23.11

(*The divergence theorem*)   *If* $\mathbf{f} = P\mathbf{i} + Q\mathbf{j} + R\mathbf{k},$ *where P, Q, and R are smooth in a region containing a closed connected surface S and its interior V, then*

$$\iint_S \mathbf{f} \cdot \mathbf{n}\, dS = \iiint_V (\nabla \cdot \mathbf{f})\, dV.$$

In component form,

$$\iint_S (P \cos \alpha + Q \cos \beta + R \cos \gamma)\, dS = \iiint_V \left(\frac{\partial P}{\partial x} + \frac{\partial Q}{\partial y} + \frac{\partial R}{\partial z} \right) dV$$

$$= \iiint_V \operatorname{div} \mathbf{f}\, dV.$$

Proof

Suppose that S is a simple surface and that S is bounded above by S_2: $z = f_2(x, y)$, below by S_1: $z = f_1(x, y)$,   and by a vertical cylindrical surface S_3 (see Figure 23.17). Let A be the projection of S upon the xy plane. Since $\cos \gamma = \cos 90° = 0$ on the vertical part S_3,

$$\iint_S R(x, y, z) \cos \gamma\, dS = \iint_{S_1} R(x, y, z) \cos \gamma\, dS + \iint_{S_2} R(x, y, z) \cos \gamma\, dS.$$

On S_2, the outward normal $\mathbf{n}_2$ points upward and

$$\iint_{S_2} R(x, y, z) \cos \gamma\, dS = \iint_A R(x, y, f_2(x, y))\, dA.$$

On S_1, the outward normal $\mathbf{n}_1$ points downward, so in the formula for surface area we must use $\cos(\pi - \gamma) = -\cos\gamma$.   Thus

$$\iint_{S_1} R(x, y, z) \cos\gamma \, dS = -\iint_A R(x, y, f_1(x, y)) \, dA.$$

Adding the integrals over S_1 and S_2, we get

$$\iint_S R(x, y, z) \cos\gamma \, dS = \iint_A R(x, y, f_2(x, y)) \, dA - \iint_A R(x, y, f_1(x, y)) \, dA$$

$$= \iint_A R(x, y, z)\Big|_{z=f_1(x, y)}^{z=f_2(x, y)} \, dA$$

$$= \iint_A \left(\int_{f_1(x, y)}^{f_2(x, y)} \frac{\partial R}{\partial z}(x, y, z) \, dz \right) dA$$

$$= \iiint_V \frac{\partial R}{\partial z} \, dV.$$

In similar fashion, by projecting onto the yz and xz planes, it can be shown that

$$\iint_S P(x, y, z) \cos\alpha \, dS = \iiint_V \frac{\partial P}{\partial x} \, dV$$

and

$$\iint_S Q(x, y, z) \cos\beta \, dS = \iiint_V \frac{\partial Q}{\partial y} \, dV.$$

We get the conclusion of the theorem by adding the three equations.

The proof can be extended to more complicated surfaces that are the sum of simple surfaces.

Note 1: $\cos\gamma \, dS = dx \, dy$, $\cos\beta \, dS = dx \, dz$,   and $\cos\alpha \, dS = dy \, dz$,   so the divergence theorem can be written

$$\iint_S P \, dy \, dz + Q \, dx \, dz + R \, dx \, dy = \iiint_V (P_x + Q_y + R_z) \, dz \, dy \, dx.$$

Note 2: The divergence theorem tells us that the flux $\iint_S \mathbf{f} \cdot \mathbf{n} \, dS$ of a velocity field $\mathbf{f}$ across the closed surface S is the integral of the divergence of $\mathbf{f}$ over the volume enclosed by S.

Example 4

Verify the divergence theorem for the solid bounded by $x = 0$, $y = 0$, $z = 0$,   and $x + y + z = 1$ and with $\mathbf{f} = x^2\mathbf{i} + 2xy\mathbf{j} + 2z\mathbf{k}$.

$$\text{div } \mathbf{f} = 2x + 2x + 2$$
$$= 4x + 2.$$

$$\iiint_V (4x + 2)\, dV = \int_0^1 \int_0^{1-x} \int_0^{1-x-y} (4x + 2)\, dz\, dy\, dx$$

$$= \int_0^1 (4x + 2) \left[\int_0^{1-x} (1 - x - y)\, dy \right] dx$$

$$= \int_0^1 (4x + 2) \left[(1 - x)y - \frac{y^2}{2} \right]\Big|_0^{1-x} dx$$

$$= \int_0^1 (2x + 1)(1 - x)^2\, dx$$

$$= x - x^3 + \frac{x^4}{2}\Big|_0^1 = \frac{1}{2}.$$

Let S_1, S_2, and S_3 be the faces in the xy, yz, and xz planes, respectively, and let S_4 be the face in the plane $x + y + z = 1$ (see Figure 23.18).

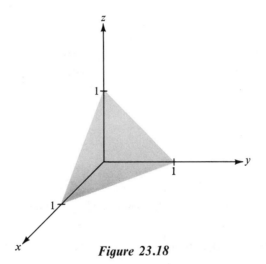

Figure 23.18

On S_1, $z = 0$, $\mathbf{n} = -\mathbf{k}$, and $\mathbf{f} \cdot \mathbf{n} = (x^2\mathbf{i} + 2xy\mathbf{j} + 0\mathbf{k}) \cdot (0\mathbf{i} + 0\mathbf{j} - \mathbf{k}) = 0$.

On S_2, $x = 0$, $\mathbf{n} = -\mathbf{i}$, and $\mathbf{f} \cdot \mathbf{n} = 0$.

On S_3, $y = 0$, $\mathbf{n} = -\mathbf{j}$, and $\mathbf{f} \cdot \mathbf{n} = 0$.

On S_4, $z = 1 - x - y$, $\mathbf{n} = (1/\sqrt{3})(\mathbf{i} + \mathbf{j} + \mathbf{k})$, and

$$\iint_{S_4} \mathbf{f} \cdot \mathbf{n}\, dS = \int_0^1 \int_0^{1-x} [x^2 + 2xy + 2(1 - x - y)] \frac{\sqrt{3}}{\sqrt{3}}\, dy\, dx$$

$$= \int_0^1 [(x^2 - 2x + 2)y + xy^2 - y^2]\Big|_0^{1-x} dx$$

$$= \int_0^1 [-(x - 1)^3 + (1 - x) + (x - 1)^3]\, dx$$

$$= x - \frac{x^2}{2}\Big|_0^1 = \frac{1}{2},$$

as before.

Example 5

Evaluate $\iint_S \mathbf{f} \cdot \mathbf{n}\, dS$, where S is the surface of the cube $\quad 0 \le x \le 1, \quad 0 \le y \le 1, \quad 0 \le z \le 1,$
and $\quad \mathbf{f} = (2z + y)\mathbf{i} + (2z - y)\mathbf{j} + (e^x + z)\mathbf{k}.$

Since $\nabla \cdot \mathbf{f} = 0 - 1 + 1 = 0,$ the divergence theorem tells us that the surface integral is zero.

Problems

A

In Problems 1–4, find the divergence of **f**.

1. $\mathbf{f} = (y^2 - z^2)\mathbf{i} + 2xyz\mathbf{j} + (z^2 - x^2)\mathbf{k}.$
2. $\mathbf{f} = ze^x \cos y\mathbf{i} - ze^x \sin y\mathbf{j} + (x^2 + y^2)\mathbf{k}.$
3. $\mathbf{f} = \sqrt{1 + x^2 + y^2}\,\mathbf{i} + \sqrt{1 + x^2 + y^2}\,\mathbf{j} + z^2\mathbf{k}.$
4. $\mathbf{f} = x\mathbf{i} + y\mathbf{j} + z\mathbf{k}.$

In Problems 5–8, find the curl of **f**.

5. $\mathbf{f} = xy^2\mathbf{i} + x^2y\mathbf{j} + z^2\mathbf{k}.$
6. $\mathbf{f} = (x + 2y - z)\mathbf{i} + (2x - y + z)\mathbf{j} + (x + y + 3z)\mathbf{k}.$
7. $\mathbf{f} = \sin y\mathbf{i} + \sin z\mathbf{j} + \sin x\mathbf{k}.$
8. $\mathbf{f} = (x^2 + y^2)\mathbf{i} + 2xy\mathbf{j}.$

In Problems 9–12, verify Stokes' theorem.

9. $\mathbf{f} = -y\mathbf{i} + x\mathbf{j} + z\mathbf{k};\quad S\colon\ x^2 + y^2 \le a^2,\quad z = 0.$
10. $\mathbf{f} = (2y + z)\mathbf{i} + (x - z)\mathbf{j} + (y - x)\mathbf{k};\quad S$ the part of $\ x + y + z = 1\ $ in the first octant.
11. $\mathbf{f} = z^2\mathbf{i} + x^2\mathbf{j} + y^2\mathbf{k};\quad S$ the part of $\ x + y + z = 2\ $ in the first octant.
12. $\mathbf{f} = y^2\mathbf{i} + 3x\mathbf{j} + 2y\mathbf{k};\quad S\colon\ x^2 + y^2 + z^2 = 4,\quad z \ge 0.$

In Problems 13–16, verify the divergence theorem.

13. $\mathbf{f} = 2y\mathbf{i} + (z - y)\mathbf{j} + (xy + z)\mathbf{k};\quad V$ is the parallelepiped $\ 0 \le x \le 2,\ \ 0 \le y \le 3,\ $ and $0 \le z \le 1.$
14. $\mathbf{f} = x^2\mathbf{i} + y^2\mathbf{j} + z^2\mathbf{k};\quad V$ is the unit cube $\ 0 \le x \le 1,\ \ 0 \le y \le 1,\ $ and $\ 0 \le z \le 1.$
15. $\mathbf{f} = z\mathbf{i} + x\mathbf{j} + 2x\mathbf{k};\quad V$ is the tetrahedron with vertices $(0, 0, 0), (1, 0, 0), (0, 1, 0),$ and $(0, 0, 1).$
16. $\mathbf{f} = 2x\mathbf{i} + (y - x)\mathbf{j} + 2z\mathbf{k};\quad V$ is the solid bounded by $\ x^2 + z^2 = 9,\ \ y = 0,\ $ and $\ y = 4.$

B

In Problems 17–21, verify the identities.

17. $\nabla \cdot \nabla \times \mathbf{f} = 0.$
18. $\nabla \times \nabla F = \mathbf{0}.$
19. $\nabla \cdot (F\mathbf{f}) = F\nabla \cdot \mathbf{f} + \mathbf{f} \cdot \nabla F.$
20. $\nabla \cdot \mathbf{r} = 3,\quad \mathbf{r} = x\mathbf{i} + y\mathbf{j} + z\mathbf{k}.$
21. $\nabla \times \mathbf{r} = \mathbf{0},\quad \mathbf{r} = x\mathbf{i} + y\mathbf{j} + z\mathbf{k}.$

22. For what values of n is $\ \nabla^2 r^n = 0\ \ (\nabla^2 = \nabla \cdot \nabla, r = |\mathbf{r}|)$?

23. What is $\displaystyle\iint_S \mathbf{n} \cdot \nabla \times \mathbf{f}\, dS$ if S is a closed surface?

24. Let u and v be scalar functions of x, y, and z and set $\mathbf{f} = u\nabla v$. Show that $\nabla \cdot \mathbf{f} = \nabla u \cdot \nabla v + u\nabla^2 v$ and

$$\iint_S \mathbf{n} \cdot u\nabla v \, dS = \iiint_V (\nabla u \cdot \nabla v + u\nabla^2 v) \, dV,$$

the first form of Green's identity.

25. Interchange the roles of u and v in Problem 24 and take the difference of the results to show that

$$\iint_S \mathbf{n} \cdot (u\nabla v - v\nabla u) \, dS = \iiint_V (u\nabla^2 v - v\nabla^2 u) \, dV,$$

the second (or symmetric) form of Green's identity.

Review Problems

A

In Problems 1–4, evaluate the line integral.

1. $\displaystyle\int_C (x^2 - y^2) \, dx - 2xy \, dy;$ $C:$ $x = 3 \cos t,$ $y = 3 \sin t,$ $0 \le t \le \pi/2.$

2. $\displaystyle\int_C x^2 \, dx + y^2 \, dy;$ C consists of the semicircle $x^2 + y^2 = 1,$ $y \ge 0,$ from $(1, 0)$ to $(-1, 0)$ followed by the segment from $(-1, 0)$ to $(1, 0)$.

3. $\displaystyle\int_C xy \, dx - x^2 \, dy$ around the square with vertices $(1, 1), (-1, 1), (-1, -1),$ and $(1, -1)$, in order.

4. $\displaystyle\int_C (x^2 + y^2) \, ds;$ $C:$ $x = t \sin t,$ $y = t \cos t,$ $0 \le t \le \pi.$

In Problems 5–8, use Green's theorem to evaluate the line integral where C is the boundary of the region bounded by $y = 3x^2,$ $y = 0,$ and $x = 1$, taken in the positive sense.

5. $\displaystyle\int_C y^2 \, dx - x^2 \, dy.$

6. $\displaystyle\int_C [\ln (1 + x^2) + y] \, dx + (\sin y^2) \, dy.$

7. $\displaystyle\int_C (x^3 - 3xy^2) \, dx + (y^2 - 3x^2y) \, dy.$

8. $\displaystyle\int_C \left(e^x + x \cos \frac{\pi y}{2}\right) dx + \left(e^x + x^2 \sin \frac{\pi y}{2}\right) dy.$

B

In Problems 9–12, test the given differential expression for exactness. If it is exact, find a function $F(x, y)$ for which it is the differential.

9. $(x^2 + xy^2) \, dx + (y^2 + x^2y) \, dy.$

10. $(\tan x + \sec y) \, dx + (x \sec y \tan y + \cos y) \, dy.$

11. $(x^3y + 3xy^2 + 4) \, dx + \left(\dfrac{3}{4} x^4 - 3xy\right) dy.$

12. $(2xe^y - y^2) \, dx + (y^2 + x^2 e^y - 2xy) \, dy.$

In Problems 13–16, find the net work done in moving a particle along the given path under the given forces.

13. Along $y = 3x + 2$, $0 \le x \le 2$; $\mathbf{f} = (1 + y)\mathbf{i} + (x - y)\mathbf{j}.$
14. Along $y = 4 - x^2$, $0 \le x \le 2$; $\mathbf{f} = 2\mathbf{i} + 3x\mathbf{j}.$
15. Along $y = 4 - x^2$, $0 \le x \le 2$; $\mathbf{f} = y\mathbf{i} - 3x\mathbf{j}$, coefficient of friction $\mu = 0.1.$
16. Along $x = y^2$, $0 \le y \le 2$; $\mathbf{f} = (x - 1)\mathbf{i} + y\mathbf{j}$, coefficient of friction $\mu.$

17. Show exact and find a potential for $(x^2 + 2xz) \, dx + (z - y^2) \, dy + (x^2 + y + z^2) \, dz.$
18. Find the area of the part of $z = x^2$ bounded by $x = 0$, $x = 1$, $y = 0$, and $y = 2.$
19. Find the area of the part of $y = x^2$ in the first octant bounded by $y = z$ and $y = 4.$
20. Find the flux across the first octant part of $z = 4 - x^2 - y^2$ if $\mathbf{f} = x\mathbf{i} + y\mathbf{j} - \mathbf{k}.$

21. Use the divergence theorem to evaluate $\displaystyle\iint_S \mathbf{f} \cdot \mathbf{n} \, dS$, where $\mathbf{f} = x^2\mathbf{i} + y^2\mathbf{j} - 2z\mathbf{k}$ and

   S is the surface of the cylindrical solid bounded by $x^2 + y^2 = 4$, $z = 0$, and $z = 5.$

C

22. Verify the divergence theorem for $\mathbf{f} = 2x\mathbf{i} - 3y\mathbf{j} + z^2\mathbf{k}$ if V is the cube $0 \le x, y, z \le 2.$
23. Verify Stokes' theorem for $\mathbf{f} = y\mathbf{i} + z\mathbf{j} + x\mathbf{k}$ and S the part of $z = 4 - x^2 - y^2$ above the xy plane.
24. Use the divergence theorem to find the total flux out of the solid $0 \le x^2 + y^2 \le 4$, $0 \le z \le 2$, if $\mathbf{f} = 3x\mathbf{i} + y^2\mathbf{j} + 3z^2\mathbf{k}.$

Appendix

A.1

Inequalities

Definition

If a and b are real numbers, $a < b$ means there is a positive number c such that $a + c = b$; $a > b$ means $b < a$.

Thus $3 < 4$, since $3 + 1 = 4$; $-2 < 6$, since $-2 + 8 = 6$; $-4 < -1$, since $-4 + 3 = -1$.

Theorem A.1

If $a < b$ and $b < c$, then $a < c$.

Proof

If $a < b$ and $b < c$, then there are positive numbers d and e such that $a + d = b$ and $b + e = c$. Thus,

$$a + d + e = b + e = c.$$

Since d and e are both positive, $d + e$ is positive and $a < c$.

Theorem A.2

If $a < b$, then $a + c < b + c$.

Proof

If $a < b$, there is a positive number d such that

$$a + d = b.$$

Then

$$a + c + d = b + c \quad \text{or} \quad a + c < b + c.$$

Theorem A.3

If $a < b$ and $c > 0$, then $ac < bc$; if $a < b$ and $c < 0$, then $ac > bc$.

Proof

If $a < b$, then there is a positive number d such that

$$a + d = b.$$

Thus,

$$ac + dc = bc.$$

If c is positive, then dc is positive and

$$ac < bc.$$

If c is negative, then dc is negative and

$$ac = bc + (-dc) \quad \text{and} \quad ac > bc.$$

Similar theorems can be stated using $a > b$. The foregoing theorems allow us to solve inequalities much as we do equations.

Example 1

Solve $3x - 2 < x + 4$.

$$3x - 2 < x + 4$$
$$2x - 2 < 4 \qquad \text{(adding } -x \text{ to both sides)}$$
$$2x < 6 \qquad \text{(adding 2 to both sides)}$$
$$x < 3. \qquad \text{(multiplying both sides by 1/2)}$$

Example 2

Solve $-x - 3 < 2x + 6$.

$$-x - 3 < 2x + 6$$
$$-3x - 3 < 6$$
$$-3x < 9$$
$$x > -3.$$

Note that the last step involved multiplying both sides by $-1/3$, which reversed the inequality.

Definition

If a and b are numbers, $a \leq b$ means $a < b$ or $a = b$; $a \geq b$ means $a > b$ or $a = b$.

The solution of inequalities involving $\leq$ or $\geq$ is basically the same as for strict inequalities or equations.

Example 3

Solve $4x - 3 \leq x - 6$.

$$4x - 3 \leq x - 6$$
$$3x - 3 \leq -6$$
$$3x \leq -3$$
$$x \leq -1.$$

Definition

If a, b, and c are real numbers, a < b < c means a < b and b < c.

With this definition, two inequalities can be solved at once in some cases.

Example 4

Solve $1 < 2x + 5 < 7.$

$$1 < 2x + 5 < 7$$
$$-4 < 2x < 2$$
$$-2 < x < 1.$$

Sometimes these double inequalities must be solved as two separate problems.

Example 5

Solve $x - 5 < 3x + 1 < -x + 2.$

$$
\begin{array}{ll}
x - 5 < 3x + 1 & \quad 3x + 1 < -x + 2 \\
-2x - 5 < 1 & \quad 4x + 1 < 2 \\
\quad -2x < 6 & \quad 4x < 1 \\
\quad\quad x > -3 & \quad x < \dfrac{1}{4}; \\
\quad -3 < x; &
\end{array}
$$

$$-3 < x < \frac{1}{4}.$$

Example 6

Solve $\dfrac{2x + 1}{x - 1} < 1.$

We have a special problem here that did not occur in the previous cases. We should like to multiply both sides by $x - 1$, but we do not know whether $x - 1$ is positive or negative. Let us consider two cases.

Case I $x - 1 > 0$ or $x > 1$:

$$\frac{2x + 1}{x - 1} < 1$$
$$2x + 1 < x - 1$$
$$x + 1 < -1$$
$$x < -2.$$

Now we have $x < -2$ provided $x > 1$. Of course, x cannot satisfy both of these inequalities. Case I gives no solution.

Case II $\quad x - 1 < 0 \quad$ or $\quad x < 1$:

$$\frac{2x+1}{x-1} < 1$$
$$2x + 1 > x - 1$$
$$x > -2.$$

Now we have $x > -2$, provided $x < 1$. Thus the solution is

$$-2 < x < 1.$$

Problems

In Problems 1–24, solve the given inequalities.

1. $2x + 5 < 3.$
2. $4x + 1 < 2x.$
3. $3x - 5 < 4.$
4. $4x - 2 < x + 1.$
5. $2x + 2 \leq x - 4.$
6. $3x + 1 \leq 4x + 4.$
7. $4x - 3 > 2x + 2.$
8. $3x - 5 > x - 2.$
9. $3x + 1 \geq 2x + 2.$
10. $2x - 4 \geq -x + 2.$
11. $-2 < x + 1 < 2.$
12. $-3 < x - 4 < 3.$
13. $1 \leq 2x + 1 \leq 4.$
14. $3 \leq 4x - 3 \leq 5.$
15. $2x - 8 < x - 1 < 2x - 4.$
16. $3x - 5 < 2x - 3 < 4x - 3.$
17. $2x - 1 \leq x + 4 \leq 3x + 1.$
18. $3x \leq x + 4 \leq 2x + 5.$

19. $\dfrac{2x-5}{x-2} < 1.$
20. $\dfrac{x-4}{x+1} < 2.$
21. $\dfrac{2x+1}{x-4} \leq 1.$

22. $\dfrac{4x-5}{x+3} \leq 1.$
23. $\dfrac{x-1}{x+1} < 1.$
24. $\dfrac{1}{x} < \dfrac{1}{x+1}.$

25. Show that if $0 < a < b$, then $1/a > 1/b$.
26. Show that if $a < b < 0$, then $1/a > 1/b$.
27. Show that if $a < 0 < b$, then $1/a < 1/b$.
28. Show that if $a < b$ and $c < d$, then $a + c < b + d$.

A.2

Absolute Values

The absolute value of a number is defined in the following way.

Definition

$$|x| = \begin{cases} x & if\ x \geq 0, \\ -x & if\ x < 0. \end{cases}$$

Thus $|5| = |-5| = 5$, or $|x| = 5$, implies that $x = \pm 5$.

Example 1

Solve $|x-1|=3$.

$$|x-1|=3$$
$$x-1=\pm3$$
$$x=1\pm3=4 \quad \text{or} \quad -2.$$

Example 2

Solve $|x+1|=|2x+3|$.

$$|x+1|=|2x+3|$$

$$
\begin{array}{ll}
x+1=2x+3 & \quad x+1=-2x-3 \\
-x=2 & \quad 3x=-4 \\
x=-2. & \quad x=-\dfrac{4}{3}.
\end{array}
$$

Let us now consider the inequality $|x|<a$, where a is a positive number. If $0\le x<a$, then $|x|=x<a$. If $-a<x<0$, then $|x|=-x<a$. Thus $|x|<a$ if $-a<x<a$. Furthermore, if $x\ge a$ or $x\le -a$, $|x| \not< a$. Thus, $|x|<a$ if and only if $-a<x<a$.

Example 3

Solve $|x-1|<1$.

$$-1<x-1<1$$
$$0<x<2.$$

Example 4

Solve $|x-1|>2$.

$$
\begin{array}{lll}
x-1>2 & \text{or} & x-1<-2 \\
x>3 & \text{or} & x<-1.
\end{array}
$$

The solution *cannot* be put into the form $-1>x>3$. The latter means $x<-1$ *and* $x>3$. Our solution is $x<-1$ *or* $x>3$.

Example 5

If $|x-2|<1$ and $f(x)=2x-1$, what can be said about $|f(x)-3|$?

$$
\begin{aligned}
|f(x)-3| &= |(2x-1)-3| \\
&= |2x-4| \\
&= 2|x-2| \\
&< 2\cdot1=2.
\end{aligned}
$$

Thus,

$$|f(x)-3|<2.$$

Problems

In Problems 1–24, solve for x.

1. $|x+2| = 5$.
2. $|2x-1| = 3$.
3. $|4x-4| = 1$.
4. $|4-2x| = 3$.
5. $|x-2| = 0$.
6. $|2x-5| = 3$.
7. $|2x+1| = |x-3|$.
8. $|x-4| = |2x-5|$.
9. $\left|\dfrac{x-2}{x+1}\right| = 3$.
10. $\left|\dfrac{4x+2}{x-3}\right| = 1$.
11. $|x+1| = |x-3|$.
12. $|x-5| = |x+2|$.
13. $|x+3| < 1$.
14. $|x-4| < 2$.
15. $|2x+3| < 3$.
16. $|3x-1| < 2$.
17. $|2x-5| \le 4$.
18. $|x-3| \le 5$.
19. $|x-4| > 1$.
20. $|2x-3| > 2$.
21. $|3x+1| \ge 4$.
22. $|2x+5| \ge 1$.
23. $|x-2| \le 0$.
24. $|2x+1| \ge 0$.

25. If $|x-2| < 1$ and $f(x) = x+1$, what can be said about $|f(x)-3|$?
26. If $|x+1| < 2$ and $f(x) = x+4$, what can be said about $|f(x)-3|$?
27. If $|x-1| < \delta$ and $f(x) = 2x+1$, what can be said about $|f(x)-3|$?
28. If $|x-3| < \delta$ and $f(x) = 3x-1$, what can be said about $|f(x)-8|$?

A.3

Absolute Values and Inequalities

Let us consider absolute values of sums, differences, products, and quotients.

Theorem A.4

If a and b are real numbers, then

$$|ab| = |a| \cdot |b|$$

$$\left|\frac{a}{b}\right| = \frac{|a|}{|b|} \quad (b \ne 0).$$

This theorem is easily proved by considering several cases ($a \ge 0$ or $a < 0$, $b \ge 0$ or $b < 0$).

Theorem A.5

If a and b are numbers, then

$$|a+b| \le |a| + |b|,$$
$$|a-b| \le |a| + |b|.$$

864   *Appendix*

Proof

By the definition of absolute value,

$$-|a| \le a \le |a| \quad \text{and} \quad -|b| \le b \le |b|.$$

Thus

$$-(|a| + |b|) \le a + b \le |a| + |b| \quad \text{(see Problem 28 in Section A.1)}$$

By the discussion following Example 2 in Section A.2,

$$|a + b| \le |a| + |b|.$$

Finally,

$$\begin{aligned} |a - b| &= |a + (-b)| \\ &\le |a| + |-b| \\ &= |a| + |b|. \end{aligned}$$

Example 1

If $|x| < 1$, what can be said about $|2x + 1|$?

$$\begin{aligned} |2x + 1| &\le |2x| + |1| \\ &= 2|x| + 1 \\ &< 2 \cdot 1 + 1 = 3. \end{aligned}$$

Thus, $|2x + 1| < 3$.

Example 2

If $|x| \le 3$, what can be said about $|x^2 - x - 1|$?

$$\begin{aligned} |x^2 - x - 1| &\le |x^2| + |-x| + |-1| \\ &= |x|^2 + |x| + 1 \\ &\le 3^2 + 3 + 1 \\ &= 13. \end{aligned}$$

Thus $|x^2 - x - 1| \le 13$.

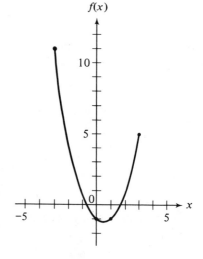

Figure A.1

If we consider the function

$$f(x) = x^2 - x - 1, \quad -3 \le x \le 3,$$

we see that

$$f'(x) = 2x - 1,$$

giving an absolute minimum $(1/2, -5/4)$. Since $f(3) = 9 - 3 - 1 = 5$ and $f(-3) = 9 + 3 - 1 = 11$, the graph of this function is as illustrated in Figure A.1. Thus we see that $|x^2 - x - 1| \le 11$ if $|x| \le 3$. This illustrates that the result found using inequalities is not necessarily the best possible result. In other words, while 13 is an upper bound for $|f(x)|$, it is not the least upper bound. More sophisticated methods are needed to find the least upper bound.

Example 3

Show that if $|x - 1| < \delta$ and $\delta \le 1$, then $|x^2 - 1| < 3\delta$.

$$|x^2 - 1| = |(x + 1)(x - 1)|$$
$$= |x + 1||x - 1|.$$

We know that $|x - 1| < \delta$. Let us consider $|x + 1|$. Since $|x - 1| < \delta$ and $\delta \le 1$,

$$|x - 1| < 1$$
$$-1 < x - 1 < 1$$
$$1 < x + 1 < 3$$
$$|x + 1| < 3.$$

Thus $|x^2 - 1| = |x + 1||x - 1| < 3\delta$.

Example 4

Show that if $|x - 1| < \delta$ and $\delta \le 1/2$, then $\left|\dfrac{1}{x} - 1\right| < 2\delta$.

$$\left|\frac{1}{x} - 1\right| = \left|\frac{1 - x}{x}\right| = \frac{1}{|x|}|x - 1|.$$

Again we know that $|x - 1| < \delta$. Let us consider $1/|x|$. Since $|x - 1| < \delta$ and $\delta \le 1/2$,

$$|x - 1| < \frac{1}{2}$$

$$-\frac{1}{2} < x - 1 < \frac{1}{2}$$

$$\frac{1}{2} < x < \frac{3}{2}$$

$$2 > \frac{1}{x} > \frac{2}{3} \qquad \text{(See Problem 25 in Section A.1)}$$

$$\frac{1}{|x|} = \left|\frac{1}{x}\right| < 2.$$

Thus $\left|\dfrac{1}{x} - 1\right| = \dfrac{1}{|x|}|x - 1| < 2\delta$.

Problems

1. If $|x| < 2$, what can be said about $|x^2 + 1|$?
2. If $|x| < 3$, what can be said about $|x^2 - 2x - 3|$?
3. If $|x| < 1$, what can be said about $|x^2 + 2x + 2|$?
4. If $|x| < 2$, what can be said about $|x^3 + x|$?
5. If $|x| < 5$, what can be said about $|x^4 + 4|$?
6. If $|x| < 1$, what can be said about $|x^4 - 2x^3 + x^2 + x|$?
7. If $-1 < x < 2$, what can be said about $|x^2 + x + 1|$?

8. If $-2 < x < 4$, what can be said about $|x^3 + x^2 + x + 1|$?

9. If $-5 < x < 1$, what can be said about $|x^2 - 3|$?

10. If $-3 < x < 2$, what can be said about $|x^3 - x + 1|$?

11. For $|x| < \delta$, show that $|x^2 + 1| < 1 + \delta^2$.

12. For $|x - 2| < \delta$ and $\delta \leq 1$, show that $|x^2 - 4| < 5\delta$.

13. For $|x - 1| < \delta$ and $\delta \leq 1$, show that $|x^2 + x - 2| < 4\delta$.

14. For $|x + 2| < \delta$ and $\delta \leq 1$, show that $|x^2 + x - 2| < 4\delta$.

15. For $|x - 1| < \delta$ and $\delta \leq 2$, show that $|x^3 - 1| < 13\delta$.

16. For $|x - 1| < \delta$ and $\delta \leq 1$, show that $|x^3 - 1| < 7\delta$.

17. For $|x - 1| < \delta$ and $\delta \leq \dfrac{3}{4}$, show that $\left| \dfrac{1}{x} - 1 \right| < 4\delta$.

18. For $|x - 2| < \delta$ and $\delta \leq 1$, show that $\left| \dfrac{1}{x} - \dfrac{1}{2} \right| < \dfrac{\delta}{2}$.

19. Prove that $|a + b| \geq \big| |a| - |b| \big|$.

B Mean-Value Theorem; Taylor's Series in Several Variables

The proofs of several important theorems were omitted in the text proper because they require the mean-value theorem. These proofs are given here. First, we reiterate the mean-value theorem upon which the proofs are based.

Theorem 17.8

(*Mean-value theorem*) *If a and b are numbers* ($a < b$) *and f is a function such that*
(a) *f is continuous on* [a, b], *and*
(b) *f'(x) exists for all x in* (a, b), *then there is a number* x_0 *between a and b such that*

$$f'(x_0) = \frac{f(b) - f(a)}{b - a}.$$

B.1

Theorem 6.2

Theorem 6.2

(*See page* 170) *If f and g are functions such that* $f'(x) = g'(x)$ *for all x, then* $f(x) - g(x)$ *is a constant.*

Proof

Let F be the function such that $F(x) = f(x) - g(x)$. Assume that F is not a constant; that is, assume there are two numbers a and b ($a < b$) such that $F(a) \neq F(b)$. Since $f'(x)$ and $g'(x)$ exist for all x, f and g are continuous for all x by Theorem 10.10 (page 308). Thus F is continuous for all x in [a, b] by Theorem 10.7 (page 305), and $F'(x) = f'(x) - g'(x)$ exists for all x in (a, b). By the mean-value theorem, there is a number x_0 between a and b such that

$$F'(x_0) = \frac{F(b) - F(a)}{b - a}.$$

Since $F(a) \neq F(b)$,

$$F'(x_0) \neq 0.$$

Thus

$$f'(x_0) - g'(x_0) \neq 0 \quad \text{and} \quad f'(x_0) \neq g'(x_0),$$

which contradicts our hypothesis. Thus $F(x) = f(x) - g(x)$ is constant.

B.2

Theorem 5.5

Theorem 5.5

(*Page* 122)   *If f is a function such that $f'(x) > 0$ for every x on (a, b), then f is increasing on (a, b); if $f'(x) < 0$ for every x on (a, b), then f is decreasing on (a, b).*

Proof

Suppose $f'(x) > 0$ for every x on (a, b). Assume that f is not increasing on (a, b); that is, there is a pair of numbers c and d such that $a < c < d < b$ and $f(c) \geq f(d)$. Since $f'(x)$ exists for all x in $[c, d]$, f is continuous on $[c, d]$. We are given that $f'(x)$ exists for all x in (c, d). By the mean-value theorem, there is a number x_0 between c and d (and thus between a and b) such that

$$f'(x_0) = \frac{f(d) - f(c)}{d - c}.$$

Since $c < d$,

$$d - c > 0;$$

since $f(c) \geq f(d)$,

$$f(d) - f(c) \leq 0.$$

Thus

$$f'(x_0) = \frac{f(d) - f(c)}{d - c} \leq 0,$$

which contradicts the hypothesis that $f'(x) > 0$ for every x on (a, b). Thus, f is increasing on (a, b). A similar argument proves the second part of the theorem.

B.3

Theorem 5.6

Before proving Theorem 5.6, let us define our terms more carefully.

Definition

The graph of $y = f(x)$ is concave upward on (a, b) means that if $a < x_1 < x_2 < x_3 < b$, then $P_2 = (x_2, f(x_2))$ is below the line determined by $P_1(x_1, f(x_1))$ and $P_3(x_3, f(x_3))$; it is concave downward if P_2 is above the line determined by P_1 and P_3.

Theorem B.1

If the graph of $y = f(x)$ is concave upward on (a, b) and $a < x_1 < x_2 < x_3 < b$, then

$$\frac{f(x_2) - f(x_1)}{x_2 - x_1} < \frac{f(x_3) - f(x_2)}{x_3 - x_2};$$

if it is concave downward, then

$$\frac{f(x_2) - f(x_1)}{x_2 - x_1} > \frac{f(x_3) - f(x_2)}{x_3 - x_2}.$$

Proof

Suppose the graph of $y = f(x)$ is concave upward on (a, b) and $a < x_1 < x_2 < x_3 < b$. Let $P(x_2, y)$ be the point on $P_1 P_3$ with abscissa x_2. The slope of $P_1 P$ is equal to the slope of PP_3. Thus

$$\frac{y - f(x_1)}{x_2 - x_1} = \frac{f(x_3) - y}{x_3 - x_2}.$$

Since the graph is concave upward on (a, b), P_2 is below P, or $f(x_2) < y$. Thus,

$$\frac{f(x_2) - f(x_1)}{x_2 - x_1} < \frac{y - f(x_1)}{x_2 - x_1} = \frac{f(x_3) - y}{x_3 - x_2} < \frac{f(x_3) - f(x_2)}{x_3 - x_2},$$

which is the desired result. A similar argument proves the second part.

Theorem 5.6

(Page 129) *If f is a function such that $f''(x)$ is positive (negative) for all x on (a, b), then the graph of f is concave upward (downward) on (a, b).*

Proof

Suppose $f''(x)$ is positive for all x on (a, b). By Theorem 5.5, f' is increasing on (a, b). Suppose $a < x_1 < x_2 < x_3 < b$. The conditions of the mean-value theorem are satisfied for the function f on $[x_1, x_2]$ as well as for f on $[x_2, x_3]$. Thus there is a number x_4 between x_1 and x_2 such that

$$f'(x_4) = \frac{f(x_2) - f(x_1)}{x_2 - x_1}.$$

Similarly there is a number x_5 between x_2 and x_3 such that

$$f'(x_5) = \frac{f(x_3) - f(x_2)}{x_3 - x_2}.$$

Since $x_1 < x_4 < x_2$ and $x_2 < x_5 < x_3$, it follows that $x_4 < x_5$. But f' is an increasing function on (a, b). Thus

$$f'(x_4) < f'(x_5)$$

or

$$\frac{f(x_2) - f(x_1)}{x_2 - x_1} < \frac{f(x_3) - f(x_2)}{x_3 - x_2}.$$

By the previous theorem, f cannot be concave downward; f cannot be linear, or the above expressions would be equal. Thus, f must be concave upward on (a, b). A similar argument proves that, if $f''(x)$ is negative for all x on (a, b), then the graph is concave downward on (a, b).

B.4

The Fundamental Theorem of Integral Calculus

In the proof of the fundamental theorem of integral calculus given on page 198 we used an area function F; however the existence of the function F is not established in that proof. The following argument, based upon the mean-value theorem, is a more rigorous proof of the fundamental theorem.

Theorem 7.5

(*Fundamental theorem of integral calculus*)   *If f is continuous on the interval $[a, b]$ and F is any function such that $F'(x) = f(x)$ for all x in $[a, b]$, then*

$$\int_a^b f(x)\, dx = F(b) - F(a).$$

Proof

Suppose the interval $[a, b]$ is subdivided by

$$a = x_0, x_1, x_2, \ldots, x_n = b.$$

Now let us consider the ith subinterval $[x_{i-1}, x_i]$. Since $F'(x)$ exists for all x in $[x_{i-1}, x_i]$, then, by Theorem 10.10 (page 308), F is continuous on $[x_{i-1}, x_i]$. The conditions of the mean-value theorem are satisfied by F. Thus, there is a number x_i^* between x_{i-1} and x_i such that

$$F'(x_i^*) = \frac{F(x_i) - F(x_{i-1})}{x_i - x_{i-1}}.$$

But $F'(x_i^*) = f(x_i^*)$ and $x_i - x_{i-1} = \Delta x_i$. Thus,

$$f(x_i^*)\,\Delta x_i = F(x_i) - F(x_{i-1}),$$

$$\sum_{i=1}^n f(x_i^*)\,\Delta x_i = \sum_{i=1}^n [F(x_i) - F(x_{i-1})]$$

$$= F(x_n) - F(x_0)$$

$$= F(b) - F(a),$$

and

$$\int_a^b f(x)\, dx = \lim_{\|S\| \to 0} \sum_{i=1}^n f(x_i^*)\,\Delta x_i$$

$$= \lim_{\|S\| \to 0} [F(b) - F(a)]$$

$$= F(b) - F(a).$$

This argument assumes that the value of the integral is independent of the choice of x_i^*. A proof of this independence rests upon the notion of uniform continuity, which we have not considered. Thus we shall *assume* the integral to be independent of the choice of the x_i^*.

B.5

Taylor's Series for Functions of Several Variables

In order to expand $f(x, y)$ in powers of $x - a$ and $y - b$, let

$$u(t) = f(a + ht, b + kt),$$

where $h = x - a$ and $k = y - b$. Then $u(0) = f(a, b)$ and $u(1) = f(x, y)$. The Taylor polynomial of degree n for $u(t)$, $t > 0$, is

$$u(t) = u(0) + u'(0)t + \frac{u''(0)}{2!} t^2 + \cdots + \frac{u^{(n)}(t_1)}{n!} t^n, \quad 0 < t_1 < t.$$

Calculating the derivatives of $u(t)$ by the chain rule,

$$u(t) = f(a, b) + \left(\frac{\partial f}{\partial x} h + \frac{\partial f}{\partial y} k \right) t + \frac{1}{2} \left(\frac{\partial^2 f}{\partial x^2} h^2 + 2 \frac{\partial^2 f}{\partial y \, \partial x} hk + \frac{\partial^2 f}{\partial y^2} k^2 \right) t^2 + \cdots,$$

where the various derivatives are evaluated at (a, b). Setting $t = 1$, we get

$$f(x, y) = f(a, b) + \frac{\partial f(a, b)}{\partial x} (x - a) + \frac{\partial f(a, b)}{\partial y} (y - b)$$

$$+ \frac{1}{2!} \left[\frac{\partial^2 f(a, b)}{\partial x^2} (x - a)^2 + 2 \frac{\partial^2 f(a, b)}{\partial x \, \partial y} (x - a)(y - b) + \frac{\partial^2 f(a, b)}{\partial y^2} (y - b)^2 \right]$$

$$+ \cdots + \frac{1}{n!} \left[\frac{\partial^n f(x_1, y_1)}{\partial x^n} (x - a)^n + n \frac{\partial^n f(x_1, y_1)}{\partial y \, \partial^{n-1} x} (x - a)^{n-1} (y - b) \right.$$

$$+ \cdots + \left. \frac{\partial^n f(x_1, y_1)}{\partial y^n} (y - b)^n \right],$$

where $x_1 = a + (x - a)t_1$ and $y_1 = b + (y - b)t_1$.

If we define the symbolic operator

$$\left(h \frac{\partial}{\partial x} + k \frac{\partial}{\partial y} \right) f(x, y) = h \frac{\partial f(x, y)}{\partial x} + k \frac{\partial f(x, y)}{\partial y},$$

the series may be written

$$f(x, y) = f(a, b) + \left[(x - a) \frac{\partial}{\partial x} + (y - b) \frac{\partial}{\partial y} \right] f(a, b)$$

$$+ \frac{1}{2!} \left[(x - a) \frac{\partial}{\partial x} + (y - b) \frac{\partial}{\partial y} \right]^2 f(a, b) + \cdots$$

$$+ \frac{1}{n!} \left[(x - a) \frac{\partial}{\partial x} + (y - b) \frac{\partial}{\partial y} \right]^n f(x_1, y_1).$$

Example 1

Find the Taylor's series about $x = 1$, $y = 2$ for $f(x, y) = x^3 + 3xy^2 + y^2 - x + 4$.

$$f(1, 2) = 20$$

$$f_x = 3x^2 + 3y^2 - 1 \qquad\qquad f_x(1, 2) = 14$$

$$f_y = 6xy + 2y \qquad\qquad f_y(1, 2) = 16$$

$$f_{xx} = 6x \qquad\qquad f_{xx}(1, 2) = 6$$

$$f_{xy} = 6y \qquad\qquad f_{xy}(1, 2) = 12$$

$$f_{yy} = 6x + 2 \qquad\qquad f_{yy}(1, 2) = 8$$

$$f_{xxx} = 6, \quad f_{xxy} = 0, \quad f_{xyy} = 6, \quad f_{yyy} = 0.$$

$$f(x, y) = 20 + 14(x - 1) + 16(y - 2) + \frac{1}{2}\left[6(x - 1)^2 + 24(x - 1)(y - 2) + 8(y - 2)^2\right]$$

$$+ \frac{1}{6}\left[6(x - 1)^3 + 0(x - 1)^2(y - 2) + 18(x - 1)(y - 2)^2 + 0(y - 2)^3\right]$$

$$= 20 + 14(x - 1) + 16(y - 2) + 3(x - 1)^2 + 12(x - 1)(y - 2) + 4(y - 2)^2$$

$$+ (x - 1)^3 + 3(x - 1)(y - 2)^2.$$

Taylor's series for functions of three or more variables can be obtained in a similar fashion. For example, let us define the operator

$$\left(h\frac{\partial}{\partial x} + k\frac{\partial}{\partial y} + l\frac{\partial}{\partial z}\right) f(x, y, z) = h\frac{\partial f(x, y, z)}{\partial x} + k\frac{\partial f(x, y, z)}{\partial y} + l\frac{\partial f(x, y, z)}{\partial z}.$$

Then the *Taylor polynomial* for $f(x, y, z)$ about $x = a$, $y = b$, $z = c$ is

$$f(x, y, z) = f(a, b, c) + \left[(x - a)\frac{\partial}{\partial x} + (y - b)\frac{\partial}{\partial y} + (z - c)\frac{\partial}{\partial z}\right] f(a, b, c)$$

$$+ \frac{1}{2!}\left[(x - a)\frac{\partial}{\partial x} + (y - b)\frac{\partial}{\partial y} + (z - c)\frac{\partial}{\partial z}\right]^2 f(a, b, c) + \cdots$$

$$+ \frac{1}{n!}\left[(x - a)\frac{\partial}{\partial x} + (y - b)\frac{\partial}{\partial y} + (z - c)\frac{\partial}{\partial z}\right]^n f(x_1, y_1, z_1),$$

where (x_1, y_1, z_1) lies on the segment between (a, b, c) and (x, y, z).

Problems

A

Find the Taylor's series of the given function about the given point.

1. $f(x, y) = 2x^2 - xy - 3y^2 + 2x + 4$;   about $(1, 0)$.
2. $f(x, y) = y^3 + 3xy^2 - x^2y + 2x^2 + 4y + 1$;   about $(0, -1)$.
3. $f(x, y, z) = x^2 + 2xz - y^2 - z^2 + y - 2z$;   about $(1, 2, 1)$.
4. $f(x, y) = \sin(x + y)$;   about $(0, \pi/2)$.
5. $f(x, y) = e^x \cos y$;   about $(0, 0)$.

B

Find the Taylor's series of the given function about the given point.

6. $f(x, y) = \text{Tan}^{-1}(y/x)$; about $(1, 1)$.
7. $f(x, y) = \ln(x + y^2)$; about $(2, 1)$.
8. $f(x, y) = \sqrt{4x + y}$; about $(2, 1)$.

C

9. Expand each of the roots of the equation $x^2 + ax + b = 0$ about $a = 0, b = -1$.
10. Expand the real root of $x^3 + ax + b = 0$ about $a = 0, b = 1$.

C.1

Elementary Algebra

---

(1) Quadratic formula

The solutions of the quadratic equation $ax^2 + bx + c = 0$ $(a \neq 0)$ are

$$x = \frac{-b \pm \sqrt{b^2 - 4ac}}{2a}.$$

(2) Binomial theorem (n a positive integer)

$$(a + b)^n = a^n + na^{n-1} b + \frac{n(n - 1)}{2!} a^{n-2} b^2 + \frac{n(n - 1)(n - 2)}{3!} a^{n-3} b^3 + \cdots$$

$$+ \frac{n(n - 1)(n - 2) \cdots (n - r + 1)}{r!} a^{n-r} b^r + \cdots + nab^{n-1} + b^n.$$

(3) Exponents

$$a^m \cdot a^n = a^{m+n}.$$

$$\frac{a^m}{a^n} = a^{m-n} \; (a \neq 0).$$

$$(a^m)^n = a^{mn}.$$

$$(ab)^n = a^n b^n.$$

$$a^{-n} = \frac{1}{a^n} \quad (a \neq 0).$$

$$a^{p/q} = \sqrt[q]{a^p} = (\sqrt[q]{a})^p \quad (a > 0).$$

(4) Logarithms

$$\log_a x + \log_a y = \log_a xy.$$

$$\log_a x - \log_a y = \log_a \frac{x}{y}.$$

$$n \log_a x = \log_a x^n.$$

$$\frac{1}{n} \log_a x = \log_a \sqrt[n]{x}.$$

$$\log_a x = \frac{\log_b x}{\log_b a}.$$

$$\log_a 1 = 0, \quad \log_a a = 1.$$

$$a^{\log_a x} = x.$$

(5) Determinants

$$\begin{vmatrix} a & b \\ c & d \end{vmatrix} = ad - bc.$$

$$\begin{vmatrix} a_1 & b_1 & c_1 \\ a_2 & b_2 & c_2 \\ a_3 & b_3 & c_3 \end{vmatrix} = a_1 \begin{vmatrix} b_2 & c_2 \\ b_3 & c_3 \end{vmatrix} - b_1 \begin{vmatrix} a_2 & c_2 \\ a_3 & c_3 \end{vmatrix} + c_1 \begin{vmatrix} a_2 & b_2 \\ a_3 & b_3 \end{vmatrix}.$$

C.2

Trigonometry

(1) Trigonometric identities

$$\tan x = \frac{\sin x}{\cos x}, \quad \cot x = \frac{\cos x}{\sin x},$$

$$\sec x = \frac{1}{\cos x}, \quad \csc x = \frac{1}{\sin x}.$$

$$\sin^2 x + \cos^2 x = 1.$$

$$\tan^2 x + 1 = \sec^2 x.$$

$$1 + \cot^2 x = \csc^2 x.$$

$$\sin (x + y) = \sin x \cos y + \cos x \sin y.$$

$$\sin (x - y) = \sin x \cos y - \cos x \sin y.$$

$$\cos (x + y) = \cos x \cos y - \sin x \sin y.$$

$$\cos (x - y) = \cos x \cos y + \sin x \sin y.$$

$$\tan (x + y) = \frac{\tan x + \tan y}{1 - \tan x \tan y}.$$

$$\tan (x - y) = \frac{\tan x - \tan y}{1 + \tan x \tan y}.$$

$$\sin 2x = 2 \sin x \cos x.$$

$$\cos 2x = \cos^2 x - \sin^2 x$$
$$= 1 - 2 \sin^2 x$$
$$= 2 \cos^2 x - 1.$$

$$\tan 2x = \frac{2 \tan x}{1 - \tan^2 x}.$$

$$\sin \frac{x}{2} = \pm \sqrt{\frac{1 - \cos x}{2}}.$$

$$\cos \frac{x}{2} = \pm \sqrt{\frac{1 + \cos x}{2}}.$$

$$\tan \frac{x}{2} = \pm \sqrt{\frac{1 - \cos x}{1 + \cos x}} = \frac{1 - \cos x}{\sin x} = \frac{\sin x}{1 + \cos x}.$$

$$\sin x + \sin y = 2 \sin \frac{x + y}{2} \cos \frac{x - y}{2}.$$

$$\sin x - \sin y = 2 \cos \frac{x + y}{2} \sin \frac{x - y}{2}.$$

$$\cos x + \cos y = 2 \cos \frac{x+y}{2} \cos \frac{x-y}{2}.$$

$$\cos x - \cos y = 2 \sin \frac{x+y}{2} \sin \frac{x-y}{2}.$$

(2) Triangles

Law of sines: $\dfrac{a}{\sin A} = \dfrac{b}{\sin B} = \dfrac{c}{\sin C}.$

Law of cosines: $c^2 = a^2 + b^2 - 2ab \cos C.$

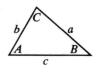

(3) Reduction formulas

$$\sin \alpha = -\cos (\alpha + \pi/2) = -\sin (\alpha + \pi) = +\cos (\alpha + 3\pi/2).$$
$$\sin \alpha = +\cos (\alpha - \pi/2) = -\sin (\alpha - \pi) = -\cos (\alpha - 3\pi/2).$$
$$\cos \alpha = +\sin (\alpha + \pi/2) = -\cos (\alpha + \pi) = -\sin (\alpha + 3\pi/2).$$
$$\cos \alpha = -\sin (\alpha - \pi/2) = -\cos (\alpha - \pi) = +\sin (\alpha - 3\pi/2).$$

C.3

Mensuration Formulas

(1) Triangle:

Area $= \dfrac{1}{2} bh$

(2) Parallelogram:

Area $= bh$

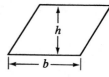

(3) Trapezoid:

Area $= \dfrac{1}{2} h(a + b)$

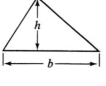

(4) Circle:

Area $= \pi r^2$

Circumference $= 2\pi r$

(5) Sector:

Area $= \dfrac{1}{2} r^2 \theta$

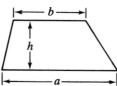

(6) Ellipse:

Area $= \pi ab$

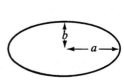

(7) Right circular cylinder:

Volume $= \pi r^2 h$

Lateral surface $= 2\pi r h$

Total surface $= 2\pi r(r + h)$

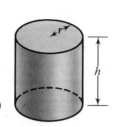

(8) Right circular cone:

Volume $= \dfrac{1}{3}\pi r^2 h$

Lateral surface $= \pi r s$

Total surface $= \pi r(r + s)$

(9) Sphere:

Volume $= \dfrac{4}{3}\pi r^3$

Surface $= 4\pi r^2$

(10) Frustum of a right circular cone:

Volume $= \dfrac{1}{3}\pi h(r_1^2 + r_1 r_2 + r_2^2)$

Lateral surface $= \pi s(r_1 + r_2)$

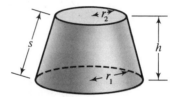

D   Tables

Table 1   Squares, Square Roots, and Prime Factors

No.	Sq.	Sq. Rt.	Factors	No.	Sq.	Sq. Rt.	Factors
1	1	1.000		51	2,601	7.141	$3 \cdot 17$
2	4	1.414	2	52	2,704	7.211	$2^2 \cdot 13$
3	9	1.732	3	53	2,809	7.280	53
4	16	2.000	2^2	54	2,916	7.348	$2 \cdot 3^3$
5	25	2.236	5	55	3,025	7.416	$5 \cdot 11$
6	36	2.449	$2 \cdot 3$	56	3,136	7.483	$2^3 \cdot 7$
7	49	2.646	7	57	3,249	7.550	$3 \cdot 19$
8	64	2.828	2^3	58	3,364	7.616	$2 \cdot 29$
9	81	3.000	3^2	59	3,481	7.681	59
10	100	3.162	$2 \cdot 5$	60	3,600	7.746	$2^2 \cdot 3 \cdot 5$
11	121	3.317	11	61	3,721	7.810	61
12	144	3.464	$2^2 \cdot 3$	62	3,844	7.874	$2 \cdot 31$
13	169	3.606	13	63	3,969	7.937	$3^2 \cdot 7$
14	196	3.742	$2 \cdot 7$	64	4,096	8.000	2^6
15	225	3.873	$3 \cdot 5$	65	4,225	8.062	$5 \cdot 13$
16	256	4.000	2^4	66	4,356	8.124	$2 \cdot 3 \cdot 11$
17	289	4.123	17	67	4,489	8.185	67
18	324	4.243	$2 \cdot 3^2$	68	4,624	8.246	$2^2 \cdot 17$
19	361	4.359	19	69	4,761	8.307	$3 \cdot 23$
20	400	4.472	$2^2 \cdot 5$	70	4,900	8.367	$2 \cdot 5 \cdot 7$
21	441	4.583	$3 \cdot 7$	71	5,041	8.426	71
22	484	4.690	$2 \cdot 11$	72	5,184	8.485	$2^3 \cdot 3^2$
23	529	4.796	23	73	5,329	8.544	73
24	576	4.899	$2^3 \cdot 3$	74	5,476	8.602	$2 \cdot 37$
25	625	5.000	5^2	75	5,625	8.660	$3 \cdot 5^2$
26	676	5.099	$2 \cdot 13$	76	5,776	8.718	$2^2 \cdot 19$
27	729	5.196	3^3	77	5,929	8.775	$7 \cdot 11$
28	784	5.292	$2^2 \cdot 7$	78	6,084	8.832	$2 \cdot 3 \cdot 13$
29	841	5.385	29	79	6,241	8.888	79
30	900	5.477	$2 \cdot 3 \cdot 5$	80	6,400	8.944	$2^4 \cdot 5$
31	961	5.568	31	81	6,561	9.000	3^4
32	1,024	5.657	2^5	82	6,724	9.055	$2 \cdot 41$
33	1,089	5.745	$3 \cdot 11$	83	6,889	9.110	83
34	1,156	5.831	$2 \cdot 17$	84	7,056	9.165	$2^2 \cdot 3 \cdot 7$
35	1,225	5.916	$5 \cdot 7$	85	7,225	9.220	$5 \cdot 17$
36	1,296	6.000	$2^2 \cdot 3^2$	86	7,396	9.274	$2 \cdot 43$
37	1,369	6.083	37	87	7,569	9.327	$3 \cdot 29$
38	1,444	6.164	$2 \cdot 19$	88	7,744	9.381	$2^3 \cdot 11$
39	1,521	6.245	$3 \cdot 13$	89	7,921	9.434	89
40	1,600	6.325	$2^3 \cdot 5$	90	8,100	9.487	$2 \cdot 3^2 \cdot 5$
41	1,681	6.403	41	91	8,281	9.539	$7 \cdot 13$
42	1,764	6.481	$2 \cdot 3 \cdot 7$	92	8,464	9.592	$2^2 \cdot 23$
43	1,849	6.557	43	93	8,649	9.644	$3 \cdot 31$
44	1,936	6.633	$2^2 \cdot 11$	94	8,836	9.695	$2 \cdot 47$
45	2,025	6.708	$3^2 \cdot 5$	95	9,025	9.747	$5 \cdot 19$
46	2,116	6.782	$2 \cdot 23$	96	9,216	9.798	$2^5 \cdot 3$
47	2,209	6.856	47	97	9,409	9.849	97
48	2,304	6.928	$2^4 \cdot 3$	98	9,604	9.899	$2 \cdot 7^2$
49	2,401	7.000	7^2	99	9,801	9.950	$3^2 \cdot 11$
50	2,500	7.071	$2 \cdot 5^2$	100	10,000	10.000	$2^2 \cdot 5^2$

Table 2 **Common Logarithms**

x	0	1	2	3	4	5	6	7	8	9
1.0	.0000	.0043	.0086	.0128	.0170	.0212	.0253	.0294	.0334	.0374
1.1	.0414	.0453	.0492	.0531	.0569	.0607	.0645	.0682	.0719	.0755
1.2	.0792	.0828	.0864	.0899	.0934	.0969	.1004	.1038	.1072	.1106
1.3	.1139	.1173	.1206	.1239	.1271	.1303	.1335	.1367	.1399	.1430
1.4	.1461	.1492	.1523	.1553	.1584	.1614	.1644	.1673	.1703	.1732
1.5	.1761	.1790	.1818	.1847	.1875	.1903	.1931	.1959	.1987	.2014
1.6	.2041	.2068	.2095	.2122	.2148	.2175	.2201	.2227	.2253	.2279
1.7	.2304	.2330	.2355	.2380	.2405	.2430	.2455	.2480	.2504	.2529
1.8	.2553	.2577	.2601	.2625	.2648	.2672	.2695	.2718	.2742	.2765
1.9	.2788	.2810	.2833	.2856	.2878	.2900	.2923	.2945	.2967	.2989
2.0	.3010	.3032	.3054	.3075	.3096	.3118	.3139	.3160	.3181	.3201
2.1	.3222	.3243	.3263	.3284	.3304	.3324	.3345	.3365	.3385	.3404
2.2	.3424	.3444	.3464	.3483	.3502	.3522	.3541	.3560	.3579	.3598
2.3	.3617	.3636	.3655	.3674	.3692	.3711	.3729	.3747	.3766	.3784
2.4	.3802	.3820	.3838	.3856	.3874	.3892	.3909	.3927	.3945	.3962
2.5	.3979	.3997	.4014	.4031	.4048	.4065	.4082	.4099	.4116	.4133
2.6	.4150	.4166	.4183	.4200	.4216	.4232	.4249	.4265	.4281	.4298
2.7	.4314	.4330	.4346	.4362	.4378	.4393	.4409	.4425	.4440	.4456
2.8	.4472	.4487	.4502	.4518	.4533	.4548	.4564	.4579	.4594	.4609
2.9	.4624	.4639	.4654	.4669	.4683	.4698	.4713	.4728	.4742	.4757
3.0	.4771	.4786	.4800	.4814	.4829	.4843	.4857	.4871	.4886	.4900
3.1	.4914	.4928	.4942	.4955	.4969	.4983	.4997	.5011	.5024	.5038
3.2	.5051	.5065	.5079	.5092	.5105	.5119	.5132	.5145	.5159	.5172
3.3	.5185	.5198	.5211	.5224	.5237	.5250	.5263	.5276	.5289	.5302
3.4	.5315	.5328	.5340	.5353	.5366	.5378	.5391	.5403	.5416	.5428
3.5	.5441	.5453	.5465	.5478	.5490	.5502	.5514	.5527	.5539	.5551
3.6	.5563	.5575	.5587	.5599	.5611	.5623	.5635	.5647	.5658	.5670
3.7	.5682	.5694	.5705	.5717	.5729	.5740	.5752	.5763	.5775	.5786
3.8	.5798	.5809	.5821	.5832	.5843	.5855	.5866	.5877	.5888	.5899
3.9	.5911	.5922	.5933	.5944	.5955	.5966	.5977	.5988	.5999	.6010
4.0	.6021	.6031	.6042	.6053	.6064	.6075	.6085	.6096	.6107	.6117
4.1	.6128	.6138	.6149	.6160	.6170	.6180	.6191	.6201	.6212	.6222
4.2	.6232	.6243	.6253	.6263	.6274	.6284	.6294	.6304	.6314	.6325
4.3	.6335	.6345	.6355	.6365	.6375	.6385	.6395	.6405	.6415	.6425
4.4	.6435	.6444	.6454	.6464	.6474	.6484	.6493	.6503	.6513	.6522
4.5	.6532	.6542	.6551	.6561	.6571	.6580	.6590	.6599	.6609	.6618
4.6	.6628	.6637	.6646	.6656	.6665	.6675	.6684	.6693	.6702	.6712
4.7	.6721	.6730	.6739	.6749	.6758	.6767	.6776	.6785	.6794	.6803
4.8	.6812	.6821	.6830	.6839	.6848	.6857	.6866	.6875	.6884	.6893
4.9	.6902	.6911	.6920	.6928	.6937	.6946	.6955	.6964	.6972	.6981
5.0	.6990	.6998	.7007	.7016	.7024	.7033	.7042	.7050	.7059	.7067
5.1	.7076	.7084	.7093	.7101	.7110	.7118	.7126	.7135	.7143	.7152
5.2	.7160	.7168	.7177	.7185	.7193	.7202	.7210	.7218	.7226	.7235
5.3	.7243	.7251	.7259	.7267	.7275	.7284	.7292	.7300	.7308	.7316
5.4	.7324	.7332	.7340	.7348	.7356	.7364	.7372	.7380	.7388	.7396
x	0	1	2	3	4	5	6	7	8	9

Table 2   (*Continued*)

x	0	1	2	3	4	5	6	7	8	9
5.5	.7404	.7412	.7419	.7427	.7435	.7443	.7451	.7459	.7466	.7474
5.6	.7482	.7490	.7497	.7505	.7513	.7520	.7528	.7536	.7543	.7551
5.7	.7559	.7566	.7574	.7582	.7589	.7597	.7604	.7612	.7619	.7627
5.8	.7634	.7642	.7649	.7657	.7664	.7672	.7679	.7686	.7694	.7701
5.9	.7709	.7716	.7723	.7731	.7738	.7745	.7752	.7760	.7767	.7774
6.0	.7782	.7789	.7796	.7803	.7810	.7818	.7825	.7832	.7839	.7846
6.1	.7853	.7860	.7868	.7875	.7882	.7889	.7896	.7903	.7910	.7917
6.2	.7924	.7931	.7938	.7945	.7952	.7959	.7966	.7973	.7980	.7987
6.3	.7993	.8000	.8007	.8014	.8021	.8028	.8035	.8041	.8048	.8055
6.4	.8062	.8069	.8075	.8082	.8089	.8096	.8102	.8109	.8116	.8122
6.5	.8129	.8136	.8142	.8149	.8156	.8162	.8169	.8176	.8182	.8189
6.6	.8195	.8202	.8209	.8215	.8222	.8228	.8235	.8241	.8248	.8254
6.7	.8261	.8267	.8274	.8280	.8287	.8293	.8299	.8306	.8312	.8319
6.8	.8325	.8331	.8338	.8344	.8351	.8357	.8363	.8370	.8376	.8382
6.9	.8388	.8395	.8401	.8407	.8414	.8420	.8426	.8432	.8439	.8445
7.0	.8451	.8457	.8463	.8470	.8476	.8482	.8488	.8494	.8500	.8506
7.1	.8513	.8519	.8525	.8531	.8537	.8543	.8549	.8555	.8561	.8567
7.2	.8573	.8579	.8585	.8591	.8597	.8603	.8609	.8615	.8621	.8627
7.3	.8633	.8639	.8645	.8651	.8657	.8663	.8669	.8675	.8681	.8686
7.4	.8692	.8698	.8704	.8710	.8716	.8722	.8727	.8733	.8739	.8745
7.5	.8751	.8756	.8762	.8768	.8774	.8779	.8785	.8791	.8797	.8802
7.6	.8808	.8814	.8820	.8825	.8831	.8837	.8842	.8848	.8854	.8859
7.7	.8865	.8871	.8876	.8882	.8887	.8893	.8899	.8904	.8910	.8915
7.8	.8921	.8927	.8932	.8938	.8943	.8949	.8954	.8960	.8965	.8971
7.9	.8976	.8982	.8987	.8993	.8998	.9004	.9009	.9015	.9020	.9025
8.0	.9031	.9036	.9042	.9047	.9053	.9058	.9063	.9069	.9074	.9079
8.1	.9085	.9090	.9096	.9101	.9106	.9112	.9117	.9122	.9128	.9133
8.2	.9138	.9143	.9149	.9154	.9159	.9165	.9170	.9175	.9180	.9186
8.3	.9191	.9196	.9201	.9206	.9212	.9217	.9222	.9227	.9232	.9238
8.4	.9243	.9248	.9253	.9258	.9263	.9269	.9274	.9279	.9284	.9289
8.5	.9294	.9299	.9304	.9309	.9315	.9320	.9325	.9330	.9335	.9340
8.6	.9345	.9350	.9355	.9360	.9365	.9370	.9375	.9380	.9385	.9390
8.7	.9395	.9400	.9405	.9410	.9415	.9420	.9425	.9430	.9435	.9440
8.8	.9445	.9450	.9455	.9460	.9465	.9469	.9474	.9479	.9484	.9489
8.9	.9494	.9499	.9504	.9509	.9513	.9518	.9523	.9528	.9533	.9538
9.0	.9542	.9547	.9552	.9557	.9562	.9566	.9571	.9576	.9581	.9586
9.1	.9590	.9595	.9600	.9605	.9609	.9614	.9619	.9624	.9628	.9633
9.2	.9638	.9643	.9647	.9652	.9657	.9661	.9666	.9671	.9675	.9680
9.3	.9685	.9689	.9694	.9699	.9703	.9708	.9713	.9717	.9722	.9727
9.4	.9731	.9736	.9741	.9745	.9750	.9754	.9759	.9763	.9768	.9773
9.5	.9777	.9782	.9786	.9791	.9795	.9800	.9805	.9809	.9814	.9818
9.6	.9823	.9827	.9832	.9836	.9841	.9845	.9850	.9854	.9859	.9863
9.7	.9868	.9872	.9877	.9881	.9886	.9890	.9894	.9899	.9903	.9908
9.8	.9912	.9917	.9921	.9926	.9930	.9934	.9939	.9943	.9948	.9952
9.9	.9956	.9961	.9965	.9969	.9974	.9978	.9983	.9987	.9991	.9996
x	0	1	2	3	4	5	6	7	8	9

Table 3   Natural Logarithms of Numbers

n	$\log_e n$	n	$\log_e n$	n	$\log_e n$
	*	4.5	1.5041	9.0	2.1972
0.1	7.6974	4.6	1.5261	9.1	2.2083
0.2	8.3906	4.7	1.5476	9.2	2.2192
0.3	8.7960	4.8	1.5686	9.3	2.2300
0.4	9.0837	4.9	1.5892	9.4	2.2407
0.5	9.3069	5.0	1.6094	9.5	2.2513
0.6	9.4892	5.1	1.6292	9.6	2.2618
0.7	9.6433	5.2	1.6487	9.7	2.2721
0.8	9.7769	5.3	1.6677	9.8	2.2824
0.9	9.8946	5.4	1.6864	9.9	2.2925
1.0	0.0000	5.5	1.7047	10	2.3026
1.1	0.0953	5.6	1.7228	11	2.3979
1.2	0.1823	5.7	1.7405	12	2.4849
1.3	0.2624	5.8	1.7579	13	2.5649
1.4	0.3365	5.9	1.7750	14	2.6391
1.5	0.4055	6.0	1.7918	15	2.7081
1.6	0.4700	6.1	1.8083	16	2.7726
1.7	0.5306	6.2	1.8245	17	2.8332
1.8	0.5878	6.3	1.8405	18	2.8904
1.9	0.6419	6.4	1.8563	19	2.9444
2.0	0.6931	6.5	1.8718	20	2.9957
2.1	0.7419	6.6	1.8871	25	3.2189
2.2	0.7885	6.7	1.9021	30	3.4012
2.3	0.8329	6.8	1.9169	35	3.5553
2.4	0.8755	6.9	1.9315	40	3.6889
2.5	0.9163	7.0	1.9459	45	3.8067
2.6	0.9555	7.1	1.9601	50	3.9120
2.7	0.9933	7.2	1.9741	55	4.0073
2.8	1.0296	7.3	1.9879	60	4.0943
2.9	1.0647	7.4	2.0015	65	4.1744
3.0	1.0986	7.5	2.0149	70	4.2485
3.1	1.1314	7.6	2.0281	75	4.3175
3.2	1.1632	7.7	2.0412	80	4.3820
3.3	1.1939	7.8	2.0541	85	4.4427
3.4	1.2238	7.9	2.0669	90	4.4998
3.5	1.2528	8.0	2.0794	100	4.6052
3.6	1.2809	8.1	2.0919	110	4.7005
3.7	1.3083	8.2	2.1041	120	4.7875
3.8	1.3350	8.3	2.1163	130	4.8676
3.9	1.3610	8.4	2.1282	140	4.9416
4.0	1.3863	8.5	2.1401	150	5.0106
4.1	1.4110	8.6	2.1518	160	5.0752
4.2	1.4351	8.7	2.1633	170	5.1358
4.3	1.4586	8.8	2.1748	180	5.1930
4.4	1.4816	8.9	2.1861	190	5.2470

* Subtract 10 for $n < 1$.   Thus $\log_e 0.1 = 7.6974 - 10 = -2.3026$.

Table 4   Exponential Functions

x	e^x	e^{-x}	x	e^x	e^{-x}
0.00	1.0000	1.0000	1.5	4.4817	0.2231
0.01	1.0101	0.9901	1.6	4.9530	0.2019
0.02	1.0202	0.9802	1.7	5.4739	0.1827
0.03	1.0305	0.9705	1.8	6.0496	0.1653
0.04	1.0408	0.9608	1.9	6.6859	0.1496
0.05	1.0513	0.9512	2.0	7.3891	0.1353
0.06	1.0618	0.9418	2.1	8.1662	0.1225
0.07	1.0725	0.9324	2.2	9.0250	0.1108
0.08	1.0833	0.9331	2.3	9.9742	0.1003
0.09	1.0942	0.9139	2.4	11.023	0.0907
0.10	1.1052	0.9048	2.5	12.182	0.0821
0.11	1.1163	0.8958	2.6	13.464	0.0743
0.12	1.1275	0.8869	2.7	14.880	0.0672
0.13	1.1388	0.8781	2.8	16.445	0.0608
0.14	1.1503	0.8694	2.9	18.174	0.0550
0.15	1.1618	0.8607	3.0	20.086	0.0498
0.16	1.1735	0.8521	3.1	22.198	0.0450
0.17	1.1853	0.8437	3.2	24.533	0.0408
0.18	1.1972	0.8353	3.3	27.113	0.0369
0.19	1.2092	0.8270	3.4	29.964	0.0334
0.20	1.2214	0.8187	3.5	33.115	0.0302
0.21	1.2337	0.8106	3.6	36.598	0.0273
0.22	1.2461	0.8025	3.7	40.447	0.0247
0.23	1.2586	0.7945	3.8	44.701	0.0224
0.24	1.2712	0.7866	3.9	49.402	0.0202
0.25	1.2840	0.7788	4.0	54.598	0.0183
0.30	1.3499	0.7408	4.1	60.340	0.0166
0.35	1.4191	0.7047	4.2	66.686	0.0150
0.40	1.4918	0.6703	4.3	73.700	0.0136
0.45	1.5683	0.6376	4.4	81.451	0.0123
0.50	1.6487	0.6065	4.5	90.017	0.0111
0.55	1.7333	0.5769	4.6	99.484	0.0101
0.60	1.8221	0.5488	4.7	109.95	0.0091
0.65	1.9155	0.5220	4.8	121.51	0.0082
0.70	2.0138	0.4966	4.9	134.29	0.0074
0.75	2.1170	0.4724	5.0	148.41	0.0067
0.80	2.2255	0.4493	5.5	244.69	0.0041
0.85	2.3396	0.4274	6.0	403.43	0.0025
0.90	2.4596	0.4066	6.5	665.14	0.0015
0.95	2.5857	0.3867	7.0	1096.6	0.0009
1.0	2.7183	0.3679	7.5	1808.0	0.0006
1.1	3.0042	0.3329	8.0	2981.0	0.0003
1.2	3.3201	0.3012	8.5	4914.8	0.0002
1.3	3.6693	0.2725	9.0	8103.1	0.0001
1.4	4.0552	0.2466	10.0	22026	0.00005

Table 5   *Trigonometric Functions*

Degrees	Radians	sin	cos	tan	cot		
0	0.0000	0.0000	1.0000	0.0000		1.5708	90
1	0.0175	0.0175	0.9998	0.0175	57.290	1.5533	89
2	0.0349	0.0349	0.9994	0.0349	28.636	1.5359	88
3	0.0524	0.0523	0.9986	0.0524	19.081	1.5184	87
4	0.0698	0.0698	0.9976	0.0699	14.301	1.5010	86
5	0.0873	0.0872	0.9962	0.0875	11.430	1.4835	85
6	0.1047	0.1045	0.9945	0.1051	9.5144	1.4661	84
7	0.1222	0.1219	0.9925	0.1228	8.1443	1.4486	83
8	0.1396	0.1392	0.9903	0.1405	7.1154	1.4312	82
9	0.1571	0.1564	0.9877	0.1584	6.3138	1.4137	81
10	0.1745	0.1736	0.9848	0.1763	5.6713	1.3963	80
11	0.1920	0.1908	0.9816	0.1944	5.1446	1.3788	79
12	0.2094	0.2079	0.9781	0.2126	4.7046	1.3614	78
13	0.2269	0.2250	0.9744	0.2309	4.3315	1.3439	77
14	0.2443	0.2419	0.9703	0.2493	4.0108	1.3265	76
15	0.2618	0.2588	0.9659	0.2679	3.7321	1.3090	75
16	0.2793	0.2/56	0.9613	0.2867	3.4874	1.2915	74
17	0.2967	0.2924	0.9563	0.3057	3.2709	1.2741	73
18	0.3142	0.3090	0.9511	0.3249	3.0777	1.2566	72
19	0.3316	0.3256	0.9455	0.3443	2.9042	1.2392	71
20	0.3491	0.3420	0.9397	0.3640	2.7475	1.2217	70
21	0.3665	0.3584	0.9336	0.3839	2.6051	1.2043	69
22	0.3840	0.3746	0.9272	0.4040	2.4751	1.1868	68
23	0.4014	0.3907	0.9205	0.4245	2.3559	1.1694	67
24	0.4189	0.4067	0.9135	0.4452	2.2460	1.1519	66
25	0.4363	0.4226	0.9063	0.4663	2.1445	1.1345	65
26	0.4538	0.4384	0.8988	0.4877	2.0503	1.1170	64
27	0.4712	0.4540	0.8910	0.5095	1.9626	1.0996	63
28	0.4887	0.4695	0.8829	0.5317	1.8807	1.0821	62
29	0.5061	0.4848	0.8746	0.5543	1.8040	1.0647	61
30	0.5236	0.5000	0.8660	0.5774	1.7321	1.0472	60
31	0.5411	0.5150	0.8572	0.6009	1.6643	1.0297	59
32	0.5585	0.5299	0.8480	0.6249	1.6003	1.0123	58
33	0.5760	0.5446	0.8387	0.6494	1.5399	0.9948	57
34	0.5934	0.5592	0.8290	0.6745	1.4826	0.9774	56
35	0.6109	0.5736	0.8192	0.7002	1.4281	0.9599	55
36	0.6283	0.5878	0.8090	0.7265	1.3764	0.9425	54
37	0.6458	0.6018	0.7986	0.7536	1.3270	0.9250	53
38	0.6632	0.6157	0.7880	0.7813	1.2799	0.9076	52
39	0.6807	0.6293	0.7771	0.8098	1.2349	0.8901	51
40	0.6981	0.6428	0.7660	0.8391	1.1918	0.8727	50
41	0.7156	0.6561	0.7547	0.8693	1.1504	0.8552	49
42	0.7330	0.6691	0.7431	0.9004	1.1106	0.8378	48
43	0.7505	0.6820	0.7314	0.9325	1.0724	0.8203	47
44	0.7679	0.6947	0.7193	0.9657	1.0355	0.8029	46
45	0.7854	0.7071	0.7071	1.0000	1.0000	0.7854	45
		cos	sin	cot	tan	Radians	Degrees

Answers to Selected Problems

Section 1.1

1. $\sqrt{65}$. **3.** 5/2. **5.** (7/2, 1). **7.** 1, 9. **9.** 2, 3. **11.** Not collinear.
13. Collinear. **15.** Collinear. **17.** Right triangle. **19.** (6, −7) **23.** (1/2, 5), 13/2.
29. 7, −9. **31.** (8, 1), (−2, 5), (4, −7).

Section 1.2

1. 5/3, 59°. **3.** 2/3, 34°. **5.** No slope, 90°. **7.** 1, 45°. **9.** Parallel.
11. Perpendicular. **13.** Coincident. **15.** None. **17.** 14/3. **19.** 2, 7/3. **21.** 9.

Section 1.3

1.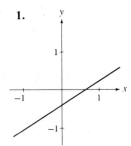

3.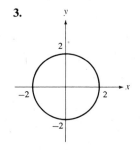

5.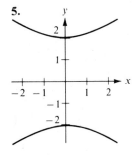

7.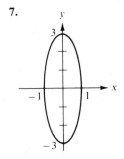

9.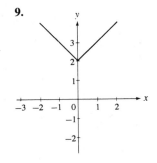

11.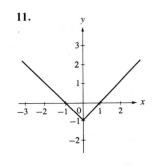

13.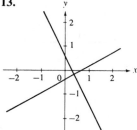

15. (2, −1), (−1, 2).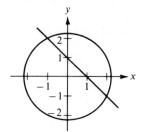

17.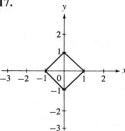

19.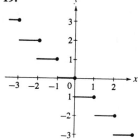

21.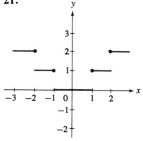

23.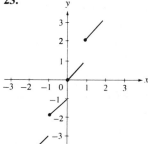

25.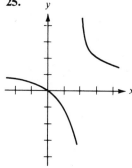

27. (0, 1), (0, −1).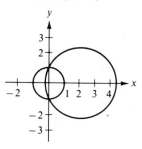

29.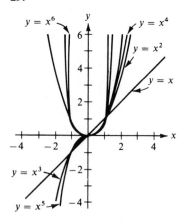

31. $y = 15 + 13\,[\![x]\!]$, $x > 0$.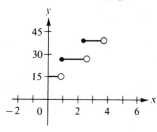

Review Problems

1. Noncollinear.

3. $\sqrt{221}/2$, $\sqrt{41}$, $\sqrt{185}/2$.

5. $(22/7, -9/7)$.

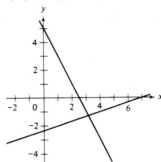

7. $-8/5$, $-1/2$, 5.

11.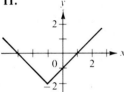

13.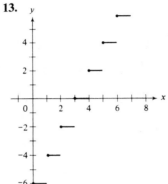

15. $(3, -2)$.   **17.** $(1, 4)$, $(4, 6)$.

Section 2.1

1. $2x + y = 0$.

3. $4x + y - 21 = 0$.

5. $y = x$.

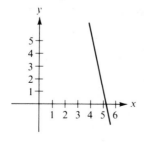

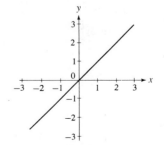

7. $y - 1 = 0.$

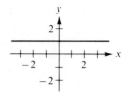

9. $x - 2y + 7 = 0.$

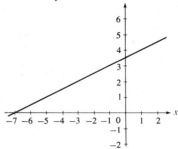

11. $x - 5y + 22 = 0.$

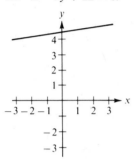

13. $5x - y = 0.$

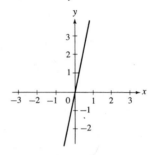

15. $x - 4 = 0.$

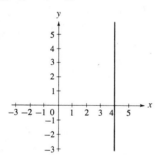

17. $2x + y - 6 = 0, x - 2y - 3 = 0, 3x - y + 1 = 0.$

19. $x - 2y - 3 = 0, 2x + y - 6 = 0, x + 3y - 3 = 0.$

21. $5x - 6y + 2 = 0, 2x + 9y - 3 = 0, 7x + 3y - 1 = 0.$

23. $2x - y + 5 = 0.$ **25.** $2x + y - 4 = 0.$

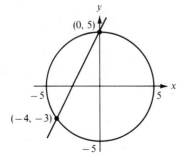

27. The given circles do not intersect.   **29.** $x - 3y + 3 = 0.$
31. $7x - y - 20 = 0, 7x - y - 20 = 0.$ **33.** $9C - 5F + 160 = 0.$

Section 2.2

1.

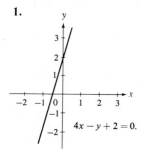

$4x - y + 2 = 0.$

3. $9x - 3y + 2 = 0.$

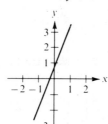

5.

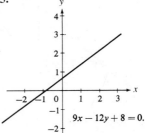

$9x - 12y + 8 = 0.$

7. $3x - y - 12 = 0.$

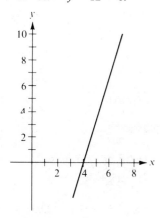

9.

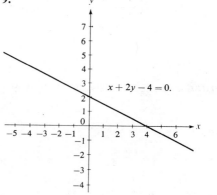

$x + 2y - 4 = 0.$

11. $12x + y - 4 = 0.$

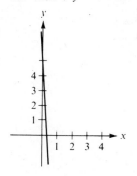

13.

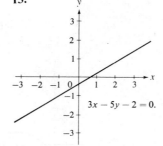

$3x - 5y - 2 = 0.$

15. $x - 5y = 0.$

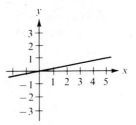

17.

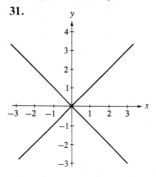

$x - 4 = 0.$

19. $3x - 2y - 13 = 0.$ **21.** $2x - y - 7 = 0.$ **23.** $(15/14, -5/14).$ **25.** $(119/41, 99/41).$
27. $2/5.$ **29.** $2/5.$

31.

33.

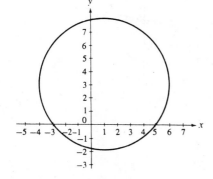

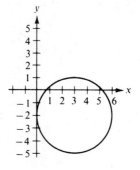

35. $0.0709 \text{ cal/mole}^{-1}.$ **39.** $(-1, 4), (4, 6), (-3, -1).$

Section 2.3

1. $\sqrt{2}.$ **3.** $1/5.$ **5.** $2/5.$ **7.** $32/\sqrt{29}.$ **9.** $10/3.$ **11.** $13/\sqrt{10}, 26/\sqrt{29}, 26/5.$
13. $7x - 7y + 2 = 0.$ **15.** $3x + y - 2 = 0.$ **17.** $2(\sqrt{2} + 1)x + 2y - (4 + 3\sqrt{2}) = 0.$
19. $3/\sqrt{29}.$ **21.** $7/2\sqrt{5}.$ **23.** $2\sqrt{5}.$ **25.** $13.$
27. $(21/10, 31/10), (-9/4, -5/4), (79/31, 31/10), (-9/4, -53/4).$ **29.** $\pm 3/\sqrt{7}.$ **31.** $\pm 20/3.$
33. $2(\sqrt{3} - 1).$

Section 2.4

1. $(x - 1)^2 + (y - 3)^2 = 25,$
    $x^2 + y^2 - 2x - 6y - 15 = 0.$

3. $(x - 3)^2 + (y + 2)^2 = 9,$
    $x^2 + y^2 - 6x + 4y + 4 = 0.$

5. $(x - 1/2)^2 + (y + 3/2)^2 = 4,$
  $2x^2 + 2y^2 - 2x + 6y - 3 = 0.$

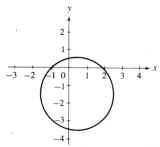

7. $(x - 5)^2 + (y - 1)^2 = 25,$
  $x^2 + y^2 - 10x - 2y + 1 = 0.$

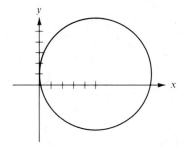

9. $x^2 + (y + 3/2)^2 = 25/4,$
  $x^2 + y^2 + 3y - 4 = 0.$

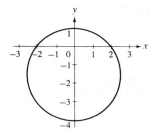

11. $(x + 5)^2 + (y - 5)^2 = 25,$
  $x^2 + y^2 + 10x - 10y + 25 = 0.$

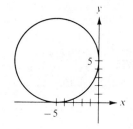

13. $(x - 1)^2 + (y - 2)^2 = 4.$

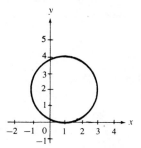

15. $(x + 1)^2 + (y - 3)^2 = 0.$

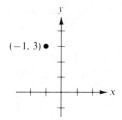

17. $(x - 4)^2 + (y - 1)^2 = 4,$
  $x^2 + y^2 - 8x - 2y + 13 = 0.$

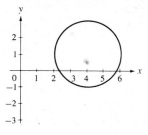

19. $(x + 3)^2 + (y - 3)^2 = 9,$
  $x^2 + y^2 + 6x - 6y + 9 = 0.$

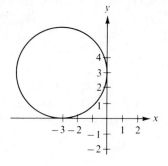
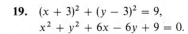

21. $(x - 1/2)^2 + (y - 3/2)^2 = 9/4$.

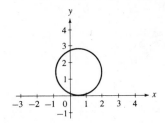

23. $(x - 1/2)^2 + (y + 3/2)^2 = 0$.

25. $(x - 1/3)^2 + (y + 1)^2 = -1/9$.   **27.** $(x - 1/2)^2 + (y + 1/4)^2 = -1$.

29. $(2, -1), (3, 3)$.   **31.** $(-1, 2), (3, 4)$.   **35.** $4x - 4y - 15 = 0$.

39. $x^2 + (y - 1)^2 = 4, x^2 + (y + 1)^2 = 4$.

Review Problems

1. (a) $2x - 3y + 13 = 0$, (b) $2x - y - 6 = 0$, (c) $3x + 3y - 1 = 0$, (d) $x - 2 = 0$.

3. $x + y - 1 = 0$.   **5.** $x - 1 = 0, 2x + 9y - 23 = 0, 10x + 9y - 31 = 0$.

7. (a) Circle with center $(5, -2)$ and radius 4, (b) Point $(1/3, -3/2)$.

9. $12/\sqrt{13}, 8/\sqrt{5}, 24/\sqrt{13}$.   **11.** $3x - y + 4 = 0$.   **13.** $2x + 6y - 11 = 0$.

15. $(x - 2)^2 + (y + 1)^2 = 17$.   **19.** $5x - 4y + 3 = 0$.

Section 3.1

1. Function.   **3.** No function.   **5.** No function.   **7.** Function.   **9.** No function.

11. $\{x | x \text{ real}, x \neq 2\}$.   **13.** $\{x | x \text{ real}, x \neq \pm 1\}$.   **15.** $\{x | x \text{ real}, x \leq -3 \text{ or } x \geq 2\}$.

17. $\{x | x \text{ real}, x < 0 \text{ or } x \geq 1\}$.   **19.** $\{x | x \text{ real}, x \geq 0\}, \{y | y \text{ real}, y \geq 1\}$.

21. $\{x | x \text{ real}, x \geq 2\}, \{y | y \text{ real}, y \geq 0\}$.

23. $\{x | x \text{ a positive integer}\}, \{y | y \text{ a positive integer}\}$.

25. $\{x | x \text{ real}, x > 0\}, \{y | y \text{ real}, y > 0\}$.   **27.** $\{x | x \text{ real}\}, \{y | y \text{ real}, y < 0 \text{ or } y = 1\}$.

29. $\{x | x \text{ real}, x \neq 0\}, \{y | y \text{ real}, y \geq 1 \text{ or } y < 0\}$.

31. $f(x) = x^3, \{x | x \text{ real}, 0 \leq x \leq 1\}, \{y | y \text{ real}, 0 \leq y \leq 1\}$.

33. $f(x) = x - 3, \{x | x \text{ real}, 0 < x \leq 6\}, \{y | y \text{ real}, -3 < y \leq 3\}$.

35. $f(x) = x, \{0, 1, 2, 3, 4, 5\}, \{0, 1, 2, 3, 4, 5\}$.

37. $f(x) = 2x, \{x | x \text{ a positive integer}\}, \{y | y \text{ a positive even integer}\}$.   **39.** $2, 5, 17, -7$.

41. -1, does not exist, $1/2, 1/(x + 1)$.

43. Does not exist, $\sqrt{3}, \sqrt{x^2 - 1}, \sqrt{x + h - 1}$.   **45.** $y^2 + 1, x^2 + 2hx + h^2 + 1$.

47. $0, 1, 2$.   **49.** $A = \pi r^2, C = 2\pi r$.   **51.** $A = s\sqrt{4R^2 - s^2}$.   **53.** $V = 2\pi r^3/3$.

55. Symmetric in the origin, symmetric in the y axis.   **57.** $4, -2$.   **59.** $x \geq 3 \text{ or } x \leq -4$.

Section 3.2

1. $2x + h$.   **3.** $2x + 2 + h$.   **5.** $28 + 4h$.   **7.** $x + a$.   **9.** $-1/x$.

11. $2 + h$ if $h \neq -1, 0$ if $h = -1$.   **13.** $\sin (1/h)$.   **15.** $3y^2(x^2 - 1)$.   **17.** $xy/(x + y)$.

19. $(f + g)(x) = x^2 + 1/x^2, (f - g)(x) = x^2 - 1/x^2, (fg)(x) = 1, (f/g)(x) = x^4, (g \circ f)(x) = 1/x^4$
$(x \neq 0)$.

21. $(f + g)(x) = \sqrt{x} + \sqrt{1 - x}$ $(0 \le x \le 1)$, $(f - g)(x) = \sqrt{x} - \sqrt{1 - x}$, $(fg)(x) = \sqrt{x - x^2}$,
$(f/g)(x) = \sqrt{x/(1 - x)}$ $(0 \le x < 1)$, $(g \circ f)(x) = \sqrt{1 - \sqrt{x}}$ $(0 \le x \le 1)$.

23. $(f + g)(x) = 4x - 2$ $(-3 < x < 4)$, $(f - g)(x) = 2x + 4$, $(fg)(x) = 3x^2 - 8x - 3$, $(f/g)(x) =$
$(3x + 1)/(x - 3)$ $(-3 < x < 4, x \ne 3)$, $(g \circ f)(x) = 3x - 2$ $(-3 < x < 1)$.

25. $(f + g)(x) = x^3 + x + 4$ $(0 \le x < 2)$, $(f - g)(x) = x^3 - x - 4$, $(fg)(x) = x^4 + 4x^3$,
$(f/g)(x) = x^3/(x + 4)$, $(g \circ f)(x) = x^3 + 4$.

27. $g(x) = 2x - 3$.   **29.** $g(x) = x^2$.   **31.** $a = 0, b = 0$ or 1.

33. (a) Even, (b) Odd, (c) Even, (d) Odd, (e) Neither, (f) Even, (g) Odd, (h) Even, (i) Even.

Section 3.3

1. 1.   **3.** -1.   **5.** 3.   **7.** No limit.   **9.** 2.   **11.** No limit.   **13.** 2.   **15.** 3.
17. $2x + 2$.   **19.** $-1/(x - 1)^2$.   **21.** $1/2\sqrt{x}$.   **23.** No limit.   **25.** -1.   **27.** $4a^3$.
29. $1/2\sqrt{a}$.   **31.** 2.

Section 3.4

1. 0, continuous.   **3.** 0, continuous.   **5.** Discontinuous.   **7.** Discontinuous.
9. 0, continuous.   **11.** Discontinuous.   **13.** 0.   **15.** 0.   **17.** 5.   **19.** 2.
21. Discontinuous.   **23.** 0, continuous.   **25.** $-1/4$.   **27.** $-2/x^3$.

Review Problems

1. (a) $\{x | x \text{ real}\}$, $\{y | y \text{ real}, y \ge 2\}$, (b) $\{x | x \text{ real}, x \ge 4\}$, $\{y | y \text{ real}, y \ge 0\}$, (c) $\{x | x \text{ real}, x \ne -5\}$,
$\{y | y \text{ real}, y \ne 0\}$.

2. (a) $\{1, 3, 4, 6, 10\}$, $\{-2, 1, 2, 5\}$.

3. (a) 0, (b) -2, (c) $-5/4$, (d) $(x + y)(x + y - 3)$, (e) $2x + h - 3$, (f) $x + a - 3$.

5. (a) 0, (b) 2, (c) Does not exist, (d) $|x|$, (e) $1/(\sqrt{1 + h} + 1)$.

7. $(f + g)(x) = x^2 + x + 2$ $(0 \le x < 4)$, $(f - g)(x) = x^2 - x - 2$ $(0 \le x < 4)$, $(fg)(x) =$
$x^3 + 2x^2$ $(0 \le x < 4)$, $(f/g)(x) = x^2/(x + 2)$ $(0 \le x < 4)$.

9. $(f \circ g) = -2x + 3$ $(x \le 0)$, $g \circ f$ does not exist.   **10.** (a) 2, (b) No limit, (d) $-1/9$.

11. (a) 9, continuous, (b) Discontinuous, (c) 3, continuous.

Section 4.1

1. -2.   **3.** 12.   **5.** 7.   **7.** -5.   **9.** $4x - y - 4 = 0$, $x + 4y - 18 = 0$.
11. $12x - y - 20 = 0$, $x + 12y - 50 = 0$.   **13.** -2.   **15.** Does not exist, $h > 0$.
17. $x - 3y - 2 = 0$, $3x + y - 6 = 0$.   **19.** $1, 1, 1, -1, -1, -1$, no slope.
21. $(2, -4)$.   **23.** $4, 4, 4$.

Section 4.2

1. $4x - 4$.   **3.** $-3/x^4$.   **5.** $4s^3 + 4s$.   **7.** 1.   **9.** $5, 5/9$.   **11.** $3x - y + 2 = 0$.
13. $19, 9$.   **15.** $5, 0$.   **17.** $15, 32$.   **19.** $20, 21$.   **21.** $144, 168$.   **23.** $(3, -7)$.

25. $(1, 2), (-1, -2)$.   **27.** $(1/\sqrt{3}, 1/3\sqrt{3}), (-1/\sqrt{3}, -1/3\sqrt{3})$.

29. $4x - y + 1 = 0$.   **31.** 36 ft, 2 sec, 48 ft/sec.

33. $(B^2 + 4AC)/4A$ ft, $(B + \sqrt{B^2 + 4AC})/2A$ sec, $\sqrt{B^2 + 4AC}$ ft/sec.

35. $f'(x)$

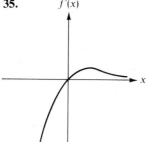

37. $f'(x)$

39. $22,000, $90; $36,000, $200.

41. $18,000, $60, increase production; $19,000, $-$50, decrease production.

Section 4.3

1. $6x + 5$.   **3.** $7x^6 + 42x^5$.   **5.** $35x^4 + 15x^2 + 1$.   **7.** $15t^2 + 4t - 6$.   **9.** 1.

11. 10.   **13.** $4(x^3 - 2x + 1)$.   **15.** $4q^3 + 10q$.   **17.** 4.

19. $(0, 0), (-4/3, -4/27)$.   **21.** $(3, -1), (5, -5)$.   **23.** $y - 2 = 0$.

25. 5 ft, 2 ft/sec, 2 ft/sec; -7 ft, -19 ft/sec, 19 ft/sec.   **29.** $3x^2 + 12x + 11$.

31. $4x^3 + 30x^2 + 70x + 50$.

Section 4.4

1. $-2/x^3$.   **3.** $-2/x^3$.   **5.** $(2/3)x^{-1/3}$.   **7.** $(2x^3 - 3x^2)/(x - 1)^2$.

9. $(2t + 1)/3t^{4/3}$.   **11.** $(-4v^2 + 8v)/(v^2 - 2v + 2)^2$.

13. $2x/\sqrt{u}$.   **15.** $9u^2 + 3$.   **17.** 0.   **19.** $-52/49$.   **21.** -3.

23. $2a^{1/3}/[3q^{2/3}(q^{1/3} + a^{1/3})^2]$.   **25.** $(4x^3 - 3x^2 + 3)/x^2$.

27. $(x + 2)/\sqrt{x^2 + 4x - 3}$.   **29.** $(2, 5), (-1, -4)$.   **31.** $(0, 0)$.   **33.** $x - 4y + 4 = 0$.

35. (a) $\sec^2 x$, (b) $-\csc^2 x$, (c) $\tan x \sec x$, (d) $-\cot x \csc x$.   **37.** -0.1 liter/min.

Section 4.5

1. $4(x + 1)^3$.   **3.** $20(4x + 3)^4$.   **5.** $2/\sqrt{4x + 2}$.   **7.** $2q^2/\sqrt[3]{q^3 - 8}$.

9. $-1/(x + 1)^{1/2}(x - 1)^{3/2}$.   **11.** $-(v + 3)/3(v + 1)^{2/3}(v - 1)^{5/3}$.

13. $8/3(2x + 1)^{5/3}(2x - 1)^{1/3}$.   **15.** $1 + 1/\sqrt{2x + 1}$.   **17.** 64.   **19.** 28.

21. $3x^5(2 - x^3)/(1 + x^3)^4$.   **23.** $2/(1 - x)^2$.

25. $-(x^3 + 2x^2 + 12x + 8)/[2\sqrt{x + 1}(x^2 - 4)^2]$.   **27.** 3/4.   **29.** 2, 4.   **31.** 2, 3, 13/5.

33. $6x + y - 15 = 0$.   **35.** $x/|x|, 1, 1, 1, -1, -1, -1$, no derivative.

Section 4.6

1. -1.   **3.** $1/4$.   **5.** $-x^2/y^2$.   **7.** $-(x/y)^{1/3}$.   **9.** $(y/x)^{2/3}$.   **11.** $-y/2x$.

13. $y/(y-x)$.   **15.** 0.   **17.** Does not exist.

19. $[2(x-y+1)-3(x+y)^2]/[2(x-y+1)+3(x+y)^2]$.   **21.** $-(x^2+2xy)/(x^2+y^2)$.

23. $[2x(1-x^2+y^2)]/[y(2y^2-2x^2-1)]$.   **25.** $[x(x-y)^2+y]/[x-y(x-y)^2]$.

27. $(-2/3, 3+2\sqrt{21}/3)$, $(-2/3, 3-2\sqrt{21}/3)$; $((-2+2\sqrt{7})/3, 3)$, $((-2-2\sqrt{7})/3, 3)$.

29. $3x-4y-18=0$.   **33.** $3x-4y+24=0, 4x+3y+7=0$.   **35.** $-4x/9y$.

37. $2, -1, 2, -1$, does not exist.

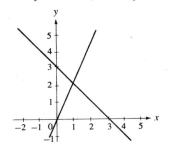

Section 4.7

1. 0.   **3.** $48v(v^2+1)(7v^2+3)$.   **5.** $(6x^2+2)/(x^2-1)^3$.   **7.** $(y^2-2xy)/(y-x)^3$.

9. $1/3x^{4/3}y^{1/3}$.   **11.** $-5/y^3$.   **13.** $3, 6, 8$.   **15.** $2, 3/4, -9/32$.   **17.** $3/4, 25/64$.

19. $-1, 4$.

21.

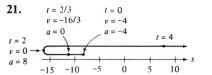

23.

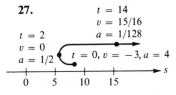

25.

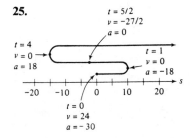

27.

29.

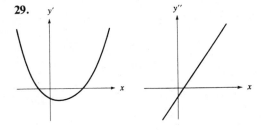

31.

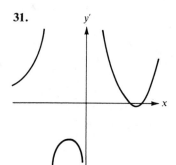 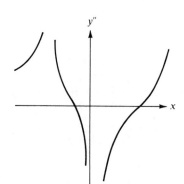

Review Problems

1. $-12/x^4$.   3. $12x^2(x^3 - 2)^3$.

5. $-1/x^2\sqrt{x^2 + 1}$.   7. $18(2x - 3)^{1/2}/(2x + 3)^{5/2}$.   9. $(2x^2 + 16x - 17)/(x + 4)^2$.

11. $(x + 2y)/(3y - 2x)$.   13. $2(x + 1)(x - 2)/(2x - 1)^2$.

15. $(3x - 4x^2 - 24)/(2x + 3)^4\sqrt{x^2 + 4}$.   17. $2u, u(1 + 2\sqrt{x + 1})/\sqrt{x + 1}$.

19. $(12y + x)/2x^2$.   21. $-5/(2x - 3)^2$.   23.

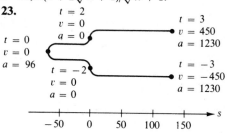

25. $(\pm 1, 0)$.   27. $3x^2 + 3y^2 + 24x - 16y - 14 = 0$.

Section 5.1

1. $(0, 0), (-3, 0)$.   3. $(-3, 0), (-2, 0), (2, 0), (0, -12)$.

5. $(-1/4, 0), (2, 0), (-3/2, 0), (0, -18)$.   7. $(-2, 0), (0, 8)$.

9. $(-1/2, 0), (1/3, 0), (0, 8)$.   11. $(0, 0)$.   13. $(3, 0), (0, 9)$.   15. None.   17. None.

19. $x = 3, y = 0$.   21. $x = -3, y = 1$.   23. $x = 1, y = 3$.   25. $x = -1, x = 3, y = 0$.

27. $y = 2$.   29. $x = -3/2, x = -1, y = 0$.   31. $y = 0$.   33. $y = 0$.

35. $x = 0, y = 4x$.   37. $x = 0, y = x + 1$.   39. $x = 0, y = x - 6$.   41. $80,000, 9$.

43. (a) $y = 0$, (b) $y = a_m/b_m$, (c) None.

Section 5.2

1. y axis.   3. y axis.   5. x axis.   7. x axis.   9. Origin.

11.

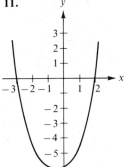

13.

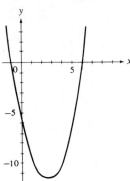

15.

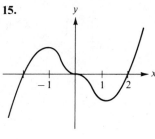

17.

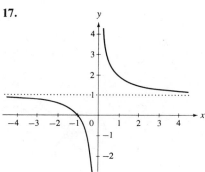

19.

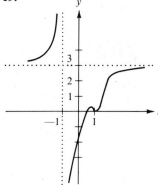

21.

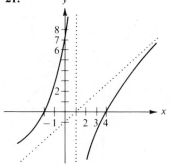

23.

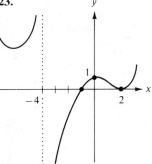

25.

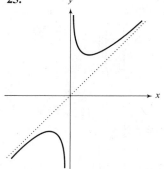

27.

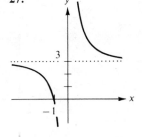

29.

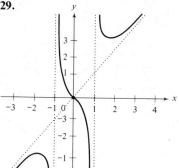

33. No.   **37.** Yes.

Section 5.3

1.

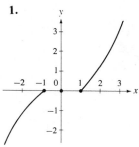

3.

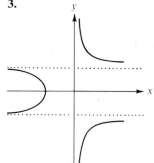

5.

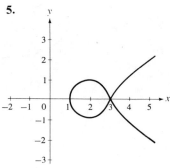

7.

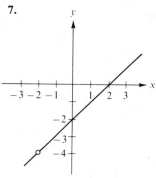

9.

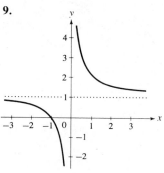

11.

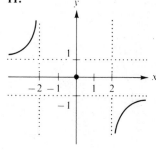

13.

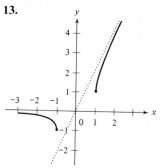

15.

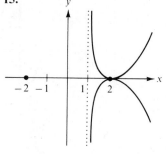

17.

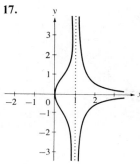

19.

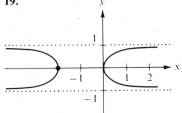

21.

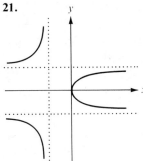

23.

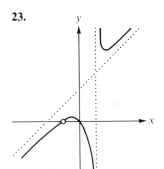

25.

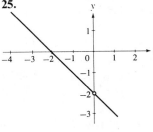

27.

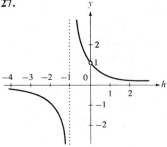

29.

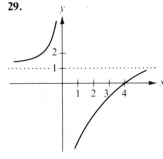

31.

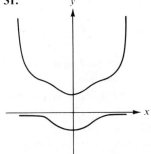

Section 5.4

1. $(-1, -4)$ absolute min.   **3.** $(4, 18)$ absolute max.

5. $(0, 1)$ relative max, $(2, -3)$ relative min.   **7.** $(2, -14)$ relative min, $(-2, 18)$ relative max.

9. $(-1, -1)$ absolute min, $(0, 0)$ neither.   **11.** $(-2, 48)$ neither, $(0, 32)$ absolute min.

13. None.   **15.** None.   **17.** $(0, -1)$ relative max.

19. $(-2, -4/27)$ relative min, $(0, 0)$ relative max.   **21.** $(-1, 0)$ relative max, $(1, -4)$ relative min.

23. $(-4, 0)$ relative max, $(-4/7, -(5832)(331{,}776)/7^7)$ relative min, $(2, 0)$ neither.

25. $(0, 0)$ neither.   **27.** $(0, -1)$ absolute min.

29. $(1, 0)$ neither, $(-2, 0)$ relative max, $(0, -\sqrt[3]{4})$ relative min.

31. $(3, 0)$ end point min.

33. $(0, 0)$ relative max, $(\pm 1/2\sqrt{2}, -1/4)$ absolute min.

37. $-B/2A$, $A < 0$ max, $A > 0$ min.

Section 5.5

1. $(1/2, -25/4)$ absolute min.   **3.** $(2, -2)$ relative min, $(-1, 25)$ relative max.

5. $(0, 24)$ relative max, $(-1, 19)$ relative min, $(2, -8)$ absolute min.

7. $(0, 0)$ relative max, $(1, -1)$ relative min.   **9.** Concave upward everywhere.

11. $(1, -1)$; concave downward for $x < 1$; concave upward for $x > 1$.

13. $(0, 0), (3, -81)$; concave downward for $0 < x < 3$; concave upward for $x < 0$ and $x > 3$.

15. $(1, 2)$ relative min, $(-1, -2)$ relative max.   **17.** $(0, 0)$ relative min, $(-2, -4)$ relative max.

19. None; concave downward for $x < -2$; concave upward for $x > -2$.

21. $(0, 0), (\sqrt{3}, \sqrt{3}/4), (-\sqrt{3}, -\sqrt{3}/4)$; concave upward for $-\sqrt{3} < x < 0$ and $x > \sqrt{3}$; concave downward for $x < -\sqrt{3}$ and $0 < x < \sqrt{3}$.

23. $(0, 0)$ relative max, $(2, -4)$ relative min, $(1, -2)$ point of inflection.

25. $(-1/2, 59/4)$ relative max, $(3, -71)$ relative min, $(5/4, -225/8)$ point of inflection.

27. $(1/2, -3/8)$ absolute min.

29. $(0, 0)$ relative max, $(4, -256)$ relative min, $(3, -162)$ point of inflection.

31. $(2, 1/4)$ absolute max, $(3, 2/9)$ point of inflection.

Section 5.6

1. $(-1, -1)$ relative max, $(0, -2)$ absolute min, $(2, 2)$ absolute max.

3. $(-2, -44)$ absolute min, $(-1/2, 13/4)$ absolute max, $(1, -17)$ relative min.

5. $(\pm 1, 1)$ absolute min.

7. $(\pm 1, 0)$ absolute min, $(3, 64)$ absolute max, $(0, 1)$ relative max, $(-2, 9)$ relative max.

9. $(0, 0)$ absolute min.   **11.** $(0, 0)$ absolute min.

13. $(-\sqrt{3}, 0)$ relative min, $(-1, 2)$ absolute max, $(1, -2)$ absolute min, $(2, 2)$ absolute max.

15. $(\pm 1, 0)$ absolute min, $(0, 1)$ absolute max.   **17.** $(0, 1)$ relative max, $(\pm 1, 0)$ absolute min.

19. $(\pm 1, -1)$ absolute min, $(0, 0)$ relative max.

21. $(-2, 2)$ absolute max, $(0, 0)$ relative min, $(1, 1)$ relative max.   **23.** $750, 11,375$.

Section 5.7

1. $(0, 0)$ min, $(-4, 32)$ max, $(-2, 16)$ point of inflection. y

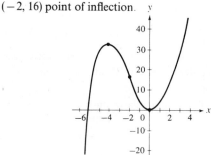

3. $(1, 1)$ max, $(3, -3)$ min, $(2, -1)$ point of inflection. y

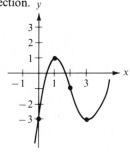

5. $(-1, 0)$ max, $(3/5, -35)$ min,
$(3, 0)$, $(1.6, -18.5)$,
$(-0.4, -14.1)$ points of inflection.

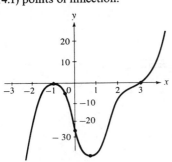

7. $(-1, 25)$ max, $(2, -2)$ min,
$(1/2, 23/2)$ point of inflection.

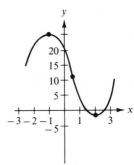

9. $(-3, 0)$ max, $(-1/3, -4\sqrt[3]{4}/3)$ min,
$(1, 0)$ point of inflection.

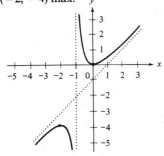

11. $(1, -2)$ min, $(0, 0)$ max.

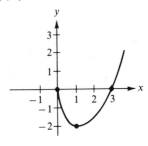

13. $(0, 0)$ min, $(-2, -4)$ max.

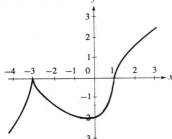

15. $(-1, -2)$ max, $(1, 2)$ min.

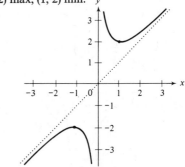

17. None.

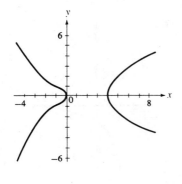

19. $(4, 4\sqrt{2})$ max, $(4, -4\sqrt{2})$ min.

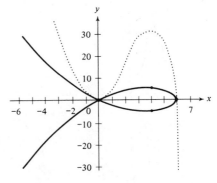

21. $(3/2, 2/3\sqrt{3})$ max, $(3/2, -2/3\sqrt{3})$ min.

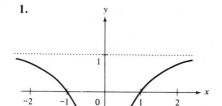

23. $(-1, 1/\sqrt{3})$ max, $(-1, -1/\sqrt{3})$ min.

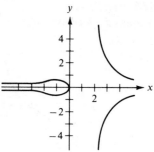

Review Problems

1.

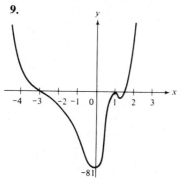

3.

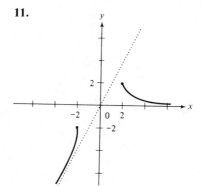

5. $(-1, 15)$ relative max, $(2, -12)$ relative min.

7. $(-\sqrt{3}, -9)$ absolute min, $(0, 0)$ relative max, $(\sqrt{3}, -9)$ absolute min, $(\pm 1, -5)$ points of inflection.

9.

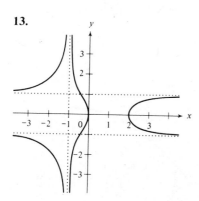

11.

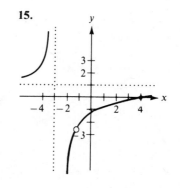

13.

15.

17. (0, 0) neither, (1/5, 4/3125) relative max, (1/3, 0) relative min.

19. $(-1, 2)$ relative max, $(1, -2)$ relative min, $(0, 0)$ neither.

21. (0, 81) relative max, $(\pm 3, 0)$ absolute min, $(\pm\sqrt{3}, 36)$ points of inflection.

23. (0, 17) relative max,   **25.** (0, 0) relative max,   **27.** $(-2, 4\sqrt{2})$ relative max,

     $(4, -15)$ relative min,   no point of inflection. $(-2, -4\sqrt{2})$ relative min,

     (2, 1) point of inflection.   (0, 0) neither.

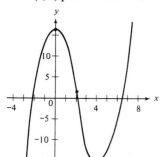 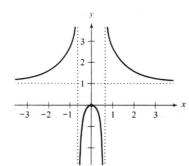 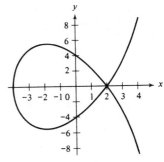

29. (a) $f(x) < 0, f'(x) > 0, f''(x) < 0$; (b) $f(x) > 0, f'(x) = 0, f''(x) < 0$;

     (c) $f(x) < 0, f'(x) < 0, f''(x) > 0$; (d) $f(x) > 0, f'(x)$ and $f''(x)$ undefined;

     (e) $f(x), f'(x),$ and $f''(x)$ all undefined; (f) $f(x) > 0, f'(x) = 0, f''(x) > 0$;

     (g) $f(x) > 0, f'(x) > 0, f''(x) = 0$; (h) $f(x) > 0, f'(x) > 0, f''(x) = 0$.

Section 6.1

1. $x = y = 24$.   **3.** 200 ft of \$1.00 fence, 800 ft of \$0.50 fence, \$600.   **5.** 2400 ft.

7. Plan b, $1800\sqrt{2}$ ft.   **9.** Base: 20×20 in.; height: 10 in.

11. Base: $2\sqrt{3}$; height: 6.   **13.** Base: $\sqrt{3}R$; height: $3R/2$.

15. Bases: $2R$ and R; height: $\sqrt{3}R/2$.   **17.** 12π.   **19.** $r = R/\sqrt{2}, h = R\sqrt{2}$.

21. Radius: 1 in.; height: 2 in.   **23.** (1, 3).   **27.** 6.25 in.².   **29.** 1768.

31. 3 mi. down the coast.   **33.** 16/3 mi. down the coast.

35. 10 mi. down the coast.   **39.** $V/2$.   **41.** \$1.00.

43. $|Ax_1 + By_1 + C|/\sqrt{A^2 + B^2}$.   **45.** $2k/3$.   **47.** 12.8 ft.

Section 6.2

1. 1400 in.³/min, 560 in.²/min.   **3.** 1/4 in./min.   **5.** 65/12 ft/sec.

7. 3π ft²/min, 2π ft²/min.   **9.** 15/16 ft/sec.   **11.** -0.1 liter/min.   **13.** 4 lb/in.²/min.

15. $1/\sqrt[4]{3}$ in./min.   **17.** 3/2 units/min, 3/4 unit/min.

19. $-1/2$ unit/min, $-1/16$ unit/min.   **21.** 2/3 in./min.

23. -17 mph, $114/\sqrt{39}$ mph.

Section 6.3

1. $(3x - 2)x\, dx$.   **3.** $-11\, dx/(2x - 3)^2$.   **5.** $8x(x^2 + 1)^3\, dx$.   **7.** $-x\, dx/4y$.

9. $(x - y)^2\, dx/x(x - 2y)$ **11.** 2.4 in.².   **13.** $\pm 0.03\pi, \pm 0.1107\pi$.   **15.** 5/9°.

17. 30 ± 0.12 ohm.   **19.** 9.111.   **21.** 2.926.   **23.** 2.03125.   **25.** 0.192.
29. 1% **31.** $2/3\%$.

Section 6.4

1. 20,000   **3.** $2.00.   **5.** $47.50.   **7.** $20, $-$0.38.   **9.** $50; 200, $3000.
11. $50.50; 195, $2802.50.   **13.** $1860.

Section 6.5

1. $3x^4 - 2x^3 + 3x + C$.   **3.** $(2/5)x^{5/2} + C$.   **5.** $(3/2)x^{2/3} + 1/2$.
7. $2x^4 + 2x^2 - 3x - 15$.   **9.** $-1/x + C$.
11. $(4/5)x^{5/2} + (8/3)x^{3/2} + 2\sqrt{x} + C$.   **13.** $(x - 1)^3/3 + C$.   **15.** $2\sqrt{x} - 1$.
17. $6x^2 - 9x + 7$.   **19.** $-x^3 + 9x - 5$.   **21.** $x^3 + x - 2$.   **23.** $s = -16t^2 + 80t$.
25. $s = 2t^3/3 + 16t + 4$.   **27.** $s = 2t^2 + t + 2$.   **29.** $s = -16t^2 + v_0 t + s_0$.
31. 9/2 ft, 81/32 ft.   **33.** 44/225, 352 ft/sec (240 mph).   **35.** 36 ft, 3 sec.
37. $x^3 - 2x^2 + 5x + 150$.   **39.** -31.148 ft/sec/sec, -31.988 ft/sec/sec.

Review Problems

1. $(x^4 + 5x - 3)/3x$.   **3.** 400.   **5.** $5/36\pi$ ft/min.   **7.** 0.208.   **9.** 40.
11. $h = 6$ in., $r = 3\sqrt{2}$ in.   **13.** $2P/(4 + \pi) \times 2P/(4 + \pi)$.
15. $10\sqrt{17}$ ft/sec.   **17.** 0.7 mi.   **19.** $2\sqrt{5}/3$ ft/sec.   **21.** $y = 1/3x + K$.

Section 7.1

1. $1 + 8 + 27 + 64 + 125 + 216 + 343$.   **3.** $-3 - 1 + 1 + 3 + 5 + 7$.
5. $1 + 3 + 5 + 7 + 9 + 11 + 13 + 15$.   **7.** $2 + 9 + 28 + \cdots + (n^3 - 3n^2 + 3n)$.
9. $3 + 5 + 7 + \cdots + (2n + 1)$.   **11.** $\sum_{i=1}^{15} i$.   **13.** $\sum_{i=1}^{11} 2i$.
15. $\sum_{i=1}^{11} (i^2 - i + 2)/2$.   **17.** $\sum_{i=1}^{n} (2i - 1)$.   **19.** $\sum_{i=3}^{n} (i^2 + 1)$.   **21.** Equal.
23. Equal.   **25.** Equal.   **27.** Not equal.   **29.** $(n + 1)(2n + 1)/6n^2$.
31. $(2n + 1)(7n + 1)/6n^2$.   **33.** $(n - 1)(2n - 1)/6n^2$.   **35.** $(38n^2 + 15n + 1)/6n^2$.
43. $n(n + 1), n^2$.

Section 7.2

5. 1/2.   **7.** 1/5.   **9.** 7/3.   **11.** 15/4   **13.** 1.   **15.** 23/6.   **17.** 1/6.   **19.** 1/3.

Section 7.3

1. 6.   **3.** 2/5.   **5.** 1/4.   **7.** 4/3.   **9.** $-1/6$.   **11.** 4.   **13.** 1.   **15.** 9/2.
17. 1/12.   **19.** 8.

Section 7.4

1. 8/3.   3. 3/4.   5. 10.   7. $-3/4$.   9. 17/12.   11. 43/2.   13. 45/4.
15. 6.   17. 1/2.   19. 10/3.   21. 4/3.   23. 16/15.   25. 1/3.   27. 2.
29. 16/3.   33. $3/\sqrt{82}, -2/\sqrt{17}$.

Section 7.5

1. $(2x-3)^3/3 + C$.   3. $(x^3-x)^2/2 + C$.   5. $(4x-3)^{3/2}/6 + C$.
7. $(x^2+2)^3/6 + C$.   9. $x^8/8 + 4x^6/3 + 4x^4 + C$.   11. $(x^3+3)^3/9 + C$.
13. $(x^{1/3}-1)^6/2 + C$.   15. $(x^{-2}+3x)^6/6 + C$.   17. $(3x^2-5)^{4/3}/8 + C$.
19. $(x^2-4)/x + C$.   21. $x^2/2 + 4x - 4/(x+2) + C$.   23. 28/6.   25. 72.   27. 670/9.
29. $-15/8$.   31. $x^2 + 2x$.

Section 7.6

1. 32/3.   3. 9/8.   5. 23/3.   7. 125/6.   9. 9/2.   11. 1/3.   13. 1/2.
15. 37/12.   17. 125/24.   19. 1/12.   21. 4/3.   23. 4/3.   25. 9.

Section 7.7

1. 9/2 ft-lb.   3. 3/4 ft-lb.   5. 30 ft-lb.   7. 1030/3.   9. 225,800 ft-lb.
11. 1046 ft-lb, 3137 ft-lb.   13. 1250 ft-lb.   15. 625 ft-lb.   17. 1514 ft-lb.
19. 1,957,000 ft-lb.; 1,731,000 ft-lb.   21. 892 ft-lb.

Section 7.8

1. 3.344, 3.333, 3.333.   3. 2.3438, 2.3333, 2.3333.   5. 1.218, 1.219, 1.219
7. 1.8772, 1.8692, 1.8692.   9. 1.509, 1.500, 1.500.   11. 1.275, 0.007, 1.268–1.282.
13. 3.110, 0.053, 3.057–3.163.   15. 11.983, E_T cannot be determined, since $f''(4)$ does not exist.
17. 0.771, E_S cannot be determined, since $f^{(4)}(1)$ does not exist.
19. 11.003, 0.00006, 11.00294–11.00306.   21. 0.307, 0.0005, 0.306–0.307.
23. 158.9, 154.8.   25. 2.897, 2.901.   27. 0.819, 0.830.   29. 7.

Review Problems

1. $2(n+1)(2-5n)/3n^2$.   3. 206/15.   5. 1/3.   7. 0.   9. 275/4.   11. 9/2.
13. 1/20.   15. 13/8.   17. 3 ft-lb; 9 ft-lb.   19. 1,008,800 ft-lb.   21. 14.237, 13.962.
23. 8.888, 0.025, 8.863–8.913.   25. 2.

Section 8.1

1. $(-6, -2)$.   3. $(4, -1)$.   5. $(3/\sqrt{10}, -1/\sqrt{10})$.   7. $(4/5)\mathbf{i} - (3/5)\mathbf{j}$.   9. $(4, 3)$.
11. $(5, 23)$.   13. $B = (4, 3)$.   15. $A = (-1, 1)$.   17. $B = (7, -3)$.   19. $A = (2, 2)$.
21. $4\mathbf{i} + \mathbf{j}$.   23. $-2\mathbf{i} - 7\mathbf{j}$.   25. $A = (5/2, -3/2), B = (11/2, 7/2)$.

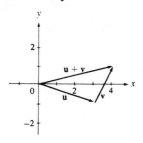
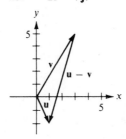

27. $A = (7/2, -3/2), B = (13/2, -5/2)$.   29. $A = (0, -7)$.   31. $A = (23/3, -1/3)$.

Section 8.2

1. $\text{Arccos } 1/5\sqrt{2} = 82°$.   3. $90°$.   5. $90°$.   7. $\text{Arccos } 1/\sqrt{5} = 63°$.
9. 1, not orthogonal.   11. 0, orthogonal.   13. -1, not orthogonal.
15. 9, not orthogonal.   17. $(1/2)\mathbf{i} + (1/2)\mathbf{j}$.   19. $(-3/25)\mathbf{i} + (21/25)\mathbf{j}$.   21. $(2/5)\mathbf{i} + (1/5)\mathbf{j}$.
23. $3\mathbf{i} - 2\mathbf{j}$.   25. 3.   27. -8.   29. -1.   31. $-1/3$.
33. $4 + 2\sqrt{3}$.   35. $8 - 13/\sqrt{3}$.   37. $-(2/17)\mathbf{i} - (8/17)\mathbf{j}, (-19/17)\mathbf{i} - (76/17)\mathbf{j}$.
39. $(-45/29)\mathbf{i} + (18/29)\mathbf{j}, (100/29)\mathbf{i} - (40/29)\mathbf{j}$.

Section 8.3

1. $6\mathbf{i} + 2\mathbf{j}$.   3. $\sqrt{65}$ lb to the right and inclined upward at an angle of $60°$ with the horizontal.
5. $\sqrt{25 - 12\sqrt{2}}$ lb to the right and inclined downward at an angle of $3°$ with the horizontal.
7. $-4\mathbf{i} + 2\mathbf{j}$.
9. $\sqrt{29}$ lb to the right and inclined downward at an angle of $68°$ with the horizontal.
11. $2\sqrt{10} + 3\sqrt{2}$ lb to the left and inclined downward at an angle of $34°$ with the horizontal.
21. $y = x^2/288$; 156,000 lb.
23. $y = x^2/720$; $240,000\sqrt{10}$ lb.   25. 1000 lb/ft.   27. 60 ft.

Review Problems

1. $(2/\sqrt{13}, -3/\sqrt{13})$.   3. $(-9/2, 5/2), (1/2, -1/2)$.   5. 8.   7. $\text{Arccos } (-\sqrt{2}/2) = 135°$.
9. $(3/2)\mathbf{i} - (3/2)\mathbf{j}$.   11. 3/4.   15. $10\sqrt{3}$ lb, 10 lb.
17. $\mathbf{v}_1 = (48/25)\mathbf{i} + (64/25)\mathbf{j}, \mathbf{v}_2 = (252/25)\mathbf{i} - (189/25)\mathbf{j}$.

Section 9.1

1. Axis: x axis, $V(0, 0)$, $F(4, 0)$,   **3.** Axis: y axis, $V(0, 0)$, $F(0, 2)$,   **5.** Axis: x axis, $V(0, 0)$, $F(5/2, 0)$,
    D: $x = -4$, lr $= 16$.   D: $y = -2$, lr $= 8$.   D: $x = -5/2$, lr $= 10$.

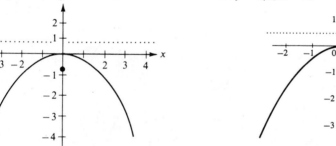

7. Axis: y axis, $V(0, 0)$, $F(0, -3/4)$,   **9.** Axis: y axis, $V(0, 0)$, $F(0, -1/2)$,
    D: $y = 3/4$, lr $= 3$.   D: $y = 1/2$, lr $= 2$.

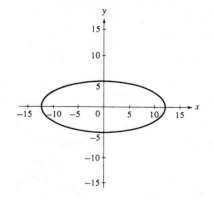

11. $y^2 = -4x/3$.   **13.** $y^2 = 5x$, $y^2 = -5x$.   **15.** $y^2 = -20x$.   **17.** $x^2 = 4y/3$.
21. $2x + y - 5 = 0$.   **23.** $2x - y - 6 = 0$.   **25.** $y = 0, 2x + y - 8 = 0$.
29. $72(3\sqrt{3} + 2)/23$ hr.

Section 9.2

1. $C(0, 0)$, $V(\pm 13, 0)$, $CV(0, \pm 5)$,   **3.** $C(0, 0)$, $V(0, \pm 6)$, $CV(\pm 3, 0)$,
    $F(\pm 12, 0)$, lr $= 50/13$. $F(0, \pm 3\sqrt{3})$, lr $= 3$.

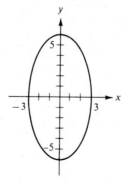

5. $C(0, 0)$, $V(0, \pm 7)$,
$CV(\pm 5, 0)$,
$F(0, \pm 2\sqrt{6})$, lr $= 50/7$.

7. $C(0, 0)$, $V(0, \pm 3)$,
$CV(\pm\sqrt{5}, 0)$,
$F(0, \pm 2)$, lr $= 10/3$.

9. $C(0, 0)$, $V(0, \pm 4)$, $CV(\pm 3, 0)$,
$F(0, \pm\sqrt{7})$, lr $= 9/2$.

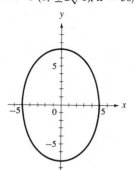

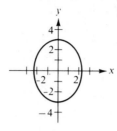

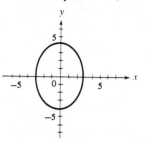

11. $x^2/25 + y^2/169 = 1$.   **13.** $x^2/25 + y^2/10 = 1$.   **15.** $x^2/16 + y^2/12 = 1$.

17. $x^2/100 + y^2/64 = 1$.   **21.** $4x + 3y - 11 = 0$.

23. $x - 4y + 14 = 0$, $11x + 12y - 70 = 0$.   **29.** 0.0167.

Section 9.3

1. $C(0, 0)$, $V(\pm 4, 0)$, $F(\pm 5, 0)$,
$A: y = \pm 3x/4$, lr $= 9/2$.

3. $C(0, 0)$, $V(0, \pm 3)$, $F(0, \pm\sqrt{10})$,
$A: y = \pm 3x$, lr $= 2/3$.

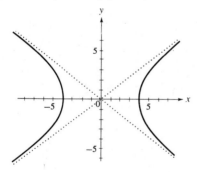

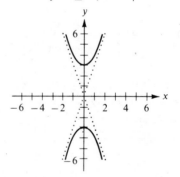

5. $C(0, 0)$, $V(\pm 12, 0)$, $F(\pm 13, 0)$,
$A: y = \pm 5x/12$, lr $= 25/6$.

7. $C(0, 0)$, $V(0, \pm 6)$, $F(0, \pm 3\sqrt{5})$,
$A: y = \pm 2x$, lr $= 3$.

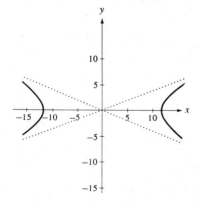

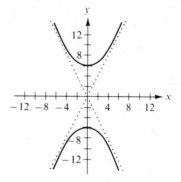

9. $C(0, 0), V(\pm 1, 0), F(\pm\sqrt{5}, 0),$
A: $y = \pm 2x$, lr $= 8$.

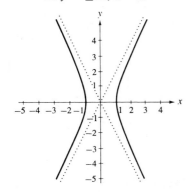

11. $C(0, 0), V(\pm 4, 0), F(\pm 4\sqrt{2}, 0),$
A: $y = \pm x$, lr $= 8$.

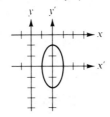

13. $C(0, 0), V(0, \pm 5/2), F(0, \pm\sqrt{34}/2),$
A: $y = \pm 5x/3$, lr $= 9/5$.

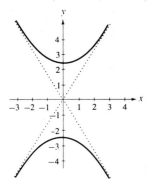

15. $x^2/4 - y^2/21 = 1.$ **17.** $x^2/36 - y^2/16 = 1.$ **19.** $16x^2 - 9y^2 = 288.$ **21.** None.
23. $y^2/25 - x^2/144 = 1.$ **25.** $x^2/9 - y^2/16 = 1.$ **27.** $5x + y - 9 = 0.$
29. $5x - 4y - 9 = 0.$ **37.** $279x^2 - 121y^2 = 84{,}397{,}500.$

Section 9.4

1. $y'^2 = 4x'.$

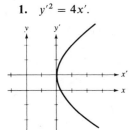

3. $x'^2 + y'^2/4 = 1.$

5. $9x'^2 - 4y'^2 = 36.$

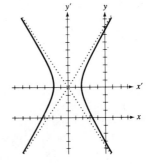

7. $x'^2/4 + y'^2/9 = 1$.

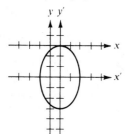

9. $x'^2 = y'$.

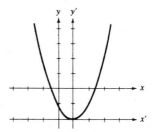

11. Axis: $x = 3$, $V(3, -1)$, $F(3, 1/2)$,
D: $y = -5/2$, lr $= 6$.

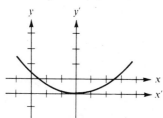

13. $C(0, 3)$, $V(\pm 3, 3)$, $F(\pm 5, 3)$,
A: $y = 3 \pm 4x/3$, lr $= 32/3$.

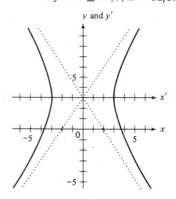

15. $C(-3, 1)$, $V(-3 \pm 3, 1)$, $CV(-3, 1 \pm 2\sqrt{2})$,
$F(-3 \pm 1, 1)$, lr $= 16/3$.

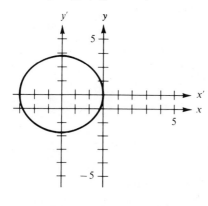

17. $C(1/3, -2/3)$, $V(1/3 \pm 2, -2/3)$,
$F(1/3 \pm 2\sqrt{2}, -2/3)$,
A: $y = -2/3 \pm (x - 1/3)$, lr $= 4$.

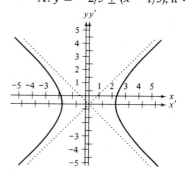

19. $25x^2 + 9y^2 - 100x - 18y - 116 = 0$. **21.** $x^2 - 15y^2 + 4x + 30y + 4 = 0$.
23. $x^2 + 4y^2 - 8x - 8y + 16 = 0$. **25.** $x'^2 - 2x'y' + 4y'^2 - 5 = 0$.
27. $x'y' + 1 = 0$. **29.** $y' = x'^3 + 3x'^2$, $y' = x'^3 - 3x'^2$. **31.** $y' = x'^5 + 2x'^3 + x'$.

Section 9.5

1. $\sqrt{13}x' = 6$.

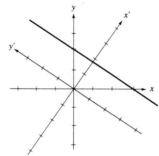

3. $x'^2 - y'^2 = 14$.

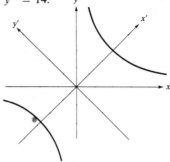

5. $9x'^2 + 4y'^2 = 36$.

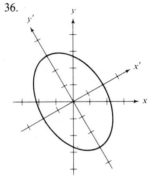

7. $y'^2 = 16x'$.

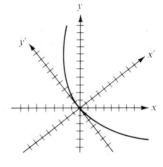

9. $3x'^2 + y'^2 - 16y' = 0$.

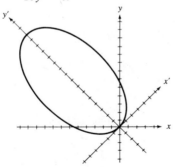

11. $x'^2/7 - y'^2/6 = 1$.

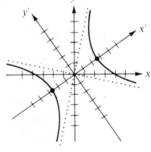

13. $4x'^2 + 9y'^2 - 36 = 0$.

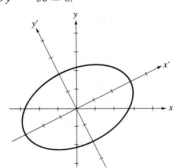

15. $x'^2 + 4x' - 5 = 0$.

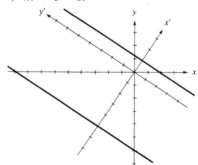

17. $y'^2 - 12x' = 0$.

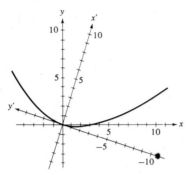

19.

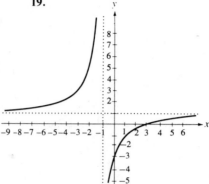

21.

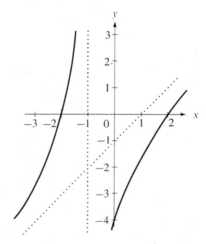

Review Problems

1. Axis: x axis; $V(0, 0)$, $F(4, 0)$,
D: $x = -4$, lr $= 16$.

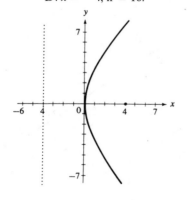

3. $C(0, 0)$, $V(0, \pm 3)$, $F(0, \pm 3\sqrt{2})$,
A: $y = \pm x$, lr $= 6$.

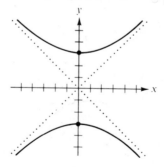

5. Axis: $y = -1$, $V(3, -1)$, $F(13/4, -1)$,
D: $x = 11/4$, lr $= 1$.

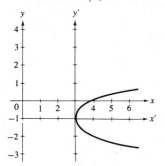

7. $C(-2, -4)$, $V(-2 \pm 4, -4)$,
$F(-2 \pm 5, -4)$,
A: $y + 4 = \pm(3/4)(x + 2)$, lr $= 9/2$.

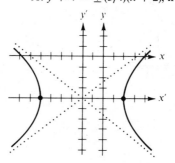

9. $C(-1, -1/2)$, $V(-1 \pm 4, -1/2)$,
$F(-1 \pm 5, -1/2)$,
A: $y = -1/2 \pm 3(x + 1)/4$, lr $= 9/2$.

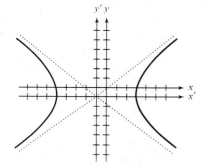

11.

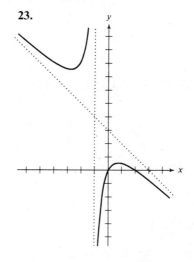

13. $y^2 = 8x$.   **15.** $16y^2 - 9x^2 = 576$.   **17.** $y^2 - 8x - 10y + 33 = 0$.
19. $8x + 3y - 8 = 0$.
21. $y' = x'^3 - 9x'$.

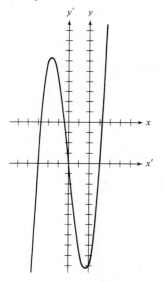

23.

25. $x'^2 + 5y'^2 = 20$.

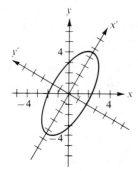

27.

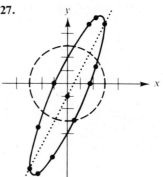

Section 10.1

1. $P = A = (0, 0)$.   **3.** $P = (2, 3)$, $A = (2, 1)$.   **5.** $P = (2, 4)$, no A.   **7.** $P = A = (2, 5)$.
9. $P = A = (0, 0)$.   **11.** Limit point.   **13.** Not a limit point.   **15.** Limit point.
17. Limit point.   **19.** Limit point.   **21.** Not a limit point.   **23.** Limit point.
25. Limit point.

Section 10.2

1. Not satisfied.   **3.** Not satisfied.   **5.** Satisfied.

Section 10.3

7. True.   **9.** True.   **11.** True.   **13.** False.   **15.** True.

Section 10.4

9. 2, 0, does not exist.   **11.** Does not exist, 0, 0.   **13.** Does not exist, 0, does not exist.

Section 10.5

15. False.   **17.** True.   **19.** True.

Section 10.6

1. Continuous.   **3.** Discontinuous.   **5.** Continuous.   **7.** Continuous.
9. Continuous.   **11.** Continuous.
13. Not continuous on reals, $x = 3$; continuous on domain.
15. Discontinuous on reals, $x = 1$; discontinuous on domain, $x = 1$.

17. Discontinuous on reals, $x = 1$; discontinuous on domain, $x = 1$.   **19.** Discontinuous.
21. Continuous.   **23.** Continuous.   **25.** Discontinuous everywhere.   **27.** $1/2$.

Section 10.7

1. -1, Theorems 10.4 and 10.3.   **3.** 2, Theorems 10.4 and 10.3.
5. 4, Theorems 10.4 and 10.3.   **7.** $-1/4$, Theorems 10.4 and 10.3.
9. $1/(x + 1)^2$, Theorems 10.4 and 10.3.

Review Problems

1. Continuous: $x \le -2$, $-1 \le x < 0$, $0 < x < 1$, $x > 1$; discontinuous: $x = 0$, $x = 1$.
3. False.   **5.** True.   **7.** True.   **9.** True.   **11.** True.   **13.** True.   **15.** False.
17. Continuous.

Section 11.1

1. $\pi/4$, $-7\pi/6$, $3\pi/2$, $\pi/6$.   **2.** $60°$, $180°$, $135°$, $-90°$.
3. $\sin \theta = -5/13$, $\cos \theta = -12/13$, $\tan \theta = 5/12$, $\csc \theta = -13/5$, $\sec \theta = -13/12$, $\cot \theta = 12/5$.
5. 1.   **7.** $-\sqrt{3}/2$.   **9.** $1 + \sin 2u$.   **11.** $\cos x(1 - \sin^2 x)$.
13. $\sec \theta \tan \theta (\sec^4 \theta - \sec^2 \theta)$.   **17.** $(\sec^2 \theta)/(2 - \sec^2 \theta)$.
19. $(x + \sqrt{9 + x^2})/3$.   **21.** $(1 - \cos 4x)/8$.
23.   **25.**   **27.**

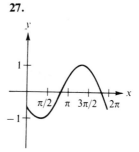

29.   **31.**

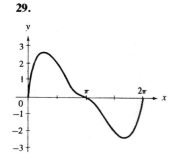

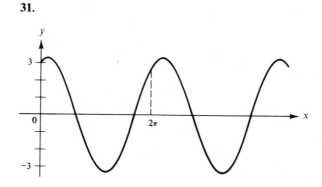

33.

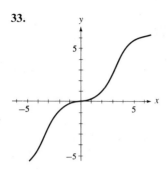

Section 11.2

1. $2 \cos 2x$.   **3.** $-3 \sin 3x$.   **5.** $6 \sin 3x \cos 3x$.   **7.** $6(\sec 4x)^{3/2} (\tan 4x)$.
9. $\sec^3 x + \sec x \tan^2 x$.   **11.** $2(\sin^2 x - \cos^2 x)$.   **13.** 0.   **15.** $1 + \sec x \tan x$.
17. $(2x \sec^2 2x - \tan 2x)/2x^2$.   **19.** $-\sqrt{2}/2$.   **21.** No value.
23. $-2 \sin x/(1 - \cos x)^2$.   **25.** $\sin x \sin (\cos x)$.   **27.** $\cos x \cos^2 y$.
29. $-\csc^2 (x + y)$.   **31.** $\sec^2 (x - y) + \tan^2 (x - y)$.
33. $(\pi/2, 2\sqrt{3} - 1)$ min, $(3\pi/2, -2\sqrt{3} - 1)$ min,   **35.** $(2n\pi + \text{Arctan } 1/\sqrt{2}, \sqrt{3})$ max,
$(\pi/3, 5/2)$ max, $(2\pi/3, 5/2)$ max, $[(2n + 1)\pi + \text{Arctan } 1/\sqrt{2}, -\sqrt{3}]$ min,
$(73°, 2.483)$, $(107°, 2.483)$, $(211°30', -1.358)$, $(n\pi - \text{Arctan } \sqrt{2}, 0)$ point of inflection.
$(328°30', -1.358)$ points of inflection.

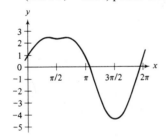

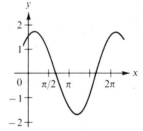

37. $3x - 6y + (3\sqrt{3} - \pi) = 0$.   **41.** 48π mi/min.
45. $\lim_{x \to 0} y = 0$, $\lim_{x \to 0} y'$ does not exist.

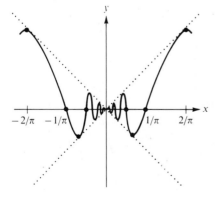

Section 11.3

1. $-(\cos 2\theta)/2 + C$.   3. $-(\csc 4x)/4 + C$.   5. $(\sec x^3)/3 + C$.
7. $2 - \sqrt{2}$.   9. $2 - \pi/2$.   11. $5x/2 - 2\sin^2 x + (3\sin 2x)/4 + C$.
13. $-(\sin 2x)/2 + C$.   15. $(\sin^5 3\theta)/15 + C$.   17. $-(\cos^3 x)/3 + C$.
19. $5\tan x + 4\sec x - 4x + C$.   21. $\theta/2 + (\sin 2\theta)/4 + C$.
23. $9x + 6\csc x - (\cot^3 x)/3 + C$.   25. $-(2\cot^{3/2} x)/3 + C$.   27. 3π.   29. 2.

Section 11.4

1. 0.   3. π.   5. $2/\sqrt{1 - 4x^2}$.   7. $-1/(x\sqrt{16x^2 - 1})$.   9. $2/(1 + 4x^2)$.
11. $(1 - x)/\sqrt{1 - x^2}$.   13. $x^2/(1 + x^2)$.   15. $(2\,\text{Arcsin}\,x)/\sqrt{1 - x^2}$.   17.   No value.
19. $-4/\sqrt{6}$.   21. $2x^2/\sqrt{1 - x^2}$.   23. $-1/(x^2\sqrt{1 - x^2})$.   25. $-1/|x|\sqrt{x^2 - 1}$.
27. $x^2/(1 - x^2)^{3/2}$.   29. $2x - 4y + (2 - \pi) = 0$.   31. $(\pi - 2)x + 4y + 2 - 2\pi = 0$.
33.   35.   37. None.

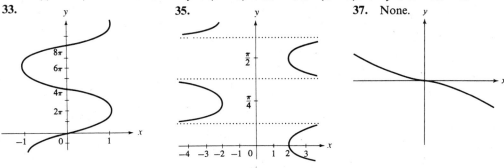

43. $\sqrt{10}$ ft.

Section 11.5

1. $\sqrt{1 - x^2} + C$.   3. $(1/2)\,\text{Arctan}\,x^2 + C$.   5. $(1/3)\,\text{Arcsin}\,3x + C$.
7. $(1/2)\,\text{Arctan}\,2x + C$.   9. $(1/2)\,\text{Arcsin}\,(2x + 1) + C$.   11. $\pi/6$.   13. $\pi/4$.
15. $(1/2)\,[\text{Arctan}\,(x - 1)/2] + C$.   17. $\text{Arcsec}\,(x - 2) + C$.   19. $\text{Arctan}\,(2x + 3) + C$.
21. $\text{Arctan}\,\sin x + C$.   23. $-\sqrt{1 - \tan^2 x} + C$.   25. $(1/2)\,\text{Arctan}^2 x + C$.   27. $\pi/6$.
29. $(1/a)\,\text{Arctan}\,(u/a) + C$.   31. $(1/a)\,\text{Arcsec}\,(u/a) + C$.
33. $2\sqrt{3} - 2\pi/3$.   35. $\pi/8$.   37. $\text{Arctan}\,k, \pi/2$.

Review Problems

1. $(1 + \sin x)/\cos^2 x$.   3. $8x\sec^2 (x^2 + 1)\tan^3 (x^2 + 1)$.   5. $9\cot^4 3x$.
7. $2\sqrt{2x - x^2}$.   9. $-4(\text{Arccos}\,2x)/\sqrt{1 - 4x^2}$.   11. $(\sin^3 x)/3 + C$.
13. $-2(\cos^{5/2} 3x)/15 + C$.   15. $-\sqrt{16 - x^2} + C$.   17. $\text{Arcsec}\,(x + 3) + C$.
19. $2\sqrt{10x} - 4y + (4 - 3\pi) = 0$.   21. $\text{Arctan}\,2$.   23. 0.534.

25. $(\pi/6 + 2n\pi, (\pi + 6\sqrt{3})/12)$ relative max,
$(5\pi/6 + 2n\pi, (5\pi - 6\sqrt{3})/12)$ relative min,
$(\pi/2 + n\pi, \pi/4 + n\pi/2)$ points of inflection.

27. $-11\pi/360$ radians/min, $-11\pi/150$ in./min.

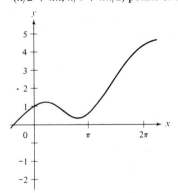

Section 12.1

1.
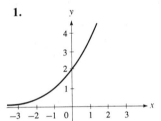

3.
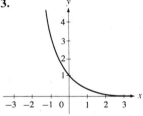

5.
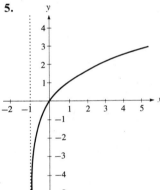

7.
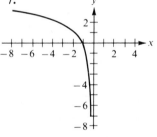

9. $\ln y$.   **11.** $5^y + 1$.
13. $2 \log x + \log (x + 2)$.

15.
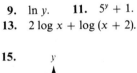

17.
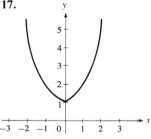

19.
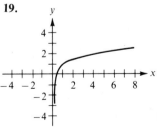

21.
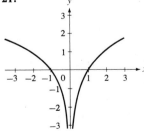

23. $\log_4 y + 2$.   **25.** y.   **27.** 2.   **29.** $4\sqrt{5}/9$.   **31.** 9.
33. $2 \log_5 x + \log_5(x + 3) - \log_5(x - 1)$.
35. $(3/5) \ln (x - 1) + (1/2) \ln (x + 2) - (1/3) \ln (x + 3)$.
41. $F = Pe^r$, \$1060.00, \$1060.90, \$1061.80.

Section 12.2

1. $(\log e)/x$.   **3.** $2x/(x^2 + 4)$.   **5.** $-1/x$.   **7.** $4/(3 + 4x - 4x^2)$.
9. $(4 \log_5 e)/(1 - 9x^2)$.   **11.** $x + 2x \ln x$.   **13.** $(1 - \ln x)/x^2$.   **15.** 1.   **17.** 1.
19. $-\tan x$.   **21.** $(\cos \ln x)/x$.   **23.** $1/\sqrt{x^2 - 1}$.   **25.** $1/(x \ln x)$.
27. $(2 \ln x + 1)/(x \ln x)$.   **29.** $y \cos x$.   **31.** $-[y \sin (x + y)]/(1 + y[\sin (x + y)])$.
33. $(1 - 2x^2 - 2xy)/(2xy + 2y^2 - 1)$.   **35.** $\dfrac{y}{2}\left(\dfrac{1}{x} + \dfrac{1}{x + 4} - \dfrac{1}{x - 2}\right)$.
37. $\dfrac{y}{3}\left(\dfrac{2}{x} + \dfrac{4}{x + 1} - \dfrac{1}{x - 5}\right)$.   **39.** $x - y - 1 = 0$.   **41.** $4x + 4y + (2 \ln 2 - \pi) = 0$.
43. $(1/e, -1/e)$ min.   **45.** $1/\sqrt{e}$.

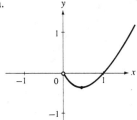

Section 12.3

1. $3^x \ln 3$.   **3.** $(3^{\sqrt{x}} \ln 3)/(2\sqrt{x})$.   **5.** $2e^{2x+2}$.   **7.** $2^{x^2+3} (2x) \ln 2$.   **9.** $1 + e^x$.
11. $3x^2 e^{x^3}(x^3 + 1)$.   **13.** $3^x[3x^2 + (x^3 - 1) \ln 3]$.   **15.** $(2 - 1/x^2)e^{x^2}$.
17. $e^{\tan x} \sec^2 x$.   **19.** $(3 \ln x + 1)x^{x^3+2}$.   **21.** $(\sin x)^x(x \cot x + \ln \sin x)$.   **23.** $2xye^{x^2}$.
25. $1/(e^y x \ln x)$.   **27.** $4e^4$.   **29.** 0.   **31.** $e^2x - y - e^2 = 0$.   **33.** $y - e = 0$.
35. $(-1, -1/e)$ min.   **37.** $(\ln 2n\pi, 1)$ max, $(\ln (2n + 1)\pi, -1)$ min, $n = 1, 2, 3, \ldots$.

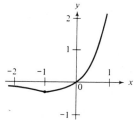

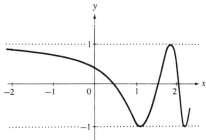

Section 12.4

1. $(1/2) \ln|2x + 1| + C$.   **3.** $-e^{-2x}/2 + C$.   **5.** $-\ln|\cos x| + C$.
7. $\ln (3/2)$.   **9.** $x^2/2 + 4x + 6 \ln |x - 1| + C$.   **11.** $2 \ln (e^x + 1) + C$.

13. $2\sqrt{x^2 + x} + C.$ **15.** $3^x/\ln 3 + C.$ **17.** $e^x + e^{-x} + C.$
19. $4x^3/3 - 3x^2/2 + x - 5 \ln |x| + C.$ **21.** $e^{\tan x} + C.$ **23.** $e^x + 3x - 2e^{-x} + C.$
25. $x^2/2 + x + 2 \ln|x + 1| + C.$ **27.** $11/2 + 9 \ln 2.$ **29.** $\ln 2.$ **31.** $1 - e^{-2}.$
33. $2 \ln 2 - 1.$ **35.** $1 - 1/e^k, 1.$ **37.** $x - \ln (e^x + 1) + C.$

Section 12.5

1. $y = ke^{x^2/2} - 1.$ **3.** $\sin x + \cot y = C.$ **5.** $e^x + e^{-y} = C.$
7. $y = 1/[1 - \ln (2 + x)].$ **9.** $y = 2 - \frac{1}{2} \ln |1 - x^2|.$ **11.** 4 hr 11.5 min.
13. 5,381,000,000.   **15.** 216 hr.   **17.** 12.4 yr.   **19.** 45 min.   **21.** 80 min.
23. 12 min.   **25.** $x = p(1 - e^{-kt})/k,$ where $k > 0; x \to p/k.$

Section 12.6

1. 2214, 2015.   **3.** 2013, 1993.   **5.** 2020.   **9.** 1000; 5000; 1000.
11. 5000, 4000, 512.   **13.** 4000.   **15.** 10,775.   **17.** 37.5%.

Section 12.7

1. $\cosh x = 5/3, \tanh x = 4/5, \coth x = 5/4, \operatorname{sech} x = 3/5, \operatorname{csch} x = 3/4.$
3. $\cosh x = 13/5, \tanh x = -12/13, \coth x = -13/12, \operatorname{sech} x = 5/13, \operatorname{csch} x = -5/12.$
5. $\sinh x = -5/12, \cosh x = 13/12, \coth x = -13/5, \operatorname{sech} x = 12/13, \operatorname{csch} x = -12/5.$
7. $3 \operatorname{sech}^2 (3x - 2).$ **9.** $2x \cosh x^2.$ **11.** $-4 \operatorname{csch}^2 x \coth x.$
13. $(-\operatorname{sech} \sqrt{x} \tanh \sqrt{x})/2\sqrt{x}.$ **15.** $\operatorname{sech}^2 xe^{\tanh x}.$ **17.** $4 \sinh x \cosh x.$
19. $\sinh x \cos (\cosh x).$ **21.** $x - y = 0.$ **23.** $3x - y = 0.$ **41.**

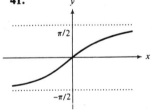

Section 12.8

1. $(1/2) \sinh (2x + 1) + C.$ **3.** $-\tanh (1/x) + C.$ **5.** $(\sinh^4 x)/4 + C.$
7. $-(1/5) \operatorname{csch}^5 x + C.$ **9.** Arctan $\cosh x + C.$
11. $2 \tanh \sqrt{x} + C.$ **13.** $\sinh 1.$ **15.** $(\sinh 2x)/4 - x/2 + C.$

Section 12.9

1. 1.818.   **3.** 0.881.   **5.** 0.347.   **7.** $3/\sqrt{9x^2 + 6x}.$ **9.** $1/2\sqrt{x + x^2}.$
11. $1/(x - x^3) - (\coth^{-1} x)/x^2.$ **13.** $e^x/(1 - e^{2x}).$ **15.** $(1 + x)/\sqrt{1 + x^2}.$
17. $(2 \sinh^{-1} x)/\sqrt{1 + x^2}.$ **19.** $2(1 + \tanh^{-1} x)/(1 - x^2).$
31. $v = 80 \tanh (-2t/5), -80$ ft/sec, -20 ft/sec.   **33.** $(\tanh^{-1} x)^2/2 + C.$

Review Problems

1. $(\sin x - \cos x)e^x/\sin^2 x.$ 3. $(e^{2x^2+1})/4 + C.$ 5. $(5x^2 + 3x - 4)/2x(x^2 - 1).$
7. $\sinh 3x (6x \cosh 3x + \sinh 3x).$ 9. $-e^x \operatorname{sech} e^x \tanh e^x.$
11. $(\ln x)^{\sin x}[\cos x \ln (\ln x) + (\sin x)/(x \ln x)].$ 13. $x - 2 \ln (x^2 + 4) + C.$
15. $\ln (x^2 + 8x + 20) + C.$ 17. $2x - ey - e = 0.$ 19. $y = 0.$
21. $(\sqrt{e}, 1/2e)$ relative max,   23. $(-1, e^{-1})$, relative min,
    $(e^{5/6}, 5/6e^{5/3})$ point of inflection.   no point of inflection.

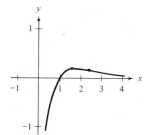
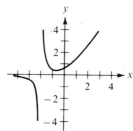

25. $(e - 1)^2/2.$ 27. $1/2.$ 31. $y = ke^{-e^{-x}}.$ 33. $330{,}000{,}000; 44{,}000{,}000{,}000.$
35. 32 min.   37. $2020, 2000.$ 39. $4000, 500.$ 41. $1400, 140.$

Section 13.1

1. $y^2 - x - 2y - 1 = 0.$ 3. $y = x^2 - 1.$ 5. $x^2 - 2xy + y^2 - 2x - 2y = 0.$

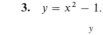
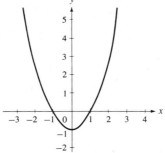
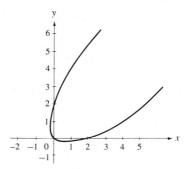

7. $y^3 = x^2.$ 9. $(x - 1)^2 + (y + 2)^2 = 1.$

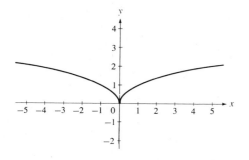
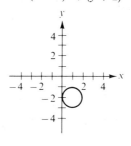

11. $(y - 1)^2 - (x - 2)^2 = 1$.

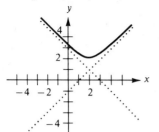

13. $y = (x - 2)^2$.

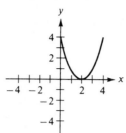

15. $y^2 = x^3$.

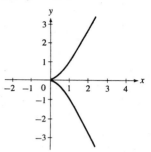

17. $x = 6r + 1, y = -2r + 5$. **19.** $x = -6r + 4, y = -2r + 3$. **21.** $x = 3r - 1, y = 3$.

23. $y - 1 = (x - 2)^3$.

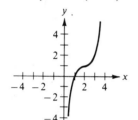

25.

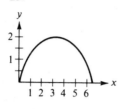

27.

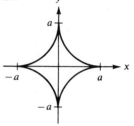

29.

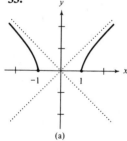

33.

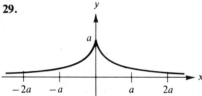

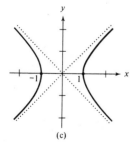

(a)

(b)

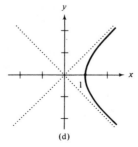

(c)

(d)

Section 13.2

1. $x = 1250t$, $y = -16t^2 + 1250\sqrt{3}t$.   3. $x = 16\sqrt{3}t$, $y = -16t^2 + 16t$; $(16\sqrt{3}, 0)$.
5. $x = (v_0 \cos \theta)t$, $y = -16t^2 + (v_0 \sin \theta)t$.   7. $x = a \sec \theta$, $y = b \tan \theta$; $x^2/a^2 - y^2/b^2 = 1$.
9. $x = a\theta - b \sin \theta$, $y = a - b \cos \theta$.   11. $x = a(1 \pm \sin \theta)$, $y = a(1 \pm \sin \theta) \tan \theta$.
13.

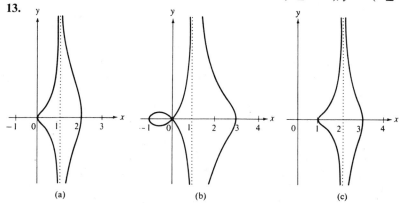

(a)   (b)   (c)

15. $x = 3a \cos \theta + a \cos 3\theta$, $y = 3a \sin \theta - a \sin 3\theta$.
17. $x = 2a \cos \theta - a \cos 2\theta$, $y = 2a \sin \theta - a \sin 2\theta$.

Section 13.3

1. $1/(2t + 1)$, $-2/(2t + 1)^3$.   3. $1/(2t - 5)$, $-2/(2t - 5)^3$.
5. $(2t - 1)/(2t + 1)$, $4/(2t + 1)^3$.   7. $-\tan t$, $-\sec^3 t$.   9. $\coth t$, $-\operatorname{csch}^3 t$.
11. $4, 2$.   13. $-2/9$, $-2/729$.   15. $0, 2$.   17. $3x - y - 1 = 0$.
19. $3x + 2y - 6\sqrt{2} = 0$.   21. $(3, 0)$, $(5, -4)$.   23. $(3, 0)$, $(0, 5)$, $(-3, 0)$, $(0, -5)$.
25. $(2n\pi, 0)$, $((2n + 1)\pi, 2)$.   27. $(1, 0)$, $(0, 1)$, $(-1, 0)$, $(0, -1)$.   29. $(0, 1)$.

Section 13.4

1. $3/2$.   3. $(5\sqrt{5} - 1)/2$.   5. $56/27$.   7. $123/32$.   9. $\sinh 1$.
11. $-1 + \ln (e^2 + 1)$.   13. $2\sqrt{2} - 2$.   15. $4\sqrt{3}$.   17. 5.   23. $\sqrt{2}/32$.
25. $3a|\sin 2t|/2$.   27. $(1, -1)$.

Section 13.5

1. 1575 lb, 1767 lb, $y = [1575 \cosh (2x/1575)]/2$.   3. 120 lb, 156 lb, $y = 120 \cosh (x/120)$.
5. 5 ft.   7. $T_0 \to +\infty$.

Review Problems

1. $y = x^2 + 2x - 5$.

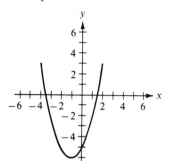

3. $x^2 - 2xy + y^2 - 2x - 2y + 1 = 0$.

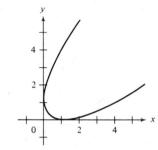

5. $y^2 = 4x^2(1 - x^2)$.

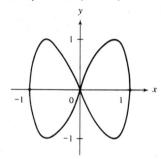

7. $x = 2r + 2, y = -8r + 3$.   **9.** $3t/4, 3/16t$.   **11.** $3x - y + 2 = 0$.
13. $x = 16t, y = -16t^2 + 16\sqrt{3}t + 64; (8(\sqrt{3} + \sqrt{19}), 0), t = (\sqrt{3} + \sqrt{19})/2$ sec.
15. $(\pi/2 + 2n\pi, 0), (3\pi/2 + 2n\pi, 2)$.
17. $3/8 + \ln 2$.   **19.** $v_0 \cos \theta (v_0 \sin \theta + \sqrt{v_0^2 \sin^2 \theta + 64H})/32$.
21. $x = 2a(\cot \theta - \sin \theta \cos \theta), y = 2a(1 - \sin^2 \theta)$.

Section 14.2

2. $(-4, 150°), (-2, 60°), (-1, 30°)$.   **3.** $(4, 5\pi/3), (3, 5\pi/3), (0, 0)$.
5.

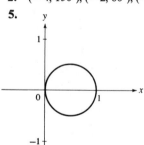

7.

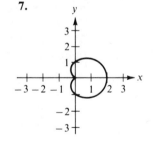

9.

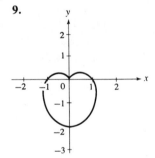

11.

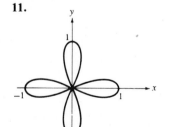

13.

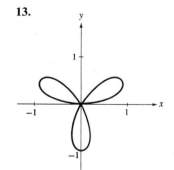

15.

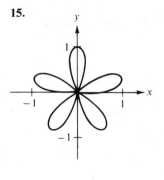

17.
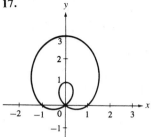

19.
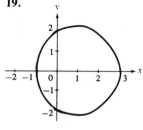

21.
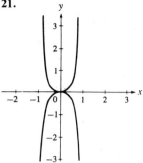

23.
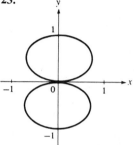

25.
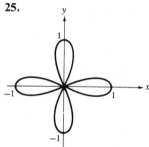

27.
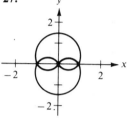

29.
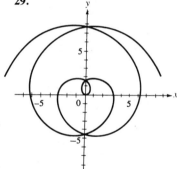

31.
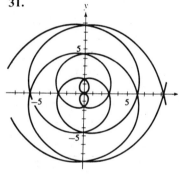

33.
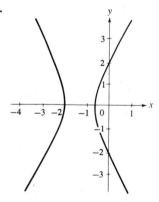

Section 14.3

1. $(\sqrt{2}, 45°), (\sqrt{2}, 315°)$.   **3.** $(2, 90°), (2, 270°)$.   **5.** $(1/2, 60°), (1/2, 300°), (0, 90°) = (0, 0°)$.
7. $(\sqrt{3}/2, 60°), (-\sqrt{3}/2, 300°), (0, 0°)$.   **9.** $(\sqrt{2}, 45°)$.   **11.** $(3, 120°), (3, 240°)$.
13. $(1, 0°), (1, 180°)$.   **15.** $(0, 0°), (1, 90°)$.   **17.** $(1 - 1/\sqrt{2}, 45°), (1 + 1/\sqrt{2}, 225°), (0, 90°) = (0, 0°)$.
19. Coincident graphs.

Section 14.4

1. $(0, 1), (3/2, \sqrt{3}/2), (2, 0), (-1, 1)$.   **2.** $(2, 7\pi/4), (2, 2\pi/3), (4, 0), (\sqrt{2}, 5\pi/4), (2, 3\pi/2)$.
3. $r = 3 \sec \theta$.   **5.** $r = 2$.   **7.** $r = 3 \csc \theta \cot \theta$.
9. $r = (\cos \theta - \sin \theta)/(1 + 2 \sin \theta \cos \theta)$.   **11.** $\tan \theta = 4$.   **13.** $r = 3/(\sin \theta - 2 \cos \theta)$.

15. $r^2 + 4r \cos\theta - 6r \sin\theta + 1 = 0.$ **17.** $r^2 = \sec\theta \csc\theta.$ **19.** $x^2 + y^2 = a^2.$
21. $x - \sqrt{3}y = 0.$ **23.** $x^2 + y^2 - 4y = 0.$ **25.** $(x^2 + y^2)^3 = (x^2 - y^2)^2.$
27. $x^2 + y^2 - 6y - 16 + 9y^2/(x^2 + y^2) = 0$
29. $(x^2 + y^2)(x^2 + y^2 - 1)^2 = y^2.$ **31.** $y^2 = 2x + 1.$ **33.** $x^2 + y^2 = 2x + 3y.$

Section 14.5

1. hyperbola, $(0, 0)$, $x = 2, 3.$ **3.** ellipse, $(0, 0)$, $y = 4, 1/3.$
5. parabola, $(0, 0)$, $y = 4, 1.$ **7.** ellipse, $(0, 0)$, $y = 3, 2/3.$ **9.** $r = 8/(3 + 2\cos\theta).$
11. $r = 3/(1 + \sin\theta).$ **13.** $r = 49/(4 + 7\cos\theta).$
15.

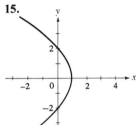

17.

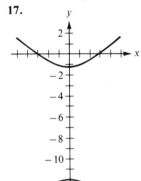

19. $r = 26/(13 + 12\sin\theta)$

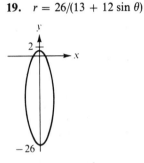

21.

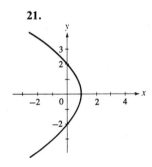

Section 14.6

1. $-(\cos\theta + \cos 2\theta)/(\sin\theta + \sin 2\theta).$ **3.** $0.$
5. $(3\sin\theta \sin 3\theta - \cos\theta \cos 3\theta)/(3\cos\theta \sin 3\theta + \sin\theta \cos 3\theta).$
7. $(\sin\theta + \theta \cos\theta)/(\cos\theta - \theta \sin\theta).$ **9.** $-\sqrt{3}.$ **11.** $-1 - \sqrt{2}.$ **13.** $-\sqrt{3}.$
15. $x - y + 1 = 0.$ **17.** $y + 1 = 0.$
19. $3\sqrt{3}x + 5y + 4 = 0.$ **21.** $-\cos\theta/(1 + \sin\theta).$ **23.** $(3/2, \pi/3)$ max, $(3/2, 5\pi/3)$ min.
25. $(2\sqrt{2}/3, 54.8°), (-2\sqrt{2}/3, 125.2°), (2\sqrt{2}/3, 234.8°), (-2\sqrt{2}/3, 305.2°).$
27. $((3 + \sqrt{33})/4, 53.6°), ((3 + \sqrt{33})/4, 306.4°), ((3 - \sqrt{33})/4, 147.0°), ((3 - \sqrt{33})/4, 213.0°).$
31. $(8 + 5\sqrt{3})/11.$ **33.** $60°, 120°.$

Section 14.7

1. 4π.   **3.** $3\pi/2$.   **5.** 1.   **7.** $19\pi/2$.   **9.** 9.   **11.** $5\pi/4 - 2$.
13. $3\pi/2 - 2\sqrt{2}$.   **15.** $(8 + \pi)/4$.   **17.** $\pi/4 - 3\sqrt{3}/16$.   **19.** $\pi/2$.   **23.** $\sqrt{2}(e^2 - 1)$.
25. $(2\pi - 3\sqrt{3})/4$.

Review Problems

1.   **3.**   **5.**

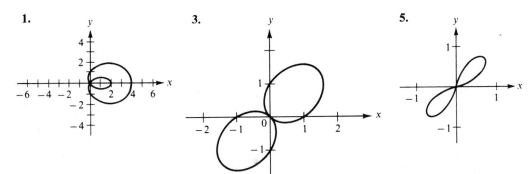

7. $(1, \pi/2), (1, 3\pi/2)$.   **9.** $(-1, 3\pi/2), (0, 0) = (0, 7\pi/6)$.   **11.** $(0, \pi/2), (1, 0), (-1, \pi)$.
13. (a) $(0, -1), (-2, 2)$; (b) $(2\sqrt{2}, 3\pi/4), (5, \pi)$.   **15.** $r = 6/(3 + 2 \sin \theta)$.   **17.** $9\pi/2$.
19. $(x^2 + y^2 - y)^2 = x^2 + y^2$.   **21.** $F(0, 0)$, D: $y = -2, e = 1$.

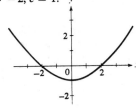

23. $x - y + 1 = 0$.   **25.** $(1.307, 3\pi/8)$ max, $(-0.541, 7\pi/8)$ min.   **27.** $\pi + 3\sqrt{3}$.

Section 15.1

1. $x^2/2 + 2x - 6 \ln |x| + C$.   **3.** $x^2 - 2x + (5 \ln |2x + 1|)/2 + C$.
5. $u^3/3 + u^2/2 + 2u + 2 \ln|u - 1| + C$.   **7.** $(2/9)(3x + 1)^{3/2} + C$.
9. $(3/10)(x^2 + 2x)^{5/3} + C$.   **11.** $e^{x^2}/2 + C$.   **13.** $-\ln |1 - \sin x| + C$.
15. $x - e^{-x} + C$.   **17.** $\ln |1 + \tan u| + C$.   **19.** $-\cos^2 3\theta/6 - \cos 3\theta/3 + C$.
21. $\ln |1 - \coth x| + C$.   **23.** $(1/2)$ Arcsec $(x/2) + C$.   **25.** Arcsin $(x/\sqrt{2}) + C$.
27. $\tan \theta - 2 \ln |\cos \theta| + C$.   **29.** $x - 4x^{3/2}/3 + x^2/2 + C$.
31. $-(4 - \ln x)^2/2 + C$.   **33.** Arctan $e^x + C$.   **35.** $-(1/2) \ln (\cos^2\theta + 1) + C$.
37. $-e^{1/x} + C$.   **39.** $-2(1 - \sqrt{x})^3/3 + C$.   **41.** $-x + 2 \ln(e^x + 1) + C$.
45. $-\ln |\csc u + \cot u| + C$.   **47.** 2 Arctan $e^u + C$.

Section 15.2

1. $(2/5)(x + 3)^{3/2}(x - 2) + C.$ 3. $(2/105)(x + 2)^{3/2}(15x^2 - 24x + 32) + C.$
5. $(1/15)(2x + 1)^{3/2}(3x - 1) + C.$ 7. $(2x/3 + 4)\sqrt{x - 3} + C.$
9. $2\sqrt{x + 2}\,(3x^2 + 8x + 32/15 + C.$ 11. $272/15.$ 13. $(30\sqrt{3} - 26)/3.$
15. $928/315.$ 17. $6.$ 19. $(10x^3 + 24x^2 + 64x + 256)(\sqrt{x - 2})/35 + C.$
21. $(3/28)(x + 1)^{4/3}(4x - 3) + C.$ 23. $4(x - 1)^{5/4}(5x + 4)/45 + C.$
25. $(2x + 1)^{3/2}(6x + 8)/15 + C.$ 27. $(2/3)(2x + 11)\sqrt{x - 2} + C.$
29. $(550\sqrt{5} - 38)/35.$

Section 15.3

1. $(1/3)\sin^3 x + C.$ 3. $(1/4)\sin^4 \theta + C.$ 5. $(3\cos^5 3x - 5\cos^3 3x)/45 + C.$
7. $16/105.$ 9. $(8x - \sin 8x)/64 + C.$
11. $(5/16)\theta + (1/4)\sin 2\theta + (3/64)\sin 4\theta - (1/48)\sin^3 2\theta + C.$ 13. $3\pi/32.$
15. $-\csc x - \sin x + C.$ 17. $(10\cos^{9/2} x - 18\cos^{5/2} x)/45 + C.$
19. $(5\sin^3 x^2 - 3\sin^5 x^2)/30 + C.$ 21. $(\cos^2 x)/2 - \ln|\cos x| + C.$
23. $(2\pi + 3\sqrt{3})/96.$ 25. $(14\sin^{3/2} t - 6\sin^{7/2} t)/21 + C.$
27. $(1/2)\sin^2 x + C_1 = -(1/2)\cos^2 x + C_2 = (-1/4)\cos 2x + C_3.$

Section 15.4

1. $(1/3)\tan^3 x + C.$ 3. $-(1/5)\cot^5 x - (1/7)\cot^7 x + C.$
5. $\csc \theta - (1/3)\csc^3 \theta + C.$ 7. $(1/3)\sec^3 \theta + C.$
9. $\tan x + (2/3)\tan^3 x + (1/5)\tan^5 x + C.$ 11. $(1/2)\sec^2 \theta - \ln|\sec \theta| + C.$
13. $\sqrt{3} - (\pi/3).$ 15. $-\cot^3 \theta/3 + C.$ 17. $(\tan^3 3x - 3\tan 3x + 9x)/9 + C.$
19. $(-\cot^3 (2x + 1) + 3\cot (2x + 1) + 6x)/6 + C.$ 21. $-(1/4)\ln|\cos 2\theta| + C.$
23. $\theta + \sin \theta + C.$ 25. $(1/4)\ln|\cos 2\theta| + C.$ 27. $\tan \theta - \cot \theta + C.$
29. $(1/3)\sin^3 x + C.$ 31. $(1/2)\tan^2 x + C_1 = (1/2)\sec^2 x + C_2 = 1/(2\cos^2 x) + C_3.$

Section 15.5

1. $-\sqrt{1 - x^2}/x - \text{Arcsin } x + C.$ 3. $\ln|x + \sqrt{x^2 + 16}| + C.$
5. $(-1/3)(x^2 + 18)\sqrt{9 - x^2} + C.$ 7. $-x/5\sqrt{x^2 - 5} + C.$
9. $\sqrt{4 - x^2} + 2\ln\left|\dfrac{2 - \sqrt{4 - x^2}}{x}\right| + C.$ 11. $\ln\left|\dfrac{1 - \sqrt{1 - x^2}}{x}\right| + C.$
13. $(243/8)\text{Arcsin }(x/3) + [(45x/8) - (x^3/4)]\sqrt{9 - x^2} + C.$

15. $\dfrac{2}{3}\text{Arcsec }\dfrac{2x}{3} - \dfrac{\sqrt{4x^2 - 9}}{2x^2} + C.$ 17. $x/(2\sqrt{3x^2 + 2}) + C.$

19. $81\pi/128.$ 21. $\ln\dfrac{\sqrt{e} - \sqrt{e - 1}}{e - \sqrt{e^2 - 1}}.$

Section 15.6

1. $\text{Arctan }(x + 1) + C.$ 3. $\text{Arcsin }(4x + 4) + C.$
5. $\text{Arcsin }(x - 1) + (x - 1)\sqrt{2x - x^2} + C.$

7. $(13/4) \operatorname{Arctan} [(x - 5)/4] + \ln (x^2 - 10x + 41) + C.$
9. $\sqrt{-x^2 + 10x - 21} + 2 \ln |(2 - \sqrt{-x^2 + 10x - 21})/(x - 5)| + C.$
11. $13 \operatorname{Arcsin} [(x - 5)/2] - 2\sqrt{-x^2 + 10x - 21} + C.$
13. $x^2/2 - 2x + \ln (x^2 + 2x + 2) + 3 \operatorname{Arctan} (x + 1) + C.$
15. $(5x + 9)/\sqrt{-x^2 - 6x - 8} - \operatorname{Arcsin} (x + 3) + C.$ 17. $\pi/4.$ 19. $1/\sqrt{2}.$

Section 15.7

1. $x \sin x + \cos x + C.$ 3. $xe^x - e^x + C.$ 5. $x \ln x^2 - 2x + C.$
7. $-\ln x/x - 1/x + C.$ 9. $x \operatorname{Arcsin} x + \sqrt{1 - x^2} + C.$
11. $x \operatorname{Arctan} 2x - (1/4) \ln |1 + 4x^2| + C.$ 13. $x \sinh^{-1} x - \sqrt{x^2 + 1} + C.$
15. $(-x^2/3)(9 - x^2)^{3/2} - (2/15)(9 - x^2)^{5/2} + C.$
17. $(2/9)x^3(x^3 + 3)^{3/2} - (4/45)(x^3 + 3)^{5/2} + C.$
19. $(x + 1) \operatorname{Arccot} \sqrt{x} + \sqrt{x} + C.$ 21. $(2x^n/3n)(x^n + 3)^{3/2} - (4/15n)(x^n + 3)^{5/2} + C.$
23. $\ln \left| \dfrac{x - 1}{x} \right| - \dfrac{\ln x}{x - 1} + C.$
25. $(-x^4 + 12x^2 - 24) \cos x + (4x^3 - 24x) \sin x + C.$
27. $e^{3x}(9x^3 - 9x^2 + 6x - 2)/27 + C.$ 29. $(1/32)x^4(8 \ln^2 x - 4 \ln x + 1) + C.$
31. $(1/2)e^x(\sin x + \cos x) + C.$
33. $(1/8)(2 \sec^3 \theta \tan \theta + 3 \sec \theta \tan \theta + 3 \ln|\sec \theta + \tan \theta|) + C.$
35. $e^x(\sin 3x - 3 \cos 3x)/10.$ 37. $(\cos x \sin 5x - 5 \sin x \cos 5x)/24 + C.$
41. $(1/2)(\theta - \sin \theta \cos \theta) + C.$ 43. $(1/8) \sin 4x - (1/12) \sin 6x + C.$

Section 15.8

1. $\ln \left| \dfrac{x}{x + 1} \right| + C.$ 3. $(1/3) \ln |(x + 2)^2(x - 1)| + C.$
5. $x + \ln \left| \dfrac{x}{(x + 3)^4} \right| + C.$ 7. $\ln \dfrac{\sqrt{x^2 - 1}}{|x|} + C.$ 9. $x + \ln \left| \dfrac{x^4(x - 2)^8}{(x - 1)^9} \right| + C.$
11. $\ln |x + 2| + \dfrac{2}{x + 2} + C.$ 13. $x - \dfrac{1}{x - 1} + 2 \ln |x - 1| + C.$ 15. $\dfrac{1}{x} + \ln \left| \dfrac{x - 1}{x} \right| + C.$
17. $(2/3x) - (7/9) \ln |x| + (1/36) \ln |x + 3| + (3/4) \ln |x - 1| + C.$
19. $-\dfrac{1}{2}\left(\ln \dfrac{|x^2 - 1|}{x^2} + \dfrac{1}{x^2 - 1} \right) + C.$ 21. $\ln \left| \dfrac{1 + \sin x}{\sin x} \right| - \csc x + C.$

Section 15.9

1. $-1/x - \operatorname{Arctan} x + C.$ 3. $\ln \dfrac{(x - 1)^2}{|x^2 + x + 1|} + C.$ 5. $\dfrac{3}{4}\left[\ln \dfrac{(x + 1)^2}{x^2 + 1} + 2 \operatorname{Arctan} x \right] + C.$
7. $\dfrac{1}{32} \ln \dfrac{(x + 2)^2}{x^2 + 4} - \dfrac{1}{8} \dfrac{1}{x + 2} + C.$ 9. $\dfrac{1}{2} \ln \dfrac{x^2 + 1}{(x + 1)^2} + \operatorname{Arctan} x + C.$
11. $\dfrac{3}{2} \operatorname{Arctan} x + \dfrac{x - 2x^2}{2(x^2 + 1)} + C.$ 13. $\dfrac{1}{2} \ln \dfrac{x^2}{x^2 + 4} + \dfrac{2}{x^2 + 4} + C.$

15. $-\dfrac{1}{3}\dfrac{1}{x-2}+\dfrac{2}{3\sqrt{3}}\,\text{Arctan}\,\dfrac{x+1}{\sqrt{3}}+C.$

17. $\dfrac{1}{2}\ln\dfrac{x^2+1}{x^2+4}+\text{Arctan}\,x-\dfrac{1}{2}\,\text{Arctan}\,\dfrac{x}{2}+\dfrac{3}{2(x^2+4)}+C.$

19. $\dfrac{1}{4}\ln\dfrac{x^2}{x^2+2x+2}-\dfrac{1}{2}\,\text{Arctan}\,(x+1)+C.$

Section 15.10

1. $x-2\sqrt{x+2}-\ln\left(\sqrt{x+2}+1\right)+C.$ **3.** $\sqrt{x}-(1/2)\ln\left(1+2\sqrt{x}\right)+C.$

5. $2\sqrt{x}+4\sqrt[4]{x}+4\ln\left(\sqrt[4]{x}-1\right)+C.$ **7.** $-4x^{1/4}-6x^{1/6}-12x^{1/12}+12\ln\left|1-x^{1/12}\right|+C.$

9. $x+3x^{2/3}+6x^{1/3}+6\ln\left|x^{1/3}-1\right|+C.$

11. $\ln\left|\dfrac{\tan(\theta/2)}{1+\tan(\theta/2)}\right|+C.$

13. $\dfrac{1}{3}\ln\left|\dfrac{(1+\sqrt{x+2})^2}{(x+3)-\sqrt{x+2}}\right|+\dfrac{2}{\sqrt{3}}\,\text{Arctan}\left[\dfrac{2}{\sqrt{3}}\left(\sqrt{x+2}-\dfrac{1}{2}\right)\right]+C.$

15. $3\sqrt[3]{2x+1}-\ln\left|\sqrt[3]{2x+1}-1\right|+\dfrac{1}{2}\ln\left[(2x+1)^{2/3}+\sqrt[3]{2x+1}+1\right]$

$+\sqrt{3}\,\text{Arctan}\left[\dfrac{2}{\sqrt{3}}(\sqrt[3]{2x+1}+1)\right]+C.$

17. $\dfrac{1}{2\sqrt{3}}\ln\left|\dfrac{\sqrt{3}+\tan(\theta/2)+1}{\sqrt{3}-\tan(\theta/2)-1}\right|+C.$ **19.** $\dfrac{1}{\sqrt{2}}\ln\dfrac{[\tan(\theta/2)-1+\sqrt{2}]^2}{|-\tan^2(\theta/2)+2\tan(\theta/2)+1|}+C.$

21. $\dfrac{1}{15}(x^2-4)^{3/2}(3x^2+8)+C.$

Section 15.11

1. $(1/3)(x-5)\sqrt{2x+5}+C.$ **3.** $-(9+x^4)^{1/2}/18x^2+C.$

5. $-e^{-6x}(3\sin 8x+4\cos 8x)/50+C.$

7. $(-1/2)(x+9)\sqrt{6x-x^2}+(27/2)\,\text{Arccos}\,[(3-x)/3]+C.$

9. $\dfrac{1}{18}\cos^5 3x\sin 3x+\dfrac{5}{72}\cos^3 3x\sin 3x+\dfrac{5}{16}x+\dfrac{5}{96}\sin 6x+C.$

11. $(5x^3-9x^2+18x-162)\sqrt{3+2x}/35+C.$ **13.** $2x\cos x+(x^2-2)\sin x+C.$

15. $(x^3/3)\,\text{Arcsin}\,x+(x^2/3)(1-x^2)^{1/2}+(2/9)(1-x^2)^{3/2}+C.$

17. $(1/2)\ln\left|2x-2+\sqrt{4x^2-8x-21}\right|+C.$

21. $(1/2)(x-1)\sin 2x+[x-(1/2)x^2-(11/4)]\cos 2x+C.$

Review Problems

1. $(1/2)\tan 2\theta-\ln\left|\cos 2\theta\right|+C.$ **3.** $(1/8)\sec^8 x-(1/6)\sec^6 x+C.$

5. $\sqrt{2}-1-\pi/4+\ln(2-\sqrt{2}).$ **7.** $(2/7)\tan^{7/2}\theta+(2/11)\tan^{11/2}\theta+C.$

9. $[2x^3\,\text{Arctan}\,x-x^2+\ln(x^2+1)]/6+C.$ **11.** $3x/2-2\cos x-(\sin 2x)/4+C.$

13. $\pi/2.$ **15.** $(2x-16)/\sqrt{x-4}+C.$ **17.** $326/135.$ **19.** $-(2\ln x+1)/4x^2+C.$

21. $\ln|1 + \tan x| + C.$ **23.** $27\pi/16.$
25. $\sqrt{x} + 9\sqrt[3]{x}/4 + 27\sqrt[6]{x}/4 + (81\ln|2\sqrt[6]{x} - 3|)/8 + C.$ **27.** $(16 - 8\sqrt{2})/243.$
29. $3\ln\left|\dfrac{x}{x - 2}\right| - \dfrac{2}{x} - \dfrac{4}{x - 2} + \dfrac{2}{(x - 2)^2} + C.$
31. $(3\sin 2x \sin 3x + 2\cos 2x \cos 3x)/5 + C.$
33. $-x + (6/\sqrt{5})\ln|x + 2 + \sqrt{5}| - [(3 - \sqrt{5})/\sqrt{5}]\ln|1 - 4x - x^2| + C.$
35. $\dfrac{x}{10(5 - x^2)} + \dfrac{\sqrt{5}}{100}\ln\left|\dfrac{\sqrt{5} + x}{\sqrt{5} - x}\right| + C.$ **37.** $\dfrac{1}{2}\ln(x^2 + 4) + \dfrac{1}{2}\text{Arctan}\dfrac{x}{2} + \dfrac{2x^2 + 5}{(x^2 + 4)^2} + C.$
39. $9 - 2e.$ **41.** $e^2(15e^2 - 1)/4.$ **43.** $\sqrt{37}/2 + (1/12)\ln(\sqrt{37} + 6).$

Section 16.1

1. $1/8.$ **3.** $1/3.$ **5.** $+\infty.$ **7.** $+\infty.$ **9.** $2.$ **11.** $0.$ **13.** $12.$ **15.** $+\infty.$
17. $+\infty.$ **19.** $1.$ **21.** $1/2.$ **23.** $3.$ **25.** $-\infty.$ **27.** $-\infty + \infty:$ diverges.

Section 16.2

1. $\pi/7.$ **3.** $\pi/30.$ **5.** $\pi^2/2.$ **7.** $\pi(e - 1)/e.$ **9.** $4\pi/21.$ **11.** $2\pi/3.$
13. $2\pi/5.$ **15.** $4\pi/15.$ **17.** $\pi(\pi - 2).$ **19.** $9\pi/10.$ **21.** $\pi(8 - \pi)/4.$ **23.** $\pi/6.$
25. $\pi/6.$ **27.** $(2 + \pi)\pi.$ **29.** $4\pi^2.$ **31.** $4\pi r^3/3.$ **33.** $+\infty, \pi.$

Section 16.3

1. $4\pi/5.$ **3.** $64\pi/5.$ **5.** $\pi/6.$ **7.** $\pi/6.$ **9.** $27\pi/2.$ **11.** $5\pi/6.$
13. $128\pi/105.$ **15.** $\pi(e^2 + 1)/2.$ **17.** $\pi.$ **19.** $+\infty.$ **21.** $144\pi/35.$
23. $\pi(e^4 - 13)/4e^4.$ **25.** $\pi/3.$

Section 16.4

1. $\pi.$ **3.** $8/3.$ **5.** $2\pi/3.$ **7.** $32.$ **9.** $64/3.$ **11.** $16.$ **13.** $4\pi.$ **15.** $192.$
17. $48\pi.$ **19.** $96.$ **21.** $24\pi.$ **23.** $2/(3\sqrt{3}).$ **25.** $128/3.$

Section 16.5

1. $6\pi a^2/5.$ **3.** $1505\pi/18.$ **5.** $2\pi(1 + \sinh 1 \cosh 1).$ **7.** $3\pi.$
9. $\pi(17\sqrt{17} - 1)/6.$ **11.** $2\pi(1 + \sinh 1 - \cosh 1).$ **13.** $\pi[9\sqrt{82} + \ln(\sqrt{82} + 9)]/2.$
15. $4\pi.$ **17.** $4\pi.$ **19.** $128(125\sqrt{10} + 1)/1215.$ **21.** $2\sqrt{2}\pi + 2\pi\ln(1 + \sqrt{2}).$
23. $2\pi\left(\dfrac{2261}{80} - \dfrac{\ln^2 4}{8} - \dfrac{16}{3}\ln 4\right).$ **25.** $\pi r\sqrt{r^2 + h^2}.$
27. $4\sqrt{2}\pi(e^{2\pi} + e^\pi - 1)/5.$ **29.** Infinite.   **31.** $2\pi a^2 + \dfrac{2\pi ab^2}{\sqrt{a^2 - b^2}}\ln\dfrac{a + \sqrt{a^2 - b^2}}{b}.$

Section 16.6

1. $(1/4, 0)$.   **3.** $(9/2, 0)$.   **5.** $(3/5, 7/5)$.   **7.** $(3/5, 17/10)$.   **9.** $(19/10, -7/10)$.
11. $(-26/3, 0)$.   **13.** $(3, 0)$.   **15.** $(-1/2, -5)$.   **17.** $(-5, -4)$.
19. 21/5 to the right and 19/10 above the lower left-hand corner.
21. Half-way between the two sides and a distance $(5 + 2\sqrt{3})/(4 + \sqrt{3})$ up from the bottom.
23. Half-way between the two sides and a distance $\sqrt{3}$ up from the bottom.
25. Half-way between the two sides and a distance $(2 + 3\pi)/(2 + \pi)$ up from the bottom.

Section 16.7

1. $(3/2, 6/5)$.   **3.** $(5/6, 5/18)$.   **5.** $(3/8, 3/5)$.   **7.** $(3/5, -2/35)$.
9. $(17/27, 5/9)$.   **11.** $(\pi/2, \pi/8)$.   **13.** $(5/9, 10/27)$.   **15.** $(0, 1/4)$.
17. $(0, 17/5)$.   **19.** $(4a/3\pi, 4b/3\pi)$.   **21.** $(a/5, a/5)$.   **23.** $(0, 4/\pi)$.   **25.** $4\pi/5$.

Section 16.8

1. $\bar{x} = 7/8$.   **3.** $\bar{y} = 1/5$.   **5.** $\bar{y} = 1/2$.   **7.** $\bar{x} = 3/8$.   **9.** $\bar{x} = 3a/8$.   **11.** $\bar{x} = 12/5$.
13. $\bar{y} = -1/10$.   **15.** $\bar{x} = (\pi^2 + 4)/4\pi$.   **17.** $\bar{x} = 1/2$.   **19.** $\bar{x} = -8/5$.

Section 16.9

1. $(2a/5, 2a/5)$.   **3.** $(15/7 + (3/28) \ln 3, 52/21)$.
5. $((8\sqrt{2} - 4)/3(\sqrt{2} + \ln (1 + \sqrt{2}), (\sqrt{2} + 5 \ln (1 + \sqrt{2})/4(\sqrt{2} + \ln (1 + \sqrt{2}))$.   **7.** $\bar{x} = 1/2$.
9. $\bar{x} = 0$.   **11.** $\bar{x} = 53,941/21,070$.   **13.** $\bar{y} = 2/\pi$.   **15.** $(\pi, 2/3)$.
17. $\left(\dfrac{6\sqrt{2} - 2 \ln (\sqrt{2} + 1)}{2\sqrt{2} + 2 \ln (\sqrt{2} + 1)}, 0\right)$.   **19.** $\bar{y} = (2 \sinh 1 \cosh 1 - \sinh^2 1 + 1)/4(\sinh 1 - \cosh 1 + 1)$.

Section 16.10

1. $34, \sqrt{17/3}$.   **3.** $66, \sqrt{22/3}$.   **5.** $4/5, \sqrt{1/5}$.   **7.** $4/15, \sqrt{2/15}$.   **9.** $8/5, \sqrt{6/5}$.
11. $38/105, \sqrt{19/35}$.   **13.** $3/4, 1/2\sqrt{2} \ln 2$.   **15.** $2\pi^2 - 8, \sqrt{\pi^2/2 - 2}$.
17. $7(2 - 5/e), \sqrt{\dfrac{2e - 5}{e - 1}}$.   **19.** $\pi/2, 1/2$.   **21.** $4\pi, 1$.   **23.** The pipe.

Section 16.11

1. 39,000 lb.   **3.** 3182.4 lb.   **5.** 12.0 lb.   **7.** 216 lb.   **9.** 1872 lb.   **11.** 15,160 lb.
13. 17.2 ft above the top of the gate.   **15.** 41.6 lb.   **17.** $62.4(1 + h)\pi$.   **19.** 8236 lb.
21. 31,350 lb.   **23.** 36,715 lb.

Review Problems

1. 0.   **3.** Diverges.   **5.** $\pi(e^2 + 1)/2$.   **7.** $32\pi/3$.   **9.** $32\pi/15$.   **11.** $\pi a^2(2 + \sinh 2)$.
13. $(1/2, -2/5)$.   **15.** $(2, 37/20)$.   **17.** $(3/2, 27/5)$.   **19.** $\bar{x} = 5/4$.   **21.** $\bar{x} = 27/16$.
23. $1/7, \sqrt{3}/14$.   **25.** $1/2$.   **27.** $4\pi\sqrt{3}$.   **29.** $191\pi/160$.
31. $\left(\dfrac{15}{3(24 + \ln 2)}, \dfrac{1008 - 64\ln 2 - \ln^2 2}{16(24 + \ln 2)} \right)$.   **33.** $\bar{x} = 15\pi/256$.   **35.** 1,065,000 lb., 520,000 lb.
37. $\left(\dfrac{2(n + 1)}{3(n + 2)}, \dfrac{2(n + 1)}{3(2n + 1)} \right)$, $(2/3, 1/3)$.   **39.** 22,800 lb.

Section 17.1

1. Limit point.   **3.** Limit point.   **5.** Not a limit point.   **7.** Limit point.
9. Limit point.   **11.** $\delta = \min\{\epsilon/5, 3\}$.   **13.** $\delta = \epsilon$.   **15.** $\delta = \epsilon$.   **17.** $\delta = \sqrt{\epsilon}$.
19. $\delta = \min\{\epsilon/3, 1\}$.   **21.** $\delta = \min\{\epsilon/2, 1\}$.   **23.** $\delta = \epsilon^2$.   **25.** $\delta = \min\{\epsilon/19, 1\}$.
27. $\delta = \min\{\epsilon/2, 1/2\}$.   **29.** $\delta = \min\{\epsilon/10, 1/2\}$.

Section 17.2

1. Choose $\epsilon \leq 2$.   **3.** Choose $\epsilon \leq 1$.   **5.** Choose $\epsilon \leq 1$.   **7.** Choose $\epsilon \leq 1$.
9. Choose any ϵ.   **11.** True.   **13.** False.   **15.** Choose any ϵ.   **17.** Choose $\epsilon \leq 2$.

Section 17.3

1. $\delta = \epsilon$.   **3.** If $N > 0$, $\delta = 1/\sqrt[4]{N}$; if $n \leq 0$, $\delta = $ anything.
5. If $N < 0$, $\delta = 1/\sqrt{-N}$; if $N \geq 0$, $\delta = $ anything.
7. If $N > 0$, $\delta = 1/N$; if $N \leq 0$, $\delta = $ anything.   **9.** $N = 1/\sqrt{\epsilon}$.   **11.** $N = 1/\epsilon$.
13. $N = \ln \epsilon$.   **15.** Choose any ϵ.   **17.** Choose any N.   **19.** Choose $N > 0$.
21. Choose any ϵ.

Section 17.4

1. $\delta = \epsilon$.   **3.** $\delta = \epsilon/2$.   **5.** $\delta = \epsilon$.   **7.** $\delta = \min\{\epsilon/7, 1\}$.   **9.** $\delta < 1$.
11. Choose $\epsilon < 1$.   **13.** $\delta = \min\{\epsilon/2, 1/2\}$.   **15.** $\delta = \min\{\epsilon/10, 1/2\}$.
17. Choose $\epsilon < 1$.   **19.** Choose $\epsilon < 1$.   **21.** 3.   **27.** Choose $\epsilon < 1$.

Section 17.5

11. Yes, there is a number c between 0 and 1 such that $f'(c) = 0$.
13. No, there is no derivative at $x = 0$.   **15.** No, $f = 0$ only at $x = 1$.
17. No, there is no derivative at $x = 1$ or $x = 2$.
19. No, f is not defined at $x = 0$ nor does $f'(0)$ exist.

Section 17.6

11. Yes, there is a number, x_0, between 0 and 1 such that $f'(x_0) = 1$.
13. No, there is no derivative at $x = 0$.
19. No, f is not defined at $x = 0$ nor does $f'(0)$ exist.

Review Problems

1. True.   **3.** False.   **5.** True.   **7.** False.   **9.** True.   **11.** False.
15. Choose $\delta = \min\{\epsilon/3, 1\}$.   **17.** -6.   **19.** No, $f(0)$ is not defined.
21. Yes, there is a number, x_0, between 0 and 8 such that $f'(x_0) = 1/2$.

Section 18.1

1. 2.   **3.** -3.   **5.** $(1/3)a^{-2/3}$.   **7.** 0.   **9.** 0.   **11.** Does not exist.   **13.** 0.
15. 0.   **17.** 1/2.   **19.** $\ln a - \ln b$.   **21.** 1.   **23.** 0.   **25.** 2.   **27.** $-1/6$.
29. 0.   **31.** $a = \pm 2, b = -1$.   **33.** 1.

Section 18.2

1. 0.   **3.** 0.   **5.** 5.   **7.** 7/3.   **9.** 0.   **11.** -1.   **13.** a.   **15.** 1/4.
17. 1/2.   **19.** 1/2.   **21.** $-1/2$.   **23.** $-\infty$.   **27.** 1/2.   **29.** $+\infty$.

Section 18.3

1. 1.   **3.** 1.   **5.** e.   **7.** 1.   **9.** 1.   **11.** e^2.   **13.** $1/e$.   **15.** 1.
17. $1/e$.   **19.** 1.   **21.** 1.   **23.** 1.   **25.** e^{ab}.
29.

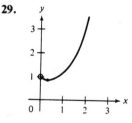

Review Problems

1. 0.   **3.** 1.   **5.** $-1/3$.   **7.** 0.   **9.** 0.   **11.** e^2.   **13.** 1.   **15.** 0.
17. 0.   **19.** 1.   **21.** 1.   **23.** $a = \pm 2, b = 0, c = 2, d = 0, e = -1$.

Section 19.1

1. 1, 4, 9, 16.   3. 2, 0, 2, 0.   5. $1/\ln 2, -1/\ln 3, 1/\ln 4, -1/\ln 5$.
7. 1, 3/2, 7/4, 15/8.   9. $s_n = n/(n + 1), n = 1, 2, 3, \dots$.
11. $s_n = (-1)^n(2^n - 1)/2^n, n = 1, 2, 3, \dots$.   13. $s_n = (2n - 1)(2n + 1), n = 1, 2, 3, \dots$.
15. $s_n = (-1)^{n+1}(n)(2^n), n = 1, 2, 3, \dots$.   17. Converges to 0.   19. Converges to 0.
21. Converges to 1.   23. Diverges.   25. Converges to 0.   27. Converges to 0.

Section 19.2

1. $1 + 1/4 + 1/9 + 1/16 + \cdots$.   3. $3/2 + 2/3 + 5/12 + 3/10 + \cdots$.
5. $-1 + 1/2 - 1/6 + 1/24 - \cdots$.   7. $\sum_{n=1}^{\infty} (1/10)^n$.   9. $\sum_{n=1}^{\infty} 2^{n-1}$.
11. $\sum_{n=1}^{\infty} (-1)^{n+1} n(n + 2)$.   13. 2/9.   15. 134/999.   17. 1.   19. 134/99.
21. Converges to 1/3.   23. Diverges.   25. Diverges.   27. Diverges.
29. Converges to 1.   31. Diverges.   33. Converges to 1/2.

Section 19.3

1. Converges.   3. Diverges.   5. Diverges.   7. Converges.   9. Converges.
11. Diverges.   13. Diverges.   15. Converges.   17. Diverges.   19. Converges.
21. Converges.   23. Converges.   25. Converges.   27. Converges.   31. Diverges.

Section 19.4

1. Converges.   3. Converges.   5. Converges.   7. Diverges.   9. Diverges.
11. Converges.   13. Diverges.   15. Converges.   17. Diverges.   19. Diverges.
21. Diverges.   23. Converges.   25. Converges.   27. Diverges.   29. Diverges.
31. Converges.
35. $(1 + 1/2 + 1/3 + \cdots + 1/8) + (1/10 + 1/11 + \cdots + 1/88) + (1/100 + \cdots + 1/888) + \cdots$
     $< 1 \cdot 8 + (1/10) \cdot 8 \cdot 9 + (1/100) \cdot 8 \cdot 9^2 + \cdots$. Converges.

Section 19.5

1. Converges conditionally.   3. Converges absolutely.   5. Converges conditionally.
7. Converges conditionally.   9. Converges absolutely.   11. Converges absolutely.
13. Diverges.   15. Converges conditionally.   17. Converges conditionally.
19. Converges absolutely.   21. 0.9.   23. -0.4.

Section 19.6

1. $-1 < x < 1$.   3. $-1 \le x < 1$.   5. $-2 < x < 2$.   7. $-3/2 < x < 3/2$.
9. $-1/2 < x < 1/2$.   11. $-3/2 < x < -1/2$.   13. $-5 < x < -1$.

15. $-5 \le x \le -1$.   **17.** $1 \le x < 2$.   **19.** All x.   **21.** $-4 \le x \le -2$.
23. $x > 1$ or $x \le -1$.   **25.** $-\pi/6 + n\pi < x < \pi/6 + n\pi$.   **27.** All real x.

Section 19.7

1. $e + e(x - 1) + (e/2!)(x - 1)^2 + (e/3!)(x - 1)^3 + \cdots$, converges for all x.
3. $x - x^3/3! + x^5/5! - x^7/7! + \cdots$, converges for all x.

5. $\dfrac{1}{\sqrt{2}}\left[\dfrac{1}{0!} + \dfrac{1}{1!}\left(x - \dfrac{\pi}{4}\right) - \dfrac{1}{2!}\left(x - \dfrac{\pi}{4}\right)^2 - \dfrac{1}{3!}\left(x - \dfrac{\pi}{4}\right)^3 + \cdots\right]$, converges for all x.

7. $1 + x + x^2 + x^3 + \cdots$, $-1 < x < 1$.   **9.** $1 + x^2/2! + x^4/4! + x^6/6! + \cdots$, converges for all x.
11. $-4 - 2(x - 1) + 3(x - 1)^2 + 4(x - 1)^3 + (x - 1)^4$, converges for all x.
13. $x - x^3/3 + x^5/5 - x^7/7 + \cdots$, converges for $-1 \le x \le 1$.
15. $2x - 8x^3/3! + 32x^5/5! - 128x^7/7! + \cdots$.

17. $1 - \dfrac{1}{2}(x - 1) + \dfrac{1 \cdot 3}{2^2 \cdot 2!}(x - 1)^2 - \dfrac{1 \cdot 3 \cdot 5}{2^3 \cdot 3!}(x - 1)^3 + \cdots$.

19. $1 - 2x^2/2! + 12x^4/4! - 120x^6/6! + \cdots$.

27. $1 - \dfrac{1}{2}x^2 - \dfrac{1}{2^2}\dfrac{x^4}{2!} - \dfrac{1 \cdot 3}{2^3}\dfrac{x^6}{3!} - \dfrac{1 \cdot 3 \cdot 5}{2^4}\dfrac{x^8}{4!} - \cdots$.

29. $2 + \dfrac{2}{3} \cdot \dfrac{x}{8} - \dfrac{2 \cdot 2}{3^2 \cdot 2!} \cdot \dfrac{x^2}{8^2} + \dfrac{2 \cdot 2 \cdot 5}{3^3 \cdot 3!} \cdot \dfrac{x^3}{8^3} - \dfrac{2 \cdot 2 \cdot 5 \cdot 8}{3^4 \cdot 4!} \cdot \dfrac{x^4}{8^4} + \cdots$.

31. $\sum_{n=0}^{\infty}\left[\dfrac{(-1)^n}{2^{n+1}} - 2\right](x - 1)^n, 0 < x < 2$.

Section 19.8

1. $2.5, 0.5$.   **3.** $0.1045, 0.0000001$.   **5.** $0.9945, 0.000000002$.   **7.** $1.17, 0.02$.
9. $0.1051, 0.000002$.   **11.** 2.7183.   **13.** 0.1736.   **15.** 0.8191.   **17.** 0.57358.
19. 0.0997.   **21.** -1.20.   **23.** 4.12.

Section 19.9

1. $1 + 2x + 3x^2 + 4x^3 + \cdots$.   **3.** $x - (x^3/3!) + (x^5/5!) - (x^7/7!) + \cdots$.
5. $x - (x^3/3) + (x^5/5) - (x^7/7) + \cdots$.   **7.** $2x + \dfrac{2x^3}{3} + \dfrac{2x^5}{5} + \dfrac{2x^7}{7} + \cdots$.

9. $1 - \dfrac{2x^2}{2!} + \dfrac{2^3 x^4}{4!} - \dfrac{2^5 x^6}{6!} + \cdots$.   **11.** $C + x - \dfrac{x^3}{3 \cdot 3!} + \dfrac{x^5}{5 \cdot 5!} - \cdots$.

13. $C + x - \dfrac{x^3}{3} + \dfrac{x^5}{5 \cdot 2!} - \dfrac{x^7}{7 \cdot 3!} + \cdots$.   **15.** 0.4994.   **17.** 0.

19. $x - \dfrac{1}{2}\dfrac{x^3}{3} + \dfrac{1 \cdot 3}{2 \cdot 4}\dfrac{x^5}{5} - \dfrac{1 \cdot 3 \cdot 5}{2^3 \cdot 6}\dfrac{x^7}{7} + \cdots$.

21. $C + x + \dfrac{1}{2}\dfrac{x^4}{4} - \dfrac{1}{2^2 \cdot 2!}\dfrac{x^7}{7} + \dfrac{1 \cdot 3}{2^3 \cdot 3!}\dfrac{x^{10}}{10} - \dfrac{1 \cdot 3 \cdot 5}{2^4 \cdot 4!}\dfrac{x^{13}}{13} + \cdots$.

23. 0.515.   **25.** 1.   **27.** $1000, 4, 3$.   **29.** 0.08.

Review Problems

1. $s_n = n2^{n-1}, n = 1, 2, 3, \dots$.   3. Converges to 2/3.   5. Diverges.
7. Converges to 1/5.   9 1.   11. Diverges.   13. Converges.   15. Converges.
17. Converges absolutely.   19. Converges absolutely.   21. Diverges.
23. Converges.   25. Diverges.   27. Converges conditionally.   29. Diverges.
31. $-3 < x < 3$.   33. $x = 0$.   35. Converges for all x.   37. $1 < x < 3$.
39. $0.1736, (9.8)(10^{-10})$.   41. $0.9781, (1.2)(10^{-7})$.   43. $x^2 - 2x^5 + 3x^8 - 4x^{11} + \cdots$.
45. $C + x - x^5/5 + x^9/9 - x^{13}/13 + \cdots$.   47. 1/120.   49. Diverges.
51. Yes, harmonic series diverges.

Section 20.1

1. 6.   3. $3\sqrt{2}$.
5. $2\mathbf{i} + 6\mathbf{j} - 5\mathbf{k}$.   7. $8\mathbf{j} - 6\mathbf{k}$.

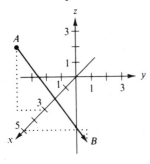
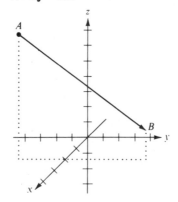

9. $(1/3, -2/3, 2/3)$.   11. $\dfrac{3}{5}\mathbf{i} - \dfrac{4}{5}\mathbf{k}$.   13. $(1, 1, 7/2)$.   15. $(4, 11/2, -2)$.
17. $(1, 4, 1)$.   19. $(-9, 7, -1)$.   21. $(5, 7, -1), (1, -5, -7), 0$; orthogonal.
23. $5\mathbf{i} + 5\mathbf{j} + 2\mathbf{k}, 3\mathbf{i} + \mathbf{j} - 4\mathbf{k}, 7$; not orthogonal.   25. $60°$.   27. $98°$.
29. $(8, 4, -6)$.   31. $(-2, 4, -3)$.   33. $5, -3$.   35. ± 5.   37. $5, -11$.
39. $A = (0, 6, -3/2), B = (4, 4, -1/2)$.

Section 20.2

1. $\dfrac{16}{13}\mathbf{j} + \dfrac{24}{13}\mathbf{k}$.   3. $\dfrac{1}{5}\mathbf{j} - \dfrac{2}{5}\mathbf{k}$.   5. $132°, 71°, 132°$.
7. $122°, 74°, 37°$.   9. $135°, 45°, 90°$.   11. $\{2, -1, 2\}$.   13. $\{1, 1, -1\}$.
15. $\{5, 2, -3\}$.   17. None.   19. Parallel.   21. Perpendicular.
23. Coincident.   25. Parallel.   27. x axis: $\{0°, 90°, 90°\}, \{1, 0, 0\}$.
31. $\mathbf{u} = (1, 0, 0), \mathbf{v}_2 = (0, 1/\sqrt{2}, 1/\sqrt{2}), \mathbf{w}_2 = (0, -1/\sqrt{2}, 1/\sqrt{2})$.

Section 20.3

1. $x = 2 + 2t, y = -4 + 3t, z = 2 + t; \dfrac{x-2}{2} = \dfrac{y+4}{3} = \dfrac{z-2}{1}.$

3. $x = 2 + 4t, y = -t, z = 3 + 3t; \dfrac{x-2}{4} = \dfrac{y}{-1} = \dfrac{z-3}{3}.$

5. $x = 1 + 3t, y = t, z = 5; \dfrac{x-1}{3} = \dfrac{y}{1}, z = 5.$

7. $x = 3 + t, y = 1, z = 2; y = 1, z = 2.$

9. $x = 3 - t, y = 3 + 3t, z = 1 - t; \dfrac{x-3}{-1} = \dfrac{y-3}{3} = \dfrac{z-1}{-1}.$

11. $x = -4 - 7t, y = 2 + t, z = -2t; \dfrac{x+4}{-7} = \dfrac{y-2}{1} = \dfrac{z}{-2}.$

13. $x = 2 + t, y = 2, z = 4 - 3t; \dfrac{x-2}{1} = \dfrac{z-4}{-3}, y = 2.$

15. $x = 2 + t, y = 4, z = -5; y = 4, z = -5.$
17. Do not intersect.   **19.** (4, 2, 3).   **21.** Do not intersect.
23. Do not intersect.   **25.** Parallel.   **27.** Coincident.   **29.** None.
31. x axis: $y = 0, z = 0.$

Section 20.4

1. $-3\mathbf{i} + 6\mathbf{j} - 3\mathbf{k}.$ **3.** $-10\mathbf{k}.$ **5.** $2\mathbf{i} - 5\mathbf{j} - \mathbf{k}.$ **7.** $\{-4, 1, 7\}.$
9. $\{1, 4, 6\}.$ **11.** $\{1, 2, 0\}.$ **13.** $x = 4 + 3t, y = -1 + t, z = -t.$
15. $2/3\sqrt{3}.$ **17.** $26/\sqrt{62}.$ **19.** 3.   **21.** $x = 3t, y = 4 - t, z = -2 + 2t.$
23. $x = 1 + 2t, y = 1 - 11t, z = 2 + 6t.$ **27.** $\sqrt{893}/2.$ **29.** 13/2.   **33.** 25.

Section 20.5

1.

3.

5.

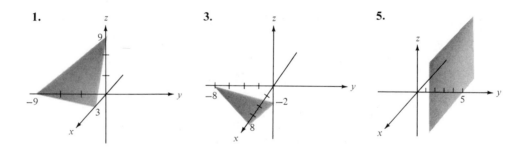

7. $2x - 5y - z + 5 = 0.$ **9.** $x + 3y + z - 14 = 0.$ **11.** $x + y - 2z + 1 = 0.$
13. $5x + 2y + z - 4 = 0.$ **15.** $x - 2y + z + 3 = 0.$ **17.** $x - 5y - 3z + 5 = 0.$
19. $4x - y - 6z - 13 = 0.$ **21.** $x = 4 + 3t, y = -2 + 2t, z = 3 - t.$
23. None.   **25.** Perpendicular.   **27.** Coincident.   **29.** $-14x + 7y + 9z + 11 = 0.$
31. $7x + 3y + z - 17 = 0.$ **33.** $x = 4 + t, y = t, z = 5 + 3t.$
37. $\{5/\sqrt{62}, 6/\sqrt{62}, -1/\sqrt{62}\}.$ **39.** $(9, -5, 12).$ **41.** $x + y - 2z - 4 = 0.$
43. $x - z = 0$ or $y - z = 0.$

Section 20.6

1. 5/3.   **3.** $3\sqrt{6}/2$.   **5.** $2/\sqrt{5}$.   **7.** 10.   **9.** 9/2, 0.   **11.** $\sqrt{186}/3$.   **13.** $3\sqrt{5}$.
15. 0.   **17.** 9.   **19.** $5/\sqrt{21}$.   **21.** $1/2\sqrt{2}$.   **23.** $11/2\sqrt{3}$.
25. 98°.   **27.** 118°.   **29.** 19°.   **31.** 118°.

Section 20.7

1.   **3.**   **5.**

7.

9.

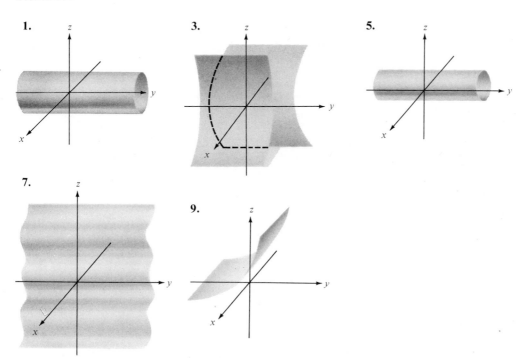

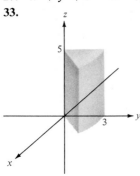

11. Sphere: $(-3, 5, -1)$, 4.   **13.** Sphere: $(-3, 4, 1)$, 2.
15. Sphere: $(1/2, -1/2, 5/2)$, 1/2.   **17.** Point: $(-2/3, 1/3, 4/3)$.
19. Sphere: $(1/2, 1/3, 1/4)$, $\sqrt{61}/12$.   **21.** $x^2 + y^2 + z^2 - 6x - 2y - 2z = 0$.
23. $x^2 + y^2 + z^2 - 8x - 2y + 6z + 17 = 0$.   **25.** $x^2 + y^2 + z^2 - 4y - 4z - 1 = 0$.
27. $x^2 + y^2 + z^2 - x - 3y + 11z - 10 = 0$.
33.   **35.**   **37.**

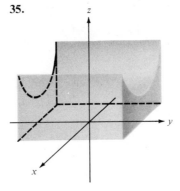

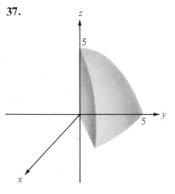

39. $6y = 9 - x^2$.

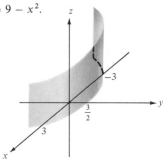

Section 20.8

1. Ellipsoid (oblate spheroid).   **3.** Hyperboloid of two sheets.   **5.** Hyberbolic paraboloid.

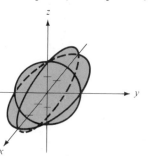

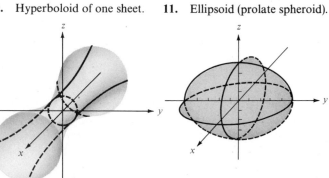

7. Circular cone.   **9.** Hyperboloid of one sheet.   **11.** Ellipsoid (prolate spheroid).

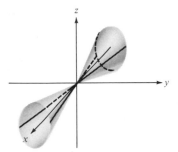

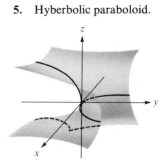

13. Circular paraboloid.   **15.** Hyperbolic paraboloid.   **17.** Circular paraboloid.

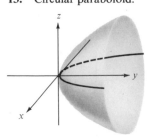

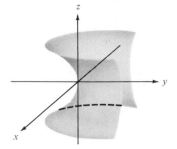

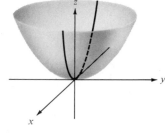

19. Ellipsoid (oblate spheroid).   **21.** Hyperboloid of two sheets.   **23.** Circular paraboloid.

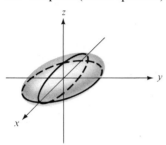 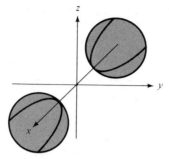

25. Hyperboloid of one sheet.   **27.**

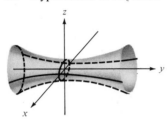

29. $\dfrac{x'^2}{36} + \dfrac{y'^2}{9} + \dfrac{z'^2}{4} = 1.$

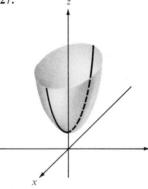 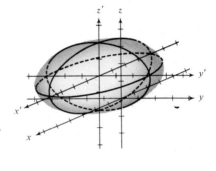

31.

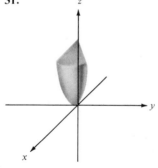

33.

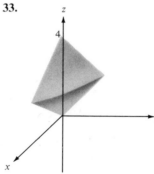

35.

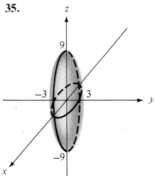

37. (a) $16z = x^2 + y^2$, (b) $x^2 + y^2 - 3z^2 - 40z - 48 = 0$, (c) $4x^2 + 4y^2 + 3z^2 - 40z + 48 = 0$.

Section 20.9

1. (a) $(\sqrt{2}, \sqrt{2}, 2)$, (b) $(-3/2, 3\sqrt{3}/2, -2)$.
2. (a) $(\sqrt{2}, 45°, 5)$, (b) $(2, 90°, -2)$.
3. (a) $(\sqrt{2}, \sqrt{2}, 2\sqrt{3})$, (b) $(0, 0, 1)$.
4. (a) $(2\sqrt{2}, 45°, 90°)$, (b) $(3, \text{Arccos } (2/\sqrt{5}), \text{Arccos } (-2/3))$.
5. (a) $(2\sqrt{2}, \pi/4, 3\pi/4)$, (b) $(5, 30°, \text{Arccos } (4/5))$.

6. (a) $(2, 45°, 2\sqrt{3})$, (b) $(2, 2\pi/3, 0)$.   **7.** $r^2 + z^2 = 4, \rho = 2$.

9. $r^2 = z \sec 2\theta, \rho = \csc \phi \cot \phi \sec 2\theta$.

11. $r^2 = (1 - z^2) \sec 2\theta, \rho^2(\sin^2 \phi \cos 2\theta + \cos^2 \phi) = 1$.

13. $r^2(4 \cos^2 \theta + 9 \sin^2 \theta) = 1 - 9z^2, \rho^2(4 \sin^2 \phi \cos^2 \theta + 9 \sin^2 \phi \sin^2 \theta + 9 \cos^2 \phi) = 1$.

15. $z = x^2 - y^2$.   **17.** $x^2 + 2yz - z^2 = 0$.   **19.** $x^2 + y^2 = 1$.   **21.** $(x^2 + y^2)z^2 = 1$.

23. Ray: $x = t, y = t, z = t, 0 \le t < \infty$.   **25.** Curve spirals around the cone;
$z = 2r$ or $4x^2 + 4y^2 - z^2 = 0$.

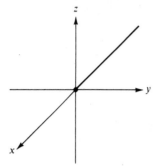

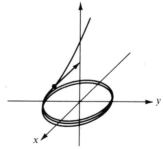

27. Spherical shell, centered at the origin; inner radius 2, outer radius 3: $4 \le x^2 + y^2 + z^2 \le 9$.

29.

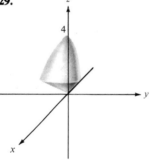

Section 20.10

1. $12\mathbf{i} + 4\mathbf{j}$.   **3.** $\mathbf{j}$.   **5.** $\mathbf{j} + \mathbf{k}$.

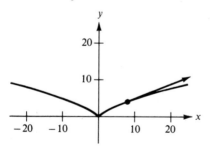

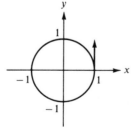

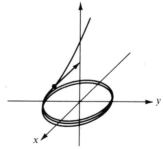

7. $2t\mathbf{i} + 2t\mathbf{j} + 2t\mathbf{k}, 2\mathbf{i} + 2\mathbf{j} + 2\mathbf{k}.$ **9.** $2t\mathbf{i} + 3t^2\mathbf{j} + \mathbf{k}, 2\mathbf{i} + 6t\mathbf{j}.$

11. $\mathbf{i} + \cos t\mathbf{j} - \sin t\mathbf{k}, -\sin t\mathbf{j} - \cos t\mathbf{k}.$

13. $\mathbf{i}, 2\mathbf{j}, 1, 0.$

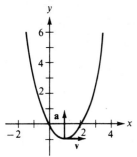

15. $4\mathbf{i} + 5\mathbf{j}, 2\mathbf{i} + 2\mathbf{j}, \sqrt{41}, 18/\sqrt{41}.$

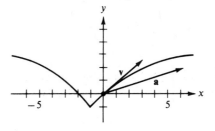

17. $\mathbf{i}, 0, 1, 0.$

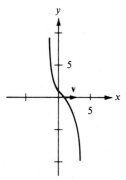

19. $3\mathbf{i} + 2\mathbf{j}, 6\mathbf{i} + 2\mathbf{j}, \sqrt{13}, 22/\sqrt{13}.$

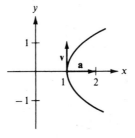

21. $2\mathbf{i} - 2\mathbf{j}, 2\mathbf{i} + 6\mathbf{j}, 0, 2\sqrt{2}, -2\sqrt{2}.$

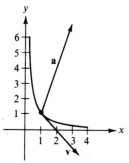

23. $\mathbf{j}, \mathbf{i}, 1, 0.$

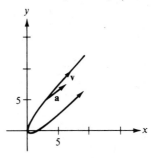

25. $\mathbf{i} + \mathbf{j}, -\mathbf{j}, \sqrt{2}, -1/\sqrt{2}.$

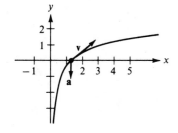

27. $\mathbf{j} + 2t\mathbf{k}, 2\mathbf{k}, \sqrt{1 + 4t^2}, 4t/\sqrt{1 + 4t^2}$.

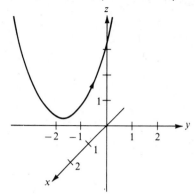

29. $\sqrt{2}\mathbf{i} + (1/2\sqrt{t})\mathbf{j} + 2\sqrt{t}\mathbf{k},$
$-(1/4t^{3/2})\mathbf{j} + (1/\sqrt{t})\mathbf{k}, 2\sqrt{t} + 1/2\sqrt{t},$
$(4t - 1)/4t^{3/2}.$

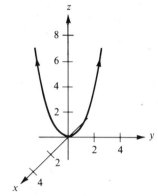

31. $e^{2t}\mathbf{i} + \sqrt{2}e^t\mathbf{j} + \mathbf{k}, 2e^{2t}\mathbf{i} + \sqrt{2}e^t\mathbf{j},$
$e^{2t} + 1, 2e^{2t}.$

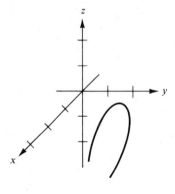

33. $t\mathbf{i} + \sqrt{2}\mathbf{j} + (1/t)\mathbf{k}, \mathbf{i} - (1/t^2)\mathbf{k},$
$t + 1/t, 1 - 1/t^2.$

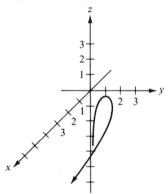

Section 20.11

1. $(-\sin t\mathbf{i} + \cos t\mathbf{j} + \mathbf{k})/\sqrt{2}, -\cos t\mathbf{i} - \sin t\mathbf{j}, (\sin t\mathbf{i} - \cos t\mathbf{j} + \mathbf{k})/)/\sqrt{2}.$

3. $(2\mathbf{i} + 2t\mathbf{j} - \mathbf{k})/\sqrt{5 + 4t^2}, (-4t\mathbf{i} + 5\mathbf{j} + 2t\mathbf{k})/\sqrt{25 + 20t^2}, (\mathbf{i} + 2\mathbf{k})/\sqrt{5}.$

5. 1. **7.** $2\sqrt{5}/(5 + 4t^2)^{3/2}.$ **9.** $2t(2 + 9t^2)/(|t|\sqrt{4 + 9t^2}), 6t^2/(|t|\sqrt{4 + 9t^2}.$

11. $2\sqrt{3} (t > 0), -2\sqrt{3} (t < 0); 0.$

Review Problems

1. $(-1, 2, 2).$ **3.** $(60°, 120°, 45°).$ **5.** Parallel.

7. $x = 2 + 3t, y = -3 - 2t, z = 5 + t; \dfrac{x - 2}{3} = \dfrac{y + 3}{-2} = \dfrac{z - 5}{1}.$ **9.** $(1, 5, -2).$

11. $x - 2y - 4 = 0$.   **13.** $1/3$.   **15.** $2\sqrt{3}/3$.   **17.** $68°$.   **19.** Point: $(1/2, 1, -3/2)$.

21. Hyperbolic cylinder.   **23.** Hyperbolic paraboloid.

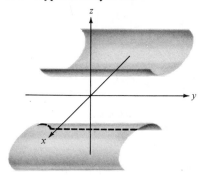   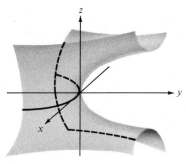

25. $z = x$.   **27.** $x^2 + y^2 + z^2 - 2x + y = 0$.   **29.** $-2, -10$.

31. $\mathbf{p} = -(5/18)\mathbf{i} - (10/9)\mathbf{j} + (5/18)\mathbf{k}, \mathbf{q} = (41/18)\mathbf{i} + (1/9)\mathbf{j} + (49/18)\mathbf{k}$.

33. $x = 4, y = 2 + 2t, z = 3 + t$.   **37.**

35. $x = 3 + 11t, y = 1 + t, z = -1 - 7t$.

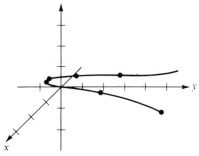

39. $4\mathbf{i} + 4\mathbf{j} + 2\mathbf{k}, 2\mathbf{i} + 2\mathbf{j}, 6, 8/3$.   **41.** $4t/\sqrt{5 + 4t^2}, \sqrt{20}/\sqrt{5 + 4t^2}$.

Section 21.2

1. $2x, 2y$.   **3.** $1/y, -x/y^2$.   **5.** $y/2\sqrt{xy}, x/2\sqrt{xy}$.

7. $3(x - 2y)^2, -6(x - 2y)^2$.   **9.** $\cos x \cos y, -\sin x \sin y$.   **11.** $-3, 4$.

13. $-1, -1$.   **15.** $1, 0$.   **17.** $\sin yz, xz \cos yz, xy \cos yz$.

19. $2x/(y^2 + z^2), 2y(z^2 - x^2)/(y^2 + z^2)^2, 2z(x^2 + y^2)/(y^2 + z^2)^2$.

21. $e^{y/z}, (x/z)e^{y/z}, -(xy/z^2)e^{y/z}$.   **23.** $-y/x^2, 1/x; 2y/x^3, -1/x^2, -1/x^2, 0$.

25. $\cos x \cos y, -\sin x \sin y; -\sin x \cos y, -\cos x \sin y, -\cos x \sin y, -\sin x \cos y$.

27. $3x^2/(x^3 + y^3), 3y^2/(x^3 + y^3); 3x(2y^3 - x^3)/(x^3 + y^3)^2, -9x^2y^2/(x^3 + y^3)^2$,

$-9x^2y^2/(x^3 + y^3)^2, 3y(2x^3 - y^3)/(x^3 + y^3)^2$.

39. $\{(x, y) \mid x + y - 4 > 0\}$.

Section 21.3

1. $4xyt + x^2 e^t$.   3. $3x^2 e^t(t+1) - \cos t$.   5. $(2xz + 4yzt - 2x^2 t - 2y^2 t)/z^2$.

7. $2xze^t - ze^{-t} + x^2 + y$.   9. $(y \sin v + xv \sin u)/y^2$, $(yu \cos v - x \cos u)/y^2$.

11. $4xy - x^2$, $2xy + 2x^2$.   13. $yz + xz + 2xy$, $yz - xz + 3xy$.

15. $3x + 2y + 3z$, $-x - y + 2z$.   17. $2t, 6u, 0$.

19. $2t \cos t + t \sin t + 2 \sin t - \cos t + (t \cos t - 3t \sin t - \sin t - 3 \cos t)/t^2$.

21. $y + xe^x (\sin x + \cos x)$.   23. $(y - x - x \cos x)/y^2$.

27. $(\partial w/\partial x) \sin \phi \cos \theta + (\partial w/\partial y) \sin \phi \sin \theta + (\partial w/\partial z) \cos \phi$, $(\partial w/\partial x) \rho \cos \phi \cos \theta$
$+ (\partial w/\partial y) \rho \cos \phi \sin \theta - (\partial w/\partial z) \rho \sin \phi$, $-(\partial w/\partial x) \rho \sin \phi \sin \theta + (\partial w/\partial y) \rho \sin \phi \cos \theta$.

31. 65π in.^{3}/min.   33. 23 in.^{2}/min.   37. $12t^2 - 8$.

Section 21.4

1. $x + 2y - z - 2 = 0$; $\dfrac{x-2}{1} = \dfrac{y-1}{2} = \dfrac{z-2}{-1}$.

3. $3x - 2y - z - 2 = 0$; $\dfrac{x-2}{3} = \dfrac{y-1}{-2} = \dfrac{z-2}{-1}$.

5. $x - y - z = 0$; $\dfrac{x-1}{1} = \dfrac{y-2}{-1} = \dfrac{z+1}{-1}$.

7. $x + 4y - z - 4 = 0$; $\dfrac{x-2}{1} = \dfrac{y-1}{4} = \dfrac{z-2}{-1}$.

9. $3x + 4y - 5z = 0$; $\dfrac{x-3}{3} = \dfrac{y-4}{4} = \dfrac{z-5}{-5}$.

11. $3x - 2y + 2z - 3 = 0$; $\dfrac{x-9}{3} = \dfrac{y-6}{-2} = \dfrac{z+6}{2}$.

13. $3x - 3y + z - 11 = 0$; $\dfrac{x-1}{-3} = \dfrac{y+2}{3} = \dfrac{z-2}{-1}$.

15. $-10x + y + 8z + 9 = 0$; $\dfrac{x-5}{-10} = \dfrac{y-9}{1} = \dfrac{z-4}{8}$.

17. $x = -4s$, $y = 3$, $z = \pi/2 + s$, $4x - z + \pi/2 = 0$.

19. $x = 5 + 4s$, $y = 3 + 4s$, $z = 2 + s$, $4x + 4y + z - 34 = 0$.

21. $x = 8 + 6s$, $y = 28 + 27s$, $z = 6 + 2s$, $6x + 27y + 2z - 816 = 0$.

23. $x = e + es$, $y = s$, $z = 1 + s$, $ex + y + z - e^2 - 1 = 0$.

27. $(2, 2, 0)$, $(-2, -2, 0)$.   29. $x = 2 + 8t$, $y = -2 + 7t$, $z = 2 - t$.

Section 21.5

1. $(3\sqrt{3}x^2 - 3y^2 - x + \sqrt{3}y)/2$.   3. $-(x^2 + 3y^2)/(\sqrt{10}x^2 y^2)$.

5. $e^{xy}(y - \sqrt{3}x)/2$.   7. $(yz - 2xz + 2xy)/3$.

9. $(-2y^2 + yz - \sqrt{2}x^2 + \sqrt{2}xz + \sqrt{5}xy)/(2\sqrt{2}x^2 y^2)$.   11. $2\sqrt{2}$.   13. 0.

15. $7/\sqrt{17}$.   17. $8/\sqrt{6}$.   19. $-(1 + 2\sqrt{6})/4$.

21. $x^2\sqrt{9y^2 + x^2}$, $(3yi + xj)/\sqrt{9y^2 + x^2}$.   23. $\sqrt{x^2 + y^2}/y^2$, $(yi - xj)/\sqrt{x^2 + y^2}$.

25. $\sqrt{y^2 z^2 + x^2 z^2 + x^2 y^2}$, $(yzi + xzj + xyk)/\sqrt{y^2 z^2 + x^2 z^2 + x^2 y^2}$.

27. $\sqrt{2}$, $(i + j)/\sqrt{2}$.   29. $\sqrt{6}$, $(i + 2j - k)/\sqrt{6}$.   31. $36°$, $-4i - 9j$.

33. $(119\sqrt{2} - 21\sqrt{13})/5$.

Section 21.6

1. $\dfrac{x-y}{x+y}$.   3. $-4x(x^2+y)/(2x^2+2y+1)$.   5. $-ye^{xy}/(xe^{xy}+2y)$.

7. $(2x-yz)/xz, (2z-xy)/xz$.   9. $y^2/x^2, z^2/x^2$.

11. $-2(x+y)/[2(x+y)-3(y-z)^2], -3(y-z)^2/[2(x+y)-3(y-z)^2]$.

13. $(x+2y-z)/(y-z)$.   15. $(x+y+z)/(u+v), (x+y+z)/(u+v)$.   17. $-z/y$.

19. $(x^2-y^2)/y^2$.   21. $-2z$.   23. x/y.   25. $r\tan\theta, -r\cot\theta$.

27. $-y(x-w)(x-z)/x(y-w)(y-z)$.   29. $(x+y)/2$.

33. $y/(2z-3y), (x+3z)/(2z-3y), -2y^2/(2z-3y)^3, (4z^2-12yz-2xy)/(2z-3y)^3,$
    $(18z^2-18xy-54yz-2x^2)/(2z-3y)^3$.

Section 21.7

1. $(2xy+y^2)\,dx+(x^2+2xy)\,dy$.   3. $(1/y)\,dx+(-x/y^2+1/y)\,dy$.

5. $3(x+\tan xy)^2[(1+y\sec^2 xy)\,dx+x\sec^2 xy\,dy]$.

7. $2xyz\,dx+(x^2z-2yz)\,dy+(x^2y-y^2+2z)\,dz$.

9. $2xy\,dx+(x^2+z/\sqrt{1-y^2})\,dy+\text{Arcsin }y\,dz$.   11. 142.83.   13. 0.487.

15. 156.0 ± 3.5.   17. 0.54.   19. 7.21 ± 0.03.   21. 0.178.

Section 21.8

1. $(-1/2, -1, -13/4)$ relative min.   3. $(0, 0, 0)$, neither.   5. $(-1, 2, -2)$ relative min.

7. $(3, 0, -27)$, neither; $(3, 4, -59)$ relative min; $(-1, 0, 5)$ relative max; $(-1, 4, -27)$, neither.

9. $(2, 1, -6)$ relative min; $(2, -2/3, -37/27)$, neither.

11. $(0, 2, -2)$ relative min; $(4, -6, 30)$, neither; $(-1, 3/2, -5/4)$, neither.

13. $(1/\sqrt{6}, 3/\sqrt{6}, 2/\sqrt{6})$, neither; $(-1/\sqrt{6}, -3/\sqrt{6}, -2/\sqrt{6})$, neither.

15. $(1/2, 4, 6)$ relative min.   17. $x=10, y=10, z=5$.

19. $16/\sqrt{3}$.   21. $32/27$.   23. $(1/3, 10/3, 11/3)$.   25. $x=75, y=100$.

Section 21.9

1. 3.   3. $18/7$.   5. $2/39$.   7. $5184/(4+3\sqrt[3]{12}+12\sqrt[3]{3})^3$.   9. $2\sqrt{2}/3\sqrt{3}$.

11. 8.   13. $37/13$.   15. $3/32$.   17. $(0, 2, 0)$.   19. $16/\sqrt{3}$.   21. $4/27$.

Review Problems

1. $2x+y, x-z, -y$.   3. $9x^2-5y^2, 9x^2-5y^2-20x^2y+40xy$.

5. $2u(x+y+5z)+6x+4y, 2v(-x+y+z)+9x+6y$.

7. $(4yz-3x^2)/(6xy+5z^2), (2y^2-10xz)/(6xy+5z^2)$.   9. $-z/2$.

11. $(y\cos xy+\cos y)\,dx+(x\cos xy-x\sin y)\,dy$.   13. $-16/5$.

15. $6x-12y-z+4=0; x=1+6t, y=1-12t, z=-2-t$.

17. $2x + y + 9z - 18 = 0; \dfrac{x-1}{2} = \dfrac{y+2}{1} = \dfrac{z-2}{9}.$

19. $x = 1/2 + \sqrt{3}s/2, y = 1/2 - \sqrt{3}s, z = \pi/6 + s.$ **21.** $288.675 \pm 1.26.$

23. $(0, 0, 0),$ neither; $(-6, 2, 8),$ neither.   **25.** $-3/4\sqrt[3]{2}.$ **27.** 4.

Section 22.1

1. 6.

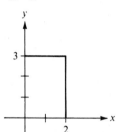

3. 16.

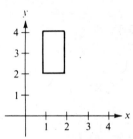

5. 32/105.

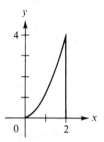

7. 5/24.

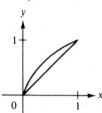

9. $4, \displaystyle\int_{-1}^{1}\int_{-1}^{1} dx\,dy.$

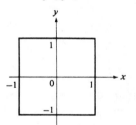

11. 88/105,

$$\int_{0}^{1}\int_{-\sqrt{y}}^{\sqrt{y}} (x^2 + y^2)\,dx\,dy.$$

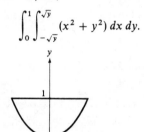

13. $(e - 1)^2/2, \displaystyle\int_{0}^{1}\int_{x}^{1} e^{x+y}\,dy\,dx.$

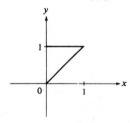

15. $-1/6.$ **17.** $-45/364.$ **19.** 9.   **21.** 31/120.   **23.** 243/40.   **25.** 52/9.

27. $(e^4 - 1)/2e^4.$

Section 22.2

1. 1/6.   **3.** 16/3.   **5.** 32/3.   **7.** 16/3.   **9.** 4/3.   **11.** 9/2.   **13.** 1.   **15.** 8/35.

17. 3/35.   **19.** $5\pi/2.$ **21.** $\pi.$ **23.** $4\pi R^3/3.$ **25.** $2\pi.$ **27.** $\pi r^2 h/3.$

Section 22.3

1. $3\pi/4$.   **3.** $5\pi/32$.   **5.** $11\pi/16$.   **7.** $8/9$.   **9.** $5\pi/3$.   **11.** $3\pi/32$.   **13.** $8/9$.
15. $\pi/8$.   **17.** $\pi/2$.   **19.** $\pi(2 - \sqrt{2})/3$.   **21.** $64/9$.   **23.** $7\pi/12$.   **27.** $\ln(\sqrt{2} + 1)$.
29. $7k/24$.   **31.** $515k\pi/16$.

Section 22.4

1. $(4/7, 3/4)$.   **3.** $(385/456, 161/456)$.   **5.** $(35/48, 35/54)$.   **7.** $(0, 1947/1733)$.
9. $25/72, 5\sqrt{5}/3\sqrt{26}$.   **11.** $1/56, \sqrt{2/7}$.   **13.** $25/126, 5\sqrt{5}/3\sqrt{19}$.   **15.** $(0, 9/14, 23/28)$.
17. $1/30, 1/\sqrt{5}$.   **19.** $80/189, 5/3\sqrt{7}$.   **21.** $5\pi/6, \sqrt{5/6}$.   **23.** $(\pi/2, \pi/8)$.   **25.** $(-1/2, 0)$.
27. $(512/105\pi, 512/105\pi)$.   **29.** $(b^4 - a^4)\pi/4$.   **31.** $\pi a^4 h/10$.

Section 22.5

1. 6.   **3.** 1.   **5.** 18.   **7.** $9/2$.   **9.** 6.   **11.** $26/105$.   **13.** $3/2$.   **15.** 2.
17. $1/12$.   **19.** π.   **21.** $3\pi/2$.   **23.** $7387/15,120$.   **25.** $4\pi abc/3$.

Section 22.6

1. $(2/3, 2/3, 2/3)$.   **3.** $(6/7, 9/14, 3/8)$.   **5.** $(0, 0, 2/3)$.   **7.** $(0, 0, 5/14)$.   **9.** $1/8, 1$.
11. $7/288, \sqrt{7}/2\sqrt{3}$.   **13.** $7/18, \sqrt{7}/3\sqrt{2}$.   **15.** $64/9, 4/3$.

Section 22.7

1. $8k\pi/3$.   **3.** $\pi/8$.   **5.** $(0, 0, 8/15)$.   **7.** $4\pi/3$.   **9.** $80\pi/3$.   **11.** $\pi(R^4 - r^4)$.
13. $(0, 0, 9(2 + \sqrt{2})/50)$.   **15.** $(0, 0, 248/225)$.   **17.** $ka^4\pi h/10, a\sqrt{30}/10$.   **19.** $(0, 0, 29/15)$.

Review Problems

1. $621/20$.

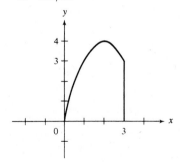

3. $9/2$.

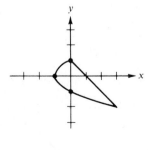

5. 16.   **7.** 3.   **9.** 160/27.   **11.** π.   **13.** $2\pi(b^3 - a^3)/3$.
15. $2{,}375/7, \sqrt{57}/2\sqrt{7}$.   **17.** $23{,}552/315, 2\sqrt{23}/\sqrt{21}$.   **19.** $-7/3$.   **21.** 832/21.
23. (30/13, 9/13, 9/13).   **25.** $2\pi(3 + 4\sqrt{2})/3$.   **27.** 24π.   **29.** 8π.

Section 23.1

1. $-20, y = (3/2)x$.

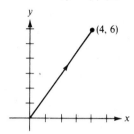

3. $2/3, x^2 + y^2 = 1$.

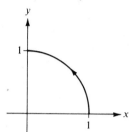

5. $-1/2, (y - 1)^2 = x$.

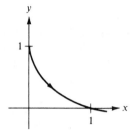

7. 16.

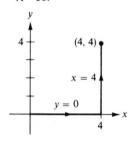

9. 31/3.

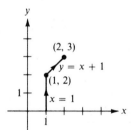

11. 0.

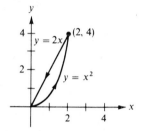

13. $16/15, y = 1 - x^2$.

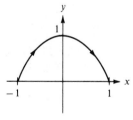

15. $8(10\sqrt{10} - 1)/27, y = x^{2/3}$.

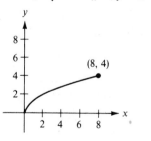

17. 1.

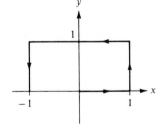

19. $-\pi/2, x^{2/3} + y^{2/3} = 1$.

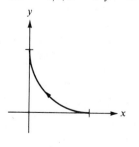

21. $0, x^2 + y^2 = 9$.

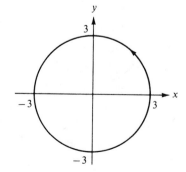

Section 23.2

1. -5π.　　3. -36.　　5. 0.　　13. 9/2.

Section 23.3

1. $xy^2 + x^3 + C$.　　3. Not exact.　　5. $\ln|xy| + y(x^2 + y^2) + C$.
7. $4x^2y + 3x + 2y^2 = C$.　　9. Not exact.　　11. $xe^{xy} - 2x^2 + \cos y + C$.
13. $e^{xy} + 2x^2y + x^2 - 2x + y = C$.　　15. $x^2 + x/y + y = C$.　　17. $x \tan y = C$.
19. $\pi\sqrt{\pi}/3$.

Section 23.4

1. $-46/3$.　　3. 3π.　　5. $e^4/2 + 25/6$.　　7. $-32/3$.　　9. 0.　　11. $3x^2/2 + y^2 + K$.
13. $\ln(2/\sqrt{3})$.　　15. 1.575.

Section 23.5

1. 41/60.　　3. $-5/2$.　　5. $x^2y + xz^2 - 2y + z + K$.　　7. $e^x + x^2y + \cos y + z^4 + K$.
9. $8\sqrt{3}$.　　11. $(17\sqrt{17} - 1)\pi/24$.　　13. $2a^2$.
15. $(0, 0, (391\sqrt{17} + 1)/[(17\sqrt{17} - 1)(10)]), (391\sqrt{17} + 1)\pi/60$.　　17. 0.　　19. $-5/6$.
21. 0.

Section 23.6

1. $2z(x + 1)$.　　3. $(x + y)/\sqrt{1 + x^2 + y^2} + 2z$.　　5. 0.
7. $-\cos z\mathbf{i} - \cos x\mathbf{j} - \cos y\mathbf{k}$.　　9. $2\pi a^2$.

Review Problems

1. -9.　　3. 0.　　5. $-33/10$.　　7. 0.　　9. $(x^3 + y^3)/3 + x^2y^2/2 + K$.
11. Not exact.　　13. -12.　　15. 2611/120.　　17. $(x^3 - y^3 + z^3)/3 + x^2z + yz + C$.
19. $(33\sqrt{17}/16 - \ln(33 + 8\sqrt{17})/128$.　　21. -40π.

Section A.1

1. $x < -1$.　　3. $x < 3$.　　5. $x \le -6$.　　7. $x > 5/2$.　　9. $x \ge 1$.　　11. $-3 < x < 1$.
13. $0 \le x \le 3/2$.　　15. $3 < x < 7$.　　17. $3/2 \le x \le 5$.　　19. $2 < x < 3$.
21. $-5 \le x < 4$.　　23. $x > -1$.

Section A.2

1. $3, -7$.   3. $5/4, 3/4$.   5. 2.   7. $-4, 2/3$.   9. $-5/2, -1/4$.   11. 1.
13. $-4 < x < -2$.   15. $-3 < x < 0$.   17. $1/2 \le x \le 9/2$.   19. $x > 5$ or $x < 3$.
21. $x \ge 1$ or $x \le -5/3$.   23. $x = 2$.   25. <1.   27. $<2\delta$.

Section A.3

1. <5.   3. <5.   5. <629.   7. <7.   9. <28.

Section B.5

1. $8 + 6(x - 1) - y + 2(x - 1)^2 - (x - 1)y - 3y^2$.
3. $-2 + 4(x - 1) - 3(y - 2) - 2(z - 1) + (x - 1)^2 + 2(x - 1)(z - 1) - (y - 2)^2 - (z - 1)^2$
5. $1 + x + x^2/2 - y^2/2 + x^3/6 - xy^2/2 + x^4/24 - x^2y^2/4 + y^4/24 + \cdots$.
7. $\ln 3 + (x - 2)/3 + 2(y - 1)/3 - (x - 2)^2/18 - 2(x - 2)(y - 1)/9 + (y - 1)^2/9$
    $+ (x - 2)^3/81 + 2(x - 2)^2(y - 1)/27 + (x - 2)(y - 1)^2/27 - 10(y - 1)^3/81 + \cdots$.
9. $x(a, b) = 1 - a/2 - (b + 1)/2 + a^2/8 - (b + 1)^2/8 + a^2(b + 1)/16 - (b + 1)^3/16 + \cdots$.

Index

What did you most like about the book?

---------------------------------FOLD HERE ---------------------------------

What did you like least about the book?

Name _____ College _____

Home address (optional) _____

---------------------------------FOLD HERE ---------------------------------

First Class
PERMIT NO. 34
Belmont, CA

BUSINESS REPLY MAIL
No postage necessary if mailed in United States

Postage will be paid by
WADSWORTH PUBLISHING COMPANY, INC.
10 Davis Drive
Belmont, California 94002
          ATTN: Rich Jones, Mathematics Editor

Your chance to rate
Calculus and Analytic Geometry, Third Edition

In order to make this text even more responsive to your needs, it would help us to know what you, the student, thought of *Calculus and Analytic Geometry, Third Edition*. We would appreciate it if you would answer the following questions. Then, cut out the page, fold, seal, and mail it; no postage is required. Thank you for your help.

What Chapters in the text did you skip, or *not* cover in class? (circle)

1 2 3 4 5 6 7 8 9 10 11 12
13 14 15 16 17 18 19 20 21 22 23

Were any figures or answers in the back of the book unclear? Which ones?

Prior to taking Calculus, which courses in mathematics had you previously taken?

College Algebra _____   Analytic Geometry _____
Trigonometry _____   Algebra and Trigonometry _____
Precalculus _____   High School Algebra _____ (1 or 2 terms? _____)

Prior to taking Calculus, how long ago did you take your last Algebra course?

Within last 2 years _____   3–5 years ago _____   Over 5 years ago _____

What is your major course of study?

Engineering _____   Chemistry _____   Physics _____
Biology _____   Mathematics _____   Computer Science _____
Business _____   Social Science _____   Other _____

Would you rather see more applications to your major in college, or more detailed examples and answers?

Applications _____   More detail _____

FORMS CONTAINING $\sqrt{a^2 - u^2}$

56. $\displaystyle \int \sqrt{a^2 - u^2}\, du = \frac{u}{2}\sqrt{a^2 - u^2} + \frac{a^2}{2}\operatorname{Sin}^{-1}\frac{u}{a} + C$

57. $\displaystyle \int u^2\sqrt{a^2 - u^2}\, du = \frac{u}{8}(2u^2 - a^2)\sqrt{a^2 - u^2} + \frac{a^4}{8}\operatorname{Sin}^{-1}\left(\frac{u}{a}\right) + C$

58. $\displaystyle \int \frac{\sqrt{a^2 - u^2}}{u}\, du = \sqrt{a^2 - u^2} - a\ln\left|\frac{a + \sqrt{a^2 - u^2}}{u}\right| + C$

59. $\displaystyle \int \frac{\sqrt{a^2 - u^2}}{u^2}\, du = -\frac{\sqrt{a^2 - u^2}}{u} - \operatorname{Sin}^{-1}\frac{u}{a} + C$

60. $\displaystyle \int \frac{u^2}{\sqrt{a^2 - u^2}}\, du = -\frac{u}{2}\sqrt{a^2 - u^2} + \frac{a^2}{2}\operatorname{Sin}^{-1}\frac{u}{a} + C$

61. $\displaystyle \int \frac{du}{u\sqrt{a^2 - u^2}} = -\frac{1}{a}\ln\left|\frac{a + \sqrt{a^2 - u^2}}{u}\right| + C$

62. $\displaystyle \int \frac{du}{u^2\sqrt{a^2 - u^2}} = -\frac{\sqrt{a^2 - u^2}}{a^2 u} + C$

63. $\displaystyle \int (a^2 - u^2)^{3/2}\, du = \frac{u}{4}(a^2 - u^2)^{3/2} + \frac{3a^2 u}{8}\sqrt{a^2 - u^2} + \frac{3a^4}{8}\operatorname{Sin}^{-1}\frac{u}{a} + C$

64. $\displaystyle \int \frac{du}{(a^2 - u^2)^{3/2}} = \frac{u}{a^2\sqrt{a^2 - u^2}} + C$

FORMS INVOLVING $\sqrt{2au - u^2}$

65. $\displaystyle \int \sqrt{2au - u^2}\, du = \frac{u - a}{2}\sqrt{2au - u^2} + \frac{a^2}{2}\operatorname{Cos}^{-1}\left(\frac{a - u}{a}\right) + C$

66. $\displaystyle \int u\sqrt{2au - u^2}\, du = \frac{2u^2 - au - 3a^2}{6}\sqrt{2au - u^2} + \frac{a^3}{2}\operatorname{Cos}^{-1}\left(\frac{a - u}{a}\right) + C$

67. $\displaystyle \int \frac{\sqrt{2au - u^2}}{u}\, du = \sqrt{2au - u^2} + a\operatorname{Cos}^{-1}\left(\frac{a - u}{a}\right) + C$

68. $\displaystyle \int \frac{\sqrt{2au - u^2}}{u^2}\, du = -\frac{2\sqrt{2au - u^2}}{u} - \operatorname{Cos}^{-1}\left(\frac{a - u}{a}\right) + C$

69. $\displaystyle \int \frac{du}{\sqrt{2au - u^2}} = \operatorname{Cos}^{-1}\left(\frac{a - u}{a}\right) + C$

70. $\displaystyle \int \frac{u\, du}{\sqrt{2au - u^2}} = -\sqrt{2au - u^2} + a\operatorname{Cos}^{-1}\left(\frac{a - u}{a}\right) + C$

71. $\displaystyle \int \frac{u^2\, du}{\sqrt{2au - u^2}} = -\frac{(u + 3a)}{2}\sqrt{2au - u^2} + \frac{3a^2}{2}\operatorname{Cos}^{-1}\left(\frac{a - u}{a}\right) + C$

72. $\displaystyle \int \frac{du}{u\sqrt{2au - u^2}} = -\frac{\sqrt{2au - u^2}}{au} + C$

73. $\displaystyle \int \frac{\sqrt{2au - u^2}}{u^n}\, du = \frac{(2au - u^2)^{3/2}}{(3 - 2n)au^n} + \frac{n - 3}{(2n - 3)a}\int \frac{\sqrt{2au - u^2}}{u^{n-1}}\, du, \quad n \neq \frac{3}{2}$

74. $\displaystyle \int \frac{u^n\, du}{\sqrt{2au - u^2}} = -\frac{u^{n-1}\sqrt{2au - u^2}}{n} + \frac{a(2n - 1)}{n}\int \frac{u^{n-1}}{\sqrt{2au - u^2}}\, du$

75. $\displaystyle \int \frac{du}{u^n\sqrt{2au - u^2}} = \frac{\sqrt{2au - u^2}}{a(1 - 2n)u^n} + \frac{n - 1}{(2n - 1)a}\int \frac{du}{u^{n-1}\sqrt{2au - u^2}}$

76. $\displaystyle \int \frac{du}{(2au - u^2)^{3/2}} = \frac{u - a}{a^2\sqrt{2au - u^2}} + C$

77. $\displaystyle \int \frac{u\, du}{(2au - u^2)^{3/2}} = \frac{u}{a\sqrt{2au - u^2}} + C$

78. $\int \sin^2 u \, du = \dfrac{u}{2} - \dfrac{\sin 2u}{4} + C$

79. $\int \cos^2 u \, du = \dfrac{u}{2} + \dfrac{\sin 2u}{4} + C$

80. $\int \tan^2 u \, du = \tan u - u + C$

81. $\int \cot^2 u \, du = -\cot u - u + C$

82. $\int \sec^3 u \, du = \frac{1}{2} \sec u \tan u + \frac{1}{2} \ln |\sec u + \tan u| + C$

83. $\int \csc^3 u \, du = -\frac{1}{2} \csc u \cot u + \frac{1}{2} \ln |\csc u - \cot u| + C$

84. $\int \sin^n u \, du = -\dfrac{1}{n} \sin^{n-1} u \cos u + \dfrac{n-1}{n} \int \sin^{n-2} u \, du$

85. $\int \cos^n u \, du = \dfrac{1}{n} \cos^{n-1} u \sin u + \dfrac{n-1}{n} \int \cos^{n-2} u \, du$

86. $\int \tan^n u \, du = \dfrac{1}{n-1} \tan^{n-1} u - \int \tan^{n-2} u \, du$

87. $\int \cot^n u \, du = \dfrac{-1}{n-1} \cot^{n-1} u - \int \cot^{n-2} u \, du$

88. $\int \sec^n u \, du = \dfrac{1}{n-1} \tan u \sec^{n-2} u + \dfrac{n-2}{n-1} \int \sec^{n-2} u \, du$

89. $\int \csc^n u \, du = \dfrac{-1}{n-1} \cot u \csc^{n-2} u + \dfrac{n-2}{n-1} \int \csc^{n-2} u \, du$

90. $\int \sin mu \sin nu \, du = -\dfrac{\sin (m+n)u}{2(m+n)} + \dfrac{\sin (m-n)u}{2(m-n)} + C, \quad m^2 \neq n^2$

91. $\int \cos mu \cos nu \, du = \dfrac{\sin (m+n)u}{2(m+n)} + \dfrac{\sin (m-n)u}{2(m-n)} + C, \quad m^2 \neq n^2$

92. $\int \sin mu \cos nu \, du = -\dfrac{\cos (m+n)u}{2(m+n)} - \dfrac{\cos (m-n)u}{2(m-n)} + C, \quad m^2 \neq n^2$

93. $\int u \sin u \, du = \sin u - u \cos u + C$

94. $\int u \cos u \, du = \cos u + u \sin u + C$

95. $\int u^2 \sin u \, du = 2u \sin u + (2 - u^2) \cos u + C$

96. $\int u^2 \cos u \, du = 2u \cos u + (u^2 - 2) \sin u + C$

97. $\int u^n \sin u \, du = -u^n \cos u + n \int u^{n-1} \cos u \, du$

98. $\int u^n \cos u \, du = u^n \sin u - n \int u^{n-1} \sin u \, du$

99. $\int \sin^m u \cos^n u \, du = -\dfrac{\sin^{m-1} u \cos^{n+1} u}{m+n} + \dfrac{m-1}{m+n} \int \sin^{m-2} u \cos^n u \, du$

$\qquad = \dfrac{\sin^{m+1} u \cos^{n-1} u}{m+n} + \dfrac{n-1}{m+n} \int \sin^m u \cos^{n-2} u \, du$

(If $m = -n$, use formula 86 or 87.)

100. $\int \text{Sin}^{-1} u \, du = u \, \text{Sin}^{-1} u + \sqrt{1 - u^2} + C$

101. $\int \text{Cos}^{-1} u \, du = u \, \text{Cos}^{-1} u - \sqrt{1 - u^2} + C$

102. $\int \text{Tan}^{-1} u \, du = u \, \text{Tan}^{-1} u - \frac{1}{2} \ln (1 + u^2) + C$

103. $\int u \, \text{Sin}^{-1} u \, du = \dfrac{2u^2 - 1}{4} \text{Sin}^{-1} u + \dfrac{u\sqrt{1 - u^2}}{4} + C$

104. $\int u \, \text{Cos}^{-1} u \, du = \dfrac{2u^2 - 1}{4} \text{Cos}^{-1} u - \dfrac{u\sqrt{1 - u^2}}{4} + C$

105. $\int u \, \text{Tan}^{-1} u \, du = \dfrac{u^2 + 1}{2} \text{Tan}^{-1} u - \dfrac{u}{2} + C$

106. $\int u^n \, \text{Sin}^{-1} u \, du = \dfrac{1}{n+1}\left(u^{n+1} \text{Sin}^{-1} u - \int \dfrac{u^{n+1} \, du}{\sqrt{1 - u^2}}\right), \quad n \neq -1$

107. $\int u^n \, \text{Cos}^{-1} u \, du = \dfrac{1}{n+1}\left(u^{n+1} \text{Cos}^{-1} u + \int \dfrac{u^{n+1} \, du}{\sqrt{1 - u^2}}\right), \quad n \neq -1$

108. $\int u^n \, \text{Tan}^{-1} u \, du = \dfrac{1}{n+1}\left(u^{n+1} \text{Tan}^{-1} u - \int \dfrac{u^{n+1} \, du}{1 + u^2}\right), \quad n \neq -1$